# Customer Support Information

## Plunkett's Biotech & Genetics Industry Almanac 2010

### Please register your book immediately...

if you did not purchase it directly from Plunkett Research, Ltd. This will enable us to fulfill your replacement request if you have a damaged product, or your requests for assistance. Also it will enable us to notify you of future editions, so that you may purchase them from the source of your choice.

## If you are an actual purchaser but did not receive a FREE CD-ROM version with your book...

you may request it by returning this form. (Proof of purchase may be required.)

_____ YES, please register me as a purchaser of the book.
I did not buy it directly from Plunkett Research, Ltd.

_____ YES, please send me a free CD-ROM version of the book.
I am an actual purchaser, but I did not receive one with my book.

Customer Name _____

Title_____

Organization _____

Address_____

City_____State_____Zip_____

Country (if other than USA) _____

Phone_____Fax _____

E-mail _____

**Mail or Fax to:** # Plunkett Research, Ltd.

Attn: FREE CD-ROM and/or Registration
P.O. Drawer 541737, Houston, TX 77254-1737 USA
713.932.0000 · Fax 713.932.7080 · www.plunkettresearch.com

# PLUNKETT'S BIOTECH & GENETICS INDUSTRY ALMANAC 2010

## The only comprehensive
## guide to biotechnology and genetics companies
## and trends

# Jack W. Plunkett

Published by:
Plunkett Research, Ltd., Houston, Texas
www.plunkettresearch.com

# PLUNKETT'S BIOTECH & GENETICS INDUSTRY ALMANAC 2010

**Editor and Publisher:**
Jack W. Plunkett

**Executive Editor and Database Manager:**
Martha Burgher Plunkett

**Senior Editors and Researchers:**
Brandon Brison
Addie K. FryeWeaver
Christie Manck

**Editors, Researchers and Assistants:**
Kalonji Bobb
Elizabeth Braddock
Michelle Dotter
Michael Esterheld
Kathi Mestousis
Lindsey Meyn
Andrew Olsen
Jill Steinberg
Kyle Wark
Suzanne Zarosky

**E-Commerce Managers:**
Mark Cassells
Emily Hurley

**Information Technology Manager:**
Wenping Guo

**Cover Design:**
Kim Paxson, Just Graphics
*Junction, TX*

**Special Thanks to:**
Burrill & Company
Centers for Medicare & Medicaid Services (CMS)
Ernst & Young
IMS Health
ISAAA
National Science Foundation
Pharmaceutical Research & Manufacturers
Association (PhRMA)
Tufts Center for the Study of Drug Development
U.S. Department of Labor, Bureau of Labor Statistics
U.S. Food & Drug Administration (FDA)
U.S. Institutes of Health
U.S. National Science Foundation
U.S. Patent & Trademark Office

**Plunkett Research, Ltd.**
P. O. Drawer 541737, Houston, Texas 77254 USA
Phone: 713.932.0000  Fax: 713.932.7080
www.plunkettresearch.com

## Published by:
**Plunkett Research, Ltd.**
**P. O. Drawer 541737**
**Houston, Texas 77254-1737**

**Phone: 713.932.0000**
**Fax: 713.932.7080**
**Internet: www.plunkettresearch.com**

**ISBN13 # 978-1-59392-150-7**

## Disclaimer of liability
## for use and results of use:

# PLUNKETT'S BIOTECH & GENETICS INDUSTRY ALMANAC 2010

## CONTENTS

*Continued on next page*

*Continued from previous page*

# A Short Biotech & Genetics Industry Glossary

**10-K:** An annual report filed by publicly held companies. It provides a comprehensive overview of the company's business and its finances. By law, it must contain specific information and follow a given form, the "Annual Report on Form 10-K." The U.S. Securities and Exchange Commission requires that it be filed within 90 days after fiscal year end. However, these reports are often filed late due to extenuating circumstances. Variations of a 10-K are often filed to indicate amendments and changes. Most publicly held companies also publish an "annual report" that is not on Form 10-K. These annual reports are more informal and are frequently used by a company to enhance its image with customers, investors and industry peers.

**510 K:** An application filed with the FDA for a new medical device to show that the apparatus is "substantially equivalent" to one that is already marketed.

**Abbreviated New Drug Application (ANDA):** An application filed with the FDA showing that a substance is the same as an existing, previously approved drug (i.e., a generic version).

**Absorption, Distribution, Metabolism and Excretion (ADME):** In clinical trials, the bodily processes studied to determine the extent and duration of systemic exposure to a drug.

**Accelerated Approval:** A process at the FDA for reducing the clinical trial length for drugs designed for certain serious or life-threatening diseases.

**ADME:** See "Absorption, Distribution, Metabolism and Excretion (ADME)."

**Adverse Event (AE):** In clinical trials, a condition not observed at baseline or worsened if present at baseline. Sometimes called Treatment Emergent Signs and Symptoms (TESS).

**AE:** See "Adverse Event (AE)."

**Agricultural Biotechnology (AgriBio):** The application of biotechnology methods to enhance agricultural plants and animals.

**Amino Acid:** Any of a class of 20 molecules that combine to form proteins.

**ANDA:** See "Abbreviated New Drug Application (ANDA)."

**Angiogenesis:** Blood vessel formation, typically in the growth of malignant tissue.

**Angioplasty:** The re-opening of a blood vessel by non-surgical techniques such as balloon dilation or laser, or through surgery.

**Antibody:** A protein produced by white blood cells in response to a foreign substance. Each antibody can bind only to one specific antigen. See "Antigen."

**Antigen:** A foreign substance that causes the immune system to create an antibody. See "Antibody."

**Antisense Technology:** The use of RNA-like oligonucleotides that bind to RNA and inhibit the expression of a gene.

**APAC:** Asia Pacific Advisory Committee. A multi-country committee representing the Asia and Pacific region.

**Apoptosis:** A normal cellular process leading to the termination of a cell's life.

**Applied Research:** The application of compounds, processes, materials or other items discovered during basic research to practical uses. The goal is to move discoveries along to the final development phase.

**Array:** An orderly arrangement, such as a rectangular matrix of data. In some laboratory systems, such as microarrays, multiple detectors (probes) are positioned in an array in order to best perform research. See "Microarray."

**Artificial Life (AL):** See "Synthetic Biology."

**Assay:** A laboratory test to identify and/or measure the amount of a particular substance in a sample. Types of assays include endpoint assays, in which a single measurement is made at a fixed time; kinetic assays, in which increasing amounts of a product are formed with time and are monitored at multiple points; microbiological assays, which measure the concentration of antimicrobials in biological material;

and immunological assays, in which analysis or measurement is based on antigen-antibody reactions.

**Astrobiology:** A field focused on the study of the origin, evolution and future of life on Earth and beyond. Participants include scientists in physics, chemistry, astronomy, biology, geology and computer science.

**Baby Boomer:** Generally refers to people born in the U.S. and Western Europe from 1946 to 1964. In the U.S., the total number of Baby Boomers is about 78 million--one of the largest and most affluent demographic groups. The term evolved to include the children of soldiers and war industry workers who were involved in World War II.

**Baseline:** A set of data used in clinical studies, or other types of research, for control or comparison.

**Basic Research:** Attempts to discover compounds, materials, processes or other items that may be largely or entirely new and/or unique. Basic research may start with a theoretical concept that has yet to be proven. The goal is to create discoveries that can be moved along to applied research. Basic research is sometimes referred to as "blue sky" research.

**Big Pharma:** The top tier of pharmaceutical companies in terms of sales and profits (e.g., Pfizer, Merck, Johnson & Johnson).

**Bioaccumulation:** A process in which chemicals are retained in fatty body tissue and increased in concentration over time.

**Bioavailability:** In pharmaceuticals, the rate and extent to which a drug is absorbed or is otherwise available to the treatment site in the body.

**Biochemical Engineering:** A sector of chemical engineering that deals with biological structures and processes. Biochemical engineers may be found in the pharmaceutical, biotechnology and environmental fields, among others.

**Biochemicals:** Chemicals that either naturally occur or are identical to naturally occurring substances. Examples include hormones, pheromones and enzymes. Biochemicals function as pesticides through non-toxic, non-lethal modes of action, disrupting insect mating patterns, regulating growth or acting as repellants. They tend to be

environmentally desirable, and may be produced by industry from organic sources such as plant waste (biomass). Biochemicals also may be referred to as bio-based chemicals, green chemicals or plant-based chemicals.

**Biodiesel:** A fuel derived when glycerin is separated from vegetable oils or animal fats. The resulting byproducts are methyl esters (the chemical name for biodiesel) and glycerin which can be used in soaps and cleaning products. It has lower emissions than petroleum diesel and is currently used as an additive to that fuel since it helps with lubricity.

**Bioengineering:** Engineering principles applied when working in biology and pharmaceuticals.

**Bioequivalence:** In pharmaceuticals, the demonstration that a drug's rate and extent of absorption are not significantly different from those of an existing drug that is already approved by the FDA. This is the basis upon which generic and brand name drugs are compared.

**Bioethanol:** A fuel produced by the fermentation of plant matter such as corn. Fermentation is enhanced through the use of enzymes that are created through biotechnology. Also, see "Ethanol."

**Biogeneric:** See "Follow-on Biologics."

**Biogenerics:** Genetic versions of drugs that have been created via biotechnology. Also, see "Follow-on Biologics."

**Bioinformatics:** Research, development or application of computational tools and approaches for expanding the use of biological, medical, behavioral or health data, including those to acquire, store, organize, archive, analyze or visualize such data. Bioinformatics is often applied to the study of genetic data. It applies principles of information sciences and technologies to make vast, diverse and complex life sciences data more understandable and useful.

**Biologics:** Drugs that are synthesized from living organisms. That is, drugs created using biotechnology, sometimes referred to as biopharmaceuticals. Specifically, biologics may be any virus, therapeutic serum, toxin, antitoxin, vaccine, blood, blood component or derivative, allergenic or analogous product, or arsphenamine or one of its derivatives used for the prevention, treatment or cure of disease. Also, see "Biologics License Application (BLA)," "Follow-on Biologics," and "Biopharmaceuticals."

**Biologics License Application (BLA):** An application to be submitted to the FDA when a firm wants to obtain permission to market a novel, new biological drug product. Specifically, these are drugs created through the use of biotechnology. It was formerly known as Product License Application (PLA). Also see "Biologics."

**Biomagnification:** The increase of tissue accumulation in species higher in the natural food chain as contaminated food species are eaten.

**Biomass:** Organic, non-fossil material of biological origin constituting a renewable energy source. The biomass can be burnt as fuel in a system that creates steam to turn a turbine, generating electricity. For example, biomass can include wood chips and agricultural crops.

**BioMEMS:** MEMS used in medicine. See "Micro Electro Mechanical Systems (MEMS)."

**Biomimetic:** Mimicking, imitating, copying or learning from nature.

**Biopharmaceuticals:** That portion of the pharmaceutical industry focused on the use of biotechnology to create new drugs. A biopharmaceutical can be any biological compound that is intended to be used as a therapeutic drug, including recombinant proteins, monoclonal and polyclonal antibodies, antisense oligonucleotides, therapeutic genes, and recombinant and DNA vaccines. Also, see "Biologics."

**Biopolymeroptoelectromechanical Systems (BioPOEMS):** Combination of MEMS and optics used in biological applications.

**Biorefinery:** A refinery that produces fuels from biomass. These fuels may include bioethanol (produced from corn or other plant matter) or biodiesel (produced from plant or animal matter).

**Biosensor:** A sensor based on the use of biological materials or that targets biological analytes.

**Biosimilar:** See "Follow-on Biologics."

**Biotechnology:** A set of powerful tools that employ living organisms (or parts of organisms) to make or modify products, improve plants or animals (including humans) or develop microorganisms for specific uses. Biotechnology is most commonly thought of to include the development of human medical therapies and processes using recombinant DNA, cell fusion, other genetic techniques and bioremediation.

**BLA:** See "Biologics License Application (BLA)."

**Blastocyst:** A fertilized embryo, aged four to 11 days, which consists of multiplying cells both outside and inside a cavity. It is the blastocyst that embeds itself in the uterine wall and ultimately develops into a fetus. Blastocysts are utilized outside the womb during the process of stem cell cultivation.

**BPO:** See "Business Process Outsourcing (BPO)."

**Branding:** A marketing strategy that places a focus on the brand name of a product, service or firm in order to increase the brand's market share, increase sales, establish credibility, improve satisfaction, raise the profile of the firm and increase profits.

**BRIC:** An acronym representing Brazil, Russia, India and China. The economies of these four countries are seen as some of the fastest growing in the world. A 2003 report by investment bank Goldman Sachs is often credited for popularizing the term; the report suggested that by 2050, BRIC economies will likely outshine those countries which are currently the richest in the world.

**B-to-B, or B2B:** See "Business-to-Business."

**B-to-C, or B2C:** See "Business-to-Consumer."

**Business Process Outsourcing (BPO):** The process of hiring another company to handle business activities. BPO is one of the fastest-growing segments in the offshoring sector. Services include human resources management, billing and purchasing and call centers, as well as many types of customer service or marketing activities, depending on the industry involved. Also, see "Knowledge Process Outsourcing (KPO)."

**Business-to-Business:** An organization focused on selling products, services or data to commercial customers rather than individual consumers. Also known as B2B.

**Business-to-Consumer:** An organization focused on selling products, services or data to individual consumers rather than commercial customers. Also known as B2C.

**CANDA:** See "Computer-Assisted New Drug Application (CANDA)."

**Captive Offshoring:** Used to describe a company-owned offshore operation. For example, Microsoft owns and operates significant captive offshore research and development centers in China and elsewhere that are offshore from Microsoft's U.S. home base. Also see "Offshoring."

**Carcinogen:** A substance capable of causing cancer. A suspected carcinogen is a substance that may cause cancer in humans or animals but for which the evidence is not conclusive.

**Case Report Form (CRF):** In clinical trials, a standard document used by clinicians to record and report subject data pertinent to the study protocol.

**Case Report Tabulation (CRT):** In clinical trials, a tabular listing of all data collected on study case report forms.

**CBER:** See "Center for Biologics Evaluation and Research (CBER)."

**CDC:** See "Centers for Disease Control and Prevention (CDC)."

**Center for Biologics Evaluation and Research (CBER):** The branch of the FDA responsible for the regulation of biological products, including blood, vaccines, therapeutics and related drugs and devices, to ensure purity, potency, safety, availability and effectiveness. www.fda.gov/cber

**Center for Devices and Radiological Health (CDRH):** The branch of the FDA responsible for the regulation of medical devices. www.fda.gov/cdrh

**Center for Drug Evaluation and Research (CDER):** The branch of the FDA responsible for the regulation of drug products. www.fda.gov/cder

**Centers for Disease Control and Prevention (CDC):** The federal agency charged with protecting the public health of the nation by providing leadership and direction in the prevention and control of diseases and other preventable conditions and responding to public health emergencies. Headquartered in Atlanta, it was established as an operating health agency within the U.S. Public Health Service on July 1, 1973. See www.cdc.gov.

**Centers for Medicare and Medicaid Services (CMS):** A federal agency responsible for administering Medicare and monitoring the states' operations of Medicaid. See www.cms.hhs.gov.

**Chromosome:** A structure in the nucleus of a cell that contains genes. Chromosomes are found in pairs.

**CIS:** See "Commonwealth of Independent States (CIS)."

**Class I Device:** An FDA classification of medical devices for which general controls are sufficient to ensure safety and efficacy.

**Class II Device:** An FDA classification of medical devices for which performance standards and special controls are sufficient to ensure safety and efficacy.

**Class III Device:** An FDA classification of medical devices for which pre-market approval is required to ensure safety and efficacy, unless the device is substantially equivalent to a currently marketed device. See "510 K."

**Climate Change (Greenhouse Effect):** A theory that assumes an increasing mean global surface temperature of the Earth caused by gases in the atmosphere (including carbon dioxide, methane, nitrous oxide, ozone and chlorofluorocarbons). The greenhouse effect allows solar radiation to penetrate the Earth's atmosphere but absorbs the infrared radiation returning to space.

**Clinical Research Associate (CRA):** An individual responsible for monitoring clinical trial data to ensure compliance with study protocol and FDA GCP regulations.

**Clinical Trial:** See "Phase I Clinical Trials," along with definitions for Phase II, Phase III and Phase IV.

**Clone:** A group of identical genes, cells or organisms derived from one ancestor. A clone is an identical copy. "Dolly" the sheep is a famous case of a clone of an animal. Also see "Cloning (Reproductive)" and "Cloning (Therapeutic)."

**Cloning (Reproductive):** A method of reproducing an exact copy of an animal or, potentially, an exact copy of a human being. A scientist removes the nucleus from a donor's unfertilized egg, inserts a nucleus from the animal to be copied and then stimulates the nucleus to begin dividing to form an embryo. In the case of a mammal, such as a human, the embryo would then be implanted in the uterus of a host female. Also see "Cloning (Therapeutic)."

**Cloning (Therapeutic):** A method of reproducing exact copies of cells needed for research or for the development of replacement tissue or organs. A scientist removes the nucleus from a donor's unfertilized egg, inserts a nucleus from the animal whose cells are to be copied and then stimulates the nucleus to begin dividing to form an embryo. However, the embryo is never allowed to grow to any significant stage of development. Instead, it is allowed to grow for a few hours or days, and stem cells are then removed from it for use in regenerating tissue. Also see "Cloning (Reproductive)."

**CMS:** See "Centers for Medicare and Medicaid Services (CMS)."

**Combinatorial Chemistry:** A chemistry in which molecules are found that control a pre-determined protein. This advanced computer technique enables scientists to use automatic fluid handlers to mix chemicals under specific test conditions at extremely high speed. Combinatorial chemistry can generate thousands of chemical compound variations in a few hours. Previously, traditional chemistry methods could have required several weeks to do the same work.

**Committee on Proprietary Medicinal Products (CPMP):** A committee, composed of two people from each EU Member State (see "European Union (EU)"), that is responsible for the scientific evaluation and assessment of marketing applications for medicinal products in the EU. The CPMP is the major body involved in the harmonization of pharmaceutical regulations within the EU and receives administrative support from the European

Medicines Evaluation Agency. See "European Medicines Evaluation Agency (EMEA)."

**Commonwealth of Independent States (CIS):** An organization consisting of 11 former members of the Soviet Union: Russia, Ukraine, Armenia, Moldova, Georgia, Belarus, Kazakhstan, Uzbekistan, Azerbaijan, Kyrgyzstan and Tajikistan. It was created in 1991. Turkmenistan recently left the Commonwealth as a permanent member, but remained as an associate member. The Commonwealth seeks to coordinate a variety of economic and social policies, including taxation, pricing, customs and economic regulation, as well as to promote the free movement of capital, goods, services and labor.

**Complementary-DNA (cDNA):** A sequence acquired by copying a messenger RNA (mRNA) molecule back into DNA. In contrast to the original DNA, mRNA codes for an expressed protein without non-coding DNA sequences (introns). Therefore, a cDNA probe can also be used to find the specific gene in a complex DNA sample from another organism with different non-coding sequences. Also called Copy DNA.

**Computational Biology:** The development and application of data-analytical and theoretical methods, mathematical modeling and computational simulation techniques to the study of biological, behavioral, and social systems. Computational biology uses mathematical and computational approaches to address theoretical and experimental questions in biology.

**Computer-Assisted New Drug Application (CANDA):** An electronic submission of a new drug application (NDA) to the FDA.

**Consumer Price Index (CPI):** A measure of the average change in consumer prices over time in a fixed market basket of goods and services, such as food, clothing and housing. The CPI is calculated by the U.S. Federal Government and is considered to be one measure of inflation.

**Contract Manufacturer:** A company that manufactures products that will be sold under the brand names of its client companies. For example, a large number of consumer electronics, such as laptop computers, are manufactured by contract manufacturers for leading brand-name computer companies such as Dell. Many other types of products are made under contract manufacturing, from apparel to pharmaceuticals. Also see "Original Equipment Manufacturer (OEM)" and "Original Design Manufacturer (ODM)."

**Contract Research Organization (CRO):** An independent organization that contracts with a client to conduct part of the work on a study or research project. For example, drug and medical device makers frequently outsource clinical trials and other research work to CROs.

**Coordinator:** In clinical trials, the person at an investigative site who handles the administrative responsibilities of the trial, acts as a liaison between the investigative site and the sponsor, and reviews data and records during a monitoring visit.

**CPMP:** See "Committee on Proprietary Medicinal Products (CPMP)."

**CRA:** See "Clinical Research Associate (CRA)."

**CRF:** See "Case Report Form (CRF)."

**CT:** See "Computed Tomography (CT)."

**Data and Safety Monitoring Board (DSMB):** See "Data Monitoring Board (DMB)."

**Data Monitoring Board (DMB):** A committee that monitors the progress of a clinical trial and carefully observes the safety data.

**Deoxyribonucleic Acid (DNA):** The carrier of the genetic information that cells need to replicate and to produce proteins.

**Development:** The phase of research and development (R&D) in which researchers attempt to create new products from the results of discoveries and applications created during basic and applied research.

**Device:** In medical products, an instrument, apparatus, implement, machine, contrivance, implant, in vitro reagent or other similar or related article, including any component, part or accessory, that 1) is recognized in the official National Formulary or United States Pharmacopoeia or any supplement to them, 2) is intended for use in the diagnosis of disease or other conditions, or in the cure, mitigation, treatment or prevention of disease, in man or animals or 3) is intended to affect the structure of the body of man or animals and does not achieve any of its principal intended purposes through chemical action within or on the body of man or animals and is not dependent upon being metabolized for the achievement of any of its principal intended purposes.

**Diagnostic Radioisotope Facility:** A medical facility in which radioactive isotopes (radiopharmaceuticals) are used as tracers or indicators to detect an abnormal condition or disease in the body.

**Distributor:** An individual or business involved in marketing, warehousing and/or shipping of products manufactured by others to a specific group of end users. Distributors do not sell to the general public. In order to develop a competitive advantage, distributors often focus on serving one industry or one set of niche clients. For example, within the medical industry, there are major distributors that focus on providing pharmaceuticals, surgical supplies or dental supplies to clinics and hospitals.

**DMB:** See "Data Monitoring Board (DMB)."

**DNA:** See "Deoxyribonucleic Acid (DNA)."

**DNA Chip:** A revolutionary tool used to identify mutations in genes like BRCA1 and BRCA2. The chip, which consists of a small glass plate encased in plastic, is manufactured using a process similar to the one used to make computer microchips. On the surface, each chip contains synthetic single-stranded DNA sequences identical to a normal gene.

**Drug Utilization Review:** A quantitative assessment of patient drug use and physicians' patterns of prescribing drugs in an effort to determine the usefulness of drug therapy.

**DSMB:** See "Data and Safety Monitoring Board (DSMB)."

**Ecology:** The study of relationships among all living organisms and the environment, especially the totality or pattern of interactions; a view that includes all plant and animal species and their unique contributions to a particular habitat.

**Efficacy:** A drug or medical product's ability to effectively produce beneficial results within a patient.

**EFGCP:** See European Forum for Good Clinical Practices (EFGCP)."

**ELA:** See "Establishment License Application (ELA)."

**Electronic Data Interchange (EDI):** An accepted standard format for the exchange of data between various companies' networks. EDI allows for the transfer of e-mail as well as orders, invoices and other files from one company to another.

**Electroporation:** A health care technology that uses short pulses of electric current (DC) to create openings (pores) in the membranes of cancerous cells, thus leading to death of the cells. It has potential as a treatment for prostate cancer. In the laboratory, it is a means of introducing foreign proteins or DNA into living cells.

**EMEA:** The region comprised of Europe, the Middle East and Africa.

**Endpoint:** A clinical or laboratory measurement used to assess safety, efficacy or other trial objectives of a test article in a clinical trial.

**Enterprise Resource Planning (ERP):** An integrated information system that helps manage all aspects of a business, including accounting, ordering and human resources, typically across all locations of a major corporation or organization. ERP is considered to be a critical tool for management of large organizations. Suppliers of ERP tools include SAP and Oracle.

**Enzyme:** A protein that acts as a catalyst, affecting the chemical reactions in cells.

**Epigenetics:** A relatively new branch of biology focused on gene "silencers." Scientists involved in epigenetics are studying the function within a gene that regulates whether a gene is operating a full capacity or is toned down to a lower level. The level of operation of a given gene may lead to a higher risk of disease, such as certain types of cancer, within a patient. Epigenetics may be very effective at correcting abnormal gene expressions that cause cancer.

**ERP:** See "Enterprise Resource Planning (ERP)."

**EST:** See "Expressed Sequence Tags (EST)."

**Establishment License Application (ELA):** Required for the approval of a biologic (see "Biologic"). It permits a specific facility to manufacture a biological product for commercial purposes. Compare to "Product License Agreement (PLA)."

**ESWL:** See "Extracorporeal Shock Wave Lithotripter (ESWL)."

**Ethanol:** A clear, colorless, flammable, oxygenated hydrocarbon, also called ethyl alcohol. In the U.S., it is used as a gasoline octane enhancer and oxygenate in a 10% blend called E10. Ethanol can be used in higher concentrations (such as an 85% blend called E85) in vehicles designed for its use. It is typically produced chemically from ethylene or biologically from fermentation of various sugars from carbohydrates found in agricultural crops and cellulose residues from crops or wood. Grain ethanol production is typically based on corn or sugarcane. Cellulosic ethanol production is based on agricultural waste, such as wheat stalks, that has been treated with enzymes to break the waste down into component sugars.

**Etiology:** The study of the causes or origins of diseases.

**EU:** See "European Union (EU)."

**EU Competence:** The jurisdiction in which the EU can take legal action.

**European Community (EC):** See "European Union (EU)."

**European Forum for Good Clinical Practices (EFGCP):** The organization dedicated to finding common ground in Europe on the implementation of Good Clinical Practices. See "Good Clinical Practices (GCP)." www.efgcp.org

**European Medicines Evaluation Agency (EMEA):** The European agency responsible for supervising and coordinating applications for marketing medicinal products in the European Union (see "European Union (EU)" and "Committee on Proprietary Medicinal Products (CPMP)"). The EMEA is headquartered in the U.K. www.eudraportal.eudra.org

**European Union (EU):** A consolidation of European countries (member states) functioning as one body to facilitate trade. Previously known as the European Community (EC), the EU expanded to include much of Eastern Europe in 2004, raising the total number of member states to 25. In 2002, the EU launched a unified currency, the Euro. See europa.eu.int.

**Expressed Sequence Tags (EST):** Small pieces of DNA sequence (usually 200 to 500 nucleotides long) that are generated by sequencing either one or both ends of an expressed gene. The idea is to sequence bits of DNA that represent genes expressed in certain cells, tissues or organs from different organisms and use these tags to fish a gene out of a portion of chromosomal DNA by matching base pairs. See "Gene Expression."

**Fast Track Development:** An enhanced process at the FDA for rapid approval of drugs that treat certain life-threatening or extremely serious conditions. Some Fast Track drugs come to market in very short periods of time. The Fast Track designation is intended for drugs that address an unmet medical need, but is independent of Priority Review and Accelerated Approval. The benefits of Fast Track include scheduled meetings to seek FDA input into development as well as the option of submitting a New Drug Application in sections rather than all components at once.

**FD&C Act:** See "Federal Food Drug and Cosmetic Act (FD&C Act)."

**FDA:** See "Food and Drug Administration (FDA)."

**Federal Food, Drug and Cosmetic Act (FD&C Act):** A set of laws passed by the U.S. Congress, which controls, among other things, residues in food and feed.

**Follow-on Biologics:** A term used to describe generic versions of drugs that have been created using biotechnology. Because biotech drugs ("biologics") are made from living cells, a generic version of a drug probably won't be biochemically identical to the original branded version of the drug. Consequently, they are described as "follow-on" biologics to set them apart. Since these drugs won't be exactly the same as the originals, there are concerns that they may not be as safe or effective unless they go through clinical trials for proof of

quality. In Europe, these drugs are referred to as "biosimilars." See "Biologics."

**Food and Drug Administration (FDA):** The U.S. government agency responsible for the enforcement of the Federal Food, Drug and Cosmetic Act, ensuring industry compliance with laws regulating products in commerce. The FDA's mission is to protect the public from harm and encourage technological advances that hold the promise of benefiting society. www.fda.gov

**Formulary:** A preferred list of drug products that typically limits the number of drugs available within a therapeutic class for purposes of drug purchasing, dispensing and/or reimbursement. A government body, third-party insurer or health plan, or an institution may compile a formulary. Some institutions or health plans develop closed (i.e. restricted) formularies where only those drug products listed can be dispensed in that institution or reimbursed by the health plan. Other formularies may have no restrictions (open formulary) or may have certain restrictions such as higher patient cost-sharing requirements for off-formulary drugs.

**Functional Foods:** Food products that contain nutrients, such as vitamins, associated with certain health benefits. The nutrients may occur naturally, or the foods may have been enhanced with them.

**Functional Genomics:** The process of attempting to convert the molecular information represented by DNA into an understanding of gene functions and effects. To address gene function and expression specifically, the recovery and identification of mutant and over-expressed phenotypes can be employed. Functional genomics also entails research on the protein function (proteomics) or, even more broadly, the whole metabolism (metabolomics) of an organism.

**Functional Imaging:** The uses of PET scan, MRI and other advanced imaging technology to see how an area of the body is functioning and responding. For example, brain activity can be viewed, and the reaction of cancer tumors to therapies can be judged using functional imaging.

**Functional Proteomics:** The study of the function of all the proteins encoded by an organism's entire genome.

**GCP:** See "Good Clinical Practices (GCP)."

**GDP:** See "Gross Domestic Product (GDP)."

**Gene:** A working subunit of DNA; the carrier of inheritable traits.

**Gene Chip:** See "DNA Chip."

**Gene Expression:** The term used to describe the transcription of the information contained within the DNA (the repository of genetic information) into messenger RNA (mRNA) molecules that are then translated into the proteins that perform most of the critical functions of cells. Scientists study the kinds and amounts of mRNA produced by a cell to learn which genes are expressed, which in turn provides insights into how the cell responds to its changing needs.

**Gene Knock-Out:** The inhibition of gene expression through various scientific methods.

**Gene Therapy:** Treatment based on the alteration or replacement of existing genes. Genetic therapy involves splicing desired genes insolated from one patient into a second patient's cells in order to compensate for that patient's inherited genetic defect, or to enable that patient's body to better fight a specific disease.

**Genetic Code:** The sequence of nucleotides, determining the sequence of amino acids in protein synthesis.

**Genetically-Modified (GM):** See "Genetically Modified (GM) Foods" and "Genetically Modified Organism (GMO)."

**Genetically Modified (GM) Foods:** Food crops that are bioengineered to resist herbicides, diseases or insects; have higher nutritional value than non-engineered plants; produce a higher yield per acre; and/or last longer on the shelf. Additional traits may include resistance to temperature and moisture extremes. Agricultural animals also may be genetically modified organisms.

**Genetically Modified Organism (GMO):** An organism that has undergone genome modification by the insertion of a foreign gene. The genetic material of a GMO is not found through mating or natural recombination.

**Genetics:** The study of the process of heredity.

**Genome:** The genetic material (composed of DNA) in the chromosomes of a living organism.

**Genomics:** The study of genes, their role in diseases and our ability to manipulate them.

**Genotype:** The genetic constitution of an organism.

**Global Warming:** An increase in the near-surface temperature of the Earth. Global warming has occurred in the distant past as the result of natural influences, but the term is most often used to refer to a theory that warming occurs as a result of increased use of hydrocarbon fuels by man. See "Climate Change (Greenhouse Effect)."

**Globalization:** The increased mobility of goods, services, labor, technology and capital throughout the world. Although globalization is not a new development, its pace has increased with the advent of new technologies, especially in the areas of telecommunications, finance and shipping.

**GLP:** See "Good Laboratory Practices (GLP)."

**GM:** See "Genetically-Modified (GM)."

**GMO:** See "Genetically Modified Organism (GMO)."

**GMP:** See "Good Manufacturing Practices (GMP)."

**Good Clinical Practices (GCP):** FDA regulations and guidelines that define the responsibilities of the key figures involved in a clinical trial, including the sponsor, the investigator, the monitor and the Institutional Review Board. See "Institutional Review Board (IRB)."

**Good Laboratory Practices (GLP):** A collection of regulations and guidelines to be used in laboratories where research is conducted on drugs, biologics or devices that are intended for submission to the FDA.

**Good Manufacturing Practices (GMP):** A collection of regulations and guidelines to be used in manufacturing drugs, biologics and medical devices.

**Gross Domestic Product (GDP):** The total value of a nation's output, income and expenditures produced with a nation's physical borders.

**Gross National Product (GNP):** A country's total output of goods and services from all forms of economic activity measured at market prices for one calendar year. It differs from Gross Domestic Product (GDP) in that GNP includes income from investments made in foreign nations.

**HESC:** See "Human Embryonic Stem Cell (HESC)."

**HHS:** See "U.S. Department of Health and Human Services (HHS)."

**High-Throughput Screening (HTP):** Makes use of techniques that allow for a fast and simple test on the presence or absence of a desirable structure, such as a specific DNA sequence. HTP screening often uses DNA chips or microarrays and automated data processing for large-scale screening, for instance, to identify new targets for drug development.

**Human Embryonic Stem Cell (HESC):** See "Stem Cells."

**IEEE:** The Institute of Electrical and Electronic Engineers. The IEEE sets global technical standards and acts as an authority in technical areas including computer engineering, biomedical technology, telecommunications, electric power, aerospace and consumer electronics, among others. www.ieee.org.

**Imaging:** In medicine, the viewing of the body's organs through external, high-tech means. This reduces the need for broad exploratory surgery. These advances, along with new types of surgical instruments, have made minimally invasive surgery possible. Imaging includes MRI (magnetic resonance imaging), CT (computed tomography or CAT scan), MEG (magnetoencephalography), improved x-ray technology, mammography, ultrasound and angiography.

**Imaging Contrast Agent:** A molecule or molecular complex that increases the intensity of the signal detected by an imaging technique, including MRI and ultrasound. An MRI contrast agent, for example, might contain gadolinium attached to a targeting antibody. The antibody would bind to a specific target, a metastatic melanoma cell for example, while the gadolinium would increase the magnetic signal detected by the MRI scanner.

**Immunoassay:** An immunological assay. Types include agglutination, complement-fixation,

precipitation, immunodiffusion and electrophoretic assays. Each type of assay utilizes either a particular type of antibody or a specific support medium (such as a gel) to determine the amount of antigen present.

**In Vitro:** Laboratory experiments conducted in the test tube, or otherwise, without using live animals and/or humans.

**In Vivo:** Laboratory experiments conducted with live animals and/or humans.

**IND:** See "Investigational New Drug Application (IND)."

**Indication:** Refers to a specific disease, illness or condition for which a drug is approved as a treatment. Typically, a new drug is first approved for one indication. Then, an application to the FDA is later made for approval of additional indications.

**Induced Pluripotent State Cell (IPSC):** A human stem cell produced without human cloning or the use of human embryos or eggs. Adult cells are drawn from a skin biopsy and treated with four reprogramming factors, rendering cells that can produce all human cell types and grow indefinitely.

**Industrial Biotechnology:** The application of biotechnology to serve industrial needs. This is a rapidly growing field on a global basis. The current focus on industrial biotechnology is primarily on enzymes and other substances for renewable energy such as biofuels; chemicals such as pharmaceuticals, food additives, solvents and colorants; and bioplastics. Industrial biotech attempts to create synergies between biochemistry, genetics and microbiology in order to develop exciting new substances.

**Informatics:** See "Bioinformatics."

**Informed Consent:** Must be obtained in writing from people who agree to be clinical trial subjects prior to their enrollment in the study. The document must explain the risks associated with the study and treatment and describe alternative therapy available to the patient. A copy of the document must also be provided to the patient.

**Initial Public Offering (IPO):** A company's first effort to sell its stock to investors (the public). Investors in an up-trending market eagerly seek stocks offered in many IPOs because the stocks of newly public companies that seem to have great promise may appreciate very rapidly in price, reaping great profits for those who were able to get the stock at the first offering. In the United States, IPOs are regulated by the SEC (U.S. Securities Exchange Commission) and by the state-level regulatory agencies of the states in which the IPO shares are offered.

**Insertion Mutants:** Mutants of genes that are obtained by inserting DNA, for instance through mobile DNA sequences. In plant research, the capacity of the bacterium Agrobacterium to introduce DNA into the plant genome is employed to induce mutants. In both cases, mutations lead to lacking or changing gene functions that are revealed by aberrant phenotypes. Insertion mutant isolation, and subsequent identification and analysis are employed in model plants such as Arabiopsis and in crop plants such as maize and rice.

**Institutional Review Board (IRB):** A group of individuals usually found in medical institutions that is responsible for reviewing protocols for ethical consideration (to ensure the rights of the patients). An IRB also evaluates the benefit-to-risk ratio of a new drug to see that the risk is acceptable for patient exposure. Responsibilities of an IRB are defined in FDA regulations.

**Interactomics (Interactome):** The study of the interactions between genes, RNA, proteins and metabolites within the cell.

**Interferon:** A type of biological response modifier (a substance that can improve the body's natural response to disease).

**Investigational New Device Exemption (IDE):** A document that must be filed with the FDA prior to initiating clinical trials of medical devices considered to pose a significant risk to human subjects.

**Investigational New Drug Application (IND):** A document that must be filed with the FDA prior to initiating clinical trials of drugs or biologics.

**Investigator:** In clinical trials, a clinician who agrees to supervise the use of an investigational drug, device or biologic in humans. Responsibilities of the investigator, as defined in FDA regulations, include administering the drug, observing and testing the patient, collecting data and monitoring the care and welfare of the patient.

**Iontophoresis:** The transfer of ions of medicine through the skin using a local electric current.

**IRB:** See "Institutional Review Board (IRB)."

**ISO 9000, 9001, 9002, 9003:** Standards set by the International Organization for Standardization. ISO 9000, 9001, 9002 and 9003 are the highest quality certifications awarded to organizations that meet exacting standards in their operating practices and procedures.

**IT:** See "Information Technology (IT)."

**IT-Enabled Services (ITES):** The portion of the Information Technology industry focused on providing business services, such as call centers, insurance claims processing and medical records transcription, by utilizing the power of IT, especially the Internet. Most ITES functions are considered to be back-office procedures. Also, see "Business Process Outsourcing (BPO)."

**ITES:** See "IT-Enabled Services (ITES)."

**Kinase (Protein Kinase):** Enzymes that influence certain basic functions within cells, such as cell division. Kinases catalyze the transfer of phosphates from ATP to proteins, thus causing changes in protein function. Defective kinases can lead to diseases such as cancer.

**Knowledge Process Outsourcing (KPO):** The use of outsourced and/or offshore workers to perform business tasks that require judgment and analysis. Examples include such professional tasks as patent research, legal research, architecture, design, engineering, market research, scientific research, accounting and tax return preparation. Also, see "Business Process Outsourcing (BPO)."

**LAC:** An acronym for Latin America and the Caribbean.

**LDCs:** See "Least Developed Countries (LDCs)."

**Least Developed Countries (LDCs):** Nations determined by the U.N. Economic and Social Council to be the poorest and weakest members of the international community. There are currently 50

LDCs, of which 34 are in Africa, 15 are in Asia Pacific and the remaining one (Haiti) is in Latin America. The top 10 on the LDC list, in descending order from top to 10th, are Afghanistan, Angola, Bangladesh, Benin, Bhutan, Burkina Faso, Burundi, Cambodia, Cape Verde and the Central African Republic. Sixteen of the LDCs are also Landlocked Least Developed Countries (LLDCs) which present them with additional difficulties often due to the high cost of transporting trade goods. Eleven of the LDCs are Small Island Developing States (SIDS), which are often at risk of extreme weather phenomenon (hurricanes, typhoons, Tsunami); have fragile ecosystems; are often dependent on foreign energy sources; can have high disease rates for HIV/AIDS and malaria; and can have poor market access and trade terms.

**Lifestyle Drug:** Lifestyle drugs target a variety of medical conditions ranging from the painful to the inconvenient, including obesity, impotence, memory loss and depression, rather than illness or disease. Drug companies continue to develop lifestyle treatments for hair loss and skin wrinkles in an effort to capture their share of the huge anti-aging market aimed at the baby-boomer generation.

**Ligand:** Any atom or molecule attached to a central atom in a complex compound.

**Liposome:** A micro or nanoscale lipid or phospholipid layer enclosing a liquid core used for transport for particular molecules or biological structures or as a model for membranes.

**M3 (Measurement):** Cubic meters.

**Marketing:** Includes all planning and management activities and expenses associated with the promotion of a product or service. Marketing can encompass advertising, customer surveys, public relations and many other disciplines. Marketing is distinct from selling, which is the process of sell-through to the end user.

**Mass Spectrometry:** Usage of analytical devices that can determine the mass (or molecular weight) of proteins and nucleic acids, the sequence of protein molecules, the chemical organization of almost all substances and the identification of gram-negative and gram-positive microorganisms.

**Medical Device:** See "Device."

**MEMS:** See "Micro Electro Mechanical Systems (MEMS)."

**Metabolomics:** The study of low-molecular-weight materials produced during genomic expression within a cell. Such studies can lead to a better understanding of how changes within genes and proteins affect the function of cells.

**Metagenomics:** An advanced form of genomics that increases the understanding of complex microbial systems.

**Micro Electro Mechanical Systems (MEMS):** Micron scale structures that transduce signals between electronic and mechanical forms.

**Microarray:** A DNA analysis tool consisting of a microscopic ordered array that enables parallel analysis of complex biochemical samples; used to analyze how large numbers of genes interact with each other; used for genotyping, mapping, sequencing, sequence detection; usually constructed by applying biomolecules onto a slide or chip and then scanning with microscope or other imaging equipment.

**Microfluidics:** Refers to the manipulation of microscopic amounts of fluid, generally for analysis in microarrays using high throughput screening. The volume of the liquid involved is on the nanolitre scale.

**Molecular Imaging:** An emerging field in which advanced biology on the molecular level is combined with noninvasive imaging to determine the presence of certain proteins and other important genetic material.

**Monoclonal Antibodies (mAb, Human Monoclonal Antibody):** Antibodies that have been cloned from a single antibody and massed produced as a therapy or diagnostic test. An example is an antibody specific to a certain protein found in cancer cells.

**Nanocantilever:** The simplest micro-electro-mechanical system (MEMS) that can be easily machined and mass-produced via the same techniques used to make computer chips. The ability to detect extremely small displacements make nanocantilever beams an ideal device for detecting extremely small forces, stresses and masses.

Nanocantilevers coated with antibodies, for example, will bend from the mass added when substrate binds to its antibody, providing a detector capable of sensing the presence of single molecules of clinical importance.

**Nanoparticle:** A nanoscale spherical or capsule-shaped structure. Most, though not all, nanoparticles are hollow, which provides a central reservoir that can be filled with anticancer drugs, detection agents, or chemicals, known as reporters, that can signal if a drug is having a therapeutic effect. The surface of a nanoparticle can also be adorned with various targeting agents, such as antibodies, drugs, imaging agents, and reporters. Most nanoparticles are constructed to be small enough to pass through blood capillaries and enter cells.

**Nanopharmaceuticals:** Nanoscale particles that modulate drug transport in drug uptake and delivery applications.

**Nanoshell:** A nanoparticle composed of a metallic shell surrounding a semiconductor. When nanoshells reach a target cancer cell, they can be irradiated with near-infrared light or excited with a magnetic field, either of which will cause the nanoshell to become hot, killing the cancer cell.

**Nanotechnology:** The science of designing, building or utilizing unique structures that are smaller than 100 nanometers (a nanometer is one billionth of a meter). This involves microscopic structures that are no larger than the width of some cell membranes.

**Nanowires:** A nanometer-scale wire made of metal atoms, silicon, or other materials that conduct electricity. Nanowires are built atom by atom on a solid surface, often as part of a microfluidic device. They can be coated with molecules such as antibodies that will bind to proteins and other substances of interest to researchers and clinicians. By the very nature of their nanoscale size, nanowires are incredibly sensitive to such binding events and respond by altering the electrical current flowing through them, and thus can form the basis of ultra sensitive molecular detectors.

**National Drug Code (NDC):** An identifying drug number maintained by the FDA.

**National Institutes of Health (NIH):** A branch of the U.S. Public Health Service that conducts biomedical research. www.nih.gov

**NCE:** See "New Chemical Entity (NCE)."

**NDA:** See "New Drug Application (NDA)."

**New Chemical Entity (NCE):** See "New Molecular Entity (NME)."

**New Drug Application (NDA):** An application requesting FDA approval, after completion of the all-important Phase III Clinical Trials, to market a new drug for human use in the U.S. The drug may contain chemical compounds that were previously approved by the FDA as distinct molecular entities suitable for use in drug trials. See "New Molecular Entity (NME)."

**New Molecular Entity (NME):** Defined by the FDA as a medication containing chemical compound that has never before been approved for marketing in any form in the U.S. An NME is sometimes referred to as a New Chemical Entity (NCE). Also, see "New Drug Application (NDA)."

**NIH:** See "National Institutes of Health (NIH)."

**NME:** See "New Molecular Entity (NME)."

**Nonclinical Studies:** In vitro (laboratory) or in vivo (animal) pharmacology, toxicology and pharmacokinetic studies that support the testing of a product in humans. Usually at least two species are evaluated prior to Phase I clinical trials. Nonclinical studies continue throughout all phases of research to evaluate long-term safety issues.

**Nucleic Acid:** A large molecule composed of nucleotides. Nucleic acids include RNA, DNA and antisense oligonucleotides.

**Nutraceutical:** Nutrient + pharmaceutical – a food or part of a food that has been isolated and sold in a medicinal form and claims to offer benefits such as the treatment or prevention of disease.

**Nutraceuticals:** Food products and dietary supplements that may have certain health benefits. Nutraceuticals may offer specific vitamins or minerals. Also see "Functional Foods."

**Nutrigenomics:** The study of how food interacts with genes.

**ODM:** See "Original Design Manufacturer (ODM)."

**OECD:** See "Organisation for Economic Co-operation and Development (OECD)."

**OEM:** See "Original Equipment Manufacturer (OEM)."

**Offshoring:** The rapidly growing tendency among U.S., Japanese and Western European firms to send knowledge-based and manufacturing work overseas. The intent is to take advantage of lower wages and operating costs in such nations as China, India, Hungary and Russia. The choice of a nation for offshore work may be influenced by such factors as language and education of the local workforce, transportation systems or natural resources. For example, China and India are graduating high numbers of skilled engineers and scientists from their universities. Also, some nations are noted for large numbers of workers skilled in the English language, such as the Philippines and India. Also see "Captive Offshoring" and "Outsourcing."

**Oncogene:** A unit of DNA that normally directs cell growth, but which can also promote or allow the uncontrolled growth of cancer.

**Oncology:** The diagnosis, study and treatment of cancer.

**Organisation for Economic Co-operation and Development (OECD):** A group of 30 countries that are strongly committed to the market economy and democracy. Some of the OECD members include Japan, the U.S., Spain, Germany, Australia, Korea, the U.K., Canada and Mexico. Although not members, Chile, Estonia, Israel, Russia and Slovenia are invited to member talks; and Brazil, China, India, Indonesia and South Africa have enhanced engagement policies with the OECD. The Organisation provides statistics, as well as social and economic data; and researches social changes, including patterns in evolving fiscal policy, agriculture, technology, trade, the environment and other areas. It publishes over 250 titles annually; publishes a corporate magazine, the OECD Observer; has radio and TV studios; and has centers in Tokyo, Washington, D.C., Berlin and Mexico City that distributed the Organisation's work and organizes events.

**Original Design Manufacturer (ODM):** A contract manufacturer that offers complete, end-to-end design, engineering and manufacturing services. ODMs design and build products, such as consumer electronics, that client companies can then brand and sell as their own. For example, a large percentage of laptop computers, cell phones and PDAs are made by ODMs. Also see "Original Equipment Manufacturer (OEM)" and "Contract Manufacturer."

**Original Equipment Manufacturer (OEM):** A company that manufactures a product or component for sale to a customer that will integrate the component into a final product or assembly. The OEM's customer will distribute the end product or resell it to an end user. For example, a personal computer made under a brand name by a given company may contain various components, such as hard drives, graphics cards or speakers, manufactured by several different OEM "vendors," but the firm doing the final assembly/manufacturing process is the final manufacturer. Also see "Original Design Manufacturer (ODM)" and "Contract Manufacturer."

**Orphan Drug:** A drug or biologic designated by the FDA as providing therapeutic benefit for a rare disease affecting less than 200,000 people in the U.S. Companies that market orphan drugs are granted a period of market exclusivity in return for the limited commercial potential of the drug.

**OTC:** See "Over-the-Counter Drugs (OTC)."

**Outsourcing:** The hiring of an outside company to perform a task otherwise performed internally by the company, generally with the goal of lowering costs and/or streamlining work flow. Outsourcing contracts are generally several years in length. Companies that hire outsourced services providers often prefer to focus on their core strengths while sending more routine tasks outside for others to perform. Typical outsourced services include the running of human resources departments, telephone call centers and computer departments. When outsourcing is performed overseas, it may be referred to as offshoring. Also see "Offshoring."

**Over-the-Counter Drugs (OTC):** FDA-regulated products that do not require a physician's prescription. Some examples are aspirin, sunscreen, nasal spray and sunglasses.

**Patent:** A property right granted by the U.S. government to an inventor to exclude others from making, using, offering for sale, or selling the invention throughout the U.S. or importing the invention into the U.S. for a limited time in exchange for public disclosure of the invention when the patent is granted.

**Pathogen:** Any microorganism (e.g., fungus, virus, bacteria or parasite) that causes a disease.

**PCR:** See "Polymerase Chain Reaction (PCR)."

**Peer Review:** The process used by the scientific community, whereby review of a paper, project or report is obtained through comments of independent colleagues in the same field.

**Pharmacodynamics (PD):** The study of reactions between drugs and living systems. It can be thought of as the study of what a drug does to the body.

**Pharmacoeconomics:** The study of the costs and benefits associated with various drug treatments.

**Pharmacogenetics:** The investigation of the different reactions of human beings to drugs and the underlying genetic predispositions. The differences in reaction are mainly caused by mutations in certain enzymes responsible for drug metabolization. As a result, the degradation of the active substance can lead to harmful by-products, or the drug might have no effect at all.

**Pharmacogenomics:** The use of the knowledge of DNA sequences for the development of new drugs.

**Pharmacokinetics (PK):** The study of the processes of bodily absorption, distribution, metabolism and excretion of compounds and medicines. It can be thought of as the study of what the body does to a drug. See "Absorption, Distribution, Metabolism and Excretion (ADME)."

**Pharmacology:** The science of drugs, their characteristics and their interactions with living organisms.

**Pharmacy Benefit Manager (PBM):** An organization that provides administrative services in processing and analyzing prescription claims for pharmacy benefit and coverage programs. Many PBMs also operate mail order pharmacies or have arrangements to include prescription availability through mail order pharmacies.

**Phase I Clinical Trials:** Studies in this phase include initial introduction of an investigational drug into humans. These studies are closely monitored and are usually conducted in healthy volunteers. Phase I trials are conducted after the completion of extensive nonclinical or pre-clinical trials not involving humans. Phase I studies include the determination of clinical pharmacology, bioavailability, drug interactions and side effects associated with increasing doses of the drug.

**Phase II Clinical Trials:** Include randomized, masked, controlled clinical studies conducted to evaluate the effectiveness of a drug for a particular indication(s). During Phase II trials, the minimum effective dose and dosing intervals should be determined.

**Phase III Clinical Trials:** Consist of controlled and uncontrolled trials that are performed after preliminary evidence of effectiveness of a drug has been established. They are conducted to document the safety and efficacy of the drug, as well as to determine adequate directions (labeling) for use by the physician. A specific patient population needs to be clearly identified from the results of these studies. Trials during Phase III are conducted using a large number of patients to determine the frequency of adverse events and to obtain data regarding intolerance.

**Phase IV Clinical Trials:** Conducted after approval of a drug has been obtained to gather data supporting new or revised labeling, marketing or advertising claims.

**Phenomics:** The study of how an organism's structure responds to such things as toxins or drugs.

**Phenotype:** Observable characteristics of an organism produced by the organism's genotype interacting with the environment.

**Phylogenetic Systematics:** The field of biology that deals with identifying and understanding the evolutionary relationships among the many different kinds of life on earth, both living (extant) and dead (extinct).

**Pivotal Studies:** In clinical trials, a Phase III trial that is designed specifically to support approval of a product. These studies are well-controlled (usually by placebo) and are generally designed with input from the FDA so that they will provide data that is adequate to support approval of the product. Two pivotal studies are required for drug product approval, but usually only one study is required for biologics.

**PLA:** See "Product License Agreement (PLA)."

**Plant Patent:** A plant patent may be granted by the U.S. Patent and Trademark Office to anyone who invents or discovers and asexually reproduces any distinct and new variety of plant.

**Platform or Technology Platform Companies:** Firms hoping to profit by providing information systems, software, databases and related support to biopharmaceutical companies.

**PMA:** See "Pre-Market Approval (PMA)."

**PMCs:** Postmarketing study commitments. PMCs are clinical studies that are not required for FDA initial approval of a drug, but the FDA nonetheless feels these studies will provide important data on a newly marketed drug. Consequently, the drug firm makes a commitment for continuing studies.

**Polymerase Chain Reaction (PCR):** In molecular biology, PCR is a technique used to reproduce or amplify small, selected sections of DNA or RNA for analysis. It enables researchers to create multiple copies of a given sequence.

**Positional Cloning:** The identification and cloning of a specific gene, with chromosomal location as the only source of information about the gene.

**Post-Marketing Surveillance:** The FDA's ongoing safety monitoring of marketed drugs.

**PPP:** See "Purchasing Power Parity (PPP) or Point-to-Point Protocol (PPP)."

**Preclinical Studies:** See "Nonclinical Studies."

**Pre-Market Approval (PMA):** Required for the approval of a new medical device or a device that is to be used for life-sustaining or life-supporting purposes, is implanted in the human body or presents potential risk of illness or injury.

**Priority Reviews:** The FDA places some drug applications that appear to promise "significant improvements" over existing drugs for priority approval, with a goal of returning approval within six months.

**Product License Agreement (PLA):** See "Biologics License Application (BLA)."

**Protein:** A large, complex molecule made up of amino acids.

**Proteome:** The genetic material (composed of amino acids) in the chromosomes of a living organism.

**Proteomics:** The study of gene expression at the protein level, by the identification and characterization of proteins present in a biological sample.

**Public Health Service (PHS):** May stand for the Public Health Service Act, a law passed by the U.S. Congress in 1944. PHS also may stand for the Public Health Service itself, a U.S. government agency established by an act of Congress in July 1798, originally authorizing hospitals for the care of American merchant seamen. Today, the Public Health Service sets national health policy; conducts medical and biomedical research; sponsors programs for disease control and mental health; and enforces laws to assure the safety and efficacy of drugs, foods, cosmetics and medical devices. The FDA ( Food and Drug Administration) is part of the Public Health Service, as are the Centers for Disease Control and Prevention (CDC).

**Purchasing Power Parity (PPP):** Currency conversion rates that attempt to reflect the actual purchasing power of a currency in its home market, as opposed to examining price levels and comparing an exchange rate. PPPs are always given in the national currency units per U.S. dollar.

**Qdots:** See "Quantum Dots (Qdots)."

**QOL:** See "Quality of Life (QOL)."

**Quality of Life (QOL):** In medicine, an endpoint of therapeutic assessment used to adjust measures of effectiveness for clinical decision-making. Typically, QOL endpoints measure the improvement of a patient's day-to-day living as a result of specific therapy.

**Quantum Dots (Qdots):** Nanometer sized semiconductor particles, made of cadmium selenide (CdSe), cadmium sulfide (CdS) or cadmium telluride (CdTe) with an inert polymer coating. The semiconductor material used for the core is chosen based upon the emission wavelength range being targeted: CdS for UV-blue, CdSe for the bulk of the visible spectrum, CdTe for the far red and near-infrared, with the particle's size determining the exact color of a given quantum dot. The polymer coating safeguards cells from cadmium toxicity but also affords the opportunity to attach any variety targeting molecules, including monoclonal antibodies directed to tumor-specific biomarkers. Because of their small size, quantum dots can function as cell- and even molecule-specific markers that will not interfere with the normal workings of a cell. In addition, the availability of quantum dots of different colors provides a powerful tool for following the actions of multiple cells and molecules simultaneously.

**R&D:** Research and development. Also see "Applied Research" and "Basic Research."

**Radioisotope:** An object that has varying properties that allows it to penetrate other objects at different rates. For example, a sheet of paper can stop an alpha particle, a beta particle can penetrate tissues in the body and a gamma ray can penetrate concrete. The varying penetration capabilities allow radioisotopes to be used in different ways. (Also called radioactive isotope or radionuclide.)

**Receptor:** Proteins in or on a cell that selectively bind a specific substance called a ligand. See "Ligand."

**Recombination:** The natural process of breaking and rejoining DNA strands to produce new combinations of genes.

**Reporter Gene:** A gene that is inserted into DNA by researchers in order to indicate when a linked gene is successfully expressed or when signal transduction has taken place in a cell.

**Return on Investment (ROI):** A measure of a company's profitability, expressed in percentage as net profit (after taxes) divided by total dollar investment.

**RNA (Ribonucleic Acid):** A macromolecule found in the nucleus and cytoplasm of cells; vital in protein synthesis.

**RNAi (RNA interference):** A biological occurrence where double-stranded RNA is used to silence genes.

**Safe Medical Devices Act (SMDA):** An act that amends the Food, Drug and Cosmetic Act to impose additional regulations on medical devices. The act became law in 1990.

**Semiconductor:** A generic term for a device that controls electrical signals. It specifically refers to a material (such as silicon, germanium or gallium arsenide) that can be altered either to conduct electrical current or to block its passage. Carbon nanotubes may eventually be used as semiconductors. Semiconductors are partly responsible for the miniaturization of modern electronic devices, as they are vital components in computer memory and processor chips. The manufacture of semiconductors is carried out by small firms, and by industry giants such as Intel and Advanced Micro Devices.

**Single Nucleotide Polymorphisms (SNPs):** Stable mutations consisting of a change at a single base in a DNA molecule. SNPs can be detected by HTP analyses, such as gene chips, and they are then mapped by DNA sequencing. They are the most common type of genetic variation.

**SMDA:** See "Safe Medical Devices Act (SMDA)."

**SNP:** See "Single-Nucleotide Polymorphisms (SNPs)."

**Sponsor:** The individual or company that assumes responsibility for the investigation of a new drug, including compliance with the FD&C Act and regulations. The sponsor may be an individual, partnership, corporation or governmental agency and may be a manufacturer, scientific institution or investigator regularly and lawfully engaged in the investigation of new drugs. The sponsor assumes most of the legal and financial responsibility of the clinical trial.

**Standard Drugs:** NCEs (New Chemical Entities) that the FDA feels offer few advantages over existing drugs that are therefore given lower status for review.

**Stem Cells:** Cells found in human bone marrow, the blood stream and the umbilical cord that can be replicated indefinitely and can turn into any type of mature blood cell, including platelets, white blood cells or red blood cells. Also referred to as pluripotent cells.

**Study Coordinator:** See "Coordinator."

**Subsidiary, Wholly-Owned:** A company that is wholly controlled by another company through stock ownership.

**Supply Chain:** The complete set of suppliers of goods and services required for a company to operate its business. For example, a manufacturer's supply chain may include providers of raw materials, components, custom-made parts and packaging materials.

**Synthetic Biology:** Synthetic biology can be defined as the design and construction of new entities, including enzymes and cells, or the reformatting of existing biological systems. This science capitalizes on previous advances in molecular biology and systems biology, by applying a focus on the design and construction of unique core components that can be integrated into larger systems in order to solve specific problems.

**Systems Biology:** The use of combinations of advanced computer hardware, software and database technologies to take a systemic approach to biological research. Advanced technologies will enable scientists to view genetic predisposition by integrating information about entire biological systems, from DNA to proteins to cells to tissues.

**T Cell (T-Cell):** A white blood cell that carries out immune system responses. The T cell originates in the bone marrow and matures in the thymus.

**Targets:** The proteins involved in a specific disease. Drug compounds concentrate on specific targets in order to have the greatest positive effect and cut down on the incidence of side effects.

**Taste Masking:** The creation of a barrier between a drug molecule and taste receptors so the drug is easier to take. It masks bitter or unpleasant tastes.

**TESS:** See "Adverse Event (AE)."

**Toxicogenomics:** The study of the relationship between responses to toxic substances and the resulting genetic changes.

**Trial Coordinator:** See "Coordinator."

**U.S. Department of Health and Human Services (HHS):** This agency has more than 300 major programs related to human health and welfare, the largest of which is Medicare. See www.hhs.gov

**Utility Patent:** A utility patent may be granted by the U.S. Patent and Trademark Office to anyone who invents or discovers any new, useful, and non-obvious process, machine, article of manufacture, or composition of matter, or any new and useful improvement thereof.

**Validation of Data:** The procedure carried out to ensure that the data contained in a final clinical trial report match the original observations.

**Value Added Tax (VAT):** A tax that imposes a levy on businesses at every stage of manufacturing based on the value it adds to a product. Each business in the supply chain pays its own VAT and is subsequently repaid by the next link down the chain; hence, a VAT is ultimately paid by the consumer, being the last link in the supply chain, making it comparable to a sales tax. Generally, VAT only applies to goods bought for consumption within a given country; export goods are exempt from VAT, and purchasers from other countries taking goods back home may apply for a VAT refund.

**World Health Organization (WHO):** A United Nations agency that assists governments in strengthening health services, furnishing technical assistance and aid in emergencies, working on the prevention and control of epidemics and promoting cooperation among different countries to improve nutrition, housing, sanitation, recreation and other aspects of environmental hygiene. Any country that is a member of the United Nations may become a member of the WHO by accepting its constitution. The WHO currently has 191 member states.

**World Trade Organization (WTO):** One of the only globally active international organizations dealing with the trade rules between nations. Its goal is to assist the free flow of trade goods, ensuring a smooth, predictable supply of goods to help raise the quality of life of member citizens. Members form consensus decisions that are then ratified by their respective parliaments. The WTO's conflict resolution process generally emphasizes interpreting existing commitments and agreements, and discovers how to ensure trade policies to conform to those agreements, with the ultimate aim of avoiding military or political conflict.

**WTO:** See "World Trade Organization (WTO)."

**Xenotransplantation:** The science of transplanting organs such as kidneys, hearts or livers into humans from other mammals, such as pigs or other agricultural animals grown with specific traits for this purpose.

**Zinc Fingers:** Naturally occurring proteins that bind to DNA to produce desired genetic effects. For example, zinc fingers used in a plant can alter its yield, taste or resistance to drought or insects. They afford very precise changes to DNA which translates into better control when modifying plants and quicker development times compared to typical genetic modification.

**Zoonosis:** An animal disease that can be transferred to man.

# INTRODUCTION

PLUNKETT'S BIOTECH & GENETICS INDUSTRY ALMANAC, the eighth edition of our guide to the biotech and genetics field, is designed to be used as a general source for researchers of all types.

The data and areas of interest covered are intentionally broad, ranging from the ethical questions facing biotechnology, to emerging technology, to an in-depth look at the major for-profit firms (which we call "THE BIOTECH 400") within the many industry sectors that make up the biotechnology and genetics arena.

This reference book is designed to be a general source for researchers. It is especially intended to assist with market research, strategic planning, employment searches, contact or prospect list creation (be sure to see the export capabilities of the accompanying CD-ROM that is available to book and eBook buyers) and financial research, and as a data resource for executives and students of all types.

PLUNKETT'S BIOTECH & GENETICS INDUSTRY ALMANAC takes a rounded approach for the general reader. This book presents a complete overview of the entire biotechnology and genetics system (see "How To Use This Book"). For example, you will find trends in the biopharmaceuticals market, along with easy-to-use charts and tables on all facets of biotechnology in general: from the sales and profits of the major drug companies to the amounts of time required in the various stages of drug approval.

THE BIOTECH 400 is our unique grouping of the biggest, most successful corporations in all segments of the global biotechnology and genetics industry. Tens of thousands of pieces of information, gathered from a wide variety of sources, have been researched and are presented in a unique form that can be easily understood. This section includes thorough indexes to THE BIOTECH 400, by geography, industry, sales, brand names, subsidiary names and many other topics. (See Chapter 4.)

Especially helpful is the way in which PLUNKETT'S BIOTECH & GENETICS INDUSTRY ALMANAC enables readers who have no business or scientific background to readily compare the financial records and growth plans of large biotech companies and major industry groups. You'll see the mid-term financial record of each firm, along with the impact of earnings, sales and strategic plans on each company's potential to fuel growth, to serve new markets and to provide investment and employment opportunities.

No other source provides this book's easy-to-understand comparisons of growth, expenditures, technologies, corporations and many other items of great importance to people of all types who may be studying this, one of the most exciting industries in the world today.

By scanning the data groups and the unique indexes, you can find the best information to fit your personal research needs. The major growth companies in biotechnology and genetics are profiled and then ranked using several different groups of specific criteria. Which firms are the biggest employers? Which companies earn the most profits? These things and much more are easy to find.

In addition to individual company profiles, an overview of biotechnology markets and trends is provided. This book's job is to help you sort through easy-to-understand summaries of today's trends in a quick and effective manner.

Whatever your purpose for researching the biotechnology and genetics field, you'll find this book to be a valuable guide. Nonetheless, as is true with all resources, this volume has limitations that the reader should be aware of:

- Financial data and other corporate information can change quickly. A book of this type can be no more current than the data that was available as of the time of editing. Consequently, the financial picture, management and ownership of the firm(s) you are studying may have changed since the date of this book. For example, this almanac includes the most up-to-date sales figures and profits available to the editors as of mid 2008. That means that we have typically used corporate financial data as of late-2008.

- Corporate mergers, acquisitions and downsizing are occurring at a very rapid rate. Such events may have created significant change, subsequent to the publishing of this book, within a company you are studying.

- Some of the companies in THE BIOTECH 400 are so large in scope and in variety of business endeavors conducted within a parent organization, that we have been unable to completely list all subsidiaries, affiliations, divisions and activities within a firm's corporate structure.

- This volume is intended to be a general guide to a vast industry. That means that researchers should look to this book for an overview and, when

conducting in-depth research, should contact the specific corporations or industry associations in question for the very latest changes and data. Where possible, we have listed contact names, toll-free telephone numbers and Internet site addresses for the companies, government agencies and industry associations involved so that the reader may get further details without unnecessary delay.

- Tables of industry data and statistics used in this book include the latest numbers available at the time of printing, generally through late-2008. In a few cases, the only complete data available was for earlier years.

- We have used exhaustive efforts to locate and fairly present accurate and complete data. However, when using this book or any other source for business and industry information, the reader should use caution and diligence by conducting further research where it seems appropriate. We wish you success in your endeavors, and we trust that your experience with this book will be both satisfactory and productive.

Jack W. Plunkett
Houston, Texas
August 2009

# HOW TO USE THIS BOOK

The two primary sections of this book are devoted first to the biotechnology and genetics industry as a whole and then to the "Individual Data Listings" for THE BIOTECH 400. If time permits, you should begin your research in the front chapters of this book. Also, you will find lengthy indexes in Chapter 4 and in the back of the book.

## THE BIOTECH AND GENETICS INDUSTRY

**Glossary:** A short list of biotech and genetics industry terms.

**Chapter 1: Major Trends Affecting the Biotech & Genetics Industry.** This chapter presents an encapsulated view of the major trends and technologies that are creating rapid changes in the biotechnology and genetics industry today.

**Chapter 2: Biotech & Genetics Industry Statistics.** This chapter presents in-depth statistics on biotechnology markets, spending, research, pharmaceuticals and more.

**Chapter 3: Important Biotech & Genetics Industry Contacts – Addresses, Telephone Numbers and Internet Sites.** This chapter covers contacts for important government agencies, biotech organizations and trade groups. Included are numerous important Internet sites.

## THE BIOTECH 400

**Chapter 4: THE BIOTECH 400: Who They Are and How They Were Chosen.** The companies compared in this book were carefully selected from the biotech and genetics industry, largely in the United States. 77 of the firms are based outside the U.S. For a complete description, see THE BIOTECH 400 indexes in this chapter.
  **Individual Data Listings:**
  Look at one of the companies in THE BIOTECH 400's Individual Data Listings. You'll find the following information fields:
  **Company Name:**
  The company profiles are in alphabetical order by company name. If you don't find the company you are seeking, it may be a subsidiary or division of one of the firms covered in this book. Try looking it up in the Index by Subsidiaries, Brand Names and Selected Affiliations in the back of the book.
  **Ranks:**
  Industry Group Code: An NAIC code used to group companies within like segments. (See Chapter 4 for a list of codes.)

Ranks Within This Company's Industry Group:
Ranks, within this firm's segment only, for annual
sales and annual profits, with 1 being the highest
rank.

**Business Activities:**
A grid arranged into six major industry categories
and several sub-categories. A "Y" indicates that the
firm operates within the sub-category. A complete
Index by Industry is included in the beginning of
Chapter 4.

**Types of Business:**
A listing of the primary types of business
specialties conducted by the firm.

**Brands/Divisions/Affiliations:**
Major brand names, operating divisions or
subsidiaries of the firm, as well as major corporate
affiliations—such as another firm that owns a
significant portion of the company's stock. A
complete Index by Subsidiaries, Brand Names and
Selected Affiliations is in the back of the book.

**Contacts:**
The names and titles up to 27 top officers of the
company are listed, including human resources
contacts.

**Address:**
The firm's full headquarters address, the
headquarters telephone, plus toll-free and fax
numbers where available. Also provided is the World
Wide Web site address.

**Financials:**
Annual Sales (2008 or the latest fiscal year
available to the editors, plus up to four previous
years): These are stated in thousands of dollars (add
three zeros if you want the full number). This figure
represents consolidated worldwide sales from all
operations. 2008 figures may be estimates or may be
for only part of the year—partial year figures are
appropriately footnoted.

Annual Profits (2008 or the latest fiscal year
available to the editors, plus up to four previous
years): These are stated in thousands of dollars (add
three zeros if you want the full number). This figure
represents consolidated, after-tax net profit from all
operations. 2008 figures may be estimates or may be
for only part of the year—partial year figures are
appropriately footnoted.

Stock Ticker, International Exchange, Parent
Company: When available, the unique stock market
symbol used to identify this firm's common stock for
trading and tracking purposes is indicated. Where
appropriate, this field may contain "private" or
"subsidiary" rather than a ticker symbol. If the firm is
a publicly-held company headquartered outside of the

U.S., its international ticker and exchange are given.
If the firm is a subsidiary, its parent company is
listed.

Total Number of Employees: The approximate
total number of employees, worldwide, as of the end
of 2008 (or the latest data available to the editors).

**Apparent Salaries/Benefits:**
(The following descriptions generally apply to
U.S. employers only.)
A "Y" in appropriate fields indicates "Yes."
Due to wide variations in the manner in which
corporations report benefits to the U.S. Government's
regulatory bodies, not all plans will have been
uncovered or correctly evaluated during our effort to
research this data. Also, the availability to employees
of such plans will vary according to the qualifications
that employees must meet to become eligible. For
example, some benefit plans may be available only to
salaried workers—others only to employees who
work more than 1,000 hours yearly. Benefits that are
available to employees of the main or parent
company may not be available to employees of the
subsidiaries. In addition, employers frequently alter
the nature and terms of plans offered.

NOTE: Generally, employees covered by wealth-
building benefit plans do not *fully* own ("vest in")
funds contributed on their behalf by the employer
until as many as five years of service with that
employer have passed. All pension plans are
voluntary—that is, employers are not obligated to
offer pensions.

Pension Plan: The firm offers a pension plan to
qualified employees. In this case, in order for a "Y"
to appear, the editors believe that the employer offers
a defined benefit or cash balance pension plan (see
discussions below).The type and generosity of these
plans vary widely from firm to firm. Caution: Some
employers refer to plans as "pension" or "retirement"
plans when they are actually 401(k) savings plans
that require a contribution by the employee.

- Defined Benefit Pension Plans: Pension plans
  that do not require a contribution from the
  employee are infrequently offered. However, a
  few companies, particularly larger employers in
  high-profit-margin industries, offer defined
  benefit pension plans where the employee is
  guaranteed to receive a set pension benefit upon
  retirement. The amount of the benefit is
  determined by the years of service with the
  company and the employee's salary during the
  later years of employment. The longer a person
  works for the employer, the higher the retirement
  benefit. These defined benefit plans are funded

entirely by the employer. The benefits, up to a reasonable limit, are guaranteed by the Federal Government's Pension Benefit Guaranty Corporation. These plans are not portable—if you leave the company, you cannot transfer your benefits into a different plan. Instead, upon retirement you will receive the benefits that vested during your service with the company. If your employer offers a pension plan, it must give you a summary plan description within 90 days of the date you join the plan. You can also request a summary annual report of the plan, and once every 12 months you may request an individual benefit statement accounting of your interest in the plan.

- Defined Contribution Plans: These are quite different. They do not guarantee a certain amount of pension benefit. Instead, they set out circumstances under which the employer will make a contribution to a plan on your behalf. The most common example is the 401(k) savings plan. Pension benefits are not guaranteed under these plans.

- Cash Balance Pension Plans: These plans were recently invented. These are hybrid plans—part defined benefit and part defined contribution. Many employers have converted their older defined benefit plans into cash balance plans. The employer makes deposits (or credits a given amount of money) on the employee's behalf, usually based on a percentage of pay. Employee accounts grow based on a predetermined interest benchmark, such as the interest rate on Treasury Bonds. There are some advantages to these plans, particularly for younger workers: a) The benefits, up to a reasonable limit, are guaranteed by the Pension Benefit Guaranty Corporation. b) Benefits are portable—they can be moved to another plan when the employee changes companies. c) Younger workers and those who spend a shorter number of years with an employer may receive higher benefits than they would under a traditional defined benefit plan.

ESOP Stock Plan (Employees' Stock Ownership Plan): This type of plan is in wide use. Typically, the plan borrows money from a bank and uses those funds to purchase a large block of the corporation's stock. The corporation makes contributions to the plan over a period of time, and the stock purchase loan is eventually paid off. The value of the plan grows significantly as long as the market price of the stock holds up. Qualified employees are allocated a share of the plan based on their length of service and their level of salary. Under federal regulations, participants in ESOPs are allowed to diversify their account holdings in set percentages that rise as the employee ages and gains years of service with the company. In this manner, not all of the employee's assets are tied up in the employer's stock.

Savings Plan, 401(k): Under this type of plan, employees make a tax-deferred deposit into an account. In the best plans, the company makes annual matching donations to the employees' accounts, typically in some proportion to deposits made by the employees themselves. A good plan will match one-half of employee deposits of up to 6% of wages. For example, an employee earning $30,000 yearly might deposit $1,800 (6%) into the plan. The company will match one-half of the employee's deposit, or $900. The plan grows on a tax-deferred basis, similar to an IRA. A very generous plan will match 100% of employee deposits. However, some plans do not call for the employer to make a matching deposit at all. Other plans call for a matching contribution to be made at the discretion of the firm's board of directors. Actual terms of these plans vary widely from firm to firm. Generally, these savings plans allow employees to deposit as much as 15% of salary into the plan on a tax-deferred basis. However, the portion that the company uses to calculate its matching deposit is generally limited to a maximum of 6%. Employees should take care to diversify the holdings in their 401(k) accounts, and most people should seek professional guidance or investment management for their accounts.

Stock Purchase Plan: Qualified employees may purchase the company's common stock at a price below its market value under a specific plan. Typically, the employee is limited to investing a small percentage of wages in this plan. The discount may range from 5 to 15%. Some of these plans allow for deposits to be made through regular monthly payroll deductions. However, new accounting rules for corporations, along with other factors, are leading many companies to curtail these plans—dropping the discount allowed, cutting the maximum yearly stock purchase or otherwise making the plans less generous or appealing.

Profit Sharing: Qualified employees are awarded an annual amount equal to some portion of a company's profits. In a very generous plan, the pool of money awarded to employees would be 15% of profits. Typically, this money is deposited into a long-term retirement account. Caution: Some employers refer to plans as "profit sharing" when they are actually 401(k) savings plans. True profit sharing plans are rarely offered.

Highest Executive Salary: The highest executive salary paid, typically a 2008 amount (or the latest year available to the editors) and typically paid to the Chief Executive Officer.

Highest Executive Bonus: The apparent bonus, if any, paid to the above person.

Second Highest Executive Salary: The next-highest executive salary paid, typically a 2008 amount (or the latest year available to the editors) and typically paid to the President or Chief Operating Officer.

Second Highest Executive Bonus: The apparent bonus, if any, paid to the above person.

**Other Thoughts:**

Apparent Women Officers or Directors: It is difficult to obtain this information on an exact basis, and employers generally do not disclose the data in a public way. However, we have indicated what our best efforts reveal to be the apparent number of women who either are in the posts of corporate officers or sit on the board of directors. There is a wide variance from company to company.

Hot Spot for Advancement for Women/Minorities: A "Y" in appropriate fields indicates "Yes." These are firms that appear either to have posted a substantial number of women and/or minorities to high posts or that appear to have a good record of going out of their way to recruit, train, promote and retain women or minorities. (See the Index of Hot Spots For Women and Minorities in the back of the book.) This information may change frequently and can be difficult to obtain and verify. Consequently, the reader should use caution and conduct further investigation where appropriate.

**Growth Plans/ Special Features:**

Listed here are observations regarding the firm's strategy, hiring plans, plans for growth and product development, along with general information regarding a company's business and prospects.

**Locations:**

A "Y" in the appropriate field indicates "Yes."

Primary locations outside of the headquarters, categorized by regions of the United States and by international locations. A complete index by locations is also in the front of this chapter.

# Chapter 1

# MAJOR TRENDS AFFECTING THE BIOTECH & GENETICS INDUSTRY

**Trends Affecting the Biotechnology & Genetics Industry:**

1) The State of the Biotechnology Industry Today
2) A Short History of Biotechnology
3) Ethanol Production Soared, But a Market Glut May Slow Expansion
4) New Money Pours into Biotech Firms/Mergers and Acquisitions Surge Ahead
5) Major Drug Companies Bet on Partnerships With Smaller Biotech Research Firms
6) From India to Singapore to Australia, Nations Compete Fiercely in Biotech Development
7) Medical Trials Conducted Abroad Spark Concerns
8) Gene Therapies and Patients' Genetic Profiles Promise a Personalized Approach to Medicine
9) Breakthrough Drugs for Cancer Treatment—Many More Will Follow
10) Few New Blockbusters: Major Drug Patents Expire While Generic Sales Growth Continues
11) Biotech and Orphan Drugs Pick Up the Slack as Blockbuster Mainstream Drugs Age
12) Biogenerics (Follow-on Biologics) are in Limbo in the U.S.
13) Breakthrough Drug Delivery Systems Evolve
14) Stem Cells—Multiple Sources Stem from New Technologies
15) U.S. Government Reverses Ban on Funding for New Stem Cell Research

16) Stem Cells—Therapeutic Cloning Techniques Advance
17) Stem Cells—A New Era of Regenerative Medicine Takes Shape
18) Nanotechnology Converges with Biotech
19) Agricultural Biotechnology Scores Breakthroughs but Causes Controversy/Selective Breeding Offers a Compromise
20) Focus on Vaccines
21) Ethical Issues Abound
22) Technology Discussion—Genomics
23) Technology Discussion—Proteomics
24) Technology Discussion—Microarrays
25) Technology Discussion—DNA Chips
26) Technology Discussion—SNPs ("Snips")
27) Technology Discussion—Combinatorial Chemistry
28) Technology Discussion—Synthetic Biology
29) Technology Discussion—Recombinant DNA
30) Technology Discussion—Polymerase Chain Reaction (PCR)

**1) The State of the Biotechnology Industry Today**

Analysts at global accounting firm Ernst & Young estimate global biotech industry revenues for publicly-held companies at $89.7 billion for 2008, a 12% increase from the previous year. The firm also estimates that revenues of publicly-held biotech companies in the U.S. rose 8% to $66.1 billion.

Genetically-engineered drugs, or "biotech" drugs, represent part of the total prescription drugs market. The U.S. Centers for Medicare & Medicaid Services (CMS) forecast calls for prescription drug purchases in the U.S. to total about $244.8 billion during 2009, representing about $800 per capita. That projected total is up from $235.4 billion in 2008 and a mere $40 billion in 1990. Estimates of the size of this market vary by source. Analysts at the widely respected firm IMS Health place U.S. domestic prescription drug sales at $291 billion for 2008. Drug industry association PhRMA estimates that 72% of its members' sales are generic drugs. In 2016, U.S. government estimates show that American drug purchases may reach $384.9 billion, thanks to a rapidly aging U.S. population and the continued introduction of expensive new drugs.

Analysts at the noted investment bank Burrill & Company estimate that global research and development (R&D) expenses at all pharmaceutical companies were $65.2 billion in 2008.

IMS Health estimates Canadian retail drug sales for the twelve months ended May 2008 at $16.7 billion, Japan at $63.2 billion (including sales at hospitals), Brazil at $11.6 billion and Mexico at $8.7 billion. In Europe's top five markets, IMS ranked Germany first, during the same twelve-month period, at $34.4 billion, followed by France at $30.8 billion, the UK at $17.1 billion, Italy at $16.9 billion and Spain at $14.8 billion. (See www.imshealth.com.)

As of early 2009, there were more than 2,900 medicines in development in the U.S. Advanced generations of drugs developed through biotechnology continue to enter the marketplace. The results may be very promising for patients, as a technology tipping point of medical care is approaching, where drugs that target specific genes and proteins may eventually become widespread. However, it continues to become more difficult and more expensive to introduce a new drug in the U.S. For example, during 2008, the FDA (Food and Drug Administration) approved only three new biologics (new biotechnology-based drugs, based on living organisms, that have never been marketed in the U.S. in any form) along with 21 new molecular entities or "NMEs" (medications containing chemical compounds that have never before been approved for marketing in the U.S.). This is up from the 18 approved in 2007, but down from the 22 in 2006 (there were 20 approved during 2005 and 36 in 2004).

These NMEs and biologics are novel new active substances that are categorized differently from "NDAs" or New Drug Applications. NDAs may seek approval for drugs based on combinations of substances that have been approved in the past. During 2008, 98 NDAs were approved by the FDA, (compared to 94 NDAs in 2007, 93 in 2006 and 79 in 2005).

---

**New Drug Application Categories**

Applications for drug approval by the FDA fall under the following categories:

**BLA (Biologics License Application)**: An application for approval of a drug synthesized from living organisms. That is, drugs created using biotechnology. Such drugs are sometimes referred to as biopharmaceuticals.

**NME (New Molecular Entity)**: A new chemical compound that has never before been approved for marketing in any form in the U.S.

**NDA (New Drug Application):** An application requesting FDA approval, after completion of the all-important Phase III Clinical Trials, to market a new drug for human use in the U.S. The drug may contain active ingredients that were previously approved by the FDA.

**Follow-On Biologics**: A term used to describe generic versions of drugs that have been created using biotechnology. Because biotech drugs ("biologics") are made from living cells, a generic version of a drug probably won't be biochemically identical to the original branded version of the drug. Consequently, they are described as "follow-on" drugs to set them apart. Since these drugs won't be exactly the same as the originals, there are concerns that they may not be as safe or effective unless they go through clinical trials for proof of quality. In Europe, these drugs are referred to as "biosimilars."

**Priority Reviews**: The FDA places some drug applications that appear to promise "significant improvements" over existing drugs for priority approval, with a goal of returning approval within six months.

**Accelerated Approval**: A process at the FDA for reducing the clinical trial length for drugs designed for certain serious or life-threatening diseases.

**Fast Track Development**: An enhanced process for rapid approval of drugs that treat certain life-threatening or extremely serious conditions. Fast Track is independent of Priority Review and Accelerated Approval.

---

Developing a new drug is an excruciatingly slow and expensive endeavor. According to PhRMA, the average time required for the drug discovery, development and clinical trials process is 16 years. The good news is that the median FDA approval time for a "priority" NME is down to about six months since 2003, compared to 16.3 months in 2002. As for for "standard" NMEs, approval time was 12.9 months in 2007, down from 23.0 in 2005.

The promising era of personalized medicine is slowly, slowly moving closer to fruition. Dozens of exciting new drugs for the treatment of dire diseases such as cancer, AIDS, Parkinson's and Alzheimer's are either on the market or are very close to regulatory approval.

Stem cell research is moving ahead briskly on a global basis, despite the previously negative effect of restrictive research funding rules of the U.S. Federal Government. This should change over the near term due to President Obama's relaxing of the limits on federal funding of stem cell research that were established by his predecessor, President Bush. In 2009, the National Institutes of Health set new guidelines for funding that will dramatically expand the number of stem cell lines that qualify for research funds from a previous 21 to as many as 700. However, research into certain extremely controversial stem cells, such as those developed via cloning, will not be funded with federal dollars.

Stem cell breakthroughs are occurring rapidly. There is truly exciting evidence of the potential for stem cells to treat many problems, from cardiovascular disease to neurological disorders. Menlo Park, California-based Geron Corporation, for example, has published the results of its experiments that show that when certain cells (called OPCs) derived from stem cells were injected in rats that had spinal cord injuries, the rats quickly recovered. According to the company, "Rats transplanted seven days after injury showed improved walking ability compared to animals receiving a control transplant. The OPC-treated animals showed improved hind limb-forelimb coordination and weight bearing capacity, increased stride length, and better paw placement compared to control-treated animals."

Despite exponential advances in biopharmaceutical knowledge and technology, biotech companies enduring the task of getting new drugs to market continue to face long timeframes, daunting costs and immense risks. Although the number of NDAs submitted to the FDA has grown dramatically since 1996, the number of new drugs receiving final approval remains relatively small. On average, of every 1,000 experimental drug compounds in some form of pre-clinical testing, only

one actually makes it to clinical trials. Then, only one in five of those drugs makes it to market. Of the drugs that get to market, only one in three recover their costs. Meanwhile, the patent expiration clock is ticking—soon enough, manufacturers of generic alternatives steal market share from the firms that invested all that time and money in the development of the original drug.

---

**Global Factors Boosting Biotech Today:**
1) A rapid aging of the population base of nations in the E.U., as well as Japan and the U.S., including the 78 million Baby Boomers in America who are entering senior years in rising numbers and needing a growing level of health care
2) A renewed, global focus on developing effective vaccines
3) Vast research investments by major pharmaceuticals firms
4) A growing global dependence on genetically-engineered agricultural seeds ("Agribio")
5) Aggressive investment in biotechnology research in Singapore, China and India, often with government sponsorship—for example, Singapore's massive Biopolis project
6) A government-subsidized emphasis on renewable energy such as bioethanol and other biofuels as substitutes for petroleum
7) Promising research into synthetic biology
8) Continuing computer-related progress in biotech areas such as gene sequencing
*Source: Plunkett Research, Ltd.*

---

*Internet Research Tip:*
You can review current and historical drug approval reports at the following page at the FDA.
www.fda.gov/Drugs/InformationOnDrugs/default.htm

The FDA regulates biologic products for use in humans. It is a source of a broad variety of data on drugs, including vaccines, blood products, counterfeit drugs, exports, drug shortages, recalls and drug safety.
www.fda.gov/BiologicsBloodVaccines/default.htm

---

According to a study released in 2001 by the Tufts Center for the Study of Drug Development, the cost of developing a new drug and getting it to market averaged $802 million, up from about $500 million in 1996. (Averaged into these figures are the costs of developing and testing drugs that never reach the market.) Expanding on the study to include post-approval research (Phase IV clinical studies), Tufts increased the number to $897 million. Tufts estimated the average cost to develop a new biologic at $1.2 billion in 2006. Even more pessimistic is research released in 2003 by Bain & Co., a consulting firm, which states that the cost is more on the order of $1.7 billion, including such factors as marketing and advertising expenses.

The typical time elapsed from the synthesis of a new chemical compound to its introduction to the market remains 12 to 20 years. Considering that the patent for a new compound only lasts about 20 years, a limited amount of time is available to reclaim the considerable investments in research, development, trials and marketing. As a result of these costs and the lengthy time-to-market, young biotech companies encounter a harsh financial reality: commercial profits take years and years to emerge from promising beginnings in the laboratory.

However, advances in systems biology (the use of a combination of state-of-the-art technologies, such as molecular diagnostics, advanced computers and extremely deep, efficient genetic databases) may eventually lead to more efficient, faster drug development at reduced costs. Much of this advance will stem from the use of technology to efficiently target the genetic causes of, and develop novel cures for, niche diseases.

The FDA is attempting to help the drug industry bring the most vital drugs to market in shorter time with three programs: Fast Track, Priority Review and Accelerated Approval. The benefits of Fast Track include scheduled meetings to seek FDA input into development as well as the option of submitting a New Drug Application in sections rather than all components at once. The Fast Track designation is intended for drugs that address an unmet medical need, but is independent of Priority Review and Accelerated Approval. Priority drugs are those considered by the FDA to offer improvements over existing drugs or to offer high therapeutic value. The priority program, along with increased budget and staffing at the FDA, are having a positive effect on total approval times for new drugs.

For example, the FDA quickly approved Novartis' new drug Gleevec (a revolutionary and highly effective treatment for patients suffering from chronic myeloid leukemia). After priority review and Fast Track status, it required only two and one-half months in the approval process (compared to a more typical six months). This rapid approval, which enabled the drug to promptly begin saving lives, was possible because of two factors aside from the FDA's

cooperation. One, Novartis mounted a targeted approach to this niche disease. Its research determined that a specific genetic malfunction causes the disease, and its drug specifically blocks the protein that causes the genetic malfunction. Two, thanks to its use of advanced genetic research techniques, Novartis was so convinced of the effectiveness of this drug that it invested heavily and quickly in its development.

---

**Key Food & Drug Administration (FDA) terms relating to human clinical trials:**

**Phase I**—Small-scale human trials to determine safety. Typically include 20 to 60 patients and are six months to one year in length.
**Phase II**—Preliminary trials on a drug's safety/efficacy. Typically include 100 to 500 patients and are one and a half to two years in length.
**Phase III**—Large-scale controlled trials for efficacy/safety; also the last stage before a request for approval for commercial distribution is made to the FDA. Typically include 1,000 to 7,500 patients and are three to five years in length.
**Phase IV**—Follow-up trials after a drug is released to the public.

---

Generally, Fast Track approval is reserved for life-threatening diseases such as rare forms of cancer, but new policies are setting the stage for accelerated approval for less deadly but more pervasive conditions such as diabetes and obesity. Approval is also being made easier by the use of genetic testing to determine a drug's efficacy, as well as the practice of drug companies working closely with federal organizations. Examples of these new policies are exemplified in the approval of Iressa, which helps fight cancer in only 10% of patients but is associated with a genetic marker that can help predict a patient's receptivity; and VELCADE, a cancer drug that received initial approval in only four months because the company that makes it, Millennium Pharmaceuticals, worked closely with the National Cancer Institute to review trials.

*Internet Research Tip:*
For extensive commentary and analysis on the development and approval of new drugs see:

**Tufts Center for the Study of Drug Development**
csdd.tufts.edu
Note: This web site gives you the opportunity to download the latest annual edition of the "Outlook", an excellent summary review of trends in drug development.

Small- to mid-size biotech firms continue to look to mature, global pharmaceutical companies for cash, marketing muscle, distribution channels and regulatory expertise. Meanwhile, major projects are underway, backed by diverse sponsors, to add to or build from scratch massive databases of genetic data on a scale not before imagined. In Iceland, for example, a group called DeCode Genetics has amassed a database of essentially all of the DNA in Iceland's unique, isolated population. Stanford University engineer Stephen R. Quake announced in mid 2009 that his new technology for decoding DNA enabled him to make an analysis of his own genome for less than $50,000 using a unique sequencer that is about the size of a household refrigerator.

With progress comes setbacks, including a massive award for damages (more than $250 million) that occurred in a small-town Texas court in August 2005. The award was made to the widow of a patient who allegedly had a fatal reaction to Merck & Co.'s Vioxx pain medication (which had previously been removed from the market due to safety concerns). Texas laws capping medical case awards reduced the damages significantly. Nonetheless, recent drug safety issues and a proliferation of lawsuits such as this may accelerate changes in the business models of drug development firms, discouraging them from risking funds on long-shot drugs intended to benefit the mass market. Meanwhile, drug makers will continue to alter marketing methods and greatly reduce consumer advertising. Virtually all drugs have significant side effect risks for certain types of patients. While drug makers have long practiced a high level of disclosure, those risks will be more clearly communicated in the future.

Global trends are affecting the biotech industry in a big way. Post 9/11, an emphasis was placed by government agencies on the prevention of bioterror risks, such as attacks by the spread of anthrax. This factor, combined with global concern about the possible spread of avian flu, has been a significant boost to vaccine research and production. At the same time, the rapid rise of offshoring and globalization is contributing to the movement of research, development and clinical trials away from the U.S., U.K. and France into lower cost technology centers in India and elsewhere. In fact, biotech firms are rising rapidly in India, China, Singapore and South Korea that will provide serious future competition to older companies in the West.

Likewise, retail drug markets have tremendous potential in emerging nations over the mid term. For example, consultants at McKinsey estimated that the drug market in India will grow from $6.3 billion in 2005 to $20 billion in 2015. China offers similar opportunities. This means that major international drug makers will be expanding their presence in these nations. However, it also means that local drug manufacturers have tremendous incentive to expand their research, product lines and marketing within their own nations.

Global panic over quickly rising food prices in 2007 and part of 2008 finally gave the genetically modified seed industry the boost it needs. Agribio (agricultural biotechnology) will become a top agenda item in government and corporate research budgets, and consumer acceptance of genetically modified food products will grow quickly.

## 2) A Short History of Biotechnology

While the 1900s will be remembered by industrial historians as the Information Technology Era and the Physics Era, the 2000s will most likely be marked by many as the Biotechnology Era because rapid advances in biotechnology will completely revolutionize many aspects of life in coming decades. However, the field of biotechnology can trace its true birth back to the dawn of civilization, when early man discovered the ability to ferment grains to make alcoholic beverages, and learned of the usefulness of cross-pollinating crops in order to create new hybrid strains—the earliest form of genetic engineering. In ancient China, people are thought to have harvested mold from soybean curd to use as an antibiotic as early as 500 B.C.

Robert Hooke first described cells as a concept in 1663 A.D., and in the late 1800s, Gregor Mendel conducted experiments that became the basis of modern theories about heredity. Alexander Fleming discovered the first commercial antibiotic, penicillin, in 1928.

The modern, more common concept of "biotech" could reasonably be said to have its beginnings shortly after World War II. In 1953, scientists James Watson and Francis Crick conceived the "double helix" model of DNA, and thus encouraged a spate of scientists to consider the further implications of human DNA. The Watson/Crick three-dimensional model began to unlock the mysteries of heredity and the methods by which replication of genetic material takes place within cells.

Significant steps toward biotech drugs occurred in the early 1970s. In 1973, Dr. Stanley N. Cohen, a Stanford University genetics professor, and Dr. Herb Boyer, a biochemist, genetic engineer and educator at UC-San Francisco, introduced the concept of gene-splicing and created the first form of recombinant DNA. In 1974, Cesar Milstein and Georges Kohler created monoclonal antibodies, cells that clone over and over again to create large quantities of a specific antibody. Many of today's top biotech drugs are monoclonal antibodies. These two discoveries (recombinant DNA and monoclonal antibodies) created the building blocks of the first modern commercial biotech drugs.

Boyer and Cohen's gene-splicing technique enabled scientists to cut genetic material from the cells of one organism and paste it into another organism. This was an important discovery because the genetic material they moved from one place to another instructs a cell as to how to make a particular protein. The organism on the receiving end of the gene-splicing technique is then able to make that protein. Over time, scientists have perfected the technique of splicing material that enables cells to create proteins that control the creation of insulin, the level of blood pressure and many other human functions. Such genetic engineering enabled, for the first time, the creation of massive vats of isolated proteins grown in bacteria or in cells harvested from mammals—in quantities large enough for the commercial production of new drugs. (In fact, Boyer and Cohen's early experiments involved inserting a gene from an African clawed toad into bacterial DNA for duplication.)

In 1975, the first human gene was isolated, opening the door to gene therapy and creating the interest that led to the beginning of the massive, publicly funded Human Genome Project in 1990. In 1976, Bob Swanson of the now-famous Silicon Valley venture capital firm Kleiner Perkins formed a new business, Genentech, in conjunction with Herb Boyer (see above). Other early biotech firms arrived soon after, generally funded by venture capital firms, angel investors and corporate venture partners. These early biotech startups included many

companies that grew into today's super-successful biopharma corporations: Amgen, Chiron, Biogen Idec and Genzyme. The creation of these startups, focused on the development of new drugs, was particularly noteworthy because it was the first time in decades that new drug companies were launched in significant numbers. In fact, most major drug companies in existence at the beginning of the 1970s could trace their histories back to the early 1900s or before.

### 3) Ethanol Production Soared, But a Market Glut May Slow Expansion

High gasoline prices, effective lobbying by agricultural and industrial interests, and a growing interest in cutting reliance on imported oil put a high national focus on bioethanol in America in recent years. Corn and other organic materials, including agricultural waste, can be converted into ethanol through the use of engineered bacteria and superenzymes manufactured by biotechnology firms. This trend has given a boost to the biotech, agriculture and alternative energy sectors. At present, corn is almost the exclusive source for bioethanol in America. This is a shift of a crop from use in the food chain to use in the energy chain that is unprecedented in all of agricultural history—a shift that is having profound effects on prices for consumers, livestock growers (where corn has long been a traditional animal feed) and food processors.

In addition to the use of ethanol in cars and trucks, the chemicals industry, faced with daunting increases in petrochemicals costs, has a new appetite for bioethanol. In fact, bioethanol can be used to create plastics—an area that consumes vast quantities of oil in America and around the globe. Archer Daniels Midland is constructing a plant in Clinton, Iowa that will produce 50,000 tons of plastic per year through the use of biotechnology to convert corn into polymers.

Ethanol is an alcohol produced by a distilling process similar to that used to produce liquors. A small amount of ethanol is added to about 30% of the gasoline sold in America, and most U.S. autos are capable of burning "E10," a gasoline blend that contains 10% ethanol. E85 is an 85% ethanol blend that may grow in popularity due to a shift in automotive manufacturing. Although only about 2,200 or so of the 170,000 U.S. service stations sold E85 as of the middle of 2009, there may be an increase in demand for ethanol in the U.S. due to mandates by the U.S. government calling for reduced dependence on oil.

Yet despite the millions of vehicles on the road that can run on E85 and billions of dollars in federal subsidies to participating refiners, many oil companies seem unenthusiastic about the adoption of the higher ethanol mix. E85 requires separate gasoline pumps, trucks and storage tanks, as well as substantial cost to the oil companies (the pumps can alone cost about $200,000 per gas station to install). The plants needed to create ethanol cost $500 million or more to build. Many drivers who have tried filling up with E85 once revert to regular unleaded when they find as much as a 25% loss in fuel economy when burning the blend.

Ethanol is a very popular fuel source in Brazil. In fact, Brazil is one of the world's largest producers of ethanol, which provides a significant amount of the fuel used in Brazil's cars. This is due to a concerted effort by the government to reduce dependency on petroleum product imports. After getting an initial boost due to government subsidies and fuel tax strategies beginning in 1975, Brazilian producers have developed methods (typically using sugar cane) that enable them to produce ethanol at moderate cost. The fact that Brazil's climate is ideally suited for sugarcane is a great asset. Also, sugar cane can be converted with one less step than corn, which is the primary source for American ethanol. Brazilian automobiles are typically equipped with engines that can burn pure ethanol or a blend of gasoline and ethanol. Brazilian car manufacturing plants operated by Ford, GM and Volkswagen all make such cars.

In America, partly in response to the energy crisis of the 1970s, Congress instituted federal ethanol production subsidies in 1979. Corn-based grain ethanol production picked up quickly, and federal subsidies have amounted to several billion dollars. The size of these subsidies and environmental concerns about the production of grain ethanol produced a steady howl of protest from observers through the years. Nonetheless, the Clean Air Act of 1990 further boosted ethanol production by increasing the use of ethanol as an additive to gasoline. Meanwhile, the largest producers of ethanol, such as Archer Daniels Midland (ADM), have reaped significant amounts of subsidies from Washington for their output.

The U.S. Energy Act of 2005 specifically requires that oil refiners mix 7.5 billion gallons of renewable fuels such as ethanol in the nation's gasoline supply by 2012. Ethanol production it the U.S. was fast approaching that level in 2007, as capacity had doubled since 2005. Although grain farmers and ethanol producers enjoyed high prices at the onset, a glut of ethanol supply eventually caused market prices to plummet. Next, the Energy Independence and Security Act of 2007 called for even more ethanol production, with a goal of 36 billion gallons per year by 2022 including 21 billion gallons to come from cellulosic and advanced biofuel sources. However, environmental concerns, the sizeable investments needed to construct ethanol refineries and questions about the advisability of using a food grain as a source for fuel may make these goals very unattainable using existing technologies. In addition, the automobile industry expects a significant amount of market share to slowly shift to electric or hybrid-electric vehicles over the long term, which will reduce dependency on liquid fuels, such as gasoline and ethanol.

More recently, some of the largest ethanol production companies have suffered severe financial problems. Notably, VeraSun filed for bankruptcy protection in late 2008, citing high corn prices and difficulty in obtaining trade credit. The Iowa-based company operated 14 ethanol plants in the Midwestern U.S. The 2008-2009 plummet in the price of crude oil made ethanol look much less attractive from a cost point-of-view. Meanwhile, ethanol factories have generally encountered great difficulty when seeking profitability in the U.S., despite immense federal government support. New plant construction projects have been cancelled or put on hold, and it is looking very unlikely that the industry can meet the production goals set by congressional mandates.

Traditional grain ethanol is typically made from corn or sugarcane. In contrast to grain ethanol, "cellulosic" ethanol is typically made from agricultural waste like corncobs, wheat husks, stems, stalks and leaves, which are treated with specially engineered enzymes to break the waste down into its component sugars. The sugars (or sucrose) are used to make ethanol. Since agricultural waste is plentiful, turning it into energy seemed a good strategy. Cellulosic ethanol can also be made from certain types of plants and grasses.

The trick to cellulosic ethanol production is the creation of efficient enzymes to treat the agricultural waste. The U.S. Department of Energy is investing heavily in research, along with major chemical companies such as Dow Chemical, DuPont and Cargill. Another challenge lies in efficient collection and delivery of cellulosic material to the refinery. It may be more costly to make cellulosic ethanol than to make it from corn. In any event, the U.S. remains far behind Brazil in cost-efficiency, as Brazil's use of sugar cane refined in smaller, nearby biorefineries creates ethanol at much lower costs per gallon.

Iogen, a Canadian biotechnology company, makes just such an enzyme and has been operating a test plant to determine how economical the process may be. The company hopes to construct a $300-million, large-scale biorefinery with a potential output of 50 million gallons per year. Its pilot plant in Ottawa utilizes wheat straw and corn stalks. In mid-2009, a Shell gasoline station in Ottawa, Canada became the first retail outlet in that nation to sell a blend of gasoline that features 10% cellulosic ethanol.

In the U.S., the Department of Energy has selected six proposed new cellulosic ethanol refineries to receive a total of $385 million in federal funding. If completed, these six refineries are expected to produce 130 million gallons of ethanol yearly. Iogen's technology will be used in one of the refineries, to be located in Shelley, Idaho. Partners in the refinery include Royal Dutch Shell.

Meanwhile, the Canadian government plans to support the Canadian biofuel industry with up to 500 million Canadian dollars for construction of next-generation plants. Iogen is expected to receive part of those funds for construction of a commercial scale cellulosic ethanol plant.

In the U.S., BP and Verenium announced plans in February 2009 to break ground on a commercial scale cellulosic ethanol plant in Highlands County, Florida as a joint venture. The plant is expected to cost $300 million and have the capacity to produce 36 million gallons of ethanol yearly from agricultural waste.

Construction of new ethanol production plants pushed total production capacity in the U.S. to about 5.4 billion gallons by the end of 2006 (about 3.4% of total U.S. gasoline consumption), up from 3.9 billion as of June 2005. Production capacity soared to 10.3 billion gallons by January 2009. Iowa, Illinois, Nebraska, Minnesota and South Dakota are the biggest producers, in that order. However, increased capacity nationwide and high corn prices are slowing the expansion trend, and many projects have recently been shelved.

Other companies, such as Syngenta, DuPont and Ceres, are genetically engineering crops so that they can be more easily converted to ethanol or other energy producing products. Syngenta, for example, is testing an engineered corn that contains the enzyme amylase. Amylase breaks down the corn's starch into sugar, which is then fermented into

ethanol. The refining methods currently used with traditional corn crops add amylase to begin the process.

Environmentalists are concerned that genetically engineering crops for use in energy-related yields will endanger the food supply through cross-pollination with traditional plants. Monsanto is focusing on conventional breeding of plants with naturally higher fermentable starch contents as an alternative to genetic engineering.

Another concern relating to ethanol use is that its production is not as energy efficient as that of biodiesel made from soybeans. According to a study at the University of Minnesota, the farming and processing of corn grain for ethanol yields only 25% more energy than it consumes, compared to 93% for biodiesel. Likewise, greenhouse gas emissions savings are greater for biodiesel. According to one estimate, producing and burning ethanol results in 12% less greenhouse gas emissions than burning gasoline, while producing and burning biodiesel results in a 41% reduction compared to making and burning regular diesel fuel. A 2009 vote by Illinois' Air Resources Board requires the use of "lower carbon intensity" fuels starting in 2011, which may have a negative long term effect on the use of ethanol.

Global warming concerns were heightened in 2009 by a report by the International Council for Science (ICSU) that concluded that the production of biofuels, including ethanol, has hurt rather than helped the fight against climate change. The report cites findings by a scientist at the Max Planck Institute for Chemistry in Germany that expand the harmful effects of a gas called nitrous oxide, which may be 300 times worse for global warming that carbon dioxide. The amounts of nitrous oxide released when farming biofuel crops such as corn may negate any advantage gained by reduced carbon dioxide emissions.

In addition, ethanol production requires enormous amounts of water. To produce one gallon of ethanol, up to four gallons of water are consumed by ethanol refineries. Add in the water needed to grow the corn in the first place, and the number grows to as much as 1,700 gallons of water for each gallon of ethanol.

Other concerns regarding the use of corn to manufacture ethanol include the fact that a great deal of energy is consumed in planting, reaping and transporting the corn in trucks. Also, high demand for corn for use in biorefineries has, from time-to-time, dramatically driven up the cost per bushel dramatically, creating burdens on consumers.

As if mid-2009, new technology was being tested that would produce ethanol from corn cobs that have been stripped of edible kernels. Poet, a South Dakota based producer of ethanol that operates 26 plants in seven U.S. states, is constructing a plant in Emmetsburg, Iowa that could be the first in the U.S. to produce ethanol on a large scale using a non-food source. The plant is slated for completion in 2011.

---

**SPOTLIGHT: Biofuels**

Corn is far from the only source of cellulose for creating biofuels.

**Municipal/Agricultural Waste**: Cheaply produced, but in limited supply compared to the billions of gallons of fuel needed in the market place.
**Wood**: Easily harvested and in somewhat healthy supply, however cellulose is more difficult to extract from wood than from other biosources.
**Algae**: The slimy green stuff does have the potential for high yields per acre, but the process for distilling cellulose is complex, requiring a source of carbon dioxide to permeate the algae.
**Grasses/Wheat**: Including switchgrass, miscanthus, sugar cane and wheat straw, the supply would be almost limitless. The challenge here is creating efficient methods for harvesting and infrastructure for delivering it to biorefineries.
**Vegetable Oils**: Including soybean, canola, sunflower, rapeseed, palm or hemp. It is difficult to keep production costs of these oils low.

---

**SPOTLIGHT: Algae Draws Major Investment**

Algae's potential as a source of biofuel got a big boost from an unlikely source in 2009. ExxonMobil announced plans to invest $300 million or more in San Diego, California-based Synthetic Genomics, a company headed by genome pioneer Craig Venter. Dr. Venter is studying ways in which an ideal species of algae can be developed for a unique culturing process. This process induces algae to release their oil (naturally stored as a foodstuff for the organisms), which can then be manipulated so that the oxygen molecules in the oil are disposed of, leaving a pure hydrocarbon suitable for use as biofuel. Another plus to Venter's process is that carbon dioxide claimed from industrial plant exhaust is used in the culturing process and then released in the atmosphere. This does not make algae biofuel production carbon neutral, but it does utilize carbon dioxide twice before it's released. Should the study go well, ExxonMobile has pledged an additional $300 million in funding to further develop the process to an industrial scale.

---

4) **New Money Pours Into Biotech Firms/Mergers and Acquisitions Surge Ahead**

Restructuring, mergers and bankruptcies are regular occurrences at smaller biotech firms, as their cash hoards from IPOs or initial rounds of venture capital dry up. Many of these companies are unable to achieve meaningful levels of revenues, and some of them incur immense annual losses due to the costs of research, development and regulatory requirements.

Much of the financial success that has been achieved by biotech firms lies with only a handful of companies such as Amgen and Genentech. These companies' initial investors have benefited greatly from growing sales and profits along with soaring stock market values, making some investors willing to take the enormous risk for the chance at a huge return.

New drugs and profitable drug companies take a long, long time to develop. The years of operation required to reach profitability at biopharmaceutical firms ranges from six years for Amgen to 10 years for Chiron and 15 years for Cephalon. These timeframes do not fit the requirements of many investors.

Meanwhile, mergers and acquisitions have reached new heights among biotechnology companies. The biggest news is Roche's early 2009 acquisition of biotech leader Genentech. Eli Lilly acquired ImClone in 2008. Japan-based Takeda

Pharmaceutical Company acquired Millennium Pharmaceuticals in 2008 also. One of the more important recent deals was the acquisition of the European biotech giant Serono by Merck KGaA. Investment analysts at Burrill & Co. estimate that over $60 billion had been invested in biotech company acquisitions by pharmaceuticals manufacturers from 2005 through mid 2008, not including the Genentech and ImClone deals.

The financial crisis had a deep effect on the biotech industry. Ernst & Young estimates that 2008 saw announcements of $16 billion in total biotech company financings worldwide, including $6 billion in venture capital. This was a dramatic, 46% decline from $30 billion in 2007. Initial public offerings of stock fell to nearly zero.

### 5) Major Drug Companies Bet on Partnerships With Smaller Biotech Research Firms

At pharmaceutical firms both large and small, profits are under constant pressure because blockbuster drugs that have made immense profits for many years quickly lose their patent protection and face vast competition from generic versions. In the U.S., generic drugs now hold between a sixty and seventy percent market share by volume. This puts pressure on large research-based drug firms to develop new avenues for profits. One such avenue is partnerships with and investments in young biotech companies, but profits from such ventures will, in most cases, be slow to appear. Meanwhile, the major, global drug firms are investing billions in-house on biotech research and development projects, but new blockbusters are elusive.

For example, Pfizer invests more than $7.5 billion yearly on R&D. That money is invested in carefully designed research programs with specific goals. In 2007, its goal was to triple the number of late-stage compounds in its research portfolio by 2009. Also in 2007, it had 11 drug programs in final testing stages, up from eight the previous year, and 47 programs in mid-stage testing, up from 32.

Much of future success for the world's major drug companies will lie in harnessing their immense financial power along with their legions of salespeople and marketing specialists to license and sell innovative new drugs that are developed by smaller companies. There are dozens of exciting, smaller biotech companies that are focused on state-of-the-art research that lack the marketing muscle needed to effectively distribute new drugs in the global marketplace. To a large degree, these companies rely on contracts and partnerships with the world's largest drug manufacturers. In addition to money to finance research and salespeople to promote new drugs to doctors, the major drug makers can offer expertise in guiding new drugs through the intricacies of the regulatory process. While these arrangements may not lead to blockbuster drugs that will sell billions of pills yearly to treat mass market diseases, they can and often do lead to very exciting targeted drugs that can produce $300 million to $1 billion in yearly revenues once they are commercialized. A string of these mid-level revenue drugs can add up to a significant amount of yearly income.

A good example is the relationship between Pfizer, one of the world's leading drug companies, and Medarex (www.medarex.com), a relative newcomer. In late 2004, Medarex announced a major partnership with Pfizer that included an initial cash infusion to Medarex of $80 million and a purchase of $30 million in Medarex stock by Pfizer. Their long-term goal was for Medarex to use its unique UltiMAb antibody technology to create up to 50 new drugs targeted at diseases specified by Pfizer. Eventually, Medarex may earn an additional $400 million in payments under the deal. As of early 2008, Medarex had more than 35 partnerships with pharmaceutical and biotech companies. Deals change quickly in the biotech world—in July 2009, Medarex announced that it intends to merge with Bristol-Myers Squibb.

### 6) From India to Singapore to Australia, Nations Compete Fiercely in Biotech Development

While pharmaceutical companies based in the nations with the largest economies, such as the U.S., U.K. and Japan, struggle to discover the next important drug, companies and government agencies in many other countries are enhancing their positions on the biotech playing field, building their own educational and technological infrastructures. Not surprisingly, countries such as India, Singapore and China, which have already made deep inroads into other technology-based industries, are making major efforts in biotechnology, which is very much an information-based science. Firms that manufacture generics and provide contract research, development and clinical trials services are already common in such nations (in India alone, clinical trials spending may reach as much as $1 billion annually by 2010). In most cases, this is just a beginning, with original drug and technology development the ultimate goal.

The government of Singapore, for example, has made biotechnology one of its top priorities for development, vowing to make it one of the staples of its economy. Its $286-million "Biopolis," (www.one-north.sg/hubs_biopolis.aspx ) a research and development center opened in 2004 that encompasses 2 million square feet of laboratories and offices, is only a small part of its $2.3-billion initiative to foster biotech in Singapore. An additional 400,000 square feet of space in the lab complex opened in late 2006. Eventually, Biopolis may house more than 2,000 scientists. Biopolis is part of a master planned science and technology park called One-North. The complex is recognized as a center for stem cell and cell therapy research. It is a melting pot of scientists and corporations from all over the world. This status is fostered by the fact that Singapore is centrally-located, has direct airline access to all of the world's major cities and has a well-educated, largely English-speaking population. For example, the Novartis Institute for Tropical Diseases has more than 100 researchers from 18 different nations.

Outsourcing of biotech tasks to India is growing at a very high annual rate. (India has dozens of drug manufacturing plants that meet FDA specifications.) Plunkett Research estimates that India's total pharmaceuticals industry revenues, largely from the manufacturing of generics, approached $11.8 billion in 2008.

India already has hundreds of firms involved in biotechnology and related support services. In 2005, the nation tightened its intellectual property laws in order to provide strong patent protection to the drug industry. As a result, drug development activity by pharma firms from around the world has increased in Indian locations in recent years. Drug sector firms to watch that are based in India include Piramal Healthcare Limited (formerly Nicholas Piramal, India Ltd.), Ranbaxy Laboratories Ltd. and Dr. Reddy's Laboratories Ltd. Services firms and research labs in India will continue to have a growing business in outsourced services for the biotech sector. The costs of developing a new drug in India can be a small fraction of those in the U.S., although drugs developed in India still are required to go through the lengthy and expensive U.S. FDA approval process before they can be sold to American patients. By 2008, India's biopharma market had reached $1 billion and ranked 11th in the world, according to BioPlan Associates, Inc. Venture capital and private equity funding of biotech firms in India reached an estimated $120 million in 2008, up from less than $10 million in 2006. In addition, India has its own

robust biotech parks, including the well-established S. P. Biotech Park in Hyderabad.

Stem cell (and cloning) research activity has been brisk in a number of nations outside the U.S. as well. To begin with, certain institutions around the world have stem cell lines in place, and some make them available for purchase. Groups that own existing lines include the National University of Singapore, Monash University in Australia and Hadassah Medical Centre in Israel. Sweden has also stepped onto the stage as a major player in stem cell research, with 40 companies focused on the field, including rising stars such as Cellartis AB, which has one of the largest lines of stem cells in the world, and NeuroNova AB, which is focusing on regenerating nerve tissue.

More importantly, several Asian nations, including Singapore, South Korea, Japan and China, are investing intensely in biotech research centered on cloning and the development of stem cell therapies. The global lead in the development of stem cell therapies may eventually pass to China, where the Chinese Ministry of Science and Technology readily sees the commercial potential and is enthusiastically funding research. On top of funding from the Chinese government, investments in labs and research are being backed by Chinese universities, private companies, venture capitalists and Hong Kong-based investors.

In late 2007, the FDA approved a Chinese drug used to fight AIDS, the first time a Chinese firm (specifically, Zhejiang Huahai Pharmaceutical Co.) has won the right to export a drug to the U.S. China is the world's largest producer of raw materials for drugs, although it falls far behind India in the export of finished generic drugs.

Meanwhile, leading biotech firms, including Roche, Pfizer and Eli Lilly, are taking advantage of China's very high quality education systems and low operating costs to establish R&D centers there. In this manner, offshore research can be complimented by offshore clinical trials.

Taiwan has opened the second of four planned biotech research parks. Vietnam has plans to open six biotech research labs. Australia also has a rapidly developing biotechnology industry.

South Korea is a world leader in research and development in a wide variety of technical sectors, and it is pushing ahead boldly into biotechnology. Initiatives include the Korea Research Institute of Bioscience and Biotechnology. The combination of government backing and extensive private capital in

Korea could make this nation a biotech powerhouse. One area of emphasis there is stem cell research.

In addition to fewer restrictions, many countries outside of the U.S. have lower labor costs, even for highly educated professionals such as doctors and scientists.

## 7) Medical Trials Conducted Abroad Spark Concerns

Another growing trend in health care offshoring is conducting clinical trials in countries outside the U.S. Researchers at Duke University reported in The New England Journal of Medicine that in November 2007, 20 of the largest U.S. drug makers conducted trials in 13,521 non-U.S. sites (compared to 10,685 U.S. sites). The study also reported that the number of countries conducting clinical drug tests has doubled in the last decade. Another study conducted by the Tufts Center for the Study of Drug Development reported that in 2007, only 54% of the roughly 26,000 FDA-regulated chief scientists conducting trials were based in the U.S. (compared to 86% in 1997). The U.S. National Institutes of Health report that in 2008, about 700 clinical trials were conducted in India, China and Russia, up from fewer than 50 in 2003.

While drug companies save significantly on trial costs overseas, there are a number of ethical concerns in this trend. On one hand, patients participating in these trials tend to be poorer than those in the U.S. and have less access to health care. These participants may view participation in trials as a form of free medical care, when in fact a therapy offered in a trial may be risky or completely ineffective. Another concern is that clinical tests performed outside the U.S. (and away from the scrutiny of the FDA) may not be properly monitored. In late 2008, a 350-participant study in India of a vaccine made by U.S. manufacturer Wyeth was stopped by Indian authorities after an infant died. The Drug Controller General of India's findings on the matter concluded that supervisory shortcomings were at fault. In 2007, two elderly patients in Poland died in a study of a bird-flu vaccine developed by Novartis. It was revealed that the patients should have been excluded from the trial due to their ages.

### Quick Tips: Genetics

A gene is a segment of DNA. Genes carry "instructions" for the construction and operation of an organism.

DNA is a nucleic acid containing genetic instructions. DNA stands for deoxyribonucleic acid.

Genetics is the study of traits inherited via specific components of genes passed from one generation to the next. The field may include the study of mutations to specific genes.

The genome is the way in which genes and DNA are arranged in an organism. A complete genome for an organism represents its full DNA sequence.

Genomics is the study of a complex set of genes in order to identify their sequence and how they are expressed.

Genetic engineering is an attempt by science to alter an organism by changing, adding or deleting genes as part of a gene therapy.

## 8) Gene Therapies and Patients' Genetic Profiles Promise a Personalized Approach to Medicine

Scientists now believe that almost all diseases have some genetic component. For example, some people have a genetic predisposition for breast cancer or heart disease. The understanding of human genetics is hoped to lead to breakthroughs in gene therapy for many ailments. Organizations ranging from the Mayo Clinic to drug giant GlaxoSmithKline are experimenting with personalized drugs that are designed to provide appropriate therapies based on a patient's personal genetic makeup or their lack of specific genes. Genetic therapy involves splicing desired genes taken from one patient into a second patient's cells in order to compensate for that patient's inherited genetic defect, or to enable that patient's body to better fight a specific disease.

For example, drugs that target the genetic origins of tumors may offer more effective, longer-lasting and far less toxic alternatives to conventional chemotherapy and radiation. One of the most noted drugs that target specific genetic action is Herceptin, a monoclonal antibody that was developed by Genentech. Approved by the FDA in 1998, Herceptin, when used in conjunction with chemotherapy, shows great promise in significantly reducing breast cancer for certain patients who are known to overproduce the HER2 gene. (A simple test is used to determine if this gene is present in the patient.) Herceptin, which works by blocking genetic signals, thus preventing the growth of cancerous cells, may show potential in treating other types of cancer, such as ovarian, pancreatic or prostate cancer.

Another genetic test is marketed by Genomic Health, based in Redwood City, California, (www.genomichealth.com). Its Oncotype DX test provides breast cancer patients with an assessment of the likelihood of the recurrence of their cancer based on the expression of 21 different genes in a tumor. The test enables patients to evaluate the results they may expect from post-operative therapies such as Tamoxifen or chemotherapy. The test costs about $3,500, and to-date has generally not been covered by health insurance, despite its potential to save treatment costs. The firm is also doing full-scale clinical development on a test to predict the likelihood of recurrence of colon cancer. Such tests will be standard preventative treatment in coming decades.

By mid-2008, as many as 1,400 genetic tests were available from various suppliers, including DeCode's and 23andMe's microarrays that test human saliva for up to 1 million genetic variations linked to disease and indicate risk factors (about $1,000 per test) and Navigenics' similar microarrays that detect predisposition to diseases such as Alzheimer's and type 2 diabetes ($2,500). On the top end is Knome's $350,000 profile, which analyzes an individual's entire genome and provides genetic variation data that can't be detected by microarrays. This is a huge drop in price from the $3 billion spent in 2003 by the Human Genome Project to sequence the first human genome. Prices should fall even more thanks to a $10 million award offered by the X Prize Foundation (www.xprize.org) to the first scientific team to sequence 100 human genomes in 10 days for less than $10,000 each. Likely contenders for the prize include Illumina, Inc., Applied Biosystems (a unit of Applera Corp.), 454 Life Sciences (owned by Roche Holding AG) and Helicos BioSciences Corp. A dark horse may be emerging in the race for the prize. A startup called Pacific Biosciences claims that it will sequence DNA in 15 minutes for less than $1,000 by 2013.

The states of California and New York issued "cease and desist" letters in mid-2008 to dozens of companies that offer genetic tests, responding to concerns about test validity, and potential patient harm. The companies providing the tests argue that the tests are personal genetic information services as opposed to medical tests. Meanwhile, the Federal Trade Commission has begun investigations into possibly deceptive advertising. Genetic testing is posing a number of issues that are unlikely to be resolved in the near term.

In 2009, a number of prominent scientists expressed doubts about the benefits of genome wide association studies, which compare the genomes of sick patients and healthy individuals in order to pinpoint DNA changes that are likely to cause certain diseases. The genetic variants are proving to be far more numerous, and more complex than first thought. Many geneticists are questioning the efficacy of genome wide studies (which cost millions of dollars to perform) and suggesting alternatives such as decoding entire genomes for individual patients.

Despite these concerns, genetic testing continues to evolve. In addition to significant strides in cutting costs, the time necessary to sequence DNA is shrinking as well thanks to improved chemicals and faster computers. New technology is simplifying the process which formerly compared multiple copies of DNA sequences to single-molecule sequencing. The new single-molecule version skips the need to compare multiple copies entirely while providing a more complete picture of the genome.

The scientific community's improving knowledge of genes and the role they play in disease is leading to several different tracks for improved treatment results. One track is to profile a patient's genetic makeup for a better understanding of a) which drugs a patient may respond to effectively, and b) whether certain defective genes reside in a patient and are causing a patient's disease or illness. Yet another application of gene testing is to study how a patient's liver is able to metabolize medication, which could help significantly when deciding upon proper dosage. Since today's widely used drugs often produce desired results in only about 50% of patients who receive them, the use of specific medications based on a patient's genetic profile could greatly boost treatment results while cutting costs. Each year, 2.2 million Americans suffer side effects from prescription drugs. Of those, more than 100,000 die, making adverse drug reaction a leading cause of death in the U.S. A Journal of the American Medical Association study states that the annual cost of treating these drug reactions totals $4 billion each year.

A second track for use of genetic knowledge is to attack, and attempt to alter, specific defective genes—this approach is sometimes referred to as "gene therapy." Generally, pure gene therapy attempts to target defective genes within a patient by introducing new copies of normal genes. These new genes may be introduced through the use of viruses or proteins that carry them into the patient's body.

Since 1989, more than 1,400 clinical trials of some form of gene therapy have been conducted with only 47 reaching phase III and none making it to regulatory approval in the U.S. China-based Shenzhen Sibiono GeneTech achieved the world's first pure gene therapy to be approved for wide commercial use in October 2003. The drug is sold under the brand name Gendicine as a treatment for a head and neck cancer known as squamous cell carcinoma (HNSC). The drug has proven highly effective in trials, and the company is testing its technology for several other types of cancer, including esophageal, gastric, colon, liver and rectum. The drug became available to Chinese patients at the Haidian Hospital in Beijing and to foreign patients at a facility in Shanghai in 2007, and was slated for approval for sale in India in 2008.

Other applications of gene therapy are already in research in the U.S. and elsewhere for treatment of a wide variety of diseases. For example, gene therapy may be highly effective in the treatment of rare immune system disorders, melanoma (a malignant skin cancer for which there is currently no effective cure once the disease has spread to other organs) and cystic fibrosis. In 2006, two male patients suffering from a rare malady called chronic granulomatous disease (CGD), which makes patients terribly vulnerable to infections, were able to cease taking daily antibiotics due to a gene therapy that introduced healthy genes to replace defective ones in their bloodstreams.

In 2008, a team of scientists led by a group at the University of Pennsylvania successfully introduced a healthy gene in six patients suffering from Leber's congenital amaurosis, a condition that leads to blindness. Four of the six patients' vision improved.

Roche Pharmaceuticals and Affymetrix, Inc. have developed a lab-on-a-chip (the AmpliChip CYP450) that can detect more than two dozen variations of two different genes. These genes are important to a patient's reaction to and use of drug therapies because the genes regulate the way in which the liver metabolizes a large number of common pharmaceuticals, such as beta blockers and antidepressants. A quick analysis of a tiny bit of a patient's blood can lead to much more effective use of prescriptions. Additional chips for specific types of genetic analysis will follow.

Another gene discovered by scientists appears to play a major role in widespread forms of the more than 30 different types of congenital heart defects, the most common of all human birth defects. Genes that are used by bacteria to trigger the infection process have also been identified, which could lead to powerful vaccines and antibiotics against life-threatening bacteria such as salmonella.

In the area of heart disease, about 300,000 patients yearly undergo heart bypass surgery in an effort to deliver increased blood flow to and from the heart. Clogged arteries are bypassed with arteries moved from the leg or elsewhere in the patient's body. Genetic experts have now developed the biobypass. That is, they have determined which genes and human proteins create a condition known as angiogenesis, which is the growth of new blood vessels that can increase blood flow without traumatic bypass surgery. This technique may become widespread in the near future.

Gene therapy is still in its infancy, and is not without its failures. Many genetic experts believe that the genetic testings will not progress to the extent where it can make a clinical difference in treating, diagnosing or predicting illness for at least another decade. However, the potential for gene therapy and genetic testing used for the choice and dosage of medications is almost limitless. Watch for major investment by pharmaceutical companies over the mid-term in further study and test development.

## 9) Breakthrough Drugs for Cancer Treatment— Many More Will Follow

Multi-kinase inhibitors are now on the leading edge of new drug development. This exciting class of drugs has the potential to target a wide variety of defective proteins. Kinases are enzymes that influence certain basic functions within cells, such as cell division. (This enzyme catalyzes the transfer of phosphates from Adenosine triphosphate (ATP) to proteins, thus causing changes in protein function.) Defective kinases can lead to diseases such as cancer. Kinase-blocking drugs are able to inhibit cell activity that can cause both tumor growth and blood vessel creation—thus they are able to shut down the blood vessels that enable tumors to grow and thrive. A good example is PTK/ZK, a drug developed by Novartis and Bayer Schering AG for the treatment of cancer of the colon, brain and other organs.

Earlier, in 2001, Novartis launched the drug Gleevec for the treatment of a blood cell cancer known as chronic myeloid leukemia. Gleevec operates by destroying diseased blood cells without harming normal cells. Patient results have been excellent. Additional projects include Genentech's highly effective colorectal and lung cancer drug Avastin, and the lung and pancreatic cancer drug Tarceva, developed by OSI Pharmaceuticals, Genentech and Roche Group.

Dozens of new kinase-blocking drugs are in various stages of development and testing. This could become a tremendous breakthrough in treatment of cancer patients and other groups. From a business standpoint, the market for such drugs is growing rapidly. By 2010, the global market may top $11.8 billion.

Another promising drug for cancer treatment is Iressa, made by AstraZeneca PLC, which was first regarded as a failure for the treatment of lung cancer. However, oncologists noted that the medication did improve outcomes for patients who were Asian or nonsmokers, or those with a particular genetic mutation involving malignant cell growth. A recent study of 400 Asian patients taking Iressa slowed or arrested the progression of the disease for a median 9.5 months (compared to 6.3 months for chemotherapy).

In 2009, a drug called Cerepro was approved for use in France and the Netherlands, making it the first approved gene therapy to hit the market in Europe. Cerepro targets fatal brain tumors called malignant glioma, and is manufactured by Ark Therapeutics in London. The drug, administered by multiple injections following the surgical removal of a malignant tumor of the brain, helps healthy tissue avoid the growth of a new tumor.

Scientists are working on ways to deliver gene therapies so that they will be accepted by healthy cells rather than fought by the body's immune system. Gene therapies are often introduced as viruses, which can be neutralized in this way. A company called Hybrid Systems Limited (www.hybridsystems.co.uk) has a portfolio of patents relating to the development of polymer coatings which protect these viruses from attack by immune systems.

## 10) Few New Blockbusters: Major Drug Patents Expire While Generic Sales Growth Continues

Historically, the drug industry has been one of the world's most profitable business sectors. The U.S. government estimated that $244 billion would be spent on prescription drugs in America during 2009. Nearly 20% of pharmaceutical company revenues are invested in discovering new medicines, well above the average of 4% reinvested in research and development in most industries. Advanced technology has allowed drug companies to saturate their development programs with smarter, more promising drugs. R&D (research and development) budgets are staggering. For example, Pfizer invests more than $7.5 billion yearly in R&D. Total spending worldwide on pharmaceutical research totaled $65.2 billion in 2008, according to estimates by Burrill & Company.

Once the patent on an existing blockbuster drug such as cholesterol treatment Zocor expires, competing drug companies are allowed to market cheaper generic brands which are identical chemical compounds. This takes huge revenues away from the maker of the original drug. According to a Kaiser Family Foundation study, the average branded drug costs three times as much as the average generic drug. When a generic equivalent of a drug hits the market, it is frequently 50% to 75% cheaper than the branded drug.

Drug makers are challenged to price their branded drugs in a way that will earn a good return on investment prior to the expiration of patent protection. The current U.S. patent policy grants drug manufacturers the normal 20 years protection from the date of the original patent (which is most likely filed very early in the research state), plus a period of 14 years after FDA approval. Since typical drugs take 10 to 15 years to research, prove in trials and bring to market, this effectively gives the patent holder only 19 to 24 years before low-price competition from generic manufacturers begins. Branded drugs tend to promptly lose 15% to 30% of market share when generic versions come on the market, eventually losing as much as 75% to 90% after a few years.

Another boost to generic drug sales is Wal-Mart's landmark move in 2006 when it began offering a large number of generic drugs for a flat $4 per month. Other discount retailers including Target, Walgreens and Kmart jumped on the generic bandwagon with their own deep discounts. In late 2007, Publix, a grocery store chain in the Southeast U.S. with more than 700 pharmacies, launched a free generic antibiotic program for prescriptions of several common drugs, for up to 14 days. The trend continued in 2008 when Wal-Mart further expanded its generic program. It offers 90-day supplies of more than 300 generic drugs for $10 in an effort to undercut mail-order pharmacy businesses. Wal-Mart is also targeting women's health by adding $9 generic prescriptions for up to a 30-day supply of generics that treat osteoporosis, breast cancer, menopause and hormone deficiencies.

Then there is the issue of "follow-on" drugs. These are drugs that hold their own patent for therapies similar to or competing with the original breakthrough patent. Examples include Zantac and Prilosec, drugs which were introduced by major firms to compete with the huge success of a pioneering drug, Tagamet. Follow-on drugs tend to get through regulatory approval much faster, and often are brought to market at much lower prices. Both factors create a significant competitive advantage for follow-on patent owners.

Major drug companies are trying to get on the generic gravy train by quietly creating their own generic drug subsidiaries. Pfizer, for example, has a division called Greenstone, Ltd., which produces generic versions of its blockbuster drugs including Zoloft, an antidepressant that brought in upwards of $2 billion in 2006 sales, at which time its patent expired. Greenstone, as of late 2008, offered 37 generic compounds. Likewise, Schering-Plough created generic subsidiary Warrick Pharmaceuticals. Naturally, independent generic drug companies oppose the practice.

Generic drugs accounted for about 63.7% of the total U.S. pharma market volume in the 12 months ending September 2008, according to market research firm IMS Health. This is down from 67% during 2007. Sales of generic drugs in the U.S. and internationally continue to grow, but growth in 2008 slowed a bit from 2007. Overall global growth in generic pharmaceuticals sales fell from 11.4% for the 12 months ending in September 2007 to 3.6% for the same period in 2008. Analysts at IMS Health attribute this slowing to declining prices and relatively fewer blockbuster drugs losing patent protection. Plunkett Research also attributes this downward trend to the overall global economic downturn. The AARP reported in 2008 that some elderly consumers were putting off filling prescriptions and delaying visits to doctors due to economic stress. However, generic sales do continue to slow, due to the fact that they offer significantly cost effective alternatives to brand name drugs.

From 2007 through 2012, patents will expire on branded drugs in the U.S. with about $60 billion in combined annual sales. For example, by 2011 a generic version of cholesterol-controlling Lipitor is expected to be on the market. Lipitor has long been the world's best selling branded drug (selling as much as $13 billion yearly at one time), but sales were declining in 2007-2008 thanks to a generic

substitute for a competing drug named Zocor; the generic of Zocor costs about 60% less than Lipitor.

Pharmaceutical company profits and research budgets are under pressure due to a wide range of additional challenges. Blockbuster drugs that might provide the high returns needed to bring a drug to market are increasingly difficult to develop. While research and development remains extremely expensive, blockbuster results are simply not emerging at the rate that they did in the past.

Meanwhile, health coverage payors are fighting to reduce the immense amounts that they spend on prescription drugs, and future revenues for drug makers will be impacted by cost-control measures at the individual level. In addition, the U.S. Government's Medicare drug benefit is putting even greater pressure on drug companies to sell drugs at lower prices. Drug makers are reacting to these pressures as best they can.

Highly effective new drug therapies can be incredibly expensive for patients. Take Iressa, a breakthrough treatment for lung cancer. It costs about $1,800 monthly per patient. Likewise, the groundbreaking colon cancer treatment Erbitux can cost as much a $30,000 for a seven-week treatment. A standard treatment of the latest drug regimens for some diseases can run up a $250,000 or higher bill over a couple of years.

The good news is that as the demand for new and improved treatments intensifies, so do the abilities of modern technology. In addition to expediting the process and lowering the costs of drug discovery and development, advanced pharmaceutical technology promises to increase the number of diseases that are treatable with drugs, enhance the effectiveness of those drugs and increase the ability to predict disease, not just the ability to react to it.

Consumers' voracious need for drugs will continue, thanks in part to the rapidly aging populations of such nations as the U.S. (where Baby Boomers make up a market of 78 million people), Japan, Italy and the U.K. Insurance companies may raise co-payments for drugs, strike deals with drug companies and employ pharmacy benefit management tactics in an effort to fend off rising pharmaceutical costs.

One trend is creating a new category for blockbuster drugs, based on vanity, convenience or personal choices. Until recently, pharmaceutical research was focused primarily on curing life-threatening or severely debilitating illnesses. But a current generation of drugs, commonly referred to as "lifestyle" drugs, is transforming the pharmaceutical industry. Lifestyle drugs target a variety of human conditions, ranging from the painful to the inconvenient, including obesity, impotence, memory loss, urinary urgency and depression. Drug companies also continue to develop lifestyle treatments for hair loss and skin wrinkles in an effort to capture their share of the huge anti-aging market aimed at the Baby Boomer generation. The use of lifestyle drugs dramatically increases the total annual consumer intake of pharmaceuticals, and creates a great deal of controversy over which drugs should be covered by managed care and which should be paid for by the consumer alone.

In coming years, taming pharmaceutical costs will be one of the biggest challenges facing the health care system. Prescription drug costs already account for about 10% of all health care expenditures in the U.S. Managed care must be able to determine which promising new drugs can deliver meaningful clinical benefits proportionate to their monetary costs.

Several developments are fueling the fire under the controversy over drug costs in the U.S. To begin with, it has become common knowledge that pharmaceutical firms tend to price their drugs at vastly lower prices outside the U.S. market. The result is that U.S. consumers and their managed care payors are bearing a disproportionate share of the costs of developing new drugs. In 2009, the Obama administration hopes to have a sweeping health care reform bill passed which may greatly impact health care costs and the way in which health care overall is managed. If such legislation is enacted, it could have very large effects on the pharmaceuticals industry. For example, extending coverage to a large number of people under some sort of universal care may significantly boost total drug sales in terms of unit volume. However, the heavy hand of government as an ever-larger buyer of health care could lead to demands for reduced drug pricing. If profits are restrained, it may discourage research or investment in biotechnology projects.

---

**Factors leading to high drug expenditures in the American health care system:**

- 78 million Baby Boomers are beginning to enter their senior years. The lifespan of Americans is increasing, and chronic illnesses are increasing as the population ages.
- Medicare Part D, the component of Medicare that pays for prescription drugs for seniors, made affordable drugs available to millions of Americans, fueling overall drug sales.
- The drug industry intensified its sales effort over recent years. Direct-to-consumer advertising and legions of sales professionals calling on physicians increased demand for the newest, most expensive drugs.
- Convoluted and uncoordinated lists of drug "formularies" (available drugs, their uses and their interactions) require increased administrative work to sort through, thus forcing costs upward.
- Physicians sometimes prescribe name-brand drugs when a generic equivalent may be available at a fraction of the cost.
- Research budgets are immense and very costly. Breakthroughs in research and development are creating significant new drug therapies, allowing a wide range of popular treatments that were not previously available. An excellent example is the rampant use of antidepressants such as Prozac and Zoloft. Meanwhile, major drug companies face the loss of patent protection on dozens of leading drugs. They are counting on research, partnerships and acquisitions to replace those marquis drugs.
- "Lifestyle" drug use is increasing, as shown by the popularity of such drugs as Viagra (for the treatment of sexual dysfunction), Propecia (for the treatment of male baldness) and Botox (for the treatment of facial wrinkles).

*Source: Plunkett Research, Ltd.*

---

### 11) Biotech and Orphan Drugs Pick Up the Slack as Blockbuster Mainstream Drugs Age

While Big Pharma traditionally concentrated on mass-market drugs that treat a broad spectrum of ailments, biotech companies have often developed treatments for rare disorders or maladies such as certain cancers that only affect a small portion of the population. For example, biotech pioneers Genentech and Biogen Idec developed Rituxan for the treatment of non-Hodgkin's lymphoma, an important but relatively small market.

Drugs such as Rituxan are commonly referred to as "orphan drugs," which means that they treat illnesses that no other drug on the market addresses. Technically, a drug designated by the FDA with orphan status provides therapeutic benefit for a disease or condition that affects less than 200,000 people in the U.S. Almost half of all drugs produced by biotech companies are for orphan diseases. These drugs enjoy a unique status due to the Orphan Drug Act of 1983, which gives pharmaceutical companies a seven-year monopoly on the drug without having to file for patent protection, plus a 50% tax credit for research and development costs. As of 2008 there 1,850 drugs designated "orphan" by the FDA, including a few hundred on the market and the balance pending approval.

As the wellspring of mainstream Big Pharma blockbusters begins to dry, many big pharmaceutical companies are rushing to develop their own orphan drugs. Pfizer released Sutent, a drug used to treat cancerous kidney and stomach-lining tumors in 2006, long after biotech firm Genentech had four drugs used for cancer on the market. Genentech's colon cancer treatment, Avastin, is expected to reach $6.2 billion in sales in 2009, while Pfizer's Sutent is projected to bring in only $570 million that same year.

---

**Commentary: The challenges facing the biopharmaceuticals industry**

- Working with governments to develop methods to safely and effectively speed approval of new drugs.
- Working with the investment community to build confidence and foster patience in the investors for the lengthy timeframe required for commercialization of promising new compounds.
- Working with civic, government, religious and academic leaders to deal with ethical questions centered on stem cells and other new technologies in a manner that will enable research and development to move forward.
- Overcoming, through research, the technical obstacles to therapeutic cloning.
- Enhancing sales and distribution channels so that they educate patients, payors and physicians about new drugs in a cost-effective manner.

---

- Emphasizing fair and appropriate pricing models that will enable payors (both private and public) to afford new drugs and diagnostics while providing ample profit incentives to the industry.
- Developing appropriate standards that fully realize the potential of systems biology (that is, the use of advanced information technology and the resources of genetic databases) in a manner that will create the synergies necessary to accelerate and lower the total cost of new drug development.
- Fostering payor acceptance, diagnostic practices and physician practices that will harness the full potential of genetically targeted, personalized medicine when a large base of new biopharma drugs becomes available.

*Source: Plunkett Research, Ltd.*

---

## 12) Biogenerics (Follow-on Biologics) are in Limbo in the U.S.

Patents on the first biotech-based drugs began expiring in 2001. Nonetheless, the FDA has no formal path for enabling drug makers to obtain approval of a generic version of biotechnology-based drugs without extensive clinical trials. Consequently, biotech drug makers as of yet haven't faced the type of generic competition that constantly challenges makers of traditional drugs, despite the fact that generic drug makers report that the biologics industry could reach $100 billion in sales by 2011. However, this situation is likely to change in the near future as more and more biotech drugs go off-patent.

Because biotech drugs ("biologics") are made from living cells, a generic version of a drug probably won't be biochemically identical to the original branded version of the drug. Therefore, they are described as "follow-on biologics" to set them apart. There are concerns that follow-on biologics may not be as safe or effective as the originals unless they go through clinical trials for proof of quality.

In addition, the development of generic versions of biologics is very costly. Industry analysts estimate that initial development for one compound runs between $100 million and $150 million, compared to $5 million to $10 million for small-molecule drugs. However, the economic potential for the marketing of these generic compounds is staggering, which makes the cost easier to swallow. Merck & Co. set up a business unit in 2009 to develop generic biologics, and is planning to invest $1.5 billion in research and development in this area by 2015.

In Europe, these drugs are referred to as "biosimilars." In the European Union, the first biosimilars were approved in April 2006 by the EMEA (European Agency for the Evaluation of Medicinal Products), the regulatory body responsible for new drugs. The first approved biosimilar was Omnitrope, a generic substitute for growth hormone Genotropin. The second was Valtropin.

FDA regulations for approval of generics, written in 1984, allow companies to achieve approval of generic versions of chemically-based, traditionally-manufactured pharmaceuticals with expired patents in a relatively short period of time. However, the law doesn't address drugs developed through biotechnology. The question remains whether the FDA will approve such generic biologics without lengthy and expensive clinical trials. It's a whole new game in the generics sector. Would-be generics makers also face the fact that biotech drugs can be vastly more complicated to manufacture than chemicals-based drugs.

Like the Europeans (EMEA), the FDA did approve a simple generic biotech drug, a human growth hormone, in May 2006. However, there are no published guidelines for getting more complex biogenerics to market. The FDA is under mounting pressure from the U.S. House of Representatives and from a number of state governors who are petitioning for these drugs.

In order to obtain approval of a biogeneric under existing regulations, a manufacturer would be forced to approve that the new drug contains the same ingredient as the original drug and that the two drugs are bioequivalent. This may be impossible without clinical trials.

India-based Dr. Reddy's Laboratories, one of that nation's leading drug firms, plans to create at least one follow-on biologic yearly over the mid-term. It already sells two such biogenerics in India, versions of Roche's Rituxan and Amgen's Neupogen. Other Indian biogeneric manufacturers, including Reliance Life Sciences and Ranbaxy, are also marketing products. Ranbaxy began marketing biogenerics in India as early as 2003. In 2008, Reliance Life Sciences launched three biosimilars (ReliPoietin, ReliFeron and ReliGrast) for sale in South Asia, South East Asia and several countries in Latin America.

In Europe, the EMEA has issued guidelines for approval of biosimilars. These include:

1) Comparability studies are required. The proof or lack of proof of comparability to the original drug will dictate how many new clinical studies may be required.
2) Clinical studies may be required to prove the biosimilar's safety and efficacy.
3) Nonclinical studies may be required as well.
4) After the biosimilar is approved and brought to market, continuing safety and efficacy study commitments will be required.

In 2008, the Biologics Price Competition and Innovation Act of 2007 regarding FDA regulation of follow-on biologics stalled in Congress and never became law. If the act had passed, FDA guidelines for biogenerics would have closely mirror those of the EMEA. In 2009, the Obama administration's sweeping health care reform initiative included no legislation regarding guidelines for the approval of biosimilars. The reform plan does include a new bill that would grant 13.5 years of intellectual property protection to biologic drugs, which would significantly extend the life of branded biotech-based drugs. This would protect the original maker of a biologic, because a generic competitor would not be able to utilize data from the clinical trials conducted by the original maker until the 13.5 years had passed and the original patent had also expired.

**13) Breakthrough Drug Delivery Systems Evolve**

Controlling how drugs are delivered is a huge business. Sales of drugs using new drug delivery systems were expected to balloon.

Until the biotech age, drugs were generally comprised of small chemical molecules capable of being absorbed by the stomach and passed into the blood stream—drugs that were swallowed as pills or liquids. However, many new biotech drugs require injection (or some other form of delivery) directly into the bloodstream, because they are based on larger molecules that cannot be absorbed by the stomach. A number of new drug delivery techniques that provide an alternative to needles are in development.

In the near future, there may be an implantable microchip, controlled by a miniature computer, capable of releasing variable doses of multiple potent medications over an extended period, potentially up to one year. The miniscule silicon chips will bear a series of tiny wells, sealed with membranes that dissolve and release the contents when a command is received by the computer. Chips that can receive commands beamed through the skin are also theoretically possible. This technology would help treatment of conditions such as Parkinson's disease or cancer, where doctors need to vary medications and dosages. The technology must first be tested in animals and then in humans to ensure that the chips are biocompatible. The chips will more likely be used first in external applications, which may facilitate laboratory testing and drug development.

Other potential needle-free drug delivery systems include synthetic molecules attached to a drug, making it harder for the stomach to render the medicine useless before it reaches the blood. High-tech inhalers, which force medicine through the lungs, are also in the works. For the patient, this means less pain and the promise of better outcomes. Needle-free systems may also make toxic drugs safer and give older drugs new life. For example, the painkiller Fentanyl is available in a lozenge form.

Some firms are developing techniques to encapsulate or rearrange drug molecules into more sturdy compounds that release steady, even doses over a prolonged period. A patch being developed by Alza Corp., now a subsidiary of Johnson & Johnson, has a network of microscopic needles that penetrate painlessly into the first layer of the skin. In addition, Alza is in competition with Vyteris to develop a patch that delivers drugs via a small electric current, activated by a button on the patch itself. Vyteris announced an agreement granting Laboratory Corporation of America Holdings rights to represent LidoSite, the first FDA-approved active transdermal patch using electric current for pain relief from blood draws and venipunctures. Another device, made by Echo Therapeutics (formerly Sontra Medical Corp.), uses ultrasound and gel to agitate and open temporary pores in the skin that can then receive a drug. The same technology can provide continuous, transdermal glucose monitoring. Yet another potential drug delivery system is edible film; quickly dissolving films treated with medication that melt on the tongue. Already in use as a breath freshener (Listerine brand PocketPaks, made by Pfizer), edible film is now the delivery method of choice for Novartis' Triaminic, Theraflu Thin Strips and Gas-X Thin Strips, as well as Enlyten's PediaStrips that replenish electrolytes in children and Electrolyte SportStrips for adults.

## 14) Stem Cells—Multiple Sources Stem from New Technologies

During the 1980s, a biologist at Stanford University, Irving L. Weissman, was the first to isolate the stem cell that builds human blood (the mammalian hematopoietic cell). Later, Weissman isolated a stem cell in a laboratory mouse and went on to co-found SysTemix, Inc. (now part of drug giant Novartis) and StemCells, Inc. to continue this work in a commercial manner.

In November 1998, two different university-based groups of researchers announced that they had accomplished the first isolation and characterization of the human embryonic stem cell (HESC). One group was led by James A. Thomson at the University of Wisconsin at Madison. The second was led by John D. Gearhart at the Johns Hopkins University School of Medicine at Baltimore. The HESC is among the most versatile basic building blocks in the human body. Embryos, when first conceived, begin creating small numbers of HESCs, and these cells eventually differentiate and develop into the more than 200 cell types that make up the distinct tissues and organs of the human body. If scientists can reproduce and then guide the development of these basic HESCs, then they could theoretically grow replacement organs and tissues in the laboratory—even such complicated tissue as brain cells or heart cells.

Ethical and regulatory difficulties have arisen from the fact that the only source for human "embryonic" stem cells is, logically enough, human embryos. A laboratory can obtain these cells in one of three ways: 1) by inserting a patient's DNA into an egg, thus producing a blastocyst that is a clone of the patient—which is then destroyed after only a few days of development; 2) by harvesting stem cells from aborted fetuses; or 3) by harvesting stem cells from embryos that are left over and unused after an in vitro fertilization of a hopeful mother. (Artificial in vitro fertilization requires the creation of a large number of test tube embryos per instance, but only one of these embryos is used in the final process.)

A rich source of similar but "non-embryonic" stem cells is bone marrow. Doctors have been performing bone marrow transplants in humans for years. This procedure essentially harnesses the healing power of stem cells, which proliferate to create healthy new blood cells in the recipient. Several other non-embryonic stem cell sources have great promise (see "Potential methods of developing 'post-embryonic' stem cells without the use of human embryos" below).

In the fall of 2001, a small biotech company called Advanced Cell Technology, Inc. announced the first cloning of a human embryo. The announcement set off yet another firestorm of rhetoric and debate on the scientific and ethical questions that cloning and related stem cell technology inspire. While medical researchers laud the seemingly infinite possibilities stem cells promise for fighting disease and the aging process, conservative theologians, many government policy makers, certain ethics organizations and pro-life groups decry the harvest of cells from aborted fetuses and the possibility of cloning as an ethical and moral abomination.

Stem cell research has been underway for years at biotech companies including Stem Cells, Inc., Geron and ViaCord (formerly ViaCell which was acquired by PerkinElmer in 2007). One company, Osiris Therapeutics, has been at work long enough to have two clinical trial programs in progress. Osiris derives its stem cells from the bone marrow of healthy adults between the ages of 18 and 30 who are volunteers. Prior to harvesting the stem cells, Osiris screens blood samples of the donors to make sure that they are free of diseases such as HIV and hepatitis. Osiris believes that this approach to gathering stem cells places them outside of the embryonic source controversy. In early 2009, Geron announced that it had received FDA clearance to begin trials of a stem cell-based therapy for acute spinal cord injuries.

The potential benefits of stem cell-based therapies are staggering. Neurological disorders might be aided with the growth of healthy cells in the brain. Injured cells in the spinal column might be regenerated. Damaged organs such as hearts, livers and kidneys might be infused with healthy cells. In China, physicians in more than 100 hospitals are already injecting stem cells into damaged spinal cords with varying levels of success.

---

**Potential methods of developing "post-embryonic" stem cells without the use of human embryos:**

- Adult Skin Cells—Exposure of harvested adult skin cells to viruses that carry specific genes, capable of reprogramming the skin cells so that they act as stem cells.
- Parthenogenesis—manipulation of unfertilized eggs.
- Other Adult Cells—Harvesting adult stem cells from bone marrow or brain tissue.
- Other Cells—harvesting of stem cells from human umbilical cords, placentas or other cells.
- De-Differentiation—use of the nucleus of an existing cell, such as a skin cell, that is altered by an egg that has had its own nucleus removed.
- Transdifferentiation—making a skin cell de-differentiate back to its primordial state so that it can then morph into a useable organ cell, such as heart tissue.

---

- Pluripotent state cells (iPSCs). Adult cells are drawn from a skin biopsy and treated with reprogramming factors.
- Most recently, researchers have found it possible to harvest stem cells from a wide variety of tissue. The looming question, however, is whether these cells can be successfully reprogrammed.

---

HESCs (typically harvested from five-day-old human embryos which are destroyed during the process) are used because of their ability to evolve into any cell or tissue in the body. Their versatility is undeniable, yet the implications of the death of the embryo are at the heart of the ethical and moral concerns.

Meanwhile, scientists have discovered that there are stem cells in existence in many diverse places in the adult human body, and they are thus succeeding in creating stem cells without embryos, by utilizing "post-embryonic" cells, such as cells from marrow. Such cells are already showing the ability to differentiate and function in animal and human recipients. Best of all, these types of stem cells may not be plagued by problems found in the use of HESCs, such as the tendency for HESCs to form tumors when they develop into differentiated cells. However, Thomas Okarma, CEO and President of Geron, argues that stem cells derived from bone marrow and other adult sources are fundamentally limited in their application, because it is nearly impossible to get them to produce anything besides blood cells. This makes it very difficult to harvest organ or nerve cells.

Studies were published in 2006 and 2007 regarding the reprogramming of adult mouse skin cells into stem cells. Initially, Shinya Yamanaka of Kyoto (Japan) University announced that he and coworkers had exposed skin cells harvested from adult mice to viruses carrying four specific genes. Yamanaka had determined that these four genes are apparently responsible for a stem cell's ability to develop into virtually any type of tissue. The technique is easy to replicate, and it was confirmed by additional studies in the U.S. published in 2007. This may be a tremendous breakthrough in stem cell research. However, scientists are still a long way from overcoming daunting technical challenges and adapting this technique for use in humans.

In late 2007, two scientific papers were published regarding induced pluripotent state cells (iPSCs). These are human stem cells produced without human cloning or the use of human embryos or eggs. Adult cells are drawn from a skin biopsy and treated with four reprogramming factors, rendering cells that can produce all human cell types and grow indefinitely. iPSCs are patient-specific, making them superior to HESCs because cell rejection by a patient's immune system is avoided. However, the risk of tumor formation appears to be higher for iPSCs due to the reprogramming genes which remain in the cell once the process is complete. Lawrence Goldstein, a researcher at the University of California at San Diego is already using iPSCs to study cells from patients with Alzheimer's in order to develop new drug therapies. Although further development is necessary in order to make iPSCs viable for general clinical use, they are a profound breakthrough in stem cell research that may resolve the ongoing controversy regarding embryo-derived cells.

Additional stem cell sources continue to be identified. In 2008, researchers at Children's Hospital of Pittsburgh found that stem cells can be harvested from the walls of small blood vessels such as capillaries. In 2009, the *Journal of Translational Medicine* reported that stem cells taken from human fallopian tubes have the potential to become a number of different cell types.

### 15) U.S. Government Reverses Ban on Funding for New Stem Cell Research

Shortly after taking office, U.S. President Barack Obama reversed an eight-year ban on the use of federal funding for embryonic stem cell research. Specifically, Obama issued Executive Order 13505, entitled "Removing Barriers to Responsible Scientific Research Involving Human Stem Cells." This executive order, dated March 9, 2009, charged the National Institutes of Health (NIH) with issuing new guidelines for stem cell research which became effective in July 2009. The order further authorized the NIH to "support and conduct responsible, scientifically worthy human stem cell research, including human embryonic stem cell research, to the extent permitted by law." The important words here are "human embryonic," since the harvesting of stem cells from discarded human embryos is what started the stem cell funding controversy in the first place. Further, this wording clearly eliminates the possibility of funding research projects involving stem cells that result from cloning. Under previous U.S. regulations, federal research funds were granted only for work with 21 specific lines of stem cells that existed in 2001. Harvesting and developing new embryonic lines did not qualify.

By mid-April 2009, the NIH had issued a new policy statement and posted it for a 30-day public comment period. The issue of funding remains politically charged. The NIH is taking a middle road. Its guidelines state that embryos donated for such research must be given voluntarily and without financial inducement. (Such embryos typically are donated by couples who have completed fertility treatments and have no need for remaining, redundant embryos. This is a common practice in seeking laboratory-aided pregnancies.)

Also of note: the NIH received $10 billion in new funds from recent economic stimulus legislation. It is possible, but not required, that NIH could use a part of that funding to boost stem cell efforts.

Once a stem cell starts to replicate, a large colony, or line, of self-replenishing cells can theoretically continue to reproduce forever. Unfortunately, only about a dozen of the stem cell lines existing at the time were considered to be useful, and some scientists believe that these lines were getting tired.

The use of non-federal funding, however, was not restricted during the eight-year ban, although many groups did want to see further state or federal level restrictions on stem cell research or usage. A major confrontation continues between American groups that advocate the potential health benefits of stem cell therapies and groups that decry the use of stem cells on ethical or religious terms. Meanwhile, stem cell development forged ahead in other technologically advanced nations.

In November 2004, voters in California approved a unique measure that provides $3 billion in state funding for stem cell research. Connecticut, Massachusetts and New Jersey also passed legislation that permits embryonic-stem cell research. California already has a massive biotech industry, spread about San Diego and San Francisco in particular. As approved, California's Proposition 71 created an oversight committee that determines how and where grants will be made, and an organization, the California Institute for Regenerative Medicine (www.cirm.ca.gov), to issue bonds for funding and to manage the entire program. The money is being invested in research at a rate of about $295 million yearly over 10 years.

In June 2007, the California Institute for Regenerative Medicine (CIRM) approved grants totaling more than $50 million to finance construction of shared research laboratories at 17 academic and non-profit institutions. These facilities are scheduled to be complete and available to researchers within six months to two years of the grant awards. By April 2009, CIRM had approved 279 grants totaling more than $693 million. The grants are funding dedicated laboratory space for the culture of human embryonic stem cells (HESCs), particularly those that fall outside federal guidelines. In early 2008, the first clinical trial began using CIRM funds at the University of California, San Diego for a therapy to treat a blood disorder.

In the private sector, funding for stem cell research was generous up until the global economic crisis beginning in 2008. For example, the Juvenile Diabetes Research Foundation has an $8 million stem cell research program underway and Stanford University has used a $12 million donation to create a research initiative. Likewise, major, privately funded efforts have been launched at Harvard and at the University of California at San Francisco.

Corporate investment in stem cells was also strong. AstraZeneca Pharmaceuticals invested $77 million in a startup firm in San Diego called BrainCells, Inc. to study how antidepressants might be used to spur brain cell growth. GlaxoSmithkline agreed to pay OncoMed Pharmaceuticals up to $1.4 billion in late 2007 for four radical new drugs that target cancer stem cells. In 2009, Geron Corporation began an FDA-approved trial of embryonic stem cells in recent spinal cord injuries in eight to ten patients.

President Obama's reversal of the stem cell funding ban comes at a time when funding from private and state sources is drying up. This is good news for further research, but it may be a bit late due to recent breakthroughs in stem cell research. In 2007, Shinya Yamanaka, a research scientist in Japan, discovered that adult cells can be reprogrammed to an embryonic state using a relatively easy process. This discovery opens to door to an almost unlimited supply of stem cells in the future and lifts most ethical arguments against using the cells for research.

## 16) Stem Cells—Therapeutic Cloning Techniques Advance

For scientists, the biggest challenge at present may be to discover the exact process by which stem cells are signaled to differentiate. Another big challenge lies in the fact that broad use of therapeutic cloning may require immense numbers of human eggs in which to grow blastocysts.

The clearest path to "therapeutic" cloning may lie in "autologous transplantation." In this method, a tiny amount of a patient's muscle or other tissue would be harvested. This sample's genetic material would then be de-differentiated; that is, reduced to a simple, unprogrammed state. The patient's DNA sample would then be inserted into an egg to grow a blastocyst. The blastocyst would be manipulated so that its stem cells would differentiate into the desired type of tissue, such as heart tissue. That newly grown tissue would then be transplanted to the patient's body. Many obstacles must be overcome before such a transplant can become commonplace, but the potential is definitely there to completely revolutionize healing through such regenerative, stem cell-based processes. One type of bone marrow stem cell, recently discovered by scientists at the University of Minnesota, appears to have a wide range of differentiation capability.

It is instructive to note that there are two distinct types of embryonic cloning: "reproductive" cloning and "therapeutic" cloning. While they have similar beginnings, the desired end results are vastly different.

"Reproductive" cloning is a method of reproducing an exact copy of an animal—or potentially an exact copy of a human being. A scientist would remove the nucleus from a donor's unfertilized egg, insert a nucleus from the animal, or human, to be copied, and then stimulate the nucleus to begin dividing to form an embryo. In the case of a mammal, such as a human, the embryo would then be implanted in the uterus of a host female for gestation and birth. The successful birth of a cloned human baby doesn't necessarily mean that a healthy adult human will result. To date, cloned animals have tended to develop severe health problems. For example, a U.S. firm, Advanced Cell Technology, reports that it has engineered the birth of cloned cows that appeared healthy at first but developed severe health problems after a few years. Nonetheless, successful cloning of animals is progressing at labs in many nations.

On the other hand, "therapeutic" cloning is a method of reproducing exact copies of cells needed for research or for the development of replacement tissue. In this case, once again a scientist removes the nucleus from a donor's unfertilized egg, inserts a nucleus from the animal, or human, whose cells are to be copied, and then stimulates the nucleus to begin dividing to form an embryo. However, in therapeutic use, the embryo would never be allowed to grow to any significant stage of development. Instead, it would be allowed to grow for a few hours or days, and stem cells would then be removed from it for use in regenerating tissue.

Because it can provide a source of stem cells, cloning has uses in regenerative medicine that can be vital in treating many types of disease. The main differences between stem cells derived from clones and those derived from aborted fetuses or fertility specimens is that a) they are made from only one source of genes, rather than by mixing sperm and eggs; and b) they are made specifically for scientific purposes, rather than being existing specimens, putting them up to more intense ethical discussions. Cloned stem cells have the added advantage of being 100% compatible with their donors, because they share the same genes, and so would provide the best possible source for replacement organs and tissues. Although the use of cloning for regeneration has stirred heated debate as well, it has not resulted in universal rejection. Most of the industrialized countries, including Canada, Russia, most of Western Europe and most of Asia, have made some government-sanctioned allowances for research into this area.

As a result of government sanction of research and development, some countries have already made progress in the field of regenerative cloning. In an important development in August 2004, scientists at Newcastle University in the U.K. announced that they were granted permission by the Human Fertilisation and Embryology Authority (HFEA), a unit of the British Government, to create human embryos as a source of stem cells for certain therapeutic purposes. Specifically, researchers will clone early-stage embryos in search of new treatments for such degenerative diseases as Parkinson's disease, Alzheimer's and diabetes. The embryos will be destroyed before they are two weeks old and will therefore not develop beyond a tiny cluster of cells.

A blow to therapeutic cloning occurred when South Korean scientist Hwang Woo Suk of Seoul National University announced that he and his team of researchers had created patient-specific stem cells; that is, cells taken from adult patients and then cloned and further manipulated to become stem cells. Backed by the South Korean government, the breakthrough promised the dawn of a new era of stem cell research with Hwang as the star. However, by December 2005, it became clear that the Korean claims were fraudulent. There were, in fact, no cloned stem cell lines at all.

The good news is that several other cloning methods are on the horizon. Markus Grompe, director of the Oregon Stem Cell Center at Oregon Health and Science University in Portland is working on research similar to that at Newcastle University. Adult donor cells are forced to create a protein called nanog, which is only found in stem cells, yet the process is altered in a way that keeps the cells from forming into embryos. In 2008, researchers at MIT's Picower Institute for Learning and Memory pinpointed stem cells within the spinal cord that may lead to a new non-surgical treatment for spinal cord injuries.

In 2009, researchers at the Chinese Academy of Sciences in Beijing made great strides in stem cell cloning or in this case "reprogramming" when they took mature mouse skin cells and returned them to an embryonic-like state. The reprogrammed cells were injected into early-stage mouse embryos. The result of the study was that out of 37 stem cell lines created by reprogramming, three lines were the genesis of 27 live mouse births, one of which was able to sire offspring of his own. Similar experiments were successfully performed at the National Institute of Biological Sciences in Beijing.

Reprogramming has evolved recently to use proteins to manipulate cells to return to embryonic states. In the past, reprogramming was accomplished by using a virus to carry genes into a mature cell. This is problematic since the virus can cause cancer or spark changes in the target cell that are undesirable. New technology uses a wash made of four proteins that are associated with the genes. The wash is absorbed in the target cell where the proteins trigger additional protein changes, causing the reversion to a primitive state. The research that led to this process was done at Scripps Research Institute in La Jolla, California. Two biotech startups in California, Fate Therapeutics and Stemgent, Inc. are partnering to produce reprogrammed tissue to drug discovery firms. Another startup, iZumi Bio, Inc. which is also in California, is teaming up with Kyoto University to further research in reprogramming.

## 17) Stem Cells—A New Era of Regenerative Medicine Takes Shape

Many firms are conducting product development and research in the areas of skin replacement, vascular tissue replacement and bone grafting or regeneration. Stem cells, as well as transgenic organs harvested from pigs, are under study for use in humans. At its highest and most promising level, regenerative medicine may eventually utilize human stem cells to create virtually any type of replacement organ or tissue.

In one recent, exciting experiment, doctors took stem cells from bone marrow and injected them into the hearts of patients undergoing bypass surgery. The study showed that the bypass patients who received the stem cells were pumping blood 24% better than patients who had not received them.

In another experiment, conducted by Dr. Mark Keating at Harvard, the first evidence was shown that stem cells may be used for regenerating lost limbs and organs. The regenerative abilities of amphibians have long been known, but exactly how they do it, or how it could be applied to mammals, has been little understood. Much of the regenerative challenge lies in differentiation, or the development of stem cells into different types of adult tissue such as muscle and bone. Creatures such as amphibians have the ability to turn their complex cells back into stem cells in order to regenerate lost parts. In the experiment, Dr. Keating made a serum from the regenerating nub (stem cells) of a newt's leg and applied it to adult mouse cells in a petri dish. He observed the mouse cells to "de-differentiate," or turn into stem cells. In a later experiment, de-differentiated cells were turned back into muscle, bone and fat. These experiments could be the first steps to true human regeneration. Keating is continuing to make exciting breakthroughs in regenerative research.

The potential of the relatively young science of tissue engineering appears to be unlimited. Transgenics (the use of organs and tissues grown in laboratory animals for transplantation to humans) is considered by many to have great future potential, and improvements in immune system suppression will eventually make it possible for the human body to tolerate foreign tissue instead of rejecting it. There is also increasing theoretical evidence that malfunctioning or defective vital organs such as livers, bladders and kidneys could be replaced with perfectly functioning "neo-organs" (like spare parts) grown in the laboratory from the patient's own stem cells, with minimal risk of rejection.

The ability of most human tissue to repair itself is a result of the activity of these cells. The potential that cultured stem cells have for transplant medicine and basic developmental biology is enormous.

Diabetics who are forced to cope with daily insulin injection treatments could also benefit from engineered tissues. If they could receive a fully functioning replacement pancreas, diabetics might be able to throw away their hypodermic needles once and for all. This could also save the health care system immense sums, since diabetics tend to suffer from many ailments that require hospitalization and intensive treatment, including blindness, organ failure, diabetic coma and circulatory diseases.

Elsewhere, the harvesting of replacement cartilage, which does not require the growth of new blood vessels, is being used to repair damaged joints and treat urological disorders. Genzyme Corp. won FDA approval for its replacement cartilage product Carticel, the first biologic cell therapy to become licensed. Genzyme's process involves harvesting the patient's own cartilage-forming cells, and, from those cells, re-growing new cartilage in the laboratory. The physician then injects the new cartilage into the damaged area. Full regeneration of the replacement cartilage is expected to take up to 18 months. The Genzyme process can cost up to $30,000, compared to $10,000 for typical cartilage surgery. Other companies are exploring alternative methods that may be less expensive and therefore more attractive to payors.

In 2007, U.S. Army medical doctors were treating five soldiers with an extracellular matrix derived from pig bladders. The material is found in all animals and shows promise for healing and regenerating tissue (it is already in use by veterinarians to help repair torn ligaments in horses). The five soldiers in the test lost fingers in the war in Iraq and the experiment is to see how the matrix affects the wounds and if regeneration of any kind occurs.

---

***Companies to Watch:*** StemCells, Inc., in Palo Alto, California (www.stemcellsinc.com), is focusing on the use of stem cells to treat damage to major organs such as the liver, pancreas and central nervous system. ViaCord (formerly ViaCell, Inc. before its acquisition by PerkinElmer), in Boston, Massachusetts (www.viacord.com), develops therapies using umbilical cord stems. Also, their ViaCord product enables families to preserve their baby's umbilical cord at the time of birth for possible future use in treating over 40 diseases and genetic disorders. As of mid-2009, more than 185,000 units of cord blood stem cells had been banked.

---

***Internet Research Tip:***
For an excellent primer on genetics and basic biotechnology techniques, see:

**National Center for Biotechnology Information**
www.ncbi.nlm.nih.gov

---

## 18) Nanotechnology Converges with Biotech

Because of their small size, nanoscale devices can readily interact with biomolecules on both the surface and the inside of cells. By gaining access to so many areas of the body, they have the potential to detect disease and deliver treatment in unique ways. Nanotechnology will create "smart drugs" that are more targeted and have fewer side effects than traditional drugs.

Current applications of nanotechnology in health care include immunosuppressants, hormone therapies, drugs for cholesterol control, and drugs for appetite enhancement, as well as advances in imaging, diagnostics and bone replacement. For example, the NanoCrystal technology developed by Elan, a major biotechnology company, enhances drug delivery in the form of tiny particles, typically less than 2,000 nanometers in diameter. The technology can be used to provide more effective delivery of drugs in tablet form, capsules, powders and liquid dispersions. Abbot Laboratories uses Elan's technology to improve results in its cholesterol drug Tricor. Par Pharmaceutical Companies uses NanoCrystal in its Megace ES drug for the improvement of appetite in people with anorexia.

Since biological processes, including events that lead to cancer, occur at the nanoscale at and inside cells, nanotechnology offers a wealth of tools that are providing cancer researchers with new and innovative ways to diagnose and treat cancer. In America, the National Cancer Institute has established the Alliance for Nanotechnology in Cancer (http://nano.cancer.gov) in order to foster breakthrough research.

Nanoscale devices have the potential to radically change cancer therapy for the better and to dramatically increase the number of effective therapeutic agents. These devices can serve as customizable, targeted drug delivery vehicles capable of ferrying large doses of chemotherapeutic agents or therapeutic genes into malignant cells while sparing healthy cells, greatly reducing or eliminating the often unpalatable side effects that accompany many current cancer therapies.

At the University of Michigan at Ann Arbor, Dr. James Baker is working with molecules known as dendrimers to create new cancer diagnostics and therapies, thanks to grants from the National Institutes of Health and other funds. This is part of a major effort named the Michigan Nanotechnology Institute for Medicine and Biological Sciences.

A dendrimer is a spherical molecule of uniform size (five to 100 nanometers) and well-defined chemical structure. Dr. Baker's lab is able to build a nanodevice with four or five attached dendrimers. To deliver cancer-fighting drugs directly to cancer cells, Dr. Baker loads some dendrimers on the device with

folic acid, while loading others with drugs that fight cancer. Since folic acid is a vitamin, many proteins in the body will bind with it, including proteins on cancer cells. When a cancer cell binds to and absorbs the folic acid on the nanodevice, it also absorbs the anticancer drug. For use in diagnostics, Dr. Baker is able to load a dendrimer with molecules that are visible to an MRI. When the dendrimer, due to its folic acid, binds with a cancer cell, the location of that cancer cell is shown on the MRI. Each of these nanodevices may be developed to the point that they are able to perform several advanced functions at once, including cancer cell recognition, drug delivery, diagnosis of the cause of a cancer cell, cancer cell location information and reporting of cancer cell death. Universities that are working on the leading edge of cancer drug delivery and diagnostics using nanotechnology include MIT and Harvard, as well as Rice University and the University of Michigan.

Meanwhile, at the University of Washington, a research group led by Babak A. Parviz is investigating manufacturing methods that resemble plants and other natural organisms by "self-assembly." If man-made machines could be designed to assemble themselves, it could revolutionize manufacturing, especially on the nanoscale level. Researchers are studying ways to program the assembly process by sparking chemical synthesis of nanoscale parts such as quantum dots or molecules which then bind to other parts through DNA hybridization or protein interactions. The group led by Professor Parviz is attempting to produce self-assembled high performance silicon circuits on plastic. It is conceivable that integrated circuits, biomedical sensors or displays could be "grown" at rates exponentially faster than current processes.

## 19) Agricultural Biotechnology Scores Breakthroughs but Causes Controversy/Selective Breeding Offers a Compromise

Global panic over quickly rising food prices in 2008, coupled with the global economic crisis, gave the genetically modified seed industry the boost it needed. Agribio (agricultural biotechnology) has become a top agenda item in government and corporate research budgets, and consumer acceptance of genetically modified food products will grow quickly.

The biotech-era technology of "molecular farming" will soon lead to broad commercialization of human drug therapies that are grown via agricultural methods. For example, by inserting human genes into plants, scientists can manipulate them so they grow certain human proteins instead of natural plant proteins. The growth in plants of transgenic protein therapies for humans may become widespread. Such drug development methods may prove to be extremely cost-effective. At the same time, hundreds of antibodies produced in farm animals for use in human drug therapies are currently under development or in clinical trials.

Meanwhile, genetically modified foods (frequently referred to as "GM" for genetically modified, or "GMO" for genetically modified organisms) offer tremendous promise in agriculture—particularly in high-population nations like China and India. A study completed by the Agriculture Policy Research Center at the Chinese Academy of Sciences in Beijing in mid-2005 found immense potential in the use of genetically modified rice in China. The center's director estimates that Chinese farmers could increase their total annual income by $4 billion through planting GM rice, which would result in higher yields per acre. China has made massive investments in agricultural biotechnology research, and as of 2008, had committed and additional $3.5 billion over 12 years for continued study. However, the planting of GM rice seeds is currently authorized only in experimental plots within China.

Agricultural biotechnology became a significant commercial industry during the 1980s. It was fostered both by startups and by large chemical or seed companies. All of these players were focused on developing genetically modified seeds and plants that had higher yields, better nutritional qualities and/or resistance to diseases or insects. Additional traits of GM plants may include resistance to temperature and moisture extremes. According to the International Service for the Acquisition of Agri-biotech Applications (ISAAA), global acreage of GM crops rose 9% to reach 308.8 million acres in 2008, up from 282.4 million acres in 2007. This amounts to about 8% of the world's agricultural acreage. This is mostly in the U.S., but large amounts were also planted in Argentina, Canada, Mexico, India, Romania, Uruguay and South Africa. Meanwhile, GM seeds have the potential to create vast benefits in low-income nations where reliance on small farms or gardens is high and food is scarce.

At the same time, researchers are modifying the structural makeup of some plants in order to alter leaves, stems, branches, roots or seed structures. The ability to modify the nutritional makeup of plants can have highly desirable effects. For example, Mycogen, an affiliate of Dow AgroSciences (www.dowagro.com/mycogen), has developed sunflower seeds with higher levels of oleic and linoleic acids—acids with exceptional nutritional value. In 2009, Mycogen added 45 new grain corn hybrids, bringing its total grain corn lineup to 132 hybrids which are high-yield and insect and disease resistant. Seven additional silage hybrids, 29 corn hybrids and six new sunflower hybrids will be added to Mycogen's portfolio in 2010.

There are currently dozens of agribio food products on the market, including a range of fruits, vegetables and nuts. There is significant potential for rapid development of new products, thanks to the same technologies that are pushing development of human gene therapies in the pharmaceutical industry.

U.S. farmers have enjoyed greatly increased crop yields and crop quality thanks to GM seeds, and by some estimates as much as 70% of U.S. food may contain ingredients that have been grown with GM methods. In particular, U.S. farmers are reaping tremendous crops of GM soybeans (89% of the U.S. market), cotton (83%) and corn (61%). These crops eventually become ingredients in everything from baked goods to soft drinks to clothing.

Although scientists have been able to engineer highly desirable traits in GM seeds for crops (such as disease-resistance and insect-resistance), and the scientific community has given GM foods a clean bill of health for years, such modified foods have faced stiff resistance among many consumers, particularly in Europe. While many areas of biotechnology are controversial, agricultural biotech has been one of the largest targets for consumer backlash and government intervention in the marketplace. Consumer resistance to food products containing material grown in this manner is sometimes fierce.

Consumers in Europe have exhibited a strong fear of GM foods. It may stem in part from a cultural preference for locally grown, natural foods. Basic European grocery shopping habits and food preparation habits vary from those of U.S. consumers. For example, Europeans tend to shop today for tonight's meal, rather than stocking up on several days' worth of food as many Americans do. Europeans also suffered mightily from the outbreak of mad cow disease in 1996, and in subsequent outbreaks such as the one in the U.K. in 2003, which contributed to their concerns about food sources.

However, with food prices a major concern, and instances of riots in a number of third world countries due to short supplies and very high prices for staples such as rice, GM foods are becoming more acceptable around the world. For example, in Japan and South Korea, a number of manufacturers have begun using genetically engineered corn in soft drinks, snacks and other foods. This is a first, but the manufacturers cannot afford corn starch and corn syrup made from conventionally grown crops. According to Yoon Chang-gyu, director of the Korean Corn Processing Industry Association, non-engineered corn cost Korean millers about $450 a metric ton in early 2008, up from $143 in 2006. Prices for GM corn were considerably less at about $350 per ton.

Syngenta (www.syngenta.com), the result of the merger between the agricultural divisions of AstraZeneca and Novartis, is focused on seeds, crop protection products, insecticides and other agricultural products. With this focus, Syngenta is in a position to make some of the best research, development and marketing decisions. The firm's annual investment in research and development is substantial, at about 10% of revenues. In 2008, its sales of seeds of all types totaled about $2.4 billion, up from $2.0 billion in 2007.

Meanwhile, Monsanto, a major competitor to Syngenta, has invested heavily in biotech seed research with terrific results. From a 2002 loss of $1.7 billion, Monsanto has evolved to a 2008 operating profit of about $2.02 billion (on sales of about $11.3 billion). The company accomplished the turnaround by continuing to invest in genetic engineering and market its products despite protests and controversy.

A particular concern among farmers in many parts of the world is that GM crops may infest neighboring plants when they pollinate, thus triggering unintended modification of plant DNA. In any event, there is a vast distrust of GM foods in certain locales. U.S. food growers and processors face significant difficulty exporting to the European Union (EU) because of the reliance that American farmers place on GM seeds.

The European Union, as well as specific nations in Europe, has kept many regulations in place that make the use of GM seeds or the import of GM food products a difficult to impossible task. These restrictions remain a hot topic of debate at the World Trade Organization and elsewhere. Meanwhile, a handful of localities in the U.S. have banned or restricted the planting of GM seeds, hoping to protect traditional crops that local growers are widely-known for. A typical restriction is to require that GM seeds

be planted at least a certain distance away from non-GM crops.

Some anti-GM activists have arguments with big business—particularly with the giant corporations like Monsanto that make GM seeds. Some people have accused Monsanto of persecuting farmers who appear to be using Monsanto-developed seeds without paying for them. The company has also received criticism for its history of manufacturing chemicals that have risen to varying levels of infamy, such as SBCs, DDT and Agent Orange. Unfortunately, protestors are sometimes violent or destructive.

While concern in many European countries continues, the number of acres sown with genetically modified corn (which is the only transgenic seed allowed by EU rules) is slowly growing due to the fact that some farmers can no longer ignore the cost savings and improved crop yields. In France, for example, the number of acres planted with GM corn (specifically, a strain produced by Monsanto) in 2006 was 12,350, up from 1,215 in 2005 according to the European Association for Bioindustries (www.abeurope.info). However, in early 2008, the French government banned the use of the Monsanto strain, sparking outrage from French farmers and delighting environmentalists.

Meanwhile, the U.S. Food and Drug Administration (FDA) declared food derived from cloned cows, pigs and goats to be safe for consumption. The European Food Safety Authority has also declared cloned animal produce to be safe. However, a number of food companies, including Smithfield Foods, Inc., Kraft Foods, Inc. and Tyson Foods, Inc., have pledged not to use milk or meat from cloned livestock.

A landmark compromise may be on the horizon thanks to a new selective breeding technique that introduces no foreign DNA such as that used in GM seeds. The technology uses old-school practices in which plants with desirable characteristics such as longer shelf life or resistance to insects are crossbred to create new, hardier specimens. The new twist to the old technique is the use of genetic markers, which make it much easier to isolate plants with a positive trait and the gene that causes it. New plants can also be quickly tested for the presence of the isolated gene. The technology cuts traditional selective breeding time in half.

A number of companies are utilizing gene markers in their breeding programs. Arcadia Biosciences is hoping to develop seeds for wheat that can be eaten by people with the intestinal disorder called Celiac Disease, which affects 1% of Americans and 4% of Europeans. Arcadia (www.arcadiabio.com ) has also developed technologies that enable crops to utilize nitrogen (part of common fertilizers) more efficiently, thus reducing the amount of fertilizer needed overall. It is working to develop plants that use water more efficiently, thus producing high crop yields in low water conditions. It has even developed technology that enables plants to be irrigated with saltwater. Monsanto is working on soybeans using the markers to make veggie burgers that taste more like beef. In late 2008, Syngenta announced the release of technology that makes soybeans resistant to aphids.

Genetic markers are not new, but the ability to use them in a cost effective manner is relatively recent thanks to falling costs since the year 2000. Where it once took several dollars to conduct a plant scan, the same test can now be conducted for pennies, making testing on a large scale possible. Look for crop biotechnology companies including DuPont, Monsanto and Syngenta to invest millions of dollars in selective breeding assisted by gene markers over the near- to mid-term.

Yet another technology may radically impact the world's food supply. Food scientists at Sangamo Biosciences (www.sangamo.com) have developed naturally occurring proteins that bind to DNA called zinc fingers. The fingers (so called because of their shape) can be used to genetically modify cells to produce desired effects such as crop yield, taste or drought resistance in plants. They afford very precise changes to DNA which translates into better control when modifying plants and quicker development times compared to typical genetic modification. Zinc fingers may also be far more acceptable to groups who are against GM foods, because the elements of the zinc finger do not remain in a plant for more than a few days. Dow AgroSciences (a subsidiary of Dow Chemical that is focused on crop production) has invested $20 million in Sangamo, hoping to compete with Monsanto and Syngenta's agribio success.

Nanotechnology is affecting foods as well. As of mid-2009, there were three nano-engineered foods on the market according to The Project on Emerging Nanotechnologies. They were Canola Active Oil, which contains an additive called nanodrops that carry vitamins, minerals and phytochemicals; Nanotea, which is formulated for better taste and increase its selenium supplement qualities; and Nanoceuticals Slim Shake Chocolate, a chocolate-flavored diet shake that uses nanoclusters to improve

taste and health benefits without the need for added sugar.

---

**Company to Watch:**

Sacramento, California-based Ventria, www.ventria.com, has received approval from the USDA every year since 1999 to produce crops for use in biopharmaceuticals. Ventria plants self-pollinating rice or barley specifically because they produce large quantities of proteins by nature and because they are not pollinated by wind or insect activity. Thus, they theoretically should have no effect on nearby traditional plantings.

---

## 20) Focus on Vaccines

Vaccines in general are experiencing a renaissance as drug companies are once again investing heavily in technologies to protect against bioterrorism and a growing number of medical diseases. Not since the 1950s, when Jonas Salk introduced his world-changing vaccine that virtually wiped out polio, have so many vaccines made such an impact on general health and the health care market. Starting in 2006, for example, Merck licensed three new vaccines, including Gardasil for cervical cancer and another treatment to prevent shingles. Merck posted revenues in vaccines alone in 2008 of $4.2 billion (up from $2.2 billion in 2006). By mid 2009, Merck's Vaccines unit offered 11 vaccines. Wyeth Pharmaceuticals' Prevnar vaccine alone sold and estimated $2.8 billion in 2008.

In May 2004, the U.S. Congress passed a bill with great bipartisan support that provided for $5.6 billion in funding over ten years for stockpiling vaccines and other medicines in defense of possible bioterror attacks on U.S. population centers. The BioShield bill further enhances the possibility of fast track research in the event of a national emergency, and allows government officials to distribute certain treatments even if they have not yet been approved by the FDA. The intent is to create effective responses to attacks from chemical or biological weapons. Among the greatest concerns are vaccines against and treatments for anthrax and smallpox. Over 100 biotechnology firms are producing products that could be candidates for BioShield contracts.

The biggest single bioterror threat to U.S. population centers is considered to be anthrax. A small parcel of anthrax sprayed over a major city could kill hundreds of thousands of people, and it is relatively easy to manufacture and hard to combat. If the release is detected or the first cases are rapidly diagnosed, quick action could save many lives.

Providing the exposed population with antibiotics followed by vaccination could be lifesaving for persons who would otherwise become ill with untreatable inhalation anthrax in the subsequent few weeks. Prophylactic antibiotics alone will prevent disease in persons exposed to antibiotic-susceptible organisms, but incorporating vaccination into the treatment regime can greatly reduce the length of treatment with antibiotics. Without vaccination, antibiotics must be continued for 60 days. However, if effective vaccination can be provided, antibiotic treatment can be reduced to 30 days.

Smallpox is also considered to be a major threat. The national stockpile (fewer than 7 million doses of vaccinia virus vaccine) is insufficient to meet national and international needs in the event of a major bioterror attack. The stockpile is also deteriorating and has a finite life span. The vaccine was made using the traditional method of scarifying and infecting the flanks and bellies of calves and harvesting the infected lymph. No manufacturer exists today with the capability to manufacture calf lymph vaccine by the traditional method. Replacing the stockpile will require the development and licensure of a new vaccine using modern cell-culture methods. This development program, which will include process development, validation of a new manufacturing process, and extensive clinical testing, will be expensive and may take several years.

Dozens of firms, particularly startups, are working on vaccines and antidotes for biological weapons with hopes of obtaining BioShield funds. Overall, the NIH grants $500 to $600 million per year (approximately one-third of the U.S. annual biodefense budget) into product development. The Department of Defense Joint Vaccine Acquisition Program has several experimental vaccines in development.

A total of seven major new, high-security, infectious disease research centers have been funded in the United States, accomplished largely with the assistance of government grants, including military funds. Much of the focus of these labs will be on biodefense, including vaccines. These centers include a 15,000 square foot lab at the CDC in Atlanta, Georgia. Additional new labs include three facilities in Fort Detrick, Maryland; as well as one each in Galveston, Texas (at the University of Texas Medical Branch-UTMB); Hamilton, Montana (Rocky Mountain Laboratories); and Boston, Massachusetts (Boston University Medical Center).

## 21) Ethical Issues Abound

Significant ethical issues face the biotech industry as it moves forward. They include, for example, the ability to determine an individual's likelihood to develop a disease in the future, based on his or her genetic makeup today; the potential to harvest replacement organs and tissues from animals or from cloned human genetic material; and the ability to alter genetically the basic foods that we eat. These are only a handful of the powers of biotechnology that must be dealt with by society. Watch for intense, impassioned discussion of such issues and a raft of governmental regulation as new technologies and therapies emerge.

The biggest single issue may be privacy. Who should have access to your personal genetic records? Where should they be stored? How should they be accessed? Can you be denied employment or insurance coverage due to your genetic makeup?

---

*Internet Research Tip:*
For the latest biotech developments check out www.biospace.com, a private sector portal for the biotech community, and www.bio.org, the web site of the highly regarded Biotechnology Industry Organization.

---

## 22) Technology Discussion—Genomics

The study of genes as a resource for the commercial development of new drugs received a significant boost from computer technology in the late 1970s. Frederick Sanger, a chemist, and Walter Gilbert, a biochemist, developed what is known as DNA sequencing technology, receiving a Nobel Prize in 1980 for their effort. In the same way that computer technology enabled the rapid growth of the Internet industry, computerization has been the catalyst for the booming biotech industry. For example, Sanger and Gilbert's computerized DNA sequencing technology enables scientists to collect massive amounts of data on human genes at high speed, analyzing how certain genes are connected with specific diseases. Using this technology, the gene responsible for Parkinson's disease was mapped in only nine days—a stark contrast to the nine years required to determine the gene connected with cystic fibrosis using traditional methods. Genomics, the mapping and analysis of genes and their uses, is the basic building block of the biopharmaceuticals business. Pharmacogenomics is the study of genomics for the purpose of creating new pharmaceuticals.

Genes are made up of DNA and reside on densely packed fibers called chromosomes. The genetic information encoded in a gene is copied into RNA (ribonucleic acid—closely related to DNA) and then used to assemble proteins. Think of DNA as a blueprint and RNA as the builder.

The human body contains about 75 trillion cells. Each cell contains 46 chromosomes, arranged in 23 pairs. Each chromosome is a strand of DNA. Each strand of DNA is composed of thousands of segments representing different genes. The sum total of the DNA contained in all 46 chromosomes is called the human genome. Until recently, it was estimated that the number of different genes in the human body would be in the 100,000 range, due to the complexity of systems within human beings. However, as genome mapping neared completion, the number appears to be a surprisingly small 20,000 to 25,000.

In October 2006, the X Prize Foundation announced a prize of $10 million for the first group to accurately produce complete genomes of 100 humans in 10 days or less and while spending less than $10,000 per genome. One of the genomics industry's goals is to able to market sub-$1,000 genomic studies to people who want a complete map of their DNA. This prize will be a big boost and as of mid 2009 was still up for grabs.

---

*Internet Research Tip:*
For a superior, easy-to-understand, illustrated resource on genetics and molecular biology, see: www.dnaftb.org/dnaftb/15/concept/.

---

## 23) Technology Discussion—Proteomics

Proteomics is the study of the proteins that a gene produces. A complete set of genetic information is contained in each cell, and this information provides a specific set of instructions to the body. The body carries out these instructions via proteins. Genes encode the genetic information for proteins.

All living organisms are composed largely of proteins. Each protein is a large, complex molecule composed of amino acids. Proteins have three main cellular functions: 1) they act as enzymes, hormones and antibodies, 2) they provide cell structure, and 3) they are involved in cell signaling and cell communication functions.

Proteins are important to researchers because they are the link between genes and pharmaceutical development. They indicate which genes are expressed or are being used. They are important for understanding gene function. They also have unique shapes or structures. Understanding these structures

and how potential pharmaceuticals will bind to them is a key element in drug design. Proteomic researchers seek to determine the unique role of each of the hundreds of thousands of proteins in the human body, as well as the relationships that such proteins have with each other and with various diseases. Microarrays enable the high-speed analysis of these proteins and the discovery of SNPs (see below).

## 24) Technology Discussion—Microarrays

In the laboratory, scientists use microarrays to map the DNA of a patient's tissues. For example, a physician may take a small biopsy of a cancer patient's tumor. That biopsy would be placed into microarray equipment. There, the microarray would arrange hundreds or thousands of microscopic dots of tumor material onto glass laboratory slides. Lasers would scan the slides, and computer software would compare the contents of the slides to vast databases, in order to determine the exact genes in the tumor. Such information may be extremely useful in treating the patient. Genetic information scanned from microarrays has been used to create giant bioinformatics databases. Ideally, such databases could analyze the genetic blueprint of a patient's tumor in order to assist a physician in determining the drugs that would best treat the specific genes that have mutated into cancerous material, and thereby have the best opportunity to cure the patient.

## 25) Technology Discussion—DNA Chips

Several chip platforms have been developed to facilitate molecular biology research. For example, industry leader Affymetrix (www.affymetrix.com) has developed chips programmed to act like living cells. These chips can run thousands of tests in short order to help scientists determine a specific gene's expression.

Scientists know that a mutation—or alteration—in a particular gene's DNA often results in a certain disease. However, it can be very difficult to develop a test to detect these mutations, because most large genes have many regions where mutations can occur. For example, researchers believe that mutations in the genes BRCA1 and BRCA2 cause as many as 60 percent of all cases of hereditary breast and ovarian cancers. But there is not one specific mutation responsible for all of these cases. Researchers have already discovered over 800 different mutations in BRCA1 alone.

The DNA chip is a tool used to identify mutations in genes like BRCA1 and BRCA2. The chip, which consists of a small glass plate encased in plastic, is manufactured somewhat like a computer microchip. On the surface, each chip contains thousands of short, synthetic, single-stranded DNA sequences, which together, add up to the normal gene in question. To determine whether an individual possesses a mutation for BRCA1 or BRCA2, a scientist first obtains a sample of DNA from the patient's blood as well as a control sample—one that does not contain a mutation in either gene.

The researcher then denatures the DNA in the samples, a process that separates the two complementary strands of DNA into single-stranded molecules. The next step is to cut the long strands of DNA into smaller, more manageable fragments and then to label each fragment by attaching a fluorescent dye. The individual's DNA is labeled with green dye and the control, or normal, DNA is labeled with red dye. Both sets of labeled DNA are then inserted into the chip and allowed to hybridize (bind) to the synthetic BRCA1 or BRCA2 DNA on the chip. If the individual does not have a mutation for the gene, both the red and green samples will bind to the sequences on the chip.

If the individual does possess a mutation, the individual's DNA will not bind properly in the region where the mutation is located. The scientist can then examine this area more closely to confirm that a mutation is present. (Explanation provided courtesy of National Institutes of Health.)

### 26) Technology Discussion—SNPs ("Snips")

The holy grail of genomics is the search for single nucleotide polymorphisms (SNPs). These are DNA sequence variations that occur when a single nucleotide (A, T, C or G) in the genome sequence is altered. For example an SNP might change the DNA sequence AAGGCTAA to ATGGCTAA. SNPs occur in every 100 to 1,000 bases along the 3 billion-base human genome. SNPs can occur in both coding (gene) and noncoding regions of the genome. Many SNPs have no effect on cell function, but scientists believe others could predispose people to diseases or influence their response to a drug.

Variations in DNA sequence can have a major impact on how humans react to disease; environmental factors such as bacteria, viruses, toxins and chemicals; and drug therapies. This makes SNPs of great value for biomedical research and for developing pharmaceutical products or medical diagnostics. Scientists believe SNP maps will help them identify the multiple genes associated with such complex diseases as cancer, diabetes, vascular disease and some forms of mental illness.

### 27) Technology Discussion—Combinatorial Chemistry

After researchers determine which protein is involved in a specific disease, they use combinatorial chemistry to find the molecule that controls the protein. High-throughput screening techniques allow scientists to use automatic fluid handlers to mix chemicals under specific test conditions at extremely high speed. Combinatorial chemistry can generate thousands of chemical compound variations in a few hours. Previously, traditional chemistry methods could have required several weeks or even months to do the same work. High-throughput screening enables automated, computerized machines to screen as many as 100,000 chemical compounds daily, seeking potential molecules as drug candidates.

One company, CombinatoRx (www.combinatorx.com), uses a combinatorial array to mix together drugs already on the market to find combinations that could have new and unforeseen uses. The company found hundreds of potential uses for combinations of drugs whose patents had expired and has patented these precise mixtures for new uses. It has received rapid FDA approval in many cases because all the drugs being combined have proven records of safety. In one particularly bizarre instance, the firm found that by combining a particular sedative with an antibiotic, it had made an effective cancer-fighting agent.

### 28) Technology Discussion—Synthetic Biology

Scientists have followed up on the task of mapping genomes by attempting to directly alter them. This effort has gone past the point of injecting a single gene into a plant cell in order to provide a single trait, as in many agricultural biotech efforts. There are now several projects underway to create entirely new versions of life forms, such as bacteria, with genetic material inserted in the desired combination in the laboratory. These include the BioBricks Foundation, the Registry of Standard Biological Parts and the Synthetic Genomics Study, all of which are part of the scientific community at MIT.

Synthetic biology can be defined as the design and construction of new entities, including enzymes and cells, or the reformatting of existing biological systems. This science capitalizes on previous advances in molecular biology and systems biology, by applying a focus on the design and construction of unique core components that can be integrated into larger systems in order to solve specific problems.

In June 2004, the first international meeting on synthetic biology was held at MIT. Called Synthetic Biology 1.0, the conference brought together researchers who are working to design and build biological parts, devices and integrated biological systems; develop technologies that enable such work; and place this research within its current and future social context. Synthetic Biology 2.0 was held in May 2006 at UC Berkeley. In fact, the National Science Foundation agreed, in mid 2006, to invest $16 million in a five-year grant to fund a new Synthetic Biology Engineering Research Center ("SynBERC) at UC Berkeley. An additional $4 million for the project has been raised from other sources, and the NSF is offering the possibility of a further five-year grant. Synthetic Biology 3.0 was held in Zurich in June 2007, and SB4.0 took place in October 2008 in Hong Kong.

Leading proponents of synthetic biology include Dr. Craig Venter, well known for his efforts in sequencing the human genome. Elsewhere, at MIT, Dr. Drew Endy and colleague Tom Knight are working on a concept called BioBricks, which are strands of DNA with connectors at each end. They can be assembled into higher-level components. Meanwhile, Dr. James J. Collins is working on using synthetic biology to program bacteria to fight disease at Boston University.

## 29) Technology Discussion—Recombinant DNA

In today's recombinant DNA therapies, the implantation of selected strands of DNA into bacteria allows large-scale production of hormones, antibodies and other exciting new drugs. Additions to the medical arsenal developed from this technique include interferons, growth factor hormone, anticoagulants, human insulin (as opposed to the cow and pig insulin that diabetics have used for 60 years) and more effective immunosuppressive drugs. An anticoagulant proven highly effective for some heart attack patients, TPA, may dramatically reduce heart damage during an attack and cut hospital stays. These drugs are often expensive, but they may greatly reduce the duration of hospital confinement and, in some cases, the need for any hospitalization at all. Long-term, potential applications of this technology include genetically-modified seeds, vaccines, production of insulin, production of blood clotting factors and the production of recombinant pharmaceuticals.

## 30) Technology Discussion—Polymerase Chain Reaction (PCR)

PCR, also called "molecular photocopying", is a fast, inexpensive method of replicating genetic material in the form of small segments of DNA. The technique was created by Kary B. Mullis of La Jolla, California. His efforts were awarded one-half of the Nobel Prize in Chemistry in 1993, which he shared with another leading DNA researcher, Michael Smith of the University of British Columbia, Canada.

PCR creates sufficient amounts of DNA to allow detailed analysis, which can be vital to researchers. This may lead to better diagnosis and treatment of diseases such as AIDS, Lyme disease, tuberculosis, viral meningitis, cystic fibrosis, Huntington's chorea, sickle cell anemia and many others. PCR also shows promise in providing cancer susceptibility warnings.

Once amplified, the DNA produced by PCR can be used in many different laboratory procedures. For example, most mapping techniques in the Human Genome Project (HGP) rely on PCR.

PCR is also valuable in a number of emerging laboratory and clinical techniques, including DNA fingerprinting, detection of bacteria or viruses (particularly AIDS), and diagnosis of genetic disorders.

# Chapter 2

# BIOTECH & GENETICS INDUSTRY STATISTICS

# Biotech Industry Overview

| Global | Amount | Units | Year | Source |
|---|---|---|---|---|
| Public & Private Biotech Companies | 4,717 | Companies | 2008 | E&Y |
| Public Biotech Companies | 776 | Companies | 2008 | E&Y |
| Number of Employees | 200.7 | Thousand | 2008 | E&Y |
| Biotech Revenues | 89.7 | Bil. US$ | 2008 | E&Y |
| Increase from 2007 | 12 | % | 2008 | E&Y |
| R&D Expenses | 31.7 | Bil. US$ | 2008 | E&Y |
| Net Income | -1.4 | Bil. US$ | 2008 | E&Y |
| Total Biotech Company Financing | 16.0 | Bil. US$ | 2008 | E&Y |
| Venture Financing | 6.0 | Bil. US$ | 2008 | E&Y |
| Total Pharmaceutical Sales, Worldwide | 773.1 | Bil. US$ | 2008 | IMS |
| Total R&D Expenses, PhRMA Member Companies | 47.9 | Bil. US$ | 2008 | PhRMA |
| Total R&D Expenses, All Pharmaceutical Companies | 65.2 | Bil. US$ | 2008 | Burrill |

| U.S. | | | | |
|---|---|---|---|---|
| Public & Private Biotech Companies | 1,754 | Companies | 2008 | E&Y |
| Public Biotech Companies | 371 | Companies | 2008 | E&Y |
| Revenues | 66.1 | Bil. US$ | 2008 | E&Y |
| % Increase | 8 | % | 2008 | E&Y |
| R&D Expenses | 25.3 | Bil. US$ | 2008 | E&Y |
| Net Income | 0.4 | Bil. US$ | 2008 | E&Y |
| Number of Employees | 128.2 | Thousand | 2008 | E&Y |
| Venture Financing | 4.5 | Bil. US$ | 2008 | E&Y |
| Area of Biotech Crops | 62.5 | Mil. Hectares | 2008 | ISAAA |
| Prescription Drug Sales, Domestic | 291.0 | Bil. US$ | 2008 | IMS |
| Number of FDA Approvals for New Drugs (NDAs) | 88 | Approvals | 2008 | FDA |
| Number of Approvals for New Molecular Entities (NMEs) | 24 | Approvals | 2008 | FDA |
| Patents Granted for "Multicellular Living Organisms & Unmodified Parts Thereof & Related Processes" | 762 | Patents | 2008 | USPTO |
| Pharmaceutical Industry Funding, NIH | 30.4 | Bil. US$ | 2009 | NIH |
| Total Requested Budget for Biological Science Research, U.S. (NSF) | 733.0 | Mil. US$ | 2010 | NSF |
| Mean Annual Salary for Biochemists & Biophysicists | 88,450 | US$ | May-08 | BLS |
| Average Cost of Developing a Biologic Drug | 1.2 | Bil. US$ | 2005 | Tufts |

| PhRMA[1] Member Statistics, U.S. | | | | |
|---|---|---|---|---|
| Pharmaceutical Sales, Domestic | 189.3 | Bil. US$ | 2008 | PhRMA |
| % Generic (by volume) | 72 | % | 2008 | PhRMA |
| Pharmaceutical Sales, Foreign[2] | 99.0 | Bil. US$ | 2008 | PhRMA |
| Pharmaceutical R&D Spending, Domestic | 38.4 | Bil. US$ | 2008 | PhRMA |
| as a Percentage of All Sales | 20.3 | % | 2008 | PhRMA |
| Share of R&D Spending by Function: | | | | |
| Prehuman/Preclinical | 27.3 | % | 2007 | PhRMA |
| Phase I | 7.4 | % | 2007 | PhRMA |
| Phase II | 13.0 | % | 2007 | PhRMA |
| Phase III | 28.5 | % | 2007 | PhRMA |
| Approval | 5.0 | % | 2007 | PhRMA |
| Phase IV | 13.4 | % | 2007 | PhRMA |
| Uncategorized | 5.2 | % | 2007 | PhRMA |

[1] PhRMA = Pharmaceutical Research and Manufacturers Association, a group of leading pharmaceutical and biotechnology companies in the U.S.  [2] Not including foreign divisions of foreign companies.

E&Y = Ernst & Young; IMS = IMS Health; Burrill = Burrill & Company; ISAAA = International Service for the Acquisition of Agri-Biotech Applications; Tufts = Tufts Center for the Study of Drug Development; FDA = U.S. Food & Drug Administration; USPTO = U.S. Patent & Trademark Office; NIH = U.S. National Institutes of Health; NSF = U.S. National Science Foundation; BLS = U.S. Bureau of Labor Statistics

# Global Biotechnology at a Glance: 2008

| (Public Company Data) | (Revenues, Expenses & Losses in Millions of US$) | | | | |
|---|---|---|---|---|---|
| | Global | U.S. | Europe | Canada | Asia-Pacific |
| Revenues | 89,648 | 66,127 | 16,515 | 2,041 | 4,965 |
| R&D Expense | 31,745 | 25,270 | 5,171 | 703 | 601 |
| Net Loss | -1,443 | 417 | -702 | -1,143 | -14 |
| Number of Employees | 200,760 | 128,200 | 49,060 | 7,970 | 15,530 |

| (Number of Companies) | | | | | |
|---|---|---|---|---|---|
| Public companies | 776 | 371 | 178 | 72 | 155 |
| Public & private companies | 4,717 | 1,754 | 1,836 | 358 | 769 |

Numbers may appear inconsistent because of rounding. Employment totals rounded to the nearest hundred in the U.S., and to the nearest 10 in other regions.

Source: Ernst & Young, Beyond Borders: Global Biotechnology Report 2009

Plunkett Research, Ltd.

www.plunkettresearch.com

# Growth in Global Biotechnology:
# 2007-2008

*(Revenues, Expenses & Losses in Millions of US$)*

| *(Public Company Data)* | 2008 | 2007 | Change |
|---|---|---|---|
| Revenues | 89,648 | 80,344 | 12% |
| R&D Expense | 31,745 | 26,881 | 18% |
| Net Loss | -1,443 | -3,055 | -53% |
| Number of Employees | 200,760 | 201,690 | -0.5% |

| *(Number of Companies)* | | | |
|---|---|---|---|
| Public Companies | 776 | 815 | -5% |
| Public & Private Companies | 4,717 | 4,799 | -2% |

2008 financials largely represent data from 1 January 2008 through 31 December 2008.  2006 financials largely represent data from 1 January 2007 through 31 December 2007.  Numbers may appear inconsistent because of rounding.

Source: Ernst & Young, Beyond Borders: Global Biotechnology Report 2009

Plunkett Research, Ltd.

www.plunkettresearch.com

# U.S. Biotechnology at a Glance: 2007-2008

*(In Billions of US$)*

| | Public Companies | | | Industry Total | | |
|---|---|---|---|---|---|---|
| | 2008 | 2007 | % Change | 2008 | 2007 | % Change |
| Product Sales | 54.1 | 49.9 | 8.4% | 57.0 | 52.7 | 8.0% |
| Revenues | 66.1 | 61.0 | 8.4% | 70.1 | 64.9 | 8.0% |
| R&D Expense | 25.3 | 21.0 | 20.5% | 30.4 | 26.1 | 16.8% |
| Net Loss | 0.4 | -0.1 | -430.7% | -3.7 | -4.2 | -11.2% |
| | | | | | | |
| Market Capitalization | 343.8 | 369.2 | -6.9% | — | — | — |
| Total Financings | 8.6 | 15.9 | -46.3% | 13.0 | 21.4 | -39.2% |
| Number of IPOs | 1 | 22 | -95.5% | 1 | 22 | -95.5% |
| Number of Companies | 371 | 395 | -6.1% | 1,754 | 1,758 | -0.2% |
| Employees | 128,200 | 131,300 | -2.4% | 190,400 | 192,600 | -1.1% |

Data were generally derived from year-end information (31 December). 2008 data are estimates based on January–September quarterly filings and preliminary annual financial performance data for some companies. 2008 employee data are obtained from 10-Ks at time of publishing and include a combination of 2007 and 2008 employee data. The 2007 estimates have been revised for compatibility with 2008 data. Numbers may appear inconsistent because of rounding.

Source: Ernst & Young, Beyond Borders: Global Biotechnology Report 2009

Plunkett Research, Ltd.

www.plunkettresearch.com

# Biotechnology Financing, U.S., Europe & Canada: 2007-2008

*(In Millions of US$)*

| 2008 | U.S. | Europe | Canada |
|---|---|---|---|
| IPO | 6 | 111 | 0 |
| Follow-on & other offerings | 8,547 | 1,115 | 271 |
| Venture financing | 4,445 | 1,369 | 207 |
| **Total** | **12,998** | **2,595** | **478** |

| 2007 | U.S. | Europe | Canada |
|---|---|---|---|
| IPO | 1,238 | 1,010 | 5 |
| Follow-on & other offerings | 14,689 | 4,880 | 702 |
| Venture financing | 5,464 | 1,604 | 353 |
| **Total** | **21,391** | **7,494** | **1,060** |

| Change | U.S. | Europe | Canada |
|---|---|---|---|
| IPO | -99.5% | -89% | -100% |
| Follow-on & other offerings | -42% | -77% | -61% |
| Venture financing | -19% | -15% | -41% |
| **Total** | **-39%** | **-66%** | **-55%** |

Numbers may appear inconsistent because of rounding. European numbers are based on a conversion from Euros to U.S. dollars. Percent changes are calculations by Plunkett Research.

Source: Ernst & Young, Beyond Borders: Global Biotechnology Report 2009

Plunkett Research, Ltd.

www.plunkettresearch.com

# Quarterly Breakdown of Biotechnology Financings, U.S. & Canada: 2008

*(In Millions of US$)*

| | 1st Quarter | | 2nd Quarter | | 3rd Quarter | | 4th Quarter | | Total | |
|---|---|---|---|---|---|---|---|---|---|---|
| | U.S. | Canada | U.S. | Canada | U.S. | Canada | U.S. | Canada | U.S. | Canada |
| **IPO** | 6 | 0 | 0 | 0 | 0 | 0 | 0 | 0 | 6 | 0 |
| # of Financings | 1 | 0 | 0 | 0 | 0 | 0 | 0 | 0 | 1 | 0 |
| **Follow-on** | 606 | 41 | 278 | 11 | 831 | 28 | 0 | 0 | 1,715 | 80 |
| # of Financings | 8 | 3 | 4 | 2 | 8 | 3 | 0 | 0 | 20 | 8 |
| **Venture** | 1,202 | 22 | 1,192 | 48 | 1,177 | 120 | 875 | 17 | 4,445 | 207 |
| # of Financings | 97 | 8 | 65 | 5 | 77 | 6 | 59 | 6 | 298 | 25 |
| **Other** | 2,120 | 41 | 879 | 48 | 3,427 | 55 | 406 | 0 | 6,832 | 191 |
| # of Financings | 41 | 12 | 48 | 15 | 43 | 10 | 36 | 12 | 168 | 49 |
| **Total** | 3,934 | 104 | 2,348 | 107 | 5,435 | 203 | 1,281 | 64 | 12,998 | 478 |
| # of Financings | 147 | 23 | 117 | 22 | 128 | 19 | 95 | 18 | 487 | 82 |

Numbers may appear inconsistent because of rounding.

Source: Ernst & Young, Beyond Borders: Global Biotechnology Report 2009

Plunkett Research, Ltd.

www.plunkettresearch.com

# Ernst & Young Survival Index, U.S. & Canadian Biotechnology Companies: 2007-2008

| | U.S. | | | |
| | 2008 | | 2007 | |
|---|---|---|---|---|
| | Number of Companies | Percent of Total | Number of Companies | Percent of Total |
| More than 5 years of cash | 76 | 20% | 124 | 31% |
| 3–5 years of cash | 18 | 5% | 53 | 13% |
| 2–3 years of cash | 41 | 11% | 39 | 10% |
| 1–2 years of cash | 74 | 20% | 81 | 21% |
| Less than 1 year of cash | 162 | 44% | 98 | 25% |
| **Total public companies** | **371** | | **395** | |

| | Canada | | | |
| | 2008 | | 2007 | |
|---|---|---|---|---|
| | Number of Companies | Percent of Total | Number of Companies | Percent of Total |
| More than 5 years of cash | 14 | 19% | 37 | 43% |
| 3-5 years of cash | 3 | 4% | 1 | 1% |
| 2-3 years of cash | 0 | 0% | 4 | 4% |
| 1-2 years of cash | 14 | 19% | 10 | 12% |
| Less than 1 year of cash | 41 | 57% | 32 | 39% |
| **Total public companies** | **72** | | **84** | |

Numbers may appear inconsistent because of rounding.

Source: Ernst & Young, Beyond Borders: Global Biotechnology Report 2009

Plunkett Research, Ltd.

www.plunkettresearch.com

# The U.S. Drug Discovery & Approval Process

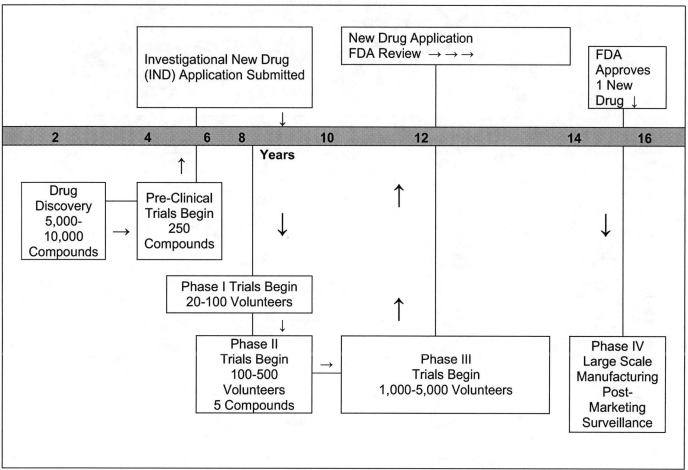

Source: Pharmaceutical Research and Manufacturers of America (PhRMA), *PhRMA Annual Membership Survey*, 2009

Plunkett Research, Ltd.

www.plunkettresearch.com

# U.S. FDA Approval Times for New Drugs: 1993-2008

| Calendar Year | Priority | | | Standard | | |
| --- | --- | --- | --- | --- | --- | --- |
| | Number Approved | Median FDA Review Time (months) | Median Total Approval Time (months) | Number Approved | Median FDA Review Time (months) | Median Total Approval Time (months) |
| 1993 | 19 | 16.3 | 20.5 | 51 | 20.8 | 26.9 |
| 1994 | 16 | 13.9 | 14.0 | 45 | 16.8 | 21.0 |
| 1995 | 16 | 7.9 | 7.9 | 67 | 16.2 | 18.7 |
| 1996 | 29 | 7.8 | 7.8 | 102 | 15.1 | 17.8 |
| 1997 | 20 | 6.3 | 6.4 | 101 | 14.7 | 15.0 |
| 1998 | 25 | 6.2 | 6.4 | 65 | 12.0 | 12.0 |
| 1999 | 28 | 6.1 | 6.1 | 55 | 12.0 | 13.8 |
| 2000 | 20 | 6.0 | 6.0 | 78 | 12.0 | 12.0 |
| 2001 | 10 | 6.0 | 6.0 | 56 | 12.0 | 14.0 |
| 2002 | 11 | 13.8 | 19.1 | 67 | 12.7 | 15.3 |
| 2003 | 14 | 7.7 | 7.7 | 58 | 11.9 | 15.4 |
| 2004* | 29 | 6.0 | 6.0 | 89 | 11.8 | 12.7 |
| 2005* | 22 | 6.0 | 6.0 | 59 | 11.8 | 13.1 |
| 2006* | 21 | 6.0 | 6.0 | 80 | 12.0 | 13.0 |
| 2007* | 23 | 6.0 | 6.0 | 55 | 10.2 | 10.4 |
| 2008* | 18 | 6 | 6 | 70 | 13 | 13 |

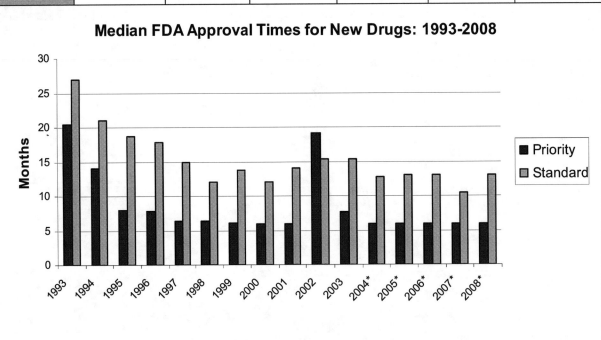

**Median FDA Approval Times for New Drugs: 1993-2008**

* Beginning in 2004, these figures include new Biologic License Applications (BLAs) for therapeutic biologic products transferred from the Center for Biologics Evaluation & Research (CBER) to the Center for Drug Evaluation & Research (CDER).

Source: U.S. Food & Drug Administration (FDA)

Plunkett Research, Ltd.

www.plunkettresearch.com

# U.S. FDA Approval Times for New Molecular Entities (NMEs): 1993-2008

| Calendar Year | Priority | | | Standard | | |
|---|---|---|---|---|---|---|
| | Number Approved | Median FDA Review Time (months) | Median Total Approval Time (months) | Number Approved | Median FDA Review Time (months) | Median Total Approval Time (months) |
| 1993 | 13 | 13.9 | 14.9 | 12 | 27.2 | 27.2 |
| 1994 | 12 | 13.9 | 14.0 | 9 | 22.2 | 23.7 |
| 1995 | 10 | 7.9 | 7.9 | 19 | 15.9 | 17.8 |
| 1996 | 18 | 7.7 | 9.6 | 35 | 14.6 | 15.1 |
| 1997 | 9 | 6.4 | 6.7 | 30 | 14.4 | 15.0 |
| 1998 | 16 | 6.2 | 6.2 | 14 | 12.3 | 13.4 |
| 1999 | 19 | 6.3 | 6.9 | 16 | 14.0 | 16.3 |
| 2000 | 9 | 6.0 | 6.0 | 18 | 15.4 | 19.9 |
| 2001 | 7 | 6.0 | 6.0 | 17 | 15.7 | 19.0 |
| 2002 | 7 | 13.8 | 16.3 | 10 | 12.5 | 15.9 |
| 2003 | 9 | 6.7 | 6.7 | 12 | 13.8 | 23.1 |
| 2004* | 21 | 6.0 | 6.0 | 15 | 16.0 | 24.7 |
| 2005* | 15 | 6.0 | 6.0 | 5 | 15.8 | 23.0 |
| 2006* | 10 | 6.0 | 6.0 | 12 | 12.5 | 13.7 |
| 2007* | 8 | 6.0 | 6.0 | 10 | 12.9 | 12.9 |
| 2008* | 9 | 6 | 6 | 15 | 13 | 13 |

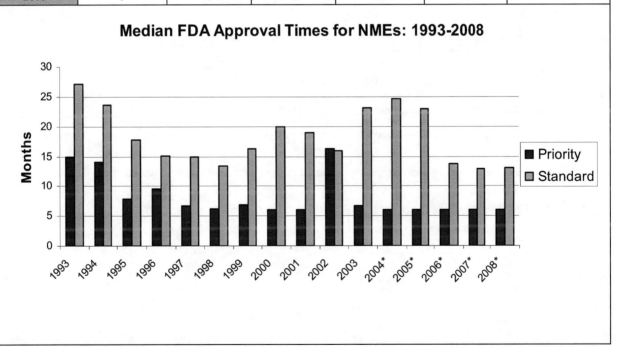

**Median FDA Approval Times for NMEs: 1993-2008**

An NME is a medication containing an active substance that has never before been approved for marketing in any form in the U.S.
* Beginning in 2004, these figures include new Biologic License Applications (BLAs) for therapeutic biologic products transferred from the Center for Biologics Evaluation & Research (CBER) to the Center for Drug Evaluation & Research (CDER).

Source: U.S. Food & Drug Administration (FDA)

Plunkett Research, Ltd.

www.plunkettresearch.com

# U.S. Pharmaceutical R&D Spending Versus the Number of New Molecular Entity (NME) Approvals: 1993-2008

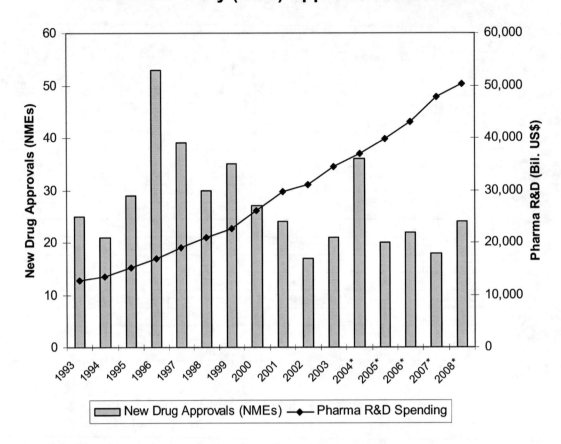

* Beginning in 2004, these figures include new BLAs for therapeutic biologic products transferred from CBER to CDER.

Notes: The FDA defines a New Molecular Entity (NME) as a medication containing an active substance that has never before been approved for marketing in any form in the U.S. Pharmaceutical R&D Spending includes expenditures inside and outside the U.S. by U.S.-owned PhRMA member companies and R&D conducted inside and outside the U.S. by the U.S. divisions of foreign-owned PhRMA member companies. R&D performed by the foreign divisions of foreign-owned PhRMA member companies is excluded.

Source: Pharmaceutical Research and Manufacturers Association (PhRMA); U.S. Food and Drug Administration

Plunkett Research, Ltd.

www.plunkettresearch.com

# Employment in Life & Physical Science Occupations by Business Type, U.S.: May 2008

| (Wage & Salary in US$) | Employment[1] | Median Hourly Wage | Mean Hourly Wage | Mean Annual Salary[2] | Mean RSE[3] (%) |
|---|---|---|---|---|---|
| Animal Scientists | 2,760 | 26.94 | 29.64 | 61,640 | 0.018 |
| Food Scientists & Technologists | 10,510 | 28.61 | 31.06 | 64,610 | 0.0 |
| Soil & Plant Scientists | 10,790 | 28.07 | 30.82 | 64,110 | 0.0 |
| Biochemists & Biophysicists | 22,230 | 39.83 | 42.53 | 88,450 | 0.0 |
| Microbiologists | 15,750 | 30.94 | 33.73 | 70,150 | 0.0 |
| Zoologists & Wildlife Biologists | 17,780 | 26.58 | 28.28 | 58,820 | 0.0 |
| Biological Scientists, All Other | 28,290 | 31.29 | 32.71 | 68,030 | 0.0 |
| Conservation Scientists | 15,830 | 28.23 | 28.93 | 60,170 | 0.0 |
| Foresters | 10,160 | 25.84 | 26.46 | 55,040 | 0.0 |
| Epidemiologists | 4,370 | 29.50 | 31.01 | 64,500 | 0.0 |
| Medical Scientists, Except Epidemiologists | 99,750 | 34.90 | 39.36 | 81,870 | 0.0 |
| Life Scientists, All Other | 12,030 | 29.55 | 33.18 | 69,020 | 0.0 |
| Astronomers | 1,280 | 48.70 | 47.95 | 99,730 | 0.0 |
| Physicists | 14,810 | 49.47 | 51.17 | 106,440 | 0.0 |
| Atmospheric & Space Scientists | 8,860 | 39.08 | 39.46 | 82,080 | 0.0 |
| Chemists | 83,080 | 31.84 | 34.17 | 71,070 | 0.0 |
| Materials Scientists | 9,650 | 38.57 | 39.23 | 81,600 | 0.0 |
| Environmental Scientists & Specialists, Including Health | 80,120 | 28.72 | 31.39 | 65,280 | 0.0 |
| Geoscientists, Except Hydrologists & Geographers | 31,260 | 38.06 | 42.93 | 89,300 | 0.0 |
| Hydrologists | 7,590 | 34.35 | 35.36 | 73,540 | 0.0 |
| Physical Scientists, All Other | 22,900 | 43.99 | 44.16 | 91,850 | 0.0 |
| Agricultural & Food Science Technicians | 18,930 | 16.34 | 17.53 | 36,470 | 0.0 |
| Biological Technicians | 72,200 | 18.46 | 19.67 | 40,900 | 0.0 |
| Chemical Technicians | 65,830 | 20.25 | 21.02 | 43,710 | 0.0 |
| Geological & Petroleum Technicians | 14,570 | 25.65 | 27.44 | 57,080 | 0.0 |
| Nuclear Technicians | 6,360 | 32.64 | 32.17 | 66,910 | 0.0 |
| Environmental Science & Protection Technicians, Including Health | 33,370 | 19.34 | 20.76 | 43,180 | 0.0 |
| Forensic Science Technicians | 11,990 | 23.97 | 25.46 | 52,960 | 0.0 |
| Forest & Conservation Technicians | 30,850 | 15.39 | 16.98 | 35,320 | 0.0 |

[1] Estimates for detailed occupations do not sum to the totals because the totals include occupations not shown separately. Estimates do not include self-employed workers.

[2] Annual wages have been calculated by multiplying the hourly mean wage by a "year-round, full-time" hours figure of 2,080 hours; for those occupations where there is not an hourly mean wage published, the annual wage has been directly calculated from the reported survey data.

[3] The relative standard error (RSE) is a measure of the reliability of a survey statistic. The smaller the relative standard error, the more precise the estimate.

Source: U.S. Bureau of Labor Statistics

Plunkett Research, Ltd.

www.plunkettresearch.com

# Federal R&D & R&D Plant Funding for General Science & Basic Research, U.S.: Fiscal Years 2007-2009

### (In Millions of US$)

| Funding Category and Agency | 2007 Actual | 2008 Prelim. | 2009 Proposed | % Change (08-09) |
|---|---|---|---|---|
| **Total** | 8,712 | 8,744 | 10,225 | 16.9 |
| National Science Foundation (NSF) | 4,440 | 4,479 | 5,175 | 15.5 |
| Biological sciences | 609 | 612 | 675 | 10.3 |
| Computer and information science and engineering | 527 | 535 | 639 | 19.5 |
| Education and human resources | 56 | 59 | 66 | 11.9 |
| Engineering | 630 | 637 | 759 | 19.2 |
| Geosciences | 746 | 753 | 849 | 12.8 |
| Integrative activities | 220 | 232 | 276 | 18.8 |
| Major research equipment and facilities | 191 | 206 | 148 | -28.2 |
| Mathematical and physical sciences | 1,151 | 1,167 | 1,403 | 20.2 |
| Office of cyberinfrastructure | 182 | 185 | 220 | 18.8 |
| Social, behavioral, and economic sciences | 215 | 215 | 233 | 8.5 |
| U.S. polar research programs | 438 | 443 | 491 | 10.9 |
| Budget authority adjustment[1] | -524 | -564 | -584 | NA |
| Department of Energy (DOE) | 3,560 | 3,574 | 4,314 | 20.7 |
| Advanced scientific computing research | 276 | 351 | 369 | 5.0 |
| Basic energy sciences | 1,221 | 1,270 | 1,568 | 23.5 |
| Biological and environmental research | 480 | 544 | 569 | 4.4 |
| Human genome research | 70 | 73 | 70 | -3.8 |
| All other biological and environmental research | 410 | 472 | 499 | 5.7 |
| Fusion energy sciences | 312 | 287 | 493 | 72.1 |
| High energy physics | 732 | 689 | 805 | 16.8 |
| Nuclear physics | 412 | 433 | 510 | 17.9 |
| Small business innovation research[2] | 126 | 0 | 0 | NA |
| Department of Homeland Security (DHS)[3] | 712 | 692 | 737 | 7 |

Notes: Detail may not add to total because of rounding. Percent change derived from unrounded data. Not all federally sponsored basic research is categorized in subfunction 251. Data derived from agencies' submission to Office of Management and Budget, Circular No. A-11, Max Schedule C; budget justification documents; and supplemental data from agencies' budget offices.

[1] Budget authority adjustment subtracts costs for research facilities, major equipment support, and other non-R&D from total NSF budget authority.
[2] DOE treats this activity as a budget execution program (i.e., funds are collected from existing appropriations and are not allocated until three-quarters into the fiscal year).
[3] In FY 2007 DHS changed its R&D portfolio to reclassify funding in defense and most administration of justice as general science and basic research.
NA = Not applicable.

Source: U.S. National Science Foundation

Plunkett Research, Ltd.

www.plunkettresearch.com

# National Health Expenditure Amounts by Type of Expenditure, U.S.: Selected Calendar Years, 2003-2018[1]

*(In Billions of US$)*

| Type of Expenditure | 2003 | 2008 | 2009 | 2010 | 2011 | 2016 | 2017 | 2018 |
|---|---|---|---|---|---|---|---|---|
| National Health Expenditures | 1734.9 | 2378.6 | 2509.5 | 2624.4 | 2770.3 | 3790.2 | 4061.7 | 4353.2 |
| Health Services & Supplies | 1623.1 | 2226.6 | 2350.1 | 2457.8 | 2595.5 | 3556.1 | 3811.4 | 4086.2 |
| Personal Health Care | 1447.5 | 1992.6 | 2099.0 | 2191.3 | 2312.0 | 3169.0 | 3395.6 | 3639.2 |
| Hospital Care | 527.4 | 746.5 | 789.4 | 829.7 | 877.4 | 1201.0 | 1284.6 | 1374.1 |
| Professional Services | 543.0 | 744.7 | 785.8 | 812.9 | 855.2 | 1167.6 | 1250.1 | 1338.1 |
| Physician & Clinical Services | 366.7 | 508.5 | 539.1 | 551.6 | 577.0 | 765.9 | 814.6 | 865.2 |
| Other Professional Services | 49.0 | 65.8 | 68.7 | 71.5 | 75.4 | 102.7 | 109.5 | 116.8 |
| Dental Services | 76.9 | 99.9 | 101.9 | 106.3 | 110.9 | 144.2 | 152.5 | 161.4 |
| Other Personal Health Care | 50.4 | 70.5 | 76.1 | 83.5 | 91.9 | 154.8 | 173.4 | 194.7 |
| Nursing Home & Home Health | 148.5 | 201.8 | 213.6 | 225.8 | 238.8 | 328.1 | 351.0 | 375.8 |
| Home Health Care | 38.0 | 64.4 | 69.7 | 74.6 | 79.7 | 115.8 | 124.9 | 134.9 |
| Nursing Home Care | 110.5 | 137.4 | 143.9 | 151.2 | 159.2 | 212.3 | 226.0 | 240.9 |
| Retail Outlet Sales of Medical Products | 228.6 | 299.6 | 310.2 | 322.9 | 340.5 | 472.3 | 509.9 | 551.3 |
| Prescription Drugs | 174.2 | 235.4 | 244.8 | 255.9 | 271.6 | 384.9 | 417.6 | 453.7 |
| Other Medical Products | 54.5 | 64.2 | 65.4 | 67.1 | 69.0 | 87.4 | 92.3 | 97.6 |
| Durable Medical Equipment | 22.4 | 25.2 | 25.2 | 25.9 | 26.6 | 34.1 | 36.0 | 38.1 |
| Other Non-Durable Medical Products | 32.1 | 39.0 | 40.2 | 41.2 | 42.4 | 53.3 | 56.3 | 59.5 |
| Program Administration & Net Cost of Private Health Insurance | 121.9 | 165.6 | 178.8 | 190.1 | 202.4 | 273.1 | 293.1 | 315.0 |
| Government Public Health Activities | 53.7 | 68.3 | 72.3 | 76.4 | 81.2 | 114.1 | 122.7 | 132.0 |
| Investment | 111.8 | 152.0 | 159.4 | 166.5 | 174.8 | 234.1 | 250.3 | 267.0 |
| Research[2] | 35.5 | 43.6 | 44.5 | 45.6 | 47.2 | 62.6 | 66.3 | 70.2 |
| Structures & Equipment | 76.3 | 108.4 | 114.9 | 120.9 | 127.6 | 171.5 | 183.9 | 196.8 |

Note: Numbers may not add to totals because of rounding.

[1] The health spending projections were based on the 2007 version of the National Health Expenditures (NHE) released in January 2009.

[2] Research and development expenditures of drug companies and other manufacturers and providers of medical equipment and supplies are excluded from research expenditures. These research expenditures are implicitly included in the expenditure class in which the product falls, in that they are covered by the payment received for that product.

Source: Centers for Medicare & Medicaid Services (CMS), Office of the Actuary

Plunkett Research, Ltd.

www.plunkettresearch.com

# U.S. Exports & Imports of Pharmaceutical Products: 2003-1st Quarter 2009

*(In Thousands of $US)*

## Exports

| Partner | 2003 | 2004 | 2005 | 2006 | 2007 | 2008 | 1Q 2008 | 1Q 2009 |
|---|---|---|---|---|---|---|---|---|
| World Total | 15,942,871 | 19,598,014 | 21,717,180 | 25,344,854 | 29,242,543 | 34,211,517 | 7,915,128 | 10,339,283 |
| Germany | 671,650 | 637,450 | 801,867 | 1,357,499 | 3,396,957 | 4,666,004 | 961,925 | 1,557,416 |
| Netherlands | 1,670,291 | 3,222,095 | 3,496,556 | 4,212,693 | 3,626,803 | 4,235,029 | 1,092,020 | 1,327,408 |
| United Kingdom | 1,779,311 | 2,195,886 | 2,771,895 | 3,604,489 | 3,683,829 | 3,936,675 | 866,239 | 1,280,737 |
| Canada | 2,546,590 | 2,712,189 | 2,794,759 | 3,341,662 | 3,336,081 | 3,118,117 | 800,240 | 985,594 |
| Switzerland | 1,013,457 | 1,308,197 | 1,433,356 | 1,804,761 | 2,243,737 | 2,449,143 | 537,152 | 503,915 |
| Belgium | 1,489,483 | 1,732,008 | 1,583,283 | 1,870,620 | 2,001,761 | 2,331,565 | 485,947 | 521,741 |
| Spain | 267,202 | 467,105 | 253,035 | 231,037 | 1,004,131 | 2,136,847 | 512,174 | 855,928 |
| Japan | 916,756 | 1,046,341 | 1,174,992 | 1,379,970 | 1,421,708 | 1,630,790 | 368,813 | 504,740 |
| France | 1,089,821 | 1,148,826 | 1,223,086 | 1,266,271 | 1,206,051 | 1,386,563 | 338,635 | 384,726 |
| Italy | 856,311 | 760,363 | 889,433 | 910,333 | 985,751 | 1,141,380 | 237,303 | 260,356 |
| Mexico | 421,991 | 549,599 | 612,695 | 697,923 | 609,542 | 863,997 | 191,557 | 222,751 |
| Ireland | 532,986 | 634,065 | 687,655 | 488,215 | 815,909 | 816,766 | 242,827 | 215,361 |
| Australia | 447,330 | 537,924 | 615,396 | 635,707 | 808,656 | 794,229 | 196,193 | 354,040 |
| Brazil | 257,236 | 316,599 | 415,832 | 526,903 | 576,222 | 711,690 | 147,912 | 170,943 |
| China | 88,304 | 98,864 | 176,227 | 238,479 | 355,058 | 392,487 | 85,233 | 114,183 |

## Imports

| Partner | 2003 | 2004 | 2005 | 2006 | 2007 | 2008 | 1Q 2008 | 1Q 2009 |
|---|---|---|---|---|---|---|---|---|
| World Total | 27,784,766 | 31,219,198 | 35,442,464 | 42,205,187 | 48,947,754 | 52,342,604 | 12,685,050 | 12,564,302 |
| United Kingdom | 4,628,850 | 4,803,592 | 4,097,241 | 4,894,358 | 6,451,320 | 7,576,662 | 1,851,873 | 1,652,448 |
| Ireland | 5,603,009 | 5,820,476 | 6,108,618 | 5,861,418 | 6,549,044 | 6,586,462 | 1,706,043 | 1,916,006 |
| Germany | 3,577,923 | 4,555,697 | 4,894,638 | 6,167,540 | 7,920,946 | 6,457,606 | 1,585,895 | 1,683,822 |
| France | 2,506,730 | 3,265,000 | 3,832,103 | 3,939,319 | 4,407,889 | 5,251,394 | 1,319,834 | 966,005 |
| Canada | 1,851,368 | 2,118,910 | 2,377,053 | 3,379,409 | 4,759,670 | 4,618,081 | 888,419 | 1,028,404 |
| Israel | 717,389 | 861,138 | 1,448,237 | 2,493,233 | 2,593,992 | 3,738,582 | 639,588 | 1,010,994 |
| Switzerland | 1,344,424 | 1,276,921 | 1,498,193 | 2,203,900 | 2,305,030 | 2,985,806 | 574,785 | 657,163 |
| Italy | 644,174 | 919,878 | 1,446,768 | 1,320,151 | 1,412,807 | 2,026,219 | 526,877 | 388,995 |
| Singapore | 9,311 | 91,128 | 1,158,317 | 2,427,199 | 3,043,012 | 1,912,644 | 809,805 | 542,546 |
| Belgium | 671,373 | 615,692 | 802,852 | 1,108,698 | 1,085,042 | 1,536,084 | 376,398 | 292,642 |
| India | 359,648 | 260,816 | 283,034 | 438,401 | 895,972 | 1,425,841 | 524,189 | 294,222 |
| Japan | 2,098,590 | 2,028,192 | 1,777,711 | 1,299,907 | 1,385,745 | 1,407,955 | 334,101 | 364,738 |
| Sweden | 1,394,607 | 1,659,667 | 1,576,642 | 2,303,653 | 1,360,010 | 1,300,479 | 342,840 | 386,709 |
| Denmark | 425,842 | 524,258 | 1,087,358 | 1,005,309 | 987,529 | 1,037,855 | 231,967 | 265,642 |
| Spain | 294,475 | 435,750 | 961,918 | 694,261 | 865,519 | 923,608 | 228,836 | 194,714 |

Note: "Pharmaceutical Products" refers to HS (Harmonized Commodity Description and Coding System) Code 30.

Source: Foreign Trade Division, U.S. Census Bureau

Plunkett Research, Ltd.

www.plunkettresearch.com

# U.S. Prescription Drug Expenditures, Aggregate & Per Capita Amounts, Percent Distribution: Selected Calendar Years, 2003-2018

*(By Source of Funds)*

| Year | Total | Out-of-Pocket Payments | Third-Party Payments | | Public | | | Medicare[2] | Medicaid[3] |
| | | | Total | Private Health Insurance | Total | Federal[1] | State & Local[1] | | |
|---|---|---|---|---|---|---|---|---|---|
| **Historical Estimates** *(In Billions of US$)* | | | | | | | | | |
| 2003 | 174.2 | 44.1 | 130.1 | 83.4 | 46.6 | 27.8 | 18.9 | 2.5 | 32.5 |
| 2004 | 188.8 | 46.2 | 142.6 | 90.0 | 52.6 | 31.4 | 21.2 | 3.4 | 36.3 |
| 2005 | 199.7 | 48.7 | 151.0 | 95.8 | 55.2 | 32.7 | 22.5 | 4.0 | 37.2 |
| 2006 | 216.8 | 46.7 | 170.1 | 96.2 | 73.9 | 58.7 | 15.1 | 39.5 | 19.1 |
| 2007 | 227.5 | 47.6 | 179.9 | 99.1 | 80.8 | 66.5 | 14.3 | 47.0 | 18.8 |
| **Projected** | | | | | | | | | |
| 2008 | 235.4 | 47.7 | 187.8 | 99.6 | 88.2 | 72.7 | 15.5 | 51.6 | 20.7 |
| 2009 | 244.8 | 47.5 | 197.3 | 100.1 | 97.3 | 80.7 | 16.6 | 57.6 | 23.3 |
| 2010 | 255.9 | 48.1 | 207.8 | 102.1 | 105.6 | 87.7 | 17.9 | 62.8 | 25.6 |
| 2011 | 271.6 | 50.0 | 221.6 | 106.7 | 114.9 | 95.6 | 19.3 | 68.9 | 27.6 |
| 2012 | 288.8 | 52.2 | 236.6 | 111.3 | 125.3 | 104.7 | 20.7 | 76.1 | 29.6 |
| 2013 | 307.8 | 54.6 | 253.2 | 116.8 | 136.4 | 114.2 | 22.2 | 83.4 | 31.8 |
| 2014 | 329.8 | 57.7 | 272.1 | 123.9 | 148.2 | 124.4 | 23.8 | 91.3 | 34.2 |
| 2015 | 355.8 | 61.5 | 294.2 | 132.4 | 161.9 | 136.3 | 25.6 | 100.7 | 36.7 |
| 2016 | 384.9 | 65.9 | 319.0 | 142.1 | 176.9 | 149.5 | 27.4 | 111.2 | 39.4 |
| 2017 | 417.6 | 70.8 | 346.8 | 153.0 | 193.8 | 164.4 | 29.4 | 123.3 | 42.0 |
| 2018 | 453.7 | 76.2 | 377.5 | 165.1 | 212.4 | 181.0 | 31.3 | 137.1 | 44.7 |
| **Historical Estimates** *(Per Capita Amount in US$)* | | | | | | | | | |
| 2003 | 599 | 152 | 447 | 287 | 160 | 96 | 65 | (5) | (5) |
| 2004 | 643 | 157 | 486 | 307 | 179 | 107 | 72 | (5) | (5) |
| 2005 | 674 | 164 | 510 | 324 | 186 | 110 | 76 | (5) | (5) |
| 2006 | 725 | 156 | 569 | 322 | 247 | 196 | 51 | (5) | (5) |
| 2007 | 753 | 157 | 596 | 328 | 268 | 220 | 47 | (5) | (5) |
| **Projected** | | | | | | | | | |
| 2008 | 772 | 156 | 616 | 327 | 289 | 238 | 51 | (5) | (5) |
| 2009 | 796 | 154 | 642 | 325 | 316 | 262 | 54 | (5) | (5) |
| 2010 | 825 | 155 | 670 | 329 | 340 | 283 | 58 | (5) | (5) |
| 2011 | 868 | 160 | 708 | 341 | 367 | 306 | 62 | (5) | (5) |
| 2012 | 915 | 165 | 749 | 352 | 397 | 332 | 65 | (5) | (5) |
| 2013 | 966 | 171 | 795 | 367 | 428 | 358 | 70 | (5) | (5) |
| 2014 | 1,027 | 180 | 847 | 386 | 461 | 387 | 74 | (5) | (5) |
| 2015 | 1,098 | 190 | 908 | 409 | 500 | 421 | 79 | (5) | (5) |
| 2016 | 1,178 | 202 | 976 | 435 | 541 | 457 | 84 | (5) | (5) |
| 2017 | 1,267 | 215 | 1,052 | 464 | 588 | 499 | 89 | (5) | (5) |
| 2018 | 1,365 | 229 | 1,136 | 497 | 639 | 545 | 94 | (5) | (5) |

Notes: Per capita amounts based on July 1 Census resident-based population estimates. Numbers and percents may not add to totals because of rounding. The health spending projections were based on the 2007 version of the National Health Expenditures (NHE) released in January 2009.

[1] Includes Medicaid SCHIP Expansion and SCHIP. [2] Subset of Federal funds. [3] Subset of Federal and State and Local Funds.

Source: Centers for Medicare & Medicaid Services (CMS), Office of the Actuary
Plunkett Research, Ltd.
www.plunkettresearch.com

# Prescription Drug Expenditures, U.S.: 1965-2018

*(In Millions of US$)*

| Year | TOTAL | Private | | | Public | | | | | | | |
| | | Total Private | Out-of-Pocket | Insurance | Total Public | Federal | | | | State & Local | | |
| | | | | | | Total Fed | Medicare | Medicaid | Other | Total S&L | Medicaid | Other |
|---|---|---|---|---|---|---|---|---|---|---|---|---|
| 1965 | 3,715 | 3,571 | 3,441 | 130 | 144 | 58 | 0 | 0 | 58 | 85 | 0 | 85 |
| 1966 | 3,985 | 3,776 | 3,594 | 182 | 209 | 95 | 0 | 50 | 45 | 115 | 53 | 62 |
| 1967 | 4,227 | 3,950 | 3,712 | 238 | 277 | 127 | 0 | 99 | 29 | 150 | 106 | 44 |
| 1968 | 4,742 | 4,437 | 4,120 | 317 | 305 | 141 | 0 | 114 | 27 | 164 | 106 | 59 |
| 1969 | 5,149 | 4,761 | 4,362 | 399 | 388 | 191 | 0 | 165 | 26 | 197 | 135 | 62 |
| 1970 | 5,497 | 5,015 | 4,531 | 484 | 482 | 237 | 0 | 224 | 13 | 245 | 193 | 53 |
| 1971 | 5,877 | 5,309 | 4,752 | 558 | 568 | 297 | 0 | 281 | 16 | 271 | 214 | 57 |
| 1972 | 6,324 | 5,678 | 5,035 | 644 | 646 | 328 | 0 | 308 | 20 | 318 | 255 | 63 |
| 1973 | 6,817 | 6,083 | 5,341 | 742 | 734 | 355 | 0 | 336 | 19 | 379 | 308 | 71 |
| 1974 | 7,422 | 6,567 | 5,714 | 854 | 855 | 443 | 0 | 419 | 24 | 412 | 321 | 91 |
| 1975 | 8,052 | 7,032 | 6,068 | 963 | 1,020 | 508 | 0 | 478 | 30 | 512 | 392 | 120 |
| 1976 | 8,722 | 7,565 | 6,476 | 1,089 | 1,157 | 610 | 0 | 576 | 35 | 547 | 394 | 153 |
| 1977 | 9,196 | 7,946 | 6,754 | 1,192 | 1,250 | 624 | 0 | 587 | 37 | 626 | 449 | 176 |
| 1978 | 9,891 | 8,531 | 7,115 | 1,416 | 1,360 | 655 | 0 | 613 | 41 | 706 | 493 | 212 |
| 1979 | 10,744 | 9,178 | 7,471 | 1,707 | 1,566 | 743 | 0 | 701 | 42 | 823 | 536 | 288 |
| 1980 | 12,049 | 10,249 | 8,466 | 1,783 | 1,800 | 857 | 0 | 813 | 44 | 943 | 595 | 347 |
| 1981 | 13,398 | 11,338 | 8,844 | 2,494 | 2,060 | 979 | 0 | 932 | 47 | 1,081 | 679 | 402 |
| 1982 | 15,029 | 12,840 | 10,272 | 2,568 | 2,189 | 990 | 0 | 934 | 56 | 1,198 | 735 | 464 |
| 1983 | 17,323 | 14,808 | 11,254 | 3,554 | 2,515 | 1,133 | 0 | 1,067 | 66 | 1,382 | 841 | 541 |
| 1984 | 19,617 | 16,670 | 12,503 | 4,168 | 2,947 | 1,311 | 2 | 1,238 | 71 | 1,636 | 982 | 655 |
| 1985 | 21,795 | 18,566 | 13,609 | 4,957 | 3,229 | 1,421 | 21 | 1,323 | 78 | 1,808 | 1,009 | 799 |
| 1986 | 24,290 | 20,198 | 15,451 | 4,746 | 4,092 | 1,834 | 44 | 1,707 | 83 | 2,259 | 1,279 | 979 |
| 1987 | 26,888 | 22,261 | 16,407 | 5,855 | 4,627 | 2,077 | 75 | 1,910 | 92 | 2,550 | 1,486 | 1,064 |
| 1988 | 30,646 | 25,325 | 18,336 | 6,990 | 5,320 | 2,398 | 102 | 2,184 | 111 | 2,923 | 1,630 | 1,293 |
| 1989 | 34,757 | 28,831 | 20,153 | 8,678 | 5,926 | 2,617 | 139 | 2,353 | 124 | 3,309 | 1,729 | 1,580 |
| 1990 | 40,290 | 33,003 | 22,376 | 10,627 | 7,288 | 3,247 | 185 | 2,915 | 147 | 4,040 | 2,162 | 1,878 |
| 1991 | 44,381 | 35,947 | 23,047 | 12,899 | 8,435 | 3,837 | 231 | 3,437 | 169 | 4,597 | 2,607 | 1,991 |
| 1992 | 47,573 | 38,060 | 23,423 | 14,637 | 9,513 | 4,476 | 293 | 3,990 | 193 | 5,037 | 2,894 | 2,144 |
| 1993 | 50,991 | 40,456 | 24,100 | 16,356 | 10,535 | 5,142 | 372 | 4,540 | 230 | 5,392 | 3,181 | 2,211 |
| 1994 | 54,301 | 42,625 | 23,390 | 19,234 | 11,677 | 5,768 | 496 | 4,990 | 282 | 5,909 | 3,638 | 2,271 |
| 1995 | 60,876 | 47,747 | 23,360 | 24,386 | 13,129 | 6,656 | 721 | 5,557 | 378 | 6,473 | 4,143 | 2,330 |
| 1996 | 68,485 | 53,764 | 24,181 | 29,583 | 14,721 | 7,878 | 1011 | 6,308 | 559 | 6,843 | 4,517 | 2,326 |
| 1997 | 77,617 | 61,138 | 25,641 | 35,496 | 16,480 | 9,080 | 1,259 | 7,025 | 796 | 7,399 | 5,077 | 2,322 |
| 1998 | 88,541 | 69,474 | 27,545 | 41,929 | 19,067 | 10,733 | 1,561 | 8,096 | 1,076 | 8,334 | 5,971 | 2,363 |
| 1999 | 104,606 | 81,409 | 30,406 | 51,003 | 23,196 | 13,213 | 1,919 | 9,756 | 1,537 | 9,984 | 7,238 | 2,746 |
| 2000 | 120,580 | 92,886 | 33,401 | 59,486 | 27,694 | 15,882 | 2,096 | 11,636 | 2,150 | 11,811 | 8,541 | 3,270 |

*(Continued on next page)*

# Prescription Drug Expenditures, U.S.: 1965-2018 (cont.)

*(In Millions of US$)*

| Year | TOTAL | Private | | | Public | | | | | | | | |
|------|-------|---------|---|---|--------|---|---|---|---|---|---|---|---|
| | | Total Private | Out-of-Pocket | Insurance | Total Public | Federal | | | | State & Local | | |
| | | | | | | Total Fed | Medicare | Medicaid | Other | Total S&L | Medicaid | Other |
| 2001 | 138,337 | 105,244 | 36,027 | 69,217 | 33,093 | 19,210 | 2,447 | 13,807 | 2,956 | 13,883 | 10,038 | 3,845 |
| 2002 | 157,638 | 118,298 | 40,353 | 77,945 | 39,340 | 23,101 | 2,472 | 16,222 | 4,407 | 16,239 | 11,589 | 4,650 |
| 2003 | 174,163 | 127,527 | 44,109 | 83,418 | 46,636 | 27,767 | 2,477 | 19,393 | 5,897 | 18,869 | 13,128 | 5,741 |
| 2004 | 188,787 | 136,217 | 46,206 | 90,011 | 52,570 | 31,417 | 3,392 | 21,282 | 6,744 | 21,152 | 15,007 | 6,145 |
| 2005 | 199,698 | 144,505 | 48,657 | 95,848 | 55,194 | 32,659 | 3,982 | 21,278 | 7,399 | 22,535 | 15,909 | 6,626 |
| 2006 | 216,837 | 142,975 | 46,731 | 96,244 | 73,862 | 58,722 | 39,516 | 10,985 | 8,221 | 15,140 | 8,123 | 7,017 |
| 2007 | 227,453 | 146,647 | 47,567 | 99,079 | 80,807 | 66,503 | 47,019 | 10,843 | 8,641 | 14,303 | 7,913 | 6,390 |
| 2008 | 235,432 | 147,246 | 47,671 | 99,574 | 88,187 | 72,686 | 51,576 | 12,003 | 9,107 | 15,501 | 8,720 | 6,780 |
| 2009 | 244,810 | 147,545 | 47,485 | 100,060 | 97,265 | 80,657 | 57,628 | 13,683 | 9,346 | 16,607 | 9,597 | 7,010 |
| 2010 | 255,870 | 150,235 | 48,114 | 102,120 | 105,636 | 87,747 | 62,799 | 15,127 | 9,821 | 17,889 | 10,455 | 7,434 |
| 2011 | 271,559 | 156,663 | 50,003 | 106,660 | 114,896 | 95,622 | 68,884 | 16,322 | 10,417 | 19,273 | 11,284 | 7,989 |
| 2012 | 288,770 | 163,423 | 52,161 | 111,262 | 125,347 | 104,682 | 76,064 | 17,508 | 11,110 | 20,665 | 12,100 | 8,565 |
| 2013 | 307,774 | 171,406 | 54,570 | 116,835 | 136,368 | 114,184 | 83,440 | 18,804 | 11,940 | 22,184 | 12,989 | 9,195 |
| 2014 | 329,792 | 181,598 | 57,713 | 123,884 | 148,194 | 124,363 | 91,298 | 20,210 | 12,854 | 23,831 | 13,955 | 9,876 |
| 2015 | 355,764 | 193,905 | 61,519 | 132,386 | 161,859 | 136,260 | 100,671 | 21,719 | 13,870 | 25,598 | 14,992 | 10,606 |
| 2016 | 384,886 | 207,977 | 65,869 | 142,108 | 176,909 | 149,459 | 111,196 | 23,287 | 14,976 | 27,449 | 16,069 | 11,380 |
| 2017 | 417,575 | 223,806 | 70,765 | 153,041 | 193,769 | 164,405 | 123,340 | 24,873 | 16,192 | 29,364 | 17,159 | 12,205 |
| 2018 | 453,664 | 241,269 | 76,171 | 165,098 | 212,395 | 181,049 | 137,065 | 26,472 | 17,512 | 31,346 | 18,258 | 13,088 |

Notes:  Federal and State and Local "Other" funds include Medicaid SCHIP Expansion and SCHIP.  The health spending projections were based on the 2007 version of the NHE released in January 2009.  2008-2018 are projections.

Source: Centers for Medicare & Medicaid Services (CMS), Office of the Actuary

Plunkett Research, Ltd.

www.plunkettresearch.com

# Total U.S. Biotechnology Patents Granted per Year by Patent Class: 1977-2008

*(Original & Cross-Reference Classifications; Duplicates Eliminated)*

| Year | Patent Class* | | | | | | |
|---|---|---|---|---|---|---|---|
| | 47 | 71 | 119 | 426 | 435 | 800 | 930 |
| 1977-87 | 1,853 | 938 | 2,271 | 7,582 | 7,668 | 66 | 1,573 |
| 1988 | 183 | 61 | 234 | 793 | 1,070 | 23 | 317 |
| 1989 | 221 | 80 | 265 | 1,157 | 1,432 | 22 | 255 |
| 1990 | 235 | 60 | 311 | 1,030 | 1,405 | 13 | 6 |
| 1991 | 244 | 68 | 391 | 1,003 | 1,561 | 29 | 12 |
| 1992 | 241 | 94 | 399 | 908 | 1,969 | 48 | 58 |
| 1993 | 226 | 73 | 341 | 899 | 2,257 | 38 | 76 |
| 1994 | 273 | 78 | 355 | 823 | 2,175 | 99 | 66 |
| 1995 | 272 | 93 | 313 | 884 | 2,250 | 91 | 38 |
| 1996 | 345 | 75 | 348 | 902 | 3,084 | 253 | 59 |
| 1997 | 296 | 100 | 362 | 814 | 4,140 | 282 | 85 |
| 1998 | 469 | 89 | 553 | 980 | 6,130 | 498 | 79 |
| 1999 | 325 | 87 | 501 | 1,261 | 6,215 | 670 | 79 |
| 2000 | 351 | 64 | 469 | 1,267 | 5,604 | 631 | 48 |
| 2001 | 305 | 97 | 447 | 1,226 | 6,275 | 666 | 46 |
| 2002 | 306 | 112 | 540 | 1,064 | 5,733 | 555 | 33 |
| 2003 | 283 | 70 | 532 | 1,015 | 5,297 | 525 | 31 |
| 2004 | 186 | 60 | 396 | 874 | 4,620 | 614 | 40 |
| 2005 | 159 | 46 | 407 | 519 | 4,132 | 521 | 29 |
| 2006 | 157 | 37 | 442 | 636 | 5,277 | 778 | 32 |
| 2007 | 124 | 32 | 380 | 524 | 5,195 | 904 | 23 |
| 2008 | 97 | 33 | 285 | 420 | 4,789 | 762 | 18 |
| Total | 7,151 | 2,447 | 10,542 | 26,581 | 88,278 | 8,088 | 3,003 |

* The patent classes are as follows:
47: Plant Husbandry
71: Chemical Fertilizers
119: Animal Husbandry
426: Food or Edible Material: Processes, Compositions and Products
435: Chemistry: Molecular Biology & Microbiology
800: Multicellular Living Organisms and Unmodified Parts Thereof and Related Processes
930: Peptide or Protein Sequence
The agricultural classes 47, 71, 119 and 426 include only a small portion of patents that are related to biotechnology. All patents in classes 435, 800 and 930 are biotechnology-related.

Source: U.S. Patent & Trademark Office (USPTO)

Plunkett Research, Ltd.

www.plunkettresearch.com

# Research Funding for Biological Sciences, U.S. National Science Foundation: Fiscal Year 2008-2010

*(In Millions of US$)*

| | FY 2008 Actual | FY 2009 Current Plan | FY 2010 Requested | Change over 2009 Request | |
|---|---|---|---|---|---|
| | | | | Amount | Percent |
| Molecular and Cellular Biosciences (MCB) | $112.28 | $121.26 | $128.83 | $7.57 | 6.2% |
| Integrative Organismal Systems (IOS) | 200.04 | 211.62 | 221.84 | 10.22 | 4.8% |
| Environmental Biology (EB) | 110.71 | 120.38 | 133.92 | 13.54 | 11.2% |
| Biological Infrastructure (BI)[1] | 109.86 | 116.80 | 130.14 | 13.34 | 11.4% |
| Emerging Frontiers (EF)[2] | 82.73 | 85.75 | 118.27 | 32.52 | 37.9% |
| **Total Biological Sciences Activity** | **$615.62** | **$655.81** | **$733.00** | **$77.19** | **11.8%** |

Totals may not add due to rounding.

[1] Funding for the Science of Learning Center (SLC) within the Division for Biological Infrastructure is included for all years for comparability. SLC will be cofunded with the Directorate for Social, Behavioral and Economic Sciences beginning in FY 2010.
[2] In FY 2010, all centers are shifted from Emerging Frontiers to Biological Infrastructure. Funding is included in Biological Infrastructure for all years for comparability.

Source: U.S. National Science Foundation
Plunkett Research, Ltd.
www.plunkettresearch.com

# Global Area of Biotech Crops by Country: 2008

*(In Millions of Hectares)*

| Rank | Country | Area | Biotech Crops |
|------|---------|------|---------------|
| 1* | USA | 62.5 | Soybean, maize, cotton, canola, squash, papaya, alfalfa, sugarbeet |
| 2* | Argentina | 21.0 | Soybean, maize, cotton |
| 3* | Brazil | 15.8 | Soybean, maise, cotton |
| 4* | India | 7.6 | Cotton |
| 5* | Canada | 7.6 | Canola, maize, soybean, sugarbeet |
| 6* | China | 3.8 | Cotton, tomato, poplar, petunia, papaya, sweet pepper, soybean |
| 7* | Paraguay | 2.7 | Soybean |
| 8* | South Africa | 1.8 | Maize, soybean, cotton |
| 9* | Uruguay | 0.7 | Soybean, maize |
| 10* | Bolivia | 0.6 | Soybean |
| 11* | Philippines | 0.4 | Maize |
| 12* | Australia | 0.2 | Cotton, canola, carnation |
| 13* | Mexico | 0.1 | Cotton, soybean |
| 14* | Spain | 0.1 | Maize |
| 15 | Chile | <0.1 | Maize, soybean, canola |
| 16 | Colombia | <0.1 | Cotton, carnation |
| 17 | Honduras | <0.1 | Maize |
| 18 | Burkina Faso | <0.1 | Cotton |
| 19 | Czech Republic | <0.1 | Maize |
| 20 | Romania | <0.1 | Maize |
| 21 | Portugal | <0.1 | Maize |
| 22 | Germany | <0.1 | Maize |
| 23 | Poland | <0.1 | Maize |
| 24 | Slovakia | <0.1 | Maize |
| 25 | Egypt | <0.1 | Maize |

* 14 biotech mega-countries growing 50,000 hectares or more of biotech crops.

Source: Clive James, ISAAA

Plunkett Research, Ltd.

www.plunkettresearch.com

# R&D as a Percentage of U.S. Biopharmaceutical Sales, PhRMA Member Companies: 1970-2008

| Year | Domestic R&D as a % of Domestic Sales | Total R&D as a % of Total Sales |
|---|---|---|
| 2008[1] | 20.3 | 17.4 |
| 2007 | 19.8 | 17.5 |
| 2006 | 19.4 | 17.1 |
| 2005 | 18.6 | 16.9 |
| 2004 | 18.4 | 16.1[2] |
| 2003 | 18.3 | 16.5[2] |
| 2002 | 18.4 | 16.1 |
| 2001 | 18.0 | 16.7 |
| 2000 | 18.4 | 16.2 |
| 1999 | 18.2 | 15.5 |
| 1998 | 21.1 | 16.8 |
| 1997 | 21.6 | 17.1 |
| 1996 | 21.0 | 16.6 |
| 1995 | 20.8 | 16.7 |
| 1994 | 21.9 | 17.3 |
| 1993 | 21.6 | 17.0 |
| 1992 | 19.4 | 15.5 |
| 1991 | 17.9 | 14.6 |
| 1990 | 17.7 | 14.4 |
| 1989 | 18.4 | 14.8 |
| 1988 | 18.3 | 14.1 |
| 1987 | 17.4 | 13.4 |
| 1986 | 16.4 | 12.9 |
| 1985 | 16.3 | 12.9 |
| 1984 | 15.7 | 12.1 |
| 1983 | 15.9 | 11.8 |
| 1982 | 15.4 | 10.9 |
| 1981 | 14.8 | 10.0 |
| 1980 | 13.1 | 8.9 |
| 1979 | 12.5 | 8.6 |
| 1978 | 12.2 | 8.5 |
| 1977 | 12.4 | 9.0 |
| 1976 | 12.4 | 8.9 |
| 1975 | 12.7 | 9.0 |
| 1974 | 11.8 | 9.1 |
| 1973 | 12.5 | 9.3 |
| 1972 | 12.6 | 9.2 |
| 1971 | 12.2 | 9.0 |
| 1970 | 12.4 | 9.3 |

[1] Estimated
[2] Revised in 2007 to reflect updated data.

Source: Pharmaceutical Research and Manufacturers of America (PhRMA), *PhRMA Annual Membership Survey*, 2009
Plunkett Research, Ltd.
www.plunkettresearch.com

# Biologics & Biotechnology R&D, PhRMA Member Companies: 2007

### (In Millions of US$)

| | Amount | Share |
|---|---|---|
| Biotechnology-Derived Therapeutic Proteins | 10,075.7 | 21.0 |
| Vaccines | 1,159.9 | 2.4 |
| CellorGene Therapy | 95.3 | 0.2 |
| All Other Biologics | 796.5 | 1.7 |
| | | |
| Total Biologics/Biotechnology R&D | 12,127.4 | 25.3 |
| Non-Biologics/Biotechnology R&D | 32,178.3 | 67.2 |
| Uncategorized R&D | 3,597.4 | 7.5 |
| Total R&D | 47,903.1 | 100.0 |

Note: All figures include company-financed R&D only. Total values may be affected by rounding.

Source: Pharmaceutical Research and Manufacturers of America (PhRMA), *PhRMA Annual Membership Survey*, 2009

Plunkett Research, Ltd.

www.plunkettresearch.com

# Domestic & Foreign Pharmaceutical Sales, PhRMA Member Companies: 1970-2008

*(In Millions of US$)*

| Year | Domestic Sales | APC | Sales Abroad[1] | APC | Total Sales | APC |
|------|---------------|-----|-----------------|-----|-------------|-----|
| 2008[2] | 189,260.5 | 2.2% | 99,025.0 | 12.3% | 288,285.5 | 5.4% |
| 2007 | 185,209.2 | 4.2% | 88,213.4 | 14.8% | 273,422.6 | 7.4% |
| 2006 | 177,736.3 | 7.0 | 76,870.2 | 10.0 | 254,606.4 | 7.9 |
| 2005 | 166,155.5 | 3.4 | 69,881.0 | 0.1 | 236,036.5 | 2.4 |
| 2004[3] | 160,751.0 | 8.6 | 69,806.9 | 14.6 | 230,557.9 | 10.3 |
| 2003[3] | 148,038.6 | 6.4 | 60,914.4 | 13.4 | 208,953.0 | 8.4 |
| 2002 | 139,136.4 | 6.4 | 53,697.4 | 12.1 | 192,833.8 | 8.0 |
| 2001 | 130,715.9 | 12.8 | 47,886.9 | 5.9 | 178,602.8 | 10.9 |
| 2000 | 115,881.8 | 14.2 | 45,199.5 | 1.6 | 161,081.3 | 10.4 |
| 1999 | 101,461.8 | 24.8 | 44,496.6 | 2.7 | 145,958.4 | 17.1 |
| 1998 | 81,289.2 | 13.3 | 43,320.1 | 10.8 | 124,609.4 | 12.4 |
| 1997 | 71,761.9 | 10.8 | 39,086.2 | 6.1 | 110,848.1 | 9.1 |
| 1996 | 64,741.4 | 13.3 | 36,838.7 | 8.7 | 101,580.1 | 11.6 |
| 1995 | 57,145.5 | 12.6 | 33,893.5 | (4) | 91,039.0 | (4) |
| 1994 | 50,740.4 | 4.4 | 26,870.7 | 1.5 | 77,611.1 | 3.4 |
| 1993 | 48,590.9 | 1.0 | 26,467.3 | 2.8 | 75,058.2 | 1.7 |
| 1992 | 48,095.5 | 8.6 | 25,744.2 | 15.8 | 73,839.7 | 11.0 |
| 1991 | 44,304.5 | 15.1 | 22,231.1 | 12.1 | 66,535.6 | 14.1 |
| 1990 | 38,486.7 | 17.7 | 19,838.3 | 18.0 | 58,325.0 | 17.8 |
| 1989 | 32,706.6 | 14.4 | 16,817.9 | -4.7 | 49,524.5 | 7.1 |
| 1988 | 28,582.6 | 10.4 | 17,649.3 | 17.1 | 46,231.9 | 12.9 |
| 1987 | 25,879.1 | 9.4 | 15,068.4 | 15.6 | 40,947.5 | 11.6 |
| 1986 | 23,658.8 | 14.1 | 13,030.5 | 19.9 | 36,689.3 | 16.1 |
| 1985 | 20,742.5 | 9.0 | 10,872.3 | 4.0 | 31,614.8 | 7.3 |
| 1984 | 19,026.1 | 13.2 | 10,450.9 | 0.4 | 29,477.0 | 8.3 |
| 1983 | 16,805.0 | 14.0 | 10,411.2 | -2.4 | 27,216.2 | 7.1 |
| 1982 | 14,743.9 | 16.4 | 10,667.4 | 0.1 | 25,411.3 | 9.0 |
| 1981 | 12,665.0 | 7.4 | 10,658.3 | 1.4 | 23,323.3 | 4.6 |
| 1980 | 11,788.6 | 10.7 | 10,515.4 | 26.9 | 22,304.0 | 17.8 |
| 1979 | 10,651.3 | 11.2 | 8,287.8 | 21.0 | 18,939.1 | 15.3 |
| 1978 | 9,580.5 | 12.0 | 6,850.4 | 22.2 | 16,430.9 | 16.1 |
| 1977 | 8,550.4 | 7.5 | 5,605.0 | 10.2 | 14,155.4 | 8.6 |
| 1976 | 7,951.0 | 11.4 | 5,084.3 | 9.7 | 13,035.3 | 10.8 |
| 1975 | 7,135.7 | 10.3 | 4,633.3 | 19.1 | 11,769.0 | 13.6 |
| 1974 | 6,740.4 | 13.8 | 3,891.0 | 23.4 | 10,361.4 | 17.2 |
| 1973 | 5,686.5 | 9.1 | 3,152.5 | 15.9 | 8,839.0 | 11.5 |
| 1972 | 5,210.1 | 1.3 | 2,720.2 | 10.6 | 7,930.3 | 4.3 |
| 1971 | 5,144.9 | 13.0 | 2,459.7 | 18.0 | 7,604.6 | 14.6 |
| 1970 | 4,552.5 | ------ | 2,084.0 | ------ | 6,636.5 | ------ |
| Average | | 10.4% | | 10.6% | | 10.4% |

Notes: Total values may be affected by rounding. APC = Annual Percent Change.
[1] Sales abroad includes sales generated outside the United States by U.S.-owned PhRMA member companies and sales generated abroad by the U.S. divisions of foreign-owned PhRMA member companies. Sales generated abroad by the foreign divisions of foreign-owned PhRMA member companies are excluded. Domestic sales, however, includes sales generated within the United States by all PhRMA member companies. [2] Estimated. [3] Revised in 2007 to reflect updated data. [4] Sales Abroad affected by merger and acquisition activity.

Source: Pharmaceutical Research and Manufacturers of America (PhRMA), *PhRMA Annual Membership Survey*, 2009
Plunkett Research, Ltd.
www.plunkettresearch.com

# Sales By Geographic Area, PhRMA Member Companies: 2007

### *(In Millions of US$; Latest Year Available)*

| Geographic Area* | Amount | Share |
|---|---|---|
| **Africa** | 1,246.6 | 0.5% |
| **Americas** | | |
| United States | 185,209.2 | 67.7 |
| Canada | 6,693.0 | 2.4 |
| Mexico | 2,987.1 | 1.1 |
| Brazil | 2,438.7 | 0.9 |
| Other Latin America[1] | 3,463.6 | 1.3 |
| **Asia-Pacific** | | |
| Japan | 9,089.4 | 3.3 |
| China | 1,586.0 | 0.6 |
| India | 589.4 | 0.2 |
| Other Asia-Pacific | 4,348.6 | 1.6 |
| **Australia & New Zealand** | 3,284.2 | 1.2 |
| **Europe** | | |
| France | 8,923.3 | 3.3 |
| Germany | 6,774.4 | 2.5 |
| Italy | 6,206.6 | 2.3 |
| Spain | 5,567.0 | 2.0 |
| United Kingdom | 5,607.4 | 2.1 |
| Other Western European nations | 10,584.7 | 3.9 |
| Turkey | 1,449.6 | 0.5 |
| Russia | 925.2 | 0.3 |
| **Other Eastern European nations[2]** | 3,755.5 | 1.4 |
| **Middle East[3]** | 1,643.7 | 0.6 |
| **Uncategorized** | 1,049.6 | 0.4 |
| **Total Sales** | 273,422.6 | 100.0 |

Note: Total values may be affected by rounding.

\* Sales abroad includes sales generated outside the United States by U.S.-owned PhRMA member companies and sales generated abroad by the U.S. divisions of foreign-owned PhRMA member companies. Sales generated abroad by the foreign divisions of foreign-owned PhRMA member companies are excluded. Domestic sales, however, includes sales generated within the United States by all PhRMA member companies.

[1] Other South American, Central American, and all Caribbean nations.
[2] Cyprus, Czech Republic, Estonia, Hungary, Poland, Slovenia, Bulgaria, Lithuania, Latvia, Romania, Slovakia, Malta and the Newly Independent States.
[3] Saudi Arabia, Yemen, United Arab Emirates, Iraq, Iran, Kuwait, Israel, Jordan, Syria, Afghanistan and Qatar.

Source: Pharmaceutical Research and Manufacturers of America (PhRMA), *PhRMA Annual Membership Survey*, 2009

Plunkett Research, Ltd.

www.plunkettresearch.com

# Domestic Biopharmaceutical R&D Breakdown, PhRMA Member Companies: 2007

*(In Millions of US$; Latest Year Available)*

| R&D Expenditures for Human-Use Pharmaceuticals | Dollars | Share (%) |
|---|---|---|
| Domestic | 36,178.0 | 75.5 |
| Abroad* | 11,006.4 | 23 |
| Total Human-Use R&D | 47,184.7 | 98.5 |

| R&D Expenditures for Veterinary-Use Pharmaceuticals | | |
|---|---|---|
| Domestic | 430.0 | 0.9 |
| Abroad* | 288.4 | 0.6 |
| Total Vet-Use R&D | 718.4 | 1.5 |
| **Total R&D** | **47,903.1** | **100.0** |

| Domestic R&D by Source | | |
|---|---|---|
| Licensed-In | 6,294.2 | 17.2 |
| Self-Originated | 27,126.9 | 74.1 |
| Uncategorized | 3,187.3 | 8.7 |
| **Total R&D** | **36,608.4** | **100.0** |

| R&D By Function | | |
|---|---|---|
| Prehuman/Preclinical | 13,087.4 | 27.3 |
| Phase I | 3,547.7 | 7.4 |
| Phase II | 6,251.0 | 13.0 |
| Phase III | 13,664.7 | 28.5 |
| Approval | 2,413.8 | 5.0 |
| Phase IV | 6,439.9 | 13.4 |
| Uncategorized | 2,498.6 | 5.2 |
| **Total R&D** | **47,903.1** | **100.0** |

Note: All figures include company-financed R&D only. Total values may be affected by rounding.

* R&D Abroad includes expenditures outside the United States by U.S.-owned PhRMA member companies and R&D conducted abroad by the U.S. divisions of foreign-owned PhRMA member companies. R&D performed abroad by the foreign divisions of foreign-owned PhRMA member companies is excluded. Domestic R&D, however, includes R&D expenditures within the United States by all PhRMA member companies.

Source: Pharmaceutical Research and Manufacturers of America (PhRMA), *PhRMA Annual Membership Survey*, 2009

Plunkett Research, Ltd.

www.plunkettresearch.com

# R&D by Global Geographic Area, PhRMA Member Companies: 2007

*(In Millions of US$; Latest Year Available)*

| Geographic Area* | Dollars | Share |
|---|---|---|
| **Africa** | 28.6 | 0.1 |
| **Americas** | | |
| United States | 36,608.4 | 76.4 |
| Canada | 612.4 | 1.3 |
| Mexico | 63.0 | 0.1 |
| Brazil | 81.2 | 0.2 |
| Other Latin America[1] | 217.9 | 0.5 |
| **Asia-Pacific** | | |
| Japan | 954.2 | 2.0 |
| China | 62.9 | 0.1 |
| India | 33.3 | 0.1 |
| Other Asia-Pacific | 191.8 | 0.4 |
| **Australia & New Zealand** | 161.0 | 0.3 |
| **Europe** | | |
| France | 521.8 | 1.1 |
| Germany | 714.7 | 1.5 |
| Italy | 240.1 | 0.5 |
| Spain | 235.5 | 0.5 |
| United Kingdom | 2,892.9 | 6.0 |
| Other Western European nations | 3,568.6 | 7.4 |
| Turkey | 39.0 | 0.1 |
| Russia | 40.1 | 0.1 |
| Central & Eastern European nations[2] | 481.8 | 1.0 |
| **Middle East[3]** | 29.7 | 0.1 |
| **Uncategorized** | 124.2 | 0.3 |
| **Total R&D** | 47,903.1 | 100.0 |

Note: All figures include company-financed R&D only. Total values may be affected by rounding.

* R&D Abroad includes expenditures outside the United States by U.S.-owned PhRMA member companies and R&D conducted abroad by the U.S. divisions of foreign-owned PhRMA member companies. R&D performed abroad by the foreign divisions of foreign-owned PhRMA member companies is excluded. Domestic R&D, however, includes R&D expenditures within the United States by all PhRMA member companies.

[1] Other South American, Central American, and all Caribbean nations.

[2] Cyprus, Czech Republic, Estonia, Hungary, Poland, Slovenia, Bulgaria, Lithuania, Latvia, Romania, Slovakia, Malta and the Newly Independent States.

[3] Saudi Arabia, Yemen, United Arab Emirates, Iraq, Iran, Kuwait, Israel, Jordan, Syria, Afghanistan and Qatar.

Source: Pharmaceutical Research and Manufacturers of America (PhRMA), *PhRMA Annual Membership Survey*, 2009

Plunkett Research, Ltd.

www.plunkettresearch.com

# Domestic Biopharmaceutical R&D Scientific, Professional & Technical Personnel by Function, PhRMA Member Companies: 2007

*(Latest Year Available)*

| Function | Personnel | Share (%) |
|---|---|---|
| Prehuman/Preclinical | 30,023 | 31.1 |
| Phase I | 6,117 | 6.3 |
| Phase II | 10,098 | 10.5 |
| Phase III | 18,579 | 19.3 |
| Approval | 4,108 | 4.3 |
| Phase IV | 13,332 | 13.8 |
| Uncategorized | 3,613 | 3.7 |
| Total R&D Staff | 85,870 | 89.0 |
| Supported R&D Nonstaff | 10,616 | 11.0 |
| TOTAL R&D PERSONNEL | 96,486 | 100.0 |

Source: Pharmaceutical Research and Manufacturers of America (PhRMA), *PhRMA Annual Membership Survey*, 2009

Plunkett Research, Ltd.

www.plunkettresearch.com

# Domestic U.S. Biopharmaceutical R&D & R&D Abroad, PhRMA Member Companies: 1970-2008

*(In Millions of US$)*

| Year | Domestic R&D | Annual % Chg. | R&D Abroad[2] | Annual % Chg. | Total R&D | Annual % Chg. |
|------|------|------|------|------|------|------|
| 2008[1] | 38,427.8 | 5.0 | 11,825.7 | 4.7 | 50,253.6 | 4.9 |
| 2007 | 36,608.4 | 7.8 | 11,294.8 | 25.4 | 47,903.1 | 11.5 |
| 2006 | 33,967.9 | 9.7 | 9,005.6 | 1.3 | 42,973.5 | 7.8 |
| 2005 | 30,969.0 | 4.8 | 8,888.9 | 19.1 | 39,857.9 | 7.7 |
| 2004 | 29,555.5 | 9.2 | 7,462.6 | 1.0 | 37,018.1 | 7.4 |
| 2003 | 27,064.9 | 5.5 | 7,388.4 | 37.9 | 34,453.3 | 11.1 |
| 2002 | 25,655.1 | 9.2 | 5,357.2 | -13.9 | 31,012.2 | 4.2 |
| 2001 | 23,502.0 | 10.0 | 6,220.6 | 33.3 | 29,772.7 | 14.4 |
| 2000 | 21,363.7 | 15.7 | 4,667.1 | 10.6 | 26,030.8 | 14.7 |
| 1999 | 18,471.1 | 7.4 | 4,219.6 | 9.9 | 22,690.7 | 8.2 |
| 1998 | 17,127.9 | 11.0 | 3,839.0 | 9.9 | 20,966.9 | 10.8 |
| 1997 | 15,466.0 | 13.9 | 3,492.1 | 6.5 | 18,958.1 | 12.4 |
| 1996 | 13,627.1 | 14.8 | 3,278.5 | -1.6 | 16,905.6 | 11.2 |
| 1995 | 11,874.0 | 7.0 | 3,333.5 | (3) | 15,207.4 | (3) |
| 1994 | 11,101.6 | 6.0 | 2,347.8 | 3.8 | 13,449.4 | 5.6 |
| 1993 | 10,477.1 | 12.5 | 2,262.9 | 5.0 | 12,740.0 | 11.1 |
| 1992 | 9,312.1 | 17.4 | 2,155.8 | 21.3 | 11,467.9 | 18.2 |
| 1991 | 7,928.6 | 16.5 | 1,776.8 | 9.9 | 9,705.4 | 15.3 |
| 1990 | 6,802.9 | 13.0 | 1,617.4 | 23.6 | 8,420.3 | 14.9 |
| 1989 | 6,021.4 | 15.0 | 1,308.6 | 0.4 | 7,330.0 | 12.1 |
| 1988 | 5,233.9 | 16.2 | 1,303.6 | 30.6 | 6,537.5 | 18.8 |
| 1987 | 4,504.1 | 16.2 | 998.1 | 15.4 | 5,502.2 | 16.1 |
| 1986 | 3,875.0 | 14.7 | 865.1 | 23.8 | 4,740.1 | 16.2 |
| 1985 | 3,378.7 | 13.3 | 698.9 | 17.2 | 4,077.6 | 13.9 |
| 1984 | 2,982.4 | 11.6 | 596.4 | 9.2 | 3,578.8 | 11.2 |
| 1983 | 2,671.3 | 17.7 | 546.3 | 8.2 | 3,217.6 | 16.0 |
| 1982 | 2,268.7 | 21.3 | 505.0 | 7.7 | 2,773.7 | 18.6 |
| 1981 | 1,870.4 | 20.7 | 469.1 | 9.7 | 2,339.5 | 18.4 |
| 1980 | 1,549.2 | 16.7 | 427.5 | 42.8 | 1,976.7 | 21.5 |
| 1979 | 1,327.4 | 13.8 | 299.4 | 25.9 | 1,626.8 | 15.9 |
| 1978 | 1,166.1 | 9.7 | 237.9 | 11.6 | 1,404.0 | 10.0 |
| 1977 | 1,063.0 | 8.1 | 213.1 | 18.2 | 1,276.1 | 9.7 |
| 1976 | 983.4 | 8.8 | 180.3 | 14.1 | 1,163.7 | 9.6 |
| 1975 | 903.5 | 13.9 | 158.0 | 7.0 | 1,061.5 | 12.8 |
| 1974 | 793.1 | 12.0 | 147.7 | 26.3 | 940.8 | 14.0 |
| 1973 | 708.1 | 8.1 | 116.9 | 64.0 | 825.0 | 13.6 |
| 1972 | 654.8 | 4.5 | 71.3 | 24.9 | 726.1 | 6.2 |
| 1971 | 626.7 | 10.7 | 57.1 | 9.2 | 683.8 | 10.6 |
| 1970 | 566.2 | ----- | 52.3 | ----- | 618.5 | ----- |
| Average | | 11.8% | | 15.5% | | 12.3% |

Note: All figures include company-financed R&D only. Total values may be affected by rounding.
[1] Estimated. [2] R&D Abroad includes expenditures outside the United States by U.S.-owned PhRMA member companies and R&D conducted abroad by the U.S. divisions of foreign-owned PhRMA member companies. R&D performed abroad by the foreign divisions of foreign-owned PhRMA member companies is excluded. Domestic R&D, however, includes R&D expenditures within the United States by all PhRMA member companies. [3] R&D Abroad affected by merger and acquisition activity.

Source: Pharmaceutical Research and Manufacturers of America (PhRMA), *PhRMA Annual Membership Survey*, 2009

Plunkett Research, Ltd.

www.plunkettresearch.com

# Chapter 3

# IMPORTANT BIOTECH & GENETICS INDUSTRY CONTACTS

**Contents:**

## I. Agricultural Biotechnology Industry Associations

**Agricultural Biotechnology in Europe (ABE)**
Ave. de l'Armée 6

Brussels, 1040 Belguim
Phone: 32-2-735-03-13
Fax: 32-2-735-49-60
E-mail Address: *info@abeurope.info*
Web Address: www.abeurope.info
Agricultural Biotechnology in Europe (ABE) publishes science-based information about agricultural biotechnology, including country-by-country statistics and news. Members of ABE are companies that have pioneered the development of agriculture biotechnology and that strongly believe that biotechnology has the potential to be of great value to Europe's agricultural and food industries.

### International Service for the Acquisition of Agri-Biotech Applications (ISAAA)
417 Bradfield Hall
Cornell University
Ithaca, NY 14853 US
Phone: 607-255-1724
Fax: 607-255-1215
E-mail Address: *americenter@isaaa.org*
Web Address: www.isaaa.org
The International Service for the Acquisition of Agri-Biotech Applications (ISAAA) is a not-for-profit organization that provides bioengineered seeds to poor and developing countries. In general, such seeds will enhance production per acre due to resistance to drought, insects and disease, and will offer additional crop enhancements.

### Plant Biotechnology Institute (PBI)
110 Gymnasium Pl.
Saskatoon, SK S7N 0W9 Canada
Phone: 306-975-5248
Fax: 306-975-4839
E-mail Address: *PBI-Info@nrc-cnrc.gc.ca*
Web Address: pbi-ibp.nrc-cnrc.gc.ca
The Plant Biotechnology Institute (PBI) is a member of Canada's National Research Council and is engaged in research regarding the genomics, metabolic pathways, gene expression, genetic transformation and structured biology of plants and crops.

## II.    Agricultural Biotechnology Resources

### Ag BioTech Infonet
Phone: 208-263-5236
E-mail Address: *karen@hillnet.com*
Web Address: www.biotech-info.net

Ag BioTech Infonet provides information on the application of biotechnology and genetic engineering in agricultural production, food processing and marketing.

### Biotechnology Knowledge Center
Monsanto Company
800 N. Lindbergh Blvd.
St. Louis, MO 63167 US
Phone: 314-694-1000
Web Address: www.biotechknowledge.com
The Biotechnology Knowledge Center, sponsored by Monsanto Company, is an online resource containing news, technical documents, a discussion board and a glossary.

### UK Agricultural Biodiversity Coalition (UKabc)
UK Food Group
94 White Lion St.
London, N1 9PF UK
Phone: 44-20-7713-5813
Fax: 44-207-837-1141
E-mail Address: *ukabc@ukabc.org*
Web Address: www.ukabc.org
The UK Agricultural Biodiversity Coalition (UKabc) provides links to life science and seed companies, databases and information resources, publicly funded research bodies and industry associations. UKabc is an activity of the UK Food Group (UKFG).

## III.    Agriculture Industry Associations

### U.S. Grains Council
1400 K St. NW, Ste. 1200
Washington, DC 20005 US
Phone: 202-789-0789
Fax: 202-898-0522
Web Address: www.grains.org
The U.S. Grains Council develops export markets for U.S. barley, corn, grain sorghum and related products. Founded in 1960, the Council is a private, nonprofit corporation with nine international offices and programs in more than 50 countries. Membership funds trigger matching market development funds from the U.S. government and support from cooperating groups in foreign countries to produce an annual development program valued at more than $25 million. The organization's website contains extensive grain production statistics and a special resource center on biotechnology.

## IV.    Alternative Energy-Ethanol

**Renewable Fuels Association (RFA)**
1 Massachusetts Ave. NW, Ste. 820
Washington, DC 20001 US
Phone: 202-289-3835
E-mail Address: *info@ethanolrfa.org*
Web Address: www.ethanolrfa.org
The Renewable Fuels Association (RFA) is a trade
organization representing the ethanol industry. It
publishes a wealth of useful information, including a
listing of biorefineries and monthly U.S. fuel ethanol
production and demand.

## V.    Biology-Synthetic

**BioBricks Foundation (BBF)**
One Kendall Sq.
PMB 126
Cambridge, MA 02139 US
E-mail Address: *info@biobricks.org*
Web Address: www.biobricks.org
The BioBricks Foundation (BBF) is a not-for-profit
organization founded by engineers and scientists
from MIT, Harvard, and UCSF with significant
experience in both nonprofit and commercial
biotechnology research. BBF encourages the
development and responsible use of technologies
based on BioBrick standard DNA parts that encode
basic biological functions.

**Synthetic Biology**
Web Address: syntheticbiology.org
Synthetic Biology is a consortium of individuals, labs
and groups working together to advance the
development of biological engineering. Its members
include 19 labs from 11 different universities,
including MIT, Princeton and Harvard. The site's
FAQ includes discussions of synthetic biology, its
applications and the difference between synthetic and
systems biology. Synthetic Biology does not maintain
a headquarters, instead allowing members to update
and edit the web site in order to disseminate
knowledge.

**Syntheticbiology.org**
E-mail Address: *synbioadmin@openwetware.org*
Web Address: www.syntheticbiology.org
An open source web site dedicated to disseminating
news about synthetic biology. It was originally
founded by a group of students, faculty and staff
from MIT and Harvard.

## VI.    Biotechnology and Biological Industry Associations

**All India Biotech Association (AIBA)**
Web Address: www.aibaonline.com
All India Biotech Association (AIBA) was
established in 1994 as a nonprofit group to represent
India's biotechnology industry.

**American Association of Immunologists**
9650 Rockville Pike
Bethesda, MD 20814 US
Phone: 301-634-7178
Fax: 301-634-7887
E-mail Address: *infoaai@aai.org*
Web Address: www.aai.org
The American Association of Immunologists is a
nonprofit organization that represents professionals in
the immunology field.

**American College of Medical Genetics (ACMG)**
9650 Rockville Pike
Bethesda, MD 20814-3998 US
Phone: 301-634-7127
Fax: 301-634-7275
E-mail Address: *acmg@acmg.net*
Web Address: www.acmg.net
The American College of Medical Genetics (ACMG)
provides education, resources and a voice for the
medical genetics profession. The ACMG promotes
the development and implementation of methods to
diagnose, treat and prevent genetic disease.

**American Peptide Society (APS)**
7668 El Camino Real, Ste. 104-140
La Costa, CA 92009-7932 US
Phone: 858-455-4752
Fax: 775-667-5332
E-mail Address:
*aps_member@americanpeptidesociety.org*
Web Address: www.americanpeptidesociety.org
The American Peptide Society (APS) is a nonprofit
organization for the advancement and promotion of
knowledge and research in the field of peptide
chemistry and biology.

**American Society for Microbiology (ASM)**
1752 N St. NW
Washington, DC 20036 US
Phone: 202-737-3600
Web Address: www.asm.org
The American Society for Microbiology (ASM) is a
life science membership organization that specializes

in the advancement of the study of bacteria, viruses, rickettsiae, mycoplasma, fungi, algae and protozoa.

**American Society of Gene Therapy (ASGT)**
555 East Wells St., Ste. 1100
Milwaukee, WI 53202 US
Phone: 414-278-1341
Fax: 414-276-3349
E-mail Address: *info@asgt.org*
Web Address: www.asgt.org
The American Society of Gene Therapy (ASGT) is a nonprofit medical and professional organization that represents researchers and scientists devoted to the discovery of new gene therapies. ASGT was established in 1996 by Dr. George Stamatoyannopoulos, professor of medicine at the University of Washington 's School of Medicine and a group of the country's leading researchers in gene therapy. With more than 2,000 members in the United States and worldwide, ASGT is the largest association of individuals involved in gene therapeutics.

**Association of Biotechnology Led Enterprises (ABLE)**
123/C, 16th Main Rd., 5th Cross, 4th Block
Near Sony World showroom / Headstart school
Koramangala, Bangalore 560034 India
Fax: 91-80-2553-3938
E-mail Address: *info@ableindia.org*
Web Address: www.ableindia.org
The Association of Biotechnology Led Enterprises (ABLE) is an organization focused on accelerating the pace of Biotechnology in India by enabling strategic alliances between researchers, the government and the global Biotech industry.

**AusBiotech**
322 Glenferrie Rd., Level 1
Malvern, VIC 3144 Australia
Phone: 03-9828-1400
Fax: 03-9824-5188
E-mail Address: *admin@ausbiotech.org*
Web Address: www.ausbiotech.org
AusBiotech is a professional organization for the biotech industry in Australia, with members in the human health, agricultural, medical device, bioinformatics, environmental and industrial sectors.

**BioAlberta**
314 Capital Pl., Ste. 9707-110
Edmonton, AB T5K 2L9 Canada
Phone: 780-425-3804

Fax: 780-409-9263
E-mail Address: *info@bioalberta.com*
Web Address: www.bioalberta.com
BioAlberta is a private, nonprofit industry association representing Alberta, Canada's biotech industry.

**Biochemical Society**
16 Procter St.
London, WC1V 6NX UK
Phone: 020-7280-4100
Fax: 020-7280-4170
E-mail Address: *genadmin@biochemistry.org*
Web Address: www.biochemsoc.org
The Biochemical Society is a professional association that promotes the advancement of the science of biochemistry, viewing cellular and molecular life sciences as a seamless continuum.

**BIOCOM**
4510 Executive Dr., Plz. 1
San Diego, CA 92121 US
Phone: 858-455-0300
Fax: 858-455-0022
E-mail Address: *inquiries@biocom.org*
Web Address: www.biocom.org
BIOCOM is a trade association for the life science industry in San Diego and Southern California.

**BioIndustry Association**
14/15 Belgrave Sq.
London, SW1X 8PS UK
Phone: 44-20-7565-7190
Fax: 44-20-7565-7191
E-mail Address: *admin@bioindustry.org*
Web Address: www.bioindustry.org
The BioIndustry Association promotes bioscience development in the U.K. The organization operates a public affairs program, a conference and seminar program, trade missions and publications for internal and external audiences.

**Bioindustry Association of Korea (BAK)**
97-3, Guro 6-dong, Guro-gu
DaeLim Office Valley, 6th Fl.
Seoul, 152-841 Korea
Phone: 82-2-855-7993
Fax: 82-2-855-7006
E-mail Address: *bioindus@chol.com*
Web Address: www.bak.or.kr
The Bioindustry Association of Korea (BAK) was established in 1991 to promote and facilitate growth and development of the bioindustry within Korea.

**Biomedical Engineering Society**
8401 Corporate Dr., Ste. 140
Landover, MD 20785-2224 US
Phone: (301) 459-1999
Fax: (301) 459-2444
E-mail Address: *info@bmes.org*
Web Address: www.bmes.org
The Biomedical Engineering Society (BMES)
members are the in biomedical engineering and
bioengineering industry.

**Biomedical Engineering Society of India**
Sree Chitra Tirunal Institute for Medical Sciences
and Technology
Satelmond Palace Campus, Poojapura
Thiruvananthapuram, 695 012 India
Phone: 91-471-2520-282
Fax: 91-471-2341-814
E-mail Address: *niranjan@sctimst.ac.in*
Web Address: www.bmesi.org.in
The Biomedical Engineering Society of India is an all
India association which seeks to advance
interdisciplinary cooperation among scientists,
engineers, and medical doctors for the growth of
teaching, research and practices of biomedical
engineering.

**BioQuebec**
381, rue Notre-Dame Ouest, Bureau 300
Montreal, QB H2Y 1V2 Canada
Phone: 514 733-8411
Fax: 514 733-8272
E-mail Address: *reception@bioquebec.com*
Web Address: www.bioquebec.com
BIOQuebec is a biotechnology and life science
industry association representing more than 175
member companies and R&D centers in Quebec. The
association works to create a positive influence on
the growth of the life science industry, including
capital access, public policy, workforce development
and international promotion.

**BioSingapore**
P.O. Box 950
Rocher Post Office
911837 Singapore
Phone: 65 6336 5371
Fax: 65 6336 5316
E-mail Address: *enquiries@biosingapore.org.sg*
Web Address: www.biosingapore.org
BioSingapore is an industry association for life
sciences business in Singapore, helping local

businesses network with financial institutions and
other important sectors.

**BIOTECanada**
130 Albert St., Ste. 420
Ottawa, ON K1P 5G4 Canada
Phone: 613-230-5585
Fax: 613-563-8850
E-mail Address: *info@biotech.ca*
Web Address: www.biotech.ca
BIOTECanada is a trade organization that promotes
the Canadian biotech industry.

**Biotechnology and Biological Sciences Research
Council (BBSRC)**
N. Star Ave.
Polaris House
Swindon, SN2 1UH UK
Phone: 44-0-1793-413200
Fax: 44-0-1793-413201
E-mail Address: *webmaster@bbsrc.ac.uk*
Web Address: www.bbsrc.ac.uk
The Biotechnology and Biological Sciences Research
Council (BBSRC) provides funding for biotech
research in the U.K.

**Biotechnology Association of Alabama**
500 Beacon Pkwy. W.
Birmingham, AL 35209 US
Phone: 205-290-9593
E-mail Address: *knugentphd@yahoo.com*
Web Address: www.bioalabama.com
The Biotechnology Association of Alabama is a
nonprofit association that represents the biotech
industry in Alabama.

**Biotechnology Industry Organization (BIO)**
1201 Maryland Ave. SW, Ste. 900
Washington, DC 20024 US
Phone: 202-962-9200
Fax: 202-488-6301
E-mail Address: *info@bio.org*
Web Address: www.bio.org
The Biotechnology Industry Organization (BIO) is
involved in the research and development of health
care, agricultural, industrial and environmental
biotechnology products. BIO has both small and
large member organizations.

**Biotechnology Research Institute of the National
Research Council Canada (NRC-BRI)**
6100 Royalmount Ave.
Montréal, QC H4P 2R2 Canada

Phone: 514-496-6100
E-mail Address: *bri-info@cnrc-nrc.gc.ca*
Web Address: www.bri.nrc-cnrc.gc.ca
The Biotechnology Research Institute of the National
Research Council Canada (NRC-BRI) is one of the
largest Canadian research facilities that focuses
solely on biotechnology.

### Brazilian Association of Biotechnology Companies (ABRABI)
Av. José Cândido da Silveira, 2.100
Horto - Belo Horizonte
MG, CEP 31170-000 Brazil
Phone: 55 31 3486-1733
E-mail Address: *abrabi@abrabi.org.br*
Web Address: http://www.abrabi.org.br/whats-
abrabi.htm
Established in 1986, ABRABI is the national entity
that represents the Brazilian Biotechnology sector.
ABRABI congregates companies, in all areas, that
are part of the biotech chain, namely small innovative
companies, national and international pharmaceutical
and diagnosis companies, agribusinesses and
industrial biotech companies, bioinformatics and
nanobiotechnology.

### California Healthcare Institute (CHI)
1020 Prospect St., Ste. 310
La Jolla, CA 92037 US
Phone: 858-551-6677
Fax: 858-551-6688
E-mail Address: *chi@chi.org*
Web Address: www.chi.org
California Healthcare Institute (CHI) is an
independent organization that represents the biotech
industry in California.

### Connecticut BioScience Cluster (CURE)
300 George St., Ste. 561
New Haven, CT 06511 US
Phone: 203-777-8747
Fax: 203-777-8754
E-mail Address: *twallace@curenet.org*
Web Address: www.curenet.org
Connecticut BioScience Cluster (CURE) is a
partnership with the Department of Economic and
Community Development to promote Connecticut's
biotech clusters.

### Council for Biotechnology Information (CBI)
1201 Maryland Ave. SW, Ste. 900
Washington, DC 20024 US
Phone: 202-962-9200

Web Address: www.whybiotech.com
The Council for Biotechnology Information (CBI) is
a trade organization dedicated to promoting biotech.

### Environmental Mutagen Society (EMS)
1821 Michael Faraday Dr., Ste. 300
Reston, VA 20190 US
Phone: 703-438-8220
Fax: 703-438-3113
E-mail Address: *emshq@ems-us.org*
Web Address: www.ems-us.org
Environmental Mutagen Society (EMS) provides
information on the process of biological mutagenesis,
and the application of this process in the field of
genetic toxicology.

### EuropaBio
Ave. de l'Armée 6
Brussels, 1040 Belgium
Phone: 32 2 735 03 13
Fax: 32 2 735 49 60
E-mail Address: *info@europabio.org*
Web Address: www.europabio.org
EuropaBio's mission is to promote an innovative and
dynamic biotechnology-based industry in Europe.
EuropaBio, (the European Association for
Bioindustries), has 79 corporate and 5 associate
members operating worldwide, 5 Bioregions and 25
national biotechnology associations representing
some 1,800 small and medium sized enterprises.

### European Society of Human Genetics (ESHG)
Vienna Medical Academy
Alser Strasse 4
Vienna, 1090 Austria
Phone: 43-1-405-13-83-20
Fax: 43-1-407-82-74
E-mail Address: *office@eshg.org*
Web Address: www.eshg.org
The European Society of Human Genetics (ESHG) is
a organization that promotes the sharing of
information among genetics societies in Europe.

### Federation of American Societies for Experimental Biology (FASEB)
9650 Rockville Pike
Bethesda, MD 20814 US
Phone: 301-634-7000
Fax: 301-634-7001
E-mail Address: *webmaster@faseb.org*
Web Address: www.faseb.org
The Federation of American Societies for
Experimental Biology (FASEB) works with its 21

member societies to advance biological science through collaborative advocacy for research policies that promote scientific progress leading to improvements in human health.

**Genetics Society of America (GSA)**
9650 Rockville Pike
Bethesda, MD 20814-3998 US
Phone: 301-634-7300
Fax: 301-634-7079
Toll Free: 866-486-4343
E-mail Address: *society@genetics-gsa.org*
Web Address: www.genetics-gsa.org
The Genetics Society of America (GSA) includes over 4,000 scientists and educators interested in the field of genetics. The society promotes the communication of advances in genetics through publication of the journal GENETICS, and by sponsoring scientific meetings focused on key organisms widely used in genetic research.

**German Medical Technology Association (BVMed)**
Bundesverband Medizintechnologie e.V.
Reinhardtstr. 29b
Berlin, 10117 Germany
Phone: 49 (0) 30 246 255-0
Fax: 49 (0) 30 246 255-99
E-mail Address: *info@bvmed.de*
Web Address: www.bvmed.de
The German Medical Technology Association (BVMed) represents about 200 manufacturers and service providers of medical devices.

**International Biometric Society (IBS)**
1444 I St. NW, Ste. 700
Washington, DC 20005 US
Phone: 202-712-9049
Fax: 202-216-9646
E-mail Address: *ibs@bostrom.com*
Web Address: www.tibs.org
The International Biometric Society (IBS) is an association that is devoted to the development and application of statistical and mathematical theory and methods in the biosciences.

**International Society for Clinical Biostatistics (ISCB)**
Teglgaarden 24
Birkerod, DK-3460 Denmark
Phone: 45-4567-2279
Fax: 45-7022-1571
E-mail Address: *office@iscb.info*

Web Address: www.iscb.info
The International Society for Clinical Biostatistics (ISCB) is a professional organization that aims to stimulate research on the biostatistical principles and methodology used in clinical research, to increase the relevance of statistical theory to clinical medicine and to promote better understanding of the use and interpretation of biostatistics by the general public.

**International Society for Stem Cell Research (ISSCR)**
111 Deer Lake Rd., Ste. 100
Deerfield, IL 60015 US
Phone: 847-509-1944
Fax: 847-480-9282
E-mail Address: *isscr@isscr.org*
Web Address: www.isscr.org
The International Society for Stem Cell Research is an independent, nonprofit organization established to promote the exchange and dissemination of information and ideas relating to stem cells, to encourage the general field of research involving stem cells and to promote professional and public education in all areas of stem cell research and application.

**Iowa Biotechnology Association**
900 E. Des Moines St.
Des Moines, IA 50309 US
Phone: 515-327-9156
Fax: 515-327-1407
E-mail Address: *dgetter@netins.net*
Web Address: www.iowabiotech.com
The Iowa Biotech Association promotes research and education in biotechnology in Iowa.

**Japan Bioindustry Association**
2-26-9 Hatchobori, Chuo-ku
Grande Bldg. 8F
Tokyo, 104-0032 Japan
Phone: 03-5541-2731
Fax: 03-5541-2737
Web Address: www.jba.or.jp
JBA is a nonprofit organization dedicated to the promotion of bioscience, biotechnology and bioindustry in both Japan and the rest of the world. Established through the support and cooperation of industry, academia, and government, JBA is the only organization of its kind in Japan.

**Korea Bio Venture Association (KOBIOVEN)**
F8th 703- 5 Seo-il Plz.
Yoeksam-dong, Kangnam-gu

Seoul, 135-513 Korea
Phone: 82-2-552-4771
Fax: 82-2-552-4840
E-mail Address: *kobioven@kobioven.or.kr*
Web Address: eng.kobioven.or.kr
Korea Bio Venture Association (KOBIOVEN) was
created to provide support to the biotech industry in
Korea. It is affiliated with the Ministry of Commerce,
Industry and Energy.

**Massachusetts Biotechnology Council (MBC)**
1 Cambridge Ctr., 9th Fl.
Cambridge, MA 02142 US
Phone: 617-674-5100
Fax: 617-674-5101
E-mail Address: *info_request@massbio.org*
Web Address: www.massbio.org
The Massachusetts Biotechnology Council (MBC) is
a nonprofit organization that promotes the
Massachusetts biotech industry.

**MdBio**
9713 Key West Ave.
Rockville, MD 20850 US
Phone: 240-243-4026
E-mail Address: *rzakour@techcouncilmd.com*
Web Address: www.mdbio.org
MdBio is a nonprofit organization that promotes
biotech in Maryland. Areas of emphasis include
corporate and business development, networking and
community building, education and workforce
development and communications.

**Michigan Biotechnology Association**
3520 Green Ct., Ste. 450
Ann Arbor, MI 48105-1579 US
Phone: 734-527-9150
Fax: 734-302-4933
E-mail Address: *info@michbio.org*
Web Address: www.michbio.org
The Michigan Biotechnology Association is a
nonprofit organization dedicated to promoting the
biotech industry in Michigan.

**Missouri Biotechnology Association (MBA)**
428 E. Capitiol
P.O. Box 148
Jefferson City, MO 65102-0148 US
Phone: 573-761-7600
Fax: 573-761-7601
E-mail Address: *gillespie@mobio.org*
Web Address: www.mobio.org

The Missouri Biotechnology Association (MBA) is
an industry organization that promotes the biotech
industry in Missouri.

**National Association for Biomedical Research
(NABR)**
818 Connecticut Ave. NW, Ste. 900
Washington, DC 20006 US
Phone: 202-857-0540
Fax: 202-659-1902
E-mail Address: *info@nabr.org*
Web Address: www.nabr.org
The National Association for Biomedical Research
(NABR) is an nonprofit group that advocates for
sound public policy that recognizes the vital role of
humane animal use in biomedical research, higher
education, and product safety testing.

**New York Biotechnology Association (NYBA)**
25 Health Sciences Dr., Ste. 203
Stony Brook, NY 11790 US
Phone: 631-444-8895
Fax: 631-444-8896
E-mail Address: *info@nyba.org*
Web Address: www.nyba.org
The New York Biotechnology Association (NYBA)
is a not-for-profit trade association dedicated to the
development and growth of biotechnology-related
industries and institutions in New York State.

**Organibio**
28 rue St. Dominique
Paris, F-75007 France
Phone: 33-147 53 09 12
Fax: 33-147 53 73 76
Web Address: www.organibio.org
ORGANIBIO represents all of the segments of the
French bioscience industry, particularly
biotechnology with applications in pharmaceuticals
and agribio.

**Pennsylvania Bioscience Association (PBA)**
7 Great Valley Pkwy., Ste. 290
Malvern, PA 19355 US
Phone: 610-578-9220
Fax: 610-578-9219
E-mail Address: *info@pennsylvaniabio.org*
Web Address: www.pennsylvaniabio.com
The Pennsylvania Bioscience Association (PBA) is a
nonprofit organization promoting bioscience in
Pennsylvania.

**Scotland Biotechnology**
5 Atlantic Quay
150 Broomielaw
Glasgow, G2 8LU UK
Phone: 0141-248-2700
E-mail Address: *enquiries@scotent.co.uk*
Web Address: www.scottish-
enterprise.com/sedotcom_home/sig/life-sciences.htm
Scotland Lifesciences promotes the biotech industry
of Scotland, as well provides information on medical
devices and pharmaceuticals.

**Society for Biomaterials**
15000 Commerce Pkwy., Ste. C
Mt. Laurel, NJ 08054 US
Phone: 856-439-0826
Fax: 856-439-0525
E-mail Address: *info@biomaterials.org*
Web Address: www.biomaterials.org
The Society for Biomaterials is a professional society
that promotes advances in all phases of materials
research and development by encouraging
cooperative educational programs, clinical
applications and professional standards in the
biomaterials field.

**Society for In Vitro Biology (SIVB)**
514 Daniels St., Ste. 411
Raleigh, NC 27605 US
Phone: 919-420-7940
Fax: 919-420-7939
Toll Free: 888-588-1923
E-mail Address: *sivb@sivb.org*
Web Address: www.sivb.org
The Society for In Vitro Biology (SIVB) is an
association of scientists working to foster the
exchange of knowledge in the field of in vitro
biology. The group maintains a variety of
publications, national and local conferences,
meetings and workshops.

**Society for Industrial Microbiology (SIM)**
3929 Old Lee Hwy., Ste. 92A
Fairfax, VA 22030-2421 US
Phone: 703-691-3357
Fax: 703-691-7991
E-mail Address: *simhq@simhq.org*
Web Address: www.simhq.org
The Society for Industrial Microbiology (SIM) is a
nonprofit professional association that works for the
advancement of microbiological sciences as they
apply to industrial products, biotechnology, materials
and processes.

**Tennessee Biotechnology Association (TBA)**
111 10th Ave. S., Ste. 110
Nashville, TN 37203 US
Phone: 615-255-6270
Fax: 615-255-0094
E-mail Address: *jrolwing@tnbio.org*
Web Address: www.tnbio.org
The Tennessee Biotechnology Association (TBA) is
a statewide clearinghouse that supports the biotech
industry in Tennessee.

**The German Association of Biotechnology
Industries (DIB)**
Deutsche Industrievereinigung Biotechnologie
HallerstraBe 6
Berlin, 10587 Germany
Phone: 49 (30) 343816-0
Fax: 49 (30) 343819-28
E-mail Address: *post@lv-no.vci.de*
Web Address: www.nordostchemie.de
The German Association of Biotechnology Industries
(DIB) represents biotechnology companies and their
interests.

**VBU Association of German Biotech Companies**
c/o DECHEMA e.V.
Theodor-Heuss-Allee 25
Frankfurt am Main, 60486 Germany
Phone: 49 (0)69 - 7564-124
Fax: 49 (0)69 - 7564-169
E-mail Address: *vbu@dechema.de*
Web Address: www.v-b-u.org
VBU Association of German Biotech Companies was
founded in October 1996. It now has over 210
corporate members, making it the largest and oldest
industrial biotechnology association in Germany.
Since its foundation VBU has devoted itself to the
promotion of science and technology and the transfer
of research findings into innovations.

**Virginia Biotechnology Association**
800 E. Leigh St., Ste. 14
Richmond, VA 23219-1534 US
Phone: 804-643-6360
Fax: 804-643-6361
E-mail Address: *questions@vabio.org*
Web Address: www.vabio.org
The Virginia Biotechnology Association is a
nonprofit organization that promotes biotechnology
in Virginia.

**Wisconsin Biotechnology and Medical Devices Association (WBMDA)**
455 Science Dr., Ste. 160
Madison, WI 53711 US
Phone: 608-236-4693
Fax: 608-236-4695
E-mail Address: *info@wisbiomed.org*
Web Address: www.wisconsinbiotech.org
The Wisconsin Biotechnology and Medical Devices Association (WBMDA) is a professional organization devoted to the promotion of the biotech industry in Wisconsin.

## VII.    Biotechnology Investing

**BioTech Stock Report**
P.O. Box 7274
Beaverton, OR 97007-7274 US
Phone: 503-649-1355
Fax: 503-649-4490
E-mail Address: *info@biotechnav.com*
Web Address: www.biotechnav.com
The BioTech Stock Report is a monthly newsletter that provides analysis, commentary, news and company developments for biotechnology investors.

**Burrill & Company**
1 Embarcadero Ctr., Ste. 2700
San Francisco, CA 94111 US
Phone: 415-591-5400
Fax: 415-591-5401
E-mail Address: *burrill@b-c.com*
Web Address: www.burrillandco.com
Burrill & Company is a leading private merchant bank concentrated on companies in the life sciences industries: biotechnology, pharmaceuticals, medical technologies, agricultural technologies, animal health and nutraceuticals.

**Medical Technology Stock Letter**
P.O. Box 40460
Berkeley, CA 94704 US
Phone: 510-843-1857
Fax: 510-843-0901
E-mail Address: *mtsl@bioinvest.com*
Web Address: www.bioinvest.com
The Medical Technology Stock Letter is a newsletter that provides financial advice about investing in biotechnology. It is distributed by mail and electronically.

## VIII.    Biotechnology Resources

**About Biotech**
Web Address: biotech.about.com
About Biotech provides news and information on the biotech industry, expertly compiled and edited by Theresa Phillips, PhD.

**BioAbility**
3200 Chapel Hill/Nelson Blvd., Ste. 201
Research Triangle Park, NC 27709-4569 US
Phone: 919-544-5111
Fax: 919-544-5401
E-mail Address: *info@bioability.com*
Web Address: www.bioability.com
BioAbility provides strategic business information to the biotechnology, pharmaceutical and life science industries.

**BioBasics**
E-mail Address: *info@biotech.gc.ca*
Web Address: biobasics.gc.ca
BioBasics is a Canadian web site that offers information and links related to gene therapy, genetic testing and xenotransplantation. It also contains information on food, health, industrial biotechnology, natural resources and sustainable development.

**Bioengineering Industry Links**
Upenn, Dept. of Bioengineering, School of Eng. & Applied Science
210 S. 33rd St., Rm. 240 Skirkanich Hall
Philadelphia, PA 19104 US
Phone: 215-898-8501
Fax: 215-573-2071
E-mail Address: *beoffice@seas.upenn.edu*
Web Address:
www.seas.upenn.edu/be/misc/bmelink/cell.html
Bioengineering Industry Links is a web site provided by the University of Pennsylvania's Department of Bioengineering. This site features links to companies involved in cell and tissue engineering.

**Biofind**
E-mail Address: *info@biofind.com*
Web Address: www.biofind.com
Biofind offers a biotech news directory, job search, chat room and event announcements, as well as a place to post announcements about biotech innovations.

**Bioinformatics Institute (A*STAR Singapore)**
30 Biopolis St.

07-01 Matrix
138671 Singapore
Phone: 65-6478-8298
E-mail Address: *enquiry@bii.a-star.edu.sg*
Web Address: www.bii.a-star.edu.sg
The Bioinformatics Institute has developed and
deployed analytical tools and computational
techniques for biology research both in-house and
through close collaboration with experimental and
clinical groups within and outside the Biopolis and
Singapore. The site includes information on current
research as well as links, databases, and events.

**BioMed Central**
34-42 Cleveland St.
Middlesex House
London, W1T 4LB UK
Phone: 44-20-7323-0323
Fax: 44-20-7631-9926
E-mail Address: *info@biomedcentral.com*
Web Address: www.biomedcentral.com
BioMed Central is an independent publishing house
that prints approximately 160 peer-reviewed journals
for the medical industry. Its web site provides free,
open access to all of its research.

**Biospace, Inc.**
Ramshorn Executive Ctr.
2399 Hwy. 34, Bldg. A-5
Manasquan, NJ 08736-1528 US
Phone: 732-528-3688
Fax: 732-528-3668
Toll Free: 888-246-7722
E-mail Address: *support@biospace.com*
Web Address: www.biospace.com
Biospace.com offers information, news and profiles
on biotech companies. It also provides an outlet for
business and scientific leaders in bioscience to
communicate with each other.

**BioTech**
The Ellington Lab, University of Texas at Austin
Chem & Biochem Dept., 1 University Station A4800
Austin, TX 78712-0165 US
E-mail Address: *feedback@biotech.icmb.utexas.edu*
Web Address: biotech.icmb.utexas.edu
The BioTech web site offers a comprehensive
dictionary of biotech terms, plus extensive research
data regarding biotechnology. BioTech is located in
The Ellington Lab at the University of Texas at
Austin.

**Biotech Rumor Mill**
E-mail Address: *info@biofind.com*
Web Address: www.biofind.com/rumor-mill
The Biotech Rumor Mill is an online discussion
forum that attracts participants from many biotech
disciplines. Rumor Mill service is provided by
Biofind Limited.

**Biotechnology Information Directory Section**
Cato Research Ltd.
4364 S. Alston Ave.
Durham, NC 27713-2280 US
Phone: 919-361-2286
Fax: 919-361-2290
E-mail Address: *info@cato.com*
Web Address: www.cato.com/pub.shtml
The Biotechnology Information Directory Section
contains links to companies, research institutes,
universities, sources of information and other
directories related to biotechnology, pharmaceutical
development and similar fields. The directory is a
service of Cato Research Ltd.

**Biotechterms.org**
E-mail Address: *webmaster@biotechterms.org*
Web Address: biotechterms.org
Biotechterms.org is an online version of Technomic
Publishing's Glossary of Biotechnology Terms by
Kimball R. Nill.

**BioWorld Online**
3525 Piedmont Rd.
Bldg. 6, Ste. 400
Atlanta, GA 30305 US
Toll Free: 800-688-2421
E-mail Address: *customerservice@bioworld.com*
Web Address: www.bioworld.com
BioWorld Online is a news and information site that
offers in-depth resources about the biotech industry
and leading companies.

**Centre for Cellular and Molecular Biology
(CCMB)**
Council of Scientific & Industrial Research
Uppal Rd.
Hyderabad, 500 007 India
Phone: 91-40-27160222
Fax: 91-040-27160591
Web Address: www.ccmb.res.in
Centre for Cellular and Molecular Biology (CCMB)
is one of the constituent Indian national laboratories
of the Council of Scientific and Industrial Research
(CSIR), the multidisciplinary research and

development organization of the Government of
India.

**Biotech-U**
Web Address: www.bikotech-u.com
Biotech-U provides a web site offering Internet-based
courses on the basics of biotechnology.

**Electronic Journal of Biotechnology**
Av. Brasil 2950
P.O. Box 4059
Valparaíso, Chile
E-mail Address: *edbiotec@ecv.cl*
Web Address: www.ejbiotechnology.info
The Electronic Journal of Biotechnology is an online
journal that publishes information about the biotech
industry.

**Genetic Education Center: Human Genome
Project Resources (GEC)**
3901 Rainbow Blvd.
Kansas City, KS 66160 US
Phone: 913-588-5000
E-mail Address: *dcollins@kumc.edu*
Web Address: www.kumc.edu/gec/prof/hgc.html
The Genetic Education Center: Human Genome
Project Resources (GEC) provides a list of links
related to the human genome including human
genome centers, information sites and news articles.

**Genetic Engineering & Biotechnology News**
140 Huguenot St., 3rd Fl.
New Rochelle, NY 10801-5215 US
Phone: 914-740-2100
Fax: 914-740-2201
Toll Free: 800-799-9436
E-mail Address: *info@liebertpub.com*
Web Address: www.genengnews.com
Genetic Engineering News is a widely read magazine
that offers weekly news on topics in biotechnology,
bioregulation, bioprocess, bioresearch and
technology transfer. It is published by Mary Ann
Liebert, Inc.

**Genome Institute of Singapore**
60 Biopolis St.
02-01 Genome
138672 Singapore
Phone: 65-6478-8000
E-mail Address: *genome@gis.a-star.edu.sg*
Web Address: www.gis.a-star.edu.sg
The Genome Institute of Singapore (GIS) is a
Singaporean national initiative with a global vision.

GIS pursues the integration of technology, genetics
and biology towards the goal of individualized
medicine.

**GrantsNet**
1200 New York Ave. NW
Washington, DC 20005 US
Phone: 202-326-6550
Web Address: www.grantsnet.org
GrantsNet is a free online service to locate funding
for training in the biomedical science industry and
undergraduate science education, provided through
ScienceCareers.org.

**Institute for Biological Sciences (IBS)**
1200 Montreal Rd.
Ottawa, Ontario K1A 0R6 Canada
Phone: 613-990-6111
Toll Free: 800-267-0441
E-mail Address: *Micha.Hage-Badr@nrc-cnrc.gc.ca*
Web Address: ibs-isb.nrc-cnrc.gc.ca
The Institute for Biological Sciences (IBS) is a
branch of Canada's National Research Council and
focuses its research and development programs on
life sciences operations designed to fight age-related
and infectious diseases.

**Institute of Bioinformatics and Applied
Biotechnology (IBAB)**
G-05, Tech Park Mall, Int'l Tech. Pk. Bangalore
Whitefield Rd.
Bangalore, 560 066 India
Phone: 91-80-2841-0029
Fax: 91-80-2841-2761
E-mail Address: *info@ibab.ac.in*
Web Address: www.ibab.ac.in
Institute of Bioinformatics and Applied
Biotechnology's (IBAB) is a joint venture of the
corporate sector and Karnataka State Government in
India. Its serves as the advisory body to Karnataka
and provides educational programs in related
bioinformatics and biotechnology industries.

**Institute of Biological Engineering (IBE)**
1020 Monarch St., Ste. 300B
Lexington, KY 40512 US
Phone: 859-977-7450
Fax: 859-977-7441
E-mail Address: *sclements@ibe.org*
Web Address: www.ibeweb.org
The Institute of Biological Engineering (IBE) is a
professional organization encouraging inquiry and

interest in biological engineering and professional development for its members.

## International Communication Forum in Human Genetics (ICFHG)
E-mail Address: *admin@hum-molgen.de*
Web Address: www.hum-molgen.de
The International Communication Forum in Human Genetics (ICFHG) contains news articles, a bulletin board and a variety of other services related to human molecular genetics.

## Korea Institute of Bioscience & Biotechnology (KRIBB)
111 Gwahangno
Yuseong-gu
Daejeon, 305-806 Korea
Phone: 042-860-4114
Fax: 042-861-1759
Web Address: www.kribb.re.kr
KRIBB is a Korean government research institute dedicated to biotechnology research across a broad span of expertise, from basic studies for the fundamental understanding of life phenomena to applied studies such as new drug discovery, novel biomaterials, integrated biotechnology, and bioinformation.

## LifeSciences World
E-mail Address: *info@lifesciencesworld.com*
Web Address: www.biotechfind.com
LifeSciences World is a directory of life science news, jobs, events, articles, reports and links to information on biotechnology, pharmaceuticals and medical devices.

## Medical Biochemistry Subject List
E-mail Address: *miking@iupui.edu*
Web Address: http://themedicalbiochemistrypage.org
The Medical Biochemistry Subject List, produced by Indiana State University, is a text-based introduction to biochemistry.

## MedWeb: Biomedical Internet Resources
Web Address: www.medweb.emory.edu/medweb
MedWeb: Biomedical Internet Resources is a web site operated by Emory University, which lists resources by medical field, and allows users to search for articles by topic or date.

## Microbiology Network
150 Parkway Dr.
N. Chili, NY 14514 US

Phone: 585-594-8273
Fax: 585-594-3338
E-mail Address: *scott.sutton@microbiol.org*
Web Address: www.microbiol.org
The Microbiology Network is a communication starting point for microbiologists, this site includes a discussion forum, user groups and extensive file libraries.

## National Human Genome Research Institute (NHGRI)
31 Center Dr.
Bldg. 31, Rm. 4B09
Bethesda, MD 20892-2152 US
Phone: 301-402-0911
Fax: 301-402-2218
Web Address: www.nhgri.nih.gov
The National Human Genome Research Institute (NHGRI) led the human genome project until its completion in April 2003. The agency, a division of the National Institutes of Health, now provides research news and information about the field of human genetics.

## Recombinant Capital
2033 N. Main St., Ste. 1050
Walnut Creek, CA 94596-3722 US
Phone: 925-952-3870
Fax: 925-952-3871
E-mail Address: *info@recap.com*
Web Address: www.recap.com
Recombinant Capital provides databases of public information, including press releases, clinical trial references and a library of analyzed collaborations. It hopes to facilitate alliances between biotechnology companies.

## Signals
2033 N. Main St., Ste. 1050
Walnut Creek, CA 94596-3722 US
Phone: 925-952-3870
Fax: 925-952-3871
E-mail Address: *signals_edit@recap.com*
Web Address: www.signalsmag.com
Signals is an online magazine of analysis for the biotechnology industry. The site provides trends dealing with biotechnology, medical device and pharmaceutical companies involved in the development of medical products.

## The Microbiology Network
150 Parkway Dr.
N. Chili, NY 14514 US

Phone: 585-594-8273
Fax: 585-594-3338
E-mail Address: *scott.sutton@microbiol.org*
Web Address: www.microbiol.org
The Microbiology Network is a virtual library
containing lists of organizations and associations in
the fields of microbiology, biology and general
science. Users can also submit new links.

**University of California at Davis Biotechnology
Program**
One Shields Ave.
301 Life Sciences
Davis, CA 95616 US
Phone: 530-752-3260
Fax: 530-752-4125
E-mail Address: *biotechprogram@ucdavis.edu*
Web Address: www.biotech.ucdavis.edu
The University of California at Davis Biotechnology
Program provides useful biotech information and
links, as well as the administrative home for UC
Davis' Biotechnology Program.

**University of Pennsylvania's Center for Bioethics**
3401 Market St., Ste. 320
Philadelphia, PA 19104 US
Phone: 215-898-7136
Web Address: www.bioethics.upenn.edu
The University of Pennsylvania's Center for
Bioethics is a world-renowned resource. It
incorporates the work of more than twenty people
from the university's schools of law, medicine,
business, philosophy, public policy and religious
studies, as well as other departments. Resources
include the PennBioethics newsletter.

**Windhover Information**
10 Hoyt St.
Norwalk, CT 06851 US
Phone: 203-838-4401
Fax: 203-838-3214
E-mail Address:
*fdcwindhover.custcare@elsevier.com*
Web Address: www.windhoverinfo.com
Windhover Information provides analysis and
commentary on healthcare and biotech business
strategy.

## IX.    Careers-Biotech

**Biotechemployment.com**
Phone: 561-630-5201
E-mail Address: *jobs@Biotechemployment.com*

Web Address: www.biotechemployment.com
Biotechemployment.com is an online resource for job
seekers in biotechnology. The site's features includes
resume posting, job search agents and employer
profiles. It is part of the eJobstores.com, Inc., which
includes the Health Care Job Store sites.

**Chase Group (The)**
10955 Lowell Ave., Ste. 500
Overland Park, KS 66210 US
Phone: 913-663-3100
Fax: 913-663-3131
E-mail Address: *chase@chasegroup.com*
Web Address: www.chasegroup.com
The Chase Group is an executive search firm
specializing in biomedical and pharmaceutical
placement.

## X.    Careers-First Time Jobs/New Grads

**Black Collegian Online (The)**
140 Carondelet St.
New Orleans, LA 70130 US
Phone: 504-523-0154
Web Address: www.black-collegian.com
The Black Collegian Online features listings for job
and internship opportunities, as well as other tools for
students of color; it is the web site of The Black
Collegian Magazine, published by IMDiversity, Inc.
The site includes a list of the top 100 minority
corporate employers and an assessment of job
opportunities.

**Collegegrad.com, Inc.**
234 E. College Ave., Ste. 200
State College, PA 16801 US
Phone: 262-375-6700
Toll Free: 1-800-991-4642
Web Address: www.collegegrad.com
Collegegrad.com, Inc. offers in-depth resources for
college students and recent grads seeking entry-level
jobs.

**Job Web**
Nat'l Association of Colleges & Employers (NACE)
62 Highland Ave.
Bethlehem, PA 18017-9085 US
Phone: 610-868-1421
Fax: 610-868-0208
Toll Free: 800-544-5272
E-mail Address: *editors@jobweb.com*
Web Address: www.jobweb.com

Job Web, owned and sponsored by National Association of Colleges and Employers (NACE), displays job openings and employer descriptions. The site also offers a database of career fairs, searchable by state or keyword, with contact information.

## MBAjobs.net
Fax: 413-556-8849
E-mail Address: *contact@mbajobs.net*
Web Address: www.mbajobs.net
MBAjobs.net is a unique international service for MBA students and graduates, employers, recruiters and business schools. The MBAjobs.net service is provided by WebInfoCo.

## MonsterTRAK
11845 W. Olympic Blvd., Ste. 500
Los Angeles, CA 90064 US
Toll Free: 800-999-8725
E-mail Address: *trakstudent@monster.com*
Web Address: www.monstertrak.monster.com
MonsterTRAK provides information about internships and entry-level jobs.

## National Association of Colleges and Employers (NACE)
62 Highland Ave.
Bethlehem, PA 18017-9085 US
Phone: 610-868-1421
Fax: 610-868-0208
Toll Free: 800-544-5272
E-mail Address: *mcollins@naceweb.org*
Web Address: www.naceweb.org
The National Association of Colleges and Employers (NACE) is a premier U.S. organization representing college placement offices and corporate recruiters who focus on hiring new grads.

## XI.  Careers-General Job Listings

## Career Exposure, Inc.
805 SW Broadway, Ste. 2250
Portland, OR 97205 US
Phone: 503-221-7779
Fax: 503-221-7780
E-mail Address: *lisam@mackenzie-marketing.com*
Web Address: www.careerexposure.com
Career Exposure, Inc. is an online career center and job placement service, with resources for employers, recruiters and job seekers.

## CareerBuilder, Inc.
200 N. LaSalle St., Ste. 1100
Chicago, IL 60601 US
Phone: 773-527-3600
Toll Free: 800-638-4212
Web Address: www.careerbuilder.com
CareerBuilder, Inc. focuses on the needs of companies and also provides a database of job openings. The site has 1.5 million jobs posted by 300,000 employers, and receives an average 23 million unique visitors monthly. The company also operates online career centers for 150 newspapers, 1,000 partners and other online portals such as America Online. Resumes are sent directly to the company, and applicants can set up a special e-mail account for job-seeking purposes. CareerBuilder is primarily a joint venture between three newspaper giants: The McClatchy Company (which recently acquired former partner Knight Ridder), Gannett Co., Inc. and Tribune Company. In 2007, Microsoft acquired a minority interest in CareerBuilder, allowing the site to ally itself with MSN.

## CareerOneStop
Toll Free: 877-348-0502
E-mail Address: *info@careeronestop.org*
Web Address: www.careeronestop.org
CareerOneStop is operated by the employment commissions of various state agencies. It contains job listings in both the private sector and in government. CareerOneStop is sponsored by the U.S. Department of Labor. It includes a wide variety of useful career resources and workforce information.

## JobCentral
DirectEmployers Association, Inc.
9002 N. Purdue Rd., Quad III, Ste. 100
Indianapolis, IN 46268 US
Phone: 317-874-9000
Fax: 317-874-9100
Toll Free: 866-268-6206
E-mail Address: *info@jobcentral.com*
Web Address: www.jobcentral.com
JobCentral, operated by the nonprofit DirectEmployers Association, Inc., links users directly to hundreds of thousands of job opportunities posted on the sites of participating employers, thus bypassing the usual job search sites. This saves employers money and allows job seekers to access many more job opportunities.

## LaborMarketInfo
Employment Dev. Dept., Labor Market Info. Div.
800 Capitol Mall, MIC 83
Sacramento, CA 95814 US

Phone: 916-262-2162
Fax: 916-262-2352
Toll Free: 800-480-3287
Web Address: www.labormarketinfo.edd.ca.gov
LaborMarketInfo, formerly the California
Cooperative Occupational Information System, is
geared to providing job seekers and employers a wide
range of resources, namely the ability to find, access
and use labor market information and services. It
provides demographical statistics for employment on
both a local and regional level, as well as career
searching tools for California residents. The web site
is sponsored by California's Employment
Development Office.

**Recruiters Online Network**
947 Essex Ln.
Medina, OH 44256 US
Phone: 888-364-4667
Fax: 888-237-8686
E-mail Address: *info@recruitersonline.com*
Web Address: www.recruitersonline.com
The Recruiters Online Network provides job postings
from thousands of recruiters, Careers Online
Magazine, a resume database, as well as other career
resources.

**True Careers, Inc.**
Web Address: www.truecareers.com
True Careers, Inc. offers job listings and provides an
array of career resources. The company also offers a
search of over 2 million scholarships. It is partnered
with CareerBuilder.com, which powers its career
information and resume posting functions.

**USAJOBS**
U.S. Office of Personnel Management
1900 E St. NW
Washington, DC 20415 US
Phone: 202-606-1800
Web Address: usajobs.opm.gov
USAJOBS, a program of the U.S. Office of Personnel
Management, is the official job site for the U.S.
Federal Government. It provides a comprehensive list
of U.S. government jobs, allowing users to search for
employment by location; agency; type of work, using
the Federal Government's numerical identification
code, the General Schedule (GS) Series; or by senior
executive positions. It also has a special veterans'
employment section; an information center, offering
resume and interview tips and other useful
information such as hiring trends and a glossary of

Federal terms; and allows users to create a profile and
post a resume.

**Wall Street Journal - CareerJournal**
Wall Street Journal
200 Liberty St.
New York, NY 10281 US
Phone: 212-416-2000
Toll Free: 800-568-7625
E-mail Address: *onelinejournal@wsj.com*
Web Address: www.online.wsj.com/careers
The Wall Street Journal's CareerJournal, an executive
career site, features a job database with thousands of
available positions; career news and employment
related articles; and advice regarding resume writing,
interviews, networking, office life and job hunting.

**Yahoo! HotJobs**
45 W. 18th St., 6th Fl.
New York, NY 10011 US
Phone: 646-351-5300
Web Address: www.hotjobs.yahoo.com
Yahoo! HotJobs, designed for experienced
professionals, employers and job seekers, is a Yahoo-
owned site that provides company profiles, a resume
posting service and a resume workshop. The site
allows posters to block resumes from being viewed
by certain companies and provides a notification
service of new jobs.

## XII.  Careers-Health Care

**Health Care Source**
8 Winchester Pl.
Winchester, MA 01890 US
Phone: 781-368-1033
Fax: 800-829-6600
Toll Free: 800-869-5200
E-mail Address: *info@healthcaresource.com*
Web Address: www.healthcaresource.com
Health Care Source offers career-related information
and job finding tools for health care professionals.

**Medzilla, Inc.**
P.O. Box 1710
Marysville, WA 98270 US
Phone: 360-657-5681
Fax: 775-514-9440
E-mail Address: *mgroutage@medzilla.com*
Web Address: www.medzilla.com
Medzilla, Inc.'s web site offers job searches, salary
surveys, a search agent and information on health
care employment.

**Monster Career Advice-Healthcare**
Monster Worldwide, Inc.
622 3rd Ave.
New York, NY 10017 US
Phone: 212-351-7000
Toll Free: 800-666-7837
Web Address: career-advice.monster.com/get-the-job/healthcare/home.aspx
Monster Career Advice-Healthcare, a service of Monster Worldwide, Inc., provides job listings, job searches and search agents for the medical field.

**PracticeLink**
P.O. Box 100
415 2nd Ave.
Hinton, WV 25951 US
Toll Free: 800-776-8383
Web Address: www.practicelink.com
PracticeLink is one of the largest physician employment web sites. It is a free service used by more than 18,000 practice-seeking physicians annually to quickly search and locate potential physician practice opportunities. PracticeLink is financially supported by more than 700 hospitals, medical groups, private practices and health care systems that advertise more than 5,000 opportunities.

**RPh on the Go USA, Inc.**
5510 Howard St.
Skokie, IL 60077-2620 US
Phone: 847-588-7170
Fax: 847-588-7060
Toll Free: 800-553-7359
E-mail Address: lbalaguer@rphonthego.com
Web Address: www.rphonthego.com
RPh on the Go USA, Inc. places temporary and permanent qualified professionals in the pharmacy community.

## XIII. Careers-Job Reference Tools

**NewsVoyager**
4401 Wilson Blvd., Ste. 900
Arlington, VA 22203-1867 US
Phone: 571-366-1000
Fax: 571-366-1195
E-mail Address: sally.clarke@naa.org
Web Address: www.newsvoyager.com
NewsVoyager, a service of the Newspaper Association of America (NAA), links individuals to local, national and international newspapers. Job seekers can search through thousands of classified sections.

**Vault.com, Inc.**
75 Varick St., 8th Fl.
New York, NY 10013 US
Phone: 212-366-4212
E-mail Address: feedback@staff.vault.com
Web Address: www.vault.com
Vault.com, Inc. is a comprehensive career web site for employers and employees, with job postings and valuable information on a wide variety of industries. Vault gears many of its features toward MBAs. The site has been recognized by Forbes and Fortune Magazines.

## XIV. Careers-Science

**Chem Jobs**
ChemIndustry.com
730 E. Cypress Ave.
Monrovia, CA 91016 US
Phone: 626-930-0808
Fax: 626-930-0102
E-mail Address: info@chemindustry.com
Web Address: www.chemjobs.net
Chem Jobs is a leading Internet site for job seekers in chemistry and related fields, aimed at chemists, biochemists, pharmaceutical scientists and chemical engineers looking for work. The web site is powered by Chemindustry.com.

**Employment Links for the Biomedical Scientist**
E-mail Address: graeme@his.com
Web Address: www.his.com/~graeme/employ.html
Employment Links for the Biomedical Scientist offers a list of links for job seekers using the Internet to locate work in the biomedical field. It is compiled by a fellow scientist, Graeme Eisenhofer.

**New Scientist Jobs**
New Scientist, Lacon House
84 Theobald's Rd.
London, WC1X 8NS UK
Phone: 44-20-7611-1200
Fax: 44-20-7611-1250
E-mail Address: nssales@elsevier.com.
Web Address: www.newscientistjobs.com
New Scientist Jobs is a web site produced by the publishers of New Scientist Magazine, which helps jobseekers and employers in the bioscience fields find each other. The site includes a job search engine and a free-of-charge e-mail job alert service. New Scientist Jobs is owned by Reed Business Information Ltd.

**Science Jobs**
E-mail Address: *help@science-jobs.org*
Web Address: www.science-jobs.org/index.htm
Science Jobs is a web site that contains many useful
categories of links, including employment
newsgroups, scientific journals and placement
agencies. It also links to sites containing information
regarding internship and fellowship opportunities for
high school students and undergrads.

## XV.    Chinese Government Agencies-Science & Technology

**China Ministry of Science and Technology (MOST)**
15B Fuxing Rd.
Beijing, 100862 China
Web Address: www.most.gov.cn
The China Ministry of Science and Technology
(MOST) has information and links to its various
departments including the Departments of Personnel;
Social Development; Rural Science and Technology;
Basic Research; and High and New Technology
Development and Industrialization.

## XVI.    Clinical Trials

**Clinical Trials**
8600 Rockville Pike
Bethesda, MD 20894 US
Phone: 301-594-5983
Fax: 301-402-1384
Toll Free: 888-346-3656
Web Address: www.clinicaltrials.gov
Clinical Trials offers up-to-date information for
locating federally and privately supported clinical
trials for a wide range of diseases and conditions. It is
a service of the National Institutes of Health (NIH).

**Institute of Clinical Research (ICR)**
Institute House
Boston Dr.
Bourne End, Buckinghamshire SL8 5YS UK
Phone: 44-1628-536960
Fax: 44-1628-530641
E-mail Address: *info@icr-global.org*
Web Address: www.icr-global.org
The Institute of Clinical Research (ICR) is a
professional organization for clinical researchers in
the pharmaceutical industry in the U.K.

**Office of Biotechnology Activities, NIH**
6705 Rockledge Dr., Ste. 750, MSC 7985
Bethesda, MD 20892-7985 US
Phone: 301-496-9838
Fax: 301-496-9839
E-mail Address: *oba@od.nih.gov*
Web Address: http://www4.od.nih.gov/oba/
This unit of the U.S. National Institutes of Health
operates a web site with links to clinical research in
recombinant DNA and gene transfer, along with
information on the National Science Advisory Board
for Biosecurity.

## XVII.   Communications Professional Associations

**Health and Science Communications Association (HeSCA)**
39 Wedgewood Dr., Ste. A
Jewett City, CT 06351 US
Phone: 860-376-5915
Fax: 860-376-6621
E-mail Address: *hesca@hesca.org*
Web Address: www.hesca.org
The Health and Science Communications Association
(HeSCA) is an association of communications
professionals committed to sharing knowledge and
resources in the health sciences arena.

## XVIII.  Consulting Industry Associations

**American Society For Quality (ASQ)**
600 N. Plankinton Ave.
Milwaukee, IL 53203 US
Fax: 414-272-1734
Toll Free: 800-248-1946
E-mail Address: *help@asq.org*
Web Address: www.asq.org
The American Society For Quality (ASQ) is a
professional association that works to advance
organizational management techniques in order to
improve business results and to create better
workplaces and communities worldwide.

## XIX.    Corporate Information Resources

**bizjournals.com**
120 W. Morehead St., Ste. 400
Charlotte, NC 28202 US
Web Address: www.bizjournals.com
Bizjournals.com is the online media division of
American City Business Journals, the publisher of

dozens of leading city business journals nationwide. It provides access to research into the latest news regarding companies small and large.

## Business Wire
44 Montgomery St., 39th Fl.
San Francisco, CA 94104 US
Phone: 415-986-4422
Fax: 415-788-5335
Toll Free: 800-227-0845
Web Address: www.businesswire.com
Business Wire offers news releases, industry- and company-specific news, top headlines, conference calls, IPOs on the Internet, media services and access to tradeshownews.com and BW Connect On-line through its informative and continuously updated web site.

## Edgar Online, Inc.
50 Washington St., 11th Fl.
Norwalk, CT 06854 US
Phone: 203-852-5666
Fax: 203-852-5667
Toll Free: 800-416-6651
Web Address: www.edgar-online.com
Edgar Online, Inc. is a gateway and search tool for viewing corporate documents, such as annual reports on Form 10-K, filed with the U.S. Securities and Exchange Commission.

## PR Newswire Association LLC
810 7th Ave., 32nd Fl.
New York, NY 10019 US
Phone: 201-360-6700
Toll Free: 800-832-5522
E-mail Address: *information@prnewswire.com*
Web Address: www.prnewswire.com
PR Newswire Association LLC provides comprehensive communications services for public relations and investor relations professionals ranging from information distribution and market intelligence to the creation of online multimedia content and investor relations web sites. Users can also view recent corporate press releases. The Association is owned by United Business Media plc.

## Silicon Investor
100 W. Main
P.O. Box 29
Freeman, MO 64746 US
E-mail Address: *admin_dave@techstocks.com*
Web Address: www.siliconinvestor.advfn.com

Silicon Investor is focused on providing information about technology companies. The company's web site serves as a financial discussion forum and offers quotes, profiles and charts.

| XX. | Economic Data & Research |
|---|---|

## Eurostat
Phone: 32-2-299-9696
Toll Free: 80-0-6789-1011
Web Address: www.epp.eurostat.ec.europa.eu
Eurostat is the European Union's service that publishes a wide variety of comprehensive statistics on European industries, populations, trade, agriculture, technology, environment and other matters.

## STAT-USA/Internet
STAT-USA, HCHB, U.S. Dept. of Commerce
Rm. 4885
Washington, DC 20230 US
Phone: 202-482-1986
Fax: 202-482-2164
Toll Free: 800-782-8872
E-mail Address: *statmail@esa.doc.gov*
Web Address: www.stat-usa.gov
STAT-USA/Internet offers daily economic news, statistical releases and databases relating to export and trade, as well as the domestic economy. It is provided by STAT-USA, which is an agency in the Economics & Statistics Administration of the U.S. Department of Commerce. The site mainly consists of two main databases, the State of the Nation (SOTN), which focuses on the current state of the U.S. economy; and the Global Business Opportunities (GLOBUS) & the National Trade Data Bank (NTDB), which deals with U.S. export opportunities, global political/socio-economic conditions and other world economic issues.

| XXI. | Engineering, Research & Scientific Associations |
|---|---|

## Agency For Science, Technology And Research (A*STAR)
1 Fusionopolis Way
20-10 Connexis N. Twr.
138632 Singapore
Phone: 65-6826-6111
Fax: 65-6777-1711
Web Address: www.a-star.edu.sg

The Agency For Science, Technology And Research (A*STAR) of Singapore comprises the Biomedical Research Council (BMRC), the Science and Engineering Research Council (SERC), Exploit Technologies Pte Ltd (ETPL), the A*STAR Graduate Academy (A*GA) and the Corporate Planning and Administration Division (CPAD). Both Councils fund the A*STAR public research institutes which conducts research in specific niche areas in science, engineering and biomedical science.

**American Association for the Advancement of Science (AAAS)**
1200 New York Ave. NW
Washington, DC 20005 US
Phone: 202-326-6400
E-mail Address: *webmaster@aaas.org*
Web Address: www.aaas.org
The American Association for the Advancement of Science (AAAS) is the world's largest scientific society and the publisher of Science magazine. It is an international nonprofit organization dedicating to advancing science.

**American National Standards Institute (ANSI)**
1819 L St. NW, 6th Fl.
Washington, DC 20036 US
Phone: 202-293-8020
Fax: 202-293-9287
E-mail Address: *info@ansi.org*
Web Address: www.ansi.org
The American National Standards Institute (ANSI) is a private, nonprofit organization that administers and coordinates the U.S. voluntary standardization and conformity assessment system. Its mission is to enhance both the global competitiveness of U.S. business and the quality of life by promoting and facilitating voluntary consensus standards and conformity assessment systems and safeguarding their integrity.

**American Physical Society (APS)**
1 Physics Ellipse
College Park, MD 20740-3844 US
Phone: 301-209-3100
Fax: 301-209-0865
Web Address: www.aps.org
The American Physical Society (APS) develops and implements effective programs in physics education and outreach.

**American Society for Healthcare Engineering (ASHE)**
1 N. Franklin, 28th Fl.
Chicago, IL 60606 US
Phone: 312-422-3800
Fax: 312-422-4571
E-mail Address: *ashe@aha.org*
Web Address: www.ashe.org
The American Society for Healthcare Engineering (ASHE) is the advocate and resource for continuous improvement in the health care engineering and facilities management professions.

**American Society of Agricultural and Biological Engineers (ASABE)**
2950 Niles Rd.
St. Joseph, MI 49085 US
Phone: 269-429-0300
Fax: 269-429-3852
E-mail Address: *hq@asabe.org*
Web Address: www.asabe.org
The American Society of Agricultural and Biological Engineers (ASABE) is a nonprofit professional and technical organization interested in engineering knowledge and technology for food and agriculture and associated industries.

**Association for Electrical, Electronic & Information Technologies (VDE)**
Stresemannallee 15
Frankfurt am Main, 60596 Germany
Phone: 49(0)69 6308 284
Fax: 49(0)69 96315215
E-mail Address: *presse@vde.com*
Web Address: www.vde.com
The Association for Electrical, Electronic & Information Technologies (VDE) is an organization with roughly 34,000 members representing one of the largest technical associations in Europe.

**Association of Official Analytical Chemists (AOAC)**
481 N. Frederick Ave., Ste. 500
Gaithersburg, MD 20877-2417 US
Phone: 301-924-7077
Fax: 301-924-7089
Toll Free: 800-379-2622
E-mail Address: *aoac@aoac.org*
Web Address: www.aoac.org
The Association of Official Analytical Chemists (AOAC) is a nonprofit scientific association committed to worldwide confidence in analytical results.

**China Association for Science and Technology (CAST)**
3 Fuxing Rd.
Beijing, 100863 China
Phone: 8610-68571898
Fax: 8610-68571897
E-mail Address: *english@cast.org.cn*
Web Address: english.cast.org.cn
The China Association for Science and Technology (CAST) is the largest national non-governmental organization of scientific and technological workers in China. The association has 167 member organizations in the field of engineering, science and technology.

**Chinese Academy of Sciences (CAS)**
52 Sanlihe Rd.
Beijing, 100864 China
Phone: 86-10-68597289
Fax: 86-10-68512458
E-mail Address: *bulletin@mail.casipm.ac.cn*
Web Address: english.cas.ac.cn
By 2010, the Chinese Academy of Sciences (CAS) plans on maintaining roughly 80 national institutes for science and technological innovation each with distinctive features with roughly thirty planned for internationally acknowledged, high-level research.

**DECHEMA (Society for Chemical Engineering and Biotechnology)**
Theodor-Heuss-Allee 25
Frankfurt am Main, 60486 Germany
Phone: 0049 69 7564-0
Fax: 0049 69 7564-201
Web Address: http://dechema.de/
The DECHEMA (Society for Chemical Engineering and Biotechnology) is a nonprofit scientific and technical society based in Frankfurt on Main. It was founded in 1926. Today it has over 5000 private and institutional members. Its aim is to promote research and technical advances in the areas of chemical engineering, biotechnology and environmental protection. The group's work is interdisciplinary, with scientists, engineers, and technologists working together under one roof. Experts from science, business, and government departments cooperate in working parties and subject divisions.

**Federation of Technology Industries (FHI)**
Federatie Van Technologiebranches
Dodeweg 6B
AK LEUSDEN, 3832 Netherlands
Phone: (033) 465 75 07

Fax: (033) 461 66 38
E-mail Address: *info@fhi.nl*
Web Address: federatie.fhi.nl
The Federation of Technology Industries (FHI) is the Dutch trade organization representing industrial electronics, industry automation, laboratory technology and medical technolgy.

**German Association of High-Tech Industries (SPECTARIS)**
SPECTARIS, Verband der Hightech-Industrie
Kirchweg 2
Koln, 50858 Germany
Phone: 0221/948628-0
Fax: 0221/948628-80
E-mail Address: *info@spectaris.de*
Web Address: www.spectaris.de/
German Association of High-Tech Industries (SPECTARIS) is the trade association for technology and research in the consumer optics, photonics, biotech and medical technology.

**Industrial Research Institute (IRI)**
2200 Clarendon Blvd., Ste. 1102
Arlington, VA 22201 US
Phone: 703-647-2580
Fax: 703-647-2581
E-mail Address: *information@iriinc.org*
Web Address: www.iriinc.org
The Industrial Research Institute (IRI) is a nonprofit organization of over 200 leading industrial companies, representing industries such as aerospace, automotive, chemical, computers and electronics, which carry out industrial research efforts in the U.S. manufacturing sector. IRI helps members improve research and development capabilities.

**Institute of Bioengineering and Nanotechnology, Singapore**
31 Biopolis Way
The Nanos 04-01
138669 Singapore
Phone: 65-6824-7000
Fax: 65-6478-9080
E-mail Address: *enquiry@ibn.a-star.edu.sg*
Web Address: www.ibn.a-star.edu.sg
As a scientific research institute, Institute of Bioengineering and Nanotechnology focuses its activities on developing the following key areas; developing a critical knowledge base in bioengineering and nanotechnology; generating new biomaterials, devices and processes; and producing and publishiing high-quality scientific research.

**Institute of Electrical and Electronics Engineers (IEEE)**
3 Park Ave., 17th Fl.
New York, NY 10016-5997 US
Phone: 212-419-7900
Fax: 212-752-4929
E-mail Address: *ieeeusa@ieee.org*
Web Address: www.ieee.org
The Institute of Electrical and Electronics Engineers (IEEE) is a nonprofit, technical professional association of more than 375,000 individual members in approximately 160 countries. The IEEE sets global technical standards and acts as an authority in technical areas ranging from computer engineering, biomedical technology and telecommunications, to electric power, aerospace and consumer electronics.

**International Federation for Medical and Biological Engineering**
Faculty of Electrical Engineering and Computing
Univ. of Zagreb, Unska 3
Zagreb, HR 10000 Croatia
Phone: 385-1-6129-938
Fax: 385-1-6129-652
E-mail Address: *ratko.magjarevic@fer.hr*
Web Address: www.ipem.ac.uk
The International Federation for Medical and Biological Engineering (IFMBE) is a federation of national and transnational organizations that represent national interests in medical and biological engineering. The objectives of the IFMBE are scientific, technological, literary, and educational.

**International Standards Organization (ISO)**
1 ch. de la Voie-Creuse
Case Postale 56
Geneva 20, CH-1211 Switzerland
Phone: 41-22-749-01-11
Fax: 41-22-733-34-30
E-mail Address: *central@iso.org*
Web Address: www.iso.org
The International Standards Organization (ISO) is a global consortium of national standards institutes from 157 countries. The established International Standards are designed to make products and services more efficient, safe and clean.

**Japan Science and Technology Agency (JST)**
Kawaguchi Ctr. Bldg.
4-1-8 Honcho, Kawaguchi-shi
Saitama, 332-0012 Japan
Phone: 81-48-226-5601
Fax: 81-48-226-5651

E-mail Address: *www-admin@tokyo.jst.go.jp*
Web Address: www.jst.go.jp/EN
The Japan Science and Technology Agency (JST) aims to act as a core organization for implementation of the nation's science and technology policies by conducting research and development with particular emphasis on new technological needs.

**Materials Research Society (MRS)**
506 Keystone Dr.
Warrendale, PA 15086-7573 US
Phone: 724-779-3003
Fax: 724-779-8313
E-mail Address: *info@mrs.org*
Web Address: www.mrs.org
The Materials Research Society (MRS) is dedicated to basic and applied research on materials of technological importance. MRS emphasizes an interdisciplinary approach to materials science and engineering.

**National Academy of Science (NAS)**
500 5th St. NW
Washington, DC 20001 US
Phone: 202-334-2000
E-mail Address: *worldwidewebfeedback@nas.edu*
Web Address: www.nationalacademies.org
The National Academies are private, nonprofit, self-perpetuating societies of scholars engaged in scientific and engineering research dedicated to the furtherance of science and technology and to their use for the general welfare. Four organizations comprise the Academies: The National Academy of Engineering, the National Research Council, the National Academy of Sciences and the Institute of Medicine.

**National Medical Research Council (NMRC)**
11 Biopolis Way
Helios 09-10/11
138667 Singapore
Phone: 65-6325-8130
Fax: 65-6324-3735
E-mail Address: *MOH_NMRC@MOH.GOV.SG*
Web Address: www.nmrc.gov.sg
National Medical Research Council (NMRC) oversees the development and advancement of medical research in Singapore.

**Netherlands Organization for Applied Scientific Research (TNO)**
Schoemakerstraat 97, Bldg. A
Delft, NL-2628 VK The Netherlands

Phone: 31-15-269-69-00
Fax: 31-15-261-24-03
E-mail Address: *infodesk@tno.nl*
Web Address: www.tno.nl
The Netherlands Organization for Applied Scientific Research (TNO) is a contract research organization that provides a link between fundamental research and practical application.

**Research in Germany, German Academic Exchange Service (DAAD)**
Kennedyallee 50
Bonn, 53175 Germany
Phone: 49(0)228 882 - 0
Fax: 49(0)228 882 - 660
Web Address: www.research-in-germany.de
The Research in Germany portal, German Academic Exchange Service (DAAD),
is an information platform and contact point for those looking to find out more about Germany's research landscape and its latest research achievements. The portal is an initiative of the Federal Ministry of Education and Research.

**Royal Society (The)**
6-9 Carlton House Ter.
London, SW1Y 5AG UK
Phone: 44-20-7451-2500
Fax: 44-20-7930-2170
E-mail Address: *info@royalsociety.org*
Web Address: www.royalsoc.ac.uk
The Royal Society is the UK's leading scientific organization. It operates as a national academy of science, supporting scientists, engineers, technologists and research. On its website, you will find a wealth of data about the research and development initiatives of its fellows and foreign members.

**Royal Society of Chemistry (RSC)**
Burlington House, Piccadilly
London, W1J 0BA UK
Phone: 44-20-7437-8656
Fax: 44-20-7437-8883
Web Address: www.rsc.org
The Royal Society of Chemistry (RSC) is one of Europe's largest organizations for advancing the chemical sciences.

**Society for Chemical Engineering and Biotechnology (DECHEMA)**
Theodor-Heuss-Allee 25
Frankfurt am Main, 60486 Germany

Phone: (069) 7564-0
Fax: (069) 7564-201
Web Address: www.dechema.de
The Society for Chemical Engineering and Biotechnology (DECHEMA) is an interdisciplinary organization representing private firms and institutions in Germany. DECHEMA cooperates with government, science and business to promote chemical engineering, biotechnology and environmental protection.

**VDI/VDE Innovation + Technik GmbH (VDI/VDE-IT)**
VDI/VDE Innovation + Technik GmbH
Steinplatz 1
Berlin, 10623 Germany
Phone: 49 (0) 3328 435-0
Fax: 49 (0) 3328 435-141
E-mail Address: *vdivde-it@vdivde-it.de*
Web Address: www.vdivde-it.de
The VDI/VDE Innovation + Technik GmbH (VDI/VDE-IT) is a management consulting subsidiary of two of the largest engineering firms in Europe, VDI GmbH and VDE (Association for Electrical, Electronic & Information Technologies), as well as a partner of federal government and the EU. VDI/VDE-IT develops and organizes state supported programs and provides consulting and support for technology-oriented companies, their banks, and investors.

| XXII. Engineering, Research & Scientific Resources |
| --- |

**Steacie Institute for Molecular Sciences (SIMS)**
100 Sussex Dr., Rm. 1151
Ottawa, ON K1A 0R6 Canada
Phone: 613-991-5419
Fax: 613-954-5242
E-mail Address: *Helene.Letourneau@nrc-cnrc.gc.ca*
Web Address: steacie.nrc-cnrc.gc.ca
The Steacie Institute for Molecular Sciences (SIMS) was created to facilitate the collaboration of scientific communities researching molecular sciences both within Canada and internationally.

| XXIII. Government Agencies-Singapore |
| --- |

**Singapore Government Online (SINGOV)**
140 Hill St., MICA Bldg., 5th Fl.
179369 Singapore
E-mail Address: *singov_webmaster@mica.gov.sg*

Web Address: www.gov.sg
Singapore Government Online (SINGOV) is the
default homepage for the Singapore Government and
is a portal for governmenal information. The website
lists governmental agencies, news, information,
policies and inititives.

## XXIV.    Health Care Business & Professional Associations

### American Medical Technologists (AMT)
10700 W. Higgins Rd., Ste. 150
Rosemont, IL 60018 US
Phone: 847-823-5169
Fax: 847-823-0458
Toll Free: 800-275-1268
E-mail Address: *membership@amt1.com*
Web Address: www.amt1.com
American Medical Technologists (AMT) is a
nonprofit certification agency and professional
membership association representing individuals in
health care.

### American Society of Clinical Oncology (ASCO)
2318 Mill Rd., Ste. 800
Alexandria, VA 22314 US
Phone: 571-483-1300
Fax: 703-299-1044
Web Address: www.asco.org
The American Society of Clinical Oncology (ASCO)
is a nonprofit organization, founded in 1964, with
overarching goals of improving cancer care and
prevention and ensuring that all patients with cancer
receive care of the highest quality. Nearly 25,000
oncology practitioners belong to ASCO, representing
all oncology disciplines.

### Association of Clinical Research Professionals (ACRP)
500 Montgomery St., Ste. 800
Alexandria, VA 22314 US
Phone: 703-254-8100
Fax: 703-254-8101
E-mail Address: *office@acrpnet.org*
Web Address: www.acrpnet.org
The Association of Clinical Research Professionals
(ACRP) is an organization for professionals in the
pharmaceutical, biotechnology and medical device
industries, as well as those in hospital, academic
medical centers and physician office settings.

### Health Industry Distributors Association (HIDA)
310 Montgomery St.

Alexandria, VA 22314-1516 US
Phone: 703-549-4432
Fax: 703-549-6495
E-mail Address: *sandler@hida.org*
Web Address: www.hida.org
The Health Industry Distributors Association (HIDA)
is the international trade association representing
medical products distributors.

### Hong Kong Medical Association
15 Hennessy Rd.
5/F Duke of Windsor Social Service Bldg.
Wanchai, Hong Kong
Phone: 2527-8285
E-mail Address: *hkma@hkma.org*
Web Address: www.hkma.org
The Hong Kong Medical Association's objective is to
promote the welfare of the medical profession and
the health of the public of Hong Kong.

### Medical Device Manufacturers Association (MDMA)
1350 I St. NW, Ste. 540
Washington, DC 20005 US
Phone: 202-354-7171
Web Address: www.medicaldevices.org
The Medical Device Manufacturers Association
(MDMA) is a national trade association that
represents independent manufacturers of medical
devices, diagnostic products and health care
information systems.

### Michigan Medical Device Association (MMDA)
P.O. Box 170
Howell, MI 48844 US
Fax: 517-546-3356
Toll Free: 800-930-5698
E-mail Address: *info@mmda.org*
Web Address: www.mmda.org
The Michigan Medical Device Association (MMDA)
sponsors educational seminars and informational
programs; is active in the areas of government
relations, networking and business development; and
acts as a source for the dissemination of matters of
interest to its members.

### Society for Pharmaceutical and Medical Device Professionals (ISPE)
3109 W. Dr. Martin Luther King, Jr. Blvd., Ste. 250
Tampa, FL 33607 US
Phone: 813-960-2105
Fax: 813-264-2816
E-mail Address: *ASK@ispe.org*

Web Address: www.ispe.org
The Society for Pharmaceutical and Medical Device Professionals (ISPE) works with its members by providing extensive education, training, technical publications, conferences and networking opportunities.

## Society of Clinical Research Associates (SOCRA)

530 W. Butler Ave., Ste. 109
Chalfont, PA 18914 US
Phone: 215-822-8644
Fax: 215-822-8633
Toll Free: 800-7627292
E-mail Address: *socramail@aol.com*
Web Address: www.socra.org
The Society of Clinical Research Associates (SOCRA) works to provide training and continuing education and to establish and maintain an international certification program for clinical research professionals.

## Society of Toxicology (SOT)

1821 Michael Faraday Dr., Ste. 300
Reston, VA 20190 US
Phone: 703-438-3115
Fax: 703-438-3113
E-mail Address: *sothq@toxicology.org*
Web Address: www.toxicology.org
The Society of Toxicology (SOT) is an association that works to advance the science of enhancing human, animal and environmental health through the understanding of toxicology.

## XXV. Health Care Resources

### Access Excellence

1350 Connecticut Ave. NW, 5th Fl.
Washington, DC 20036 US
Phone: 650-712-1723
Web Address: www.accessexcellence.org
Access Excellence provides information for high school biology and life science teachers. It is produced by the National Health Museum.

### Singapore Medical Council (SMC)

16 College Rd.
01-01 College of Medicine Bldg.
169854 Singapore
Phone: 65-6372-3065
Fax: 65-6221-0558
E-mail Address: *moh_smc@moh.gov.sg*
Web Address: www.smc.gov.sg

The Singapore Medical Council (SMC), a statutory board under the Ministry of Health, maintains the Register of Medical Practitioners in Singapore, administers the compulsory continuing medical education program and also governs and regulates the professional conduct and ethics of registered medical practitioners.

## XXVI. Health Facts-Global

### Organisation for Economic Co-Operation and Development (OECD) - Health Statistics

2 rue André Pascal
F-75775
Paris, Cedex 16 France
Phone: 33-145-24-8200
Fax: 33-145-24-8500
E-mail Address: *health.contact@oecd.org*
Web Address: www.oecd.org
The Organisation for Economic Co-Operation and Development (OECD) offers extensive health statistics on a country-by-country basis. Data ranges from health expenditures per capita to health expenditures as percent of GDP for the 30 nations with the world's largest economies.

## XXVII. Immunization Resources

### CDC National Immunization Information Hotline (NIIH)

1600 Clifton Rd.
Atlanta, GA 30333 US
Toll Free: 800-232-4636
The CDC National Immunization Information Hotline (NIIH) offers up-to-date immunization information, including vaccine schedules, side effects, contraindications, recommendations and more.

## XXVIII. Industry Research/Market Research

### Forrester Research

400 Technology Sq.
Cambridge, MA 02139 US
Phone: 617-613-6000
Fax: 617-613-5200
Toll Free: 866-367-7378
Web Address: www.forrester.com
Forrester Research identifies and analyzes emerging trends in technology and their impact on business. Among the firm's specialties are the financial

services, retail, health care, entertainment, automotive and information technology industries.

## Marketresearch.com

11200 Rockville Pike, Ste. 504
Rockville, MD 20852 US
Phone: 240-747-3000
Fax: 240-747-3004
Toll Free: 800-298-5699
E-mail Address:
*customerservice@marketresearch.com*
Web Address: www.marketresearch.com
Marketresearch.com is a leading broker for professional market research and industry analysis. Users are able to search the company's database of research publications including data on global industries, companies, products and trends.

## Plunkett Research, Ltd.

P.O. Drawer 541737
Houston, TX 77254-1737 US
Phone: 713-932-0000
Fax: 713-932-7080
E-mail Address:
*customersupport@plunkettresearch.com*
Web Address: www.plunkettresearch.com
Plunkett Research, Ltd. is a leading provider of market research, industry trends analysis and business statistics. Since 1985, it has served clients worldwide, including corporations, universities, libraries, consultants and government agencies. At the firm's web site, visitors can view product information and pricing and access a great deal of basic market information on industries such as financial services, infotech, e-commerce, health care and biotech.

## XXIX.    Libraries-Medical Data

### Library of the National Medical Society

Web Address: www.medical-library.org
The Library of the National Medical Society provides a free resource of medical information for both health care consumers and medical professionals.

### Weill Cornell Medical Library

Weill Cornell Medical College, C.V. Starr
Biomedical Information Center
1300 York Ave.
New York, NY 10021-4896 US
Phone: 212-746-6055
E-mail Address: *infodesk@med.cornell.edu*
Web Address: library.med.cornell.edu

The Weill Cornell Medical Library houses information on the biomedical sciences, as well as performing data retrieval, management and evaluation.

## XXX.    MBA Resources

### MBA Depot

Phone: 512-499-8728
Web Address: www.mbadepot.com
MBA Depot is an online community for MBA professionals.

## XXXI.    Nanotechnology Associations

### IndiaNano

Range Hills Rd., Shivaji Nagar
IndiaCo Ctr., 4th Fl., Symphony, S. 210 A/1
Pune, Maharashtra 411 020 India
Phone: 91-20-25513254
Fax: 91-20-25513243
E-mail Address: *info@indianano.com*
Web Address: www.indianano.com
IndiaNano is a nonprofit organization located in India and supported by academic and industry experts. It aims to develop collaboration in order to advance technologies, including nanotechnology.

### Nano Science and Technology Institute (NSTI)

1 Kendall Sq.
PMB 308
Cambridge, MA 02139 US
Phone: 508-357-2925
Fax: 925-886-8461
E-mail Address: *mlaudon@nsti.org*
Web Address: www.nsti.org
The Nano Science and Technology Institute (NSTI) is engaged in the promotion and integration of nano and other advanced technologies through education, technology and business development. NSTI offers consulting services, continuing education programs, scientific and business publishing and community outreach.

### NCI Alliance for Nanotechnology in Cancer

c/o Nat'l Cancer Institute, Office of Tech. & Industrial Rel.
Bldg. 21, Rm. 10A49, 31 Center Dr., MSC 2580
Bethesda, MD 20892-2580 US
Phone: 301-496-1550
Fax: 301-496-7807
E-mail Address: *cancer.nano@mail.nih.gov*

Web Address: nano.cancer.gov
The NCI Alliance for Nanotechnology in Cancer, a service of the National Cancer Institute, is dedicated to using nanotechnology to advance the ways we prevent, treat and diagnose cancer. It especially seeks advances in the near and medium terms and to lower the barriers for those advances to be handed off to the private sector for commercial development. The Alliance focuses on translational research and development work in six major challenge areas, where nanotechnology can have the biggest and fastest impact on cancer treatment.

## XXXII.   Online Health Data

**Medscape**
76 Ninth Ave., Ste. 719
New York, NY 10011 US
Phone: 212-624-3700
Toll Free: 888-506-6098
E-mail Address:
*medscapecustomersupport@medscape.net*
Web Address: www.medscape.com
Medscape, an online resource for better patient care, provides links to journal articles, health care-related sites and health care information.

**PubMed**
E-mail Address: *pubmedcentral@nih.gov*
Web Address: www.ncbi.nlm.nih.gov/entrez/query
PubMed provides access to over 17 million MEDLINE citations dating back to the mid-1960s and additional life science journals. PubMed includes links to open access full text articles.

## XXXIII.   Patent Organizations

**European Patent Office**
Av. de Cortenbergh 60
Brussels, 1000 Belgium
Phone: 32-2-274-15-90
Web Address: www.epo.org
The European Patent Office (EPO) provides a uniform application procedure for individual inventors and companies seeking patent protection in up to 38 European countries. It is the executive arm of the European Patent Organization and is supervised by the Administrative Council.

**U.S. Patent and Trademark Office (PTO)**
U.S. Patent and Trademark Office
Office of Public Affairs, P. O. Box 1450

Alexandria, VA 22313-1450 US
Phone: 571-272-1000
Fax: 571-273-8300
Toll Free: 800-786-9199
E-mail Address: *usptoinfo@uspto.gov*
Web Address: www.uspto.gov
The U.S. Patent and Trademark Office (PTO) administers patent and trademark laws for the U.S. and enables registration of patents and trademarks.

## XXXIV.   Patent Organizations-Global

**World Intellectual Property Organization (WIPO)**
34 chemin des Colombettes
Geneva 20, CH-1211 Switzerland
Phone: 41-22-338-9111
Fax: 41-22-733-54-28
E-mail Address: *publicinf@wipo.int*
Web Address: www.wipo.int
The WIPO has a United Nations mandate to assist organizations and companies in filing patents and other intellectual property data on a global basis. At its website, you can download free copies of its WIPO magazine, and you can search its international patent applications.

## XXXV.   Patent Resources

**Patent Docs**
E-mail Address: *patentdocs@gmail.com*
Web Address: patentdocs.typepad.com
Patent Docs is an excellent blog about patent law and patent news in the fields of biotechnology and pharmaceuticals.

**Patent Law for Non-Lawyers**
E-mail Address: *info@thinkbiotech.com*
Web Address: www.dnapatent.com/law
Patent Law for Non-Lawyers is an informative site detailing the patent process in the fields of biotechnology and engineering. The site assumes a working knowledge of the industry.

## XXXVI.   Pharmaceutical Industry Associations (Drug Industry)

**Academy of Pharmaceutical Physicians and Investigators (APPI)**
500 Montgomery St., Ste. 800
Alexandria, VA 22314 US
Phone: 703-254-8100
Fax: 703-254-8101

Toll Free: 866-225-2779
E-mail Address: *office@acrpnet.org*
Web Address: www.appinet.org
The Academy of Pharmaceutical Physicians and
Investigators (APPI) is an association that arose when
the American Academy of Pharmaceutical Physicians
and the Association of Clinical Research
Professionals merged. It is a membership
organization that provides scientific and educational
activities on issues concerning pharmaceutical
medicine.

### Accreditation Council for Pharmacy Education (ACPE)

20 N. Clark St., Ste. 2500
Chicago, IL 60602-5109 US
Phone: 312-664-3575
Fax: 312-664-4652
E-mail Address: *info@acpe-accredit.org*
Web Address: www.acpe-accredit.org
The Accreditation Council for Pharmacy Education
(ACPE) provides accreditation for pharmaceutical
programs.

### American Association of Pharmaceutical Sciences (AAPS)

2107 Wilson Blvd., Ste. 700
Arlington, VA 22201-3042 US
Phone: 703-243-2800
Fax: 703-243-9650
Web Address: www.aapspharmaceutica.com
The American Association of Pharmaceutical
Scientists (AAPS) represents scientists in the
pharmaceutical field. Members are given access to
international forum, scientific programs, ongoing
education, opportunities for networking and
professional development.

### American Association of Pharmacy Technicians (AAPT)

P.O. Box 1447
Greensboro, NC 27402 US
Phone: 336-333-9356
Fax: 336-333-9068
Toll Free: 877-368-4771
E-mail Address: *secretary@pharmacytechnician.com*
Web Address: www.pharmacytechnician.com
The American Association of Pharmacy Technicians
(AAPT) provides leadership and represents the
interests of pharmacy technicians in the United
States. The group also offers continuing education
programs and services to its members.

### American Pharmaceutical Association (APhA)

1100 15th St. NW, Ste. 400
Washington, DC 20005-1707 US
Phone: 202-628-4410
Fax: 202-783-2351
Toll Free: 800-237-2742
E-mail Address: *infocenter@aphanet.org*
Web Address: www.aphanet.org
American Pharmaceutical Association (APhA) is a
national professional society that provides news and
information to pharmacists.

### American Society for Automation in Pharmacy (ASAP)

492 Norristown Rd., Ste. 160
Blue Bell, PA 19422 US
Phone: 610-825-7783
Fax: 610-825-7641
E-mail Address: *will@computertalk.com*
Web Address: www.asapnet.org
The American Society for Automation in Pharmacy
(ASAP) is a nonprofit organization that seeks to
advance the application of computer technology in
the pharmacists role as caregiver and in the efficient
operation and management of a pharmacy.

### American Society for Clinical Pharmacology and Therapeutics (ASCPT)

528 N. Washington St.
Alexandria, VA 22314 US
Phone: 703-836-6981
Fax: 703-836-5223
E-mail Address: *info@ascpt.org*
Web Address: www.ascpt.org
The American Society for Clinical Pharmacology and
Therapeutics (ASCPT) is a nonprofit organization
that is devoted to the discovery, development,
regulation, and use of safe and effective medications
necessary for the prevention and treatment of illness.

### American Society for Pharmacology and Experimental Therapeutics (ASPET)

9650 Rockville Pike
Bethesda, MD 20814-3995 US
Phone: 301-634-7060
Fax: 301-634-7061
E-mail Address: *info@aspet.org*
Web Address: www.aspet.org
The American Society for Pharmacology and
Experimental Therapeutics (ASPET) is a scientific
society whose members conduct basic and clinical
pharmacological research in academia, industry and
the government.

**American Society of Consultant Pharmacists (ASCP)**
1321 Duke St.
Alexandria, VA 22314-3563 US
Phone: 703-739-1300
Fax: 800-220-1321
Toll Free: 800-355-2727
E-mail Address: *info@ascp.com*
Web Address: www.ascp.com
The American Society of Consultant Pharmacists (ASCP) is an international professional association that provides leadership, education, advocacy and resources to advance the practice of consultant and senior care pharmacy.

**American Society of Pharmacognosy (ASP)**
3149 Dundee Rd., Ste. 270
Northbrook, IL 60062 US
Phone: 623-202-3500
Fax: 847-656-2800
E-mail Address: *asphocog@aol.com*
Web Address: www.phcog.org
The American Society of Pharmacognosy (ASP) is a volunteer organization that promotes the growth and development of pharmacognosy, the study of the physical, chemical, biochemical and biological properties of drugs of natural origin and drugs from natural sources.

**Association of the British Pharmaceutical Industry (ABPI)**
12 Whitehall
London, SW1A 2DY UK
Phone: 44-870-890-4333
Fax: 44-20-7747-1414
Web Address: www.abpi.org.uk
The Association of the British Pharmaceutical Industry (ABPI) is a trade association that provides research and information for the British pharmaceuticals industry.

**Canadian Pharmacists Association (CPHA)**
1785 Alta Vista Dr.
Ottawa, ON K1G 3Y6 Canada
Phone: 613-523-7877
Fax: 613-523-0445
Toll Free: 800-917-9489
E-mail Address: *info@pharmacists.ca*
Web Address: www.pharmacists.ca
The Canadian Pharmacists Association (CPHA) is a professional organization providing drug information, pharmacy practice support material, patient information and news about the pharmacy industry.

**Canadian Research-Based Pharmaceutical Companies Association (Rx&D)**
55 Metcalfe St., Ste. 1220
Ottawa, ON K1P 6L5 Canada
Phone: 613-236-0455
Fax: 613-236-6756
E-mail Address: *info@canadapharma.org*
Web Address: www.canadapharma.org
The Canadian Research-Based Pharmaceutical Companies Association (Rx&D) is a trade organization providing news and information to the Canadian biotech industry.

**Canadian Society for Pharmaceutical Sciences (CSPS)**
3126 Dentistry/Pharmacy Ctr.
University of Alberta Campus
Edmonton, Alberta T6G 2N8 Canada
Phone: 780-492-0950
Fax: 780-492-0951
E-mail Address: *csps@cspscanada.org*
Web Address: www.cspscanada.org
The Canadian Society for Pharmaceutical Sciences (CSPS) is a nonprofit organization that works to advance pharmaceutical research. CSPS also maintains the Journal of Pharmacy and Pharmaceutical Sciences, an international online publication.

**Controlled Release Society (CRS)**
3340 Pilot Knob Rd.
St. Paul, MN 55421 US
Phone: 651-454-7250
Fax: 651-454-0766
E-mail Address: *crs@scisoc.org*
Web Address: www.controlledrelease.org
The Controlled Release Society (CRS) is an organization that promotes the science of the controlled delivery of bioactive substances.

**Drug, Chemical & Associated Technologies Association (DCAT)**
1 Washington Blvd., Ste. 7
Robbinsville, NJ 08691 US
Phone: 609-448-1000
Fax: 609-448-1944
Toll Free: 800-640-3228
E-mail Address: *mtimony@dcat.org*
Web Address: www.dcat.org
The Drug, Chemical & Associated Technologies Association (DCAT) is a business development association whose membership is made up of companies that manufacture, distribute or provide

services to the pharmaceutical, chemical, nutritional and related industries.

**Generic Pharmaceutical Association (GPhA)**
2300 Clarendon Blvd., Ste. 400
Arlington, VA 22201 US
Phone: 703-647-2480
Web Address: www.gphaonline.org
The Generic Pharmaceutical Association (GPhA) represents the manufacturers and distributors of finished generic pharmaceutical products, manufacturers and distributors of bulk active pharmaceutical chemicals, and suppliers of other goods and services to the generic pharmaceutical industry.

**International Academy of Compounding Pharmacists (IACP)**
4638 Riverstone Blvd.
Missouri City, TX 77459 US
Phone: 281-933-8400
Fax: 281-495-0602
Toll Free: 800-927-4227
E-mail Address: *iacpinfo@iacprx.org*
Web Address: www.iacprx.org
The International Academy of Compounding Pharmacists (IACP) is a nonprofit association that seeks to protect, promote and advance the art the customizing, compounding pharmacy profession.

**International Association for Pharmaceutical Technology (APV)**
Kurfurstenstrasse 59
Mainz, 55118 Germany
Phone: 49-6131-9769
Fax: 49-6131-97-6969
E-mail Address: *apv@apv-mainz.de*
Web Address: www.apv-mainz.de
The International Association for Pharmaceutical Technology (APV) is a nonprofit scientific association that is located in Mainz, Germany, and publishes its own scientific journal.

**International Federation of Pharmaceutical Manufacturers Associations (IFPMA)**
15 Ch. Louis-Dunant
P.O. Box 195, 1211
Geneva, 20 Switzerland
Phone: 41-22-338-32-00
Fax: 41-22-338-32-99
E-mail Address: *admin@ifpma.org*
Web Address: www.ifpma.org

The International Federation of Pharmaceutical Manufacturers Associations (IFPMA) is a nonprofit organization that represents the world's research-based pharmaceutical and biotech companies.

**International Federation of Pharmaceutical Wholesalers (IFPW)**
10569 Crestwood Dr.
Manassas, VA 20109 US
Phone: 703-331-3714
Fax: 703-331-3715
E-mail Address: *info@ifpw.com*
Web Address: www.ifpw.com
The International Federation of Pharmaceutical Wholesalers (IFPW) collaborates with its six member associations to promote the pharmaceutical manufacturing industry.

**International Pharmaceutical Excipients Council of the Americas (IPEC-Americas)**
1655 N. Fort Myer Dr., Ste. 700
Arlington, VA 22209 US
Phone: 703-875-2127
Fax: 703-525-5157
E-mail Address: *info@ipecamericas.org*
Web Address: www.ipecamericas.org
International Pharmaceutical Excipients Council of the Americas (IPEC-Americas) is a trade organization that promotes standardized approval criteria for drug inert ingredients, or excipients, among different nations. The organization also works to promote safe and useful excipients in the U.S.

**International Pharmaceutical Federation (FIP)**
Andries Bickerweg 5
P.O. Box 84200
The Hague, AE 2508 The Netherlands
Phone: 31-70-3021970
Fax: 31-70-3021999
E-mail Address: *fip@fip.org*
Web Address: www.fip.org
The International Pharmaceutical Federation (FIP) is a worldwide organization of pharmaceutical professional and scientific associations.

**International Pharmaceutical Students Federation (IPSF)**
P.O. Box 84200
Den Haag, 2506 AE The Netherlands
Phone: 31-70-302-1992
Fax: 31-70-302-1999
E-mail Address: *president@ipsf.org*
Web Address: www.ipsf.org

The International Pharmaceutical Students Federation (IPSF) is an organization that aims to promote the interests of pharmacy students and encourage international co-operation amongst them.

## International Society for Pharmacoepidemiology (ISPE)

5272 River Rd., Ste. 630
Bethesda, MD 20816 US
Phone: 301-718-6500
Fax: 301-656-0989
E-mail Address: *ispe@paimgmt.com*
Web Address: www.pharmacoepi.org
The International Society for Pharmacoepidemiology (ISPE) is an international organization dedicated to the health of the public by advancing the study of the effects and determinants of pharmacology on epidemic diseases and to help provide risk benefit assessments on drugs with large scale distributions.

## International Society of Pharmaceutical Engineers (ISPE)

3109 W. Dr. Martin Luther King, Jr. Blvd., Ste. 250
Tampa, FL 33607 US
Phone: 813-960-2105
Fax: 813-264-2816
E-mail Address: *ASK@ispe.org*
Web Address: www.ispe.org
The International Society of Pharmaceutical Engineers (ISPE) is a worldwide nonprofit society dedicated to educating and advancing pharmaceutical manufacturing professionals and the pharmaceutical industry.

## International Society of Regulatory Toxicology & Pharmacology (ISRTP)

6546 Belleview Dr.
Columbia, MD 21046-1054 US
Phone: 410-992-9083
Fax: 410-740-9181
E-mail Address: *c.carr65@comcast.net*
Web Address: www.isrtp.org
The International Society of Regulatory Toxicology & Pharmacology (ISRTP) is an association of professionals that mediates between policy makers and scientists in order to promote sound toxicologic and pharmacologic science as a basis for regulation affecting the environment and human safety and health.

## International Union of Basic and Clinical Pharmacology (IUPHAR)

3901 Rainbow Blvd.

Mail Stop 4016
Kansas City, KS 66160 US
Phone: 913-588-7533
Fax: 913-588-7373
E-mail Address: *iuphar@kumc.edu*
Web Address: www.iuphar.org
The International Union of Basic and Clinical Pharmacology (IUPHAR) is a nonprofit association representing the interests of pharmacologists around the world.

## Korean Research-based Pharmaceutical Industry Association (KRPIA)

201-6 Guui-Dong, Kwangjin-Gu
5th Fl., Dae Han Bldg.
Seoul, 143-200 Korea
Phone: 82-2-456-8553
Fax: 82-2-456-8320
E-mail Address: *krpia@krpia*
Web Address: www.krpia.or.kr
The Korean Research-based Pharmaceutical Industry Association (KRPIA) is an association of 24 research-based pharmaceutical companies operating in Korea.

## LEEM (French Pharmaceutical Companies Association)

88 Rue de la Faisanderie
Paris, 75016 France
Phone: 01 45 03 88 88
Web Address: www.leem.org
LEEM (Les Entreprises du Médicament) represents the 335 pharmaceutical companies operating in France, engaged in the research and/or development of medicines for human use.

## National Association of Boards of Pharmacy (NABP)

1600 Feehanville Dr.
Mount Prospect, IL 60056 US
Phone: 847-391-4406
Fax: 847-391-4502
E-mail Address: *exec-office@nabp.net*
Web Address: www.nabp.net
The National Association of Boards of Pharmacy (NABP) is the association of the member boards and jurisdictions in the field of pharmacy for the development, implementation, and enforcement of uniform standards for the purpose of protecting the public health.

**National Association of Pharmaceutical Manufacturers (NAPM)**
6 De Veer Ln.
Arcadia, Pretoria 0007 South Africa
Phone: 012-323-7529
Fax: 086-529-4245
E-mail Address: *napm@mweb.co.za*
Web Address: www.napm.co.za
The National Association of Pharmaceutical Manufacturers (NAPM) is a nonprofit trade association consisting of South African, generic-based pharmaceutical manufacturers and distributors.

**Parenteral Drug Association (PDA)**
4350 E. West Hwy., Ste. 200
Bethesda, MD 20814 US
Phone: 301-656-5900
Fax: 301-986-1093
E-mail Address: *swan@pda.org*
Web Address: www.pda.org
The Parenteral Drug Association (PDA) is a global provider of science, technology and regulatory information and education for the pharmaceutical and biopharmaceutical community.

**Pharmaceutical Information and Pharmacovigilance Association (PIPA)**
P.O. Box 254
Haslemere, Surrey GU27 9AF UK
Phone: 07531-899537
E-mail Address: *pipa@pipaonline.org*
Web Address: www.aiopi.org.uk
The Pharmaceutical Information and Pharmacovigilance Association (PIPA) is a professional organization that promotes the advancement of information in the pharmaceutical industry in the U.K.

**Pharmaceutical Research and Manufacturers of America (PhRMA)**
950 F St. NW, Ste. 300
Washington, DC 20004 US
Phone: 202-835-3400
Fax: 202-835-3414
Web Address: www.phrma.org
Pharmaceutical Research and Manufacturers of America (PhRMA) represents the nation's leading research-based pharmaceutical and biotechnology companies.

**Pharmacy Council of India**
Kotla Rd., Aiwan-E-Ghalib, Marg
Combined Councils' Bldg.

New Delhi, 110-002 India
Phone: 011-23239184
Fax: 011-23239184
E-mail Address: *pci@ndb.vsnl.net.in*
Web Address: www.pci.nic.in
The Pharmacy Council of India provides regulation of pharmacists under the Pharmacy Act and is a statutory body working under the Ministry of Health Family Welfare, Government of India.

**PharmaSUG**
421 New Parkside Dr.
Chapel Hill, NC 27516 US
E-mail Address: *syamala.ponnapalli@gmail.com*
Web Address: www.pharmasug.org
PharmaSUG promotes biotech information technology using SAS software.

**Royal Pharmaceutical Society of Great Britain (RPSGB)**
1 Lambeth High St.
London, SE1 7JN UK
Phone: 020-7735-9141
Fax: 020-7735-7629
E-mail Address: *enquiries@rpsgb.org*
Web Address: www.rpsgb.org.uk
The Royal Pharmaceutical Society of Great Britain (RPSGB) is the regulatory agency for pharmacists in England, Wales and Scotland.

**Singapore Association of Pharmaceutical Industries (SAPI)**
151 Chin Swee Rd.
02-13A/14 Manhattan House
169876 Singapore
Phone: 65-6738-0966
Fax: 65-6738-0977
E-mail Address: *FokTaiHung@sapi.org.sg*
Web Address: www.sapi.org.sg
Singapore Association of Pharmaceutical Industries (SAPI) represents a wide spectrum of pharmaceutical related businesses, namely the trading houses, manufacturers, representative offices and pharmacies.

**Society of Infectious Diseases Pharmacists (SIDP)**
823 Congress Ave., Ste. 230
Austin, TX 78701 US
Phone: 512-479-0425
Fax: 512-495-9031
E-mail Address: *sidp@eami.com*
Web Address: www.sidp.org
The Society of Infectious Diseases Pharmacists (SIDP) is an association of health professionals

dedicated to promoting the appropriate use of antimicrobials, providing its members with education, advocacy and leadership in all aspects of the treatment of infectious diseases.

## XXXVII. Pharmaceutical Industry Resources (Drug Industry)

### American Institute of the History of Pharmacy
777 Highland Ave.
Madison, WI 53705-2222 US
Phone: 608-262-5378
E-mail Address: *AIHP@AIHP.org*
Web Address: www.pharmacy.wisc.edu/aihp/
The American Institute of the History of Pharmacy is a nonprofit national organization that works to advance knowledge of the role of pharmacy in history through its programs and print publications.

### Board of Pharmaceutical Specialties (BPS)
1100 15th St. NW, Ste. 400
Washington, DC 20005-1707 US
Phone: 202-429-7591
Fax: 202-429-6304
E-mail Address: *bps@aphanet.org*
Web Address: www.bpsweb.org
The Board of Pharmaceutical Specialties (BPS) operates a certification program for specialized clinical pharmacists.

### European Medicines Agency (EMEA)
7 Westferry Circus
Canary Wharf
London, E14 4HB UK
Phone: 44-2074188400
Fax: 44-2074188416
E-mail Address: *info@emea.europa.eu*
Web Address: www.emea.europa.eu
The European Medicines Agency (EMEA) is the European agency charged with approving new drugs and monitoring the efficacy of existing drugs.

### Tufts Center for the Study of Drug Development
75 Kneeland St., Ste. 1100
Boston, MA 02111 US
Phone: 617-636-2170
Fax: 617-636-2425
E-mail Address: *csdd@tufts.edu*
Web Address: csdd.tufts.edu
The Tufts Center for the Study of Drug Development, an affiliate of Tuft's University, provides analyses and commentary on pharmaceutical issues. Its mission is to improve the quality and efficiency of pharmaceutical development, research and utilization. It is famous, among other things, for its analysis of the true total cost of developing and commercializing a new drug. Tuft's Center conducts research in areas of drug development, public policy and regulation, and biotechnology.

### Pharmaportal.com
Web Address: www.pharmaportal.com
Pharmaportal.com is a pharmaceutical portal containing information about the Pharmaceutical Magazine, as well as links for executives in the industry to meet each other.

### United States Pharmacopeia
12601 Twinbrook Pkwy.
Rockville, MD 20852-1790 US
Phone: 800-227-8772
Fax: 301-816-8148
Toll Free: 301-881-0666
E-mail Address: *custsvc@usp.org*
Web Address: www.usp.org
The United States Pharmacopeia is the official public standards-setting authority for all over-the-counter and prescription medicines, dietary supplements and other healthcare products manufactured and sold in the United States.

## XXXVIII. Research & Development, Laboratories

### Battelle Memorial Institute
505 King Ave.
Columbus, OH 43201-2693 US
Phone: 614-424-5853
Toll Free: 800-201-2011
Web Address: www.battelle.org
Battelle Memorial Institute serves commercial and governmental customers in developing new technologies and products. The institute adds technology to systems and processes for manufacturers; pharmaceutical and agrochemical industries; trade associations; and government agencies supporting energy, the environment, health, national security and transportation.

### Commonwealth Scientific and Industrial Research Organization (CSRIO)
CSIRO Enquiries
Bag 10
Clayton South, Victoria 3169 Australia
Phone: 61-3-9545-2176
Fax: 61-3-9545-2175

E-mail Address: *enquiries@csiro.au*
Web Address: www.csiro.au
The Commonwealth Scientific and Industrial
Research Organization (CSRIO) is Australia's
national science agency and a leading international
research agency. CSRIO performs research in
Australia over a broad range of areas including
agriculture, minerals and energy, manufacturing,
communications, construction, health and the
environment.

**Computational Neurobiology Laboratory**
The Salk Institute
10010 N. Torrey Pines Rd.
La Jolla, CA 92037 US
Phone: 858-453-4100
E-mail Address: *sejnowski@salk.edu*
Web Address: www.cnl.salk.edu
The Computational Neurobiology Laboratory at The
Salk Institute strives to understand the computational
resources of the brain from the biophysical to the
systems levels.

**Council of Scientific & Industrial Research
(CSIR)**
2 Rafi Marg
Anusandhan Bhawan
New Delhi, 110 001 India
Phone: 011-23710618
Fax: 011-23713011
E-mail Address: *itweb@csir.res.in*
Web Address: www.csir.res.in
The Council of Scientific & Industrial Research
(CSIR) is a government-funded organization that
promotes research and development initiatives in
India. It operates in the fields of energy,
biotechnology, space, science and technology.

**National Research Council Canada (NRC)**
NRC Communications & Corp. Rel.
1200 Montreal Rd., Bldg. M-58
Ottawa, ON K1A 0R6 Canada
Phone: 613-993-9101
Fax: 613-952-9907
Toll Free: 877-672-2672
E-mail Address: *info@nrc-cnrc.gc.ca*
Web Address: www.nrc-cnrc.gc.ca
National Research Council Canada (NRC) is a
government organization of 20 research institutes that
carry out multidisciplinary research with partners in
industries and sectors key to Canada's economic
development.

**SRI International**
333 Ravenswood Ave.
Menlo Park, CA 94025-3493 US
Phone: 650-859-2000
E-mail Address: *ellie.javadi@sri.com*
Web Address: www.sri.com
SRI International is a nonprofit organization offering
a wide range of services, including engineering
services, information technology, pure and applied
physical sciences, product development,
pharmaceutical discovery, biopharmaceutical
discovery and policy issues. SRI conducts research
for commercial and governmental customers.

| XXXIX.    Robotics Associations |
|---|

**Laboratory Robotics Interest Group (LRIG)**
1730 W. Circle Dr.
Martinsville, NJ 08836-2147 US
Phone: 732-302-1038
Fax: 732-875-0270
E-mail Address: *andy.zaayenga@lab-robotics.org*
Web Address: www.lab-robotics.org
Laboratory Robotics Interest Group (LRIG) is a
membership group focused on the application of
robotics in the laboratory.

| XL.   Technology Transfer Associations |
|---|

**Association of University Technology Managers
(AUTM)**
111 Deer Lake Rd., Ste. 100
Deerfield, IL 60015 US
Phone: 847-559-0846
Fax: 847-480-9282
E-mail Address: *info@autm.net*
Web Address: www.autm.net
The Association of University Technology Managers
(AUTM) is a nonprofit professional association with
membership of more than 3,600 intellectual property
managers and business executives from 45 countries.
The association's mission is to advance the field of
technology transfer, and enhance our ability to bring
academic and nonprofit research to people around the
world.

| XLI.    Trade Associations-General |
|---|

**BUSINESSEUROPE**
168 Ave. de Cortenbergh
Brussels, 1000 Belgium
Phone: 32-0-2-237-65-11

Fax: 32-0-2-231-14-45
E-mail Address: *main@businesseurope.eu*
Web Address: www.businesseurope.eu
BUSINESSEUROPE is a major European trade
federation that operates in a manner similar to a
chamber of commerce. Its members are the central
national business federations of the 34 countries
throughout Europe from which they come.
Companies cannot become direct members of
BUSINESSEUROPE, though there is a support group
which offers the opportunity for firms to encourage
BUSINESSEUROPE objectives in various ways.

## XLII.   Trade Associations-Global

**World Trade Organization (WTO)**
Centre William Rappard
Rue de Lausanne 154
Geneva 21, CH-1211 Switzerland
Phone: 41-22-739-51-11
Fax: 41-22-731-42-06
E-mail Address: *enquiries@wto.og*
Web Address: www.wto.org
The World Trade Organization (WTO) is a global
organization dealing with the rules of trade between
nations. To become a member, nations must agree to
abide by certain guidelines. Membership increases a
nation's ability to import and export efficiently.

## XLIII.   U.S. Government Agencies

**Bureau of Economic Analysis (BEA)**
1441 L St. NW
Washington, DC 20230 US
Phone: 202-606-9900
E-mail Address: *customerservice@bea.gov*
Web Address: www.bea.gov
The Bureau of Economic Analysis (BEA), an agency
of the U.S. Department of Commerce, is the nation's
economic accountant, preparing estimates that
illuminate key national, international and regional
aspects of the U.S. economy.

**Bureau of Labor Statistics (BLS)**
2 Massachusetts Ave. NE
Washington, DC 20212-0001 US
Phone: 202-691-5200
Web Address: stats.bls.gov
The Bureau of Labor Statistics (BLS) is the principal
fact-finding agency for the Federal Government in
the field of labor economics and statistics. It is an
independent national statistical agency that collects,
processes, analyzes and disseminates statistical data
to the American public, U.S. Congress, other federal
agencies, state and local governments, business and
labor. The BLS also serves as a statistical resource to
the Department of Labor.

**Center for Biologics Evaluation and Research
(CBER)**
1401 Rockville Pike, Ste. 200N
Rockville, MD 20852-1448 US
Phone: 301-827-1800
Toll Free: 800-835-4709
E-mail Address: *octma@fda.hhs.gov*
Web Address: www.fda.gov/Cber
The Center for Biologics Evaluation and Research
(CBER) regulates biologic products for use in
humans. It is a source for a broad variety of data on
drugs, including blood products, counterfeit drugs,
exports, drug shortages, recalls and drug safety.

**Center for Devices and Radiological Health
(CDRH)**
1350 Piccard Dr.
HFZ-210
Rockville, MD 20850 US
Phone: 301-276-3103
Fax: 301-443-8818
Toll Free: 800-638-2041
E-mail Address: *dsmica@cdrh.fda.gov*
Web Address: www.fda.gov/cdrh
The Center for Devices and Radiological Health
(CDRH) is a unit of the FDA that regulates medical
devices and radiation-emitting products.

**Center for Drug Evaluation and Research
(CDER)**
5600 Fishers Ln.
HFD-240
Rockville, MD 20852-9787 US
Phone: 301-796-3400
Toll Free: 888-463-6332
E-mail Address: *druginfo@cder.fea.gov*
Web Address: www.fda.gov/cder
The Center for Drug Evaluation and Research
(CDER) is a division of the FDA that offers a wealth
of information on new drug approval statistics and
the approval process.

**Center for Food Safety and Applied Nutrition-
Biotechnology (CFSAN)**
5100 Paint Branch Pkwy.
HFS-555
College Park, MD 20740-3835 US

Toll Free: 888-723-3366
Web Address: www.cfsan.fda.gov/list.html
The Center for Food Safety and Applied Nutrition-
Biotechnology (CFSAN) is an FDA site that provides
information about genetically engineered food
products.

**Centers for Disease Control and Prevention
(CDC)**
1600 Clifton Rd.
Atlanta, GA 30333 US
Phone: 404-639-3311
Toll Free: 800-232-4636
E-mail Address: *cdcinfo@cdc.gov*
Web Address: www.cdc.gov
The Centers for Disease Control and Prevention
(CDC), headquartered in Atlanta and established as
an operating health agency within the U.S. Public
Health Service, is the federal agency charged with
protecting the public health of the nation by
providing leadership and direction in the prevention
and control of diseases and other preventable
conditions and responding to public health
emergencies.

**Government Printing Office (GPO)**
732 N. Capitol St. NW
Washington, DC 20401 US
Phone: 202-512-0000
Fax: 202-512-2104
Toll Free: 866.512.1800
E-mail Address: *contactcenter@gpo.gov*
Web Address: www.gpo.gov
The U.S. Government Printing Office (GPO) is the
primary information source concerning the activities
of Federal agencies. GPO gathers, catalogues,
produces, provides, authenticates and preserves
published information.

**National Cancer Institute (NCI)**
6116 Executive Blvd., Ste. 3036A
Bethesda, MD 20892-8322 US
Toll Free: 800-422-6237
E-mail Address: *ncergovstaff@mail.nih.gov.*
Web Address: www.cancer.gov
The National Cancer Institute (NCI) is the Federal
Government's principal agency for cancer research
and training.

**National Center for Biotechnology Information
(NCBI)**
U.S. National Library of Medicine
National Institues of Health

8600 Rockville Pike, Bldg. 38A
Bethesda, MD 20894 US
Phone: 301-496-2475
Fax: 301-480-9241
E-mail Address: *info@ncbi.nlm.nih.gov*
Web Address: www.ncbi.nlm.nih.gov
The National Center for Biotechnology Information
(NCBI) creates public databases, conducts research in
computational biology, develops software for
analyzing genome data and disseminates biomedical
information. It is part of the U.S. National Library of
Medicine (NLM), which is located on the campus of
the National Institutes of Health (NIH).

**National Center for Research Resources (NCRR)**
6701 Democracy Blvd.
MSC 4874, 1 Democracy Plz., 9th Fl.
Bethesda, MD 20892-4874 US
Phone: 301-435-0888
Fax: 301-480-3558
E-mail Address: *info@ncrr.nih.gov*
Web Address: www.ncrr.nih.gov
The National Center for Research Resources (NCRR)
supports primary health and life sciences research to
create and develop critical resources, models and
technologies.

**National Center for Toxicological Research**
3900 NCTR Rd.
Jefferson, AR 72079 US
Phone: 870-543-7130
Toll Free: 800-216-7331
E-mail Address: *rhuber@fda.hhs.gov*
Web Address: www.fda.gov/nctr
The mission of the National Center for Toxicological
Research is to conduct peer-reviewed scientific
research that supports and anticipates the FDA's
current and future regulatory needs.

**National Heart, Lung, and Blood Institute
(NHLBI)**
P.O. Box 30105
Bethesda, MD 20824-0105 US
Phone: 301-592-8573
Fax: 240-629-3246
E-mail Address: *nhlbiinfo@nhlbi.nih.gov*
Web Address: www.nhlbi.nih.gov
The National Heart, Lung, and Blood Institute
(NHLBI) provides leadership for a national program
in diseases of the heart, blood vessels, lung and
blood; blood resources; and sleep disorders.

## National Institute of Standards and Technology (NIST)

100 Bureau Dr.
Stop 1070
Gaithersburg, MD 20899-1070 US
Phone: 301-975-6478
E-mail Address: *inquiries@nist.gov*
Web Address: www.nist.gov
The National Institute of Standards and Technology (NIST) is an agency of the U.S. Department of Commerce's Technology Administration. It works with various industries to develop and apply technology, measurements and standards.

## National Institutes of Health (NIH)

9000 Rockville Pike
Bethesda, MD 20892 US
Phone: 301-496-4000
Toll Free: 800-411-1222
E-mail Address: *nihinfo@od.nih.gov*
Web Address: www.nih.gov
The National Institutes of Health (NIH) is the leader of medical and behavioral research for the nation, with over 15 institutes ranging from the National Cancer Institute to the National Institute of Mental Health.

## National Science Foundation (NSF)

4201 Wilson Blvd.
Arlington, VA 22230 US
Phone: 703-292-5111
Toll Free: 800-877-8339
E-mail Address: *info@nsf.gov*
Web Address: www.nsf.gov
The National Science Foundation (NSF) is an independent U.S. government agency responsible for promoting science and engineering. The foundation provides grants and funding for research.

## U.S. Census Bureau

4600 Silver Hill Rd.
Washington, DC 20233-8800 US
Phone: 301-763-4636
Fax: 301-457-3670
Toll Free: 800-923-8282
E-mail Address: *pio@census.gov*
Web Address: www.census.gov
The U.S. Census Bureau is the official collector of data about the people and economy of the U.S. Founded in 1790, it provides official social, demographic and economic information.

## U.S. Department of Commerce (DOC)

1401 Constitution Ave. NW
Washington, DC 20230 US
Phone: 202-482-2000
E-mail Address: *cgutierrez@doc.gov*
Web Address: www.doc.gov
The U.S. Department of Commerce (DOC) regulates trade and provides valuable economic analysis of the economy.

## U.S. Department of Labor (DOL)

Frances Perkins Bldg.
200 Constitution Ave. NW
Washington, DC 20210 US
Toll Free: 866-487-2365
Web Address: www.dol.gov
The U.S. Department of Labor (DOL) is the government agency responsible for labor regulations. This site provides tools to help citizens find out whether companies are complying with family and medical-leave requirements.

## U.S. Food and Drug Administration (FDA)

5600 Fishers Ln.
Rockville, MD 20857 US
Toll Free: 888-463-6332
Web Address: www.fda.gov
The U.S. Food and Drug Administration (FDA) promotes and protects the public health by helping safe and effective products reach the market in a timely way and by monitoring products for continued safety after they are in use. It regulates both prescription and over-the-counter drugs as well as medical devices and food products.

## U.S. Securities and Exchange Commission (SEC)

100 F St. NE
Washington, DC 20549 US
Phone: 202-551-6000
Toll Free: 888-732-6585
E-mail Address: *publicinfo@sec.gov*
Web Address: www.sec.gov
The U.S. Securities and Exchange Commission (SEC) is a nonpartisan, quasi-judicial regulatory agency responsible for administering federal securities laws. These laws are designed to protect investors in securities markets and ensure that they have access to disclosure of all material information concerning publicly traded securities. Visitors to the web site can access the EDGAR database of corporate financial and business information.

# Chapter 4

# THE BIOTECH 400:
# WHO THEY ARE AND HOW THEY WERE CHOSEN

## Includes Indexes by Company Name, Industry & Location, And a Complete Table of Sales, Profits and Ranks

The companies chosen to be listed in PLUNKETT'S BIOTECH & GENETICS INDUSTRY ALMANAC comprise a unique list. THE BIOTECH 400 (the actual count is 378 companies) were chosen specifically for their dominance in the many facets of biotechnology and genetics in which they operate. Complete information about each firm can be found in the "Individual Profiles," beginning at the end of this chapter. These profiles are in alphabetical order by company name.

THE BIOTECH 400 includes leading companies from all parts of the United States as well as many other nations, and from all biotech and genetics related industry segments: pharmaceuticals; diagnostics; research and development; and support services.

Simply stated, the list contains 378 of the largest, most successful, fastest growing firms in biotech and related industries in the world. To be included in our list, the firms had to meet the following criteria:

1) Generally, these are corporations based in the U.S., however, the headquarters of 77 firms are located in other nations.

2) Prominence, or a significant presence, in biotech, genetics and supporting fields. (See the following Industry Codes section for a complete list of types of businesses that are covered).

3) The companies in THE BIOTECH 400 do not have to be exclusively in the biotech and genetics field.

4) Financial data and vital statistics must have been available to the editors of this book, either directly from the company being written about or from outside sources deemed reliable and accurate by the editors. A small number of companies that we would like to have included are not listed because of a lack of sufficient, objective data.

## INDUSTRY LIST, WITH CODES

**This book refers to the following list of unique
industry codes, based on the 2007 NAIC code
system (NAIC is used by many analysts as a
replacement for older SIC codes because NAIC is
more specific to today's industry sectors, see
www.census.gov/NAICS). Companies profiled in
this book are given a primary NAIC code,
reflecting the main line of business of each firm.**

### Agriculture

*Agriculture*
11511       Agricultural Crop Production Support, Seeds,
              Fertilizers

### Financial Services

*Stocks & Investments*
523110      Investment Banking

### Health Care

*Health Products, Manufacturing*
325411      Medicinals & Botanicals, Manufacturing
325412      Drugs (Pharmaceuticals), Discovery &
              Manufacturing
325412A     Drug Delivery Systems
325412B     Veterinary Products Manufacturing
325413      Diagnostic Services and Substances
              Manufacturing
325414      Biological Products, Manufacturing
33911       Medical/Dental/Surgical Equipment &
              Supplies, Manufacturing
*Health Products, Wholesale Distribution*
423450      Medical/Dental/Surgical Equipment &
              Supplies, Distribution
*Health Care-Clinics, Labs and
Organizations*
6215        Laboratories & Diagnostic Services--Medical

### InfoTech

*Computers & Electronics Manufacturing*
3345        Instrument Manufacturing, including
              Measurement, Control, Test & Navigational
*Software*
511210D     Computer Software, Healthcare &
              Biotechnology

### Manufacturing

*Textiles Manufacturing*
313         Textiles, Fabrics, Sheets/Towels,
              Manufacturing
*Chemicals*
325         Chemicals, Manufacturing

### Nanotechnology

*Nanotechnology*
541711      Nanotechnology-Biotech/Health

### Services

*Consulting & Professional Services*
541690      Consulting--Scientific & Technical
541712      Research & Development-Physical,
              Engineering & Life Sciences
541910      Market Research

# INDEX OF RANKINGS WITHIN INDUSTRY GROUPS

| Company | Industry Code | 2008 Sales (U.S. $ thousands) | Sales Rank | 2008 Profits (U.S. $ thousands) | Profits Rank |
|---|---|---|---|---|---|
| **Agricultural Crop Production Support, Seeds, Fertilizers** | | | | | |
| ARCADIA BIOSCIENCES INC | 11511 | | | | |
| DOW AGROSCIENCES LLC | 11511 | | | | |
| DUPONT AGRICULTURE & NUTRITION | 11511 | | | | |
| MONSANTO CO | 11511 | 11,365,000 | 2 | 2,024,000 | 1 |
| SYNGENTA AG | 11511 | 11,624,000 | 1 | 1,385,000 | 2 |
| **Biological Products, Manufacturing** | | | | | |
| ALPHARMA ANIMAL HEALTH | 325414 | | | | |
| ANIKA THERAPEUTICS INC | 325414 | 33,055 | 5 | 3,629 | 3 |
| BIOHEART INC | 325414 | | | | |
| CELLARTIS AB | 325414 | | | | |
| CELLULAR DYNAMICS INTERNATIONAL | 325414 | | | | |
| CSL LIMITED | 325414 | 2,698,750 | 2 | 499,350 | 1 |
| EISAI CO LTD | 325414 | 7,304,620 | 1 | -169,110 | 8 |
| GENENCOR INTERNATIONAL | 325414 | | | | |
| GTC BIOTHERAPEUTICS INC | 325414 | 16,656 | 6 | -22,665 | 6 |
| JUBILANT BIOSYS LTD | 325414 | | | | |
| LIFECELL CORPORATION | 325414 | | | | |
| NOVOZYMES | 325414 | 1,453,610 | 3 | 189,510 | 2 |
| ORGANOGENESIS INC | 325414 | | | | |
| POLYDEX PHARMACEUTICALS | 325414 | 5,735 | 7 | -885 | 4 |
| SERACARE LIFE SCIENCES INC | 325414 | 48,967 | 4 | -11,963 | 5 |
| STEMCELLS INC | 325414 | 232 | 8 | -29,087 | 7 |
| VIACELL INC | 325414 | | | | |
| **Chemicals, Manufacturing** | | | | | |
| AKZO NOBEL NV | 325 | 20,387,300 | 4 | 981,340 | 4 |
| BASF AG | 325 | 83,990,800 | 1 | 3,925,610 | 1 |
| BAYER AG | 325 | 43,536,000 | 2 | 2,273,480 | 2 |
| BAYER CORP | 325 | | | | |
| E I DU PONT DE NEMOURS & CO (DUPONT) | 325 | 30,529,000 | 3 | 2,007,000 | 3 |
| INTERNATIONAL ISOTOPES INC | 325 | 5,602 | 7 | -2,167 | 7 |
| LONZA GROUP | 325 | 2,712,290 | 5 | 386,940 | 5 |
| SIGMA-ALDRICH CORP | 325 | 2,200,700 | 6 | 341,500 | 6 |
| **Computer Software, Healthcare & Biotechnology** | | | | | |
| ACCELRYS INC | 511210D | 79,739 | 3 | 1,321 | 3 |
| CEGEDIM SA | 511210D | 1,121,330 | 1 | 44,250 | 1 |
| DECODE GENETICS INC | 511210D | 58,095 | 4 | -80,947 | 4 |
| ERESEARCH TECHNOLOGY INC | 511210D | 133,140 | 2 | 25,002 | 2 |
| MEDIDATA SOLUTIONS INC | 511210D | | | | |
| TRIPOS INTERNATIONAL | 511210D | | | | |
| **Consulting--Scientific & Technical** | | | | | |
| CANREG INC | 541690 | | | | |
| **Diagnostic Services and Substances Manufacturing** | | | | | |
| AFFYMETRIX INC | 325413 | 410,249 | 5 | -307,919 | 22 |

| Company | Industry Code | 2008 Sales (U.S. $ thousands) | Sales Rank | 2008 Profits (U.S. $ thousands) | Profits Rank |
|---|---|---|---|---|---|
| AVIVA BIOSCIENCES CORP | 325413 | | | | |
| BIOSITE INC | 325413 | | | | |
| CALIPER LIFE SCIENCES | 325413 | 134,054 | 10 | -68,292 | 20 |
| CELSIS INTERNATIONAL PLC | 325413 | 52,900 | 15 | 6,700 | 8 |
| CEPHEID | 325413 | 169,627 | 7 | -27,713 | 15 |
| EPIX PHARMACEUTICALS INC | 325413 | 28,628 | 18 | -36,671 | 17 |
| GENOMIC HEALTH INC | 325413 | 110,579 | 11 | -16,089 | 13 |
| GEN-PROBE INC | 325413 | 472,695 | 4 | 59,498 | 4 |
| HEMAGEN DIAGNOSTICS INC | 325413 | 6,375 | 20 | 427 | 10 |
| HYCOR BIOMEDICAL INC | 325413 | | | | |
| IDEXX LABORATORIES INC | 325413 | 1,024,030 | 2 | 116,169 | 1 |
| IMMUNOMEDICS INC | 325413 | 3,651 | 21 | -22,909 | 14 |
| LIFE TECHNOLOGIES CORP | 325413 | 1,620,323 | 1 | 31,321 | 5 |
| LUMINEX CORPORATION | 325413 | 104,447 | 12 | 3,057 | 9 |
| MALLINCKRODT INC | 325413 | | | | |
| MERIDIAN BIOSCIENCE INC | 325413 | 139,639 | 9 | 30,202 | 6 |
| NANOGEN INC | 325413 | 46,920 | 17 | -38,107 | 18 |
| NEOGEN CORPORATION | 325413 | 102,418 | 13 | 12,098 | 7 |
| NEOPROBE CORPORATION | 325413 | 7,886 | 19 | -5,166 | 12 |
| ORASURE TECHNOLOGIES INC | 325413 | 71,104 | 14 | -31,275 | 16 |
| PROMEGA CORP | 325413 | | | | |
| QIAGEN NV | 325413 | 892,975 | 3 | 89,033 | 3 |
| SEQUENOM INC | 325413 | 47,149 | 16 | -44,154 | 19 |
| SIEMENS HEALTHCARE DIAGNOSTICS | 325413 | | | | |
| SPECTRAL DIAGNOSTICS INC | 325413 | 2,800 | 22 | -1,400 | 11 |
| STRATAGENE CORP | 325413 | | | | |
| TECHNE CORP | 325413 | 257,420 | 6 | 103,558 | 2 |
| THERMO SCIENTIFIC | 325413 | | | | |
| THIRD WAVE TECHNOLOGIES | 325413 | | | | |
| TRINITY BIOTECH PLC | 325413 | 140,139 | 8 | -77,778 | 21 |
| VERMILLION INC | 325413 | | | | |
| VYSIS INC | 325413 | | | | |
| **Drug Delivery Systems** | | | | | |
| ALKERMES INC | 325412A | 240,717 | 10 | 166,979 | 4 |
| ALZA CORP | 325412A | | | | |
| ARADIGM CORPORATION | 325412A | 251 | 27 | -22,608 | 19 |
| BARR PHARMACEUTICALS INC | 325412A | | | | |
| BIOVAIL CORPORATION | 325412A | 757,178 | 6 | 199,904 | 3 |
| CARACO PHARMACEUTICAL LABORATORIES | 325412A | 350,367 | 9 | 35,388 | 7 |
| DEPOMED INC | 325412A | 34,842 | 18 | -15,843 | 17 |
| DOR BIOPHARMA INC | 325412A | 2,310 | 24 | -3,422 | 11 |
| DURECT CORP | 325412A | 27,101 | 19 | -43,907 | 26 |
| EMISPHERE TECHNOLOGIES | 325412A | 251 | 26 | -24,388 | 20 |
| FLAMEL TECHNOLOGIES SA | 325412A | 38,619 | 17 | -12,084 | 16 |
| GENEREX BIOTECHNOLOGY | 325412A | 125 | 28 | -36,229 | 24 |
| HI-TECH PHARMACAL CO INC | 325412A | 62,017 | 16 | -5,098 | 13 |
| IMPAX LABORATORIES INC | 325412A | 210,071 | 11 | 18,700 | 9 |
| INSITE VISION INC | 325412A | 13,706 | 20 | -21,310 | 18 |

| Company | Industry Code | 2008 Sales (U.S. $ thousands) | Sales Rank | 2008 Profits (U.S. $ thousands) | Profits Rank |
|---|---|---|---|---|---|
| KV PHARMACEUTICAL CO | 325412A | 601,896 | 7 | 88,354 | 6 |
| LANNETT COMPANY INC | 325412A | 72,403 | 15 | -2,318 | 10 |
| MDRNA INC | 325412A | 2,609 | 23 | -59,220 | 28 |
| MYLAN INC | 325412A | 5,137,585 | 2 | -181,215 | 29 |
| NANOBIO CORPORATION | 325412A | | | | |
| NEKTAR THERAPEUTICS | 325412A | 90,185 | 14 | -34,336 | 22 |
| NEOPHARM INC | 325412A | 0 | | -8,218 | 15 |
| NEXMED INC | 325412A | 5,957 | 22 | -5,171 | 14 |
| NOVAVAX INC | 325412A | 1,064 | 25 | -36,049 | 23 |
| NOVEN PHARMACEUTICALS | 325412A | 108,175 | 12 | 21,412 | 8 |
| ONCOGENEX PHARMACEUTICALS | 325412A | 0 | | -4,204 | 12 |
| PAR PHARMACEUTICAL COMPANIES INC | 325412A | 578,115 | 8 | -47,768 | 27 |
| PENWEST PHARMACEUTICALS | 325412A | 8,534 | 21 | -26,734 | 21 |
| PERRIGO CO | 325412A | 1,822,131 | 4 | 135,773 | 5 |
| RANBAXY LABORATORIES LIMITED | 325412A | 1,531,700 | 5 | -196,320 | 30 |
| SKYEPHARMA PLC | 325412A | 94,710 | 13 | -43,700 | 25 |
| TARO PHARMACEUTICAL INDUSTRIES | 325412A | | | | |
| TEVA PHARMACEUTICAL INDUSTRIES | 325412A | 11,085,000 | 1 | 635,000 | 1 |
| WATSON PHARMACEUTICALS | 325412A | 2,535,501 | 3 | 238,379 | 2 |
| **Drugs (Pharmaceuticals), Discovery & Manufacturing** | | | | | |
| 3SBIO INC | 325412 | 35,653 | 95 | 5,795 | 54 |
| 4SC AG | 325412 | 4,220 | 130 | -16,830 | 97 |
| ABBOTT LABORATORIES | 325412 | 29,527,600 | 8 | 4,880,700 | 10 |
| ACAMBIS PLC | 325412 | | | | |
| ACCESS PHARMACEUTICALS | 325412 | 291 | 152 | -20,573 | 103 |
| ACTELION LTD | 325412 | 1,254,960 | 37 | 273,810 | 28 |
| ADAMIS PHARMACEUTICALS | 325412 | 0 | | -1,534 | 62 |
| ADOLOR CORP | 325412 | 48,208 | 85 | -30,122 | 123 |
| AEOLUS PHARMACEUTICALS | 325412 | 0 | | -2,973 | 64 |
| AETERNA ZENTARIS INC | 325412 | 38,478 | 92 | -59,817 | 144 |
| AKORN INC | 325412 | 93,598 | 70 | -7,939 | 73 |
| ALCON INC | 325412 | 6,294,000 | 22 | 2,046,000 | 15 |
| ALEXION PHARMACEUTICALS | 325412 | 259,099 | 53 | 33,149 | 42 |
| ALFACELL CORPORATION | 325412 | 0 | | -12,321 | 86 |
| ALIZYME PLC | 325412 | 3,060 | 136 | -19,350 | 101 |
| ALLERGAN INC | 325412 | 4,339,700 | 27 | 578,600 | 23 |
| ALLIANCE PHARMACEUTICAL | 325412 | | | | |
| ALLOS THERAPEUTICS INC | 325412 | 0 | | -51,730 | 140 |
| ALSERES PHARMACEUTICALS | 325412 | 0 | | -20,847 | 105 |
| ALTANA AG | 325412 | 1,774,430 | 33 | 136,760 | 33 |
| AMARIN CORPORATION PLC | 325412 | | | | |
| AMGEN INC | 325412 | 15,003,000 | 14 | 4,196,000 | 12 |
| AMYLIN PHARMACEUTICALS | 325412 | 840,109 | 40 | -315,405 | 169 |
| ANGIOTECH PHARMACEUTICALS | 325412 | 283,272 | 51 | -741,176 | 173 |
| ANTIGENICS INC | 325412 | 2,651 | 139 | -28,488 | 120 |
| APP PHARMACEUTICALS INC | 325412 | | | | |
| ARENA PHARMACEUTICALS INC | 325412 | 9,809 | 115 | -237,573 | 167 |
| ARIAD PHARMACEUTICALS INC | 325412 | 7,082 | 124 | -71,052 | 151 |

| Company | Industry Code | 2008 Sales (U.S. $ thousands) | Sales Rank | 2008 Profits (U.S. $ thousands) | Profits Rank |
|---|---|---|---|---|---|
| ARQULE INC | 325412 | 14,141 | 110 | -50,864 | 138 |
| ASTELLAS PHARMA INC | 325412 | 9,726,000 | 19 | 1,774,000 | 18 |
| ASTRAZENECA CANADA | 325412 | | | | |
| ASTRAZENECA PLC | 325412 | 31,601,000 | 7 | 6,130,000 | 7 |
| AUTOIMMUNE INC | 325412 | 318 | 150 | -345 | 59 |
| AVANIR PHARMACEUTICALS | 325412 | 6,829 | 125 | -17,496 | 98 |
| AVAX TECHNOLOGIES INC | 325412 | | | | |
| AVI BIOPHARMA INC | 325412 | 21,258 | 102 | -23,953 | 112 |
| AVIGEN INC | 325412 | 7,100 | 123 | -25,099 | 116 |
| AXCAN PHARMA INC | 325412 | | | | |
| BAYER SCHERING PHARMA AG | 325412 | 14,243,900 | 15 | | |
| BIOCRYST PHARMACEUTICALS | 325412 | 56,561 | 81 | -24,732 | 115 |
| BIOGEN IDEC INC | 325412 | 4,097,500 | 28 | 783,200 | 22 |
| BIOMARIN PHARMACEUTICAL | 325412 | 296,493 | 49 | 30,831 | 43 |
| BIOPURE CORPORATION | 325412 | 3,128 | 135 | -20,282 | 102 |
| BIOTECH HOLDINGS LTD | 325412 | 340 | 148 | -1,300 | 61 |
| BIOTIME INC | 325412 | 1,504 | 142 | -3,781 | 66 |
| BRISTOL-MYERS SQUIBB CO | 325412 | 20,597,000 | 11 | 5,247,000 | 9 |
| CAMBREX CORP | 325412 | 249,618 | 54 | 7,929 | 52 |
| CANGENE CORP | 325412 | 152,800 | 62 | 27,300 | 46 |
| CARDIOME PHARMA CORP | 325412 | 1,500 | 143 | -55,700 | 142 |
| CARDIUM THERAPEUTICS INC | 325412 | 2,417 | 140 | -24,598 | 114 |
| CELGENE CORP | 325412 | 2,254,781 | 31 | -1,533,653 | 174 |
| CELL GENESYS INC | 325412 | 94,571 | 69 | -46,975 | 134 |
| CELL THERAPEUTICS INC | 325412 | 11,432 | 114 | -180,029 | 165 |
| CELLEDEX THERAPEUTICS INC | 325412 | 7,456 | 122 | -47,501 | 136 |
| CEL-SCI CORPORATION | 325412 | 5 | 159 | -7,703 | 72 |
| CEPHALON INC | 325412 | 1,974,554 | 32 | 222,548 | 30 |
| CERUS CORPORATION | 325412 | 16,507 | 107 | -29,181 | 121 |
| CHIRON CORP | 325412 | | | | |
| COLUMBIA LABORATORIES INC | 325412 | 36,340 | 94 | -14,077 | 89 |
| COMPUGEN LTD | 325412 | 338 | 149 | -12,527 | 87 |
| CORTEX PHARMACEUTICALS | 325412 | 0 | | -14,596 | 91 |
| CRUCELL NV | 325412 | 379,440 | 45 | 20,720 | 48 |
| CUBIST PHARMACEUTICALS INC | 325412 | 433,641 | 44 | 169,819 | 31 |
| CURAGEN CORPORATION | 325412 | 1,174 | 145 | 24,781 | 47 |
| CURIS INC | 325412 | 8,367 | 118 | -12,123 | 84 |
| CYPRESS BIOSCIENCE INC | 325412 | 17,159 | 106 | -18,226 | 99 |
| CYTRX CORPORATION | 325412 | 6,266 | 126 | -27,803 | 119 |
| DAIICHI SANKYO CO LTD | 325412 | 9,282,450 | 20 | 1,030,000 | 20 |
| DENDREON CORPORATION | 325412 | 111 | 155 | -71,644 | 152 |
| DISCOVERY LABORATORIES | 325412 | 4,600 | 129 | -39,106 | 130 |
| DIVI'S LABORATORIES LIMITED | 325412 | | | | |
| DR REDDY'S LABORATORIES LIMITED | 325412 | 1,230,100 | 38 | 116,900 | 35 |
| DSM PHARMACEUTICALS INC | 325412 | | | | |
| DUSA PHARMACEUTICALS INC | 325412 | 29,545 | 98 | -6,250 | 71 |
| DYAX CORP | 325412 | 43,429 | 88 | -66,468 | 149 |
| ELAN CORP PLC | 325412 | 1,000,200 | 39 | -71,000 | 150 |

| Company | Industry Code | 2008 Sales (U.S. $ thousands) | Sales Rank | 2008 Profits (U.S. $ thousands) | Profits Rank |
|---|---|---|---|---|---|
| ELI LILLY & COMPANY | 325412 | 20,378,000 | 12 | -2,071,900 | 175 |
| ENDO PHARMACEUTICALS HOLDINGS INC | 325412 | 1,260,536 | 36 | 261,741 | 29 |
| ENTREMED INC | 325412 | 7,477 | 121 | -23,862 | 111 |
| ENZO BIOCHEM INC | 325412 | 77,795 | 73 | -10,653 | 80 |
| ENZON PHARMACEUTICALS INC | 325412 | 196,938 | 58 | -2,715 | 63 |
| EUSA PHARMA (USA) INC | 325412 | | | | |
| EXELIXIS INC | 325412 | 117,859 | 64 | -162,854 | 163 |
| FOREST LABORATORIES INC | 325412 | 3,501,802 | 29 | 967,933 | 21 |
| GENENTECH INC | 325412 | 13,418,000 | 17 | 3,427,000 | 14 |
| GENTA INC | 325412 | 363 | 147 | -505,838 | 172 |
| GENVEC INC | 325412 | 15,121 | 108 | -26,063 | 118 |
| GENZYME CORP | 325412 | 4,605,039 | 25 | 421,081 | 27 |
| GENZYME ONCOLOGY | 325412 | | | | |
| GERON CORPORATION | 325412 | 2,803 | 138 | -62,021 | 145 |
| GILEAD SCIENCES INC | 325412 | 5,335,750 | 23 | 2,011,154 | 16 |
| GLAXOSMITHKLINE PLC | 325412 | 36,127,200 | 6 | 10,594,000 | 2 |
| HEMISPHERX BIOPHARMA INC | 325412 | 265 | 153 | -12,219 | 85 |
| HOLLIS-EDEN PHARMACEUTICALS | 325412 | 0 | | -21,565 | 107 |
| HUMAN GENOME SCIENCES | 325412 | 48,422 | 84 | -244,915 | 168 |
| IDERA PHARMACEUTICALS INC | 325412 | 26,376 | 101 | 1,509 | 57 |
| IDM PHARMA INC | 325412 | 3,147 | 134 | -18,606 | 100 |
| IMCLONE SYSTEMS INC | 325412 | | | | |
| IMMTECH PHARMACEUTICALS | 325412 | 9,717 | 116 | -10,513 | 79 |
| IMMUNOGEN INC | 325412 | 40,249 | 90 | -32,020 | 126 |
| INCYTE CORP | 325412 | 3,919 | 132 | -178,920 | 164 |
| INSMED INCORPORATED | 325412 | 11,699 | 112 | -15,667 | 94 |
| INSPIRE PHARMACEUTICALS | 325412 | 70,498 | 75 | -51,603 | 139 |
| INTERMUNE INC | 325412 | 48,152 | 86 | -97,739 | 158 |
| ISIS PHARMACEUTICALS INC | 325412 | 107,190 | 65 | -11,963 | 83 |
| JAZZ PHARMACEUTICALS | 325412 | 67,514 | 78 | -184,339 | 166 |
| JOHNSON & JOHNSON | 325412 | 63,747,000 | 1 | 12,949,000 | 1 |
| KENDLE INTERNATIONAL INC | 325412 | 678,581 | 41 | 29,397 | 45 |
| KERYX BIOPHARMACEUTICALS | 325412 | 1,283 | 144 | -52,881 | 141 |
| KING PHARMACEUTICALS INC | 325412 | 1,565,061 | 34 | -333,063 | 170 |
| LA JOLLA PHARMACEUTICAL | 325412 | 0 | | -62,854 | 147 |
| LEXICON PHARMACEUTICALS | 325412 | 32,321 | 97 | -76,860 | 153 |
| LIGAND PHARMACEUTICALS | 325412 | 27,315 | 100 | -98,114 | 159 |
| LORUS THERAPEUTICS INC | 325412 | 40 | 157 | -11,680 | 82 |
| MANHATTAN PHARMACEUTICALS INC | 325412 | 0 | | -4,269 | 67 |
| MAXYGEN INC | 325412 | 100,709 | 67 | 30,325 | 44 |
| MEDAREX INC | 325412 | 52,292 | 83 | -38,465 | 129 |
| MEDICINES CO (THE) | 325412 | 348,157 | 47 | -8,504 | 75 |
| MEDICIS PHARMACEUTICAL | 325412 | 517,750 | 43 | 10,276 | 50 |
| MEDIMMUNE INC | 325412 | | | | |
| MERCK & CO INC | 325412 | 23,850,300 | 9 | 7,808,400 | 5 |
| MERCK KGAA | 325412 | 10,000,300 | 18 | 501,600 | 25 |
| MERCK SERONO SA | 325412 | | | | |
| MGI PHARMA INC | 325412 | | | | |

| Company | Industry Code | 2008 Sales (U.S. $ thousands) | Sales Rank | 2008 Profits (U.S. $ thousands) | Profits Rank |
|---|---|---|---|---|---|
| MIGENIX INC | 325412 | 180 | 154 | -520 | 60 |
| MILLENNIUM PHARMACEUTICALS INC | 325412 | | | | |
| MYRIAD GENETICS INC | 325412 | 334,000 | 48 | 47,845 | 39 |
| NABI BIOPHARMACEUTICALS | 325412 | 0 | | -10,716 | 81 |
| NEUROBIOLOGICAL TECHNOLOGIES INC | 325412 | 14,760 | 109 | -16,322 | 95 |
| NEUROCRINE BIOSCIENCES | 325412 | 3,975 | 131 | -88,613 | 157 |
| NEUROGEN CORP | 325412 | 3,000 | 137 | -34,514 | 127 |
| NOVARTIS AG | 325412 | 41,459,000 | 4 | 8,233,000 | 3 |
| NOVO-NORDISK AS | 325412 | 8,239,030 | 21 | 1,744,460 | 19 |
| NPS PHARMACEUTICALS INC | 325412 | 102,279 | 66 | -31,726 | 124 |
| NYCOMED | 325412 | 4,540,310 | 26 | -110,860 | 160 |
| ONCOTHYREON INC | 325412 | 39,998 | 91 | 7,125 | 53 |
| ONYX PHARMACEUTICALS INC | 325412 | 194,343 | 59 | 1,948 | 56 |
| OSCIENT PHARMACEUTICALS CORPORATION | 325412 | 86,325 | 72 | -64,755 | 148 |
| OSI PHARMACEUTICALS INC | 325412 | 379,388 | 46 | 471,485 | 26 |
| OXIGENE INC | 325412 | 12 | 158 | -21,401 | 106 |
| PAIN THERAPEUTICS INC | 325412 | 63,725 | 80 | 15,347 | 49 |
| PALATIN TECHNOLOGIES INC | 325412 | 11,483 | 113 | -14,384 | 90 |
| PDL BIOPHARMA | 325412 | 294,270 | 50 | 68,387 | 36 |
| PEREGRINE PHARMACEUTICALS INC | 325412 | 6,093 | 127 | -23,176 | 109 |
| PFIZER INC | 325412 | 48,296,000 | 2 | 8,104,000 | 4 |
| PHARMACYCLICS INC | 325412 | 0 | | -24,298 | 113 |
| PHARMOS CORP | 325412 | 0 | | -10,089 | 78 |
| PONIARD PHARMACEUTICALS | 325412 | 0 | | -48,565 | 137 |
| POZEN INC | 325412 | 66,133 | 79 | -5,976 | 69 |
| PROGENICS PHARMACEUTICALS | 325412 | 67,671 | 77 | -44,672 | 132 |
| QLT INC | 325412 | 124,140 | 63 | 134,891 | 34 |
| QUESTCOR PHARMACEUTICALS | 325412 | 95,248 | 68 | 40,532 | 40 |
| REGENERON PHARMACEUTICALS INC | 325412 | 238,457 | 55 | -82,710 | 154 |
| REPLIGEN CORPORATION | 325412 | 19,296 | 104 | 37,107 | 41 |
| REPROS THERAPEUTICS INC | 325412 | 433 | 146 | -25,202 | 117 |
| RIGEL PHARMACEUTICALS INC | 325412 | 0 | | -132,435 | 162 |
| ROCHE HOLDING LTD | 325412 | 41,676,500 | 3 | 7,803,000 | 6 |
| SALIX PHARMACEUTICALS | 325412 | 178,766 | 60 | -47,037 | 135 |
| SANOFI-AVENTIS SA | 325412 | 36,751,700 | 5 | 5,721,790 | 8 |
| SAVIENT PHARMACEUTICALS | 325412 | 3,181 | 133 | -84,169 | 155 |
| SCHERING-PLOUGH CORP | 325412 | 18,502,000 | 13 | 1,903,000 | 17 |
| SCICLONE PHARMACEUTICALS | 325412 | 54,113 | 82 | -8,348 | 74 |
| SCIELE PHARMA INC | 325412 | | | | |
| SCIOS INC | 325412 | | | | |
| SEATTLE GENETICS | 325412 | 35,236 | 96 | -85,501 | 156 |
| SENETEK PLC | 325412 | 1,845 | 141 | -3,759 | 65 |
| SEPRACOR INC | 325412 | 1,292,289 | 35 | 515,110 | 24 |
| SHIRE BIOCHEM INC | 325412 | | | | |
| SHIRE PLC | 325412 | 3,022,200 | 30 | 156,000 | 32 |
| SIGA TECHNOLOGIES INC | 325412 | 8,066 | 119 | -8,599 | 76 |
| SIMCERE PHARMACEUTICAL GROUP | 325412 | 225,206 | 57 | 51,323 | 38 |

| Company | Industry Code | 2008 Sales (U.S. $ thousands) | Sales Rank | 2008 Profits (U.S. $ thousands) | Profits Rank |
|---|---|---|---|---|---|
| SPECTRUM PHARMACEUTICALS | 325412 | 28,725 | 99 | -15,467 | 93 |
| STIEFEL LABORATORIES INC | 325412 | | | | |
| SUPERGEN INC | 325412 | 38,422 | 93 | -9,111 | 77 |
| SYNVISTA THERAPEUTICS INC | 325412 | | | | |
| TAKEDA PHARMACEUTICAL COMPANY LTD | 325412 | 13,748,020 | 16 | 3,554,540 | 13 |
| TARGETED GENETICS CORP | 325412 | 8,718 | 117 | -20,720 | 104 |
| TELIK INC | 325412 | 0 | | -31,763 | 125 |
| TEVA PARENTERAL MEDICINES | 325412 | | | | |
| TITAN PHARMACEUTICALS | 325412 | | | | |
| TORREYPINES THERAPEUTICS | 325412 | 6,071 | 128 | -22,785 | 108 |
| TRIMERIS INC | 325412 | 20,613 | 103 | 8,009 | 51 |
| UCB SA | 325412 | 4,860,380 | 24 | -58,040 | 143 |
| UNIGENE LABORATORIES | 325412 | 19,229 | 105 | -6,078 | 70 |
| UNITED THERAPEUTICS CORP | 325412 | 281,497 | 52 | -42,789 | 131 |
| UNITED-GUARDIAN INC | 325412 | 12,292 | 111 | 3,163 | 55 |
| URIGEN PHARMACEUTICALS | 325412 | 0 | | -4,844 | 68 |
| VALEANT PHARMACEUTICALS INTERNATIONAL | 325412 | 656,977 | 42 | -23,714 | 110 |
| VASOGEN INC | 325412 | 0 | | -14,800 | 92 |
| VAXGEN INC | 325412 | 293 | 151 | -12,563 | 88 |
| VERNALIS PLC | 325412 | 89,900 | 71 | -100 | 58 |
| VERTEX PHARMACEUTICALS | 325412 | 175,504 | 61 | -459,851 | 171 |
| VICAL INC | 325412 | 7,956 | 120 | -36,896 | 128 |
| VION PHARMACEUTICALS INC | 325412 | 43 | 156 | -29,848 | 122 |
| VIROPHARMA INC | 325412 | 232,307 | 56 | 67,617 | 37 |
| WARNER CHILCOTT PLC | 325412 | | | | |
| WYETH | 325412 | 22,833,908 | 10 | 4,417,833 | 11 |
| XECHEM INTERNATIONAL | 325412 | | | | |
| XENOPORT INC | 325412 | 41,996 | 89 | -62,540 | 146 |
| XOMA LTD | 325412 | 67,987 | 76 | -45,245 | 133 |
| ZILA INC | 325412 | 45,061 | 87 | -16,378 | 96 |
| ZYMOGENETICS INC | 325412 | 73,989 | 74 | -116,241 | 161 |
| **Instrument Manufacturing, including Measurement, Control, Test & Navigational** | | | | | |
| AGILENT TECHNOLOGIES INC | 3345 | 5,774,000 | 1 | 693,000 | 1 |
| HARVARD BIOSCIENCE INC | 3345 | 88,049 | 4 | 1,673 | 4 |
| ILLUMINA INC | 3345 | 573,225 | 3 | 50,477 | 3 |
| MILLIPORE CORP | 3345 | 1,602,138 | 2 | 145,801 | 2 |
| WHATMAN PLC | 3345 | | | | |
| **Investment Banking** | | | | | |
| BURRILL & COMPANY | 523110 | | | | |
| **Laboratories & Diagnostic Services--Medical** | | | | | |
| 23ANDME | 6215 | | | | |
| BIORELIANCE CORP | 6215 | | | | |
| CML HEALTHCARE INCOME FUND | 6215 | 425,500 | 3 | 93,400 | 2 |
| MDS INC | 6215 | 1,315,000 | 2 | -553,000 | 5 |
| MEDTOX SCIENTIFIC INC | 6215 | 85,813 | 4 | 5,572 | 3 |
| ORCHID CELLMARK INC | 6215 | 57,595 | 5 | -4,481 | 4 |
| QUEST DIAGNOSTICS INC | 6215 | 7,249,447 | 1 | 581,490 | 1 |

| Company | Industry Code | 2008 Sales (U.S. $ thousands) | Sales Rank | 2008 Profits (U.S. $ thousands) | Profits Rank |
|---|---|---|---|---|---|
| SPECIALTY LABORATORIES INC | 6215 | | | | |
| **Market Research** | | | | | |
| IMS HEALTH INC | 541910 | 2,329,528 | 1 | 311,250 | 1 |
| **Medical/Dental/Surgical Equipment & Supplies, Distribution** | | | | | |
| THERMO FISHER SCIENTIFIC | 423450 | 10,498,000 | 1 | 994,200 | 1 |
| **Medical/Dental/Surgical Equipment & Supplies, Manufacturing** | | | | | |
| AASTROM BIOSCIENCES INC | 33911 | 522 | 17 | -20,133 | 16 |
| ABAXIS INC | 33911 | 100,551 | 10 | 12,503 | 9 |
| ADVANCED BIONICS CORP | 33911 | | | | |
| ADVANSOURCE BIOMATERIALS CORPORATION | 33911 | 1,950 | 16 | -6,090 | 14 |
| APPLIED BIOSYSTEMS INC | 33911 | 2,361,484 | 6 | 213,808 | 6 |
| BAUSCH & LOMB INC | 33911 | | | | |
| BAXTER INTERNATIONAL INC | 33911 | 12,348,000 | 2 | 2,014,000 | 2 |
| BIO RAD LABORATORIES INC | 33911 | 1,764,365 | 7 | 89,510 | 7 |
| BIOCOMPATIBLES INTERNATIONAL PLC | 33911 | 29,050 | 14 | -750 | 12 |
| BIOMET INC | 33911 | 2,383,300 | 5 | 1,018,800 | 4 |
| BIOSPHERE MEDICAL INC | 33911 | 29,258 | 13 | -5,492 | 13 |
| COVIDIEN PLC | 33911 | 9,910,000 | 3 | 1,361,000 | 3 |
| GE HEALTHCARE | 33911 | 17,392,000 | 1 | 2,851,000 | 1 |
| GENZYME BIOSURGERY | 33911 | | | | |
| HOSPIRA INC | 33911 | 3,629,500 | 4 | 320,900 | 5 |
| INSTITUT STRAUMANN AG | 33911 | 726,850 | 8 | 7,650 | 11 |
| INTEGRA LIFESCIENCES HOLDINGS CORP | 33911 | 654,604 | 9 | 34,933 | 8 |
| LIFECORE BIOMEDICAL INC | 33911 | | | | |
| NORTH AMERICAN SCIENTIFIC | 33911 | 13,927 | 15 | -10,945 | 15 |
| SYNOVIS LIFE TECHNOLOGIES | 33911 | 49,800 | 12 | 11,485 | 10 |
| THERAGENICS CORP | 33911 | 67,358 | 11 | -58,540 | 17 |
| **Medicinals & Botanicals, Manufacturing** | | | | | |
| GALDERMA PHARMA SA | 325411 | | | | |
| PROTEIN POLYMER TECHNOLOGIES | 325411 | 25 | 1 | -2,938 | 1 |
| VENTRIA BIOSCIENCE | 325411 | | | | |
| **Nanotechnology-Biotech/Health** | | | | | |
| ARRYX INC | 541711 | | | | |
| BIONANOMATRIX | 541711 | | | | |
| **Research & Development--Physical, Engineering & Life Sciences** | | | | | |
| ADVANCED CELL TECHNOLOGY | 541712 | 787 | 20 | -33,904 | 14 |
| ALBANY MOLECULAR RESEARCH INC | 541712 | 229,260 | 8 | 20,560 | 6 |
| APPLIED MOLECULAR EVOLUTION INC | 541712 | | | | |
| ARRAY BIOPHARMA INC | 541712 | 28,808 | 15 | -96,288 | 15 |
| BIOANALYTICAL SYSTEMS INC | 541712 | 41,697 | 13 | -1,489 | 9 |
| CELERA CORPORATION | 541712 | 138,700 | 9 | -104,100 | 16 |
| CHARLES RIVER LABORATORIES INTERNATIONAL INC | 541712 | 1,343,493 | 3 | -521,843 | 20 |
| CHESAPEAKE BIOLOGICAL LABORATORIES INC | 541712 | | | | |
| COMMONWEALTH BIOTECHNOLOGIES INC | 541712 | 9,435 | 17 | -9,863 | 10 |
| COVANCE INC | 541712 | 1,827,067 | 1 | 196,760 | 1 |

| Company | Industry Code | 2008 Sales (U.S. $ thousands) | Sales Rank | 2008 Profits (U.S. $ thousands) | Profits Rank |
|---|---|---|---|---|---|
| ENCORIUM GROUP INC | 541712 | 35,912 | 14 | -21,073 | 11 |
| EVOTEC AG | 541712 | 56,300 | 12 | -110,400 | 17 |
| ICON PLC | 541712 | 865,248 | 5 | 64,676 | 3 |
| INFINITY PHARMACEUTICALS | 541712 | 83,441 | 10 | 23,654 | 5 |
| LIFE SCIENCES RESEARCH INC | 541712 | 242,422 | 7 | 10,418 | 7 |
| ORE PHARMACEUTICALS INC | 541712 | 1,950 | 19 | -22,461 | 12 |
| PACIFIC BIOMETRICS INC | 541712 | 8,265 | 18 | -571 | 8 |
| PAREXEL INTERNATIONAL | 541712 | 964,283 | 4 | 64,640 | 4 |
| PHARMACEUTICAL PRODUCT DEVELOPMENT INC | 541712 | 1,569,901 | 2 | 187,519 | 2 |
| PHARMANET DEVELOPMENT GROUP INC | 541712 | 451,453 | 6 | -251,093 | 19 |
| PRA INTERNATIONAL | 541712 | | | | |
| QUINTILES TRANSNATIONAL | 541712 | | | | |
| ROSETTA INPHARMATICS LLC | 541712 | | | | |
| SANGAMO BIOSCIENCES INC | 541712 | 16,186 | 16 | -24,302 | 13 |
| VERENIUM CORPORATION | 541712 | 69,659 | 11 | -185,542 | 18 |
| **Textiles, Fabrics, Sheets/Towels, Manufacturing** | | | | | |
| TOYOBO CO LTD | 313 | 4,305,989 | 1 | 46,891 | 1 |
| **Veterinary Products Manufacturing** | | | | | |
| HESKA CORP | 325412B | 81,653 | 1 | -850 | 1 |
| IMMUCELL CORPORATION | 325412B | 4,634 | 2 | -469 | 2 |
| SYNBIOTICS CORP | 325412B | | | | |
| VIRBAC CORP | 325412B | | | | |

# ALPHABETICAL INDEX

COVANCE INC
COVIDIEN PLC
CRUCELL NV
CSL LIMITED
CUBIST PHARMACEUTICALS INC
CURAGEN CORPORATION
CURIS INC
CYPRESS BIOSCIENCE INC
CYTRX CORPORATION
DAIICHI SANKYO CO LTD
DECODE GENETICS INC
DENDREON CORPORATION
DEPOMED INC
DISCOVERY LABORATORIES INC
DIVI'S LABORATORIES LIMITED
DOR BIOPHARMA INC
DOW AGROSCIENCES LLC
DR REDDY'S LABORATORIES LIMITED
DSM PHARMACEUTICALS INC
DUPONT AGRICULTURE & NUTRITION
DURECT CORP
DUSA PHARMACEUTICALS INC
DYAX CORP
E I DU PONT DE NEMOURS & CO (DUPONT)
EISAI CO LTD
ELAN CORP PLC
ELI LILLY & COMPANY
EMISPHERE TECHNOLOGIES INC
ENCORIUM GROUP INC
ENDO PHARMACEUTICALS HOLDINGS INC
ENTREMED INC
ENZO BIOCHEM INC
ENZON PHARMACEUTICALS INC
EPIX PHARMACEUTICALS INC
ERESEARCH TECHNOLOGY INC
EUSA PHARMA (USA) INC
EVOTEC AG
EXELIXIS INC
FLAMEL TECHNOLOGIES SA
FOREST LABORATORIES INC
GALDERMA PHARMA SA
GE HEALTHCARE
GENENCOR INTERNATIONAL INC
GENENTECH INC
GENEREX BIOTECHNOLOGY
GENOMIC HEALTH INC
GEN-PROBE INC
GENTA INC
GENVEC INC
GENZYME BIOSURGERY
GENZYME CORP
GENZYME ONCOLOGY
GERON CORPORATION
GILEAD SCIENCES INC
GLAXOSMITHKLINE PLC
GTC BIOTHERAPEUTICS INC
HARVARD BIOSCIENCE INC
HEMAGEN DIAGNOSTICS INC

HEMISPHERX BIOPHARMA INC
HESKA CORP
HI-TECH PHARMACAL CO INC
HOLLIS-EDEN PHARMACEUTICALS
HOSPIRA INC
HUMAN GENOME SCIENCES INC
HYCOR BIOMEDICAL INC
ICON PLC
IDERA PHARMACEUTICALS INC
IDEXX LABORATORIES INC
IDM PHARMA INC
ILLUMINA INC
IMCLONE SYSTEMS INC
IMMTECH PHARMACEUTICALS INC
IMMUCELL CORPORATION
IMMUNOGEN INC
IMMUNOMEDICS INC
IMPAX LABORATORIES INC
IMS HEALTH INC
INCYTE CORP
INFINITY PHARMACEUTICALS INC
INSITE VISION INC
INSMED INCORPORATED
INSPIRE PHARMACEUTICALS INC
INSTITUT STRAUMANN AG
INTEGRA LIFESCIENCES HOLDINGS CORP
INTERMUNE INC
INTERNATIONAL ISOTOPES INC
ISIS PHARMACEUTICALS INC
JAZZ PHARMACEUTICALS
JOHNSON & JOHNSON
JUBILANT BIOSYS LTD
KENDLE INTERNATIONAL INC
KERYX BIOPHARMACEUTICALS INC
KING PHARMACEUTICALS INC
KV PHARMACEUTICAL CO
LA JOLLA PHARMACEUTICAL
LANNETT COMPANY INC
LEXICON PHARMACEUTICALS INC
LIFE SCIENCES RESEARCH INC
LIFE TECHNOLOGIES CORP
LIFECELL CORPORATION
LIFECORE BIOMEDICAL INC
LIGAND PHARMACEUTICALS INC
LONZA GROUP
LORUS THERAPEUTICS INC
LUMINEX CORPORATION
MALLINCKRODT INC
MANHATTAN PHARMACEUTICALS INC
MAXYGEN INC
MDRNA INC
MDS INC
MEDAREX INC
MEDICINES CO (THE)
MEDICIS PHARMACEUTICAL CORP
MEDIDATA SOLUTIONS INC
MEDIMMUNE INC
MEDTOX SCIENTIFIC INC

MERCK & CO INC
MERCK KGAA
MERCK SERONO SA
MERIDIAN BIOSCIENCE INC
MGI PHARMA INC
MIGENIX INC
MILLENNIUM PHARMACEUTICALS INC
MILLIPORE CORP
MONSANTO CO
MYLAN INC
MYRIAD GENETICS INC
NABI BIOPHARMACEUTICALS
NANOBIO CORPORATION
NANOGEN INC
NEKTAR THERAPEUTICS
NEOGEN CORPORATION
NEOPHARM INC
NEOPROBE CORPORATION
NEUROBIOLOGICAL TECHNOLOGIES INC
NEUROCRINE BIOSCIENCES INC
NEUROGEN CORP
NEXMED INC
NORTH AMERICAN SCIENTIFIC
NOVARTIS AG
NOVAVAX INC
NOVEN PHARMACEUTICALS
NOVO-NORDISK AS
NOVOZYMES
NPS PHARMACEUTICALS INC
NYCOMED
ONCOGENEX PHARMACEUTICALS
ONCOTHYREON INC
ONYX PHARMACEUTICALS INC
ORASURE TECHNOLOGIES INC
ORCHID CELLMARK INC
ORE PHARMACEUTICALS INC
ORGANOGENESIS INC
OSCIENT PHARMACEUTICALS CORPORATION
OSI PHARMACEUTICALS INC
OXIGENE INC
PACIFIC BIOMETRICS INC
PAIN THERAPEUTICS INC
PALATIN TECHNOLOGIES INC
PAR PHARMACEUTICAL COMPANIES INC
PAREXEL INTERNATIONAL CORP
PDL BIOPHARMA
PENWEST PHARMACEUTICALS CO
PEREGRINE PHARMACEUTICALS INC
PERRIGO CO
PFIZER INC
PHARMACEUTICAL PRODUCT DEVELOPMENT INC
PHARMACYCLICS INC
PHARMANET DEVELOPMENT GROUP INC
PHARMOS CORP
POLYDEX PHARMACEUTICALS
PONIARD PHARMACEUTICALS INC
POZEN INC
PRA INTERNATIONAL

PROGENICS PHARMACEUTICALS
PROMEGA CORP
PROTEIN POLYMER TECHNOLOGIES
QIAGEN NV
QLT INC
QUEST DIAGNOSTICS INC
QUESTCOR PHARMACEUTICALS
QUINTILES TRANSNATIONAL CORP
RANBAXY LABORATORIES LIMITED
REGENERON PHARMACEUTICALS INC
REPLIGEN CORPORATION
REPROS THERAPEUTICS INC
RIGEL PHARMACEUTICALS INC
ROCHE HOLDING LTD
ROSETTA INPHARMATICS LLC
SALIX PHARMACEUTICALS
SANGAMO BIOSCIENCES INC
SANOFI-AVENTIS SA
SAVIENT PHARMACEUTICALS INC
SCHERING-PLOUGH CORP
SCICLONE PHARMACEUTICALS
SCIELE PHARMA INC
SCIOS INC
SEATTLE GENETICS
SENETEK PLC
SEPRACOR INC
SEQUENOM INC
SERACARE LIFE SCIENCES INC
SHIRE BIOCHEM INC
SHIRE PLC
SIEMENS HEALTHCARE DIAGNOSTICS
SIGA TECHNOLOGIES INC
SIGMA-ALDRICH CORP
SIMCERE PHARMACEUTICAL GROUP
SKYEPHARMA PLC
SPECIALTY LABORATORIES INC
SPECTRAL DIAGNOSTICS INC
SPECTRUM PHARMACEUTICALS INC
STEMCELLS INC
STIEFEL LABORATORIES INC
STRATAGENE CORP
SUPERGEN INC
SYNBIOTICS CORP
SYNGENTA AG
SYNOVIS LIFE TECHNOLOGIES INC
SYNVISTA THERAPEUTICS INC
TAKEDA PHARMACEUTICAL COMPANY LTD
TARGETED GENETICS CORP
TARO PHARMACEUTICAL INDUSTRIES
TECHNE CORP
TELIK INC
TEVA PARENTERAL MEDICINES INC
TEVA PHARMACEUTICAL INDUSTRIES
THERAGENICS CORP
THERMO FISHER SCIENTIFIC INC
THERMO SCIENTIFIC
THIRD WAVE TECHNOLOGIES INC
TITAN PHARMACEUTICALS

TORREYPINES THERAPEUTICS INC
TOYOBO CO LTD
TRIMERIS INC
TRINITY BIOTECH PLC
TRIPOS INTERNATIONAL
UCB SA
UNIGENE LABORATORIES
UNITED THERAPEUTICS CORP
UNITED-GUARDIAN INC
URIGEN PHARMACEUTICALS INC
VALEANT PHARMACEUTICALS INTERNATIONAL
VASOGEN INC
VAXGEN INC
VENTRIA BIOSCIENCE
VERENIUM CORPORATION
VERMILLION INC
VERNALIS PLC
VERTEX PHARMACEUTICALS INC
VIACELL INC
VICAL INC
VION PHARMACEUTICALS INC
VIRBAC CORP
VIROPHARMA INC
VYSIS INC
WARNER CHILCOTT PLC
WATSON PHARMACEUTICALS INC
WHATMAN PLC
WYETH
XECHEM INTERNATIONAL
XENOPORT INC
XOMA LTD
ZILA INC
ZYMOGENETICS INC

# INDEX OF U.S. HEADQUARTERS
# LOCATION BY STATE

To help you locate members of THE BIOTECH 400 geographically, the city and state of the headquarters of each company are in the following index.

## ALABAMA
BIOCRYST PHARMACEUTICALS INC; Birmingham

## ARIZONA
MEDICIS PHARMACEUTICAL CORP; Scottsdale
ZILA INC; Scottsdale

## CALIFORNIA
23ANDME; Mountain View
ABAXIS INC; Union City
ACCELRYS INC; San Diego
ADAMIS PHARMACEUTICALS CORPORATION; Del Mar
ADVANCED BIONICS CORPORATION; Sylmar
AEOLUS PHARMACEUTICALS INC; Mission Viejo
AFFYMETRIX INC; Santa Clara
AGILENT TECHNOLOGIES INC; Santa Clara
ALLERGAN INC; Irvine
ALLIANCE PHARMACEUTICAL CORP; San Diego
ALZA CORP; Mountain View
AMGEN INC; Thousand Oaks
AMYLIN PHARMACEUTICALS INC; San Diego
APPLIED BIOSYSTEMS INC; Foster City
APPLIED MOLECULAR EVOLUTION INC; San Diego
ARADIGM CORPORATION; Hayward
ARCADIA BIOSCIENCES INC; Davis
ARENA PHARMACEUTICALS INC; San Diego
AUTOIMMUNE INC; Pasadena
AVANIR PHARMACEUTICALS; Aliso Viejo
AVIGEN INC; Alameda
AVIVA BIOSCIENCES CORP; San Diego
BIO RAD LABORATORIES INC; Hercules
BIOMARIN PHARMACEUTICAL INC; Novato
BIOSITE INC; San Diego
BIOTIME INC; Alameda
BURRILL & COMPANY; San Francisco
CARDIUM THERAPEUTICS INC; San Diego
CELERA CORPORATION; Alameda
CELL GENESYS INC; South San Francisco
CEPHEID; Sunnyvale
CERUS CORPORATION; Concord
CHIRON CORP; Emeryville
CORTEX PHARMACEUTICALS INC; Irvine
CYPRESS BIOSCIENCE INC; San Diego
CYTRX CORPORATION; Los Angeles
DEPOMED INC; Menlo Park
DURECT CORP; Cupertino
EXELIXIS INC; South San Francisco
GENENTECH INC; South San Francisco

GENOMIC HEALTH INC; Redwood City
GEN-PROBE INC; San Diego
GERON CORPORATION; Menlo Park
GILEAD SCIENCES INC; Foster City
HOLLIS-EDEN PHARMACEUTICALS; San Diego
HYCOR BIOMEDICAL INC; Garden Grove
IDM PHARMA INC; Irvine
ILLUMINA INC; San Diego
IMPAX LABORATORIES INC; Hayward
INSITE VISION INC; Alameda
INTERMUNE INC; Brisbane
ISIS PHARMACEUTICALS INC; Carlsbad
JAZZ PHARMACEUTICALS; Palo Alto
LA JOLLA PHARMACEUTICAL; San Diego
LIFE TECHNOLOGIES CORP; Carlsbad
LIGAND PHARMACEUTICALS INC; San Diego
MAXYGEN INC; Redwood City
NANOGEN INC; San Diego
NEKTAR THERAPEUTICS; San Carlos
NEUROBIOLOGICAL TECHNOLOGIES INC;
Emeryville
NEUROCRINE BIOSCIENCES INC; San Diego
NORTH AMERICAN SCIENTIFIC; Chatsworth
ONYX PHARMACEUTICALS INC; Emeryville
PAIN THERAPEUTICS INC; San Mateo
PDL BIOPHARMA; Redwood City
PEREGRINE PHARMACEUTICALS INC; Tustin
PHARMACYCLICS INC; Sunnyvale
PONIARD PHARMACEUTICALS INC; South San
Francisco
PROTEIN POLYMER TECHNOLOGIES; San Diego
QUESTCOR PHARMACEUTICALS; Union City
RIGEL PHARMACEUTICALS INC; South San Francisco
SANGAMO BIOSCIENCES INC; Richmond
SCICLONE PHARMACEUTICALS; Foster City
SCIOS INC; Mountain View
SENETEK PLC; Napa
SEQUENOM INC; San Diego
SPECIALTY LABORATORIES INC; Valencia
SPECTRUM PHARMACEUTICALS INC; Irvine
STEMCELLS INC; Palo Alto
STRATAGENE CORP; La Jolla
SUPERGEN INC; Dublin
TELIK INC; Palo Alto
TEVA PARENTERAL MEDICINES INC; Irvine
TITAN PHARMACEUTICALS; South San Francisco
TORREYPINES THERAPEUTICS INC; La Jolla
URIGEN PHARMACEUTICALS INC; San Francisco
VALEANT PHARMACEUTICALS INTERNATIONAL;
Aliso Viejo
VAXGEN INC; South San Francisco
VENTRIA BIOSCIENCE; Sacramento
VERMILLION INC; Fremont
VICAL INC; San Diego
WATSON PHARMACEUTICALS INC; Corona
XENOPORT INC; Santa Clara
XOMA LTD; Berkeley

**COLORADO**
ALLOS THERAPEUTICS INC; Westminster
ARRAY BIOPHARMA INC; Boulder
HESKA CORP; Loveland

**CONNECTICUT**
ALEXION PHARMACEUTICALS INC; Cheshire
CURAGEN CORPORATION; Branford
IMS HEALTH INC; Norwalk
NEUROGEN CORP; Branford
PENWEST PHARMACEUTICALS CO; Danbury
VION PHARMACEUTICALS INC; New Haven

**DELAWARE**
DUPONT AGRICULTURE & NUTRITION; Wilmington
E I DU PONT DE NEMOURS & CO (DUPONT);
Wilmington
INCYTE CORP; Wilmington

**FLORIDA**
BIOHEART INC; Sunrise
NOVEN PHARMACEUTICALS; Miami
STIEFEL LABORATORIES INC; Coral Gables

**GEORGIA**
SCIELE PHARMA INC; Atlanta
THERAGENICS CORP; Buford

**IDAHO**
INTERNATIONAL ISOTOPES INC; Idaho Falls

**ILLINOIS**
ABBOTT LABORATORIES; Abbott Park
AKORN INC; Lake Forest
APP PHARMACEUTICALS INC; Schaumburg
ARRYX INC; Chicago
BAXTER INTERNATIONAL INC; Deerfield
CELSIS INTERNATIONAL PLC; Chicago
HOSPIRA INC; Lake Forest
NEOPHARM INC; Lake Bluff
VYSIS INC; Des Plaines

**INDIANA**
BIOANALYTICAL SYSTEMS INC; West Lafayette
BIOMET INC; Warsaw
DOW AGROSCIENCES LLC; Indianapolis
ELI LILLY & COMPANY; Indianapolis

**MAINE**
IDEXX LABORATORIES INC; Westbrook
IMMUCELL CORPORATION; Portland

**MARYLAND**
BIORELIANCE CORP; Rockville
CHESAPEAKE BIOLOGICAL LABORATORIES INC;
Baltimore
ENTREMED INC; Rockville

GENVEC INC; Gaithersburg
HEMAGEN DIAGNOSTICS INC; Columbia
HUMAN GENOME SCIENCES INC; Rockville
MEDIMMUNE INC; Gaithersburg
NABI BIOPHARMACEUTICALS; Rockville
NOVAVAX INC; Rockville
ORE PHARMACEUTICALS INC; Gaithersburg
UNITED THERAPEUTICS CORP; Silver Spring

## MASSACHUSETTS
ADVANCED CELL TECHNOLOGY INC; Worcester
ADVANSOURCE BIOMATERIALS CORPORATION;
Wilmington
ALKERMES INC; Cambridge
ALSERES PHARMACEUTICALS INC; Hopkinton
ANIKA THERAPEUTICS INC; Bedford
ARIAD PHARMACEUTICALS INC; Cambridge
ARQULE INC; Woburn
BIOGEN IDEC INC; Cambridge
BIOPURE CORPORATION; Cambridge
BIOSPHERE MEDICAL INC; Rockland
CALIPER LIFE SCIENCES; Hopkinton
CELLEDEX THERAPEUTICS INC; Needham
CHARLES RIVER LABORATORIES
INTERNATIONAL INC; Wilmington
CUBIST PHARMACEUTICALS INC; Lexington
CURIS INC; Cambridge
DUSA PHARMACEUTICALS INC; Wilmington
DYAX CORP; Cambridge
EPIX PHARMACEUTICALS INC; Lexington
GENZYME BIOSURGERY; Cambridge
GENZYME CORP; Cambridge
GENZYME ONCOLOGY; Cambridge
GTC BIOTHERAPEUTICS INC; Framingham
HARVARD BIOSCIENCE INC; Holliston
IDERA PHARMACEUTICALS INC; Cambridge
IMMUNOGEN INC; Waltham
INFINITY PHARMACEUTICALS INC; Cambridge
MILLENNIUM PHARMACEUTICALS INC; Cambridge
MILLIPORE CORP; Billerica
ORGANOGENESIS INC; Canton
OSCIENT PHARMACEUTICALS CORPORATION;
Waltham
OXIGENE INC; Waltham
PAREXEL INTERNATIONAL CORP; Waltham
REPLIGEN CORPORATION; Waltham
SEPRACOR INC; Marlborough
SERACARE LIFE SCIENCES INC; Milford
THERMO FISHER SCIENTIFIC INC; Waltham
VERENIUM CORPORATION; Cambridge
VERTEX PHARMACEUTICALS INC; Cambridge
VIACELL INC; Cambridge

## MICHIGAN
AASTROM BIOSCIENCES INC; Ann Arbor
CARACO PHARMACEUTICAL LABORATORIES;
Detroit
NANOBIO CORPORATION; Ann Arbor

NEOGEN CORPORATION; Lansing
PERRIGO CO; Allegan

## MINNESOTA
LIFECORE BIOMEDICAL INC; Chaska
MEDTOX SCIENTIFIC INC; St. Paul
MGI PHARMA INC; Bloomington
SYNOVIS LIFE TECHNOLOGIES INC; St. Paul
TECHNE CORP; Minneapolis

## MISSOURI
KV PHARMACEUTICAL CO; St. Louis
MALLINCKRODT INC; Hazelwood
MONSANTO CO; St. Louis
SIGMA-ALDRICH CORP; St. Louis
SYNBIOTICS CORP; Kansas City
TRIPOS INTERNATIONAL; St. Louis

## NEW JERSEY
ALFACELL CORPORATION; Somerset
ALPHARMA ANIMAL HEALTH; Bridgewater
BARR PHARMACEUTICALS INC; Montvale
CAMBREX CORP; East Rutherford
CELGENE CORP; Summit
COLUMBIA LABORATORIES INC; Livingston
COVANCE INC; Princeton
DOR BIOPHARMA INC; Princeton
EMISPHERE TECHNOLOGIES INC; Cedar Knolls
ENZON PHARMACEUTICALS INC; Bridgewater
GENTA INC; Berkeley Heights
IMMUNOMEDICS INC; Morris Plains
INTEGRA LIFESCIENCES HOLDINGS CORP;
Plainsboro
JOHNSON & JOHNSON; New Brunswick
LIFE SCIENCES RESEARCH INC; East Millstone
LIFECELL CORPORATION; Branchburg
MEDAREX INC; Princeton
MEDICINES CO (THE); Parsippany
MERCK & CO INC; Whitehouse Station
NEXMED INC; Robbinsville
NPS PHARMACEUTICALS INC; Bedminster
ORCHID CELLMARK INC; Princeton
PALATIN TECHNOLOGIES INC; Cranbury
PAR PHARMACEUTICAL COMPANIES INC;
Woodcliff Lake
PHARMANET DEVELOPMENT GROUP INC;
Princeton
PHARMOS CORP; Iselin
QUEST DIAGNOSTICS INC; Madison
SAVIENT PHARMACEUTICALS INC; E. Brunswick
SCHERING-PLOUGH CORP; Kenilworth
SYNVISTA THERAPEUTICS INC; Montvale
UNIGENE LABORATORIES; Boonton
WARNER CHILCOTT PLC; Rockaway
WYETH; Madison
XECHEM INTERNATIONAL; Edison

## NEW YORK
ALBANY MOLECULAR RESEARCH INC; Albany
ANTIGENICS INC; New York
BAUSCH & LOMB INC; Rochester
BRISTOL-MYERS SQUIBB CO; New York
ENZO BIOCHEM INC; New York
FOREST LABORATORIES INC; New York
GENENCOR INTERNATIONAL INC; Rochester
HI-TECH PHARMACAL CO INC; Amityville
IMCLONE SYSTEMS INC; New York
IMMTECH PHARMACEUTICALS INC; New York
KERYX BIOPHARMACEUTICALS INC; New York
MANHATTAN PHARMACEUTICALS INC; New York
MEDIDATA SOLUTIONS INC; New York
OSI PHARMACEUTICALS INC; Melville
PFIZER INC; New York
PROGENICS PHARMACEUTICALS; Tarrytown
REGENERON PHARMACEUTICALS INC; Tarrytown
SIEMENS HEALTHCARE DIAGNOSTICS; Tarrytown
SIGA TECHNOLOGIES INC; New York
UNITED-GUARDIAN INC; Hauppauge

## NORTH CAROLINA
DSM PHARMACEUTICALS INC; Greenville
INSPIRE PHARMACEUTICALS INC; Durham
PHARMACEUTICAL PRODUCT DEVELOPMENT
INC; Wilmington
POZEN INC; Chapel Hill
PRA INTERNATIONAL; Raleigh
QUINTILES TRANSNATIONAL CORP; Durham
SALIX PHARMACEUTICALS; Morrisville
TRIMERIS INC; Durham

## OHIO
KENDLE INTERNATIONAL INC; Cincinnati
MERIDIAN BIOSCIENCE INC; Cincinnati
NEOPROBE CORPORATION; Dublin

## OREGON
AVI BIOPHARMA INC; Corvallis

## PENNSYLVANIA
ADOLOR CORP; Exton
AVAX TECHNOLOGIES INC; Philadelphia
BAYER CORP; Pittsburgh
BIONANOMATRIX; Philadelphia
CEPHALON INC; Frazer
DISCOVERY LABORATORIES INC; Warrington
ENCORIUM GROUP INC; Wayne
ENDO PHARMACEUTICALS HOLDINGS INC; Chadds
Ford
ERESEARCH TECHNOLOGY INC; Philadelphia
EUSA PHARMA (USA) INC; Langhorne
HEMISPHERX BIOPHARMA INC; Philadelphia
LANNETT COMPANY INC; Philadelphia
MYLAN INC; Canonsburg
ORASURE TECHNOLOGIES INC; Bethlehem
VIROPHARMA INC; Exton

## TENNESSEE
KING PHARMACEUTICALS INC; Bristol

## TEXAS
ACCESS PHARMACEUTICALS INC; Dallas
LEXICON PHARMACEUTICALS INC; The Woodlands
LUMINEX CORPORATION; Austin
REPROS THERAPEUTICS INC; The Woodlands
VIRBAC CORP; Fort Worth

## UTAH
MYRIAD GENETICS INC; Salt Lake City

## VIRGINIA
CEL-SCI CORPORATION; Vienna
COMMONWEALTH BIOTECHNOLOGIES INC;
Richmond
INSMED INCORPORATED; Richmond

## WASHINGTON
CELL THERAPEUTICS INC; Seattle
DENDREON CORPORATION; Seattle
MDRNA INC; Bothell
ONCOGENEX PHARMACEUTICALS; Bothell
ONCOTHYREON INC; Seattle
PACIFIC BIOMETRICS INC; Seattle
ROSETTA INPHARMATICS LLC; Seattle
SEATTLE GENETICS; Bothell
TARGETED GENETICS CORP; Seattle
ZYMOGENETICS INC; Seattle

## WISCONSIN
CELLULAR DYNAMICS INTERNATIONAL; Madison
GE HEALTHCARE; Waukesha
PROMEGA CORP; Madison
THIRD WAVE TECHNOLOGIES INC; Madison

# INDEX OF NON-U.S. HEADQUARTERS LOCATION BY COUNTRY

## AUSTRALIA
CSL LIMITED; Parkville

## BELGIUM
UCB SA; Brussels

## BERMUDA
COVIDIEN PLC; Hamilton

## CANADA
AETERNA ZENTARIS INC; Quebec City
ANGIOTECH PHARMACEUTICALS; Vancouver
ASTRAZENECA CANADA; Mississauga
AXCAN PHARMA INC; Mont-Saint-Hilaire
BIOTECH HOLDINGS LTD; Richmond
BIOVAIL CORPORATION; Mississauga
CANGENE CORP; Winnipeg
CANREG INC; Dundas
CARDIOME PHARMA CORP; Vancouver
CML HEALTHCARE INCOME FUND; Mississauga
GENEREX BIOTECHNOLOGY; Toronto
LORUS THERAPEUTICS INC; Toronto
MDS INC; Mississauga
MIGENIX INC; Vancouver
POLYDEX PHARMACEUTICALS; Toronto
QLT INC; Vancouver
SHIRE BIOCHEM INC; Ville Saint-Laurent
SPECTRAL DIAGNOSTICS INC; Toronto
VASOGEN INC; Mississauga

## CHINA
3SBIO INC; Shenyang
SIMCERE PHARMACEUTICAL GROUP; Jiangsu

## DENMARK
NOVO-NORDISK AS; Bagsvaerd
NOVOZYMES; Bagsvaerd

## FRANCE
CEGEDIM SA; Boulogne-Billancourt
FLAMEL TECHNOLOGIES SA; Venissieux Cedex
SANOFI-AVENTIS SA; Paris

## GERMANY
4SC AG; Planegg-Martinsried
ALTANA AG; Wesel
BASF AG; Ludwigshafen
BAYER AG; Leverkusen
BAYER SCHERING PHARMA AG; Berlin
EVOTEC AG; Hamburg
MERCK KGAA; Darmstadt

## ICELAND
DECODE GENETICS INC; Reykjavik

## INDIA
DIVI'S LABORATORIES LIMITED; Hyderabad
DR REDDY'S LABORATORIES LIMITED; Hyderabad
JUBILANT BIOSYS LTD; Noida
RANBAXY LABORATORIES LIMITED; Gurgaon

## ISRAEL
COMPUGEN LTD; Tel Aviv
TARO PHARMACEUTICAL INDUSTRIES; Yakum
TEVA PHARMACEUTICAL INDUSTRIES; Petach Tikva

## JAPAN
ASTELLAS PHARMA INC; Tokyo
DAIICHI SANKYO CO LTD; Tokyo
EISAI CO LTD; Tokyo
TAKEDA PHARMACEUTICAL COMPANY LTD; Osaka
TOYOBO CO LTD; Osaka

## SWEDEN
CELLARTIS AB; Goteborg
THERMO SCIENTIFIC; Helsingborg

## SWITZERLAND
ACTELION LTD; Allschwil
ALCON INC; Hunenberg
GALDERMA PHARMA SA; Lausanne
INSTITUT STRAUMANN AG; Basel
LONZA GROUP; Basel
MERCK SERONO SA; Geneva 20
NOVARTIS AG; Basel
NYCOMED; Zurich
ROCHE HOLDING LTD; Basel
SYNGENTA AG; Basel

## THE NETHERLANDS
AKZO NOBEL NV; Amsterdam
CRUCELL NV; Leiden
QIAGEN NV; Venlo

## UNITED KINGDOM
ACAMBIS PLC; Cambridge
ALIZYME PLC; Cambridge
AMARIN CORPORATION PLC; Dublin
ASTRAZENECA PLC; London
BIOCOMPATIBLES INTERNATIONAL PLC; Farnham
ELAN CORP PLC; Dublin
GLAXOSMITHKLINE PLC; Middlesex
ICON PLC; Dublin
SHIRE PLC; Basingstoke
SKYEPHARMA PLC; London
TRINITY BIOTECH PLC; Bray
VERNALIS PLC; Winnersh
WHATMAN PLC; Maidstone

# INDEX BY REGIONS OF THE U.S.
## WHERE THE FIRMS HAVE LOCATIONS

### WEST
23ANDME
ABAXIS INC
ABBOTT LABORATORIES
ACCELRYS INC
ACTELION LTD
ADAMIS PHARMACEUTICALS CORPORATION
ADVANCED BIONICS CORPORATION
ADVANCED CELL TECHNOLOGY INC
AEOLUS PHARMACEUTICALS INC
AFFYMETRIX INC
AGILENT TECHNOLOGIES INC
ALBANY MOLECULAR RESEARCH INC
ALCON INC
ALLERGAN INC
ALLIANCE PHARMACEUTICAL CORP
ALLOS THERAPEUTICS INC
ALPHARMA ANIMAL HEALTH
ALZA CORP
AMGEN INC
AMYLIN PHARMACEUTICALS INC
ANGIOTECH PHARMACEUTICALS
APP PHARMACEUTICALS INC
APPLIED BIOSYSTEMS INC
APPLIED MOLECULAR EVOLUTION INC
ARADIGM CORPORATION
ARCADIA BIOSCIENCES INC
ARENA PHARMACEUTICALS INC
ARRAY BIOPHARMA INC
ASTELLAS PHARMA INC
ASTRAZENECA PLC
AUTOIMMUNE INC
AVANIR PHARMACEUTICALS
AVI BIOPHARMA INC
AVIGEN INC
AVIVA BIOSCIENCES CORP
BASF AG
BAUSCH & LOMB INC
BAXTER INTERNATIONAL INC
BAYER AG
BAYER CORP
BAYER SCHERING PHARMA AG
BIO RAD LABORATORIES INC
BIOANALYTICAL SYSTEMS INC
BIOGEN IDEC INC
BIOMARIN PHARMACEUTICAL INC
BIOMET INC
BIOSITE INC
BIOTIME INC
BRISTOL-MYERS SQUIBB CO
BURRILL & COMPANY
CALIPER LIFE SCIENCES
CANGENE CORP
CARDIUM THERAPEUTICS INC
CELERA CORPORATION

CELL GENESYS INC
CELL THERAPEUTICS INC
CEPHALON INC
CEPHEID
CERUS CORPORATION
CHARLES RIVER LABORATORIES
INTERNATIONAL INC
CHIRON CORP
CORTEX PHARMACEUTICALS INC
COVANCE INC
CYPRESS BIOSCIENCE INC
CYTRX CORPORATION
DECODE GENETICS INC
DENDREON CORPORATION
DEPOMED INC
DISCOVERY LABORATORIES INC
DUPONT AGRICULTURE & NUTRITION
DURECT CORP
E I DU PONT DE NEMOURS & CO (DUPONT)
ELAN CORP PLC
ELI LILLY & COMPANY
EXELIXIS INC
GENENCOR INTERNATIONAL INC
GENENTECH INC
GENOMIC HEALTH INC
GEN-PROBE INC
GENZYME CORP
GERON CORPORATION
GILEAD SCIENCES INC
GLAXOSMITHKLINE PLC
HARVARD BIOSCIENCE INC
HEMAGEN DIAGNOSTICS INC
HESKA CORP
HOLLIS-EDEN PHARMACEUTICALS
HOSPIRA INC
HYCOR BIOMEDICAL INC
ICON PLC
IDEXX LABORATORIES INC
IDM PHARMA INC
ILLUMINA INC
IMPAX LABORATORIES INC
IMS HEALTH INC
INCYTE CORP
INSITE VISION INC
INSPIRE PHARMACEUTICALS INC
INTEGRA LIFESCIENCES HOLDINGS CORP
INTERMUNE INC
INTERNATIONAL ISOTOPES INC
ISIS PHARMACEUTICALS INC
JAZZ PHARMACEUTICALS
JOHNSON & JOHNSON
KENDLE INTERNATIONAL INC
LA JOLLA PHARMACEUTICAL
LANNETT COMPANY INC
LIFE TECHNOLOGIES CORP
LIGAND PHARMACEUTICALS INC
MALLINCKRODT INC
MAXYGEN INC

MDRNA INC
MDS INC
MEDAREX INC
MEDICIS PHARMACEUTICAL CORP
MEDIMMUNE INC
MERCK & CO INC
MERCK KGAA
MILLIPORE CORP
MONSANTO CO
MYRIAD GENETICS INC
NANOGEN INC
NEKTAR THERAPEUTICS
NEUROBIOLOGICAL TECHNOLOGIES INC
NEUROCRINE BIOSCIENCES INC
NORTH AMERICAN SCIENTIFIC
NOVARTIS AG
NOVAVAX INC
NOVO-NORDISK AS
NOVOZYMES
ONCOGENEX PHARMACEUTICALS
ONCOTHYREON INC
ONYX PHARMACEUTICALS INC
OSI PHARMACEUTICALS INC
OXIGENE INC
PACIFIC BIOMETRICS INC
PAIN THERAPEUTICS INC
PAREXEL INTERNATIONAL CORP
PDL BIOPHARMA
PEREGRINE PHARMACEUTICALS INC
PFIZER INC
PHARMACEUTICAL PRODUCT DEVELOPMENT INC
PHARMACYCLICS INC
PHARMANET DEVELOPMENT GROUP INC
PONIARD PHARMACEUTICALS INC
PRA INTERNATIONAL
PROMEGA CORP
PROTEIN POLYMER TECHNOLOGIES
QIAGEN NV
QLT INC
QUEST DIAGNOSTICS INC
QUESTCOR PHARMACEUTICALS
QUINTILES TRANSNATIONAL CORP
RIGEL PHARMACEUTICALS INC
ROCHE HOLDING LTD
ROSETTA INPHARMATICS LLC
SANGAMO BIOSCIENCES INC
SANOFI-AVENTIS SA
SCHERING-PLOUGH CORP
SCICLONE PHARMACEUTICALS
SEATTLE GENETICS
SENETEK PLC
SEQUENOM INC
SIEMENS HEALTHCARE DIAGNOSTICS
SIGA TECHNOLOGIES INC
SIGMA-ALDRICH CORP
SPECIALTY LABORATORIES INC
SPECTRAL DIAGNOSTICS INC
SPECTRUM PHARMACEUTICALS INC

STEMCELLS INC
STIEFEL LABORATORIES INC
STRATAGENE CORP
SUPERGEN INC
SYNBIOTICS CORP
SYNGENTA AG
TAKEDA PHARMACEUTICAL COMPANY LTD
TARGETED GENETICS CORP
TELIK INC
TEVA PARENTERAL MEDICINES INC
TEVA PHARMACEUTICAL INDUSTRIES
THERAGENICS CORP
THERMO FISHER SCIENTIFIC INC
TITAN PHARMACEUTICALS
TORREYPINES THERAPEUTICS INC
TRINITY BIOTECH PLC
URIGEN PHARMACEUTICALS INC
VALEANT PHARMACEUTICALS INTERNATIONAL
VAXGEN INC
VENTRIA BIOSCIENCE
VERENIUM CORPORATION
VERMILLION INC
VERTEX PHARMACEUTICALS INC
VICAL INC
WATSON PHARMACEUTICALS INC
WYETH
XENOPORT INC
XOMA LTD
ZYMOGENETICS INC

**SOUTHWEST**
ABBOTT LABORATORIES
ACCESS PHARMACEUTICALS INC
ALCON INC
ARCADIA BIOSCIENCES INC
ASTRAZENECA PLC
BASF AG
BAXTER INTERNATIONAL INC
BAYER AG
BAYER CORP
BIO RAD LABORATORIES INC
BIOHEART INC
BRISTOL-MYERS SQUIBB CO
CELL THERAPEUTICS INC
CHARLES RIVER LABORATORIES
INTERNATIONAL INC
COVANCE INC
DUPONT AGRICULTURE & NUTRITION
E I DU PONT DE NEMOURS & CO (DUPONT)
ELI LILLY & COMPANY
GENZYME CORP
GLAXOSMITHKLINE PLC
HOSPIRA INC
ICON PLC
IDEXX LABORATORIES INC
JOHNSON & JOHNSON
LEXICON PHARMACEUTICALS INC
LUMINEX CORPORATION

MDS INC
MEDICIS PHARMACEUTICAL CORP
MERCK & CO INC
MILLIPORE CORP
MONSANTO CO
MYLAN INC
ONCOTHYREON INC
ORCHID CELLMARK INC
PFIZER INC
PHARMACEUTICAL PRODUCT DEVELOPMENT INC
PRA INTERNATIONAL
QUEST DIAGNOSTICS INC
QUINTILES TRANSNATIONAL CORP
REPROS THERAPEUTICS INC
ROCHE HOLDING LTD
SANOFI-AVENTIS SA
SCHERING-PLOUGH CORP
SIGMA-ALDRICH CORP
STRATAGENE CORP
SYNGENTA AG
TEVA PHARMACEUTICAL INDUSTRIES
THERAGENICS CORP
VIRBAC CORP
WYETH
ZILA INC

**MIDWEST**
AASTROM BIOSCIENCES INC
ABBOTT LABORATORIES
AKORN INC
AKZO NOBEL NV
ALKERMES INC
ALPHARMA ANIMAL HEALTH
ALTANA AG
ANGIOTECH PHARMACEUTICALS
APP PHARMACEUTICALS INC
ARRYX INC
ASTELLAS PHARMA INC
ASTRAZENECA PLC
BASF AG
BAXTER INTERNATIONAL INC
BAYER AG
BAYER CORP
BIOANALYTICAL SYSTEMS INC
BIOMET INC
BRISTOL-MYERS SQUIBB CO
CAMBREX CORP
CANREG INC
CARACO PHARMACEUTICAL LABORATORIES
CELLULAR DYNAMICS INTERNATIONAL
CELSIS INTERNATIONAL PLC
CEPHALON INC
CHARLES RIVER LABORATORIES
INTERNATIONAL INC
CHIRON CORP
COVANCE INC
CSL LIMITED
DECODE GENETICS INC

DOW AGROSCIENCES LLC
DUPONT AGRICULTURE & NUTRITION
E I DU PONT DE NEMOURS & CO (DUPONT)
EISAI CO LTD
ELI LILLY & COMPANY
ENZON PHARMACEUTICALS INC
FOREST LABORATORIES INC
GE HEALTHCARE
GENENCOR INTERNATIONAL INC
GENZYME CORP
GLAXOSMITHKLINE PLC
HOSPIRA INC
ICON PLC
IDEXX LABORATORIES INC
IMMTECH PHARMACEUTICALS INC
INTEGRA LIFESCIENCES HOLDINGS CORP
JOHNSON & JOHNSON
KENDLE INTERNATIONAL INC
KING PHARMACEUTICALS INC
KV PHARMACEUTICAL CO
LIFECORE BIOMEDICAL INC
LONZA GROUP
MALLINCKRODT INC
MDS INC
MEDICIS PHARMACEUTICAL CORP
MEDTOX SCIENTIFIC INC
MERCK & CO INC
MERCK KGAA
MERIDIAN BIOSCIENCE INC
MGI PHARMA INC
MILLIPORE CORP
MONSANTO CO
MYLAN INC
NANOBIO CORPORATION
NEOGEN CORPORATION
NEOPHARM INC
NEOPROBE CORPORATION
ORCHID CELLMARK INC
PFIZER INC
PHARMACEUTICAL PRODUCT DEVELOPMENT INC
PHARMANET DEVELOPMENT GROUP INC
PRA INTERNATIONAL
PROMEGA CORP
QUEST DIAGNOSTICS INC
QUINTILES TRANSNATIONAL CORP
ROCHE HOLDING LTD
SANOFI-AVENTIS SA
SCHERING-PLOUGH CORP
SIGMA-ALDRICH CORP
SPECTRAL DIAGNOSTICS INC
SYNGENTA AG
SYNOVIS LIFE TECHNOLOGIES INC
TAKEDA PHARMACEUTICAL COMPANY LTD
TECHNE CORP
TEVA PHARMACEUTICAL INDUSTRIES
THERMO FISHER SCIENTIFIC INC
THIRD WAVE TECHNOLOGIES INC
TRINITY BIOTECH PLC

TRIPOS INTERNATIONAL
UCB SA
VENTRIA BIOSCIENCE
VIACELL INC
VIRBAC CORP
VYSIS INC
WATSON PHARMACEUTICALS INC
WYETH

**SOUTHEAST**
AKZO NOBEL NV
ALCON INC
ALPHARMA ANIMAL HEALTH
AMGEN INC
ANGIOTECH PHARMACEUTICALS
ASTRAZENECA PLC
AXCAN PHARMA INC
BASF AG
BAUSCH & LOMB INC
BAXTER INTERNATIONAL INC
BAYER CORP
BIOCRYST PHARMACEUTICALS INC
BIOMET INC
BRISTOL-MYERS SQUIBB CO
CANGENE CORP
CHARLES RIVER LABORATORIES
INTERNATIONAL INC
COMPUGEN LTD
COVANCE INC
CRUCELL NV
CSL LIMITED
DUPONT AGRICULTURE & NUTRITION
E I DU PONT DE NEMOURS & CO (DUPONT)
ELAN CORP PLC
ELI LILLY & COMPANY
GENZYME CORP
GLAXOSMITHKLINE PLC
ICON PLC
IDEXX LABORATORIES INC
JOHNSON & JOHNSON
KING PHARMACEUTICALS INC
MEDICIS PHARMACEUTICAL CORP
MERCK & CO INC
MERCK KGAA
MERIDIAN BIOSCIENCE INC
MILLIPORE CORP
MONSANTO CO
NEKTAR THERAPEUTICS
NOVEN PHARMACEUTICALS
ORCHID CELLMARK INC
PFIZER INC
PHARMACEUTICAL PRODUCT DEVELOPMENT INC
PHARMANET DEVELOPMENT GROUP INC
QUEST DIAGNOSTICS INC
QUINTILES TRANSNATIONAL CORP
RANBAXY LABORATORIES LIMITED
ROCHE HOLDING LTD
SANOFI-AVENTIS SA

SCHERING-PLOUGH CORP
SCIELE PHARMA INC
SIGMA-ALDRICH CORP
STIEFEL LABORATORIES INC
SYNGENTA AG
SYNOVIS LIFE TECHNOLOGIES INC
TEVA PHARMACEUTICAL INDUSTRIES
THERAGENICS CORP
THERMO FISHER SCIENTIFIC INC
UCB SA
UNITED THERAPEUTICS CORP
VERENIUM CORPORATION
WATSON PHARMACEUTICALS INC
WYETH
ZILA INC

**NORTHEAST**
ABBOTT LABORATORIES
ACAMBIS PLC
ACCELRYS INC
ACTELION LTD
ADOLOR CORP
ADVANCED CELL TECHNOLOGY INC
ADVANSOURCE BIOMATERIALS CORPORATION
AETERNA ZENTARIS INC
AGILENT TECHNOLOGIES INC
AKORN INC
AKZO NOBEL NV
ALBANY MOLECULAR RESEARCH INC
ALEXION PHARMACEUTICALS INC
ALFACELL CORPORATION
ALKERMES INC
ALLOS THERAPEUTICS INC
ALPHARMA ANIMAL HEALTH
ALSERES PHARMACEUTICALS INC
ALTANA AG
AMARIN CORPORATION PLC
AMGEN INC
ANGIOTECH PHARMACEUTICALS
ANIKA THERAPEUTICS INC
ANTIGENICS INC
APP PHARMACEUTICALS INC
APPLIED BIOSYSTEMS INC
ARIAD PHARMACEUTICALS INC
ARQULE INC
ASTELLAS PHARMA INC
ASTRAZENECA PLC
AVAX TECHNOLOGIES INC
BARR PHARMACEUTICALS INC
BASF AG
BAUSCH & LOMB INC
BAXTER INTERNATIONAL INC
BAYER AG
BAYER CORP
BAYER SCHERING PHARMA AG
BIO RAD LABORATORIES INC
BIOANALYTICAL SYSTEMS INC
BIOCRYST PHARMACEUTICALS INC

BIOGEN IDEC INC
BIOMET INC
BIONANOMATRIX
BIOPURE CORPORATION
BIORELIANCE CORP
BIOSPHERE MEDICAL INC
BIOVAIL CORPORATION
BRISTOL-MYERS SQUIBB CO
BURRILL & COMPANY
CALIPER LIFE SCIENCES
CAMBREX CORP
CANGENE CORP
CEGEDIM SA
CELERA CORPORATION
CELGENE CORP
CELLEDEX THERAPEUTICS INC
CEL-SCI CORPORATION
CELSIS INTERNATIONAL PLC
CEPHALON INC
CHARLES RIVER LABORATORIES
INTERNATIONAL INC
CHESAPEAKE BIOLOGICAL LABORATORIES INC
CHIRON CORP
CML HEALTHCARE INCOME FUND
COLUMBIA LABORATORIES INC
COMMONWEALTH BIOTECHNOLOGIES INC
COVANCE INC
COVIDIEN PLC
CRUCELL NV
CSL LIMITED
CUBIST PHARMACEUTICALS INC
CURAGEN CORPORATION
CURIS INC
DAIICHI SANKYO CO LTD
DECODE GENETICS INC
DISCOVERY LABORATORIES INC
DIVI'S LABORATORIES LIMITED
DOR BIOPHARMA INC
DR REDDY'S LABORATORIES LIMITED
DSM PHARMACEUTICALS INC
DUPONT AGRICULTURE & NUTRITION
DUSA PHARMACEUTICALS INC
DYAX CORP
E I DU PONT DE NEMOURS & CO (DUPONT)
EISAI CO LTD
ELAN CORP PLC
ELI LILLY & COMPANY
EMISPHERE TECHNOLOGIES INC
ENCORIUM GROUP INC
ENDO PHARMACEUTICALS HOLDINGS INC
ENTREMED INC
ENZO BIOCHEM INC
ENZON PHARMACEUTICALS INC
EPIX PHARMACEUTICALS INC
ERESEARCH TECHNOLOGY INC
EUSA PHARMA (USA) INC
EVOTEC AG
FOREST LABORATORIES INC

GALDERMA PHARMA SA
GENENCOR INTERNATIONAL INC
GENEREX BIOTECHNOLOGY
GENTA INC
GENVEC INC
GENZYME BIOSURGERY
GENZYME CORP
GENZYME ONCOLOGY
GILEAD SCIENCES INC
GLAXOSMITHKLINE PLC
GTC BIOTHERAPEUTICS INC
HARVARD BIOSCIENCE INC
HEMAGEN DIAGNOSTICS INC
HEMISPHERX BIOPHARMA INC
HI-TECH PHARMACAL CO INC
HOSPIRA INC
HUMAN GENOME SCIENCES INC
ICON PLC
IDERA PHARMACEUTICALS INC
IDEXX LABORATORIES INC
IMCLONE SYSTEMS INC
IMMTECH PHARMACEUTICALS INC
IMMUCELL CORPORATION
IMMUNOGEN INC
IMMUNOMEDICS INC
IMPAX LABORATORIES INC
IMS HEALTH INC
INCYTE CORP
INFINITY PHARMACEUTICALS INC
INSMED INCORPORATED
INSPIRE PHARMACEUTICALS INC
INSTITUT STRAUMANN AG
INTEGRA LIFESCIENCES HOLDINGS CORP
JOHNSON & JOHNSON
JUBILANT BIOSYS LTD
KENDLE INTERNATIONAL INC
KERYX BIOPHARMACEUTICALS INC
KING PHARMACEUTICALS INC
LANNETT COMPANY INC
LEXICON PHARMACEUTICALS INC
LIFE SCIENCES RESEARCH INC
LIFECELL CORPORATION
LIGAND PHARMACEUTICALS INC
LONZA GROUP
MALLINCKRODT INC
MANHATTAN PHARMACEUTICALS INC
MDRNA INC
MDS INC
MEDAREX INC
MEDICINES CO (THE)
MEDICIS PHARMACEUTICAL CORP
MEDIDATA SOLUTIONS INC
MEDIMMUNE INC
MEDTOX SCIENTIFIC INC
MERCK & CO INC
MERCK KGAA
MERCK SERONO SA
MERIDIAN BIOSCIENCE INC

MGI PHARMA INC
MILLENNIUM PHARMACEUTICALS INC
MILLIPORE CORP
MONSANTO CO
MYLAN INC
NABI BIOPHARMACEUTICALS
NEUROBIOLOGICAL TECHNOLOGIES INC
NEUROGEN CORP
NEXMED INC
NORTH AMERICAN SCIENTIFIC
NOVARTIS AG
NOVAVAX INC
NOVO-NORDISK AS
NOVOZYMES
NPS PHARMACEUTICALS INC
ORASURE TECHNOLOGIES INC
ORCHID CELLMARK INC
ORE PHARMACEUTICALS INC
ORGANOGENESIS INC
OSCIENT PHARMACEUTICALS CORPORATION
OSI PHARMACEUTICALS INC
OXIGENE INC
PALATIN TECHNOLOGIES INC
PAR PHARMACEUTICAL COMPANIES INC
PAREXEL INTERNATIONAL CORP
PENWEST PHARMACEUTICALS CO
PERRIGO CO
PFIZER INC
PHARMACEUTICAL PRODUCT DEVELOPMENT INC
PHARMANET DEVELOPMENT GROUP INC
PHARMOS CORP
POZEN INC
PRA INTERNATIONAL
PROGENICS PHARMACEUTICALS
PROMEGA CORP
QIAGEN NV
QUEST DIAGNOSTICS INC
QUINTILES TRANSNATIONAL CORP
RANBAXY LABORATORIES LIMITED
REGENERON PHARMACEUTICALS INC
REPLIGEN CORPORATION
ROCHE HOLDING LTD
SALIX PHARMACEUTICALS
SANOFI-AVENTIS SA
SAVIENT PHARMACEUTICALS INC
SCHERING-PLOUGH CORP
SCIOS INC
SEPRACOR INC
SEQUENOM INC
SERACARE LIFE SCIENCES INC
SHIRE PLC
SIEMENS HEALTHCARE DIAGNOSTICS
SIGA TECHNOLOGIES INC
SIGMA-ALDRICH CORP
SPECTRAL DIAGNOSTICS INC
STEMCELLS INC
STIEFEL LABORATORIES INC
SYNBIOTICS CORP

SYNGENTA AG
SYNVISTA THERAPEUTICS INC
TAKEDA PHARMACEUTICAL COMPANY LTD
TARO PHARMACEUTICAL INDUSTRIES
TEVA PHARMACEUTICAL INDUSTRIES
THERMO FISHER SCIENTIFIC INC
TOYOBO CO LTD
TRIMERIS INC
TRINITY BIOTECH PLC
UCB SA
UNIGENE LABORATORIES
UNITED THERAPEUTICS CORP
UNITED-GUARDIAN INC
VERENIUM CORPORATION
VERTEX PHARMACEUTICALS INC
VIACELL INC
VION PHARMACEUTICALS INC
VIROPHARMA INC
WARNER CHILCOTT PLC
WATSON PHARMACEUTICALS INC
WHATMAN PLC
WYETH
XECHEM INTERNATIONAL

# INDEX OF FIRMS WITH
# INTERNATIONAL OPERATIONS

IMS HEALTH INC
INSTITUT STRAUMANN AG
INTEGRA LIFESCIENCES HOLDINGS CORP
JOHNSON & JOHNSON
JUBILANT BIOSYS LTD
KENDLE INTERNATIONAL INC
LA JOLLA PHARMACEUTICAL
LIFE SCIENCES RESEARCH INC
LIFE TECHNOLOGIES CORP
LIFECORE BIOMEDICAL INC
LONZA GROUP
LORUS THERAPEUTICS INC
LUMINEX CORPORATION
MALLINCKRODT INC
MAXYGEN INC
MDS INC
MEDICINES CO (THE)
MEDIDATA SOLUTIONS INC
MEDIMMUNE INC
MERCK & CO INC
MERCK KGAA
MERCK SERONO SA
MERIDIAN BIOSCIENCE INC
MIGENIX INC
MILLIPORE CORP
MONSANTO CO
MYLAN INC
NANOGEN INC
NEKTAR THERAPEUTICS
NEOGEN CORPORATION
NEXMED INC
NORTH AMERICAN SCIENTIFIC
NOVARTIS AG
NOVO-NORDISK AS
NOVOZYMES
NYCOMED
ORCHID CELLMARK INC
ORGANOGENESIS INC
OSI PHARMACEUTICALS INC
OXIGENE INC
PAREXEL INTERNATIONAL CORP
PDL BIOPHARMA
PEREGRINE PHARMACEUTICALS INC
PERRIGO CO
PFIZER INC
PHARMACEUTICAL PRODUCT DEVELOPMENT INC
PHARMANET DEVELOPMENT GROUP INC
POLYDEX PHARMACEUTICALS
PRA INTERNATIONAL
PROMEGA CORP
QIAGEN NV
QLT INC
QUEST DIAGNOSTICS INC
QUINTILES TRANSNATIONAL CORP
RANBAXY LABORATORIES LIMITED
ROCHE HOLDING LTD
SANOFI-AVENTIS SA
SCHERING-PLOUGH CORP

SCICLONE PHARMACEUTICALS
SENETEK PLC
SEPRACOR INC
SEQUENOM INC
SERACARE LIFE SCIENCES INC
SHIRE BIOCHEM INC
SHIRE PLC
SIEMENS HEALTHCARE DIAGNOSTICS
SIGMA-ALDRICH CORP
SIMCERE PHARMACEUTICAL GROUP
SKYEPHARMA PLC
SPECTRAL DIAGNOSTICS INC
SPECTRUM PHARMACEUTICALS INC
STIEFEL LABORATORIES INC
STRATAGENE CORP
SYNBIOTICS CORP
SYNGENTA AG
TAKEDA PHARMACEUTICAL COMPANY LTD
TARO PHARMACEUTICAL INDUSTRIES
TECHNE CORP
TEVA PARENTERAL MEDICINES INC
TEVA PHARMACEUTICAL INDUSTRIES
THERMO FISHER SCIENTIFIC INC
THERMO SCIENTIFIC
TORREYPINES THERAPEUTICS INC
TOYOBO CO LTD
TRINITY BIOTECH PLC
TRIPOS INTERNATIONAL
UCB SA
UNITED THERAPEUTICS CORP
VALEANT PHARMACEUTICALS INTERNATIONAL
VASOGEN INC
VERNALIS PLC
VERTEX PHARMACEUTICALS INC
VIACELL INC
VIRBAC CORP
VYSIS INC
WARNER CHILCOTT PLC
WHATMAN PLC
WYETH
XECHEM INTERNATIONAL
ZILA INC

**Individual Profiles
On Each Of
THE BIOTECH 400**

# 23ANDME

www.23andme.com

**Industry Group Code: 6215  Ranks within this company's industry group:  Sales:    Profits:**

| Drugs: | Other: | | Clinical: | | Computers: | | Services: | |
|--------|--------|--|-----------|--|-----------|--|-----------|--|
| Discovery: | AgriBio: | | Trials/Services: | Y | Hardware: | | Specialty Services: | Y |
| Licensing: | Genetic Data: | Y | Labs: | Y | Software: | Y | Consulting: | |
| Manufacturing: | Tissue Replacement: | | Equipment/Supplies: | | Arrays: | | Blood Collection: | |
| Genetics: | | | Research & Development Services: | Y | Database Management: | Y | Drug Delivery: | |
| | | | Diagnostics: | Y | | | Drug Distribution: | |

## TYPES OF BUSINESS:

Genetic Testing Services
Genetic Research

## BRANDS/DIVISIONS/AFFILIATES:

Genentech Inc
Google Inc
New Enterprise Associates

## CONTACTS: *Note: Officers with more than one job title may be intentionally listed here more than once.*

Sarah Imbach, COO
Larry Tesler, Product Fellow
Rachel G. Cohen, Mgr.-Comm.
Anne Wojcicki, Co-Founder
Linda Avey, Co-Founder

| Phone: 650-938-6300 | Fax: |
|---------------------|------|
| **Toll-Free:** | |
| **Address:** 2606 Bayshore Pkwy., Mountain View, CA 94043 US | |

## GROWTH PLANS/SPECIAL FEATURES:

23AndMe, Inc. is a start-up company based in Mountain View, California that specializes in the field of personalized genetics. Specifically, the firm offers its customers a detailed genetic profile using DNA analysis technologies and web-based interactive tools. To utilize 23AndMe's services, a customer receives a sample collection kit in the mail, along with instructions on how to collect and ship the approximately 2.5 milliliters of saliva necessary for the analysis. The kit includes a pre-paid, pre-addressed envelope along with a bar-coded specimen tube. After receiving the sample, the firm's technicians extract DNA from cells in the saliva. The DNA is then processed on an Illumina BeadChip, which reads more than 550,000 SNPs (single nucleotide polymorphisms), plus a custom designed set that analyzes more than 50,000 additional SNPs. Using information from the DNA analysis, the firm creates a genetic profile of the client, and in 4-6 weeks allows the customer to review results through an interactive web site that interprets the data into a variety of graphical formats, offering customers insight into ancestry, genetic predisposition to disease, traits attributable to genetics and other information. The firm's web site has additional features such as online surveys and social networking-style functions allowing users to share and compare their genetic data with others site users and participate in online forums and discussion groups related to genetics. The company utilizes regular security audits and multiple encryption layers to protect customers' privacy. The compiled genetic information is also used for scientific research to advance the field of genetics; research studies are conducted both by 23AndMe and by various partner organizations and companies. 23andMe's investors include Genentech, Inc.; Google, Inc.; and New Enterprise Associates, a venture capital firm. In 2008, the firm expanded its service to customers in Canada and approximately 49 European countries.

## FINANCIALS:  Sales and profits are in thousands of dollars—add 000 to get the full amount. 2008 Note: Financial information for 2008 was not available for all companies at press time.

| | | |
|--|--|--|
| 2008 Sales: $ | 2008 Profits: $ | **U.S. Stock Ticker: Private** |
| 2007 Sales: $ | 2007 Profits: $ | **Int'l Ticker:**    Int'l Exchange: |
| 2006 Sales: $ | 2006 Profits: $ | Employees: |
| 2005 Sales: $ | 2005 Profits: $ | Fiscal Year Ends: |
| 2004 Sales: $ | 2004 Profits: $ | Parent Company: |

## SALARIES/BENEFITS:

| Pension Plan: | ESOP Stock Plan: | Profit Sharing: | Top Exec. Salary: $ | Bonus: $ |
|---------------|------------------|-----------------|---------------------|----------|
| Savings Plan: | Stock Purch. Plan: | | Second Exec. Salary: $ | Bonus: $ |

## OTHER THOUGHTS:

**Apparent Women Officers or Directors**: 4
**Hot Spot for Advancement for Women/Minorities**: Y

## LOCATIONS: ("Y" = Yes)

| West: | Southwest: | Midwest: | Southeast: | Northeast: | International: |
|-------|-----------|----------|-----------|-----------|---------------|
| Y | | | | | |

# 3SBIO INC

## www.3sbio.com

**Industry Group Code: 325412  Ranks within this company's industry group: Sales: 95  Profits: 54**

| Drugs: | | Other: | Clinical: | Computers: | Services: |
|---|---|---|---|---|---|
| Discovery: | Y | AgriBio: | Trials/Services: | Hardware: | Specialty Services: |
| Licensing: | | Genetic Data: | Labs: | Software: | Consulting: |
| Manufacturing: | Y | Tissue Replacement: | Equipment/Supplies: | Arrays: | Blood Collection: |
| Genetics: | | | Research & Development Services: | Database Management: | Drug Delivery: |
| | | | Diagnostics: | | Drug Distribution: |

## TYPES OF BUSINESS:

Biopharmaceutical Manufacturing & Design
Anemia Treatments
Cancer Treatments
Exporting Biopharmaceuticals

## BRANDS/DIVISIONS/AFFILIATES:

Shenyang Sunshine Pharmaceutical Company Limited
EPIAO
TPIAO
INTEFEN
INLEUSIN
Tietai Iron Sucrose Supplement
NuPIAO
NuLeusin

## CONTACTS: Note: Officers with more than one job title may be intentionally listed here more than once.

Jing Lou, CEO
David Chen, COO/Exec. VP
Bo Tan, CFO
Bin Huang, VP-Human Resources
Yingfei Wei, Chief Scientific Officer
Dongmei Su, CTO
Ke Li, Corp. Sec.
Yingfei Wei, VP-Bus. Dev.
Dan Lou, Chmn.

| Phone: 86-24-2581-1820 | Fax: |
|---|---|
| Toll-Free: | |
| Address: No. 3 A1, Rd. 10, Shenyang Development Zone, Shenyang, 110027 China | |

## GROWTH PLANS/SPECIAL FEATURES:

3SBio, Inc. is a leading biotechnology company that researches, develops, manufactures and markets biopharmaceutical products, primarily in China. Substantially operating through wholly-owned subsidiary Shenyang Sunshine Pharmaceutical Company Limited, 3SBio currently markets four drugs. EPIAO, an injectable drug, stimulates the production of red blood cells for anemic patients, thereby reducing their need for transfusions. TPIAO is a protein-based treatment for chemotherapy-induced thrombocytopenia, a deficiency of platelets. INTEFEN is a treatment for lymphatic or hematopoietic system carcinomas, such as lymphoma and leukemia, as well as for viral infections such as hepatitis C. INLEUSIN has several applications, including the treatment of renal cell carcinoma, a common type of kidney cancer; metastatic melanoma, a skin cancer; and thoracic fluid build-up caused by cancer or tuberculosis. Lastly, Tietai Iron Sucrose Supplement is an injectable anemia treatment indicated for end-stage renal disease requiring iron replacement therapy. The company's development products, mostly related to its marketed products, include the following. NuPIAO is a second-generation EPIAO product for treating anemia following renal failure or chemotherapy. NuLeusin is a next generation INLEUSIN product, designed for similar uses. The company is adapting TPIAO for the treatment of idiopathic thrombocytopenic purpura (ITP), an immune disorder that causes the body to destroy its own platelets. A human papilloma virus (HPV) vaccine is also being developed. Finally, an anti-TNF (Tumor necrosis factor alpha) antibody is underway; TNF is responsible for regulating the inflammatory process and plays an underlying role in rheumatoid arthritis, psoriasis and other inflammatory disorders. During 2008, EPIAO sales generated approximately 63.5% of the firm's revenue; TPIAO, 27.8%; INTEFEN, 2.1%; INLEUSIN, 0.3%; and Tietai, 2.9%, while exports accounted for 3.4%. 3SBio exports its products primarily to Egypt, Pakistan, Thailand, Trinidad and Tobago, Brazil, Korea and Mexico.

## FINANCIALS: Sales and profits are in thousands of dollars—add 000 to get the full amount. 2008 Note: Financial information for 2008 was not available for all companies at press time.

| | | |
|---|---|---|
| 2008 Sales: $35,653 | 2008 Profits: $5,795 | U.S. Stock Ticker: SSRX |
| 2007 Sales: $24,700 | 2007 Profits: $11,174 | Int'l Ticker:    Int'l Exchange: |
| 2006 Sales: $16,373 | 2006 Profits: $3,907 | Employees:   483 |
| 2005 Sales: $ | 2005 Profits: $ | Fiscal Year Ends: 12/31 |
| 2004 Sales: $ | 2004 Profits: $ | Parent Company: |

## SALARIES/BENEFITS:

| Pension Plan: | ESOP Stock Plan: | Profit Sharing: | Top Exec. Salary: $ | Bonus: $ |
|---|---|---|---|---|
| Savings Plan: | Stock Purch. Plan: | | Second Exec. Salary: $ | Bonus: $ |

## OTHER THOUGHTS:

**Apparent Women Officers or Directors:** 2
**Hot Spot for Advancement for Women/Minorities:** Y

## LOCATIONS: ("Y" = Yes)

| West: | Southwest: | Midwest: | Southeast: | Northeast: | International: Y |
|---|---|---|---|---|---|

Note: Financial information, benefits and other data can change quickly and may vary from those stated here.

# 4SC AG

**www.4sc.de**

**Industry Group Code:** 325412  **Ranks within this company's industry group:** Sales: 130  Profits: 97

| Drugs: | | Other: | Clinical: | Computers: | Services: |
|---|---|---|---|---|---|
| Discovery: | Y | AgriBio: | Trials/Services: | Hardware: | Specialty Services: |
| Licensing: | Y | Genetic Data: | Labs: | Software: | Consulting: |
| Manufacturing: | | Tissue Replacement: | Equipment/Supplies: | Arrays: | Blood Collection: |
| Genetics: | | | Research & Development Services: | Database Management: | Drug Delivery: |
| | | | Diagnostics: | | Drug Distribution: |

## TYPES OF BUSINESS:

Drug Discovery & Pre-Clinical Development
Drugs-Oncology
Drugs-Inflammation
Drugs-Autoimmune Diseases
High Throughput Screening

## BRANDS/DIVISIONS/AFFILIATES:

4SCan Technology
4SC-101

## CONTACTS: Note: Officers with more than one job title may be intentionally listed here more than once.

Ulrich Dauer, CEO
Enno Spillner, CFO
Marion Fischbach, Mgr.-Human Resources
Daniel Vitt, Chief Science Officer
Bernd Hentsch, Chief Dev. Officer
Bettina von Klitzing, Mgr.-Public Rel.
Bettina von Klitzing, Mgr.-Investor Rel.
Stefan Strobl, Dir.-Discovery Research
Charlotte Herrlinger, VP-Medicine
Jorg Neermann, Chmn.
Tanja Wieber, Mgr.-Purchasing

| Phone: 49-89-700-763-0 | Fax: 49-89-700-763-29 |
|---|---|
| Toll-Free: | |
| Address: Am Klopferspitz 19a, Planegg-Martinsried, 82152 Germany | |

## GROWTH PLANS/SPECIAL FEATURES:

4SC AG is a German-based drug discovery and development company that focuses on possible treatments for cancer and inflammatory diseases. The company develops its compounds to the clinical phase, after which they will be licensed to the biopharmaceutical industry for further testing, development and commercialization. 4SC's technology platform uses a virtual High Throughput Screening technology known as 4SCan, which allows the high volume analysis of molecules based on protein structures, biological activity or homology modeling on a computer, rather than in a laboratory. Using this technology, the company has compiled a virtual library of small molecule drug candidates. It has had collaborative efforts with AiCuris, establishing a pipeline for anti-infective drug candidates; Axxima, researching human cytomegalovirus treatments using small molecule kinase inhibitors; Beohringer Ingelheim, researching drugs based on ligands; Esteve, identifying drug candidates for central nervous system-related diseases; Mutabilis S.A., researching infectious disease related drugs; ProQinase, researching anticancer drugs based on protein kinase inhibitors; Switch Biotech, developing active compounds to treat psoriasis; Wilex, researching second-generation urokinase inhibitors; and Schwarz Pharma, identifying pre-development drug candidates. It also has collaborations with Recordati, Sanofi-Aventis, Sanwa Kagaku Kenkyusho and Schering, working on undisclosed targets. Currently, the firm has six compounds in development: two for oncology, two for rheumatoid arthritis, one for multiple sclerosis and one for viral infection. In 2008, the company acquired eight oncology projects from Nycomed. In March 2009, 4SC began Phase IIa clinical trials of 4sc-101 for the treatment of Crohn's disease.

## FINANCIALS: Sales and profits are in thousands of dollars—add 000 to get the full amount. 2008 Note: Financial information for 2008 was not available for all companies at press time.

| | | |
|---|---|---|
| 2008 Sales: $4,220 | 2008 Profits: $-16,830 | U.S. Stock Ticker: Private |
| 2007 Sales: $2,100 | 2007 Profits: $-12,700 | Int'l Ticker: Int'l Exchange: |
| 2006 Sales: $5,700 | 2006 Profits: $-8,600 | Employees: 80 |
| 2005 Sales: $3,200 | 2005 Profits: $-9,800 | Fiscal Year Ends: |
| 2004 Sales: $ | 2004 Profits: $ | Parent Company: |

## SALARIES/BENEFITS:

| Pension Plan: | ESOP Stock Plan: | Profit Sharing: | Top Exec. Salary: $184,952 | Bonus: $35,886 |
|---|---|---|---|---|
| Savings Plan: | Stock Purch. Plan: | | Second Exec. Salary: $182,192 | Bonus: $46,928 |

## OTHER THOUGHTS:

Apparent Women Officers or Directors: 4
Hot Spot for Advancement for Women/Minorities: Y

## LOCATIONS: ("Y" = Yes)

| West: | Southwest: | Midwest: | Southeast: | Northeast: | International: Y |
|---|---|---|---|---|---|

Note: Financial information, benefits and other data can change quickly and may vary from those stated here.

# AASTROM BIOSCIENCES INC
www.aastrom.com

**Industry Group Code: 33911 Ranks within this company's industry group: Sales: 17 Profits: 16**

| Drugs: | Other: | Clinical: | | Computers: | Services: |
|---|---|---|---|---|---|
| Discovery: | AgriBio: | Trials/Services: | Y | Hardware: | Specialty Services: |
| Licensing: | Genetic Data: | Labs: | | Software: | Consulting: |
| Manufacturing: | Tissue Replacement: Y | Equipment/Supplies: | | Arrays: | Blood Collection: |
| Genetics: | | Research & Development Services: | | Database Management: | Drug Delivery: |
| | | Diagnostics: | | | Drug Distribution: |

## TYPES OF BUSINESS:
Cell Products Development
Regenerative medicine products
Clinical & Pre-clinical Development

## BRANDS/DIVISIONS/AFFILIATES:
Tissue Repair Cells
Cardiac Repair Cells
Single-Pass Perfusion (SPP)
RESTORE-CLI
IMPACT-DCM

## CONTACTS: Note: Officers with more than one job title may be intentionally listed here more than once.
George W. Dunbar, CEO
George W. Dunbar, Pres.
George W. Dunbar, CFO
Tod Barton, Sr. Dir.-Eng. & Dev.
Sherrill M. MacKay, Sr. Dir.-Mfg.
Julie A. Caudill, Corp. Sec./Controller
Ronnda L. Bartel, Sr. Dir.-Tech. Oper.
Sheldon A. Schaffer, VP-Corp. Dev. & Intellectual Property
Kris M. Maly, Sr. Dir.-Comm. & Admin.
Frances J. Kivel, Sr. Dir.-Clinical Trials
Elmar R. Burchardt, VP-Medical Affairs
Peter C. Alain, Sr. Dir.-Quality Assurance
Nelson M. Sims, Chmn.

**Phone:** 734-930-5555 **Fax:** 734-665-0485
**Toll-Free:**
**Address:** 24 Frank Lloyd Wright Dr., Ann Arbor, MI 48105 US

## GROWTH PLANS/SPECIAL FEATURES:
Aastrom Biosciences, Inc. is a development stage company focused on the development of autologous cell products for use in regenerative medicine. The company's pre-clinical and clinical products development programs utilize patient-derived bone marrow stem and progenitor cell populations, which are being investigated for their ability to aid in the regeneration of vascular, bone, cardiac and neural tissues. The firm's primary business is to develop Tissue Repair Cell (TRC) based products using Single-Pass Perfusion (SPP) to replicate early-stage stem and progenitor cells. The TRC platform technology is based on its cell products (a unique cell mixture containing large numbers of stromal, stem and progenitor cells, produced outside the body from a small amount of bone marrow taken from a patient) and the means to produce these products in an automated process. TRC-based products have been used in over 290 patients. The pre-clinical data for the TRCs showed a substantial increase in the stem and progenitor cells that can develop into tissues such as hematopoietic (i.e., blood forming) or mesenchymal (i.e., developing into tissues characteristic of certain internal organs), as well as stromal progenitor cells that produce various growth factors. The company demonstrated in the laboratory that TRCs can progress into bone cell and blood vessel cell lineages. Based on these pre-clinical observations, the TRCs are currently in active clinical trials for bone regeneration and vascular regeneration applications. The firm developed a patented manufacturing system to produce human cells for clinical use. Aastrom currently owns 26 U.S. patents and 41 foreign patents. The company is currently focusing on its IMPACT-DCM (dilated cardiomyopathy) and RESTORE-CLI (critical limb ischemia) trials. In May 2009, the company temporarily suspended its Phase 2 IMPACT-DCM clinical trial due to the death of an involved patient.

## FINANCIALS: Sales and profits are in thousands of dollars—add 000 to get the full amount. 2008 Note: Financial information for 2008 was not available for all companies at press time.

| | | |
|---|---|---|
| 2008 Sales: $ 522 | 2008 Profits: $-20,133 | **U.S. Stock Ticker: ASTM** |
| 2007 Sales: $ 685 | 2007 Profits: $-17,594 | **Int'l Ticker:** Int'l Exchange: |
| 2006 Sales: $ 863 | 2006 Profits: $-16,475 | Employees: 67 |
| 2005 Sales: $ 909 | 2005 Profits: $-11,811 | Fiscal Year Ends: 6/30 |
| 2004 Sales: $1,300 | 2004 Profits: $-10,500 | Parent Company: |

## SALARIES/BENEFITS:
| | | | | |
|---|---|---|---|---|
| Pension Plan: | ESOP Stock Plan: | Profit Sharing: | Top Exec. Salary: $375,000 | Bonus: $42,188 |
| Savings Plan: | Stock Purch. Plan: | | Second Exec. Salary: $239,583 | Bonus: $19,721 |

## OTHER THOUGHTS:
**Apparent Women Officers or Directors:** 5
**Hot Spot for Advancement for Women/Minorities:** Y

## LOCATIONS: ("Y" = Yes)
| West: | Southwest: | Midwest: | Southeast: | Northeast: | International: |
|---|---|---|---|---|---|
| | | Y | | | |

Note: Financial information, benefits and other data can change quickly and may vary from those stated here.

# ABAXIS INC

**www.abaxis.com**

Industry Group Code: 33911  **Ranks within this company's industry group:**  Sales: 10    Profits: 9

| Drugs: | Other: | Clinical: | | Computers: | Services: |
|---|---|---|---|---|---|
| Discovery: | AgriBio: | Trials/Services: | | Hardware: | Specialty Services: |
| Licensing: | Genetic Data: | Labs: | | Software: | Consulting: |
| Manufacturing: | Tissue Replacement: | Equipment/Supplies: | Y | Arrays: | Blood Collection: |
| Genetics: | | Research & Development Services: | Y | Database Management: | Drug Delivery: |
| | | Diagnostics: | Y | | Drug Distribution: |

## TYPES OF BUSINESS:

Point-of-Care Blood Analyzer Systems Equipment
Veterinary Blood Analyzer Systems
Reagents & Supplies

## BRANDS/DIVISIONS/AFFILIATES:

VetScan VS2
VetScan Classic
VetScan Vspro
Piccolo Xpress
VetScan HM2
Orbos Discrete Lyophilization Process
VetScan HM5

## CONTACTS: *Note: Officers with more than one job title may be intentionally listed here more than once.*

Clinton H. Severson, CEO
Donald Wood, COO
Clinton H.Severson, Pres.
Al Santa Ines, CFO
Kenneth P. Aron, CTO
Al Santa Ines, VP-Finance
Martin Mulroy, VP-Mktg. & Sales, Veterinary Market
Rick Betts, Dir.-Mktg. (Veterinary)
Valerie Goodwin, Dir.-Mktg. (Medical)
Achim Henkel, Dir.-European Sales & Mktg.
Clinton H. Severson, Chmn.
Vladimir E. Ostoich, VP-Gov't Affairs & Mktg., Pacific Rim

| Phone: 510-675-6500 | Fax: 510-441-6150 |
|---|---|
| Toll-Free: | |
| Address: 3240 Whipple Rd., Union City, CA 94587 US | |

## GROWTH PLANS/SPECIAL FEATURES:

Abaxis, Inc. develops, manufactures and markets portable blood analysis systems for use in veterinary or human patient-care settings to provide clinicians with rapid blood constituent measurements.  The company provides over 90% of the chemical tests in animal and medical diagnoses. The medical market accounted for 23% of total revenue in 2008, and the veterinary market accounted for 71%.  The analysis systems are marketed as VetScan VS2, VetScan HM2 (formerly VetScan HMII),VetScan HM5, VetScan VSpro and VetScan Classic in the veterinary market and as the Piccolo Xpress in the human medical market.  Abaxis' blood analysis systems consist of a compact analyzer and a series of single-use plastic discs, called reagent discs, containing all the chemicals required to perform a panel of up to 24 diagnostic tests.  In addition to blood analysis systems, Abaxis sells the VetScan HM5 hematology analyzer that provides a 22-parameter blood count analysis, including a five-part white blood cell differential.  To produce the dry reagents used in the reagent disks, Abaxis uses its Orbos Discrete Lyophilization Process (Orbos process), a process which freeze dries reagents in small quantities, enabling efficient manufacturing of reagents in a convenient and stable format.  Abaxis licenses its Orbos process to bioMerieux, Cepheid and GE Healthcare.  In January 2009, Abaxis announced the release of two new products: the VetScan VSpro Coagulation Analyzer and the VetScan Canine Heartworm Rapid Test.

Abaxis offers vision, life, health and dental insurance; 401(K) plans and flexible spending accounts.

## FINANCIALS:  Sales and profits are in thousands of dollars—add 000 to get the full amount. 2008 Note: Financial information for 2008 was not available for all companies at press time.

| | | |
|---|---|---|
| 2008 Sales: $100,551 | 2008 Profits: $12,503 | **U.S. Stock Ticker:** ABAX |
| 2007 Sales: $86,221 | 2007 Profits: $10,073 | **Int'l Ticker:**  Int'l Exchange: |
| 2006 Sales: $68,928 | 2006 Profits: $7,475 | Employees:  339 |
| 2005 Sales: $52,758 | 2005 Profits: $4,851 | Fiscal Year Ends: 3/31 |
| 2004 Sales: $46,874 | 2004 Profits: $24,033 | Parent Company: |

## SALARIES/BENEFITS:

| Pension Plan: | ESOP Stock Plan: | Profit Sharing: | Top Exec. Salary: $336,500 | Bonus: $456,000 |
|---|---|---|---|---|
| Savings Plan: Y | Stock Purch. Plan: | | Second Exec. Salary: $202,077 | Bonus: $261,250 |

## OTHER THOUGHTS:

**Apparent Women Officers or Directors**: 1
**Hot Spot for Advancement for Women/Minorities**:

## LOCATIONS: ("Y" = Yes)

| West: | Southwest: | Midwest: | Southeast: | Northeast: | International: |
|---|---|---|---|---|---|
| Y | | | | | Y |

# ABBOTT LABORATORIES

**www.abbott.com**

Industry Group Code: 325412  Ranks within this company's industry group: Sales: 8   Profits: 10

| Drugs: | | Other: | Clinical: | | Computers: | | Services: | |
|---|---|---|---|---|---|---|---|---|
| Discovery: | Y | AgriBio: | Trials/Services: | | Hardware: | | Specialty Services: | |
| Licensing: | | Genetic Data: | Labs: | | Software: | | Consulting: | |
| Manufacturing: | Y | Tissue Replacement: | Equipment/Supplies: | | Arrays: | Y | Blood Collection: | |
| Genetics: | | | Research & Development Services: | | Database Management: | | Drug Delivery: | |
| | | | Diagnostics: | | | | Drug Distribution: | |

## TYPES OF BUSINESS:

Pharmaceuticals Manufacturing
Nutritional Products
Diagnostics
Consumer Health Products
Medical & Surgical Devices
Pharmaceutical Products
Animal Health

## BRANDS/DIVISIONS/AFFILIATES:

Abbott Medical Optics Inc
Prevacid
Lupix
Trilipix
Humira
Simcor
Glucerna
Ensure

## CONTACTS: *Note: Officers with more than one job title may be intentionally listed here more than once.*

Miles D. White, CEO
Thomas C. Freyman, CFO
Stephen R. Fussell, Sr. VP-Human Resources
John M. Leonard, Sr. VP-R&D
John C. Landgraf, Sr. VP-Global Pharmaceutical Mgmt. & Supply
Laura J. Schumacher, General Counsel/Exec. VP/Corp. Sec.
Richard W. Ashley, Exec. VP-Corp. Dev.
Melissa Brotz, VP-External Comm.
Thomas C. Freyman, Exec. VP-Finance
James L. Tyree, Exec. VP-Pharmaceutical Products Group
Holger Liepmann, Exec. VP-Global Nutrition
John M. Capek, Exec. VP-Medical Devices
Michael J. Warmuth, Sr. VP-Diagnostics
Miles D. White, Chmn.
Olivier Bohuon, Sr. VP-Int'l Oper.

| Phone: 847-937-6100 | Fax: 847-937-9555 |
|---|---|
| Toll-Free: | |
| Address: 100 Abbott Park Rd., Abbott Park, IL 60064-3500 US | |

## GROWTH PLANS/SPECIAL FEATURES:

Abbott Laboratories develops, manufactures and sells health care products and technologies ranging from pharmaceuticals to medical devices. The firm markets its products in more than 130 countries. The pharmaceutical segment deals with adult and pediatric conditions such as rheumatoid arthritis, HIV, epilepsy and manic depression. The diagnostic instruments and test segment deals with a range of medical tests to diagnose infectious diseases, cancer, diabetes and genetic conditions. Products include Humira for arthritis, Prevacid (Ogastro), a proton pump inhibitor for the short-term treatment of gastroesophageal reflux disease and Lupix for prostate cancer. The nutritional products segment offers consumer products such as Similac, Ensure, Glucerna and AdvantEdge, as well as medical nutritional products and feeding devices. The diagnostic products segment includes diagnostic systems and tests such as immunoassay, chemistry and hematology systems, which are manufactured, marketed and sold to blood banks, hospitals, commercial laboratories, physicians' offices and plasma protein therapeutic companies. The vascular products segment consists of coronary, endovascular and vessel closure devices, used in the treatment of vascular disease. These products are generally marketed and sold directly to hospitals from Abbot-owned distribution centers and public warehouses. The firm also produces a line of animal products such as anesthetics, wound care products and intravenous fluid therapy for the veterinary market. The company operates internationally in Europe, Asia, Africa, Latin and South America and the Middle East. In December 2008, Abbott received FDA approval for Trilipix, a cholesterol management medication. In February 2009, the company acquired Advanced Medical Optics which has been renamed Abbott Medical Optics Inc.

Employees are offered medical, dental and vision insurance; flexible spending accounts; child care solutions; an employee assistance program; legal referral services; a pension plan; a 401(k) plan; profit sharing; tuition assistance; travel accident insurance; long-term care insurance; and life insurance.

## FINANCIALS: Sales and profits are in thousands of dollars—add 000 to get the full amount. 2008 Note: Financial information for 2008 was not available for all companies at press time.

| | | |
|---|---|---|
| 2008 Sales: $29,527,600 | 2008 Profits: $4,880,700 | U.S. Stock Ticker: ABT |
| 2007 Sales: $25,914,200 | 2007 Profits: $3,606,300 | Int'l Ticker:    Int'l Exchange: |
| 2006 Sales: $22,476,322 | 2006 Profits: $1,716,755 | Employees: 69,000 |
| 2005 Sales: $22,337,808 | 2005 Profits: $3,372,065 | Fiscal Year Ends: 12/31 |
| 2004 Sales: $19,680,016 | 2004 Profits: $3,235,851 | Parent Company: |

## SALARIES/BENEFITS:

| Pension Plan: Y | ESOP Stock Plan: | Profit Sharing: Y | Top Exec. Salary: $1,795,471 | Bonus: $4,200,000 |
|---|---|---|---|---|
| Savings Plan: Y | Stock Purch. Plan: | | Second Exec. Salary: $886,363 | Bonus: $1,232,500 |

## OTHER THOUGHTS:

**Apparent Women Officers or Directors**: 5
**Hot Spot for Advancement for Women/Minorities**: Y

## LOCATIONS: ("Y" = Yes)

| West: | Southwest: | Midwest: | Southeast: | Northeast: | International: |
|---|---|---|---|---|---|
| Y | Y | Y | | Y | Y |

*Note: Financial information, benefits and other data can change quickly and may vary from those stated here.*

# ACAMBIS PLC

**www.acambis.com**

**Industry Group Code:** 325412 **Ranks within this company's industry group:** Sales: Profits:

| Drugs: | | Other: | | Clinical: | Computers: | Services: | |
|---|---|---|---|---|---|---|---|
| Discovery: | Y | AgriBio: | | Trials/Services: | Hardware: | Specialty Services: | |
| Licensing: | | Genetic Data: | Y | Labs: | Software: | Consulting: | |
| Manufacturing: | Y | Tissue Replacement: | | Equipment/Supplies: | Arrays: | Blood Collection: | |
| Genetics: | | | | Research & Development Services: | Database Management: | Drug Delivery: | |
| | | | | Diagnostics: | | Drug Distribution: | Y |

## TYPES OF BUSINESS:

Vaccine Development & Manufacturing
Vaccine Sales & Distribution

## BRANDS/DIVISIONS/AFFILIATES:

Sanofi-Avnetis SA
Sanofi Pasteur
C-VIG
ACAM-FLU-A
Baxter Healthcare Corporation
ChimeriVax-JE
ChimeriVax-West Nile
ChimeriVax-Dengue

## CONTACTS: *Note: Officers with more than one job title may be intentionally listed here more than once.*

Ian Garland, CEO
Elizabeth Jones, CFO
Daniel Fredian, VP-Human Resources
Michael Watson, Exec. VP-R&D
Donna Radzik, Sr. VP-Oper.
Paul Giannasca, VP-Dev.
Lyndsay Wright, VP-Corp. Comm.
Lyndsay Wright, VP-Investor Rel.
Davin Wonnacott, Sr. VP-Regulatory Affairs & Quality Systems
Harry Kleanthous, VP-Research
Mark Felton, VP-Clinical Oper. & Medical Affairs
Peter Fellner, Chmn.

| **Phone:** 44-1223-275-300 | **Fax:** 44-1223-416-300 |
|---|---|
| **Toll-Free:** | |
| **Address:** Peterhouse Technology Park, 100 Fulbourn Rd., Cambridge, CB1 9PT UK | |

## GROWTH PLANS/SPECIAL FEATURES:

Acambis plc is a major U.K.-based developer and producer of vaccines to prevent and treat infectious diseases. The company is internationally recognized as a leading producer of smallpox vaccines and a top supplier to governments wishing to build vaccine stockpiles, having supplied 200 million doses of the second-generation ACAM2000. Acambis also produces C-VIG, an antibody product developed with Cangene and used to prep recipients or to treat severe reactions to the smallpox vaccine. With Baxter Healthcare Corporation, the firm was developing a third-generation smallpox vaccine, MVA 3000, which is a weakened form of the current generation of smallpox vaccines, but Acambis is scaling down its MVA-related activities. Acambis is developing travel vaccines for Japanese encephalitis (ChimeriVax-JE, which is undergoing Phase III pediatric trials in India), and dengue fever (ChimeriVax-Dengue, a vaccine currently in Phase II clinical trials designed to protect against all dengue virus serotypes). Moreover, Acambis has one of the only vaccines in development against the antibiotic-resistant, hospital-acquired diarrhea bacteria Clostridium difficile. Acambis has also been conducting Phase I trials of ACAM-FLU-A, a universal vaccine designed to target all strains of the influenza virus, both pandemic and seasonal. In February 2008, the company began pre-clinical testing of a herpes simplex virus (HSV) vaccine. In April 2008, Acambis finalized a multi-year, warm-based manufacturing contract, with the U.S. government, for ACAM2000. The contract is expected to generate $425 million over 10 years. The company has partnered with Sanofi Pasteur, the U.S.-based vaccines division of French pharmaceutical company Sanofi-Aventis, to develop a West Nile vaccine, with Sanofi Pasteur initially financing the research and Acambis conducting the development of the vaccine. The vaccine, ChimeriVax-West Nile, is expected to generate $80 million in milestone and royalty payments from the U.S. market. In September 2008, Sanofi Pasteur acquired Acambis for approximately $450 million, subsequently taking the company private.

## FINANCIALS: Sales and profits are in thousands of dollars—add 000 to get the full amount. 2008 Note: Financial information for 2008 was not available for all companies at press time.

| | | |
|---|---|---|
| 2008 Sales: $ | 2008 Profits: $ | **U.S. Stock Ticker:** Private |
| 2007 Sales: $19,000 | 2007 Profits: $-62,600 | **Int'l Ticker:** Int'l Exchange: |
| 2006 Sales: $61,800 | 2006 Profits: $-33,000 | Employees: 206 |
| 2005 Sales: $70,400 | 2005 Profits: $-44,700 | Fiscal Year Ends: 12/31 |
| 2004 Sales: $163,900 | 2004 Profits: $38,000 | Parent Company: |

## SALARIES/BENEFITS:

| Pension Plan: | ESOP Stock Plan: | Profit Sharing: | Top Exec. Salary: $ | Bonus: $ |
|---|---|---|---|---|
| Savings Plan: | Stock Purch. Plan: | | Second Exec. Salary: $ | Bonus: $ |

## OTHER THOUGHTS:

**Apparent Women Officers or Directors:** 4
**Hot Spot for Advancement for Women/Minorities:** Y

## LOCATIONS: ("Y" = Yes)

| West: | Southwest: | Midwest: | Southeast: | Northeast: | International: |
|---|---|---|---|---|---|
| | | | | Y | Y |

# ACCELRYS INC

**www.accelrys.com**

**Industry Group Code: 511210D  Ranks within this company's industry group: Sales: 3  Profits: 3**

| Drugs: | Other: | Clinical: | Computers: | | Services: | |
|---|---|---|---|---|---|---|
| Discovery: | AgriBio: | Trials/Services: | Hardware: | | Specialty Services: | |
| Licensing: | Genetic Data: | Labs: | Software: | Y | Consulting: | |
| Manufacturing: | Tissue Replacement: | Equipment/Supplies: | Arrays: | | Blood Collection: | |
| Genetics: | | Research & Development Services: | Database Management: | | Drug Delivery: | |
| | | Diagnostics: | | | Drug Distribution: | |

## TYPES OF BUSINESS:

Software - Simulation & Informatics
Computational Nanotechnology Tools
Informatics Software
Modeling & Simulation Software

## BRANDS/DIVISIONS/AFFILIATES:

SciTegic Pipeline Pilot
Pipeline Pilot Enterprise Server 7.5

## CONTACTS: *Note: Officers with more than one job title may be intentionally listed here more than once.*

Todd Johnson, CEO
Todd Johnson, Pres.
Rick Russo, CFO/Sr. VP
Rohit Shyam, VP-Mktg.
Judith Ohrn Hicks, VP-Human Resources
Frank K. Brown, Chief Science Officer/Sr. VP
David Mersten, General Counsel/Corp. Sec./Sr. VP
Philomena Walsh, Dir.-Corp. Comm.
Matt Hahn, VP-R&D
Ilene Vogt, VP-Worldwide Sales & Svcs.
Kenneth L. Coleman, Chmn.

| Phone: 858-799-5000 | Fax: 858-799-5100 |
|---|---|
| Toll-Free: | |
| Address: 10188 Telesis Ct., Ste. 100, San Diego, CA 92121-3752 US | |

## GROWTH PLANS/SPECIAL FEATURES:

Accelrys, Inc. develops and commercializes scientific business intelligence software for biologists, chemists, materials scientists and IT professionals for product design and drug discovery and development. Its customers include leading commercial, government and academic organizations. Accelrys' SciTegic Pipeline Pilot platform allows users to aggregate, integrate and mine vast quantities of both structured and unstructured scientific data such as chemical structures, biological sequences and complex digital images. It also filters, normalizes and performs statistical analysis on this scientific data and provides interactive visual reports to both scientists and scientific managers. Accelrys offers modeling and simulation software comprised of over 100 application modules that employ fundamental scientific principles, advanced computer visualization, molecular modeling techniques and computational chemistry enabling scientists to perform computations of chemical, biological and materials properties; to simulate, visualize and analyze chemical and biological systems; and to communicate the results to other scientists. The firm also offers open access to many of its software development environments, within which customers and third-party licensees can develop, integrate and distribute their own software applications for computational chemistry, biology and materials research. Accelrys provides a bioinformatics suite of programs enabling biologists to search, edit, compare, map and align sequence data; and cheminformatics software enabling users to search, retrieve and use chemical structures, registration information and biological, chemical and experimental data. In February 2009, the company released Pipeline Pilot Enterprise Server 7.5, an integration and analysis platform that reports scientific data.

Employees are offered medical, dental and vision insurance; flexible health and dependent care spending accounts; an employee assistance program; life insurance; short-and long-term disability coverage; group travel and accident insurance; legal services plan; a retirement savings plan; educational savings programs; employee stock equity programs; tuition reimbursement; a computer loan program; and subsidized gym memberships.

## FINANCIALS: Sales and profits are in thousands of dollars—add 000 to get the full amount. 2008 Note: Financial information for 2008 was not available for all companies at press time.

| | | |
|---|---|---|
| 2008 Sales: $79,739 | 2008 Profits: $1,321 | **U.S. Stock Ticker: ACCL** |
| 2007 Sales: $80,955 | 2007 Profits: $-1,525 | **Int'l Ticker:**   Int'l Exchange: |
| 2006 Sales: $82,001 | 2006 Profits: $-7,739 | Employees:   364 |
| 2005 Sales: $79,030 | 2005 Profits: $-16,578 | Fiscal Year Ends: 3/31 |
| 2004 Sales: $86,209 | 2004 Profits: $-4,596 | Parent Company: |

## SALARIES/BENEFITS:

| Pension Plan: | ESOP Stock Plan: | Profit Sharing: | Top Exec. Salary: $400,000 | Bonus: $400,000 |
|---|---|---|---|---|
| Savings Plan: Y | Stock Purch. Plan: Y | | Second Exec. Salary: $255,000 | Bonus: $127,500 |

## OTHER THOUGHTS:

**Apparent Women Officers or Directors:** 2
**Hot Spot for Advancement for Women/Minorities:** Y

## LOCATIONS: ("Y" = Yes)

| West: | Southwest: | Midwest: | Southeast: | Northeast: | International: |
|---|---|---|---|---|---|
| Y | | | | Y | Y |

# ACCESS PHARMACEUTICALS INC                   www.accesspharma.com

**Industry Group Code: 325412  Ranks within this company's industry group:  Sales: 152    Profits: 103**

| Drugs: | | Other: | | Clinical: | Computers: | | Services: | |
|---|---|---|---|---|---|---|---|---|
| Discovery: | Y | AgriBio: | | Trials/Services: | Hardware: | | Specialty Services: | |
| Licensing: | Y | Genetic Data: | | Labs: | Software: | | Consulting: | |
| Manufacturing: | Y | Tissue Replacement: | | Equipment/Supplies: | Arrays: | | Blood Collection: | |
| Genetics: | | | | Research & Development Services: | Database Management: | | Drug Delivery: | Y |
| | | | | Diagnostics: | | | Drug Distribution: | |

## TYPES OF BUSINESS:

Pharmaceutical Development
Drug Delivery Systems
Polymer Technology
Oncology Products

## BRANDS/DIVISIONS/AFFILIATES:

MuGard
ProLindac
ProDrax
Angiolix
Alchemix
EcoNail
Somanta Pharmaceuticals, Inc.
MacroChem Corp.

## CONTACTS: *Note: Officers with more than one job title may be intentionally listed here more than once.*

Jeffrey B. Davis, CEO
Stephen B. Thompson, CFO/VP
David P. Nowotnik, Sr. VP-R&D
David P. Luci, General Counsel/VP
Phillip Wise, VP-Bus. Dev. & Strategy
Esteban Cvitkovic, Vice Chmn.
Esteban Cvitkovic, Sr. Dir.-Clinical Oncology R&D
Agamemnon Epenetos, Chief Science Officer, Europe
Steven H. Rouhandeh, Chmn.

| **Phone:** 214-905-5100 | **Fax:** 214-905-5101 |
|---|---|
| **Toll-Free:** | |
| **Address:** 2600 Stemmons Freeway, Ste. 176, Dallas, TX 75207 US | |

## GROWTH PLANS/SPECIAL FEATURES:

Access Pharmaceuticals, Inc. is a biopharmaceutical company developing products for use in the treatment of cancer, the supportive care of cancer and other disease states. Beyond its own research, Access also incorporates the work of its recent acquisitions, Somanta Pharmaceuticals and MacroChem Corp. The company has one technology approved by the FDA, MuGard, and three drug delivery technology platforms: synthetic polymer targeted delivery, which is designed to exploit enhanced permeability and retention at tumor sites to selectively accumulate drug and control drug release; Cobalamin-medicated oral delivery, which utilizes vitamin B12 to increase absorption of orally consumed medicines; and Cobalamin-medicated targeted delivery. Other preclinical drugs in development include Angiolix, a humanized monoclonal antibody that targets cancer cells; Prodrax, a prodrug activated in solid tumors; and Alchemix, a chemotherapeutic agent that combines different modes of action to combat drug resistance. Pexiganan, a topical anti-infective treatment for foot ulcers, is in the latter stages of testing. Both Thiarabine, an anti-cancer drug and EcoNail (Onychomycosis), a topical anti-fungal, have proven to be effective in Phase 1 and 2 human testing. Phenylbutyrate has shown promise in anti-cancer treatments and has performed well in Phase 2 testing. MuGard is a viscous polymer solution that coats the oral cavity and is used in the treatment of mucositis, an issue that is faced by many chemo-therapy patients. Access is also in the process of developing its drug candidate ProLindac (AP5346), which links the DACH platform to a polymer and selectively releases the active drug to the tumor based on a difference between pH, as well as blood permeability, of a tumor and healthy tissue. The drug is in Phase 2 study. In April 2009, the company launched MuGard in Europe through its partner SpePharm.

## FINANCIALS:  Sales and profits are in thousands of dollars—add 000 to get the full amount. 2008 Note: Financial information for 2008 was not available for all companies at press time.

| | | |
|---|---|---|
| 2008 Sales: $ 291 | 2008 Profits: $-20,573 | **U.S. Stock Ticker: ACCP** |
| 2007 Sales: $ 57 | 2007 Profits: $-36,652 | **Int'l Ticker:**    Int'l Exchange: |
| 2006 Sales: $ | 2006 Profits: $-12,874 | Employees:    9 |
| 2005 Sales: $ | 2005 Profits: $-1,700 | Fiscal Year Ends: 12/31 |
| 2004 Sales: $ | 2004 Profits: $-10,200 | Parent Company: |

## SALARIES/BENEFITS:

| Pension Plan: | ESOP Stock Plan: | Profit Sharing: | Top Exec. Salary: $266,076 | Bonus: $ |
|---|---|---|---|---|
| Savings Plan: Y | Stock Purch. Plan: Y | | Second Exec. Salary: $253,620 | Bonus: $ |

## OTHER THOUGHTS:

**Apparent Women Officers or Directors:**
**Hot Spot for Advancement for Women/Minorities:**

## LOCATIONS: ("Y" = Yes)

| West: | Southwest: | Midwest: | Southeast: | Northeast: | International: |
|---|---|---|---|---|---|
| | Y | | | | Y |

# ACTELION LTD

**www.actelion.com**

**Industry Group Code: 325412  Ranks within this company's industry group:** Sales: 37   Profits: 28

| Drugs: | | Other: | Clinical: | Computers: | Services: |
|---|---|---|---|---|---|
| Discovery: | Y | AgriBio: | Trials/Services: | Hardware: | Specialty Services: |
| Licensing: | | Genetic Data: | Labs: | Software: | Consulting: |
| Manufacturing: | | Tissue Replacement: | Equipment/Supplies: | Arrays: | Blood Collection: |
| Genetics: | | | Research & Development Services: | Database Management: | Drug Delivery: |
| | | | Diagnostics: | | Drug Distribution: |

## TYPES OF BUSINESS:

Drugs, Discovery & Development
Pharmaceutical Research
Cardiovascular Treatment
Genetic Disorder Treatment

## BRANDS/DIVISIONS/AFFILIATES:

G-Protein Coupled Receptors (GPCRs)
Tracleer
Zavesca
Ventavis
Bosentan
Miglustat
Macitentan
Almorexant

## CONTACTS: *Note: Officers with more than one job title may be intentionally listed here more than once.*

Jean-Paul Clozel, CEO
Andrew J. Oakley, CFO/VP
Walter Fischli, Sr. VP/Head-Molecular Biology & Biochemistry
Marian Borovsky, General Counsel/VP/Corp. Sec.
Otto Schwarz, Pres., Bus. Oper.
Simon Buckingham, Pres., Corp. & Bus. Dev.
Roland Haefeli, Head-Public Affairs
Roland Haefeli, VP/Head-Investor Rel.
Thomas Weller, VP/Head-Drug Discovery, Chemistry
Isaac Kobrin, Sr. VP/Head-Clinical Dev.
Martine Clozel, Sr. VP/Head-Pharmacology & Pre-Clinical Dev.
Robert E. Cawthorn, Chmn.

| Phone: 41-61-565-65-65 | Fax: 41-61-565-65-00 |
|---|---|
| **Toll-Free:** | |
| **Address:** Gewerbestrasse 16, Allschwil, 4123 Switzerland | |

## GROWTH PLANS/SPECIAL FEATURES:

Actelion, Ltd. is a biopharmaceutical company that focuses on the discovery, development and marketing of drugs for unaddressed medical needs. Actelion focuses its drug discovery efforts on the design and synthesis of novel low molecular weight, drug-like molecules. Additional drug discovery platforms include G-Protein Coupled Receptors (GPCRs), aspartic proteinases, anti-infectives and ion channels. Actelion's most recognized product is Tracleer, a dual endothelin receptor antagonist used for pulmonary arterial hypertension (PAH). Another popular Actelion drug, Zavesca, is one of the first approved oral drug therapy treatments for a genetic lipid metabolic disorder called Gaucher disease. The company's Ventavis is an inhaled formulation of iloprost for the treatment of PAH. Additional drugs that are currently in the developmental stage include Bosentan, Miglustat, Almorexant, Clazosentan and Macitentan. Bosentan is an orally active dual endothelin receptor antagonist for the treatment of PAH. Miglustat is a low molecular weight inhibitor of glucosylceramide synthase and glucosidase for the treatment of Gaucher disease. Almorexant is a first-in-class receptor antagonist for the treatment of sleep disorders. Clazosentan is an intravenous endothelin receptor antagonist intended for the prevention and treatment of vasospasms, a life-threatening condition that leads to neurological deficits after a patient suffers an aneurysm. Macitentan is a tissue-targeting endothelin receptor antagonist intended for treatment of the cardiovascular system. Actelion is based in Switzerland and has subsidiaries in more than 20 countries. In February 2008, Actelion agreed to collaborate with Nippon Shinyaku on the development of Nippon's prostaglandin I2 (PGI -2) receptor agonist NS-304 for the treatment of PAH. In September 2008, Actelion joined the Swiss Market Index (SMI). In February 2009, Actelion agreed to acquire an epoprostenol sodium formulation for the intravenous treatment of PAH from GeneraMedix, Inc.

## FINANCIALS: Sales and profits are in thousands of dollars—add 000 to get the full amount. 2008 Note: Financial information for 2008 was not available for all companies at press time.

| | | |
|---|---|---|
| 2008 Sales: $1,254,960 | 2008 Profits: $273,810 | **U.S. Stock Ticker:** |
| 2007 Sales: $1,291,000 | 2007 Profits: $122,100 | **Int'l Ticker: ATLN**   Int'l Exchange: Zurich-SWX |
| 2006 Sales: $926,800 | 2006 Profits: $236,300 | Employees:  400 |
| 2005 Sales: $545,168 | 2005 Profits: $103,135 | Fiscal Year Ends: 12/31 |
| 2004 Sales: $471,880 | 2004 Profits: $87,219 | Parent Company: |

## SALARIES/BENEFITS:

| Pension Plan: | ESOP Stock Plan: | Profit Sharing: | Top Exec. Salary: $ | Bonus: $ |
|---|---|---|---|---|
| Savings Plan: | Stock Purch. Plan: | | Second Exec. Salary: $ | Bonus: $ |

## OTHER THOUGHTS:

**Apparent Women Officers or Directors:** 1
**Hot Spot for Advancement for Women/Minorities:**

## LOCATIONS: ("Y" = Yes)

| West: | Southwest: | Midwest: | Southeast: | Northeast: | International: |
|---|---|---|---|---|---|
| Y | | | | Y | Y |

# ADAMIS PHARMACEUTICALS CORPORATION

**Industry Group Code: 325412　Ranks within this company's industry group:** Sales:　　Profits: 62

| Drugs: | | Other: | Clinical: | Computers: | Services: |
|---|---|---|---|---|---|
| Discovery: | Y | AgriBio: | Trials/Services: | Hardware: | Specialty Services: |
| Licensing: | Y | Genetic Data: | Labs: | Software: | Consulting: |
| Manufacturing: | | Tissue Replacement: | Equipment/Supplies: | Arrays: | Blood Collection: |
| Genetics: | | | Research & Development Services: | Database Management: | Drug Delivery: |
| | | | Diagnostics: | | Drug Distribution: |

## TYPES OF BUSINESS:

Drugs
HIV Prevention

## BRANDS/DIVISIONS/AFFILIATES:

Cellegy Pharmaceuticals, Inc.
Biosyn, Inc.
Savvy

## CONTACTS: *Note: Officers with more than one job title may be intentionally listed here more than once.*

Richard C. Williams, Interim CEO
Robert J. (Rob) Caso, CFO/VP
Richard C. Williams, Chmn.

| **Phone:** 858-401-3984 | **Fax:** |
|---|---|
| **Toll-Free:** | |
| **Address:** 2658 Del Mar Heights Rd., Ste. 555, Del Mar, CA 92014 US | |

## GROWTH PLANS/SPECIAL FEATURES:

Adamis Pharmaceuticals Corporation, formerly Cellegy Pharmaceuticals, Inc., is a specialty pharmaceutical company that, through its Biosyn, Inc. subsidiary, has intellectual property relating to a portfolio of proprietary product candidates known as microbicides. Biosyn's product candidates include Savvy, which underwent now-terminated Phase III clinical trials in Ghana and Nigeria for anti-HIV microbial efficacy and a Phase III contraception trial in the U.S. Savvy has also shown promising preliminary results in the prevention of additional STDs, including those caused by the herpes simplex virus and Chlamydia. The active compound in Savvy is C31G, a broad-spectrum compound with antiviral, antibacterial and antifungal activity. The Phase III trial of Savvy in the U.S. for contraception was completed in late 2008, and analysis of the trial is expected by the end of 2009. Adamis is not directly involved with the trial and, due to the cessation of the HIV Phase III trials, it is uncertain whether the company will ever realize revenue from Savvy. Certain Phase III trial expenses for Savvy, and certain other clinical and preclinical developments costs for the Biosyn pipeline, are funded by grant and contract commitments through agencies including the United States Agency for International Development (USAID); the National Institute for Child Health Development (NICHD); the National Institute for Allergy and Infectious Disease (NIAID); the Contraceptive Research and Development Organization (CONRAD); and other governmental and philanthropic organizations. In February 2009, Cellegy Pharmaceuticals was acquired through merger by Adamis Pharmaceuticals.

## FINANCIALS: Sales and profits are in thousands of dollars—add 000 to get the full amount. 2008 Note: Financial information for 2008 was not available for all companies at press time.

| | | |
|---|---|---|
| 2008 Sales: $ | 2008 Profits: $-1,534 | **U.S. Stock Ticker: ADMP** |
| 2007 Sales: $ | 2007 Profits: $-1,927 | **Int'l Ticker:**　Int'l Exchange: |
| 2006 Sales: $2,660 | 2006 Profits: $9,672 | Employees:　14 |
| 2005 Sales: $12,199 | 2005 Profits: $-5,008 | Fiscal Year Ends: 12/31 |
| 2004 Sales: $2,033 | 2004 Profits: $-28,154 | Parent Company: |

## SALARIES/BENEFITS:

| Pension Plan: | ESOP Stock Plan: | Profit Sharing: | Top Exec. Salary: $273,770 | Bonus: $ |
|---|---|---|---|---|
| Savings Plan: | Stock Purch. Plan: | | Second Exec. Salary: $182,103 | Bonus: $ |

## OTHER THOUGHTS:

**Apparent Women Officers or Directors:**
**Hot Spot for Advancement for Women/Minorities:**

## LOCATIONS: ("Y" = Yes)

| West: | Southwest: | Midwest: | Southeast: | Northeast: | International: |
|---|---|---|---|---|---|
| Y | | | | | |

# ADOLOR CORP

www.adolor.com

Industry Group Code: 325412 **Ranks within this company's industry group:** Sales: 85 Profits: 123

| Drugs: | | Other: | Clinical: | | Computers: | | Services: | |
|---|---|---|---|---|---|---|---|---|
| Discovery: | Y | AgriBio: | Trials/Services: | | Hardware: | | Specialty Services: | |
| Licensing: | Y | Genetic Data: | Labs: | | Software: | | Consulting: | |
| Manufacturing: | | Tissue Replacement: | Equipment/Supplies: | | Arrays: | | Blood Collection: | |
| Genetics: | | | Research & Development Services: | | Database Management: | | Drug Delivery: | |
| | | | Diagnostics: | | | | Drug Distribution: | |

## TYPES OF BUSINESS:

Drugs, Discovery & Development
Pain Management Products
Gastrointestinal Products

## BRANDS/DIVISIONS/AFFILIATES:

Entereg

## CONTACTS: Note: Officers with more than one job title may be intentionally listed here more than once.

Michael R. Dougherty, CEO
Michael R. Dougherty, Pres.
Stephen W. Webster, CFO
John P. Wilson, VP-Mktg. & Sales
Eliseo Oreste Salinas, Sr. VP-R&D/Chief Medical Officer
George R. Maurer, Sr. VP-Mfg. & Pharmaceutical Tech.
John M. Limongelli, General Counsel/Sr. VP/Sec.
Richard M. Mangano, VP-Clinical Oper.
Kevin G. Taylor, VP-Bus. Dev.
Stephen W. Webster, Sr. VP-Finance
Elizabeth V. Jobes, VP/Chief Compliance Officer
Richard M. Woodward, VP-Discovery
David M. Stephon, VP-Quality Assurance
Lee M. Techner, VP-Medical Affairs
David M. Madden, Chmn.

| Phone: 484-595-1500 | Fax: 484-595-1520 |
|---|---|
| Toll-Free: | |
| Address: 700 Pennsylvania Dr., Exton, PA 19341 US | |

## GROWTH PLANS/SPECIAL FEATURES:

Adolor Corporation is a development stage biopharmaceutical corporation specializing in the discovery and development of prescription pain management products. The company's leading product is Entereg (alvimopan), which was developed in collaboration with Glaxo Group Limited to treat gastrointestinal (GI) complications caused by post-operative ileus (POI) and OBD (opioid bowel dysfunction). These complications are typically caused by opioids such as morphine, which provide several beneficial pain-management results with the central nervous system (CNS) but can lead to serious side effects. The FDA approved the use of Entereg for POI in May 2008. Adolor is also exploring methods designed to alleviate pain resulting from inflammation, nerve damage, and damage joints. In addition to Entereg, Adolor has identified a series of novel, orally active delta agonists that selectively stimulate the delta opioid receptor, whereas all marketed opioid drugs currently interact with the mu receptors in the brain and spinal cord. The company is collaborating with Pfizer, Inc. to develop new alternatives for pain management based on this research. In December 2008, the company decided to halt testing of Entereg as a treatment of OBDfor commercial reasons, but hopes to continue research in this area with other solutions.

Adolor offers employees life, health, dental insurance; prescription drug coverage; flexible spending accounts and a 401(k) plan.

## FINANCIALS: Sales and profits are in thousands of dollars—add 000 to get the full amount. 2008 Note: Financial information for 2008 was not available for all companies at press time.

| | | |
|---|---|---|
| 2008 Sales: $48,208 | 2008 Profits: $-30,122 | U.S. Stock Ticker: ADLR |
| 2007 Sales: $9,120 | 2007 Profits: $-48,443 | Int'l Ticker: Int'l Exchange: |
| 2006 Sales: $15,087 | 2006 Profits: $-69,738 | Employees: 128 |
| 2005 Sales: $15,719 | 2005 Profits: $-56,797 | Fiscal Year Ends: 12/31 |
| 2004 Sales: $25,542 | 2004 Profits: $-43,586 | Parent Company: |

## SALARIES/BENEFITS:

| Pension Plan: | ESOP Stock Plan: | Profit Sharing: | Top Exec. Salary: $426,942 | Bonus: $151,900 |
|---|---|---|---|---|
| Savings Plan: Y | Stock Purch. Plan: Y | | Second Exec. Salary: $276,536 | Bonus: $179,300 |

## OTHER THOUGHTS:

**Apparent Women Officers or Directors**: 1
**Hot Spot for Advancement for Women/Minorities**: Y

## LOCATIONS: ("Y" = Yes)

| West: | Southwest: | Midwest: | Southeast: | Northeast: | International: |
|---|---|---|---|---|---|
| | | | | Y | |

# ADVANCED BIONICS CORPORATION

www.advancedbionics.com

**Industry Group Code:** 33911 **Ranks within this company's industry group:** Sales: Profits:

| Drugs: | Other: | Clinical: | | Computers: | | Services: | |
|---|---|---|---|---|---|---|---|
| Discovery: | AgriBio: | Trials/Services: | | Hardware: | | Specialty Services: | |
| Licensing: | Genetic Data: | Labs: | | Software: | Y | Consulting: | |
| Manufacturing: | Tissue Replacement: | Equipment/Supplies: | Y | Arrays: | | Blood Collection: | |
| Genetics: | | Research & Development Services: | | Database Management: | | Drug Delivery: | |
| | | Diagnostics: | | | | Drug Distribution: | |

## TYPES OF BUSINESS:

Medical Equipment-Manufacturing
Bionic Devices
Cochlear Implant Technology

## BRANDS/DIVISIONS/AFFILIATES:

HiRes
HiResolution Bionic Ear System
Harmony
Platinum

## CONTACTS: Note: Officers with more than one job title may be intentionally listed here more than once.

Jeff Greiner, CEO
Jeff Greiner, Pres
Russ Ramanayake, Controller
Al Mann, Chmn.

| Phone: 661-362-1400 | Fax: 661-362-1500 |
|---|---|
| Toll-Free: 800-678-2575 | |
| Address: 12740 San Fernando Rd., Sylmar, CA 91342 US | |

## GROWTH PLANS/SPECIAL FEATURES:

Advanced Bionics Corporation develops and markets bionic technologies that employ implantable neurostimulation devices such as cochlear implants for the restoration of hearing in deaf individuals. In 2008, Advanced Bionics, formerly a subsidiary of Boston Scientific, went private and subsequently acquired Boston Scientific's auditory and drug pump operations. Boston Scientific retained its pain-management business, which formerly operated under the Advanced Bionics banner. Advanced Bionics is currently the only company in the U.S. that develops cochlear implant technology. Products in the company's HiResolution Bionic Ear System include Harmony, a behind-the-ear processor that produces high-quality sound resolution; Platinum, a sound processor that is worn around a belt instead of behind the ears; and HiRes, a 90K implant that uses integrated circuit computer technology with an internal memory and Hifocus electrodes for neutral targeting. The firm currently has corporate offices in California, France and Japan, serving countries in North America, Latin and South America, Europe, Asia and the Pacific Rim.

Employees are offered medical, dental and vision insurance; life insurance; disability coverage; flexible spending accounts; and a 401(k) savings plan.

## FINANCIALS: Sales and profits are in thousands of dollars—add 000 to get the full amount. 2008 Note: Financial information for 2008 was not available for all companies at press time.

| | | | |
|---|---|---|---|
| 2008 Sales: $ | 2008 Profits: $ | **U.S. Stock Ticker:** Private | |
| 2007 Sales: $ | 2007 Profits: $ | **Int'l Ticker:** Int'l Exchange: | |
| 2006 Sales: $ | 2006 Profits: $ | Employees: 500 | |
| 2005 Sales: $ | 2005 Profits: $ | Fiscal Year Ends: 12/31 | |
| 2004 Sales: $ | 2004 Profits: $ | Parent Company: | |

## SALARIES/BENEFITS:

| Pension Plan: | ESOP Stock Plan: | Profit Sharing: | Top Exec. Salary: $ | Bonus: $ |
|---|---|---|---|---|
| Savings Plan: Y | Stock Purch. Plan: Y | | Second Exec. Salary: $ | Bonus: $ |

## OTHER THOUGHTS:

**Apparent Women Officers or Directors:**
**Hot Spot for Advancement for Women/Minorities:**

## LOCATIONS: ("Y" = Yes)

| West: | Southwest: | Midwest: | Southeast: | Northeast: | International: |
|---|---|---|---|---|---|
| Y | | | | | Y |

# ADVANCED CELL TECHNOLOGY INC

www.advancedcell.com

Industry Group Code: 541712  Ranks within this company's industry group: Sales: 20  Profits: 14

| Drugs: | | Other: | | Clinical: | | Computers: | | Services: | |
|---|---|---|---|---|---|---|---|---|---|
| Discovery: | Y | AgriBio: | | Trials/Services: | | Hardware: | | Specialty Services: | |
| Licensing: | | Genetic Data: | | Labs: | | Software: | | Consulting: | |
| Manufacturing: | | Tissue Replacement: | Y | Equipment/Supplies: | | Arrays: | | Blood Collection: | |
| Genetics: | | | | Research & Development Services: | | Database Management: | | Drug Delivery: | |
| | | | | Diagnostics: | | | | Drug Distribution: | |

## TYPES OF BUSINESS:

Human Stem Cell Research
Patent Licensing

## BRANDS/DIVISIONS/AFFILIATES:

Mytogen, Inc.

## CONTACTS: *Note: Officers with more than one job title may be intentionally listed here more than once.*

William M. Caldwell, IV, CEO
William M. Caldwell, IV, Principle Financial Officer
Robert Lanza, Chief Scientific Officer
Rita Parker, Media Contact
Jonathan H. Dinsmore, Sr. VP-Regulatory & Clinical
William M. Caldwell, IV, Chmn.

| Phone: 508-756-1212 | Fax: 508-756-4468 |
|---|---|
| Toll-Free: 800-218-4202 | |
| Address: 381 Plantation St., Worcester, MA 01605 US | |

## GROWTH PLANS/SPECIAL FEATURES:

Advanced Cell Technology, Inc. (ACT) is a biotechnology company focusing on developing and commercializing embryonic and adult stem cell technology in the emerging field of regenerative medicine. Regenerative medicine treats chronic degenerative diseases and facilitates regenerative repair of acute disease, such as trauma, infarction and burns. ACT owns or licenses over 380 patents and patent applications related to the field of stem cell therapy, including nuclear transfer, which allows the production of stem cells genetically matched to the patient, and a reduced complexity library of stem cells for acute clinical applications. It was one of the first companies to develop a technique to generate an embryonic stem cell line that does not damage the embryo's developmental potential. ACT divides its research into three categories. Cellular reprogramming involves turning stem cells into one of over 200 different human cell types that may be therapeutically relevant in treating diseased or destroyed tissue, and which are tailored to each patient's needs. The reduced complexity program uses proprietary technology to generate cell therapy products for patients with acute medical needs that do not allow time for patient-specific reprogramming of cells. Lastly, its stem cell differentiation segment controls the differentiation, culture and growth of the company's stem cells. The company is currently focused on developing eye treatments based on retinal pigment epithelium (RPE) cells and is developing preclinical treatments for cardiovascular disease, stroke and cancer based on hemangioblast (HG) cells. It also has projects working on stem-cell-based dermal treatments that would provide scar-free skin grafts. Additionally, following its acquisition of Mytogen, Inc., the firm is developing a Myoblast Program to treat heart failure, currently undergoing Phase II clinical trials. Myoblasts are stem cells derived from the patient's skeletal leg muscle.

## FINANCIALS: Sales and profits are in thousands of dollars—add 000 to get the full amount. 2008 Note: Financial information for 2008 was not available for all companies at press time.

| | | |
|---|---|---|
| 2008 Sales: $ 787 | 2008 Profits: $-33,904 | U.S. Stock Ticker: ACTC.PK |
| 2007 Sales: $ 647 | 2007 Profits: $-15,899 | Int'l Ticker:   Int'l Exchange: |
| 2006 Sales: $ 441 | 2006 Profits: $-18,720 | Employees:   12 |
| 2005 Sales: $ 395 | 2005 Profits: $-9,394 | Fiscal Year Ends: 12/31 |
| 2004 Sales: $ 3 | 2004 Profits: $- 52 | Parent Company: |

## SALARIES/BENEFITS:

| Pension Plan: | ESOP Stock Plan: | Profit Sharing: | Top Exec. Salary: $350,000 | Bonus: $ |
|---|---|---|---|---|
| Savings Plan: | Stock Purch. Plan: | | Second Exec. Salary: $290,000 | Bonus: $ |

## OTHER THOUGHTS:

**Apparent Women Officers or Directors:**
**Hot Spot for Advancement for Women/Minorities:**

## LOCATIONS: ("Y" = Yes)

| West: | Southwest: | Midwest: | Southeast: | Northeast: | International: |
|---|---|---|---|---|---|
| Y | | | | Y | |

# ADVANSOURCE BIOMATERIALS CORPORATION

www.advbiomaterials.com

Industry Group Code: 33911  Ranks within this company's industry group: Sales: 16   Profits: 14

| Drugs: | Other: | Clinical: | | Computers: | Services: | |
|---|---|---|---|---|---|---|
| Discovery: | AgriBio: | Trials/Services: | | Hardware: | Specialty Services: | |
| Licensing: | Genetic Data: | Labs: | | Software: | Consulting: | |
| Manufacturing: | Tissue Replacement: | Equipment/Supplies: | Y | Arrays: | Blood Collection: | |
| Genetics: | | Research & Development Services: | | Database Management: | Drug Delivery: | |
| | | Diagnostics: | | | Drug Distribution: | |

## TYPES OF BUSINESS:

Polymer Material Manufacturing
Medical Device Design
Polyurethane-Based Biomaterials

## BRANDS/DIVISIONS/AFFILIATES:

AdvanSource Biomaterials Corp.
CardioPass
CardioTech International, Ltd.
Catheter and Disposables Technology, Inc.
ChronoFlex
HydroThane

## CONTACTS: Note: Officers with more than one job title may be intentionally listed here more than once.

Michael F. Adams, CEO
Michael F. Adams, Pres.
David Volpe, Acting CFO
Andrew M. Reed, VP-Tech. & Science
William J. O'Neill, Jr., Chmn.

| Phone: 978-657-0075 | Fax: 978-657-0074 |
|---|---|
| Toll-Free: | |
| Address: 229 Andover St., Wilmington, MA 01887 US | |

## GROWTH PLANS/SPECIAL FEATURES:

AdvanSource Biomaterials Corp., formerly CardioTech International, Inc., develops advanced polymer materials used in the design and development of medical devices for a range of anatomical site and disease state treatment.  The firm provides products with both short and long-term implant applications, including stents, artificial hearts, catheters and vascular access ports.  The company sells its products under the ChronoFilm, ChronoFlex, ChronoThane, ChronoPrene, HydroThane and PolyBlend brand names.  ChronoFlex is a family of medical grade polyurethanes for the prevention of in-vivo formation of environmental stress cracking (ESC). ChronoThane is an aromatic ether-based product line of thermoplastic polymers.  HydroThane is a family of highly absorbent thermoplastic, hydrophilic, elastomers allowing for targeted water absorption. HydroMed is a line of hydrophilic polyurethanes used as lubricious coatings for medical devices.  The PolyBlend product line includes extremely soft, extrudable and injection moldable polyurethane elastomeric alloys.  ChronoPrene thermoplastic elastomeric materials are made of easy molding, high flow rubber with high chemical resistance to alcohols, acids and bases.  CardioPass is a synthetic coronary artery bypass graft, currently undergoing a clinical trial in Europe.  AdvanSource's services include polymer customization, including the addition of radiopaque polymers; manufacturing; university collaboration; feasibility advising; and product concept support through its Concept Center.  AdvanSource serves the intervention radiology, diabetes management, peripheral vascular, gastroenterology, ear, nose and throat, cardiovascular, drug delivery, orthopedic, endoscopy, neurology, oncology, urology and spine market segments.  In March 2008, AdvanSource sold wholly-owned subsidiary Catheter and Disposables Technology, Inc. to TACPRO, Inc for approximately $1.2 million.  In May 2008, the company formed AdvanSource Biomaterials Corporation as an operating subsidiary, and in October aligned with this subsidiary and changed its name from CardioTech International, Inc.

## FINANCIALS: Sales and profits are in thousands of dollars—add 000 to get the full amount. 2008 Note: Financial information for 2008 was not available for all companies at press time.

| | | |
|---|---|---|
| 2008 Sales: $1,950 | 2008 Profits: $-6,090 | U.S. Stock Ticker: ASB |
| 2007 Sales: $2,275 | 2007 Profits: $-2,962 | Int'l Ticker:     Int'l Exchange: |
| 2006 Sales: $22,381 | 2006 Profits: $-5,069 | Employees:    20 |
| 2005 Sales: $21,841 | 2005 Profits: $-1,595 | Fiscal Year Ends: 3/31 |
| 2004 Sales: $21,799 | 2004 Profits: $-1,515 | Parent Company: |

## SALARIES/BENEFITS:

| Pension Plan: | ESOP Stock Plan: | Profit Sharing: | Top Exec. Salary: $279,231 | Bonus: $ |
|---|---|---|---|---|
| Savings Plan: | Stock Purch. Plan: | | Second Exec. Salary: $189,615 | Bonus: $ |

## OTHER THOUGHTS:

Apparent Women Officers or Directors:
Hot Spot for Advancement for Women/Minorities:

## LOCATIONS: ("Y" = Yes)

| West: | Southwest: | Midwest: | Southeast: | Northeast: | International: |
|---|---|---|---|---|---|
| | | | | Y | |

# AEOLUS PHARMACEUTICALS INC

## www.aeoluspharma.com

Industry Group Code: 325412 Ranks within this company's industry group: Sales: Profits: 64

| Drugs: | | Other: | Clinical: | | Computers: | Services: |
|---|---|---|---|---|---|---|
| Discovery: | Y | AgriBio: | Trials/Services: | | Hardware: | Specialty Services: |
| Licensing: | | Genetic Data: | Labs: | | Software: | Consulting: |
| Manufacturing: | | Tissue Replacement: | Equipment/Supplies: | | Arrays: | Blood Collection: |
| Genetics: | | | Research & Development Services: | | Database Management: | Drug Delivery: |
| | | | Diagnostics: | | | Drug Distribution: |

## TYPES OF BUSINESS:

Pharmaceutical Development
Catalytic Antioxidants

## BRANDS/DIVISIONS/AFFILIATES:

AEOL 10150
AEOL 11207

## CONTACTS: Note: Officers with more than one job title may be intentionally listed here more than once.

John L. McManus, CEO
John L. McManus, Pres.
Michael McManus, CFO
Brian J. Day, Chief Scientific Officer
Michael McManus, Sec.
Michael McManus, Treas.
David C. Cavalier, Chmn.

| Phone: 949-481-9825 | Fax: 949-481-9829 |
|---|---|
| Toll-Free: | |
| Address: 26361 Crown Valley Pkwy., Ste. 150, Mission Viejo, CA 92691 US | |

## GROWTH PLANS/SPECIAL FEATURES:

Aeolus Pharmaceuticals, Inc. is a Southern California-based biopharmaceutical company. It is developing a new class of catalytic antioxidant compounds for disease and disorders of the central nervous system, respiratory system, autoimmune system and oncology. The company's lead drug candidate is AEOL 10150, the first drug in its class of catalytic antioxidant compounds to enter human clinical evaluation. AEOL 10150 is a small molecule catalytic antioxidant with the ability to scavenge a broad range of reactive oxygen species, or free radicals. As a catalytic antioxidant, AEOL 10150 mimics and amplifies the body's natural enzymatic systems for eliminating these damaging compounds. AEOL 10150 is thought to be useful as a treatment for amyotrophic lateral sclerosis, or ALS, also known as Lou Gehrig's disease. Aeolus boasts positive safety results from two completed Phase 1 single dose studies of AEOL 10150 in patients diagnosed with ALS, with no serious adverse effects reported. The drug has also shown promise in the field of radiation therapy and in the treatment of mustard gas exposure. Tests on mice demonstrate that AEOL 10150 protects healthy lung tissue from radiation injury delivered in a single dose or by fractionated radiation therapy doses, and that the drug does not negatively affect tumor radiotherapy. Aeolus has also selected AEOL 11207 as the company's second development candidate through the Aeolus Pipeline Initiative, an internal development initiative. Collected data suggests the compound may be useful as a potential once-every-other-day oral therapeutic treatment option for central nervous system disorders, most likely Parkinson's disease. In November 2008, the company announced that AEOL 10150 has proven to be effective in protecting against radiations and mustard gas in animals and has the potential to be fast-tracked by the FDA.

## FINANCIALS: Sales and profits are in thousands of dollars—add 000 to get the full amount. 2008 Note: Financial information for 2008 was not available for all companies at press time.

| | | | |
|---|---|---|---|
| 2008 Sales: $ | 2008 Profits: $-2,973 | U.S. Stock Ticker: AOLS | |
| 2007 Sales: $ | 2007 Profits: $-3,024 | Int'l Ticker: Int'l Exchange: | |
| 2006 Sales: $ 92 | 2006 Profits: $-5,728 | Employees: 1 | |
| 2005 Sales: $ 252 | 2005 Profits: $-6,905 | Fiscal Year Ends: 9/30 | |
| 2004 Sales: $ 305 | 2004 Profits: $-17,167 | Parent Company: | |

## SALARIES/BENEFITS:

| Pension Plan: | ESOP Stock Plan: | Profit Sharing: | Top Exec. Salary: $250,200 | Bonus: $ |
|---|---|---|---|---|
| Savings Plan: | Stock Purch. Plan: Y | | Second Exec. Salary: $33,333 | Bonus: $37,707 |

## OTHER THOUGHTS:

Apparent Women Officers or Directors:
Hot Spot for Advancement for Women/Minorities:

## LOCATIONS: ("Y" = Yes)

| West: | Southwest: | Midwest: | Southeast: | Northeast: | International: |
|---|---|---|---|---|---|
| Y | | | | | |

# AETERNA ZENTARIS INC

www.aeternazentaris.com

**Industry Group Code: 325412  Ranks within this company's industry group:** Sales: 92    Profits: 144

| Drugs: | Other: | Clinical: | Computers: | Services: |
|---|---|---|---|---|
| Discovery: | AgriBio: | Trials/Services: | Hardware: | Specialty Services: |
| Licensing: | Genetic Data: | Labs: | Software: | Consulting: |
| Manufacturing: Y | Tissue Replacement: | Equipment/Supplies: | Arrays: | Blood Collection: |
| Genetics: | | Research & Development Services: | Database Management: | Drug Delivery: |
| | | Diagnostics: | | Drug Distribution: |

*(Other: AgriBio: Y)*

## TYPES OF BUSINESS:

Drug Development
Oncology Products
Endocrine Therapy Products

## BRANDS/DIVISIONS/AFFILIATES:

Cetrotide
Cetrorelix
Ozarelix
Perifosine
Zentaris GmbH
Impavido

## CONTACTS: *Note: Officers with more than one job title may be intentionally listed here more than once.*

Jurgen Engel, CEO
Jurgen Engel, Pres.
Dennis Turpin, CFO/Sr. VP
Chantal Gravel, Human Resources Contact-Canada & US
Paul Blake, Sr. VP/Chief Medical Officer
Matthias Seeber, VP-Admin.
Matthias Seeber, VP-Legal Affairs
Chantal Gravel, Mgr.-Planning & Market Research
Paul Burroughs, Dir.-Comm.
Nicholas J. Pelliccione, Sr. VP-Regulatory Affairs & Quality Assurance
Daniel Perrissoud, Project Mgr.-Clinical Affairs
Juergen Ernst, Chmn.
Angelika Kirsch, Legal/Human Resources-Europe

| Phone: 418-652-8525 | Fax: 418-652-0881 |
|---|---|
| Toll-Free: | |
| Address: 1405 du Parc-Technologique Blvd., Quebec City, QC G1P 4P5 Canada | |

## GROWTH PLANS/SPECIAL FEATURES:

AEterna Zentaris Inc. is a Canadian biopharmaceutical company focused on endocrine therapy and oncology. The company is devoted to discovering and developing drugs for the treatment of certain forms of cancer, endocrine disorders and infectious diseases. The firm has one product on the commercial market, Cetrotide. Cetrotide was the first hormone antagonist treatment approved for in vitro fertilization. The drug is administered to women in order to prevent premature ovulation in order to increase fertility success rates. Cetrotide is approved in over 80 countries. The drug is marketed worldwide by Merck Serono, except in Japan where it is marketed by Shionogi and Nippon Kayaku. Currently, the company has two drugs in Phase I of clinical trial, three in Phase II and one in Phase III. The firm's Phase II drugs include Ozarelix, for prostate cancer; Perifosine, for multiple cancers; and AEZS-108, for ovarian and endometrial cancer. The drug in Phase III clinical trials is Cetrorelix, for benign prostatic hyperplasia and endometriosis. In addition, the firm has two products in preclinical in vitro testing and three in preclinical development. The company owns 100% of Zentaris GmbH, an integrated clinical research company. In April 2008, the company sold the rights for the manufacture, production, distribution, marketing, sale and use of its miltefosine product, under the brand name Impavido, to Paladin Labs Inc. for approximately $9.2 million. In June 2008, AEterna Zentaris conducted a sale-and-leaseback of its Quebec City office property; the firm sold the property for $7.1 million in cash and subsequently entered into a long-term lease agreement to continue occupying its current office space. In December 2008, the company completed the sale of certain royalty rights for sales of Cetrotide to Cowen Healthcare Royalty Partners, L.P., an international healthcare private equity firm, for approximately $52.5 million.

## FINANCIALS: Sales and profits are in thousands of dollars—add 000 to get the full amount. 2008 Note: Financial information for 2008 was not available for all companies at press time.

| | | |
|---|---|---|
| 2008 Sales: $38,478 | 2008 Profits: $-59,817 | **U.S. Stock Ticker:** AEZS |
| 2007 Sales: $42,100 | 2007 Profits: $-32,300 | **Int'l Ticker:** AEZ    Int'l Exchange: Toronto-TSX |
| 2006 Sales: $38,800 | 2006 Profits: $33,400 | Employees:  109 |
| 2005 Sales: $47,204 | 2005 Profits: $10,571 | Fiscal Year Ends: 12/31 |
| 2004 Sales: $ | 2004 Profits: $ | Parent Company: |

## SALARIES/BENEFITS:

| Pension Plan: | ESOP Stock Plan: | Profit Sharing: | Top Exec. Salary: $405,925 | Bonus: $248,093 |
|---|---|---|---|---|
| Savings Plan: | Stock Purch. Plan: | | Second Exec. Salary: $355,250 | Bonus: $135,000 |

## OTHER THOUGHTS:

**Apparent Women Officers or Directors:** 2
**Hot Spot for Advancement for Women/Minorities:**

## LOCATIONS: ("Y" = Yes)

| West: | Southwest: | Midwest: | Southeast: | Northeast: | International: |
|---|---|---|---|---|---|
| | | | | Y | Y |

# AFFYMETRIX INC

www.affymetrix.com

**Industry Group Code:** 325413 **Ranks within this company's industry group:** Sales: 5 Profits: 22

| Drugs: | Other: | | Clinical: | Computers: | | Services: | |
|---|---|---|---|---|---|---|---|
| Discovery: | AgriBio: | | Trials/Services: | Hardware: | Y | Specialty Services: | |
| Licensing: | Genetic Data: | Y | Labs: | Software: | | Consulting: | |
| Manufacturing: | Tissue Replacement: | | Equipment/Supplies: | Arrays: | Y | Blood Collection: | |
| Genetics: | | | Research & Development Services: | Database Management: | | Drug Delivery: | |
| | | | Diagnostics: | | | Drug Distribution: | |

## TYPES OF BUSINESS:

Chips-Genetics
DNA Array Technology
Genomics

## BRANDS/DIVISIONS/AFFILIATES:

GeneChip
CustomExpress
CustomSeq
USB Corporation
Panomics, Inc.

## CONTACTS: *Note: Officers with more than one job title may be intentionally listed here more than once.*

Kevin M. King, CEO
Kevin M. King, Pres.
John C. Batty, CFO
Rick Runkel, General Counsel
John C. Batty, Exec. VP-Finance
Stephen P. A. Fodor, Chmn.

| **Phone:** 408-731-5000 | **Fax:** 408-731-5441 |
|---|---|
| **Toll-Free:** 888-362-2447 | |
| **Address:** 3420 Central Expwy., Santa Clara, CA 95051 US | |

## GROWTH PLANS/SPECIAL FEATURES:

Affymetrix, Inc. develops, manufactures, sells and services consumables and systems for genetic analysis in the life sciences and clinical healthcare markets. The firm sells its products directly to pharmaceutical, biotechnology, agrichemical, diagnostics and consumer products companies, as well as academic research centers, government research laboratories, private foundation laboratories and clinical reference laboratories, in North America and Europe. The company also sells its products through life science supply specialists acting as authorized distributors in Latin America, India and the Middle East and Asia Pacific regions. Affymetrix markets products for two principal applications: monitoring of gene or exon expression levels and investigation of genetic variation. Its catalogue GeneChip expression arrays are available for the study of human, rat, mouse and a range of other mammalian and model organisms. Human, mouse and rat exon analysis arrays are also available. The firm's integrated GeneChip microarray platform includes disposable DNA probe arrays consisting of nucleic acid sequences set out in an ordered, high density pattern; certain reagents for use with the probe arrays; a scanner and other instruments used to process the probe arrays; and software to analyze and manage genomic or genetic information obtained from the probe arrays. Additionally, the company markets CustomExpress and CustomSeq products, which enable its customers to design their own custom GeneChip expression arrays or a sequence of arrays for organisms of interest to them. In January 2008, Affymetrix acquired USB Corporation, an Ohio-based developer of molecular biology and biochemical reagent products. In December 2008, Affymetrix acquired Panomics, Inc., a provider of assay products for a variety of low- to mid-plex genetic, protein and cellular analysis applications.

Affymetrix offers its employees a tuition assistance plan, health fitness membership discounts, a lunch program, a family resources program, domestic partner benefits, reimbursement accounts and medical, dental, vision, life and disability insurance.

## FINANCIALS: Sales and profits are in thousands of dollars—add 000 to get the full amount. 2008 Note: Financial information for 2008 was not available for all companies at press time.

| | | |
|---|---|---|
| 2008 Sales: $410,249 | 2008 Profits: $-307,919 | **U.S. Stock Ticker:** AFFX |
| 2007 Sales: $371,320 | 2007 Profits: $12,593 | **Int'l Ticker:** Int'l Exchange: |
| 2006 Sales: $355,317 | 2006 Profits: $-13,704 | Employees: 1,128 |
| 2005 Sales: $367,602 | 2005 Profits: $65,787 | Fiscal Year Ends: 12/31 |
| 2004 Sales: $345,962 | 2004 Profits: $47,608 | Parent Company: |

## SALARIES/BENEFITS:

| Pension Plan: | ESOP Stock Plan: | Profit Sharing: | Top Exec. Salary: $655,673 | Bonus: $ |
|---|---|---|---|---|
| Savings Plan: Y | Stock Purch. Plan: | | Second Exec. Salary: $491,346 | Bonus: $ |

## OTHER THOUGHTS:

**Apparent Women Officers or Directors:** 1
**Hot Spot for Advancement for Women/Minorities:**

## LOCATIONS: ("Y" = Yes)

| West: | Southwest: | Midwest: | Southeast: | Northeast: | International: |
|---|---|---|---|---|---|
| Y | | | | | Y |

# AGILENT TECHNOLOGIES INC

www.agilent.com

**Industry Group Code: 3345  Ranks within this company's industry group: Sales: 1  Profits: 1**

| Drugs: | Other: | Clinical: | | Computers: | Services: |
|---|---|---|---|---|---|
| Discovery: | AgriBio: | Trials/Services: | | Hardware: | Specialty Services: |
| Licensing: | Genetic Data: | Labs: | | Software: | Consulting: |
| Manufacturing: | Tissue Replacement: | Equipment/Supplies: | Y | Arrays: | Blood Collection: |
| Genetics: | | Research & Development Services: | | Database Management: | Drug Delivery: |
| | | Diagnostics: | | | Drug Distribution: |

## TYPES OF BUSINESS:

Test Equipment
Communications Test Equipment
Integrated Circuits Test Equipment
Optoelectronics Test Equipment
Image Sensors
Bioinstrumentation
Software Products
Informatics Products

## BRANDS/DIVISIONS/AFFILIATES:

Agilent Technologies Laboratories
Particle Sizing Systems
RVM Scientific Inc
TILL Photonics GmbH
Stratagene Corp.

## CONTACTS: *Note: Officers with more than one job title may be intentionally listed here more than once.*

William P. Sullivan, CEO
William P. Sullivan, Pres.
Adrian T. Dillon, CFO
Jean M. Halloran, Sr. VP-Human Resources
Darlene J. S. Solomon, VP-Agilent Laboratories
Darlene J. S. Solomon, CTO
Adrian T. Dillon, Exec. VP-Admin.
D. Craig Nordlund, General Counsel/Sec./Sr. VP
Amy Flores, Mgr.-Public Rel.
Rodney Gonsalves, Dir.-Investor Rel.
Adrian T. Dillon, Exec. VP-Finance
Gooi Soon Chai, VP/Gen. Mgr.-Electronic Instruments
Ron Nersesian, VP/Gen. Mgr.-Wireless
Nick Roelofs, VP/Gen. Mgr.-Life Sciences Solutions
David Churchill, VP/Gen. Mgr.-Network & Digital Solutions
James G. Cullen, Chmn.

| **Phone:** 408-345-8886 | **Fax:** 408-345-8474 |
|---|---|
| **Toll-Free:** 877-424-4536 | |
| **Address:** 5301 Stevens Creek Blvd., Santa Clara, CA 95051 US | |

## GROWTH PLANS/SPECIAL FEATURES:

Agilent Technologies, Inc. is a diversified technology company with two main business segments: Electronic Measurement and Bio-analytical Measurement. The Electronic Measurement business operates in two markets. Its products for the communications testing market include testing equipment for fiber optic networks; broadband and data networks; and wireless communications and microwave networks. It also assists in installing, activating and maintaining optical, wireless, wireline and large-company networks. Supplying the aerospace, defense, computer and semiconductor industries, its offerings for the general purpose testing market include general purpose instruments, including voltmeters and signal generators; modular instruments and test software used as reconfigurable testing platforms; digital design products, including complex high-speed servers and logic analyzers; high-frequency electronic design automation software tools used to construct computer simulations; parametric test instruments and systems for semiconductor wafers; and electronic manufacturing test products such as automated x-ray inspection and in-circuit testing products. The Bio-analytical Measurement business serves two main life sciences markets: Pharmaceuticals, biotech, contract research and contract manufacturing; and academic and government institutions. It also serves five main chemical analysis markets: petroleum and chemicals; the environment; forensics and homeland security; bio-agriculture and food safety; and materials science. Its main product categories are gas chromatography, liquid chromatography, mass spectrometry, microfluidics, microarrays, atomic force microscopy, PCR (Polymerase Chain Reaction) instrumentation, software and informatics. It also supplies consumables and related bioagents. Agilent conducts centralized research for both segments through Agilent Technologies Laboratories, based in Santa Clara, California. In 2008, the company acquired Particle Sizing Systems, a particle measuring instruments manufacturer; RVM Scientific Inc, a manufacturer of direct heating/cooling systems for gas chromatography capillary columns; and TILL Photonics GmbH, a developer and manufacturer of microscopy products.

Employees are offered medical, dental and vision insurance; life insurance; disability coverage; an employee and family assistance plan; and adoption assistance.

## FINANCIALS: Sales and profits are in thousands of dollars—add 000 to get the full amount. 2008 Note: Financial information for 2008 was not available for all companies at press time.

| | | |
|---|---|---|
| 2008 Sales: $5,774,000 | 2008 Profits: $693,000 | **U.S. Stock Ticker: A** |
| 2007 Sales: $5,420,000 | 2007 Profits: $638,000 | **Int'l Ticker:**   Int'l Exchange: |
| 2006 Sales: $4,973,000 | 2006 Profits: $3,307,000 | Employees: 19,400 |
| 2005 Sales: $4,685,000 | 2005 Profits: $327,000 | Fiscal Year Ends: 10/31 |
| 2004 Sales: $4,556,000 | 2004 Profits: $369,000 | Parent Company: |

## SALARIES/BENEFITS:

| Pension Plan: Y | ESOP Stock Plan: | Profit Sharing: | Top Exec. Salary: $986,667 | Bonus: $1,305,563 |
|---|---|---|---|---|
| Savings Plan: Y | Stock Purch. Plan: Y | | Second Exec. Salary: $699,996 | Bonus: $627,775 |

## OTHER THOUGHTS:

**Apparent Women Officers or Directors: 3**
**Hot Spot for Advancement for Women/Minorities: Y**

## LOCATIONS: ("Y" = Yes)

| West: | Southwest: | Midwest: | Southeast: | Northeast: | International: |
|---|---|---|---|---|---|
| Y | | | | Y | Y |

Note: Financial information, benefits and other data can change quickly and may vary from those stated here.

# AKORN INC

**www.akorn.com**

Industry Group Code: 325412  Ranks within this company's industry group: Sales: 70  Profits: 73

| Drugs: | | Other: | Clinical: | | Computers: | | Services: | |
|---|---|---|---|---|---|---|---|---|
| Discovery: | Y | AgriBio: | Trials/Services: | | Hardware: | | Specialty Services: | |
| Licensing: | | Genetic Data: | Labs: | | Software: | | Consulting: | |
| Manufacturing: | Y | Tissue Replacement: | Equipment/Supplies: | | Arrays: | | Blood Collection: | |
| Genetics: | Y | | Research & Development Services: | | Database Management: | | Drug Delivery: | |
| | | | Diagnostics: | | | | Drug Distribution: | |

## TYPES OF BUSINESS:

Ophthalmic & Hospital Drugs & Injectables
Contract Services
Vaccines

## BRANDS/DIVISIONS/AFFILIATES:

Akorn (New Jersey), Inc.
Akorn-Strides, LLC

## CONTACTS: *Note: Officers with more than one job title may be intentionally listed here more than once.*

Raj Rai, Interim CEO
Timothy A. Dick, CFO
John R. Sabat, Sr. VP-Global Sales & Mktg.
Neill Shanahan, VP-Human Resources
Mark M. Silverberg, Exec. VP-Tech. Svcs.
Joseph Bonaccorsi, General Counsel/Sr. VP/Corp. Sec.
Mark M. Silverberg, Exec. VP-Oper.
Abu Alam, Sr. VP-New Bus. Dev.
John R. Sabat, Sr. VP-National Acct.
Mark M. Silverberg, Exec. VP-Global Quality Assurance
Sam Boddapati, Sr. VP-Regulatory Affairs
Michael P. Stehn, Sr. VP-Oper.
John N. Kapoor, Chmn.

| | |
|---|---|
| **Phone:** 847-279-6100 | **Fax:** 800-943-3694 |
| **Toll-Free:** 800-932-5676 | |
| **Address:** 1925 W. Field Ct., Ste. 300, Lake Forest, IL 60045 US | |

## GROWTH PLANS/SPECIAL FEATURES:

Akorn, Inc. provides diagnostic and therapeutic pharmaceuticals for specialty areas such as ophthalmology, rheumatology, anesthesia and antidotal medicine. The company operates in four segments: ophthalmic; hospital drugs and injectables; biologics and vaccines; and contract services. The ophthalmic division markets a line of diagnostic and therapeutic products for eye health. Diagnostic products, primarily used in the office setting, include mydriatics, cycloplegics, anesthetics, topical stains, gonioscopic solutions and angiography dyes. Therapeutic products, sold primarily to wholesalers and other national account customers, include antibiotics, anti-infectives, steroids, glaucoma medications, decongestants /antihistamines and anti-edema medications. Non-pharmaceutical products include various artificial tear solutions, preservative-free lubricating ointments, eyelid cleansers, vitamin supplements and contact lens accessories. Through the hospital drugs and injectables segment, Akorn markets a line of specialty injectable pharmaceutical products, including antidotes, anesthesia and products used in the treatment of rheumatoid arthritis and pain management. The biologics and vaccines segment focuses on adult Tetanus-Diphtheria vaccines and is expanded into the production of flu vaccine during 2008. The contract services segment manufactures products for third-party pharmaceutical and biotechnology customers based on their specifications. Akorn (New Jersey), Inc., the firm's wholly-owned subsidiary, is involved in manufacturing, product development and administrative activities related to the ophthalmic and hospital drugs and injectables segments. The firm has manufacturing facilities in Illinois and New Jersey. Customers include physicians, optometrists, hospitals, wholesalers, group purchasing organizations and other pharmaceutical companies. Akorn currently holds seven U.S. patents and has two U.S. patents that are pending. The firm's has a joint venture with Strides Arcolab Limited, called Akorn-Strides, LLC, which focuses on developing liquid, lyophilized and dry powder formulations of generic injectable drugs. In October 2008, the firm received FDA approval for its Akten, an ophthalmic anesthetic.

Akorn offers its employees medical, dental, vision and life insurance; an employee assistance program; and tuition reimbursement.

## FINANCIALS: Sales and profits are in thousands of dollars—add 000 to get the full amount. 2008 Note: Financial information for 2008 was not available for all companies at press time.

| | | |
|---|---|---|
| 2008 Sales: $93,598 | 2008 Profits: $-7,939 | **U.S. Stock Ticker:** AKRX |
| 2007 Sales: $52,895 | 2007 Profits: $-19,168 | **Int'l Ticker:** Int'l Exchange: |
| 2006 Sales: $71,250 | 2006 Profits: $-5,960 | Employees: 351 |
| 2005 Sales: $44,484 | 2005 Profits: $-8,609 | Fiscal Year Ends: 12/31 |
| 2004 Sales: $50,708 | 2004 Profits: $-3,026 | Parent Company: |

## SALARIES/BENEFITS:

| Pension Plan: | ESOP Stock Plan: | Profit Sharing: | Top Exec. Salary: $440,000 | Bonus: $ |
|---|---|---|---|---|
| Savings Plan: Y | Stock Purch. Plan: Y | | Second Exec. Salary: $275,000 | Bonus: $ |

## OTHER THOUGHTS:

**Apparent Women Officers or Directors:**
**Hot Spot for Advancement for Women/Minorities:**

## LOCATIONS: ("Y" = Yes)

| West: | Southwest: | Midwest: | Southeast: | Northeast: | International: |
|---|---|---|---|---|---|
| | | Y | | Y | |

Note: Financial information, benefits and other data can change quickly and may vary from those stated here.

# AKZO NOBEL NV

### www.akzonobel.com

Industry Group Code: 325  Ranks within this company's industry group:  Sales: 4   Profits: 4

| Drugs: | | Other: | Clinical: | Computers: | Services: |
|---|---|---|---|---|---|
| Discovery: | Y | AgriBio: | Trials/Services: | Hardware: | Specialty Services: |
| Licensing: | | Genetic Data: | Labs: | Software: | Consulting: |
| Manufacturing: | | Tissue Replacement: | Equipment/Supplies: | Arrays: | Blood Collection: |
| Genetics: | | | Research & Development Services: | Database Management: | Drug Delivery: |
| | | | Diagnostics: | | Drug Distribution: |

## TYPES OF BUSINESS:

Specialty Chemicals
Coatings
Decorative Paints

## BRANDS/DIVISIONS/AFFILIATES:

Imperial Chemical Industries PLC
Nippon Kayaku, Kayaku Akzo Co Ltd
LII Europe
Sikkens
Dulux
Hammerite

## CONTACTS: Note: Officers with more than one job title may be intentionally listed here more than once.

Hanz Wijers, CEO
Keith Nichols, CFO
Jennifer Midura, Dir.-Corp. Strategy
Leif Darner, Mgr.-Performance Coatings
Rob Frohn, Mgr.-Specialty Chemicals
Tex Gunning, Dir.-Decorative Paints
Maarten van den Bergh, Chmn.

| Phone: 31-20-502-7555 | Fax: 31-20-502-7666 |
|---|---|
| Toll-Free: | |
| Address: Strawinskylaan 2555, Amsterdam, 1077 ZZ The Netherlands | |

## GROWTH PLANS/SPECIAL FEATURES:

Akzo Nobel N.V. produces paints, coatings and chemicals, and operates in over 80 countries. The company's operations are divided into three segments: Decorative Paints, Performance Coatings and Specialty Chemicals. The Decorative Paints division includes paint, lacquer and varnish products, as well as adhesives, and floor leveling compounds. Brands consist of Sikkens, Dulux and Hammerite. The segment has offices in the U.K., Continental Europe, the Americas and Asia. Akzo Nobel's Performance Coatings division makes a variety of chemical products including powder, wood, coil, and marine coatings; tile and wood adhesives; and a line of car refinishes. The company's Specialty Chemicals division produces pulp and paper chemicals; polymer chemicals such as metal alkyls and suspending agents; surfactants used in hair and skincare products; base chemicals such as salt and chlor-alkali products used in the manufacture of glass and plastics; and functional chemicals used in toothpaste, ice cream and flame retardants. In April 2008, the company sold its Adhesives and Electronic Materials business. In December of the same year, Akzo Nobel acquired a 75% stake in Kayaku Akzo Co. Ltd., its Japanese joint venture with Nippon Kayaku. In January 2009, the firm acquired LII Europe.

## FINANCIALS: Sales and profits are in thousands of dollars—add 000 to get the full amount. 2008 Note: Financial information for 2008 was not available for all companies at press time.

| | | |
|---|---|---|
| 2008 Sales: $20,387,300 | 2008 Profits: $981,340 | U.S. Stock Ticker: |
| 2007 Sales: $18,481,000 | 2007 Profits: $12,770,600 | Int'l Ticker: AKZA   Int'l Exchange: Amsterdam-Euronext |
| 2006 Sales: $15,836,300 | 2006 Profits: $1,821,700 | Employees:  42,600 |
| 2005 Sales: $20,540,000 | 2005 Profits: $1,518,400 | Fiscal Year Ends: 12/31 |
| 2004 Sales: $17,187,000 | 2004 Profits: $1,159,000 | Parent Company: |

## SALARIES/BENEFITS:

| Pension Plan: | ESOP Stock Plan: | Profit Sharing: | Top Exec. Salary: $ | Bonus: $ |
|---|---|---|---|---|
| Savings Plan: | Stock Purch. Plan: | | Second Exec. Salary: $ | Bonus: $ |

## OTHER THOUGHTS:

**Apparent Women Officers or Directors:** 2
**Hot Spot for Advancement for Women/Minorities:** Y

## LOCATIONS: ("Y" = Yes)

| West: | Southwest: | Midwest: | Southeast: | Northeast: | International: |
|---|---|---|---|---|---|
| | | Y | Y | Y | Y |

# ALBANY MOLECULAR RESEARCH INC

www.amriglobal.com

Industry Group Code: 541712  Ranks within this company's industry group: Sales: 8  Profits: 6

| Drugs: | Other: | Clinical: | | Computers: | Services: | |
|---|---|---|---|---|---|---|
| Discovery: | AgriBio: | Trials/Services: | | Hardware: | Specialty Services: | |
| Licensing: | Genetic Data: | Labs: | | Software: | Consulting: | |
| Manufacturing: | Tissue Replacement: | Equipment/Supplies: | | Arrays: | Blood Collection: | |
| Genetics: | | Research & Development Services: | Y | Database Management: | Drug Delivery: | |
| | | Diagnostics: | | | Drug Distribution: | |

## TYPES OF BUSINESS:

Contract Drug Discovery & Development
Custom Biotech & Genomic Research
Chemistry Research
Manufacturing Services
Consulting Services
Analytical Chemistry Services

## BRANDS/DIVISIONS/AFFILIATES:

## CONTACTS: Note: Officers with more than one job title may be intentionally listed here more than once.

Thomas E. D'Ambra, CEO
Thomas E. D'Ambra, Pres.
Mark T. Frost, CFO
W. Steven (Steve) Jennings, Sr. VP-Sales & Mktg.
Brian D. Russell, VP-Human Resources
Bruce J. Sargent, VP-Discovery R&D
Harold Meckler, VP-Science & Tech.
Steven R. Hagen, VP-Pharmaceutical Dev. & Mfg.
W. Steven (Steve) Jennings, Sr. VP-Bus. Dev.
Peter Jerome, Dir.-Investor Rel.
Mark T. Frost, Treas.
Michael P. Trova, Sr. VP-Chemistry
Richard A. Saffee, Gen. Mgr.-Large Scale Mfg.
Thomas E. D'Ambra, Chmn.
Michael D. Ironside, Dir.-Global Project Mgmt.

| Phone: 518-464-0279 | Fax: 518-512-2020 |
|---|---|

Toll-Free:

Address: 21 Corporate Cir., P.O. Box 15098, Albany, NY 12212-5098 US

## GROWTH PLANS/SPECIAL FEATURES:

Albany Molecular Research, Inc. (AMRI) is a chemistry-based drug discovery and development company, focusing on applications for new small-molecule and prescription drugs. It derives revenue from discovering then licensing new compounds with commercial potential for service fees, milestone and royalty payments. Some of the products of this research led to the development of the active ingredient (fexofenadine HCl) for a non-sedating antihistamine marketed by Sanofi-Aventis S.A. as Allegra in the U.S. and as Telfast outside the U.S. Since its launch in 1995, AMRI has earned more than $367.3 million in royalty and milestone revenue from this product. The firm has also licensed the rights to develop and commercialize two chemicals (amine neurotransmitter reuptake inhibitors) from Bristol-Myers Squibb Company (BMS). In addition to developing its own drugs, AMRI has increasingly acted as a custom research and development source for the pharmaceutical, genomic and biotechnology industries. It provides contract services across the entire product development cycle, from lead discovery to commercial manufacturing. The company's services allow pharmaceutical companies to outsource their chemistry departments in order to pursue a greater number of drug discovery and development opportunities. An integral part of these contract operations consists of several facilities in India and Singapore, which were launched as part of a strategic move to globalize its services. The firm also has domestic research facilities in Albany, Syracuse and Rensselaer, New York as well as in Bothell, Washington. In February 2008, AMRI acquired a custom pilot scale intermediaries manufacturing firm, FineKem Laboratories Pvt. Limited, based in Aurangabad, India. In June 2008, the first BMS-licensed chemical, indicated for treating depression, entered Phase I clinical trials in Canada. In September 2008, the second chemical, a novel cancer treatment, entered preclinical trials, and subsequent positive test results moved the chemical into Phase I trials by October 2008.

## FINANCIALS: Sales and profits are in thousands of dollars—add 000 to get the full amount. 2008 Note: Financial information for 2008 was not available for all companies at press time.

| | | |
|---|---|---|
| 2008 Sales: $229,260 | 2008 Profits: $20,560 | U.S. Stock Ticker: AMRI |
| 2007 Sales: $192,511 | 2007 Profits: $8,936 | Int'l Ticker:    Int'l Exchange: |
| 2006 Sales: $179,807 | 2006 Profits: $2,183 | Employees: 1,357 |
| 2005 Sales: $183,906 | 2005 Profits: $16,321 | Fiscal Year Ends: 12/31 |
| 2004 Sales: $169,527 | 2004 Profits: $-11,691 | Parent Company: |

## SALARIES/BENEFITS:

| Pension Plan: | ESOP Stock Plan: | Profit Sharing: | Top Exec. Salary: $484,615 | Bonus: $202,750 |
|---|---|---|---|---|
| Savings Plan: Y | Stock Purch. Plan: Y | | Second Exec. Salary: $309,692 | Bonus: $95,550 |

## OTHER THOUGHTS:

**Apparent Women Officers or Directors:** 2
**Hot Spot for Advancement for Women/Minorities:** Y

## LOCATIONS: ("Y" = Yes)

| West: | Southwest: | Midwest: | Southeast: | Northeast: | International: |
|---|---|---|---|---|---|
| Y | | | | Y | Y |

Note: Financial information, benefits and other data can change quickly and may vary from those stated here.

# ALCON INC

**www.alcon.com**

**Industry Group Code: 325412  Ranks within this company's industry group:** Sales: 22  Profits: 15

| Drugs: | | Other: | Clinical: | | Computers: | | Services: | |
|---|---|---|---|---|---|---|---|---|
| Discovery: | Y | AgriBio: | Trials/Services: | | Hardware: | | Specialty Services: | |
| Licensing: | | Genetic Data: | Labs: | | Software: | | Consulting: | |
| Manufacturing: | Y | Tissue Replacement: | Equipment/Supplies: | Y | Arrays: | | Blood Collection: | |
| Genetics: | | | Research & Development Services: | | Database Management: | | Drug Delivery: | |
| | | | Diagnostics: | | | | Drug Distribution: | |

## TYPES OF BUSINESS:

Eye Care Products
Ophthalmic Products & Equipment
Contact Lens Care Products
Surgical Instruments

## BRANDS/DIVISIONS/AFFILIATES:

Opti-Free
Patanol
AcrySof
Systane
WaveLight AG
Alcon Surgical
William C. Conner Research Center
Nestle Corporation

## CONTACTS: *Note: Officers with more than one job title may be intentionally listed here more than once.*

Cary Rayment, CEO
Cary Rayment, Pres.
Richard J. Croarkin, CFO/Sr. VP
Kevin J. Buehler, Chief Mktg. Officer
Sabri Markabi, Sr. VP-R&D
Ed McGough, Sr. VP-Tech.
Ed McGough, Sr. VP-Global Mfg.
Elaine E. Whitbeck, General Counsel/Chief Legal Officer
Cary Rayment, Chmn.

| **Phone:** 41-41-785-8888 | **Fax:** |
|---|---|
| **Toll-Free:** | |
| **Address:** Bosch 69, Hunenberg, CH-6331 Switzerland | |

## GROWTH PLANS/SPECIAL FEATURES:

Alcon, Inc. is one of the world's largest eye care product companies. The firm manages its business through two business segments: Alcon United States and Alcon International. Its portfolio spans three key ophthalmic categories: pharmaceutical, surgical and consumer eye care products. The divisions develop, manufacture and market ophthalmic pharmaceuticals, surgical equipment and devices, contact lens care products and other consumer eye care products that treat diseases and conditions of the eye. Alcon maintains manufacturing plants, laboratories and offices in 75 countries and offers its products and services in over 180 countries. The firm holds approximately 4,500 global patents and 3,200 pending patent applications. The company makes more than 10,000 unique products, including prescription and over-the-counter drugs, contact lens solutions, surgical instruments, intraocular lenses and office systems for ophthalmologists. Its brand names include Patanol solution for eye allergies, AcrySof intraocular lenses, Systane lubricant drops for dry eye and the Opti-Free system for contact lens care. Alcon's research and development headquarters houses the 400,000 square-foot William C. Conner Research Center in Texas. The firm also has research and development laboratories in California, Florida, Switzerland and Spain. Alcon Surgical creates implantable lenses, viscoelastics and medical tools specifically made for ocular surgeons, including instruments for cataract removal and absorbable sutures. Sales of glaucoma products account for 35.9% of total pharmaceutical sales. Nestle Corporation owns 77% of the firm. In 2008, the firm agreed to be acquired from Nestle SA by Novartis AG for $39 billion. In September 2008, Alcon announced plans to build an additional 74,000 square foot manufacturing facility in West Virginia.

Alcon matches employee contributions to a 401(k) up to 5% of total compensation. In addition, it offers employees a retirement plan and total company combined contributions to the Alcon 401(k) and the Alcon Retirement Plan can be as much as 12%.

## FINANCIALS: Sales and profits are in thousands of dollars—add 000 to get the full amount. 2008 Note: Financial information for 2008 was not available for all companies at press time.

| | | |
|---|---|---|
| 2008 Sales: $6,294,000 | 2008 Profits: $2,046,000 | **U.S. Stock Ticker: ACL** |
| 2007 Sales: $5,599,600 | 2007 Profits: $1,586,400 | **Int'l Ticker:**  Int'l Exchange: |
| 2006 Sales: $4,896,600 | 2006 Profits: $1,348,100 | Employees:  14,500 |
| 2005 Sales: $4,368,500 | 2005 Profits: $931,000 | Fiscal Year Ends: 12/31 |
| 2004 Sales: $3,913,600 | 2004 Profits: $871,800 | Parent Company: |

## SALARIES/BENEFITS:

| Pension Plan: Y | ESOP Stock Plan: | Profit Sharing: | Top Exec. Salary: $1,250,000 | Bonus: $1,375,000 |
|---|---|---|---|---|
| Savings Plan: Y | Stock Purch. Plan: | | Second Exec. Salary: $570,833 | Bonus: $390,000 |

## OTHER THOUGHTS:

**Apparent Women Officers or Directors**: 2
**Hot Spot for Advancement for Women/Minorities**: Y

## LOCATIONS: ("Y" = Yes)

| West: | Southwest: | Midwest: | Southeast: | Northeast: | International: |
|---|---|---|---|---|---|
| Y | Y | | Y | | Y |

# ALEXION PHARMACEUTICALS INC

### www.alexionpharm.com

**Industry Group Code: 325412 Ranks within this company's industry group:** Sales: 53 Profits: 42

| Drugs: | | Other: | Clinical: | Computers: | Services: |
|---|---|---|---|---|---|
| Discovery: | Y | AgriBio: | Trials/Services: | Hardware: | Specialty Services: |
| Licensing: | | Genetic Data: | Labs: | Software: | Consulting: |
| Manufacturing: | Y | Tissue Replacement: | Equipment/Supplies: | Arrays: | Blood Collection: |
| Genetics: | | | Research & Development Services: | Database Management: | Drug Delivery: |
| | | | Diagnostics: | | Drug Distribution: |

## TYPES OF BUSINESS:

Therapeutic Products
Hematologic Diseases
Neurological Diseases
Cancer
Autoimmune Disorders

## BRANDS/DIVISIONS/AFFILIATES:

Soliris
Alexion Europe SAS
ALXN6000
Eculizumab

## CONTACTS: *Note: Officers with more than one job title may be intentionally listed here more than once.*

Leonard Bell, CEO/Treas./Sec.
David W. Keiser, COO
David W. Keiser, Pres.
Vikas Sinha, CFO/Sr. VP
Glenn Melrose, VP-Human Resources
Stephen P. Squinto, Exec. VP/Head-R&D
Daniel N. Caron, Exec. Dir.-Eng.
M. Stacy Hooks, Sr. VP-Mfg. & Tech. Oper.
Thomas I.H. Dubin, General Counsel/Sr. VP
Daniel N. Caron, Exec. Dir.-Oper.
Claude Nicaise, Sr. VP-Strategic Product Dev.
Barry P. Luke, VP-Finance/Asst. Sec.
Camille Bedrosian, Sr. VP/Chief Medical Officer
David Hallal, Sr. VP-U.S. Commercial Oper.
Russell P. Rother, Sr. VP/Chief Scientific Officer
Glenn Melrose, VP-Human Resources
Max Link, Chmn.
Patrice Coissac, Gen. Mgr./Pres., Alexion Europe SAS

| | |
|---|---|
| **Phone:** 203-272-2596 | **Fax:** 203-271-8190 |
| **Toll-Free:** | |
| **Address:** 325 Knotter Dr., Cheshire, CT 06410 US | |

## GROWTH PLANS/SPECIAL FEATURES:

Alexion Pharmaceuticals, Inc. engages in the discovery, development and delivery of therapeutic products to treat patients with severe disease states, including hematologic diseases, neurologic disease, cancer and autoimmune disorders. The company devotes substantially all of its resources to drug discovery, research, and product and clinical development. The firm has one market product, Soliris (eculizumab), the first therapy approved for the treatment of paroxysmal nocturnal hemoglobinuria (PNH). PNH is a rare genetic deficiency blood disorder, in which a patient's own complement system attacks and destroys blood cells. Soliris, a genetically altered antibody known as C5 complement inhibitor, treats PNH by selectively blocking the production of inflammation-causing proteins in the complement cascade. The company's other products are currently in preclinical or phase trial stages. These products include intravenous eculizumab for the treatment of myasthenia gravis, multifocal motor neuropathy and asthma; intravitreal eculizumab for the treatment of age-related macular degeneration; anti-CD200 MAb (monoclonal antibody), which uses the working name ALXN6000, for the treatment of multiple myeloma and leukemia; and DC-SIGN MAb, a cancer vaccine. Preliminary findings have also supported the use of Soliris to fight atypical Hemolytic Uremic Syndrome (aHUS). All of the firm's products focus on anti-inflammatory therapeutics for disease in which the complement cascade is activated. Alexion also operates several subsidiaries in Europe, including Alexion Europe SAS, which supports commercial and regulatory international operations. In January 2009, Soliris was approved for treatment use in Canada.

## FINANCIALS: Sales and profits are in thousands of dollars—add 000 to get the full amount. 2008 Note: Financial information for 2008 was not available for all companies at press time.

| | | |
|---|---|---|
| 2008 Sales: $259,099 | 2008 Profits: $33,149 | **U.S. Stock Ticker: ALXN** |
| 2007 Sales: $72,041 | 2007 Profits: $-92,290 | **Int'l Ticker:** Int'l Exchange: |
| 2006 Sales: $1,558 | 2006 Profits: $-131,514 | Employees: 504 |
| 2005 Sales: $1,064 | 2005 Profits: $-108,750 | Fiscal Year Ends: 12/31 |
| 2004 Sales: $4,609 | 2004 Profits: $-74,095 | Parent Company: |

## SALARIES/BENEFITS:

| | | | | |
|---|---|---|---|---|
| Pension Plan: | ESOP Stock Plan: | Profit Sharing: | Top Exec. Salary: $622,752 | Bonus: $750,000 |
| Savings Plan: | Stock Purch. Plan: | | Second Exec. Salary: $399,538 | Bonus: $184,000 |

## OTHER THOUGHTS:

**Apparent Women Officers or Directors:**
**Hot Spot for Advancement for Women/Minorities:**

## LOCATIONS: ("Y" = Yes)

| West: | Southwest: | Midwest: | Southeast: | Northeast: | International: |
|---|---|---|---|---|---|
| | | | | Y | Y |

# ALFACELL CORPORATION

**www.alfacell.com**

**Industry Group Code: 325412  Ranks within this company's industry group:** Sales:     Profits: 86

| Drugs: | | Other: | Clinical: | Computers: | Services: |
|---|---|---|---|---|---|
| Discovery: | Y | AgriBio: | Trials/Services: | Hardware: | Specialty Services: |
| Licensing: | | Genetic Data: | Labs: | Software: | Consulting: |
| Manufacturing: | Y | Tissue Replacement: | Equipment/Supplies: | Arrays: | Blood Collection: |
| Genetics: | | | Research & Development Services: | Database Management: | Drug Delivery: |
| | | | Diagnostics: | | Drug Distribution: |

## TYPES OF BUSINESS:

Cancer & Pathological Conditions Drugs
RNase-based Drugs

## BRANDS/DIVISIONS/AFFILIATES:

Onconase
Amphinases

## CONTACTS: *Note: Officers with more than one job title may be intentionally listed here more than once.*

Kuslima Shogen, CEO
Diane Scudiery, Dir.-Clinical & Regulatory Oper.
David Sidransky, Chmn.

| **Phone:** 732-652-4525 | **Fax:** 732-652-4575 |
|---|---|
| **Toll-Free:** | |
| **Address:** 300 Atrium Dr., Somerset, NJ 08873 US | |

## GROWTH PLANS/SPECIAL FEATURES:

Alfacell Corporation is a biopharmaceutical company engaged in the research, development and commercialization of drugs for life-threatening diseases, such as malignant mesothelioma and other cancers. The company's drug discovery and development program consists of novel therapeutics that are being developed from amphibian ribonucleases (RNases). RNases are biologically active enzymes that split RNA molecules. The firm uses RNases for the development of therapeutics for cancer and other life-threatening diseases, including HIV and autoimmune diseases, that require anti-proliferative and apoptotic, or programmed cell death, properties. The company's proprietary product is Onconase, which targets solid tumors that have become resistant to other chemotherapeutic drugs. Onconase affects primarily exponentially growing malignant cells, with activity controlled through specific molecular mechanisms. The drug is being evaluated as a treatment for inoperable malignant mesothelioma, a rare cancer primarily affecting the pleura (lining of the lungs) usually caused by exposure to asbestos. The drug received orphan drug designation for malignant mesothelioma in Australia and the U.S., as well as from the European Agency for the Evaluation of Medicinal Products (EMEA). Alfacell is also evaluating Onconase in Phase I/II clinical development for applications in treating lung cancer and other solid tumors. The company also has another amphibian RNases product, known as Amphinases, in pre-clinical research and development. The firm was awarded a U.S. patent for a methodology for synthesizing gene sequences of ranpirnase, the active component of Onconase. In July 2008, the firm announced an exclusive distribution agreement with Megapharm, a leading Israeli pharmaceutical company, for the commercialization on Onconase in Israel.

## FINANCIALS: Sales and profits are in thousands of dollars—add 000 to get the full amount. 2008 Note: Financial information for 2008 was not available for all companies at press time.

| | | |
|---|---|---|
| 2008 Sales: $ | 2008 Profits: $-12,321 | **U.S. Stock Ticker: ACEL** |
| 2007 Sales: $ | 2007 Profits: $-8,755 | **Int'l Ticker:**     Int'l Exchange: |
| 2006 Sales: $ | 2006 Profits: $-7,810 | Employees:   15 |
| 2005 Sales: $ 152 | 2005 Profits: $-6,462 | Fiscal Year Ends: 7/31 |
| 2004 Sales: $ 42 | 2004 Profits: $-5,070 | Parent Company: |

## SALARIES/BENEFITS:

| Pension Plan: | ESOP Stock Plan: | Profit Sharing: | Top Exec. Salary: $278,877 | Bonus: $500,000 |
|---|---|---|---|---|
| Savings Plan: | Stock Purch. Plan: | | Second Exec. Salary: $215,231 | Bonus: $42,000 |

## OTHER THOUGHTS:

**Apparent Women Officers or Directors**: 1
**Hot Spot for Advancement for Women/Minorities**:

## LOCATIONS: ("Y" = Yes)

| West: | Southwest: | Midwest: | Southeast: | Northeast: | International: |
|---|---|---|---|---|---|
| | | | | Y | |

# ALIZYME PLC

www.alizyme.com

Industry Group Code: 325412  Ranks within this company's industry group:  Sales: 136   Profits: 101

| Drugs: | | Other: | Clinical: | | Computers: | | Services: | |
|---|---|---|---|---|---|---|---|---|
| Discovery: | Y | AgriBio: | Trials/Services: | | Hardware: | | Specialty Services: | |
| Licensing: | Y | Genetic Data: | Labs: | | Software: | | Consulting: | |
| Manufacturing: | | Tissue Replacement: | Equipment/Supplies: | | Arrays: | | Blood Collection: | |
| Genetics: | | | Research & Development Services: | | Database Management: | | Drug Delivery: | Y |
| | | | Diagnostics: | | | | Drug Distribution: | |

## TYPES OF BUSINESS:

Drugs-Gastrointestinal
Colonic Drug Delivery System
Obesity Treatments

## BRANDS/DIVISIONS/AFFILIATES:

COLAL
Renzapride
Cetilistat
ATL-104
COLAL-PRED
Norgine B.V.
Prometheus Laboratories

## CONTACTS: Note: Officers with more than one job title may be intentionally listed here more than once.

Timothy P. McCarthy, CEO
Roger I. Hickling, Dir.-R&D
Nick Blech, Corp. Sec.
Neil Fish, Dir.-Bus. Dev.
Brian Richards, Chmn.

| Phone: 44-1223-896-000 | Fax: 44-1223-896-001 |
|---|---|
| Toll-Free: | |
| Address: Granta Park, Great Abington, Cambridge,  CB21 6GX UK | |

## GROWTH PLANS/SPECIAL FEATURES:

Alizyme plc is a specialty biopharmaceutical company focused on therapy associated with metabolic disorders, gastrointestinal distress and cancer.  The firm develops prescription drugs for the treatment of obesity and related diseases as well as gastrointestinal disorders, including irritable bowel syndrome (IBS), inflammatory bowel diseases (IBD), ulcerative colitis and mucositis, a gastrointestinal side effect of chemotherapy.  The company owns the intellectual property rights for many of its key products, including the COLAL drug delivery system, which delivers drugs directly to the colon.  The firm has received FDA approval to begin Phase III clinical trials for Cetilistat (previously ATL-962) for the treatment of obesity.  Alizyme has a licensing agreement with Takeda Chemical Industries, Ltd. for Cetilistat in Japan. The company has completed Phase IIa trials for ATL-104 in the U.K. for the treatment of mucositis, and is in preparation to begin Phase III trials.  COLAL-PRED, used for the treatment of IBD, is also currently in development.  This product incorporates an anti-inflammatory steroid with the colonic drug delivery system COLAL and was developed for the management of ulcerative colitis.  COLAL-PRED is also in Pre-Phase III development for the maintenance of remission following ulcerative colitis.  In May 2008, the company's partner for COLAL-PRED development in North America, Prometheus Laboratories, commenced a Phase II study of the drug in the U.S.  In July 2008, a Phase III clinical trail of COLAL-PRED involving approximately 800 patients in Europe was completed, with the drug showing continuing good prospects with regard to safety and efficacy.  In December 2008, the firm commenced Phase III clinical trials of Cetilistat in Japan.  Also during 2008, the firm removed the drug Renzapride from its portfolio.

## FINANCIALS: Sales and profits are in thousands of dollars—add 000 to get the full amount. 2008 Note: Financial information for 2008 was not available for all companies at press time.

| | | |
|---|---|---|
| 2008 Sales: $3,060 | 2008 Profits: $-19,350 | U.S. Stock Ticker: Private |
| 2007 Sales: $ 20 | 2007 Profits: $-55,220 | Int'l Ticker: AZM    Int'l Exchange: London-LSE |
| 2006 Sales: $2,296 | 2006 Profits: $-39,820 | Employees:   20 |
| 2005 Sales: $ | 2005 Profits: $-38,260 | Fiscal Year Ends: 12/31 |
| 2004 Sales: $3,333 | 2004 Profits: $-12,360 | Parent Company: |

## SALARIES/BENEFITS:

| Pension Plan: | ESOP Stock Plan: | Profit Sharing: | Top Exec. Salary: $ | Bonus: $ |
|---|---|---|---|---|
| Savings Plan: | Stock Purch. Plan: | | Second Exec. Salary: $ | Bonus: $ |

## OTHER THOUGHTS:

Apparent Women Officers or Directors:
Hot Spot for Advancement for Women/Minorities:

## LOCATIONS: ("Y" = Yes)

| West: | Southwest: | Midwest: | Southeast: | Northeast: | International: |
|---|---|---|---|---|---|
| | | | | | Y |

# ALKERMES INC

**www.alkermes.com**

**Industry Group Code: 325412A Ranks within this company's industry group:** Sales: 10   Profits: 4

| Drugs: | | Other: | Clinical: | Computers: | Services: |
|---|---|---|---|---|---|
| Discovery: | | AgriBio: | Trials/Services: | Hardware: | Specialty Services: |
| Licensing: | | Genetic Data: | Labs: | Software: | Consulting: |
| Manufacturing: | Y | Tissue Replacement: | Equipment/Supplies: | Arrays: | Blood Collection: |
| Genetics: | | | Research & Development Services: | Database Management: | Drug Delivery: |
| | | | Diagnostics: | | Drug Distribution: |

## TYPES OF BUSINESS:

Drug Delivery Systems
Pulmonary Drug Delivery Systems
Sustained Release Injection Delivery Systems

## BRANDS/DIVISIONS/AFFILIATES:

Vivitrol
Risperdal Consta
Medisorb
Janssen Pharmaceutica, Inc
Cephalon, Inc.

## CONTACTS: Note: Officers with more than one job title may be intentionally listed here more than once.

David A. Broecker, CEO
Gordon G. Pugh, COO/Sr. VP
David A. Broecker, Pres.
James M. Frates, CFO/Sr. VP
Madeline D. Coffin, VP-Human Resources
Elliot W. Ehrich, Sr. VP-R&D/Chief Medical Officer
Kathryn L. Biberstein, General Counsel/Sr. VP/Sec.
Michael J. Landine, Sr. VP-Corp. Dev.
James M. Frates, Treas.
Kathryn L. Biberstein, Chief Compliance Officer
Dennis J. Bucceri, VP-Regulatory Affairs
Stephen King, VP-Commercial Oper.
Richard Pops, Chmn.

| Phone: 617-494-0171 | Fax: 617-494-9263 |
|---|---|
| Toll-Free: | |
| Address: 88 Sidney St., Cambridge, MA 02139 US | |

## GROWTH PLANS/SPECIAL FEATURES:

Alkermes, Inc. is a biotechnology company that specializes in the development of sophisticated drug delivery technologies. The company currently markets two commercial products: Risperdal Consta, a long-acting atypical antipsychotic medication for schizophrenia, and Vivitrol, an injectable medication for the treatment of alcohol dependence. Risperdal Consta is a long-acting formulation of risperidone, a product of Janssen Pharmaceutica, Inc. Risperdal Consta is the first long-acting, atypical antipsychotic to be approved by the U.S. Food and Drug Administration (FDA). The medication uses the firm's proprietary Medisorb technology to deliver and maintain therapeutic medication levels in the body through just one injection every two weeks. Risperdal Consta is approved in approximately 85 countries and marketed in approximately 60 countries, and Janssen continues to launch the product around the world. In May 2008, Alkermes and Janssen agreed to begin development of a four-week formulation of Risperdal Consta. Vivitrol is an extended-release formulation, for the treatment of alcohol dependence. Each injection of Vivitrol provides medication for one month and alleviates the need for patients to make daily medication dosing decisions. Cephalon, Inc. is primarily responsible for marketing Vivitrol in the U.S. Alkermes is the exclusive manufacturer of Vivitrol. In addition to its marketed products, Alkermes also develops extended-release injectable, pulmonary and oral products for the treatment of central nervous system disorders, addiction and diabetes. Additional developing products in the company's pipeline include an injectable form of Amylin's Byetta, used to treat type 2 diabetes; ALKS 27 for the treatment of chronic obstructive pulmonary disease; ALKS 29 for treatment of alcohol dependence; and AKLS 33, which may be used to treat a broad range of diseases and medical conditions. In April 2009, the firm announced plans to relocate its headquarters from Cambridge, Massachusetts to Waltham, Massachusetts. The move is expected to be complete by 2010.

## FINANCIALS: Sales and profits are in thousands of dollars—add 000 to get the full amount. 2008 Note: Financial information for 2008 was not available for all companies at press time.

| | | |
|---|---|---|
| 2008 Sales: $240,717 | 2008 Profits: $166,979 | **U.S. Stock Ticker: ALKS** |
| 2007 Sales: $239,965 | 2007 Profits: $9,445 | **Int'l Ticker:** Int'l Exchange: |
| 2006 Sales: $166,601 | 2006 Profits: $3,818 | Employees: 570 |
| 2005 Sales: $76,126 | 2005 Profits: $-73,916 | Fiscal Year Ends: 3/31 |
| 2004 Sales: $39,054 | 2004 Profits: $-102,385 | Parent Company: |

## SALARIES/BENEFITS:

| Pension Plan: | ESOP Stock Plan: Y | Profit Sharing: | Top Exec. Salary: $608,721 | Bonus: $ |
|---|---|---|---|---|
| Savings Plan: Y | Stock Purch. Plan: | | Second Exec. Salary: $472,278 | Bonus: $ |

## OTHER THOUGHTS:

**Apparent Women Officers or Directors:** 2
**Hot Spot for Advancement for Women/Minorities:** Y

## LOCATIONS: ("Y" = Yes)

| West: | Southwest: | Midwest: | Southeast: | Northeast: | International: |
|---|---|---|---|---|---|
| | | Y | | Y | |

# ALLERGAN INC

**www.allergan.com**

Industry Group Code: 325412    Ranks within this company's industry group: Sales: 27    Profits: 23

| Drugs: | | Other: | Clinical: | Computers: | Services: |
|---|---|---|---|---|---|
| Discovery: | Y | AgriBio: | Trials/Services: | Hardware: | Specialty Services: |
| Licensing: | | Genetic Data: | Labs: | Software: | Consulting: |
| Manufacturing: | Y | Tissue Replacement: | Equipment/Supplies: | Arrays: | Blood Collection: |
| Genetics: | | | Research & Development Services: | Database Management: | Drug Delivery: |
| | | | Diagnostics: | | Drug Distribution: |

## TYPES OF BUSINESS:

Pharmaceutical Development
Eye Care Supplies
Dermatological Products
Neuromodulator Products
Obesity Intervention Products
Urologic Products
Medical Aesthetics

## BRANDS/DIVISIONS/AFFILIATES:

Restasis
Lumigan
Optive
Latisse
Botox
Sanctura XR
Aczone
Spectrum Pharmaceuticals, Inc.

## CONTACTS: Note: Officers with more than one job title may be intentionally listed here more than once.

David Pyott, CEO
F. Michael Ball, Pres.
Jeffrey L. Edwards, CFO
Dianne Dyer-Bruggeman, Exec. VP-Human Resources
Scott M. Whitcup, Exec. VP-R&D
Raymond H. Diradoorian, Exec. VP-Global Tech. Oper.
Douglas S. Ingram, Chief Admin. Officer/Chief Ethics Officer
Douglas S. Ingram, General Counsel/Corp. Sec./Exec. VP
Jeffrey L. Edwards, Exec. VP-Bus. Dev.
Jeffrey L. Edwards, Exec. VP-Finance
David Pyott, Chmn.

| Phone: 714-246-4500 | Fax: 714-246-6987 |
|---|---|
| Toll-Free: 800-433-8871 | |
| Address: 2525 Dupont Dr., Irvine, CA 92612 US | |

## GROWTH PLANS/SPECIAL FEATURES:

Allergan, Inc. is a technology-driven global health care company that develops and commercializes specialty pharmaceutical products, biologics and medical devices for the ophthalmic, neurological, medical aesthetics, medical dermatology, breast aesthetics, obesity intervention, urological and other specialty markets in more than 100 countries. The company focuses on treatments for chronic dry eye, glaucoma, retinal disease, psoriasis, acne, movement disorders, neuropathic pain and genitourinary diseases. The company operates in two segments: specialty pharmaceuticals and medical devices. The specialty pharmaceuticals segment includes eye care products, such as Restasis ophthalmic emulsion, Lumigan ophthalmic solution, Optive lubricant eye drops and the Refresh line of artificial tears; Botox, used in the treatment of neuromuscular disorders, pain management, the temporary improvement of wrinkles and for certain other therapeutic and aesthetic indications; skin care products, principally tazarotene products in cream and gel formulations for the treatment of acne, facial wrinkles and psoriasis, marketed under the name Tazorac; eyelash growth products; and urologics products, including Sanctura XR, a medication for overactive bladder. The medical devices segment includes breast implants for augmentation, revision and reconstructive surgery; obesity intervention products, including the Lap-Band, an adjustable gastric banding system, and the Orbera intragastric balloon system; and facial aesthetics products, including the Juvederm line of dermal filler products. In July 2008, the company completed the acquisition of the Aczone Gel 5% product, a topical treatment for acne vulgaris, from QLT, Inc. In January 2009, the firm launched Latisse, the first and only FDA-approved prescription treatment designed for eyelash growth.

The firm offers employees benefits including a 401(k) plan; a defined benefit retirement contribution; adoption assistance; education assistance; before-tax flex dollars and flexible spending accounts; backup child care; an employee credit union; an employee assistance program; dependent scholarship awards; and U.S. savings bond deductions.

## FINANCIALS: Sales and profits are in thousands of dollars—add 000 to get the full amount. 2008 Note: Financial information for 2008 was not available for all companies at press time.

| | | |
|---|---|---|
| 2008 Sales: $4,339,700 | 2008 Profits: $578,600 | U.S. Stock Ticker: AGN |
| 2007 Sales: $3,879,000 | 2007 Profits: $499,300 | Int'l Ticker:    Int'l Exchange: |
| 2006 Sales: $3,010,100 | 2006 Profits: $-127,400 | Employees: 7,886 |
| 2005 Sales: $2,319,200 | 2005 Profits: $403,900 | Fiscal Year Ends: 12/31 |
| 2004 Sales: $2,045,600 | 2004 Profits: $377,100 | Parent Company: |

## SALARIES/BENEFITS:

| | | | | |
|---|---|---|---|---|
| Pension Plan: Y | ESOP Stock Plan: | Profit Sharing: | Top Exec. Salary: $1,350,000 | Bonus: $1,212,100 |
| Savings Plan: Y | Stock Purch. Plan: | | Second Exec. Salary: $683,308 | Bonus: $324,100 |

## OTHER THOUGHTS:

**Apparent Women Officers or Directors:** 3
**Hot Spot for Advancement for Women/Minorities:** Y

## LOCATIONS: ("Y" = Yes)

| West: | Southwest: | Midwest: | Southeast: | Northeast: | International: |
|---|---|---|---|---|---|
| Y | | | | | Y |

Note: Financial information, benefits and other data can change quickly and may vary from those stated here.

# ALLIANCE PHARMACEUTICAL CORP

www.allp.com

**Industry Group Code: 325412  Ranks within this company's industry group:  Sales:     Profits:**

| Drugs: | | Other: | Clinical: | Computers: | Services: |
|---|---|---|---|---|---|
| Discovery: | Y | AgriBio: | Trials/Services: | Hardware: | Specialty Services: |
| Licensing: | | Genetic Data: | Labs: | Software: | Consulting: |
| Manufacturing: | | Tissue Replacement: | Equipment/Supplies: | Arrays: | Blood Collection: |
| Genetics: | | | Research & Development Services: | Database Management: | Drug Delivery: |
| | | | Diagnostics: | | Drug Distribution: |

## TYPES OF BUSINESS:

Drugs-Cardiovascular & Respiratory
Blood Substitutes
Immune Disorder Therapies

## BRANDS/DIVISIONS/AFFILIATES:

Oxygent
Beijing Double-Crane Pharmaceutical Co., Ltd.
Leo Pharma A/S
Il Yang Pharm. Co., Ltd.

## CONTACTS: *Note: Officers with more than one job title may be intentionally listed here more than once.*

Duane J. Roth, CEO
B. Jack Defranco, COO
B. Jack DeFranco, Pres.
B. Jack DeFranco, CFO
Duane J. Roth, Chmn.

| Phone: 858-779-1458 | Fax: 858-427-0646 |
|---|---|
| Toll-Free: | |
| Address: 4660 La Jolla Village Dr., Ste. 825, San Diego, CA 92122 US | |

## GROWTH PLANS/SPECIAL FEATURES:

Alliance Pharmaceutical Corp. develops therapeutic and diagnostic products that utilize perfluorochemicals, which are chemical substances with high oxygen-carrying capacity. Artificial oxygen carriers are intended to help ease blood shortages and avoid
transfusion-related safety issues, such as the presence of blood transmissible diseases like Chagas disease in a blood supply.  The company's leading product candidate, Oxygent, is an intravascular oxygen carrier designed to augment oxygen delivery in surgical patients.   Oxygent is also intended to be an emulsion-based red blood cell substitute. The blood substitute is sterile and universally compatible with all blood types.  Oxygent has two indications, the first being to provide oxygen to tissues during elective surgeries, and the second  as a therapeutic for organ perfusion during major elective surgeries to protect vital organs from hypoxic injury. In clinical studies, Oxygent has been shown to correct or mean arterial pressure, heart rate and mixed venous oxygen tension.   Alliance  has  formed  many  collaborative relationships with other companies to test, manufacture and eventually distribute Oxygent in other countries, including Beijing Double-Crane Pharmaceutical Co., Ltd. in China; Leo Pharma A/S in Europe, including the EU member and applicant countries, and Canada; and Il Yang Pharm. Co., Ltd. in South Korea.

## FINANCIALS:  Sales and profits are in thousands of dollars—add 000 to get the full amount. 2008 Note: Financial information for 2008 was not available for all companies at press time.

| | | |
|---|---|---|
| 2008 Sales: $ | 2008 Profits: $ | U.S. Stock Ticker: ALLP |
| 2007 Sales: $ 573 | 2007 Profits: $-4,093 | Int'l Ticker:     Int'l Exchange: |
| 2006 Sales: $ 129 | 2006 Profits: $-9,575 | Employees:     4 |
| 2005 Sales: $1,477 | 2005 Profits: $-5,743 | Fiscal Year Ends: 6/30 |
| 2004 Sales: $ 549 | 2004 Profits: $10,016 | Parent Company: |

## SALARIES/BENEFITS:

| Pension Plan: | ESOP Stock Plan: | Profit Sharing: | Top Exec. Salary: $235,000 | Bonus: $ |
|---|---|---|---|---|
| Savings Plan: | Stock Purch. Plan: | | Second Exec. Salary: $120,000 | Bonus: $ |

## OTHER THOUGHTS:

**Apparent Women Officers or Directors:**
**Hot Spot for Advancement for Women/Minorities:**

## LOCATIONS: ("Y" = Yes)

| West: | Southwest: | Midwest: | Southeast: | Northeast: | International: |
|---|---|---|---|---|---|
| Y | | | | | |

# ALLOS THERAPEUTICS INC

**www.allos.com**

**Industry Group Code: 325412  Ranks within this company's industry group:** Sales:    Profits: 140

| Drugs: | | Other: | Clinical: | Computers: | Services: |
|---|---|---|---|---|---|
| Discovery: | Y | AgriBio: | Trials/Services: | Hardware: | Specialty Services: |
| Licensing: | Y | Genetic Data: | Labs: | Software: | Consulting: |
| Manufacturing: | Y | Tissue Replacement: | Equipment/Supplies: | Arrays: | Blood Collection: |
| Genetics: | | | Research & Development Services: | Database Management: | Drug Delivery: |
| | | | Diagnostics: | | Drug Distribution: |

## TYPES OF BUSINESS:

Cancer Treatment Drugs
Small-Molecule Therapies

## BRANDS/DIVISIONS/AFFILIATES:

Pralatrexate

## CONTACTS: Note: Officers with more than one job title may be intentionally listed here more than once.

Paul L. Berns, CEO
Paul L. Berns, Pres.
James V. Caruso, Chief Commercial Officer/Exec. VP
Pablo J. Cagnoni, Chief Medical Officer/Sr. VP
Marc H. Graboyes, General Counsel/Sr. VP
Bruce K. Bennett, VP-Pharmaceutical Oper.
Monique Greer, Corp. Comm.
Monique Greer, Investor Rel.
David C. Clark, VP-Finance
Stephen J. Hoffman, Chmn.

| **Phone:** 303-426-6262 | **Fax:** 303-426-4731 |
|---|---|

**Toll-Free:**

**Address:** 11080 CirclePoint Rd., Ste. 200, Westminster, CO 80020 US

## GROWTH PLANS/SPECIAL FEATURES:

Allos Therapeutics, Inc. focuses on developing and commercializing small molecule drugs for cancer treatment. The company currently has one product candidate: Pralatrexate. Pralatrexate is a small molecule chemotherapeutic agent that inhibits dihydrofolate reductase, a folic acid-dependent enzyme involved in several processes, including the building of nucleic acids. Pralatrexate was designed for transport into tumor cells via the reduced folate carrier and effective intracellular drug retention. The company has been conducting Phase II studies of Pralatrexate in patients with relapsed or refractory peripheral T-cell Lymphoma (the PROPEL trial) and relapsed or refractory Non-Hodgkin's Lymphoma or Hodgkin's disease (B-cell). Additional Phase II studies are underway for Pralatrexate in the areas of stage IIIB/IV non-small cell lung cancer (NSCLC) and advanced or metastatic relapsed transitional cell carcinoma (TCC) of the urinary bladder, in addition to a Phase I study of patients with previously treated (Stage IIB-IV) advanced NSCLC. In February 2009, the company announced the completion of the PROPEL trial, with Pralatrexate eliciting a response in 27% of the 109 subjects. Based on these results, in March 2009, Allos submitted a new drug application (NDA) to the U.S. Food and Drug Administration for Pralatrexate to be considered as a treatment for patients with relapsed or refractory peripheral T-cell Lymphoma.

The company offers its employees a 401(k) plan; health, dental and vision insurance; group life insurance; short- and long-term disability; educational reimbursement; an employee stock purchase plan; and flexible spending accounts.

## FINANCIALS: Sales and profits are in thousands of dollars—add 000 to get the full amount. 2008 Note: Financial information for 2008 was not available for all companies at press time.

| | | |
|---|---|---|
| 2008 Sales: $ | 2008 Profits: $-51,730 | **U.S. Stock Ticker: ALTH** |
| 2007 Sales: $ | 2007 Profits: $-39,370 | **Int'l Ticker:**    Int'l Exchange: |
| 2006 Sales: $ | 2006 Profits: $-30,212 | Employees:   81 |
| 2005 Sales: $ | 2005 Profits: $-20,137 | Fiscal Year Ends: 12/31 |
| 2004 Sales: $ | 2004 Profits: $-21,800 | Parent Company: |

## SALARIES/BENEFITS:

| Pension Plan: | ESOP Stock Plan: | Profit Sharing: | Top Exec. Salary: $497,200 | Bonus: $298,300 |
|---|---|---|---|---|
| Savings Plan: Y | Stock Purch. Plan: Y | | Second Exec. Salary: $395,600 | Bonus: $168,900 |

## OTHER THOUGHTS:

**Apparent Women Officers or Directors:**
**Hot Spot for Advancement for Women/Minorities:**

## LOCATIONS: ("Y" = Yes)

| West: | Southwest: | Midwest: | Southeast: | Northeast: | International: |
|---|---|---|---|---|---|
| Y | | | | Y | |

# ALPHARMA ANIMAL HEALTH                    www.alpharmaah.com

**Industry Group Code: 325414  Ranks within this company's industry group:** Sales:    Profits:

| Drugs: | | Other: | | Clinical: | Computers: | Services: |
|---|---|---|---|---|---|---|
| Discovery: | Y | AgriBio: | Y | Trials/Services: | Hardware: | Specialty Services: |
| Licensing: | | Genetic Data: | | Labs: | Software: | Consulting: |
| Manufacturing: | Y | Tissue Replacement: | | Equipment/Supplies: | Arrays: | Blood Collection: |
| Genetics: | | | | Research & Development Services: | Database Management: | Drug Delivery: |
| | | | | Diagnostics: | | Drug Distribution: |

## TYPES OF BUSINESS:

Drugs-Animal Health
Human Pharmaceuticals
Animal Feed Additives

## BRANDS/DIVISIONS/AFFILIATES:

BIO-COX
Histostat
BMD
Albac
3-Nitro
Deccox
Zoamix
King Pharmaceuticals Inc

## CONTACTS: *Note: Officers with more than one job title may be intentionally listed here more than once.*

Eric J. Bruce, Pres.
Jeffrey Mellinger, Global Head-Mktg. & Sales
Mark LaVorgna, Global Dir.-Tech. Svcs.
Carol A. Wrenn, Pres., Animal Health Division
Curtis Shuey, Gen. Mgr.-APAC
Will Hart, Head- Latin America
John Langemeier, Dir.-North American Livestock & Canada
Brian A. Markison, Chmn.-King Pharmaceuticals
Karl DeBruyne, Dir.-Sales & Mktg., EMEA & South Asia

| **Phone:** 908-566-3800 | **Fax:** 908-566-4137 |
|---|---|
| **Toll-Free:** 866-322-2525 | |
| **Address:** 440 U.S. Highway 22 East, Bridgewater, NJ 08807 US | |

## GROWTH PLANS/SPECIAL FEATURES:

Alpharma Animal Health, formerly Alpharma, Inc., is a multinational pharmaceutical company that manufactures and markets pharmaceutical products for farm animals. The company has operations in the U.S., Canada, Asia Pacific, Europe, the Middle East and Africa. The firm divides its operations based on these geographic locations; in addition, the U.S. group is divided into poultry and livestock. Alpharma markets over 100 animal feed additives and water soluble therapeutic products, including BMD, a feed additive that promotes growth and feed efficiency, and prevents/treats diseases in poultry and swine; Albac, a feed additive for poultry, swine and calves; BIO-COX, which prevents coccidiosis in poultry; and 3-Nitro, Zoamix and Histostat feed grade anticoccidials. Additional product names include Aureomycin, Beta Mos and Deccox. Alpharma Animal Health manufactures its products at facilities in Maryland, Colorado, West Virginia, Illinois, Arkansas and China. In December 2008, Alpharma Inc., was acquired by King Pharmaceuticals Inc., through its subsidiary Albert Acquisition Corp., for roughly $1.6 billion; Alpharma now acts as a wholly-owned subsidiary (Alpharma Animal Health) of King Pharmaceuticals, Inc.

## FINANCIALS: Sales and profits are in thousands of dollars—add 000 to get the full amount. 2008 Note: Financial information for 2008 was not available for all companies at press time.

| | | |
|---|---|---|
| 2008 Sales: $ | 2008 Profits: $ | **U.S. Stock Ticker: Subsidiary** |
| 2007 Sales: $722,425 | 2007 Profits: $-13,581 | **Int'l Ticker:**    Int'l Exchange: |
| 2006 Sales: $653,828 | 2006 Profits: $82,544 | Employees:  2,000 |
| 2005 Sales: $553,617 | 2005 Profits: $133,769 | Fiscal Year Ends: 12/31 |
| 2004 Sales: $513,329 | 2004 Profits: $-314,737 | Parent Company: KING PHARMACEUTICALS INC |

## SALARIES/BENEFITS:

| Pension Plan: | ESOP Stock Plan: | Profit Sharing: | Top Exec. Salary: $646,923 | Bonus: $845,000 |
|---|---|---|---|---|
| Savings Plan: Y | Stock Purch. Plan: Y | | Second Exec. Salary: $436,762 | Bonus: $212,000 |

## OTHER THOUGHTS:

**Apparent Women Officers or Directors:** 3
**Hot Spot for Advancement for Women/Minorities:** Y

## LOCATIONS: ("Y" = Yes)

| West: | Southwest: | Midwest: | Southeast: | Northeast: | International: |
|---|---|---|---|---|---|
| Y | | Y | Y | Y | Y |

# ALSERES PHARMACEUTICALS INC

www.alseres.com

**Industry Group Code:** 325412 **Ranks within this company's industry group:** Sales: Profits: 105

| Drugs: | | Other: | Clinical: | Computers: | Services: |
|---|---|---|---|---|---|
| Discovery: | | AgriBio: | Trials/Services: | Hardware: | Specialty Services: |
| Licensing: | Y | Genetic Data: | Labs: | Software: | Consulting: |
| Manufacturing: | | Tissue Replacement: | Equipment/Supplies: | Arrays: | Blood Collection: |
| Genetics: | | | Research & Development Services: | Database Management: | Drug Delivery: |
| | | | Diagnostics: | | Drug Distribution: |

## TYPES OF BUSINESS:

Pharmaceutical Discovery & Development
Patent & Development Rights
CNS Disorder Treatments

## BRANDS/DIVISIONS/AFFILIATES:

Oncomodulin
Inosine
Cethrin
DAT Blocker
Altropane
Rho Inhibitor
ALSE-100

## CONTACTS: Note: Officers with more than one job title may be intentionally listed here more than once.

Peter G. Savas, CEO
Mark J. Pykett, COO
Mark J. Pykett, Pres.
Kenneth L. Rice, Jr., CFO
Noel J. Cusack, Sr. VP-Preclinical Dev.
Kennith L. Rice, Jr., Exec. VP-Admin.
Kenneth L. Rice, Jr., In-House Counsel
Richard M. Thorn, Sr. VP-Program Oper.
Kenneth L. Rice, Jr., Exec. VP-Finance
James R. Weston, Sr. VP-Regulatory Affairs & Quality
Mark Hurtt, Chief Medical Officer
Susan M. Flint, Sr. VP-Drug Dev.
Peter G. Savas, Chmn.

| Phone: 508-497-2360 | Fax: 508-497-9964 |
|---|---|
| Toll-Free: | |
| Address: 239 South St., Hopkinton, MA 01748 US | |

## GROWTH PLANS/SPECIAL FEATURES:

Alseres Pharmaceuticals, Inc. is a biotechnology company engaged in the research and development of diagnostic and therapeutic products primarily for central nervous system (CNS) disorders. The company's research and development is based on three technology platforms: the regenerative therapeutics program, which focuses on nerve repair, restoring movement and returning sensory activity in patients with loss of CNS function from traumas or degenerative diseases; the molecular imaging program, which focuses on the diagnosis of Parkinsonian Syndromes, including Parkinson's Disease (PD), Dementia and Attention Deficit Hyperactivity Disorder (ADHD); and the neurodegenerative disease program, which focuses on treating the symptoms of PD and slowing or stopping the progression of PD. The regenerative therapeutics program has four drugs under development: Cethrin for the treatment of acute spinal cord injuries (SCIs); Inosine for the treatment of SCIs, TBIs (traumatic brain injuries) and stroke; Oncomodulin for the treatment of optic nerve injuries and Glaucoma and the Rho Inhibitor (which uses the working name ALSE-100) for bone repair. Within the molecular imaging program, the company has two products in the pre-clinical stages of development, as well as Altropane for PD in Phase 3 clinical trials and for Dementia with Lewy Bodies (DLB) in Phase 2 of clinical trials. In the neurodegenerative program, the company is in the preclinical developmental stages for a novel dopamine transporter blocker (DAT) treatment for PD and ADHD. The firm owns or has licensed 41 issued U.S. patents and has 29 pending U.S. patent applications. In April 2009, the company was delisted from the NASDAQ due to its loss of market value and announced that it must raise cash or be acquired in order to remain in business.

Alseres offers employees health, dental, vision, life and disability insurance as well as a 401(K) savings plan.

## FINANCIALS: Sales and profits are in thousands of dollars—add 000 to get the full amount. 2008 Note: Financial information for 2008 was not available for all companies at press time.

| | | |
|---|---|---|
| 2008 Sales: $ | 2008 Profits: $-20,847 | U.S. Stock Ticker: ALSE |
| 2007 Sales: $ | 2007 Profits: $-19,548 | Int'l Ticker: Int'l Exchange: |
| 2006 Sales: $ | 2006 Profits: $-26,355 | Employees: 27 |
| 2005 Sales: $ | 2005 Profits: $-11,501 | Fiscal Year Ends: 12/31 |
| 2004 Sales: $ | 2004 Profits: $-11,251 | Parent Company: |

## SALARIES/BENEFITS:

| Pension Plan: | ESOP Stock Plan: | Profit Sharing: | Top Exec. Salary: $400,000 | Bonus: $ |
|---|---|---|---|---|
| Savings Plan: Y | Stock Purch. Plan: | | Second Exec. Salary: $340,000 | Bonus: $ |

## OTHER THOUGHTS:

**Apparent Women Officers or Directors:** 1
**Hot Spot for Advancement for Women/Minorities:**

## LOCATIONS: ("Y" = Yes)

| West: | Southwest: | Midwest: | Southeast: | Northeast: | International: |
|---|---|---|---|---|---|
| | | | | Y | |

# ALTANA AG                                                    www.altana.com

**Industry Group Code: 325412  Ranks within this company's industry group:  Sales: 33    Profits: 33**

| Drugs: | Other: | Clinical: | Computers: | Services: |
|---|---|---|---|---|
| Discovery: | AgriBio: | Trials/Services: | Hardware: | Specialty Services: |
| Licensing: | Genetic Data: | Labs: | Software: | Consulting: |
| Manufacturing:    Y | Tissue Replacement: | Equipment/Supplies:    Y | Arrays: | Blood Collection: |
| Genetics: | | Research & Development Services: | Database Management: | Drug Delivery: |
| | | Diagnostics: | | Drug Distribution: |

## TYPES OF BUSINESS:

Specialty Chemical Manufacturing
Imaging Products
Electrical Insulation
Coatings

## BRANDS/DIVISIONS/AFFILIATES:

BYK Additives & Instruments
ECKART Effect Pigments
ELANTAS Electrical Insulation
ACTEGA Coatings & Sealants
BYK-Chemie
BYK-Gardner
Dick Peters BV
BYK USA Inc

## CONTACTS: *Note: Officers with more than one job title may be intentionally listed here more than once.*

Matthias L. Wolfgruber, CEO
Martin Babilas, CFO
Roland Peter, Pres., Additives & Instruments Div.
Christoph Schlunken, Pres., Effect Pigments Div.
Wolfgang Schutt, Pres., Electrical Insulation Div.
Guido Forstbach, Pres., Coatings & Sealants Div.
Fritz Frohlich, Chmn.

| Phone: 49-281-670-8 | Fax: 49-281-670-376 |
|---|---|
| Toll-Free: | |
| Address: Abelstrasse 43, Wesel,  46483 Germany | |

## GROWTH PLANS/SPECIAL FEATURES:

Altana AG is an international chemicals company that develops, manufactures and markets products for a range of targeted, highly specialized applications.   The company serves customers in the coatings, paint and plastics, printing, cosmetics, electrical and electronics industries.   The firm operates through four divisions: BYK Additives & Instruments; ECKART Effect Pigments; ELANTAS Electrical Insulation; and ACTEGA Coatings & Sealants.   BYK Additives & Instruments offers a range of chemical additives, produced by subsidiary BYK-Chemie, that help to improve and regulate the quality and processability of coatings and plastics.   This division also offers testing and measuring equipment, produced by subsidiary BYK-Gardner, allowing manufacturers to predetermine the color, gloss and other physical properties of paints and plastics products. ECKART Effect Pigments develops and produces metallic effect and pearlescent pigments, as well as gold-bronze and zinc pigments, used to produce certain optical effects in paints, inks, cosmetics and coatings. ELANTAS Electrical Insulation produces insulating materials used in electrical and electronics applications, including electric motors, household appliances, cars, generators, transformers, capacitors, televisions, computers, wind mills, circuit boards and sensors.   ACTEGA Coatings & Sealants develops and produces specialty coatings and sealants used primarily by the graphic arts and packaging industries. These products, which include sealants for glass bottles and metal cans as well as coatings for flexible packaging, help to regulate the physical properties of packaging, preserve the freshness of contents and contribute to the overall appearance of packaged goods.   The full Altana group consists of approximately 42 operational companies and 46 research laboratories worldwide. Sales outside of Germany account for approximately 83% of annual revenues, with products sold in over 100 countries.   In November 2008, Altana announced that it would acquire Dick Peters BV, the wax additives business of Swiss chemicals firm Clariant, for approximately $22.4 million.

## FINANCIALS: Sales and profits are in thousands of dollars—add 000 to get the full amount. 2008 Note: Financial information for 2008 was not available for all companies at press time.

| | | |
|---|---|---|
| 2008 Sales: $1,774,430 | 2008 Profits: $136,760 | **U.S. Stock Ticker:** |
| 2007 Sales: $1,825,660 | 2007 Profits: $171,790 | **Int'l Ticker: ALT**    Int'l Exchange: Frankfurt-Euronext |
| 2006 Sales: $2,039,470 | 2006 Profits: $89,500 | Employees:  4,646 |
| 2005 Sales: $1,405,400 | 2005 Profits: $679,100 | Fiscal Year Ends: 12/31 |
| 2004 Sales: $4,013,500 | 2004 Profits: $529,200 | Parent Company: |

## SALARIES/BENEFITS:

| | | | | |
|---|---|---|---|---|
| Pension Plan: | ESOP Stock Plan: | Profit Sharing: | Top Exec. Salary: $ | Bonus: $ |
| Savings Plan: | Stock Purch. Plan: | | Second Exec. Salary: $ | Bonus: $ |

## OTHER THOUGHTS:

**Apparent Women Officers or Directors**: 1
**Hot Spot for Advancement for Women/Minorities**:

## LOCATIONS: ("Y" = Yes)

| West: | Southwest: | Midwest: | Southeast: | Northeast: | International: |
|---|---|---|---|---|---|
| | | Y | | Y | Y |

# ALZA CORP

www.jnj.com

**Industry Group Code: 325412A  Ranks within this company's industry group:  Sales:    Profits:**

| Drugs: | | Other: | | Clinical: | Computers: | | Services: | |
|---|---|---|---|---|---|---|---|---|
| Discovery: | Y | AgriBio: | | Trials/Services: | Hardware: | | Specialty Services: | Y |
| Licensing: | | Genetic Data: | | Labs: | Software: | | Consulting: | |
| Manufacturing: | Y | Tissue Replacement: | | Equipment/Supplies: | Arrays: | | Blood Collection: | |
| Genetics: | | | | Research & Development Services: | Database Management: | | Drug Delivery: | Y |
| | | | | Diagnostics: | | | Drug Distribution: | |

## TYPES OF BUSINESS:

Drug Delivery Systems
Pharmaceutical Development & Marketing

## BRANDS/DIVISIONS/AFFILIATES:

Johnson & Johnson
OROS
L-OROS
D-TRANS
STEALTH
DUROS
E-TRANS
IONSYS

## CONTACTS: *Note: Officers with more than one job title may be intentionally listed here more than once.*

Michael R. Jackson, Pres.
Erik Wiberg, VP-Bus. Dev.
Ravi Kiron, Exec. Dir.- New Tech. Assessment & Planning

| Phone: 650-564-5000 | Fax: 732-342-9819 |
|---|---|
| Toll-Free: | |
| Address: 1900 Charleston Rd., Mountain View, CA 94042 US | |

## GROWTH PLANS/SPECIAL FEATURES:

ALZA Corp., a subsidiary of Johnson & Johnson, develops and markets pharmaceutical products involving advanced drug delivery technologies.  The company focuses on several therapeutic areas, including oncology, AIDS, pain management, endocrinology and urology.  ALZA partners with other companies to combine new pharmaceutical compounds with its oral, transdermal, implantable and liposomal delivery systems.  It has five key brand name technologies.  OROS oral delivery technology, currently incorporated into 13 different products worldwide including Sudafed 24 Hour, uses osmosis to provide controlled drug delivery for up to 24 hours, with L-OROS available for liquid formulations.  D-TRANS transdermal technology is a patch drug delivery system currently used for seven different products, including Nicoderm CQ.  Its STEALTH liposomal technology consists of microscopic lipid particles that incorporate a polyethylene glycol coating, which allows cancer therapeutics and gene therapy vectors to evade detection by the immune system, targeting specific areas of disease within the body.  DUROS implants are miniature titanium cylinders which deliver therapeutics including small drugs, proteins, DNA and other bioactive macromolecules at a continuous rate for up to one year.  E-TRANS is an electron transport technology which actively transports drugs through intact skin, both locally and systemically, using low-level electrical energy.  ALZA Corp.'s drug delivery technology has been incorporated in approximately 30 commercialized products sold worldwide.

ALZA provides its employees with a wide range of benefits including pre-tax accounts for health and dependant care; medical, dental, life and accident insurance; adoption assistance; tuition reimbursement; short- and long-term disability care; daycare discounts; exercise programs; and a charitable matching gift program.

## FINANCIALS:  Sales and profits are in thousands of dollars—add 000 to get the full amount. 2008 Note: Financial information for 2008 was not available for all companies at press time.

| 2008 Sales: $ | 2008 Profits: $ | U.S. Stock Ticker: Subsidiary |
|---|---|---|
| 2007 Sales: $198,400 | 2007 Profits: $ | Int'l Ticker:    Int'l Exchange: |
| 2006 Sales: $ | 2006 Profits: $ | Employees:  1,845 |
| 2005 Sales: $ | 2005 Profits: $ | Fiscal Year Ends: 12/31 |
| 2004 Sales: $ | 2004 Profits: $ | Parent Company: JOHNSON & JOHNSON |

## SALARIES/BENEFITS:

| Pension Plan: Y | ESOP Stock Plan: | Profit Sharing: | Top Exec. Salary: $ | Bonus: $ |
|---|---|---|---|---|
| Savings Plan: Y | Stock Purch. Plan: | | Second Exec. Salary: $ | Bonus: $ |

## OTHER THOUGHTS:

**Apparent Women Officers or Directors:**
**Hot Spot for Advancement for Women/Minorities:**

## LOCATIONS: ("Y" = Yes)

| West: | Southwest: | Midwest: | Southeast: | Northeast: | International: |
|---|---|---|---|---|---|
| Y | | | | | |

# AMARIN CORPORATION PLC

**www.amarincorp.com**

Industry Group Code: 325412  **Ranks within this company's industry group:**  Sales:     Profits:

| Drugs: | | Other: | | Clinical: | | Computers: | | Services: | |
|---|---|---|---|---|---|---|---|---|---|
| Discovery: | Y | AgriBio: | | Trials/Services: | | Hardware: | | Specialty Services: | |
| Licensing: | Y | Genetic Data: | | Labs: | | Software: | | Consulting: | |
| Manufacturing: | Y | Tissue Replacement: | | Equipment/Supplies: | | Arrays: | | Blood Collection: | |
| Genetics: | | | | Research & Development Services: | | Database Management: | | Drug Delivery: | |
| | | | | Diagnostics: | | | | Drug Distribution: | |

## TYPES OF BUSINESS:

Drugs, Neurology
Drugs, Huntington's Disease
Drugs, Depression
Drugs, Parkinson's Disease
Drugs, CNS Disorders
Drugs, Multiple Sclerosis Fatigue

## BRANDS/DIVISIONS/AFFILIATES:

Miraxion
EN101
AMR101
AMR103

## CONTACTS: Note: Officers with more than one job title may be intentionally listed here more than once.

Thomas G. Lynch, CEO
Alan Cooke, COO
Alan Cooke, Pres.
Declan Doogan, Head-R&D
Tom Maher, General Counsel/Corp. Sec.
Stuart Sedlack, Exec. VP-Corp. Dev.
Darren Cunningham, Exec. VP-Investor Rel.
Conor Dalton, VP-Finance
Darren Cunningham, Exec. VP-Strategic Dev.
David Boal, VP-Bus. Dev.
Mehar Manku, Chief Scientist
Thomas G. Lynch, Chmn.

| Phone: 353-01-6699-020 | Fax: 353-01-6699-028 |
|---|---|
| Toll-Free: | |
| Address: First Fl., Block 3, Shellbourne Rd., Ballsbridge, Dublin, Ireland 4 UK | |

## GROWTH PLANS/SPECIAL FEATURES:

Amarin Corporation plc is a specialty pharmaceutical company focused on cardiovascular disorders and diseases of the central nervous system. The company's platforms utilize the cardiovascular system's innate quality of being affected by polyunsaturated fatty acids. Amarin's lipophilic drugs are fat-soluble, allowing easy transportation across the blood-brain barrier. The company is also utilizing a combinational lipid platform, where bioactive lipids are attached either to other drugs or other lipids. Amarin's current leading development project is its prescription grade Omega-3 fatty acid AMR101 (previously Miraxion) for hypertriglyceridemia and entering Phase III trials in 2009. Hypertriglyceridemia is a condition in which patients exhibit high blood levels of triglycerides, a known risk indicator for cardiovascular disease, which AMR101 may help to combat. Other Amarin pipeline projects are focused on disorders of the central nervous system, and include AMR101 for Huntington's disease, a treatment in which the drug acts in part by replenishing lipid bi-layers and allowing distressed neurons to regain functionality; EN101, an AChE-R mRNA inhibitor currently in Phase II trials and designed to treat the debilitating neuromuscular disease Myasthenia Gravis; a sublingual apomorphine treatment currently in Phase I trials and indicated for Parkinson's Disease; a novel nazal lorazapem formulation in the pre-clinical stage intended for the treatment of epileptic seizures; and AMR103 for the treatment of Parkinson's, also currently in the pre-clinical stage. The company's AMR101 for Huntington's has already completed Phase III trials involving over 600 patients in North America and Europe, showing signs of efficacy with long-term use. The company's strategy is eventually to commercialize its AMR101 for hypertriglyceridemia with a larger pharmaceutical firm. Amarin also seeks to partner with other firms for the trials and development of its pipeline of drugs for disorders of the central nervous system. In December 2008, Amarin opened a research and development office in Mystic, Connecticut.

## FINANCIALS: Sales and profits are in thousands of dollars—add 000 to get the full amount. 2008 Note: Financial information for 2008 was not available for all companies at press time.

| | | |
|---|---|---|
| 2008 Sales: $ | 2008 Profits: $ | **U.S. Stock Ticker: AMRN** |
| 2007 Sales: $ | 2007 Profits: $-38,197 | **Int'l Ticker:**     Int'l Exchange: |
| 2006 Sales: $ 111 | 2006 Profits: $-23,707 | Employees:   28 |
| 2005 Sales: $ 500 | 2005 Profits: $-19,630 | Fiscal Year Ends: 8/31 |
| 2004 Sales: $1,000 | 2004 Profits: $4,700 | Parent Company: |

## SALARIES/BENEFITS:

| Pension Plan: Y | ESOP Stock Plan: | Profit Sharing: | Top Exec. Salary: $515,000 | Bonus: $291,000 |
|---|---|---|---|---|
| Savings Plan: | Stock Purch. Plan: | | Second Exec. Salary: $482,000 | Bonus: $ |

## OTHER THOUGHTS:

**Apparent Women Officers or Directors:**
**Hot Spot for Advancement for Women/Minorities:**

## LOCATIONS: ("Y" = Yes)

| West: | Southwest: | Midwest: | Southeast: | Northeast: | International: |
|---|---|---|---|---|---|
| | | | | Y | Y |

Note: Financial information, benefits and other data can change quickly and may vary from those stated here.

# AMGEN INC
**www.amgen.com**

Industry Group Code: 325412  Ranks within this company's industry group: Sales: 14   Profits: 12

| Drugs: | | Other: | | Clinical: | | Computers: | | Services: | |
|---|---|---|---|---|---|---|---|---|---|
| Discovery: | Y | AgriBio: | | Trials/Services: | | Hardware: | | Specialty Services: | |
| Licensing: | | Genetic Data: | | Labs: | | Software: | | Consulting: | |
| Manufacturing: | Y | Tissue Replacement: | | Equipment/Supplies: | | Arrays: | | Blood Collection: | |
| Genetics: | | | | Research & Development Services: | | Database Management: | | Drug Delivery: | |
| | | | | Diagnostics: | | | | Drug Distribution: | |

## TYPES OF BUSINESS:
Drugs-Diversified
Oncology Drugs
Nephrology Drugs
Inflammation Drugs
Neurology Drugs
Metabolic Drugs

## BRANDS/DIVISIONS/AFFILIATES:
Aranesp
EPOGEN
Neulasta
NEUPOGEN
Enbrel

## CONTACTS: *Note: Officers with more than one job title may be intentionally listed here more than once.*
Kevin W. Sharer, CEO
Kevin W. Sharer, Pres.
Robert A. Bradway, CFO/Exec. VP
Brian McNamee, Sr. VP-Human Resources
Roger M. Perlmutter, Exec. VP-R&D
Thomas J. (Tom) Flanagan, CIO/Sr. VP
David J. Scott, General Counsel/Sr. VP/Corp. Sec.
Fabrizio Bonanni, Exec. VP-Oper.
David Beier, Sr. VP-Corp. Affairs & Global Gov't
Anna Richo, Sr. VP/Chief Compliance Officer
Kevin W. Sharer, Chmn.
George J. Morrow, Exec. VP-Global Commercial Oper.

| Phone: 805-447-1000 | Fax: 805-447-1010 |
|---|---|
| Toll-Free: | |
| Address: 1 Amgen Center Dr., Thousand Oaks, CA 91320-1799 US | |

## GROWTH PLANS/SPECIAL FEATURES:

Amgen, Inc. is a global biotechnology company that develops, manufactures and markets human therapeutics based on cellular and molecular biology. Its products are used for treatment in the fields of supportive cancer care, nephrology and inflammation. Amgen's primary products include Aranesp, EPOGEN, Neulasta, NEUPOGEN and Enbrel, which together represent 94% of the company's sales. Aranesp and EPOGEN stimulate the production of red blood cells to treat anemia and belong to a class of drugs referred to as erythropoiesis-stimulating agents. Aranesp is used for the treatment of anemia both in supportive cancer care and in nephrology. EPOGEN is used to treat anemia associated with chronic renal failure. Neulasta and NEUPOGEN selectively stimulate the production of neutrophils, one type of white blood cell that helps the body fight infections. ENBREL inhibits tumor necrosis factor (TNF), a substance induced in response to inflammatory and immunological responses, such as rheumatoid arthritis and psoriasis. Amgen maintains sales and marketing forces primarily in the U.S., Europe and Canada, and markets its products to healthcare providers including physicians, dialysis centers, hospitals and pharmacies. Amgen focuses its research and development efforts in the core areas of oncology, inflammation, bone, metabolic disorders and neuroscience, taking a modality-independent approach to drug discovery by choosing the best possible approach to block a specific disease process before considering the type of drug that may be required to pursue that approach. In September 2008, Biovitrum AB acquired the marketed biologic therapeutic products Kepivance and Stemgen from Amgen and obtained a worldwide exclusive license from the firm for Kineret.

Amgen offers its employees an education reimbursement plan, a Long Term Incentive program and medical, prescription, vision and dental benefits.

## FINANCIALS: Sales and profits are in thousands of dollars—add 000 to get the full amount. 2008 Note: Financial information for 2008 was not available for all companies at press time.

| | | |
|---|---|---|
| 2008 Sales: $15,003,000 | 2008 Profits: $4,196,000 | U.S. Stock Ticker: AMGN |
| 2007 Sales: $14,771,000 | 2007 Profits: $3,166,000 | Int'l Ticker:   Int'l Exchange: |
| 2006 Sales: $14,268,000 | 2006 Profits: $2,950,000 | Employees: 17,500 |
| 2005 Sales: $12,430,000 | 2005 Profits: $3,674,000 | Fiscal Year Ends: 12/31 |
| 2004 Sales: $10,550,000 | 2004 Profits: $2,363,000 | Parent Company: |

## SALARIES/BENEFITS:

| Pension Plan: | ESOP Stock Plan: | Profit Sharing: | Top Exec. Salary: $1,561,923 | Bonus: $3,875,000 |
|---|---|---|---|---|
| Savings Plan: Y | Stock Purch. Plan: Y | | Second Exec. Salary: $970,408 | Bonus: $1,290,000 |

## OTHER THOUGHTS:
**Apparent Women Officers or Directors**: 2
**Hot Spot for Advancement for Women/Minorities**: Y

## LOCATIONS: ("Y" = Yes)

| West: | Southwest: | Midwest: | Southeast: | Northeast: | International: |
|---|---|---|---|---|---|
| Y | | | Y | Y | Y |

Note: Financial information, benefits and other data can change quickly and may vary from those stated here.

# AMYLIN PHARMACEUTICALS INC                          www.amylin.com

**Industry Group Code: 325412  Ranks within this company's industry group:** Sales: 40   Profits: 169

| Drugs: | | Other: | Clinical: | Computers: | Services: |
|--------|---|--------|-----------|------------|-----------|
| Discovery: | Y | AgriBio: | Trials/Services: | Hardware: | Specialty Services: |
| Licensing: | Y | Genetic Data: | Labs: | Software: | Consulting: |
| Manufacturing: | | Tissue Replacement: | Equipment/Supplies: | Arrays: | Blood Collection: |
| Genetics: | | | Research & Development Services: | Database Management: | Drug Delivery: |
| | | | Diagnostics: | | Drug Distribution: |

## TYPES OF BUSINESS:

Pharmaceutical Discovery & Development
Drugs, Obesity
Drugs, Diabetes

## BRANDS/DIVISIONS/AFFILIATES:

SYMLIN
BYETTA
Exenatide LAR
Integrated Neurohormonal Therapy for Obesity
Pramlintide/Metreleptin
Davalintide

## CONTACTS: Note: Officers with more than one job title may be intentionally listed here more than once.

Daniel M. Bradbury, CEO
Daniel M. Bradbury, Pres.
Mark G. Foletta, CFO
Vincent P. Mihalik, Sr. VP-Sales & Mktg.
Roger Marchetti, Sr. VP-Human Resources
Orville G. Kolterman, Sr. VP-R&D
Roger Marchetti, Sr. VP-Info. Mgmt.
Marcea B. Lloyd, General Counsel/Sr. VP-Gov't Affairs
Paul Marshall, Sr. VP-Oper.
Mark J. Gergen, Sr. VP-Corp. Dev.
Marcea B. Lloyd, Sr. VP-Corp. Affairs
Mark G. Foletta, Sr. VP-Finance
Vincent P. Mihalik, Chief Commercial Officer
Joseph C. Cook, Jr., Chmn.

| Phone: 858-552-2200 | Fax: 858-552-2212 |
|---|---|
| Toll-Free: | |
| Address: 9360 Towne Centre Dr., San Diego, CA 92121 US | |

## GROWTH PLANS/SPECIAL FEATURES:

Amylin Pharmaceuticals, Inc. is engaged in the discovery, development and commercialization of drug candidates for the treatment of diabetes, obesity and other diseases. Amylin's research process is focused on identifying potentially useful peptide hormones, experimenting to discover their therapeutic applications and developing new treatments. The firm has amassed a polypeptide hormone library encompassing over 1,000 potentially useful biologics. Amylin is currently marketing two medicines to treat diabetes: BYETTA (exenatide) injection and SYMLIN (pramlintide acetate) injection. Additionally, the company is in Phase III of developing exenatide once weekly and in Phase I for the development of exenatide nasal, both drugs targeted for the treatment of diabetes. Amylin's Integrated Neurohormonal Therapy for Obesity (INTO) program studies the safety and efficacy of multiple neurohormones when used in combination with pramlintide to treat obesity. The company's obesity drugs, which are primarily a part of Amylin's INTO program, include Pramlintide/Metreleptin, currently in Phase II, and the second generation product Davalintide, also in Phase II of development. Amylin has strategic alliance partnerships with Alkermes, Inc. and Eli Lilly for the development of the exenatide once weekly diabetes treatment.

Amylin provides its employees with medical, dental and voluntary vision plans with domestic partner coverage; life and disability insurance; flexible spending accounts; a 401(k) plan; a 529 college savings plan; an employee stock purchase plan; an employee assistance program; education assistance programs; and discounted gym memberships and online concierge services.

## FINANCIALS: Sales and profits are in thousands of dollars—add 000 to get the full amount. 2008 Note: Financial information for 2008 was not available for all companies at press time.

| | | |
|---|---|---|
| 2008 Sales: $840,109 | 2008 Profits: $-315,405 | **U.S. Stock Ticker: AMLN** |
| 2007 Sales: $780,997 | 2007 Profits: $-211,136 | **Int'l Ticker:** Int'l Exchange: |
| 2006 Sales: $510,875 | 2006 Profits: $-218,856 | Employees: 1,800 |
| 2005 Sales: $140,474 | 2005 Profits: $-206,832 | Fiscal Year Ends: 12/31 |
| 2004 Sales: $34,268 | 2004 Profits: $-157,157 | Parent Company: |

## SALARIES/BENEFITS:

| Pension Plan: | ESOP Stock Plan: | Profit Sharing: | Top Exec. Salary: $655,769 | Bonus: $ |
|---|---|---|---|---|
| Savings Plan: Y | Stock Purch. Plan: | | Second Exec. Salary: $432,308 | Bonus: $ |

## OTHER THOUGHTS:

**Apparent Women Officers or Directors**: 4
**Hot Spot for Advancement for Women/Minorities**: Y

## LOCATIONS: ("Y" = Yes)

| West: | Southwest: | Midwest: | Southeast: | Northeast: | International: |
|---|---|---|---|---|---|
| Y | | | | | Y |

# ANGIOTECH PHARMACEUTICALS
## www.angiotech.com

**Industry Group Code: 325412 Ranks within this company's industry group: Sales: 51 Profits: 173**

| Drugs: | | Other: | Clinical: | | Computers: | | Services: | |
|---|---|---|---|---|---|---|---|---|
| Discovery: | Y | AgriBio: | Trials/Services: | Y | Hardware: | | Specialty Services: | |
| Licensing: | Y | Genetic Data: | Labs: | | Software: | | Consulting: | |
| Manufacturing: | Y | Tissue Replacement: | Equipment/Supplies: | Y | Arrays: | | Blood Collection: | |
| Genetics: | | | Research & Development Services: | | Database Management: | | Drug Delivery: | |
| | | | Diagnostics: | | | | Drug Distribution: | |

## TYPES OF BUSINESS:

Medical Device Coatings
Drugs, Inflammatory Disease
Surgical Equipment & Technology
Dialysis Catheters

## BRANDS/DIVISIONS/AFFILIATES:

TAXUS
Vascular Wrap
CoSeal
Vitagel Surgical Hemostat
BioSeal
MultiStem
Quill SRS
Option Vena Cava Filter

## CONTACTS: Note: Officers with more than one job title may be intentionally listed here more than once.

William L. Hunter, CEO
William L. Hunter, Pres.
K. Thomas Bailey, CFO
Tammy Neske, Sr. VP-Human Resources
Jeffery M. Gross, Sr. VP-R&D
Victor Diaz, Sr. VP-Global Mfg.
David D. McMasters, General Counsel/Sr. VP-Legal
Jonathan W. Chen, Sr. VP-Bus. Dev. & Financial Strategy
Jay Dent, Sr. VP-Finance
Rui Avelar, Chief Medical Officer
David T. Howard, Chmn.

| Phone: 604-221-7676 | Fax: 604-221-2330 |
|---|---|
| Toll-Free: | |
| Address: 1618 Station St., Vancouver, BC V6A 1B6 Canada | |

## GROWTH PLANS/SPECIAL FEATURES:

Angiotech Pharmaceuticals, Inc. is a Canadian pharmaceutical company that develops medical products for complications associated with medical device implants, wound closure, surgical interventions and acute injury. The company consists of two segments: pharmaceutical technologies and medical products. The pharmaceutical technologies segment develops, licenses and sells technologies that improve the performances of medical devices and the outcomes of surgical procedures. The medical products segment establishes development and marketing partnerships with medical device, pharmaceutical and biomaterials companies. The firm's products address restenosis treatment, surgical adhesions, surgical sealants, surgical hemostats, systemic programs and clinical programs. Angiotech's flagship products are the Quill SRS (Self-Retaining System) suture, which does not require knots, and TAXUS, a polymeric formulation that delivers paclitaxel as a stent coating for restenosis. CoSeal is the firm's surgical sealant, which aids in the healing of suture lines and synthetic grafts during surgery. Hemostatic devices, like Vitagel Surgical Hemostat, control bleeding during surgery and help prevent complications including blood loss and ineffective closure. The firm also has various products in clinical trials, including the HemoStream central venous catheters to administer fluid and nourishment to critically ill patients; Vascular Wrap, a paclitaxel-eluting mesh surgical implant to treat complications with vascular graft implants; BioSeal, a biopsy track plug; the Option Vena Cava filter and the MultiStem stem cell therapy program. In February 2009, Canada approved Angiotech's license to market the Quill SRS. In May 2009, the company announced U.S. FDA approval of its TAXUS Liberte Atom and Element stents.

## FINANCIALS: Sales and profits are in thousands of dollars—add 000 to get the full amount. 2008 Note: Financial information for 2008 was not available for all companies at press time.

| | | |
|---|---|---|
| 2008 Sales: $283,272 | 2008 Profits: $-741,176 | U.S. Stock Ticker: ANPI |
| 2007 Sales: $287,694 | 2007 Profits: $-65,940 | Int'l Ticker: ANP Int'l Exchange: Toronto-TSX |
| 2006 Sales: $315,075 | 2006 Profits: $10,714 | Employees: 1,450 |
| 2005 Sales: $199,600 | 2005 Profits: $-1,200 | Fiscal Year Ends: 12/31 |
| 2004 Sales: $130,800 | 2004 Profits: $52,500 | Parent Company: |

## SALARIES/BENEFITS:

| Pension Plan: | ESOP Stock Plan: | Profit Sharing: | Top Exec. Salary: $741,088 | Bonus: $ |
|---|---|---|---|---|
| Savings Plan: | Stock Purch. Plan: | | Second Exec. Salary: $483,643 | Bonus: $196,825 |

## OTHER THOUGHTS:

**Apparent Women Officers or Directors: 1**
**Hot Spot for Advancement for Women/Minorities:**

## LOCATIONS: ("Y" = Yes)

| West: | Southwest: | Midwest: | Southeast: | Northeast: | International: |
|---|---|---|---|---|---|
| Y | | Y | Y | Y | Y |

# ANIKA THERAPEUTICS INC

**www.anikatherapeutics.com**

Industry Group Code: 325414  **Ranks within this company's industry group:** Sales: 5    Profits: 3

| Drugs: | | Other: | | Clinical: | Computers: | Services: |
|---|---|---|---|---|---|---|
| Discovery: | Y | AgriBio: | | Trials/Services: | Hardware: | Specialty Services: |
| Licensing: | Y | Genetic Data: | | Labs: | Software: | Consulting: |
| Manufacturing: | Y | Tissue Replacement: | | Equipment/Supplies: | Arrays: | Blood Collection: |
| Genetics: | | | | Research & Development Services: | Database Management: | Drug Delivery: |
| | | | | Diagnostics: | | Drug Distribution: |

## TYPES OF BUSINESS:

Tissue Protection, Healing & Repair Drugs
Hyaluronic Acid Based Drugs
Tissue Augmentation Products
Osteoarthritis Pain Relief
Post-Surgical Anti-Adhesion

## BRANDS/DIVISIONS/AFFILIATES:

Orthovisc
Elevess
Amvisc
Staarvisc-II
ShellGel
Incert
Hyvisc
Monovisc

## CONTACTS: Note: Officers with more than one job title may be intentionally listed here more than once.

Charles H. Sherwood, CEO
Frank J. Luppino, COO
Charles H. Sherwood, Pres.
Kevin W. Quinlan, CFO
William J. Mrachek, VP-Human Resources
Andrew J. Carter, CTO
Randall W. Wilhoite, VP-Oper.
Gregory T. Fulton, Chief Commercial Officer
Irina B. Kulinets, VP-Regulatory & Clinical Affairs

| **Phone:** 781-457-9000 | **Fax:** 781-305-9720 |
|---|---|
| **Toll-Free:** | |
| **Address:** 32 Wiggins Ave., Bedford, MA 01730 US | |

## GROWTH PLANS/SPECIAL FEATURES:

Anika Therapeutics, Inc. develops, manufactures and commercializes therapeutic products for tissue protection, healing and repair. These products are based on hyaluronic acid (HA), a naturally occurring, biocompatible polymer found throughout the body that plays an important role in a number of physiological functions such as the protection and lubrication of soft tissues and joints, the maintenance of the structural integrity of tissues and the transport of molecules to and within cells. The firm's currently marketed products consist of Orthovisc, an HA product which provides lubrication for the knee and helps cushion the knee joint; Monovisc, a single-injection drug designed to alleviate arthritis pain; Amvisc, Amvisc Plus, Staarvisc-II and ShellGel, each an injectable ophthalmic viscoelastic HA product; Hyvisc, an HA product used for treatment of joint dysfunction in horses; Incert-S, an HA-based anti-adhesive for surgical applications and Elevess, an HA-based injectable dermal filler used for cosmetic tissue applications and products associated with joint health. Another osteoarthritis solution, Cingal, is currently in pre-clinical development. Anika is working to produce more Elevess products in the future, as there is a growing demand for soft tissue fillers for facial wrinkles, scar remediation and lip augmentation. Orthovisc is marketed in the U.S. by DePuy Mitek, a subsidiary of Johnson & Johnson, and outside the U.S. by distributors in roughly 20 countries. Hyvisc is marketed in the U.S. through Boehringer Ingelheim Vetmedica, Inc. In June and July of 2008, Monovisc and Elevess were approved in Europe and the United States, respectively. In October 2008, the company began distribution of Orthovisc in South America, South Korea and Canada.

The company offers its employees benefits that include health, dental, life and disability insurance; prescription drug coverage and a 401(k) plan.

## FINANCIALS: Sales and profits are in thousands of dollars—add 000 to get the full amount. 2008 Note: Financial information for 2008 was not available for all companies at press time.

| | | |
|---|---|---|
| 2008 Sales: $33,055 | 2008 Profits: $3,629 | **U.S. Stock Ticker:** ANIK |
| 2007 Sales: $30,830 | 2007 Profits: $6,035 | **Int'l Ticker:**      Int'l Exchange: |
| 2006 Sales: $26,841 | 2006 Profits: $4,604 | Employees:    84 |
| 2005 Sales: $29,835 | 2005 Profits: $5,893 | Fiscal Year Ends: 12/31 |
| 2004 Sales: $26,466 | 2004 Profits: $11,190 | Parent Company: |

## SALARIES/BENEFITS:

| Pension Plan: | ESOP Stock Plan: | Profit Sharing: | Top Exec. Salary: $428,480 | Bonus: $192,816 |
|---|---|---|---|---|
| Savings Plan: Y | Stock Purch. Plan: | | Second Exec. Salary: $261,888 | Bonus: $62,583 |

## OTHER THOUGHTS:

**Apparent Women Officers or Directors**: 1
**Hot Spot for Advancement for Women/Minorities**:

## LOCATIONS: ("Y" = Yes)

| West: | Southwest: | Midwest: | Southeast: | Northeast: | International: |
|---|---|---|---|---|---|
| | | | | Y | |

# ANTIGENICS INC

www.antigenics.com

Industry Group Code: 325412  Ranks within this company's industry group: Sales: 139  Profits: 120

| Drugs: | | Other: | Clinical: | Computers: | Services: |
|---|---|---|---|---|---|
| Discovery: | Y | AgriBio: | Trials/Services: | Hardware: | Specialty Services: |
| Licensing: | | Genetic Data: | Labs: | Software: | Consulting: |
| Manufacturing: | Y | Tissue Replacement: | Equipment/Supplies: | Arrays: | Blood Collection: |
| Genetics: | | | Research & Development Services: | Database Management: | Drug Delivery: |
| | | | Diagnostics: | | Drug Distribution: |

## TYPES OF BUSINESS:

Cancers & Infectious Diseases Products & Technologies

## BRANDS/DIVISIONS/AFFILIATES:

Oncophage
Araplatin
AG-707
QS-21 Stimulon
Quilimmune-P
Atra-IV

## CONTACTS: Note: Officers with more than one job title may be intentionally listed here more than once.

Garo H. Armen, CEO
Shalini Sharp, CFO/VP
John Cerio, VP-Human Resources
Stephen Monks, VP-Mfg.
Karen Higgins Valentine, General Counsel/VP/Sec.
Sunny Uberoi, VP-Corp. Comm.
Christine M. Klaskin, VP-Finance
Kerry A. Wentworth, VP-Clinical Oper. & Regulatory Affairs
Garo H. Armen, Chmn.

| Phone: 212-994-8200 | Fax: 212-994-8299 |
|---|---|
| Toll-Free: | |
| Address: 162 5th Ave., Ste. 900, New York, NY 10010 US | |

## GROWTH PLANS/SPECIAL FEATURES:

Antigenics, Inc. develops technologies and products to treat cancers and infectious diseases, primarily based on immunological approaches. The company's most advanced product candidate is Oncophage (vitespen), a personalized therapeutic cancer vaccine candidate based on a heat shock protein called gp96 that is currently tested in several cancer indications, including in Phase III clinical trials for the treatment of renal cell carcinoma, the most common type of kidney cancer, and for metastatic melanoma. Oncophage is also being tested in Phase I and Phase II clinical trials in a range of indications, including brain cancer, and has received fast-track designation from the FDA. The firm's product candidate portfolio also includes QS-21 Stimulon, an adjuvant used in numerous vaccines under development for diseases including hepatitis, human immunodeficiency virus (HIV), influenza, cancer, Alzheimer's disease, malaria, and tuberculosis; AG-707, a therapeutic vaccine program in Phase I clinical trial for the treatment of genital herpes; Quilimmune-P, a vaccine to prevent Streptococcus pneumoniae infections in elderly patients; Atra-IV, an intravenous treatment for promyelocytic leukemia and Araplatin, a liposomal chemotherapeutic in Phase I clinical trial for the treatment of solid tumors and B-cell lymphomas. Antigenics is based in New York City with research, development and manufacturing facilities in Lexington, Massachusetts. The firm has over 175 issued patents and more than 99 patents pending worldwide. In April 2008, Oncophage was approved for use in treating kidney cancer in Russia. In December 2008, the company was delisted from the NASDAQ due to minimum asset and revenue requirements, but regained its listing in April 2009.

Antigenics offers its employees medical, dental, life, AD&D and travel accident insurance; short and long-term disability coverage; flexible spending accounts; a 401(k) plan; an educational assistance program; an employee stock purchase program and an employee assistance program.

## FINANCIALS: Sales and profits are in thousands of dollars—add 000 to get the full amount. 2008 Note: Financial information for 2008 was not available for all companies at press time.

| | | |
|---|---|---|
| 2008 Sales: $2,651 | 2008 Profits: $-28,488 | U.S. Stock Ticker: AGEN |
| 2007 Sales: $5,552 | 2007 Profits: $-36,795 | Int'l Ticker:   Int'l Exchange: |
| 2006 Sales: $ 692 | 2006 Profits: $-51,881 | Employees:   80 |
| 2005 Sales: $ 630 | 2005 Profits: $-74,104 | Fiscal Year Ends: 12/31 |
| 2004 Sales: $ 707 | 2004 Profits: $-56,162 | Parent Company: |

## SALARIES/BENEFITS:

| Pension Plan: | ESOP Stock Plan: | Profit Sharing: | Top Exec. Salary: $440,000 | Bonus: $ |
|---|---|---|---|---|
| Savings Plan: Y | Stock Purch. Plan: Y | | Second Exec. Salary: $240,000 | Bonus: $ |

## OTHER THOUGHTS:

Apparent Women Officers or Directors: 4
Hot Spot for Advancement for Women/Minorities: Y

## LOCATIONS: ("Y" = Yes)

| West: | Southwest: | Midwest: | Southeast: | Northeast: | International: |
|---|---|---|---|---|---|
| | | | | Y | |

# APP PHARMACEUTICALS INC

**www.apppharma.com**

Industry Group Code: 325412 Ranks within this company's industry group: Sales: Profits:

| Drugs: | | Other: | Clinical: | Computers: | Services: |
|---|---|---|---|---|---|
| Discovery: | Y | AgriBio: | Trials/Services: | Hardware: | Specialty Services: |
| Licensing: | Y | Genetic Data: | Labs: | Software: | Consulting: |
| Manufacturing: | Y | Tissue Replacement: | Equipment/Supplies: | Arrays: | Blood Collection: |
| Genetics: | | | Research & Development Services: | Database Management: | Drug Delivery: |
| | | | Diagnostics: | | Drug Distribution: |

## TYPES OF BUSINESS:

Pharmaceuticals Manufacturing
Injectable Oncology Drugs
Anti-Infective Drugs
Critical Care Drugs

## BRANDS/DIVISIONS/AFFILIATES:

Fresenius Kabi Pharmaceuticals Holding Inc
Cefotan

## CONTACTS: Note: Officers with more than one job title may be intentionally listed here more than once.

Thomas H. Silberg, CEO
Frank Harmon, COO/Exec. VP
Thomas H. Silberg, Pres.
Richard J. Tajak, CFO/Exec. VP
James W. Callanan, VP-Human Resources
Richard E. Maroun, Chief Admin. Officer
Richard E. Maroun, General Counsel/Corp. Sec.
Katherine Gregory, VP-Business Dev.
Debra Lynn Ross, Dir.-Comm.

| Phone: 847-413-2073 | Fax: 800-743-7082 |
|---|---|
| Toll-Free: 888-386-1300 | |
| Address: 1501 E. Woodfield Rd., Ste. 300 E., Schaumburg, IL 60173-5837 US | |

## GROWTH PLANS/SPECIAL FEATURES:

APP Pharmaceuticals, Inc., a subsidiary of Fresenius Kabi Pharmaceuticals Holding, Inc., is a fully integrated biopharmaceutical company that develops, manufactures and markets injectable pharmaceutical products, focusing on the injectable critical care, oncology and anti-infective markets. The firm offers injectable products in each of the three basic forms: Liquid, powder and lyophilized (freeze-dried). APP manufactures and markets more than 60 injectable critical care products, including heparin, used to treat and prevent blood clotting. The firm currently manufactures 17 injectable oncology products in 38 dosages and formulations, including the recently launched fludabarine, used to treat adult B-cell chronic lymphocytic leukemia patients who have not responded to other treatments. The company also offers 21 injectable anti-infective products, including 11 classes of antimicrobials; the company's newest product, a second-generation cephalosporin called cefotetan, is used for surgical prophylaxis and has the longest half-life of any cephalosporin. APP is currently the only company marketing cefotetan in the U.S., which it does under the brand name Cefotan. APP's products are used in hospitals, long-term care facilities, alternate care sites and clinics. In March 2008, the company launched a cefepime hydrochloride injection, to be used as a treatment for treatment of urinary tract infections, skin and skin structure infections and intra-abdominal infections. In September of the same year, APP was acquired by Fresenius Kabi Pharmaceuticals Holding, Inc. In January 2009, the firm launched a rocuronium bromide injection, the generic version of the medication Zemuron, which is used as a muscle relaxant.

Employees are offered medical, dental and vision insurance; disability and life insurance; flexible spending accounts; legal assistance; dependent spouse and child life insurance; a 401(k) plan; a tuition assistance program; and an employee assistance plan.

## FINANCIALS: Sales and profits are in thousands of dollars—add 000 to get the full amount. 2008 Note: Financial information for 2008 was not available for all companies at press time.

| | | |
|---|---|---|
| 2008 Sales: $ | 2008 Profits: $ | U.S. Stock Ticker: Subsidiary |
| 2007 Sales: $647,374 | 2007 Profits: $34,358 | Int'l Ticker: Int'l Exchange: |
| 2006 Sales: $583,201 | 2006 Profits: $-46,897 | Employees: 1,375 |
| 2005 Sales: $385,082 | 2005 Profits: $17,657 | Fiscal Year Ends: 12/31 |
| 2004 Sales: $405,247 | 2004 Profits: $18,221 | Parent Company: FRESENIUS KABI PHARMACEUTICALS HOLDING INC |

## SALARIES/BENEFITS:

| Pension Plan: | ESOP Stock Plan: | Profit Sharing: | Top Exec. Salary: $692,885 | Bonus: $100,000 |
|---|---|---|---|---|
| Savings Plan: Y | Stock Purch. Plan: | | Second Exec. Salary: $623,077 | Bonus: $275,000 |

## OTHER THOUGHTS:

**Apparent Women Officers or Directors**: 2
**Hot Spot for Advancement for Women/Minorities**: Y

## LOCATIONS: ("Y" = Yes)

| West: | Southwest: | Midwest: | Southeast: | Northeast: | International: |
|---|---|---|---|---|---|
| Y | | Y | | Y | Y |

# APPLIED BIOSYSTEMS INC

**www.applera.com**

Industry Group Code: 33911  **Ranks within this company's industry group:** Sales: 6   Profits: 6

| Drugs: | | Other: | | Clinical: | | Computers: | | Services: | |
|---|---|---|---|---|---|---|---|---|---|
| Discovery: | Y | AgriBio: | | Trials/Services: | | Hardware: | | Specialty Services: | |
| Licensing: | | Genetic Data: | Y | Labs: | | Software: | Y | Consulting: | |
| Manufacturing: | | Tissue Replacement: | | Equipment/Supplies: | Y | Arrays: | | Blood Collection: | |
| Genetics: | | | | Research & Development Services: | | Database Management: | Y | Drug Delivery: | |
| | | | | Diagnostics: | Y | | | Drug Distribution: | |

## TYPES OF BUSINESS:

Equipment-Life Sciences & Genomics
Genetic Database Management
Proteomics
Medical Software
DNA Sequencing
Drug Discovery & Development
Diagnostics

## BRANDS/DIVISIONS/AFFILIATES:

InteRNA Genomics B.V.
Biotique Systems, Inc.
SOLiD System

## CONTACTS: *Note: Officers with more than one job title may be intentionally listed here more than once.*

Gregory T. Lucier, CEO
Mark Stevenson, COO
Mark Stevenson, Pres.
David F. Hoffmeister, CFO/Sr. VP
Peter Leddy, Sr. VP-Global Human Resources
Brian Pollok, Chief Scientific Officer/Head-Global R&D
Joe Beery, CIO/Sr. VP
John Cottingham, Chief Legal Officer
Mark O'Donnell, Sr. VP-Global Oper. & Svcs.
Paul Grossman, Sr. VP-Strategy & Corp. Dev.
Amanda Clardy, VP-Corp. Comm.
Amanda Clardy, VP-Investor Rel.
Kelli A. Richard, Chief Acct. Officer/VP-Finance
Nicolas M. Barthelemy, Pres., Cell Systems
Peter Dansky, Pres., Molecular Biology Systems
Laura Lauman, Pres., Mass Spectrometry
Bernd Brust, Chief Commercial Oper. Officer/Pres.
Gregory T. Lucier, Chmn.

| Phone: 650-638-5800 | Fax: 650-638-5998 |
|---|---|
| **Toll-Free:** 800-327-3002 | |
| **Address:** 850 Lincoln Centre Dr., Foster City, CA 94404 US | |

## GROWTH PLANS/SPECIAL FEATURES:

Applied Biosystems Inc., formerly Applera Corporation, serves the life science industry and research community by developing and marketing instrument-based systems, consumables, software and services. The company focuses on the following markets: basic research, commercial research (pharmaceutical and biotechnology) and standardized testing, including forensic human identification, paternity testing and food testing. The firm is a global leader in the life sciences industry, operating from an installed base of approximately 180,000 instrument systems in nearly 100 countries. Its products are used for synthesis, amplification, purification, isolation, analysis and sequencing of DNA, RNA, proteins and other biological molecules. Customers use these products for applications including research, pharmaceutical discovery and development, biosecurity, food and environmental testing, analysis of infectious diseases, human identification and forensic DNA analysis. In 2008, the firm, under its previous name, Appelra Corp., separated itself from its previously held subsidiary, Celera Genomics Group. Now Celera Corporation; a company that develops and manufactures molecular diagnostic products by hospitals and other clinical laboratories to detect and treat disease; the company operates as a separate, independent, publicly-traded company. Also in July 2008, the company took on the name of its remaining subsidiary, Applied Biosystems, Inc., to accurately reflect the company's services following its separation from Celera. Additionally, Applied Biosystems recently announced the formation of two new partnerships with InteRNA Genomics B.V. and Biotique Systems, Inc. for the purpose of incorporating the commercial software solutions offered by these companies to advance data analysis and management for researchers utilizing Applied Biosystems' SOLiD System, the company's ultra-high-throughput platform for DNA sequencing and genomic analysis research.

Applied Biosystems offers its employees health coverage, an educational assistance plan, employee assistance programs, gift matching programs and adoption assistance.

## FINANCIALS: Sales and profits are in thousands of dollars—add 000 to get the full amount. 2008 Note: Financial information for 2008 was not available for all companies at press time.

| | | |
|---|---|---|
| 2008 Sales: $2,361,484 | 2008 Profits: $213,808 | **U.S. Stock Ticker:** Private |
| 2007 Sales: $2,132,493 | 2007 Profits: $150,771 | **Int'l Ticker:**   Int'l Exchange: |
| 2006 Sales: $1,949,390 | 2006 Profits: $212,492 | Employees:  5,160 |
| 2005 Sales: $1,845,140 | 2005 Profits: $159,795 | Fiscal Year Ends: 6/30 |
| 2004 Sales: $1,825,200 | 2004 Profits: $114,953 | Parent Company: |

## SALARIES/BENEFITS:

| Pension Plan: | ESOP Stock Plan: | Profit Sharing: | Top Exec. Salary: $1,096,154 | Bonus: $717,978 |
|---|---|---|---|---|
| Savings Plan: Y | Stock Purch. Plan: Y | | Second Exec. Salary: $571,154 | Bonus: $717,978 |

## OTHER THOUGHTS:

**Apparent Women Officers or Directors:** 3
**Hot Spot for Advancement for Women/Minorities:** Y

## LOCATIONS: ("Y" = Yes)

| West: | Southwest: | Midwest: | Southeast: | Northeast: | International: |
|---|---|---|---|---|---|
| Y | | | | Y | Y |

Note: Financial information, benefits and other data can change quickly and may vary from those stated here.

# APPLIED MOLECULAR EVOLUTION INC                        www.amevolution.com

**Industry Group Code: 541712  Ranks within this company's industry group: Sales:    Profits:**

| Drugs: | | Other: | | Clinical: | Computers: | Services: |
|---|---|---|---|---|---|---|
| Discovery: | Y | AgriBio: | | Trials/Services: | Hardware: | Specialty Services: |
| Licensing: | | Genetic Data: | Y | Labs: | Software: | Consulting: |
| Manufacturing: | | Tissue Replacement: | | Equipment/Supplies: | Arrays: | Blood Collection: |
| Genetics: | | | | Research & Development Services: | Database Management: | Drug Delivery: |
| | | | | Diagnostics: | | Drug Distribution: |

## TYPES OF BUSINESS:

Research-Directed Molecular Evolution
Human Biotherapeutics
Small-Molecule Drugs

## BRANDS/DIVISIONS/AFFILIATES:

AME System
DirectAME
ExpressAME
ScreenAME

## CONTACTS: *Note: Officers with more than one job title may be intentionally listed here more than once.*

Thomas F. Bumol, Pres.
Cheryl C. Gabele, Dir.-Admin.
Cheryl C. Gabele, Dir.-Finance

| **Phone:** 858-597-4990 | **Fax:** 858-597-4950 |
|---|---|
| **Toll-Free:** | |
| **Address:** 3520 Dunhill St., San Diego, CA 92121 US | |

## GROWTH PLANS/SPECIAL FEATURES:

Applied Molecular Evolution, Inc. (AME), a subsidiary of Eli Lilly, is a drug development company and a leader in the application of directed molecular evolution to the development of biotherapeutics. The company uses its proprietary AMEsystem technology to develop improved versions of currently marketed, U.S. FDA-approved biopharmaceuticals as well as novel human biotherapeutics. It does so by adjusting amino acids in individual proteins to create positive characteristics. There are three components to the AMEsystem technology. DirectAME is a gene synthesis process that enables the rapid production of a library of variant genes based on an initial gene. ExpressAME consists of gene expression systems that produce proteins from the genes generated using DirectAME. ScreenAME is a series of tests that facilitate the selection and identification of proteins with the desired commercial properties from the protein libraries produced using ExpressAME. While AME had collaborative projects in the past, all of them were completed before its acquisition by Eli Lilly, and its employees now devote themselves entirely to working on projects for Eli Lilly.

The firm offers employees health and dental insurance; a prescription drug plan; a 401(k) plan; a retirement plan; flexible spending accounts; financial planning services; educational assistance; domestic partner benefits; on-site employee health services; an employee assistance program; on-site food services; a business casual dress code; banking, postal and dry cleaning services; free gym membership; maternity and paternity leave; nursing mothers program; adoption assistance; flextime and flex week; on-site child development centers; telecommuting; education leave; and discounted event tickets.

## FINANCIALS: Sales and profits are in thousands of dollars—add 000 to get the full amount. 2008 Note: Financial information for 2008 was not available for all companies at press time.

| | | | |
|---|---|---|---|
| 2008 Sales: $ | 2008 Profits: $ | **U.S. Stock Ticker: Subsidiary** | |
| 2007 Sales: $10,200 | 2007 Profits: $ | **Int'l Ticker:** Int'l Exchange: | |
| 2006 Sales: $ | 2006 Profits: $ | Employees: 106 | |
| 2005 Sales: $ | 2005 Profits: $ | Fiscal Year Ends: 12/31 | |
| 2004 Sales: $ | 2004 Profits: $ | Parent Company: ELI LILLY AND COMPANY | |

## SALARIES/BENEFITS:

| Pension Plan: Y | ESOP Stock Plan: | Profit Sharing: | Top Exec. Salary: $350,000 | Bonus: $70,000 |
|---|---|---|---|---|
| Savings Plan: Y | Stock Purch. Plan: | | Second Exec. Salary: $300,000 | Bonus: $81,000 |

## OTHER THOUGHTS:

**Apparent Women Officers or Directors**: 1
**Hot Spot for Advancement for Women/Minorities**:

## LOCATIONS: ("Y" = Yes)

| West: | Southwest: | Midwest: | Southeast: | Northeast: | International: |
|---|---|---|---|---|---|
| Y | | | | | |

# ARADIGM CORPORATION

**www.aradigm.com**

Industry Group Code: 325412A **Ranks within this company's industry group:** Sales: 27    Profits: 19

| Drugs: | | Other: | | Clinical: | | Computers: | | Services: | |
|---|---|---|---|---|---|---|---|---|---|
| Discovery: | | AgriBio: | | Trials/Services: | | Hardware: | | Specialty Services: | |
| Licensing: | | Genetic Data: | | Labs: | | Software: | | Consulting: | |
| Manufacturing: | Y | Tissue Replacement: | | Equipment/Supplies: | Y | Arrays: | | Blood Collection: | |
| Genetics: | | | | Research & Development Services: | | Database Management: | | Drug Delivery: | Y |
| | | | | Diagnostics: | | | | Drug Distribution: | |

## TYPES OF BUSINESS:

Drug Delivery Systems
Pulmonary Drug Delivery Systems

## BRANDS/DIVISIONS/AFFILIATES:

AERx
AERx iDMS
AERx Essence
Novo Nordisk

## CONTACTS: *Note: Officers with more than one job title may be intentionally listed here more than once.*

Igor Gonda, CEO
Igor Gonda, Pres.
Nancy Pecota, CFO
D. Jeffery Grimes, General Counsel/VP-Legal Affairs/Sec.
Virgil D. Thompson, Chmn.

| **Phone:** 510-265-9000 | **Fax:** 510-265-0277 |
|---|---|
| **Toll-Free:** | |
| **Address:** 3929 Point Eden Way, Hayward, CA 94545 US | |

## GROWTH PLANS/SPECIAL FEATURES:

Aradigm Corporation is an emerging specialty pharmaceutical company focused on the development and commercialization of a selection of drugs delivered by inhalation for the treatment of severe respiratory disease by pulmonologists. Aradigm's operations are centered on its hand-held AERx pulmonary drug delivery system, which creates aerosols from liquid drug formulations. This system is marketed as a replacement for medical devices such as nebulizers, metered-dose inhalers and dry powder inhalers. It is particularly suitable for drugs where highly efficient and precise delivery to the respiratory tract is advantageous or essential. Aradigm is focusing on product development for the treatment of respiratory disease, including identifying opportunities that it can develop and commercialize in the U.S. without a partner. In selecting its development programs, the company primarily seeks drugs approved by the U.S. FDA that can be reformulated for existing and new indications in respiratory disease. Aradigm is developing liposomal ciprofloxacin for the treatment of cystic fibrosis, currently in Phase II development; bronchiectasis, in Phase II development; inhalation anthrax, in preclinical development and smoking cessation, in Phase I development. In collaboration with other companies, the firm is additionally developing treatments for pulmonary arterial hypertension, asthma and other chronic obstructive diseases of airways. One of the company's most notable collaborations has been with Novo Nordisk on the AERx insulin diabetes management system (AERx iDMS and AERx Essence) for the treatment of Type I and Type II diabetes, which has included 15 Phase III clinical trials. In January 2008, Novo Nordisk terminated all pending clinical trials for fast-acting inhaled insulin delivered using the AERx system due to poor sales of a rival product and gave Aradigm the patent portfolio for the technology in October.

Aradigm offers its employees medical, dental, life, and disability insurance; a 401(k) program; a 529 college savings plan; tuition reimbursement and an educational rewards program.

## FINANCIALS: Sales and profits are in thousands of dollars—add 000 to get the full amount. 2008 Note: Financial information for 2008 was not available for all companies at press time.

| | | |
|---|---|---|
| 2008 Sales: $ 251 | 2008 Profits: $-22,608 | **U.S. Stock Ticker:** ARDM |
| 2007 Sales: $ 961 | 2007 Profits: $-24,201 | **Int'l Ticker:**    Int'l Exchange: |
| 2006 Sales: $4,814 | 2006 Profits: $-13,027 | Employees:    38 |
| 2005 Sales: $10,507 | 2005 Profits: $-29,215 | Fiscal Year Ends: 12/31 |
| 2004 Sales: $28,045 | 2004 Profits: $-30,189 | Parent Company: |

## SALARIES/BENEFITS:

| Pension Plan: | ESOP Stock Plan: | Profit Sharing: | Top Exec. Salary: $380,000 | Bonus: $ |
|---|---|---|---|---|
| Savings Plan: Y | Stock Purch. Plan: Y | | Second Exec. Salary: $322,000 | Bonus: $ |

## OTHER THOUGHTS:

**Apparent Women Officers or Directors:** 1
**Hot Spot for Advancement for Women/Minorities:**

## LOCATIONS: ("Y" = Yes)

| West: | Southwest: | Midwest: | Southeast: | Northeast: | International: |
|---|---|---|---|---|---|
| Y | | | | | |

# ARCADIA BIOSCIENCES INC

## www.arcadiabio.com

Industry Group Code: 11511 Ranks within this company's industry group: Sales:    Profits:

| Drugs: | Other: | | Clinical: | Computers: | Services: |
|---|---|---|---|---|---|
| Discovery: | AgriBio: | Y | Trials/Services: | Hardware: | Specialty Services: |
| Licensing: | Genetic Data: | | Labs: | Software: | Consulting: |
| Manufacturing: | Tissue Replacement: | | Equipment/Supplies: | Arrays: | Blood Collection: |
| Genetics: | | | Research & Development Services: | Database Management: | Drug Delivery: |
| | | | Diagnostics: | | Drug Distribution: |

## TYPES OF BUSINESS:

Agricultural-Based Technologies
Environment Health Technologies
Human Health Technologies

## BRANDS/DIVISIONS/AFFILIATES:

Washington State University

## CONTACTS: Note: Officers with more than one job title may be intentionally listed here more than once.

Eric Rey, CEO
Eric Rey, Pres.
Vic Knauf, Chief Scientific Officer
Steve Brandwein, VP-Admin.
Wendy Neal, Chief Legal Officer/VP
Roger Salameh, VP-Bus. Dev., Agriculture
Steve Brandwein, VP-Finance
Frank Flider, VP-Bus. Dev., Nutrition
Don Emlay, Dir.-Regulatory Affairs & Compliance

| Phone: 530-756-7077 | Fax: 530-756-7027 |
|---|---|
| Toll-Free: | |
| Address: 202 Cousteau Place, Ste. 200, Davis, CA 95618 US | |

## GROWTH PLANS/SPECIAL FEATURES:

Arcadia Biosciences, Inc. specializes in developing and commercializing technologies based in agriculture that target the environment as well as human health. The company uses advanced breeding techniques, genetic screening and genetic engineering to develop its product portfolio. Its current environmental health areas of focus include Nitrogen Use Efficient (NUE) crops and salt tolerant plants. Its NUE project seeks to minimize the amount of nitrogen fertilizer required to produce crops, in some cases utilizing up to 66% less nitrogen than conventional fertilizers to produce an equivalent yield. Its salt tolerance project aims to develop plants able to produce normal quality and yields in high saline conditions, with a variety of crop applications including corn, rice, alfalfa, soybeans, turf, wheat and vegetables. These plants not only produce normal results in high saline areas, they are also engineered to bind excess salt into the plant, thus reducing an area's saline levels over time. The company's human health projects also have two current areas of focus: GLA safflower oil and extended shelf-life produce. The GLA safflower oil project aims to breed new varieties of safflower whose seeds will have as much as 40% GLA (gamma linolenic acid), an omega-6 fatty acid believed to have therapeutic benefits, which could be utilized to manufacture supplements, functional foods and nutraceuticals. Its extended shelf-life produce project also uses the firm's TILLING breeding technology to seek new genetic varieties of tomatoes, lettuce, melons and strawberries. The firm is also currently conducting research, in partnership with Washington State University, focused on the production of wheat for individuals with Celiac disease (an autoimmune disorder triggered by glutens found in wheat), with grant funding from the National Institutes of Health.

## FINANCIALS: Sales and profits are in thousands of dollars—add 000 to get the full amount. 2008 Note: Financial information for 2008 was not available for all companies at press time.

| | | |
|---|---|---|
| 2008 Sales: $ | 2008 Profits: $ | U.S. Stock Ticker: Private |
| 2007 Sales: $4,000 | 2007 Profits: $ | Int'l Ticker:    Int'l Exchange: |
| 2006 Sales: $ | 2006 Profits: $ | Employees:    68 |
| 2005 Sales: $ | 2005 Profits: $ | Fiscal Year Ends: |
| 2004 Sales: $ | 2004 Profits: $ | Parent Company: |

## SALARIES/BENEFITS:

| Pension Plan: | ESOP Stock Plan: | Profit Sharing: | Top Exec. Salary: $ | Bonus: $ |
|---|---|---|---|---|
| Savings Plan: Y | Stock Purch. Plan: | | Second Exec. Salary: $ | Bonus: $ |

## OTHER THOUGHTS:

Apparent Women Officers or Directors: 1
Hot Spot for Advancement for Women/Minorities:

## LOCATIONS: ("Y" = Yes)

| West: | Southwest: | Midwest: | Southeast: | Northeast: | International: |
|---|---|---|---|---|---|
| Y | Y | | | | |

# ARENA PHARMACEUTICALS INC

## www.arenapharm.com

Industry Group Code: 325412  Ranks within this company's industry group: Sales: 115  Profits: 167

| Drugs: | | Other: | Clinical: | | Computers: | Services: |
|---|---|---|---|---|---|---|
| Discovery: | Y | AgriBio: | Trials/Services: | | Hardware: | Specialty Services: |
| Licensing: | | Genetic Data: | Labs: | | Software: | Consulting: |
| Manufacturing: | | Tissue Replacement: | Equipment/Supplies: | | Arrays: | Blood Collection: |
| Genetics: | | | Research & Development Services: | Y | Database Management: | Drug Delivery: |
| | | | Diagnostics: | | | Drug Distribution: |

## TYPES OF BUSINESS:
Oral Drugs Discovery, Development & Commercialization

## BRANDS/DIVISIONS/AFFILIATES:
Lorcaserin Hydrochloride
APD 916
APD 791
APD 811
APD 597
Niacin Receptor Agonist

## CONTACTS: *Note: Officers with more than one job title may be intentionally listed here more than once.*
Jack Lief, CEO
Jack Lief, Pres.
Robert E. Hoffman, CFO
Dominic P. Behan, Chief Scientific Officer/Sr. VP
Steven W. Spector, General Counsel/Sr. VP/Sec.
Robert E. Hoffman, VP-Finance
K.A. Ajit-Simh, VP-Quality Systems
William R. Shanahan, VP/Chief Medical Officer
Jack Lief, Chmn.

| Phone: 858-453-7200 | Fax: 858-453-7210 |
|---|---|
| Toll-Free: | |
| Address: 6166 Nancy Ridge Dr., San Diego, CA 92121 US | |

## GROWTH PLANS/SPECIAL FEATURES:

Arena Pharmaceuticals, Inc. is a clinical-stage biopharmaceutical company that focuses on the discovery, development and commercialization of oral drugs in four major therapeutic areas: cardiovascular, central nervous system, inflammatory and metabolic diseases. The company's most advanced drug candidate, lorcaserin hydrochloride, is currently in a Phase III clinical trial program for the treatment of obesity. Arena has initiated three planned Phase III trials evaluating the efficacy and safety of lorcaserin. The first trial is known as BLOOM (Behavioral modification and Lorcaserin for Overweight and Obesity Management), the second as BLOSSOM (Behavioral modification and Lorcaserin Second Study for Obesity Management) and the third as BLOOM-DM (Behavioral modification and Lorcaserin for Overweight and Obesity Management in Diabetes Mellitus). In addition to lorcaserin, the company's internal development programs include APD916 for narcolepsy and cataplexy; APD811for pulmonary arterial hypertension and APD791 for arterial thrombosis. Arena completed a Phase I clinical trial of APD791, an orally available drug candidate the treatment and prevention of diseases such as acute coronary syndrome. In addition to internal programs, the firm is developing drugs with two pharmaceutical companies: APD597 with Ortho-McNeil, focusing on investigating receptor agonists of an orphan G protein-coupled receptors, the glucose-dependent insulinotropic receptor for treatment of Type II diabetes, and a Niacin Receptor Agonist with Merck, focusing on treatments for atherosclerosis and other disorders. In May 2009, the company announced that it will lay off 31% of its workforce, or about 130 employees.

Arena offers employees medical, dental and vision insurance; life, AD&D and disability insurance; flexible spending accounts; a 401(k) plan; an employee assistance program; stock options and a stock purchase plan.

## FINANCIALS: Sales and profits are in thousands of dollars—add 000 to get the full amount. 2008 Note: Financial information for 2008 was not available for all companies at press time.

| | | |
|---|---|---|
| 2008 Sales: $9,809 | 2008 Profits: $-237,573 | U.S. Stock Ticker: ARNA |
| 2007 Sales: $19,332 | 2007 Profits: $-143,166 | Int'l Ticker:    Int'l Exchange: |
| 2006 Sales: $30,569 | 2006 Profits: $-86,248 | Employees:   499 |
| 2005 Sales: $23,233 | 2005 Profits: $-67,901 | Fiscal Year Ends: 12/31 |
| 2004 Sales: $13,686 | 2004 Profits: $-57,992 | Parent Company: |

## SALARIES/BENEFITS:

| Pension Plan: | ESOP Stock Plan: | Profit Sharing: | Top Exec. Salary: $668,021 | Bonus: $300,000 |
|---|---|---|---|---|
| Savings Plan: Y | Stock Purch. Plan: Y | | Second Exec. Salary: $387,813 | Bonus: $125,000 |

## OTHER THOUGHTS:

Apparent Women Officers or Directors:
Hot Spot for Advancement for Women/Minorities:

## LOCATIONS: ("Y" = Yes)

| West: | Southwest: | Midwest: | Southeast: | Northeast: | International: |
|---|---|---|---|---|---|
| Y | | | | | |

# ARIAD PHARMACEUTICALS INC

**www.ariad.com**

Industry Group Code: 325412  **Ranks within this company's industry group:** Sales: 124   Profits: 151

| Drugs: | | Other: | Clinical: | | Computers: | | Services: | |
|---|---|---|---|---|---|---|---|---|
| Discovery: | Y | AgriBio: | Trials/Services: | | Hardware: | | Specialty Services: | |
| Licensing: | Y | Genetic Data: | Labs: | | Software: | | Consulting: | |
| Manufacturing: | | Tissue Replacement: | Equipment/Supplies: | | Arrays: | | Blood Collection: | |
| Genetics: | | | Research & Development Services: | Y | Database Management: | | Drug Delivery: | |
| | | | Diagnostics: | | | | Drug Distribution: | |

## TYPES OF BUSINESS:

Signaling Inhibitors Drugs

## BRANDS/DIVISIONS/AFFILIATES:

Ariad Gene Therapeutics, Inc.
AP24534
Ridaforolimus
AP24334
Ariad Gene Therapeutics, Inc.
ARGENT Technology

## CONTACTS: *Note: Officers with more than one job title may be intentionally listed here more than once.*

Harvey J. Berger, CEO
Harvey J. Berger, Pres.
Edward M. Fitzgerald, CFO/Sr. VP/Treas.
Virginia R. Dean, VP-Human Resources
Timothy P. Clackson, Chief Scientific Officer/Sr. VP
Kelly M. Schmitz, VP-IT & Oper.
David C. Dalgarno, VP-Research Tech.
Andreas Woppmann, VP-Mfg. Oper.
Raymond T. Keane, General Counsel/Sr. VP/Sec.
Matthew E. Ros, Sr. VP-Commercial Oper.
John D. Iuliucci, Sr. VP-Dev./Chief Dev. Officer
Maria E. Cantor, VP-Corp. Comm.
Maria E. Cantor, VP-Investor Rel.
Joseph Bratica, VP-Finance/Controller
Pierre F. Dodion, Sr. VP/Chief Medical Officer
Shirish Hirani, VP-Program/Alliance Mgr.
David L. Berstein, Sr. VP/Chief Intellectual Property Officer
Daniel M. Bollag, Sr. VP-Regulatory Affairs & Quality
Harvey J. Berger, Chmn.

| Phone: 617-494-0400 | Fax: 617-494-8144 |
|---|---|
| Toll-Free: | |
| Address: 26 Landsdowne St., Cambridge, MA 02139-4234 US | |

## GROWTH PLANS/SPECIAL FEATURES:

Ariad Pharmaceuticals, Inc. is an oncology company that is engaged in the discovery and development of its ARGENT technology, which revolves around medicines used to treat cancers by regulating cell signaling with small molecules. The company's lead product candidate is ridaforolimus, an inhibitor of the protein mTOR, which is a master switch in cancers. Blocking mTOR creates a starvation-like effect in cancer cells by interfering with cell growth, division, metabolism and angiogenesis. It is currently being developed by Ariad in partnership with Merck & Co., Inc. The firm has tested ridaforolimus in Phase III trials for patients with metastatic soft-tissue and bone sarcomas and is planning to initiate multiple clinical trials of ridaforolimus, including Phase II clinical trials in endometrial, breast and prostate cancers. Ariad's second product candidate, AP24534 is an orally active kinase inhibitor, which is in development for the treatment of various forms of leukemia and blood-based cancers. AP24334 has shown to be a potent inhibition of Bcr-Abl, a kinase that causes chronic myeloid leukemia, and of Flt3, a kinase involved in acute myeloid leukemia and the progression of multiple solid tumors. The firm licenses both the intellectual property related to the ARGENT technology and the product candidates resulting from the application of this technology, including mTOR inhibitors. In September 2008, the company announced it would merge with Ariad Gene Therapeutics, Inc., a subsidiary in which it already held an 80% stake.

Ariad offers employees medical, dental, life and short and long-term disability insurance; a 401(k) plan; stock options; an employee stock purchase plan; tuition reimbursement and a parking or public transportation pass.

## FINANCIALS:  Sales and profits are in thousands of dollars—add 000 to get the full amount. 2008 Note: Financial information for 2008 was not available for all companies at press time.

| | | |
|---|---|---|
| 2008 Sales: $7,082 | 2008 Profits: $-71,052 | **U.S. Stock Ticker:** ARIA |
| 2007 Sales: $3,583 | 2007 Profits: $-58,522 | **Int'l Ticker:**   Int'l Exchange: |
| 2006 Sales: $ 896 | 2006 Profits: $-61,928 | Employees:   150 |
| 2005 Sales: $1,217 | 2005 Profits: $-55,482 | Fiscal Year Ends: 12/31 |
| 2004 Sales: $ 742 | 2004 Profits: $-35,573 | Parent Company: |

## SALARIES/BENEFITS:

| Pension Plan: | ESOP Stock Plan: | Profit Sharing: | Top Exec. Salary: $607,500 | Bonus: $ |
|---|---|---|---|---|
| Savings Plan: Y | Stock Purch. Plan: Y | | Second Exec. Salary: $350,000 | Bonus: $158,000 |

## OTHER THOUGHTS:

**Apparent Women Officers or Directors**: 2
**Hot Spot for Advancement for Women/Minorities**: Y

## LOCATIONS: ("Y" = Yes)

| West: | Southwest: | Midwest: | Southeast: | Northeast: | International: |
|---|---|---|---|---|---|
| | | | | Y | |

Note: Financial information, benefits and other data can change quickly and may vary from those stated here.

# ARQULE INC

**www.arqule.com**

Industry Group Code: 325412  Ranks within this company's industry group: Sales: 110  Profits: 138

| Drugs: | | Other: | | Clinical: | | Computers: | | Services: | |
|---|---|---|---|---|---|---|---|---|---|
| Discovery: | Y | AgriBio: | | Trials/Services: | | Hardware: | Y | Specialty Services: | Y |
| Licensing: | | Genetic Data: | Y | Labs: | | Software: | Y | Consulting: | |
| Manufacturing: | | Tissue Replacement: | | Equipment/Supplies: | | Arrays: | | Blood Collection: | |
| Genetics: | | | | Research & Development Services: | Y | Database Management: | Y | Drug Delivery: | |
| | | | | Diagnostics: | | | | Drug Distribution: | |

## TYPES OF BUSINESS:

Research-Drug Discovery
Small-Molecule Compounds
Systems & Software
Predictive Modeling

## BRANDS/DIVISIONS/AFFILIATES:

Activated Checkpoint Therapy (ACT)
ARQ 197
ARQ 501
ARQ 171
ARQ 761
ARQ 621

## CONTACTS: *Note: Officers with more than one job title may be intentionally listed here more than once.*

Paolo Pucci, CEO
Peter S. Lawrence, COO
Peter S. Lawrence, Pres.
Anthony S. Messina, VP-Human Dev.
Thomas C.K. Chan, Chief Scientific Officer
Brian Schwartz, Chief Medical Officer
Patrick J. Zenner, Chmn.

| Phone: 781-994-0300 | Fax: 781-376-6019 |
|---|---|
| Toll-Free: | |
| Address: 19 Presidential Way, Woburn, MA 01801-5140 US | |

## GROWTH PLANS/SPECIAL FEATURES:

ArQule, Inc. is a clinical-stage biotechnology company engaged in the research and development of cancer therapeutics that act selectively against cancer cells, target multiple tumor types and are well tolerated by patients. The company's lead approach to cancer therapy development is based on its Activated Checkpoint Therapy (ACT) platform, which restores the cells' internal damage response mechanism, which then triggers programmed cell death. Because cancer cells invariably contain both damaged DNA and de-activated damage response checkpoints, ACT selectively induces the self-destruction of cancer cells while leaving normal cells undamaged. ArQule's lead products consist of ARQ 197, an orally-administered inhibitor of the c-Met receptor tyrosine kinase; ARQ 501, an intravenously administered novel activator of the cell's damage response mechanism mediated by the E2F-1 transcription factor; and ARQ 171, an intravenously administered second generation activator of E2F-1. Early-stage clinical trial results, which are available for ARQ 197 and ARQ 501, have demonstrated anti-cancer activity across multiple types of tumors. ArQule is also developing ARQ 761, a new chemical entity based on ARQ 501, for which it has filed an Investigational New Drug Application (IND). The company has been focusing a significant portion of its drug discovery efforts on its newly acquired understanding of the way ARQ 197 binds to c-Met. Insights into this binding mechanism form the basis of a discovery platform that ArQule plans to leverage to generate new types of selective kinase inhibitors. In December 2008, the U.S. Food and Drug Administration (FDA) accepted an IND for ARQ 621, a new product focused on inhibition of the Eg5 kinesin spindle protein.

ArQule offers its employees tuition reimbursement; credit union membership; an employee assistance program; an on-site fitness center; discounted theme park tickets, movie tickets and ski vouchers; flexible spending accounts; and medical, dental, vision and disability insurance.

## FINANCIALS: Sales and profits are in thousands of dollars—add 000 to get the full amount. 2008 Note: Financial information for 2008 was not available for all companies at press time.

| | | |
|---|---|---|
| 2008 Sales: $14,141 | 2008 Profits: $-50,864 | U.S. Stock Ticker: ARQL |
| 2007 Sales: $9,165 | 2007 Profits: $-53,372 | Int'l Ticker:   Int'l Exchange: |
| 2006 Sales: $6,626 | 2006 Profits: $-31,440 | Employees:  107 |
| 2005 Sales: $6,628 | 2005 Profits: $-7,520 | Fiscal Year Ends: 12/31 |
| 2004 Sales: $5,012 | 2004 Profits: $-4,921 | Parent Company: |

## SALARIES/BENEFITS:

| Pension Plan: | ESOP Stock Plan: | Profit Sharing: | Top Exec. Salary: $373,846 | Bonus: $150,000 |
|---|---|---|---|---|
| Savings Plan: Y | Stock Purch. Plan: Y | | Second Exec. Salary: $287,808 | Bonus: $73,267 |

## OTHER THOUGHTS:

Apparent Women Officers or Directors: 1
Hot Spot for Advancement for Women/Minorities:

## LOCATIONS: ("Y" = Yes)

| West: | Southwest: | Midwest: | Southeast: | Northeast: | International: |
|---|---|---|---|---|---|
| | | | | Y | |

Note: Financial information, benefits and other data can change quickly and may vary from those stated here.

# ARRAY BIOPHARMA INC

www.arraybiopharma.com

Industry Group Code: 541712 Ranks within this company's industry group: Sales: 15 Profits: 15

| Drugs: | | Other: | Clinical: | | Computers: | | Services: | |
|---|---|---|---|---|---|---|---|---|
| Discovery: | Y | AgriBio: | Trials/Services: | | Hardware: | | Specialty Services: | |
| Licensing: | Y | Genetic Data: | Labs: | | Software: | | Consulting: | |
| Manufacturing: | | Tissue Replacement: | Equipment/Supplies: | | Arrays: | Y | Blood Collection: | |
| Genetics: | | | Research & Development Services: | Y | Database Management: | | Drug Delivery: | |
| | | | Diagnostics: | | | | Drug Distribution: | |

## TYPES OF BUSINESS:

Drug Development & Research Services
Small-Molecule Drugs
Arrays

## BRANDS/DIVISIONS/AFFILIATES:

Array Discovery Platform
MEK
Optimer Building Blocks
ARRY-162
ARRY-797

## CONTACTS: Note: Officers with more than one job title may be intentionally listed here more than once.

Robert E. Conway, CEO
David L. Snitman, COO
Kevin Koch, Pres.
R. Michael Carruthers, CFO
Kevin Koch, Chief Scientific Officer
John R. Moore, General Counsel/VP
David L. Snitman, VP-Bus. Dev.
John A. Josey, VP-Discovery Chem.
James D. Winkler, VP.-Discovery Biology
Gary M. Clark, VP-Biostatistics & Data Mgmt.
John Yates, Chief Medical Officer
Kyle A. Lefkoff, Chmn.

| Phone: 303-381-6600 | Fax: 303-386-1390 |
|---|---|
| Toll-Free: 877-633-2436 | |
| Address: 3200 Walnut St., Boulder, CO 80301 US | |

## GROWTH PLANS/SPECIAL FEATURES:

Array BioPharma, Inc. is a drug discovery company creating small-molecule drugs through the integration of chemistry, biology and informatics. The firm's scientists use its Array Discovery Platform, an integrated set of drug discovery technologies including predictive informatics and high throughput screening, to invent novel small-molecule drugs in collaboration with leading pharmaceutical and biotechnology companies, as well as for its own pipeline of proprietary drugs. The company holds collaborative partnerships with AstraZeneca; Genentech; InterMune; Ono Pharmaceutical Co., Ltd.; Amgen Inc.; Eli Lilly and Company; Japan Tobacco, Inc.; and Takeda Pharmaceutical Company, Ltd. Array's prime research focuses are cancer and inflammatory diseases. The company has four cancer drugs and four inflammation drugs in its advanced pipeline. The clinical candidate ARRY-162, or MEK for Cancer, is an orally active drug that is currently in Phase II of study, with additional plans for other Phase II studies by Astra Zeneca. Other cancer products in development target EGFR, Erb-2, VEGF and PDGF receptors that are over expressed in tumors. The inflammatory and metabolic drugs aim to regulate biological factors contributing to those conditions. The company also sells its Optimer building blocks, which are the starting materials used to create more complex chemical compounds in the drug discovery process, on a per-compound basis without any restrictions on use. In June 2008, the firm announced successful Phase I results for ARRY-162, a small molecule MEK inhibitor, and ARRY-797, a pan-cytokine inhibitor. In January 2009, the company announced several cost-cutting measures, including layoffs of 10% of its workforce.

Array offers medical, dental, vision and prescription coverage; life insurance and disability coverage; an employee assistance program; flexible spending accounts; a 401(k) plan; and educational assistance.

## FINANCIALS: Sales and profits are in thousands of dollars—add 000 to get the full amount. 2008 Note: Financial information for 2008 was not available for all companies at press time.

| | | |
|---|---|---|
| 2008 Sales: $28,808 | 2008 Profits: $-96,288 | U.S. Stock Ticker: ARRY |
| 2007 Sales: $36,970 | 2007 Profits: $-55,443 | Int'l Ticker: Int'l Exchange: |
| 2006 Sales: $45,003 | 2006 Profits: $39,614 | Employees: 386 |
| 2005 Sales: $45,505 | 2005 Profits: $-23,244 | Fiscal Year Ends: 6/30 |
| 2004 Sales: $34,831 | 2004 Profits: $-25,504 | Parent Company: |

## SALARIES/BENEFITS:

| Pension Plan: | ESOP Stock Plan: | Profit Sharing: | Top Exec. Salary: $477,500 | Bonus: $235,125 |
|---|---|---|---|---|
| Savings Plan: Y | Stock Purch. Plan: Y | | Second Exec. Salary: $429,905 | Bonus: $ |

## OTHER THOUGHTS:

Apparent Women Officers or Directors:
Hot Spot for Advancement for Women/Minorities:

## LOCATIONS: ("Y" = Yes)

| West: | Southwest: | Midwest: | Southeast: | Northeast: | International: |
|---|---|---|---|---|---|
| Y | | | | | |

# ARRYX INC
## www.arryx.com

**Industry Group Code: 541711  Ranks within this company's industry group:** Sales:  Profits:

| Drugs: | Other: | Clinical: | Computers: | | Services: |
|---|---|---|---|---|---|
| Discovery: | AgriBio: | Trials/Services: | Hardware: | Y | Specialty Services: |
| Licensing: | Genetic Data: | Labs: | Software: | | Consulting: |
| Manufacturing: | Tissue Replacement: | Equipment/Supplies: | Arrays: | | Blood Collection: |
| Genetics: | | Research & Development Services: | Database Management: | | Drug Delivery: |
| | | Diagnostics: | | | Drug Distribution: |

## TYPES OF BUSINESS:

Holographic Laser Steering
Optics
Nanomaterial Manipulation Technology

## BRANDS/DIVISIONS/AFFILIATES:

BioRyx 200
University of Chicago
Laser Tweezers
HOTkit Complete Holo-Tweezers Solution

## CONTACTS: *Note: Officers with more than one job title may be intentionally listed here more than once.*

Michael Reese, CFO
Daniel Mueth, Lead Scientist
Daniel Mueth, CTO
Nicole S. Williams, Chmn.

| **Phone:** 312-726-6675 | **Fax:** 312-726-6652 |
|---|---|
| **Toll-Free:** | |
| **Address:** 316 N. Michigan Ave., Ste. CL20, Chicago, IL 60601 US | |

## GROWTH PLANS/SPECIAL FEATURES:

Arryx, Inc., a subsidiary of Haemonetics Corporation, develops advanced systems and products for the optoelectronics and biotechnology industries. The company created the BioRyx 200 system, which employs holographic laser steering to improve productivity and profitability for manufacturing and processing in industries ranging from pharmaceuticals to integrated circuit manufacturing. The system integrates proprietary laser steering technology with easy-to-use software to give researchers the ability to manipulate multiple microscopic objects independently and simultaneously in three dimensions. The firm is the exclusive licensee of holographic optical trapping (HOT) technology, originally developed at the University of Chicago. HOT uses a holographic device, such as a spatial light modulator, to operate beams of light to capture and manipulate microscopic and nanoscopic objects such as carbon nanotubes in optical traps. The optical traps are also called Laser Tweezers. The system can grab, move, spin, assemble, stretch, join, separate and otherwise control materials ranging in size from 1/1000th the diameter of a human hair to the size of human cells. HOT uses non-standard beam profiles such as Bessel beams, which are non-diffracting, and optical vortices, which are optical fields with phase singularities. Applications include isolating valuable cells or molecules from other cells, tissues and contaminants; detecting and measuring the presence of materials to increase test sophistication and sensitivity; and manufacturing sensors to detect biohazards, chemical hazards and other contaminants on a universal basis for the rapidly growing homeland security industry. Clients of the firm have included universities and government institutions, such as the National Institute of Standards and Technology. In February 2008, Arryx launched the HOTkit Holo-Tweezers Solution, a holographic optical trapping add-on allowing scientists to incorporate event manipulation into existing microscope or imaging technologies.

Arryx offers its employees medical, dental and vision plans; life and disability insurance; a 401(k) plan; flex time; and a stock purchase plan.

## FINANCIALS: Sales and profits are in thousands of dollars—add 000 to get the full amount. 2008 Note: Financial information for 2008 was not available for all companies at press time.

| | | |
|---|---|---|
| 2008 Sales: $ | 2008 Profits: $ | **U.S. Stock Ticker:** Subsidiary |
| 2007 Sales: $ | 2007 Profits: $ | **Int'l Ticker:**   Int'l Exchange: |
| 2006 Sales: $ | 2006 Profits: $ | Employees: |
| 2005 Sales: $ | 2005 Profits: $ | Fiscal Year Ends: 12/31 |
| 2004 Sales: $ | 2004 Profits: $ | Parent Company: HAEMONETICS CORPORATION |

## SALARIES/BENEFITS:

| | | | | |
|---|---|---|---|---|
| Pension Plan: | ESOP Stock Plan: | Profit Sharing: | Top Exec. Salary: $ | Bonus: $ |
| Savings Plan: Y | Stock Purch. Plan: | | Second Exec. Salary: $ | Bonus: $ |

## OTHER THOUGHTS:

**Apparent Women Officers or Directors:** 1
**Hot Spot for Advancement for Women/Minorities:**

## LOCATIONS: ("Y" = Yes)

| West: | Southwest: | Midwest: | Southeast: | Northeast: | International: |
|---|---|---|---|---|---|
| | | Y | | | |

Note: Financial information, benefits and other data can change quickly and may vary from those stated here.

# ASTELLAS PHARMA INC                                     www.astellas.com

**Industry Group Code: 325412  Ranks within this company's industry group:** Sales: 19   Profits: 18

| Drugs: | | Other: | | Clinical: | | Computers: | | Services: | |
|---|---|---|---|---|---|---|---|---|---|
| Discovery: | Y | AgriBio: | | Trials/Services: | | Hardware: | | Specialty Services: | |
| Licensing: | | Genetic Data: | Y | Labs: | | Software: | | Consulting: | |
| Manufacturing: | Y | Tissue Replacement: | | Equipment/Supplies: | Y | Arrays: | | Blood Collection: | |
| Genetics: | | | | Research & Development Services: | Y | Database Management: | | Drug Delivery: | |
| | | | | Diagnostics: | | | | Drug Distribution: | |

## TYPES OF BUSINESS:

Drugs, Manufacturing
Immunological Pharmaceuticals
Over-the-Counter Products
Reagents
Genomic Research
Venture Capital
Drug Licensing

## BRANDS/DIVISIONS/AFFILIATES:

Yamanouchi Pharmaceutical Co., Ltd.
Fujisawa Pharmaceutical Co., Ltd.
Prograf
Lipitor
Harnal
Flomax
Micardis
Astellas Venture Management, LLC

## CONTACTS: Note: Officers with more than one job title may be intentionally listed here more than once.

Masafumi Nogimori, CEO
Masafumi Nogimori, Pres.
Yasuo Ishii, Chief Sales & Mktg. Officer/Exec. VP
Hirofumi Onosaka, Sr. Corp. Officer
Hitoshi Ohta, Sr. Corp. Officer
Iwaki Miyazaki, Sr. Corp. Officer
Katsuro Yamada, Sr. Corp. Officer
Toichi Takenaka, Chmn.
Yoshihiko Hatanaka, CEO/Pres., Astellas Pharma U.S., Inc.

| **Phone:** 81-3-3244-3000 | **Fax:** 81-3-3244-3272 |
|---|---|
| **Toll-Free:** | |
| **Address:** 2-3-11 Nihonbashi-Honcho, Chuo-ku, Tokyo, 103-8411 Japan | |

## GROWTH PLANS/SPECIAL FEATURES:

Astellas Pharma, Inc. is one of the largest pharmaceuticals manufacturers in Japan. It was formed from the recent merger of Yamanouchi Pharmaceutical Co., Ltd. and Fujisawa Pharmaceutical Co., Ltd. In addition to developing its own pharmaceuticals, Astellas pursues in-licensing and co-promotion agreements with biotechnology firms and a host of other pharmaceutical companies such as Pfizer, Inc. Nearly all of Astellas's sales relate to pharmaceuticals, led by Prograf, which is used as an immunosuppressant in conjunction with organ transplantation. Other products target needs in dermatology, urology, gastrointestinal disorders, immunology, infectious diseases, psychiatry and cardiology. Some of the firm's main products include Lipitor (developed by Pfizer) for high cholesterol; Micardis (co-promoted with Nippon Boehringer Ingelheim), a hypertension treatment; Myslee, an insomnia treatment co-promoted with sanofi-aventis S.A.; Seroquel, an antipsychotic; Gaster, for peptic ulcers and gastritis; fungal infection treatments AmBisome and Mycamine; overactive bladder treatment VESIcare; and Harnal for symptoms caused by enlarged prostates. Harnal is marketed by Boehringer Ingelheim Pharmaceuticals, Inc. in the U.S. under the name Flomax. Besides developing its own drugs or marketing drugs developed by others, the firm maintains Los Altos, CA-based subsidiary Astellas Venture Management, LLC, which is engaged in investing in biotechnology companies, starting with $30 million in initial capitalization. Astellas has 24 subsidiaries and affiliates in Europe; 11 in North America; seven in Asia; and three manufacturing subsidiaries in Japan. Geographically, Japan accounted for 51% of 2008 sales; Europe, 25%; North America, 22%; and Asia, 2%. In January 2009, the firm announced its intention to acquired CV Therapeutics, Inc. for $1.1 billion.

## FINANCIALS: Sales and profits are in thousands of dollars—add 000 to get the full amount. 2008 Note: Financial information for 2008 was not available for all companies at press time.

| | | |
|---|---|---|
| 2008 Sales: $9,726,000 | 2008 Profits: $1,774,000 | **U.S. Stock Ticker: ALPMF.PK** |
| 2007 Sales: $9,114,200 | 2007 Profits: $1,299,700 | **Int'l Ticker: 4503**  Int'l Exchange: Tokyo-TSE |
| 2006 Sales: $8,705,700 | 2006 Profits: $1,026,200 | Employees: 7,453 |
| 2005 Sales: $4,425,800 | 2005 Profits: $333,800 | Fiscal Year Ends: 3/31 |
| 2004 Sales: $4,839,100 | 2004 Profits: $568,500 | Parent Company: |

## SALARIES/BENEFITS:

| Pension Plan: Y | ESOP Stock Plan: | Profit Sharing: | Top Exec. Salary: $ | Bonus: $ |
|---|---|---|---|---|
| Savings Plan: | Stock Purch. Plan: | | Second Exec. Salary: $ | Bonus: $ |

## OTHER THOUGHTS:

**Apparent Women Officers or Directors**: 1
**Hot Spot for Advancement for Women/Minorities**:

## LOCATIONS: ("Y" = Yes)

| West: | Southwest: | Midwest: | Southeast: | Northeast: | International: |
|---|---|---|---|---|---|
| Y | | Y | | Y | Y |

# ASTRAZENECA CANADA

www.astrazeneca.ca

**Industry Group Code: 325412 Ranks within this company's industry group: Sales: Profits:**

| Drugs: | | Other: | Clinical: | Computers: | Services: |
|---|---|---|---|---|---|
| Discovery: | Y | AgriBio: | Trials/Services: | Hardware: | Specialty Services: |
| Licensing: | | Genetic Data: | Labs: | Software: | Consulting: |
| Manufacturing: | Y | Tissue Replacement: | Equipment/Supplies: | Arrays: | Blood Collection: |
| Genetics: | | | Research & Development Services: | Database Management: | Drug Delivery: |
| | | | Diagnostics: | | Drug Distribution: |

## TYPES OF BUSINESS:

Pharmaceutical Research & Development
Oncology Research
Cardiovascular Research
Neuroscience Research

## BRANDS/DIVISIONS/AFFILIATES:

AstraZeneca PLC

## CONTACTS: Note: Officers with more than one job title may be intentionally listed here more than once.

Mark Jones, CEO
Mark Jones, Pres.

**Phone:** 905-277-7111  **Fax:** 905-270-3248
**Toll-Free:** 800-565-5877
**Address:** 1004 Middlegate Rd., Mississauga, ON L4Y 1M4 Canada

## GROWTH PLANS/SPECIAL FEATURES:

AstraZeneca Canada, a subsidiary of pharmaceutical giant AstraZeneca PLC, is a pharmaceutical research and development company based in Ontario, Canada. AstraZeneca Canada focuses on creating therapeutic interventions in the key areas of cardiovascular, gastrointestinal, oncology, respiratory, neuroscience and infection. The company spends over $2 million every week on research initiatives. A significant portion of this funding is invested in the company's Montreal research facility, where the firm employs more than 120 scientists to conduct research in pain treatment. The majority of the company's employees work at its high-tech headquarters in Mississauga, Ontario, where the company engages in the packaging, clinical research, sales, marketing and distribution portion of its business. The firm also sponsors several research chairs and fellowships at many Canadian universities, including the Universities of Toronto, Alberta, Montreal, and Saskatchewan.

In 2008, AstraZeneca Canada was ranked second on Canada's 30 Best Pension and Benefits Plans according to Benefits Canada, the country's leading benefit and pension publication.

## FINANCIALS: Sales and profits are in thousands of dollars—add 000 to get the full amount. 2008 Note: Financial information for 2008 was not available for all companies at press time.

| | | |
|---|---|---|
| 2008 Sales: $ | 2008 Profits: $ | U.S. Stock Ticker: Subsidiary |
| 2007 Sales: $ | 2007 Profits: $ | Int'l Ticker:   Int'l Exchange: |
| 2006 Sales: $ | 2006 Profits: $ | Employees: 1,400 |
| 2005 Sales: $ | 2005 Profits: $ | Fiscal Year Ends: |
| 2004 Sales: $ | 2004 Profits: $ | Parent Company: ASTRAZENECA PLC |

## SALARIES/BENEFITS:

| | | | | |
|---|---|---|---|---|
| Pension Plan: Y | ESOP Stock Plan: | Profit Sharing: | Top Exec. Salary: $ | Bonus: $ |
| Savings Plan: | Stock Purch. Plan: | | Second Exec. Salary: $ | Bonus: $ |

## OTHER THOUGHTS:

**Apparent Women Officers or Directors:**
**Hot Spot for Advancement for Women/Minorities:**

## LOCATIONS: ("Y" = Yes)

| West: | Southwest: | Midwest: | Southeast: | Northeast: | International: Y |
|---|---|---|---|---|---|

# ASTRAZENECA PLC
### www.astrazeneca.com

Industry Group Code: 325412  Ranks within this company's industry group:  Sales: 7    Profits: 7

| Drugs: | | Other: | Clinical: | Computers: | Services: |
|---|---|---|---|---|---|
| Discovery: | Y | AgriBio: | Trials/Services: | Hardware: | Specialty Services: |
| Licensing: | | Genetic Data: | Labs: | Software: | Consulting: |
| Manufacturing: | Y | Tissue Replacement: | Equipment/Supplies: | Arrays: | Blood Collection: |
| Genetics: | | | Research & Development Services: | Database Management: | Drug Delivery: |
| | | | Diagnostics: | | Drug Distribution: |

## TYPES OF BUSINESS:
Drugs-Diversified
Pharmaceutical Research & Development

## BRANDS/DIVISIONS/AFFILIATES:
Nexium
Seroquel
Crestor
Arimidex
Symbicort
Pulmicort
AstraTech
Aptium Oncology

## CONTACTS: Note: Officers with more than one job title may be intentionally listed here more than once.
David R. Brennan, CEO
Simon Lowth, CFO
Tony Zook, Exec. VP-Global Mktg.
Lynn Tetrault, Exec. VP-Human Resources & Corp. Affairs
Jan Lundberg, Exec. VP-Discovery Research
Jeff Pott, General Counsel
David Smith, Exec. VP-Oper.
Anders Ekblom, Exec. VP-Dev.
Tony Zook, Pres., MedImmune
Tony Zook, CEO-North America
Louis Schweitzer, Chmn.
Bruno Angelici, Exec. VP-Europe/Japan/Asia Pacific/Rest of World

| Phone: 44-20-7304-5000 | Fax: 44-20-7304-5151 |
|---|---|
| Toll-Free: | |
| Address: 15 Stanhope Gate, London,  W1Y 6LN UK | |

## GROWTH PLANS/SPECIAL FEATURES:
AstraZeneca plc is a leading global pharmaceutical company that discovers, develops, manufactures and markets prescription pharmaceuticals, biologics and vaccines for the treatment or prevention of diseases in such areas of healthcare as cardiovascular, gastrointestinal, infection, neuroscience, oncology and respiratory and inflammation. The company is the result of the merger of the Zeneca Group with Astra. The firm invests over $5 billion annually in research and development and enjoys sales in over 100 countries. It operates 26 manufacturing sites in 18 countries and 17 major research centers in eight countries. AstraZeneca's cardiovascular products include Seloken ZOK, Crestor, Atacand, Plendil, Zestril and Tenormin. The firm's gastrointestinal products include Nexium, Entocort and Prilosec. Merrem, its primary infection product, is an antibiotic for serious hospital-acquired infections. AstraZeneca's neuroscience offering includes Zomig, a migraine treatment; anesthetics Diprivan and Xylocaine; Naropin, a long-acting anesthetic; and Seroquel for the treatment of schizophrenia and bipolar mania. AstraZeneca's cancer treatments include Casodex for prostate cancer; Zoladex; Armidex and Faslodex for breast cancer; Iressa for lung cancer; and Nolvadex. The firm's respiratory and inflammation brands include Pulmicort, Symbicort, Rhinocort, Accolate and Oxis. Nexium, Seroquel, Crestor, Arimidex, Symbicort, Pulmicort, Casodex, Atacand, Synagis, Prilosec and Zoladex all have sales in excess of $1 billion. Subsidiary AstraTech is engaged in the research, development, manufacture and marketing of medical devices and implants. Another subsidiary, Aptium Oncology, is a leading provider of outpatient oncology management and consulting services in the U.S., with full-service outpatient comprehensive cancer centers in California, Florida and New York. In February 2008, AstraZeneca formed Albireo, a joint venture with Nomura Phase4 Ventures, for the development of treatments for gastrointestinal disorders. In November 2008, the firm sold its Nordic over-the-counter portfolio to GlaxoSmithKline.

## FINANCIALS: Sales and profits are in thousands of dollars—add 000 to get the full amount. 2008 Note: Financial information for 2008 was not available for all companies at press time.

| | | |
|---|---|---|
| 2008 Sales: $31,601,000 | 2008 Profits: $6,130,000 | U.S. Stock Ticker: AZN |
| 2007 Sales: $29,559,000 | 2007 Profits: $5,627,000 | Int'l Ticker: AZN    Int'l Exchange: London-LSE |
| 2006 Sales: $26,475,000 | 2006 Profits: $6,063,000 | Employees:  65,000 |
| 2005 Sales: $23,950,000 | 2005 Profits: $3,881,000 | Fiscal Year Ends: 12/31 |
| 2004 Sales: $21,426,000 | 2004 Profits: $3,813,000 | Parent Company: |

## SALARIES/BENEFITS:
| Pension Plan: | ESOP Stock Plan: | Profit Sharing: | Top Exec. Salary: $1,191,000 | Bonus: $588,000 |
|---|---|---|---|---|
| Savings Plan: | Stock Purch. Plan: | | Second Exec. Salary: $732,000 | Bonus: $347,000 |

## OTHER THOUGHTS:
**Apparent Women Officers or Directors**: 4
**Hot Spot for Advancement for Women/Minorities**: Y

## LOCATIONS: ("Y" = Yes)
| West: | Southwest: | Midwest: | Southeast: | Northeast: | International: |
|---|---|---|---|---|---|
| Y | Y | Y | Y | Y | Y |

Note: Financial information, benefits and other data can change quickly and may vary from those stated here.

# AUTOIMMUNE INC

www.autoimmuneinc.com

**Industry Group Code: 325412  Ranks within this company's industry group: Sales: 150  Profits: 59**

| Drugs: | | Other: | | Clinical: | Computers: | Services: |
|---|---|---|---|---|---|---|
| Discovery: | Y | AgriBio: | | Trials/Services: | Hardware: | Specialty Services: |
| Licensing: | Y | Genetic Data: | | Labs: | Software: | Consulting: |
| Manufacturing: | | Tissue Replacement: | | Equipment/Supplies: | Arrays: | Blood Collection: |
| Genetics: | | | | Research & Development Services: | Database Management: | Drug Delivery: |
| | | | | Diagnostics: | | Drug Distribution: |

## TYPES OF BUSINESS:

Drugs-Immune System & Inflammatory Disease
Drug Development & Licensing
Nutraceuticals

## BRANDS/DIVISIONS/AFFILIATES:

Colloral
Teva Pharmaceutical Industries
BioMS Medical Corporation
Coral
AI 401
Eli Lilly & Company
Copaxone
MBP8298

## CONTACTS: Note: Officers with more than one job title may be intentionally listed here more than once.

Robert C. Bishop, CEO
Robert C. Bishop, Pres.
Diane M. McClintock, Dir.-Finance
Diane M. McClintock, Treas.
Robert C. Bishop, Chmn.

| Phone: 626-792-1235 | Fax: 626-792-1236 |
|---|---|
| Toll-Free: | |
| Address: 1199 Madia St, Pasadena, CA 91103 US | |

## GROWTH PLANS/SPECIAL FEATURES:

AutoImmune, Inc. is a healthcare company that owns and licenses rights to technology primarily based upon the principles of mucosal intolerance used in the development of products for the treatment of autoimmune and other cell-mediated inflammatory diseases and conditions. AutoImmune is currently attempting to demonstrate to the FDA that Colloral, its main product initially developed as a rheumatoid arthritis medication, meets the statutory definition of a dietary supplement, as it failed to gain FDA approval as a pharmaceutical product. AutoImmune has license agreements with Teva Pharmaceutical Industries, Ltd. and BioMS Medical Corporation. The firm has a license agreement with Teva Pharmaceutical Industries Ltd. for technology pertaining to the development of an oral formulation of Copaxone, Teva's injectable multiple sclerosis drug, under the name of Coral. Although Teva ceased development of Coral, it does continue to maintain its license to AutoImmune's technology by paying a portion of its patent expenses. AutoImmune's dirucotide product (formerly known as MBP8298) for secondary progressive multiple sclerosis was licensed to BioMS and is currently in Phase II/III trials. AutoImmune licenses AI 401 to Eli Lilly & Company for the Diabetes Prevention Trial conducted by the National Health Institute, determining if it can delay or prevent the onset of Type I diabetes. The company holds 201 US and foreign patents, with three more pending.

## FINANCIALS: Sales and profits are in thousands of dollars—add 000 to get the full amount. 2008 Note: Financial information for 2008 was not available for all companies at press time.

| | | |
|---|---|---|
| 2008 Sales: $ 318 | 2008 Profits: $- 345 | **U.S. Stock Ticker: AIMM.PK** |
| 2007 Sales: $ 294 | 2007 Profits: $- 118 | **Int'l Ticker:** Int'l Exchange: |
| 2006 Sales: $ 401 | 2006 Profits: $- 481 | Employees: 2 |
| 2005 Sales: $ 179 | 2005 Profits: $- 666 | Fiscal Year Ends: 12/31 |
| 2004 Sales: $ 130 | 2004 Profits: $- 761 | Parent Company: |

## SALARIES/BENEFITS:

| Pension Plan: | ESOP Stock Plan: | Profit Sharing: | Top Exec. Salary: $75,000 | Bonus: $ |
|---|---|---|---|---|
| Savings Plan: | Stock Purch. Plan: | | Second Exec. Salary: $60,500 | Bonus: $ |

## OTHER THOUGHTS:

**Apparent Women Officers or Directors:** 1
**Hot Spot for Advancement for Women/Minorities:** Y

## LOCATIONS: ("Y" = Yes)

| West: | Southwest: | Midwest: | Southeast: | Northeast: | International: |
|---|---|---|---|---|---|
| Y | | | | | |

Note: Financial information, benefits and other data can change quickly and may vary from those stated here.

# AVANIR PHARMACEUTICALS

www.avanir.com

**Industry Group Code: 325412  Ranks within this company's industry group:** Sales: 125  Profits: 98

| Drugs: | | Other: | | Clinical: | Computers: | Services: |
|---|---|---|---|---|---|---|
| Discovery: | Y | AgriBio: | | Trials/Services: | Hardware: | Specialty Services: |
| Licensing: | Y | Genetic Data: | | Labs: | Software: | Consulting: |
| Manufacturing: | | Tissue Replacement: | | Equipment/Supplies: | Arrays: | Blood Collection: |
| Genetics: | | | | Research & Development Services: | Database Management: | Drug Delivery: |
| | | | | Diagnostics: | | Drug Distribution: |

## TYPES OF BUSINESS:

Pharmaceutical Discovery & Development
Human Antibody Technology Research
Central Nervous System Research
Allergy & Asthma Drugs
Antibody Generation
Drugs - HSV1 Treatment

## BRANDS/DIVISIONS/AFFILIATES:

Xenerex
Abreva
Zenvia

## CONTACTS: *Note: Officers with more than one job title may be intentionally listed here more than once.*

Keith A. Katkin, CEO
Keith A. Katkin, Pres.
Gregory J. Flesher, VP-Bus. Dev.
Eric S. Benevich, VP-Comm.
Christine G. Ocamp, VP-Finance
Randall E. Kaye, Sr. VP-Clinical Research & Medical Affairs
Randall E. Kaye, Chief Medical Officer

| Phone: 949-389-6700 | Fax: 949-643-6800 |
|---|---|
| Toll-Free: | |
| Address: 101 Enterprise, Ste. 300, Aliso Viejo, CA 92656 US | |

## GROWTH PLANS/SPECIAL FEATURES:

Avanir Pharmaceuticals is a biopharmaceutical company engaged in research, development, commercialization, licensing and sales of innovative drug products and antibody generation services. The firm developed Abreva (docosonal 10% cream), the only over-the-counter, FDA-approved treatment for Type-1 Herpes Simplex (HSV1, more commonly known as cold sores or fever blisters). GlaxoSmithKline is the company's marketing partner for Abreva in the U.S. and Canada. Avanir's lead drug candidate, Zenvia, is in Phase III clinical trials for the treatment of diabetic peripheral neuropathic pain and pseudobulbar affect (PBA). PBA is characterized by unprovoked and uncontrollable episodes of crying or laughing, and afflicts patients with neurological disorders such as Lou Gehrig's disease, Alzheimer's disease, MS, stroke and traumatic brain injury. The company is collaborating with Novartis to develop a treatment that targets macrophage migration factor (MIF) for use with inflammatory diseases. The company is also engaged in small-molecule research to develop treatments for central nervous system disorder and inflammatory diseases. Another area of development includes testing the company's patented Xenerex antibody technology for discovery of fully human monoclonal antibodies in treating inhaled anthrax. In June 2008, Avanir received a European patent for continued commercial exclusivity of Zenvia through 2023.

Avanir offers major medical, dental, vision and disability insurance; life and AD&D insurance; and a 401(k) plan.

## FINANCIALS: Sales and profits are in thousands of dollars—add 000 to get the full amount. 2008 Note: Financial information for 2008 was not available for all companies at press time.

| | | |
|---|---|---|
| 2008 Sales: $6,829 | 2008 Profits: $-17,496 | U.S. Stock Ticker: AVNR |
| 2007 Sales: $9,153 | 2007 Profits: $-20,933 | Int'l Ticker:    Int'l Exchange: |
| 2006 Sales: $15,186 | 2006 Profits: $-62,553 | Employees:    20 |
| 2005 Sales: $16,691 | 2005 Profits: $-30,607 | Fiscal Year Ends: 9/30 |
| 2004 Sales: $3,589 | 2004 Profits: $-28,155 | Parent Company: |

## SALARIES/BENEFITS:

| Pension Plan: | ESOP Stock Plan: | Profit Sharing: | Top Exec. Salary: $343,544 | Bonus: $210,568 |
|---|---|---|---|---|
| Savings Plan: Y | Stock Purch. Plan: | | Second Exec. Salary: $317,117 | Bonus: $155,496 |

## OTHER THOUGHTS:

**Apparent Women Officers or Directors:** 1
**Hot Spot for Advancement for Women/Minorities:**

## LOCATIONS: ("Y" = Yes)

| West: | Southwest: | Midwest: | Southeast: | Northeast: | International: |
|---|---|---|---|---|---|
| Y | | | | | |

# AVAX TECHNOLOGIES INC

### www.avax-tech.com

Industry Group Code: 325412 **Ranks within this company's industry group:** Sales: Profits:

| Drugs: | | Other: | Clinical: | Computers: | Services: |
|---|---|---|---|---|---|
| Discovery: | Y | AgriBio: | Trials/Services: | Hardware: | Specialty Services: |
| Licensing: | Y | Genetic Data: | Labs: | Software: | Consulting: |
| Manufacturing: | Y | Tissue Replacement: | Equipment/Supplies: | Arrays: | Blood Collection: |
| Genetics: | | | Research & Development Services: | Database Management: | Drug Delivery: |
| | | | Diagnostics: | | Drug Distribution: |

## TYPES OF BUSINESS:

Drugs-Cancer
Melanoma Treatment
Non-Small Cell Lung Cancer Treatment
Ovarian Cancer Treatment
Vaccine Therapies

## BRANDS/DIVISIONS/AFFILIATES:

AC Vaccine
Mvax
Ovax
LungVax

## CONTACTS: *Note: Officers with more than one job title may be intentionally listed here more than once.*

Francois R. Martelet, CEO
Francois R. Martelet, Pres.
Henry E. Schea, III, Dir.-Global Quality & Regulatory Affairs
Isabelle Fourthin, Gen. Mgr.-Genopoietic
Jean-Louis Misset, Chmn.-Scientific & Medical Advisory Board
John K. A. Prendergast, Chmn.
Isabelle Fourthin, Chief Medical Officer-EMEA

| Phone: 215-241-9760 | Fax: 215-241-9684 |
|---|---|
| Toll-Free: | |
| Address: 2000 Hamilton St., Ste. 204, Philadelphia, PA 19130 US | |

## GROWTH PLANS/SPECIAL FEATURES:

AVAX Technologies, Inc. is a development stage biotechnology company that specializes in the development and commercialization of individualized vaccine therapies and other technologies for the treatment of cancer. The company's vaccine consists of autologous (the patient's own) cancer cells that have been treated with a chemical (haptenized) to make them more visible to the patient's immune system. AVAX refers to its cancer vaccine technology as autologous cell vaccine immunotherapy and to the vaccine as AC Vaccine. The firm's previous clinical trials have focused on melanoma, ovarian carcinoma and non-small cell lung cancer. AVAX's AC Vaccine candidates are MVax, currently in Phase III trials for the treatment of melanoma; LungVax, in Phase II trials for the treatment of non-small cell lung cancer; and OVax, in Phase I-II trials for the treatment of ovarian cancer. The company's leading AC Vaccine is MVax, which is designed as an immunotherapy for the post-surgical treatment of late stage (stages three and four) melanoma. AVAX believes that MVax is the first immunotherapy to show a substantial increase in the survival rate for patients with this type of melanoma. Of 214 stage-three melanoma patients treated with MVax, mature studies (in which all surviving patients completed the five-year follow-up) evidenced a five-year overall survival rate of 44%, as opposed to the historical post-surgical survival rates of approximately 22-32%. In total studies of over 600 patients, no serious adverse side effects have yet been reported. Subsidiary Genopoietic, based in Lyons, France, oversees the company's European activities, including a drug manufacturing facility in France.

## FINANCIALS: Sales and profits are in thousands of dollars—add 000 to get the full amount. 2008 Note: Financial information for 2008 was not available for all companies at press time.

| | | |
|---|---|---|
| 2008 Sales: $ | 2008 Profits: $ | U.S. Stock Ticker: AVXT |
| 2007 Sales: $ 617 | 2007 Profits: $-6,414 | Int'l Ticker: Int'l Exchange: |
| 2006 Sales: $ 735 | 2006 Profits: $-5,356 | Employees: 29 |
| 2005 Sales: $1,624 | 2005 Profits: $-3,704 | Fiscal Year Ends: 12/31 |
| 2004 Sales: $1,691 | 2004 Profits: $-3,457 | Parent Company: |

## SALARIES/BENEFITS:

| Pension Plan: | ESOP Stock Plan: | Profit Sharing: | Top Exec. Salary: $1,339,339 | Bonus: $2,708,940 |
|---|---|---|---|---|
| Savings Plan: | Stock Purch. Plan: | | Second Exec. Salary: $554,769 | Bonus: $866,320 |

## OTHER THOUGHTS:

Apparent Women Officers or Directors: 1
Hot Spot for Advancement for Women/Minorities:

## LOCATIONS: ("Y" = Yes)

| West: | Southwest: | Midwest: | Southeast: | Northeast: | International: |
|---|---|---|---|---|---|
| | | | | Y | Y |

# AVI BIOPHARMA INC

**www.avibio.com**

**Industry Group Code: 325412  Ranks within this company's industry group:**  Sales: 102    Profits: 112

| Drugs: | | Other: | Clinical: | Computers: | Services: |
|---|---|---|---|---|---|
| Discovery: | Y | AgriBio: | Trials/Services: | Hardware: | Specialty Services: |
| Licensing: | | Genetic Data: | Labs: | Software: | Consulting: |
| Manufacturing: | | Tissue Replacement: | Equipment/Supplies: | Arrays: | Blood Collection: |
| Genetics: | | | Research & Development Services: | Database Management: | Drug Delivery: |
| | | | Diagnostics: | | Drug Distribution: |

## TYPES OF BUSINESS:

Gene-Targeted Pharmaceuticals
Drugs - Cardiovascular Disease
Drugs - Cancer
Drugs - Infectious Disease

## BRANDS/DIVISIONS/AFFILIATES:

Translation Suppressing Oligomers (TSO)
Splice Switching Oligomers (SSO)
AVI-4658
Ercole Biotechnology Inc
AVI-6002
AVI-6003

## CONTACTS: *Note: Officers with more than one job title may be intentionally listed here more than once.*

Leslie Hudson, CEO
Leslie Hudson, Pres.
J. David Boyle II, CFO/Sr. VP
Ryszard Kole, Sr. VP-Discovery Research & Dev.
Dwight D. Weller, Sr. VP-Mfg. & Chemistry
Shirley J. Leow, VP-Clinical Oper. & Project Mgmt.
Paul Medeiros, Sr. VP-Bus. Dev./Chief Business Officer
Steve Shrewsbury, Chief Medical Officer
Steve Shrewsbury, Sr. VP-Preclinical, Clinical & Reg. Affairs
Patrick L. Iversen, Sr. VP-Strategic Alliances
Michael D. Casey, Chmn.

| Phone: 541-753-3635 | Fax: 503-227-0751 |
|---|---|
| Toll-Free: | |
| Address: 4575 SW Research Way, Corvallis, OR 97333 US | |

## GROWTH PLANS/SPECIAL FEATURES:

AVI BioPharma, Inc. is a biopharmaceutical company that develops drug treatments principally based on RNA therapeutics. Proprietary technologies include Translation Suppressing Oligomers (TSO), which are antisense compounds that are designed to suppress protein translation, and Splice Switching Oligomers (SSO), which block disease-related pathways in protein production. The company's principal products in development target life-threatening diseases, including cardiovascular, genetic and infectious diseases. These products have also been tested in preclinical trials and some clinical studies for the treatment of Duchenne muscular dystrophy (DMD), prevention of Restenosis and the Ebola and Marburg viruses. AVI's leading drug candidate, AVI-4658, currently in Phase I clinical trials, aims to reverse protein deletions that contribute to DMD. The firm's biodefense program is developing two antisense drugs, AVI-6002 and AVI-6003, which treat the Ebola and Margburg hemorrhagic fever viruses, respectively. This research is being supported by the US Department of Defense. The company is also working with Global Therapeutics to develop a cardiovascular restenosis drug for use in bare metal stents. AVI is also affiliated with a number of other institutions, including the US Army Medical Research Institute of Infectious Diseases (AMRIID) to find solutions for infections such as the avian influenza virus; Eleos, Inc. for cancer treatment and the Imperial College of London for muscular dystrophy. The firm owns 222 issued or licensed patent worldwide and 196 pending patent applications. In 2008, AVI acquired Ercole Biotechnology Inc., the firm's former collaboration partner in the development of drugs which splice messenger RNA in order to treat a variety of genetic and acquired diseases. In May 2009, the company announced it had received a $2.5 million grant from the Department of Defense to accelerate the development of its AVI-4658 drug for DMD.

AVI offers employees medical and dental insurance, a 401(k) savings plan and employee stock option plans.

## FINANCIALS:  Sales and profits are in thousands of dollars—add 000 to get the full amount. 2008 Note: Financial information for 2008 was not available for all companies at press time.

| | | | |
|---|---|---|---|
| 2008 Sales: $21,258 | 2008 Profits: $-23,953 | **U.S. Stock Ticker: AVII** | |
| 2007 Sales: $10,985 | 2007 Profits: $-27,168 | **Int'l Ticker:**    Int'l Exchange: | |
| 2006 Sales: $ 115 | 2006 Profits: $-31,073 | Employees:    83 | |
| 2005 Sales: $4,783 | 2005 Profits: $-16,676 | Fiscal Year Ends: 12/31 | |
| 2004 Sales: $ 430 | 2004 Profits: $-24,778 | Parent Company: | |

## SALARIES/BENEFITS:

| Pension Plan: | ESOP Stock Plan: | Profit Sharing: | Top Exec. Salary: $411,551 | Bonus: $196,020 |
|---|---|---|---|---|
| Savings Plan: Y | Stock Purch. Plan: Y | | Second Exec. Salary: $269,533 | Bonus: $ |

## OTHER THOUGHTS:

**Apparent Women Officers or Directors**: 1
**Hot Spot for Advancement for Women/Minorities**:

## LOCATIONS: ("Y" = Yes)

| West: | Southwest: | Midwest: | Southeast: | Northeast: | International: |
|---|---|---|---|---|---|
| Y | | | | | |

# AVIGEN INC

www.avigen.com

**Industry Group Code: 325412  Ranks within this company's industry group: Sales: 123  Profits: 116**

| Drugs: | Other: | | Clinical: | Computers: | Services: |
|---|---|---|---|---|---|
| Discovery: | AgriBio: | | Trials/Services: | Hardware: | Specialty Services: |
| Licensing: | Genetic Data: | Y | Labs: | Software: | Consulting: |
| Manufacturing: | Tissue Replacement: | | Equipment/Supplies: | Arrays: | Blood Collection: |
| Genetics: | | | Research & Development Services: | Database Management: | Drug Delivery: |
| | | | Diagnostics: | | Drug Distribution: |

## TYPES OF BUSINESS:

Neurological & Neuromuscular Therapeutics

## BRANDS/DIVISIONS/AFFILIATES:

AV650
AV411
AV333

## CONTACTS: *Note: Officers with more than one job title may be intentionally listed here more than once.*

Andrew A. Sauter, Acting CEO
Andrew A. Sauter, Acting Pres.
Andrew A. Sauter, CFO
Kirk W. Johnson, VP-R&D
Zola Horovitz, Chmn.

| Phone: 510-748-7150 | Fax: 510-748-7155 |
|---|---|
| Toll-Free: | |
| Address: 1301 Harbor Bay Pkwy., Alameda, CA 94502 US | |

## GROWTH PLANS/SPECIAL FEATURES:

Avigen, Inc. is a biopharmaceutical company that develops and commercializes differentiated products to treat serious disorders. The company's current lead product candidate is AV411 (ibudilast), a glial attenuator for neuropathic pain and opioid withdrawal and methamphetamine addiction. Major products in development include AV333, a plasmid treatment for chronic neuropathic pain through glial cell activation. The firm maintains a small ongoing preclinical research effort to identify additional opportunities to expand its product development pipeline. The efforts primarily focus on additional treatments for neuropathic pain. In October 2008, Avigen announced that one of its leading product candidates, AV650, a treatment for disabling neuromuscular spasticity and spasm, did not achieve statistically significant results in its Phase IIb clinical trial, and that its development was being terminated. In connection with this termination, the company announced a restructuring plan that included workforce reduction of over 70%. In December 2008, the firm sold the rights to its early-stage blood coagulation compound, AV513, to Baxter Healthcare Corporation for $7 million.

The company offers its employees medical, dental, vision, life and short- and long-term disability insurance; a 401(k) plan; an employee assistance program; stock options; and flexible spending accounts.

**FINANCIALS:**  Sales and profits are in thousands of dollars—add 000 to get the full amount. 2008 Note: Financial information for 2008 was not available for all companies at press time.

| | | | |
|---|---|---|---|
| 2008 Sales: $7,100 | 2008 Profits: $-25,099 | U.S. Stock Ticker: AVGN | |
| 2007 Sales: $ | 2007 Profits: $-25,164 | Int'l Ticker:  Int'l Exchange: | |
| 2006 Sales: $ 103 | 2006 Profits: $-24,256 | Employees:  12 | |
| 2005 Sales: $12,026 | 2005 Profits: $-14,696 | Fiscal Year Ends: 12/31 | |
| 2004 Sales: $2,195 | 2004 Profits: $-23,923 | Parent Company: | |

## SALARIES/BENEFITS:

| Pension Plan: | ESOP Stock Plan: | Profit Sharing: | Top Exec. Salary: $443,251 | Bonus: $80,892 |
|---|---|---|---|---|
| Savings Plan: Y | Stock Purch. Plan: Y | | Second Exec. Salary: $313,903 | Bonus: $50,978 |

## OTHER THOUGHTS:

**Apparent Women Officers or Directors:**
**Hot Spot for Advancement for Women/Minorities:**

## LOCATIONS: ("Y" = Yes)

| West: | Southwest: | Midwest: | Southeast: | Northeast: | International: |
|---|---|---|---|---|---|
| Y | | | | | |

# AVIVA BIOSCIENCES CORP

## www.avivabio.com

Industry Group Code: 325413  Ranks within this company's industry group: Sales:     Profits:

| Drugs: | Other: | Clinical: | Computers: | | Services: | |
|---|---|---|---|---|---|---|
| Discovery: | AgriBio: | Trials/Services: | Hardware: | Y | Specialty Services: | Y |
| Licensing: | Genetic Data: | Labs: | Software: | | Consulting: | |
| Manufacturing: | Tissue Replacement: | Equipment/Supplies: | Arrays: | | Blood Collection: | |
| Genetics: | | Research & Development Services: | Database Management: | | Drug Delivery: | |
| | | Diagnostics: | | | Drug Distribution: | |

## TYPES OF BUSINESS:

Cellular Biology Equipment
Cancer Cell Isolation Technology
Drug Candidate Screening Technology
Automated Patch Clamp Electrophysiology
Rare Cell Enrichment
Multiple Force Biochips

## BRANDS/DIVISIONS/AFFILIATES:

Sealchip
Electrophysiology on Demand
Fetal Cell Enrichment Kit
hERGexpress
China Development Industrial Bank
CapityalBio Corporation
Pac-Link
WI Harper Group

## CONTACTS: Note: Officers with more than one job title may be intentionally listed here more than once.

Julian Yuan, CEO
Jia Xu, VP-R&D
Lei Wu, VP-Mfg.
Lei Wu, VP-Oper.
Vytas P. Ambutas, Chmn.

| Phone: 858-522-0888 | Fax: 858-522-9040 |
|---|---|
| Toll-Free: 888-284-8224 | |
| Address: 11045 Roselle St., Ste. 100, San Diego, CA 92121 US | |

## GROWTH PLANS/SPECIAL FEATURES:

AVIVA Biosciences Corp. develops drug technologies for the research and development of drugs. The firm's key development areas include the integrating of biochips for electrophysiology research, ion channel drug screening and rare-cell isolation. In addition, the company has developed proprietary surface chemistries, microbeads and reagent compositions as solutions to cell biology and bioassay applications. AVIVA's introduction of improved and automated cell manipulation platforms has enabled greater productivity of cell-based assays for use in the biotechnological development of medicinal products. The firm offers Electrophysiology on Demand (EPOD), a drug discovery service that utilizes ion channel drug screening in order to provide clients with a full line of automated electrophysiology instruments, experienced personnel and customer service. The main product line within the EPOD service is Sealchip, a single-use disposable cartridge designed for use in high fidelity ion channel measurements and higher throughput screening for drug discovery customers. Optimized and validated ion channel cell lines are also provided for patch clamp electrophysiology experiments. Additionally, AVIVA develops cancer cell isolation systems that reliably detect targeted tumor cells from blood samples. Through a proprietary depletion approach, the system removes non-relevant cells and exposes target cells that can then be analyzed and quantified. This deletion process could be used in prenatal care in the future, such as detecting Down syndrome in the first trimester of pregnancy. In addition to its product lines, the firm offers hERGexpress, a screening service for medicinal chemists and toxicologists that provides high quality data as guidance in assessing the cardiac safety of certain pharmaceutical compounds. AVIVA is owned by four investors: CapitalBio Corporation (its primary shareholder), China Development Industrial Bank, WI Harper Group and Pac-Link.

## FINANCIALS: Sales and profits are in thousands of dollars—add 000 to get the full amount. 2008 Note: Financial information for 2008 was not available for all companies at press time.

| | | |
|---|---|---|
| 2008 Sales: $ | 2008 Profits: $ | U.S. Stock Ticker: Private |
| 2007 Sales: $ | 2007 Profits: $ | Int'l Ticker:     Int'l Exchange: |
| 2006 Sales: $ | 2006 Profits: $ | Employees:   26 |
| 2005 Sales: $ | 2005 Profits: $ | Fiscal Year Ends: |
| 2004 Sales: $ | 2004 Profits: $ | Parent Company: |

## SALARIES/BENEFITS:

| Pension Plan: | ESOP Stock Plan: | Profit Sharing: | Top Exec. Salary: $ | Bonus: $ |
|---|---|---|---|---|
| Savings Plan: | Stock Purch. Plan: | | Second Exec. Salary: $ | Bonus: $ |

## OTHER THOUGHTS:

Apparent Women Officers or Directors:
Hot Spot for Advancement for Women/Minorities:

## LOCATIONS: ("Y" = Yes)

| West: | Southwest: | Midwest: | Southeast: | Northeast: | International: |
|---|---|---|---|---|---|
| Y | | | | | |

Note: Financial information, benefits and other data can change quickly and may vary from those stated here.

# AXCAN PHARMA INC

**www.axcan.com**

Industry Group Code: 325412  **Ranks within this company's industry group:** Sales:    Profits:

| Drugs: | | Other: | Clinical: | Computers: | Services: | |
|---|---|---|---|---|---|---|
| Discovery: | | AgriBio: | Trials/Services: | Hardware: | Specialty Services: | |
| Licensing: | | Genetic Data: | Labs: | Software: | Consulting: | |
| Manufacturing: | Y | Tissue Replacement: | Equipment/Supplies: | Arrays: | Blood Collection: | |
| Genetics: | | | Research & Development Services: | Database Management: | Drug Delivery: | |
| | | | Diagnostics: | | Drug Distribution: | |

## TYPES OF BUSINESS:

Pharmaceutical Manufacturing
Gastroenterology Treatment Products

## GROWTH PLANS/SPECIAL FEATURES:

Axcan Pharma, Inc. is a leading specialty pharmaceutical company concentrating in the field of gastroenterology, with operations in North America and Europe.  Axcan markets and sells pharmaceutical products used in the treatment of a variety of gastrointestinal diseases and disorders, including inflammatory bowel disease, cholestatic liver diseases, irritable bowel syndrome and complications related to pancreatic insufficiency.  For the treatment of inflammatory bowel diseases, such as ulcerative colitis and ulcerative proctitis, Axcan markets mesalamine-based products Salofalk and Canasa.  Axcan is currently developing products for the prevention and treatment of colorectal cancer, including SUDCA, which is currently in Phase I trial for the prevention of the recurrence of colorectal polyps.  For the treatment of the cholestatic liver disease Primary Biliary Cirrhosis (PBC), a condition that causes the slow destruction of bile ducts in the liver, Axcan markets URSO 250 and URSO Forte.  For the treatment of both PBC and the cholestatic liver disease Primary Sclerosing Cholangitis (PSC), a condition that narrows the bile ducts inside and outside of the liver through inflammation and scarring, Axcan markets Urso, Urso DS and Delursan.  For the treatment of pancreatic insufficiency, Axcan markets Ultrase, Panzytrat and Viokase.  For the treatment of duodenal ulcers, Axcan markets Carafate and Sulcrate.  Recently, Axcan released Pylera, a therapy for the eradication of Helicobacter pylori, a bacterium recognized as the main cause of gastric and duodenal ulcers.  In February 2008, Axcan was acquired by an affiliate of private investment firm TPG (Texas Pacific Group) for approximately $1.3 billion.

## BRANDS/DIVISIONS/AFFILIATES:

TPG (Texas Pacific Group)
Salofalk
Canasa
Sudca
Urso
Delursan
Ultrase
Viokase

## CONTACTS: *Note: Officers with more than one job title may be intentionally listed here more than once.*

Frank A. G. M. Verwiel, CEO
David W. Mims, COO/Exec. VP
Frank A. G. M. Verwiel, Pres.
Steve Gannon, CFO/Sr. VP
Jean-Francois Hebert, Sr. VP-Mfg. Oper.
Martha Donze, VP-Corp. Admin.
Richard Tarte, General Counsel
Richard Tarte, VP-Corp. Dev.
Isabelle Adjahi, Sr. Dir.-Comm.
Isabelle Adjahi, Sr. Dir.-Investor Rel.
Darcy Toms, VP-Bus. Dev.
Nicholas Franco, Sr. VP- Int'l Commercial Oper.

| | |
|---|---|
| **Phone:** 450-467-5138 | **Fax:** 450-464-9979 |
| **Toll-Free:** 800-809-4950 | |
| **Address:** 597 Laurier Blvd., Mont-Saint-Hilaire, QC J3H 6CA8 Canada | |

## FINANCIALS: Sales and profits are in thousands of dollars—add 000 to get the full amount. 2008 Note: Financial information for 2008 was not available for all companies at press time.

| | | |
|---|---|---|
| 2008 Sales: $ | 2008 Profits: $ | **U.S. Stock Ticker:** Private |
| 2007 Sales: $349,000 | 2007 Profits: $ | **Int'l Ticker:** AXP    Int'l Exchange: Toronto-TSX |
| 2006 Sales: $292,320 | 2006 Profits: $39,120 | Employees:   480 |
| 2005 Sales: $251,300 | 2005 Profits: $26,400 | Fiscal Year Ends: 9/30 |
| 2004 Sales: $243,800 | 2004 Profits: $44,500 | Parent Company: TPG (TEXAS PACIFIC GROUP) |

## SALARIES/BENEFITS:

| | | | | |
|---|---|---|---|---|
| Pension Plan: | ESOP Stock Plan: | Profit Sharing: Y | Top Exec. Salary: $ | Bonus: $ |
| Savings Plan: Y | Stock Purch. Plan: | | Second Exec. Salary: $ | Bonus: $ |

## OTHER THOUGHTS:

**Apparent Women Officers or Directors**: 2
**Hot Spot for Advancement for Women/Minorities**: Y

## LOCATIONS: ("Y" = Yes)

| West: | Southwest: | Midwest: | Southeast: | Northeast: | International: |
|---|---|---|---|---|---|
| | | | Y | | Y |

# BARR PHARMACEUTICALS INC

www.barrlabs.com

Industry Group Code: 325412A  Ranks within this company's industry group: Sales:    Profits:

| Drugs: | | Other: | Clinical: | Computers: | Services: |
|---|---|---|---|---|---|
| Discovery: | Y | AgriBio: | Trials/Services: | Hardware: | Specialty Services: |
| Licensing: | | Genetic Data: | Labs: | Software: | Consulting: |
| Manufacturing: | Y | Tissue Replacement: | Equipment/Supplies: | Arrays: | Blood Collection: |
| Genetics: | Y | | Research & Development Services: | Database Management: | Drug Delivery: |
| | | | Diagnostics: | | Drug Distribution: |

## TYPES OF BUSINESS:

Drugs-Generic Pharmaceuticals
Contraceptives
Hormone Therapy Drugs
Female Healthcare Pharmaceuticals

## BRANDS/DIVISIONS/AFFILIATES:

Barr Laboratories
PLIVA
Duramed Pharmaceuticals
SEASONIQUE
PLAN B
ENJUVIA

## CONTACTS: Note: Officers with more than one job title may be intentionally listed here more than once.

William T. McKee, CFO/Exec. VP
Jane F. Greenman, Exec. VP-Human Resources
Fredrick J. Killion, General Counsel/Exec. VP
Carol A. Cox, Sr. VP-Corp. Comm.
Carol A. Cox, Sr. VP-Global Investor Rel.
Christine A. Mundkur, CEO-Barr Laboratories, Inc.
Michael J. Bogda, COO/Pres., Barr Laboratories, Inc.
G. Frederick Wilkinson, CEO-Duramed Pharmaceuticals, Inc.
Zeljko Covic, COO/Pres., PLIVA d.d.
Bruce L. Downey, Chmn.

| Phone: 201-930-3300 | Fax: 201-930-3330 |
|---|---|
| Toll-Free: 800-227-7522 | |
| Address: 225 Summit Ave., Montvale, NJ 07645 US | |

## GROWTH PLANS/SPECIAL FEATURES:

Barr Pharmaceuticals, Inc. engages in the development, manufacture and marketing of generic and proprietary prescription pharmaceuticals in more than 30 countries. Its operations are based primarily in North America and Europe, with its key markets located in the U.S., Croatia, Germany, Poland and Russia. The firm conducts its generics business in North America principally through its Barr Laboratories subsidiary and internationally through its PLIVA subsidiary. Barr markets in the U.S. approximately 245 different dosage forms and strengths of approximately 120 different generic pharmaceutical products, including 25 oral contraceptive products. Internationally, Barr markets approximately 255 different molecules, representing 1,025 generic pharmaceutical products in approximately 2,790 different presentations. The firm markets and sells approximately 27 proprietary pharmaceutical products primarily in the U.S., largely concentrated in the area of female healthcare. Its proprietary business is conducted through its Duramed Pharmaceuticals subsidiary. Proprietary products include SEASONIQUE, an extended-cycle oral contraceptive product; PLAN B, a dual-label, over-the-counter emergency contraceptive; and ENJUVIA, a line of hormone therapy products. Barr has several generic biologics products in various stages of development in the U.S. and European markets, including granulocyte colony stimulating factor (G-CSF), a protein that stimulates the growth of certain white blood cells. In March 2008, the firm received FDA approval to market a generic version of Taxol Injection. In May 2008, Barr received FDA approval to market a generic version of Yasmin, an oral contraceptive product. In December 2008, the firm was acquired by Teva Pharmaceutical Industries Ltd. for approximately $7.5 billion; subsequent to the acquisition, Barr was delisted from the New York Stock Exchange.

## FINANCIALS: Sales and profits are in thousands of dollars—add 000 to get the full amount. 2008 Note: Financial information for 2008 was not available for all companies at press time.

| | | |
|---|---|---|
| 2008 Sales: $ | 2008 Profits: $ | U.S. Stock Ticker: Subsidiary |
| 2007 Sales: $2,500,582 | 2007 Profits: $128,350 | Int'l Ticker:    Int'l Exchange: |
| 2006 Sales: $1,314,465 | 2006 Profits: $336,477 | Employees: 8,900 |
| 2005 Sales: $1,047,399 | 2005 Profits: $214,988 | Fiscal Year Ends: 6/30 |
| 2004 Sales: $1,309,088 | 2004 Profits: $123,103 | Parent Company: TEVA PHARMACEUTICAL INDUSTRIES |

## SALARIES/BENEFITS:

| Pension Plan: | ESOP Stock Plan: | Profit Sharing: | Top Exec. Salary: $1,198,654 | Bonus: $331,391 |
|---|---|---|---|---|
| Savings Plan: Y | Stock Purch. Plan: | | Second Exec. Salary: $662,782 | Bonus: $331,391 |

## OTHER THOUGHTS:

Apparent Women Officers or Directors: 5
Hot Spot for Advancement for Women/Minorities: Y

## LOCATIONS: ("Y" = Yes)

| West: | Southwest: | Midwest: | Southeast: | Northeast: | International: |
|---|---|---|---|---|---|
| | | | | Y | |

# BASF AG

**www.basf.com**

Industry Group Code: 325   Ranks within this company's industry group: Sales: 1   Profits: 1

| Drugs: | Other: | Clinical: | | Computers: | Services: | |
|--------|--------|-----------|--|------------|-----------|--|
| Discovery: | AgriBio: | Trials/Services: | | Hardware: | Specialty Services: | Y |
| Licensing: | Genetic Data: | Labs: | | Software: | Consulting: | |
| Manufacturing: | Tissue Replacement: | Equipment/Supplies: | Y | Arrays: | Blood Collection: | |
| Genetics: | | Research & Development Services: | Y | Database Management: | Drug Delivery: | |
| | | Diagnostics: | | | Drug Distribution: | |

## TYPES OF BUSINESS:

Chemicals Manufacturing
Agricultural Products
Oil & Gas Production
Plastics
Coatings
Nanotechnology Research
Nutritional Products
Agricultural Biotechnology

## BRANDS/DIVISIONS/AFFILIATES:

Wintershall AG
Orgamol SA
BASF Catalysts LLC
Johnson Polymer
CropDesign
Hansa Chemie International
Sorex Holdings Ltd

## CONTACTS: Note: Officers with more than one job title may be intentionally listed here more than once.

Jurgen Hambrect, CEO
Kurt W. Bock, CFO
Harald Schwager, Exec. Dir.-Human Resources
Stefan Marcinowski, Exec. Dir.-Research Planning
Kurt W. Bock, Exec. Dir.-Info. Svcs.
Stefan Marcinowski, Exec. Dir.-Corp. Eng.
Magdalena Moll, Sr. VP-Investor Rel.
Kurt W. Bock, Exec. Dir.-Finance
Andreas Kreimeyer, Exec. Dir.-Performance Chemicals
John Feldmann, Exec. Dir.-Oil & Gas
Stefan Marcinowski, Exec. Dir.-Inorganics & Petrochemicals
Peter Oakley, Dir.-Agricultural Prod.
Juergen Hambrecht, Chmn.
Martin Brudermueller, Exec. Dir.-APAC
Hans-Ulrich Engel, Exec. Dir.-Procurement & Logistics

| Phone: 49-621-60-0 | Fax: 49-621-60-42525 |
|---|---|
| Toll-Free: | |
| Address: 38 Carl-Bosch St., Ludwigshafen, 67056 Germany | |

## GROWTH PLANS/SPECIAL FEATURES:

BASF AG is a chemical manufacturing company that operates 330 production facilities in 38 countries and serves customers in more than 170 countries. Around 21% of BASF sales are made to North American industries. The firm operates in six business segments: chemicals; plastics; performance products; agricultural solutions; functional solutions; and oil and gas. The chemicals segment manufactures inorganic, petrochemical and intermediate chemicals for the pharmaceutical, construction, textile and automotive industries. The plastics segment primarily manufactures polystyrene, styrenics and performance polymers for the manufacturing and packaging industries. The performance polymers segment produces pigments, inks, printing supplies, coatings and polymers for the automotive, oil, packaging, textile, detergent, sanitary care, construction and chemical industries. The firm's agricultural solutions segment produces and markets genetically engineered plants, nutritional supplements, herbicides, fungicides and insecticides for use in agriculture, public health and pest control. The functional solutions segment develops automotive and industrial catalysts; construction chemicals; and coatings and refinishes for automotive and construction markets. The oil and gas segment is operated through BASF subsidiary Wintershall AG, which focuses on petroleum and natural gas exploration and production in North America, Asia, Europe, the Middle East and Africa. BASF also employs chemical nanotechnology in pigments that are used to color coatings, paints and plastics; and sunscreen. The company is one of the world's leading R&D firms, with 8,000 employees working in research in 70 sites worldwide, employing a research budget of $1.3 billion Euros yearly. In September 2008, the firm agreed to acquire specialty chemicals maker Ciba Holding AG. In December of the same year, the company acquired Sorex Holdings Ltd., a manufacturer of branded chemical and other products for pest management. In 2009, BASF plans to eliminate 2,000 jobs and shorten hours for another 3,000 workers.

U.S. employees are offered medical, dental and vision insurance; life insurance; disability coverage; an employee savings plan; tuition reimbursement; auto and home insurance; adoption assistance; and a preferred supplier discount.

## FINANCIALS: Sales and profits are in thousands of dollars—add 000 to get the full amount. 2008 Note: Financial information for 2008 was not available for all companies at press time.

| | | |
|---|---|---|
| 2008 Sales: $83,990,800 | 2008 Profits: $3,925,610 | **U.S. Stock Ticker: BF** |
| 2007 Sales: $78,122,600 | 2007 Profits: $5,479,950 | **Int'l Ticker: BAS**   Int'l Exchange: Frankfurt-Euronext |
| 2006 Sales: $69,448,400 | 2006 Profits: $4,575,330 | Employees: 96,924 |
| 2005 Sales: $52,080,500 | 2005 Profits: $3,663,700 | Fiscal Year Ends: 12/31 |
| 2004 Sales: $51,572,600 | 2004 Profits: $2,550,700 | Parent Company: |

## SALARIES/BENEFITS:

| Pension Plan: | ESOP Stock Plan: | Profit Sharing: | Top Exec. Salary: $ | Bonus: $ |
|---|---|---|---|---|
| Savings Plan: | Stock Purch. Plan: | | Second Exec. Salary: $ | Bonus: $ |

## OTHER THOUGHTS:

**Apparent Women Officers or Directors: 1**
**Hot Spot for Advancement for Women/Minorities:**

## LOCATIONS: ("Y" = Yes)

| West: | Southwest: | Midwest: | Southeast: | Northeast: | International: |
|---|---|---|---|---|---|
| Y | Y | Y | Y | Y | Y |

Note: Financial information, benefits and other data can change quickly and may vary from those stated here.

# BAUSCH & LOMB INC

## www.bausch.com

**Industry Group Code: 33911　Ranks within this company's industry group:**　Sales:　Profits:

| Drugs: | Other: | Clinical: | Computers: | Services: |
|---|---|---|---|---|
| Discovery: | AgriBio: | Trials/Services: | Hardware: | Specialty Services: |
| Licensing: | Genetic Data: | Labs: | Software: | Consulting: |
| Manufacturing: Y | Tissue Replacement: | Equipment/Supplies: Y | Arrays: | Blood Collection: |
| Genetics: Y | | Research & Development Services: | Database Management: | Drug Delivery: |
| | | Diagnostics: | | Drug Distribution: |

## TYPES OF BUSINESS:

Supplies-Eye Care
Contact Lens Products
Ophthalmic Pharmaceuticals
Surgical Products

## BRANDS/DIVISIONS/AFFILIATES:

Alrex
Warburg Pincus LLC
Eyeonics Inc
Ocuvite
Lotemax
Alrex
PreserVision
Zyoptic

## CONTACTS: *Note: Officers with more than one job title may be intentionally listed here more than once.*

Gerald M. Ostrov, CEO
Brian J. Harris, CFO/Corp. VP
Paul H. Sartori, Corp. VP- Human Resources
Alan H. Farnsworth, CIO/Sr. VP-IT/Corp. VP
John W. Sheets, Jr., CTO/Corp. VP
A Robert D. Bailey, General Counsel/Corp. VP
Michael Gowen, Exec. VP-Global Bus. Oper. & Process Excellence
Paul H. Sartori, Corp. VP-Public Affairs
J. Andy Corley, Global Pres./Corp. VP-Surgical Products
David N. Edwards, Pres., Asia Pacific Region/Corp. VP
Stuart Heap, Global Pres., Vision Care/Corp. VP
Flemming Ornskov, Global Pres., Pharmaceuticals/Corp. VP
Gerald M. Ostrov, Chmn.
John H. Brown, Pres., EMEA/Corp. VP

| Phone: 585-338-6000 | Fax: |
|---|---|
| **Toll-Free:** 800-344-8815 | |
| **Address:** 1 Bausch & Lomb Pl., Rochester, NY 14604-2701 US | |

## GROWTH PLANS/SPECIAL FEATURES:

Bausch & Lomb, Inc. (B&L), a subsidiary of Warburg Pincus LLC, is a world leader in the development, marketing and manufacturing of eye care products. The firm's products are marketed in more than 100 countries and in five categories: contact lenses; lens care; pharmaceuticals; cataract and vitreoretinal surgery; and refractive surgery. In its contact lens category, B&L's product portfolio includes traditional, planned replacement disposable, daily disposable, continuous wear, toric soft contact lenses and rigid gas-permeable materials. The firm's lens care products include multi-purpose solutions, enzyme cleaners and saline solutions. The firm's pharmaceuticals include generic and branded prescription ophthalmic pharmaceuticals, ocular vitamins, over-the-counter medications and vision accessories. Key pharmaceutical trademarks of the firm are Bausch & Lomb, Alrex, Liposic, Lotemax, Ocuvite, PreserVision and Zylet. B&L's cataract and vitreoretinal division offers a broad line of intraocular lenses and delivery systems, as well as the Millennium and Stellaris lines of phacoemulsification equipment used in the extraction of the patient's natural lens during cataract surgery. The company's refractive surgery products include lasers and diagnostic equipment used in the LASIK surgical procedure under the brand Zyoptic. B&L's global operations include research and development units on six continents and operating offices in over 43 countries. In February 2008, B&L completed the acquisition of Eyeonics, Inc., a company specializing in ophthalmic medical devices based in Aliso Viejo, California.

The firm offers employees medical and dental coverage; various work/life programs; a vacation buy/sell program; domestic partner benefits; flexible spending accounts; and education reimbursement.

## FINANCIALS: Sales and profits are in thousands of dollars—add 000 to get the full amount. 2008 Note: Financial information for 2008 was not available for all companies at press time.

| | | |
|---|---|---|
| 2008 Sales: $ | 2008 Profits: $ | **U.S. Stock Ticker: Private** |
| 2007 Sales: $ | 2007 Profits: $ | **Int'l Ticker:**　Int'l Exchange: |
| 2006 Sales: $2,293,400 | 2006 Profits: $ | Employees: 13,700 |
| 2005 Sales: $2,353,800 | 2005 Profits: $19,200 | Fiscal Year Ends: 12/31 |
| 2004 Sales: $2,233,500 | 2004 Profits: $153,900 | Parent Company: WARBURG PINCUS LLC |

## SALARIES/BENEFITS:

| Pension Plan: | ESOP Stock Plan: | Profit Sharing: | Top Exec. Salary: $1,100,000 | Bonus: $ |
|---|---|---|---|---|
| Savings Plan: | Stock Purch. Plan: | | Second Exec. Salary: $410,001 | Bonus: $295,000 |

## OTHER THOUGHTS:

**Apparent Women Officers or Directors:** 1
**Hot Spot for Advancement for Women/Minorities:** Y

## LOCATIONS: ("Y" = Yes)

| West: | Southwest: | Midwest: | Southeast: | Northeast: | International: |
|---|---|---|---|---|---|
| Y | | | Y | Y | Y |

# BAXTER INTERNATIONAL INC

**www.baxter.com**

Industry Group Code: 33911 **Ranks within this company's industry group:** Sales: 2 Profits: 2

| Drugs: | Other: | Clinical: | | Computers: | Services: |
|---|---|---|---|---|---|
| Discovery: | AgriBio: | Trials/Services: | | Hardware: | Specialty Services: |
| Licensing: | Genetic Data: | Labs: | | Software: | Consulting: |
| Manufacturing: | Tissue Replacement: | Equipment/Supplies: | Y | Arrays: | Blood Collection: |
| Genetics: | | Research & Development Services: | | Database Management: | Drug Delivery: |
| | | Diagnostics: | | | Drug Distribution: |

## TYPES OF BUSINESS:

Medical Equipment Manufacturing
Supplies-Intravenous & Renal Dialysis Systems
Medication Delivery Products & IV Fluids
Biopharmaceutical Products
Plasma Collection & Processing
Vaccines
Software
Contract Research

## BRANDS/DIVISIONS/AFFILIATES:

Medication Delivery
BioScience
Renal
Colleague CX
Enlightened
ADVATE
RenalSoft HD
ARTISS

## CONTACTS: Note: Officers with more than one job title may be intentionally listed here more than once.

Robert L. Parkinson, Jr., CEO
Robert L. Parkinson, Jr., Pres.
Robert M. Davis, CFO/VP
Jeanne K. Mason, VP-Human Resources
Norbert G. Riedel, Chief Scientific Officer/VP
Karenann Terrell, CIO/VP
J. Michael Gatling, VP-Mfg.
Susan R. Lichtenstein, General Counsel/VP
Michael J. Baughman, Controller/VP
Joy A. Amundson, VP/Pres., Bioscience
Bruce McGillivray, Pres., Renal/VP
Peter J. Arduini, Pres., Medication Delivery/VP
Gerald Lema, VP/Pres., Asia Pacific
Robert L. Parkinson, Jr., Chmn.
John J. Greisch, VP/Pres., Int'l

| Phone: 847-948-2000 | Fax: 847-948-3642 |
|---|---|
| Toll-Free: 800-422-9837 | |
| Address: 1 Baxter Pkwy., Deerfield, IL 60015-4625 US | |

## GROWTH PLANS/SPECIAL FEATURES:

Baxter International, Inc. manufactures and markets products for the treatment of hemophilia, immune disorders, cancer, infectious diseases, kidney disease, trauma and other chronic and acute medical conditions, offering expertise in medical devices, pharmaceuticals and biotechnology. Baxter markets its offerings to hospitals; clinical and medical research labs; blood and blood dialysis centers; rehab facilities; nursing homes; doctor's offices; and patients undergoing supervised home care. The firm has manufacturing facilities in 26 countries and offers products and services in 100 countries. Baxter operates in three segments: Medication Delivery, its largest sector, which provides a range of intravenous solutions and specialty products that are used in combination for fluid replenishment, nutrition therapy, pain management, antibiotic therapy and chemotherapy; BioScience, which develops biopharmaceuticals, biosurgery products, vaccines, blood collection, processing and storage products and technologies; and Renal, which develops products and provides services to treat end-stage kidney disease. Products include the Colleague CX infusion pump; the Enlightened bar-coding system for flexible IV containers; ADVATE, a coagulant for hemophilia patients; and RenalSoft HD, a software module for the management of prescription, therapy and monitoring information relating to patients suffering from kidney failure. In addition, the company provides the following services: BioLife Plasma Services, a plasma collection and processing business; BioPharma Solutions, biotechnology; Global Technical Services, providing instrument service and support for devices manufactured and marketed by Baxter; Renal Clinical Helpline; Renal Services, an education and research operation; and Training and Education, a portfolio of interactive clinical web sites. In March 2008, the company received FDA approval of ARTISS, a slow-setting fibrin sealant for the use of adhering skin grafts in burn patients.

Employees are offered medical and dental insurance; vision care discounts; health and dependent care reimbursement accounts; an educational assistance program; credit union membership; adoption reimbursement; an employee assistance program; a 401(k) plan; and a stock purchase plan.

## FINANCIALS: Sales and profits are in thousands of dollars—add 000 to get the full amount. 2008 Note: Financial information for 2008 was not available for all companies at press time.

| | | |
|---|---|---|
| 2008 Sales: $12,348,000 | 2008 Profits: $2,014,000 | **U.S. Stock Ticker:** BAX |
| 2007 Sales: $11,263,000 | 2007 Profits: $1,707,000 | **Int'l Ticker:** Int'l Exchange: |
| 2006 Sales: $10,378,000 | 2006 Profits: $1,397,000 | Employees: 48,500 |
| 2005 Sales: $9,849,000 | 2005 Profits: $956,000 | Fiscal Year Ends: 12/31 |
| 2004 Sales: $9,509,000 | 2004 Profits: $388,000 | Parent Company: |

## SALARIES/BENEFITS:

| Pension Plan: | ESOP Stock Plan: | Profit Sharing: | Top Exec. Salary: $1,296,153 | Bonus: $3,000,000 |
|---|---|---|---|---|
| Savings Plan: Y | Stock Purch. Plan: | | Second Exec. Salary: $595,165 | Bonus: $1,089,050 |

## OTHER THOUGHTS:

**Apparent Women Officers or Directors:** 6
**Hot Spot for Advancement for Women/Minorities:** Y

## LOCATIONS: ("Y" = Yes)

| West: | Southwest: | Midwest: | Southeast: | Northeast: | International: |
|---|---|---|---|---|---|
| Y | Y | Y | Y | Y | Y |

Note: Financial information, benefits and other data can change quickly and may vary from those stated here.

# BAYER AG

### www.bayer.com

**Industry Group Code: 325  Ranks within this company's industry group:** Sales: 2  Profits: 2

| Drugs: | Other: | | Clinical: | | Computers: | Services: | |
|---|---|---|---|---|---|---|---|
| Discovery: | AgriBio: | Y | Trials/Services: | | Hardware: | Specialty Services: | Y |
| Licensing: | Genetic Data: | | Labs: | | Software: | Consulting: | |
| Manufacturing: Y | Tissue Replacement: | | Equipment/Supplies: | Y | Arrays: | Blood Collection: | |
| Genetics: | | | Research & Development Services: | | Database Management: | Drug Delivery: | |
| | | | Diagnostics: | | | Drug Distribution: | |

## TYPES OF BUSINESS:

Chemicals Manufacturing
Pharmaceuticals
Animal Health Products
Synthetic Materials
Crop Science
Plant Biotechnology
Health Care Products

## BRANDS/DIVISIONS/AFFILIATES:

Bayer Corp
Bayer CropScience
Bayer HealthCare
Bayer MaterialScience
Bayer Business Services
Bayer Technology Services
Currenta GmbH & Co.
Bayer Schering Pharma AG

## CONTACTS: Note: Officers with more than one job title may be intentionally listed here more than once.

Werner Wenning, Chmn.-Mgmt. Board
Klaus Kuhn, Dir.-Finance
Richard Pott, Dir.-Human Resources
Wolfgang Plischke, Dir.-Innovation
Wolfgang Plischke, Dir.-Tech.
Richard Pott, Dir.-Strategy
Michael Schade, Head-Comm.
Alexander Rosar, Head-Investor Rel.
Wolfgang Plischke, Dir.-Environment
A.J. Higgins, Chmn.-Bayer Health Care
F. Berschauer, Chmn.-Bayer Crop Sciences
P. Thomas, Chmn.-Bayer Material Science
Manfred Schneider, Chmn.-Supervisory Board
Klaus Kuhn, Dir.-EMEA

| **Phone:** 49-214-30-1 | **Fax:** 49-214-30-66328 |
|---|---|
| **Toll-Free:** 800-269-2377 | |
| **Address:** Bayerwerk Gebaeude W11, Leverkusen, D-51368 Germany | |

## GROWTH PLANS/SPECIAL FEATURES:

Bayer AG is a German holding company encompassing over 300 consolidated subsidiaries on five continents. The company has five business segments: Bayer HealthCare, Bayer CropScience, Bayer MaterialScience, Bayer Business Services and Bayer Technology Services. The Bayer HealthCare segment develops, produces and markets products for the prevention, diagnosis and treatment of human and animal diseases. Bayer CropScience is active in the areas of chemical crop protection and seed treatment, non-agricultural pest and weed control and plant biotechnology. Bayer MaterialScience develops, manufactures and markets polyurethane, polycarbonate, cellulose derivatives and special metals products. Bayer Business Services offers IT infrastructure and applications, procurement and logistics, human resources and management services. Bayer Technology Services offers process development, process and plant engineering, construction and optimization services. Bayer also operates the Currenta GmbH & Co. joint venture with Lanxess AG. Currenta offers utility supply, waste management, infrastructure, safety, security, analytics and vocational training services to the chemical industry. In March 2008, the firm acquired the over-the-counter (OTC) brand portfolio of Sagmel, Inc. In April 2008, the company acquired the remaining shares of BaySystems BUFA Polyurethane GmbH & Co. In July 2008, Bayer acquired Maxygen's hemophilia program assets. Also in July 2008, Bayer acquired the Western OTC cough and cold portfolio of Topsun Science and Technology Qidong Gaitianli Pharmaceutical Co., Ltd. In September 2008, the company acquired all outstanding shares of Bayer Schering Pharma AG, making it a wholly-owned subsidiary of Bayer. In March 2009, the company sold its Thermoplastics Testing Center to Underwriters Laboratories, Inc.

Bayer offers its employees deferred compensation, a defined benefit pension fund, sports amenities, flexible work schedules and a varied program of cultural events.

## FINANCIALS: Sales and profits are in thousands of dollars—add 000 to get the full amount. 2008 Note: Financial information for 2008 was not available for all companies at press time.

| | | |
|---|---|---|
| 2008 Sales: $43,536,000 | 2008 Profits: $2,273,480 | **U.S. Stock Ticker: BAY** |
| 2007 Sales: $42,831,100 | 2007 Profits: $6,230,580 | **Int'l Ticker: BAY GR**  Int'l Exchange: Frankfurt-Euronext |
| 2006 Sales: $38,710,400 | 2006 Profits: $2,249,950 | Employees: 108,600 |
| 2005 Sales: $32,662,374 | 2005 Profits: $1,902,517 | Fiscal Year Ends: 12/31 |
| 2004 Sales: $27,731,937 | 2004 Profits: $816,045 | Parent Company: |

## SALARIES/BENEFITS:

| Pension Plan: Y | ESOP Stock Plan: | Profit Sharing: | Top Exec. Salary: $1,700,721 | Bonus: $2,985,500 |
|---|---|---|---|---|
| Savings Plan: Y | Stock Purch. Plan: Y | | Second Exec. Salary: $1,147,526 | Bonus: $1,851,035 |

## OTHER THOUGHTS:

**Apparent Women Officers or Directors:**
**Hot Spot for Advancement for Women/Minorities:**

## LOCATIONS: ("Y" = Yes)

| West: | Southwest: | Midwest: | Southeast: | Northeast: | International: |
|---|---|---|---|---|---|
| Y | Y | Y | | Y | Y |

# BAYER CORP

**www.bayerus.com**

Industry Group Code: 325  Ranks within this company's industry group: Sales:   Profits:

| Drugs: | | Other: | | Clinical: | | Computers: | | Services: | |
|---|---|---|---|---|---|---|---|---|---|
| Discovery: | Y | AgriBio: | Y | Trials/Services: | | Hardware: | | Specialty Services: | |
| Licensing: | | Genetic Data: | | Labs: | | Software: | | Consulting: | |
| Manufacturing: | Y | Tissue Replacement: | | Equipment/Supplies: | Y | Arrays: | | Blood Collection: | |
| Genetics: | | | | Research & Development Services: | | Database Management: | | Drug Delivery: | |
| | | | | Diagnostics: | Y | | | Drug Distribution: | |

## TYPES OF BUSINESS:

Chemicals Manufacturing
Animal Health Products
Over-the-Counter Drugs
Diagnostic Products
Coatings, Adhesives & Sealants
Polyurethanes & Plastics
Herbicides, Fungicides & Insecticides

## BRANDS/DIVISIONS/AFFILIATES:

Bayer
Bayer HealthCare AG
Bayer MaterialSciences, LLC
Bayer CropScience, LP
Aleve
BREEZE
BaySystems
Alka-Seltzer Plus

## CONTACTS: Note: Officers with more than one job title may be intentionally listed here more than once.

Gregory S. Babe, CEO
Gregory S. Babe, Pres.
Willy Scherf, CFO
Joyce Burgess, Dir.-Human Resources
Claudio Abreu, CIO
George J. Lykos, Chief Legal Officer
Mark Ryan, Chief Comm. Officer
Andreas Beier, Chief Acct. Officer
Arthur Higgins, Chmn.-Bayer HealthCare AG
Gregory Babe, Bayer MaterialScience LLC
William Buckner, CEO/Pres., Bayer CropScience, LP
Timothy Roseberry, Chief Procurement Officer/VP-Corp. Materials Mgmt.

| Phone: 412-777-2000 | Fax: 412-777-2034 |
|---|---|
| Toll-Free: | |
| Address: 100 Bayer Rd., Pittsburgh, PA 15205-9741 US | |

## GROWTH PLANS/SPECIAL FEATURES:

Bayer Corporation is the U.S. subsidiary of chemical and pharmaceutical giant Bayer AG. The company operates through four subsidiaries: Bayer HealthCare; Bayer MaterialScience; Bayer Corporate and Business Services; and Bayer CropScience. Bayer HealthCare operates through five divisions: pharmaceuticals, consumer care, diagnostics, diabetes care and animal health. Its animal health products include vaccines and other preventative measures for farm and domestic animals. Its consumer care products include analgesics (Aleve and Bayer); cold and cough treatments (Alka-Seltzer Plus and Talcio); digestive relief products (Alka-Mints and Phillips' Milk of Magnesia); topical skin preparations (Domeboro and Bactine); and vitamins (One-A-Day and Flintstones). The diabetes care division is a leader in self-test blood glucose diagnostic systems, and has recently released the BREEZE product family that offers alternate site testing and automatic coding and requires smaller blood samples. Bayer HealthCare's diagnostics division, now called Siemens Medical Solutions Diagnostics, produces diagnostic systems for critical and intensive care, hematology, urinalysis, immunology, clinical chemistry and molecular testing. The Advia Centaur system is used for the diagnosis of diseases like cancer, cardiovascular diseases, allergies and infections; the Versant and Trugent brands of assays are used for the detection of HIV and hepatitis virus. Bayer's MaterialScience segment produces coatings, adhesives and sealant raw materials; polyurethanes; and plastics. Bayer CropScience makes products directed toward crop protection, environmental science and bioscience, which include herbicides, fungicides and insecticides. Bayer Corporate and Business Services provides business services to the aforementioned Bayer subsidiaries, such as administration, technology services, mergers/acquisitions and internal auditing. In March 2008, Bayer MaterialScience introduced Makrolon LED2245, a polycarbonate resin designed specifically for high-brightness LED applications.

## FINANCIALS: Sales and profits are in thousands of dollars—add 000 to get the full amount. 2008 Note: Financial information for 2008 was not available for all companies at press time.

| | | | |
|---|---|---|---|
| 2008 Sales: $ | 2008 Profits: $ | **U.S. Stock Ticker: Subsidiary** | |
| 2007 Sales: $ | 2007 Profits: $ | **Int'l Ticker:** Int'l Exchange: | |
| 2006 Sales: $10,262,800 | 2006 Profits: $ | Employees: 17,200 | |
| 2005 Sales: $8,747,200 | 2005 Profits: $ | Fiscal Year Ends: 12/31 | |
| 2004 Sales: $11,504,000 | 2004 Profits: $ | Parent Company: BAYER AG | |

## SALARIES/BENEFITS:

| Pension Plan: | ESOP Stock Plan: | Profit Sharing: | Top Exec. Salary: $ | Bonus: $ |
|---|---|---|---|---|
| Savings Plan: Y | Stock Purch. Plan: Y | | Second Exec. Salary: $ | Bonus: $ |

## OTHER THOUGHTS:

**Apparent Women Officers or Directors**: 1
**Hot Spot for Advancement for Women/Minorities**:

## LOCATIONS: ("Y" = Yes)

| West: | Southwest: | Midwest: | Southeast: | Northeast: | International: |
|---|---|---|---|---|---|
| Y | Y | Y | Y | Y | |

Note: Financial information, benefits and other data can change quickly and may vary from those stated here.

# BAYER SCHERING PHARMA AG

**www.bayerscheringpharma.de**

Industry Group Code: 325412  Ranks within this company's industry group:  Sales: 15   Profits:

| Drugs: | | Other: | | Clinical: | | Computers: | | Services: | |
|---|---|---|---|---|---|---|---|---|---|
| Discovery: | | AgriBio: | | Trials/Services: | | Hardware: | | Specialty Services: | |
| Licensing: | | Genetic Data: | Y | Labs: | | Software: | | Consulting: | |
| Manufacturing: | Y | Tissue Replacement: | | Equipment/Supplies: | | Arrays: | | Blood Collection: | |
| Genetics: | | | | Research & Development Services: | | Database Management: | | Drug Delivery: | |
| | | | | Diagnostics: | Y | | | Drug Distribution: | |

## TYPES OF BUSINESS:

Pharmaceuticals Discovery, Development & Manufacturing
Gynecology & Andrology Treatments
Contraceptives
Cancer Treatments
Multiple Sclerosis Treatments
Circulatory Disorder Treatments
Diagnostic & Radiopharmaceutical Agents
Proteomics

## BRANDS/DIVISIONS/AFFILIATES:

Schering AG
Angeliq
Yasmin
Menostar
Betaferon
Fludara
Leukine
Zevalin

## CONTACTS: *Note: Officers with more than one job title may be intentionally listed here more than once.*

Andreas Busch, Chmn.-Exec. Board
Werner Baumann, Member-Exec. Board, Human Resources
Ulrich Koestlin, Member-Exec. Board
Kemal Malik, Member-Exec. Board
Gunnar Riemann, Member-Exec. Board
Andreas Busch, Member-Exec. Board
Werner Wenning, Chmn.-Supervisory Board

| Phone: 49-30-468-1111 | Fax: 49-30-468-15305 |
|---|---|
| Toll-Free: | |
| Address: Mullerstrasse 178, Berlin, 13353 Germany | |

## GROWTH PLANS/SPECIAL FEATURES:

Bayer Schering Pharma AG, formerly Schering AG, is a major global research-based pharmaceutical company that operates through subsidiaries in more than 100 countries. The firm concentrates its activities on four business areas: women's healthcare, specialty medicine, general medicine and diagnostic imaging. Schering's women's health products include birth control pills (Yasmin), hormone therapy (Angeliq and Menostar) and other contraceptives for women (Mirena). The firm's specialty medicine unit focuses on cancer, central nervous system disease and age related eye disease treatments. The division recently introduced the drug Fludara to provide treatment for chronic lymphocytic leukemia, a variety of leukemia. Another product, Leukine, is a drug administered to treat the immune system weakened by chemotherapy. Zevalin is a readioimmunotherapy for follicular B-cell non-Hodgkin's lymphoma approved for use in E.U. countries, and MabCampath/Campath is a chemotherapy drug often used on those patients who do not respond to traditional chemotherapy. Schering has contributed to the body of research on multiple sclerosis (MS) through its Beyond and Benefit studies. Its Betaferon drug reduces the frequency of MS episodes significantly. The general medicine segment focuses on anti-infective treatments, men's health care and cardiovascular, metabolic and thromboembolic diseases. Schering's diagnostics imaging products include a range of contrast media, such as Magnevist, a general MRI contrast agent and Gadovist, a central nervous system MRI contrast agent. The company is a subsidiary of Bayer A.G. In December 2008, the company signed an agreement with ProStrakan Group plc to market the ProStrakan's testosterone gel, Tostrex, in Canadam Latin America, Africa, Asia and the Middle East. In February 2009, the company announced plans to invest in the formation of a research and development facility in Beijing.

## FINANCIALS: Sales and profits are in thousands of dollars—add 000 to get the full amount. 2008 Note: Financial information for 2008 was not available for all companies at press time.

| | | |
|---|---|---|
| 2008 Sales: $14,243,900 | 2008 Profits: $ | **U.S. Stock Ticker: Subsidiary** |
| 2007 Sales: $5,806,500 | 2007 Profits: $1,980,100 | **Int'l Ticker:** Int'l Exchange: |
| 2006 Sales: $8,471,600 | 2006 Profits: $3,495,700 | Employees: 25,000 |
| 2005 Sales: $7,802,800 | 2005 Profits: $909,900 | Fiscal Year Ends: 12/31 |
| 2004 Sales: $6,647,000 | 2004 Profits: $677,000 | Parent Company: BAYER AG |

## SALARIES/BENEFITS:

| Pension Plan: Y | ESOP Stock Plan: | Profit Sharing: | Top Exec. Salary: $ | Bonus: $ |
|---|---|---|---|---|
| Savings Plan: | Stock Purch. Plan: | | Second Exec. Salary: $ | Bonus: $ |

## OTHER THOUGHTS:

**Apparent Women Officers or Directors:**
**Hot Spot for Advancement for Women/Minorities:**

## LOCATIONS: ("Y" = Yes)

| West: | Southwest: | Midwest: | Southeast: | Northeast: | International: |
|---|---|---|---|---|---|
| Y | | | | Y | Y |

# BIO RAD LABORATORIES INC

## www.bio-rad.com

**Industry Group Code: 33911  Ranks within this company's industry group:** Sales: 7    Profits: 7

| Drugs: | Other: | | Clinical: | | Computers: | | Services: | |
|---|---|---|---|---|---|---|---|---|
| Discovery: | AgriBio: | Y | Trials/Services: | | Hardware: | | Specialty Services: | |
| Licensing: | Genetic Data: | Y | Labs: | Y | Software: | Y | Consulting: | |
| Manufacturing: | Tissue Replacement: | | Equipment/Supplies: | Y | Arrays: | | Blood Collection: | |
| Genetics: | | | Research & Development Services: | | Database Management: | | Drug Delivery: | |
| | | | Diagnostics: | Y | | | Drug Distribution: | |

## TYPES OF BUSINESS:

Equipment-Life Sciences Research
Clinical Diagnostics Products
Analytical Instruments
Laboratory Devices
Biomaterials
Imaging Products
Assays
Software

## BRANDS/DIVISIONS/AFFILIATES:

DiaMed Holding AG
BioPlex 2200
iScript
ProteOn XPR36

## CONTACTS: Note: Officers with more than one job title may be intentionally listed here more than once.

Norman Schwartz, CEO
Norman Schwartz, Pres.
Christine Tsingos, CFO/VP
Colleen Corey, Dir.-Corp. Human Resources
Sanford S. Wadler, General Counsel/VP/Sec.
Tina Cuccia, Mgr.-Corp. Comm.
James R. Stark, Corp. Controller
Ronald W. Hutton, Treas.
Brad Crutchfield, VP/Group Mgr.-Life Science
John Goetz, VP/Group Mgr.-Clinical Diagnostics
David Schwartz, Chmn.
Giovanni Magni, VP/Mgr.-Int'l Sales

| **Phone:** 510-724-7000 | **Fax:** 510-741-5815 |
|---|---|
| **Toll-Free:** 800-424-6723 | |
| **Address:** 1000 Alfred Nobel Dr., Hercules, CA 94547 US | |

## GROWTH PLANS/SPECIAL FEATURES:

Bio-Rad Laboratories, Inc., supplies the life science research, health care and analytical chemistry markets with a broad range of products and systems. These are used to separate complex chemical and biological materials and to identify, analyze and purify components. The company operates in two industry segments: life science and clinical diagnostics. The firm's life science division develops products for applications including electrophoresis, image analysis, molecular detection, chromatography, gene transfer, sample preparation and amplification. Products include a range of laboratory instruments, apparatus and consumables used for research in genomics, proteomics and food safety. The Bio-Rad life science division provides its services to universities; medical schools; pharmaceutical manufacturers; industrial research organizations; food testing laboratories; government agencies; and biotechnology researchers. The clinical diagnostics division encompasses an array of technologies incorporated into a variety of tests used to detect, identify and quantify substances in blood or other body fluids and tissues. The test results are used as aids for medical diagnosis, detection, evaluation, monitoring and treatment of diseases and other conditions. This division is known for diabetes monitoring products, quality control systems, blood virus testing, blood typing, toxicology, genetic disorders products, molecular pathology and Internet-based software. Bio-Rad is also an international provider of bovine spongiform encephalopathy (mad cow disease) tests. Bio-Rad's brand name systems include the BioPlex 2200 multiplex testing platform; iScript, reverse transcription reagent kits; and ProteOn XPR36, a protein interaction array system. In 2008, Bio-Rad acquired additional shares of DiaMed Holding AG, which develops and markets products used in blood typing/screening; the firm now owns 93.46% of DiaMed Holdings. In March 2009, the firm signed an exclusive marketing and development agreement with Bruker Corporation, a manufacturer of mass spectrometry instruments; the companies will work together to develop new products to identify intact peptides and proteins under 30 kilodaltons.

## FINANCIALS: Sales and profits are in thousands of dollars—add 000 to get the full amount. 2008 Note: Financial information for 2008 was not available for all companies at press time.

| | | |
|---|---|---|
| 2008 Sales: $1,764,365 | 2008 Profits: $89,510 | **U.S. Stock Ticker: BIO** |
| 2007 Sales: $1,461,052 | 2007 Profits: $92,994 | **Int'l Ticker:**    Int'l Exchange: |
| 2006 Sales: $1,273,930 | 2006 Profits: $103,263 | Employees:  6,400 |
| 2005 Sales: $1,180,985 | 2005 Profits: $81,553 | Fiscal Year Ends: 12/31 |
| 2004 Sales: $1,090,012 | 2004 Profits: $68,242 | Parent Company: |

## SALARIES/BENEFITS:

| Pension Plan: | ESOP Stock Plan: | Profit Sharing: | Top Exec. Salary: $690,450 | Bonus: $559,212 |
|---|---|---|---|---|
| Savings Plan: Y | Stock Purch. Plan: | | Second Exec. Salary: $520,065 | Bonus: $208,000 |

## OTHER THOUGHTS:

**Apparent Women Officers or Directors:** 3
**Hot Spot for Advancement for Women/Minorities:** Y

## LOCATIONS: ("Y" = Yes)

| West: | Southwest: | Midwest: | Southeast: | Northeast: | International: |
|---|---|---|---|---|---|
| Y | Y | | | Y | Y |

Note: Financial information, benefits and other data can change quickly and may vary from those stated here.

# BIOANALYTICAL SYSTEMS INC

### www.bioanalytical.com

**Industry Group Code: 541712  Ranks within this company's industry group: Sales: 13    Profits: 9**

| Drugs: | Other: | Clinical: | | Computers: | Services: | |
|--------|--------|-----------|---|-----------|-----------|---|
| Discovery: | AgriBio: | Trials/Services: | Y | Hardware: | Specialty Services: | Y |
| Licensing: | Genetic Data: | Labs: | Y | Software: | Consulting: | Y |
| Manufacturing: | Tissue Replacement: | Equipment/Supplies: | Y | Arrays: | Blood Collection: | |
| Genetics: | | Research & Development Services: | Y | Database Management: | Drug Delivery: | |
| | | Diagnostics: | | | Drug Distribution: | |

## TYPES OF BUSINESS:

Contract Research Services
Screening & Testing Services
Bioanalytical Instruments
Immunochemistry
Toxicology Testing
Formulation Development
Testing Equipment
Phase I Clinical Trials

## BRANDS/DIVISIONS/AFFILIATES:

Culex APS
Vetronics
epsilon
Baltimore Clinical Pharmacology Research

## CONTACTS: Note: Officers with more than one job title may be intentionally listed here more than once.

Richard M. Shepperd, CEO
Richard M. Shepperd, Pres.
Michael M. Cox, CFO
Craig S. Bruntlett, VP-Sales
Lina L. Reeves-Kerner, VP-Human Resources
Ronald E. Shoup, Chief Scientific Officer
Ed Burrow, Dir.-Mfg.
Michael M. Cox, VP-Admin.
Edward M. Chait, Chief Bus. Officer
Michael M. Cox, VP-Finance
Emilio Cordova, VP-Bus. Dev.
Johan Reinhoudt, VP/Gen. Mgr.-Clinical Pharmacology Research
Mark Gehrke, Gen. Mgr.-West Lafayette Contract Research
Lori Payne, Gen. Mgr.-Northwest Laboratories
William E. Baitinger, Chmn.
Mark Wareing, Managing Dir.-UK & Europe

| **Phone:** 765-463-4527 | **Fax:** 765-497-1102 |
|---|---|
| **Toll-Free:** 800-845-4246 | |
| **Address:** Purdue Research Park, 2701 Kent Ave., West Lafayette, IN 47906 US | |

## GROWTH PLANS/SPECIAL FEATURES:

Bioanalytical Systems, Inc. (BASi) is a drug development firm that serves leading pharmaceutical, medical device and biotechnology companies and institutions worldwide.  Its principal clients consist of scientists engaged in analytical chemistry, drug metabolism studies, clinical trials, pharmacokinetics and basic neuroscience research at major pharmaceutical organizations.  BASi operates two business segments, both of which address bioanalytical, preclinical and clinical research needs of drug developers.  The Contract Research Services segment provides product characterization, method development and validation services, including determining the purity, potency, chemical composition, structure and physical properties of a drug compound; bioanalytical testing, mainly targeting drug and metabolite concentrations in complex biological matrices; and stability testing, mainly analyzing product purity, potency and shelf life.  It also offers in vivo pharmacology, that is, analyzing drug reactions in living specimens (often animals), as opposed to in vitro (or in the glass) laboratory testing, which takes place outside an organism; preclinical and pathology services, including pharmacokinetic and safety testing; sand Phase I clinical trials.  The Research Products segment offers robotic sampling systems and accessories; in vivo microdialysis collection systems; physiology monitoring tools; and liquid chromatography and electrochemical instruments.  Its products include the patented Culex APS robotic pharmacology system, an automated program used during bioanalytical testing; Vetronics, a small animal electro-cardiogram and vital signs monitor mainly used by veterinarians; and epsilon, a single liquid chromatography and electrochemistry instrument used for separation systems and chemical analysis.  The firm operates four labs in the U.S. and one in the U.K.  In July 2008, the company sold its Baltimore Clinical Pharmacology Research subsidiary.

Employees of BASi receive medical, dental, vision and life insurance, as well as disability benefits.

## FINANCIALS:  Sales and profits are in thousands of dollars—add 000 to get the full amount. 2008 Note: Financial information for 2008 was not available for all companies at press time.

| | | |
|---|---|---|
| 2008 Sales: $41,697 | 2008 Profits: $-1,489 | **U.S. Stock Ticker: BASI** |
| 2007 Sales: $39,753 | 2007 Profits: $ 926 | **Int'l Ticker:**    Int'l Exchange: |
| 2006 Sales: $43,048 | 2006 Profits: $-2,670 | Employees:   311 |
| 2005 Sales: $42,395 | 2005 Profits: $-  80 | Fiscal Year Ends: 9/30 |
| 2004 Sales: $37,152 | 2004 Profits: $- 203 | Parent Company: |

## SALARIES/BENEFITS:

| Pension Plan: | ESOP Stock Plan: | Profit Sharing: | Top Exec. Salary: $420,000 | Bonus: $ |
|---|---|---|---|---|
| Savings Plan: Y | Stock Purch. Plan: | | Second Exec. Salary: $165,000 | Bonus: $25,000 |

## OTHER THOUGHTS:

**Apparent Women Officers or Directors:** 2
**Hot Spot for Advancement for Women/Minorities:** Y

## LOCATIONS: ("Y" = Yes)

| West: | Southwest: | Midwest: | Southeast: | Northeast: | International: |
|---|---|---|---|---|---|
| Y | | Y | | Y | Y |

Note: Financial information, benefits and other data can change quickly and may vary from those stated here.

# BIOCOMPATIBLES INTERNATIONAL PLC    www.biocompatibles.com

Industry Group Code: 33911  Ranks within this company's industry group:  Sales: 14   Profits: 12

| Drugs: | | Other: | | Clinical: | | Computers: | | Services: | |
|---|---|---|---|---|---|---|---|---|---|
| Discovery: | | AgriBio: | | Trials/Services: | | Hardware: | | Specialty Services: | |
| Licensing: | | Genetic Data: | | Labs: | | Software: | | Consulting: | |
| Manufacturing: | Y | Tissue Replacement: | | Equipment/Supplies: | | Arrays: | Y | Blood Collection: | |
| Genetics: | | | | Research & Development Services: | | Database Management: | | Drug Delivery: | Y |
| | | | | Diagnostics: | | | | Drug Distribution: | |

## TYPES OF BUSINESS:
Medical Implant Technology
Supplies-Phosphorycholine Coatings
Embolisation Microspheres
Biomaterials
Cancer Treatment
Drug Delivery Platforms

## BRANDS/DIVISIONS/AFFILIATES:
CellMed
Bead Block
CellBeads
NFil Technology
PRECISION Drug Eluting Bead
Phosphorylcholine (PC) Technology
BrachySciences, Inc.
Merz Pharmaceuticals GmbH

## CONTACTS: *Note: Officers with more than one job title may be intentionally listed here more than once.*
Crispin Simon, CEO
Tim Maloney, Dir.-Mktg.
Geoff Tompsett, Dir.-Human Resources
Andy Lewis, Dir.-Research
Geoff Tompsett, Dir.-IT
Andy Lewis, Dir.-Tech
Ian Ardill, Sec.
Ian Ardill, Dir.-Finance
Mike Motion, Dir.-Sales & Mktg, Oncology Prod. Div.
Alistair Taylor, Dir.-Quality & Regulatory Affairs
Paul Baxter, Dir.-Intellectual Property
John Sylvestor, Managing Dir.-Oncology Prod. Div.
Gerry Brown, Chmn.

| Phone: 44-1252-732-732 | Fax: 44-1252-732-777 |
|---|---|
| Toll-Free: | |
| Address: Chapman House, Farnham Business Park, Weydon Ln., Farnham, Surrey GU9 8QL UK | |

## GROWTH PLANS/SPECIAL FEATURES:
Biocompatibles International plc is an international provider of medical products that combine medical devices with ancillary therapeutic drugs. The company operates through its Oncology Products division as well as two research and development businesses. The Oncology Products division conducts the marketing for Biocompatibles' Bead products. These products are used for the treatment of liver cancer and other types of cancer. Biocompatibles uses bead technology to deliver therapeutic agents in its products, as well as providing Bead Block and Alginate Bead, two bead products that contain no drug. The company's three biomedical polymer systems are its NFil technology; CellBeads, which are used in the delivery of biological agents and its Phosphorylcholine (PC) Technology, used for such applications as Medtronic's Endeavor Drug Eluting Stent. Four products comprise the company's product portfolio: Bead Block for use in embolisation therapy; DC Bead drug delivery embolisation system; LC Bead controlled embolisation system; and PRECISION Drug Eluting Bead, an embolisation system which releases a local, controlled and sustained dose of doxorubicin for the treatment of Hepatocellular Carcinoma (HCC). The Oncology Products Division also includes Brachytherapy products, which are radiation-delivering seeds that are used to treat prostate cancer. The firm's research and development division is responsible for the marketing of the Drug Delivery and Cellmed businesses, both of which are involved in product development and licensing. The Drug Delivery business is focused on developing oncology products, while the German facility, CellMed, is currently working on a Drug-Eluting Bead product for stroke treatment, and a GLP-1 analog for the treatment of obesity and diabetes. In August 2008, the company acquired BrachySciences, Inc. In March 2009, the company's cosmetic dermal filler bead was approved for the European market. It will be launched under the name Novabel, by the firm's partner Merz Pharmaceuticals GmbH.

## FINANCIALS: Sales and profits are in thousands of dollars—add 000 to get the full amount. 2008 Note: Financial information for 2008 was not available for all companies at press time.

| | | |
|---|---|---|
| 2008 Sales: $29,050 | 2008 Profits: $- 750 | U.S. Stock Ticker: |
| 2007 Sales: $18,112 | 2007 Profits: $-3,890 | Int'l Ticker: BII   Int'l Exchange: London-LSE |
| 2006 Sales: $11,864 | 2006 Profits: $9,990 | Employees:  72 |
| 2005 Sales: $6,880 | 2005 Profits: $-9,490 | Fiscal Year Ends: 12/31 |
| 2004 Sales: $5,000 | 2004 Profits: $- 300 | Parent Company: |

## SALARIES/BENEFITS:
| Pension Plan: | ESOP Stock Plan: | Profit Sharing: | Top Exec. Salary: $ | Bonus: $ |
|---|---|---|---|---|
| Savings Plan: | Stock Purch. Plan: | | Second Exec. Salary: $ | Bonus: $ |

## OTHER THOUGHTS:
Apparent Women Officers or Directors: 1
Hot Spot for Advancement for Women/Minorities:

## LOCATIONS: ("Y" = Yes)
| West: | Southwest: | Midwest: | Southeast: | Northeast: | International: |
|---|---|---|---|---|---|
| | | | | | Y |

Note: Financial information, benefits and other data can change quickly and may vary from those stated here.

# BIOCRYST PHARMACEUTICALS INC

## www.biocryst.com

Industry Group Code: 325412  Ranks within this company's industry group: Sales: 81    Profits: 115

| Drugs: | | Other: | Clinical: | Computers: | Services: |
|---|---|---|---|---|---|
| Discovery: | Y | AgriBio: | Trials/Services: | Hardware: | Specialty Services: |
| Licensing: | Y | Genetic Data: | Labs: | Software: | Consulting: |
| Manufacturing: | | Tissue Replacement: | Equipment/Supplies: | Arrays: | Blood Collection: |
| Genetics: | | | Research & Development Services: | Database Management: | Drug Delivery: |
| | | | Diagnostics: | | Drug Distribution: |

## TYPES OF BUSINESS:

Small-Molecule Pharmaceutical Products
Drugs-Immunological, Infectious & Inflammatory Disease

## BRANDS/DIVISIONS/AFFILIATES:

Fodosine HCl
BCX-4208/R3421
Peramivir

## CONTACTS: Note: Officers with more than one job title may be intentionally listed here more than once.

Jon P. Stonehouse, CEO
Jon P. Stonehouse, Pres.
Stuart Grant, CFO
Robert Stoner, VP-Human Resources
William P. Sheridan, Chief Medical Officer
Alane Barnes, General Counsel/Sec.
David S. McCullough, Sr. VP-Strategic Planning & Corp. Dev.
Yarlagadda S. Babu, VP-Drug Discovery
Elliott Berger, Sr. VP-Regulatory Affairs
Walter G. Gowan, VP-Chemistry Mfg. & Control
Zola P. Horovitz, Chmn.

| Phone: 205-444-4600 | Fax: 205-444-4640 |
|---|---|
| Toll-Free: | |
| Address: 2190 Parkway Lake Dr., Birmingham, AL 35244 US | |

## GROWTH PLANS/SPECIAL FEATURES:

BioCryst Pharmaceuticals, Inc. is a biotechnology company that designs and develops drugs to block key enzymes involved in cancer, viral infections and autoimmune diseases. The company's technology is designed to stifle the generation of malignant cells at their origin, which is the site of chemical reactions in enzymes. BioCryst integrates the disciplines of biology, crystallography, medicinal chemistry and computer modeling to design structure-based drugs and develop small-molecule pharmaceuticals. BioCryst has three main drugs in development, with several smaller ones in trials. The drugs are forodesine HCl, peramivir and BCX-4208/R3421. Forodesine HCl is in Phase II trials for the treatment of patients with cutaneous T-cell lymphoma and for patients with chronic lymphocytic leukemia. The FDA has granted orphan drug status for Forodesine in three indications, including T-cell non-Hodgkins lymphoma; chronic lymphocytic leukemia and related leukemias, including T-cell prolymphocytic, adult T-cell, hairy cell and B-cell acute lymphoblastic leukemias. Peramivir, a neuraminidase inhibitor, has gone through Phase II trials for acute uncomplicated influenza. Biocryst is planning another Phase II trial for the drug, testing both the intramuscular and intravenous form. BCX-4208/R3421 is a second-generation PNP inhibitor in development for the prevention of acute rejection in transplantation and for the treatment of autoimmune diseases, including psoriasis. In addition to these drugs, the firm also owns the rights to inhibitors of parainfluenza neuraminidase and hepatitis C, which it continues to evaluate and test in preclinical trials. The firm has corporate alliances with Roche, Mundipharma, Green Cross and Shionogi & Co., Ltd. Biocryst is currently developing peramivir for possible global influenza pandemics, including avian and swine flus, based on a four-year $102.6 million contract from a recent US Department of Health and Human Services award. In May 2009, the Japanese firm Shionogi & Co., Ltd. announced plans to market BioCryst's peramivir in the country by 2010.

BioCryst offers employees medical, dental, disability and life insurance; a 401(k) retirement plan; equity participation through a stock option plan and an employee stock purchase plan.

## FINANCIALS: Sales and profits are in thousands of dollars—add 000 to get the full amount. 2008 Note: Financial information for 2008 was not available for all companies at press time.

| | | |
|---|---|---|
| 2008 Sales: $56,561 | 2008 Profits: $-24,732 | **U.S. Stock Ticker:** BCRX |
| 2007 Sales: $71,238 | 2007 Profits: $-29,056 | **Int'l Ticker:** Int'l Exchange: |
| 2006 Sales: $6,212 | 2006 Profits: $-43,618 | Employees:    80 |
| 2005 Sales: $ 152 | 2005 Profits: $-26,099 | Fiscal Year Ends: 12/31 |
| 2004 Sales: $ 337 | 2004 Profits: $-21,104 | Parent Company: |

## SALARIES/BENEFITS:

| Pension Plan: | ESOP Stock Plan: | Profit Sharing: | Top Exec. Salary: $420,835 | Bonus: $180,700 |
|---|---|---|---|---|
| Savings Plan: Y | Stock Purch. Plan: Y | | Second Exec. Salary: $384,375 | Bonus: $115,900 |

## OTHER THOUGHTS:

**Apparent Women Officers or Directors:** 1
**Hot Spot for Advancement for Women/Minorities:**

## LOCATIONS: ("Y" = Yes)

| West: | Southwest: | Midwest: | Southeast: | Northeast: | International: |
|---|---|---|---|---|---|
| | | | Y | Y | |

# BIOGEN IDEC INC

**www.biogenidec.com**

**Industry Group Code: 325412 Ranks within this company's industry group: Sales: 28 Profits: 22**

| Drugs: | | Other: | | Clinical: | | Computers: | | Services: | |
|---|---|---|---|---|---|---|---|---|---|
| Discovery: | Y | AgriBio: | | Trials/Services: | | Hardware: | | Specialty Services: | |
| Licensing: | Y | Genetic Data: | | Labs: | Y | Software: | | Consulting: | |
| Manufacturing: | Y | Tissue Replacement: | | Equipment/Supplies: | | Arrays: | | Blood Collection: | |
| Genetics: | | | | Research & Development Services: | | Database Management: | | Drug Delivery: | |
| | | | | Diagnostics: | | | | Drug Distribution: | |

## TYPES OF BUSINESS:

Drugs-Immunology, Neurology & Oncology
Autoimmune & Inflammatory Disease Treatments
Drugs-Multiple Sclerosis
Drugs-Cancer

## BRANDS/DIVISIONS/AFFILIATES:

AVONEX
TYSABRI
RITUXAN
FUMADERM

## CONTACTS: Note: Officers with more than one job title may be intentionally listed here more than once.

James C. Mullen, CEO
Robert A. Hamm, COO
James C. Mullen, Pres.
Paul J. Clancy, CFO/Exec. VP
Craig Eric Schneier, Exec. VP-Human Resources
Cecil Pickett, Pres., R&D
Susan H. Alexander, General Counsel/Exec. VP/Corp. Sec.
Michael Lytton, Exec. VP-Corp. & Bus. Dev.
Craig Eric Schneier, Exec. VP-Corp. Comm. & Public Affairs
Michael F. MacLean, Chief Acct. Officer/Controller/Sr. VP
John M. Dunn, Exec. VP-New Ventures
Bruce R. Ross, Chmn.

| **Phone:** 617-679-2000 | **Fax:** 617-679-2617 |
|---|---|
| **Toll-Free:** | |
| **Address:** 14 Cambridge Ctr., Cambridge, MA 02142-1481 US | |

## GROWTH PLANS/SPECIAL FEATURES:

Biogen IDEC, Inc. is a biotechnology company that develops, manufactures and markets therapeutic pharmaceuticals in the fields of immunology, neurology and oncology. Biogen currently has four approved products: AVONEX, which is designed to treat relapsing forms of multiple sclerosis (MS) and is used by more than 135,000 patients in 70 countries globally; TYSABRI, which is approved for the treatment of relapsing forms of MS and in the U.S. is approved to treat moderate to severe active Crohn's disease; RITUXAN, which is globally approved for the treatment of relapsed or refractory low-grade or follicular, CD20-positive, B-cell non-Hodgkin's lymphomas (NHLs) and has had approximately 1.5 million patient exposures worldwide; and FUMADERM, which acts as an immunomodulator and is approved in Germany for the treatment of severe psoriasis. RITUXAN, in combination with methotrexate, is also approved to reduce signs and symptoms in adult patients with moderately-to-severely active rheumatoid arthritis who have had an inadequate response to one or more tumor necrosis factor antagonist therapies. The company is working with Genentech and Roche on the development of RITUXAN in additional oncology and other indications. In addition to its approved drugs, the firm currently has 22 drugs under development, including drugs in the company's core areas of focus as well as in other therapeutic areas such as cardiovascular disease and hemophilia. Biogen also generates revenue by licensing drugs it has developed to other companies, including Schering-Plough, Merck and GlaxoSmithKline. In May 2009, Biogen and Genentech submitted RITUXAN for approval in the U.S. for the treatment of adult leukemia; the indication is for RITUXAN (in conjunction with standard chemotherapy) for previously untreated or treated chronic lymphocytic leukemia.

Biogen offers employees medical, dental and vision insurance; tuition reimbursement; commuter benefits; discounts on health clubs and local events; flexible spending accounts; and an employee assistance program.

## FINANCIALS: Sales and profits are in thousands of dollars—add 000 to get the full amount. 2008 Note: Financial information for 2008 was not available for all companies at press time.

| | | |
|---|---|---|
| 2008 Sales: $4,097,500 | 2008 Profits: $783,200 | **U.S. Stock Ticker: BIIB** |
| 2007 Sales: $3,171,600 | 2007 Profits: $638,172 | **Int'l Ticker:** Int'l Exchange: |
| 2006 Sales: $2,683,049 | 2006 Profits: $217,511 | Employees: 4,700 |
| 2005 Sales: $2,422,500 | 2005 Profits: $160,711 | Fiscal Year Ends: 12/31 |
| 2004 Sales: $2,211,562 | 2004 Profits: $25,086 | Parent Company: |

## SALARIES/BENEFITS:

| Pension Plan: | ESOP Stock Plan: | Profit Sharing: | Top Exec. Salary: $1,192,308 | Bonus: $2,400,000 |
|---|---|---|---|---|
| Savings Plan: Y | Stock Purch. Plan: Y | | Second Exec. Salary: $816,923 | Bonus: $879,450 |

## OTHER THOUGHTS:

**Apparent Women Officers or Directors:** 3
**Hot Spot for Advancement for Women/Minorities:** Y

## LOCATIONS: ("Y" = Yes)

| West: | Southwest: | Midwest: | Southeast: | Northeast: | International: |
|---|---|---|---|---|---|
| Y | | | | Y | Y |

Note: Financial information, benefits and other data can change quickly and may vary from those stated here.

# BIOHEART INC
## www.bioheartinc.com
**Industry Group Code: 325414  Ranks within this company's industry group:**  Sales:     Profits:

| Drugs: | Other: | | Clinical: | Computers: | Services: | |
|---|---|---|---|---|---|---|
| Discovery: | AgriBio: | | Trials/Services: | Hardware: | Specialty Services: | Y |
| Licensing: | Genetic Data: | | Labs: | Software: | Consulting: | |
| Manufacturing: | Tissue Replacement: | Y | Equipment/Supplies: | Arrays: | Blood Collection: | |
| Genetics: | | | Research & Development Services: | Database Management: | Drug Delivery: | |
| | | | Diagnostics: | | Drug Distribution: | |

## TYPES OF BUSINESS:
Research & Development-Heart-Related Therapies
Distribution-Medical Devices

## BRANDS/DIVISIONS/AFFILIATES:
MyoCell
MyoCell SDF-1
Tissue Genesis, Inc.
TGI 1200
RTX Heathcare A/S
Bioheart 3370 Heart Failure Monitor
Monebo Technologies, Inc.
CardioBelt

## CONTACTS: *Note: Officers with more than one job title may be intentionally listed here more than once.*
Howard J. Leonhardt, CEO
Matt Fendrich, VP-Mktg. & Sales
Catherine Sulawske-Guck, VP-Human Resources
Kristin Comella, VP-Research
Howard J. Leonhardt, CTO
Catherine Sulawske-Guck, VP-Admin.
Kristin Comella, VP-Corp. Dev.
Scott Bromley, VP-Public Rel.
Howard J. Leonhardt, Chmn.

| Phone: 954-835-1500 | Fax: 954-845-9976 |
|---|---|
| Toll-Free: | |
| Address: 13794 N.W. 4th St., Ste. 212, Sunrise, FL 33325 US | |

## GROWTH PLANS/SPECIAL FEATURES:
Bioheart, Inc. discovers, develops and commercializes autologous cell therapies (that is, those derived from the patient themselves) in order to treat chronic acute heart damage and peripheral vascular disease.  Its primary product, called MyoCell, is a muscle-derived cell therapy designed to repopulate scar tissue in the patient's heart after they have experienced heart failure.  To date, MyoCell has completed a 40-patient Phase II-a test in Europe and a 20-patient Phase I test in the U.S.  The U.S. Food and Drug Administration (FDA) has approved a further 330-patient, Phase II/III multicenter test in North America and Europe (the MARVEL Trial).  Thus far, 50 patients have been enrolled in the MARVEL Trial.  Bioheart is seeking an amendment with the FDA to use mobile cardiac telemetry monitor recorders in the test.  If it receives this amendment, it will seek an additional $5 million in funding to enroll 150 more patients in the MARVEL Trial.  Should that funding not be found, the company will have to seek alternative options, including slowing down or stopping the MARVEL Trial.  Bioheart has other product candidates in the works.  The MyoCell SDF-1 product utilizes autologous cells that have been genetically modified to express various therapeutic growth proteins.  Bioheart Acute Cell Therapy is a product that offers an autologous adipose (fat) cell treatment for heart damage.  It is designed to be used with Tissue Genesis, Inc.'s TGI 1200 tissue processing system.  In April 2008, Bioheart signed a worldwide distribution agreement for Danish firm RTX Heathcare A/S's Bioheart 3370 Heart Failure Monitor, an at-home system that can be remote-monitored by a medical professional in contact with the attending physician.

## FINANCIALS:  Sales and profits are in thousands of dollars—add 000 to get the full amount. 2008 Note: Financial information for 2008 was not available for all companies at press time.

| | | |
|---|---|---|
| 2008 Sales: $ | 2008 Profits: $ | U.S. Stock Ticker: BHRT |
| 2007 Sales: $ | 2007 Profits: $ | Int'l Ticker:     Int'l Exchange: |
| 2006 Sales: $ | 2006 Profits: $ | Employees: |
| 2005 Sales: $ | 2005 Profits: $ | Fiscal Year Ends: 12/31 |
| 2004 Sales: $ | 2004 Profits: $ | Parent Company: |

## SALARIES/BENEFITS:
| Pension Plan: | ESOP Stock Plan: | Profit Sharing: | Top Exec. Salary: $164,223 | Bonus: $162,022 |
|---|---|---|---|---|
| Savings Plan: | Stock Purch. Plan: | | Second Exec. Salary: $150,000 | Bonus: $ |

## OTHER THOUGHTS:
**Apparent Women Officers or Directors**: 3
**Hot Spot for Advancement for Women/Minorities**: Y

## LOCATIONS: ("Y" = Yes)
| West: | Southwest: | Midwest: | Southeast: | Northeast: | International: |
|---|---|---|---|---|---|
| | Y | | | | Y |

Note: Financial information, benefits and other data can change quickly and may vary from those stated here.

# BIOMARIN PHARMACEUTICAL INC

**www.biomarinpharm.com**

Industry Group Code: 325412 **Ranks within this company's industry group:** Sales: 49    Profits: 43

| Drugs: | | Other: | Clinical: | Computers: | Services: | |
|--------|---|--------|-----------|------------|-----------|---|
| Discovery: | Y | AgriBio: | Trials/Services: | Hardware: | Specialty Services: | |
| Licensing: | Y | Genetic Data: | Labs: | Software: | Consulting: | |
| Manufacturing: | Y | Tissue Replacement: | Equipment/Supplies: | Arrays: | Blood Collection: | |
| Genetics: | | | Research & Development Services: | Database Management: | Drug Delivery: | Y |
| | | | Diagnostics: | | Drug Distribution: | |

## TYPES OF BUSINESS:

Biopharmaceutical Product Development
Drugs-Severe Conditions
Pediatric Disease Treatments
Asthma Treatments
Drug Delivery Technologies

## BRANDS/DIVISIONS/AFFILIATES:

Naglazyme
Kuvan
Aldurazyme
Genzyme Corporation
PEG-PAL
BH4
6R-BH4

## CONTACTS: *Note: Officers with more than one job title may be intentionally listed here more than once.*

Jean-Jacques Bienaime, CEO
Jeffrey H. Cooper, CFO/Sr. VP
Lewis Chapman, VP-Global Mktg.
Mark Wood, VP-Human Resources
Henry J. Fuchs, Chief Medical Officer/Sr. VP
Ed Von Pervieux, CIO/VP
Robert A. Baffi, Sr. VP-Tech. Oper.
Daniel P. Maher, VP-Prod. Dev.
R. Andrew Ramelmeier, VP-Mfg. & Process Dev.
G. Eric Davis, General Counsel/VP/Corp. Sec.
Jeff Ajer, VP-Commercial Oper.
Stephen Aselage, Chief Bus. Oper./Sr. VP
Amy Waterhouse, VP-Regulatory & Gov't Affairs
Stuart J. Swiedler, Sr. VP-Clinical Affairs
Victoria Sluzky, VP-Quality & Analytical Chemistry
Charles A. O'Neill, VP-Pharmacological Sciences
Pierre LaPalme, Chmn.
William E. Aliski, VP/Gen. Mgr.-European Oper.
Steven Jungles, VP-Supply Chain

| Phone: 415-506-6700 | Fax: 415-382-7889 |
|---------------------|-------------------|
| **Toll-Free:** | |
| **Address:** 105 Digital Dr., Novato, CA 94949 US | |

## GROWTH PLANS/SPECIAL FEATURES:

BioMarin Pharmaceutical, Inc. develops and commercializes biopharmaceutical products for serious diseases and medical conditions. BioMarin has three approved products: Naglazyme, for the treatment of mucopolysaccharidosis VI (MPS-VI); Kuvan, for which the firm was recently granted marketing approval in the U.S. for the treatment of phenylketonuria (PKU); and Aldurazyme, for the treatment of mucopolysaccharidosis I (MPS-I). MPS-VI is a debilitating life-threatening genetic disease for which no other drug treatment currently exists. Naglazyme has been granted orphan drug status in the U.S. and E.U., which confers market exclusivity for the treatment of MPS VI expiring in 2012 and 2016, respectively. BioMarin has been granted orphan drug status in the U.S. for Kuvan, the first drug treatment for PKU, an inherited metabolic disease that affects at least 50,000 diagnosed patients globally under the age of 40. Aldurazyme, developed through a 50/50 joint-venture with Genzyme Corporation, has been approved for marketing in the U.S., European Union (E.U.) and other countries for patients with MPS I. Aldurazyme has been granted orphan drug status in the U.S., expiring in 2010, and in the E.U., expiring in 2013. BioMarin's PEG-PAL (formerly referred to as Phenylase) is an investigational enzyme substitution therapy being developed as a subcutaneous injection for those who do not respond to Kuvan. The firm is developing BH4 for the treatment of indications associated with endothelial dysfunction. BioMarin initiated Phase II clinical trials of 6R-BH4 for peripheral arterial disease and sickle cell disease as well as preclinical studies for Duchenne Muscular Dystrophy and IgA Nephropathy. In December 2008, the firm obtained marketing rights for Kuvan in the European Union.

BioMarin offers its employees an education assistance program; a flexible spending plan; life, medical, dental and vision insurance; an employee assistance program; AD&D coverage; a 401(k) plan; short and long-term disability; and bi-weekly chair massages.

## FINANCIALS: Sales and profits are in thousands of dollars—add 000 to get the full amount. 2008 Note: Financial information for 2008 was not available for all companies at press time.

| | | |
|---|---|---|
| 2008 Sales: $296,493 | 2008 Profits: $30,831 | **U.S. Stock Ticker:** BMRN |
| 2007 Sales: $121,581 | 2007 Profits: $-15,803 | **Int'l Ticker:**    Int'l Exchange: |
| 2006 Sales: $84,209 | 2006 Profits: $-28,533 | **Employees:**   649 |
| 2005 Sales: $25,669 | 2005 Profits: $-74,270 | **Fiscal Year Ends:** 12/31 |
| 2004 Sales: $18,641 | 2004 Profits: $-187,443 | **Parent Company:** |

## SALARIES/BENEFITS:

| Pension Plan: | ESOP Stock Plan: | Profit Sharing: | Top Exec. Salary: $673,439 | Bonus: $769,500 |
|---------------|------------------|-----------------|----------------------------|-----------------|
| Savings Plan: Y | Stock Purch. Plan: Y | | Second Exec. Salary: $348,519 | Bonus: $139,650 |

## OTHER THOUGHTS:

**Apparent Women Officers or Directors**: 3
**Hot Spot for Advancement for Women/Minorities:** Y

## LOCATIONS: ("Y" = Yes)

| West: | Southwest: | Midwest: | Southeast: | Northeast: | International: |
|-------|-----------|----------|------------|------------|---------------|
| Y | | | | | Y |

*Note: Financial information, benefits and other data can change quickly and may vary from those stated here.*

# BIOMET INC

## www.biomet.com

**Industry Group Code: 33911  Ranks within this company's industry group:  Sales: 5    Profits: 4**

| Drugs: | Other: | Clinical: | | Computers: | Services: | |
|---|---|---|---|---|---|---|
| Discovery: | AgriBio: | Trials/Services: | | Hardware: | Specialty Services: | |
| Licensing: | Genetic Data: | Labs: | | Software: | Consulting: | |
| Manufacturing: | Tissue Replacement: Y | Equipment/Supplies: | Y | Arrays: | Blood Collection: | |
| Genetics: | | Research & Development Services: | | Database Management: | Drug Delivery: | |
| | | Diagnostics: | | | Drug Distribution: | |

## TYPES OF BUSINESS:

Orthopedic Supplies
Electrical Bone Growth Stimulators
Orthopedic Support Devices
Operating Room Supplies
Powered Surgical Instruments
Arthroscopy Products
Imaging Equipment
Human Bone Joint Replacement Systems

## BRANDS/DIVISIONS/AFFILIATES:

Vanguard System
Oxford Unicompartmental Knee
Alpina Unicompartmental Knee
Vanguard M System
Repicci II Unicondylar Knee System
Biomet OSS Orthopaedic Salvage System
Arthrotek, Inc.

## CONTACTS: Note: Officers with more than one job title may be intentionally listed here more than once.

Jeffrey R. Binder, CEO
Jeffrey R. Binder, Pres.
Daniel P. Florin, CFO/Sr. VP
Peggy Taylor, Sr. VP-Human Resources
Bradley J. Tandy, General Counsel/Sr. VP/Corp. Sec.
Robin T. Barney, Sr. VP-Worldwide Oper.
Glen A. Kashuba, Sr. VP/Pres., Biomet Trauma & Biomet Spine
Gregory W. Sasso, Sr. VP/Pres., Biomet SBU Oper.
Steven F. Schiess, Sr. VP/Pres., Biomet 3i LLC
Jon C. Serbousek, Sr. VP/Pres., Biomet Orthopedics LLC
Jeffrey R. Binder, Chmn.
Roger P. Van Broeck, Pres., Biomet EMEA/Sr. VP

| | |
|---|---|
| **Phone:** 574-267-6639 | **Fax:** 574-267-8137 |
| **Toll-Free:** | |
| **Address:** 56 E. Bell Dr., P.O. Box 587, Warsaw, IN 46581-0587 US | |

## GROWTH PLANS/SPECIAL FEATURES:

Biomet, Inc. designs, manufactures and markets products that are used primarily by musculoskeletal medical specialists in both surgical and non-surgical therapy. The company's product portfolio encompasses reconstructive products, fixation devices, spinal products and other products. Reconstructive products include knee, hip and extremity joint replacement systems, as well as dental reconstructive implants, bone cements and accessories. Fixation devices include electrical stimulation systems; internal and external fixation devices; craniomaxillofacial fixation systems; and bone substitution materials. Spinal products include spinal fusion stimulation systems, spinal fixation systems, spinal bone substitution materials, precision machine allograft and motion preservation products. The other product market segment includes arthroscopy products, orthopedic support products, operating room supplies, casting materials, general surgical instruments and wound care products. Biomet manufactures numerous knee systems, including the Vanguard System, the Oxford Unicompartmental Knee, the Alpina Unicompartmental Knee, the Vanguard M System, the Repicci II Unicondylar Knee System and the Biomet OSS Orthopaedic Salvage System. Biomet's Arthrotek, Inc. subsidiary manufactures arthroscopy products in five product categories: power instruments, manual instruments, visualization products, soft tissue anchors and procedure-specific instruments and implants. Biomet is owned by a private-equity group that includes affiliates of the Blackstone Group, Goldman Sachs & Co., Kohlberg Kravis Roberts & Co. L.P. and TPG Capital, who together paid roughly $11.4 billion for the firm.

## FINANCIALS: Sales and profits are in thousands of dollars—add 000 to get the full amount. 2008 Note: Financial information for 2008 was not available for all companies at press time.

| | | |
|---|---|---|
| 2008 Sales: $2,383,300 | 2008 Profits: $1,018,800 | **U.S. Stock Ticker: Private** |
| 2007 Sales: $2,107,428 | 2007 Profits: $335,892 | **Int'l Ticker:**    Int'l Exchange: |
| 2006 Sales: $2,025,739 | 2006 Profits: $405,908 | Employees:  4,177 |
| 2005 Sales: $1,879,950 | 2005 Profits: $349,373 | Fiscal Year Ends: 5/31 |
| 2004 Sales: $1,615,751 | 2004 Profits: $320,324 | Parent Company: BLACKSTONE GROUP LP (THE) |

## SALARIES/BENEFITS:

| | | | | |
|---|---|---|---|---|
| Pension Plan: | ESOP Stock Plan: | Profit Sharing: | Top Exec. Salary: $358,800 | Bonus: $289,200 |
| Savings Plan: | Stock Purch. Plan: | | Second Exec. Salary: $341,300 | Bonus: $289,200 |

## OTHER THOUGHTS:

Apparent Women Officers or Directors: 2
Hot Spot for Advancement for Women/Minorities:

## LOCATIONS: ("Y" = Yes)

| West: | Southwest: | Midwest: | Southeast: | Northeast: | International: |
|---|---|---|---|---|---|
| Y | | Y | Y | Y | Y |

Note: Financial information, benefits and other data can change quickly and may vary from those stated here.

# BIONANOMATRIX
### www.bionanomatrix.com

**Industry Group Code: 541711 Ranks within this company's industry group: Sales: Profits:**

| Drugs: | Other: | | Clinical: | | Computers: | Services: |
|---|---|---|---|---|---|---|
| Discovery: | AgriBio: | | Trials/Services: | | Hardware: | Specialty Services: |
| Licensing: | Genetic Data: | Y | Labs: | | Software: | Consulting: |
| Manufacturing: | Tissue Replacement: | | Equipment/Supplies: | Y | Arrays: | Blood Collection: |
| Genetics: | | | Research & Development Services: | | Database Management: | Drug Delivery: |
| | | | Diagnostics: | | | Drug Distribution: |

## TYPES OF BUSINESS:
Genome Imaging & Analysis

## BRANDS/DIVISIONS/AFFILIATES:
National Institutes of Health
Battelle Ventures
KT Venture Group
Ben Franklin Technology Partners
21Ventures
Complete Genomics, Inc.
National Human Genome Research Institute

## CONTACTS: *Note: Officers with more than one job title may be intentionally listed here more than once.*
Michael Boyce-Jacino, CEO
Michael Boyce-Jacino, Pres.
Lorraine LoPresti, CFO
Han Cao, Chief Scientific Officer/Founder
Michael Kochersperger, VP-Eng.
Lorraine LoPresti, VP-Admin.
Gary Zweiger, VP-Bus. Dev.
Lorraine LoPresti, VP-Finance

| Phone: 267-499-2014 | Fax: 267-499-2015 |
|---|---|
| Toll-Free: | |
| Address: 3701 Market Street, 4th Floor, Philadelphia, PA 19104 US | |

## GROWTH PLANS/SPECIAL FEATURES:
BioNanomatrix develops nanoscale whole genome imaging and analytic platforms for applications in biomedical research, genetic diagnostics and personalized medicine. The company's technology is designed to dramatically reduce the time and cost required to analyze genomic DNA. Its proprietary technology works by detecting, identifying and analyzing long strands of DNA in sequence and in a massively parallel format, which allows for the survey of the DNA of an entire genome without amplification, as opposed to current methods of analyzing short fragments of DNA. Nanochannel arrays are incorporated into a nanochip fluidics device, designed to direct long strands of DNA through the chip into the nanochannels in a linear fashion, making them easy to assess for specific information. The nanochip device's design makes it possible to conduct millions of these analyses simultaneously. BioNanomatrix's development programs are supported in part by grants from the National Institutes of Health (NIH) and an $8.8 million government award to develop, in conjunction with Complete Genomics, Inc., a platform capable of sequencing the entire human genome at a cost of $100. The company's other major investors include Battelle Ventures, KT Venture Group, Ben Franklin Technology Partners and 21Ventures. In August 2008, BioNanomatrix received a grant from the National Human Genome Research Institute (NHGRI) of NIH for the development of a nanoscale platform for single-molecule haplotyping imaging and analysis of long strands of DNA in a massively parallel format. In March 2009, the company received a Phase 2 grant from NHGRI for the continued development of its whole genome imaging and analysis platform.

## FINANCIALS: Sales and profits are in thousands of dollars—add 000 to get the full amount. 2008 Note: Financial information for 2008 was not available for all companies at press time.

| | | |
|---|---|---|
| 2008 Sales: $ | 2008 Profits: $ | U.S. Stock Ticker: Private |
| 2007 Sales: $ | 2007 Profits: $ | Int'l Ticker: Int'l Exchange: |
| 2006 Sales: $ | 2006 Profits: $ | Employees: |
| 2005 Sales: $ | 2005 Profits: $ | Fiscal Year Ends: |
| 2004 Sales: $ | 2004 Profits: $ | Parent Company: |

## SALARIES/BENEFITS:
| | | | | |
|---|---|---|---|---|
| Pension Plan: | ESOP Stock Plan: | Profit Sharing: | Top Exec. Salary: $ | Bonus: $ |
| Savings Plan: | Stock Purch. Plan: | | Second Exec. Salary: $ | Bonus: $ |

## OTHER THOUGHTS:
**Apparent Women Officers or Directors:** 2
**Hot Spot for Advancement for Women/Minorities:** Y

## LOCATIONS: ("Y" = Yes)
| West: | Southwest: | Midwest: | Southeast: | Northeast: | International: |
|---|---|---|---|---|---|
| | | | | Y | |

# BIOPURE CORPORATION

www.biopure.com

**Industry Group Code: 325412 Ranks within this company's industry group:** Sales: 135    Profits: 102

| Drugs: | | Other: | Clinical: | Computers: | Services: |
|---|---|---|---|---|---|
| Discovery: | Y | AgriBio: | Trials/Services: | Hardware: | Specialty Services: |
| Licensing: | | Genetic Data: | Labs: | Software: | Consulting: |
| Manufacturing: | Y | Tissue Replacement: | Equipment/Supplies: | Arrays: | Blood Collection: |
| Genetics: | | | Research & Development Services: | Database Management: | Drug Delivery: |
| | | | Diagnostics: | | Drug Distribution: |

## TYPES OF BUSINESS:

Blood Transfusion Products
Oxygen Therapeutics
Veterinary Drugs

## BRANDS/DIVISIONS/AFFILIATES:

Hemopure
Oxyglobin

## CONTACTS: Note: Officers with more than one job title may be intentionally listed here more than once.

Zafiris G. Zafirelis, CEO
Zafiris G. Zafirelis, Pres.
W. Richard Light, VP-Tech. Dev.
Jane Kober, General Counsel/Sr. VP/Sec.
Barry L. Scott, VP-Bus. Dev.
Tiana Gorham, Mgr.-Corp. Comm.
Tiana Gorham, Mgr.-Investor Rel.
Virginia T. Rentko, VP-Preclinical Dev.
A. Gerson Greenburg, VP-Medical Affairs
Zafiris G. Zafirelis, Chmn.

| Phone: 617-234-6500 | Fax: 617-234-6505 |
|---|---|
| Toll-Free: | |
| Address: 11 Hurley St., Cambridge, MA 02141 US | |

## GROWTH PLANS/SPECIAL FEATURES:

Biopure Corporation develops, manufactures and markets oxygen therapeutics, a class of intravenous pharmaceuticals that increase oxygen transport to the body's tissues. Products include Hemopure for human use and Oxyglobin for veterinary use. Hemopure is approved in South Africa for treating adult surgical patients who are acutely anemic and for eliminating, reducing or delaying the need for red blood cell transfusion in these patients. Biopure's current clinical development efforts for Hemopure are focused on potential indications in anemia and cardiovascular ischemia, and on supporting the U.S. Navy's government-funded efforts to develop a potential out-of-hospital trauma indication. While the treatment is approved for sale to post-surgical, acutely anemic patients in South Africa, it remains on FDA clinical hold in the U.S. The firm is also developing Hemopure for use in anemic cancer patients and end-of-life patients. Once infused into a patient, Hemopure molecules disperse throughout the plasma space and are in continuous contact with the blood vessel wall, turning plasma into an oxygen-delivering substance which can bypass partial blockages or pass through constricted vessels that impede the normal passage of red blood cells. Oxyglobin, the firm's veterinary product, is used for the treatment of anemia in dogs and is marketed and sold to veterinary hospitals and small animal veterinary practices in the U.S. and Europe. Both of these products consist of hemoglobin that has been taken out of the red blood cells of cattle and then purified, chemically cross-linked for stability and formulated in a balanced salt solution to create a solution that does not contain any cells. In May 2008, Biopure Corporation named Dechra Veterinary Products, a division of Dechra Pharmaceuticals PLC, the exclusive distributor of Oxyglobin in the U.S.

The company's employee benefits include medical and dental coverage; life and disability insurance; a 401(k); and tuition reimbursement.

## FINANCIALS: Sales and profits are in thousands of dollars—add 000 to get the full amount. 2008 Note: Financial information for 2008 was not available for all companies at press time.

| | | |
|---|---|---|
| 2008 Sales: $3,128 | 2008 Profits: $-20,282 | U.S. Stock Ticker: BPUR |
| 2007 Sales: $2,556 | 2007 Profits: $-36,282 | Int'l Ticker:    Int'l Exchange: |
| 2006 Sales: $1,715 | 2006 Profits: $-26,454 | Employees:    7 |
| 2005 Sales: $2,110 | 2005 Profits: $-28,671 | Fiscal Year Ends: 10/31 |
| 2004 Sales: $3,750 | 2004 Profits: $-41,665 | Parent Company: |

## SALARIES/BENEFITS:

| Pension Plan: | ESOP Stock Plan: | Profit Sharing: | Top Exec. Salary: $250,016 | Bonus: $ |
|---|---|---|---|---|
| Savings Plan: Y | Stock Purch. Plan: | | Second Exec. Salary: $231,036 | Bonus: $ |

## OTHER THOUGHTS:

**Apparent Women Officers or Directors:** 3
**Hot Spot for Advancement for Women/Minorities:** Y

## LOCATIONS: ("Y" = Yes)

| West: | Southwest: | Midwest: | Southeast: | Northeast: | International: |
|---|---|---|---|---|---|
| | | | | Y | |

Note: Financial information, benefits and other data can change quickly and may vary from those stated here.

# BIORELIANCE CORP

**www.bioreliance.com**

**Industry Group Code: 6215 Ranks within this company's industry group:** Sales: Profits:

| Drugs: | | Other: | | Clinical: | | Computers: | | Services: | |
|---|---|---|---|---|---|---|---|---|---|
| Discovery: | | AgriBio: | | Trials/Services: | | Hardware: | | Specialty Services: | |
| Licensing: | Y | Genetic Data: | | Labs: | Y | Software: | | Consulting: | |
| Manufacturing: | | Tissue Replacement: | | Equipment/Supplies: | | Arrays: | | Blood Collection: | |
| Genetics: | | | | Research & Development Services: | Y | Database Management: | | Drug Delivery: | |
| | | | | Diagnostics: | | | | Drug Distribution: | |

## TYPES OF BUSINESS:

Research-Nonclinical Product Testing
Contract Biologics Manufacturing
Biologics Pharmaceutical Services

## BRANDS/DIVISIONS/AFFILIATES:

Microbiological Associates
Avista Capital Partners
Invitrogen Corporation
Analytical Services
Laboratory Animal Diagnostic Services (LADS)
Clinical Trial Support Services

## CONTACTS: *Note: Officers with more than one job title may be intentionally listed here more than once.*

David A. Dodd, CEO
David A. Dodd, Pres.
David S. Walker, CFO
David E. Onions, Chief Scientific Officer
James J. Kramer, VP-Oper., U.S. Biologics
Darryl L. Goss, VP-Global Process Excellence
David A. Dodd, Chmn.
David L. Bellitt, VP-Global Comm. Oper., Biologics

| **Phone:** 301-738-1000 | **Fax:** 301-610-2590 |
|---|---|
| **Toll-Free:** 800-553-5372 | |
| **Address:** 14920 Broschart Rd., Rockville, MD 20850 US | |

## GROWTH PLANS/SPECIAL FEATURES:

BioReliance Corp., founded in 1947 as Microbiological Associates, is a global contract services organization providing biological safety testing, toxicology, viral manufacturing and laboratory animal diagnostic services to the pharmaceutical and biopharmaceutical industries. BioReliance provides services to over 600 clients annually, including most of the largest pharmaceutical and biopharmaceutical companies in the world. The company also supports early stage companies lacking the staff, expertise and financial resources to conduct many aspects of product development in-house. BioReliance provides its services throughout the product cycle, starting from early preclinical development through licensed production. Services include biologics safety testing, viral clearance studies, manufacturing, toxicology testing, Analytical Services, Laboratory Animal Diagnostic Services (LADS) and Clinical Trial Support Services. The firm provides GLP compliant genetic, in vivo and molecular toxicology testing on pharmaceutical, biopharmaceutical, medical device and pesticide products. BioReliance's Analytical Services department physiochemically characterizes biotechnology products to determine whether their molecular structures meet predefined criteria and whether their macro-molecules remain safe, stable and efficacious in the final product formulations. The company's LADS department offers a spectrum of animal diagnostic and analytical testing programs, including rodent and simian diagnostics, which are used by laboratories, breeding colonies and animal facilities worldwide for both routine monitoring and emergency situations. BioReliance's Clinical Trial Support Services department has significant expertise in manufacturing and testing viral vectors and vaccines, which may require tests beyond the standard clinical chemistry and hematology required for other clinical trials, including an analysis of patient samples for shed virus, expression of the delivered gene, presence of anti-virus neutralizing antibodies and cytokines. Recently, the company was acquired by private equity firm Avista Capital Partners from Invitrogen Corporation. In March 2008, BioReliance expanded its operations in the Asia Pacific region with the opening of a new commercial office in Tokyo.

## FINANCIALS: Sales and profits are in thousands of dollars—add 000 to get the full amount. 2008 Note: Financial information for 2008 was not available for all companies at press time.

| | | |
|---|---|---|
| 2008 Sales: $ | 2008 Profits: $ | **U.S. Stock Ticker: Private** |
| 2007 Sales: $ | 2007 Profits: $ | **Int'l Ticker:** Int'l Exchange: |
| 2006 Sales: $ | 2006 Profits: $ | Employees: 638 |
| 2005 Sales: $ | 2005 Profits: $ | Fiscal Year Ends: 12/31 |
| 2004 Sales: $ | 2004 Profits: $ | Parent Company: AVISTA CAPITAL PARTNERS |

## SALARIES/BENEFITS:

| Pension Plan: | ESOP Stock Plan: | Profit Sharing: | Top Exec. Salary: $401,923 | Bonus: $181,149 |
|---|---|---|---|---|
| Savings Plan: | Stock Purch. Plan: | | Second Exec. Salary: $229,391 | Bonus: $76,991 |

## OTHER THOUGHTS:

**Apparent Women Officers or Directors**: 1
**Hot Spot for Advancement for Women/Minorities**:

## LOCATIONS: ("Y" = Yes)

| West: | Southwest: | Midwest: | Southeast: | Northeast: | International: |
|---|---|---|---|---|---|
| | | | | Y | Y |

# BIOSITE INC

**www.biosite.com**

Industry Group Code: 325413 **Ranks within this company's industry group:** Sales: Profits:

| Drugs: | Other: | Clinical: | | Computers: | | Services: | |
|--------|--------|-----------|---|-----------|---|-----------|---|
| Discovery: | AgriBio: | Trials/Services: | | Hardware: | | Specialty Services: | Y |
| Licensing: | Genetic Data: | Labs: | | Software: | | Consulting: | Y |
| Manufacturing: | Tissue Replacement: | Equipment/Supplies: | Y | Arrays: | | Blood Collection: | |
| Genetics: | | Research & Development Services: | Y | Database Management: | | Drug Delivery: | |
| | | Diagnostics: | Y | | | Drug Distribution: | |

## TYPES OF BUSINESS:

Medical Diagnostics Products
Rapid Immunoassays
Antibody Development Services

## BRANDS/DIVISIONS/AFFILIATES:

Biosite Discovery
Triage Drugs of Abuse Panel
Triage Cardiac Panel
Triage TOX Drug Screen
Triage BNP Test
Triage Profiler Panels
Triage Parasite Panel
Biosite Encompass

## CONTACTS: Note: Officers with more than one job title may be intentionally listed here more than once.

Kim D. Blickenstaff, CEO
Kenneth F. Buechler, Pres.
Christopher J. Twomey, CFO
Kenneth F. Buechler, Chief Scientific Officer
Doug Guarino, Media Contact
Kim D. Blickenstaff, Chmn.

| **Phone:** 858-805-8378 | **Fax:** 858-455-4815 |
|---|---|
| **Toll-Free:** 888-246-7483 | |
| **Address:** 9975 Summers Ridge Rd., San Diego, CA 92121 US | |

## GROWTH PLANS/SPECIAL FEATURES:

Biosite, Inc. is a global diagnostics company dedicated to utilizing biotechnology in the development of diagnostic products. The firm, a private subsidiary of Inverness Medical Innovations, validates and patents novel protein biomarkers and panels of biomarkers; develops and markets products; conducts strategic research on its products; and educates healthcare providers about its products. Biosite markets immunoassay diagnostics in the areas of cardiovascular disease, drug overdose and infectious disease. Cardiovascular products include the Triage BNP Test, Triage Cardiac Panel, Triage Profiler Panels, Triage D-Dimer Test and Triage Stroke Panel. The Triage BNP test is used in more than 3,000 hospitals and doctor's offices, and helps in the diagnosis, and severity assessment, of heart failure. The Triage Drugs of Abuse Panel and Triage TOX Drug Screen are rapid, qualitative urine screens that test for up to nine different illicit and prescription drugs, or drug classes, and provide results in less than 15 minutes. The firm's Biosite Discovery research business seeks to identify new protein markers of diseases that lack effective diagnostic tests. Additionally, with Biosite Discovery, the company has the capacity to offer antibody development services to companies seeking high-affinity antibodies for use in drug research. In return, Biosite seeks diagnostic licenses. The firm's Encompass program provides comprehensive education and consultation programs for all Biosite customers. These programs include training and education; evaluation support; product training; clinical training; outcomes tracking; POC reimbursement; CLIA audit reports; audits on request; and consulting services. In all, Biosite's products are used in about half of all U.S. hospitals, as well as in approximately 50 international markets.

Biosite offers employees a benefits package including medical, dental, vision and life insurance; flexible spending accounts; an employee assistance program; bereavement and paternity leave; education reimbursement; a 401(k) plan; and an employee stock purchase plan.

## FINANCIALS: Sales and profits are in thousands of dollars—add 000 to get the full amount. 2008 Note: Financial information for 2008 was not available for all companies at press time.

| | | |
|---|---|---|
| 2008 Sales: $ | 2008 Profits: $ | **U.S. Stock Ticker: Subsidiary** |
| 2007 Sales: $ | 2007 Profits: $ | **Int'l Ticker:** Int'l Exchange: |
| 2006 Sales: $308,592 | 2006 Profits: $39,994 | Employees: 1,036 |
| 2005 Sales: $287,699 | 2005 Profits: $54,029 | Fiscal Year Ends: 12/31 |
| 2004 Sales: $244,900 | 2004 Profits: $41,400 | Parent Company: INVERNESS MEDICAL INNOVATIONS INC |

## SALARIES/BENEFITS:

| Pension Plan: | ESOP Stock Plan: | Profit Sharing: | Top Exec. Salary: $553,500 | Bonus: $264,424 |
|---|---|---|---|---|
| Savings Plan: Y | Stock Purch. Plan: Y | | Second Exec. Salary: $430,961 | Bonus: $203,944 |

## OTHER THOUGHTS:

**Apparent Women Officers or Directors:** 3
**Hot Spot for Advancement for Women/Minorities:** Y

## LOCATIONS: ("Y" = Yes)

| West: | Southwest: | Midwest: | Southeast: | Northeast: | International: |
|---|---|---|---|---|---|
| Y | | | | | Y |

# BIOSPHERE MEDICAL INC

**www.biospheremed.com**

**Industry Group Code: 33911  Ranks within this company's industry group:** Sales: 13  Profits: 13

| Drugs: | | Other: | | Clinical: | | Computers: | | Services: | |
|---|---|---|---|---|---|---|---|---|---|
| Discovery: | Y | AgriBio: | | Trials/Services: | | Hardware: | | Specialty Services: | |
| Licensing: | | Genetic Data: | | Labs: | | Software: | | Consulting: | |
| Manufacturing: | Y | Tissue Replacement: | | Equipment/Supplies: | Y | Arrays: | | Blood Collection: | |
| Genetics: | | | | Research & Development Services: | | Database Management: | | Drug Delivery: | |
| | | | | Diagnostics: | | | | Drug Distribution: | |

## TYPES OF BUSINESS:

Drugs-Bioengineered Microspheres
Cancer Treatments
Microsphere Delivery Systems

## BRANDS/DIVISIONS/AFFILIATES:

Embosphere
EmboGold
HepaSphere
EmboCath Plus
Sequitor Guidewire
QuatraSphere

## CONTACTS: *Note: Officers with more than one job title may be intentionally listed here more than once.*

Richard J. Faleschini, CEO
Richard J. Faleschini, Pres.
Martin J. Joyce, CFO/Exec. VP-Finance & Admin.
Joel B. Weinstein, VP-Global Mktg. & Sales
Peter Sutcliffe, VP-Mfg.
Martin J. Joyce, Exec. VP-Admin.
Willard W. Hennemann, VP-Bus. & New Prod. Dev.
Martin J. Joyce, Exec. VP-Finance
Melodie R. Domurad, VP-Regulatory, Medical Affairs & Quality Systems
Timothy J. Barberich, Chmn.

| **Phone:** 781-681-7900 | **Fax:** 781-792-2745 |
|---|---|
| **Toll-Free:** 800-394-0295 | |
| **Address:** 1050 Hingham St., Rockland, MA 02370 US | |

## GROWTH PLANS/SPECIAL FEATURES:

Biosphere Medical, Inc. develops, manufactures and markets products for medical procedures that use embolotherapy techniques. Embolotherapy, the introduction of biocompatible substances into the circulatory system, works to occlude blood vessels in order to arrest hemorrhaging or to devitalize a structure. Biosphere's core technology consists of patented bioengineered polymers, which help produce miniature spherical embolic particles called microspheres. Embosphere Microspheres and EmboGold Microspheres, the company's embolic products, are made of an acrylic co-polymer that is cross-linked with gelatin. Due to their uniform, spherical shape and soft, slippery surface, microspheres are easy to inject through microcatheters, resulting in an even distribution within the vessel network. The microspheres selectively block the target tissue's blood supply, which destroys or devitalizes its target. Biosphere provides its products in calibrated size ranges, so they can be selected to target occlusion of specific sized vessels. The company's principle focus is the treatment of symptomatic uterine fibroids, which are non-cancerous tumors growing in the uterus, using a procedure called uterine fibroid embolization (UFE). UFE is an alternative to hysterectomy and myomectomy that has been shown to allow faster recovery time and provide equivalent quality of life benefit. In addition to its microspheres, Biosphere produces two microcatheters: the Embocath Plus for general application and the Sequitor Guidewire for increased access to pelvic and visceral anatomy. Biosphere also produces the QuatraSphere and HepaSphere product lines for a number of other medical treatments, including the use of microspheres in the treatment of peripheral arteriovenous malformations and hypervascularized tumors like primary cancer of the liver. QuatraSphere is marketed in the U.S. while HepaSphere is marketed in Japan and Europe. HepaSphere is also approved for transarterial chemoembolization of liver cancer, when used in conjunction with doxorubicin, an anticancer drug.

Biosphere offers employees health, life, dental and disability insurance.

## FINANCIALS: Sales and profits are in thousands of dollars—add 000 to get the full amount. 2008 Note: Financial information for 2008 was not available for all companies at press time.

| | | |
|---|---|---|
| 2008 Sales: $29,258 | 2008 Profits: $-5,492 | **U.S. Stock Ticker: BSMD** |
| 2007 Sales: $26,900 | 2007 Profits: $-1,854 | **Int'l Ticker:**  Int'l Exchange: |
| 2006 Sales: $22,891 | 2006 Profits: $-2,324 | Employees:  84 |
| 2005 Sales: $18,484 | 2005 Profits: $-2,801 | Fiscal Year Ends: 12/31 |
| 2004 Sales: $14,158 | 2004 Profits: $-6,841 | Parent Company: |

## SALARIES/BENEFITS:

| Pension Plan: | ESOP Stock Plan: | Profit Sharing: | Top Exec. Salary: $397,800 | Bonus: $120,533 |
|---|---|---|---|---|
| Savings Plan: Y | Stock Purch. Plan: | | Second Exec. Salary: $242,481 | Bonus: $59,456 |

## OTHER THOUGHTS:

**Apparent Women Officers or Directors:** 1
**Hot Spot for Advancement for Women/Minorities:**

## LOCATIONS: ("Y" = Yes)

| West: | Southwest: | Midwest: | Southeast: | Northeast: | International: |
|---|---|---|---|---|---|
| | | | | Y | Y |

Note: Financial information, benefits and other data can change quickly and may vary from those stated here.

# BIOTECH HOLDINGS LTD

www.biotechltd.com

**Industry Group Code: 325412  Ranks within this company's industry group:  Sales: 148    Profits: 961**

| Drugs: | | Other: | Clinical: | Computers: | Services: |
|---|---|---|---|---|---|
| Discovery: | Y | AgriBio: | Trials/Services: | Hardware: | Specialty Services: |
| Licensing: | | Genetic Data: | Labs: | Software: | Consulting: |
| Manufacturing: | Y | Tissue Replacement: | Equipment/Supplies: | Arrays: | Blood Collection: |
| Genetics: | | | Research & Development Services: | Database Management: | Drug Delivery: |
| | | | Diagnostics: | | Drug Distribution: |

## TYPES OF BUSINESS:

Pharmaceuticals Development & Manufacturing
Drugs-Diabetes

## BRANDS/DIVISIONS/AFFILIATES:

Sucanon
Diab II
Glucanin

## CONTACTS: Note: Officers with more than one job title may be intentionally listed here more than once.

Robert B. Rieveley, Pres.
Lorne D. Brown, CFO
Gale Belding, Exec. VP-Admin. Affairs
Gale Belding, Exec. VP-Regulatory Affairs
Luis M. Ornelas Lopez, VP-Latin American Oper.

| **Phone:** 604-295-1119 | **Fax:** 604-295-1110 |
|---|---|
| **Toll-Free:** 888-216-1111 | |
| **Address:** 3751 Shell Rd., Ste. 160, Richmond, BC V6X 2W2 Canada | |

## GROWTH PLANS/SPECIAL FEATURES:

Biotech Holdings, Ltd. researches, develops, manufactures and markets pharmaceutical products.  The firm's only marketed drug, trademarked as Sucanon and Diab II, treats Type II diabetes mellitus, the most common form of diabetes, accounting for roughly 90% of all diabetic cases.  Sucanon treats Type II diabetes by acting as an insulin sensitizer, increasing the body's ability to absorb and use insulin, thus decreasing blood sugar levels, improving metabolism and enhancing the patient's sense of well-being.  Studies indicate Sucanon has no substantial side effects.  One of only a handful of insulin sensitizers approved anywhere in the world, Sucanon is primarily marketed in Mexico and Latin America.  In Mexico, Sucanon's distribution channels include Costco Mexico and Wal-Mart Mexico.  The drug is sold as Glucanin in Peru.  The firm has production facilities in Canada and Mexico.  Biotech maintains an e-commerce site, sucanonhealth.com, which sells the drug for personal use worldwide, including for those in the U.S. and Canada.  Through distribution agreements with private companies, the firm markets Sucanon to four Central American countries: El Salvador, Honduras, Guatemala and Nicaragua; and the United Arab Emirates, including Dubai.

## FINANCIALS:  Sales and profits are in thousands of dollars—add 000 to get the full amount. 2008 Note: Financial information for 2008 was not available for all companies at press time.

| | | |
|---|---|---|
| 2008 Sales: $ 340 | 2008 Profits: $-1,300 | **U.S. Stock Ticker: BIOHF.OB** |
| 2007 Sales: $ 310 | 2007 Profits: $-1,420 | **Int'l Ticker: BIO**    Int'l Exchange: Toronto-TSX |
| 2006 Sales: $ 450 | 2006 Profits: $-1,870 | Employees:    8 |
| 2005 Sales: $ | 2005 Profits: $-1,575 | Fiscal Year Ends: 3/31 |
| 2004 Sales: $ | 2004 Profits: $-1,800 | Parent Company: |

## SALARIES/BENEFITS:

| Pension Plan: | ESOP Stock Plan: | Profit Sharing: | Top Exec. Salary: $ | Bonus: $ |
|---|---|---|---|---|
| Savings Plan: | Stock Purch. Plan: | | Second Exec. Salary: $ | Bonus: $ |

## OTHER THOUGHTS:

**Apparent Women Officers or Directors**: 2
**Hot Spot for Advancement for Women/Minorities**: Y

## LOCATIONS: ("Y" = Yes)

| West: | Southwest: | Midwest: | Southeast: | Northeast: | International: Y |
|---|---|---|---|---|---|

# BIOTIME INC

www.biotimeinc.com

**Industry Group Code:** 325412 **Ranks within this company's industry group:** Sales: 142 Profits: 66

| Drugs: | | Other: | | Clinical: | | Computers: | | Services: | |
|---|---|---|---|---|---|---|---|---|---|
| Discovery: | Y | AgriBio: | | Trials/Services: | | Hardware: | | Specialty Services: | |
| Licensing: | Y | Genetic Data: | | Labs: | | Software: | | Consulting: | |
| Manufacturing: | | Tissue Replacement: | | Equipment/Supplies: | Y | Arrays: | | Blood Collection: | |
| Genetics: | | | | Research & Development Services: | | Database Management: | | Drug Delivery: | |
| | | | | Diagnostics: | | | | Drug Distribution: | |

## TYPES OF BUSINESS:

Drugs-Surgical
Blood Plasma Expanders
Blood Replacement Solutions
Regenerative Medicine

## BRANDS/DIVISIONS/AFFILIATES:

Hextend
PentaLyte
ACTcellerate
Enbryome Sciences, Inc.
Embryome.com Database
International Embryome Initiative
Reproductive Genetics Institute

## CONTACTS: *Note: Officers with more than one job title may be intentionally listed here more than once.*

Michael D. West, CEO
Robert W. Peabody, COO/Sr. VP
Steven A. Seinberg, CFO
Hal Sternberg, VP-Research
Harold Waitz, VP-Eng. & Regulatory Affairs
Judith Segall, Corp. Sec.
Judith Segall, VP-Oper.
Jeffrey Nickel, VP-Bus. Dev.
Judith Segall, Press Contact
Steven A. Seinberg, Treas.

| Phone: 510-521-3390 | Fax: 510-521-3389 |
|---|---|
| Toll-Free: | |
| Address: 1301 Harbor Bay Pkwy., Alameda, CA 94502 US | |

## GROWTH PLANS/SPECIAL FEATURES:

BioTime, Inc. is a development-stage company engaged in two areas of biomedical research and development: aqueous-based synthetic surgery products, and regenerative medicine through stem cell-related products. The firm's synthetic surgery formulas can be used as blood plasma volume expanders and blood replacement solutions in surgery, emergency trauma treatment and other applications. Its lead blood plasma expander product is Hextend, an intravenous solution used in the treatment of hypovolemia, a condition caused by low blood volume, often from blood loss during surgery. Hextend maintains circulatory system fluid volume and blood pressure and keeps vital organs perfused during surgery. Another product still in development is PentaLyte, now in clinical evaluation, a pentastarch-based synthetic plasma expander. The company's regenerative medicine business is operated through wholly-owned subsidiary Embryome Sciences, Inc. The regenerative medicine business focuses on the development and sale of advanced human stem cell products and technology for use by researchers at universities; companies in the bioscience and biopharmaceutical industries; and other companies providing research products to those industries. Embryome's first product is the Embryome.com Database, a database that provides a detailed map of the embryome. Embryome has collaborated with the International Stem Cell Corporation, Lifeline Cell Technology, the International Longevity Center-USA and the Wisconsin Alumni Research Foundation to jointly produce and distribute research products. In June 2008, BioTime launched Embryome.com and the International Embryome Initiative, an international collaboration to create the first systematic map of all the cell types derived from human embryonic stem cells. In July 2008, Embryome Sciences acquired Advanced Cell Technology, Inc., an exclusive license to use ACTCellerate embryonic stem cell technology and a bank of over 140 diverse progenitor cell lines derived using that technology. In February 2009, Embryome agreed to obtain the rights to market human embryonic stem cell (hES) lines derived by Reproductive Genetics Institute (RGI).

## FINANCIALS: Sales and profits are in thousands of dollars—add 000 to get the full amount. 2008 Note: Financial information for 2008 was not available for all companies at press time.

| | | | |
|---|---|---|---|
| 2008 Sales: $1,504 | 2008 Profits: $-3,781 | U.S. Stock Ticker: BTIM | |
| 2007 Sales: $1,046 | 2007 Profits: $-1,438 | Int'l Ticker: | Int'l Exchange: |
| 2006 Sales: $1,162 | 2006 Profits: $-1,864 | Employees: 11 | |
| 2005 Sales: $ 903 | 2005 Profits: $-2,074 | Fiscal Year Ends: 12/31 | |
| 2004 Sales: $ 688 | 2004 Profits: $-3,085 | Parent Company: | |

## SALARIES/BENEFITS:

| Pension Plan: | ESOP Stock Plan: | Profit Sharing: | Top Exec. Salary: $250,000 | Bonus: $ |
|---|---|---|---|---|
| Savings Plan: | Stock Purch. Plan: | | Second Exec. Salary: $160,000 | Bonus: $ |

## OTHER THOUGHTS:

**Apparent Women Officers or Directors:** 1
**Hot Spot for Advancement for Women/Minorities:**

## LOCATIONS: ("Y" = Yes)

| West: | Southwest: | Midwest: | Southeast: | Northeast: | International: |
|---|---|---|---|---|---|
| Y | | | | | |

# BIOVAIL CORPORATION

**www.biovail.com**

Industry Group Code: 325412A  Ranks within this company's industry group: Sales: 6  Profits: 3

| Drugs: | Other: | Clinical: | | Computers: | | Services: | |
|---|---|---|---|---|---|---|---|
| Discovery: | AgriBio: | Trials/Services: | Y | Hardware: | | Specialty Services: | |
| Licensing: | Genetic Data: | Labs: | Y | Software: | | Consulting: | |
| Manufacturing: Y | Tissue Replacement: | Equipment/Supplies: | | Arrays: | | Blood Collection: | |
| Genetics: Y | | Research & Development Services: | Y | Database Management: | | Drug Delivery: | Y |
| | | Diagnostics: | | | | Drug Distribution: | |

## TYPES OF BUSINESS:

Drug Delivery Systems Technologies
Generic Drugs
Drugs-Hypertension
Drugs-Antidepressants
Contract Research Services
Drug Development

## BRANDS/DIVISIONS/AFFILIATES:

BTA Pharmaceuticals, Inc.
Biovail Pharmaceuticals Canada
Biovail Laboratories International SRL
Biovail Technologies, Inc.
Prestwick Pharmaceuticals, Inc.
Xenazine
Cardizem LA
Wellbutrin XL

## CONTACTS: Note: Officers with more than one job title may be intentionally listed here more than once.

William M. Wells, CEO
Gilbert Godin, COO/Exec. VP
Peggy Mulligan, CFO/Sr. VP
Mark Durham, Sr. VP-Human Resources & Shared Svcs.
John Sebben, VP-Tech. Oper.
Wendy Kelley, General Counsel/Sr. VP/Corp. Sec.
Gregory Gubitz, Sr. VP-Corp. Dev.
Nelson F. Isabel, VP-Corp. Comm.
Nelson F. Isabel, VP-Investor Rel.
Christopher Bovaird, VP-Corp. Finance
Jennifer Tindale, VP/Associate General Counsel
Todd Zater, Controller/VP
Rick Albert, Treas./VP/
Christine Mayer, Sr. VP-Bus. Dev., BTA Pharmaceuticals, Inc.
Douglas J.P. Squires, Chmn.
Michel Chouinard, COO-Biovail Laboratories Int'l SRL

| Phone: 905-286-3000 | Fax: 905-286-3050 |
|---|---|
| Toll-Free: | |
| Address: 7150 Mississauga Rd., Mississauga, ON L5N 8M5 Canada | |

## GROWTH PLANS/SPECIAL FEATURES:

Biovail Corporation is a specialty pharmaceutical company focused on the development and commercialization of products that target specialty central nervous system (CNS) disorders without significant existing treatments. Historically, the firm's primary areas of focus include cardiovascular disease, Type II diabetes and pain management. The company has primarily employed its drug delivery technologies to provide include controlled release, graded release, enhanced absorption, rapid absorption, taste masking and oral disintegration processes. The firm's products include Wellbutrin XL, the brand name of bupropion, an anti-depressant; Cardizem LA and Tiazac, blood pressure medications; Zovirax, a topical form of acyclovir, used in the treatment of the herpes virus; Ralivia and Ultram pain medications; Retavase, for the treatment of acute myocardial infarction; and Glumetza, a diabetes medication. In addition, the company operates a contract research division that provides Biovail and other pharmaceutical companies with a broad range of Phase I/II clinical research services using pharmacokinetic studies and bioanalytical laboratory testing. Biovail Laboratories International SRL (BLS), the firm's primary operating subsidiary, manages the company's intellectual property; develops, manufactures, and sells its pharmaceutical products; and performs strategic planning. Biovail markets its products through its marketing divisions, BTA Pharmaceuticals, Inc. and Biovail Pharmaceuticals Canada, and through other strategic partners to health care professionals. The company's other subsidiary is Biovail Technologies, Inc. In September 2008, Biovail acquired Prestwick Pharmaceuticals, Inc., a pharmaceutical company that holds the licensing rights to Xenazine tetrabenazine tablets in the U.S. and Canada. In May 2009, BLS acquired the U.S. and Canadian rights to develop, manufacture and commercialize pimavanserin tartrate for treatment of Parkinson's disease and Alzheimer's, as well as the full U.S. commercialization rights to Wellbutrin XL. Another of Biovail's subsidiaries also agreed to acquire worldwide development and commercialization rights to Cambridge Laboratories (Ireland), Ltd.'s entire portfolio of tetrabenazine products, including Xenazine/Nitoman tetrabenazine tablets.

## FINANCIALS: Sales and profits are in thousands of dollars—add 000 to get the full amount. 2008 Note: Financial information for 2008 was not available for all companies at press time.

| | | |
|---|---|---|
| 2008 Sales: $757,178 | 2008 Profits: $199,904 | **U.S. Stock Ticker: BVF** |
| 2007 Sales: $842,818 | 2007 Profits: $195,539 | **Int'l Ticker: BVF**   Int'l Exchange: Toronto-TSX |
| 2006 Sales: $1,067,722 | 2006 Profits: $211,626 | Employees:  1,389 |
| 2005 Sales: $935,500 | 2005 Profits: $89,000 | Fiscal Year Ends: 10/31 |
| 2004 Sales: $886,500 | 2004 Profits: $161,000 | Parent Company: |

## SALARIES/BENEFITS:

| Pension Plan: | ESOP Stock Plan: | Profit Sharing: | Top Exec. Salary: $750,607 | Bonus: $ |
|---|---|---|---|---|
| Savings Plan: | Stock Purch. Plan: | | Second Exec. Salary: $700,000 | Bonus: $525,000 |

## OTHER THOUGHTS:

**Apparent Women Officers or Directors**: 4
**Hot Spot for Advancement for Women/Minorities**: Y

## LOCATIONS: ("Y" = Yes)

| West: | Southwest: | Midwest: | Southeast: | Northeast: | International: |
|---|---|---|---|---|---|
| | | | | Y | Y |

Note: Financial information, benefits and other data can change quickly and may vary from those stated here.

# BRISTOL-MYERS SQUIBB CO

**www.bms.com**

Industry Group Code: 325412  Ranks within this company's industry group: Sales: 11    Profits: 9

| Drugs: | | Other: | Clinical: | | Computers: | | Services: | |
|---|---|---|---|---|---|---|---|---|
| Discovery: | Y | AgriBio: | Trials/Services: | | Hardware: | | Specialty Services: | |
| Licensing: | | Genetic Data: | Labs: | | Software: | | Consulting: | |
| Manufacturing: | Y | Tissue Replacement: | Equipment/Supplies: | Y | Arrays: | | Blood Collection: | |
| Genetics: | | | Research & Development Services: | | Database Management: | | Drug Delivery: | |
| | | | Diagnostics: | | | | Drug Distribution: | |

## TYPES OF BUSINESS:

Drugs-Diversified
Medical Imaging Products
Nutritional Products

## BRANDS/DIVISIONS/AFFILIATES:

Plavix
Enfamil
Reyataz
Ixempra
Sprycel
Abilify
Medarex
Kosan Biosciences Incorporated

## CONTACTS: Note: Officers with more than one job title may be intentionally listed here more than once.

James M. Cornelius, CEO
Lamberto Andreotti, COO
Lamberto Andreotti, Pres.
Jean-Marc Huet, CFO/Exec. VP
Anthony McBride, Sr. VP-Human Resources
Elliott Sigal, Exec. VP/Chief Scientific Officer/Pres., R&D
Carlo de Notaristefani, Pres., Tech. Oper. & Global Support Functions
Sandra Leung, General Counsel/Corp. Sec./Sr. VP
Robert T. Zito, Chief Comm. Officer/Sr. VP-Corp. & Bus. Comm.
Anthony C. Hooper, Pres., Americas
John E. Celentano, Pres., Emerging Markets & Asia Pacific
Brian Daniels, Sr. VP-Global Dev. & Medical Affairs
James M. Cornelius, Chmn.
Beatrice Cazala, Pres., Europe & Global Commercialization
Quentin Roach, Chief Procurement Officer/Sr. VP

| Phone: 212-546-4000 | Fax: 212-546-4020 |
|---|---|
| **Toll-Free:** | |
| **Address:** 345 Park Ave., New York, NY 10154 US | |

## GROWTH PLANS/SPECIAL FEATURES:

Bristol-Myers Squibb Co. discovers, develops, licenses, manufactures, markets, distributes and sells pharmaceuticals and other health care related products. It operates in two segments: Pharmaceuticals and Nutritionals. The pharmaceuticals segment, accounting for 86% of net sales, manufactures drugs across multiple therapeutic classes, including cardiovascular; virology, including immunodeficiency virus infection; oncology; affective and other psychiatric disorders; and immunoscience. Products include Plavix, Avapro/Avalide, Reyataz, Sprycel and Ixempra. These products are manufactured in the U.S. and Puerto Rico and 11 foreign countries. The nutritionals segment, through Mead Johnson, manufactures, markets, distributes and sells infant formulas and other nutritional products, including the entire line of Enfamil products. Nutritional products are generally sold by wholesalers and retailers and are promoted primarily to health care professionals. In 2008, the FDA approved Orencia for the treatment of moderate-to-severe Polyarticular Juvenile Idiopathic Arthritis, as well as Abilify, for the add-on treatment to Lithium or Valproate in the acute treatment of manic episodes of Bipolar disorder. In July 2008, the company acquired Kosan Biosciences Incorporated for $190 million. In August 2008, the firm completed the divestment of ConvaTec to Nordic Capital Fund VII and Avista Capital Partners. In July 2009, the firm agreed to acquire Medarex.

Employees are offered medical and dental insurance; health care reimbursement accounts; a pension plan; a 401(k) plan; short-and long-term disability coverage; life insurance; travel accident insurance; an employee assistance plan; and adoption assistance.

## FINANCIALS: Sales and profits are in thousands of dollars—add 000 to get the full amount. 2008 Note: Financial information for 2008 was not available for all companies at press time.

| | | |
|---|---|---|
| 2008 Sales: $20,597,000 | 2008 Profits: $5,247,000 | **U.S. Stock Ticker: BMY** |
| 2007 Sales: $18,193,000 | 2007 Profits: $2,165,000 | **Int'l Ticker:**   Int'l Exchange: |
| 2006 Sales: $16,208,000 | 2006 Profits: $1,585,000 | Employees:  42,000 |
| 2005 Sales: $18,605,000 | 2005 Profits: $3,000,000 | Fiscal Year Ends: 12/31 |
| 2004 Sales: $19,380,000 | 2004 Profits: $2,388,000 | Parent Company: |

## SALARIES/BENEFITS:

| Pension Plan: Y | ESOP Stock Plan: | Profit Sharing: | Top Exec. Salary: $1,488,077 | Bonus: $4,475,000 |
|---|---|---|---|---|
| Savings Plan: Y | Stock Purch. Plan: | | Second Exec. Salary: $1,211,141 | Bonus: $2,676,668 |

## OTHER THOUGHTS:

**Apparent Women Officers or Directors**: 2
**Hot Spot for Advancement for Women/Minorities**: Y

## LOCATIONS: ("Y" = Yes)

| West: | Southwest: | Midwest: | Southeast: | Northeast: | International: |
|---|---|---|---|---|---|
| Y | Y | Y | Y | Y | Y |

Note: Financial information, benefits and other data can change quickly and may vary from those stated here.

# BURRILL & COMPANY                                    www.burrillandco.com

**Industry Group Code: 523110  Ranks within this company's industry group:** Sales:     Profits:

| Drugs: | Other: | Clinical: | Computers: | Services: | |
|---|---|---|---|---|---|
| Discovery: | AgriBio: | Trials/Services: | Hardware: | Specialty Services: | Y |
| Licensing: | Genetic Data: | Labs: | Software: | Consulting: | Y |
| Manufacturing: | Tissue Replacement: | Equipment/Supplies: | Arrays: | Blood Collection: | |
| Genetics: | | Research & Development Services: | Database Management: | Drug Delivery: | |
| | | Diagnostics: | | Drug Distribution: | |

## TYPES OF BUSINESS:

Investment Banking-Life Sciences
Strategic Partnership & Spin-Off/Outlicensing Consulting
Life Sciences Publications
Industry Conferences
Venture Capital Funds

## BRANDS/DIVISIONS/AFFILIATES:

Burrill Life Sciences Capital Fund
Burrill Biotechnology Capital Fund
Burrill Agbio Capital Fund
Biotech Meeting at Laguna Beach
Burrill Nutraceuticals Capital Fund
Indiana Life Sciences Forum
Burrill Personalized Medicine Meeting
Burrill International Group

## CONTACTS: *Note: Officers with more than one job title may be intentionally listed here more than once.*

G. Steven Burrill, CEO
Victor Herbert, Chief Admin. Officer
Victor Herbert, Chief Legal Officer
Leslie Errington, Dir.-Bus. Dev.
Peter Winter, Dir.-Comm.
Helena Sen, Controller
Ganesh Kishore, CEO-Malaysian Life Sciences Capital Fund
Tania Fernandez, Dir.-India
James D. Watson, Head-Merchant Banking
Hal Gerber, Managing Dir.-Private Equity
Ann F. Hanham, Managing Dir.-Int'l Group

| Phone: 415-591-5400 | Fax: 415-591-5401 |
|---|---|
| Toll-Free: | |
| Address: 1 Embarcadero Ctr., Ste. 2700, San Francisco, CA 94111 US | |

## GROWTH PLANS/SPECIAL FEATURES:

Burrill & Company is a life-sciences merchant bank focused exclusively on companies involved in biotechnology; pharmaceuticals; drug delivery devices; diagnostics; medical devices; human health care and related medical technologies; nutraceuticals; agricultural biotechnologies; and industrial biomaterials and bioprocesses. The company operates through several business units, including venture capital, merchant banking, private equity, publications and conferences. The venture capital unit manages and offers various funds totaling more than $950 million under management, including the Burrill Life Sciences Capital Fund; Burrill Biotechnology Capital Fund; Burrill Agbio Capital Fund; and Burrill Nutraceuticals Capital Fund. The merchant banking unit assists life science companies in identifying, negotiation and forming strategic partnerships with other companies for access to resources, technologies or collaborations. The unit also works with major life sciences companies to spin off divisions or out-license technologies. The firm's dedicated private equity team focuses on investments in small and midcap public companies, along with select spin-offs and buy-outs from larger life sciences corporations. The publications unit publishes monthly indices on biotech industry stock market performance; quarterly reports that highlight important industry developments such as advancements in science, technology breakthroughs and important business transactions and deals; articles and commentary on the biotechnology industry for various publications; and annual biotechnology industry reports. The conference unit annually hosts and sponsors various industry conferences including the Biotech Meeting at Laguna Beach, the Indiana Life Sciences Forum, the Stem Cell Meeting, the Burrill Personalized Medicine Meeting and the Ageing Meeting. Past conferences hosted by the firm have included the Burrill China Life Sciences Meeting, the Burrill India Life Sciences Meeting and the Japan Biotech Meeting. Additionally, the Burrill International Group concentrates on investment opportunities outside the U.S., with current areas of focus including China, India, Japan, Malaysia, Korea, Australia, New Zealand, Canada, Russia, the Middle East and Eastern Europe.

## FINANCIALS: Sales and profits are in thousands of dollars—add 000 to get the full amount. 2008 Note: Financial information for 2008 was not available for all companies at press time.

| | | |
|---|---|---|
| 2008 Sales: $ | 2008 Profits: $ | **U.S. Stock Ticker: Private** |
| 2007 Sales: $ | 2007 Profits: $ | **Int'l Ticker:**   Int'l Exchange: |
| 2006 Sales: $ | 2006 Profits: $ | Employees:   40 |
| 2005 Sales: $ | 2005 Profits: $ | Fiscal Year Ends: 12/31 |
| 2004 Sales: $ | 2004 Profits: $ | Parent Company: |

## SALARIES/BENEFITS:

| Pension Plan: | ESOP Stock Plan: | Profit Sharing: | Top Exec. Salary: $ | Bonus: $ |
|---|---|---|---|---|
| Savings Plan: | Stock Purch. Plan: | | Second Exec. Salary: $ | Bonus: $ |

## OTHER THOUGHTS:

**Apparent Women Officers or Directors:** 10
**Hot Spot for Advancement for Women/Minorities:** Y

## LOCATIONS: ("Y" = Yes)

| West: | Southwest: | Midwest: | Southeast: | Northeast: | International: |
|---|---|---|---|---|---|
| Y | | | | Y | Y |

# CALIPER LIFE SCIENCES

## www.calipertech.com

Industry Group Code: 325413 Ranks within this company's industry group: Sales: 10 Profits: 20

| Drugs: | Other: | Clinical: | | Computers: | | Services: | |
|--------|--------|-----------|---|------------|---|-----------|---|
| Discovery: | AgriBio: | Trials/Services: | | Hardware: | Y | Specialty Services: | |
| Licensing: | Genetic Data: | Labs: | | Software: | Y | Consulting: | |
| Manufacturing: | Tissue Replacement: | Equipment/Supplies: | Y | Arrays: | Y | Blood Collection: | |
| Genetics: | | Research & Development Services: | | Database Management: | | Drug Delivery: | |
| | | Diagnostics: | | | | Drug Distribution: | |

## TYPES OF BUSINESS:

Bioanalysis Equipment
Microfluidic Systems
High-Throughput Screening Machines
Liquid Handling Systems
Drug Discovery Platforms
Laboratory Automation Solutions
Software

## BRANDS/DIVISIONS/AFFILIATES:

Caliper Technologies
LabChip 90
Lab Chip 3000
NovaScreen Biosciences Corp.
Xenogen Corp.
Caliper Discovery Alliances & Services
Zephyr Genomics Workstation

## CONTACTS: Note: Officers with more than one job title may be intentionally listed here more than once.

E. Kevin Hrusovsky, CEO
E. Kevin Hrusovsky, Pres.
Peter F. McAree, CFO/Sr. VP
Paula J. Cassidy, VP-Human Resources
Bradley W. Rice, Sr. VP-R&D
Stephen E. Creager, General Counsel/Corp. Sec./Sr. VP
Bruce J. Bal, Sr. VP-Oper.
William C. Kruka, Sr. VP-Corp. Dev.
David M. Manyak, Exec. VP-Caliper Discovery Alliances & Svcs.
Enrique Bernal, Sr. VP-In Vitro Bus. Dev.
Mark T. Roskey, VP-Reagents & Applied Biology
Bob Bishop, Chmn.

| Phone: 508-435-9500 | Fax: 508-435-3439 |
|---------------------|---------------------|

Toll-Free: 877-522-2447

Address: 68 Elm St., Hopkinton, MA 01748 US

## GROWTH PLANS/SPECIAL FEATURES:

Caliper Life Sciences uses its core technologies of liquid handling, automation and LabChip microfluidics to foster developments in the life sciences industry. The company manufactures high-throughput screening machines, automated liquid handling machines, micro-plate management, pharmaceutical development and quality control systems. Caliper is best known for its LabChip systems, which are designed to accelerate laboratory experimentation with applicability in the pharmaceutical and diagnostics industries. The firm makes two types of LabChip systems: LabChip 90 and LabChip 3000. LabChip 90, which is designed to meet the high-throughput needs of laboratories, uses microfluidic technology to automate the analysis of proteins and DNA fragments. The LabChip 3000 drug discovery system miniaturizes, integrates and automates enzymatic and cell-based assays even when unattended. LabChip assays are separations-based, so the quality of results exceeds what is achievable in homogeneous, well-based assays. Other products include various plate management; pharmaceutical development; evaporation and solid phase extraction devices; software; and workstations. Caliper Discovery Alliances & Services (CDAS) is the firm's contract research division that combines the biosciences of its two subsidiaries, NovaScreen and Xenogen, both providers of in vitro discovery services. In January 2009, the company launched the Zephyr Genomics Workstation, an improved automated solution liquid handling designed to improve quality and consistency.

Employees are offered medical, dental and vision insurance; life insurance; short-and long-term disability coverage; a 401(k) plan; a stock purchase plan; tuition reimbursement; and an employee assistance plan.

## FINANCIALS: Sales and profits are in thousands of dollars—add 000 to get the full amount. 2008 Note: Financial information for 2008 was not available for all companies at press time.

| | | |
|---|---|---|
| 2008 Sales: $134,054 | 2008 Profits: $-68,292 | U.S. Stock Ticker: CALP |
| 2007 Sales: $140,707 | 2007 Profits: $-24,080 | Int'l Ticker: Int'l Exchange: |
| 2006 Sales: $107,871 | 2006 Profits: $-28,934 | Employees: 489 |
| 2005 Sales: $87,009 | 2005 Profits: $-14,457 | Fiscal Year Ends: 12/31 |
| 2004 Sales: $80,127 | 2004 Profits: $-31,600 | Parent Company: |

## SALARIES/BENEFITS:

| Pension Plan: | ESOP Stock Plan: | Profit Sharing: | Top Exec. Salary: $451,567 | Bonus: $ |
|---------------|------------------|-----------------|----------------------------|----------|
| Savings Plan: Y | Stock Purch. Plan: Y | | Second Exec. Salary: $261,995 | Bonus: $ |

## OTHER THOUGHTS:

Apparent Women Officers or Directors: 2
Hot Spot for Advancement for Women/Minorities: Y

## LOCATIONS: ("Y" = Yes)

| West: | Southwest: | Midwest: | Southeast: | Northeast: | International: |
|-------|-----------|----------|------------|------------|---------------|
| Y | | | | Y | Y |

Note: Financial information, benefits and other data can change quickly and may vary from those stated here.

# CAMBREX CORP

www.cambrex.com

**Industry Group Code:** 325412  **Ranks within this company's industry group:** Sales: 54  Profits: 52

| Drugs: | | Other: | Clinical: | | Computers: | | Services: | |
|---|---|---|---|---|---|---|---|---|
| Discovery: | | AgriBio: | Trials/Services: | Y | Hardware: | | Specialty Services: | |
| Licensing: | | Genetic Data: | Labs: | | Software: | | Consulting: | |
| Manufacturing: | Y | Tissue Replacement: | Equipment/Supplies: | Y | Arrays: | | Blood Collection: | |
| Genetics: | | | Research & Development Services: | Y | Database Management: | | Drug Delivery: | |
| | | | Diagnostics: | | | | Drug Distribution: | |

## TYPES OF BUSINESS:

Contract Pharmaceutical Manufacturing
Contract Research
Pharmaceutical Ingredients
Testing Products & Services
Technical Support

## BRANDS/DIVISIONS/AFFILIATES:

Cambrex Charles City Inc
Cambrex Kariskoga AB
Cambrex Profarmaco
Cambrex Tallinn AS
FlashGel Rapid Electrophoresis System
Platinum UltraPAK

## CONTACTS: *Note: Officers with more than one job title may be intentionally listed here more than once.*

Steven M. Klosk, CEO
Steven M. Klosk, Pres.
Greg Sargen, CFO/VP
Paolo Russolo, Pres., Cambrex Profarmaco
John Miller, Chmn.

| **Phone:** 201-804-3000 | **Fax:** 201-804-9852 |
|---|---|
| **Toll-Free:** | |
| **Address:** 1 Meadowlands Plz., East Rutherford, NJ 07073 US | |

## GROWTH PLANS/SPECIAL FEATURES:

Cambrex Corp. provides products and services to aid and enhance the discovery and commercialization of therapeutics. The company offers a variety of outsourcing products and services for drug discovery research and therapeutic testing. The firm manufactures products, which are sold to research organizations, pharmaceutical, biopharmaceutical and generic drug companies. Outsourcing options include bulk biologics manufacturing; development, manufacturing and commercialization services for cell-based therapeutics and pharmaceutical products; and testing services including assays for microbiology, sterility and veterinary services. Products offered for drug discovery research include bioassays; cell model systems; cell analysis stains; electrophoresis products, including the FlashGel Rapid Electrophoresis System; and protein analysis products. Cambrex offers technical support for all its research products. In addition, Cambrex offers Platinum UltraPAK, a line of flexible packaging systems that can be modified to fit specific customer needs. Therapeutic testing products include a range of endotoxin services and products, including endotoxin detection assays, removal products, testing services accessory products, instrumentation and software. The company also offers testing products using a wide range of assays. Subsidiaries of the firm include Cambrex Charles City, Inc.; Cambrex Karlskoga AB; Cambrex Profarmaco; and Cambrex Tallinn AS. In 2008, Swedish subsidiary Cambrex Kariskoga AB acquired ProSyntest AS, an Estonia-based active pharmaceutical ingredients research and development company. ProSyntest was consequently renamed Cambrex Tallinn. In January 2009, Cambrex signed a joint marketing and development agreement with Skinvisible, Inc. to develop acne products.

Employees are offered medical, dental, vision and life insurance; short-and long-term disability coverage; a 401(k) plan; and tuition reimbursement.

## FINANCIALS: Sales and profits are in thousands of dollars—add 000 to get the full amount. 2008 Note: Financial information for 2008 was not available for all companies at press time.

| | | |
|---|---|---|
| 2008 Sales: $249,618 | 2008 Profits: $7,929 | **U.S. Stock Ticker: CBM** |
| 2007 Sales: $252,574 | 2007 Profits: $209,248 | **Int'l Ticker:**  Int'l Exchange: |
| 2006 Sales: $236,659 | 2006 Profits: $-30,100 | Employees:  856 |
| 2005 Sales: $223,565 | 2005 Profits: $-110,458 | Fiscal Year Ends: 12/31 |
| 2004 Sales: $395,906 | 2004 Profits: $-26,870 | Parent Company: |

## SALARIES/BENEFITS:

| Pension Plan: | ESOP Stock Plan: | Profit Sharing: | Top Exec. Salary: $431,634 | Bonus: $ |
|---|---|---|---|---|
| Savings Plan: Y | Stock Purch. Plan: | | Second Exec. Salary: $382,538 | Bonus: $ |

## OTHER THOUGHTS:

**Apparent Women Officers or Directors:** 2
**Hot Spot for Advancement for Women/Minorities:** Y

## LOCATIONS: ("Y" = Yes)

| West: | Southwest: | Midwest: | Southeast: | Northeast: | International: |
|---|---|---|---|---|---|
| | | Y | | Y | Y |

# CANGENE CORP

www.cangene.com

Industry Group Code: 325412  Ranks within this company's industry group:  Sales: 62  Profits: 46

| Drugs: | | Other: | | Clinical: | | Computers: | | Services: | |
|---|---|---|---|---|---|---|---|---|---|
| Discovery: | Y | AgriBio: | | Trials/Services: | | Hardware: | | Specialty Services: | |
| Licensing: | | Genetic Data: | Y | Labs: | | Software: | | Consulting: | |
| Manufacturing: | Y | Tissue Replacement: | | Equipment/Supplies: | | Arrays: | | Blood Collection: | |
| Genetics: | Y | | | Research & Development Services: | Y | Database Management: | | Drug Delivery: | |
| | | | | Diagnostics: | | | | Drug Distribution: | |

## TYPES OF BUSINESS:

Drugs-Hyperimmunes
Generic Drugs
Contract Manufacturing
Vaccines

## BRANDS/DIVISIONS/AFFILIATES:

WinRho SDF
VariZIG
VIG
HepaGam B
Anthrax Immune Globulin (AIG)
Leucotropin
Accretropin

## CONTACTS: Note: Officers with more than one job title may be intentionally listed here more than once.

John M. Langstaff, CEO
John M. Langstaff, Pres.
Michael Graham, CFO
Paul Brisebois, VP-Mktg. & Sales
Grant McClarty, VP-R&D
John W. McMillan, Corp. Sec.
William Bees, Sr. VP-Oper.
Paul Brisebois, VP-Bus. Dev.
Andrew D. Storey, VP-Quality Assurance
Andrew D. Storey, VP-Clinical & Regulatory Affairs
Jack Kay, Chmn.

| Phone: 204-275-4200 | Fax: 204-269-7003 |
|---|---|

Toll-Free: 800-768-2304
Address: 155 Innovation Dr., Winnipeg, MB R3T 5Y3 Canada

## GROWTH PLANS/SPECIAL FEATURES:

Cangene Corp. is a leading global developer and manufacturer of specialty hyperimmune plasma and biotechnology products. The company offers contract research and process development services; bulk products manufacturing; and finished products manufacturing services to biopharmaceutical companies. Cangene's leading product offering is WinRho SDF, a hyperimmune used to prevent hemolytic disease in newborns and to treat immune thrombocytopenic purpura (ITP), a clotting disorder. The company's other approved hyperimmune products include VariZIG, a chicken pox vaccine, developed to prevent chicken pox in pregnant women; VIG (Vaccinia Immune Globulin), an antibody product used to prep recipients for or to treat severe reactions to the smallpox vaccine; Accretropin Injection, a recombinant human growth hormone; and HepaGam B, a specialized antibody for treatment following acute exposure to the hepatitis B virus. Cangene's products currently under development include Botulism Antitoxin, a product containing neutralizing antibodies to the seven botulinum types; Anthrax Immune Globulin (AIG), an adjunct to antibiotic therapy in critically ill patients with anthrax; and Leucotropin, a recombinant version of a protein that stimulates the production of certain white blood cells, for the treatment of acute radiation syndrome. The company offers contract manufacturing services in the areas of process development, bulk product manufacturing and finished product manufacturing. Services include fermentation and purification process development, optimization and scale-up; formulation development; lyophilization cycle development; cGMP manufacturing; labeling and packaging; quality control testing; stability studies; and regulatory support.

## FINANCIALS: Sales and profits are in thousands of dollars—add 000 to get the full amount. 2008 Note: Financial information for 2008 was not available for all companies at press time.

| | | |
|---|---|---|
| 2008 Sales: $152,800 | 2008 Profits: $27,300 | U.S. Stock Ticker: |
| 2007 Sales: $85,000 | 2007 Profits: $9,300 | Int'l Ticker: CNJ    Int'l Exchange: Toronto-TSX |
| 2006 Sales: $100,600 | 2006 Profits: $12,100 | Employees:  650 |
| 2005 Sales: $90,620 | 2005 Profits: $-13,638 | Fiscal Year Ends: 7/31 |
| 2004 Sales: $118,000 | 2004 Profits: $24,500 | Parent Company: |

## SALARIES/BENEFITS:

| Pension Plan: | ESOP Stock Plan: | Profit Sharing: | Top Exec. Salary: $ | Bonus: $ |
|---|---|---|---|---|
| Savings Plan: | Stock Purch. Plan: | | Second Exec. Salary: $ | Bonus: $ |

## OTHER THOUGHTS:

**Apparent Women Officers or Directors**: 1
**Hot Spot for Advancement for Women/Minorities**:

## LOCATIONS: ("Y" = Yes)

| West: | Southwest: | Midwest: | Southeast: | Northeast: | International: |
|---|---|---|---|---|---|
| Y | | | Y | Y | Y |

# CANREG INC

**www.canreginc.com**

Industry Group Code: 541690 **Ranks within this company's industry group:** Sales:     Profits:

| Drugs: | Other: | Clinical: | Computers: | Services: | |
|---|---|---|---|---|---|
| Discovery: | AgriBio: | Trials/Services: | Hardware: | Specialty Services: | Y |
| Licensing: | Genetic Data: | Labs: | Software: | Consulting: | Y |
| Manufacturing: | Tissue Replacement: | Equipment/Supplies: | Arrays: | Blood Collection: | |
| Genetics: | | Research & Development Services: | Database Management: | Drug Delivery: | |
| | | Diagnostics: | | Drug Distribution: | |

## TYPES OF BUSINESS:

Regulatory Affairs Consulting

## BRANDS/DIVISIONS/AFFILIATES:

## CONTACTS: Note: Officers with more than one job title may be intentionally listed here more than once.

Anne Tomalin, Pres.
Lynda Rattenbury, Exec. Dir.-Human Resources
Harold DeVenne, VP-Oper.
Darrell Ethell, Exec Dir.-Bus. Dev.
Lynda Rattenbury, Exec. Dir.-Comm. Affairs
Harold DeVenne, VP-Finance
Patricia Anderson, VP-Regulatory Svcs.
Mary Speagle, Exec. Dir.-Canadian Regulatory Affairs
Stuart Wright, Exec. Dir.-CMC & Quality
Bernard Chiasson, Exec. Dir.-U.S. Bus. & Medical Svcs.

| Phone: 905-689-3980 | Fax: 905-689-1465 |
|---|---|
| Toll-Free: 866-722-6734 | |
| Address: 4 Innovation Drive, Dundas, ON L9H 7P3 Canada | |

## GROWTH PLANS/SPECIAL FEATURES:

Canreg, Inc. is a regulatory affairs consulting company, focusing on the pharmaceutical, biotechnology and medical device industries. Other clients include venture capitalists, contract research organizations, natural health product and food companies, cosmetic manufacturers and governments. The firm's more than 100 consultants and staff serve companies at all levels in Canada, the U.S. and Europe, serving either as the global regulatory affairs department or simply augmenting internal staff. The firm's Regulatory Operations Team works with the FDA, Health Canada and European regulatory agencies in both electronic and paper form. In Canada, Canreg provides New Drug Submissions, Abbreviated New Drug Submissions, Clinical Trial Applications, DIN applications, Medical Device License Applications, Natural Health Product (NHP) Submissions, Notifiable Changes, Common Drug Review strategy and submissions, Provincial and Private Payer Formulary submissions and Patented Medicine Prices Review Board reporting, as well as a Health Canada Liaison and Canadian Agent services. For the U.S., Canreg provides Investigational New Drug Applications, New Drug Applications, Abbreviated New Drug Applications, Biologic License Applications, Orphan Drug Applications and Medical Device Applications, as well as a liaison with the FDA, U.S. Agent service and structured product labeling. In Europe, the firm offers Marketing Authorization applications (centralized, de-centralized or national), Clinical Trial Submissions and Orphan Drug applications, as well as agency liaisons and European Agent services. Globally, the company provides regulatory assessments and strategy development, quality and compliance services, clinical compliance services, pharmacovigilance, medical writing, due diligence, regulatory operations (electronic and paper) and training.

## FINANCIALS: Sales and profits are in thousands of dollars—add 000 to get the full amount. 2008 Note: Financial information for 2008 was not available for all companies at press time.

| | | |
|---|---|---|
| 2008 Sales: $ | 2008 Profits: $ | U.S. Stock Ticker: Private |
| 2007 Sales: $ | 2007 Profits: $ | Int'l Ticker:     Int'l Exchange: |
| 2006 Sales: $ | 2006 Profits: $ | Employees:   100 |
| 2005 Sales: $ | 2005 Profits: $ | Fiscal Year Ends: |
| 2004 Sales: $ | 2004 Profits: $ | Parent Company: |

## SALARIES/BENEFITS:

| | | | | |
|---|---|---|---|---|
| Pension Plan: | ESOP Stock Plan: | Profit Sharing: | Top Exec. Salary: $ | Bonus: $ |
| Savings Plan: | Stock Purch. Plan: | | Second Exec. Salary: $ | Bonus: $ |

## OTHER THOUGHTS:

**Apparent Women Officers or Directors**: 4
**Hot Spot for Advancement for Women/Minorities**: Y

## LOCATIONS: ("Y" = Yes)

| West: | Southwest: | Midwest: | Southeast: | Northeast: | International: |
|---|---|---|---|---|---|
| | | Y | | | Y |

# CARACO PHARMACEUTICAL LABORATORIES     www.caraco.com

**Industry Group Code: 325412A  Ranks within this company's industry group:** Sales: 9  Profits: 7

| Drugs: | | Other: | Clinical: | Computers: | Services: |
|---|---|---|---|---|---|
| Discovery: | Y | AgriBio: | Trials/Services: | Hardware: | Specialty Services: |
| Licensing: | | Genetic Data: | Labs: | Software: | Consulting: |
| Manufacturing: | Y | Tissue Replacement: | Equipment/Supplies: | Arrays: | Blood Collection: |
| Genetics: | Y | | Research & Development Services: | Database Management: | Drug Delivery: |
| | | | Diagnostics: | | Drug Distribution: |

## TYPES OF BUSINESS:

Drugs-Generic

## BRANDS/DIVISIONS/AFFILIATES:

Sun Pharmaceutical Industries
Allopurinol
Baclofen
Carbamazepine
Fluvoxamine
Hydrochlorothiazide
Midrin
Topiramate

## CONTACTS: *Note: Officers with more than one job title may be intentionally listed here more than once.*

Daniel H. Movens, CEO
Mukul Rathi, Interim CFO
Thomas Larkin, Dir.-Mktg.
Tammy Bitterman, Dir.-Human Resources
Kaushikkumar Gandhi, VP-Mfg.
Gurpartap Singh Sachdeva, Sr. VP-Bus. Strategies
Robert Kurkiewicz, Sr. VP-Regulatory Affairs
Jayesh Shah, Dir.-Commercial
Daniel Barone, Dir.-Quality
David Risk, Dir.-Bus. Dev.
Dilip S. Shanghvi, Chmn.

| **Phone:** 313-871-8400 | **Fax:** 313-871-8314 |
|---|---|
| **Toll-Free:** 800-818-4555 | |
| **Address:** 1150 Elijah McCoy Dr., Detroit, MI 48202 US | |

## GROWTH PLANS/SPECIAL FEATURES:

Caraco Pharmaceutical Laboratories, Ltd., develops, manufactures and markets generic and private-label drugs for prescription and over-the-counter markets. The company's product portfolio includes 52 products in 114 strengths and various package sizes. These drugs relate to a variety of therapeutic segments, including arthritis, pain control, epilepsy, diabetes, antipsychotic and neurological disorders. Pharmaceutical products that the company produces include Allopurinol, and anti-gout medication; Baclofen, a skeletal muscle relaxant; Carbamazepine, a chewable anticonvulsant; Fluvoxamine, an antidepressant; Hydrochlorothiazide, an antihypertensive; and Midrin, a vascular and migraine headache suppressant. The company also has several drugs awaiting FDA approval. The company has collaborative agreements with several companies, with the most prominent being Sun Pharmaceutical Industries (Sun Pharma), the majority stock holder of Caraco. Under these agreements, Caraco develops generic drugs for each company to market as its own brand. The firm distributes its products through wholesalers, chain drug stores, retail pharmacies, mail-order companies, managed care organizations, hospital groups and nursing homes. Some of the wholesalers that distribute Caraco's products include Amerisource-Bergen Corporation, McKesson Corporation and Cardinal Health. In early 2008, Sun Pharma Global, Inc., a wholly-owned subsidary of Sun Pharma, completed its agreement with Caraco to transfer 25 products to the firm. Also in 2008, the company released several new products on behalf of Sun Pharma, including generic Depakote delayed release tablets for seizures and other conditions; and generic Sinemet, called Carbidopa and Levodopa, for Parkinson's disease. In April and May 2009, the company launched Topiramate tablets (generic Topamax) for seizures and generic Roxicodone tablets for moderate to severe pain; these were also released on behalf of Sun Pharma.

Caraco offers employees benefits which include medical, dental, and vision care; paid time off and holiday pay; a 401(k); life and disability insurance; and health and dependent care accounts.

## FINANCIALS: Sales and profits are in thousands of dollars—add 000 to get the full amount. 2008 Note: Financial information for 2008 was not available for all companies at press time.

| | | |
|---|---|---|
| 2008 Sales: $350,367 | 2008 Profits: $35,388 | **U.S. Stock Ticker: CPD** |
| 2007 Sales: $117,027 | 2007 Profits: $26,858 | **Int'l Ticker:** Int'l Exchange: |
| 2006 Sales: $82,789 | 2006 Profits: $-10,423 | Employees: 662 |
| 2005 Sales: $64,116 | 2005 Profits: $-2,278 | Fiscal Year Ends: 3/31 |
| 2004 Sales: $60,340 | 2004 Profits: $- 199 | Parent Company: SUN PHARMACEUTICAL INDUSTRIES LTD |

## SALARIES/BENEFITS:

| Pension Plan: | ESOP Stock Plan: | Profit Sharing: | Top Exec. Salary: $427,596 | Bonus: $150,478 |
|---|---|---|---|---|
| Savings Plan: Y | Stock Purch. Plan: | | Second Exec. Salary: $224,345 | Bonus: $51,660 |

## OTHER THOUGHTS:

**Apparent Women Officers or Directors:** 1
**Hot Spot for Advancement for Women/Minorities:** Y

## LOCATIONS: ("Y" = Yes)

| West: | Southwest: | Midwest: | Southeast: | Northeast: | International: |
|---|---|---|---|---|---|
| | | Y | | | |

Note: Financial information, benefits and other data can change quickly and may vary from those stated here.

# CARDIOME PHARMA CORP

## www.cardiome.com

**Industry Group Code:** 325412   **Ranks within this company's industry group:** Sales: 143   Profits: 142

| Drugs: | | Other: | Clinical: | Computers: | Services: |
|---|---|---|---|---|---|
| Discovery: | Y | AgriBio: | Trials/Services: | Hardware: | Specialty Services: |
| Licensing: | | Genetic Data: | Labs: | Software: | Consulting: |
| Manufacturing: | | Tissue Replacement: | Equipment/Supplies: | Arrays: | Blood Collection: |
| Genetics: | | | Research & Development Services: | Database Management: | Drug Delivery: |
| | | | Diagnostics: | | Drug Distribution: |

## TYPES OF BUSINESS:

Drugs-Cardiac

## BRANDS/DIVISIONS/AFFILIATES:

Astellas Pharma US, Inc.
vernakalant (oral)
vernakalant (iv)
KYNAPID

## CONTACTS: Note: Officers with more than one job title may be intentionally listed here more than once.

Bob Rieder, CEO
Doug Janzen, Pres.
Curtis Sikorsky, CFO
Karim Lalji, Sr. VP-Commercial Affairs
Donald A. McAfee, Chief Scientific Officer
Sheila M. Grant, VP-Prod. Dev.
Taryn Boivin, VP-Mfg. & Pharmaceutical Sciences
Doug Janzen, Chief Bus. Officer
Peter Hofman, Sr. Dir.-Investor Rel.
Charles Fisher, Chief Medical Officer/Exec. VP-Clinical Affairs
Guy F. Cipriani, VP-Bus. Dev.
Gregory N. Beatch, VP-Scientific Affairs
Bob Rieder, Chmn.

| **Phone:** 604-677-6905 | **Fax:** 604-677-6915 |
|---|---|
| **Toll-Free:** 800-330-9928 | |
| **Address:** 6190 Agronomy Rd., 6th Fl., Vancouver, BC V6T 1Z3 Canada | |

## GROWTH PLANS/SPECIAL FEATURES:

Cardiome Pharma Corp. is a drug discovery and development company that is focused on the development of proprietary drugs for the treatment and prevention of cardiovascular diseases. The company currently has two late-stage clinical drug programs for atrial arrhythmia: a Phase I program for GED-aPC, an engineered analog of recombinant human activated Protein C; and a pre-clinical program that specifically focuses on the improvement of cardiovascular function. Cardiome's main product focus is on Vernakalant (formerly known as RSD1235), which was created for the acute conversion of atrial fibrillation to a normal heart rhythm. The drug is being developed in two forms: the intravenous vernakalant (iv) and an oral drug formulation, vernakalant (oral). The drug works by selectively blocking ion channels in the heart that become active during episodes of atrial fibrillation. Vernakalant is also being considered as a chronic-use oral drug for the maintenance of normal heart rhythm following atrial fibrillation. In recent years, the firm and its co-development partner, Astellas Pharma US, Inc., completed the Phase III clinical study of the efficacy and safety of vernakalant (iv) for patients with post-operative atrial arrhythmia. Currently, vernakalant (iv) is in an ongoing Phase III comparator study; KYNAPID is the proposed North American brand name for the intravenous treatment. In July 2008, the firm announced positive clinical results from the Phase 2b clinical study of vernakalant (oral) in humans. In May 2009, the firm sold the exclusive global commercialization rights to vernakalant (oral) to Merck & Co., Inc., and sold rights to the vernakalant (iv) outside of North America to Merck Sharp & Dohme (Switzerland) GmbH.

Cardiome offers employees medical and dental benefits; an employee incentive stock option program; professional development support; health and wellness resources; and activities and social events that are organized by an interdepartmental social committee.

## FINANCIALS: Sales and profits are in thousands of dollars—add 000 to get the full amount. 2008 Note: Financial information for 2008 was not available for all companies at press time.

| | | |
|---|---|---|
| 2008 Sales: $1,500 | 2008 Profits: $-55,700 | **U.S. Stock Ticker: CRME** |
| 2007 Sales: $4,500 | 2007 Profits: $-78,600 | **Int'l Ticker: COM**   Int'l Exchange: Toronto-TSX |
| 2006 Sales: $19,000 | 2006 Profits: $-33,300 | Employees:   104 |
| 2005 Sales: $15,800 | 2005 Profits: $-52,300 | Fiscal Year Ends: 12/31 |
| 2004 Sales: $22,000 | 2004 Profits: $-23,100 | Parent Company: |

## SALARIES/BENEFITS:

| Pension Plan: | ESOP Stock Plan: Y | Profit Sharing: | Top Exec. Salary: $555,023 | Bonus: $ |
|---|---|---|---|---|
| Savings Plan: | Stock Purch. Plan: | | Second Exec. Salary: $445,373 | Bonus: $ |

## OTHER THOUGHTS:

**Apparent Women Officers or Directors:** 2
**Hot Spot for Advancement for Women/Minorities:** Y

## LOCATIONS: ("Y" = Yes)

| West: | Southwest: | Midwest: | Southeast: | Northeast: | International: |
|---|---|---|---|---|---|
| | | | | | Y |

# CARDIUM THERAPEUTICS INC

**www.cardiumthx.com**

**Industry Group Code: 325412 Ranks within this company's industry group: Sales: 140 Profits: 114**

| Drugs: | Other: | | Clinical: | | Computers: | | Services: | |
|---|---|---|---|---|---|---|---|---|
| Discovery: | AgriBio: | | Trials/Services: | | Hardware: | | Specialty Services: | |
| Licensing: | Genetic Data: | Y | Labs: | | Software: | | Consulting: | |
| Manufacturing: | Tissue Replacement: | | Equipment/Supplies: | Y | Arrays: | | Blood Collection: | |
| Genetics: | | | Research & Development Services: | Y | Database Management: | | Drug Delivery: | |
| | | | Diagnostics: | | | | Drug Distribution: | |

## TYPES OF BUSINESS:
Biologic Therapeutics & Medical Devices

## BRANDS/DIVISIONS/AFFILIATES:
Generx
Corgentin
Celsius Control System
Excellerate
Cardium Biologics
InnerCool Therapeutics
Tissue Repair Co.
UroCool

## CONTACTS: Note: Officers with more than one job title may be intentionally listed here more than once.
Christopher J. Reinhard, CEO
Christopher J. Reinhard, Pres.
Dennis M. Mulroy, CFO
Gabor M. Rubanyi, Chief Scientific Officer
Ted Williams, VP-Tech. Oper.
Ted Williams, VP-Mfg.
Tyler M. Dylan, General Counsel/Exec. VP/Sec./Chief Bus. Officer
Mark McCutchen, VP-Bus. Dev.
Bonnie Ortega, Dir.-Public Rel.
Bonnie Ortega, Dir.-Investor Rel.
Christopher J. Reinhard, Treas.
Robert L. Engler, Chief Medical Advisor
Barabara K. Sosnowski, VP-Biologics Dev.
Michael L. Magers, Pres./COO-Innercool Therapies
Christopher J. Reinhard, Chmn.

| Phone: 858-436-1000 | Fax: 858-436-1001 |
|---|---|
| Toll-Free: | |
| Address: 12255 El Camino Real, Ste. 250, San Diego, CA 92130 US | |

## GROWTH PLANS/SPECIAL FEATURES:
Cardium Therapeutics, Inc. is a medical technology company primarily focused on the development and commercialization of medical devices and biologic therapeutics for the treatment of cardiovascular and ischemic diseases. The company's products are divided between three companies: Cardium Biologics, and subsidiaries InnerCool Therapeutics and Tissue Repair Company. Product candidates for Cardium Biologics include Generx, a late-stage DNA-based growth factor therapeutic that is being developed to promote and stimulate the growth of collateral circulation in the hearts of patients with ischemic conditions such as recurrent angina; and Corgentin, a DNA-based therapeutic being developed to enhance myocardial healing in and around the infarct zone when used as an adjunct to existing vascular-directed pharmacologic and interventional therapies. The FDA has granted fast track designation to Generx for the potential treatment of myocardial ischemia. InnerCool Therapies focuses on therapeutic hypothermia or patient temperature modulation. Its Celsius Control System, a rapid endovascular-based system, is used for fever reduction and the induction, maintenance and reversal of mild hypothermia in neurosurgery and cardiac surgery patients. This subsidiary also offers the RapidBlue system for high-performance endovascular temperature modulation, which includes a programmable console with an enhanced user interface (CoolBlue) and the Accutrol catheter, which is designed to quickly modulate patient temperature in association with surgery or other medical procedures. The Tissue Repair Company's lead product candidate, Excellarate, is a DNA-activated collagen gel that utilizes the company's Gene Activated Matrix Technology and is formulated with an adenovector delivery carrier; it is being developed for the treatment of chronic diabetic wounds, such as non-healing diabetic foot ulcers. In October 2008, the firm received 510(k) clearance from the U.S. FDA to market RapidBlue. In November 2008, InnerCool developed UroCool, a pelvic cooling catheter system designed to induce localized cooling during prostate cancer surgery.

## FINANCIALS: Sales and profits are in thousands of dollars—add 000 to get the full amount. 2008 Note: Financial information for 2008 was not available for all companies at press time.

| | | |
|---|---|---|
| 2008 Sales: $2,417 | 2008 Profits: $-24,598 | **U.S. Stock Ticker: CXM** |
| 2007 Sales: $1,587 | 2007 Profits: $-25,322 | **Int'l Ticker:** Int'l Exchange: |
| 2006 Sales: $ 756 | 2006 Profits: $-18,593 | Employees: 43 |
| 2005 Sales: $ | 2005 Profits: $-5,442 | Fiscal Year Ends: 12/31 |
| 2004 Sales: $ | 2004 Profits: $ | Parent Company: |

## SALARIES/BENEFITS:

| Pension Plan: | ESOP Stock Plan: | Profit Sharing: | Top Exec. Salary: $359,811 | Bonus: $ |
|---|---|---|---|---|
| Savings Plan: | Stock Purch. Plan: | | Second Exec. Salary: $334,536 | Bonus: $ |

## OTHER THOUGHTS:
**Apparent Women Officers or Directors:** 2
**Hot Spot for Advancement for Women/Minorities:** Y

## LOCATIONS: ("Y" = Yes)

| West: | Southwest: | Midwest: | Southeast: | Northeast: | International: |
|---|---|---|---|---|---|
| Y | | | | | |

Note: Financial information, benefits and other data can change quickly and may vary from those stated here.

# CEGEDIM SA

**www.cegedim.fr**

Industry Group Code: 511210D  **Ranks within this company's industry group:** Sales: 1  Profits: 1

| Drugs: | Other: | Clinical: | | Computers: | | Services: | |
|---|---|---|---|---|---|---|---|
| Discovery: | AgriBio: | Trials/Services: | Y | Hardware: | | Specialty Services: | |
| Licensing: | Genetic Data: | Labs: | | Software: | Y | Consulting: | Y |
| Manufacturing: | Tissue Replacement: | Equipment/Supplies: | | Arrays: | | Blood Collection: | |
| Genetics: | | Research & Development Services: | | Database Management: | Y | Drug Delivery: | |
| | | Diagnostics: | | | | Drug Distribution: | |

## TYPES OF BUSINESS:

Software - Development & Marketing
Databases
Pharmaceuticals Marketing & Tracking Software
Compliance Services

## BRANDS/DIVISIONS/AFFILIATES:

Cegedim Dendrite
Dendrite International Inc
Itops
Cegedim Strategic Data
Cegers
Infopharm
InfoSante
Santesurf

## CONTACTS: *Note: Officers with more than one job title may be intentionally listed here more than once.*

Jean-Claude Labrune, CEO
Pierre Marucchi, Managing Dir.
Laurent Labrune, Chmn./Managing Dir.-Cegedim SRH
Jean-Claude Labrune, Chmn.

| Phone: 33-1-49-09-22-00 | Fax: 33-1-46-03-45-95 |
|---|---|
| Toll-Free: | |
| Address: 127-137, rue d'Aguesseau, Boulogne-Billancourt, 92641 France | |

## GROWTH PLANS/SPECIAL FEATURES:

Cegedim S.A., formerly Dendrite International, Inc., is a developer of databases and software solutions for pharmaceutical companies, healthcare professionals, health insurance providers and France-based companies in virtually any market sector. The company operates through two divisions: the healthcare and strategic data division, which generates roughly 90% of its revenue; and the technologies and services division, which generates roughly 10% of its revenue. Cegedim's healthcare and strategic data division serves pharmaceutical companies in over 80 countries, healthcare professionals primarily located in Europe and health insurers. The company develops databases that provide pharmaceutical companies with information regarding where their drugs are sold, who prescribes them and why their drugs are prescribed. Additional healthcare products and services include tools for optimizing information resources; report and analysis tools for office and hospital sales forces; strategic marketing, operational marketing and competition monitoring tools and studies; performance measurement and promotional spending auditing tools; and pharmacy order-taking tools. Cegedim offers customer relationship management (CRM) services for medical representatives through Cegedim Dendrite. Sales force optimization services are offered through subsidiary Itops; market research studies are offered through subsidiary Cegedim Strategic Data; sales statistics are offered through subsidiaries Cegers, Infopharm and InfoSante; and prescription analysis is offered through Cegedim Customer Information. Software products for pharmacists are offered through subsidiaries Alliadis and Cegedim Rx. Subsidiary Santesurf provides the only free and secure French intranet exclusively for healthcare professionals. Cegedim's technologies and services division provides outsourced human resources management, Internet hosting services, medical financial leasing, corporate databases, printing services and sample management.

## FINANCIALS: Sales and profits are in thousands of dollars—add 000 to get the full amount. 2008 Note: Financial information for 2008 was not available for all companies at press time.

| | | U.S. Stock Ticker: |
|---|---|---|
| 2008 Sales: $1,121,330 | 2008 Profits: $44,250 | Int'l Ticker: CGM    Int'l Exchange: Paris-Euronext |
| 2007 Sales: $995,010 | 2007 Profits: $58,340 | Employees: 2,534 |
| 2006 Sales: $868,600 | 2006 Profits: $60,700 | Fiscal Year Ends: 12/31 |
| 2005 Sales: $437,240 | 2005 Profits: $21,447 | Parent Company: |
| 2004 Sales: $399,197 | 2004 Profits: $29,565 | |

## SALARIES/BENEFITS:

| Pension Plan: | ESOP Stock Plan: | Profit Sharing: | Top Exec. Salary: $522,917 | Bonus: $ |
|---|---|---|---|---|
| Savings Plan: | Stock Purch. Plan: | | Second Exec. Salary: $509,760 | Bonus: $ |

## OTHER THOUGHTS:

**Apparent Women Officers or Directors:**
**Hot Spot for Advancement for Women/Minorities:**

## LOCATIONS: ("Y" = Yes)

| West: | Southwest: | Midwest: | Southeast: | Northeast: | International: |
|---|---|---|---|---|---|
| | | | | Y | Y |

# CELERA CORPORATION

**www.celera.com**

Industry Group Code: 541712  Ranks within this company's industry group: Sales: 9   Profits: 16

| Drugs: | Other: | | Clinical: | | Computers: | | Services: | |
|---|---|---|---|---|---|---|---|---|
| Discovery: | AgriBio: | | Trials/Services: | | Hardware: | | Specialty Services: | |
| Licensing: | Genetic Data: | Y | Labs: | | Software: | Y | Consulting: | |
| Manufacturing: | Tissue Replacement: | Y | Equipment/Supplies: | | Arrays: | | Blood Collection: | |
| Genetics: | | | Research & Development Services: | Y | Database Management: | Y | Drug Delivery: | |
| | | | Diagnostics: | | | | Drug Distribution: | |

## TYPES OF BUSINESS:

Research-Human Genome Mapping
Information Management & Analysis Software
Consulting, Research & Development Services

## BRANDS/DIVISIONS/AFFILIATES:

Applera Corporation
Applied Biosystems Group
Abbott Laboratories
ViroSeq HIV-1 Genotyping System
Berkeley HeartLab, Inc.
Atria Genetics, Inc.
4MyHeart.com

## CONTACTS: *Note: Officers with more than one job title may be intentionally listed here more than once.*

Kathy Ordonez, Pres.
Joel Jung, CFO
Paul Arata, VP-Human Resources
Thomas White, Chief Scientific Officer
Paul Arata, VP-Admin.
Scott Milsten, General Counsel/VP/Corp. Sec.
Stacey Sias, Chief Bus. Officer
David P. Speechly, VP-Corp. Affairs
Samuel Broder, Chief Medical Officer
Russell Warnick, Chief Scientific Officer-Berkeley HeartLab, Inc.
Christopher Hall, Chief Bus. Officer-Berkeley HeartLab, Inc.
Mike Zoccoli, Gen. Mgr.-Products Bus.

| Phone: 510-749-4200 | Fax: |
|---|---|
| Toll-Free: | |
| Address: 1401 Harbor Bay Parkway, Alameda, CA 94502 US | |

## GROWTH PLANS/SPECIAL FEATURES:

Celera Corporation, formerly Celera Genomics Group, is primarily a molecular diagnostics business that uses proprietary genomics and proteomics discovery platforms to identify and validate novel diagnostic markers. In July 2008, Celera separated from Applera Corporation, becoming an independent, publicly-traded company. Celera pursues a strategy that it refers to as targeted medicine, which is based on the belief that understanding the genetic basis of biology and disease is crucial to improving the diagnosis and treatment of many common complex diseases. The company develops products that facilitate disease detection, prediction of disease predisposition, monitoring of disease progression and determination of patient responsiveness to treatments. Its subsidiary, Berkeley HeartLab, offers innovative services to predict the risk of cardiovascular disease and to optimize patient management. Through 4myheart.com, Celera offers cardiovascular patients and physicians a web-based program that makes cardiovascular health records and lab results easily accessible to physicians and patients, among other services. The company maintains a strategic alliance with Abbott Laboratories for the development and commercialization of molecular, or nucleic acid-based, diagnostic products. These products currently include ViroSeq HIV-1 Genotyping System; products used for the detection of mutations in the CFTR gene, which cause cystic fibrosis; hepatitis C virus ASRs; ASRs for the detection of mutations in the FMR-1 gene, which cause Fragile X Syndrome; and ASRs for the detection of mutations in genes known to be involved in deep vein thrombosis. Through its genomics and proteomics research efforts, Celera additionally discovers and validates therapeutic targets, and it is seeking strategic partnerships to develop therapeutic products based on these targets.

## FINANCIALS:  Sales and profits are in thousands of dollars—add 000 to get the full amount. 2008 Note: Financial information for 2008 was not available for all companies at press time.

| | | | |
|---|---|---|---|
| 2008 Sales: $138,700 | 2008 Profits: $-104,100 | U.S. Stock Ticker: CRA | |
| 2007 Sales: $43,400 | 2007 Profits: $-20,600 | Int'l Ticker: | Int'l Exchange: |
| 2006 Sales: $46,200 | 2006 Profits: $-63,600 | Employees:   554 | |
| 2005 Sales: $31,000 | 2005 Profits: $-77,100 | Fiscal Year Ends: 6/30 | |
| 2004 Sales: $60,100 | 2004 Profits: $-57,500 | Parent Company: | |

## SALARIES/BENEFITS:

| Pension Plan: | ESOP Stock Plan: | Profit Sharing: | Top Exec. Salary: $650,000 | Bonus: $338,254 |
|---|---|---|---|---|
| Savings Plan: Y | Stock Purch. Plan: Y | | Second Exec. Salary: $428,000 | Bonus: $147,802 |

## OTHER THOUGHTS:

**Apparent Women Officers or Directors:** 3
**Hot Spot for Advancement for Women/Minorities:** Y

## LOCATIONS: ("Y" = Yes)

| West: | Southwest: | Midwest: | Southeast: | Northeast: | International: |
|---|---|---|---|---|---|
| Y | | | | Y | |

# CELGENE CORP

**www.celgene.com**

**Industry Group Code:** 325412  **Ranks within this company's industry group:** Sales: 31   Profits: 174

| Drugs: | | Other: | | Clinical: | | Computers: | | Services: | |
|---|---|---|---|---|---|---|---|---|---|
| Discovery: | Y | AgriBio: | | Trials/Services: | | Hardware: | | Specialty Services: | |
| Licensing: | Y | Genetic Data: | | Labs: | | Software: | | Consulting: | |
| Manufacturing: | | Tissue Replacement: | | Equipment/Supplies: | | Arrays: | | Blood Collection: | |
| Genetics: | | | | Research & Development Services: | Y | Database Management: | | Drug Delivery: | |
| | | | | Diagnostics: | | | | Drug Distribution: | |

## TYPES OF BUSINESS:

Cancer & Immune-Inflammatory Related Diseases  Drugs

## BRANDS/DIVISIONS/AFFILIATES:

REVLIMID
THALOMID
ALKERAN
FOCALIN
FOCALIN XR
RITALIN
VIDAZA
Pharmion Corp

## CONTACTS: *Note: Officers with more than one job title may be intentionally listed here more than once.*

Sol J. Barer, CEO
Robert J. Hugin, COO
Robert J. Hugin, Pres.
David W. Gryska, CFO/Sr. VP
Graham Burton, Sr. VP-Regulatory Affairs & Pharmacovigilance
Sol J. Barer, Chmn.
Aart Brouwer, Pres., Int'l

| Phone: 908-673-9000 | Fax: 732-271-4184 |
|---|---|
| Toll-Free: | |
| Address: 86 Morris Ave., Summit, NJ 07901 US | |

## GROWTH PLANS/SPECIAL FEATURES:

Celgene Corp. is a global integrated biopharmaceutical company primarily engaged in the discovery, development and commercialization of therapies designed to treat cancer and immune-inflammatory related diseases. The company's commercial stage products are Revlimid, Thalomid and Vidaza. Revlimid has been approved by the U.S. FDA, the European Commission (EC), the Swiss Agency for Therapeutic Products (Swissmedic) and the Australian Therapeutic Goods Administration for treatment in combination with dexamethasone for multiple myeloma patients who have received at least one prior therapy. In addition, Revlimid has been approved by the FDA and the Canadian Therapeutics Directorate for treatment of patients with transfusion-dependent anemia due to low- or intermediate-1-risk myelodysplastic syndromes (MDS) associated with a deletion 5q cytogenetic abnormality with or without additional cytogenetic abnormalities. Thalomid has been approved by the FDA for treatment in combination with dexamethasone of patients with newly diagnosed multiple myeloma and is also approved for the treatment and suppression of cutaneous manifestations of erythema nodosum leprosum (ENL), an inflammatory complication of leprosy. Vidaza, which was integrated into the Celgene lineup after their 2008 acquisition of Pharmion Corp., is approved for treatment in patients with various myelodysplastic syndrome subtypes. Celgene also sells Alkeran, which it obtains through a supply and distribution agreement with GlaxoSmithKline (GSK), and Focalin, which it sells exclusively to Novartis Pharma AG. Other sources of revenue include royalties which the company receives primarily from Novartis on its sales of the entire family of Ritalin drugs and Focalin XR, in addition to revenues from collaborative agreements and licensing fees. Its portfolio of drug candidates includes IMiDs compounds, which are proprietary to the firm and have demonstrated certain immunomodulatory and other biologically important properties.

Celgene offers its employees educational assistance; travel assistance; and medical, dental, vision, life, AD&D, business travel accident and disability insurance.

## FINANCIALS: Sales and profits are in thousands of dollars—add 000 to get the full amount. 2008 Note: Financial information for 2008 was not available for all companies at press time.

| | | |
|---|---|---|
| 2008 Sales: $2,254,781 | 2008 Profits: $-1,533,653 | **U.S. Stock Ticker:** CELG |
| 2007 Sales: $1,405,820 | 2007 Profits: $226,433 | **Int'l Ticker:**   Int'l Exchange: |
| 2006 Sales: $898,873 | 2006 Profits: $68,981 | Employees:  2,441 |
| 2005 Sales: $536,941 | 2005 Profits: $63,656 | Fiscal Year Ends: 12/31 |
| 2004 Sales: $377,502 | 2004 Profits: $52,756 | Parent Company: |

## SALARIES/BENEFITS:

| Pension Plan: | ESOP Stock Plan: | Profit Sharing: | Top Exec. Salary: $939,000 | Bonus: $2,166,955 |
|---|---|---|---|---|
| Savings Plan: Y | Stock Purch. Plan: | | Second Exec. Salary: $733,333 | Bonus: $1,571,730 |

## OTHER THOUGHTS:

**Apparent Women Officers or Directors**: 1
**Hot Spot for Advancement for Women/Minorities**:

## LOCATIONS: ("Y" = Yes)

| West: | Southwest: | Midwest: | Southeast: | Northeast: | International: |
|---|---|---|---|---|---|
| | | | | Y | Y |

Note: Financial information, benefits and other data can change quickly and may vary from those stated here.

# CELL GENESYS INC

### www.cellgenesys.com

**Industry Group Code:** 325412 **Ranks within this company's industry group:** Sales: 69 Profits: 134

| Drugs: | | Other: | Clinical: | Computers: | Services: |
|---|---|---|---|---|---|
| Discovery: | Y | AgriBio: | Trials/Services: | Hardware: | Specialty Services: |
| Licensing: | Y | Genetic Data: | Labs: | Software: | Consulting: |
| Manufacturing: | Y | Tissue Replacement: | Equipment/Supplies: | Arrays: | Blood Collection: |
| Genetics: | | | Research & Development Services: | Database Management: | Drug Delivery: |
| | | | Diagnostics: | | Drug Distribution: |

## TYPES OF BUSINESS:

Cancer Immunotherapy
Oncolytic Virus Therapy Drugs

## BRANDS/DIVISIONS/AFFILIATES:

GVAX

## CONTACTS: *Note: Officers with more than one job title may be intentionally listed here more than once.*

Stephen A. Sherwin, CEO
Sharon E. Tetlow, CFO/Sr. VP
Christine B. McKinley, Sr. VP-Human Resources
Peter K. Working, Sr. VP-R&D
Robert H. Tidwell, Sr. VP-Corp. Dev.
Marc L. Belsky, Chief Acct. Officer/VP
Robert J. Dow, Sr. VP-Medical Affairs/Chief Medical Officer
Carol C. Grundfest, Sr. VP-Regulatory Affairs & Portfolio Mgmt.
Kristen M. Hege, VP-Clinical Research
Stephen A. Sherwin, Chmn.

| **Phone:** 650-266-3000 | **Fax:** 650-266-3010 |
|---|---|
| **Toll-Free:** | |
| **Address:** 400 Oyster Point Blvd., South San Francisco, CA 94080 US | |

## GROWTH PLANS/SPECIAL FEATURES:

Cell Genesys, Inc. is a biotechnology company that focuses on the development and commercialization of biological therapies for patients with cancer. The company's lead program was the GVAX cell-based immunotherapy for cancer. In August 2008, the firm decided to discontinue to clinical trials of GVAX based a recommendation from the study's Independent Data Monitoring Committee (IDMC) due to a higher number of deaths of patients being treated with the medication than those without. The company owns roughly 400 U.S. and foreign patents issued or granted to it or available based on licensing arrangements and roughly 256 U.S. and foreign applications pending in its name or available based on licensing agreements. Cell Genesys recently sold all of its assets, intellectual property and previously established licensing agreements relating to its lentiviral gene delivery technology to GDP IP, LLC for approximately $12 million. In April 2009, the firm announced it would be laying off about 95% of its workforce (a reduction of around 274 employees), terminating its major facility leases and will reduce its number of patents owned.

Cell Genesys offers its employees an employee assistance program; a flexible spending plan and medical, dental, vision, life, AD&D, travel accident and disability insurance.

## FINANCIALS: Sales and profits are in thousands of dollars—add 000 to get the full amount. 2008 Note: Financial information for 2008 was not available for all companies at press time.

| | | |
|---|---|---|
| 2008 Sales: $94,571 | 2008 Profits: $-46,975 | **U.S. Stock Ticker:** CEGE |
| 2007 Sales: $1,380 | 2007 Profits: $-99,274 | **Int'l Ticker:** Int'l Exchange: |
| 2006 Sales: $1,364 | 2006 Profits: $-82,929 | Employees: 61 |
| 2005 Sales: $4,584 | 2005 Profits: $-64,939 | Fiscal Year Ends: 12/31 |
| 2004 Sales: $11,458 | 2004 Profits: $-97,411 | Parent Company: |

## SALARIES/BENEFITS:

| Pension Plan: | ESOP Stock Plan: | Profit Sharing: | Top Exec. Salary: $602,500 | Bonus: $ |
|---|---|---|---|---|
| Savings Plan: Y | Stock Purch. Plan: Y | | Second Exec. Salary: $445,000 | Bonus: $ |

## OTHER THOUGHTS:

**Apparent Women Officers or Directors:** 5
**Hot Spot for Advancement for Women/Minorities:** Y

## LOCATIONS: ("Y" = Yes)

| West: | Southwest: | Midwest: | Southeast: | Northeast: | International: |
|---|---|---|---|---|---|
| Y | | | | | |

# CELL THERAPEUTICS INC

**www.cticseattle.com**

**Industry Group Code: 325412  Ranks within this company's industry group:** Sales: 114    Profits: 165

| Drugs: | | Other: | Clinical: | Computers: | Services: |
|---|---|---|---|---|---|
| Discovery: | Y | AgriBio: | Trials/Services: | Hardware: | Specialty Services: |
| Licensing: | Y | Genetic Data: | Labs: | Software: | Consulting: |
| Manufacturing: | | Tissue Replacement: | Equipment/Supplies: | Arrays: | Blood Collection: |
| Genetics: | | | Research & Development Services: | Database Management: | Drug Delivery: |
| | | | Diagnostics: | | Drug Distribution: |

## TYPES OF BUSINESS:

Cancer Treatment Drugs

## BRANDS/DIVISIONS/AFFILIATES:

Zevalin
Paclitaxel Poliglumex
Pixantrone
Brostallicin
Systems Medicine LLC
Aequus BioPharma, Inc.
Opaxio

## CONTACTS: *Note: Officers with more than one job title may be intentionally listed here more than once.*

James A. Bianco, CEO
Craig W. Philips, Pres.
Jack W. Singer, Chief Medical Officer
Louis A. Bianco, Exec. VP-Admin.
Dan Eramian, Exec. VP-Corp. Comm.
Louis A. Bianco, Exec. VP-Finance
Pres.-CTI Sede Secondaria, Tim Williamson
Christina Waters, Chief Bus. Officer-Systems Medicine
Phillip Nudelman, Chmn.

| **Phone:** 206-282-7100 | **Fax:** 206-284-6206 |
|---|---|
| **Toll-Free:** 800-215-2355 | |
| **Address:** 501 Elliott Ave. W., Ste. 400, Seattle, WA 98119 US | |

## GROWTH PLANS/SPECIAL FEATURES:

Cell Therapeutics, Inc. (CTI) develops, acquires and commercializes treatments for cancer. The company's research and in-licensing activities are concentrated on identifying new, personalized, less toxic and more effective ways to treat cancer. The firm is currently developing thee drugs: opaxio (paclitaxel poliglumex), pixantrone and brostallicin. Opaxio, a biologically enhanced chemotherapeutic agent, links a widely used anti-cancer agent, paclitaxel, to a polyglutamate polymer delivery system for the potential treatment of ovarian and other cancers. Pixantrone is an anthracycline derivative for the safer treatment of non-Hodgkin's lymphoma. Finally, brostallicin, which is a small molecule, anti-cancer drug with a unique mechanism of action and composition of matter patent coverage, is being tested for treatment of relapsed/refractory soft tissue sarcoma. CTI is also working on new novel bisplatinum analogues that build on the successes of treatments such as Cisplatin, a platinum-based chemotherapy drug. The company has exclusive rights to 12 issued U.S. patents and 123 U.S. and foreign pending or issued patent applications relating to its polymer drug delivery technology. CTI also works with its subsidiaries Systems Medicine LLC, which integrates cellular genomics with medicine, and Aequus BioPharma, Inc., which develops biotherapeutics based on CTI's proprietary Genetic Polymer technology. In February 2009, the company optioned to sell its 50% stake in its joint venture with Spectrum Pharmaceuticals to market Zevalin, a radioimmunotherapy cancer treatment. In May 2009, the company closed its preclinical research facility in Bresso, Italy.

CTI offers its employees medical, dental, vision, life and disability insurance; an employee assistance program; travel assistance; an employee stock purchase plan; an educational assistance program; a 401(k) plan Additional perks include wellness seminars, discounted health club memberships, flu shots and ice cream socials.

## FINANCIALS: Sales and profits are in thousands of dollars—add 000 to get the full amount. 2008 Note: Financial information for 2008 was not available for all companies at press time.

| | | |
|---|---|---|
| 2008 Sales: $11,432 | 2008 Profits: $-180,029 | **U.S. Stock Ticker:** CTIC |
| 2007 Sales: $ 127 | 2007 Profits: $-138,108 | **Int'l Ticker:**    Int'l Exchange: |
| 2006 Sales: $  80 | 2006 Profits: $-135,819 | **Employees:**   194 |
| 2005 Sales: $16,092 | 2005 Profits: $-102,505 | **Fiscal Year Ends:** 12/31 |
| 2004 Sales: $29,594 | 2004 Profits: $-252,298 | **Parent Company:** |

## SALARIES/BENEFITS:

| Pension Plan: | ESOP Stock Plan: | Profit Sharing: | Top Exec. Salary: $650,000 | Bonus: $579,438 |
|---|---|---|---|---|
| Savings Plan: Y | Stock Purch. Plan: Y | | Second Exec. Salary: $340,000 | Bonus: $153,000 |

## OTHER THOUGHTS:

**Apparent Women Officers or Directors:** 2
**Hot Spot for Advancement for Women/Minorities:** Y

## LOCATIONS: ("Y" = Yes)

| West: | Southwest: | Midwest: | Southeast: | Northeast: | International: |
|---|---|---|---|---|---|
| Y | Y | | | | Y |

Note: Financial information, benefits and other data can change quickly and may vary from those stated here.

# CELLARTIS AB

www.cellartis.com

**Industry Group Code: 325414  Ranks within this company's industry group: Sales:   Profits:**

| Drugs: | Other: | Clinical: | | Computers: | Services: | |
|---|---|---|---|---|---|---|
| Discovery: | AgriBio: | Trials/Services: | Y | Hardware: | Specialty Services: | Y |
| Licensing: | Genetic Data: | Labs: | | Software: | Consulting: | |
| Manufacturing: | Tissue Replacement: | Equipment/Supplies: | | Arrays: | Blood Collection: | |
| Genetics: | | Research & Development Services: | | Database Management: | Drug Delivery: | |
| | | Diagnostics: | | | Drug Distribution: | |

## TYPES OF BUSINESS:

Cardiac Stemcell Fabrication
Drug Development Tests

## BRANDS/DIVISIONS/AFFILIATES:

## CONTACTS: *Note: Officers with more than one job title may be intentionally listed here more than once.*

Mats Lundwall, CEO
Johan Hyllner, COO
Johan Hyllner, Chief Science Officer
Kristina runeberg, VP-Bus. Dev.

| Phone: 46-31-758-09-00 | Fax: 46-31-758-09-10 |
|---|---|

**Toll-Free:**

**Address:** Arvid Wallgrens Backe 20, Goteborg, SE-413 46 Sweden

## GROWTH PLANS/SPECIAL FEATURES:

Cellartis AB is a startup biotechnology firm focused on human embryonic stem (hES) cells and technology utilized in drug discovery research, toxicity testing and regenerative medicine. Cellartis has developed more than 30 stem cell lines, making it one of the world's largest sources of stem cells. Primarily, the firm is concerned with the development of hepatocytes (cells in the liver) and cardiomyocytes (cells in the heart) for use in drug discovery. Cellartis is testing drugs for liver treatments with AstraZeneca and birth defects with Pfizer. The company is headquartered in Sweden, with offices in Dundee in the U.K. In 2008, the firm partnered with Pfizer to develop a screening system for human toxicity; VivoMedica PLC to apply VivoMedica's DrugPrint technology for cardiomyocites from human stem cells; and Novo Nordisk and Lund University to develop insulin producing cells from stem cells. In January 2009, the company signed a licensing agreement with the Wisconsin Alumni Research Foundation (WARF) for hES products. In February 2009, Cellartis signed a license agreement with ITI life Sciences for large scale commercial hES production. Also in February, the firm extended its collaboration with AstraZeneca to test drug safety. In April 2009, Cellartis partnered with ReproCELL to allow ReproCELL to use beating hES derived cardiomyocites with its QTempo Cardiotoxicity platform.

## FINANCIALS: Sales and profits are in thousands of dollars—add 000 to get the full amount. 2008 Note: Financial information for 2008 was not available for all companies at press time.

| | | |
|---|---|---|
| 2008 Sales: $ | 2008 Profits: $ | U.S. Stock Ticker: Private |
| 2007 Sales: $ | 2007 Profits: $ | Int'l Ticker:     Int'l Exchange: |
| 2006 Sales: $ | 2006 Profits: $ | Employees: |
| 2005 Sales: $ | 2005 Profits: $ | Fiscal Year Ends: |
| 2004 Sales: $ | 2004 Profits: $ | Parent Company: |

## SALARIES/BENEFITS:

| Pension Plan: | ESOP Stock Plan: | Profit Sharing: | Top Exec. Salary: $ | Bonus: $ |
|---|---|---|---|---|
| Savings Plan: | Stock Purch. Plan: | | Second Exec. Salary: $ | Bonus: $ |

## OTHER THOUGHTS:

**Apparent Women Officers or Directors:** 2
**Hot Spot for Advancement for Women/Minorities:**

## LOCATIONS: ("Y" = Yes)

| West: | Southwest: | Midwest: | Southeast: | Northeast: | International: Y |
|---|---|---|---|---|---|

Note: Financial information, benefits and other data can change quickly and may vary from those stated here.

# CELLEDEX THERAPEUTICS INC

www.celldextherapeutics.com

**Industry Group Code: 325412  Ranks within this company's industry group:** Sales: 122    Profits: 136

| Drugs: | | Other: | Clinical: | Computers: | Services: |
|---|---|---|---|---|---|
| Discovery: | Y | AgriBio: | Trials/Services: | Hardware: | Specialty Services: |
| Licensing: | | Genetic Data: | Labs: | Software: | Consulting: |
| Manufacturing: | Y | Tissue Replacement: | Equipment/Supplies: | Arrays: | Blood Collection: |
| Genetics: | | | Research & Development Services: | Database Management: | Drug Delivery: |
| | | | Diagnostics: | | Drug Distribution: |

## TYPES OF BUSINESS:

Drugs - Vaccines & Immunotherapeutics
Drugs - Oncology, Inflammatory Disease & Infectious Disease

## BRANDS/DIVISIONS/AFFILIATES:

AVANT Immunotherapeutics, Inc.
Precision Targeted Immunotherapy Platform
Rotarix
CholeraGarde
CDX-110
CDX-1307
APC Targeting Technology
CDX-1135

## CONTACTS: *Note: Officers with more than one job title may be intentionally listed here more than once.*

Anthony S. Marucci, CEO
Anthony S. Marucci, Pres.
Avery W. (Chip) Catlin, CFO/Sr. VP
Tibor Keler, Chief Scientific Officer/Sr. VP
Thomas Davis, Chief Medical Officer/Sr. VP
Charles R. Schaller, Chmn.

| Phone: 781-433-0771 | Fax: 781-433-0262 |
|---|---|
| Toll-Free: | |
| Address: 119 4th Ave., Needham, MA 02494-2725 US | |

## GROWTH PLANS/SPECIAL FEATURES:

Celldex Therapeutics, Inc., formerly AVANT Immunotherapeutics, Inc., is a biopharmaceutical company that applies its Precision Targeted Immunotherapy Platform to generate a pipeline of candidates to treat cancer and other difficult-to-treat diseases. The company's APC Targeting Technology uses human monoclonal antibodies linked to disease associated antigens to efficiently deliver the attached antigens to immune cells known as antigen presenting cells (APC). The firm's immunotherapy platform includes a complementary portfolio of monoclonal antibodies, antibody-targeted vaccines and immunomodulators to create novel disease-specific drug candidates. Celldex currently has one product on the market and six products in clinical development. Its marketed product, Rotarix, is a rotavirus vaccine. In Phase II trials are its CholeraGarde cholera vaccine, Ty800 typhoid vaccine and CDX-110 vaccine for the tumor specific molecule EGFRvIII. In Phase I trials are its CDX-1307, its primary its primary ACP Targeting Technology product candidate, for the treatment of colorectal, pancreatic, bladder, ovarian and breast cancers; CDX-1135 for renal disease; and ETEC E. coli vaccine. In pre-clinical development are its CDX-1401 for multiple solid tumors; CDX-1127 for immuno-modulation and solid tumors; CDX-1189 for renal disease; and CDX-2401 HIV (Human Immunodeficiency Virus) vaccine. In March 2008, the company acquired Celldex Therapeutics, Inc. through a merger transaction for approximately $75 million. In October 2008, the firm changed its name to Celldex Therapeutics. In January 2009, Celldex sold its poultry vaccines business to Lohmann Animal Health International. In April 2009, the company acquired exclusive rights to immune-stimulatory molecules from Amgen. In May 2009, the firm agreed to acquire CuraGen Corporation.

Celldex offers its employees educational assistance; an employee assistance program; an incentive stock option plan; flexible spending accounts; and medical, dental, life, AD&D, disability and business travel insurance.

## FINANCIALS: Sales and profits are in thousands of dollars—add 000 to get the full amount. 2008 Note: Financial information for 2008 was not available for all companies at press time.

| | | |
|---|---|---|
| 2008 Sales: $7,456 | 2008 Profits: $-47,501 | **U.S. Stock Ticker:** CLDX |
| 2007 Sales: $1,406 | 2007 Profits: $-15,073 | **Int'l Ticker:**    Int'l Exchange: |
| 2006 Sales: $ 899 | 2006 Profits: $-17,835 | Employees:   78 |
| 2005 Sales: $3,088 | 2005 Profits: $-18,097 | Fiscal Year Ends: 12/31 |
| 2004 Sales: $6,859 | 2004 Profits: $-13,204 | Parent Company: |

## SALARIES/BENEFITS:

| Pension Plan: | ESOP Stock Plan: | Profit Sharing: | Top Exec. Salary: $302,800 | Bonus: $137,400 |
|---|---|---|---|---|
| Savings Plan: Y | Stock Purch. Plan: Y | | Second Exec. Salary: $300,000 | Bonus: $96,600 |

## OTHER THOUGHTS:

**Apparent Women Officers or Directors**: 1
**Hot Spot for Advancement for Women/Minorities**:

## LOCATIONS: ("Y" = Yes)

| West: | Southwest: | Midwest: | Southeast: | Northeast: | International: |
|---|---|---|---|---|---|
| | | | | Y | |

# CELLULAR DYNAMICS INTERNATIONAL     www.cellular-dynamics.com

**Industry Group Code: 325414  Ranks within this company's industry group:** Sales:      Profits:

| Drugs: | Other: | Clinical: | | Computers: | | Services: | |
|--------|--------|-----------|--|-----------|--|----------|--|
| Discovery: | AgriBio: | Trials/Services: | Y | Hardware: | | Specialty Services: | Y |
| Licensing: | Genetic Data: | Labs: | | Software: | | Consulting: | |
| Manufacturing: | Tissue Replacement: | Equipment/Supplies: | | Arrays: | | Blood Collection: | |
| Genetics: | | Research & Development Services: | | Database Management: | | Drug Delivery: | |
| | | Diagnostics: | | | | Drug Distribution: | |

## TYPES OF BUSINESS:
Cardiac Stemcell Fabrication
Drug Assays

## BRANDS/DIVISIONS/AFFILIATES:
GLP hERG Channel Drug Candidate Screening
Non-GLP hERG Channel Drug Candidate Screening
hERG Channel Trafficking Assay
Repolarization Assay Using Human Cardiomyocytes

## GROWTH PLANS/SPECIAL FEATURES:
Cellular Dynamics International (CDI) is a startup biotechnology firm that utilizes embryonic stem cells to create human heart cells. The cells are used to screen drug candidates for safety and effectiveness in the laboratory before performing clinical trials on human patients. CDI has four drug screening products and services: GLP hERG Channel Drug Candidate Screening; Non-GLP hERG Channel Drug Candidate Screening; hERG Channel Trafficking Assay; and the Repolarization Assay Using Human Cardiomyocytes. All products promote accelerated drug development in specially targeted therapeutic areas related to cardiac health. In 2008, CDI began tests to screen cancer drugs for toxicity to the heart. The company also signed Roche Holding Ltd. as its first customer. In November 2008, the firm merged Stem Cell Products Inc. and iPS Cells Inc. with itself as the surviving entity.

## CONTACTS: Note: Officers with more than one job title may be intentionally listed here more than once.
Robert J. Palay, CEO
Emile Nuwaysir, COO/VP
Thomas M. Palay, Pres.
David S. Snyder, CFO/VP
James A. Thomson, Chief Scientific Officer
Nicholas Seay, CTO/VP
David Sneider, Chief Bus. Officer/VP
Chris Kendrick-Parker, Chief Commercial Officer/VP

| Phone: 608-310-5100 | Fax: |
|---|---|
| **Toll-Free:** | |
| **Address:** 525 Science Dr., Madison, WI 53711 US | |

## FINANCIALS: Sales and profits are in thousands of dollars—add 000 to get the full amount. 2008 Note: Financial information for 2008 was not available for all companies at press time.

| | | |
|---|---|---|
| 2008 Sales: $ | 2008 Profits: $ | **U.S. Stock Ticker: Private** |
| 2007 Sales: $ | 2007 Profits: $ | **Int'l Ticker:**     Int'l Exchange: |
| 2006 Sales: $ | 2006 Profits: $ | Employees: |
| 2005 Sales: $ | 2005 Profits: $ | Fiscal Year Ends: |
| 2004 Sales: $ | 2004 Profits: $ | Parent Company: |

## SALARIES/BENEFITS:

| | | | | |
|---|---|---|---|---|
| Pension Plan: | ESOP Stock Plan: | Profit Sharing: | Top Exec. Salary: $ | Bonus: $ |
| Savings Plan: | Stock Purch. Plan: | | Second Exec. Salary: $ | Bonus: $ |

## OTHER THOUGHTS:
**Apparent Women Officers or Directors:**
**Hot Spot for Advancement for Women/Minorities:**

## LOCATIONS: ("Y" = Yes)

| West: | Southwest: | Midwest: | Southeast: | Northeast: | International: |
|-------|-----------|----------|-----------|-----------|---------------|
| | | Y | | | |

# CEL-SCI CORPORATION

**www.cel-sci.com**

**Industry Group Code:** 325412  **Ranks within this company's industry group:** Sales: 159   Profits: 72

| Drugs: | | Other: | Clinical: | Computers: | Services: |
|---|---|---|---|---|---|
| Discovery: | Y | AgriBio: | Trials/Services: | Hardware: | Specialty Services: |
| Licensing: | | Genetic Data: | Labs: | Software: | Consulting: |
| Manufacturing: | Y | Tissue Replacement: | Equipment/Supplies: | Arrays: | Blood Collection: |
| Genetics: | | | Research & Development Services: | Database Management: | Drug Delivery: |
| | | | Diagnostics: | | Drug Distribution: |

## TYPES OF BUSINESS:

Drugs-Cancer & Infectious Disease
Vaccines

## BRANDS/DIVISIONS/AFFILIATES:

MultiKine
LEAPS
Cel-1000
Cel-2000

## CONTACTS: *Note: Officers with more than one job title may be intentionally listed here more than once.*

Geert R. Kersten, CEO
Maximilian de Clara, Pres.
Eyal Talor, Sr. VP-Research
Eyal Talor, Sr. VP-Mfg.
Patricia B. Prichep, Sr. VP-Oper.
William (Brooke) Jones, VP-Quality Assurance

| Phone: 703-506-9460 | Fax: 703-506-9471 |
|---|---|
| Toll-Free: | |
| Address: 8229 Boone Blvd., Ste. 802, Vienna, VA 22182 US | |

## GROWTH PLANS/SPECIAL FEATURES:

CEL-SCI Corp. researches and develops immunotherapy products for the treatment of cancer and infectious diseases. Its flagship product, Multikine, is the first of a new class of cancer immunotherapy drugs called Immune SIMULATORs. It simulates the activities of a healthy person's immune system as well as killing cancer cells in a targeted fashion. The company is currently planning to conduct an international Phase III clinical study of Multikine in the treatment of advanced primary squamous cell carcinoma of the head and neck.  Previous trials have shown that Multikine is safe and non-toxic and renders cancer cells much more susceptible to radiation therapy.  Additionally, the company owns a pre-clinical T-cell modulation process technology called Ligand Epitope Antigen Presentations System (LEAPS), through which it has developed the Cel-1000 peptide.  Cel-1000 has proven to be effective in protecting animals against herpes, malaria, viral encephalitis and cancer.  The company has also tested the effectiveness of Cel-1000 against avian flu and the hepatitis B virus. LEAPS may also have applications in any disease for which antigenic epitope sequences have been identified, such as infectious diseases, autoimmune diseases, allergic asthma, allergies and select CNS diseases, such as Alzheimer's. CEL-2000 is a peptide for the prevention or retardation of tissue damage caused by rheumatoid arthritis.  In 2008, the firm announced the discovery of Cel-2000.  CEL-SHI also recently entered into exclusive licensing agreements for Multikine with Teva Pharmaceutical Industries, Ltd., Orient Europharma Co., Ltd., and Byron Biopharma; took delivery of a new manufacturing facility for Multikine; and began pre-clinical testing, using LEAPS, against H1N1 swine flu and other influenzas.

CEL-SCI offers its employees health, dental, vision, and long term disability insurance; a 401(k) plan; and flexible spending accounts that allow for pre-tax dependent care, unreimbursed medical expenses and medical premiums.

## FINANCIALS:  Sales and profits are in thousands of dollars—add 000 to get the full amount. 2008 Note: Financial information for 2008 was not available for all companies at press time.

| | | |
|---|---|---|
| 2008 Sales: $ 5 | 2008 Profits: $-7,703 | **U.S. Stock Ticker: CVM** |
| 2007 Sales: $ 57 | 2007 Profits: $-9,630 | **Int'l Ticker:**   Int'l Exchange: |
| 2006 Sales: $ 125 | 2006 Profits: $-7,939 | Employees:   30 |
| 2005 Sales: $ 270 | 2005 Profits: $-3,040 | Fiscal Year Ends: 9/30 |
| 2004 Sales: $ 325 | 2004 Profits: $-2,952 | Parent Company: |

## SALARIES/BENEFITS:

| Pension Plan: | ESOP Stock Plan: | Profit Sharing: | Top Exec. Salary: $389,637 | Bonus: $ |
|---|---|---|---|---|
| Savings Plan: Y | Stock Purch. Plan: | | Second Exec. Salary: $363,000 | Bonus: $ |

## OTHER THOUGHTS:

**Apparent Women Officers or Directors:** 1
**Hot Spot for Advancement for Women/Minorities:**

## LOCATIONS: ("Y" = Yes)

| West: | Southwest: | Midwest: | Southeast: | Northeast: | International: |
|---|---|---|---|---|---|
| | | | | Y | |

# CELSIS INTERNATIONAL PLC

www.celsis.com

**Industry Group Code: 325413  Ranks within this company's industry group: Sales: 15  Profits: 8**

| Drugs: | Other: | Clinical: | | Computers: | | Services: | |
|---|---|---|---|---|---|---|---|
| Discovery: | AgriBio: | Trials/Services: | | Hardware: | | Specialty Services: | Y |
| Licensing: | Genetic Data: | Labs: | Y | Software: | | Consulting: | |
| Manufacturing: | Tissue Replacement: | Equipment/Supplies: | Y | Arrays: | Y | Blood Collection: | |
| Genetics: | | Research & Development Services: | | Database Management: | | Drug Delivery: | |
| | | Diagnostics: | Y | | | Drug Distribution: | |

## TYPES OF BUSINESS:

Microbial Contamination Detection Instruments
Laboratory Services Outsourcing

## BRANDS/DIVISIONS/AFFILIATES:

Celsis Rapid Detection
Celsis In Vitro Technologies
Celsis Analytical Services
RapiScreen
AKuScreen
ReACT

## CONTACTS: *Note: Officers with more than one job title may be intentionally listed here more than once.*

Jay LeCoque, CEO
Christian Madrolle, Dir.-Finance
Jack Rowell, Chmn.

| Phone: 312-476-1200 | Fax: 312-476-1201 |
|---|---|
| Toll-Free: | |
| Address: 600 W. Chicago Ave., Ste. 625, Chicago, IL 60654-2822 US | |

## GROWTH PLANS/SPECIAL FEATURES:

Celsis International plc is a global provider of life sciences products and laboratory services to the pharmaceutical and consumer products industries. Its core technology involves bioluminescence-based technology, in which ATP (adenosine triphosphate), a nucleotide present in all biological matter, is catalyzed to produce light (as in the abdomen of a firefly). The firm has offices in the U.K., the U.S., France, Germany and Belgium. Sales to the Americas generated 66% of the company's 2008 revenue, while sales to Europe generated 26% and sales to other regions generated 8%. Celsis serves the pharmaceuticals, personal care, home care, consumer products and food and beverage industries. During 2008, the pharmaceutical industry generated 64% of the company's revenue, the beauty and home industries generated 21% and the beverage industry generated 15%. The firm operates through three divisions: Celsis Rapid Detection, Celsis In Vitro Technologies and Celsis Analytical Services. Celsis Rapid Detection uses proprietary enzyme technology to develop and supply screening systems, including instruments, software and reagents, for the rapid detection of microbial contamination in pharmaceutical and consumer products. Celsis's detection product lines include RapiScreen and AKuScreen assays. Celsis In Vitro Technologies uses proprietary hepatocyte (liver cell) and cryopreservation technologies to provide in vitro testing products for pharmaceutical and biotechnological companies, and is a leading provider of ADME-Tox products for drug discovery and development. Celsis Analytical Services provides outsourced laboratory testing services, including analytical chemistry and microbiological laboratory services and stability testing and storage programs, to pharmaceutical, biopharmaceutical, medical device and consumer products companies. The division can conduct up to 90% of USP (United States Pharmacopeia) compendial methods testing. Products offered by Celsis generated 61% of its 2008 revenue, while services generated 39%. In March 2009, Celsis launched its ReACT RNA-based assay for detecting objectionable organisms without the need for expensive equipment or specialized training.

## FINANCIALS: Sales and profits are in thousands of dollars—add 000 to get the full amount. 2008 Note: Financial information for 2008 was not available for all companies at press time.

| | | |
|---|---|---|
| 2008 Sales: $52,900 | 2008 Profits: $6,700 | U.S. Stock Ticker: |
| 2007 Sales: $47,441 | 2007 Profits: $5,921 | Int'l Ticker: CEL    Int'l Exchange: London-LSE |
| 2006 Sales: $33,104 | 2006 Profits: $4,603 | Employees:  230 |
| 2005 Sales: $30,397 | 2005 Profits: $7,999 | Fiscal Year Ends: 3/31 |
| 2004 Sales: $50,400 | 2004 Profits: $12,200 | Parent Company: |

## SALARIES/BENEFITS:

| | | | | |
|---|---|---|---|---|
| Pension Plan: | ESOP Stock Plan: | Profit Sharing: | Top Exec. Salary: $451,523 | Bonus: $151,235 |
| Savings Plan: Y | Stock Purch. Plan: | | Second Exec. Salary: $394,485 | Bonus: $142,078 |

## OTHER THOUGHTS:

Apparent Women Officers or Directors:
Hot Spot for Advancement for Women/Minorities:

## LOCATIONS: ("Y" = Yes)

| West: | Southwest: | Midwest: | Southeast: | Northeast: | International: |
|---|---|---|---|---|---|
| | | Y | | Y | Y |

Note: Financial information, benefits and other data can change quickly and may vary from those stated here.

# CEPHALON INC

### www.cephalon.com

**Industry Group Code: 325412  Ranks within this company's industry group: Sales: 32    Profits: 30**

| Drugs: | | Other: | | Clinical: | | Computers: | | Services: | |
|---|---|---|---|---|---|---|---|---|---|
| Discovery: | Y | AgriBio: | | Trials/Services: | | Hardware: | | Specialty Services: | |
| Licensing: | Y | Genetic Data: | | Labs: | | Software: | | Consulting: | |
| Manufacturing: | | Tissue Replacement: | | Equipment/Supplies: | | Arrays: | | Blood Collection: | |
| Genetics: | | | | Research & Development Services: | Y | Database Management: | | Drug Delivery: | |
| | | | | Diagnostics: | | | | Drug Distribution: | |

## TYPES OF BUSINESS:

Pharmaceutical Discovery & Development
Neurological Disorder Treatments
Cancer Treatments
Pain Medications
Addiction Treatment

## BRANDS/DIVISIONS/AFFILIATES:

Provigil
Actiq
Fentora
Trisenox
Vivitrol
Nuvigil
Treanda
AMRIX

## CONTACTS: *Note: Officers with more than one job title may be intentionally listed here more than once.*

Frank Baldino, Jr., CEO
J. Kevin Buchi, CFO/Exec. VP
Jeffry L. Vaught, Chief Science Officer/Exec. VP-R&D
Peter E. Grebow, Sr. VP-Worldwide Tech. Oper.
Carl A. Savini, Chief Admin. Officer/Exec. VP
Gerald J. Pappert, General Counsel/Exec. VP
Lesley Russell, Exec. VP-Worldwide Medical & Regulatory Oper.
Valli F. Baldassano, Chief Compliance Officer/Exec. VP
Frank Baldino, Jr., Chmn.
Robert P. Roche, Jr., Sr. VP-Worldwide Pharmaceutical Oper.

| Phone: 610-344-0200 | Fax: 610-738-6590 |
|---|---|
| Toll-Free: | |
| Address: 41 Moores Rd., Frazer, PA 19355 US | |

## GROWTH PLANS/SPECIAL FEATURES:

Cephalon, Inc. is a biopharmaceutical company focused on the discovery, development and marketing of products in four core areas: Central nervous system (CNS) disorders, pain, oncology and addiction.   It conducts research and development as well as marketing its products in the U.S. and Europe.  Cephalon's technology principally focuses on understanding the class of enzymes known as kinases and the role they play in cellular survival and proliferation.  The company's CNS products include its most significant product, Provigil which is designed to treat extreme sleepiness associated with narcolepsy and other sleep disorders.  The firm's other CNS medication is Gabitril, designed for partial seizures in epileptic patients.    The firm's two pain management products, Actiq and Fentora, are designed for patients who are opiod-tolerant.   Cephalon's oncology products include Trisenox (an arsenic salt), which is marketed in the U.S. and Europe to treat patients with relapsed acute promyelocytic leukemia (APL).    Other products include Amrix, a muscle relaxant (obtained from E. Claiborne Robins Co., Inc.) designed to relieve spasms due to musculoskeletal conditions; Treanda, a chemotherapy drug used for chronic lymphocytic leukemia; and Nuvigil, a treatment similar to Provigil.  Cephalon also has a number of products in development, including treatments for Alzheimer's Disease, pain management, tumors, leukemia, multiple myeloma, lymphoma and lupus.  In January 2009, the company acquired Ception Therapeutics, a private biopharmaceutical company developing drugs to treat pediatric eosinophilic esophagitis and adult eosinophilic asthma.  In February 2009, the firm obtained a majority stake in the Australian biopharmaceutical company Arana, which develops and licenses anti-inflammatory treatments.

Cephalon's benefits include medical, prescription drug, dental and disability coverage; life insurance; a 401(k) profit sharing plan; flexible spending accounts; educational reimbursements; and health advocacy services.

## FINANCIALS: Sales and profits are in thousands of dollars—add 000 to get the full amount. 2008 Note: Financial information for 2008 was not available for all companies at press time.

| | | |
|---|---|---|
| 2008 Sales: $1,974,554 | 2008 Profits: $222,548 | U.S. Stock Ticker: CEPH |
| 2007 Sales: $1,772,638 | 2007 Profits: $-191,704 | Int'l Ticker:     Int'l Exchange: |
| 2006 Sales: $1,764,069 | 2006 Profits: $144,816 | Employees:  2,780 |
| 2005 Sales: $1,211,892 | 2005 Profits: $-174,954 | Fiscal Year Ends: 12/31 |
| 2004 Sales: $1,015,400 | 2004 Profits: $-73,800 | Parent Company: |

## SALARIES/BENEFITS:

| Pension Plan: | ESOP Stock Plan: | Profit Sharing: Y | Top Exec. Salary: $1,244,600 | Bonus: $1,779,800 |
|---|---|---|---|---|
| Savings Plan: Y | Stock Purch. Plan: | | Second Exec. Salary: $567,600 | Bonus: $380,300 |

## OTHER THOUGHTS:

**Apparent Women Officers or Directors**: 1
**Hot Spot for Advancement for Women/Minorities**:

## LOCATIONS: ("Y" = Yes)

| West: | Southwest: | Midwest: | Southeast: | Northeast: | International: |
|---|---|---|---|---|---|
| Y | | Y | | Y | Y |

# CEPHEID

www.cepheid.com

**Industry Group Code: 325413  Ranks within this company's industry group:** Sales: 7  Profits: 15

| Drugs: | Other: | Clinical: | | Computers: | | Services: | |
|---|---|---|---|---|---|---|---|
| Discovery: | AgriBio: | Trials/Services: | | Hardware: | | Specialty Services: | |
| Licensing: | Genetic Data: | Labs: | | Software: | | Consulting: | |
| Manufacturing: | Tissue Replacement: | Equipment/Supplies: | Y | Arrays: | | Blood Collection: | |
| Genetics: | | Research & Development Services: | | Database Management: | | Drug Delivery: | |
| | | Diagnostics: | Y | | | Drug Distribution: | |

## TYPES OF BUSINESS:

Equipment-Biological Testing
Genetic Profiling
DNA Analysis Systems
Molecular Diagnostics

## BRANDS/DIVISIONS/AFFILIATES:

Smart Cycler
Smart Cycler II
GeneXpert
IVD
ASR
RUO
Sangtec Molecular Diagnostics AB
Xpert HemosIL

## CONTACTS: *Note: Officers with more than one job title may be intentionally listed here more than once.*

John L. Bishop, CEO
Andrew D. Miller, CFO/Sr. VP
Robert J. Koska, Sr. VP-Sales & Mktg.
Laurie King, VP-Human Resources
Peter J. Dailey, VP-R&D
Jan Steuperaert, VP-IT
David H. Persing, CTO/Exec. VP
Nicki Bowen, VP-Eng.
Joseph H. Smith, Sr. VP-Legal
Humberto Reyes, Exec. VP-Oper.
Joseph H. Smith, Sr. VP-Bus. Dev.
Jared Tipton, Dir.-Corp. Comm.
Jared Tipton, Dir.-Investor Rel.
Michael Myhre, VP-Corp. Controller
Lee Christel, VP-Research & Systems Integration
Russel K. Enns, Sr. VP-Regulatory, Clinical & Govt. Affairs
Robert J. Koska, Sr. VP-Worldwide Commercial Oper.
Sandra Finley, VP-Mktg.
Thomas L. Gutshall, Chmn.
Rika Dutau, VP/Managing Dir.-Cepeid Europe

| **Phone:** 408-541-4191 | **Fax:** 408-541-4192 |
|---|---|
| **Toll-Free:** 888-838-3222 | |
| **Address:** 904 Caribbean Dr., Sunnyvale, CA 94089 US | |

## GROWTH PLANS/SPECIAL FEATURES:

Cepheid is a molecular diagnostics company that develops, manufactures and markets fully integrated systems for the clinical genetic assessment, life sciences, industrial and biothreat markets. Cephid systems enable rapid molecular testing for organisms and genetic-based diseases by implementing automated technology to reduce the complicated and time-intensive steps that are usually involved in molecular testing. Complex biological systems are analyzed in the company's proprietary biocartridges, which eliminates any lengthy preparation, amplification and detection of targeted genes. The company's two principal system platforms are the SmartCycler and GeneXpert systems. SmartCycles integrates DNA amplification and detection in order to rapidly analyze samples while the GeneXpert system incorporates sample preparation with DNA amplification and detection. Both systems are utilized in areas of critical infectious diseases, immunocompromised transplantations, women's health and oncology. Other products include the IVD line, ASR line, RUO products line, biothreat products and life science products. The company has sold units to a wide range of customers, including the Center for Disease Control and Prevention, the U.S. Food and Drug Administration, Johns Hopkins University, Memorial Sloan-Kettering Cancer Center, the National Institute of Health, Stanford University and the U.S. Army Medical Research Institute for Infectious Disease (USAMRIID). GeneXpert has also been incorporated into the United States Postal Service's Biohazard Detection System to identify the presence of anthrax from air samples. In February 2007, the company acquired Sangtec Molecular Diagnostics AB, which added a line of products that manages infections in immunocompromised patients to Cepheid's technologies. In February 2008, in conjunction with Instrumentation Laboratory, the firm introduced the Xpert HemosIL assay product to the European clinical testing market; the device detects genetic variations associated with thrombophilia (increased risk of blood clotting).

Cepheid offers employees flexible spending accounts, income protections plans and an employee assistance program.

## FINANCIALS: Sales and profits are in thousands of dollars—add 000 to get the full amount. 2008 Note: Financial information for 2008 was not available for all companies at press time.

| | | |
|---|---|---|
| 2008 Sales: $169,627 | 2008 Profits: $-27,713 | **U.S. Stock Ticker:** CPHD |
| 2007 Sales: $129,473 | 2007 Profits: $-21,423 | **Int'l Ticker:** Int'l Exchange: |
| 2006 Sales: $87,352 | 2006 Profits: $-25,985 | Employees: 547 |
| 2005 Sales: $85,010 | 2005 Profits: $-13,594 | Fiscal Year Ends: 12/31 |
| 2004 Sales: $52,968 | 2004 Profits: $-13,800 | Parent Company: |

## SALARIES/BENEFITS:

| Pension Plan: | ESOP Stock Plan: | Profit Sharing: | Top Exec. Salary: $488,462 | Bonus: $ |
|---|---|---|---|---|
| Savings Plan: Y | Stock Purch. Plan: Y | | Second Exec. Salary: $383,850 | Bonus: $ |

## OTHER THOUGHTS:

**Apparent Women Officers or Directors:** 7
**Hot Spot for Advancement for Women/Minorities:** Y

## LOCATIONS: ("Y" = Yes)

| West: | Southwest: | Midwest: | Southeast: | Northeast: | International: |
|---|---|---|---|---|---|
| Y | | | | | Y |

# CERUS CORPORATION

**www.cerus.com**

Industry Group Code: 325412  Ranks within this company's industry group: Sales: 107  Profits: 121

| Drugs: | | Other: | Clinical: | Computers: | Services: |
|---|---|---|---|---|---|
| Discovery: | Y | AgriBio: | Trials/Services: | Hardware: | Specialty Services: |
| Licensing: | Y | Genetic Data: | Labs: | Software: | Consulting: |
| Manufacturing: | | Tissue Replacement: | Equipment/Supplies: | Arrays: | Blood Collection: |
| Genetics: | | | Research & Development Services: | Database Management: | Drug Delivery: |
| | | | Diagnostics: | | Drug Distribution: |

## TYPES OF BUSINESS:

Pharmaceutical Development
Cancer Vaccines
Blood Treatment Systems

## BRANDS/DIVISIONS/AFFILIATES:

INTERCEPT
Helinx

## CONTACTS: Note: Officers with more than one job title may be intentionally listed here more than once.

Claes Glassell, CEO
Claes Glassell, Pres.
Laurence M. Corash, Chief Scientific Officer/Sr. VP
Lori L. Roll, VP-Admin./Corp. Sec.
Howard G. Ervin, VP-Legal Affairs
William M. Greenman, Sr. VP-Bus. Dev. & Mktg.
Kevin D. Green, VP-Finance/Chief Acct. Officer
B. J. Cassin, Chmn.

| Phone: 925-288-6000 | Fax: 925-288-6001 |
|---|---|
| Toll-Free: | |
| Address: 2411 Stanwell Dr., Concord, CA 94520 US | |

## GROWTH PLANS/SPECIAL FEATURES:

Cerus Corp. develops and commercializes novel, proprietary products and technologies within the field blood safety. Cerus is developing and commercializing the INTERCEPT Blood System for platelets, plasma and red blood cells, which is based on the company's proprietary Helinx technology for controlling biological replication. The company also had an immunotherapy program, but this was recently spun off to Anza Therapeutics, Inc. in order to better focus on the blood safety program. INTERCEPT is designed to enhance the safety of donated blood components by inactivating viruses, bacteria, parasites and other pathogens, as well as potentially harmful white blood cells. Cerus' Helinx technology prevents the replication of the DNA or RNA of susceptible pathogens, such as hepatitis B, hepatitis C and HIV. INTERCEPT for platelets and plasma is marketed in the European Union and undergoing clinical trials in the US. Trials for INTERCEPT for red blood cells were temporarily halted due to unsatisfactory results in 2003, but the company believes that the problems have been corrected and a new round of testing has begun. The firm licenses its technology to other companies for international distribution, including subsidiaries for Europe and BioOne Corp. for select Asian countries. A number of distribution agreements exist for partners in various other countries, including some in South America, the Middle East and CIS countries. The company owns about 23 US and 47 foreign patents relating to INTERCEPT. In March 2009, Cerus announced it would be laying off about 30% of its workforce, or 31 employees. In May 2009, the company signed an agreement with Grifols S.A. to expand the distribution of INTERCEPT to Italy.

Cerus offers medical, dental, and vision coverage; life and AD&D insurance; tuition reimbursement; flexible spending accounts; a 401(k) plan; stock options; a health club membership discount; and an employee stock purchase plan.

## FINANCIALS: Sales and profits are in thousands of dollars—add 000 to get the full amount. 2008 Note: Financial information for 2008 was not available for all companies at press time.

| | | |
|---|---|---|
| 2008 Sales: $16,507 | 2008 Profits: $-29,181 | U.S. Stock Ticker: CERS |
| 2007 Sales: $11,044 | 2007 Profits: $-45,304 | Int'l Ticker:    Int'l Exchange: |
| 2006 Sales: $30,310 | 2006 Profits: $-4,779 | Employees:   107 |
| 2005 Sales: $13,497 | 2005 Profits: $13,064 | Fiscal Year Ends: 12/31 |
| 2004 Sales: $13,911 | 2004 Profits: $-31,153 | Parent Company: |

## SALARIES/BENEFITS:

| Pension Plan: | ESOP Stock Plan: | Profit Sharing: | Top Exec. Salary: $486,969 | Bonus: $ |
|---|---|---|---|---|
| Savings Plan: Y | Stock Purch. Plan: Y | | Second Exec. Salary: $373,190 | Bonus: $ |

## OTHER THOUGHTS:

**Apparent Women Officers or Directors**: 2
**Hot Spot for Advancement for Women/Minorities**:

## LOCATIONS: ("Y" = Yes)

| West: | Southwest: | Midwest: | Southeast: | Northeast: | International: |
|---|---|---|---|---|---|
| Y | | | | | Y |

# CHARLES RIVER LABORATORIES INTERNATIONAL INC

www.criver.com

Industry Group Code: 541712  Ranks within this company's industry group: Sales: 3  Profits: 20

| Drugs: | Other: | Clinical: | | Computers: | | Services: | |
|---|---|---|---|---|---|---|---|
| Discovery: | AgriBio: | Trials/Services: | | Hardware: | | Specialty Services: | Y |
| Licensing: | Genetic Data: | Labs: | Y | Software: | Y | Consulting: | |
| Manufacturing: | Tissue Replacement: | Equipment/Supplies: | Y | Arrays: | | Blood Collection: | |
| Genetics: | | Research & Development Services: | | Database Management: | | Drug Delivery: | |
| | | Diagnostics: | Y | | | Drug Distribution: | |

## TYPES OF BUSINESS:

Animal Research Models
Consulting
Bioactivity Software
Biosafety Testing
Contract Staffing
Laboratory Diagnostics
Intellectual Property Consulting
Analytical Testing

## BRANDS/DIVISIONS/AFFILIATES:

NewLab BioQuality AG
MIR Preclinical Services

## CONTACTS: Note: Officers with more than one job title may be intentionally listed here more than once.

James C. Foster, CEO
James C. Foster, Pres.
Thomas F. Ackerman, CFO/Exec. VP
Stephanie B. Wells, Sr. VP-Mktg./Chief Mktg. Officer
David P. Johst, Exec. VP-Human Resources
Real H. Renaud, Pres., Global Research Model Prod. & Svcs.
Nicholas Ventresca, CIO/Sr. VP-IT
David P. Johst, Chief Admin. Officer
David P. Johst, General Counsel/Exec. VP
John C. Ho, Sr. VP-Corp. Strategy
Nancy A. Gillett, Exec. VP/Pres., Global Preclinical Svcs.
Christophe Berthoux, Chief Commercial Officer
Brian Bathgate, Sr. VP/Pres., European Preclinical Svcs.
Christopher Perkin, Sr. VP/Pres., Canadian & Chinese Preclinical Svcs.
James C. Foster, Chmn.
Jorg Geller, Sr. VP-Japanese Oper. & Select Research Model Bus.

| Phone: 978-658-6000 | Fax: 978-658-7132 |
|---|---|
| Toll-Free: | |
| Address: 251 Ballardvale St., Wilmington, MA 01887 US | |

## GROWTH PLANS/SPECIAL FEATURES:

Charles River Laboratories International, Inc. (CRL), founded in 1947, is a global provider of solutions that accelerate the drug discovery and development process, including research models and outsourced preclinical services. Charles River operates roughly 70 facilities in 17 countries. The firm's customer base includes global pharmaceutical companies, a wide range of biotechnology companies, government agencies, hospitals and academic institutions. Charles River organizes its operations into two segments: research models and services (RMS); and preclinical services (PCS). RMS, which generated 49% of the company's sales in 2008, is focused on the commercial production and sale of research models; research model services, including transgenic services, research animal diagnostics, discovery services and consulting and staffing services; and other related products and services, such as vaccine support and in vitro technology. With approximately 150 different strains, the company primarily provides genetically and virally defined purpose-bred rats and mice as research models. The segment has multiple facilities located throughout North America, Europe and Japan and maintains approximately 180 barrier rooms or isolator facilities. PCS, which generated 51% of Charles River's sales in 2008, is principally engaged in the discovery and development of new drugs, devices and therapies. Charles River offers particular expertise in the design, execution and reporting of general and specialty toxicology studies, especially those dealing with innovative therapies and biologicals. Additional PCS include pathology services; bioanalysis, pharmacokinetics and drug metabolism analysis; discovery support; biopharmaceutical services; and Phase I trials in healthy, normal and special populations. The company currently provides PCS at multiple facilities in the U.S., Canada, Europe and China. As a result of increasing demand for outsourced PCS, Charles River has recently conducted significant facilities expansions at its preclinical facilities in Massachusetts and Nevada. In September 2008, the company acquired NewLab BioQuality AG, for $53 million, and MIR Preclinical Services, for $12.5 million.

## FINANCIALS: Sales and profits are in thousands of dollars—add 000 to get the full amount. 2008 Note: Financial information for 2008 was not available for all companies at press time.

| | | |
|---|---|---|
| 2008 Sales: $1,343,493 | 2008 Profits: $-521,843 | U.S. Stock Ticker: CRL |
| 2007 Sales: $1,230,626 | 2007 Profits: $154,406 | Int'l Ticker:    Int'l Exchange: |
| 2006 Sales: $1,058,385 | 2006 Profits: $-55,783 | Employees: 9,000 |
| 2005 Sales: $993,328 | 2005 Profits: $141,999 | Fiscal Year Ends: 12/31 |
| 2004 Sales: $724,221 | 2004 Profits: $89,792 | Parent Company: |

## SALARIES/BENEFITS:

| Pension Plan: | ESOP Stock Plan: | Profit Sharing: | Top Exec. Salary: $948,500 | Bonus: $817,133 |
|---|---|---|---|---|
| Savings Plan: Y | Stock Purch. Plan: | | Second Exec. Salary: $496,460 | Bonus: $386,288 |

## OTHER THOUGHTS:

Apparent Women Officers or Directors: 5
Hot Spot for Advancement for Women/Minorities: Y

## LOCATIONS: ("Y" = Yes)

| West: | Southwest: | Midwest: | Southeast: | Northeast: | International: |
|---|---|---|---|---|---|
| Y | Y | Y | Y | Y | Y |

Note: Financial information, benefits and other data can change quickly and may vary from those stated here.

# CHESAPEAKE BIOLOGICAL LABORATORIES INC    www.cblinc.com

**Industry Group Code: 541712  Ranks within this company's industry group:  Sales:    Profits:**

| Drugs: | Other: | Clinical: | | Computers: | Services: | |
|---|---|---|---|---|---|---|
| Discovery: | AgriBio: | Trials/Services: | Y | Hardware: | Specialty Services: | Y |
| Licensing: | Genetic Data: | Labs: | Y | Software: | Consulting: | Y |
| Manufacturing: Y | Tissue Replacement: | Equipment/Supplies: | | Arrays: | Blood Collection: | |
| Genetics: | | Research & Development Services: | Y | Database Management: | Drug Delivery: | |
| | | Diagnostics: | | | Drug Distribution: | |

## TYPES OF BUSINESS:

Research & Development
Contract Pharmaceutical Services
Regulatory & Compliance Consulting

## BRANDS/DIVISIONS/AFFILIATES:

Cangene Corporation

## CONTACTS: *Note: Officers with more than one job title may be intentionally listed here more than once.*

William Labossiere (Bill) Bees, VP-Oper., Cangene
Steven E. Rowan, VP-Mktg.
Vicki Wolff-Long, VP-Oper.
Steven E. Rowan, VP-Bus. Dev.

| Phone: 410-843-5000 | Fax: 410-843-4414 |
|---|---|
| Toll-Free: 800-441-4225 | |
| Address: 1111 S. Paca St., Baltimore, MD 21230-2591 US | |

## GROWTH PLANS/SPECIAL FEATURES:

Chesapeake Biological Laboratories, Inc. (CBL), a subsidiary of Cangene Corporation, provides contract pharmaceutical and biopharmaceutical product development and production services. Customers typically hire the company to produce developmental stage products for FDA clinical trials or to produce and manufacture FDA approved products for commercial sale. Since its inception, CBL has developed over 175 therapeutic products and provides a variety of services, including clinical manufacturing, commercial manufacturing, lyophilization and product stability testing. The company performs clinical manufacturing for all regulatory phases; has a large selection of pre-qualified vials, stoppers and syringes; and can fill liquid or lyophilized products in vial sizes from 3cc to 100cc, as well as most major syringe sizes. CBL has a 240 square-foot lyophilizer capable of holding roughly 60,000 3cc vials. Products that require lyophilization are filled, partially stoppered and loaded into the pre-chilled lyophilizer throughout the filling period. Its lyophilizer has both low and high speed filling options, temperature control to as low as negative 65 degrees Celsius and vacuum control between 50 and 1,000 millitorr. The time from product initiation to production in most cases can be reduced to two months by choosing one of the firm's pre-qualified combinations. CBL currently fills 15 commercial products, with several more products pending approval. The company typically performs identification testing on client supplied raw materials; in-process testing sufficient to ensure bulk formulation is done correctly; and finished product testing as required by the client. CBL's Analytical Services Department can perform Karl Fischer, Dionex IC, Viscosity, Dissolved Oxygen and Osmolality tests, among others. Its Microbiology Department can perform Enhancement & Inhibition, Endotoxin, Bioburden Method Qualification, Bacteriostasis/Fungistasis, Bulk Sterility, Final Product Sterility and ELISA tests. Recently, CBL completed the validation of an additional GMP (Good Manufacturing Practice) oven.

CBL offers its employees medical, dental, vision, life and disability insurance.

## FINANCIALS:  Sales and profits are in thousands of dollars—add 000 to get the full amount. 2008 Note: Financial information for 2008 was not available for all companies at press time.

| | | |
|---|---|---|
| 2008 Sales: $ | 2008 Profits: $ | **U.S. Stock Ticker: Subsidiary** |
| 2007 Sales: $ | 2007 Profits: $ | **Int'l Ticker:**    Int'l Exchange: |
| 2006 Sales: $ | 2006 Profits: $ | Employees:   125 |
| 2005 Sales: $ | 2005 Profits: $ | Fiscal Year Ends: 7/31 |
| 2004 Sales: $ | 2004 Profits: $ | Parent Company: CANGENE CORP |

## SALARIES/BENEFITS:

| Pension Plan: | ESOP Stock Plan: | Profit Sharing: | Top Exec. Salary: $ | Bonus: $ |
|---|---|---|---|---|
| Savings Plan: Y | Stock Purch. Plan: | | Second Exec. Salary: $ | Bonus: $ |

## OTHER THOUGHTS:

**Apparent Women Officers or Directors: 1**
**Hot Spot for Advancement for Women/Minorities:**

## LOCATIONS: ("Y" = Yes)

| West: | Southwest: | Midwest: | Southeast: | Northeast: | International: |
|---|---|---|---|---|---|
| | | | | Y | |

# CHIRON CORP

www.chiron.com

**Industry Group Code: 325412 Ranks within this company's industry group: Sales: Profits:**

| Drugs: | | Other: | | Clinical: | | Computers: | | Services: | |
|---|---|---|---|---|---|---|---|---|---|
| Discovery: | Y | AgriBio: | | Trials/Services: | | Hardware: | | Specialty Services: | |
| Licensing: | | Genetic Data: | Y | Labs: | | Software: | | Consulting: | |
| Manufacturing: | Y | Tissue Replacement: | | Equipment/Supplies: | | Arrays: | | Blood Collection: | |
| Genetics: | | | | Research & Development Services: | | Database Management: | | Drug Delivery: | |
| | | | | Diagnostics: | Y | | | Drug Distribution: | |

## TYPES OF BUSINESS:

Pharmaceuticals Discovery & Development
Biopharmaceuticals
Vaccines
Blood Screening Assays

## BRANDS/DIVISIONS/AFFILIATES:

Novartis AG
Vaccines and Diagnostics
Gen-Probe
PROCLEIX
ZymeQuest

## CONTACTS: Note: Officers with more than one job title may be intentionally listed here more than once.

Gene Walther, Pres.

| Phone: 510-655-8730 | Fax: 510-655-9910 |
|---|---|
| Toll-Free: | |
| Address: 4560 Horton St., Emeryville, CA 94608-2916 US | |

## GROWTH PLANS/SPECIAL FEATURES:

Chiron Corp., the blood screening business of Novartis AG's Vaccines and Diagnostics division, develops tools, products and services focused on preventing transfusion-transmitted infection. The company was acquired by Novartis in April 2006. Chiron's transcription-mediated amplification (TMA) products are designed to simplify nucleic acid testing (NAT) by enabling simultaneous detection of multiple viruses within a single tube. TMA technology was developed by Gen-Probe, Chiron's NAT innovation partner. TMA begins by preparing samples for testing by lysing the viruses to release genetic material, a process which involves no pretreatment or handling, thus reducing the risk of contamination. Capture probes hybridize internal control (IC) and viral nucleic acids and bind them to magnetic particles, and then the unbound material is washed away. A DNA copy of the target nucleic acid is created using reverse transcriptase and RNA is synthesized using RNA polymerase. The newly synthesized RNA can then reenter the TMA process and serve as further amplification templates. With this process, billions of copies can potentially be created in under an hour. Simultaneous detection of both IC- and viral-encoded RNA is enabled by dual kinetic assay (DKA) technology, which produces a flash of light for IC-encoded RNA, and a long glow for viral-encoded RNA. Chiron's TMA products include instruments, software and assays, sold under the PROCLEIX brand. Chiron is currently developing an enhanced immunoassay for the detection of abnormal prions (protein particles) in blood and blood products that are associated with variant Creuzfeldt-Jakob Disease, a rare, degenerative and fatal brain disorder believed to be developed by consuming cattle products contaminated with mad-cow disease (bovine spongiform encephalopathy). With its partner ZymeQuest, Chiron is also developing a system to convert A, B and AB red blood cells to enzyme-converted, universally transfusable group O red blood cells. Chiron also offers its clients NAT training and support.

## FINANCIALS: Sales and profits are in thousands of dollars—add 000 to get the full amount. 2008 Note: Financial information for 2008 was not available for all companies at press time.

| | | | U.S. Stock Ticker: Subsidiary |
|---|---|---|---|
| 2008 Sales: $ | 2008 Profits: $ | | Int'l Ticker: Int'l Exchange: |
| 2007 Sales: $ | 2007 Profits: $ | | Employees: 5,500 |
| 2006 Sales: $ | 2006 Profits: $ | | Fiscal Year Ends: 12/31 |
| 2005 Sales: $1,921,000 | 2005 Profits: $187,000 | | Parent Company: NOVARTIS AG |
| 2004 Sales: $1,723,355 | 2004 Profits: $78,917 | | |

## SALARIES/BENEFITS:

| Pension Plan: | ESOP Stock Plan: | Profit Sharing: | Top Exec. Salary: $800,000 | Bonus: $897,293 |
|---|---|---|---|---|
| Savings Plan: Y | Stock Purch. Plan: | | Second Exec. Salary: $552,461 | Bonus: $1,500,000 |

## OTHER THOUGHTS:

**Apparent Women Officers or Directors:**
**Hot Spot for Advancement for Women/Minorities:**

## LOCATIONS: ("Y" = Yes)

| West: | Southwest: | Midwest: | Southeast: | Northeast: | International: |
|---|---|---|---|---|---|
| Y | | Y | | Y | Y |

# CML HEALTHCARE INCOME FUND

www.cmlhealthcare.com

**Industry Group Code: 6215 Ranks within this company's industry group: Sales: 3 Profits: 2**

| Drugs: | Other: | Clinical: | | Computers: | Services: |
|---|---|---|---|---|---|
| Discovery: | AgriBio: | Trials/Services: | | Hardware: | Specialty Services: |
| Licensing: | Genetic Data: | Labs: | Y | Software: | Consulting: |
| Manufacturing: | Tissue Replacement: | Equipment/Supplies: | | Arrays: | Blood Collection: |
| Genetics: | | Research & Development Services: | | Database Management: | Drug Delivery: |
| | | Diagnostics: | Y | | Drug Distribution: |

## TYPES OF BUSINESS:
Medical Diagnostic Services
Laboratory Testing Services
Medical Imaging Services

## BRANDS/DIVISIONS/AFFILIATES:
CML Healthcare Inc.

## CONTACTS: *Note: Officers with more than one job title may be intentionally listed here more than once.*
Paul J. Bristow, CEO
Kent B. Nicholson, COO/Exec. VP
Paul J. Bristow, Pres.
Tom Weber, CFO/Exec. VP
Cameron Duff, VP-Corp. Dev.
Donald W. Kerr, Gen. Mgr.-Laboratory Svcs.
Kent Wentzell, Gen. Mgr.-Imaging Svcs.
Stephen R. Wiseman, Chmn.

| Phone: 905-565-0043 | Fax: 905-565-2844 |
|---|---|
| Toll-Free: 800-263-0801 | |
| Address: 6560 Kennedy Rd., Mississauga, ON L5T 2X4 Canada | |

## GROWTH PLANS/SPECIAL FEATURES:

CML HealthCare Income Fund is an open-ended trust that owns CML Healthcare, Inc., one of Canada's largest diagnostic services businesses, providing laboratory testing services in Ontario and medical imaging services across Canada. The firm operates in two divisions: laboratory services and imaging services. The laboratory services division conducts a wide range of medical tests, including hematology, biochemistry, cytology, microbiology, histology, Holter monitoring, prostate specific antigen (PSA) and HPV (human papillomavirus) testing. All of the tests are used by physicians to assist in diagnosis and treatment. The laboratory services division operates through the company's network of two licensed medical diagnostic laboratories and 125 operating licensed specimen collection centers. The imaging services division provides medical imaging services, including x-ray and fluoroscopy, ultrasound, mammography, bone densitometry, nuclear medicine, magnetic resonance imaging (MRI) and computed tomography through a network of 115 non-hospital based medical imaging clinics located in Ontario, British Columbia, Alberta, Manitoba and Quebec. In January 2008, the firm acquired Vancouver-based Croyden Radiological Services Ltd. In February 2008, CML acquired American Radiology Services, Inc. for $150.4 million. The acquisition expands the firm's business into the U.S. with 15 fixed-site multi-modality and two single-modality outpatient centers Delaware and Maryland. In April 2008, the company acquired another Vancouver area medical imaging firm, Laurel Radiology, Inc. In July 2008, CML acquired Applemed X-ray and Ultrasound Services, Inc., operating in the area of Saint Catharine's, Ontario. In September 2008, the firm completed its acquisition of General Medical Imaging Services, Inc., located in British Columbia.

## FINANCIALS: Sales and profits are in thousands of dollars—add 000 to get the full amount. 2008 Note: Financial information for 2008 was not available for all companies at press time.

| | | |
|---|---|---|
| 2008 Sales: $425,500 | 2008 Profits: $93,400 | **U.S. Stock Ticker:** |
| 2007 Sales: $287,800 | 2007 Profits: $92,300 | **Int'l Ticker: CLC.UN**   Int'l Exchange: Toronto-TSX |
| 2006 Sales: $265,800 | 2006 Profits: $90,900 | Employees: 2,000 |
| 2005 Sales: $256,753 | 2005 Profits: $74,979 | Fiscal Year Ends: 12/31 |
| 2004 Sales: $ | 2004 Profits: $ | Parent Company: |

## SALARIES/BENEFITS:

| Pension Plan: | ESOP Stock Plan: | Profit Sharing: | Top Exec. Salary: $441,047 | Bonus: $198,471 |
|---|---|---|---|---|
| Savings Plan: | Stock Purch. Plan: | | Second Exec. Salary: $310,728 | Bonus: $91,403 |

## OTHER THOUGHTS:
**Apparent Women Officers or Directors**: 1
**Hot Spot for Advancement for Women/Minorities**:

## LOCATIONS: ("Y" = Yes)

| West: | Southwest: | Midwest: | Southeast: | Northeast: | International: |
|---|---|---|---|---|---|
| | | | | Y | Y |

Note: Financial information, benefits and other data can change quickly and may vary from those stated here.

# COLUMBIA LABORATORIES INC

## www.columbialabs.com

**Industry Group Code: 325412  Ranks within this company's industry group: Sales: 94   Profits: 89**

| Drugs: | | Other: | Clinical: | Computers: | Services: | |
|---|---|---|---|---|---|---|
| Discovery: | Y | AgriBio: | Trials/Services: | Hardware: | Specialty Services: | |
| Licensing: | Y | Genetic Data: | Labs: | Software: | Consulting: | |
| Manufacturing: | Y | Tissue Replacement: | Equipment/Supplies: | Arrays: | Blood Collection: | |
| Genetics: | | | Research & Development Services: | Database Management: | Drug Delivery: | Y |
| | | | Diagnostics: | | Drug Distribution: | |

## TYPES OF BUSINESS:

Women's Healthcare Drugs
Endocrine Drugs

## BRANDS/DIVISIONS/AFFILIATES:

Prochieve
Crinone
Striant
RepHresh
Lidocaine
Bioadhesive Delivery System
Assisted Reproductive Technology

## CONTACTS: *Note: Officers with more than one job title may be intentionally listed here more than once.*

Robert S. Mills, CEO
Robert S. Mills, Pres.
James A. Meer, CFO/Sr. VP/Treas.
Carl Worrell, Head-Sales & Mktg.
George W. Creasy, VP-Clinical Research & Dev.
Michael McGrane, General Counsel/Sr. VP/Sec.
James A. Meer, Treas.
Stephen G. Kasnet, Chmn.

| Phone: 973-994-3999 | Fax: 973-994-3001 |
|---|---|
| **Toll-Free:** 866-566-5636 | |
| **Address:** 354 Eisenhower Pkwy., Plz. 1, Fl. 2, Livingston, NJ 07039 US | |

## GROWTH PLANS/SPECIAL FEATURES:

Columbia Laboratories, Inc. develops, manufactures and markets drugs that target women's healthcare and endocrine-related disorders. The company's products are used for vaginal delivery of hormones, analgesics and other drugs; buccal delivery of hormones and peptides; and for support in the reproductive market. The vaginal products adhere to the vaginal epithelium, and the buccal products adhere to the mucosal membrane of the gum and cheek. Both forms provide sustained and controlled delivery of active drug ingredients. All of the firm's products utilize its patented Bioadhesive Delivery System (BDS) technology, which consists principally of a polymer and an active ingredient. Products include CRINONE and PROCHIEVE progesterone gels, designed to deliver progesterone directly to the uterus; STRAINT, used to treat hypogonadism, a deficiency or absence of endogenous testosterone production; Replens vaginal moisturizer, which replenishes vaginal moisture on a sustained basis and relieves discomfort associated with dryness; and RepHresh vaginal gel, a feminine hygiene product that can eliminate vaginal odor by maintaining a normal pH range. PROCHIEVE is part of the firm's Assisted Reproductive Technology (ART) treatment for infertile women with progesterone deficiencies. Outside the U.S., CRINONE is approved for marketing for medical indications such as menstrual irregularities, dysmenorrheal and dysfunctional uterine bleeding. Columbia contracts its manufacturing activities to third parties in the U.K., Switzerland and Italy. Merck Serono S.A. sells CRINONE outside of the U.S. The company began Phase II testing of vaginally-administered Lidocaine, for the prevention and the treatment of dysmenorrhea, a condition that affects over 50% of all menstruating women.

Columbia offers a 401(k) plan; medical, dental, life and disability insurance; and paid vacation and holidays.

## FINANCIALS: Sales and profits are in thousands of dollars—add 000 to get the full amount. 2008 Note: Financial information for 2008 was not available for all companies at press time.

| | | |
|---|---|---|
| 2008 Sales: $36,340 | 2008 Profits: $-14,077 | **U.S. Stock Ticker: CBRX** |
| 2007 Sales: $29,627 | 2007 Profits: $-14,292 | **Int'l Ticker:**    Int'l Exchange: |
| 2006 Sales: $17,393 | 2006 Profits: $-12,485 | Employees:   62 |
| 2005 Sales: $22,041 | 2005 Profits: $-10,104 | Fiscal Year Ends: 12/31 |
| 2004 Sales: $17,860 | 2004 Profits: $-25,130 | Parent Company: |

## SALARIES/BENEFITS:

| Pension Plan: | ESOP Stock Plan: | Profit Sharing: | Top Exec. Salary: $383,933 | Bonus: $ |
|---|---|---|---|---|
| Savings Plan: Y | Stock Purch. Plan: | | Second Exec. Salary: $294,867 | Bonus: $ |

## OTHER THOUGHTS:

**Apparent Women Officers or Directors:** 1
**Hot Spot for Advancement for Women/Minorities:**

## LOCATIONS: ("Y" = Yes)

| West: | Southwest: | Midwest: | Southeast: | Northeast: | International: |
|---|---|---|---|---|---|
| | | | | Y | Y |

# COMMONWEALTH BIOTECHNOLOGIES INC          www.cbi-biotech.com

**Industry Group Code: 541712  Ranks within this company's industry group:** Sales: 17   Profits: 10

| Drugs: | Other: | | Clinical: | | Computers: | | Services: | |
|---|---|---|---|---|---|---|---|---|
| Discovery: | AgriBio: | | Trials/Services: | | Hardware: | | Specialty Services: | |
| Licensing: | Genetic Data: | Y | Labs: | Y | Software: | | Consulting: | |
| Manufacturing: | Tissue Replacement: | Y | Equipment/Supplies: | | Arrays: | | Blood Collection: | |
| Genetics: | | | Research & Development Services: | Y | Database Management: | Y | Drug Delivery: | |
| | | | Diagnostics: | Y | | | Drug Distribution: | |

## TYPES OF BUSINESS:

Contract Research-Biotech & Genetics
DNA & Genome Sequencing
Genetic Testing Services
Molecular Biology
Biophysical Analysis
Nucleic Acid Synthesis
Peptide Sequencing & Synthesis
Biodefense Services

## BRANDS/DIVISIONS/AFFILIATES:

Fairfax Identity Labs
Mimotopes Pty Ltd
Tripos Discovery Research Limited
Exelgen, Ltd.
Venturepharm-ASIA

## CONTACTS: *Note: Officers with more than one job title may be intentionally listed here more than once.*

Bill Guo, Acting CEO
Richard J. Freer, COO
Robert B. Harris, Pres.
Thomas R. Reynolds, Exec. VP-Science
Thomas R. Reynolds, Exec. VP-Tech.
Thomas R. Reynolds, Corp. Sec.
James H. Brennan, VP-Financial Oper.
Bill Guo, Chmn.

| Phone: 804-648-3820 | Fax: 804-648-2641 |
|---|---|
| Toll-Free: 800-735-9224 | |
| Address: 601 Biotech Dr., Richmond, VA 23235 US | |

## GROWTH PLANS/SPECIAL FEATURES:

Commonwealth Biotechnologies, Inc. (CBI) provides early developmental contract research and services for global biotechnology industries, academic institutions, government agencies and pharmaceutical companies. The company offers a broad array of analytical and synthetic chemistries and biophysical analysis technologies. CBI offers its services in fully-integrated research programs. Through its Fairfax Identity Labs subsidiary, CBI offers comprehensive genetic identity testing, including paternity, forensic and Combined DNA Index System (CODIS) analyses. The firm is accredited by the American Association of Blood Banks, Clinical Laboratory Improvement Amendments (CLIA) and the National Forensic Science Technology Council, and operates fully accredited Biosafety level 3 (BSL-3) laboratories. CBI has numerous specialty labs, including labs for bacteriology and virology; a DNA reference lab; calorimetry and mass spectrometry labs; cell culture and fermentation labs; high throughput DNA sequence labs; and peptide synthesis labs. CBI operates in four principal areas: bio-defense, which includes environmental and laboratory testing for pathogens; laboratory support services for on-going clinical trials; comprehensive contract projects in the private sector; and DNA reference, which includes paternity testing, forensic case-work analysis, mitochondrial DNA analysis and CODIS sequence analysis. The company views commercial and government contracts as its most important sources of revenue, and has consequently moved away from piece work for individual investigators. The company attracts customers from its presentations at national trade shows and advertisements in professional journals. In 2008, the company announced plans to establish two drug discovery service facilities in China through its joint venture with Venturepharm Laboratories Ltd., Venturepharm-ASIA. Also in 2008, the company opened an office in Tokyo.

## FINANCIALS:  Sales and profits are in thousands of dollars—add 000 to get the full amount. 2008 Note: Financial information for 2008 was not available for all companies at press time.

| | | |
|---|---|---|
| 2008 Sales: $9,435 | 2008 Profits: $-9,863 | **U.S. Stock Ticker: CBTE** |
| 2007 Sales: $9,025 | 2007 Profits: $-2,758 | **Int'l Ticker:**   Int'l Exchange: |
| 2006 Sales: $6,532 | 2006 Profits: $-1,153 | Employees:   65 |
| 2005 Sales: $7,803 | 2005 Profits: $ 79 | Fiscal Year Ends: 12/31 |
| 2004 Sales: $5,749 | 2004 Profits: $- 368 | Parent Company: |

## SALARIES/BENEFITS:

| Pension Plan: | ESOP Stock Plan: | Profit Sharing: | Top Exec. Salary: $191,500 | Bonus: $ |
|---|---|---|---|---|
| Savings Plan: | Stock Purch. Plan: | | Second Exec. Salary: $167,900 | Bonus: $ |

## OTHER THOUGHTS:

**Apparent Women Officers or Directors:**
**Hot Spot for Advancement for Women/Minorities:**

## LOCATIONS: ("Y" = Yes)

| West: | Southwest: | Midwest: | Southeast: | Northeast: | International: |
|---|---|---|---|---|---|
| | | | | Y | |

# COMPUGEN LTD

www.cgen.com

**Industry Group Code: 325412  Ranks within this company's industry group:** Sales: 149   Profits: 87

| Drugs: | | Other: | | Clinical: | | Computers: | | Services: | |
|---|---|---|---|---|---|---|---|---|---|
| Discovery: | Y | AgriBio: | Y | Trials/Services: | | Hardware: | Y | Specialty Services: | |
| Licensing: | | Genetic Data: | Y | Labs: | | Software: | Y | Consulting: | |
| Manufacturing: | | Tissue Replacement: | | Equipment/Supplies: | | Arrays: | | Blood Collection: | |
| Genetics: | | | | Research & Development Services: | | Database Management: | | Drug Delivery: | |
| | | | | Diagnostics: | Y | | | Drug Distribution: | |

## TYPES OF BUSINESS:

Drug Discovery
Biotechnology Databases & Diagnostics
Agricultural Biotechnology
Bioinformatics Platform

## BRANDS/DIVISIONS/AFFILIATES:

LEADS
Evogene Ltd.
CGEN-25007

## CONTACTS: *Note: Officers with more than one job title may be intentionally listed here more than once.*

Martin S. Gerstal, Co-CEO
Martin S. Gerstal, Pres.
Dikla Czaczkes Axselbrad, CFO
Dorit Amitay, Dir., Human Resources
Yossi Cohen, CTO
Eli Zangvil, VP-Bus. Dev.
Anat Cohen-Dayag, Co-CEO
Dov Hershberg, Chmn.

| **Phone:** 972-3-765-8585 | **Fax:** 972-3-765-8555 |
|---|---|
| **Toll-Free:** | |
| **Address:** 72 Pinchas Rosen St., Tel Aviv, 69512 Israel | |

## GROWTH PLANS/SPECIAL FEATURES:

Compugen, Ltd. engages in drug and diagnostic product candidate discovery and commercialization, largely through early stage licensing and co-development agreements. Compugen focuses on using field-focused discovery platforms to predict, select and validate therapeutic drug candidates and diagnostic biomarker candidates. Headquartered in Israel, it also works through a wholly-owned subsidiary in Maryland. The company's initial discovery platforms have focused mainly on cancer, cardiovascular and immune-related diseases. The firm is structured along the lines of three principle activities: research and discovery; therapeutics; and diagnostics. Its research and discovery activities are focused on developing field-focused discovery platforms for the prediction and selection of product candidates. Current areas of research include identification of GPCR peptide ligands, Disease Associated Conformation peptide blockers, targets for monoclonal antibodies, drug response genomic markers and pathway kinetic simulation. Therapeutic and diagnostic activities both consist of identification and development of candidates as either therapeutic or diagnostic products. Current therapeutic drug candidates are potential treatments for various types of cancer, inflammatory diseases and cardiovascular indications. Current diagnostic programs include immunoassay diagnostics, drug-induced toxicity biomarkers, genomic markers and nucleic acid diagnostics. Compugen's core product is LEADS, a bioinformatics platform for analyzing genomic and protein data. The company maintains numerous collaborations for the development and licensing of its product candidates, and currently has therapeutic, diagnostic and research agreements with Biosite, Medarex, Inc.; Merck & Co., Inc.; Ortho-Clinical Diagnostics; Roche; Siemens Healthcare Diagnostics, Inc.; and Teva Pharmaceutical Industries. Evogene, Ltd., a wholly-owned subsidiary, focuses on agricultural biotechnology, especially plant genomics. In July 2008, Compugen announced the positive results of an in vivo study of CGEN-25007, a novel peptide antagonist of the gp96 protein as well as the discovery of new targets for antibody therapy for various kinds of cancer and four biomarkers for early detection of drug-induced nephrotoxicity.

## FINANCIALS: Sales and profits are in thousands of dollars—add 000 to get the full amount. 2008 Note: Financial information for 2008 was not available for all companies at press time.

| | | |
|---|---|---|
| 2008 Sales: $ 338 | 2008 Profits: $-12,527 | **U.S. Stock Ticker: CGEN** |
| 2007 Sales: $ 180 | 2007 Profits: $-12,114 | **Int'l Ticker: CGEN**  Int'l Exchange: Tel Aviv-TASE |
| 2006 Sales: $ 215 | 2006 Profits: $-13,020 | Employees: 57 |
| 2005 Sales: $ 646 | 2005 Profits: $-13,978 | Fiscal Year Ends: 12/31 |
| 2004 Sales: $ 2,630 | 2004 Profits: $-13,722 | Parent Company: |

## SALARIES/BENEFITS:

| Pension Plan: | ESOP Stock Plan: | Profit Sharing: | Top Exec. Salary: $ | Bonus: $ |
|---|---|---|---|---|
| Savings Plan: | Stock Purch. Plan: | | Second Exec. Salary: $ | Bonus: $ |

## OTHER THOUGHTS:

**Apparent Women Officers or Directors**: 4
**Hot Spot for Advancement for Women/Minorities**: Y

## LOCATIONS: ("Y" = Yes)

| West: | Southwest: | Midwest: | Southeast: | Northeast: | International: |
|---|---|---|---|---|---|
| | | | Y | | Y |

# CORTEX PHARMACEUTICALS INC

**www.cortexpharm.com**

Industry Group Code: 325412  Ranks within this company's industry group:  Sales:       Profits: 91

| Drugs: | | Other: | | Clinical: | | Computers: | | Services: | |
|---|---|---|---|---|---|---|---|---|---|
| Discovery: | Y | AgriBio: | | Trials/Services: | | Hardware: | | Specialty Services: | |
| Licensing: | Y | Genetic Data: | | Labs: | | Software: | | Consulting: | |
| Manufacturing: | Y | Tissue Replacement: | | Equipment/Supplies: | | Arrays: | | Blood Collection: | |
| Genetics: | | | | Research & Development Services: | | Database Management: | | Drug Delivery: | |
| | | | | Diagnostics: | | | | Drug Distribution: | |

## TYPES OF BUSINESS:

Drugs-Neurological
Psychiatric Drugs

## BRANDS/DIVISIONS/AFFILIATES:

AMPAKINE
CX717
CX2007
CX1739
CX1942

## CONTACTS:
*Note: Officers with more than one job title may be intentionally listed here more than once.*

Mark A. Varney, CEO
Mark A. Varney, Pres.
Maria S. Messinger, CFO/VP
Maria S. Messinger, Corp. Sec.
James H. Coleman, Sr. VP-Bus. Dev.
Janet Vasquez, Contact-Media
Pierre V. Tran, Chief Medical Officer/VP-Clinical Dev.
Steven A. Johnson, VP-Preclinical Dev.
Roger G. Stoll, Chmn.

| Phone: 949-727-3157 | Fax: 949-727-3657 |
|---|---|
| Toll-Free: | |
| Address: 15231 Barranca Pkwy., Irvine, CA 92618 US | |

## GROWTH PLANS/SPECIAL FEATURES:

Cortex Pharmaceuticals, Inc., focuses on developing small-molecule compounds that positively modulate AMPA-type glutamate receptors, a complex of proteins involved in the communication between nerve cells in the brain. The firm's compounds, which are being developed under the brand name AMPAKINE, enhance the activity of the AMPA receptor; AMPAKINE shows promise in treating neurological and psychiatric diseases/disorders that involve the depressed functioning of pathways in the brain using glutamate as a neurotransmitter. The drugs thus have the potential to treat mental illnesses such as Alzheimer's disease; depression; mild cognitive impairment (MCI); schizophrenia; Attention Deficit Hyperactivity Disorder (ADHD); respiratory depression caused by opiate analgesics; and possibly sleep apnea. Ampakine comes in two categories: low impact and high impact compounds, defined by the site to which they bind on the AMPA receptor complex. The company's most advanced low impact Ampakine compound is CX717, which is currently in Phase II clinical development. CX717 is also in preclinical trials for respiratory depression. The firm has three other compounds in clinical or preclinical development: CX1739, currently in a Phase II sleep apnea study; CX2007, a long-lasting chemical compound for Alzheimer's disease and ADHD treatment currently in the preclinical stage; and CX1942, a possible treatment of sleep apnea and oral respiratory depression, also in the preclinical stage. Cortex has several other low impact compounds in preclinical trials for schizophrenia, depression and veterinary uses, and one high impact compound in a preclinical trial. Cortex owns or has exclusive rights to more than 180 issued or allowed U.S. and foreign patents. In March 2009, Cortex announced a restructuring plan which included laying off approximately 50% of its employees.

The company offers employees medical, dental, life and vision insurance; a 401(k) plan; a cafeteria plan; and stock options.

## FINANCIALS:
Sales and profits are in thousands of dollars—add 000 to get the full amount. 2008 Note: Financial information for 2008 was not available for all companies at press time.

| | | | |
|---|---|---|---|
| 2008 Sales: $ | 2008 Profits: $-14,596 | U.S. Stock Ticker: COR | |
| 2007 Sales: $ | 2007 Profits: $-12,969 | Int'l Ticker:    Int'l Exchange: | |
| 2006 Sales: $1,177 | 2006 Profits: $-16,055 | Employees:    27 | |
| 2005 Sales: $2,577 | 2005 Profits: $-11,605 | Fiscal Year Ends: 12/31 | |
| 2004 Sales: $6,973 | 2004 Profits: $-5,994 | Parent Company: | |

## SALARIES/BENEFITS:

| Pension Plan: | ESOP Stock Plan: | Profit Sharing: | Top Exec. Salary: $370,000 | Bonus: $ |
|---|---|---|---|---|
| Savings Plan: Y | Stock Purch. Plan: | | Second Exec. Salary: $347,277 | Bonus: $ |

## OTHER THOUGHTS:

**Apparent Women Officers or Directors:** 2
**Hot Spot for Advancement for Women/Minorities:**

## LOCATIONS: ("Y" = Yes)

| West: | Southwest: | Midwest: | Southeast: | Northeast: | International: |
|---|---|---|---|---|---|
| Y | | | | | |

# COVANCE INC

www.covance.com

Industry Group Code: 541712 **Ranks within this company's industry group:** Sales: 1 Profits: 1

| Drugs: | Other: | Clinical: | | Computers: | | Services: | |
|---|---|---|---|---|---|---|---|
| Discovery: | AgriBio: | Trials/Services: | Y | Hardware: | | Specialty Services: | Y |
| Licensing: | Genetic Data: | Labs: | | Software: | Y | Consulting: | |
| Manufacturing: | Tissue Replacement: | Equipment/Supplies: | | Arrays: | | Blood Collection: | |
| Genetics: | | Research & Development Services: | Y | Database Management: | | Drug Delivery: | |
| | | Diagnostics: | | | | Drug Distribution: | |

## TYPES OF BUSINESS:

Pharmaceutical Research & Development
Drug Preclinical/Clinical Trials
Laboratory Testing & Analysis
Approval Assistance
Health Economics & Outcomes Services
Online Tools

## BRANDS/DIVISIONS/AFFILIATES:

LabLink
Study Tracker
Trial Tracker

## CONTACTS: *Note: Officers with more than one job title may be intentionally listed here more than once.*

Joseph L. Herring, CEO
Wendel Barr, COO/Exec. VP
William Klitgaard, CFO/Sr. VP
James W. Lovett, General Counsel/Sr. VP
Richard F. Cimino, Pres., Late Stage Dev.
Joseph L. Herring, Chmn.
Anthony Cork, Pres., Early Dev. Europe

| Phone: 609-452-4440 | Fax: |
|---|---|
| **Toll-Free:** 888-268-2623 | |
| **Address:** 210 Carnegie Ctr., Princeton, NJ 08540 US | |

## GROWTH PLANS/SPECIAL FEATURES:

Covance, Inc. is a leading drug development services company and contract research organization. It provides a wide range of product development services to pharmaceutical, biotechnology and medical device industries across the globe. The company also provides laboratory testing services for clients in the chemical, agrochemical and food businesses. The firm operates two business segments: early development services and late-stage development services. Covance's early development services include preclinical services (such as toxicology, pharmaceutical development, research products and a bioanalytical testing service) and Phase I clinical services. Its late-stage development services cover clinical development and support; clinical trials; periapproval and market access; and central laboratory operations. Covance has also introduced several Internet-based products: Study Tracker, an Internet-based client access product, which permits customers of toxicology services to review study data and schedules on a near real-time basis; LabLink, a client access program that allows customers of central laboratory services to review and query lab data; and Trial Tracker, a web-enabled clinical trial project management and tracking tool intended to allow both employees and customers of its late-stage clinical business to review and manage all aspects of clinical-trial projects. In October 2008, Covance purchased an early drug development facility in Greenfield, Indiana from Eli Lilly and Company. In addition, Covance upgraded several research facilities and in June 2008 opened a 50,000 square foot research clinic in Evansville, Indiana. In September 2008 , the firm announced plans to build a preclinical facility in China, and announced in 2009 that it plans to open clinical development offices in Ukraine, Slovakia and Israel. In December 2008, Covance purchased a minority stake in Caprion Proteomics, a provider of proteomics based services to the pharmaceutical industry.

Covance offers its employees benefits such as medical, dental and vision plans; a range of insurance benefits; employee assistance; financial planning services; and tuition reimbursement.

## FINANCIALS: Sales and profits are in thousands of dollars—add 000 to get the full amount. 2008 Note: Financial information for 2008 was not available for all companies at press time.

| | | |
|---|---|---|
| 2008 Sales: $1,827,067 | 2008 Profits: $196,760 | **U.S. Stock Ticker: CVD** |
| 2007 Sales: $1,631,516 | 2007 Profits: $175,929 | **Int'l Ticker:** Int'l Exchange: |
| 2006 Sales: $1,406,058 | 2006 Profits: $144,998 | Employees: 9,600 |
| 2005 Sales: $1,250,400 | 2005 Profits: $119,600 | Fiscal Year Ends: 12/31 |
| 2004 Sales: $1,056,397 | 2004 Profits: $97,947 | Parent Company: |

## SALARIES/BENEFITS:

| Pension Plan: | ESOP Stock Plan: | Profit Sharing: | Top Exec. Salary: $729,167 | Bonus: $850,000 |
|---|---|---|---|---|
| Savings Plan: Y | Stock Purch. Plan: Y | | Second Exec. Salary: $415,000 | Bonus: $255,000 |

## OTHER THOUGHTS:

**Apparent Women Officers or Directors:** 2
**Hot Spot for Advancement for Women/Minorities:** Y

## LOCATIONS: ("Y" = Yes)

| West: | Southwest: | Midwest: | Southeast: | Northeast: | International: |
|---|---|---|---|---|---|
| Y | Y | Y | Y | Y | Y |

Note: Financial information, benefits and other data can change quickly and may vary from those stated here.

# COVIDIEN PLC

**www.covidien.com**

Industry Group Code: 33911 Ranks within this company's industry group: Sales: 3   Profits: 3

| Drugs: | Other: | Clinical: | | Computers: | Services: | |
|---|---|---|---|---|---|---|
| Discovery: | AgriBio: | Trials/Services: | | Hardware: | Specialty Services: | |
| Licensing: | Genetic Data: | Labs: | | Software: | Consulting: | |
| Manufacturing: Y | Tissue Replacement: | Equipment/Supplies: | Y | Arrays: | Blood Collection: | |
| Genetics: | | Research & Development Services: | Y | Database Management: | Drug Delivery: | Y |
| | | Diagnostics: | | | Drug Distribution: | Y |

## TYPES OF BUSINESS:

Medical Equipment & Supplies, Manufacturing
Imaging Agents
Pharmaceutical Products
Medical Devices

## BRANDS/DIVISIONS/AFFILIATES:

Kendall
Autosuture
Syneture
AbsorbaTack
VNUS Medical Technologies, Inc.
Nellcor
Puritan Bennett
TussiCaps

## CONTACTS: *Note: Officers with more than one job title may be intentionally listed here more than once.*

Richard J. Meelia, CEO
Richard J. Meelia, Pres.
Charles J. Dockendorff, CFO/Exec. VP
Karen A. Quinn-Quintin, Sr. VP-Human Resources
Steven M. McManama, CIO
John H. Masterson, General Counsel/Sr. VP
Amy A. Wendell, Sr. VP-Bus. Dev. & Strategy
Eric A. Kraus, Sr. VP-Corp. Comm.
Coleman Lannum, VP-Investor Rel.
Richard G. Brown, Jr., Chief Acct. Officer/Controller
Jose E. Almeida, Pres., Medical Devices
Timothy R. Wright, Pres., Pharmaceutical Prod. & Imaging Solutions
James C. Clemmer, Pres., Medical Supplies
Kevin G. DaSilva, VP/Treas.
Richard J. Meelia, Chmn.
James M. Muse, Sr. VP-Global Supply Chain

| Phone: 441-298-2480 | Fax: |
|---|---|
| Toll-Free: | |
| Address: 131 Front St., Hamilton, HM 12 Bermuda | |

## GROWTH PLANS/SPECIAL FEATURES:

Covidien plc, formerly Covidien Ltd., is a global healthcare products company that markets items for both clinical and home settings. The firm manufactures and distributes industry-leading brands such as Kendall, Autosuture, Syneture, Nellcor and Puritan Bennett. The company's business consists of four segments: medical devices, imaging solutions, pharmaceutical products and medical supplies. The medical devices segment offers products such as endomechanical instruments; soft tissue repair products; energy devices; oximetry and monitoring products; airway and ventilation products; vascular devices; sharpsafety products; and clinical care products. This division, which generates 68% of sales, markets these products under the Autosuture, AbsorbaTack, Syneture and Valleylab names. The imaging solutions segment (12% of sales) includes contrast agents, contrast systems and radiopharmaceuticals. The pharmaceutical products segment (10% of sales) produces active pharmaceutical ingredients, dosage pharmaceuticals and specialty chemicals. Pharmaceutical products include brand and generic pharmaceuticals; addiction treatment products; medicinal narcotics; acetaminophen; peptides; stearates; and phosphates. The medical supplies segment (10% of sales) provides various medical supplies (needles, syringes, etc.) and traditional wound care, incontinence and medical surgical products. Covidien's products are sold in over 140 countries, with roughly 55% of sales in the U.S. and 45% internationally. In May 2008, Covidien sold its European Incontinence business to a French firm and agreed to acquire certain assets of Pinyons Medical Technology Inc. In August 2008, the firm founded Covidien Ventures, a corporate venture investment company. New 2008 Covidien products include TussiCaps, a 12-hour cough suppressant; the Permacol Biologic Implant, a biologic mesh for hernia repair; and the RapidVac Smoke Evacuator System, which filters airborne contaminants from operating room environments. In May 2009, the company agreed to acquire medical device developer VNUS Medical Technologies, Inc., for approximately $440 million. In June 2009, Covidien Ltd. changed its country incorporation to Ireland and became Covidien plc.

## FINANCIALS: Sales and profits are in thousands of dollars—add 000 to get the full amount. 2008 Note: Financial information for 2008 was not available for all companies at press time.

| | | |
|---|---|---|
| 2008 Sales: $9,910,000 | 2008 Profits: $1,361,000 | U.S. Stock Ticker: COV |
| 2007 Sales: $8,895,000 | 2007 Profits: $-342,000 | Int'l Ticker:   Int'l Exchange: |
| 2006 Sales: $8,313,000 | 2006 Profits: $1,155,000 | Employees: 41,700 |
| 2005 Sales: $ | 2005 Profits: $ | Fiscal Year Ends: 9/30 |
| 2004 Sales: $ | 2004 Profits: $ | Parent Company: |

## SALARIES/BENEFITS:

| Pension Plan: | ESOP Stock Plan: | Profit Sharing: | Top Exec. Salary: $905,163 | Bonus: $514,002 |
|---|---|---|---|---|
| Savings Plan: Y | Stock Purch. Plan: | | Second Exec. Salary: $535,000 | Bonus: $514,002 |

## OTHER THOUGHTS:

**Apparent Women Officers or Directors**: 3
**Hot Spot for Advancement for Women/Minorities**: Y

## LOCATIONS: ("Y" = Yes)

| West: | Southwest: | Midwest: | Southeast: | Northeast: | International: |
|---|---|---|---|---|---|
| | | | | Y | Y |

Note: Financial information, benefits and other data can change quickly and may vary from those stated here.

# CRUCELL NV

www.crucell.com

**Industry Group Code: 325412  Ranks within this company's industry group: Sales: 45  Profits: 48**

| Drugs: | | Other: | Clinical: | | Computers: | | Services: |
|---|---|---|---|---|---|---|---|
| Discovery: | Y | AgriBio: | Trials/Services: | | Hardware: | | Specialty Services: |
| Licensing: | | Genetic Data: | Labs: | | Software: | | Consulting: |
| Manufacturing: | | Tissue Replacement: | Equipment/Supplies: | Y | Arrays: | | Blood Collection: |
| Genetics: | | | Research & Development Services: | | Database Management: | | Drug Delivery: |
| | | | Diagnostics: | | | | Drug Distribution: |

## TYPES OF BUSINESS:

Biopharmaceuticals
Biopharmaceutical Development Technologies

## BRANDS/DIVISIONS/AFFILIATES:

Quinvaxem
Hepavax-Gene
MoRu-Viraten
Epaxal
Dukoral
Inflexal V
PER.C6
AdVac

## CONTACTS: Note: Officers with more than one job title may be intentionally listed here more than once.

Ronald H. P. Brus, CEO
Cees de Jong, COO
Ronald H. P. Brus, Pres.
Leonard Kruimer, CFO
Jaap Goudsmit, Chief Scientific Officer
Rene Beukema, General Counsel/Corp. Sec.
Arthur Lahr, Exec. VP-Bus. Dev./Chief Strategy Officer
Bjorn Sjostrand, Chief Bus. Officer
Jan Pieter Oosterveld, Chmn.

| Phone: 31-71-519-91-00 | Fax: 31-71-519-98-00 |
|---|---|

**Toll-Free:**

**Address:** Archimedesweg 4-6, Leiden, 2333 CN The Netherlands

## GROWTH PLANS/SPECIAL FEATURES:

Crucell N.V. focuses on combating infectious diseases. With operations in 12 countries, Crucell sold over 100 million vaccine doses in more than 80 countries during 2008, making it one of the largest independent vaccine companies in the world. It has three categories of products. Pediatric products comprise the fully liquid Quinvaxem vaccine that protects against five childhood diseases (diphtheria, tetanus, whooping cough, hepatitis B and Haemophilus influenzae type B), aluminum-free hepatitis A vaccine Epaxal Junior (the lack of aluminum reduces pain during administration), recombinant hepatitis B vaccine Hepavax-Gene and measles and rubella vaccine MoRu-Viraten. Travel and endemic products comprise an adult version of Epaxal, typhoid fever vaccine Vivotif and drinkable cholera vaccine Dukoral. Lastly, the only respiratory product is adjuvanated flu vaccine Inflexal V. The antigen in Inflexal varies year-to-year based on World Health Organization (WHO) recommendations. Products in development include yellow fever vaccine Flavimun, currently undergoing Phase III trials, and vaccines for tuberculosis, seasonal flu strains, malaria, Ebola and HIV. The tuberculosis and flu vaccines are in Phase II testing, while the others are undergoing Phase I trials. Crucell is also testing human antibodies for the flu and rabies and a blood coagulation factor. The company develops its products through five proprietary technologies. PER.C6, the most important of these technologies, is a human designer cell line for the development and large-scale manufacturing of biopharmaceutical products. AdVac is used with PER.C6 to develop recombinant vaccines. STAR enhances the yield of recombinant human antibodies and proteins on mammalian cell lines. MAbstract is used to discover novel drug targets and identify human antibodies. Lastly, Virosome enables the use of virus antigens in the making of vaccines. Product sales generated 85% of the firm's 2008 revenues, with license revenues (11%) and service fees (4%) filling in the remainder.

## FINANCIALS: Sales and profits are in thousands of dollars—add 000 to get the full amount. 2008 Note: Financial information for 2008 was not available for all companies at press time.

| | | |
|---|---|---|
| 2008 Sales: $379,440 | 2008 Profits: $20,720 | **U.S. Stock Ticker: CRXL** |
| 2007 Sales: $289,430 | 2007 Profits: $-60,740 | **Int'l Ticker: CRX**  Int'l Exchange: Zurich-SWX |
| 2006 Sales: $186,860 | 2006 Profits: $-123,845 | Employees: |
| 2005 Sales: $ | 2005 Profits: $ | Fiscal Year Ends: 12/31 |
| 2004 Sales: $ | 2004 Profits: $ | Parent Company: |

## SALARIES/BENEFITS:

| Pension Plan: | ESOP Stock Plan: | Profit Sharing: | Top Exec. Salary: $ | Bonus: $ |
|---|---|---|---|---|
| Savings Plan: | Stock Purch. Plan: | | Second Exec. Salary: $ | Bonus: $ |

## OTHER THOUGHTS:

**Apparent Women Officers or Directors:**
**Hot Spot for Advancement for Women/Minorities:**

## LOCATIONS: ("Y" = Yes)

| West: | Southwest: | Midwest: | Southeast: | Northeast: | International: |
|---|---|---|---|---|---|
| | | | Y | Y | Y |

Note: Financial information, benefits and other data can change quickly and may vary from those stated here.

# CSL LIMITED

**www.csl.com.au**

Industry Group Code: 325414  **Ranks within this company's industry group:** Sales: 2    Profits: 1

| Drugs: | | Other: | | Clinical: | | Computers: | | Services: | |
|---|---|---|---|---|---|---|---|---|---|
| Discovery: | | AgriBio: | | Trials/Services: | | Hardware: | | Specialty Services: | Y |
| Licensing: | | Genetic Data: | | Labs: | | Software: | | Consulting: | |
| Manufacturing: | Y | Tissue Replacement: | | Equipment/Supplies: | Y | Arrays: | | Blood Collection: | Y |
| Genetics: | | | | Research & Development Services: | Y | Database Management: | | Drug Delivery: | |
| | | | | Diagnostics: | Y | | | Drug Distribution: | |

## TYPES OF BUSINESS:

Human Blood-Plasma Collection
Plasma Products
Immunohematology Products
Vaccines
Pharmaceutical Marketing
Antivenom
Drugs-Cancer

## BRANDS/DIVISIONS/AFFILIATES:

ZLB Behring
CSL Bioplasma
CSL Behring LLC
CSL Pharmaceutical
CSL Biotherapies
Zenyth Therapeutics
CytoGam
ZLB Plasma Services

## CONTACTS: *Note: Officers with more than one job title may be intentionally listed here more than once.*

Brian McNamee, CEO/Managing Dir.
Tony M. Cipa, Exec. Dir.-Finance
Kenneth J. Roberts, Dir.-Mktg.
Kenneth J. Roberts, Gen. Mgr.-Human Resources
Andrew Cuthbertson, Chief Scientific Officer
John Akehurst, Dir.-Eng.
Greg Boss, General Counsel/Sr. VP
Paul Walton, Sr. VP-Corp. Dev.
Rachel David, Dir.-Public Affairs
Elizabeth A. Alexander, Mgr.-Finance & Risk
Peter Turner, Pres., CSL Behring
Colin Armit, Pres., CSL Biotherapies
Mary Sontrop, Gen Mgr.-CSL Biotherapies, Australia & New Zealand
Edward Bailey, Australian General Counsel/Corp. Sec.
Elizabeth A. Alexander, Chmn.
Jeff Davies, Pres., Bioplasma Asia Pacific

| Phone: 61-3-9389-1911 | Fax: 61-3-9389-1434 |
|---|---|
| Toll-Free: | |
| Address: 45 Poplar Rd., Parkville, VIC 3052 Australia | |

## GROWTH PLANS/SPECIAL FEATURES:

CSL Limited develops, manufactures and markets pharmaceutical products of biological origin in 27 countries worldwide. The company operates through several subsidiaries that manufacture and distribute pharmaceuticals, vaccines and diagnostics derived from human plasma. The firm's subsidiaries include CSL Behring, CSL Limited, CSL Bioplasma, CSL Biotherapies, CSL Research & Development, ZLB Behring and Zenyth Therapeutics. CSL Behring is a world leader in the manufacture of plasma products such as hemophilia treatments, immunoglobulins and wound healing agents. CSL Limited operates one of the largest plasma collection networks in the world, named ZLB Plasma Services, which includes 65 collection centers in the U.S. and eight in Germany. CSL Bioplasma is one of the largest manufacturers of plasma products in the southern hemisphere and works with the Red Cross and government entities to supply such products in Australia, New Zealand, Singapore, Malaysia and Hong Kong. It also provides contract plasma fractionation services. CSL Biotherapies, formerly CSL Pharmaceutical, manufactures and markets vaccines for human use, including children's vaccines, travel vaccines, respiratory vaccines and antivenom. Currently, its primary focus is the manufacturing of flu vaccines. The company's research and development portfolio includes treatments for stroke, acute coronary syndromes, cervical cancer, melanoma, genital warts, papilloma viruses and hepatitis C, in addition to a method of topical delivery of drugs to the eye. ZLB Behring holds the rights to CytoGam, an intravenous drug for the prevention of antibodies against cytomegalovirus in transplant patients. Zenyth Therapeutics develops and commercializes therapeutic antibodies for cancer and inflammation. In August 2008, the company agreed to acquire Talecris Biotherapeutics, a U.S.-based biotherapeutic and biotechnology company, for $3.1 billion.

CSL Limited provides flexible work arrangements, a Global Employee Share Plan for stock purchase, ongoing training programs and study assistance.

## FINANCIALS: Sales and profits are in thousands of dollars—add 000 to get the full amount. 2008 Note: Financial information for 2008 was not available for all companies at press time.

| | | |
|---|---|---|
| 2008 Sales: $2,698,750 | 2008 Profits: $499,350 | **U.S. Stock Ticker:** |
| 2007 Sales: $2,354,470 | 2007 Profits: $383,400 | **Int'l Ticker: CSL**   Int'l Exchange: Sydney-ASX |
| 2006 Sales: $2,146,111 | 2006 Profits: $88,406 | Employees:  7,000 |
| 2005 Sales: $1,965,359 | 2005 Profits: $176,824 | Fiscal Year Ends: 6/30 |
| 2004 Sales: $1,650,196 | 2004 Profits: $219,625 | Parent Company: |

## SALARIES/BENEFITS:

| Pension Plan: | ESOP Stock Plan: | Profit Sharing: | Top Exec. Salary: $782,154 | Bonus: $273,292 |
|---|---|---|---|---|
| Savings Plan: | Stock Purch. Plan: Y | | Second Exec. Salary: $409,753 | Bonus: $273,292 |

## OTHER THOUGHTS:

**Apparent Women Officers or Directors**: 3
**Hot Spot for Advancement for Women/Minorities**: Y

## LOCATIONS: ("Y" = Yes)

| West: | Southwest: | Midwest: | Southeast: | Northeast: | International: |
|---|---|---|---|---|---|
| | | Y | Y | Y | Y |

# CUBIST PHARMACEUTICALS INC

www.cubist.com

**Industry Group Code: 325412  Ranks within this company's industry group:  Sales: 44  Profits: 31**

| Drugs: | | Other: | Clinical: | Computers: | Services: |
|---|---|---|---|---|---|
| Discovery: | Y | AgriBio: | Trials/Services: | Hardware: | Specialty Services: |
| Licensing: | | Genetic Data: | Labs: | Software: | Consulting: |
| Manufacturing: | | Tissue Replacement: | Equipment/Supplies: | Arrays: | Blood Collection: |
| Genetics: | | | Research & Development Services: | Database Management: | Drug Delivery: |
| | | | Diagnostics: | | Drug Distribution: |

## TYPES OF BUSINESS:

Drugs-Infectious Disease
Antimicrobial Drugs
Antiviral Drugs

## BRANDS/DIVISIONS/AFFILIATES:

CUBICIN

## CONTACTS: Note: Officers with more than one job title may be intentionally listed here more than once.

Michael W. Bonney, CEO
Robert J. (Rob) Perez, COO/Exec. VP
Michael W. Bonney, Pres.
David W.J. McGirr, CFO/Sr. VP
Maureen H. Powers, VP-Human Resources
Steven C. Gilman, Chief Scientific Officer
Anthony S. Murabito, CIO
Lindon M. Fellows, Sr. VP-Tech. Oper.
Steven C. Gilman, Sr. VP-Discovery & Nonclinical Dev.
Tamara L. Joseph, General Counsel/Sr. VP/Corp. Sec.
Ed Campanaro, VP-Clinical Oper.
Praveen Tipirneni, VP-Bus. Dev.
Mary C. Stack, VP-Finance
Gregory Stea, Sr. VP-Commercial Oper.
Santosh J. Vetticaden, Chief Medical Officer/Sr. VP-Clinical Dev.
Mark Battaglini, VP-Gov't Affairs
Dennis D. Keith, VP-Chemistry

| Phone: 781-860-8660 | Fax: 781-861-0566 |
|---|---|
| Toll-Free: | |
| Address: 65 Hayden Ave., Lexington, MA 02421 US | |

## GROWTH PLANS/SPECIAL FEATURES:

Cubist Pharmaceuticals, Inc. is a biopharmaceutical company focused on the research, development and commercialization of pharmaceutical products for the acute care environment.  Its one marketed product is CUBICIN, a once-daily, bactericidal, intravenous antibiotic with activity against methicillin-resistant Staphylococcus aureus (MRSA).  CUBICIN is approved in the U.S. for the treatment of complicated skin and skin structure infections caused by MRSA and certain other Gram-positive bacteria, as well as for MRSA bloodstream infections.  Cubist markets CUBICIN to over 2,000 U.S. hospitals and outpatient acute care settings.  CUBICIN has received regulatory approval in 58 countries and is marketed in 25 countries.  Novartis AG, through a subsidiary, is responsible for regulatory filings, sales, marketing and distribution costs of CUBICIN in Europe, Australia, New Zealand, India and certain Central American, South American and Middle Eastern countries.  Merck & Co. subsidiary Banyu Pharmaceutical is responsible for the development and commercialization of CUBICIN in Japan.  AstraZeneca develops and commercializes CUBICIN in China and over 100 additional countries.  Other international partners for CUBICIN include Medison Pharma for Israel; Sepreacor for Canada; TTY BioPharm for Taiwan; and Kuhnil Pharma for Korea.  Under an agreement with AstraZeneca, Cubist promotes MERREM I.V., an intravenous broad spectrum carbapenem antibiotic for the treatment of serious hospital-acquired infections.  In April 2008, the firm agreed with Dyax to the development and commercialization in North America and Europe of the intravenous form of Dyax's ecallantide compound for the prevention of blood loss during surgery.  In September 2008, the company opened a 35,000 square foot research facility, which will accommodate 100 employees.  In January 2009, Cubist agreed with Alnylam to the development and commercialization of RNAi therapeutics for the potential treatment of RSV infection.

Cubist Pharmaceuticals offers its employees tuition reimbursement, an employee assistance program, fitness club reimbursement, adoption assistance and medical, dental, life and disability insurance.

## FINANCIALS:  Sales and profits are in thousands of dollars—add 000 to get the full amount. 2008 Note: Financial information for 2008 was not available for all companies at press time.

| | | |
|---|---|---|
| 2008 Sales: $433,641 | 2008 Profits: $169,819 | **U.S. Stock Ticker:** CBST |
| 2007 Sales: $294,620 | 2007 Profits: $48,147 | **Int'l Ticker:**   Int'l Exchange: |
| 2006 Sales: $194,748 | 2006 Profits: $- 376 | Employees:  554 |
| 2005 Sales: $120,645 | 2005 Profits: $-31,852 | Fiscal Year Ends: 12/31 |
| 2004 Sales: $68,071 | 2004 Profits: $-76,512 | Parent Company: |

## SALARIES/BENEFITS:

| Pension Plan: | ESOP Stock Plan: | Profit Sharing: | Top Exec. Salary: $460,000 | Bonus: $500,480 |
|---|---|---|---|---|
| Savings Plan: Y | Stock Purch. Plan: Y | | Second Exec. Salary: $400,000 | Bonus: $272,107 |

## OTHER THOUGHTS:

**Apparent Women Officers or Directors**: 5
**Hot Spot for Advancement for Women/Minorities**: Y

## LOCATIONS: ("Y" = Yes)

| West: | Southwest: | Midwest: | Southeast: | Northeast: | International: |
|---|---|---|---|---|---|
| | | | | Y | |

Note: Financial information, benefits and other data can change quickly and may vary from those stated here.

# CURAGEN CORPORATION
www.curagen.com

Industry Group Code: 325412  Ranks within this company's industry group:  Sales: 145    Profits: 47

| Drugs: | | Other: | | Clinical: | Computers: | Services: |
|---|---|---|---|---|---|---|
| Discovery: | Y | AgriBio: | | Trials/Services: | Hardware: | Specialty Services: |
| Licensing: | | Genetic Data: | Y | Labs: | Software: | Consulting: |
| Manufacturing: | | Tissue Replacement: | | Equipment/Supplies: | Arrays: | Blood Collection: |
| Genetics: | | | | Research & Development Services: | Database Management: | Drug Delivery: |
| | | | | Diagnostics: | | Drug Distribution: |

## TYPES OF BUSINESS:
Drug Discovery

## BRANDS/DIVISIONS/AFFILIATES:
Belinostat
CR011-vcMMAE

## CONTACTS: Note: Officers with more than one job title may be intentionally listed here more than once.
Timothy M. Shannon, CEO
Timothy M. Shannon, Pres.
Sean Cassidy, CFO/VP
Ronit Simantov, VP-Medical Dev.
Henri S. Lichenstein, VP-Prod. Dev.
Paul M. Finigan, General Counsel/Exec. VP
Cyrus Karkaria, VP-Oper. & BPS
Elizabeth Crowley, VP-Dev. Oper.
Glenn Schulman, Dir.-Investor Rel.
Hans Scholl, VP-Regulatory Affairs & Quality Assurance
Robert E. Patricelli, Chmn.

Phone: 203-481-1104   Fax: 203-483-2552
Toll-Free: 888-436-6642
Address: 322 E. Main St., Branford, CT 06405 US

## GROWTH PLANS/SPECIAL FEATURES:
CuraGen Corporation is a biopharmaceutical company dedicated to developing cancer treatments. The firm's two major drug candidates are Belinostat (formerly PXD101) and CR011-vcMMAE, both developed in collaboration with other companies. Belinostat (PXD101), which is an inhibitor for the enzyme histone deactylase (HDAC), was developed with TopoTarget. HDAC inhibitors have been shown to arrest cancer cell growth; induce programmed cell death; and sensitize cancer cells to make them more susceptible to treatments. The drug being tested in both an oral and intravenous format, either as a monotherapy or in conjunction with other drugs, and is in various phases of clinical trials for the treatment of numerous cancers, including Phase II for the treatment of T-cell lymphomas and ovarian cancer; and Phase I for treating solid tumors and soft tissue sarcoma. The firm developed CR011, a fully-human monoclonal antibody drug, with Amgen Fremont (formerly Abgenix), and licensed antibody-drug conjugation (ADC) technology from Seattle Genetics to attach monomethylauristatin E (vcMMAE ) to CR011, creating CR011-vcMMAE. CR011 attaches itself to a cancerous cell and transports the ADC inside it, after which the MMAE splits off the antibody and is activated. The drug is currently in Phase I testing for treating metastatic melanoma. Early trials suggest CR011-vcMMAE can cause complete and durable tumor regression, without any notable toxicity or weight loss.
Employees of CuraGen receive medical, dental, prescription, life and AD&D insurance.

## FINANCIALS: Sales and profits are in thousands of dollars—add 000 to get the full amount. 2008 Note: Financial information for 2008 was not available for all companies at press time.

| | | |
|---|---|---|
| 2008 Sales: $1,174 | 2008 Profits: $24,781 | U.S. Stock Ticker: CRGN |
| 2007 Sales: $ 88 | 2007 Profits: $25,398 | Int'l Ticker:  Int'l Exchange: |
| 2006 Sales: $2,298 | 2006 Profits: $-59,839 | Employees: 16 |
| 2005 Sales: $4,825 | 2005 Profits: $-73,244 | Fiscal Year Ends: 12/31 |
| 2004 Sales: $6,339 | 2004 Profits: $-90,397 | Parent Company: |

## SALARIES/BENEFITS:
| Pension Plan: | ESOP Stock Plan: | Profit Sharing: | Top Exec. Salary: $375,000 | Bonus: $136,500 |
|---|---|---|---|---|
| Savings Plan: Y | Stock Purch. Plan: | | Second Exec. Salary: $315,000 | Bonus: $80,262 |

## OTHER THOUGHTS:
Apparent Women Officers or Directors: 1
Hot Spot for Advancement for Women/Minorities: Y

## LOCATIONS: ("Y" = Yes)
| West: | Southwest: | Midwest: | Southeast: | Northeast: | International: |
|---|---|---|---|---|---|
| | | | | Y | |

Note: Financial information, benefits and other data can change quickly and may vary from those stated here.

# CURIS INC

www.curis.com

**Industry Group Code: 325412** **Ranks within this company's industry group:** Sales: 118 Profits: 84

| Drugs: | | Other: | | Clinical: | | Computers: | | Services: | |
|---|---|---|---|---|---|---|---|---|---|
| Discovery: | Y | AgriBio: | | Trials/Services: | | Hardware: | | Specialty Services: | |
| Licensing: | Y | Genetic Data: | | Labs: | | Software: | | Consulting: | |
| Manufacturing: | | Tissue Replacement: | | Equipment/Supplies: | | Arrays: | | Blood Collection: | |
| Genetics: | | | | Research & Development Services: | | Database Management: | | Drug Delivery: | |
| | | | | Diagnostics: | | | | Drug Distribution: | |

## TYPES OF BUSINESS:

Drugs-Regenerative Medicine
Signaling Pathway Therapeutics

## BRANDS/DIVISIONS/AFFILIATES:

GDC-0449
BMP-7

## CONTACTS: Note: Officers with more than one job title may be intentionally listed here more than once.

Daniel R. Passeri, CEO
Michael P. Gray, COO
Daniel R. Passeri, Pres.
Michael P. Gray, CFO
Changgeng Qian, VP-Discovery & Preclinical Dev.
Mark Noel, VP-Tech. Mgmt.
James R. McNab, Jr., Chmn.

| Phone: 617-503-6500 | Fax: 617-503-6501 |
|---|---|
| Toll-Free: | |
| Address: 45 Moulton St., Cambridge, MA 02138 US | |

## GROWTH PLANS/SPECIAL FEATURES:

Curis, Inc. focuses on the discovery and development of products that use signaling pathway drug technologies. Its product development approach involves using small molecules to target components of abnormally regulated signaling pathway networks, the networks used by cells to exchange instructional messages regulating specific biological functions. The firm primarily focuses on targeting signaling pathways in the treatment cancer, but is also expanding its research to neurological disease and cardiovascular disease. Curis collaborates with other biopharmaceutical companies, including the Hedgehog antagonist program with Genentech and the recent sale of the BMP-7 program assets to Stryker Corp. The Hedgehog program targets the Hedgehog signaling pathway, which controls the development and growth of many tissues in the body. The program's leading drug candidate, GDC-0449, is being tested in Phase I and II trials for locally advanced, multi-focal or metastatic basal cell carcinoma. The company's other main program is its targeted cancer drug development platform, primarily consisting of several proprietary cancer drug programs that target multiple signaling pathways. In addition to cancer products, the firm develops drugs including a bone morphogenic protein (BMP-7), which promotes the health of the kidneys, skeletal system and vascular system, and the Hsp90 Inhibitor, for the treatment of stroke and several neurodegenerative such as Alzheimer's, Parkinson's and Huntington's Disease. Curis is also developing a Hedgehog small molecule agonist program, which targets the Hedgehog pathway for the treatment of neurological, cardiovascular and bone diseases. In February 2009, the company announced a Cooperative Research and Development Agreement (CRADA) with Genentech and the National Cancer Institute to further develop GDC-0449.

Curis offers employees health, dental and vision benefits; an employee assistance program; education assistance; short- and long-term disability benefits; a 401(k) plan; stock option grants; company-paid parking; transportation passes; and has a company sports teams.

## FINANCIALS: Sales and profits are in thousands of dollars—add 000 to get the full amount. 2008 Note: Financial information for 2008 was not available for all companies at press time.

| | | |
|---|---|---|
| 2008 Sales: $8,367 | 2008 Profits: $-12,123 | **U.S. Stock Ticker:** CRIS |
| 2007 Sales: $16,389 | 2007 Profits: $-6,964 | **Int'l Ticker:** Int'l Exchange: |
| 2006 Sales: $14,936 | 2006 Profits: $-8,829 | Employees: 34 |
| 2005 Sales: $6,002 | 2005 Profits: $-14,855 | Fiscal Year Ends: 12/31 |
| 2004 Sales: $3,699 | 2004 Profits: $-15,075 | Parent Company: |

## SALARIES/BENEFITS:

| Pension Plan: | ESOP Stock Plan: | Profit Sharing: | Top Exec. Salary: $384,615 | Bonus: $ |
|---|---|---|---|---|
| Savings Plan: Y | Stock Purch. Plan: | | Second Exec. Salary: $292,308 | Bonus: $ |

## OTHER THOUGHTS:

**Apparent Women Officers or Directors:** 1
**Hot Spot for Advancement for Women/Minorities:**

## LOCATIONS: ("Y" = Yes)

| West: | Southwest: | Midwest: | Southeast: | Northeast: | International: |
|---|---|---|---|---|---|
| | | | | Y | |

Note: Financial information, benefits and other data can change quickly and may vary from those stated here.

# CYPRESS BIOSCIENCE INC

www.cypressbio.com

Industry Group Code: 325412 Ranks within this company's industry group: Sales: 106 Profits: 99

| Drugs: | | Other: | | Clinical: | | Computers: | | Services: | |
|---|---|---|---|---|---|---|---|---|---|
| Discovery: | Y | AgriBio: | | Trials/Services: | | Hardware: | | Specialty Services: | Y |
| Licensing: | Y | Genetic Data: | | Labs: | | Software: | | Consulting: | |
| Manufacturing: | Y | Tissue Replacement: | | Equipment/Supplies: | | Arrays: | | Blood Collection: | |
| Genetics: | | | | Research & Development Services: | | Database Management: | | Drug Delivery: | |
| | | | | Diagnostics: | Y | | | Drug Distribution: | |

## TYPES OF BUSINESS:

Drugs-Manufacturing
Drugs-Fibromyalgia Syndrome
Personalized Medicine Laboratory Services

## BRANDS/DIVISIONS/AFFILIATES:

Proprius Pharmaceuticals Inc
Forest Laboratories Inc
Milnacipran
Savella
Avise PG
Avise MCV

## CONTACTS: Note: Officers with more than one job title may be intentionally listed here more than once.

Jay D. Kranzler, CEO
Sabrina Martucci Johnson, COO
Sabrina Martucci Johnson, CFO/Exec. VP
Srinivas G. Rao, Chief Scientific Officer
R. Michael Gendreau, VP-Clinical Dev./Chief Medical Officer
Denise L. Wheeler, General Counsel
Michael J. Walsh, Exec. VP/Chief Commercial Officer
Jay D. Kranzler, Chmn.

| Phone: 858-452-2323 | Fax: 858-452-1222 |
|---|---|
| Toll-Free: | |
| Address: 4350 Executive Dr., Ste. 325, San Diego, CA 92121 US | |

## GROWTH PLANS/SPECIAL FEATURES:

Cypress Bioscience, Inc. develops and markets pharmaceutical products and personalized medicine laboratory services that allow physicians to serve unmet medical needs. The company is developing Savella (milnacipran) for fibromyalgia (FM), a dual reuptake inhibitor that blocks norepinephrine with higher potency than serotonin, two neurotransmitters known to play an essential role in regulating pain and mood. The firm has exercised the right granted by its partner Forest Laboratories, Inc. to co-promote Savella and intends to detail it to rheumatologists, pain centers and physical and rehabilitation medicine specialists. With the recent $37.5 million acquisition of Proprius Pharmaceuticals, Inc. in March 2008, Cypress also provides personalized medicine laboratory services to rheumatologists. These services include Avise PG, a test for determining optimal metabolization of methotrexate in rheumatoid arthritis (RA) patients, and Avise MCV, a test which measures mutated citrullinated vimentin (MCV) to diagnose and measure the severity of RA. Cypress also has a number of Proof of Concept stage opportunities in development, including two pharmaceutical candidates acquired from Proprius: A topical NSAID and a possible treatment for RA. In January 2009, the FDA approved Savella for use in treating fibromyalgia.

## FINANCIALS: Sales and profits are in thousands of dollars—add 000 to get the full amount. 2008 Note: Financial information for 2008 was not available for all companies at press time.

| | | |
|---|---|---|
| 2008 Sales: $17,159 | 2008 Profits: $-18,226 | U.S. Stock Ticker: CYPB |
| 2007 Sales: $13,941 | 2007 Profits: $3,488 | Int'l Ticker: Int'l Exchange: |
| 2006 Sales: $4,322 | 2006 Profits: $-8,318 | Employees: 145 |
| 2005 Sales: $8,384 | 2005 Profits: $-7,650 | Fiscal Year Ends: 12/31 |
| 2004 Sales: $14,415 | 2004 Profits: $-11,215 | Parent Company: |

## SALARIES/BENEFITS:

| Pension Plan: | ESOP Stock Plan: | Profit Sharing: | Top Exec. Salary: $578,112 | Bonus: $269,921 |
|---|---|---|---|---|
| Savings Plan: | Stock Purch. Plan: | | Second Exec. Salary: $312,613 | Bonus: $54,707 |

## OTHER THOUGHTS:

**Apparent Women Officers or Directors:** 3
**Hot Spot for Advancement for Women/Minorities:** Y

## LOCATIONS: ("Y" = Yes)

| West: | Southwest: | Midwest: | Southeast: | Northeast: | International: |
|---|---|---|---|---|---|
| Y | | | | | |

Note: Financial information, benefits and other data can change quickly and may vary from those stated here.

# CYTRX CORPORATION

**cytrx.com**

**Industry Group Code: 325412  Ranks within this company's industry group:  Sales: 126  Profits: 119**

| Drugs: | | Other: | | Clinical: | Computers: | Services: |
|---|---|---|---|---|---|---|
| Discovery: | Y | AgriBio: | | Trials/Services: | Hardware: | Specialty Services: |
| Licensing: | Y | Genetic Data: | | Labs: | Software: | Consulting: |
| Manufacturing: | | Tissue Replacement: | | Equipment/Supplies: | Arrays: | Blood Collection: |
| Genetics: | | | | Research & Development Services: | Database Management: | Drug Delivery: |
| | | | | Diagnostics: | | Drug Distribution: |

## TYPES OF BUSINESS:

Pharmaceutical Research & Development
Small-Molecule Drugs

## BRANDS/DIVISIONS/AFFILIATES:

Tamibarotene
Iroxanadine
INNO-206
INNO-406
Arimoclomol

## CONTACTS: *Note: Officers with more than one job title may be intentionally listed here more than once.*

Steven A. Kriegsman, CEO
Steven A. Kriegsman, Pres.
John C. Caloz, CFO
Jack R. Barber, Chief Scientific Officer
Scott Wieland, Sr. VP-Drug Dev.
Benjamin S. Levin, General Counsel/VP-Legal Affairs/Corp. Sec.
Jaisim Shah, Chief Bus. Officer/Sr. VP-Bus. Dev.
David J. Haen, VP-Bus. Dev.
Shi Chung Ng, Sr. VP-R&D
Max Link, Chmn.

| Phone: 310-826-5648 | Fax: 310-826-6139 |
|---|---|
| **Toll-Free:** | |
| **Address:** 11726 San Vicente Blvd., Ste. 650, Los Angeles, CA 90049 US | |

## GROWTH PLANS/SPECIAL FEATURES:

CytRx Corporation is a biopharmaceutical company engaged in the development of human therapeutic products utilizing its core technology in small-molecule molecular chaperone amplification. The firm's small molecule drug candidates provide cellular protection by activating molecular chaperones that can detect and repair or degrade misfolded proteins that are believed to cause many diseases. The company is currently developing treatments for cancer, neurodegenerative disorders and diabetic complications. CytRx's leading drug candidate is tamibarotene, a treatment for acute myeloid leukemia that seeks to overcome the shortcomings of current treatments. Additional oncology works include INNO-406, a treatment for Chronic Myeloid Leukemia, and INNO-206, which can be used to combat a variety of different cancers. Another possible product is arimoclomol, which has the potential to treat amyotrophic lateral sclerosis (ALS, or Lou Gherig's Disease) and stroke recovery. The company's other product under development is iroxanadine, which has completed Phase I trials for the treatment of diabetic foot ulcers. CytRx does not manufacture any of its products, but instead relies on third parties to supply materials for its clinical studies. In 2008, the firm awarded shares of its subsidiary RXi Pharmaceuticals Corporation to its stockholders, effectively decreasing its ownership of RXi to 45%. RXi develops products based on RNA interference technology, which interfere with the expression of targeted disease-associated genes. Also in 2008, the company discovered a new series of compounds that amplify the natural cellular chaperone response to toxic misfolded proteins, providing potential treatment for cancer, cardiovascular disease, diabetes and neurodegenerative diseases. In June 2008, CytRx agreed to purchase Innovive Pharmaceuticals, Inc., a biopharmaceutical company with four clinical stage oncology drug candidates.

CytRx offers employees medical, dental and vision coverage; life and disability insurance; a 401(k) plan; flexible spending accounts; an incentive stock option plan; and tuition reimbursement.

## FINANCIALS: Sales and profits are in thousands of dollars—add 000 to get the full amount. 2008 Note: Financial information for 2008 was not available for all companies at press time.

| | | |
|---|---|---|
| 2008 Sales: $6,266 | 2008 Profits: $-27,803 | **U.S. Stock Ticker: CYTR** |
| 2007 Sales: $7,459 | 2007 Profits: $-21,890 | **Int'l Ticker:**    Int'l Exchange: |
| 2006 Sales: $2,066 | 2006 Profits: $-16,752 | Employees:  35 |
| 2005 Sales: $ 184 | 2005 Profits: $-15,093 | Fiscal Year Ends: 12/31 |
| 2004 Sales: $ 428 | 2004 Profits: $-16,392 | Parent Company: |

## SALARIES/BENEFITS:

| Pension Plan: | ESOP Stock Plan: | Profit Sharing: | Top Exec. Salary: $551,000 | Bonus: $150,000 |
|---|---|---|---|---|
| Savings Plan: Y | Stock Purch. Plan: Y | | Second Exec. Salary: $364,375 | Bonus: $55,000 |

## OTHER THOUGHTS:

**Apparent Women Officers or Directors:**
**Hot Spot for Advancement for Women/Minorities:**

## LOCATIONS: ("Y" = Yes)

| West: | Southwest: | Midwest: | Southeast: | Northeast: | International: |
|---|---|---|---|---|---|
| Y | | | | | |

# DAIICHI SANKYO CO LTD

## www.daiichisankyo.co.jp

**Industry Group Code: 325412  Ranks within this company's industry group:** Sales: 20   Profits: 20

| Drugs: | | Other: | Clinical: | Computers: | Services: |
|---|---|---|---|---|---|
| Discovery: | Y | AgriBio: | Trials/Services: | Hardware: | Specialty Services: |
| Licensing: | Y | Genetic Data: | Labs: | Software: | Consulting: |
| Manufacturing: | Y | Tissue Replacement: | Equipment/Supplies: | Arrays: | Blood Collection: |
| Genetics: | | | Research & Development Services: | Database Management: | Drug Delivery: |
| | | | Diagnostics: | | Drug Distribution: |

## TYPES OF BUSINESS:

Pharmaceuticals
Prescription Drugs
Over-the-Counter Drugs
Functional Foods

## BRANDS/DIVISIONS/AFFILIATES:

Sankyo Co., Ltd.
Daiichi Pharmaceutical Co., Ltd.
Daiichi Sankyo Healthcare Co. Ltd.
Pravastatin
Olmesartan
Lamisil AT
Daiichi Sankyo Healthcare Co. Ltd.
Ranbaxy Laboratories Limited

## CONTACTS: *Note: Officers with more than one job title may be intentionally listed here more than once.*

Takashi Shoda, CEO
Takashi Shoda, Pres.
Ryuzo Takada, Sr. Exec. Officer-Sales & Mktg.
Takeshi Ogita, Sr. Exec. Officer-Human Resources
Kazunori Hirokawa, Head-R&D Div.
Tsutomu Une, Sr. Exec. Officer-Corp. Strategy
Hitoshi Matsuda, Sr. Exec. Officer-Corp. Bus. Mgmt.
Yoshihiko Suzuki, Head-Sales & Mktg. Div.
Kazuhiko Tanzawa, Pres., Daiichi Sankyo Research Institute
Yuki Sato, Head-Pharmaceutical Tech. Div.
Kiyoshi Morita, Chmn.
George Nakayama, Gen. Mgr.-Int'l Bus. Mgmt.
Toru Kuroda, Head-Supply Chain Div.

| Phone:  81-3-6225-1111 | Fax: |
|---|---|
| Toll-Free: | |
| Address:  3-5-1, Nihonbashi-honcho, Chuo-ku, Tokyo,  103-8426 Japan | |

## GROWTH PLANS/SPECIAL FEATURES:

Daiichi Sankyo Co., Ltd., formed from the merger of the 106-year old Sankyo Co., Ltd. and the 90-year old Daiichi Pharmaceutical Co., Ltd, is primarily a pharmaceutical manufacturing firm.  Its products are distributed through roughly 5,000 representatives, about half of which are in Japan with the rest spread throughout the U.S., Europe and Asia.  Daiichi Sankyo's research and development is focused mainly on cardiovascular diseases; cancer; glucose metabolic disorders; bone/joint diseases; immunity and allergies; and infectious diseases.  The firm has seven research and development facilities in the U.S., the U.K., Germany, China and Japan.  Some of the company's major drugs are Pravastatin, an antihyperlipidemic agent that is also sold under the name Mevalotin; Levofloxacin, an oral antibacterial agent, also called Cravit; and Olmesartan, an antihypertension agent sold in the U.S. as Benicar, and sold in Japan and Europe as Olmetec.  Some of its other drugs include Omnipaque, a non-ionicity contrast agent; Loxonin, a non-steroidal analgesic and anti-inflammatory agent; Artist, an antihypertensive; and WelChol, an antihyperlipidemic agent and a treatment for Type 2 diabetes.  Its OTC (over-the-counter) drug products include Lamisil AT, an athlete's foot and ringworm treatment; and Patecs Felbinac, an external anti-inflammatory analgesic.  Besides its domestic subsidiaries, the firm has over 23 subsidiaries overseas in more than 18 countries.  Domestic subsidiaries include Daiichi Sankyo Healthcare Co. Ltd., which offers self-medication options exclusively in Japan, including OTC medicines, skin care products and medical equipment. Overseas subsidiaries include Daiichi Sankyo, Inc. in the U.S., which markets the firm's products as well as conducting clinical trials and overseeing U.S. regulatory affairs and other functions. Geographically, Japan generates around 68% of Daiichi Sankyo's sales; North America, 20%; Europe, 9%; and other regions, 3%. In November 2008, the company completed its acquisition of a majority share in India-based Ranbaxy Laboratories Limited for approximately $3.3 billion.

## FINANCIALS:  Sales and profits are in thousands of dollars—add 000 to get the full amount. 2008 Note: Financial information for 2008 was not available for all companies at press time.

| | | |
|---|---|---|
| 2008 Sales: $9,282,450 | 2008 Profits: $1,030,000 | **U.S. Stock Ticker:** |
| 2007 Sales: $7,910,000 | 2007 Profits: $670,000 | **Int'l Ticker: 4568**    Int'l Exchange: Tokyo-TSE |
| 2006 Sales: $8,703,320 | 2006 Profits: $735,030 | Employees: |
| 2005 Sales: $8,669,610 | 2005 Profits: $820,240 | Fiscal Year Ends: 3/31 |
| 2004 Sales: $ | 2004 Profits: $ | Parent Company: |

## SALARIES/BENEFITS:

| Pension Plan: | ESOP Stock Plan: | Profit Sharing: | Top Exec. Salary: $ | Bonus: $ |
|---|---|---|---|---|
| Savings Plan: | Stock Purch. Plan: | | Second Exec. Salary: $ | Bonus: $ |

## OTHER THOUGHTS:

**Apparent Women Officers or Directors:**
**Hot Spot for Advancement for Women/Minorities:**

## LOCATIONS: ("Y" = Yes)

| West: | Southwest: | Midwest: | Southeast: | Northeast: | International: |
|---|---|---|---|---|---|
| | | | | Y | Y |

# DECODE GENETICS INC

**www.decode.com**

**Industry Group Code: 511210D  Ranks within this company's industry group:** Sales: 4  Profits: 4

| Drugs: | | Other: | | Clinical: | | Computers: | | Services: | |
|---|---|---|---|---|---|---|---|---|---|
| Discovery: | Y | AgriBio: | | Trials/Services: | | Hardware: | | Specialty Services: | Y |
| Licensing: | | Genetic Data: | Y | Labs: | | Software: | Y | Consulting: | |
| Manufacturing: | | Tissue Replacement: | | Equipment/Supplies: | | Arrays: | | Blood Collection: | |
| Genetics: | | | | Research & Development Services: | Y | Database Management: | | Drug Delivery: | |
| | | | | Diagnostics: | | | | Drug Distribution: | |

## TYPES OF BUSINESS:
Bioinformatics & Medical Records Databases
Bioinformatics Software Products
Genetic Disease Research

## BRANDS/DIVISIONS/AFFILIATES:
deCODE Chemistry, Inc.
deCODE Biostructures, Inc.
deCODEme
Secure Robotized Sample Vault

## CONTACTS: Note: Officers with more than one job title may be intentionally listed here more than once.
Kari Stefansson, CEO
Axel Nielsen, COO
Kari Stefansson, Pres.
Lance Thibault, CFO
Jeffrey Gulcher, Chief Scientific Officer
Hakon Gudbjartsson, VP-Informatics
Daniel L. Hartman, Sr. VP-Prod. Dev.
Jakob Sigurdsson, Sr. VP-Corp. Dev.
Lance Thibault, Treas.
Mark Gurney, Sr. VP-Drug Discovery & Dev.
C. Augustine Kong, VP-Statistics
Kari Stefansson, Chmn.

| **Phone:** 354-570-1900 | **Fax:** 354-570-1903 |
|---|---|
| **Toll-Free:** | |
| **Address:** Sturlugata 8, Reykjavik, IS-101 Iceland | |

## GROWTH PLANS/SPECIAL FEATURES:
deCODE genetics, Inc., an Icelandic biopharmaceutical company, performs population-based genetic and medical research to identify diseased genes, drug targets and diagnostic targets.  By analyzing the genotypic and medical data from over 60% of Iceland's genetically homogenous adult population, combined with genealogical data that covers the past 1,100 years and links the entire population, deCODE gains insights into the pathogenesis of certain illnesses and the reasons why some patients respond differently to the same drugs.  The comparison of DNA samples from closely and distantly related individuals with the same disease helps expedite the process of identifying specific genes and specific markers with those genes that cause the illness.  deCODE has studied the genetics and pathology of over 50 different common diseases.  The company's lead product candidate is DG031, a heart attack prevention treatment designed to limit inflammatory activity.  The firm has a handful of other drugs in development, including DG051, a similar heart attack drug; DG041, an anti-platelet preventative for arterial thrombosis; and DG071, a compound to limit memory loss and cognitive deficiencies caused by Alzheimer's Disease.  deCODE's genetic analysis service, deCODEme, offers subscribers an expert analysis of their genome, including ancestry and risk for certain diseases.  Subscribers view their results through a secure web page.  Washington-based deCODE chemistry, Inc. provides drug discovery technology and services; while Illinois-based deCODE biostructures, Inc. analyzes the crystal structures of proteins for structure-based drug design and development.  deCODE also maintains the first Secure Robotized Sample Vault (SRSV) that archives its 530,000 clinical specimens, and the company offers several product versions of the SRSV to the research and clinical communities.  In January 2009, the company introduced two new additions to their deCODEme service, deCODEme Cardio and deCODEme Cancer, designed to evaluate an individual's risk for cardiovascular diseases or cancer.

## FINANCIALS: Sales and profits are in thousands of dollars—add 000 to get the full amount. 2008 Note: Financial information for 2008 was not available for all companies at press time.

| | | |
|---|---|---|
| 2008 Sales: $58,095 | 2008 Profits: $-80,947 | **U.S. Stock Ticker: DCGN** |
| 2007 Sales: $40,403 | 2007 Profits: $-95,526 | **Int'l Ticker:**  Int'l Exchange: |
| 2006 Sales: $40,510 | 2006 Profits: $-85,473 | Employees:  431 |
| 2005 Sales: $43,955 | 2005 Profits: $-62,750 | Fiscal Year Ends: 12/31 |
| 2004 Sales: $42,127 | 2004 Profits: $-57,255 | Parent Company: |

## SALARIES/BENEFITS:

| Pension Plan: | ESOP Stock Plan: | Profit Sharing: | Top Exec. Salary: $662,296 | Bonus: $ |
|---|---|---|---|---|
| Savings Plan: Y | Stock Purch. Plan: | | Second Exec. Salary: $365,000 | Bonus: $ |

## OTHER THOUGHTS:
**Apparent Women Officers or Directors:**
**Hot Spot for Advancement for Women/Minorities:**

## LOCATIONS: ("Y" = Yes)

| West: | Southwest: | Midwest: | Southeast: | Northeast: | International: |
|---|---|---|---|---|---|
| Y | | Y | | Y | Y |

# DENDREON CORPORATION

www.dendreon.com

**Industry Group Code: 325412  Ranks within this company's industry group:  Sales: 155    Profits: 152**

| Drugs: | | Other: | Clinical: | Computers: | Services: |
|---|---|---|---|---|---|
| Discovery: | Y | AgriBio: | Trials/Services: | Hardware: | Specialty Services: |
| Licensing: | | Genetic Data: | Labs: | Software: | Consulting: |
| Manufacturing: | | Tissue Replacement: | Equipment/Supplies: | Arrays: | Blood Collection: |
| Genetics: | | | Research & Development Services: | Database Management: | Drug Delivery: |
| | | | Diagnostics: | | Drug Distribution: |

## TYPES OF BUSINESS:

Cancer Drugs

## BRANDS/DIVISIONS/AFFILIATES:

Provenge
Neuvenge
Trp-p8
Genetech, Inc.
Amgen Fremont, Inc.
CA-9 (MN)
CEA

## CONTACTS: Note: Officers with more than one job title may be intentionally listed here more than once.

Mitchell H. Gold, CEO
Mitchell H. Gold, Pres.
Gregory T. Schiffman, CFO/Sr. VP
David L. Urdal, Chief Scientific Officer/Sr. VP
Rick Hamm, General Counsel/Sec.
Rick Hamm, Sr. VP-Corp. Dev.
Mark Frolich, Sr. VP-Clinical Affairs & Chief Medical Officer
Richard B. Brewer, Chmn.

| Phone: 206-256-4545 | Fax: 206-256-0571 |
|---|---|
| Toll-Free: | |
| Address: 3005 First Ave., Seattle, WA 98121 US | |

## GROWTH PLANS/SPECIAL FEATURES:

Dendreon Corporation develops and commercializes active cellular immunotherapy, monoclonal antibody and small molecule product candidates to treat a wide range of cancers. The most advanced product candidate is Provenge, an active cellular immunotherapy that has completed Phase III trials for the treatment of asymptomatic, metastatic, androgen-independent prostrate cancer (AIPC). The company's second candidate, Neuvenge, is the investigational active immunotherapy for the treatment of patients with breast, ovarian and other solid tumor expressing HER2/neu. The firm is currently evaluating future development plans for Neuvenge. Dendreon has a few product candidates in the research and development program, such as Trp-p8 for lung, breast, prostrate and colon cancer; CA-9 (MN) for kidney, colon and cervical cancer; and CEA for breast, lung and colon cancer. The company is in collaboration with Genentech, Inc. for the preclinical research, clinical development and commercialization of potential products derived from Trp-p8, an ion channel found in prostrate cancer cells. The firm also has preclinical collaborations with Amgen Fremont, Inc., focused on the discovery, development and commercialization of fully-human monoclonal antibodies against a membrane-bound serine protease. In April 2009, Dendreon announced that Provenge had met its main goal in the Phase III trials, prompting the company to state plans for seek FDA approval.

The company offers life, medical, dental and vision insurance; AD&D and disability coverage; a 401(k) plan; an employee assistance program; a public transportation subsidy; and a tuition subsidy program.

## FINANCIALS: Sales and profits are in thousands of dollars—add 000 to get the full amount. 2008 Note: Financial information for 2008 was not available for all companies at press time.

| | | |
|---|---|---|
| 2008 Sales: $ 111 | 2008 Profits: $-71,644 | U.S. Stock Ticker: DNDN |
| 2007 Sales: $ 743 | 2007 Profits: $-99,264 | Int'l Ticker:      Int'l Exchange: |
| 2006 Sales: $ 273 | 2006 Profits: $-91,642 | Employees:   198 |
| 2005 Sales: $ 210 | 2005 Profits: $-81,547 | Fiscal Year Ends: 12/31 |
| 2004 Sales: $5,035 | 2004 Profits: $-75,240 | Parent Company: |

## SALARIES/BENEFITS:

| Pension Plan: | ESOP Stock Plan: | Profit Sharing: | Top Exec. Salary: $500,000 | Bonus: $212,500 |
|---|---|---|---|---|
| Savings Plan: Y | Stock Purch. Plan: | | Second Exec. Salary: $391,875 | Bonus: $125,400 |

## OTHER THOUGHTS:

**Apparent Women Officers or Directors**: 1
**Hot Spot for Advancement for Women/Minorities**: Y

## LOCATIONS: ("Y" = Yes)

| West: | Southwest: | Midwest: | Southeast: | Northeast: | International: |
|---|---|---|---|---|---|
| Y | | | | | |

# DEPOMED INC

www.depomedinc.com

**Industry Group Code: 325412A  Ranks within this company's industry group: Sales: 18  Profits: 17**

| Drugs: | | Other: | | Clinical: | Computers: | | Services: | |
|---|---|---|---|---|---|---|---|---|
| Discovery: | Y | AgriBio: | | Trials/Services: | Hardware: | | Specialty Services: | |
| Licensing: | Y | Genetic Data: | | Labs: | Software: | | Consulting: | |
| Manufacturing: | Y | Tissue Replacement: | | Equipment/Supplies: | Arrays: | | Blood Collection: | |
| Genetics: | | | | Research & Development Services: | Database Management: | | Drug Delivery: | Y |
| | | | | Diagnostics: | | | Drug Distribution: | |

## TYPES OF BUSINESS:
Drug Delivery System-Based Drugs

## BRANDS/DIVISIONS/AFFILIATES:
AcuForm
Glumetza
ProQuin XR
Gabapentin GR

## CONTACTS: *Note: Officers with more than one job title may be intentionally listed here more than once.*
Carl A. Pelzel, CEO
Carl A. Pelzel, Pres.
Tammy L. Cameron, Interim Principal Acct. & Financial Officer
Shay Weisbrich, VP-Mktg. & Sales
Kera Alexander, VP-Human Resources
Michael Sweeney, VP-Prod. Dev.
Kera Alexander, VP-Admin.
Matthew Gosling, General Counsel/VP
John N. Shell, VP-Oper.
Thadd M. Vargas, Sr. VP-Bus. Dev.
Tammy L. Cameron, VP-Finance
Craig R. Smith, Chmn.

| Phone: 650-462-5900 | Fax: 650-462-9993 |
|---|---|
| Toll-Free: | |
| Address: 1360 O'Brien Dr., Menlo Park, CA 94025 US | |

## GROWTH PLANS/SPECIAL FEATURES:

Depomed, Inc. focuses on the development and commercialization of differentiated products that are based on proprietary drug delivery technologies. The company has developed two commercial products: Glumetza, a once-daily treatment for adults with Type 2 diabetes that the firm jointly commercializes in the U.S. with Santarus, Inc.; and ProQuin XR, a once-daily treatment of uncomplicated urinary tract infections that Watson Pharmaceuticals, Inc. markets in the U.S. Depomed's most advanced product candidate in development is Gabapentin GR, an extended release form of gabapentin. The company has submitted an investigational new drug application for Gabapentin GR, and is currently in Phase III studies for the treatment of menopausal hot flashes and postherpetic neuralgia, and in Phase II studies for diabetic peripheral neuropathy. In addition, the firm has other product candidates in earlier stages of development, including treatments for gastroesophageal reflux disease and Parkinson's Disease. Depomed's drugs are based on its proprietary drug delivery system, AcuForm. The AcuForm technology is a polymer-based drug delivery platform that provides targeted drug delivery solutions for a wide range of compounds. The technology embraces diffusional, erosional, bilayer and multi-drug systems that can optimize oral drug delivery for both soluble and insoluble drugs. One application of the technology allows standard-sized tablets to be retained in the stomach for 6-8 hours after administration, extending the time of drug delivery to the small intestine.

The company offers medical, dental, life and long-term disability insurance; flexible spending accounts; a 401(k) plan; and an employee stock purchase plan.

## FINANCIALS: Sales and profits are in thousands of dollars—add 000 to get the full amount. 2008 Note: Financial information for 2008 was not available for all companies at press time.

| | | |
|---|---|---|
| 2008 Sales: $34,842 | 2008 Profits: $-15,843 | U.S. Stock Ticker: DEPO |
| 2007 Sales: $65,582 | 2007 Profits: $49,219 | Int'l Ticker:     Int'l Exchange: |
| 2006 Sales: $9,551 | 2006 Profits: $-39,659 | Employees:  81 |
| 2005 Sales: $4,405 | 2005 Profits: $-24,467 | Fiscal Year Ends: 12/31 |
| 2004 Sales: $ 203 | 2004 Profits: $-26,775 | Parent Company: |

## SALARIES/BENEFITS:

| Pension Plan: | ESOP Stock Plan: | Profit Sharing: | Top Exec. Salary: $425,000 | Bonus: $275,000 |
|---|---|---|---|---|
| Savings Plan: Y | Stock Purch. Plan: Y | | Second Exec. Salary: $325,000 | Bonus: $124,000 |

## OTHER THOUGHTS:
**Apparent Women Officers or Directors**: 4
**Hot Spot for Advancement for Women/Minorities**: Y

## LOCATIONS: ("Y" = Yes)

| West: | Southwest: | Midwest: | Southeast: | Northeast: | International: |
|---|---|---|---|---|---|
| Y | | | | | |

# DISCOVERY LABORATORIES INC			www.discoverylabs.com

**Industry Group Code: 325412  Ranks within this company's industry group:  Sales: 129   Profits: 130**

| Drugs: | | Other: | Clinical: | Computers: | Services: |
|---|---|---|---|---|---|
| Discovery: | Y | AgriBio: | Trials/Services: | Hardware: | Specialty Services: |
| Licensing: | | Genetic Data: | Labs: | Software: | Consulting: |
| Manufacturing: | Y | Tissue Replacement: | Equipment/Supplies: | Arrays: | Blood Collection: |
| Genetics: | | | Research & Development Services: | Database Management: | Drug Delivery: |
| | | | Diagnostics: | | Drug Distribution: |

## TYPES OF BUSINESS:

Respiratory Disease Treatments
Pulmonary Drug Delivery Products

## BRANDS/DIVISIONS/AFFILIATES:

Surfaxin
Aerosurf
Dr. Esteve, S.A.
KL-4

## CONTACTS: Note: Officers with more than one job title may be intentionally listed here more than once.

Robert J. Capetola, CEO
Robert J. Capetola, Pres.
John G. Cooper, CFO/Exec. VP
Kathryn Cole, Sr. VP-Human Resources
Robert Segal, Sr. VP-Medical & Scientific Affairs
Charles F. Katzer, Sr. VP-Mfg. Oper.
David L. Lopez, General Counsel/Exec. VP
Gerald J. Orehostky, Sr. VP-Quality Oper.
Thomas F. Miller, Sr. VP-Corp. Dev. & Commercialization
Robert Segal, Chief Medical Officer
Mary B. Templeton, Sr. VP/Deputy General Counsel
W. Thomas Amick, Chmn.

| Phone: 215-488-9300 | Fax: 215-488-9301 |
|---|---|
| **Toll-Free:** | |
| **Address:** 2600 Kelly Road, Ste. 100, Warrington, PA 18976 US | |

## GROWTH PLANS/SPECIAL FEATURES:

Discovery Laboratories, Inc. (DLI) is a biotechnology company that develops proprietary surfactant technology as Surfactant Replacement Therapies (SRT) for respiratory disorders and diseases. Surfactants are produced naturally by the lungs and are essential for breathing. Discovery's technology produces a precision-engineered surfactant designed to closely mimic the essential properties of natural human lung surfactant. The SRT pipeline is focused primarily on the most significant respiratory conditions prevalent in the neonatal intensive care unit. The firm's lead product, Surfaxin, is used in the prevention of respiratory distress syndrome in premature infants. DLI also develops Surfaxin for the prevention and treatment of bronchopulmonary dysplasia in premature infants. Aerosurf, the company's proprietary SRT in aerosolized form administered through nasal continuous positive airway pressure, is being developed for the prevention and treatment of infants at risk for respiratory failure. DLI's technology also utilizes KL-4, a peptide also known as sinapultide, which mimics the essential properties of SP-B, a surfactant protein that lowers surface tension and promotes oxygen exchange. DLI has granted development and marketing rights of its SRT products to Dr. Esteve, S.A., one of the largest pharmaceutical companies in Southern Europe. The firm is also working to develop and commercialize aerosol SRT (previously conducted with Chrysalis Technologies, a division of Philip Morris USA, Inc.) to address a broad range of serious respiratory conditions such as neonatal respiratory failure, cystic fibrosis, chronic obstructive respiratory disorder and asthma. In April 2009, the FDA declined to approve Surfaxin for the U.S. market, citing questions about the drug's stability test, among other issues.

Discovery offers its employees medical, vision, dental and life insurance; flexible spending accounts; short- and long-term disability; a 401(k) plan; stock options; access to a credit union; employee referral bonuses; a 529 college savings plan; and an employee assistance program.

## FINANCIALS: Sales and profits are in thousands of dollars—add 000 to get the full amount. 2008 Note: Financial information for 2008 was not available for all companies at press time.

| | | |
|---|---|---|
| 2008 Sales: $4,600 | 2008 Profits: $-39,106 | **U.S. Stock Ticker:** DSCO |
| 2007 Sales: $ | 2007 Profits: $-40,005 | **Int'l Ticker:**    Int'l Exchange: |
| 2006 Sales: $ | 2006 Profits: $-46,333 | Employees:   118 |
| 2005 Sales: $ 134 | 2005 Profits: $-58,904 | Fiscal Year Ends: 12/31 |
| 2004 Sales: $1,200 | 2004 Profits: $-46,200 | Parent Company: |

## SALARIES/BENEFITS:

| Pension Plan: | ESOP Stock Plan: | Profit Sharing: | Top Exec. Salary: $470,000 | Bonus: $300,000 |
|---|---|---|---|---|
| Savings Plan: Y | Stock Purch. Plan: Y | | Second Exec. Salary: $292,000 | Bonus: $150,000 |

## OTHER THOUGHTS:

**Apparent Women Officers or Directors**: 2
**Hot Spot for Advancement for Women/Minorities**: Y

## LOCATIONS: ("Y" = Yes)

| West: | Southwest: | Midwest: | Southeast: | Northeast: | International: |
|---|---|---|---|---|---|
| Y | | | | Y | |

# DIVI'S LABORATORIES LIMITED

## www.divislabs.com

**Industry Group Code: 325412 Ranks within this company's industry group: Sales:      Profits:**

| Drugs: | Other: | | Clinical: | | Computers: | | Services: | |
|---|---|---|---|---|---|---|---|---|
| Discovery: | AgriBio: | | Trials/Services: | | Hardware: | | Specialty Services: | |
| Licensing: | Genetic Data: | | Labs: | | Software: | | Consulting: | |
| Manufacturing: Y | Tissue Replacement: | | Equipment/Supplies: | | Arrays: | | Blood Collection: | |
| Genetics: | | | Research & Development Services: Y | | Database Management: | | Drug Delivery: | |
| | | | Diagnostics: | | | | Drug Distribution: | |

## TYPES OF BUSINESS:

Pharmaceutical Manufacturing
Pharmaceutical Research & Development

## BRANDS/DIVISIONS/AFFILIATES:

Divis Laboratories (USA), Inc.
Divis Laboratories (Europe) AG

## CONTACTS: *Note: Officers with more than one job title may be intentionally listed here more than once.*

Murali K. Divi, Managing Dir.
L. Kishorebabu, CFO
P. Gundu Rao, Dir.-R&D
P.V. Lakshmi Rajani, Sec.
Kiran S. Divi, Dir.-Bus. Dev.
N. V. Ramana, Exec. Dir.
Madhusudana Rao Divi, Exec. Dir.
Murali K. Divi, Chmn.

| **Phone:** 91-40-2373-1318 | **Fax:** 91-40-2373-3242 |
|---|---|

**Toll-Free:**

**Address:** Divi Towers, 7-1-77/E/1/303 Dharam Karan Rd., Hyderabad, 500016 India

## GROWTH PLANS/SPECIAL FEATURES:

Divi's Laboratories Limited is an Indian company engaged in pharmaceutical research and development, as well as pharmaceutical manufacturing. The company produces active pharmaceutical ingredients (APIs) and intermediates for generics; peptide components; nucleotide components; carotenoids; and chiral ligands. Some of the APIs manufactured by the company include naproxen, naproxen sodium, niacin, telmisartan, gabapentin and carbidopa, as well as carotenoids such as betacarotene, lycopene, astaxanthin and canthaxanthin. The company has an installed capacity of 3,500 million tons; production in 2008 was 2902.35 million tons. The firm's services include custom synthesis and contract research, specializing in process design for new drug candidates, development, structural elucidation, impurity profile studies, process validation, process justification, process optimization, analytical methods development and validation, environment impact analysis, safety studies and time cycle studies. Divi's operates three manufacturing facilities and four R&D facilities. The firm's main manufacturing and research and development facilities are located in Andhra Pradesh, India. Divi's operates two regional subsidiaries, Divi's Laboratories (USA), Inc., located in New Jersey, and Divi's Laboratories (Europe) AG, located in Switzerland. The majority of the company's sales, 52.9%, come from North America; Europe accounts for 28.9%; India, 5.7%; Far East, 5.6%; South America, 1.8%; and Asia, 1.4%. The rest of the world accounts for the remainder.

## FINANCIALS: Sales and profits are in thousands of dollars—add 000 to get the full amount. 2008 Note: Financial information for 2008 was not available for all companies at press time.

| | | |
|---|---|---|
| 2008 Sales: $ | 2008 Profits: $ | **U.S. Stock Ticker:** |
| 2007 Sales: $ | 2007 Profits: $ | **Int'l Ticker: 532488**   Int'l Exchange: Bombay-BSE |
| 2006 Sales: $ | 2006 Profits: $ | Employees: |
| 2005 Sales: $ | 2005 Profits: $ | Fiscal Year Ends: |
| 2004 Sales: $ | 2004 Profits: $ | Parent Company: |

## SALARIES/BENEFITS:

| Pension Plan: | ESOP Stock Plan: | Profit Sharing: | Top Exec. Salary: $ | Bonus: $ |
|---|---|---|---|---|
| Savings Plan: | Stock Purch. Plan: | | Second Exec. Salary: $ | Bonus: $ |

## OTHER THOUGHTS:

**Apparent Women Officers or Directors:**
**Hot Spot for Advancement for Women/Minorities:**

## LOCATIONS: ("Y" = Yes)

| West: | Southwest: | Midwest: | Southeast: | Northeast: | International: |
|---|---|---|---|---|---|
| | | | | Y | Y |

Note: Financial information, benefits and other data can change quickly and may vary from those stated here.

# DOR BIOPHARMA INC

**www.dorbiopharma.com**

**Industry Group Code: 325412A  Ranks within this company's industry group:** Sales: 24  Profits: 11

| Drugs: | | Other: | Clinical: | Computers: | Services: |
|---|---|---|---|---|---|
| Discovery: | Y | AgriBio: | Trials/Services: | Hardware: | Specialty Services: |
| Licensing: | | Genetic Data: | Labs: | Software: | Consulting: |
| Manufacturing: | | Tissue Replacement: | Equipment/Supplies: | Arrays: | Blood Collection: |
| Genetics: | | | Research & Development Services: | Database Management: | Drug Delivery: |
| | | | Diagnostics: | | Drug Distribution: |

## TYPES OF BUSINESS:

Drugs-Drug Delivery
Oral Formulations
Biodefense Vaccines

## BRANDS/DIVISIONS/AFFILIATES:

orBec
RiVax
BT-VACC
Leuprolide
Oraprine
DOR BioPharma UK Ltd.

## CONTACTS: *Note: Officers with more than one job title may be intentionally listed here more than once.*

Christopher J. Schaber, CEO
Christopher J. Schaber, Pres.
Evan Myrianthopoulos, CFO
Robert N. Brey, Chief Scientific Officer
James Clavijo, Corp. Sec.
James Clavijo, Treas./Controller
James S. Kuo, Chmn.

| Phone: 786-425-3848 | Fax: 786-425-3853 |
|---|---|
| Toll-Free: | |
| Address: 29 Emmons Dr. Ste. C-10, Princeton, NJ 08540 US | |

## GROWTH PLANS/SPECIAL FEATURES:

DOR BioPharma, Inc. develops biodefense vaccines and oral formulations of therapeutic drugs designed to treat life-threatening diseases. The firm operates in two business units, biotherapeutics and biodefense. The company's leading product, orBec, which is part of its biotherapeutics platform, is used to treat gastrointestinal Graft-versus-Host-Disease. The firm has completed Phase III trials of orBec, which has been granted orphan fast track drug status by the FDA. Other biotherapeutic drugs in development include Leuprolide, a treatment for prostate cancer, endometriosis, and precocious puberty, and Oraprine, which is used as an immunosuppressant to inhibit rejection of transplanted organs and as a second-line treatment for severe, active rheumatoid arthritis. DOR is developing an oral formulation of Leuprolide, using a lipid polymer micelle system for enhancing the intestinal absorption of water-soluble peptides. The biodefense division, working in collaboration with the University of Texas Southwest Medical Center and Thomas Jefferson University, has vaccines under development for the treatment of two toxins, ricin and botulinum. RiVax, a vaccine for ricin, has completed Phase I trials and has received support from the NIH and the National Institute of Allergy and Infectious Diseases. The vaccine for botulinum toxin, BT-VACC, is a mucosally administered vaccine. DOR is conducting preclinical research of this vaccine under a cooperative agreement with the U.S. Army's Medical Research Institute of Infectious Diseases. The vaccine is currently in the animal testing stage of development. In September 2008, the company entered into a Named Patient Access Program (NPAP) agreement with BurnsAdler to distribute orBec in Latin America, despite its status as an investigational drug. In December 2008, the FDA cleared DOR201, a drug that uses the active ingredient in orBec, for Investigational New Drug status in treating radiation enteritis. In February 2009, the firm announced a $30 million partner agreement with Sigma-Tau Pharmaceuticals, Inc. to develop and distribute orBec in North America.

## FINANCIALS: Sales and profits are in thousands of dollars—add 000 to get the full amount. 2008 Note: Financial information for 2008 was not available for all companies at press time.

| | | |
|---|---|---|
| 2008 Sales: $2,310 | 2008 Profits: $-3,422 | **U.S. Stock Ticker:** DORB.OB |
| 2007 Sales: $1,258 | 2007 Profits: $-6,165 | **Int'l Ticker:** Int'l Exchange: |
| 2006 Sales: $2,313 | 2006 Profits: $-8,163 | Employees: 7 |
| 2005 Sales: $3,076 | 2005 Profits: $-4,720 | Fiscal Year Ends: 12/31 |
| 2004 Sales: $ 997 | 2004 Profits: $-5,872 | Parent Company: |

## SALARIES/BENEFITS:

| Pension Plan: | ESOP Stock Plan: | Profit Sharing: | Top Exec. Salary: $300,000 | Bonus: $100,000 |
|---|---|---|---|---|
| Savings Plan: | Stock Purch. Plan: | | Second Exec. Salary: $200,000 | Bonus: $50,000 |

## OTHER THOUGHTS:

Apparent Women Officers or Directors:
Hot Spot for Advancement for Women/Minorities:

## LOCATIONS: ("Y" = Yes)

| West: | Southwest: | Midwest: | Southeast: | Northeast: | International: |
|---|---|---|---|---|---|
| | | | | Y | Y |

*Note: Financial information, benefits and other data can change quickly and may vary from those stated here.*

# DOW AGROSCIENCES LLC                          www.dowagro.com

**Industry Group Code:** 11511 **Ranks within this company's industry group:** Sales:     Profits:

| Drugs: | Other: | | Clinical: | Computers: | Services: |
|---|---|---|---|---|---|
| Discovery: | AgriBio: | Y | Trials/Services: | Hardware: | Specialty Services: |
| Licensing: | Genetic Data: | Y | Labs: | Software: | Consulting: |
| Manufacturing: | Tissue Replacement: | | Equipment/Supplies: | Arrays: | Blood Collection: |
| Genetics: | | | Research & Development Services: | Database Management: | Drug Delivery: |
| | | | Diagnostics: | | Drug Distribution: |

## TYPES OF BUSINESS:

Agricultural Chemicals
Agricultural Biotechnology Products
Herbicides, Pesticides & Fungicides
Plant Genetics

## BRANDS/DIVISIONS/AFFILIATES:

Dow Chemical Company (The)
DowElanco
Herculex
WideStrike
Nexera
Agromen Technologia Ltda.
Agrigenetics, Inc.
Duo Maize

## CONTACTS: Note: Officers with more than one job title may be intentionally listed here more than once.

Jerome A. Peribere, CEO
Jerome A. Peribere, Pres.
Bill Wales, General Counsel/VP-Legal Office/Sec.
Robyn Heine, Leader-Global Public Affairs

| Phone: 317-337-3000 | Fax: |
|---|---|
| Toll-Free: | |
| Address: 9330 Zionsville Rd., Indianapolis, IN 46268 US | |

## GROWTH PLANS/SPECIAL FEATURES:

Dow AgroSciences, LLC, a wholly-owned subsidiary of Dow Chemical Co., is a global provider of pest management and biotechnology products for agricultural and specialty markets. The company, formerly DowElanco, was formed by a joint venture between Dow Chemical and Eli Lilly; Dow Chemical purchased Eli Lilly's share in 1997. The firm's products are broken into two categories: agricultural chemicals and plant genetics/ biotechnology. The agricultural chemicals unit produces herbicides, insecticides, fungicides, pest management solutions such as gas fumigants and termite detection tools and more. The plant genetics and biotechnology business develops agricultural products that protect crops against insects, boost nutritional value and increase crop yields. Products are broken into three segments: traits, which encompasses the brands Herculex and WideStrike; seeds, which includes Mycogen brand seeds, Nexera canola and sunflower seeds and PhytoGen cottonseed; and oils. The subsidiary has operations in 140 countries worldwide. In March 2008, the company acquired Triumph Seed, a planting seed company which develops, markets, and produces seeds, for an undisclosed amount. In May 2008, the firm entered a discovery research agreement with GVK Biosciences, a contract research organization to global pharmaceutical and biotech companies based in India, to generate novel molecules for synthesis and testing as insecticides and fungicides. In May 2008, the company entered a collaborative agreement with Martek Biosciences, to commercialize and develop canola seed that produces DHA, the omega-3 fatty acid. In September 2008, the firm acquired Dairyland Seed Co., specializing in hybrid alfalfa, hybrid corn and soybean plant breeding, and Bio-Plant Research Ltd., which represents more than 200 seed companies, for undisclosed amounts. In October 2008, Dow AgroSciences expanded its Brazilian productions by partnering with Coodetec, a cooperative of over 180 farmers. In December 2008, the company acquired Sudwestsaat, a German-based hybrid maize company, for an undisclosed amount.

## FINANCIALS: Sales and profits are in thousands of dollars—add 000 to get the full amount. 2008 Note: Financial information for 2008 was not available for all companies at press time.

| | | |
|---|---|---|
| 2008 Sales: $ | 2008 Profits: $ | **U.S. Stock Ticker: Subsidiary** |
| 2007 Sales: $ | 2007 Profits: $ | **Int'l Ticker:** Int'l Exchange: |
| 2006 Sales: $ | 2006 Profits: $ | Employees: 5,500 |
| 2005 Sales: $3,364,000 | 2005 Profits: $ | Fiscal Year Ends: 12/31 |
| 2004 Sales: $3,368,000 | 2004 Profits: $ | Parent Company: DOW CHEMICAL COMPANY (THE) |

## SALARIES/BENEFITS:

| Pension Plan: Y | ESOP Stock Plan: | Profit Sharing: | Top Exec. Salary: $ | Bonus: $ |
|---|---|---|---|---|
| Savings Plan: Y | Stock Purch. Plan: Y | | Second Exec. Salary: $ | Bonus: $ |

## OTHER THOUGHTS:

**Apparent Women Officers or Directors:** 1
**Hot Spot for Advancement for Women/Minorities:**

## LOCATIONS: ("Y" = Yes)

| West: | Southwest: | Midwest: | Southeast: | Northeast: | International: |
|---|---|---|---|---|---|
| | | Y | | | Y |

Note: Financial information, benefits and other data can change quickly and may vary from those stated here.

# DR REDDY'S LABORATORIES LIMITED

## www.drreddys.com

**Industry Group Code: 325412 Ranks within this company's industry group:** Sales: 38     Profits: 35

| Drugs: | | Other: | | Clinical: | Computers: | | Services: | |
|---|---|---|---|---|---|---|---|---|
| Discovery: | Y | AgriBio: | | Trials/Services: | Hardware: | | Specialty Services: | |
| Licensing: | | Genetic Data: | | Labs: | Software: | | Consulting: | |
| Manufacturing: | Y | Tissue Replacement: | | Equipment/Supplies: | Arrays: | | Blood Collection: | |
| Genetics: | Y | | | Research & Development Services: | Database Management: | | Drug Delivery: | |
| | | | | Diagnostics: | | | Drug Distribution: | Y |

## TYPES OF BUSINESS:

Pharmaceuticals
Generic Drugs
Active Ingredients
Drug Discovery & Development

## BRANDS/DIVISIONS/AFFILIATES:

Grafeel
Reditux
Dr. Reddy's Laboratories (EU) Limited

## CONTACTS: *Note: Officers with more than one job title may be intentionally listed here more than once.*

G. V. Prasad, CEO/Exec. Vice Chmn.
Satish Reddy, COO
Satish Reddy, Managing Dir.
Umang Vohra, CFO/Sr. VP
Prabir Jha, Chief Human Resources Officer/Sr. VP
Rajinder Kumar, Pres., R&D & Commercialization
K. B. Sankara Rao, Exec. VP-Integrated Prod. Dev.
Vilas Dholye, Exec. VP/Head-Formulations Mfg.
V. S. Suresh, Corp. Sec.
Saumen Chakraborty, Pres., Corp. & Global Generics Oper.
Amit Patel, Sr. VP-Strategic Bus. Planning
Mythili Mamidanna, Corp. Comm. Contact
Kedar Upadhye, Investor Rel. Contact
Abijit Mukherjee, Pres., Pharma Svcs. & Active Ingredients (PSAI)
Jeffrey Wasserstein, Exec. VP-North America Specialty
Amit Patel, Sr. VP-North America Generics
Cartikeya Reddy, Sr. VP/Head-Biologics
K. Anji Reddy, Chmn.
V. S. Vasudevan, Head/Pres., European Oper.

| Phone: 91-40-23731946 | Fax: 91-40-23731955 |
|---|---|
| Toll-Free: | |
| Address: 7-1-27, Ameerpet, Hyderabad, 500 016 India | |

## GROWTH PLANS/SPECIAL FEATURES:

Dr. Reddy's Laboratories Limited is a global pharmaceutical manufacturer. It operates three segments: Global Generics (which generated approximately 72% of 2009 revenue), Pharmaceutical Services & Active Ingredients (PSAI) (27%) and Proprietary Products and other (1%). Global Generics consists of finished pharmaceutical products ready for consumption by the patient, marketed under a brand name or as generic products. The company currently offers over 200 generic drugs in therapeutic areas including gastrointestinal, oncology, cardiovascular, pediatrics, dermatology, anti-inflammatory and pain management. Dr. Reddy's calls its generic biopharmaceuticals Biologics. Its Biologics include Grafeel, a cancer treatment that stimulates white blood cell growth in bone marrow; and a generic monoclonal antibody called Reditux, a non-Hodgkins lymphoma treatment. PSAI comprises raw active pharmaceutical ingredients transformed by other manufacturers into pharmaceutical products ready for human consumption such as a tablets, capsules or liquids using additional inactive ingredients. Dr. Reddy's manufactures and markets over 140 bulk active ingredients, the principal ingredients in the finished dosages of drugs. This segment also includes contract research services and contract manufacturing. Lastly, Proprietary Products involves discovering new chemicals for subsequent commercialization and out-licensing, as well as Dr. Reddy's specialty pharmaceuticals business, which primarily sells in-licensed and co-developed dermatology products. The company's drug discovery activities, with its primary laboratory in Hyderabad, focus on the areas of metabolic disorders, cardiovascular disorders, bacterial infections, inflammation and cancer. Geographically, Europe generated the largest share of 2009 revenues (33%), followed by North America (21%), India (13%) and the rest of the world (34%). In April 2008, subsidiary Dr. Reddy's Laboratories (EU) Limited acquired certain U.K.-based assets from The Dow Chemical Company for $32 million. In April 2008, the firm acquired a newly completed Shreveport, Louisiana-based contract manufacturing facility (now part of Global Generics) from BASF Corporation for $40 million.

## FINANCIALS: Sales and profits are in thousands of dollars—add 000 to get the full amount. 2008 Note: Financial information for 2008 was not available for all companies at press time.

| | | |
|---|---|---|
| 2008 Sales: $1,230,100 | 2008 Profits: $116,900 | **U.S. Stock Ticker: RDY** |
| 2007 Sales: $1,488,010 | 2007 Profits: $215,790 | **Int'l Ticker: 500124**   Int'l Exchange: Bombay-BSE |
| 2006 Sales: $541,304 | 2006 Profits: $36,620 | Employees: 9,575 |
| 2005 Sales: $438,000 | 2005 Profits: $4,800 | Fiscal Year Ends: 3/31 |
| 2004 Sales: $463,900 | 2004 Profits: $57,200 | Parent Company: |

## SALARIES/BENEFITS:

| Pension Plan: Y | ESOP Stock Plan: | Profit Sharing: | Top Exec. Salary: $794,352 | Bonus: $ |
|---|---|---|---|---|
| Savings Plan: Y | Stock Purch. Plan: Y | | Second Exec. Salary: $529,496 | Bonus: $ |

## OTHER THOUGHTS:

**Apparent Women Officers or Directors**: 1
**Hot Spot for Advancement for Women/Minorities**:

## LOCATIONS: ("Y" = Yes)

| West: | Southwest: | Midwest: | Southeast: | Northeast: | International: |
|---|---|---|---|---|---|
| | | | | Y | Y |

Note: Financial information, benefits and other data can change quickly and may vary from those stated here.

# DSM PHARMACEUTICALS INC

### www.dsmpharmaceuticals.com

Industry Group Code: 325412  Ranks within this company's industry group:  Sales:  Profits:

| Drugs: | | Other: | | Clinical: | | Computers: | | Services: | |
|---|---|---|---|---|---|---|---|---|---|
| Discovery: | | AgriBio: | | Trials/Services: | Y | Hardware: | | Specialty Services: | Y |
| Licensing: | | Genetic Data: | | Labs: | | Software: | | Consulting: | |
| Manufacturing: | Y | Tissue Replacement: | | Equipment/Supplies: | Y | Arrays: | | Blood Collection: | |
| Genetics: | | | | Research & Development Services: | Y | Database Management: | | Drug Delivery: | |
| | | | | Diagnostics: | | | | Drug Distribution: | |

## TYPES OF BUSINESS:

Drugs-Custom Manufacturing
Oral & Topical Formulations
Sterile Liquids
Lyophilization Services
Pharmaceutical Packaging
Product Development & Clinical Trial Services
Product Packaging

## BRANDS/DIVISIONS/AFFILIATES:

DSM
Lyo-Advantage
Liquid-Advantage
DSM Pharmaceuticals Packaging Group
DSM Pharma Chemicals
APT Pharmaceuticals, Inc.

## CONTACTS: Note: Officers with more than one job title may be intentionally listed here more than once.

Hans Engels, COO
Hans Engels, Pres.
Rolf-Dieter Schwalb, CFO
Terry Novak, Chief Mktg. Officer
Cor Visker, Sr. VP-Human Resources
Feike Sijbesma, Chmn.

| Phone: 252-758-3436 | Fax: 973-257-8481 |
|---|---|
| Toll-Free: | |
| Address: 5900 NW Greenville Blvd., Greenville, NC 27834 US | |

## GROWTH PLANS/SPECIAL FEATURES:

DSM Pharmaceuticals, Inc., a subsidiary of DSM NV, is an outsourcing partner for both large and small pharmaceutical companies that specializes in custom chemical manufacturing of synthetic drugs and biopharmaceuticals. The firm also provides products and services from early and preclinical development to initial synthesis of promising new drugs and dosage formulation and packaging. Company products and services fall under five categories: Manufacturing; orals and topicals; sterile liquids; lyophilization (freeze drying); and packaging. DSM's manufacturing services convert active pharmaceutical ingredients (APIs) and excipients processed by DSM Pharma Chemicals and other suppliers into finished dosage forms such as solid-dose creams, ointments, steriles and liquids. Its oral and topical products include creams and ointments, while services in this area include blending and granulation, coating, tablet compression, capsule filing, drying and all related packaging. Sterile liquid operations include the manufacturing, aseptic filling, terminal sterilization, cold storage and packaging of sterile liquids. The company's steriles processing facilities use state-of-the-art Liquid-Advantage and Lyo-Advantage systems, both developed in-house. These systems provide a precise level of manufacturing and environment control. The firm offers cutting edge lyophilization services, which involve freeze-drying of biological substances. The DSM Pharmaceuticals Packaging Group is capable of producing high-quality specified package engineering, graphics design and testing services. Its offerings include bottles, liners, stoppers, dosage cups and droppers and thermo films, among various other packaging items. Other services include clinical trial management and the development and manufacturing of finished dosage forms for pharmaceutical companies. In recent news, DSM Pharmaceuticals, Inc. signed a manufacturing agreement with APT Pharmaceuticals, Inc., a drug company focused primarily on inhalants for serious lung diseases, where DSM will provide sterile manufacturing services for APT, utilizing DSM's commercial facilities in Greenville, North Carolina.

DSM Pharmaceuticals offers employees a benefits plan including prepaid legal assistance and educational reimbursement.

## FINANCIALS:  Sales and profits are in thousands of dollars—add 000 to get the full amount. 2008 Note: Financial information for 2008 was not available for all companies at press time.

| | | | |
|---|---|---|---|
| 2008 Sales: $ | 2008 Profits: $ | U.S. Stock Ticker: Subsidiary | |
| 2007 Sales: $ | 2007 Profits: $ | Int'l Ticker: Int'l Exchange: | |
| 2006 Sales: $ | 2006 Profits: $ | Employees: | |
| 2005 Sales: $ | 2005 Profits: $ | Fiscal Year Ends: 12/31 | |
| 2004 Sales: $ | 2004 Profits: $ | Parent Company: DSM NV | |

## SALARIES/BENEFITS:

| Pension Plan: | ESOP Stock Plan: | Profit Sharing: | Top Exec. Salary: $265,000 | Bonus: $170,000 |
|---|---|---|---|---|
| Savings Plan: Y | Stock Purch. Plan: Y | | Second Exec. Salary: $208,000 | Bonus: $21,000 |

## OTHER THOUGHTS:

Apparent Women Officers or Directors:
Hot Spot for Advancement for Women/Minorities:

## LOCATIONS: ("Y" = Yes)

| West: | Southwest: | Midwest: | Southeast: | Northeast: | International: |
|---|---|---|---|---|---|
| | | | | Y | Y |

Note: Financial information, benefits and other data can change quickly and may vary from those stated here.

# DUPONT AGRICULTURE & NUTRITION

www.dupont.com/ag

Industry Group Code: 11511  Ranks within this company's industry group: Sales:  Profits:

| Drugs: | Other: | | Clinical: | Computers: | Services: |
|---|---|---|---|---|---|
| Discovery: | AgriBio: | Y | Trials/Services: | Hardware: | Specialty Services: |
| Licensing: | Genetic Data: | | Labs: | Software: | Consulting: |
| Manufacturing: | Tissue Replacement: | | Equipment/Supplies: | Arrays: | Blood Collection: |
| Genetics: | | | Research & Development Services: | Database Management: | Drug Delivery: |
| | | | Diagnostics: | | Drug Distribution: |

## TYPES OF BUSINESS:

Agricultural Biotechnology Products & Chemicals Manufacturing
Insecticides
Herbicides
Fungicides
Genetically Modified Plants
Soy Products
Forage & Grain Additives

## BRANDS/DIVISIONS/AFFILIATES:

Pioneer Hi-Bred International
DuPont Crop Protection
DuPont Nutrition and Health
Solae Company (The)
Supro
Agroproducts Corey, S.A. de C.V.
Nurish
Qualicon

## CONTACTS: Note: Officers with more than one job title may be intentionally listed here more than once.

John Bedbrook, VP-R&D
Don Wirth, VP-Oper.
Kathleen H. Forte, VP-Public Affairs
James C. Borel, VP-DuPont Agriculture & Nutrition
James C. Collins, Pres., DuPont Crop Protection
Craig F. Binetti, Pres., DuPont Nutrition & Health
Paul Schickler, Pres., Pioneer Hi-Bred
Charles O. Holliday, Jr., Chmn.
Donald Wirth, VP-Global Oper.

| Phone: 302-774-1000 | Fax: |
|---|---|
| Toll-Free: 888-638-7668 | |
| Address: DuPont Bldg., 1007 Market St., Wilmington, DE 19898 US | |

## GROWTH PLANS/SPECIAL FEATURES:

DuPont Agriculture & Nutrition (DPAN) is a business unit of global chemical giant DuPont. The company oversees a number of business units covering many aspects of crop protection, optimization and additives. Pioneer Hi-Bred International develops advanced plant genetics, including seeds and forage and grain additives. DuPont Crop Protection produces herbicide, fungicide and insecticide products and services. DuPont Nutrition and Health provides soy protein and soy fiber ingredients under brand names including The Solae Company, a joint venture with Bunge; Supro; SuproSoy; and Nurish. DPAN has joint ventures in the U.S. and around the world, with projects such as an Agricultural products venture with Agroproducts Corey, S.A. de C.V. in Mexico; a crop protection venture with AO Khimprom in Russia; a biofuel production partnership with British Petroleum (BP); and soy-based ventures with General Mills/PTI in Minnesota, So Good in the U.K and Syngenta in Illinois. DPAN also has a microbial testing branch, Qualicon. Recently a number of new products have been approved for registration by the Environmental Protection Agency (EPA), including the cleaning and personal care product Zemea propanediol and the herbicide Herculex. In 2008, the company announced plans to open a corn research center in India to help farmers increase grain productivity, and two new research centers, in Italy and Hungary, for the purpose of cultivating higher yielding corn and sunflower hybrids. Late in 2008, DuPont opened the DuPont Knowledge Center in Hyderabad, India, a $30 million, agricultural and biotechnology research center for the India and the Asia Pacific region. These recent expansions reflect DuPont's current focus on research and development in Europe and India, aimed at boosting farmer productivity and profitability, and fulfilling a growing global demand for agricultural products.

The company offers employees a comprehensive benefits package with dependent care spending accounts, flexible work practices and adoption assistance.

## FINANCIALS: Sales and profits are in thousands of dollars—add 000 to get the full amount. 2008 Note: Financial information for 2008 was not available for all companies at press time.

| | | |
|---|---|---|
| 2008 Sales: $ | 2008 Profits: $ | U.S. Stock Ticker: Subsidiary |
| 2007 Sales: $ | 2007 Profits: $ | Int'l Ticker:  Int'l Exchange: |
| 2006 Sales: $ | 2006 Profits: $ | Employees: |
| 2005 Sales: $ | 2005 Profits: $ | Fiscal Year Ends: 12/31 |
| 2004 Sales: $6,247,000 | 2004 Profits: $ | Parent Company: E I DU PONT DE NEMOURS & CO (DUPONT) |

## SALARIES/BENEFITS:

| Pension Plan: | ESOP Stock Plan: | Profit Sharing: | Top Exec. Salary: $ | Bonus: $ |
|---|---|---|---|---|
| Savings Plan: | Stock Purch. Plan: | | Second Exec. Salary: $ | Bonus: $ |

## OTHER THOUGHTS:

Apparent Women Officers or Directors: 1
Hot Spot for Advancement for Women/Minorities:

## LOCATIONS: ("Y" = Yes)

| West: | Southwest: | Midwest: | Southeast: | Northeast: | International: |
|---|---|---|---|---|---|
| Y | Y | Y | Y | Y | Y |

# DURECT CORP

**www.durect.com**

Industry Group Code: 325412A  **Ranks within this company's industry group:** Sales: 19  Profits: 26

| Drugs: | | Other: | | Clinical: | | Computers: | | Services: | |
|---|---|---|---|---|---|---|---|---|---|
| Discovery: | | AgriBio: | | Trials/Services: | | Hardware: | | Specialty Services: | |
| Licensing: | Y | Genetic Data: | | Labs: | | Software: | | Consulting: | |
| Manufacturing: | Y | Tissue Replacement: | | Equipment/Supplies: | | Arrays: | | Blood Collection: | |
| Genetics: | | | | Research & Development Services: | | Database Management: | | Drug Delivery: | Y |
| | | | | Diagnostics: | | | | Drug Distribution: | |

## TYPES OF BUSINESS:

Drug Delivery Systems
Biodegradable Polymer Manufacturing
Pharmaceutical Products

## BRANDS/DIVISIONS/AFFILIATES:

SABER
ORADUR
TRANSDUR
DURIN
MICRODUR
ELADUR
POSIDUR
ALZET

## CONTACTS: *Note: Officers with more than one job title may be intentionally listed here more than once.*

James E. Brown, CEO
James E. Brown, Pres.
Matthew J. Hogan, CFO
Felix Theeuwes, Chief Scientific Officer
Harry Guy, VP-Eng. & Safety
Paula Mendenhall, Exec. VP-Admin.
Jean I. Liu, General Counsel/Sr. VP/Sec.
Paula Mendenhall, Exec. VP-Oper.
Michael H. Arenberg, VP-Bus. Dev.
Jian Li, VP-Finance/Corp. Controller
Peter J. Langecker, Chief Medical Officer
Su IL Yum, Exec. VP-Pharmaceutical Systems R&D/Principal Eng.
Nacer E. Dean Abrouk, VP-Biostatistics
Andrew R. Miksztal, VP-Pharmaceutical R&D
Felix Theeuwes, Chmn.

| Phone: 408-777-1417 | Fax: 408-777-3577 |
|---|---|
| Toll-Free: | |
| Address: 2 Results Way, Cupertino, CA 95014 US | |

## GROWTH PLANS/SPECIAL FEATURES:

Durect Corp. is an emerging specialty pharmaceutical company focused on the development of pharmaceutical products based on its proprietary drug delivery technology platforms. The company's pharmaceutical systems enable optimized therapy for a given disease or patient population by controlling the rate and duration of drug administration as well as, for certain applications, placement of the drug at the intended site of action. Durect's development efforts are focused on the application of its pharmaceutical systems technologies in a variety of chronic and episodic disease areas including pain, central nervous system disorders, cardiovascular disease and other chronic diseases. The firm's delivery platforms include Saber, an injectable used for proteins, peptides, and small molecules; Oradur, for oral gel-cap administration; Transdur, a transdermal patch; and the Durin and MICRODurin biodegradable input systems. The company has two marketed products, including Alzet, an implantable osmotic pump for laboratory animal testing; and Lactel, a line of biodegradable polymers. Durect's lead development product in Remoxy, a chronic pain solution based on Oradur technology and developed in partnership with Pain Therapeutics, Inc. and King Pharmaceuticals, Inc. Other products in development include Posidur, a product for post-operative pain, which is based on Saber technology and developed in conjunction with Nycomed; Transdur Sufentanil, a chronic pain treatment; Eladur, a drug for post-herpetic neuralgia also developed with King Pharmaceuticals; and two yet-undisclosed opioids based on Oradur. In December 2008, the company received a notice from the FDA stating that they would not approve a new drug application (NDA) for Remoxy.

Durect offers its employees medical, dental and vision insurance; life, AD&D, short- and long-term disability insurance; a 401(k) plan; an employee stock purchase plan; stock options; an employee assistance program; subsidized health club memberships; flexible spending accounts; a wellness program; and annual company events.

## FINANCIALS: Sales and profits are in thousands of dollars—add 000 to get the full amount. 2008 Note: Financial information for 2008 was not available for all companies at press time.

| | | |
|---|---|---|
| 2008 Sales: $27,101 | 2008 Profits: $-43,907 | **U.S. Stock Ticker: DRRX** |
| 2007 Sales: $30,675 | 2007 Profits: $-24,339 | **Int'l Ticker:**     Int'l Exchange: |
| 2006 Sales: $21,894 | 2006 Profits: $-33,327 | Employees:   171 |
| 2005 Sales: $28,571 | 2005 Profits: $-18,128 | Fiscal Year Ends: 12/31 |
| 2004 Sales: $13,853 | 2004 Profits: $-27,637 | Parent Company: |

## SALARIES/BENEFITS:

| Pension Plan: | ESOP Stock Plan: | Profit Sharing: | Top Exec. Salary: $464,373 | Bonus: $25,076 |
|---|---|---|---|---|
| Savings Plan: Y | Stock Purch. Plan: Y | | Second Exec. Salary: $452,907 | Bonus: $24,457 |

## OTHER THOUGHTS:

**Apparent Women Officers or Directors:** 3
**Hot Spot for Advancement for Women/Minorities:** Y

## LOCATIONS: ("Y" = Yes)

| West: | Southwest: | Midwest: | Southeast: | Northeast: | International: |
|---|---|---|---|---|---|
| Y | | | | | |

# DUSA PHARMACEUTICALS INC

### www.dusapharma.com

**Industry Group Code:** 325412 **Ranks within this company's industry group:** Sales: 98    Profits: 71

| Drugs: | | Other: | Clinical: | Computers: | Services: |
|---|---|---|---|---|---|
| Discovery: | Y | AgriBio: | Trials/Services: | Hardware: | Specialty Services: |
| Licensing: | Y | Genetic Data: | Labs: | Software: | Consulting: |
| Manufacturing: | Y | Tissue Replacement: | Equipment/Supplies: | Arrays: | Blood Collection: |
| Genetics: | | | Research & Development Services: | Database Management: | Drug Delivery: |
| | | | Diagnostics: | | Drug Distribution: |

## TYPES OF BUSINESS:

Photodynamic Therapy Products

## BRANDS/DIVISIONS/AFFILIATES:

Levulan
BLU-U
Kerastick
Nicomide
Nicomide-T
ClindaReach
Sirius Labratories, Inc.
Psoriatec

## CONTACTS: *Note: Officers with more than one job title may be intentionally listed here more than once.*

Robert F. Doman, CEO
Robert F. Doman, Pres.
Richard Christopher, CFO
William F. O'Dell, Exec. VP-Sales & Mktg.
Stuart Marcus, VP-Scientific Affairs/Chief Medical Officer
Nanette W. Mantell, Sec.
Mark Carota, VP-Oper.
D. Geoffrey Shulman, Chief Strategic Officer
Richard Christopher, VP-Finance
Michael J. Todisco, VP/Controller
Scott Lundahl, VP-Regulatory Affairs & Intellectual Property
D. Geoffrey Shulman, Chmn.

| Phone: 978-657-7500 | Fax: 978-657-9193 |
|---|---|
| Toll-Free: | |
| Address: 25 Upton Dr., Wilmington, MA 01887 US | |

## GROWTH PLANS/SPECIAL FEATURES:

DUSA Pharmaceuticals, Inc. is a dermatology company that develops and markets Levulan photodynamic therapy (PDT) and other products for common skin conditions. The company's marketed products include Levulan Kerastick 20% topical solution with PDT, the BLU-U brand light source; and the newly launched product, ClindaReach. The firm's Levulan brand of aminolevulinic acid HCl is designed to be utilized when followed by exposure to light, either to treat a medical condition (Levulan PDT) or to detect a medical condition (Levulan photodetection, or Levulan PD). The Levulan Kerastick 20% topical solution with PDT and the BLU-U light source are used for the treatment of non-hyperkeratotic actinic keratoses, precancerous skin lesions caused by chronic sun exposure of the face or scalp. BLU-U is also used for the treatment of moderate inflammatory acne vulgaris and general dermatological conditions. ClindaReach provides medication for acne vulgaris as well as supplying a unique system for applying this medication to all hard-to-reach areas, including the back. Other products include Psoriatec, Nicomide-T and METED Shampoo, which are distributed by Sirius Labratories, Inc., a wholly owned subsidiary of DUSA Pharmaceuticals, Inc. DUSA has an agreement with Stiefel Laboratories to market the Levulan Kerastick in Mexico, Central and South America; with Coherent-AMT to market its products in Canada; and with Daewoong Pharmaceutical Co., Ltd. to market Levulan Kerastick in certain Asian territories. In July 2008, the company announced it would begin marketing its Nicomide line as a prescription product and seek to rebrand it as a dietary supplement. In May 2009, DUSA announced it would begin clinical trials to determine the effectiveness of Levulan PDT in treating skin cancer in immunosuppressed organ transplant recipients.

The company offers its employees a 401(k) plan; health, dental and life insurance; flexible spending accounts; tuition reimbursement; short- and long-term disability; and business travel insurance.

## FINANCIALS: Sales and profits are in thousands of dollars—add 000 to get the full amount. 2008 Note: Financial information for 2008 was not available for all companies at press time.

| | | |
|---|---|---|
| 2008 Sales: $29,545 | 2008 Profits: $-6,250 | **U.S. Stock Ticker:** DUSA |
| 2007 Sales: $27,663 | 2007 Profits: $-14,714 | **Int'l Ticker:**    Int'l Exchange: |
| 2006 Sales: $25,583 | 2006 Profits: $-31,350 | Employees:   86 |
| 2005 Sales: $11,337 | 2005 Profits: $-14,999 | Fiscal Year Ends: 12/31 |
| 2004 Sales: $7,988 | 2004 Profits: $-15,629 | Parent Company: |

## SALARIES/BENEFITS:

| Pension Plan: | ESOP Stock Plan: | Profit Sharing: | Top Exec. Salary: $417,000 | Bonus: $141,000 |
|---|---|---|---|---|
| Savings Plan: Y | Stock Purch. Plan: | | Second Exec. Salary: $329,363 | Bonus: $ |

## OTHER THOUGHTS:

**Apparent Women Officers or Directors:** 1
**Hot Spot for Advancement for Women/Minorities:**

## LOCATIONS: ("Y" = Yes)

| West: | Southwest: | Midwest: | Southeast: | Northeast: | International: |
|---|---|---|---|---|---|
| | | | | Y | Y |

Note: Financial information, benefits and other data can change quickly and may vary from those stated here.

# DYAX CORP

**www.dyax.com**

**Industry Group Code: 325412  Ranks within this company's industry group:** Sales: 88  Profits: 149

| Drugs: | | Other: | | Clinical: | | Computers: | | Services: | |
|---|---|---|---|---|---|---|---|---|---|
| Discovery: | Y | AgriBio: | | Trials/Services: | | Hardware: | | Specialty Services: | |
| Licensing: | Y | Genetic Data: | | Labs: | | Software: | Y | Consulting: | |
| Manufacturing: | | Tissue Replacement: | | Equipment/Supplies: | | Arrays: | | Blood Collection: | |
| Genetics: | | | | Research & Development Services: | Y | Database Management: | | Drug Delivery: | |
| | | | | Diagnostics: | | | | Drug Distribution: | |

## TYPES OF BUSINESS:

Pharmaceuticals Discovery & Development
Proteins, Peptides & Antibodies
Drugs-Cancer
Drugs-Anti-Inflammatory
Drugs-Blood Clotting Regulation

## BRANDS/DIVISIONS/AFFILIATES:

DX-88
DX-2400
DX-2240
WebPhage

## CONTACTS: *Note: Officers with more than one job title may be intentionally listed here more than once.*

Gustav A. Christensen, CEO
Gustav A. Christensen, Pres.
George Migausky, CFO/Exec. VP
Clive R. Wood, Exec. VP-Research/Chief Scientific Officer
Ivana Magovcevic-Liebisch, Exec. VP-Admin.
Ivana Magovcevic-Liebisch, General Counsel
William E. Pullman, Chief Dev. Officer/Exec. VP
William E. Pullman, Chief Dev. Officer/Exec. VP
Nathaniel S. Gardiner, Sec.
Henry E. Blair, Chmn.

| **Phone:** 617-225-2500 | **Fax:** 617-225-2501 |
|---|---|
| **Toll-Free:** | |
| **Address:** 300 Technology Sq., Cambridge, MA 02139 US | |

## GROWTH PLANS/SPECIAL FEATURES:

Dyax Corp. is a biopharmaceutical company principally focused on the discovery, development and commercialization of antibodies, proteins and peptides as therapeutic products, with an emphasis on cancer and inflammatory conditions. In general, the firm strives to maintain 10 active therapeutic programs in its pipeline at all times. Its lead product candidate is DX-88 (ecallantide), a recombinant form of a small protein that is being tested for applications in hereditary angioedema, a genetic disease that causes swelling of the larynx, gastrointestinal tract and extremities. DX-88 is also being tested to treat complications, including blood loss and systemic inflammation, during on-pump cardiothoracic surgery. This application passed Phase I and II trials and began a second Phase II trial. After DX-88, its most advanced candidates are DX-2240 and DX-2400, both fully human monoclonal antibodies designed to attack cancerous tumors. The company signed a license agreement granting sanofi-aventis S.A. the responsibility for the future development of DX-2240. Dyax identifies its pipeline compounds through its proprietary phage display methodology, called WebPhage. The process involves generating phage display libraries of protein variations, screening libraries to select binding compounds with high affinity and high specificity to a target and producing and evaluating the selected binding compounds. The company also allows other companies to use WebPhage display technology to discover new compounds through its Licensing and Funded Research Program. Over 70 companies licensed this technology and 12 new compounds currently undergoing clinical trials have been derived. In February 2009, Dyax agreed to license DX-88 to Fovea Pharmaceuticals, SA for the treatment of retinal diseases in Europe. In March 2009, the company announced it would be eliminating about a third of its workforce, or 60 employees.

Employees of the firm receive medical and dental benefits; short- and long-term disability coverage; life and AD&D insurance; flexible spending accounts; paid holidays; educational assistance; an employee assistance program; health club reimbursement; and a transportation subsidy.

## FINANCIALS: Sales and profits are in thousands of dollars—add 000 to get the full amount. 2008 Note: Financial information for 2008 was not available for all companies at press time.

| | | |
|---|---|---|
| 2008 Sales: $43,429 | 2008 Profits: $-66,468 | **U.S. Stock Ticker: DYAX** |
| 2007 Sales: $26,096 | 2007 Profits: $-56,309 | **Int'l Ticker:**  Int'l Exchange: |
| 2006 Sales: $12,776 | 2006 Profits: $-50,323 | Employees:  164 |
| 2005 Sales: $19,859 | 2005 Profits: $-30,944 | Fiscal Year Ends: 12/31 |
| 2004 Sales: $16,590 | 2004 Profits: $-33,114 | Parent Company: |

## SALARIES/BENEFITS:

| Pension Plan: | ESOP Stock Plan: | Profit Sharing: Y | Top Exec. Salary: $647,115 | Bonus: $300,000 |
|---|---|---|---|---|
| Savings Plan: Y | Stock Purch. Plan: Y | | Second Exec. Salary: $373,846 | Bonus: $200,000 |

## OTHER THOUGHTS:

**Apparent Women Officers or Directors:** 3
**Hot Spot for Advancement for Women/Minorities:** Y

## LOCATIONS: ("Y" = Yes)

| West: | Southwest: | Midwest: | Southeast: | Northeast: | International: |
|---|---|---|---|---|---|
| | | | | Y | Y |

Note: Financial information, benefits and other data can change quickly and may vary from those stated here.

# E I DU PONT DE NEMOURS & CO (DUPONT)

www2.dupont.com

**Industry Group Code: 325 Ranks within this company's industry group: Sales: 3 Profits: 3**

| Drugs: | Other: | Clinical: | Computers: | Services: |
|---|---|---|---|---|
| Discovery: | AgriBio: Y | Trials/Services: | Hardware: | Specialty Services: |
| Licensing: | Genetic Data: | Labs: | Software: | Consulting: |
| Manufacturing: | Tissue Replacement: | Equipment/Supplies: | Arrays: | Blood Collection: |
| Genetics: | | Research & Development Services: | Database Management: | Drug Delivery: |
| | | Diagnostics: | | Drug Distribution: |

## TYPES OF BUSINESS:

Chemicals Manufacturing
Polymers
Performance Coatings
Nutrition & Health Products
Electronics Materials
Agricultural Seeds
Fuel-Cell, Biofuels & Solar Panel Technology
Contract Research & Development

## BRANDS/DIVISIONS/AFFILIATES:

Pioneer
Teflon
Corian
Kevlar
Tyvek
Coastal Training Technologies Corporation
MapShots, Inc.

## CONTACTS: *Note: Officers with more than one job title may be intentionally listed here more than once.*

Ellen J. Kullman, CEO
Richard R. Goodmanson, COO/Exec. VP
Jeffrey L. Keefer, CFO/Exec. VP
David G. Bills, Chief Mktg. & Sales Officer
W. Donald Johnson, Sr. VP-DuPont Human Resources
John Bedbrook, VP-R&D, Agriculture & Nutrition
Phuong Tram, CIO/VP-DuPont IT
Uma Chowdhry, CTO/Chief Science Officer/Sr. VP
Thomas M. Connelly, Jr., Chief Innovation Officer/Exec. VP
Mathieu Vrijsen, Sr. VP-DuPont Eng.
Thomas L. Sager, General Counsel/Sr. VP-DuPont Legal
Mathieu Vrijsen, Sr. VP-DuPont Oper.
Peter C. Hemken, VP-Strategic Dir. & Bus. Dev., Pioneer Hi-Bred
Karen A. Fletcher, VP-DuPont Investor Rel.
Susan M. Stalnecker, VP-Finance/Treas.
Criag F. Binetti, Sr. VP-DuPont Nutrition & Health
Diane H. Gulyas, Group VP-DuPont Performance Materials
Terry Caloghiris, Group VP-DuPont Coatings & Color Tech.
Nicholas C. Fanandakis, Group VP-DuPont Applied BioSciences
Charles O. Holliday, Jr., Chmn.
Don Wirth, VP-Global Oper.
Jeffrey A. Coe, Sr. VP-DuPont Sourcing & Logistics

| Phone: 302-774-1000 | Fax: 302-773-2631 |
|---|---|
| Toll-Free: 800-441-7515 | |
| Address: 1007 Market St., Wilmington, DE 19898 US | |

## GROWTH PLANS/SPECIAL FEATURES:

E. I. du Pont de Nemours & Co. (DuPont), founded in 1802, develops and manufactures products in the biotechnology, electronics, materials science, synthetic fibers and safety and security sectors. DuPont operates in five segments: Agriculture and Nutrition (A&N); Coatings and Color Technologies (C&CT); Electronic and Communication Technologies (E&C); Performance Materials (PM); and Safety and Protection (S&P). A&N delivers Pioneer brand seed products, insecticides, fungicides, herbicides, soy-based food ingredients, food quality diagnostic testing equipment and liquid food packaging systems. The C&CT segment supplies automotive coatings, titanium dioxide white pigments and pigment and dye-based inks for ink-jet digital printing. E&C provides a range of advanced materials for the electronics industry, flexographic printing, color communication systems and a range of fluoropolymer and fluorochemical products. PM manufactures polymer-based materials, which include engineered polymers, specialized resins and films for use in food packaging, sealants, adhesives, sporting goods and laminated safety glass. The S&P segment provides protective materials and safety consulting services. Significant brands include Teflon fluoropolymers, films, fabric protectors, fibers and dispersions; Corian surfaces; Kevlar high strength material; and Tyvek protective material. Recent acquisitions include Chemtura Corporation's fluorine chemicals business in February 2008; the Industrial Apparel line of Cardinal Health's Scientific and Production Products business in April 2008; Coastal Training Technologies Corporation, a producer and marketer of training programs, in October 2008; and MapShots, Inc., an agricultural data management company, in December 2008. Recent divestitures include DuPont Super Boll and FreeFall brand cotton products in February 2008; and its 8th Continent soy milk joint venture with General Mills in February 2008. Also in 2008, DuPont opened new offices in Abu Dhabi, U.A.E. and Hyderabad, India.

DuPont offers its employees tuition assistance, ongoing training programs, flexible work practices, adoption assistance, an employee resource program, an emergency backup childcare resource and dependent care spending accounts.

## FINANCIALS: Sales and profits are in thousands of dollars—add 000 to get the full amount. 2008 Note: Financial information for 2008 was not available for all companies at press time.

| | | |
|---|---|---|
| 2008 Sales: $30,529,000 | 2008 Profits: $2,007,000 | **U.S. Stock Ticker: DD** |
| 2007 Sales: $29,378,000 | 2007 Profits: $2,988,000 | **Int'l Ticker:** Int'l Exchange: |
| 2006 Sales: $27,421,000 | 2006 Profits: $3,148,000 | Employees: 60,000 |
| 2005 Sales: $26,639,000 | 2005 Profits: $2,053,000 | Fiscal Year Ends: 12/31 |
| 2004 Sales: $27,340,000 | 2004 Profits: $1,780,000 | Parent Company: |

## SALARIES/BENEFITS:

| Pension Plan: Y | ESOP Stock Plan: | Profit Sharing: | Top Exec. Salary: $1,320,000 | Bonus: $2,207,000 |
|---|---|---|---|---|
| Savings Plan: Y | Stock Purch. Plan: | | Second Exec. Salary: $835,384 | Bonus: $918,000 |

## OTHER THOUGHTS:

**Apparent Women Officers or Directors**: 17
**Hot Spot for Advancement for Women/Minorities**: Y

## LOCATIONS: ("Y" = Yes)

| West: | Southwest: | Midwest: | Southeast: | Northeast: | International: |
|---|---|---|---|---|---|
| Y | Y | Y | Y | Y | Y |

Note: Financial information, benefits and other data can change quickly and may vary from those stated here.

# EISAI CO LTD

**www.eisai.co.jp**

**Industry Group Code:** 325414 **Ranks within this company's industry group:** Sales: 1 Profits: 8

| Drugs: | | Other: | | Clinical: | | Computers: | | Services: | |
|---|---|---|---|---|---|---|---|---|---|
| Discovery: | | AgriBio: | | Trials/Services: | | Hardware: | | Specialty Services: | |
| Licensing: | | Genetic Data: | | Labs: | | Software: | | Consulting: | |
| Manufacturing: | Y | Tissue Replacement: | | Equipment/Supplies: | | Arrays: | Y | Blood Collection: | |
| Genetics: | | | | Research & Development Services: | Y | Database Management: | | Drug Delivery: | |
| | | | | Diagnostics: | Y | | | Drug Distribution: | |

## TYPES OF BUSINESS:

Pharmaceuticals Manufacturing
Over-the-Counter Pharmaceuticals
Pharmaceutical Production Equipment
Diagnostic Products
Food Additives
Personal Health Care Products
Vitamins & Nutritional Supplements

## BRANDS/DIVISIONS/AFFILIATES:

Aricept
Aciphex/Pariet
Coretec
Myonal
Chocola
Travelmin
Juvelux
MGI Pharma Inc

## CONTACTS: *Note: Officers with more than one job title may be intentionally listed here more than once.*

Haruo Naito, CEO
Haruo Naito, Pres.
Hideaki Matsui, CFO
Soichi Matsuno, Deputy Pres., Global Human Resources
Kentaro Yoshimatsu, Sr. VP-R&D/Pres., Eisai R&D Mgmt. Co., Ltd.
Kazuo Hirai, VP-Info. System & Corp. Mgmt. Planning
Takafumi Asano, VP-Prod.
Kenta Takahashi, General Counsel/VP/Sr. Dir.-Legal Dept.
Makoto Shiina, Exec. VP-Strategy
Akira Fujiyoshi, VP-Corp. Comm. Dept.
Nobuo Deguchi, Exec. VP-Internal Control & Intellectual Property
Hajime Shimizu, Chmn/CEO-Eisai Corp. N. America/Chmn/CEO-Eisai Inc
Yutaka Tsuchiya, Chmn./Pres., Eisai Europe Ltd.
Yukio Akada, Chmn./Pres., Eisai China, Inc.
Masanori Tsuno, Pres., Eisai Medical Research, Inc.
Norihiko Tanikawa, Chmn.
Soichi Matsuno, Deputy Pres., Global Pharmaceuticals Bus.
Takafumi Asano, VP-Logistics & Transformation

| Phone: 81-3-3817-3700 | Fax: 81-3-3811-3077 |
|---|---|
| **Toll-Free:** | |
| **Address:** 4-6-10 Koishikawa, Bunkyo-ku, Tokyo, 112-8088 Japan | |

## GROWTH PLANS/SPECIAL FEATURES:

Eisai Co., Ltd. primarily develops, manufactures and distributes medical products, operating in two segments: prescription pharmaceuticals, which represented 96.9% of its 2008 net sales, and other. These other products include over-the-counter pharmaceuticals, consumer health care products, food additives, pharmaceutical production equipment and diagnostic products. Its two largest pharmaceuticals, Aricept, a treatment for Alzheimer's dementia, and proton pump inhibitor (PPI) AcipHex/Pariet, a treatment for gastroesophageal reflux disease and ulcers, accounted for 39.6% and 24% of 2008 net sales, respectively. The firm's research teams are currently investigating new indications for Aricept including dementia associated with Parkinson's disease. Other prescription pharmaceuticals include Coretec, an agent for acute heart failure; Glakay, an osteoporosis treatment; Azeptin, an antiallergic agent; and Myonal, a muscle relaxant. In the consumer health care field, the firm's products include Seabond denture adhesive (manufactured by U.S.-based Combe Laboratories, Inc.); motion sickness remedy Travelmin; Breathe Right nasal strips (manufactured by GlaxoSmithKline plc); and Juvelux natural vitamin E preparation. Eisai has a long-standing leading position in the Japanese vitamin and nutritional supplement market, focusing on synthetic and natural vitamin E products and derivatives; it also markets a full line of vitamin-enriched dietary supplements under the brand name Chocola. The firm also markets diagnostic products and manufactures pharmaceutical production systems including continuous sterilization devices, inspection systems and ampoule packing machines. Geographically, North America generated 46.2% of 2008 net sales; Japan, 42.6%; Europe, 7.4%; and Asia and other regions, 3.8%. The company maintains approximately 32 overseas subsidiaries, in Asia, Europe and North America; and 12 domestic subsidiaries. In January 2008, the firm acquired U.S.-based MGI Pharma, Inc. for $3.9 billion. Now a subsidiary of Eisai Corporation of North America, MGI is a biopharmaceutical company that focuses on oncology and acute care.

## FINANCIALS: Sales and profits are in thousands of dollars—add 000 to get the full amount. 2008 Note: Financial information for 2008 was not available for all companies at press time.

| | | |
|---|---|---|
| 2008 Sales: $7,304,620 | 2008 Profits: $-169,110 | **U.S. Stock Ticker:** ESALY |
| 2007 Sales: $6,705,770 | 2007 Profits: $702,310 | **Int'l Ticker: 4523** Int'l Exchange: Tokyo-TSE |
| 2006 Sales: $5,113,100 | 2006 Profits: $539,200 | Employees: 9,649 |
| 2005 Sales: $4,530,600 | 2005 Profits: $471,760 | Fiscal Year Ends: 3/31 |
| 2004 Sales: $4,734,600 | 2004 Profits: $474,700 | Parent Company: |

## SALARIES/BENEFITS:

| | | | | |
|---|---|---|---|---|
| Pension Plan: Y | ESOP Stock Plan: | Profit Sharing: | Top Exec. Salary: $ | Bonus: $ |
| Savings Plan: | Stock Purch. Plan: | | Second Exec. Salary: $ | Bonus: $ |

## OTHER THOUGHTS:

**Apparent Women Officers or Directors:**
**Hot Spot for Advancement for Women/Minorities:**

## LOCATIONS: ("Y" = Yes)

| West: | Southwest: | Midwest: | Southeast: | Northeast: | International: |
|---|---|---|---|---|---|
| | | Y | | Y | Y |

Note: Financial information, benefits and other data can change quickly and may vary from those stated here.

# ELAN CORP PLC

www.elan.com

Industry Group Code: 325412  Ranks within this company's industry group: Sales: 39   Profits: 150

| Drugs: | | Other: | Clinical: | Computers: | Services: |
|---|---|---|---|---|---|
| Discovery: | Y | AgriBio: | Trials/Services: | Hardware: | Specialty Services: |
| Licensing: | | Genetic Data: | Labs: | Software: | Consulting: |
| Manufacturing: | Y | Tissue Replacement: | Equipment/Supplies: | Arrays: | Blood Collection: |
| Genetics: | | | Research & Development Services: | Database Management: | Drug Delivery: |
| | | | Diagnostics: | | Drug Distribution: |

## TYPES OF BUSINESS:

Drugs-Neurology
Acute Care Drugs
Pain Management Drugs
Autoimmune Disease Drugs
Drug Delivery Technologies

## BRANDS/DIVISIONS/AFFILIATES:

TYSABRI
PRIALT
Azactam
Maxipime
NanoCrystal

## CONTACTS: Note: Officers with more than one job title may be intentionally listed here more than once.

G. Kelly Martin, CEO
Carlos Paya, Pres.
Shane Cooke, CFO/Exec. VP
Kathleen Martorano, Exec. VP-Strategic Human Resources
Dale Schenk, Chief Scientific Officer/Exec. VP
Shane Cooke, Head-Elan Drug Tech.
Richard T. Collier, General Counsel/Exec. VP
Johannes Roebers, Sr. VP/Head-Biologic Oper., Strategy & Planning
Menghis Bairu, Exec. VP/Head-Global Dev.
Mary Stutts, Exec. VP/Head-Corp. Rel.
Chris Burns, Sr. VP-Investor Rel.
Nigel Clerkin, Sr. VP-Finance/Controller
Shane Cooke, Head-Elan Drug Technologies
Allison Hulme, Exec. VP/Head-Global Clinical Relationships
Ted Yednock, Exec. VP/Head-Global Research
Karen Kim, Exec. VP-Special Projects
Kyran McLaughlin, Chmn.
Menghis Bairu, Sr. VP/Head-Int'l

| Phone: 353-1-709-4000 | Fax: 353-1-709-4108 |
|---|---|
| Toll-Free: | |
| Address: Treasury Bldg., Lower Grand Canal St., Dublin, Ireland 2 UK | |

## GROWTH PLANS/SPECIAL FEATURES:

Elan Corp. plc, a leading global specialty pharmaceutical company, focuses on the discovery, development and marketing of therapeutic products and services in neurology, acute care and pain management. Elan is divided into two segments: Biopharmaceuticals and Elan Drug Technologies (EDT). The Irish company conducts its worldwide business through subsidiaries in Ireland, the U.S., the U.K. and other countries. Nearly half of its revenue is generated from EDT, which controls its four marketed products. The biopharmaceuticals division focuses its research and development on Alzheimer's disease, Parkinson's disease, multiple sclerosis, pain management and Crohn's disease. The company uses proprietary technologies to develop, market and license drug delivery products to its pharmaceutical clients. It does not manufacture any of its products. Elan has developed three novel therapeutic approaches in its research program in Alzheimer's disease: Immunotherapy, beta secretase inhibitors and gamma secretase inhibitors. The company's four products are TYSABRI, an alpha four integrin antagonist for the treatment of recurring multiple sclerosis; PRIALT, used for the management of severe chronic pain; Azactam, which is used to treat gram-negative organism induced disorders such as urinary and lower respiratory tract infections and intra-abdominal infections; and Maxipime, which treats a number of disorders including urinary tract infections and pneumonia. Elan also owns proprietary rights to NanoCrystal technology, which it licenses to Roche and Johnson & Johnson. In the European Union, the firm's primary product is TYSABRI. The firm is currently developing two drugs used to treat Alzheimer's disease, bapineuzumab (AAB-001) and ELND005 (AZD-103), both in phase 2 of development. In July 2009, Johnson & Johnson announced that it would invest approximately $1.5 billion for an 18.4% stake in Elan. The transaction will also give Johnson & Johnson certain rights related to Elan's Alzheimer's drug development program.

## FINANCIALS: Sales and profits are in thousands of dollars—add 000 to get the full amount. 2008 Note: Financial information for 2008 was not available for all companies at press time.

| | | |
|---|---|---|
| 2008 Sales: $1,000,200 | 2008 Profits: $-71,000 | U.S. Stock Ticker: ELN |
| 2007 Sales: $759,400 | 2007 Profits: $-405,000 | Int'l Ticker: DRX    Int'l Exchange: Dublin-ISE |
| 2006 Sales: $560,400 | 2006 Profits: $-267,300 | Employees: 1,687 |
| 2005 Sales: $490,300 | 2005 Profits: $-383,600 | Fiscal Year Ends: 12/31 |
| 2004 Sales: $481,700 | 2004 Profits: $-394,700 | Parent Company: |

## SALARIES/BENEFITS:

| Pension Plan: | ESOP Stock Plan: | Profit Sharing: | Top Exec. Salary: $702,854 | Bonus: $800,000 |
|---|---|---|---|---|
| Savings Plan: Y | Stock Purch. Plan: Y | | Second Exec. Salary: $266,666 | Bonus: $150,000 |

## OTHER THOUGHTS:

Apparent Women Officers or Directors: 7
Hot Spot for Advancement for Women/Minorities: Y

## LOCATIONS: ("Y" = Yes)

| West: | Southwest: | Midwest: | Southeast: | Northeast: | International: |
|---|---|---|---|---|---|
| Y | | | Y | Y | Y |

# ELI LILLY & COMPANY                                    www.lilly.com

Industry Group Code: 325412  Ranks within this company's industry group: Sales: 12  Profits: 175

| Drugs: | | Other: | Clinical: | Computers: | Services: | |
|---|---|---|---|---|---|---|
| Discovery: | Y | AgriBio: | Trials/Services: | Hardware: | Specialty Services: | Y |
| Licensing: | | Genetic Data: | Labs: | Software: | Consulting: | |
| Manufacturing: | Y | Tissue Replacement: | Equipment/Supplies: | Arrays: | Blood Collection: | |
| Genetics: | | | Research & Development Services: | Database Management: | Drug Delivery: | |
| | | | Diagnostics: | | Drug Distribution: | |

## TYPES OF BUSINESS:

Pharmaceuticals Discovery & Development
Veterinary Products

## BRANDS/DIVISIONS/AFFILIATES:

Zyprexa
Prozac
Humalog
Gemzar
Coban
Applied Molecular Evolution Inc
Icos Corporation
ImClone Systems Inc.

## CONTACTS: Note: Officers with more than one job title may be intentionally listed here more than once.

John Lechleiter, CEO
John Lechleiter, Pres.
Derica Rice, CFO/Sr. VP
Bryce D. Carmine, Exec. VP-Global Mktg. & Sales
Steven M. Paul, Exec. VP-Science
Michael Heim, CIO/VP-IT
Steven M. Paul, Exec. VP-Tech.
Thomas Verhoeven, Pres., Global Prod. Dev.
W. Darin Moody, VP-Corp. Eng. & Continuous Improvement
Frank Deane, Pres., Mfg.
Robert A. Armitage, Co-General Counsel/Sr. VP
Gino Santini, Exec. Dir.-Corp. Strategy & Policy
Alex M. Azar II, Sr. VP-Corp. Affairs & Comm.
Thomas W. Grein, Treas./VP
Enrique Conterno, Pres., Lilly USA
Alfonso Zulueta, Pres./Gen. Mgr.-Lilly Japan
Alecia A. DeCoudreaux, Co-General Counsel/VP
Tim Garnett, Chief Medical Officer/VP-Medical
John Lechleiter, Chmn.
Karim Bitar, Pres., European Oper.

| Phone: 317-276-2000 | Fax: |
|---|---|
| Toll-Free: 800-545-5979 | |
| Address: Lilly Corporate Ctr., Indianapolis, IN 46285 US | |

## GROWTH PLANS/SPECIAL FEATURES:

Eli Lilly & Co. researches, develops, manufactures and sells pharmaceuticals designed to treat a variety of conditions. Most of Eli Lilly's products are developed by its in-house research staff, which primarily directs its research efforts towards the search for products to prevent and treat cancer and diseases of the central nervous, endocrine and cardiovascular systems. The firm's other research lies in anti-infectives and products to treat animal diseases. Major brands include neuroscience products Zyprexa, Strattera, Prozac, Cymbalta and Permax; endocrine products Humalog, Humulin and Actos; oncology products Gemzar and Alimta; animal health products Tylan, Rumensin and Coban; cardiovascular products ReoPro and Xigris; anti-infectives Ceclor and Vancocin; and Cialis, for erectile dysfunction. In the U.S., the company distributes pharmaceuticals primarily through independent wholesale distributors. The company manufactures and distributes its products through facilities in the U.S., Puerto Rico and 25 other countries, which are then sold to markets in 135 countries throughout the world. In 2008, the firm owned 15 production and distribution facilities in the U.S. and Puerto Rico. Major research and development facilities abroad are located in the U.K., Canada, Singapore and Spain. In April 2008, the firm acquired Hypnion, Inc., a neuroscience drug discovery company focused on sleep disorder research. In October 2008, the company agreed to acquire ImClone Systems, Inc. for $6.5 billion.

Eli Lilly offers its employees domestic partner benefits and an employee assistance program, as well as up to 10 weeks of paid maternity leave. The firm also offers an on-site fitness center, flexible hours or telecommuting, parenting and dependant care leaves, adoption assistance and tuition reimbursement.

## FINANCIALS: Sales and profits are in thousands of dollars—add 000 to get the full amount. 2008 Note: Financial information for 2008 was not available for all companies at press time.

| | | |
|---|---|---|
| 2008 Sales: $20,378,000 | 2008 Profits: $-2,071,900 | U.S. Stock Ticker: LLY |
| 2007 Sales: $18,633,500 | 2007 Profits: $2,953,000 | Int'l Ticker:    Int'l Exchange: |
| 2006 Sales: $15,691,000 | 2006 Profits: $2,662,700 | Employees:  40,500 |
| 2005 Sales: $14,645,300 | 2005 Profits: $1,979,600 | Fiscal Year Ends: 12/31 |
| 2004 Sales: $13,857,900 | 2004 Profits: $1,810,100 | Parent Company: |

## SALARIES/BENEFITS:

| Pension Plan: Y | ESOP Stock Plan: | Profit Sharing: | Top Exec. Salary: $1,339,125 | Bonus: $2,709,053 |
|---|---|---|---|---|
| Savings Plan: Y | Stock Purch. Plan: | | Second Exec. Salary: $1,000,250 | Bonus: $1,309,327 |

## OTHER THOUGHTS:

Apparent Women Officers or Directors: 4
Hot Spot for Advancement for Women/Minorities: Y

## LOCATIONS: ("Y" = Yes)

| West: | Southwest: | Midwest: | Southeast: | Northeast: | International: |
|---|---|---|---|---|---|
| Y | Y | Y | Y | Y | Y |

Note: Financial information, benefits and other data can change quickly and may vary from those stated here.

# EMISPHERE TECHNOLOGIES INC

www.emisphere.com

**Industry Group Code: 325412A  Ranks within this company's industry group:** Sales: 26  Profits: 20

| Drugs: | | Other: | | Clinical: | | Computers: | | Services: | |
|---|---|---|---|---|---|---|---|---|---|
| Discovery: | Y | AgriBio: | | Trials/Services: | | Hardware: | | Specialty Services: | |
| Licensing: | | Genetic Data: | | Labs: | | Software: | | Consulting: | |
| Manufacturing: | Y | Tissue Replacement: | | Equipment/Supplies: | | Arrays: | | Blood Collection: | |
| Genetics: | | | | Research & Development Services: | Y | Database Management: | | Drug Delivery: | Y |
| | | | | Diagnostics: | | | | Drug Distribution: | |

## TYPES OF BUSINESS:

Drug Delivery Systems
Cardiovascular Diseases Oral Drugs
Osteoarthritis & Osteoporosis Oral Drugs
Growth Disorders Oral Drugs
Diabetes Oral Drugs
Asthma & Allergies Oral Drugs
Obesity Oral Drugs
Infectious Diseases & Oncology Oral Drugs

## BRANDS/DIVISIONS/AFFILIATES:

Eligen Technology
Novartis Pharma AG

## CONTACTS: Note: Officers with more than one job title may be intentionally listed here more than once.

Michael V. Novinski, CEO
Michael V. Novinski, Pres.
Michael R. Garone, CFO/VP
Daria M. Palestina, Dir.-Human Resources
Laura Kragie, Chief Medical Officer/VP-Clinical Dev.
Patrick J. Osinski, General Counsel
Nicholas J. Hart, VP-Dev. & Strategy
M. Gary I. Riley, VP-Non-Clinical Dev. & Applied Biology

| Phone: 973-532-8000 | Fax: 973-532-8121 |
|---|---|
| Toll-Free: | |
| Address: 240 Cedar Knolls Rd., Ste. 200, Cedar Knolls, NJ 07927 US | |

## GROWTH PLANS/SPECIAL FEATURES:

Emisphere Technologies, Inc. is a biopharmaceutical company that focuses on the delivery of therapeutic molecules using its proprietary Eligen technology. The company's product pipeline includes product candidates for the treatment of cardiovascular diseases, osteoarthritis, osteoporosis, growth disorders, diabetes, asthma/allergies, obesity, infectious diseases and oncology. Emisphere's Eligen technology is based on the use of proprietary, synthetic chemical compounds known as Emisphere delivery agents. These delivery agents facilitate and enable the transport of therapeutic macromolecules (such as proteins, peptides and polysaccharides) and poorly absorbed small molecules across biological membranes such as the small intestine. The Eligen technology is rapidly absorbed, metabolized and eliminated from the body. It does not accumulate in the organs and tissues and is considered safe at anticipated dose and dosing regimens. The company's lead product candidate is Salmon Calcitonin for Osteoarthritis and Osteoporosis, developed in conjunction with Novartis Pharma AG and its partner, Nordic Bioscience. The drug uses Emisphere's Eligen delivery technology to create an effective oral treatment as an alternative to existing injectable and intranasal options. Another product in development combines GLP-1 and PYY with Eligen technology to treat Type II diabetes. Finally, Emisphere is developing an oral system for vitamin B12, a process that promises to be less costly and better absorbed than current injection methods. Other products in development include oral deliveries for insulin, heparin, recombinant human growth hormone, acyclovir, PYY, gallium and recombinant parathyroid hormone. In December 2008, the company announced it would shutter its Tarrytown, New York facility and lay off many of that location's employees. In April 2009, the firm announced an agreement with AAIPharma to combine its drug development services with Eligen technology.

Emisphere offers its employees tuition reimbursement; flex time; an employee assistance program; college savings plans; financial advisors; credit union membership; childcare/eldercare referral programs; and medical, dental, life and disability insurance.

## FINANCIALS: Sales and profits are in thousands of dollars—add 000 to get the full amount. 2008 Note: Financial information for 2008 was not available for all companies at press time.

| | | |
|---|---|---|
| 2008 Sales: $ 251 | 2008 Profits: $-24,388 | **U.S. Stock Ticker:** EMIS |
| 2007 Sales: $4,077 | 2007 Profits: $-16,928 | **Int'l Ticker:**  Int'l Exchange: |
| 2006 Sales: $7,259 | 2006 Profits: $-41,766 | Employees:   20 |
| 2005 Sales: $3,540 | 2005 Profits: $-18,051 | Fiscal Year Ends: 12/31 |
| 2004 Sales: $1,953 | 2004 Profits: $-37,522 | Parent Company: |

## SALARIES/BENEFITS:

| | | | | |
|---|---|---|---|---|
| Pension Plan: | ESOP Stock Plan: | Profit Sharing: | Top Exec. Salary: $554,231 | Bonus: $357,123 |
| Savings Plan: Y | Stock Purch. Plan: Y | | Second Exec. Salary: $267,039 | Bonus: $40,000 |

## OTHER THOUGHTS:

**Apparent Women Officers or Directors:** 1
**Hot Spot for Advancement for Women/Minorities:**

## LOCATIONS: ("Y" = Yes)

| West: | Southwest: | Midwest: | Southeast: | Northeast: | International: |
|---|---|---|---|---|---|
| | | | | Y | |

Note: Financial information, benefits and other data can change quickly and may vary from those stated here.

# ENCORIUM GROUP INC

**www.encorium.com**

**Industry Group Code: 541712  Ranks within this company's industry group:** Sales: 14   Profits: 11

| Drugs: | Other: | Clinical: | | Computers: | | Services: | |
|---|---|---|---|---|---|---|---|
| Discovery: | AgriBio: | Trials/Services: | Y | Hardware: | | Specialty Services: | |
| Licensing: | Genetic Data: | Labs: | | Software: | Y | Consulting: | |
| Manufacturing: | Tissue Replacement: | Equipment/Supplies: | | Arrays: | | Blood Collection: | Y |
| Genetics: | | Research & Development Services: | | Database Management: | | Drug Delivery: | |
| | | Diagnostics: | | | | Drug Distribution: | |

## TYPES OF BUSINESS:

Contract Research
Clinical Trial & Data Management
Disease Assessment Software
Biostatistical Analysis
Regulatory Affairs Services

## BRANDS/DIVISIONS/AFFILIATES:

Remedium Oy
Covalent Group, Inc.

## CONTACTS: Note: Officers with more than one job title may be intentionally listed here more than once.

David Ginsberg, CEO
Linda L. Nardone, COO/Exec. VP
Kenneth M. Borow, Pres.
Philip L. Calamia, Interim CFO
Eeva-Kaarina Koskelo, VP-Clinical Oper., Europe & Asia
Kai E. Lindevall, Chmn.
Kai E. Lindevall, Pres., Int'l Oper.

| Phone: 610-975-9533 | Fax: 610-975-9556 |
|---|---|
| Toll-Free: | |
| Address: 1 Glenhardie Corp Ctr, 1275 Drummers Ln., Ste. 100, Wayne, PA 19087 US | |

## GROWTH PLANS/SPECIAL FEATURES:

Encorium Group, Inc., formed in 2006 from a merger between Remedium Oy and Covalent Group, Inc., is a contract research organization (CRO) that designs and manages clinical trials in the pharmaceutical, biotechnology and medical device development process. Encorium offers therapeutic expertise, experienced team management and advanced technologies, with the capacity to conduct clinical trials on a global basis. The company specializes in Phase I through IV clinical trials, cost-effectiveness studies and outcomes studies for pharmaceutical companies, managed-care organizations, insurers and employers. It offers a full array of integrated services, including strategic trial planning; project management; monitoring; data management; biostatistics; pharmacovigilance; medical writing; quality assurance; outsourcing of clinical staff; and medical device certification in the European Union. The firm's clinical trials management services include assistance with case report form design, investigator recruitment, patient enrollment and study monitoring and data collection. Encorium has clinical trial experience across a wide variety of therapeutic areas, such as cardiovascular, nephrology, endocrinology/metabolism, hematology, diabetes, neurology, oncology, immunology, vaccines, infectious diseases, gastroenterology, dermatology, hepatology, rheumatology, urology, ophthalmology, women's health and respiratory medicine. Encorium's clients consist of many of the largest companies in the pharmaceutical, biotechnology and medical device industries. During 2007, it provided services to 96 different clients covering 215 separate studies, of which 77 clients and 184 studies were associated with its European operations. Roughly 37% of Encorium's revenue was generated by its operations in the U.S. during 2007 and roughly 63% by its operations in Europe. In 2007, the firm's three largest clients accounted for 38% of its net revenues. Encorium is generally awarded contracts based upon its response to requests for proposals received from pharmaceutical, biotechnology and medical device companies, and its business development and marketing strategy is based on expanding its relationships with its existing clients as well as gaining new clients.

## FINANCIALS: Sales and profits are in thousands of dollars—add 000 to get the full amount. 2008 Note: Financial information for 2008 was not available for all companies at press time.

| | | |
|---|---|---|
| 2008 Sales: $35,912 | 2008 Profits: $-21,073 | **U.S. Stock Ticker: ENCO** |
| 2007 Sales: $36,802 | 2007 Profits: $-2,751 | **Int'l Ticker:**   Int'l Exchange: |
| 2006 Sales: $17,684 | 2006 Profits: $- 494 | Employees:  226 |
| 2005 Sales: $12,727 | 2005 Profits: $-1,484 | Fiscal Year Ends: 12/31 |
| 2004 Sales: $13,590 | 2004 Profits: $-4,223 | Parent Company: |

## SALARIES/BENEFITS:

| Pension Plan: | ESOP Stock Plan: | Profit Sharing: | Top Exec. Salary: $373,628 | Bonus: $58,432 |
|---|---|---|---|---|
| Savings Plan: Y | Stock Purch. Plan: | | Second Exec. Salary: $322,714 | Bonus: $ |

## OTHER THOUGHTS:

**Apparent Women Officers or Directors**: 2
**Hot Spot for Advancement for Women/Minorities**: Y

## LOCATIONS: ("Y" = Yes)

| West: | Southwest: | Midwest: | Southeast: | Northeast: | International: |
|---|---|---|---|---|---|
| | | | | Y | Y |

Note: Financial information, benefits and other data can change quickly and may vary from those stated here.

# ENDO PHARMACEUTICALS HOLDINGS INC

**www.endo.com**

**Industry Group Code: 325412  Ranks within this company's industry group: Sales: 36   Profits: 29**

| Drugs: | | Other: | Clinical: | | Computers: | | Services: | |
|---|---|---|---|---|---|---|---|---|
| Discovery: | Y | AgriBio: | Trials/Services: | Y | Hardware: | | Specialty Services: | Y |
| Licensing: | Y | Genetic Data: | Labs: | | Software: | | Consulting: | |
| Manufacturing: | Y | Tissue Replacement: | Equipment/Supplies: | | Arrays: | | Blood Collection: | |
| Genetics: | Y | | Research & Development Services: | | Database Management: | | Drug Delivery: | |
| | | | Diagnostics: | | | | Drug Distribution: | Y |

## TYPES OF BUSINESS:

Drugs-Pain Management
Pharmaceutical Preparations

## BRANDS/DIVISIONS/AFFILIATES:

Endo Pharmaceuticals, Inc.
Lidoderm
Opana
Percocet
Frova
Valstar
Nebido
Indevus Pharmaceuticals, Inc.

## CONTACTS: Note: Officers with more than one job title may be intentionally listed here more than once.

David P. Holveck, CEO
Nancy J. Wysenski, COO
David P. Holveck, Pres.
Alan G. Levin, CFO/Exec. VP
Larry Cunningham, Sr. VP-Human Resources
Ivan Gergel, Exec. VP-R&D
Jim Maguire, CIO/Sr. VP
Caroline B. Manogue, Chief Legal Officer/Corp. Sec./Exec. VP
Robert J. Cobuzzi, VP-Corp. Dev.
Blaine T. Davis, VP-Corp. Affairs
Edward J. Sweeney, Principal Acct. Officer/VP/Controller
Colleen M. Craven, VP-Corp. Compliance & Bus. Practices
Sandeep Gupta, Sr. VP-Discovery & Early Dev.
Roger H. Kimmel, Chmn.

| **Phone:** 610-558-9800 | **Fax:** 610-558-8979 |
|---|---|
| **Toll-Free:** | |
| **Address:** 100 Endo Blvd., Chadds Ford, PA 19317 US | |

## GROWTH PLANS/SPECIAL FEATURES:

Endo Pharmaceuticals Holdings, Inc., through subsidiary Endo Pharmaceuticals, Inc., is a specialty pharmaceutical company engaged in the research, development, sale and marketing of branded and generic prescription pharmaceuticals used primarily to treat and manage pain. The company has a portfolio of branded products that includes such names as Lidoderm, a topical patch containing lidocaine; Opana and Opana ER, indicated for the relief of moderate to severe pain in patients requiring continuous opioid treatment; Percocet, indicated for the treatment of moderate to moderately severe pain; and Frova, a migraine treatment product. Branded products generate approximately 93% of Endo's net sales in 2008, with 61% of its net sales generated from Lidoderm. Its non-branded generic portfolio currently consists of products primarily focused in pain management. The firm focuses selectively on generics that have one or more barriers to market entry, such as complex formulation, regulatory or legal challenges or difficulty in raw material sourcing. Endo's product pipeline includes treatments for BCG-refractory bladder cancer (Valstar), hypogonadism (Nebido), the prevention of HIV and STDs (Pro 2000), acromegaly and carcinoid syndrome (Octreotide Implant), moderate to severe pain (Axomadol) and stuttering (Pagoclone). The firm's primary suppliers of contract manufacturing services are Novartis Consumer Health, Inc. and Teikoku Seiyaku Co., Ltd. It markets its branded products through a dedicated sales force of approximately 700 sales representatives in the U.S. to high-prescribing physicians in the pain management, neurology, surgery, anesthesiology, oncology and primary care, as well as retail pharmacies. In March 2009, Endo completed its acquisition of Indevus Pharmaceuticals, Inc.

Endo offers it's a range of benefits, including employees educational assistance; individual development plans; financial planning assistance; flexible time off; parenting benefits; an employee assistance program; flexible spending accounts; and medical, dental, vision, pharmacy and disability insurance.

## FINANCIALS: Sales and profits are in thousands of dollars—add 000 to get the full amount. 2008 Note: Financial information for 2008 was not available for all companies at press time.

| | | |
|---|---|---|
| 2008 Sales: $1,260,536 | 2008 Profits: $261,741 | **U.S. Stock Ticker: ENDP** |
| 2007 Sales: $1,085,608 | 2007 Profits: $227,440 | **Int'l Ticker:**   Int'l Exchange: |
| 2006 Sales: $909,659 | 2006 Profits: $137,839 | Employees:  1,216 |
| 2005 Sales: $820,164 | 2005 Profits: $202,295 | Fiscal Year Ends: 12/31 |
| 2004 Sales: $615,100 | 2004 Profits: $143,300 | Parent Company: |

## SALARIES/BENEFITS:

| Pension Plan: | ESOP Stock Plan: | Profit Sharing: | Top Exec. Salary: $600,000 | Bonus: $746,637 |
|---|---|---|---|---|
| Savings Plan: Y | Stock Purch. Plan: Y | | Second Exec. Salary: $465,000 | Bonus: $286,177 |

## OTHER THOUGHTS:

**Apparent Women Officers or Directors:** 4
**Hot Spot for Advancement for Women/Minorities:** Y

## LOCATIONS: ("Y" = Yes)

| West: | Southwest: | Midwest: | Southeast: | Northeast: | International: |
|---|---|---|---|---|---|
| | | | | Y | |

# ENTREMED INC

www.entremed.com

**Industry Group Code: 325412  Ranks within this company's industry group: Sales: 121  Profits: 111**

| Drugs: | | Other: | Clinical: | Computers: | Services: |
|---|---|---|---|---|---|
| Discovery: | Y | AgriBio: | Trials/Services: | Hardware: | Specialty Services: |
| Licensing: | Y | Genetic Data: | Labs: | Software: | Consulting: |
| Manufacturing: | | Tissue Replacement: | Equipment/Supplies: | Arrays: | Blood Collection: |
| Genetics: | | | Research & Development Services: | Database Management: | Drug Delivery: |
| | | | Diagnostics: | | Drug Distribution: |

## TYPES OF BUSINESS:

Drugs-Angiogenesis
Drugs-Cancer & Rheumatoid Arthritis

## BRANDS/DIVISIONS/AFFILIATES:

Panzem
Panzem NCD
Miikana Therapeutics, Inc.
MKC-1
ENMD-1198
ENMD-2076

## CONTACTS: Note: Officers with more than one job title may be intentionally listed here more than once.

Cynthia Wong Hu, COO
Mark R. Bray, VP-Research
Cynthia Wong Hu, General Counsel/VP/Corp. Sec./Chief Legal Officer
Kathy R. Wehmeir-Davis, Principal Acct. Officer/Controller
Carolyn F. Sidor, Chief Medical Officer/VP
Michael Tarnow, Chmn.

| Phone: 240-864-2600 | Fax: 240-864-2601 |
|---|---|
| Toll-Free: | |
| Address: 9640 Medical Center Dr., Rockville, MD 20850 US | |

## GROWTH PLANS/SPECIAL FEATURES:

EntreMed, Inc. is a clinical-stage biopharmaceutical company that develops multi-mechanism oncology drugs to treat cancer and inflammatory disease by targeting diseases cells and their nourishing blood vessels. The company's four clinical product candidates, Panzem NCD, MKC-1, ENMD-1198 and ENMD-2076, are aimed at controlling angiogenesis, cell cycle regulation and inflammation to prevent the progression of cancer. Panzem NCD is the company's current lead product candidate, and works by inhibiting angiogenesis, disrupting microtubule formation, down regulating hypoxia inducible factor-one alpha and inducing apoptosis. In recent years, the FDA has approved the use of Panzem NMD, which utilizes NanoCrystal technology through its licensing agreement with Elan Corporation for the treatment of glioblastoma multiforme, multiple myeloma and ovarian cancer. MKC-1 arrests cellular mitosis by inhibiting a novel intracellular target involved in cell division during cellular trafficking in metastatic breast cancer patients, hematological cancers and non-small cell lung cancer. ENMD-1198 inhibits tumor growth through the modification of the chemical structure of 2-methoxyestradiol, which increases the molecule's anti-tumor properties, anti-angiogenic properties and metabolic rate. ENMD-2076, or Aurora Kase Inhibitors, is an atimitotic agent that disables Aurora A, a mitosis regulator that is often over-expressed in cancer cases. In addition to these four products, the corporation is also investigating the possible use of Panzem, or 2ME2, for the treatment of rheumatoid arthritic and is also developing multi-kinase, tubulin and HDAS inhibitors for the treatment of cancer. MKC-1 and Pnazem NMD are in Phase II of clinical trial, and ENMD-1198 remains in Phase I. In the past year, both Panzem and the Aurora Kinase/Angiogenesis Inhibitor were accepted as Investigational New Drugs (IND) by the FDA.

EntreMed offers employees tuition reimbursement; an employee assistance plan; health, dental, and vision insurance, a 401(k) plan; company-paid life insurance and supplemental coverage of home and auto insurance; and a 529 savings plan.

## FINANCIALS: Sales and profits are in thousands of dollars—add 000 to get the full amount. 2008 Note: Financial information for 2008 was not available for all companies at press time.

| | | |
|---|---|---|
| 2008 Sales: $7,477 | 2008 Profits: $-23,862 | **U.S. Stock Ticker: ENMD** |
| 2007 Sales: $7,396 | 2007 Profits: $-22,411 | **Int'l Ticker:** Int'l Exchange: |
| 2006 Sales: $6,894 | 2006 Profits: $-49,889 | Employees: 21 |
| 2005 Sales: $5,918 | 2005 Profits: $-16,313 | Fiscal Year Ends: 12/31 |
| 2004 Sales: $ 500 | 2004 Profits: $-12,600 | Parent Company: |

## SALARIES/BENEFITS:

| Pension Plan: | ESOP Stock Plan: | Profit Sharing: | Top Exec. Salary: $430,000 | Bonus: $ |
|---|---|---|---|---|
| Savings Plan: Y | Stock Purch. Plan: | | Second Exec. Salary: $320,000 | Bonus: $ |

## OTHER THOUGHTS:

**Apparent Women Officers or Directors:** 4
**Hot Spot for Advancement for Women/Minorities:** Y

## LOCATIONS: ("Y" = Yes)

| West: | Southwest: | Midwest: | Southeast: | Northeast: | International: |
|---|---|---|---|---|---|
| | | | | Y | |

Note: Financial information, benefits and other data can change quickly and may vary from those stated here.

# ENZO BIOCHEM INC

**www.enzo.com**

Industry Group Code: 325412  **Ranks within this company's industry group:**  Sales: 73    Profits: 80

| Drugs: | | Other: | | Clinical: | | Computers: | | Services: | |
|---|---|---|---|---|---|---|---|---|---|
| Discovery: | | AgriBio: | | Trials/Services: | | Hardware: | | Specialty Services: | Y |
| Licensing: | | Genetic Data: | Y | Labs: | Y | Software: | | Consulting: | |
| Manufacturing: | Y | Tissue Replacement: | | Equipment/Supplies: | Y | Arrays: | | Blood Collection: | |
| Genetics: | | | | Research & Development Services: | | Database Management: | | Drug Delivery: | |
| | | | | Diagnostics: | Y | | | Drug Distribution: | |

## TYPES OF BUSINESS:

Pharmaceutical Discovery & Development
Genetic Analysis Products
Therapeutic Products
Diagnostic Products
Drug Development
Genetics Research
Clinical Laboratories
Biomedical Research Products & Tools

## BRANDS/DIVISIONS/AFFILIATES:

Enzo Life Sciences, Inc.
Enzo Therapeutics, Inc.
Enzo Clinical Labs, Inc.
Axxora Life Sciences
Affinity Ltd.

## CONTACTS: Note: Officers with more than one job title may be intentionally listed here more than once.

Elazar Rabbani, CEO
Barry W. Weiner, Pres.
Barry W. Weiner, CFO
Shahram K. Rabbani, Corp. Sec./Treas./Pres., Enzo Clinical Labs
Barbara E. Thalenfeld, VP-Corp. Dev. & Clinical Affairs
Andrew R. Crescenzo, Sr. VP-Finance
David C. Goldberg, VP-Bus. Dev./Sr. VP-Enzo Clinical Labs
Norman E. Kelker, Sr. VP
Carl W. Balezentis, Pres., Enzo Life Sciences
Christine T. Fischette, Pres., Enzo Therapeutics
Elazar Rabbani, Chmn.

| Phone: 212-583-0100 | Fax: |
|---|---|
| Toll-Free: | |
| Address: 527 Madison Ave., New York, NY 10022 US | |

## GROWTH PLANS/SPECIAL FEATURES:

Enzo Biochem, Inc. is a life sciences and biotechnology company that employs a variety of genetic processes to develop research tools, diagnostics and therapeutics, and serves as a diagnostic services provider to the medical community. The firm has approximately 220 patents and roughly 180 pending patent applications. The main focus of Enzo is to research the use of nucleic acids as informational molecules and to investigate the use of compounds for immune modulation. Enzo is particularly interested in the potency of gene-based diagnostics, which offers a greater advantage than immunoassay technology by allowing detection of diseases that are in the earlier stages of development. Operations are conducted under three subsidiary companies: Enzo Life Sciences, Inc.; Enzo Therapeutics, Inc.; and Enzo Clinical Labs, Inc. Enzo Life Sciences develops non-radioactive labeling and detection protocols and reagents for molecular biology research. Enzo Therapeutics works in collaboration with Enzo Life Sciences to develop multiple approaches in the areas of gastrointestinal, infectious, ophthalmic and metabolic diseases. This sector has also generated several clinical and preclinical pipelines in its effort to develop treatments for diseases and conditions where current treatment options are either ineffective, costly or cause undesired side effects, including HIV, Hepatitis B, Hepatitis C and Crohn's Disease. Enzo's therapeutic technologies deal with gene regulation, in which genes must be delivered into cells efficiently and without eliciting an unfavorable immune response. Enzo Clinical Labs, located in New York and New Jersey, offer a variety of routine clinical laboratory tests and procedures for general patient care. In Farmingdale, New York, Enzo Clinical Labs offers a full-service clinical laboratory with 19 patient service centers, a rapid response laboratory and a phlebotomy department. In May 2008, the company acquired Affinity Ltd. and the U.S.-based assets of Biomol International, LP.

## FINANCIALS: Sales and profits are in thousands of dollars—add 000 to get the full amount. 2008 Note: Financial information for 2008 was not available for all companies at press time.

| | | |
|---|---|---|
| 2008 Sales: $77,795 | 2008 Profits: $-10,653 | **U.S. Stock Ticker: ENZ** |
| 2007 Sales: $52,908 | 2007 Profits: $-13,260 | **Int'l Ticker:**    Int'l Exchange: |
| 2006 Sales: $39,826 | 2006 Profits: $-15,667 | Employees:   510 |
| 2005 Sales: $43,403 | 2005 Profits: $3,004 | Fiscal Year Ends: 7/31 |
| 2004 Sales: $41,644 | 2004 Profits: $-6,232 | Parent Company: |

## SALARIES/BENEFITS:

| Pension Plan: | ESOP Stock Plan: | Profit Sharing: | Top Exec. Salary: $510,817 | Bonus: $375,000 |
|---|---|---|---|---|
| Savings Plan: Y | Stock Purch. Plan: | | Second Exec. Salary: $453,094 | Bonus: $51,300 |

## OTHER THOUGHTS:

**Apparent Women Officers or Directors:** 3
**Hot Spot for Advancement for Women/Minorities:** Y

## LOCATIONS: ("Y" = Yes)

| West: | Southwest: | Midwest: | Southeast: | Northeast: | International: |
|---|---|---|---|---|---|
| | | | | Y | |

# ENZON PHARMACEUTICALS INC

www.enzon.com

**Industry Group Code: 325412  Ranks within this company's industry group:** Sales: 58   Profits: 63

| Drugs: | | Other: | Clinical: | Computers: | Services: |
|---|---|---|---|---|---|
| Discovery: | Y | AgriBio: | Trials/Services: | Hardware: | Specialty Services: |
| Licensing: | Y | Genetic Data: | Labs: | Software: | Consulting: |
| Manufacturing: | | Tissue Replacement: | Equipment/Supplies: | Arrays: | Blood Collection: |
| Genetics: | | | Research & Development Services: | Database Management: | Drug Delivery: |
| | | | Diagnostics: | | Drug Distribution: |

## TYPES OF BUSINESS:

Pharmaceutical Development
Drugs-Oncology & Hematology
Drugs-Transplantation
Drugs-Infectious Diseases
Single-Chain Antibody Technology
Polyethylene Glycol Technology

## BRANDS/DIVISIONS/AFFILIATES:

Abelcet
Adagen
DepoCyt
Oncaspar
Micromet, Inc.
Cimzia

## CONTACTS: *Note: Officers with more than one job title may be intentionally listed here more than once.*

Jeffrey H. Buchalter, CEO
Jeffrey H. Buchalter, Pres.
Craig Tooman, CFO
Paul Davit, Exec. VP-Human Resources
Ivan Horak, Chief Scientific Officer/Exec. VP-R&D
Ralph del Campo, Exec. VP-Tech. Oper.
Paul S. Davit, Corp. Sec.
Craig Tooman, Exec. VP-Finance
Jeffrey H. Buchalter, Chmn.

| Phone: 908-541-8600 | Fax: 908-575-9457 |
|---|---|
| Toll-Free: | |
| Address: 685 Route 202/206, Bridgewater, NJ 08807 US | |

## GROWTH PLANS/SPECIAL FEATURES:

Enzon, Inc. is a biopharmaceutical company that develops and commercializes pharmaceutical products in oncology and hematology, transplantation and infectious diseases. The firm advances its products through continued research and development of its proprietary PEG and SCA technologies. Enzon's PEG (polyethylene glycol) technology is used to improve the delivery, safety and efficacy of proteins and small molecules with known therapeutic value. The company's current portfolio includes four marketed products, as well as several others in development. Marketed products include the following: Abelcet, a lipid complex formulation of amphotericin B used to treat immuno-compromised patients with invasive fungal infections; Adagen, a PEG treatment for severe combined immunodeficiency disease (SCID); DepoCyt, an injectable chemotherapeutic agent for the treatment of lymphomatous meningitis; and Oncaspar, a PEG-enhanced version of the naturally occurring enzyme L-asparaginase that can be used in conjunction with other chemotherapeutics for the treatment of acute lymphoblastic leukemia. SCA (single-chain antibody) technology is used to discover and produce antibody-like molecules that can offer the therapeutic benefits of monoclonal antibodies while overcoming some of their limitations. Through a marketing agreement with biopharmaceutical firm Micromet, Inc., the firm licenses its SCA technology to other companies, sharing licensing revenues with Micromet. In May 2009, the U.S. Food and Drug Administration approved the use of Enzon's Cimzia for the treatment of adult patients suffering from moderate to severe rheumatoid arthritis.

Enzon offers its employees medical, dental and vision coverage, along with a prescription plan; dependent and healthcare flexible spending accounts; life and disability insurance; a 401(k) plan; tuition reimbursement; and an employee assistance plan.

## FINANCIALS: Sales and profits are in thousands of dollars—add 000 to get the full amount. 2008 Note: Financial information for 2008 was not available for all companies at press time.

| | | |
|---|---|---|
| 2008 Sales: $196,938 | 2008 Profits: $-2,715 | U.S. Stock Ticker: ENZN |
| 2007 Sales: $185,601 | 2007 Profits: $83,053 | Int'l Ticker:    Int'l Exchange: |
| 2006 Sales: $185,653 | 2006 Profits: $21,309 | Employees:   351 |
| 2005 Sales: $166,250 | 2005 Profits: $-89,606 | Fiscal Year Ends: 12/31 |
| 2004 Sales: $169,571 | 2004 Profits: $4,208 | Parent Company: |

## SALARIES/BENEFITS:

| Pension Plan: | ESOP Stock Plan: | Profit Sharing: | Top Exec. Salary: $853,271 | Bonus: $1,111,500 |
|---|---|---|---|---|
| Savings Plan: Y | Stock Purch. Plan: | | Second Exec. Salary: $532,857 | Bonus: $320,000 |

## OTHER THOUGHTS:

**Apparent Women Officers or Directors:**
**Hot Spot for Advancement for Women/Minorities:**

## LOCATIONS: ("Y" = Yes)

| West: | Southwest: | Midwest: | Southeast: | Northeast: | International: |
|---|---|---|---|---|---|
| | | Y | | Y | |

Note: Financial information, benefits and other data can change quickly and may vary from those stated here.

# EPIX PHARMACEUTICALS INC

**www.epixmed.com**

Industry Group Code: 325413  Ranks within this company's industry group: Sales: 18  Profits: 17

| Drugs: | | Other: | Clinical: | | Computers: | | Services: | |
|---|---|---|---|---|---|---|---|---|
| Discovery: | Y | AgriBio: | Trials/Services: | | Hardware: | | Specialty Services: | |
| Licensing: | Y | Genetic Data: | Labs: | | Software: | | Consulting: | |
| Manufacturing: | Y | Tissue Replacement: | Equipment/Supplies: | Y | Arrays: | | Blood Collection: | |
| Genetics: | | | Research & Development Services: | Y | Database Management: | | Drug Delivery: | |
| | | | Diagnostics: | | | | Drug Distribution: | |

## TYPES OF BUSINESS:

Medical Diagnostics Products
MRI Contrast Agents

## BRANDS/DIVISIONS/AFFILIATES:

Vasovist
PRX-08066
PRX-03140
PRX-07034
MS-325

## CONTACTS: *Note: Officers with more than one job title may be intentionally listed here more than once.*

Elkan Gamzu, CEO
Elkan Gamzu, Pres.
Kim C. Drapkin, CFO
Brenda Sousa, VP-Human Resources
Sheila DeWitt, VP-Discovery & Mfg.
Simon S. Jones, VP-Biology & ADMET
Frederick Frank, Chmn.

| Phone: 781-761-7600 | Fax: 781-761-7641 |
|---|---|
| Toll-Free: | |
| Address: 4 Maguire Rd., Lexington, MA 02421 US | |

## GROWTH PLANS/SPECIAL FEATURES:

EPIX Pharmaceuticals, Inc. develops targeted contrast agents to improve the quality of magnetic resonance imaging (MRI) as a tool for diagnosing human disease. The firm has three product candidates in clinical trials: PRX-08066, a treatment for pulmonary hypertension in chronic obstructive pulmonary disease (COPD); PRX-03140, a drug for Alzheimer's Disease; and PRX-07034, which targets obesity, Alzheimer's Disease, and cognitive impairment associated with schizophrenia. The company's lead product, MS-325 (formerly marketed as Vasovist), is an injectable intravascular contrast agent designed for multiple cardiovascular imaging applications, including peripheral vascular disease and coronary artery disease. EPIX believes that MS-325 will significantly enhance the quality of MRI images and provide physicians with a minimally-invasive and cost-effective diagnostic method. MS-325 has the potential to replace highly invasive and costly conventional X-ray angiography, over which it has a number of advantages. The agent is safer, involves no patient exposure to ionizing radiation, allows 60 minutes of arterial and venous imaging, facilitates imaging of multiple vascular areas in a single exam and provides three-dimensional data. MS-325 has been approved for marketing in the U.S. and 37 countries abroad. EPIX formerly collaborated with Schering AG for the development and commercialization of the drug. EPIX has two preclinical products in the pipeline, including a partnership with Amgen to develop an S1P1 modulator for a variety of auto-immune diseases; and the development of a cystic fibrosis therapy along with Cystic Fibrosis Therapeutics, Inc. The firm continues to maintain an Israeli headquarters in Ramat-Gan. In October 2008, the company announced a reduction of about 23% of its workforce. In April 2009, Epix sold the U.S., Canadian and Australian rights for MS-325 to Lantheus Medical Imaging, but retained the right to other markets.

## FINANCIALS: Sales and profits are in thousands of dollars—add 000 to get the full amount. 2008 Note: Financial information for 2008 was not available for all companies at press time.

| | | |
|---|---|---|
| 2008 Sales: $28,628 | 2008 Profits: $-36,671 | **U.S. Stock Ticker: EPIX** |
| 2007 Sales: $14,960 | 2007 Profits: $-62,789 | **Int'l Ticker:**  Int'l Exchange: |
| 2006 Sales: $6,041 | 2006 Profits: $-157,393 | Employees:  91 |
| 2005 Sales: $7,190 | 2005 Profits: $-21,269 | Fiscal Year Ends: 12/31 |
| 2004 Sales: $12,259 | 2004 Profits: $-22,621 | Parent Company: |

## SALARIES/BENEFITS:

| Pension Plan: | ESOP Stock Plan: | Profit Sharing: | Top Exec. Salary: $261,562 | Bonus: $65,611 |
|---|---|---|---|---|
| Savings Plan: Y | Stock Purch. Plan: Y | | Second Exec. Salary: $256,846 | Bonus: $65,000 |

## OTHER THOUGHTS:

**Apparent Women Officers or Directors:** 5
**Hot Spot for Advancement for Women/Minorities:** Y

## LOCATIONS: ("Y" = Yes)

| West: | Southwest: | Midwest: | Southeast: | Northeast: | International: |
|---|---|---|---|---|---|
| | | | | Y | Y |

# ERESEARCH TECHNOLOGY INC

www.ert.com

**Industry Group Code: 511210D  Ranks within this company's industry group:** Sales: 2    Profits: 2

| Drugs: | Other: | Clinical: | | Computers: | | Services: | |
|---|---|---|---|---|---|---|---|
| Discovery: | AgriBio: | Trials/Services: | | Hardware: | | Specialty Services: | Y |
| Licensing: | Genetic Data: | Labs: | Y | Software: | Y | Consulting: | Y |
| Manufacturing: | Tissue Replacement: | Equipment/Supplies: | | Arrays: | | Blood Collection: | |
| Genetics: | | Research & Development Services: | | Database Management: | | Drug Delivery: | |
| | | Diagnostics: | | | | Drug Distribution: | |

## TYPES OF BUSINESS:

Software-Clinical Research Technology
Technology Consulting Services
Cardiac Safety Services
Service-Business Services, NEC
Service-Testing Laboratories

## BRANDS/DIVISIONS/AFFILIATES:

EXPeRT
ePRO Solutions
My Study Portal
Electronic Data Capture

## CONTACTS: *Note: Officers with more than one job title may be intentionally listed here more than once.*

Michael J. McKelvey, CEO
Michael J. McKelvey, Pres.
Keith D. Schneck, CFO/Exec. VP
John Blakeley, Exec. VP-Mktg. & Sales
Valerie Mattern, VP-Human Resources
Joel Morganroth, Chief Scientist
Thomas P. Devine, Chief Dev. Officer/Exec. VP
Tom Devine, Chief Dev. Officer/Exec. VP
Amy Furlong, Exec. VP-Cardiac Safety Oper.
Jeffrey S. Litwin, Chief Medical Officer/Exec. VP
George Tiger, Sr. VP-Americas Sales
Robert Brown, Sr. VP-Strategic Mktg., Planning, & Partnerships
Joel Morganroth, Chmn.

| Phone: 215-972-0420 | Fax: 215-972-0414 |
|---|---|
| Toll-Free: | |
| Address: 30 S. 17th St., Philadelphia, PA 19103-4001 US | |

## GROWTH PLANS/SPECIAL FEATURES:

eResearch Technology, Inc. (ERT) is a provider of technology-based products and services that enable the pharmaceutical, biotechnology and medical device and contract resource companies to collect, interpret and distribute cardiac safety and clinical data more efficiently. The company also provides centralized electrocardiograph (ECG) services (through its EXPeRT Cardiac Safety Intelligent Data Management system) and clinical research technology and services, which include the development, marketing and support of clinical research technology. Clinical trial sponsors and clinical research organizations use the cardiac safety services during clinical trials.  ERT's clinical research technology and services include the licensing of its proprietary software products and the provision of maintenance and consulting services supporting its proprietary software products.  The service's My Study Portal supports access to data without a software footprint, as it utilizes the web rather than installed programs.  The eResearch Electronic Data Capture (EDC) provides standardize data collection and access for clinical trial studies.   The company's electronic patient reported outcomes (ePRO) business allows users to collect, monitor and store patient information in a centralized database and utilize expert consultants to extrapolate strategies.  In January 2009, eResearch Technology announced an alliance with Integrium, LLC to combine the two companies' services for enhanced data collection, analysis and support services.

eResearch Technology's benefits include medical, vision and dental coverage; spending accounts; employee stock programs; a 401(k) plan; short- and long-term disability coverage; tuition reimbursement; and transit credits.

## FINANCIALS: Sales and profits are in thousands of dollars—add 000 to get the full amount. 2008 Note: Financial information for 2008 was not available for all companies at press time.

| | | |
|---|---|---|
| 2008 Sales: $133,140 | 2008 Profits: $25,002 | **U.S. Stock Ticker: ERES** |
| 2007 Sales: $98,698 | 2007 Profits: $15,252 | **Int'l Ticker:**    Int'l Exchange: |
| 2006 Sales: $86,368 | 2006 Profits: $8,310 | Employees:   41 |
| 2005 Sales: $86,847 | 2005 Profits: $15,365 | Fiscal Year Ends: 12/31 |
| 2004 Sales: $109,393 | 2004 Profits: $29,724 | Parent Company: |

## SALARIES/BENEFITS:

| Pension Plan: | ESOP Stock Plan: | Profit Sharing: | Top Exec. Salary: $500,000 | Bonus: $414,831 |
|---|---|---|---|---|
| Savings Plan: Y | Stock Purch. Plan: Y | | Second Exec. Salary: $273,000 | Bonus: $202,017 |

## OTHER THOUGHTS:

**Apparent Women Officers or Directors**: 2
**Hot Spot for Advancement for Women/Minorities**: Y

## LOCATIONS: ("Y" = Yes)

| West: | Southwest: | Midwest: | Southeast: | Northeast: | International: |
|---|---|---|---|---|---|
| | | | | Y | Y |

# EUSA PHARMA (USA) INC                                www.eusapharma.com

**Industry Group Code:** 325412  **Ranks within this company's industry group:** Sales:    Profits:

| Drugs: | | Other: | Clinical: | | Computers: | | Services: | |
|---|---|---|---|---|---|---|---|---|
| Discovery: | Y | AgriBio: | Trials/Services: | | Hardware: | | Specialty Services: | |
| Licensing: | Y | Genetic Data: | Labs: | | Software: | | Consulting: | |
| Manufacturing: | | Tissue Replacement: | Equipment/Supplies: | | Arrays: | | Blood Collection: | |
| Genetics: | | | Research & Development Services: | | Database Management: | | Drug Delivery: | |
| | | | Diagnostics: | Y | | | Drug Distribution: | |

## TYPES OF BUSINESS:

Drugs-Oncology
Diagnostic Imaging Agents

## BRANDS/DIVISIONS/AFFILIATES:

ProstaScint
Caphosol
Quadramet
EUSA Pharma, Inc.

## CONTACTS: Note: Officers with more than one job title may be intentionally listed here more than once.

Kevin G. Lokay, Pres.
Stephen A. Ross, Sr. VP-Sales & Mktg.
Rita Auld, VP-Human Resources
Michael J. Manyak, VP-U.S. Medical Science
Rita Auld, VP-Admin.
William F. Goeckeler, Sr. VP-Oper.
Kevin J. Bratton, Sr. VP-Finance
Thu Dang, VP-Finance
Rolf Stahel, Chmn.

| Phone: 215-867-4900 | Fax: 215-579-0384 |
|---|---|
| Toll-Free: 800-833-3533 | |
| Address: 1717 Langhorne-Newtown Rd. Ste. 201, Langhorne, PA 19047 US | |

## GROWTH PLANS/SPECIAL FEATURES:

EUSA Pharma (USA) Inc., formerly Cytogen Corp., develops and commercializes oncology products. Its FDA-approved and marketed products include ProstaScint, a monoclonal antibody-based agent used to image the extent and spread of prostate cancer; QuadraMet, a therapeutic agent marketed for the relief of pain in prostate, breast and other cancers that have spread to the bone; and Caphosol, an advanced electrolyte solution for the treatment of oral mucositis and dry mouth. While ProstaScint and QuadraMet are marketed and sold exclusively in the U.S., Caphosol is available in the U.S. and Europe. The company is also developing CYT-500, a third-generation radio-labeled antibody to treat prostate cancer. In May 2008, Cytogen was acquired by EUSA Pharma, Inc., an international specialty pharmaceutical company focused on licensing, developing and marketing late-stage oncology, pain control and critical care products. Cytogen became EUSA Pharma (USA) Inc., and now operates as a wholly-owned subsidiary of EUSA Pharma.

## FINANCIALS: Sales and profits are in thousands of dollars—add 000 to get the full amount. 2008 Note: Financial information for 2008 was not available for all companies at press time.

| | | |
|---|---|---|
| 2008 Sales: $ | 2008 Profits: $ | U.S. Stock Ticker: Subsidiary |
| 2007 Sales: $ | 2007 Profits: $ | Int'l Ticker:    Int'l Exchange: |
| 2006 Sales: $17,307 | 2006 Profits: $-15,103 | Employees:    95 |
| 2005 Sales: $15,946 | 2005 Profits: $-26,289 | Fiscal Year Ends: 12/31 |
| 2004 Sales: $14,619 | 2004 Profits: $-20,540 | Parent Company: EUSA PHARMA INC |

## SALARIES/BENEFITS:

| | | | | |
|---|---|---|---|---|
| Pension Plan: | ESOP Stock Plan: | Profit Sharing: | Top Exec. Salary: $361,166 | Bonus: $ |
| Savings Plan: Y | Stock Purch. Plan: Y | | Second Exec. Salary: $276,855 | Bonus: $ |

## OTHER THOUGHTS:

Apparent Women Officers or Directors: 1
Hot Spot for Advancement for Women/Minorities:

## LOCATIONS: ("Y" = Yes)

| West: | Southwest: | Midwest: | Southeast: | Northeast: | International: |
|---|---|---|---|---|---|
| | | | | Y | |

# EVOTEC AG

www.evotecoai.com

**Industry Group Code: 541712  Ranks within this company's industry group:** Sales: 12   Profits: 17

| Drugs: | | Other: | | Clinical: | | Computers: | | Services: | |
|---|---|---|---|---|---|---|---|---|---|
| Discovery: | Y | AgriBio: | | Trials/Services: | | Hardware: | Y | Specialty Services: | Y |
| Licensing: | Y | Genetic Data: | | Labs: | Y | Software: | Y | Consulting: | |
| Manufacturing: | Y | Tissue Replacement: | | Equipment/Supplies: | | Arrays: | | Blood Collection: | |
| Genetics: | | | | Research & Development Services: | Y | Database Management: | Y | Drug Delivery: | |
| | | | | Diagnostics: | | | | Drug Distribution: | |

## TYPES OF BUSINESS:

Drug Discovery & Development Services
Assay Development
High-Throughput Screening
Chemical Compound Libraries
Medicinal Chemistry
Manufacturing Services
Drug Discovery Hardware & Software

## BRANDS/DIVISIONS/AFFILIATES:

Zebrafish
EVT 101
EVT 103
EVT 302
EVT 201
VR1

## CONTACTS: Note: Officers with more than one job title may be intentionally listed here more than once.

Werner Lanthaler, CEO
Mario Polywka, COO
Klaus Maleck, CFO
Charmion Gillmore, Sr. VP-Human Resources & Internal Comm.
John Kemp, Chief R&D Officer
Mark Ashton, Exec. VP-Bus. Dev.
Anne Hennecke, Sr. VP-Corp. Comm.
Anne Hennecke, Sr. VP-Investor Rel.
David Brister, Chief Business Officer
Flemming Ornskov, Chmn.

| Phone: 49-40-5-60-81-0 | Fax: 49-40-5-60-81-222 |
|---|---|
| **Toll-Free:** | |
| **Address:** Schnackenburgallee 114, Hamburg,  22525 Germany | |

## GROWTH PLANS/SPECIAL FEATURES:

Evotec AG provides investigational new drug programs for its biotechnology and pharmaceutical company clients.  The company has specific expertise in the area of Central Nervous System (CNS) related diseases.  Evotec's collaborative research division provides contract research and development services through its Innovation Centers, with capabilities including assay development; high-throughput screening; compound libraries; medicinal chemistry; chemical and pharmaceutical development; and formulation and manufacture.  Assay development identifies chemical compounds that interact with a selected biological target.  High-throughput screening takes the assay format and uses it to test up to 100,000 chemical compounds per day in Evotec's, and its clients' chemical libraries.  Primary screens determine whether a compound shows biological activity worth exploring.   The company fine-tunes its chemical library findings into pre-clinical candidates, and then develops processes for manufacturing enough of the test drug for Phase I-III clinical trials.  Evotec's Pipeline division conducts internal research to discover compounds for the treatment of major CNS related disorders including sleep disorders and Alzheimer's disease, as well as inflammation and pain.  For the treatment of insomnia, Evotec has developed EVT 201, which has completed phase II trials with positive results.  EVT 101, in Phase II, and EVT 103, in Phase I are the company's Alzheimer's disease and neuropathic pain products.   The P2X7 antagonist, for Rheumatoid Arthritis began is also currently in Phase I. Evotec has an MAO-B inhibitor EVT 302, for the purpose of smoking cessation, which began Phase II clinical trials in 2008.  In August 2008, Evotec began Phase I trials, in conjunction with Pfizer Inc., for VR1, a vanilloid receptor antagonist thought to be helpful in treating pain, urinary incontinence, and other diseases.   In May 2009, the company acquired the Zebrafish drug screening technology from Summit Corporation plc.

## FINANCIALS: Sales and profits are in thousands of dollars—add 000 to get the full amount. 2008 Note: Financial information for 2008 was not available for all companies at press time.

| | | |
|---|---|---|
| 2008 Sales: $56,300 | 2008 Profits: $-110,400 | **U.S. Stock Ticker: EVT** |
| 2007 Sales: $46,700 | 2007 Profits: $-15,700 | **Int'l Ticker: EVT**    Int'l Exchange: Frankfurt-Euronext |
| 2006 Sales: $57,600 | 2006 Profits: $-39,100 | Employees:   386 |
| 2005 Sales: $101,630 | 2005 Profits: $-42,780 | Fiscal Year Ends: 12/31 |
| 2004 Sales: $99,200 | 2004 Profits: $-114,900 | Parent Company: |

## SALARIES/BENEFITS:

| Pension Plan: | ESOP Stock Plan: | Profit Sharing: | Top Exec. Salary: $ | Bonus: $ |
|---|---|---|---|---|
| Savings Plan: | Stock Purch. Plan: | | Second Exec. Salary: $ | Bonus: $ |

## OTHER THOUGHTS:

**Apparent Women Officers or Directors**: 1
**Hot Spot for Advancement for Women/Minorities**:

## LOCATIONS: ("Y" = Yes)

| West: | Southwest: | Midwest: | Southeast: | Northeast: | International: |
|---|---|---|---|---|---|
| | | | | Y | Y |

Note: Financial information, benefits and other data can change quickly and may vary from those stated here.

# EXELIXIS INC

**www.exelixis.com**

**Industry Group Code: 325412  Ranks within this company's industry group:** Sales: 64  Profits: 163

| Drugs: | | Other: | | Clinical: | Computers: | Services: |
|---|---|---|---|---|---|---|
| Discovery: | Y | AgriBio: | Y | Trials/Services: | Hardware: | Specialty Services: |
| Licensing: | Y | Genetic Data: | Y | Labs: | Software: | Consulting: |
| Manufacturing: | | Tissue Replacement: | Y | Equipment/Supplies: | Arrays: | Blood Collection: |
| Genetics: | | | | Research & Development Services: | Database Management: | Drug Delivery: |
| | | | | Diagnostics: | | Drug Distribution: |

## TYPES OF BUSINESS:

Genetic Research & Drug Development
Crop Protection Products
Genomics
Anti-Cancer Compounds

## BRANDS/DIVISIONS/AFFILIATES:

Exelixis Plant Sciences Inc

## CONTACTS: *Note: Officers with more than one job title may be intentionally listed here more than once.*

George A. Scangos, CEO
George A. Scangos, Pres.
Frank Karbe, CFO/Exec. VP
Michael Morrissey, Pres., R&D
Pamela A. Simonton, General Counsel/Exec. VP
Lupe M. Rivera, Sr. VP-Oper.
Fran Heller, Exec. VP-Bus. Dev.
D. Ry Wagner, VP-Research, Exelixis Plant Sciences
Gisela M. Schwab, Chief Medical Officer/Exec. VP
Peter Lamb, Chief Scientific Officer/Sr. VP-Discovery Research
Stelios Papadopoulos, Chmn.

| Phone: 650-837-7000 | Fax: 650-837-8300 |
|---|---|
| Toll-Free: | |
| Address: 210 E. Grand Ave., South San Francisco, CA 94083 US | |

## GROWTH PLANS/SPECIAL FEATURES:

Exelixis, Inc. is a biotechnology company that focuses on the discovery and development of potential new drug therapies for cancer and other life-threatening diseases. The company has developed an integrated research and discovery platform utilizing proprietary technologies such as medicinal chemistry, bioinformatics, structural biology and early in vivo testing to provide an efficient and cost-effective process in gene analysis and drug development. Exelixis' translational research group employs knowledge generated in the discovery process to identify targeted patient populations for possible gene mutations or gene variants that impact response to therapy. The company's clinical program then conducts clinical trials to move candidate compounds through clinical registration phases in order to market newly discovered treatments. Exelixis Plant Sciences, a subsidiary of Exelixis, is working in collaboration with agricultural companies in plant biotechnology and crop protection. Research areas in crop protection focus on chemical products such as herbicides, insecticides and nematides designed specifically to target implicated crops. The plant biotechnology sector aims to develop crops with higher yields and improved nutritional profiles in oil content and protein composition. Exelixis has entered into research collaborations with Bristol-Myers Squibb to identify novel targets for new drugs in the fields of oncology and cardiovascular disease. The firm collaborates its research with GlaxoSmithKline in therapeutic areas such as vascular biology, inflammatory disease and oncology.

Employees are offered medical, dental and vision insurance; domestic partner benefits; health and dependent care flexible spending accounts; life and AD&D insurance; an employee assistance program; a 401(k) plan; long-term disability coverage; business travel accident insurance; a college savings plan; a group legal plan; an adoption assistance program; an infertility assistance program; onsite programs such as dry-cleaning and massage; pet insurance; back up dependent care; a tuition reimbursement program; a stock purchase plan; discounts on entertainment events; subsidized cafeteria meals; and subsidies for fitness center memberships.

## FINANCIALS:  Sales and profits are in thousands of dollars—add 000 to get the full amount. 2008 Note: Financial information for 2008 was not available for all companies at press time.

| | | | |
|---|---|---|---|
| 2008 Sales: $117,859 | 2008 Profits: $-162,854 | **U.S. Stock Ticker: EXEL** | |
| 2007 Sales: $113,470 | 2007 Profits: $-86,381 | **Int'l Ticker:** Int'l Exchange: | |
| 2006 Sales: $98,670 | 2006 Profits: $-101,492 | Employees:   676 | |
| 2005 Sales: $75,961 | 2005 Profits: $-84,404 | Fiscal Year Ends: 12/31 | |
| 2004 Sales: $52,857 | 2004 Profits: $-137,245 | Parent Company: | |

## SALARIES/BENEFITS:

| Pension Plan: | ESOP Stock Plan: | Profit Sharing: | Top Exec. Salary: $848,731 | Bonus: $255,000 |
|---|---|---|---|---|
| Savings Plan: Y | Stock Purch. Plan: Y | | Second Exec. Salary: $483,612 | Bonus: $121,157 |

## OTHER THOUGHTS:

**Apparent Women Officers or Directors**: 4
**Hot Spot for Advancement for Women/Minorities**: Y

## LOCATIONS: ("Y" = Yes)

| West: | Southwest: | Midwest: | Southeast: | Northeast: | International: |
|---|---|---|---|---|---|
| Y | | | | | Y |

Note: Financial information, benefits and other data can change quickly and may vary from those stated here.

# FLAMEL TECHNOLOGIES SA

**www.flamel-technologies.fr**

**Industry Group Code:** 325412A **Ranks within this company's industry group:** Sales: 17 Profits: 16

| Drugs: | | Other: | Clinical: | | Computers: | Services: | |
|---|---|---|---|---|---|---|---|
| Discovery: | | AgriBio: | Trials/Services: | | Hardware: | Specialty Services: | |
| Licensing: | Y | Genetic Data: | Labs: | | Software: | Consulting: | |
| Manufacturing: | Y | Tissue Replacement: | Equipment/Supplies: | Y | Arrays: | Blood Collection: | |
| Genetics: | | | Research & Development Services: | | Database Management: | Drug Delivery: | Y |
| | | | Diagnostics: | | | Drug Distribution: | |

## TYPES OF BUSINESS:

Drug Delivery Systems
Extended-Release Formulations
Collagen-Based Biomaterials
Long-Acting Insulin

## BRANDS/DIVISIONS/AFFILIATES:

Medusa
Micropump
Asacard
Genvir
Colcys
Basulin
Trigger-Lock
Metformin XL

## CONTACTS: *Note: Officers with more than one job title may be intentionally listed here more than once.*

Stephen H. Willard, CEO
Raphael Jorda, COO/Exec. VP
Sian Crouzet, Principal Financial Officer
Remi Meyrueix, Dir.-Scientific
Rafael Jorda, Dir.-Mfg. & Dev.
Charles Marlio, Dir.-Strategic Planning
Charles Marlio, Dir.-Investor Rel.
Christian Kalita, Chief Pharmacist
Catherine Castan, Dir.-R&D Head-Micropump Team
Roger Kravtzoff, Dir.-Preclinical & Early Clinical Dev.
Katherine Hanras, Dir.-Analytics
Elie Vannier, Chmn.
Nigel McWilliam, Dir.-Bus. Dev., U.S.
David Weber, Dir.-Purchasing Oper.

| Phone: 33-472-783-434 | Fax: 33-472-783-435 |
|---|---|

**Toll-Free:**

**Address:** 33 Ave. du Docteur Georges Levy, Venissieux Cedex, 69693 France

## GROWTH PLANS/SPECIAL FEATURES:

Flamel Technologies S.A. is a drug delivery company that is engaged in the development of small molecule and protein therapeutic products for the biotechnology and pharmaceutical industries. Named after a 14th Century French alchemist, the company operates in a 60,000 square foot pharmaceutical production facility in Pessac, France and produces over 2 billion tablets and capsules every year. Flamel currently holds two patented drug delivery platforms. Medusa consists of an injectable 20-50 nanometer diameter self-assembled poly-amino-acid nanoparticles system with no side effects and long duration, which enables the controlled delivery of fully human and non-denatured proteins. Medusa technology is currently being tested for FT-105, a long-acting basal insulin for the treatment of type II diabetes; and in IFN alpha-2b XL for the treatment of hepatitis B, hepatitis C and some forms of cancers. Micropump, a small-sized tablet geared specifically towards pediatric and geriatric customers, encapsulates 5,000-10,000 individual 200-500 micrometer in diameter microparticles per tablet, releasing drugs through osmotic pressure at an adjustable rate (controlled for Micropump I, delayed for II). Since the active ingredients are enclosed within each microparticle, drugs can easily be delivered in suspensions or syrups that completely mask any taste of the active pharmaceutical compound. Micropump is being tested for numerous products, including Genvir, a treatment for acute genital herpes; Metformin XL, a treatment for type II diabetes; Augmentin SR, an antibiotic; and ACE and protein pump inhibitors. In addition, the company owns ColCys, which consists of a family of proprietary collagen-based implantable biomaterials for post-surgical adhesion prevention. Flamel operates primarily through licensing agreements and collaborations with other pharmaceutical companies such as GlaxoSmithKline (GSK), Merck and Servier Monde.

## FINANCIALS: Sales and profits are in thousands of dollars—add 000 to get the full amount. 2008 Note: Financial information for 2008 was not available for all companies at press time.

| | | | |
|---|---|---|---|
| 2008 Sales: $38,619 | 2008 Profits: $-12,084 | **U.S. Stock Ticker:** FLML | |
| 2007 Sales: $36,654 | 2007 Profits: $-37,737 | **Int'l Ticker:** FL3 | **Int'l Exchange:** Frankfurt-Euronext |
| 2006 Sales: $23,020 | 2006 Profits: $-35,201 | **Employees:** 302 | |
| 2005 Sales: $23,598 | 2005 Profits: $-27,377 | **Fiscal Year Ends:** 12/31 | |
| 2004 Sales: $55,410 | 2004 Profits: $12,499 | **Parent Company:** | |

## SALARIES/BENEFITS:

| Pension Plan: Y | ESOP Stock Plan: | Profit Sharing: | Top Exec. Salary: $ | Bonus: $ |
|---|---|---|---|---|
| Savings Plan: | Stock Purch. Plan: | | Second Exec. Salary: $ | Bonus: $ |

## OTHER THOUGHTS:

**Apparent Women Officers or Directors:** 3
**Hot Spot for Advancement for Women/Minorities:** Y

## LOCATIONS: ("Y" = Yes)

| West: | Southwest: | Midwest: | Southeast: | Northeast: | International: |
|---|---|---|---|---|---|
| | | | | | Y |

# FOREST LABORATORIES INC

**www.frx.com**

**Industry Group Code: 325412  Ranks within this company's industry group:** Sales: 29  Profits: 21

| Drugs: | | Other: | Clinical: | Computers: | Services: |
|---|---|---|---|---|---|
| Discovery: | Y | AgriBio: | Trials/Services: | Hardware: | Specialty Services: |
| Licensing: | | Genetic Data: | Labs: | Software: | Consulting: |
| Manufacturing: | Y | Tissue Replacement: | Equipment/Supplies: | Arrays: | Blood Collection: |
| Genetics: | | | Research & Development Services: | Database Management: | Drug Delivery: |
| | | | Diagnostics: | | Drug Distribution: |

## TYPES OF BUSINESS:

Drugs, Manufacturing
Over-the-Counter Pharmaceuticals
Generic Pharmaceuticals
Antidepressants
Asthma Medications
Cardiovascular Products
OB/Gyn Products
Endocrinology

## BRANDS/DIVISIONS/AFFILIATES:

Lexapro
Namenda
Benicar
Forest Research Institute
Forest Pharmaceuticals, Inc.
Forest Laboratories Europe
Inwood Laboratories
Cerexa, Inc.

## CONTACTS: *Note: Officers with more than one job title may be intentionally listed here more than once.*

Howard Solomon, CEO
Lawrence S. Olanoff, COO
Lawrence S. Olanoff, Pres.
Francis I. Perier, Jr., CFO
Elaine Hochberg, Sr. VP-Mktg./Chief Commercial Officer
William J. Candee III, Sec.
Frank Murdolo, VP-Investor Rel.
Francis I. Perier, Jr., Sr. VP-Finance
Howard Solomon, Chmn.

| Phone: 212-421-7850 | Fax: |
|---|---|
| **Toll-Free:** 800-947-5227 | |
| **Address:** 909 3rd Ave., New York, NY 10022 US | |

## GROWTH PLANS/SPECIAL FEATURES:

Forest Laboratories, Inc. develops, delivers and sells pharmaceutical products. It currently covers six therapeutic areas, developing treatments for respiratory, pain management, ob/gyn, endocrinology, central nervous system and cardiovascular conditions. Forest's four principal brands are Lexapro, an antidepressant; Benicar, a hypertension treatment; Namenda, a therapy for moderate or severe Alzheimer's disease; and Campral, which helps reduce withdrawals for those seeking to eliminate alcohol dependence. Other products include Aerobid, an asthma medication; AeroChamber Plus, an inhalant delivery system for asthma medications; Infasurf, used to prevent respiratory distress syndrome (RDS), a condition caused by a lack of surfactant, found mainly in premature infants; Armour Thyroid, Levothroid and Thyrolar, for treating hypothyroidism; Celexa, an antidepressant; Cervidil, used to prepare the cervix before inducing labor; and Combunox, a pain medication combining both opioids and non-steroidal anti-inflammatory drugs. Forest markets directly to physicians who have the most potential for growth and are agreeable to the introduction of new products, as well as to pharmacies, hospitals, managed care and other healthcare organizations. Forest Research Institute, Forest's scientific division, maintains labs on Long Island and in New Jersey. Subsidiary Forest Pharmaceuticals, Inc. manufactures and distributes Forest's branded prescription products in the U.S. Subsidiary Forest Laboratories Europe has two manufacturing sites in Dublin, Ireland and one in Bexley, Kent, and distributes prescription and over-the-counter drugs in Europe, the Middle East, Australia and Asia. Subsidiary Inwood Laboratories manufactures and supplies generic versions of Forest's medications. Cerexa, Inc., acquired in 2007, develops and commercializes treatments for life-threatening infections. In January 2009, the company's Savella, a selective serotonin and norepinephrine inhibitor, was approved by the FDA for the management of fibromyalgia, a chronic pain condition.

Employees at Forest receive financial assistance for adoption and fertility treatments; medical, dental and life insurance; flexible spending accounts; child-care resources; and a commuter benefit program.

## FINANCIALS: Sales and profits are in thousands of dollars—add 000 to get the full amount. 2008 Note: Financial information for 2008 was not available for all companies at press time.

| | | |
|---|---|---|
| 2008 Sales: $3,501,802 | 2008 Profits: $967,933 | **U.S. Stock Ticker: FRX** |
| 2007 Sales: $3,183,324 | 2007 Profits: $454,103 | **Int'l Ticker:** Int'l Exchange: |
| 2006 Sales: $2,793,934 | 2006 Profits: $708,514 | Employees: 5,225 |
| 2005 Sales: $3,052,408 | 2005 Profits: $838,805 | Fiscal Year Ends: 3/31 |
| 2004 Sales: $2,650,432 | 2004 Profits: $735,874 | Parent Company: |

## SALARIES/BENEFITS:

| Pension Plan: | ESOP Stock Plan: | Profit Sharing: Y | Top Exec. Salary: $1,162,500 | Bonus: $635,000 |
|---|---|---|---|---|
| Savings Plan: Y | Stock Purch. Plan: | | Second Exec. Salary: $758,750 | Bonus: $400,000 |

## OTHER THOUGHTS:

**Apparent Women Officers or Directors**: 1
**Hot Spot for Advancement for Women/Minorities**: Y

## LOCATIONS: ("Y" = Yes)

| West: | Southwest: | Midwest: | Southeast: | Northeast: | International: |
|---|---|---|---|---|---|
| | | Y | | Y | Y |

*Note: Financial information, benefits and other data can change quickly and may vary from those stated here.*

# GALDERMA PHARMA SA

**www.galderma.com**

**Industry Group Code: 325411  Ranks within this company's industry group: Sales:    Profits:**

| Drugs: | | Other: | | Clinical: | | Computers: | | Services: | |
|---|---|---|---|---|---|---|---|---|---|
| Discovery: | | AgriBio: | Y | Trials/Services: | | Hardware: | | Specialty Services: | |
| Licensing: | | Genetic Data: | | Labs: | Y | Software: | | Consulting: | |
| Manufacturing: | Y | Tissue Replacement: | | Equipment/Supplies: | | Arrays: | | Blood Collection: | |
| Genetics: | | | | Research & Development Services: | | Database Management: | | Drug Delivery: | |
| | | | | Diagnostics: | | | | Drug Distribution: | |

## TYPES OF BUSINESS:

Dermatological Pharmaceuticals
Dermatological Product Research and Development

## BRANDS/DIVISIONS/AFFILIATES:

Nestle SA
Loreal SA
Cetaphil
Clobex
Differin
Epiduo
Loceryl
Rozex

## CONTACTS: *Note: Officers with more than one job title may be intentionally listed here more than once.*

Humberto C. Antunes, CEO
Humberto C. Antunes, Pres.

| Phone: 41-21-641-1151 | Fax: 41-21-641-1161 |
|---|---|
| Toll-Free: | |
| Address: Avenue de Gratta-Paille 1, Case Postale 453, Lausanne,  CH-1000 Switzerland | |

## GROWTH PLANS/SPECIAL FEATURES:

Galderma Pharma S.A. is engaged in the research and development, manufacture and distribution of dermatological products worldwide.  The firm, a joint venture between Nestle and L'Oreal, has a portfolio that includes treatment for skin conditions such as acne, dry skin, fungal nail infections, pigmentary disorders, skin cancers, rosacea, psoriasis/SRD (Steroid-Responsive Dermatoses) and skin senescence.  Galderma's products vary from topical and oral treatments to photodynamic therapy and include key brands such as Cetaphil, Clobex, Differin, Epiduo, Loceryl, Rozex, Metvix, Oracea, Silkis and Tri-Luma.    The company's primary research and development department, located in Sophia Antipolis, France, uses technologies such as human genome analysis, robotic screening, pharmaco-dynamic modeling, computer-assisted imaging and miniaturized human biological assaying.  Galderma also houses research and development teams in Princeton, New Jersey and Tokyo, Japan, as well as manufacturing facilities in Alby-sur-Cheran, France; Montreal, Quebec; and Hortolandia, Brazil.  The company holds operating subsidiaries around the globe, including Galderma Australia (PTY) Ltd. in Australia; Galderma Laboratories L.P. in the U.S.; Galderma Canada, Inc.; Galderma Argentina S.A.; Galderma Brasil LTDA; and Galderma (Japan) KK, among many others.  The company's products are distributed in over 65 countries, and the firm allocates approximately 20% of revenues for research and development of new products.  The most recent products from Galderma's pipeline are Vectical, a vitamin D3 product for plaque psoriasis, and Epiduo Gel, a topical acne treatment.  In April 2008, Galderma acquired Collagenex Pharmaceuticals, Inc., a specialty pharmaceutical company focused on dermatological therapies.  CollaGenex's chief products include Oracea for the treatment of inflammatory lesions of rosacea in adults; Pandel, used for the treatment of dermatitis and psoriasis; Alcortin, a mild dermatosis gel with combined anti-fungal, anti-bacterial and anti-inflammatory effects; and Novacort, a topical corticosteroid with anti-inflammatory and anesthetic treatment for certain dermatoses.

## FINANCIALS: Sales and profits are in thousands of dollars—add 000 to get the full amount. 2008 Note: Financial information for 2008 was not available for all companies at press time.

| | | | |
|---|---|---|---|
| 2008 Sales: $ | 2008 Profits: $ | **U.S. Stock Ticker: Joint Venture** | |
| 2007 Sales: $ | 2007 Profits: $ | **Int'l Ticker:**    Int'l Exchange: | |
| 2006 Sales: $ | 2006 Profits: $ | Employees: | |
| 2005 Sales: $ | 2005 Profits: $ | Fiscal Year Ends: | |
| 2004 Sales: $ | 2004 Profits: $ | Parent Company: | |

## SALARIES/BENEFITS:

| Pension Plan: | ESOP Stock Plan: | Profit Sharing: | Top Exec. Salary: $ | Bonus: $ |
|---|---|---|---|---|
| Savings Plan: | Stock Purch. Plan: | | Second Exec. Salary: $ | Bonus: $ |

## OTHER THOUGHTS:

**Apparent Women Officers or Directors:**
**Hot Spot for Advancement for Women/Minorities:**

## LOCATIONS: ("Y" = Yes)

| West: | Southwest: | Midwest: | Southeast: | Northeast: | International: |
|---|---|---|---|---|---|
| | | | | Y | Y |

# GE HEALTHCARE

**www.gehealthcare.com**

Industry Group Code: 33911  Ranks within this company's industry group:  Sales: 1    Profits: 1

| Drugs: | | Other: | Clinical: | | Computers: | | Services: | |
|---|---|---|---|---|---|---|---|---|
| Discovery: | Y | AgriBio: | Trials/Services: | | Hardware: | | Specialty Services: | Y |
| Licensing: | | Genetic Data: | Labs: | | Software: | Y | Consulting: | |
| Manufacturing: | | Tissue Replacement: | Equipment/Supplies: | Y | Arrays: | Y | Blood Collection: | |
| Genetics: | | | Research & Development Services: | Y | Database Management: | | Drug Delivery: | |
| | | | Diagnostics: | Y | | | Drug Distribution: | |

## TYPES OF BUSINESS:

Medical Imaging & Information Technology
Magnetic Resonance Imaging Systems
Patient Monitoring Systems
Clinical Information Systems
Nuclear Medicine
Surgery & Vascular Imaging
X-Ray & Ultrasound Bone Densitometers
Clinical & Business Services

## BRANDS/DIVISIONS/AFFILIATES:

General Electric Co (GE)
GE Healthcare Bio-Sciences
GE Healthcare Technologies
GE Healthcare Information Technologies
GE Technology Infrastructure
Innova
Vital Signs
Whatman plc

## CONTACTS: *Note: Officers with more than one job title may be intentionally listed here more than once.*

John Dineen, CEO
John Dineen, Pres.
Frank Schulkes, CFO/Exec. VP
Jean-Michel Cossery, Chief Mktg. Officer
Mike Hanley, VP-Human Resources
Russel P. Meyer, CIO
Michael J. Barber, CTO
Peter Y. Solmssen, General Counsel/Exec. VP
Ralph Strosin, Gen. Mgr.-Oper.
Michael A. Jones, Exec. VP-Bus. Dev.
Lynne Gailey, Exec. VP-Global Comm.
Pete McCabe, CEO-Surgery
Peter Ehrenheim, CEO/Pres., Life Sciences
Vishal Wanchoo, CEO/Pres., Integrated IT Solutions
Reinaldo Garcia, CEO/Pres., Int'l
Brian Masterson, VP-Supply Chain

| Phone: 414-721-2407 | Fax: |
|---|---|
| Toll-Free: | |
| Address: 3000 N. Grandview Blvd., Waukesha, WI 53188 US | |

## GROWTH PLANS/SPECIAL FEATURES:

GE Healthcare, a subsidiary of General Electric Co. (GE) and a part of GE's Technology Infrastructure division, is a global leader in medical imaging and information technologies, patient monitoring systems and health care services. The company operates through seven divisions, including diagnostic imaging; global services; clinical systems; life systems; medical diagnostics; integrated information technology solutions; and interventional, cardiology and surgery. The diagnostic imaging business provides X-ray, digital mammography, computed tomography, magnetic resonance and molecular imaging technologies. GE Healthcare's global services business provides maintenance of a wide range of medical systems and devices. The clinical systems business provides technologies and services for clinicians and healthcare administrators. GE Healthcare's life sciences segment offers drug discovery, biopharmaceutical manufacturing and cellular technologies, enabling scientists and specialists around the world to discover new ways to predict, diagnose and treat disease earlier. The segment also makes systems and equipment for the purification of biopharmaceuticals. The firm's medical diagnostics business researches, manufactures and markets agents used during medical scanning procedures to highlight organs, tissue and functions inside the human body. The integrated information technology (IT) solutions business provides clinical and financial information technology solutions including enterprise and departmental IT products, revenue cycle management and practice applications, to help customers streamline healthcare costs and improve the quality of care. GE Healthcare's interventional, cardiology and surgery (ICS) business provides tools and technologies for fully integrated cardiac, surgical and interventional care. Recent acquisitions include Vital Signs, a global provider of medical products; Whatman plc, a supplier of filtration products and technologies; VersaMed Corporation, a provider of portable critical care ventilators; and Image Diagnost International GmbH, a developer or innovative products optimized for the needs of diagnostic and screening mammography.

## FINANCIALS: Sales and profits are in thousands of dollars—add 000 to get the full amount. 2008 Note: Financial information for 2008 was not available for all companies at press time.

| | | |
|---|---|---|
| 2008 Sales: $17,392,000 | 2008 Profits: $2,851,000 | U.S. Stock Ticker: Subsidiary |
| 2007 Sales: $16,997,000 | 2007 Profits: $3,056,000 | Int'l Ticker:    Int'l Exchange: |
| 2006 Sales: $16,560,000 | 2006 Profits: $3,142,000 | Employees: 47,000 |
| 2005 Sales: $15,016,000 | 2005 Profits: $2,601,000 | Fiscal Year Ends: 12/31 |
| 2004 Sales: $13,411,000 | 2004 Profits: $2,263,000 | Parent Company: GENERAL ELECTRIC CO (GE) |

## SALARIES/BENEFITS:

| | | | | |
|---|---|---|---|---|
| Pension Plan: Y | ESOP Stock Plan: | Profit Sharing: | Top Exec. Salary: $ | Bonus: $ |
| Savings Plan: Y | Stock Purch. Plan: | | Second Exec. Salary: $ | Bonus: $ |

## OTHER THOUGHTS:

**Apparent Women Officers or Directors**: 3
**Hot Spot for Advancement for Women/Minorities**: Y

## LOCATIONS: ("Y" = Yes)

| West: | Southwest: | Midwest: | Southeast: | Northeast: | International: |
|---|---|---|---|---|---|
| | | Y | | | Y |

# GENENCOR INTERNATIONAL INC

**www.genencor.com**

**Industry Group Code: 325414  Ranks within this company's industry group: Sales:    Profits:**

| Drugs: | | Other: | | Clinical: | Computers: | | Services: | |
|---|---|---|---|---|---|---|---|---|
| Discovery: | Y | AgriBio: | | Trials/Services: | Hardware: | | Specialty Services: | Y |
| Licensing: | | Genetic Data: | Y | Labs: | Software: | | Consulting: | |
| Manufacturing: | Y | Tissue Replacement: | Y | Equipment/Supplies: | Arrays: | | Blood Collection: | |
| Genetics: | | | | Research & Development Services: | Database Management: | | Drug Delivery: | |
| | | | | Diagnostics: | | | Drug Distribution: | |

## TYPES OF BUSINESS:

Biological Manufacturing
Biotech Research & Discovery
Enzyme-Based Products
Protein-Based Products
Bioethanol Technology Research & Development
Enzymes for Ethanol Production

## BRANDS/DIVISIONS/AFFILIATES:

Danisco A/S
Multifect Protex
Accellerase
Optimax
Gensweet
Agtech Products Inc.

## CONTACTS: Note: Officers with more than one job title may be intentionally listed here more than once.

Tjerk De Ruiter, CEO
Jim Sjoerdsma, VP-Human Resources
Michael V. Arbige, Exec. VP-R&D
Michael V. Arbige, Exec. VP-Tech.
Soonhee Jang, VP-Intellectual Property/Chief IP Counsel
Philippe Lavielle, Exec. VP-Bus. Dev.
Jennifer Hutchins, Media
Andrew Ashworth, VP-Finance
James Laughton, Exec. VP-Animal Nutrition/Food & Beverage Enzymes
Glenn Nedwin, Exec. VP-Technical Enzymes
Philippe Lavielle, Exec. VP-Bus. Dev., Biomass
Ken Herfert, Sr. VP-Supply

| **Phone:** 585-256-5200 | **Fax:** 585-256-6952 |
|---|---|
| **Toll-Free:** | |
| **Address:** 200 Meridian Centre Blvd., Rochester, NY 14618-3916 US | |

## GROWTH PLANS/SPECIAL FEATURES:

Genencor International, Inc., the biotechnology division of Danish firm Danisco A/S, discovers, develops and sells biocatalysts and other biochemicals. It delivers 250 products to customers in 80 countries throughout the world. The firm does so through five R&D centers located in the USA, Denmark, Finland, the Netherlands and China; and 10 manufacturing centers throughout world. Its products serve the following sectors: Agri Processing, Industrial Processing and Consumer Products. Agri Processing products serve customers who process agricultural raw materials, including plant proteins, to produce animal feeds, food, food ingredients and renewable fuels. Specific enzymes include Multifect Protex, used in rice and soybean processing; Accellerase, used to enhance the production of ethanol from biomass; Optimax, which converts starches, such as wheat, into glucose, a non-sweet sugar; and Gensweet, which coverts glucose into fructose, a sweetener. Industrial Processing products mainly include enzymes used in everything from denim finishing in the textiles industry and as processing agents in the wastewater treatment and oil and gas production sectors to medical instrument cleaning and even as decontamination agents in chemical and biological warfare attack situations. Consumer Products applications include enzymes used in fabric and household care products to remove recalcitrant stains more efficiently than soaps and detergents alone; and in personal care items such as contact lens cleaner, whitening toothpaste and diabetic test kits. The firm recently joined forces with BRAIN, a leading metagenomic enzyme and screening technology company, to produce biobased chemicals from renewable feedstock. In 2008, Genencor launched a new biosafety program that focuses on biodefense, bioremediation and prion disinfection in primarily the military, medical and dental sectors. The firm's parent company, Danisco, recently furthered its biotechnology efforts with the acquisition of Agtech Products Inc., a US-based agricultural biotechnology company dedicated to producing microbial-based products for animal nutrition.

## FINANCIALS: Sales and profits are in thousands of dollars—add 000 to get the full amount. 2008 Note: Financial information for 2008 was not available for all companies at press time.

| | | | |
|---|---|---|---|
| 2008 Sales: $ | 2008 Profits: $ | **U.S. Stock Ticker: Subsidiary** | |
| 2007 Sales: $90,700 | 2007 Profits: $ | **Int'l Ticker:** Int'l Exchange: | |
| 2006 Sales: $ | 2006 Profits: $ | Employees: 1,098 | |
| 2005 Sales: $ | 2005 Profits: $ | Fiscal Year Ends: 12/31 | |
| 2004 Sales: $410,417 | 2004 Profits: $26,178 | Parent Company: DANISCO A/S | |

## SALARIES/BENEFITS:

| Pension Plan: | ESOP Stock Plan: | Profit Sharing: | Top Exec. Salary: $540,346 | Bonus: $370,900 |
|---|---|---|---|---|
| Savings Plan: Y | Stock Purch. Plan: | | Second Exec. Salary: $337,888 | Bonus: $172,800 |

## OTHER THOUGHTS:

**Apparent Women Officers or Directors:** 2
**Hot Spot for Advancement for Women/Minorities:**

## LOCATIONS: ("Y" = Yes)

| West: | Southwest: | Midwest: | Southeast: | Northeast: | International: |
|---|---|---|---|---|---|
| Y | | Y | | Y | Y |

Note: Financial information, benefits and other data can change quickly and may vary from those stated here.

# GENENTECH INC

**www.gene.com**

**Industry Group Code:** 325412 **Ranks within this company's industry group:** Sales: 17 Profits: 14

| Drugs: | | Other: | Clinical: | Computers: | Services: |
|---|---|---|---|---|---|
| Discovery: | Y | AgriBio: | Trials/Services: | Hardware: | Specialty Services: |
| Licensing: | | Genetic Data: | Labs: | Software: | Consulting: |
| Manufacturing: | Y | Tissue Replacement: | Equipment/Supplies: | Arrays: | Blood Collection: |
| Genetics: | | | Research & Development Services: | Database Management: | Drug Delivery: |
| | | | Diagnostics: | | Drug Distribution: |

## TYPES OF BUSINESS:

Drug Development & Manufacturing
Genetically Engineered Drugs

## BRANDS/DIVISIONS/AFFILIATES:

Avastin
TNKase
Herceptin
Rituxan
Activase
Pulmozyme
Nutropin

## CONTACTS: Note: Officers with more than one job title may be intentionally listed here more than once.

Arthur D. Levinson, CEO
David A. Ebersman, CFO/Exec. VP
Richard H. Scheller, Exec. VP-Research
Susan Desmond-Hellmann, Pres., Prod. Dev.
Stephen G. Juelsgaard, Exec. VP/Corp. Sec.
Ian T. Clark, Exec. VP-Comm. Oper.
Robert E. Andreatta, Chief Acct. Officer/Controller
Stephen G. Juelsgaard, Chief Compliance Officer
Patrick Y. Yang, Exec. VP-Prod. Oper.
Arthur D. Levinson, Chmn.

| **Phone:** 650-225-1000 | **Fax:** 650-225-6000 |
|---|---|
| **Toll-Free:** | |
| **Address:** 1 DNA Way, South San Francisco, CA 94080 US | |

## GROWTH PLANS/SPECIAL FEATURES:

Genentech, Inc. makes medicines by splicing genes into fast-growing bacteria that then produce therapeutic proteins and combat diseases on a molecular level. Genentech uses cutting-edge technologies such as computer visualization of molecules, micro arrays and sensitive assaying techniques to develop, manufacture and market pharmaceuticals for unmet medical needs. Genentech's research is directed toward the oncology, immunology and vascular biology fields. The company's products consist of a variety of cardio-centric medications, as well as cancer, growth hormone deficiency (GHD) and cystic fibrosis treatments. Biotechnology products offered by Genentech include Herceptin, used to treat metastatic breast cancers; Avastin, used to inhibit angiogenesis of solid-tumor cancers; Nutropin, a growth hormone for the treatment of GHD in children and adults; TNKase, for the treatment of acute myocardial infarction; and Pulmozyme, for the treatment of cystic fibrosis. The company also produces the Rituxan antibody, used for the treatment of patients with non-Hodgkin's lymphoma. Through its long-standing Genentech Access to Care Foundation, Genentech assists those without sufficient health insurance to receive its medicines. In 2008, sales to Genentech's three major distributors, AmerisourceBergen, McKesson and Cardinal Health, represented 86% of its total U.S. net product sales. There are three manufacturing sites in California, with an additional facility planned for 2010 licensure in Hillsboro, Oregon. In addition, it expects FDA licensure of a bulk drug substance manufacturing site in Singapore in 2010. The firm recently completed the acquisition of Tanox, a firm that focuses on monoclonal antibody technology and development partner for its Xolair asthma product. In March 2009, Roche Group completed its acquisition of the company.

For the last ten years, the company has been named to Fortune Magazine's 100 Best Companies to Work For. Every Friday evening, Genentech hosts socials called Ho-Hos, providing free food, beverages and a chance to socialize with co-workers.

## FINANCIALS: Sales and profits are in thousands of dollars—add 000 to get the full amount. 2008 Note: Financial information for 2008 was not available for all companies at press time.

| | | |
|---|---|---|
| 2008 Sales: $13,418,000 | 2008 Profits: $3,427,000 | **U.S. Stock Ticker: Subsidiary** |
| 2007 Sales: $11,724,000 | 2007 Profits: $2,769,000 | **Int'l Ticker:** Int'l Exchange: |
| 2006 Sales: $9,284,000 | 2006 Profits: $2,113,000 | Employees: 11,186 |
| 2005 Sales: $6,633,372 | 2005 Profits: $1,278,991 | Fiscal Year Ends: 12/31 |
| 2004 Sales: $4,621,157 | 2004 Profits: $784,816 | Parent Company: ROCHE GROUP |

## SALARIES/BENEFITS:

| Pension Plan: | ESOP Stock Plan: | Profit Sharing: | Top Exec. Salary: $995,000 | Bonus: $2,725,000 |
|---|---|---|---|---|
| Savings Plan: Y | Stock Purch. Plan: Y | | Second Exec. Salary: $503,833 | Bonus: $920,000 |

## OTHER THOUGHTS:

**Apparent Women Officers or Directors:** 2
**Hot Spot for Advancement for Women/Minorities:** Y

## LOCATIONS: ("Y" = Yes)

| West: | Southwest: | Midwest: | Southeast: | Northeast: | International: |
|---|---|---|---|---|---|
| Y | | | | | Y |

Note: Financial information, benefits and other data can change quickly and may vary from those stated here.

# GENEREX BIOTECHNOLOGY

**www.generex.com**

**Industry Group Code: 325412A  Ranks within this company's industry group: Sales: 28   Profits: 24**

| Drugs: | | Other: | Clinical: | Computers: | Services: | |
|---|---|---|---|---|---|---|
| Discovery: | Y | AgriBio: | Trials/Services: | Hardware: | Specialty Services: | |
| Licensing: | | Genetic Data: | Labs: | Software: | Consulting: | |
| Manufacturing: | | Tissue Replacement: | Equipment/Supplies: | Arrays: | Blood Collection: | |
| Genetics: | | | Research & Development Services: | Database Management: | Drug Delivery: | Y |
| | | | Diagnostics: | | Drug Distribution: | |

## TYPES OF BUSINESS:

Drug Delivery Systems
Buccal Drug Delivery Systems
Diabetes Treatment-Insulin
Infectious & Autoimmune Disease Treatments
Large-Molecule Drug Delivery Systems

## BRANDS/DIVISIONS/AFFILIATES:

RapidMist
Oral-Lyn
Glucose RapidSpray
Antigen Express, Inc.
BaBOOM! Energy Spray
Crave-NX 7-Day Diet Aid Spray

## CONTACTS: *Note: Officers with more than one job title may be intentionally listed here more than once.*

Anna E. Gluskin, CEO
Rose C. Perri, COO
Anna E. Gluskin, Pres.
Rose C. Perri, CFO/Corp. Sec.
Gerald Bernstein, VP-Medical Affairs
Mark Fletcher, General Counsel/Exec. VP
William D. Abajian, Sr. Exec. Advisor-Global Bus. Dev. & Alliances
Rose C. Perri, Treas.
Eric von Hofe, VP/Pres., Antigen Express
Anna E. Gluskin, Chmn.

| **Phone:** 416-364-2551 | **Fax:** 416-364-9363 |
|---|---|
| **Toll-Free:** | |
| **Address:** 33 Harbour Square, Ste. 202, Toronto, ON M5J 2G2 Canada | |

## GROWTH PLANS/SPECIAL FEATURES:

Generex Biotechnology Corporation primarily researches and develops drug delivery systems and technologies. It currently focuses on formulations that use its RapidMist hand-held aerosol applicator to administer large-molecule drugs through the buccal mucosa, primarily the inner cheek walls. For instance, Oral-Lyn is a formulation of human insulin designed for buccal delivery; it is commercially approved in Ecuador and is undergoing Phase III studies in the U.S. and Canada. The firm's Glucose RapidSpray product offers a fat-free, low-calorie glucose supplement for those who require extra glucose in their diet. The company has also begun development on buccal formulations of morphine and of the synthetic opioid analgesic fentanyl, which is around 80 times stronger than morphine. However, these projects are on hold until the late stage trials for its oral insulin products are completed for the U.S., Canada and Europe. In general, these oral formulations are designed to ease patients onto medication without invasive procedures, such as injections, and with the added benefit of being self-administered. Generex also owns Antigen Express, Inc., a Massachusetts-based company that develops treatments, primarily vaccine formulations, for malignant, infectious, autoimmune and allergic diseases. Diseases currently being studied include melanoma, breast cancer, prostate cancer, HIV, influenza, avian influenza (or bird flu) and Type-1 diabetes. Antigen's product candidates are in pre-clinical stages of development. Additional Generex products include BaBOOM! Energy Spray, designed to enhance energy levels for sports, work, travel and overall fatigue, with ingredients including glucose, caffeine, ginseng and Vitamins B and C. In February 2009, the company launched Crave-NX 7-Day Diet Aid Spray, designed to help control cravings and support customers' weight loss efforts. In May 2009, Generex announced the commercial launch of Oral-Lyn in Lebanon, where the product has been approved by the Ministry of Public Health for the treatment of Type-1 and Type-2 diabetes.

## FINANCIALS: Sales and profits are in thousands of dollars—add 000 to get the full amount. 2008 Note: Financial information for 2008 was not available for all companies at press time.

| | | |
|---|---|---|
| 2008 Sales: $ 125 | 2008 Profits: $-36,229 | **U.S. Stock Ticker: GNBT** |
| 2007 Sales: $ 180 | 2007 Profits: $-23,505 | **Int'l Ticker:**   Int'l Exchange: |
| 2006 Sales: $ 175 | 2006 Profits: $-67,967 | Employees:   27 |
| 2005 Sales: $ 392 | 2005 Profits: $-24,002 | Fiscal Year Ends: 7/31 |
| 2004 Sales: $ 627 | 2004 Profits: $-18,363 | Parent Company: |

## SALARIES/BENEFITS:

| Pension Plan: | ESOP Stock Plan: | Profit Sharing: | Top Exec. Salary: $514,583 | Bonus: $215,000 |
|---|---|---|---|---|
| Savings Plan: | Stock Purch. Plan: | | Second Exec. Salary: $411,667 | Bonus: $165,000 |

## OTHER THOUGHTS:

**Apparent Women Officers or Directors**: 3
**Hot Spot for Advancement for Women/Minorities**: Y

## LOCATIONS: ("Y" = Yes)

| West: | Southwest: | Midwest: | Southeast: | Northeast: | International: |
|---|---|---|---|---|---|
| | | | | Y | Y |

Note: Financial information, benefits and other data can change quickly and may vary from those stated here.

# GENOMIC HEALTH INC

## www.genomichealth.com

**Industry Group Code:** 325413 **Ranks within this company's industry group:** Sales: 11    Profits: 13

| Drugs: | Other: | | Clinical: | | Computers: | Services: | |
|---|---|---|---|---|---|---|---|
| Discovery: | AgriBio: | | Trials/Services: | | Hardware: | Specialty Services: | Y |
| Licensing: | Genetic Data: | Y | Labs: | Y | Software: | Consulting: | |
| Manufacturing: | Tissue Replacement: | | Equipment/Supplies: | | Arrays: | Blood Collection: | |
| Genetics: | | | Research & Development Services: | | Database Management: | Drug Delivery: | |
| | | | Diagnostics: | Y | | Drug Distribution: | |

## TYPES OF BUSINESS:

Genomic-Based Cancer Diagnostic Test Development

## BRANDS/DIVISIONS/AFFILIATES:

Oncotype DX

## CONTACTS: *Note: Officers with more than one job title may be intentionally listed here more than once.*

Kim Popovits, CEO
Brad Cole, COO
Kim Popovits, Pres.
Brad Cole, CFO
Tricia Tomlinson, Sr. VP-Human Resources
Joffre B. Baker, Chief Scientific Officer
Laura Leber, VP-Corp. Comm.
Steven Shak, Chief Medical Officer
Randal W. Scott, Chmn.
David Logan, Sr. VP-Worldwide Commercialization

| **Phone:** 650-556-9300 | **Fax:** 650-556-1132 |
|---|---|
| **Toll-Free:** 866-662-6897 | |
| **Address:** 301 Penobscot Dr., Redwood City, CA 94063 US | |

## GROWTH PLANS/SPECIAL FEATURES:

Genomic Health, Inc. focuses on the development and commercialization of genomic-based clinical diagnostic tests for cancer. Oncotype DX, its first and only marketed test product, is designed to help physicians determine the most effective treatment type for patients with node-negative, estrogen-receptor positive early stage breast cancer. Unlike comparable test assay systems that use fresh or frozen samples, Oncotype DX uses tissue samples that are chemically preserved and sealed in paraffin wax, thus allowing for easier handling and transportation. Hence, unlike other tests that may require a frozen sample shipped on dry ice, the Oncotype DX sample can be sent via regular overnight mail. All samples are processed at the firm's laboratory in Redwood City, California, and most physicians receive the results of their test within 10-14 days. Oncotype DX is offered as a laboratory service that tests 21 specific genes in a tumor sample to provide tumor-specific information (or the oncotype of the tumor), mainly to calculate the probability of recurrence and the potential efficacy of chemotherapy. This technology compares the genetic makeup of the tissue sample with archived information from other cancer patients, in order to facilitate a correlation between the patient's condition and known clinical outcomes. It expresses the resulting profile as a single quantitative score, called the Recurrence Score, which ranges from 0-100. Higher Recurrence Scores indicate a more aggressive tumor, lower scores indicate less aggressive tumors. The company is also developing tests to detect colon, prostate, melanoma, renal and non-small cell lung cancers. Since its inception in January 2004, Oncotype DX has been used in over 100,000 tests.

Genomic employees receiver medical, dental, vision, and disability coverage; a 401(k) plan; reimbursement accounts; and discounted gym memberships.

## FINANCIALS: Sales and profits are in thousands of dollars—add 000 to get the full amount. 2008 Note: Financial information for 2008 was not available for all companies at press time.

| | | |
|---|---|---|
| 2008 Sales: $110,579 | 2008 Profits: $-16,089 | **U.S. Stock Ticker:** GHDX |
| 2007 Sales: $64,027 | 2007 Profits: $-27,292 | **Int'l Ticker:**     Int'l Exchange: |
| 2006 Sales: $29,174 | 2006 Profits: $-28,920 | Employees:   288 |
| 2005 Sales: $5,202 | 2005 Profits: $-31,361 | Fiscal Year Ends: 12/31 |
| 2004 Sales: $ | 2004 Profits: $ | Parent Company: |

## SALARIES/BENEFITS:

| Pension Plan: | ESOP Stock Plan: | Profit Sharing: | Top Exec. Salary: $380,000 | Bonus: $35,350 |
|---|---|---|---|---|
| Savings Plan: Y | Stock Purch. Plan: Y | | Second Exec. Salary: $350,000 | Bonus: $32,550 |

## OTHER THOUGHTS:

**Apparent Women Officers or Directors:** 4
**Hot Spot for Advancement for Women/Minorities:** Y

## LOCATIONS: ("Y" = Yes)

| West: | Southwest: | Midwest: | Southeast: | Northeast: | International: |
|---|---|---|---|---|---|
| Y | | | | | |

# GEN-PROBE INC

www.gen-probe.com

**Industry Group Code: 325413 Ranks within this company's industry group:** Sales: 4    Profits: 4

| Drugs: | Other: | Clinical: | | Computers: | Services: |
|---|---|---|---|---|---|
| Discovery: | AgriBio: | Trials/Services: | | Hardware: | Specialty Services: |
| Licensing: | Genetic Data: | Labs: | | Software: | Consulting: |
| Manufacturing: | Tissue Replacement: | Equipment/Supplies: | | Arrays: | Blood Collection: |
| Genetics: | | Research & Development Services: | | Database Management: | Drug Delivery: |
| | | Diagnostics: | Y | | Drug Distribution: |

## TYPES OF BUSINESS:

Medical Diagnostics Products
Diagnostic Tests
Blood Screening Assays
Services-Commercial Physical Research
Services-Commercial Biological Research

## BRANDS/DIVISIONS/AFFILIATES:

TIGRIS
Tepnel Life Sciences, plc

## CONTACTS: Note: Officers with more than one job title may be intentionally listed here more than once.

Carl W. Hull, CEO
Carl W. Hull, Pres.
Herm Rosenman, CFO/Sr. VP-Finance
Stephen J. Kondor, Sr. VP-Mktg. & Sales
Diana De Walt, Sr. VP-Human Resources
Daniel L. Kacian, Chief Scientist/Exec. VP
Brad Phillips, CIO/VP
R. William Bowen, General Counsel/Sr. VP/Sec.
Jorgine Ellerbrock, Sr. VP-Oper.
Paul Gargan, Sr. VP-Bus. Dev.
Michael Watts, VP-Corp. Comm.
Michael Watts, VP-Investor Rel.
Kevin Herde, VP-Finance/Corp. Controller
Eric Lai, Sr. VP-R&D
Eric Tardif, Sr. VP-Corp. Strategy
Christina Yang, Sr. VP-Clinical, Regulatory & Quality
Tammy J. Brach, VP-Program Mgmt.
Henry L. Nordhoff, Chmn.
Brian B. Hansen, VP-North American Sales

| Phone: 858-410-8000 | Fax: 800-288-3141 |
|---|---|
| Toll-Free: 800-523-5001 | |
| Address: 10210 Genetic Center Dr., San Diego, CA 92121-4362 US | |

## GROWTH PLANS/SPECIAL FEATURES:

Gen-Probe, Inc. develops, manufactures and markets nucleic acid testing (NAT) products for clinical diagnosis of human diseases and screening of human blood donations. Gen-Probe has received U.S. Food and Drug Administration (FDA) approval for over 60 products to detect infectious microorganisms causing sexually transmitted diseases, tuberculosis, strep throat, pneumonia and fungal infections. It has developed and commercialized one of the first fully automated, integrated, high-throughput NAT instrument systems, the TIGRIS instrument.   The company also developed, and now manufactures, the only FDA-approved blood screening assay for the simultaneous detection of HIV-1 and HCV (hepatitis C virus), presently marketed by Novartis Corporation and used by blood collection agencies like the American Red Cross and America's Blood Centers. The firm has also designed and developed, often with outside vendors, a range of instruments for use with its assays.   Its clinical diagnostic products are marketed to clinical laboratories, public health institutions and hospitals in the U.S., Canada and certain countries in Europe.   Gen-Probe's blood screening products are marketed and distributed worldwide by Novartis.   In addition, the firm has agreements with Siemens, bioMerieux and Fujirebio to market products in various overseas markets. The company is currently developing NAT assays and instruments for the detection of harmful pathogens in the environment and biopharmaceutical and beverage manufacturing processes. In April 2009, Gen-Probe acquired Tepnel Life Sciences plc, a U.K.-based molecular diagnostics and pharmaceutical services company.

Gen-Probe employee benefits include income protection, an employee assistance plan, flexible spending accounts and an on-site cafeteria.   In addition, the company has a tuition assistance program, many in-house training opportunities and the Gen-Probe E-Learning program, through which employees can take over 100 online courses to develop skills at their own pace.

## FINANCIALS: Sales and profits are in thousands of dollars—add 000 to get the full amount. 2008 Note: Financial information for 2008 was not available for all companies at press time.

| | | |
|---|---|---|
| 2008 Sales: $472,695 | 2008 Profits: $59,498 | **U.S. Stock Ticker: GPRO** |
| 2007 Sales: $403,014 | 2007 Profits: $86,140 | **Int'l Ticker:**    Int'l Exchange: |
| 2006 Sales: $354,764 | 2006 Profits: $59,498 | Employees:   991 |
| 2005 Sales: $305,965 | 2005 Profits: $60,089 | Fiscal Year Ends: 12/31 |
| 2004 Sales: $269,707 | 2004 Profits: $54,575 | Parent Company: |

## SALARIES/BENEFITS:

| Pension Plan: | ESOP Stock Plan: | Profit Sharing: | Top Exec. Salary: $710,000 | Bonus: $798,750 |
|---|---|---|---|---|
| Savings Plan: Y | Stock Purch. Plan: Y | | Second Exec. Salary: $482,293 | Bonus: $441,788 |

## OTHER THOUGHTS:

**Apparent Women Officers or Directors:** 6
**Hot Spot for Advancement for Women/Minorities:** Y

## LOCATIONS: ("Y" = Yes)

| West: | Southwest: | Midwest: | Southeast: | Northeast: | International: |
|---|---|---|---|---|---|
| Y | | | | | |

Note: Financial information, benefits and other data can change quickly and may vary from those stated here.

# GENTA INC

**www.genta.com**

**Industry Group Code: 325412  Ranks within this company's industry group:  Sales: 147    Profits: 172**

| Drugs: | | Other: | Clinical: | Computers: | Services: |
|---|---|---|---|---|---|
| Discovery: | Y | AgriBio: | Trials/Services: | Hardware: | Specialty Services: |
| Licensing: | | Genetic Data: | Labs: | Software: | Consulting: |
| Manufacturing: | | Tissue Replacement: | Equipment/Supplies: | Arrays: | Blood Collection: |
| Genetics: | | | Research & Development Services: | Database Management: | Drug Delivery: |
| | | | Diagnostics: | | Drug Distribution: |

## TYPES OF BUSINESS:

Anticancer & Related Diseases Drugs
Antisense Drugs

## BRANDS/DIVISIONS/AFFILIATES:

Genasense
Ganite
Tesetaxel

## CONTACTS: *Note: Officers with more than one job title may be intentionally listed here more than once.*

Raymond P. Warrell, Jr., CEO
W. Lloyd Sanders, COO/Sr. VP
Gary Siegel, VP-Finance
Loretta M. Itri, Chief Medical Officer/Pres., Pharmaceutical Dev.
Raymond P. Warrell, Jr., Chmn.
Michael M. Yoshitsu, VP-Global Bus. Dev.

| | |
|---|---|
| **Phone:** 908-286-9800 | **Fax:** 908-464-1701 |
| **Toll-Free:** | |
| **Address:** 200 Connell Dr., Berkeley Heights, NJ 07922 US | |

## GROWTH PLANS/SPECIAL FEATURES:

Genta, Inc. is a biopharmaceutical company engaged in pharmaceutical research and development of drugs for the treatment of cancer and related diseases. The company's research portfolio consists of two major programs: DNA/RNA medicines and small molecules. The DNA/RNA medicines program includes drugs that are based on using modifications of either DNA or RNA as drugs that can be used to treat disease. The program includes technologies such as antisense, decoys and small interfering or micro RNAs. The lead drug from this program is an investigational antisense compound known as Genasense. Genasense is designed to block the production of Bcl-2, a protein that fortifies cancer cells against treatment, and is being developed primarily as a means of amplifying the cytotoxic effects of other anticancer treatments. The small molecules program includes drugs that are based on gallium-containing compounds. The lead drug from this program is Ganite, which is FDA approved for the treatment of patients with symptomatic cancer-related hypercalcemia that is resistant to hydration. The firm is engaged in developing new formulations of gallium-containing compounds that may be orally absorbed. The company is also developing Tesetaxel, which is an oral version of the prominent anti-cancer taxane, currently available only in intravenous forms. Genta owns 65 patents, and has 66 pending patents applications in the U.S. and foreign countries. In December 2008, the FDA declined to approve Genasense for the use of chronic lymphocytic leukemia, citing the need for an additional clinical study. Also in December 2008, the FDA granted Tesetaxel orphan drug status.

Genta offers a 401(k) plan; medical, dental and life insurance; and a stock options plan.

## FINANCIALS:  Sales and profits are in thousands of dollars—add 000 to get the full amount. 2008 Note: Financial information for 2008 was not available for all companies at press time.

| | | |
|---|---|---|
| 2008 Sales: $ 363 | 2008 Profits: $-505,838 | **U.S. Stock Ticker: GNTAD** |
| 2007 Sales: $ 580 | 2007 Profits: $-23,320 | **Int'l Ticker:**    Int'l Exchange: |
| 2006 Sales: $ 708 | 2006 Profits: $-56,781 | Employees:   25 |
| 2005 Sales: $26,585 | 2005 Profits: $-2,203 | Fiscal Year Ends: 12/31 |
| 2004 Sales: $14,615 | 2004 Profits: $-32,685 | Parent Company: |

## SALARIES/BENEFITS:

| | | | | |
|---|---|---|---|---|
| Pension Plan: | ESOP Stock Plan: | Profit Sharing: | Top Exec. Salary: $467,500 | Bonus: $ |
| Savings Plan: Y | Stock Purch. Plan: | | Second Exec. Salary: $409,662 | Bonus: $ |

## OTHER THOUGHTS:

**Apparent Women Officers or Directors:** 1
**Hot Spot for Advancement for Women/Minorities:** Y

## LOCATIONS: ("Y" = Yes)

| West: | Southwest: | Midwest: | Southeast: | Northeast: | International: |
|---|---|---|---|---|---|
| | | | | Y | |

# GENVEC INC

**www.genvec.com**

**Industry Group Code: 325412  Ranks within this company's industry group:** Sales: 108  Profits: 118

| Drugs: | | Other: | | Clinical: | Computers: | Services: |
|---|---|---|---|---|---|---|
| Discovery: | Y | AgriBio: | | Trials/Services: | Hardware: | Specialty Services: |
| Licensing: | | Genetic Data: | Y | Labs: | Software: | Consulting: |
| Manufacturing: | Y | Tissue Replacement: | | Equipment/Supplies: | Arrays: | Blood Collection: |
| Genetics: | | | | Research & Development Services: | Database Management: | Drug Delivery: |
| | | | | Diagnostics: | | Drug Distribution: |

## TYPES OF BUSINESS:

Gene-Based Therapeutic Drugs & Vaccines
Cancer Drugs

## BRANDS/DIVISIONS/AFFILIATES:

TNFerade
AdPEDF
TherAtoh

## CONTACTS: Note: Officers with more than one job title may be intentionally listed here more than once.

Paul H. Fischer, CEO
Paul H. Fischer, Pres.
Douglas J. Swirsky, CFO/Sr. VP
C. Richter King, Sr. VP-Research
Mark O. Thornton, Sr. VP-Prod. Dev.
Douglas J. Swirsky, Corp. Sec.
Douglas J. Swirsky, Treas.
Milan Kovacevic, VP-Clinical Oper.
Bryan T. Butman, Sr. VP-Vector Oper.
Zola P. Horovitz, Chmn.

**Phone:** 240-632-0740   **Fax:** 240-632-0735
**Toll-Free:** 877-943-6832
**Address:** 65 W. Watkins Mill Rd., Gaithersburg, MD 20878 US

## GROWTH PLANS/SPECIAL FEATURES:

GenVec, Inc. is a biopharmaceutical company that develops gene-based therapeutic drugs and vaccines. The company's lead product candidate, TNFerade biologic, is being developed for use in the treatment of cancer. The drug is an adenovector, or DNA carrier, and is administered directly into tumors. TNFerade is currently the subject of Phase III clinical trials to assess the effectiveness of using TNFerade in combination with standard care treatment for patients in advanced stages of pancreatic cancer. TNFerade is also in Phase I/II trials for the treatment of rectal cancer and melanoma, and in Phase I trials for the treatment of head and neck cancer. GenVec is also developing TherAtoh, a therapy for delivering the human atonal gene to trigger the production of therapeutic proteins by cells in the inner ear. GenVec and its collaborators also have multiple vaccines in development, which use the firm's adenovector technology. The company has collaborations with the National Institute of Allergy and Infectious Diseases (NIAID) to develop vaccines for HIV and influenza viruses; the U.S. Naval Medical Research Center and the PATH Malaria Vaccine Initiative to develop vaccines for malaria; and with the U.S. Department of Homeland Security and the U.S. Department of Agriculture to develop a vaccine for food-and-mouth diseases. The firm has access to over 274 issued, allowed or pending patents worldwide. In January 2009, the company eliminated 22 positions in order to cut costs.

The company offers its employees medical, dental and vision insurance; life, AD&D, short- and long-term disability insurance; a 401(k); an employee assistance program; tuition reimbursement; and an employee stock purchase program. Additional benefits include car washes, dry cleaning services and credit union membership.

## FINANCIALS: Sales and profits are in thousands of dollars—add 000 to get the full amount. 2008 Note: Financial information for 2008 was not available for all companies at press time.

| | | |
|---|---|---|
| 2008 Sales: $15,121 | 2008 Profits: $-26,063 | **U.S. Stock Ticker: GNVC** |
| 2007 Sales: $14,047 | 2007 Profits: $-18,708 | **Int'l Ticker:** Int'l Exchange: |
| 2006 Sales: $18,923 | 2006 Profits: $-19,272 | Employees: 98 |
| 2005 Sales: $26,554 | 2005 Profits: $-13,992 | Fiscal Year Ends: 12/31 |
| 2004 Sales: $11,853 | 2004 Profits: $-18,894 | Parent Company: |

## SALARIES/BENEFITS:

| Pension Plan: | ESOP Stock Plan: | Profit Sharing: | Top Exec. Salary: $410,000 | Bonus: $ |
|---|---|---|---|---|
| Savings Plan: Y | Stock Purch. Plan: Y | | Second Exec. Salary: $304,950 | Bonus: $ |

## OTHER THOUGHTS:

**Apparent Women Officers or Directors:**
**Hot Spot for Advancement for Women/Minorities:**

## LOCATIONS: ("Y" = Yes)

| West: | Southwest: | Midwest: | Southeast: | Northeast: | International: |
|---|---|---|---|---|---|
| | | | | Y | |

# GENZYME BIOSURGERY

## www.genzymebiosurgery.com

**Industry Group Code: 33911  Ranks within this company's industry group:**  Sales:      Profits:

| Drugs: | Other: | Clinical: | | Computers: | Services: |
|--------|--------|-----------|---|-----------|-----------|
| Discovery: | AgriBio: | Trials/Services: | | Hardware: | Specialty Services: |
| Licensing: | Genetic Data: | Labs: | | Software: | Consulting: |
| Manufacturing: | Tissue Replacement: Y | Equipment/Supplies: | Y | Arrays: | Blood Collection: |
| Genetics: | | Research & Development Services: | | Database Management: | Drug Delivery: |
| | | Diagnostics: | | | Drug Distribution: |

## TYPES OF BUSINESS:

Equipment-Surgery & Orthopedic Products
Burn Treatment Products
Biomaterials & Biotherapeutics

## BRANDS/DIVISIONS/AFFILIATES:

Genzyme Corp.
Synvisc
Seprafilm
Seprapack
Sepragel Sinus
Carticel
Epicel
HyluMed

## CONTACTS: Note: *Officers with more than one job title may be intentionally listed here more than once.*

C. Ann Merrifield, Pres.
Caren Arnstein, VP-Corp. Comm.
Henri A. Termeer, Chmn./CEO/Pres., Genzyme Corp.

| Phone: 617-252-7500 | Fax:  617-252-7368 |
|---------------------|---------------------|
| Toll-Free: | |
| Address:  55 Cambridge Pkwy., Cambridge, MA 02142 US | |

## GROWTH PLANS/SPECIAL FEATURES:

Genzyme Biosurgery (GB), a division of Genzyme Corp., develops and markets a portfolio of devices, biomaterials and biotherapeutics primarily for the general surgery market. GB's products are used for osteoarthritis relief, adhesion prevention, hernia repair, cartilage repair, burn treatment and bulk sodium hyaluronate (HA) powder.  Synvisc, a treatment for the pain and immobility associated with osteoarthritis of the knee, is an elastic viscous fluid that acts as a shock absorber and lubricant for the knee joint when injected. Genzyme's adhesion prevention products include Seprafilm adhesion barrier, used in abdominal and pelvic surgery; Sepramesh Biosurgical Composite, used in hernia repair surgery; and Seprapack and Sepragel Sinus, which are used after nasal or sinus surgery.  Carticel, which is used to treat damaged articular cartilage in the knee, is an injection of a patient's cultured cartilage cells into a knee that has had an inadequate response to a prior arthroscopic or other surgical repair procedure.  Epicel, made from cultured epidermal autografts grown from the patient's own skin cells in approximately 16 days, is a permanent skin replacement product for severe burn victims.  HyluMed is GB's sterile and medical grade HA powder, which is useful in helping the body adapt to medical implants and to accept medications. GB's latest product, Synvisc, is a joint lubrication designed to ease knee pain due to osteoarthritis; the product is administered via injection.   The company markets its products directly to physicians and hospital administrators throughout the U.S. and Europe.  It also uses Genzyme Corp.'s network of distributors to sell certain products in the U.S., Europe, Asia and Latin America.

## FINANCIALS:  Sales and profits are in thousands of dollars—add 000 to get the full amount. 2008 Note: Financial information for 2008 was not available for all companies at press time.

| | | |
|---|---|---|
| 2008 Sales: $ | 2008 Profits: $ | **U.S. Stock Ticker: Subsidiary** |
| 2007 Sales: $ | 2007 Profits: $ | **Int'l Ticker:**    Int'l Exchange: |
| 2006 Sales: $ | 2006 Profits: $ | Employees:  3,800 |
| 2005 Sales: $ | 2005 Profits: $ | Fiscal Year Ends: 12/31 |
| 2004 Sales: $ | 2004 Profits: $ | Parent Company: GENZYME CORP |

## SALARIES/BENEFITS:

| | | | | |
|---|---|---|---|---|
| Pension Plan: | ESOP Stock Plan: | Profit Sharing: | Top Exec. Salary: $1,300,000 | Bonus: $1,770,000 |
| Savings Plan: Y | Stock Purch. Plan: Y | | Second Exec. Salary: $625,000 | Bonus: $455,000 |

## OTHER THOUGHTS:

**Apparent Women Officers or Directors:** 2
**Hot Spot for Advancement for Women/Minorities:** Y

## LOCATIONS: ("Y" = Yes)

| West: | Southwest: | Midwest: | Southeast: | Northeast: | International: |
|-------|-----------|----------|-----------|-----------|---------------|
| | | | | Y | |

# GENZYME CORP

www.genzyme.com

**Industry Group Code: 325412 Ranks within this company's industry group: Sales: 25   Profits: 27**

| Drugs: | | Other: | | Clinical: | | Computers: | | Services: | |
|---|---|---|---|---|---|---|---|---|---|
| Discovery: | Y | AgriBio: | | Trials/Services: | | Hardware: | | Specialty Services: | |
| Licensing: | | Genetic Data: | Y | Labs: | | Software: | | Consulting: | |
| Manufacturing: | Y | Tissue Replacement: | | Equipment/Supplies: | Y | Arrays: | | Blood Collection: | |
| Genetics: | | | | Research & Development Services: | Y | Database Management: | | Drug Delivery: | |
| | | | | Diagnostics: | Y | | | Drug Distribution: | |

## TYPES OF BUSINESS:

Pharmaceuticals Discovery & Development
Genetic Disease Treatments
Surgical Products
Diagnostic Products
Genetic Testing Services
Oncology Products
Biomaterials
Medical Devices

## BRANDS/DIVISIONS/AFFILIATES:

Renagel
Cerezyme
Fabrazyme
Mozobil
Thyrogen
MACI
Clolar
Synvisc-One

## CONTACTS: *Note: Officers with more than one job title may be intentionally listed here more than once.*

Henri A. Termeer, CEO
Henri A. Termeer, Pres.
Michael S. Wyzga, CFO
Zoltan Csimma, Chief Human Resources Officer/Sr. VP
Alan E. Smith, Chief Scientific Officer/Sr. VP-Research
Thomas J. DesRosier, Chief Legal Officer/General Counsel/Sr. VP
Mark R. Bamforth, Sr. VP-Corp. Oper. & Pharmaceuticals
Richard H. Douglas, Sr. VP-Corp. Dev.
Mary McGrane, Sr. VP-Gov't Rel.
Michael S. Wyzga, Exec. VP-Finance
Mark J. Enyedy, Sr. VP/Pres., Oncology & Multiple Sclerosis
John Butler, Sr. VP/Pres., Cardiometabolic & Renal
C. Ann Merrifield, Sr. VP/Pres., Genzyme & Biosurgery
Donald E. Pogorzelski, VP/Pres., Diagnostic Prod.
Henri A. Termeer, Chmn.
Sandford D. Smith, Pres., Int'l Group

| Phone: 617-252-7500 | Fax: 617-252-7600 |
|---|---|
| Toll-Free: | |
| Address: 500 Kendall St., Cambridge, MA 02142 US | |

## GROWTH PLANS/SPECIAL FEATURES:

Genzyme Corporation is a major biotech drug manufacturer operating through four major units: Genetic Diseases; Cardiometabolic and Renal; Biosurgery; and Hematologic Oncology. The Genetic Diseases segment develops therapeutic products to treat patients suffering from genetic and other chronic debilitating diseases including lysosomal disorders (LSDs). Products include Cerezyme, an enzyme replacement treatment for Type 1 Gaucher disease and Fabrazyme for Fabry disease. The Cardiometabolic and Renal unit produces products for patients suffering from renal disease, including chronic renal failure, and endocrine and cardiovascular diseases. Products include Renagel, a calcium-free, metal-free phosphate binder for patients with Chronic Kidney disease on hemodialysis; and Thyrogen, an injection used as a diagnostic in follow-up screenings of cancer patients with thyroid cancer. The Biosurgery division develops biotherapeutics and biomaterial-based products for the orthopaedics sector and broader surgical areas. Its main products are Synvisc, a lubricant and pain reducer for the knee joint in patients with osteoarthritic knees; and the MACI implant, which uses the patient's own culture cartilage cells to repair a damaged knee joint. The Hematologic Oncology segment focuses on the treatment of cancer. Its main products consists of Mozobil, an injection used to mobilize stem cells in the blood stream for collection and for transplantation in patients with non-Hodgkin's lymphoma and multiple myeloma; and Clolar, a treatment for children with acute lymphoblastic leukemia. In February 2009, the FDA approved Synvisc-One, an injection designed for pain relief in osteoarthritic knees.

Employees are offered medical and dental insurance; life and dependent life insurance; business travel accident insurance; flexible spending accounts; short-and long-term disability coverage; a 401(k) plan; a stock purchase plan; financial education programs; tuition reimbursement; a college savings plan; and an employee discount program.

## FINANCIALS: Sales and profits are in thousands of dollars—add 000 to get the full amount. 2008 Note: Financial information for 2008 was not available for all companies at press time.

| | | |
|---|---|---|
| 2008 Sales: $4,605,039 | 2008 Profits: $421,081 | **U.S. Stock Ticker: GENZ** |
| 2007 Sales: $3,813,519 | 2007 Profits: $480,193 | **Int'l Ticker:**   Int'l Exchange: |
| 2006 Sales: $3,187,013 | 2006 Profits: $-16,797 | Employees: 11,000 |
| 2005 Sales: $2,734,842 | 2005 Profits: $441,489 | Fiscal Year Ends: 12/31 |
| 2004 Sales: $2,201,145 | 2004 Profits: $86,527 | Parent Company: |

## SALARIES/BENEFITS:

| Pension Plan: | ESOP Stock Plan: | Profit Sharing: | Top Exec. Salary: $1,578,514 | Bonus: $1,962,725 |
|---|---|---|---|---|
| Savings Plan: Y | Stock Purch. Plan: Y | | Second Exec. Salary: $734,331 | Bonus: $489,500 |

## OTHER THOUGHTS:

**Apparent Women Officers or Directors: 6**
**Hot Spot for Advancement for Women/Minorities: Y**

## LOCATIONS: ("Y" = Yes)

| West: | Southwest: | Midwest: | Southeast: | Northeast: | International: |
|---|---|---|---|---|---|
| Y | Y | Y | Y | Y | Y |

Note: Financial information, benefits and other data can change quickly and may vary from those stated here.

# GENZYME ONCOLOGY

www.genzymeoncology.com

**Industry Group Code: 325412  Ranks within this company's industry group:**  Sales:    Profits:

| Drugs: | | Other: | | Clinical: | Computers: | Services: |
|---|---|---|---|---|---|---|
| Discovery: | Y | AgriBio: | | Trials/Services: | Hardware: | Specialty Services: |
| Licensing: | | Genetic Data: | Y | Labs: | Software: | Consulting: |
| Manufacturing: | Y | Tissue Replacement: | | Equipment/Supplies: | Arrays: | Blood Collection: |
| Genetics: | | | | Research & Development Services: | Database Management: | Drug Delivery: |
| | | | | Diagnostics: | | Drug Distribution: |

## TYPES OF BUSINESS:

Drugs-Gene-Based
Cancer Vaccines
Angiogenesis Inhibitors
Drug Discovery Platforms

## BRANDS/DIVISIONS/AFFILIATES:

Genzyme Corp.
SAGE
Tasidotin Hydrochloride
Tumor Endothelial Marker (TEM)
Campath
Clolar
ILEX Oncology, Inc.
DENSPM Melan-AMART

## CONTACTS: Note: Officers with more than one job title may be intentionally listed here more than once.

Mark J. Enyedy, Pres.
Terry L. Murdock, Sr. VP/Gen. Mgr.-Prod.
Frederic J. Vinick, Sr VP-Drug Discovery

| Phone: 617-761-8777 | Fax: 617-761-8918 |
|---|---|
| Toll-Free: | |
| Address: 55 Cambridge Pkwy., Cambridge, MA 02142 US | |

## GROWTH PLANS/SPECIAL FEATURES:

Genzyme Oncology, a division of Genzyme Corp., creates cancer vaccines and angiogenesis inhibitors through the integration of its genomics, gene and cell therapy; small-molecule drug discovery; and protein therapeutic capabilities. The company's research consists of Tasidotin Hydrochloride, GC1008 and Topoisomeric I Inhibitor. Tasidotin Hydrochloride is a synthetic dolastatin analog currently being studied in Phase II metastatic melanoma, non-small cell lung cancer and prostate cancer clinical trials. GC1008 is a human antibody which combats all forms of Transforming Growth Factor, which is itself responsible for a patient's inability to produce an immune response to cancer tumors. The Topoisomeric I Inhibitor, currently in preclinical development, prevents Topoisomeric I enzymes from replicating DNA. The company's acquisition of ILEX Oncology, Inc. gave the firm its marketed cancer drugs Campath and Clolar. Campath is indicated for the treatment of B-cell chronic lymphocytic leukemia, while Clolar is used for the treatment of children with refractory or relapsed acute lymphoblastic leukemia. The company is collaborating with researchers at Johns Hopkins University in its attempt to expand its portfolio of TEMs by applying SAGE technology to endothelial cell samples. SAGE facilitates rapid, accurate analysis of gene expression patterns with the ability to identify previously unknown genes, tumor antigens and anti-angiogenic factors, analyzing the effects of drugs on human tissue and gaining insight into disease pathways. Genzyme Oncology licenses SAGE to pharmaceutical, biotechnology and genomics companies. The firm has a number of products in the pipeline predominantly focused on antibody and small molecule therapies. Genzyme is working to develop therapeutics that target antiangiogenesis, cell death, proliferation and metastasis.

## FINANCIALS: Sales and profits are in thousands of dollars—add 000 to get the full amount. 2008 Note: Financial information for 2008 was not available for all companies at press time.

| | | |
|---|---|---|
| 2008 Sales: $ | 2008 Profits: $ | U.S. Stock Ticker: Subsidiary |
| 2007 Sales: $ | 2007 Profits: $ | Int'l Ticker:    Int'l Exchange: |
| 2006 Sales: $ | 2006 Profits: $ | Employees: 5,200 |
| 2005 Sales: $ | 2005 Profits: $ | Fiscal Year Ends: 12/31 |
| 2004 Sales: $ | 2004 Profits: $ | Parent Company: GENZYME CORP |

## SALARIES/BENEFITS:

| Pension Plan: | ESOP Stock Plan: | Profit Sharing: | Top Exec. Salary: $ | Bonus: $ |
|---|---|---|---|---|
| Savings Plan: Y | Stock Purch. Plan: Y | | Second Exec. Salary: $ | Bonus: $ |

## OTHER THOUGHTS:

**Apparent Women Officers or Directors:**
**Hot Spot for Advancement for Women/Minorities:**

## LOCATIONS: ("Y" = Yes)

| West: | Southwest: | Midwest: | Southeast: | Northeast: | International: |
|---|---|---|---|---|---|
| | | | | Y | |

# GERON CORPORATION

www.geron.com

**Industry Group Code: 325412  Ranks within this company's industry group: Sales: 138  Profits: 145**

| Drugs: | | Other: | | Clinical: | Computers: | Services: | |
|---|---|---|---|---|---|---|---|
| Discovery: | Y | AgriBio: | | Trials/Services: | Hardware: | Specialty Services: | |
| Licensing: | Y | Genetic Data: | | Labs: | Software: | Consulting: | |
| Manufacturing: | | Tissue Replacement: | Y | Equipment/Supplies: | Arrays: | Blood Collection: | |
| Genetics: | | | | Research & Development Services: | Database Management: | Drug Delivery: | |
| | | | | Diagnostics: | | Drug Distribution: | |

## TYPES OF BUSINESS:

Drug Discovery & Development
Telomerase Technologies
Human Stem Cell Technologies

## BRANDS/DIVISIONS/AFFILIATES:

Start Licensing, Inc.
GRN163L
GRNVAC1
GRNOPC1
GRNCM1

## CONTACTS: Note: Officers with more than one job title may be intentionally listed here more than once.

Thomas B. Okarma, CEO
Thomas B. Okarma, Pres.
David L. Greenwood, CFO/Exec. VP
Calvin B. Harley, Chief Scientific Officer
David L. Greenwood, Sec.
David J. Earp, Sr. VP-Bus. Dev./Chief Patent Counsel
David L. Greenwood, Treas./Sec.
Fabio M. Benedetti, Sr. VP-Oncology
Jane S. Lebkowski, Sr. VP-Regenerative Medicine
Melissa A. Kelly Behrs, Sr. VP-Therapeutic Dev. & Oncology
David J. Earp, Chief Patent Counsel
Alexander E. Barkas, Chmn.

| Phone: 650-473-7700 | Fax: 650-473-7750 |
|---|---|
| Toll-Free: | |
| Address: 230 Constitution Dr., Menlo Park, CA 94025 US | |

## GROWTH PLANS/SPECIAL FEATURES:

Geron Corp. develops biopharmaceuticals for the treatment of cancer and chronic degenerative diseases such as spinal cord injuries, heart failure and diabetes. The company's therapies are based on telomerase and human embryonic stem cell technologies. The company is advancing telomerase targeted therapies, which enable cell division, protect chromosomes from degradation and act as molecular clocks for cellular aging. The enzyme telomerase restores telomere length, which shortens as cells multiply, extending a cell's ability to replicate. The company seeks to use human embryonic stem cells (hESC) as a potential source for manufacturing replacement cells and tissues for organ repair. The firm is now testing hESC-derived therapeutic cell types for oncology and regenerative applications. The company has two oncology candidates in clinical trials: GRN163L, a telomerase inhibitor drug, in patients with advanced non-small cell lung cancer; and GRNVAC1, a developmental vaccine for leukemia. In regenerative medicine, the Geron is testing GRNOPC1, a targeted treatment for spinal cord injuries, and has multiple others in developmental stages, including GRNCM1, a treatment for patients with myocardial disease; GRNIC1 for diabetes, Osteoblasts and Chondrocytes for osteoporosis; and research towards a better understanding of hESC treatments in humans. The firm owns or licenses over 170 U.S. and 340 foreign patents, with more than 360 pending applications worldwide. Geron formed Start Licensing, Inc., a joint venture with Exeter Life Sciences, to manage and license a portfolio of animal reproductive and cloning technologies. In August 2008, Start Licensing merged with Viagen Inc., an Exeter Life Sciences subsidiary. In January 2009, Geron received FDA clearance to test GRNOPC1 for acute spinal cord injuries, marking the world's first human clinical trials using embryonic stem cells.

Geron Corp. offers its employees medical, dental, vision, life and AD&D insurance; short- and long-term disability; 401(k) plan; and employee stock purchase plan; and flexible spending accounts.

## FINANCIALS: Sales and profits are in thousands of dollars—add 000 to get the full amount. 2008 Note: Financial information for 2008 was not available for all companies at press time.

| | | |
|---|---|---|
| 2008 Sales: $2,803 | 2008 Profits: $-62,021 | **U.S. Stock Ticker: GERN** |
| 2007 Sales: $7,622 | 2007 Profits: $-36,697 | **Int'l Ticker:** Int'l Exchange: |
| 2006 Sales: $3,277 | 2006 Profits: $-31,365 | Employees: 159 |
| 2005 Sales: $6,158 | 2005 Profits: $-33,689 | Fiscal Year Ends: 12/31 |
| 2004 Sales: $1,053 | 2004 Profits: $-79,558 | Parent Company: |

## SALARIES/BENEFITS:

| Pension Plan: | ESOP Stock Plan: | Profit Sharing: | Top Exec. Salary: $535,000 | Bonus: $280,900 |
|---|---|---|---|---|
| Savings Plan: Y | Stock Purch. Plan: Y | | Second Exec. Salary: $415,000 | Bonus: $163,400 |

## OTHER THOUGHTS:

**Apparent Women Officers or Directors**: 2
**Hot Spot for Advancement for Women/Minorities**: Y

## LOCATIONS: ("Y" = Yes)

| West: | Southwest: | Midwest: | Southeast: | Northeast: | International: |
|---|---|---|---|---|---|
| Y | | | | | |

Note: Financial information, benefits and other data can change quickly and may vary from those stated here.

# GILEAD SCIENCES INC

**www.gilead.com**

Industry Group Code: 325412  Ranks within this company's industry group: Sales: 23    Profits: 16

| Drugs: | | Other: | | Clinical: | | Computers: | | Services: | |
|---|---|---|---|---|---|---|---|---|---|
| Discovery: | Y | AgriBio: | | Trials/Services: | | Hardware: | | Specialty Services: | |
| Licensing: | Y | Genetic Data: | | Labs: | | Software: | | Consulting: | |
| Manufacturing: | Y | Tissue Replacement: | | Equipment/Supplies: | | Arrays: | | Blood Collection: | |
| Genetics: | | | | Research & Development Services: | Y | Database Management: | | Drug Delivery: | |
| | | | | Diagnostics: | | | | Drug Distribution: | |

## TYPES OF BUSINESS:

Viral & Bacterial Infections Drugs
Respiratory & Cardiopulmonary Diseases Drugs

## BRANDS/DIVISIONS/AFFILIATES:

Vistide
Truvada
Emtriva
Atripla
Hepsera
Viread
Lexiscan
Ranexa

## CONTACTS: Note: Officers with more than one job title may be intentionally listed here more than once.

John C. Martin, CEO
John F. Milligan, COO
John F. Milligan, Pres.
Robin Washington, CFO/Sr. VP
Kristen M. Metza, Sr. VP-Human Resources
Norbert W. Bischofberger, Chief Scientific Officer/Exec. VP-R&D
Anthony D. Caracciolo, Sr. VP-Mfg.
Gregg H. Alton, General Counsel/Sr. VP
Anthony D. Caracciolo, Sr. VP-Oper.
John Toole, Sr. VP-Corp. Dev.
Kevin Young, Exec. VP-Commercial Oper.
A. Bruce Montgomery, Sr. VP/Head-Respiratory Therapeutics
Seigo Izump, Sr. VP-Cardiovascular Therapeutics
William A. Lee, Sr. VP-Research
John C. Martin, Chmn.
Paul Carter, Sr. VP-Int'l Commercial Oper.

| Phone: 650-574-3000 | Fax: 650-578-9264 |
|---|---|
| Toll-Free: 800-445-3235 | |
| Address: 333 Lakeside Dr., Foster City, CA 94404 US | |

## GROWTH PLANS/SPECIAL FEATURES:

Gilead Sciences, Inc. is a biopharmaceutical company that discovers, develops and commercializes therapeutics for the treatment of life-threatening diseases such as viral and bacterial infections.  The company expanded its efforts to include respiratory and cardiopulmonary diseases.  The firm maintains research, development, manufacturing, sales and marketing facilities in the U.S., Europe and Australia and operates marketing subsidiaries in another 12 countries. Gilead currently has 12 products on the market: Viread, Truvada and Emtriva, which are oral medicines used as part of a combination therapy to treat HIV; Atripla, an oral formulation for treatment of HIV; Hespera, an oral medication used for treatment of Hepatitis B; AmBisome, an antifungal agent to treat serious invasive fungal infections; Vistide, an antiviral medication for the treatment of cytomegalovirus retinitis in patients with AIDS; Ranexa, a treatment for chronic angina; Letairis, for the treatment of pulmonary arterial hypertension; and Lexiscan, which is used as a pharmacologic stress agent.  The firm also has a number of products in development, including treatments for cystic fibrosis, hypertension, heart failure, hepatitis C, HIV/AIDS, and others.  The company also derives revenues from licensing agreements for Macugen, a macular degeneration treatment developed by OSI Pharmaceuticals, Inc.; and Tamiflu, an influenza medication sold by F. Hoffman-LaRoche.  In August 2008, Gilead entered into an agreement with Merck & Co., Inc. to distribute Atripla in 12 countries, mostly in Latin America and Asia.  Also in August 2008, the FDA approved Viread for treatment of chronic hepatitis B in adults.  In April 2009, the firm acquired CV Therapeutics.

The company offers its employees medical, vision, dental, life and AD&D insurance; short- and long-term disability coverage; a 401(k) plan; a stock purchase plan; an employee assistance plan; and tuition reimbursement.

## FINANCIALS:  Sales and profits are in thousands of dollars—add 000 to get the full amount. 2008 Note: Financial information for 2008 was not available for all companies at press time.

| | | |
|---|---|---|
| 2008 Sales: $5,335,750 | 2008 Profits: $2,011,154 | **U.S. Stock Ticker: GILD** |
| 2007 Sales: $4,230,045 | 2007 Profits: $1,615,298 | **Int'l Ticker:**    Int'l Exchange: |
| 2006 Sales: $3,026,139 | 2006 Profits: $-1,189,957 | Employees:  3,441 |
| 2005 Sales: $2,028,400 | 2005 Profits: $813,914 | Fiscal Year Ends: 12/31 |
| 2004 Sales: $1,324,621 | 2004 Profits: $449,371 | Parent Company: |

## SALARIES/BENEFITS:

| | | | | |
|---|---|---|---|---|
| Pension Plan: | ESOP Stock Plan: | Profit Sharing: | Top Exec. Salary: $1,146,261 | Bonus: $1,651,650 |
| Savings Plan: Y | Stock Purch. Plan: Y | | Second Exec. Salary: $727,988 | Bonus: $693,589 |

## OTHER THOUGHTS:

Apparent Women Officers or Directors: 4
Hot Spot for Advancement for Women/Minorities: Y

## LOCATIONS: ("Y" = Yes)

| West: | Southwest: | Midwest: | Southeast: | Northeast: | International: |
|---|---|---|---|---|---|
| Y | | | | Y | Y |

---

Note: Financial information, benefits and other data can change quickly and may vary from those stated here.

# GLAXOSMITHKLINE PLC

**www.gsk.com**

**Industry Group Code:** 325412 **Ranks within this company's industry group:** Sales: 6 Profits: 2

| Drugs: | | Other: | Clinical: | Computers: | Services: |
|---|---|---|---|---|---|
| Discovery: | Y | AgriBio: | Trials/Services: | Hardware: | Specialty Services: |
| Licensing: | | Genetic Data: | Labs: | Software: | Consulting: |
| Manufacturing: | Y | Tissue Replacement: | Equipment/Supplies: | Arrays: | Blood Collection: |
| Genetics: | | | Research & Development Services: | Database Management: | Drug Delivery: |
| | | | Diagnostics: | | Drug Distribution: |

## TYPES OF BUSINESS:

Prescription Medications
Asthma Drugs
Respiratory Drugs
Antibiotics
Antivirals
Dermatological Drugs
Over-the-Counter & Nutritional Products

## BRANDS/DIVISIONS/AFFILIATES:

Lanoxin
Lamictal
Adoair
Alvedon
Zantac
Nicorette
Ceravix
Genelabs Technologies

## CONTACTS: Note: Officers with more than one job title may be intentionally listed here more than once.

Andrew Witty, CEO
Julian Heslop, CFO
Daniel Phelan, Chief of Staff
Moncef Slaoui, Chmn.-R&D
Bill Louv, CIO
David Pulman, Pres., Global Mfg. & Supply
Simon Bicknell, Company Sec./Compliance Officer/Sr. VP
David Redfern, Chief Strategy Officer
Duncan Learmouth, Sr. VP-Corp. Comm. & Comm. Partnerships
Marc Dunoyer, Pres., Asia Pacific & Japan
Deirdre Connelly, Pres., North American Pharmaceuticals
Eddie Gray, Pres., Pharmaceuticals Europe
John Clarke, Pres., Consumer Healthcare
Christopher Gent, Chmn.

| | |
|---|---|
| **Phone:** 44-20-8047-5000 | **Fax:** 44-20-8047-7807 |
| **Toll-Free:** 888-825-5249 | |
| **Address:** 980 Great West Rd., Brentford, Middlesex, TW8 9GS UK | |

## GROWTH PLANS/SPECIAL FEATURES:

GlaxoSmithKline (GSK) is a leading research-based pharmaceutical company formed from the merger of Glaxo Wellcome and SmithKline Beecham. Its subsidiaries consist of global drug and health companies engaged in the creation, discovery, development, manufacturing and marketing of pharmaceuticals and other consumer health products. GSK operates in two segments: pharmaceuticals, and consumer health care. The pharmaceuticals segment includes prescription medications and vaccines. GSK designs prescription medications for the treatment many conditions including heart and circulatory conditions, cancer, and malaria. Major prescription medication approvals in 2008 included Lamictal and Adoair for use in Japan. GSK's vaccines are designed to treat life-threatening illnesses such as hepatitis A, diphtheria, influenza and bacterial meningitis. GSK received several drug approvals by the Food and Drug Administration in 2008 including Ceravix, a cervical cancer vaccine. The consumer health care division is divided into three segments: over-the-counter, oral healthcare and nutritional healthcare. Products from the consumer health care segment include over-the-counter medications such as Citrucel and Nicorette; oral care products such as Aquafresh; and nutritional products such as Boost. Research and development operations take place at 17 sites in four countries. Its research areas include neuroscience, oncology, infectious diseases, cardiovascular/metabolic respiratory, musculoskeletal/inflammation and gastrointestinal/urology. In 2008, GSK agreed to divest four products, Eltroxin, Lanoxin, Imuran and Zyloric, to Aspen Global Incorporated. GSK announced in October 2008 that it agreed to acquire the Egyptian products business of Bristol Myers Squibb and the U.S. company Genelabs Technologies. In November 2008, it entered into an agreement with AstraZeneca to acquire a number of leading over-the-counter (OTC) medicines predominantly sold in Sweden, including Alvedon.

## FINANCIALS: Sales and profits are in thousands of dollars—add 000 to get the full amount. 2008 Note: Financial information for 2008 was not available for all companies at press time.

| | | |
|---|---|---|
| 2008 Sales: $36,127,200 | 2008 Profits: $10,594,000 | **U.S. Stock Ticker: GSK** |
| 2007 Sales: $33,700,100 | 2007 Profits: $11,264,500 | **Int'l Ticker: GSK** Int'l Exchange: London-LSE |
| 2006 Sales: $45,595,800 | 2006 Profits: $10,793,000 | Employees: 110,000 |
| 2005 Sales: $37,783,631 | 2005 Profits: $8,400,952 | Fiscal Year Ends: 12/31 |
| 2004 Sales: $34,863,347 | 2004 Profits: $7,015,930 | Parent Company: |

## SALARIES/BENEFITS:

| | | | | |
|---|---|---|---|---|
| Pension Plan: Y | ESOP Stock Plan: Y | Profit Sharing: | Top Exec. Salary: $1,523,000 | Bonus: $2,250,000 |
| Savings Plan: Y | Stock Purch. Plan: | | Second Exec. Salary: $949,136 | Bonus: $ |

## OTHER THOUGHTS:

**Apparent Women Officers or Directors:** 3
**Hot Spot for Advancement for Women/Minorities:** Y

## LOCATIONS: ("Y" = Yes)

| West: | Southwest: | Midwest: | Southeast: | Northeast: | International: |
|---|---|---|---|---|---|
| Y | Y | Y | Y | Y | Y |

Note: Financial information, benefits and other data can change quickly and may vary from those stated here.

# GTC BIOTHERAPEUTICS INC

**www.gtc-bio.com**

Industry Group Code: 325414  Ranks within this company's industry group: Sales: 6    Profits: 6

| Drugs: | | Other: | | Clinical: | Computers: | Services: |
|---|---|---|---|---|---|---|
| Discovery: | Y | AgriBio: | Y | Trials/Services: | Hardware: | Specialty Services: |
| Licensing: | Y | Genetic Data: | Y | Labs: | Software: | Consulting: |
| Manufacturing: | Y | Tissue Replacement: | | Equipment/Supplies: | Arrays: | Blood Collection: |
| Genetics: | | | | Research & Development Services: | Database Management: | Drug Delivery: |
| | | | | Diagnostics: | | Drug Distribution: |

## TYPES OF BUSINESS:

Recombinant Proteins
Drugs-Anticoagulants
Transgenic Animals

## BRANDS/DIVISIONS/AFFILIATES:

Atryn
LFB Biotechnologies

## CONTACTS: *Note: Officers with more than one job title may be intentionally listed here more than once.*

Geoffrey F. Cox, CEO
Geoffrey F. Cox, Pres.
John B. Green, CFO
Harry M. Meade, Sr. VP-R&D
Daniel S. Woloshen, General Counsel/Sr. VP
Gregory Liposky, Sr. VP-Oper.
Ashley Lawton, VP-Bus. Dev.
Thomas E. Newberry, VP-Corp. Comm.
John B. Green, Sr. VP-Finance
Carol A. Ziomek, VP-Dev.
Suzanne Groet, VP-Therapeutic Protein Dev.
Richard A. Scotland, Sr. VP-Regulatory Affairs
Geoffrey F. Cox, Chmn.

| Phone: 508-620-9700 | Fax: 508-370-3797 |
|---|---|
| Toll-Free: | |
| Address: 175 Crossing Blvd., Framingham, MA 01702 US | |

## GROWTH PLANS/SPECIAL FEATURES:

GTC Biotherapeutics, Inc. (GTC) applies transgenic technology to develop recombinant proteins for human therapeutic uses. The company uses transgenic animals that express specific recombinant proteins in their milk. The firm generates transgenic animals through microinjection and nuclear transfer. GTC uses goats in most of its commercial development programs due to the relatively short gestation times and relatively high milk production volume of the animals. The company's leading product is ATryn for patients with hereditary antithrombin deficiency undergoing surgical procedures. Antithrombin is an important protein found in the bloodstream with anticoagulant and anti-inflammatory properties. The drug is in clinical trials for disseminated intravascular coagulation, which is an acquired deficiency of antithrombin that occurs in sepsis. The company is also using transgenic methods to produce monoclonal antibodies (MAbs), including potential therapeutic and follow-on biologics. This line of research could lead to treatments for cancer and autoimmune diseases. Other transgenic projects currently under development include a recombinant human coagulation factor for the treatment of hemophilia; a second recombinant human coagulation factor for type B hemophilia; and an elastase inhibitor to treat emphysema and several other respiratory disorders. The company works with LFB Biotechnologies to develop selected recombinant plasma proteins and monoclonal antibodies with GTC's protein process. In February 2009, the FDA approved ATryn for hereditary antithrombin deficient patients.

GTC offers its employees tuition reimbursement, heath, dental and life insurance; stock options; tuition reimbursement; and a 401(k) investment plan.

## FINANCIALS: Sales and profits are in thousands of dollars—add 000 to get the full amount. 2008 Note: Financial information for 2008 was not available for all companies at press time.

| | | | |
|---|---|---|---|
| 2008 Sales: $16,656 | 2008 Profits: $-22,665 | **U.S. Stock Ticker: GTCB** | |
| 2007 Sales: $13,896 | 2007 Profits: $-36,321 | **Int'l Ticker:** Int'l Exchange: | |
| 2006 Sales: $6,128 | 2006 Profits: $-33,345 | Employees: 159 | |
| 2005 Sales: $4,152 | 2005 Profits: $-30,112 | Fiscal Year Ends: 12/31 | |
| 2004 Sales: $6,626 | 2004 Profits: $-29,493 | Parent Company: | |

## SALARIES/BENEFITS:

| Pension Plan: | ESOP Stock Plan: | Profit Sharing: | Top Exec. Salary: $480,000 | Bonus: $44,447 |
|---|---|---|---|---|
| Savings Plan: Y | Stock Purch. Plan: Y | | Second Exec. Salary: $306,342 | Bonus: $20,673 |

## OTHER THOUGHTS:

**Apparent Women Officers or Directors:** 4
**Hot Spot for Advancement for Women/Minorities:** Y

## LOCATIONS: ("Y" = Yes)

| West: | Southwest: | Midwest: | Southeast: | Northeast: | International: |
|---|---|---|---|---|---|
| | | | | Y | |

# HARVARD BIOSCIENCE INC

### www.harvardbioscience.com

**Industry Group Code: 3345  Ranks within this company's industry group:  Sales: 4    Profits: 4**

| Drugs: | Other: | Clinical: | | Computers: | Services: |
|---|---|---|---|---|---|
| Discovery: | AgriBio: | Trials/Services: | | Hardware: | Specialty Services: |
| Licensing: | Genetic Data: | Labs: | | Software: | Consulting: |
| Manufacturing: | Tissue Replacement: | Equipment/Supplies: | Y | Arrays: | Blood Collection: |
| Genetics: | | Research & Development Services: | Y | Database Management: | Drug Delivery: |
| | | Diagnostics: | | | Drug Distribution: |

## TYPES OF BUSINESS:
Apparatus & Scientific Instruments

## BRANDS/DIVISIONS/AFFILIATES:
Hoefer, Inc.
BTX Molecular Delivery Systems
GE Healthcare
Biochrom Ltd.
GeneMachines
Warner Instruments

## CONTACTS: Note: Officers with more than one job title may be intentionally listed here more than once.
Chane Graziano, CEO
Susan Luscinski, COO
David Green, Pres.
Thomas McNaughton, CFO
Chane Graziano, Chmn.

| Phone: 508-893-8999 | Fax: 508-429-5732 |
|---|---|
| Toll-Free: 800-272-2775 | |
| Address: 84 October Hill Rd., Holliston, MA 01746 US | |

## GROWTH PLANS/SPECIAL FEATURES:

Harvard Bioscience, Inc. is a provider of life science research that specializes in broad product lines that are inexpensive and geared toward niche markets. Products are targeted toward two major areas of application: ADMET screening and molecular biology. ADMET screening is used to identify compounds that have toxic side effects or undesirable physiological or pharmacological properties. These pharmacological properties consist of absorption, distribution, metabolism and elimination. The company's products in this area include absorption diffusion chambers, 96 well equilibrium dialysis plate, organ testing systems, precision infusion pumps, cell injection systems, ventilators and electroporation products. The molecular biology products are mainly scientific instruments such as spectrophotometers and plate readers that analyze light to detect and quantify a wide range of molecular and cellular processes or apparatus such as gel electrophoresis units. These products can quantify the amount of DNA, RNA or protein in a sample; can use chromatography to separate the amino acids in a sample; and can use the method of gel electrophoresis to separate and purify DNA, RNA and proteins. Most of these molecular biology products are sold through GE Healthcare (formerly Amersham Biosciences). Harvard Bioscience operates though its 21 wholly-owned subsidiaries; which have the rights to numerous trade names and trade marks, notably Hoefer, Warner, BTX, Biochrom, and GeneMachines. Harvard Bioscience has manufacturing operations in the U.S., the U.K., Germany, Spain and Austria with sales facilities in France and Canada. The company sells its products to thousands of researches in over 100 countries primarily through its 900 page catalog, and secondarily through its web site and through other distributors. Customers primarily include research scientists at pharmaceutical and biotechnology companies, universities and government laboratories, including the U.S. National Institutes of Health.

## FINANCIALS: Sales and profits are in thousands of dollars—add 000 to get the full amount. 2008 Note: Financial information for 2008 was not available for all companies at press time.

| | | |
|---|---|---|
| 2008 Sales: $88,049 | 2008 Profits: $1,673 | **U.S. Stock Ticker: HBIO** |
| 2007 Sales: $83,407 | 2007 Profits: $-1,354 | **Int'l Ticker:**    Int'l Exchange: |
| 2006 Sales: $76,181 | 2006 Profits: $-2,341 | Employees:   315 |
| 2005 Sales: $67,431 | 2005 Profits: $-31,877 | Fiscal Year Ends: 12/31 |
| 2004 Sales: $64,745 | 2004 Profits: $-2,329 | Parent Company: |

## SALARIES/BENEFITS:

| Pension Plan: | ESOP Stock Plan: | Profit Sharing: | Top Exec. Salary: $535,500 | Bonus: $ |
|---|---|---|---|---|
| Savings Plan: Y | Stock Purch. Plan: Y | | Second Exec. Salary: $441,000 | Bonus: $ |

## OTHER THOUGHTS:
**Apparent Women Officers or Directors**: 1
**Hot Spot for Advancement for Women/Minorities:**

## LOCATIONS: ("Y" = Yes)

| West: | Southwest: | Midwest: | Southeast: | Northeast: | International: |
|---|---|---|---|---|---|
| Y | | | | Y | Y |

Note: Financial information, benefits and other data can change quickly and may vary from those stated here.

# HEMAGEN DIAGNOSTICS INC

www.hemagen.com

**Industry Group Code:** 325413 **Ranks within this company's industry group:** Sales: 20　Profits: 10

| Drugs: | | Other: | | Clinical: | | Computers: | | Services: | |
|---|---|---|---|---|---|---|---|---|---|
| Discovery: | | AgriBio: | | Trials/Services: | | Hardware: | | Specialty Services: | |
| Licensing: | Y | Genetic Data: | | Labs: | | Software: | | Consulting: | |
| Manufacturing: | Y | Tissue Replacement: | | Equipment/Supplies: | Y | Arrays: | | Blood Collection: | |
| Genetics: | | | | Research & Development Services: | | Database Management: | | Drug Delivery: | |
| | | | | Diagnostics: | Y | | | Drug Distribution: | |

## TYPES OF BUSINESS:

Medical Diagnostics Products
Clinical Chemical Analysis Products
Clinical Chemistry Reagents

## BRANDS/DIVISIONS/AFFILIATES:

Hemagen Diagnosticos Comercio, Ltd.
EasyLyte
EasySampler
Analyst
Virgo

## CONTACTS: Note: Officers with more than one job title may be intentionally listed here more than once.

William P. Hales, CEO
William P. Hales, Pres.
Catherine M. Davidson, CFO
Catherine M. Davidson, Controller
William P. Hales, Chmn.

| **Phone:** 443-367-5500 | **Fax:** 443-367-5527 |
|---|---|
| **Toll-Free:** 800-436-2436 | |
| **Address:** 9033 Red Branch Rd., Columbia, MD 21045 US | |

## GROWTH PLANS/SPECIAL FEATURES:

Hemagen Diagnostics, Inc. develops, manufactures and markets proprietary medical diagnostic test kits for both human and veterinary subjects. Hemagen has two different product lines. The Virgo product line of diagnostic test kits is used to aid in the diagnosis of certain autoimmune and infectious diseases, using enzyme-linked immunosorbent assay (ELISA), immunoflourescence and hemaglutination technology. In addition, Hemagen also manufactures and sells the Analyst, an FDA-cleared clinical chemistry analyzer used to measure important constituents in human and animal blood. Hemagen currently offers more than 150 test kits that have been cleared by the FDA for sale in the U.S. Brand-name products are sold directly to physicians, veterinarians, laboratories and blood banks within the U.S. Products are also made available through private-label agreements with global distributors of medical supplies. A network of international distributors represents the company's various product lines in overseas markets. In total, more than 1,000 customers use Hemagen products worldwide. In South America, Hemagen's operations are managed by the company's majority-owned Brazilian subsidiary, Hemagen Diagnosticos Comercio, Importacao e Exportacao, Ltd. Hemagen has increased its holding significantly by finalizing the purchase of this Brazilian subsidiary, which enables the company to prioritize expanding its South and Central American client base. Medical devices sold by Hemagen include the EasySampler for humans; and the Analyst III, Hematology Anaylzer, EasyLyte, and EasySampler units for veterinary use.

Hemagen Diagnostics, Inc. offers its employees a 401(k) plan; medical, dental and disability coverage; paid holidays and comprehensive leave; and an ESOP stock ownership plan.

## FINANCIALS: Sales and profits are in thousands of dollars—add 000 to get the full amount. 2008 Note: Financial information for 2008 was not available for all companies at press time.

| | | |
|---|---|---|
| 2008 Sales: $6,375 | 2008 Profits: $ 427 | **U.S. Stock Ticker: HMGN.OB** |
| 2007 Sales: $4,487 | 2007 Profits: $- 850 | **Int'l Ticker:**　Int'l Exchange: |
| 2006 Sales: $7,250 | 2006 Profits: $ 313 | Employees:　32 |
| 2005 Sales: $7,586 | 2005 Profits: $-1,337 | Fiscal Year Ends: 9/30 |
| 2004 Sales: $7,471 | 2004 Profits: $-3,599 | Parent Company: |

## SALARIES/BENEFITS:

| Pension Plan: | ESOP Stock Plan: Y | Profit Sharing: | Top Exec. Salary: $172,500 | Bonus: $ |
|---|---|---|---|---|
| Savings Plan: Y | Stock Purch. Plan: | | Second Exec. Salary: $115,000 | Bonus: $ |

## OTHER THOUGHTS:

**Apparent Women Officers or Directors:** 1
**Hot Spot for Advancement for Women/Minorities:**

## LOCATIONS: ("Y" = Yes)

| West: | Southwest: | Midwest: | Southeast: | Northeast: | International: |
|---|---|---|---|---|---|
| Y | | | | Y | Y |

Note: Financial information, benefits and other data can change quickly and may vary from those stated here.

# HEMISPHERX BIOPHARMA INC

## www.hemispherx.net

**Industry Group Code: 325412  Ranks within this company's industry group:** Sales: 153    Profits: 85

| Drugs: | | Other: | | Clinical: | Computers: | | Services: | |
|---|---|---|---|---|---|---|---|---|
| Discovery: | Y | AgriBio: | | Trials/Services: | Hardware: | | Specialty Services: | |
| Licensing: | | Genetic Data: | Y | Labs: | Software: | | Consulting: | |
| Manufacturing: | Y | Tissue Replacement: | | Equipment/Supplies: | Arrays: | | Blood Collection: | |
| Genetics: | | | | Research & Development Services: | Database Management: | | Drug Delivery: | |
| | | | | Diagnostics: | | | Drug Distribution: | |

## TYPES OF BUSINESS:

RNA-Related Drugs
HIV Treatments
Antivirals

## BRANDS/DIVISIONS/AFFILIATES:

Oragen
Ampligen
Alferon N
Alferon LDO

## CONTACTS: *Note: Officers with more than one job title may be intentionally listed here more than once.*

William A. Carter, CEO
William A. Carter, Pres.
Charles T. Bernhardt, CFO
Carol A. Smith, VP-Mfg. Quality Assurance
Ransom W. Etheridge, General Counsel
Wayne S. Springate, VP-Oper.
Carol A. Smith, VP-Process Dev.
David R. Strayer, Medical Dir.-Regulatory Affairs
Katalin Ferencz-Biro, Sr. VP-Regulatory Affairs
Russel Lander, VP-Quality Assurance
William A. Carter, Chmn.

| **Phone:** 215-988-0080 | **Fax:** 215-988-1739 |
|---|---|
| **Toll-Free:** | |
| **Address:** 1 Penn Ctr., 1617 JFK Blvd., 6th Fl., Philadelphia, PA 19103 US | |

## GROWTH PLANS/SPECIAL FEATURES:

Hemispherx Biopharma, Inc. is a biopharmaceutical company that develops and manufactures nucleic acid drugs for the treatment of viral and immune-based disorders. The company's proprietary drug technology uses specifically configured RNA. Hemispherx's flagship products include the antiviral/immunotherapeutic drugs Ampligen and Oragen, as well its Alferon N Injection. Ampligen is a synthetic double-stranded RNA configured to address a variety of chronic diseases and viral disorders such as HIV and chronic fatigue syndrome. Oragen drugs are being evaluated for use in treating chronic viral and immunological disorders. Alferon, in either its injection form (N) or its low dose oral form (LDO), is a purified, natural source alpha interferon (interferon alfa-n3, human leukocyte derived). Alferon N is also in development for treating avian flu, West Nile Virus and other diseases. Alferon LDO is currently in early stage development targeting influenza and viral diseases. In addition to these flagship drugs and projects, Hemispherx has a patent portfolio of over 50 patents issued worldwide. The company is engaged in ongoing experimental studies in partnership with Biovail; Defence R&D Canada; the St. Vincent's Hospital Clinical Trial Centre in Australia; Hollister-Stier Laboratories LLC; the National Institutes of Health in the U.S.; and the National Institute of Infectious Diseases in Tokyo, Japan. In November 2008, the firm entered into a contract with the Lovelace Respiratory Institute to expand research based on Ampligen's cellular functions.

## FINANCIALS: Sales and profits are in thousands of dollars—add 000 to get the full amount. 2008 Note: Financial information for 2008 was not available for all companies at press time.

| | | | |
|---|---|---|---|
| 2008 Sales: $ 265 | 2008 Profits: $-12,219 | **U.S. Stock Ticker:** HEB | |
| 2007 Sales: $1,059 | 2007 Profits: $-18,139 | **Int'l Ticker:** | **Int'l Exchange:** |
| 2006 Sales: $ 933 | 2006 Profits: $-19,399 | **Employees:** 32 | |
| 2005 Sales: $1,083 | 2005 Profits: $-13,213 | **Fiscal Year Ends:** 12/31 | |
| 2004 Sales: $1,229 | 2004 Profits: $-24,140 | **Parent Company:** | |

## SALARIES/BENEFITS:

| Pension Plan: | ESOP Stock Plan: | Profit Sharing: | Top Exec. Salary: $664,624 | Bonus: $ |
|---|---|---|---|---|
| Savings Plan: Y | Stock Purch. Plan: | | Second Exec. Salary: $259,164 | Bonus: $ |

## OTHER THOUGHTS:

**Apparent Women Officers or Directors:** 1
**Hot Spot for Advancement for Women/Minorities:** Y

## LOCATIONS: ("Y" = Yes)

| West: | Southwest: | Midwest: | Southeast: | Northeast: | International: |
|---|---|---|---|---|---|
| | | | | Y | Y |

Note: Financial information, benefits and other data can change quickly and may vary from those stated here.

# HESKA CORP

**www.heska.com**

Industry Group Code: 325412B  **Ranks within this company's industry group:**  Sales: 1   Profits: 1

| Drugs: | | Other: | | Clinical: | | Computers: | | Services: | |
|---|---|---|---|---|---|---|---|---|---|
| Discovery: | Y | AgriBio: | Y | Trials/Services: | | Hardware: | | Specialty Services: | |
| Licensing: | | Genetic Data: | | Labs: | Y | Software: | | Consulting: | |
| Manufacturing: | Y | Tissue Replacement: | | Equipment/Supplies: | Y | Arrays: | | Blood Collection: | |
| Genetics: | | | | Research & Development Services: | | Database Management: | | Drug Delivery: | |
| | | | | Diagnostics: | Y | | | Drug Distribution: | |

## TYPES OF BUSINESS:

Drugs-Animal Health & Pet Care
Veterinary Diagnostics Products & Services
Animal Dietary Supplements
Veterinary Vaccines

## BRANDS/DIVISIONS/AFFILIATES:

Core Companion Animal Health
HESKA Feline UltraNasal FVRCP Vaccine
i-STAT 1 Handheld Clinical Analyzer
DRI-CHEM Veterinary Chemistry Analyzer
HEMATRUE Veterinary Hematology Analyzer
VET/IV 2.2 Infusion Pump
ERD - Healthscreen Urine Test
ALLERCEPT

## CONTACTS:  *Note: Officers with more than one job title may be intentionally listed here more than once.*

Robert B. Grieve, CEO
Robert B. Grieve, Pres.
Jason A. Napolitano, CFO/Exec. VP
Nancy Wisnewski, VP-Prod. Dev.
John R. Flanders, General Counsel/VP/Corp. Sec.
Michael J. McGinley, VP-Global Oper.
Michael A. Bent, Controller/Principle Acct. Officer
Kenneth Schrader, VP-Bus. Oper.
G. Lynn Snodgrass, VP-Sales
Nancy Wisnewski, VP-Tech. Customer Svc.
Donald L. Wassom, Managing Dir.-Heska AG
Robert B. Grieve, Chmn.

| Phone: 970-493-7272 | Fax: 970-619-3003 |
|---|---|
| Toll-Free: 800-464-3752 | |
| Address: 3760 Rocky Mountain Ave., Loveland, CO 80538 US | |

## GROWTH PLANS/SPECIAL FEATURES:

Heska Corp. focuses on the discovery, development, marketing and support of animal health care products. The company uses biotechnology to create a broad range of diagnostic, therapeutic and vaccine products for dogs, cats and other animals.   The business is divided into two segments: Core Companion Animal Health (CCA) and Other Vaccines, Pharmaceuticals and Products (OVP), previously Diamond Animal Health.   The CCA segment includes diagnostic and monitoring instruments and supplies, single-use diagnostic tests, vaccines, pharmaceuticals and nutritional supplements, primarily for canine and feline use. Its line of veterinary diagnostic instruments includes the i-STAT 1 Handheld Clinical Analyzer, for electrolyte, blood gas, chemistry and basic hematology analysis; the DRI-CHEM Veterinary Chemistry Analyzer, which analyzes blood chemistry and electrolytes; the HEMATRUE Veterinary Hematology Analyzer, a blood analyzer focused on white blood cell count, red blood cell count, platelet count and hemoglobin levels; and the VET/IV 2.2 infusion pump for regulated infusion of fluids, drugs or nutritional products. Heska's diagnostic tests include tests for heartworms, allergy tests and early renal damage detection products.   The E.R.D.-Healthscreen Urine Tests can identify dogs and cats at risk for kidney disease before the majority of kidney function is lost.  The company also sells the HESKA Feline Ultranasal FVRCP Vaccine, a three-way modified live vaccine to prevent disease caused by respiratory viruses, and more general ALLERCEPT Allergy Treatment Sets for animals with positive allergy results.  The OVP business manufactures private label vaccines and pharmaceutical products that are marketed and distributed by third parties primarily for cattle, although some products are also manufactured for small mammals and fish.  In addition, OVP manufactures certain companion animal health products for marketing and sale by Heska.   Other products include heartworm prevention tablets, a fatty acid supplement and a chewable thyroid supplement.

## FINANCIALS:  Sales and profits are in thousands of dollars—add 000 to get the full amount. 2008 Note: Financial information for 2008 was not available for all companies at press time.

| | | |
|---|---|---|
| 2008 Sales: $81,653 | 2008 Profits: $- 850 | **U.S. Stock Ticker: HSKA** |
| 2007 Sales: $82,335 | 2007 Profits: $34,808 | Int'l Ticker:     Int'l Exchange: |
| 2006 Sales: $75,060 | 2006 Profits: $1,828 | Employees:   312 |
| 2005 Sales: $69,437 | 2005 Profits: $ 282 | Fiscal Year Ends: 12/31 |
| 2004 Sales: $67,691 | 2004 Profits: $-4,815 | Parent Company: |

## SALARIES/BENEFITS:

| | | | | |
|---|---|---|---|---|
| Pension Plan: | ESOP Stock Plan: | Profit Sharing: | Top Exec. Salary: $416,666 | Bonus: $ |
| Savings Plan: | Stock Purch. Plan: Y | | Second Exec. Salary: $241,263 | Bonus: $ |

## OTHER THOUGHTS:

**Apparent Women Officers or Directors:** 3
**Hot Spot for Advancement for Women/Minorities:** Y

## LOCATIONS: ("Y" = Yes)

| West: | Southwest: | Midwest: | Southeast: | Northeast: | International: |
|---|---|---|---|---|---|
| Y | | | | | Y |

# HI-TECH PHARMACAL CO INC

## www.hitechpharm.com

**Industry Group Code: 325412A Ranks within this company's industry group:** Sales: 16 Profits: 13

| Drugs: | | Other: | Clinical: | Computers: | Services: | |
|---|---|---|---|---|---|---|
| Discovery: | | AgriBio: | Trials/Services: | Hardware: | Specialty Services: | Y |
| Licensing: | | Genetic Data: | Labs: | Software: | Consulting: | |
| Manufacturing: | Y | Tissue Replacement: | Equipment/Supplies: | Arrays: | Blood Collection: | |
| Genetics: | Y | | Research & Development Services: | Database Management: | Drug Delivery: | |
| | | | Diagnostics: | | Drug Distribution: | |

## TYPES OF BUSINESS:

Drugs-Generic
Nutritional Products
Over-the-Counter Products
Ophthalmic Products
Manufacturing Contract Services
Inhalation Products
Diabetes Products
Generic Drugs

## BRANDS/DIVISIONS/AFFILIATES:

Diabetic Tussin
DiabetiSweet
DiabetDerm
Multi-Betic
Zostrix
Midlothian Laboratories, LLC
ECR Pharmaceuticals

## CONTACTS: *Note: Officers with more than one job title may be intentionally listed here more than once.*

David S. Seltzer, CEO
David S. Seltzer, Pres.
William Peters, CFO/VP
Edwin A. Berrios, VP-Mktg. & Sales
Polireddy Dondeti, Sr. Dir.-R&D
James P. Tracy, VP-Info. Systems
David S. Seltzer, Sec.
Eyal Mares, VP-Oper.
Christopher LoSardo, VP-Corp. Dev.
Margaret Santorufo, Controller/VP
Gary M. April, Pres., Health Care Prod. Div./Divisional VP-Sales
Joanne Curri, Dir.-Regulatory Affairs
Pudpong Poolsuk, Sr. Dir.-Science
Tanya Akimova, Dir.-New Bus. Dev.
David S. Seltzer, Chmn.

| Phone: 631-789-8228 | Fax: 631-789-8429 |
|---|---|
| Toll-Free: | |
| Address: 369 Bayview Ave., Amityville, NY 11701-2802 US | |

## GROWTH PLANS/SPECIAL FEATURES:

Hi-Tech Pharmacal Co., Inc. is a manufacturer and marketer of prescription, over-the-counter (OTC) and nutritional products that are sold in liquid and cream forms. A wide range of products are produced for various disease states, including asthma, bronchial disorders, dermatological disorders, allergies, pain, stomach, oral care, neurological disorders and other conditions. The firm divides its products into two main lines: generic and OTC brands. The generic products division primarily includes prescription items such as oral solutions and suspensions, topical creams and ointments as well as nasal sprays. This division also manufactures ophthalmic, optic and inhalation products and provides sterile manufacturing contract services. Hi-Tech's top five selling generic products are sulfamethoxazole and trimethoprim; promethazine products; chlorhexadine gluconate; pediatric multivitamins; and urea based creams, lotions, gels and nail sticks. The line was expanded due to the company's recent acquisition of Midlothian Labratories, LLC. The firm's OTC brands division, also named the health care products division, develops and markets a line of branded products primarily for people with diabetes, including Diabetic Tussin, a line of cough medications; DiabetiSweet, sugar substitutes that can be used for baking and cooking; Multi-betic, a daily multi-vitamin; and DiabetDerm, a diabetic skin care line. This division also sells Zostrix, a brand of capsaisin products for pain and arthritis. In 2008, sales of generic pharmaceuticals represented 81% of total sales and sales of the health care products line of over-the-counter products accounted for 19% of total sales. The company's customers consist of generic distributors, drug wholesalers, chain drug stores, mass merchandise chains, certain government agencies and mail-order pharmacies. Business has mainly depended upon the following customers: McKesson Corporation; Walgreens; Cardinal Health, Inc.; CVS; AmeriSourceBergen Corporation; CVS; and Wal-Mart. In March 2009, the Hi-Tech acquired the assets of ECR Pharmaceuticals, which specializes in allergy, dermal and analgesic products.

## FINANCIALS: Sales and profits are in thousands of dollars—add 000 to get the full amount. 2008 Note: Financial information for 2008 was not available for all companies at press time.

| | | |
|---|---|---|
| 2008 Sales: $62,017 | 2008 Profits: $-5,098 | **U.S. Stock Ticker:** HITK |
| 2007 Sales: $58,898 | 2007 Profits: $-2,036 | **Int'l Ticker:** Int'l Exchange: |
| 2006 Sales: $78,020 | 2006 Profits: $11,453 | Employees: 366 |
| 2005 Sales: $67,683 | 2005 Profits: $8,288 | Fiscal Year Ends: 4/30 |
| 2004 Sales: $56,366 | 2004 Profits: $6,592 | Parent Company: |

## SALARIES/BENEFITS:

| Pension Plan: | ESOP Stock Plan: | Profit Sharing: | Top Exec. Salary: $421,000 | Bonus: $ |
|---|---|---|---|---|
| Savings Plan: Y | Stock Purch. Plan: | | Second Exec. Salary: $237,000 | Bonus: $ |

## OTHER THOUGHTS:

**Apparent Women Officers or Directors:** 3
**Hot Spot for Advancement for Women/Minorities:** Y

## LOCATIONS: ("Y" = Yes)

| West: | Southwest: | Midwest: | Southeast: | Northeast: | International: |
|---|---|---|---|---|---|
| | | | | Y | |

Note: Financial information, benefits and other data can change quickly and may vary from those stated here.

# HOLLIS-EDEN PHARMACEUTICALS

**www.holliseden.com**

Industry Group Code: 325412  Ranks within this company's industry group: Sales:  Profits: 107

| Drugs: | | Other: | Clinical: | Computers: | Services: |
|---|---|---|---|---|---|
| Discovery: | Y | AgriBio: | Trials/Services: | Hardware: | Specialty Services: |
| Licensing: | | Genetic Data: | Labs: | Software: | Consulting: |
| Manufacturing: | | Tissue Replacement: | Equipment/Supplies: | Arrays: | Blood Collection: |
| Genetics: | | | Research & Development Services: | Database Management: | Drug Delivery: |
| | | | Diagnostics: | | Drug Distribution: |

## TYPES OF BUSINESS:

Drugs-Immune System Regulation
Drugs-Infectious Diseases
Drugs-Hormonal Imbalances

## BRANDS/DIVISIONS/AFFILIATES:

Triolex
Apoptone
Hormonal Signaling Technology Platform

## CONTACTS: Note: Officers with more than one job title may be intentionally listed here more than once.

James M. Frincke, Interim CEO
James M. Frincke, COO
Richard B. Hollis, Pres.
Robert L. Marsella, Sr. VP-Mktg.
Christopher L. Reading, Chief Scientific Officer
Robert L. Marsella, Sr. VP-Bus. Dev.
Robert W. Weber, Chief Acct. Officer/VP/Controller
Dwight R. Stickney, Chief Medical Officer
Salvatore J. Zizza, Chmn.

| Phone: 858-587-9333 | Fax: 858-558-6470 |
|---|---|
| Toll-Free: | |
| Address: 4435 Eastgate Mall, Ste. 400, San Diego, CA 92121 US | |

## GROWTH PLANS/SPECIAL FEATURES:

Hollis-Eden Pharmaceuticals, Inc., a development-stage pharmaceutical company, is engaged in the discovery, development and commercialization of products for the treatment of diseases and disorders against which the body is unable to mount an appropriate immune response. It has focused its initial technology efforts on a series of adrenal steroid hormones and hormone analogs, derived from its Hormonal Signaling Technology Platform. The compounds are safe, cost-effective to manufacture and unlikely to produce resistance. The firm is currently focused on the development of two clinical drug development candidates: Triolex (HE3286) and Apoptone (HE3235). Triolex is a next-generation compound currently in a multi-dose Phase I/II clinical trial for the treatment of metabolic and autoimmune disorders and is cleared for clinical trials for type II diabetes, rheumatoid arthritis and ulcerative colitis, with developments for cystic fibrosis in preclinical stages. Apoptone is a next-generation compound selected for clinical development for prostate cancer. In preclinical models of prostate and breast cancer, Apoptone has been shown to reduce the incidence, growth and progression of tumors. In February 2009, Hollis-Eden announced several cost-cutting measures, including layoffs of 33% of its workforce, salary freezes and bonus suspensions.

## FINANCIALS: Sales and profits are in thousands of dollars—add 000 to get the full amount. 2008 Note: Financial information for 2008 was not available for all companies at press time.

| | | |
|---|---|---|
| 2008 Sales: $ | 2008 Profits: $-21,565 | U.S. Stock Ticker: HEPH |
| 2007 Sales: $ 645 | 2007 Profits: $-23,121 | Int'l Ticker:  Int'l Exchange: |
| 2006 Sales: $ 444 | 2006 Profits: $-30,231 | Employees:  42 |
| 2005 Sales: $ 56 | 2005 Profits: $-29,441 | Fiscal Year Ends: 12/31 |
| 2004 Sales: $ 63 | 2004 Profits: $-24,757 | Parent Company: |

## SALARIES/BENEFITS:

| Pension Plan: | ESOP Stock Plan: | Profit Sharing: | Top Exec. Salary: $553,779 | Bonus: $ |
|---|---|---|---|---|
| Savings Plan: Y | Stock Purch. Plan: | | Second Exec. Salary: $371,394 | Bonus: $ |

## OTHER THOUGHTS:

Apparent Women Officers or Directors:
Hot Spot for Advancement for Women/Minorities:

## LOCATIONS: ("Y" = Yes)

| West: | Southwest: | Midwest: | Southeast: | Northeast: | International: |
|---|---|---|---|---|---|
| Y | | | | | |

# HOSPIRA INC

**www.hospira.com**

Industry Group Code: 33911  Ranks within this company's industry group: Sales: 4  Profits: 5

| Drugs: | | Other: | Clinical: | | Computers: | | Services: | |
|---|---|---|---|---|---|---|---|---|
| Discovery: | Y | AgriBio: | Trials/Services: | | Hardware: | Y | Specialty Services: | |
| Licensing: | Y | Genetic Data: | Labs: | | Software: | Y | Consulting: | |
| Manufacturing: | Y | Tissue Replacement: | Equipment/Supplies: | Y | Arrays: | | Blood Collection: | |
| Genetics: | | | Research & Development Services: | | Database Management: | | Drug Delivery: | |
| | | | Diagnostics: | | | | Drug Distribution: | |

## TYPES OF BUSINESS:

Pharmaceutical Development
Generic Pharmaceuticals
Medication Delivery Systems
Anesthetics
Injectable Medications
Diagnostic Imaging Agents
Contract Manufacturing

## BRANDS/DIVISIONS/AFFILIATES:

One 2 One
Sculptor Developmental Technologies
EndoTool

## CONTACTS: Note: Officers with more than one job title may be intentionally listed here more than once.

Christopher B. Begley, CEO
Terrence C. Kearney, COO
Thomas E. Werner, CFO
Ron Squarer, Sr. VP-Global Mktg.
Ken Meyers, Sr. VP-People Dev.
Sumant Ramachandra, Sr. VP-R&D/Chief Science Officer
Brian J. Smith, General Counsel/Sec./Sr. VP
Ron Squarer, Sr. VP-Corp. Dev.
Thomas E. Werner, Sr. VP-Finance
Ken Meyers, Sr. VP-Organizational Transformation
Sumant Ramachandra, Sr. VP-Medical Affairs
Christopher B. Begley, Chmn.

| | |
|---|---|
| **Phone:** 224-212-2000 | **Fax:** |
| **Toll-Free:** 877-946-7747 | |
| **Address:** 275 N. Field Dr., Lake Forest, IL 60045 US | |

## GROWTH PLANS/SPECIAL FEATURES:

Hospira, Inc. is a global specialty pharmaceutical and medication delivery company. The company's primary operations involve the research and development of generic pharmaceuticals, pharmaceuticals based on proprietary pharmaceuticals whose patents have expired. The company's activities include the development, manufacture and marketing of generic acute-care and oncology injectables, as well as integrated infusion therapy and medication management systems. Hospira divides its products into four categories: specialty injectable pharmaceuticals, other pharmaceuticals, medication management systems and other devices. The specialty injectable pharmaceuticals segment consists of approximately 200 injectable generic drugs; the proprietary sedation drug Precedex; and Retacrit, a biogeneric version of erythropoietin. The other pharmaceuticals segment consists primarily of large volume I.V. solutions and nutritional products. Through its One 2 One manufacturing services group, this segment also offers contract manufacturing services to proprietary pharmaceutical and biotechnology companies for formulation development, filling and finishing of injectable and oral drugs worldwide. The medication management systems segment's products include electronic drug delivery pumps, safety software, administration sets that are used to deliver I.V. fluids and medications and other related services. The other devices segment includes gravity administration sets; critical care devices; needlestick safety products; and other products. The firm operates under three reportable business segments: Americas; Europe, Middle East and Africa; and Asia Pacific. In February 2008, the company introduced an irinotecan hydrochloride injection for use with patients with colon or rectal cancer. In April 2008, Hospira acquired Sculptor Developmental Technologies, a software engineering company and subsidiary of St. Clair Health Corporation. In October 2008, the firm acquired the EndoTool glucose management system from MD Scientific LLC.

Hospira offers its employees medical, dental and vision plans, flexible spending accounts, tuition reimbursement, adoption assistance, an employee assistance program and a 401(k) plan.

## FINANCIALS: Sales and profits are in thousands of dollars—add 000 to get the full amount. 2008 Note: Financial information for 2008 was not available for all companies at press time.

| | | |
|---|---|---|
| 2008 Sales: $3,629,500 | 2008 Profits: $320,900 | **U.S. Stock Ticker:** HSP |
| 2007 Sales: $3,436,238 | 2007 Profits: $136,758 | **Int'l Ticker:**  Int'l Exchange: |
| 2006 Sales: $2,688,505 | 2006 Profits: $237,679 | Employees: 14,500 |
| 2005 Sales: $2,626,696 | 2005 Profits: $235,638 | Fiscal Year Ends: 12/31 |
| 2004 Sales: $2,645,036 | 2004 Profits: $301,552 | Parent Company: |

## SALARIES/BENEFITS:

| | | | | |
|---|---|---|---|---|
| Pension Plan: Y | ESOP Stock Plan: | Profit Sharing: | Top Exec. Salary: $1,029,231 | Bonus: $976,740 |
| Savings Plan: Y | Stock Purch. Plan: | | Second Exec. Salary: $613,461 | Bonus: $465,617 |

## OTHER THOUGHTS:

**Apparent Women Officers or Directors:** 2
**Hot Spot for Advancement for Women/Minorities:**

## LOCATIONS: ("Y" = Yes)

| West: | Southwest: | Midwest: | Southeast: | Northeast: | International: |
|---|---|---|---|---|---|
| Y | Y | Y | | Y | Y |

Note: Financial information, benefits and other data can change quickly and may vary from those stated here.

# HUMAN GENOME SCIENCES INC

www.hgsi.com

Industry Group Code: 325412  Ranks within this company's industry group: Sales: 84   Profits: 168

| Drugs: | | Other: | | Clinical: | Computers: | | Services: | |
|---|---|---|---|---|---|---|---|---|
| Discovery: | Y | AgriBio: | | Trials/Services: | Hardware: | | Specialty Services: | |
| Licensing: | | Genetic Data: | Y | Labs: | Software: | | Consulting: | |
| Manufacturing: | Y | Tissue Replacement: | | Equipment/Supplies: | Arrays: | | Blood Collection: | |
| Genetics: | | | | Research & Development Services: | Database Management: | | Drug Delivery: | |
| | | | | Diagnostics: | | | Drug Distribution: | |

## TYPES OF BUSINESS:

Oncology, Immunology & Infectious Diseases Drugs

## BRANDS/DIVISIONS/AFFILIATES:

LymphoStat-B
Albuferon
Abthrax
Darapladib
Syncria
TRAIL Receptor Antibodies

## CONTACTS: Note: Officers with more than one job title may be intentionally listed here more than once.

H. Thomas Watkins, CEO
H. Thomas Watkins, Pres.
Timothy C. Barabe, CFO/Sr. VP
Susan Bateson, Sr. VP-Human Resources
David C. Stump, Exec. VP-R&D
Joseph A. Morin, VP-Eng.
Randy Maddox, VP-Mfg. Oper.
James H. Davis, General Counsel/Exec. VP/Sec.
Curran M. Simpson, Sr. VP-Oper.
Sally D. Bolmer, Sr. VP-Dev. & Regulatory Affairs
Jerry Parrott, VP-Corp. Comm. & Public Policy
Barry A. Labinger, Chief Commercial Officer/Exec. VP
Ann L. Wang, VP-Clinical Oper.
Giles Gallant, VP-Clinical Research, Oncology
Daniel J. Odenheimer, VP-Clinical Research, Gen. Medicine
Argeris N. Karabelas, Chmn.

| Phone: 301-309-8504 | Fax: 301-309-8512 |
|---|---|
| Toll-Free: | |
| Address: 14200 Shady Grove Rd., Rockville, MD 20850 US | |

## GROWTH PLANS/SPECIAL FEATURES:

Human Genome Sciences, Inc. (HGS) is a commercially focused drug development company with five products in late-stage clinical development: Albuferon for chronic hepatitis, LymphoStat-B for systemic lupus erythematosus, ABthrax for anthrax disease, Darapladib for heart disease and Syncria for diabetes. The company also has a pipeline of novel compounds in earlier stages of clinical development in oncology and immunology. The firm's partners conduct clinical trials of additional drugs to treat cardiovascular, metabolic and central nervous system diseases and advanced a number of products derived from the company's technology to clinical development. The Albuferon collaborator is Novartis and the LymphoStat-B collaborator is GlaxoSmithKline. ABthrax is being developed under a contract with the U.S. government based on highly successfully efficacy studies. HGS began manufacturing on schedule and by January 2009, the company had delivered 20,000 doses to the Strategic National Stockpile. HGS's leading products still in early to mid development are based on its TRAIL receptor antibodies for cancer. In the areas of research and development, the company also specializes in human-antibody and albumin-fusion technology, which is buffeted by a collaboration agreement with Xencor, Inc. and their XmAb research. HGS has over 500 U.S. patents covering genes and proteins. In September 2008, HGS announced a process development and manufacturing agreement with Hospira, Inc., wherein Hospira products would be produced using excess HGS manufacturing resources. In March 2009, the company announced a collaboration agreement with Morphotek, Inc., to develop monoclonal antibodies for oncology and immunology.

The company offers its employees medical, dental and vision insurance; flexible spending accounts; life and AD&D insurance; short- and long-term disability; a 401(k) plan; health and dependent care flexible spending accounts; an employee stock purchase plan; education assistance; and ongoing training through employee development programs.

## FINANCIALS: Sales and profits are in thousands of dollars—add 000 to get the full amount. 2008 Note: Financial information for 2008 was not available for all companies at press time.

| | | |
|---|---|---|
| 2008 Sales: $48,422 | 2008 Profits: $-244,915 | **U.S. Stock Ticker: HGSI** |
| 2007 Sales: $41,851 | 2007 Profits: $-262,448 | **Int'l Ticker:** Int'l Exchange: |
| 2006 Sales: $25,755 | 2006 Profits: $-251,173 | Employees: 880 |
| 2005 Sales: $19,113 | 2005 Profits: $-239,439 | Fiscal Year Ends: 12/31 |
| 2004 Sales: $3,831 | 2004 Profits: $-242,898 | Parent Company: |

## SALARIES/BENEFITS:

| Pension Plan: | ESOP Stock Plan: Y | Profit Sharing: | Top Exec. Salary: $696,154 | Bonus: $450,000 |
|---|---|---|---|---|
| Savings Plan: Y | Stock Purch. Plan: Y | | Second Exec. Salary: $474,638 | Bonus: $190,000 |

## OTHER THOUGHTS:

Apparent Women Officers or Directors: 5
Hot Spot for Advancement for Women/Minorities: Y

## LOCATIONS: ("Y" = Yes)

| West: | Southwest: | Midwest: | Southeast: | Northeast: | International: |
|---|---|---|---|---|---|
| | | | | Y | |

# HYCOR BIOMEDICAL INC

www.hycorbiomedical.com

Industry Group Code: 325413 Ranks within this company's industry group: Sales: Profits:

| Drugs: | Other: | Clinical: | | Computers: | | Services: | |
|--------|--------|-----------|---|------------|---|-----------|---|
| Discovery: | AgriBio: | Trials/Services: | | Hardware: | Y | Specialty Services: | |
| Licensing: | Genetic Data: | Labs: | | Software: | | Consulting: | |
| Manufacturing: | Tissue Replacement: | Equipment/Supplies: | Y | Arrays: | | Blood Collection: | |
| Genetics: | | Research & Development Services: | | Database Management: | | Drug Delivery: | |
| | | Diagnostics: | Y | | | Drug Distribution: | |

## TYPES OF BUSINESS:

Medical Diagnostics Products
Allergy & Autoimmune Diagnostics
Urinalysis Products

## BRANDS/DIVISIONS/AFFILIATES:

Stratagene Corp.
Agilent Technologies, Inc.
Hycor Biomedical, Ltd.
Hycor Biomedical GmbH
KOVA Urinalysis
KOVA-Trol
KOVA Refractrol SP
HY-TEC

## CONTACTS: Note: Officers with more than one job title may be intentionally listed here more than once.

Nelson F. Thune, Gen. Mgr.
Cheryl Graham, Dir.-Human Resources
Nelson F. Thune, Sr. VP-Oper.
Vance Mitchell, Dir.-Oper.

| Phone: 714-933-3000 | Fax: 714-933-3222 |
|---|---|
| Toll-Free: 800-382-2527 | |
| Address: 7272 Chapman Ave., Garden Grove, CA 92841 US | |

## GROWTH PLANS/SPECIAL FEATURES:

Hycor Biomedical, Inc., a subsidiary of Agilent Technologies, researches, develops and manufactures medical diagnostic products for clinical laboratories and specialty physicians. The firm has a particular focus on allergy and autoimmune testing and urinalysis products. Hycor operates two foreign subsidiaries, Hycor Europe, based in the Netherlands, and Hycor Biomedical, Ltd., in the U.K. The company's KOVA family of urinalysis products include KOVA Urinalysis, the market leader in Standardized Microscopic Urinalysis; KOVA-Trol tri-level, lyophilized urine dipstick chemistry control, which is stable for 5-7 days at room temperature and for four months when frozen; KOVA Liqua-Trol, a ready-to-use bi-level liquid control, useable with most brands of urine chemistry dipsticks; and KOVA Refractrol SP, a tri-level control product with a 24-month shelf life designed for use with temperature compensated and non-compensated refractometers. Hycor's allergy diagnostic product line, which includes HY-TEC specific IgE tests, features radio-immunoassays and enzymatic immunoassays that test for reactions to more than 1,000 allergens. The company's autoimmune products division manufactures devices (branded as Autostat II) that test for disorders such as systemic lupus erythematosus and rheumatoid arthritis. Hycor also produces HY-TEC 288, an automated consolidated workstation capable of storing up to 50 allergy and 100 autoimmune samples, conducting 288 tests per run, providing bar-coded sample identification, high precision robotic liquid handling and real-time incubation control. The firm recently announced the release of the Extended Range Calibrator Set for the HY-TEC 288 Allergy System.

## FINANCIALS: Sales and profits are in thousands of dollars—add 000 to get the full amount. 2008 Note: Financial information for 2008 was not available for all companies at press time.

| | | | |
|---|---|---|---|
| 2008 Sales: $ | 2008 Profits: $ | U.S. Stock Ticker: Subsidiary | |
| 2007 Sales: $19,800 | 2007 Profits: $ | Int'l Ticker: Int'l Exchange: | |
| 2006 Sales: $ | 2006 Profits: $ | Employees: 100 | |
| 2005 Sales: $ | 2005 Profits: $ | Fiscal Year Ends: 12/31 | |
| 2004 Sales: $ | 2004 Profits: $ | Parent Company: AGILENT TECHNOLOGIES INC | |

## SALARIES/BENEFITS:

| Pension Plan: Y | ESOP Stock Plan: | Profit Sharing: | Top Exec. Salary: $255,131 | Bonus: $51,026 |
|---|---|---|---|---|
| Savings Plan: Y | Stock Purch. Plan: Y | | Second Exec. Salary: $237,083 | Bonus: $47,417 |

## OTHER THOUGHTS:

Apparent Women Officers or Directors: 1
Hot Spot for Advancement for Women/Minorities:

## LOCATIONS: ("Y" = Yes)

| West: | Southwest: | Midwest: | Southeast: | Northeast: | International: |
|-------|-----------|----------|-----------|-----------|---------------|
| Y | | | | | Y |

Note: Financial information, benefits and other data can change quickly and may vary from those stated here.

# ICON PLC

**www.iconclinical.com**

Industry Group Code: 541712  Ranks within this company's industry group: Sales: 5  Profits: 3

| Drugs: | Other: | Clinical: | | Computers: | | Services: | |
|---|---|---|---|---|---|---|---|
| Discovery: | AgriBio: | Trials/Services: | Y | Hardware: | Y | Specialty Services: | |
| Licensing: | Genetic Data: | Labs: | Y | Software: | | Consulting: | Y |
| Manufacturing: | Tissue Replacement: | Equipment/Supplies: | | Arrays: | | Blood Collection: | |
| Genetics: | | Research & Development Services: | Y | Database Management: | | Drug Delivery: | |
| | | Diagnostics: | | | | Drug Distribution: | Y |

## TYPES OF BUSINESS:

Clinical Trial Research Services
Data Management & Analysis
Laboratory Services
Regulatory Affairs Consulting
Interactive Voice Response Systems
Strategic Consulting & Marketing

## BRANDS/DIVISIONS/AFFILIATES:

ICON Central Laboratores
ICON Clinical Research
ICON Contracting Solutions
ICON Development Solutions
ICON Medical Imaging
Prevalere Life Sciences Inc
Beacon Bioscience Inc

## CONTACTS: Note: Officers with more than one job title may be intentionally listed here more than once.

Peter Gray, CEO
Ciaran Murray, CFO
Simon Holmes, VP-Mktg. & Market Dev.
Eimear Kenny, VP-Strategic Human Resources
Peter Sowood, Chief Scientific Officer-ICON Clinical Research
Michael McGrath, VP-IT
Malcolm Burgess, COO-ICON U.S.
Bill Taaffe, Pres., Corp. Dev.
Ciaran Murray, Head-Investor Rel.
Sean Leech, Pres., ICON Contracting Solutions
Bob Scott Edwards, Pres., Central Labs
John Hubbard, Pres., Clinical Research
Ted Gastineau, CEO-Medical Imaging
John Climax, Exec. Chmn.

| Phone: 353-291-2000 | Fax: 353-291-2700 |
|---|---|
| Toll-Free: | |
| Address: S. County Business Park, Leopardstown, Dublin, Ireland 18 UK | |

## GROWTH PLANS/SPECIAL FEATURES:

ICON plc is a contract research organization serving the pharmaceutical, medical device and biotechnology industries. The company has 67 locations in 36 countries and it operates in five divisions. ICON Clinical Research, ICON's core service, specializes in planning management, execution and analysis of Phase I-IV clinical trials. It has conducted over 1,000 clinical studies in over 31,000 centers worldwide in areas such as the central nervous system (CNS), oncology, cardiology, transplant dermatology, pediatrics, respiratory and urology. ICON Central Laboratories offers services ranging from complex microbiology and chemistry to simple safety tests. Research focuses include clinical chemistry, hematology, toxicology, infectious disease, immunology/serology, flow cytometry, pharmacogenomics and molecular diagnostics. ICON Contracting Solutions offers temporary or permanent research personnel, supplementing staff shortages, unpredicted monitoring issues and elevated study initiatives. ICON Development Solutions offers strategy, management and execution of products and early phase clinical planning development, particularly in the areas of regulatory affairs and pharmacokinetics/biopharmaceutics. Lastly, ICON Medical Imaging provides image based product development services. ICON also offers technological products which include ICOPhone, an Interactive Voice Recognition (IVR) system used to increase accuracy, efficiency and cost effectiveness of clinical trials; ICOnet, an online tool offering access to case report form (CRF) pages, status reports and project management documentation; and Electronic Data Capture (EDC) Technology, which provides electronic data collection and data management for clinical trials. In November 2008, the company acquired Prevalere Life Sciences, Inc. for approximately $35 million. In January 2009, ICON completed its acquisition of the remaining 30% of Beacon Bioscience, Inc. which it did not already own. Beacon Bioscience was previously renamed ICON Medical Imaging.

ICON offers its employees benefits including medical, dental and vision insurance; flexible spending accounts; short and long term disability coverage; life and AD&D insurance; a 401(k) plan with company match; and paid vacation time.

## FINANCIALS: Sales and profits are in thousands of dollars—add 000 to get the full amount. 2008 Note: Financial information for 2008 was not available for all companies at press time.

| | | |
|---|---|---|
| 2008 Sales: $865,248 | 2008 Profits: $64,676 | **U.S. Stock Ticker:** ICLR |
| 2007 Sales: $630,722 | 2007 Profits: $55,142 | **Int'l Ticker:** IJF  **Int'l Exchange:** Dublin-ISE |
| 2006 Sales: $649,826 | 2006 Profits: $38,304 | Employees: 6,552 |
| 2005 Sales: $326,658 | 2005 Profits: $13,545 | Fiscal Year Ends: 12/31 |
| 2004 Sales: $443,875 | 2004 Profits: $25,742 | Parent Company: |

## SALARIES/BENEFITS:

| Pension Plan: | ESOP Stock Plan: | Profit Sharing: | Top Exec. Salary: $856,193 | Bonus: $599,025 |
|---|---|---|---|---|
| Savings Plan: Y | Stock Purch. Plan: | | Second Exec. Salary: $633,888 | Bonus: $507,147 |

## OTHER THOUGHTS:

Apparent Women Officers or Directors: 11
Hot Spot for Advancement for Women/Minorities: Y

## LOCATIONS: ("Y" = Yes)

| West: | Southwest: | Midwest: | Southeast: | Northeast: | International: |
|---|---|---|---|---|---|
| Y | Y | Y | Y | Y | Y |

Note: Financial information, benefits and other data can change quickly and may vary from those stated here.

# IDERA PHARMACEUTICALS INC                     www.iderapharma.com

**Industry Group Code: 325412  Ranks within this company's industry group: Sales: 101     Profits: 57**

| Drugs: | | Other: | | Clinical: | Computers: | | Services: | |
|---|---|---|---|---|---|---|---|---|
| Discovery: | Y | AgriBio: | | Trials/Services: | Hardware: | | Specialty Services: | |
| Licensing: | Y | Genetic Data: | Y | Labs: | Software: | | Consulting: | |
| Manufacturing: | | Tissue Replacement: | | Equipment/Supplies: | Arrays: | | Blood Collection: | |
| Genetics: | | | | Research & Development Services: | Database Management: | | Drug Delivery: | |
| | | | | Diagnostics: | | | Drug Distribution: | |

## TYPES OF BUSINESS:

Drugs-Targeted Immune Therapies
Drugs-Cancer
Drugs-Infectious Diseases
Drugs-Respiratory Diseases
Drugs-Autoimmune Diseases

## BRANDS/DIVISIONS/AFFILIATES:

IMO-2125
IMO-2055
IMO-3100
QAX935
IMO-2134

## CONTACTS: Note: Officers with more than one job title may be intentionally listed here more than once.

Sudhir Agrawal, CEO
Sudhir Agrawal, Pres.
Lou Arcudi, CFO/Treas.
Sudhir Agrawal, Chief Scientific Officer
Lou Arcudi, Sec.
David M. Lough, Dir.-Bus. Dev. & Alliance Mgmt.
Frank Whalen, Controller
Alice Bexon, VP-Clinical Dev.
Timothy M. Sullivan, VP-Dev. Programs
Ekambar R. Kandimalla, VP-Discovery
Steven J. Ritter, VP-Intellectual Property & Contracts
James B. Wyngaarden, Chmn.

| Phone: 617-679-5500 | Fax: 617-679-5592 |
|---|---|
| Toll-Free: | |
| Address: 167 Sidney St., Cambridge, MA 02139 US | |

## GROWTH PLANS/SPECIAL FEATURES:

Idera Pharmaceuticals, Inc. is a biotechnology company engaged in the discovery and development of therapeutics that modulate immune responses through Toll-like receptors (TLR) for the treatment of multiple diseases, including cancer, infectious diseases, asthma/allergies, autoimmune diseases and vaccine adjuvants. The firm focuses on DNA- and RNA-based drug candidates, which are used to stimulate or block an immune response through the targeted TLR. Idera's research aims to modulate the immune system by using DNA and RNA compounds to interact with the TLR 7, 8 and 9 receptors. The firm develops this technology on multiple fronts by entering into collaboration agreements with other companies, allowing a variety of different research programs to be conducted simultaneously. The company's lead drug candidate, IMO-2055, is a cancer treatment being developed in conjunction with Merck Pharmaceuticals. Other drugs in development include QAX935 (IMO-2134), a treatment for asthma and allergies being developed with Novartis; IMO-2125, a treatment for hepatitis C; IMO-3100, a treatment for autoimmune diseases such as lupus, rheumatoid arthritis, multiple sclerosis, colitis and psoriasis; vaccines for cancer, infectious diseases, and Alzheimer's Disease (also being developed with Merck); and other research-stage drugs for cancer and infectious diseases. The company claims 61 U.S. patents or patent applications and 204 corresponding worldwide patents. In addition to other pharmaceutical companies, Idera also works with academic institutions, including Boston University, Duke University and the University of North Carolina. In December 2008, Massachusetts Life Science Center granted the firm a cooperative research grant with Massachusetts General Hospital to develop new therapies for lupus based on TLR targeting.

## FINANCIALS: Sales and profits are in thousands of dollars—add 000 to get the full amount. 2008 Note: Financial information for 2008 was not available for all companies at press time.

| | | |
|---|---|---|
| 2008 Sales: $26,376 | 2008 Profits: $1,509 | U.S. Stock Ticker: IDRA |
| 2007 Sales: $7,981 | 2007 Profits: $-13,208 | Int'l Ticker:     Int'l Exchange: |
| 2006 Sales: $2,421 | 2006 Profits: $-16,525 | Employees:    37 |
| 2005 Sales: $2,467 | 2005 Profits: $-13,706 | Fiscal Year Ends: 12/31 |
| 2004 Sales: $ 942 | 2004 Profits: $-12,735 | Parent Company: |

## SALARIES/BENEFITS:

| Pension Plan: | ESOP Stock Plan: | Profit Sharing: | Top Exec. Salary: $485,000 | Bonus: $340,000 |
|---|---|---|---|---|
| Savings Plan: Y | Stock Purch. Plan: Y | | Second Exec. Salary: $265,000 | Bonus: $55,000 |

## OTHER THOUGHTS:

**Apparent Women Officers or Directors: 2**
**Hot Spot for Advancement for Women/Minorities:**

## LOCATIONS: ("Y" = Yes)

| West: | Southwest: | Midwest: | Southeast: | Northeast: | International: |
|---|---|---|---|---|---|
| | | | | Y | |

Note: Financial information, benefits and other data can change quickly and may vary from those stated here.

# IDEXX LABORATORIES INC

www.idexx.com

Industry Group Code: 325413  Ranks within this company's industry group:  Sales: 2   Profits: 1

| Drugs: | Other: | Clinical: | | Computers: | | Services: | |
|---|---|---|---|---|---|---|---|
| Discovery: | AgriBio: | Trials/Services: | | Hardware: | | Specialty Services: | |
| Licensing: | Genetic Data: | Labs: | Y | Software: | Y | Consulting: | Y |
| Manufacturing: | Tissue Replacement: | Equipment/Supplies: | Y | Arrays: | | Blood Collection: | |
| Genetics: | | Research & Development Services: | | Database Management: | | Drug Delivery: | |
| | | Diagnostics: | Y | | | Drug Distribution: | Y |

## TYPES OF BUSINESS:

Veterinary Laboratory Testing & Consulting
Point-of-Care Diagnostic Products
Veterinary Pharmaceuticals
Information Management Software
Food & Water Testing Products

## BRANDS/DIVISIONS/AFFILIATES:

VetTest
VetLyte
VetStat
LaserCyte
SNAPshot DX
Coag Dx
Colisure
Parallux

## CONTACTS: Note: Officers with more than one job title may be intentionally listed here more than once.

Jonathan W. Ayers, CEO
Jonathan W. Ayers, Pres.
Merilee Raines, CFO/VP
William C. Wallen, Chief Scientific Officer/Sr. VP
S. Sam Fratoni, VP-Computer Systems
William E. Brown, III, VP-Mfg. & Instrument R&D
Conan R. Deady, General Counsel/Sec./VP
Irene C. Kerr, VP-Worldwide Oper.
Merilee Raines, Treas.
James Polewaczyck, VP-Rapid Assay & Digital Radiography
Michael Williams, VP-Instrument Diagnostics
Thomas J. Dupree, VP-Companion Animal Group
Johnny D. Powers, VP-IDEXX Reference Laboratories
Jonathan W. Ayers, Chmn.
Ali Naqui, VP-Int'l

| Phone: 207-556-0300 | Fax: 207-556-4286 |
|---|---|
| Toll-Free: 800-548-6733 | |
| Address: 1 Idexx Dr., Westbrook, ME 04092-2041 US | |

## GROWTH PLANS/SPECIAL FEATURES:

IDEXX Laboratories, Inc. develops, manufactures and distributes products and provides services for the veterinary and the food and water testing markets. The company operates in two business segments: The Companion Animal Group, which provides products and services for the veterinary market, and the Production Animal Segment, which provides products for production animal health. The company also operated two smaller segments: Dairy, comprising products for dairy quality, and OPTI Medical, comprising products for the human medical diagnostic market. Its primary business focus is on animal health. IDEXX currently markets an integrated and flexible suite of in-house laboratory analyzers for use in veterinary practices, which is referred to as the VetLab suite of analyzers. The suite includes several instrument systems, as well as associated proprietary consumable products such as VetTest, VetLyte, VetStat, and LaserCyte analyzers, the IDEXX SNAPshot Dx and the Coag Dx Analyzer, among other offerings. In addition, it also provides assay kits, software and instrumentation for accurate assessment of infectious disease in production animals, such as cattle, swine and poultry. The company currently offers commercial veterinary laboratory and consulting services throughout the U.S. The water quality segment's products include Colilert-18 and Colisure tests, which simultaneously detect total coliforms and E. coli in water. IDEXX's two principal products for use in testing for antibiotic residue in milk are the SNAP Beta-lactam test and the Parallux system. In March 2009, IDEXX announced new Vector-Borne Disease Panels, its next-generation in infectious disease testing, which allow for more comprehensive diagnoses.

IDEXX offers its employee health, dental and life insurance; a 401(k) plan; short- and long-term disability programs; flexible spending accounts; an employee stock purchase plan; and employee assistance programs.

## FINANCIALS: Sales and profits are in thousands of dollars—add 000 to get the full amount. 2008 Note: Financial information for 2008 was not available for all companies at press time.

| | | |
|---|---|---|
| 2008 Sales: $1,024,030 | 2008 Profits: $116,169 | U.S. Stock Ticker: IDXX |
| 2007 Sales: $922,555 | 2007 Profits: $94,014 | Int'l Ticker:   Int'l Exchange: |
| 2006 Sales: $739,117 | 2006 Profits: $93,678 | Employees: 4,700 |
| 2005 Sales: $638,095 | 2005 Profits: $78,254 | Fiscal Year Ends: 12/31 |
| 2004 Sales: $549,181 | 2004 Profits: $78,332 | Parent Company: |

## SALARIES/BENEFITS:

| Pension Plan: | ESOP Stock Plan: | Profit Sharing: | Top Exec. Salary: $700,000 | Bonus: $675,000 |
|---|---|---|---|---|
| Savings Plan: Y | Stock Purch. Plan: Y | | Second Exec. Salary: $375,000 | Bonus: $200,000 |

## OTHER THOUGHTS:

Apparent Women Officers or Directors: 2
Hot Spot for Advancement for Women/Minorities: Y

## LOCATIONS: ("Y" = Yes)

| West: | Southwest: | Midwest: | Southeast: | Northeast: | International: |
|---|---|---|---|---|---|
| Y | Y | Y | Y | Y | Y |

# IDM PHARMA INC

**www.idm-biotech.com**

Industry Group Code: 325412  **Ranks within this company's industry group:** Sales: 134    Profits: 100

| Drugs: | | Other: | Clinical: | Computers: | Services: |
|---|---|---|---|---|---|
| Discovery: | Y | AgriBio: | Trials/Services: | Hardware: | Specialty Services: |
| Licensing: | | Genetic Data: | Labs: | Software: | Consulting: |
| Manufacturing: | | Tissue Replacement: | Equipment/Supplies: | Arrays: | Blood Collection: |
| Genetics: | | | Research & Development Services: | Database Management: | Drug Delivery: |
| | | | Diagnostics: | | Drug Distribution: |

## TYPES OF BUSINESS:

Drugs-Cancer
Cancer Cell Destroying Products
Tumor Recurrence Prevention
Drugs-Vaccines

## BRANDS/DIVISIONS/AFFILIATES:

MEPACT
BEXIDEM
UVIDEM
COLLIDEM
IDM-2101
Takeda Pharmaceutical Company Ltd

## CONTACTS: *Note: Officers with more than one job title may be intentionally listed here more than once.*

Timothy P. Walbert, CEO
Timothy P. Walbert, Pres.
Robert J. De Vaere, CFO
Robert W. Metz, Commercial Oper.
Jeffrey Sherman, Chief Medical Officer/Sr. VP-R&D
Robert J. De Vaere, Sr. VP-Admin.
Timothy C. Melkus, Sr. VP-Oper.
Timothy C. Melkus, Sr. VP-Bus. Dev.
Robert J. De Vaere, Sr. VP-Finance
Bonnie Mills, VP-Clinical Oper.
Michael G. Grey, Chmn.
Herve Duchesne de Lamotte, Gen. Mgr.-Europe

| Phone: 949-470-4751 | Fax: 949-470-6470 |
|---|---|
| Toll-Free: | |
| Address: 9 Parker, Ste. 100, Irvine, CA 92618-1605 US | |

## GROWTH PLANS/SPECIAL FEATURES:

IDM Pharma Inc. is a biopharmaceutical company focused on developing products to treat and control cancer by targeting the immune system to destroy cancer cells or prevent tumor recurrence. The firm's lead product, mifamurtide (L-MTP-PE), is being developed for the treatment of patients with non-metastatic resectable osteosarcoma, or bone cancer. L-MTP-PE, known as Mepact in Europe, activates certain immune cells called macrophages in vivo (inside the body), in order to enhance their ability to destroy cancer cells. IDM has completed Phase III trials of this drug and has submitted marketing applications for the U.S. and has been approved in Europe. The other product that the firm is developing to destroy cancer cells is Bexidem, a drug based on MAK (Monocyte-derived Activated Killer) cells, which are produced by activating the patient's own white blood cells. Bexidem is in Phase II/III clinical development for the treatment of bladder cancer. Products designed to prevent tumor recurrence include IDM-2101 (a synthetic vaccine in clinical development for the treatment of non-small cell lung cancer), Uvidem and Collidem (for the treatment of melanoma and colorectal cancer, respectively). Uvidem and Collidem are both dendritophages, which are specialized immune cells derived from a patient's white blood cells, exposed to tumor cell antigens, and then reinjected into the patient in order to stimulate the immune system to recognize and kill tumor cells that display these antigens on their surface. The company holds 27 issued patents and 7 patent applications, with rights to 135 licensed patents. In May 2009, IDM was acquired by Takeda America Holdings, Ltd., a subsidiary of Takeda Pharmaceutical Company, Ltd. IDM's primary pharmaceutical asset, mifamurtide, will transfer over to Millenium: The Takeda Oncology Company.

## FINANCIALS: Sales and profits are in thousands of dollars—add 000 to get the full amount. 2008 Note: Financial information for 2008 was not available for all companies at press time.

| | | |
|---|---|---|
| 2008 Sales: $3,147 | 2008 Profits: $-18,606 | **U.S. Stock Ticker:** Subsidiary |
| 2007 Sales: $14,630 | 2007 Profits: $-18,350 | **Int'l Ticker:**    Int'l Exchange: |
| 2006 Sales: $11,286 | 2006 Profits: $-23,455 | Employees:   16 |
| 2005 Sales: $8,539 | 2005 Profits: $-39,209 | Fiscal Year Ends: 12/31 |
| 2004 Sales: $5,805 | 2004 Profits: $-31,657 | Parent Company: TAKEDA PHARMACEUTICAL COMPANY LTD |

## SALARIES/BENEFITS:

| Pension Plan: | ESOP Stock Plan: | Profit Sharing: | Top Exec. Salary: $268,910 | Bonus: $ |
|---|---|---|---|---|
| Savings Plan: Y | Stock Purch. Plan: Y | | Second Exec. Salary: $231,789 | Bonus: $ |

## OTHER THOUGHTS:

**Apparent Women Officers or Directors**: 1
**Hot Spot for Advancement for Women/Minorities:**

## LOCATIONS: ("Y" = Yes)

| West: | Southwest: | Midwest: | Southeast: | Northeast: | International: |
|---|---|---|---|---|---|
| Y | | | | | Y |

# ILLUMINA INC

**www.illumina.com**

**Industry Group Code: 3345 Ranks within this company's industry group: Sales: 3 Profits: 3**

| Drugs: | Other: | | Clinical: | | Computers: | | Services: | |
|---|---|---|---|---|---|---|---|---|
| Discovery: | AgriBio: | | Trials/Services: | | Hardware: | Y | Specialty Services: | Y |
| Licensing: | Genetic Data: | Y | Labs: | | Software: | Y | Consulting: | |
| Manufacturing: | Tissue Replacement: | | Equipment/Supplies: | Y | Arrays: | Y | Blood Collection: | |
| Genetics: | | | Research & Development Services: | | Database Management: | | Drug Delivery: | |
| | | | Diagnostics: | | | | Drug Distribution: | |

## TYPES OF BUSINESS:

Instruments-Genetic Variation Measurement
Array Technology
Digital Microbead Technology
Software
Genotyping Services

## BRANDS/DIVISIONS/AFFILIATES:

BeadArray Technology
Genome Analyzer
iScan System
Oligator
VeraCode
Illumina Sequencing
BeadXpress Reader
Solexa, Inc.

## CONTACTS: Note: Officers with more than one job title may be intentionally listed here more than once.

Jay T. Flatley, CEO
Jay T. Flatley, Pres.
Christian Henry, CFO/Sr. VP
Mark Lewis, Sr. VP-Product Dev.
Mostafa Ronaghi, CTO/Sr. VP
Christian G. Cabou, General Counsel/Sr. VP
Bill Bonnar, Sr. VP-Oper.
Joel McComb, Sr. VP/Gen. Mgr.-Life Sciences
John West, Sr. VP/Gen. Mgr.-DNA Sequencing
Tristan Orpin, Sr. VP-Commercial Oper.
Gregory F. Heath, Sr. VP/Gen. Mgr.-Diagnostics
William H. Rastetter, Chmn.

| Phone: 858-202-4500 | Fax: 858-202-4545 |
|---|---|
| Toll-Free: 800-809-4566 | |
| Address: 9885 Towne Centre Dr., San Diego, CA 92121-1975 US | |

## GROWTH PLANS/SPECIAL FEATURES:

Illumina, Inc. develops tools for the large-scale analysis of genetic variation and function. The firm's tools provide genomic information that can be used to improve drugs and therapies, customize diagnoses and treatments and cure disease. The company deploys three primary systems for analyzing genetic variation and function: the Genome Analyzer, based on Illumina Sequencing technology, provides DNA sequencing services; the iScan System, based on BeadArray technology, performs multiple assays simultaneously; and the BeadXpress Reader, based on VeraCode technology acquired from the CyVera Corporation, is similar to the BeadArray but is used in lower multiplex projects. These use fiber optics to achieve a level of array miniaturization that allows experimentation to be easily scaled up. The firm also markets a series of sequencing kits and BeadChips for use in their instrumentation. Illumina arranges its arrays in patterns that match the wells of industry-standard microtiter plates, allowing higher throughput than other technologies. The company's other complementary technology, Oligator, permits parallel synthesis of the millions of pieces of DNA necessary to perform large-scale genetic analysis. Illumina additionally provides genotyping services for other companies, as well as software, benchtop and production systems, installation and certain warranty services for its products. Illumina is one of five U.S. research groups, participating in the International HapMap Project, a global consortium aimed at creating a detailed map of genetic variation. The firm is also one of three participating in the 1000 Genomes Project, which will build on HapMap.

Illumina offers its employees medical, dental, vision and prescription drug plans; flexible spending accounts; an employee stock purchase plan; an educational assistance program; training opportunities; disability coverage; an on-site gym; wellness benefits; and many company activities.

## FINANCIALS: Sales and profits are in thousands of dollars—add 000 to get the full amount. 2008 Note: Financial information for 2008 was not available for all companies at press time.

| | | |
|---|---|---|
| 2008 Sales: $573,225 | 2008 Profits: $50,477 | **U.S. Stock Ticker: ILMN** |
| 2007 Sales: $366,799 | 2007 Profits: $-278,359 | **Int'l Ticker:** Int'l Exchange: |
| 2006 Sales: $184,586 | 2006 Profits: $39,968 | Employees: 1,041 |
| 2005 Sales: $73,501 | 2005 Profits: $-20,874 | Fiscal Year Ends: 12/31 |
| 2004 Sales: $50,583 | 2004 Profits: $-6,225 | Parent Company: |

## SALARIES/BENEFITS:

| Pension Plan: | ESOP Stock Plan: | Profit Sharing: | Top Exec. Salary: $647,038 | Bonus: $604,500 |
|---|---|---|---|---|
| Savings Plan: Y | Stock Purch. Plan: Y | | Second Exec. Salary: $356,118 | Bonus: $205,601 |

## OTHER THOUGHTS:

**Apparent Women Officers or Directors:** 1
**Hot Spot for Advancement for Women/Minorities:**

## LOCATIONS: ("Y" = Yes)

| West: | Southwest: | Midwest: | Southeast: | Northeast: | International: |
|---|---|---|---|---|---|
| Y | | | | | Y |

# IMCLONE SYSTEMS INC

www.imclone.com

**Industry Group Code: 325412 Ranks within this company's industry group:** Sales: Profits:

| Drugs: | | Other: | Clinical: | Computers: | Services: |
|---|---|---|---|---|---|
| Discovery: | Y | AgriBio: | Trials/Services: | Hardware: | Specialty Services: |
| Licensing: | | Genetic Data: | Labs: | Software: | Consulting: |
| Manufacturing: | Y | Tissue Replacement: | Equipment/Supplies: | Arrays: | Blood Collection: |
| Genetics: | | | Research & Development Services: | Database Management: | Drug Delivery: |
| | | | Diagnostics: | | Drug Distribution: |

## TYPES OF BUSINESS:

Drugs-Cancer

## BRANDS/DIVISIONS/AFFILIATES:

Erbitux
Eli Lilly & Co.
Bristol-Myers Squibb Company,

## CONTACTS: Note: Officers with more than one job title may be intentionally listed here more than once.

John H. Johnson, CEO
Kenneth J. Zuerblis, CFO/Sr. VP
Greg Reynolds, VP-Human Resources
Larry Witte, Sr. VP-Research
Greg Mayes, Deputy General Counsel/VP/Chief Compliance Officer
Joseph I. DePinto, VP-Commercial Oper.
Joan Connolly, Associate VP-Strategy & Oper.
Tracy Henrikson, Head-Corp. Comm.
Richard P. Crowley, Sr. VP-Biopharmaceutical Oper.
Eric K. Rowinsky, Chief Medical Officer/Exec. VP
Bernhard Ehmer, Sr. VP/Managing Dir.-Int'l Oper.

| Phone: 212-645-1405 | Fax: 212-645-2054 |
|---|---|
| Toll-Free: | |
| Address: 180 Varick St., 6th Fl., New York, NY 10014 US | |

## GROWTH PLANS/SPECIAL FEATURES:

ImClone Systems, Inc. is a biopharmaceutical company that is engaged in the research and development of cancer treatments through growth factor blockers and angiogenesis inhibitors. The firm's commercially available product is Erbitux (Cetuximab), a growth factor blocker that has been approved by the U.S. Food and Drug Administration (FDA) for the treatment of metastatic colorectal cancer and of squamous cell carcinoma of the head and neck, or head and neck cancer. Erbitux works by binding specifically to EGFR on both normal and tumor cells, and competitively inhibits the binding of the epidermal growth factor and other growth stimulatory ligands, such as transforming growth factor-alpha. ImClone is currently working in collaboration with its partner, Bristol-Myers Squibb Company, in development trials of Erbitux for broader use in colorectal, head and neck, lung, pancreatic, esophageal, stomach, prostate, bladder and other cancers. In addition to the development and commercialization of Erbitux, the company has developed several investigational agents that are in various stages of clinical development, and the firm is also developing monoclonal antibodies to inhibit angiogenesis. In 2008, Erbitux received marketing authorization in Japan for use in treating patients with advanced or metastatic colorectal cancer. In November 2008, the company was acquired by Eli Lilly & Co., a global pharmaceutical company, for approximately $6.5 billion. ImClone Systems is now a wholly-owned subsidiary Eli Lilly.

## FINANCIALS: Sales and profits are in thousands of dollars—add 000 to get the full amount. 2008 Note: Financial information for 2008 was not available for all companies at press time.

| | | |
|---|---|---|
| 2008 Sales: $ | 2008 Profits: $ | U.S. Stock Ticker: Subsidiary |
| 2007 Sales: $590,833 | 2007 Profits: $39,799 | Int'l Ticker: Int'l Exchange: |
| 2006 Sales: $677,847 | 2006 Profits: $370,674 | Employees: 1,128 |
| 2005 Sales: $383,673 | 2005 Profits: $86,496 | Fiscal Year Ends: 12/31 |
| 2004 Sales: $388,690 | 2004 Profits: $113,653 | Parent Company: ELI LILLY & COMPANY |

## SALARIES/BENEFITS:

| Pension Plan: | ESOP Stock Plan: | Profit Sharing: | Top Exec. Salary: $415,000 | Bonus: $275,643 |
|---|---|---|---|---|
| Savings Plan: | Stock Purch. Plan: | | Second Exec. Salary: $325,000 | Bonus: $218,200 |

## OTHER THOUGHTS:

**Apparent Women Officers or Directors:** 1
**Hot Spot for Advancement for Women/Minorities:** Y

## LOCATIONS: ("Y" = Yes)

| West: | Southwest: | Midwest: | Southeast: | Northeast: | International: |
|---|---|---|---|---|---|
| | | | | Y | |

Note: Financial information, benefits and other data can change quickly and may vary from those stated here.

# IMMTECH PHARMACEUTICALS INC

**www.immtechpharma.com**

**Industry Group Code: 325412  Ranks within this company's industry group: Sales: 116   Profits: 79**

| Drugs: | | Other: | Clinical: | Computers: | Services: | |
|---|---|---|---|---|---|---|
| Discovery: | Y | AgriBio: | Trials/Services: | Hardware: | Specialty Services: | Y |
| Licensing: | | Genetic Data: | Labs: | Software: | Consulting: | |
| Manufacturing: | | Tissue Replacement: | Equipment/Supplies: | Arrays: | Blood Collection: | |
| Genetics: | | | Research & Development Services: | Database Management: | Drug Delivery: | |
| | | | Diagnostics: | | Drug Distribution: | |

## TYPES OF BUSINESS:

Pharmaceuticals Discovery & Development
Drugs-Antifungal
Drugs-Infectious Diseases
Drugs-Antibacterial

## BRANDS/DIVISIONS/AFFILIATES:

Immtech Hong Kong Limited
Immtech Therapeutics Limited
Super Insight Limited
Immtech Life Science Limited
Gold Avenue Ltd.

## CONTACTS: *Note: Officers with more than one job title may be intentionally listed here more than once.*

Eric L. Sorkin, CEO
Gary C. Parks, CFO
Gary C. Parks, Sec.
Gary C. Parks, Treas.
Cecilia Chan, Vice Chmn.
Norman A. Abood, Sr. VP-Discovery Programs
Eric L. Sorkin, Chmn.

| **Phone:** 212-791-2911 | **Fax:** 212-791-2917 |
|---|---|
| **Toll-Free:** 877-898-8038 | |
| **Address:** 1 North End Ave., New York, NY 10282 US | |

## GROWTH PLANS/SPECIAL FEATURES:

Immtech Pharmaceuticals, Inc., focuses on the discovery and commercialization of therapeutics for the treatment of infectious diseases. In order to capitalize on the opportunities arising in the global healthcare market, the firm emphasizes the development of operations in China. The company's drug candidates are produced using its proprietary library of aromatic cation compounds. These compounds have at least one positively charged end and at least one benzene ring in their structure and many of them demonstrate the potential to prevent growth in pathogens. Immtech focuses on the composition of cations with links of different length, shape, flexibility and curvature, in order to target DNA and other receptors. The company's drug development section currently has programs researching treatments in three primary areas: Hepatitis C (HCV), bacterial infections and fungal diseases. Immtech has also generated data indicating that a subset of compounds generated for the HCV program show potential activity against the West Nile virus. The company owns 148 issued domestic U.S. and foreign patents that cover many classes of novel chemical compounds. The firm has three subsidiaries: Immtech Hong Kong Limited, Immtech Therapeutics Limited and Super Insight Limited. In 2008, the firm's drug development program for pafuramidine was discontinued due to findings of renal and liver adverse events among participants in a study of healthy volunteers conducted in South Africa. Pafuramidine was being tested for the treatment of African sleeping sickness, malaria and pneumocystis pneumonia indications. In June 2008, Immtech agreed to provide contract research services in China in collaboration with Beijing Capital Medical University with a primary focus on pre-clinical drug development. In April 2009, the firm agreed to sell its Chinese subsidiary, Immtech Life Science Limited, for approximately $2 million. Also in April 2009, the company invested in Gold Avenue Ltd., a firm involved in Chinese tin production.

## FINANCIALS: Sales and profits are in thousands of dollars—add 000 to get the full amount. 2008 Note: Financial information for 2008 was not available for all companies at press time.

| | | |
|---|---|---|
| 2008 Sales: $9,717 | 2008 Profits: $-10,513 | **U.S. Stock Ticker: IMMP** |
| 2007 Sales: $4,318 | 2007 Profits: $-11,133 | **Int'l Ticker:**   Int'l Exchange: |
| 2006 Sales: $3,575 | 2006 Profits: $-15,526 | Employees:  24 |
| 2005 Sales: $5,931 | 2005 Profits: $-13,433 | Fiscal Year Ends: 3/31 |
| 2004 Sales: $2,416 | 2004 Profits: $-12,846 | Parent Company: |

## SALARIES/BENEFITS:

| Pension Plan: | ESOP Stock Plan: | Profit Sharing: | Top Exec. Salary: $375,000 | Bonus: $ |
|---|---|---|---|---|
| Savings Plan: Y | Stock Purch. Plan: | | Second Exec. Salary: $235,000 | Bonus: $ |

## OTHER THOUGHTS:

**Apparent Women Officers or Directors:** 2
**Hot Spot for Advancement for Women/Minorities:** Y

## LOCATIONS: ("Y" = Yes)

| West: | Southwest: | Midwest: | Southeast: | Northeast: | International: |
|---|---|---|---|---|---|
| | | Y | | Y | |

*Note: Financial information, benefits and other data can change quickly and may vary from those stated here.*

# IMMUCELL CORPORATION                 www.immucell.com

**Industry Group Code: 325412B  Ranks within this company's industry group:  Sales: 2   Profits: 2**

| Drugs: | | Other: | | Clinical: | | Computers: | | Services: | |
|---|---|---|---|---|---|---|---|---|---|
| Discovery: | Y | AgriBio: | | Trials/Services: | | Hardware: | | Specialty Services: | |
| Licensing: | Y | Genetic Data: | | Labs: | | Software: | | Consulting: | |
| Manufacturing: | Y | Tissue Replacement: | | Equipment/Supplies: | Y | Arrays: | | Blood Collection: | |
| Genetics: | | | | Research & Development Services: | Y | Database Management: | | Drug Delivery: | |
| | | | | Diagnostics: | Y | | | Drug Distribution: | |

## TYPES OF BUSINESS:

Diagnostic Tests & Products
Animal Health Drugs

## BRANDS/DIVISIONS/AFFILIATES:

Wipe Out Dairy Wipes
California Mastitis Test
Mastik
MastOut
First Defense
RJT

## CONTACTS: Note: Officers with more than one job title may be intentionally listed here more than once.

Michael F. Brigham, CEO
Michael F. Brigham, Pres.
Joseph H. Crabb, Chief Scientific Officer/VP
Michael F. Brigham, Sec.
Michael F. Brigham, Treas.

| **Phone:** 207-878-2770 | **Fax:** 207-878-2117 |
|---|---|
| **Toll-Free:** 800-466-8235 | |
| **Address:** 56 Evergreen Dr., Portland, ME 04103 US | |

## GROWTH PLANS/SPECIAL FEATURES:

ImmuCell Corp. is a biotechnology company serving veterinarians and producers in the dairy and beef industries with products that improve animal health and productivity. The company sells diagnostic tests and products for therapeutic and preventive use against certain infectious diseases in animals and humans.    The firm's leading product, First Defense, is an oral medicine manufactured from cows' colostrum using the proprietary vaccine and milk protein purification technologies.    The drug targets calf diseases, which cause diarrhea and dehydration in newborn calves and often leads to serious sickness or death. ImmuCell also sells products designed to aid in the management of mastitis (inflammation of the mammary gland) caused by bacterial infections.    The company's second leading source of product sales is Wipe Out Dairy Wipes, which consist of pre-moistened, biodegradable towelettes that are impregnated with Nisin, a natural antibacterial peptide, to prepare the teat area of a cow in advance of milking.    The firm also offers MASTiK (Mastitis Antibiotic Susceptibility Test Kit), which helps veterinarians and producers select the antibiotic most likely to be effective in the treatment of individual cases of mastitis; California Mastitis Test (CMT), which can be performed at cow side for early detection of mastitis; and RJT (Rapid Johne's Test) that can identify cattle with symptomatic Johne's disease. The company is currently developing a treatment called Mast Out, designed to treat Mastitis.

## FINANCIALS: Sales and profits are in thousands of dollars—add 000 to get the full amount. 2008 Note: Financial information for 2008 was not available for all companies at press time.

| | | | |
|---|---|---|---|
| 2008 Sales: $4,634 | 2008 Profits: $- 469 | **U.S. Stock Ticker: ICCC** | |
| 2007 Sales: $6,069 | 2007 Profits: $ 662 | **Int'l Ticker:**    Int'l Exchange: | |
| 2006 Sales: $4,801 | 2006 Profits: $ 647 | Employees:    30 | |
| 2005 Sales: $4,983 | 2005 Profits: $ 708 | Fiscal Year Ends: 12/31 | |
| 2004 Sales: $3,696 | 2004 Profits: $ 144 | Parent Company: | |

## SALARIES/BENEFITS:

| Pension Plan: | ESOP Stock Plan: | Profit Sharing: | Top Exec. Salary: $183,932 | Bonus: $ |
|---|---|---|---|---|
| Savings Plan: Y | Stock Purch. Plan: | | Second Exec. Salary: $183,932 | Bonus: $ |

## OTHER THOUGHTS:

**Apparent Women Officers or Directors**: 1
**Hot Spot for Advancement for Women/Minorities**:

## LOCATIONS: ("Y" = Yes)

| West: | Southwest: | Midwest: | Southeast: | Northeast: | International: |
|---|---|---|---|---|---|
| | | | | Y | |

# IMMUNOGEN INC

**www.immunogen.com**

**Industry Group Code: 325412  Ranks within this company's industry group:** Sales: 90  Profits: 126

| Drugs: | | Other: | Clinical: | Computers: | Services: |
|---|---|---|---|---|---|
| Discovery: | Y | AgriBio: | Trials/Services: | Hardware: | Specialty Services: |
| Licensing: | Y | Genetic Data: | Labs: | Software: | Consulting: |
| Manufacturing: | | Tissue Replacement: | Equipment/Supplies: | Arrays: | Blood Collection: |
| Genetics: | | | Research & Development Services: | Database Management: | Drug Delivery: |
| | | | Diagnostics: | | Drug Distribution: |

## TYPES OF BUSINESS:

Anticancer Biopharmaceuticals
Tumor-Activated Drugs

## BRANDS/DIVISIONS/AFFILIATES:

IMGN901
IMGN388
BIIB015
Transtuzumab-DM1
SAR3419
BT-062
Tumor-Activated Prodrug
AVE1642

## CONTACTS: *Note: Officers with more than one job title may be intentionally listed here more than once.*

Daniel M. Junius, CEO
Daniel M. Junius, Pres.
Gregory Perry, CFO/Sr. VP
John M. Lambert, Chief Science Officer/Exec. VP-R&D
Craig Barrows, General Counsel/VP/Sec.
John Tagliamonte, VP-Bus. Dev.
Daniel M. Junius, Treas.
James O'Leary, Chief Medical Officer/VP
Mitchel Sayare, Chmn.

| Phone: 781-895-0600 | Fax: 781-895-0611 |
|---|---|
| Toll-Free: | |
| Address: 830 Winter St., Waltham, MA 02451-1477 US | |

## GROWTH PLANS/SPECIAL FEATURES:

ImmunoGen develops targeted therapeutics for the treatment of cancer. The company's Tumor-Activated Prodrug (TAP) technology uses antibodies to deliver a cytotoxic agent specifically to cancer cells. The TAP technology is designed to enable the creation of potent well-tolerated anticancer products. TAP compounds are designed to bind to and destroy certain targets using monoclonal antibodies, for instance antigens on cancer cell. These cell-killing agents (CKAs) are believed to be far more powerful and potent than traditional chemotherapy techniques. The company then licenses this technology to a variety of pharmaceutical companies. ImmunoGen's collaborative partners include Amgen, Inc.; Biogen Idec, Inc.; Centocor, Inc.; Genetech, Inc.; the sanofi-aventis Group; and Biotest AG. The company's lead collaborative product in development is trastuzumab-DM1 (T-DM1), a CKA that targets HER2. The drug is currently in testing for indications of metastatic breast cancer. The firm has a licensing agreement for T-DM1 with Roche Group outside the U.S. and with its subsidiary Genentech for American sales. Other products in development include AVE9633, AVE1642 and SAR3419, in development for cancer by sanofi-aventis; IMGN901, IMGN242 and IMGN388 for myeloma and other cancers, which are wholly owned by ImmunoGen; BIIB015, in development for solid tumors by Biogen Idec; and BT-062, in development for myeloma by Biotest. All are TAP compounds. The firm is also developing a handful of preclinical and research projects, including SAR566658 and SAR650984 with sanofi-aventis. In December 2008, the company licensed a fifth TAP target to Genentech. In June 2009, ImmunoGen opted to halt development of its IMGN242 prospect and make it a possible out-license candidate due to changing priorities within the company.

The company offers its employees health and dental coverage; life and AD&D insurance; a 401(k) plan; tuition reimbursement; and stock options.

## FINANCIALS: Sales and profits are in thousands of dollars—add 000 to get the full amount. 2008 Note: Financial information for 2008 was not available for all companies at press time.

| | | |
|---|---|---|
| 2008 Sales: $40,249 | 2008 Profits: $-32,020 | **U.S. Stock Ticker: IMGN** |
| 2007 Sales: $38,212 | 2007 Profits: $-18,987 | **Int'l Ticker:**　Int'l Exchange: |
| 2006 Sales: $32,088 | 2006 Profits: $-17,834 | Employees:　210 |
| 2005 Sales: $35,718 | 2005 Profits: $-10,951 | Fiscal Year Ends: 6/30 |
| 2004 Sales: $25,956 | 2004 Profits: $-5,917 | Parent Company: |

## SALARIES/BENEFITS:

| Pension Plan: | ESOP Stock Plan: | Profit Sharing: | Top Exec. Salary: $463,680 | Bonus: $179,096 |
|---|---|---|---|---|
| Savings Plan: Y | Stock Purch. Plan: | | Second Exec. Salary: $346,500 | Bonus: $93,684 |

## OTHER THOUGHTS:

**Apparent Women Officers or Directors**: 1
**Hot Spot for Advancement for Women/Minorities**: Y

## LOCATIONS: ("Y" = Yes)

| West: | Southwest: | Midwest: | Southeast: | Northeast: | International: |
|---|---|---|---|---|---|
| | | | | Y | |

# IMMUNOMEDICS INC

## www.immunomedics.com

**Industry Group Code: 325413** Ranks within this company's industry group: Sales: 21  Profits: 14

| Drugs: | | Other: | | Clinical: | | Computers: | | Services: | |
|---|---|---|---|---|---|---|---|---|---|
| Discovery: | Y | AgriBio: | | Trials/Services: | | Hardware: | | Specialty Services: | |
| Licensing: | Y | Genetic Data: | | Labs: | | Software: | | Consulting: | |
| Manufacturing: | Y | Tissue Replacement: | | Equipment/Supplies: | | Arrays: | | Blood Collection: | |
| Genetics: | | | | Research & Development Services: | | Database Management: | | Drug Delivery: | |
| | | | | Diagnostics: | Y | | | Drug Distribution: | |

## TYPES OF BUSINESS:

Monoclonal Antibody-Based Products
Cancer & Autoimmune  Products

## BRANDS/DIVISIONS/AFFILIATES:

Epratuzumab
CEA-Scan
LeukoScan
LymphoScan
AFP-Scan
Veltuzumab
Milatuzumab
Labetuzumab

## CONTACTS: Note: Officers with more than one job title may be intentionally listed here more than once.

Cynthia L. Sullivan, CEO
Cynthia L. Sullivan, Pres.
Gerald G. Gorman, CFO
David M. Goldenberg, Chief Medical Officer/Chief Scientific Officer
Gerard G. Gorman, Sr. VP-Bus. Dev.
Gerard G. Gorman, Sr. VP-Finance
David M. Goldenberg, Chmn.

| Phone: 973-605-8200 | Fax: 973-605-8282 |
|---|---|
| Toll-Free: | |
| Address:  300 American Rd., Morris Plains, NJ 07950 US | |

## GROWTH PLANS/SPECIAL FEATURES:

Immunomedics, Inc. is a biopharmaceutical company focused on the development of monoclonal antibody-based products for the targeted treatment of cancer, autoimmune diseases and other indications.  The company creates humanized antibodies that can be used either alone in unlabeled form or conjugated with radioactive isotopes, chemotherapeutics or toxins to create targeted agents. Diagnostic imaging products manufactured by Immunomedics include CEA-Scan, for the detection of colorectal, lung and breast cancers, and LeukoScan, for the detection of bone infections.  CEA-Scan has been approved in the U.S., Canada and Europe, and LeukoScan has been approved in Canada, Europe and Australia.  The company also offers ImmuSTRIP, a laboratory test kit for the detection of human anti-murine antibodies (HAMA).  Immunomedics has a number of therapeutic product candidates in clinical trials including epratuzumab (LymphoScan), veltuzumab, Y-90 hPAM4, milatuzumab and labetuzumab, most of which target autoimmune diseases like systemic lupus erythematosus, Sjogren's syndrome, B-cell non-Hodgkins lymphoma (NHL) and various solid tumors.  In 2008, the firm entered a license and collaboration agreement with Nycomed to develop, manufacture and commercialize veltuzumab, an anti-CD22 humanized antibody in the subcutaneous formulation similar to epretuzumab, for treating all non-cancer indications.  Veltuzumab is currently in Phase II clinical trials, having shown positive results in earlier testing on follicular lymphoma.  The company's preclinical developments include RNASe-Conjugate, used mainly for solid tumors and NHL; DNL (Dock-and-Lock), a method for creating proteins and antibodies for specific diseases; and a Dendritic Cell Vaccine to use antibodies to create a preventative measure against tumors.  The firm's majority-owned subsidiary, IBC Pharmaceuticals, Inc., develops radioimmunotherapeutics with target-specific antibodies.  In February 2009, the FDA granted Immunomedics' hPAm4 a fast-track designation for use with pancreatic cancer.

The company offers its employees medical, dental and vision insurance; flexible spending accounts; life insurance; a 401(k) plan; and tuition reimbursement.

## FINANCIALS: Sales and profits are in thousands of dollars—add 000 to get the full amount. 2008 Note: Financial information for 2008 was not available for all companies at press time.

| | | |
|---|---|---|
| 2008 Sales: $3,651 | 2008 Profits: $-22,909 | **U.S. Stock Ticker: IMMU** |
| 2007 Sales: $8,506 | 2007 Profits: $-16,655 | **Int'l Ticker:**  Int'l Exchange: |
| 2006 Sales: $4,353 | 2006 Profits: $-28,764 | Employees:   117 |
| 2005 Sales: $3,813 | 2005 Profits: $-26,758 | Fiscal Year Ends: 6/30 |
| 2004 Sales: $4,306 | 2004 Profits: $-22,355 | Parent Company: |

## SALARIES/BENEFITS:

| Pension Plan: | ESOP Stock Plan: | Profit Sharing: | Top Exec. Salary: $532,000 | Bonus: $239,400 |
|---|---|---|---|---|
| Savings Plan: Y | Stock Purch. Plan: | | Second Exec. Salary: $500,000 | Bonus: $225,000 |

## OTHER THOUGHTS:

**Apparent Women Officers or Directors**: 2
**Hot Spot for Advancement for Women/Minorities**: Y

## LOCATIONS: ("Y" = Yes)

| West: | Southwest: | Midwest: | Southeast: | Northeast: | International: |
|---|---|---|---|---|---|
| | | | | Y | Y |

Note: Financial information, benefits and other data can change quickly and may vary from those stated here.

# IMPAX LABORATORIES INC

**www.impaxlabs.com**

Industry Group Code: 325412A **Ranks within this company's industry group:** Sales: 11  Profits: 9

| Drugs: | Other: | Clinical: | Computers: | Services: |
|---|---|---|---|---|
| Discovery: | AgriBio: | Trials/Services: | Hardware: | Specialty Services: |
| Licensing: | Genetic Data: | Labs: | Software: | Consulting: |
| Manufacturing: Y | Tissue Replacement: | Equipment/Supplies: | Arrays: | Blood Collection: |
| Genetics: Y | | Research & Development Services: | Database Management: | Drug Delivery: Y |
| | | Diagnostics: | | Drug Distribution: Y |

## TYPES OF BUSINESS:

Drug Delivery Systems
Drugs-Generic
Pharmaceutical Discovery & Development
Drugs-Central Nervous System

## BRANDS/DIVISIONS/AFFILIATES:

Global Pharmaceuticals
IMPAX Pharmaceuticals

## CONTACTS: *Note: Officers with more than one job title may be intentionally listed here more than once.*

Larry Hsu, CEO
Larry Hsu, Pres.
Arthur A. Koch, Jr., CFO
Charles V. Hildenbrand, Sr. VP-Oper.
Arthur A. Koch, Jr., Sr. VP-Finance
Michael J. Nestor, Pres., IMPAX Pharmaceuticals
Robert L. Burr, Chmn.
Christopher Mengler, Pres., Global Pharmaceuticals

| Phone: 510-476-2000 | Fax: 510-471-3200 |
|---|---|
| Toll-Free: | |
| Address: 30831 Huntwood Ave., Hayward, CA 94544 US | |

## GROWTH PLANS/SPECIAL FEATURES:

IMPAX Laboratories, Inc. is a technology-based, specialty pharmaceutical company focused on the development and commercialization of generic and brand-name pharmaceuticals. In the generic pharmaceuticals market, the firm focuses on controlled-release generic versions of brand-name pharmaceuticals covering a range of therapeutic areas with technically challenging drug-delivery mechanisms or limited competition. The company also develops specialty generic pharmaceuticals that present barriers to entry by competitors, such as difficulty in raw materials sourcing; complex formulation or development characteristics; or special handling requirements. In the brand-name pharmaceuticals market, IMPAX develops products for the treatment of central nervous system disorders. The company markets generic products through its Global Pharmaceuticals division and intends to market branded products through its IMPAX Pharmaceuticals division. The firm markets 70 generic pharmaceuticals representing dosage variations of 24 different pharmaceutical compounds through its Global Pharmaceuticals division and another 16 products representing dosage variations of four different pharmaceutical compounds through its alliance agreements' partners. It has 53 abbreviated new drug applications (ANDAs) approved by the U.S. Food and Drug Administration (FDA), including generic versions of Brethine, Floriner, Minocine, Claritin-D 12-hour, Claritin-D 24-hour, Wellbutrin SR and Prilosec; 24 applications pending at the FDA, including two tentatively approved; and 40 products in various stages of development. IMPAX also has one branded pharmaceutical product for which it has completed a Phase III clinical study, a second product in Phase II and four programs in the early exploratory stage. In May 2009, the firm received final FDA approval for its generic versions of Depakote and Precose Tablets in 25 mg (milligram), 50 mg and 100 mg strengths. In March 2009, IMPAX began trading on The NASDAQ Global Market.

IMPAX offers its employees flexible spending accounts; an employee assistance program; health maintenance organization (HMO) and preferred provider organization (PPO) medical coverage; and dental, life and disability insurance.

## FINANCIALS: Sales and profits are in thousands of dollars—add 000 to get the full amount. 2008 Note: Financial information for 2008 was not available for all companies at press time.

| | | |
|---|---|---|
| 2008 Sales: $210,071 | 2008 Profits: $18,700 | **U.S. Stock Ticker:** IPXL |
| 2007 Sales: $273,753 | 2007 Profits: $125,925 | **Int'l Ticker:**   Int'l Exchange: |
| 2006 Sales: $135,246 | 2006 Profits: $-12,044 | Employees:   768 |
| 2005 Sales: $ | 2005 Profits: $ | Fiscal Year Ends: 12/31 |
| 2004 Sales: $ | 2004 Profits: $ | Parent Company: |

## SALARIES/BENEFITS:

| Pension Plan: | ESOP Stock Plan: | Profit Sharing: | Top Exec. Salary: $375,000 | Bonus: $1,233,141 |
|---|---|---|---|---|
| Savings Plan: Y | Stock Purch. Plan: Y | | Second Exec. Salary: $302,404 | Bonus: $218,902 |

## OTHER THOUGHTS:

**Apparent Women Officers or Directors:** 1
**Hot Spot for Advancement for Women/Minorities:**

## LOCATIONS: ("Y" = Yes)

| West: | Southwest: | Midwest: | Southeast: | Northeast: | International: |
|---|---|---|---|---|---|
| Y | | | | Y | |

Note: Financial information, benefits and other data can change quickly and may vary from those stated here.

# IMS HEALTH INC

**www.imshealth.com**

**Industry Group Code: 541910 Ranks within this company's industry group: Sales: 1 Profits: 1**

| Drugs: | Other: | Clinical: | Computers: | | Services: | |
|---|---|---|---|---|---|---|
| Discovery: | AgriBio: | Trials/Services: | Hardware: | | Specialty Services: | Y |
| Licensing: | Genetic Data: | Labs: | Software: | Y | Consulting: | Y |
| Manufacturing: | Tissue Replacement: | Equipment/Supplies: | Arrays: | | Blood Collection: | |
| Genetics: | | Research & Development Services: | Database Management: | Y | Drug Delivery: | |
| | | Diagnostics: | | | Drug Distribution: | |

## TYPES OF BUSINESS:

Market Research - Pharmaceuticals
Pharmaceutical Sales Tracking
Health Care Databases
Software-Sales Management & Market Research
Physician Profiling
Industry Audits
Prescription Tracking Reporting Services

## BRANDS/DIVISIONS/AFFILIATES:

MIDAS
Market Research Publications
Pharmaceutical World Review
ValueMedics Research LLC
MIHS Holdings, Inc.
HIS
Health Services Research Network
RMBC

## CONTACTS: *Note: Officers with more than one job title may be intentionally listed here more than once.*

David R. Carlucci, CEO
Giles V. J. Pajot, COO
Leslye G. Katz, CFO/Sr. VP
Karla L. Packer, Sr. VP-Human Resources
Robert H. Steinfeld, General Counsel/Sr. VP/Corp. Sec.
John R. Walsh, Sr. VP-Strategy & Bus. Dev.
Jeffrey J. Ford, Treas./VP
Murray L. Aitken, Sr. VP-Healthcare Insight
Tatsuyuki Saeki, Pres., Japan
Kevin Knightly, Sr. VP-Bus Line Mgmt.
William J. Nelligan, Pres., Americas
David R. Carlucci, Chmn.
Adel Al-Saleh, Pres., EMEA

| Phone: 203-845-5200 | Fax: |
|---|---|
| Toll-Free: | |
| Address: 901 Main Ave., Ste. 612, Norwalk, CT 06851-1187 US | |

## GROWTH PLANS/SPECIAL FEATURES:

IMS Health, Inc. is a leading global provider of market intelligence to the pharmaceutical and health care industries. IMS offers such products and services as portfolio optimization capabilities; launch and brand management solutions; sales force effectiveness innovations; managed markets and consumer health offerings; and consulting and services solutions that improve ROI and the delivery of quality healthcare worldwide. The company's information products use data secured from a worldwide network of suppliers in over 100 countries. The firm's sales force effectiveness products generated roughly 45% of its worldwide revenue in 2008 and include sales territory reporting services; prescription tracking reporting services; and sales and account management services. Portfolio optimization products generated roughly 28% of IMS's revenue, and include pharmaceutical, medical, hospital and prescription audits; MIDAS, its online multinational integrated data analysis tool used to assess global pharmaceutical information and trends; other portfolio optimization reports, including personal care reports, reports of bulk chemical shipments and Market Research Publications such as the Pharmaceutical World Review; and consulting and services. Launch, brand management and other offerings comprised roughly 27% of the company's 2007 revenue. Launch and brand management offerings combine information, analytical tools and consulting and services to address client needs relevant to the management of each stage of the lifecycle of their pharmaceutical brands. Additional products offered by IMS include information to quantify the effects of managed markets on the pharmaceutical and healthcare industries and product movement, market share and pricing information for over-the-counter, personal care, patient care and nutritional products. In April 2008, the company joined with 10 academic researchers in medicine, public health and health economics to establish the Health Services Research Network, a consortium that will address healthcare issues. In August of the same year, the firm acquired RMBC, a Russia-based provider of pharmaceutical market intelligence and analytics.

## FINANCIALS: Sales and profits are in thousands of dollars—add 000 to get the full amount. 2008 Note: Financial information for 2008 was not available for all companies at press time.

| | | |
|---|---|---|
| 2008 Sales: $2,329,528 | 2008 Profits: $311,250 | **U.S. Stock Ticker: RX** |
| 2007 Sales: $2,192,571 | 2007 Profits: $234,040 | **Int'l Ticker:** Int'l Exchange: |
| 2006 Sales: $1,958,588 | 2006 Profits: $315,511 | Employees: 7,500 |
| 2005 Sales: $1,754,791 | 2005 Profits: $284,091 | Fiscal Year Ends: 12/31 |
| 2004 Sales: $1,569,045 | 2004 Profits: $285,422 | Parent Company: |

## SALARIES/BENEFITS:

| Pension Plan: | ESOP Stock Plan: | Profit Sharing: | Top Exec. Salary: $850,000 | Bonus: $567,500 |
|---|---|---|---|---|
| Savings Plan: | Stock Purch. Plan: | | Second Exec. Salary: $725,000 | Bonus: $364,400 |

## OTHER THOUGHTS:

**Apparent Women Officers or Directors**: 4
**Hot Spot for Advancement for Women/Minorities**: Y

## LOCATIONS: ("Y" = Yes)

| West: | Southwest: | Midwest: | Southeast: | Northeast: | International: |
|---|---|---|---|---|---|
| Y | | | | Y | Y |

Note: Financial information, benefits and other data can change quickly and may vary from those stated here.

# INCYTE CORP

**www.incyte.com**

Industry Group Code: 325412  **Ranks within this company's industry group:** Sales: 132   Profits: 164

| Drugs: | | Other: | | Clinical: | Computers: | Services: |
|---|---|---|---|---|---|---|
| Discovery: | Y | AgriBio: | | Trials/Services: | Hardware: | Specialty Services: |
| Licensing: | Y | Genetic Data: | Y | Labs: | Software: | Consulting: |
| Manufacturing: | | Tissue Replacement: | | Equipment/Supplies: | Arrays: | Blood Collection: |
| Genetics: | | | | Research & Development Services: | Database Management: | Drug Delivery: |
| | | | | Diagnostics: | | Drug Distribution: |

## TYPES OF BUSINESS:

Drug Discovery & Development
Genetic Information Research
Gene Expression Services

## BRANDS/DIVISIONS/AFFILIATES:

INCB18424
INCB28050
INCB13739
INCB7839

## CONTACTS: *Note: Officers with more than one job title may be intentionally listed here more than once.*

Paul A. Friedman, CEO
Paul A. Friedman, Pres.
David C. Hastings, CFO/Exec. VP
Paula J. Swain, Exec. VP-Human Resources
Brian W. Metcalf, Chief Drug Discovery Scientist/Exec. VP
Patricia A. Schreck, General Counsel/Exec. VP
Patricia S. Andrews, Chief Commercial Officer/Exec. VP
Richard U. DeSchutter, Chmn.

| Phone: 302-498-6700 | Fax: 302-425-2750 |
|---|---|
| Toll-Free: | |
| Address: Rte. 141 & Henry Clay Rd., Bldg. E336, Wilmington, DE 19880 US | |

## GROWTH PLANS/SPECIAL FEATURES:

Incyte Corp., formerly Incyte Genomics, discovers and develops small molecule drugs for the treatment of HIV, diabetes, oncology and inflammation. The company currently has multiple products in its pipeline, all within preclinical, Phase I and Phase II clinical studies. The firm has three active, two pending and four inactive programs, with the pending and inactive programs awaiting clinical trials or external partners. Targets for the active programs include three target inhibitors: Janus Kinase (JAK), HSD1 and Sheddase. The JAK program has been given the highest priority, and products in development are INCB18424 (oral) for myelofibrosis, polycythemia vera/essential thrombocythemia and rheumatoid arthritis; INCB18424 (topical) for psoriasis; and INCB28050 for rheumatoid arthritis. The drugs in development are designed to influence inflammatory cytokines that lead to the aforementioned conditions. HSD1 product INCB 13739 is being developed for type II diabetes and is designed to reduce insulin resistance caused by cortisol. The Sheddase program, INCB7839, aims to treat breast cancer by inhibiting HER pathway signaling. Pending programs target CMET and IDO, and are geared towards solid cancers and general oncology. The inactive programs plan to treat multiple myeloma, prostate cancer, type II diabetes, multiple sclerosis and HIV. In these programs, Incyte is developing CCR5 Antagonists, which belong to a new class of drugs called HIV Entry Inhibitors. The product works by blocking HIV before the virus enters the cell and begins the replication process. In May 2009, the company announced plans to initiate Phase III trials for INCB18424, the developmental treatment for myelofibrosis.

Incyte offers its employees medical, vision and dental plans; life insurance; disability coverage; a 401(k) plan; tuition reimbursement; employee discounts; referral and spot bonuses; and variable compensation plans.

## FINANCIALS: Sales and profits are in thousands of dollars—add 000 to get the full amount. 2008 Note: Financial information for 2008 was not available for all companies at press time.

| | | |
|---|---|---|
| 2008 Sales: $3,919 | 2008 Profits: $-178,920 | **U.S. Stock Ticker:** INCY |
| 2007 Sales: $34,440 | 2007 Profits: $-86,881 | **Int'l Ticker:** Int'l Exchange: |
| 2006 Sales: $27,643 | 2006 Profits: $-74,166 | Employees: 196 |
| 2005 Sales: $7,846 | 2005 Profits: $-103,043 | Fiscal Year Ends: 12/31 |
| 2004 Sales: $14,146 | 2004 Profits: $-164,817 | Parent Company: |

## SALARIES/BENEFITS:

| Pension Plan: | ESOP Stock Plan: | Profit Sharing: | Top Exec. Salary: $587,933 | Bonus: $431,842 |
|---|---|---|---|---|
| Savings Plan: Y | Stock Purch. Plan: Y | | Second Exec. Salary: $397,789 | Bonus: $194,786 |

## OTHER THOUGHTS:

**Apparent Women Officers or Directors:** 2
**Hot Spot for Advancement for Women/Minorities:** Y

## LOCATIONS: ("Y" = Yes)

| West: | Southwest: | Midwest: | Southeast: | Northeast: | International: |
|---|---|---|---|---|---|
| Y | | | | Y | |

# INFINITY PHARMACEUTICALS INC

www.infi.com

Industry Group Code: 541712 **Ranks within this company's industry group:** Sales: 10  Profits: 5

| Drugs: | | Other: | Clinical: | Computers: | Services: | |
|---|---|---|---|---|---|---|
| Discovery: | Y | AgriBio: | Trials/Services: | Hardware: | Specialty Services: | |
| Licensing: | | Genetic Data: | Labs: | Software: | Consulting: | |
| Manufacturing: | | Tissue Replacement: | Equipment/Supplies: | Arrays: | Blood Collection: | |
| Genetics: | | | Research & Development Services: | Database Management: | Drug Delivery: | Y |
| | | | Diagnostics: | | Drug Distribution: | |

## TYPES OF BUSINESS:

Research & Development Services
Drug Discovery Services
Drugs-Cancer

## BRANDS/DIVISIONS/AFFILIATES:

IPI-504
IPI-493
IPI-926

## CONTACTS: Note: Officers with more than one job title may be intentionally listed here more than once.

Steven H. Holtzman, CEO
Adelene Q. Perkins, Pres.
Jeanette W. Kohlbrenner, Sr. Dir.-Human Resources
Julian Adams, Chief Scientific Officer/Pres., R&D
John J. Keilty, Sr. Dir.-IT & Informatics
Jeffrey K. Tong, VP-Corp. & Prod. Dev.
Gerald E. Quirk, General Counsel/VP
John F. McPherson, VP-Oper. & Facilities
Adelene Q. Perkins, Chief Bus. Officer
Steven J. Kafka, VP-Finance
David S. Grayzel, VP-Clinical Dev. & Medical Affairs
Vito J. Palombella, VP-Drug Discovery
Michael S. Curtis, VP-Pharmaceutical Dev.
Steven H. Holtzman, Chmn.

| Phone: 617-453-1000 | Fax: 617-453-1001 |
|---|---|
| Toll-Free: | |
| Address: 780 Memorial Dr., Cambridge, MA 02139 US | |

## GROWTH PLANS/SPECIAL FEATURES:

Infinity Pharmaceuticals, Inc. (IPI) discovers, develops and delivers medicines for the treatment of cancer and related conditions. IPI has a pipeline of small molecule drug technologies for multiple cancer indications. The company's lead product candidate, IPI-504, developed in collaboration with AstraZeneca/MedImmune (AZ/MI), is currently undergoing a Phase I clinical trial in patients with Gleevec-refractory gastrointestinal stromal tumors, as well as a Phase I/II clinical trial in patients with advanced non-small cell lung cancer. IPI-504 inhibits heat shock protein 90 (hsp90), which inhibition is believed to have broad therapeutic potential for patients with solid and hematological tumors. IPI-504 is also in Phase Ib clinical trials in combination with Taxotere (docetaxel) in patients with advanced solid tumors. The company's next most advanced program, IPI-926, is also developed in collaboration with AZ/MI and is directed against the Hedgehog cell pathway, which normally regulates tissue and organ formation during embryonic development. When abnormally activated during adulthood, the Hedgehog pathway is believed to play a central role in the proliferation and survival of certain cancer-causing cells, and is implicated in many of the most deadly cancers. The drug is a derivative of the plant Veratrum californicum. The oral application of IPI-504, IPI-493, is being evaluated for Phase I trials. In December 2008, Infinity reacquired commercialization rights to its hsp90 program from AstraZeneca's MedImmune.

Infinity offers its employees health, dental, AD&D and life insurance; short- and long-term disability coverage; a 401(k) plan; profit sharing; stock options; flexible spending accounts; a first time homebuyer assistance program; tuition reimbursement; fitness room access; and transportation assistance.

## FINANCIALS: Sales and profits are in thousands of dollars—add 000 to get the full amount. 2008 Note: Financial information for 2008 was not available for all companies at press time.

| | | |
|---|---|---|
| 2008 Sales: $83,441 | 2008 Profits: $23,654 | U.S. Stock Ticker: INFI |
| 2007 Sales: $24,536 | 2007 Profits: $-16,898 | Int'l Ticker:    Int'l Exchange: |
| 2006 Sales: $18,495 | 2006 Profits: $-28,448 | Employees:   161 |
| 2005 Sales: $ 522 | 2005 Profits: $-36,369 | Fiscal Year Ends: 12/31 |
| 2004 Sales: $44,268 | 2004 Profits: $3,903 | Parent Company: |

## SALARIES/BENEFITS:

| Pension Plan: | ESOP Stock Plan: | Profit Sharing: Y | Top Exec. Salary: $480,000 | Bonus: $288,000 |
|---|---|---|---|---|
| Savings Plan: Y | Stock Purch. Plan: Y | | Second Exec. Salary: $390,000 | Bonus: $195,000 |

## OTHER THOUGHTS:

**Apparent Women Officers or Directors:** 3
**Hot Spot for Advancement for Women/Minorities:** Y

## LOCATIONS: ("Y" = Yes)

| West: | Southwest: | Midwest: | Southeast: | Northeast: | International: |
|---|---|---|---|---|---|
| | | | | Y | |

Note: Financial information, benefits and other data can change quickly and may vary from those stated here.

# INSITE VISION INC

www.insitevision.com

**Industry Group Code: 325412A  Ranks within this company's industry group:** Sales: 20    Profits: 18

| Drugs: | | Other: | Clinical: | Computers: | Services: |
|---|---|---|---|---|---|
| Discovery: | Y | AgriBio: | Trials/Services: | Hardware: | Specialty Services: |
| Licensing: | Y | Genetic Data: | Labs: | Software: | Consulting: |
| Manufacturing: | Y | Tissue Replacement: | Equipment/Supplies: | Arrays: | Blood Collection: |
| Genetics: | | | Research & Development Services: | Database Management: | Drug Delivery: |
| | | | Diagnostics: | | Drug Distribution: |

## TYPES OF BUSINESS:

Drug Delivery Systems
Drugs-Ophthalmic
Eye Disease Diagnostics
Ear Infection Diagnostics

## BRANDS/DIVISIONS/AFFILIATES:

DuraSite
OcuGene
AzaSite
AzaSite Plus
AzaSite Xtra
ISV-403
AquaSite
ISV-502

## CONTACTS: Note: Officers with more than one job title may be intentionally listed here more than once.

Louis Drapeau, Interim CEO
Louis Drapeau, CFO/VP
Kamran Hosseini, Chief Medical Officer
Surendra Patel, VP-Oper.
Lyle M. Bowman, VP-Dev.
David F. Heniges, VP/Gen. Mgr.-Commercial Opportunities
Kamran Hosseini, VP-Clinical Affairs
Surendra Patel, VP-Quality
Evan Melrose, Chmn.

| Phone: 510-865-8800 | Fax: 510-865-5700 |
|---|---|
| Toll-Free: | |
| Address: 965 Atlantic Ave., Alameda, CA 94501 US | |

## GROWTH PLANS/SPECIAL FEATURES:

InSite Vision, Inc. is an ophthalmic product development company focusing on therapies that treat ocular infections, diseases and glaucoma. Products are based on proprietary DuraSite eye drop drug delivery technology, a patented system that allows medication to stay in the eye for several hours rather than just a few minutes, reducing the number of doses needed and lowering the potential for adverse side effects. InSite's lead product, AzaSite (a DuraSite formulation of azithromycin, a broad spectrum antibiotic) has been approved by the FDA for the treatment of bacterial conjunctivitis. InSite has entered into an initial agreement with Inspire Pharmaceuticals to commercialize AzaSite in the U.S. and Canada. Two other products are also in development for ocular infections in the AzaSite family: AzaSite Plus, currently in Phase III trials; and AzaSite Xtra, currently in preclinical trails. InSite's first product utilizing the DuraSite technology, AquaSite dry eye treatment, was launched as an over-the-counter medication by CIBA Vision. OcuGene, the company's first genetic tool, is commercially available to those with glaucoma. ISV-403, a treatment for bacterial infection, was sold to Bausch and Lomb, Inc. and is nearing FDA approval. In March 2009, InSite received regulatory approval to market Azasite in Canada. Also in March 2009, the firm announced a broad restructuring program in which it eliminated about 52% of its workforce. In June 2009, the company announced FDA approval for a Bausch and Lomb product based on DuraSite technology. The firm stands to gain royalties from the sale of this new product.

InSite offers its employees medical, dental, vision, life and health insurance; short- and long-term disability coverage; an employee assistance program; travel assistance; a 401(k) plan; stock options; an employee stock purchase plan; and tuition reimbursement.

## FINANCIALS: Sales and profits are in thousands of dollars—add 000 to get the full amount. 2008 Note: Financial information for 2008 was not available for all companies at press time.

| | | |
|---|---|---|
| 2008 Sales: $13,706 | 2008 Profits: $-21,310 | **U.S. Stock Ticker: ISV** |
| 2007 Sales: $23,761 | 2007 Profits: $5,535 | **Int'l Ticker:**    Int'l Exchange: |
| 2006 Sales: $  2 | 2006 Profits: $-16,611 | Employees:    36 |
| 2005 Sales: $  4 | 2005 Profits: $-15,215 | Fiscal Year Ends: 12/31 |
| 2004 Sales: $ 542 | 2004 Profits: $-5,514 | Parent Company: |

## SALARIES/BENEFITS:

| Pension Plan: | ESOP Stock Plan: | Profit Sharing: | Top Exec. Salary: $390,000 | Bonus: $200,000 |
|---|---|---|---|---|
| Savings Plan: Y | Stock Purch. Plan: Y | | Second Exec. Salary: $235,000 | Bonus: $ |

## OTHER THOUGHTS:

**Apparent Women Officers or Directors:** 2
**Hot Spot for Advancement for Women/Minorities:** Y

## LOCATIONS: ("Y" = Yes)

| West: | Southwest: | Midwest: | Southeast: | Northeast: | International: |
|---|---|---|---|---|---|
| Y | | | | | |

# INSMED INCORPORATED

**www.insmed.com**

**Industry Group Code: 325412  Ranks within this company's industry group: Sales: 112  Profits: 94**

| Drugs: | | Other: | | Clinical: | | Computers: | | Services: | |
|---|---|---|---|---|---|---|---|---|---|
| Discovery: | Y | AgriBio: | | Trials/Services: | | Hardware: | | Specialty Services: | |
| Licensing: | Y | Genetic Data: | | Labs: | | Software: | | Consulting: | |
| Manufacturing: | Y | Tissue Replacement: | | Equipment/Supplies: | | Arrays: | | Blood Collection: | |
| Genetics: | | | | Research & Development Services: | | Database Management: | | Drug Delivery: | |
| | | | | Diagnostics: | | | | Drug Distribution: | |

## TYPES OF BUSINESS:

Drugs-Metabolic & Endocrine Diseases
Drugs-Cancer
Drugs-Myotonic Muscular Dystrophy
Drugs-Amyotrophic Lateral Sclerosis

## BRANDS/DIVISIONS/AFFILIATES:

INSM-18
IPLEX
rhIGFBP-3

## CONTACTS: Note: Officers with more than one job title may be intentionally listed here more than once.

Geoffrey Allan, CEO
Geoffrey Allan, Pres.
Kevin P. Tully, CFO/Exec. VP
Steve Glover, Chief Business Officer
Doug Farrar, VP-Insmed Therapeutic Proteins
Geoffrey Allan, Chmn.

| **Phone:** 804-565-3000 | **Fax:** 804-565-3500 |
|---|---|
| **Toll-Free:** | |
| **Address:** 8720 Stony Point Pkwy., Ste. 200, Richmond, VA 23235 US | |

## GROWTH PLANS/SPECIAL FEATURES:

Insmed, Inc. is a biologics manufacturing and development stage company with expertise in recombinant protein drug development. In the proprietary protein field, the firm's lead product, IPLEX is FDA-approved. It is a composition of recombinant human IGF-1 and IGF binding protein 3. IPLEX is now being studied as a treatment for several medical conditions including Myotonic Muscular Dystrophy (MMD) and ALS (Lou Gehrig's Disease) The company has acquired non-exclusive patent rights to the use of an insulin-like growth factor (IGF-I) therapy for the treatment of extreme or severe insulin-resistant diabetes from Fujisawa Pharmaceutical Co., Ltd. Under the terms of the agreement, Insmed will obtain worldwide rights in territories, excluding Japan and including the U.S. and Europe. The company has two oncology compounds, rhIGFBP-3 and INSM-18, in development. Preclinical models show that one or both treatments interact with the IGF-1 system to reduce tumor growth in models of breast, prostate, lung, colorectal and head and neck cancers. In November 2008, the firm obtained royalty-free worldwide rights to administer compassionate treatments to patients with ALS despite the drug not having been cleared for that particular indication. In March 2009, Insmed sold its Follow-On Biologics (FOB) program to Merck for $130 million, including its Boulder, Colorado facility and offered positions to the staff assigned to it. Through this deal, Merck acquired the company's FOB developmental assets, including INS-19 and INS-20, which aimed to treat anemia, neutropenia and autoimmune diseases with recombinant human colony-stimulating factors.

## FINANCIALS: Sales and profits are in thousands of dollars—add 000 to get the full amount. 2008 Note: Financial information for 2008 was not available for all companies at press time.

| | | |
|---|---|---|
| 2008 Sales: $11,699 | 2008 Profits: $-15,667 | **U.S. Stock Ticker: INSM** |
| 2007 Sales: $7,581 | 2007 Profits: $-19,962 | **Int'l Ticker:**  Int'l Exchange: |
| 2006 Sales: $1,025 | 2006 Profits: $-56,139 | Employees:  97 |
| 2005 Sales: $ 131 | 2005 Profits: $-40,929 | Fiscal Year Ends: 12/31 |
| 2004 Sales: $ 137 | 2004 Profits: $-27,203 | Parent Company: |

## SALARIES/BENEFITS:

| Pension Plan: | ESOP Stock Plan: | Profit Sharing: | Top Exec. Salary: $438,408 | Bonus: $225,000 |
|---|---|---|---|---|
| Savings Plan: Y | Stock Purch. Plan: Y | | Second Exec. Salary: $311,827 | Bonus: $110,250 |

## OTHER THOUGHTS:

Apparent Women Officers or Directors:
Hot Spot for Advancement for Women/Minorities:

## LOCATIONS: ("Y" = Yes)

| West: | Southwest: | Midwest: | Southeast: | Northeast: | International: |
|---|---|---|---|---|---|
| | | | | Y | |

# INSPIRE PHARMACEUTICALS INC

## www.inspirepharm.com

**Industry Group Code: 325412  Ranks within this company's industry group:  Sales: 75    Profits: 139**

| Drugs: | | Other: | Clinical: | Computers: | Services: |
|---|---|---|---|---|---|
| Discovery: | Y | AgriBio: | Trials/Services: | Hardware: | Specialty Services: |
| Licensing: | Y | Genetic Data: | Labs: | Software: | Consulting: |
| Manufacturing: | Y | Tissue Replacement: | Equipment/Supplies: | Arrays: | Blood Collection: |
| Genetics: | | | Research & Development Services: | Database Management: | Drug Delivery: |
| | | | Diagnostics: | | Drug Distribution: |

## TYPES OF BUSINESS:

Drugs-Respiratory & Ocular
Cystic Fibrosis Treatment
Retinal Disease Treatment
Dry Eye Treatment
Allergic Conjunctivitis
Cardiovascular Disease Treatment

## BRANDS/DIVISIONS/AFFILIATES:

Restasis
Elestat
AzaSite
Prolacria
Bilastine
Denufosol Tetrasodium

## CONTACTS: Note: Officers with more than one job title may be intentionally listed here more than once.

Christy L. Shaffer, CEO
Christy L. Shaffer, Pres.
Thomas R. Staab, II, CFO
Benjamin R. Yerxa, Chief R&D Officer/Exec. VP
Joseph M. Spagnardi, General Counsel/Sr. VP/Sec.
Joseph K. Schachle, Chief Commercial & Corp. Oper. Officer/Exec. VP
R. Kim Brazzell, Head-Opthalmology Bus./Exec. VP
Kenneth B. Lee, Jr., Chmn.

| **Phone:** 919-941-9777 | **Fax:** 919-941-9797 |
|---|---|
| **Toll-Free:** 877-800-4536 | |
| **Address:** 4222 Emperor Blvd., Ste. 200, Durham, NC 27703-8466 US | |

## GROWTH PLANS/SPECIAL FEATURES:

Inspire Pharmaceuticals, Inc. is a biopharmaceutical company dedicated to discovering, developing and commercializing prescription pharmaceutical products in the opthamalic and respiratory/allergies areas. The firm also focuses on the design and synthesis of P2 receptor related agonists, where it has significant expertise. Inspire's ophthalmic products and product candidates are concentrated in the allergic conjunctivitis, dry eye disease, cystic fibrosis, acute cardiac care, seasonal allergic rhinitis and glaucoma indications. The firm's product portfolio includes Elestat for allergic conjunctivitis, Restasis for dry eye disease and AzaSite for bacterial conjunctivitis. Elestat and Restasis are licensed to Allergan and currently marketed in the U.S. AzaSite was developed, as a collaborative project, with InSite Vision. Products in clinical development include Prolacria (diquafosol tetrasodium) for dry eye, which has completed Phase III testing and has received two approvable letters from the FDA; denufosol tetrasodium for cystic fibrosis, currently in Phase III trials; AzaSite for blephartis, currently in Phase II; and INS115644 and INS117548 for glaucoma, both currently in Phase I. The company had been developing Epinastine for the treatment and prevention of seasonal allergic rhinitis with Boehringer Ingelheim International, but discontinued the drug due to unmet endpoints in Phase III trials. In October 2008, Inspire was issued a patent for Elestat for allergic conjunctivitis. In May 2009, the company announced it would lay off about 12% of its workforce and shift focus away from early drug development.

Inspire offers its employees medical, dental, life and AD&D insurance; short- and long-term disability coverage; a 401(k) plan; flexible spending accounts; and employee assistance program; and wellness benefits program.

## FINANCIALS:  Sales and profits are in thousands of dollars—add 000 to get the full amount. 2008 Note: Financial information for 2008 was not available for all companies at press time.

| | | |
|---|---|---|
| 2008 Sales: $70,498 | 2008 Profits: $-51,603 | **U.S. Stock Ticker: ISPH** |
| 2007 Sales: $48,665 | 2007 Profits: $-63,740 | **Int'l Ticker:**    Int'l Exchange: |
| 2006 Sales: $37,059 | 2006 Profits: $-42,115 | Employees:   250 |
| 2005 Sales: $23,266 | 2005 Profits: $-31,847 | Fiscal Year Ends: 12/31 |
| 2004 Sales: $11,068 | 2004 Profits: $-44,069 | Parent Company: |

## SALARIES/BENEFITS:

| | | | | |
|---|---|---|---|---|
| Pension Plan: | ESOP Stock Plan: | Profit Sharing: | Top Exec. Salary: $466,444 | Bonus: $185,306 |
| Savings Plan: Y | Stock Purch. Plan: | | Second Exec. Salary: $317,000 | Bonus: $107,250 |

## OTHER THOUGHTS:

**Apparent Women Officers or Directors:** 2
**Hot Spot for Advancement for Women/Minorities:** Y

## LOCATIONS: ("Y" = Yes)

| West: | Southwest: | Midwest: | Southeast: | Northeast: | International: |
|---|---|---|---|---|---|
| Y | | | | Y | |

# INSTITUT STRAUMANN AG                                    www.straumann.com

**Industry Group Code: 33911  Ranks within this company's industry group:** Sales: 8   Profits: 11

| Drugs: | Other: | Clinical: | Computers: | Services: |
|---|---|---|---|---|
| Discovery: | AgriBio: | Trials/Services: | Hardware: | Specialty Services: |
| Licensing: | Genetic Data: | Labs: | Software: Y | Consulting: |
| Manufacturing: | Tissue Replacement: Y | Equipment/Supplies: Y | Arrays: | Blood Collection: |
| Genetics: | | Research & Development Services: | Database Management: | Drug Delivery: |
| | | Diagnostics: | | Drug Distribution: |

## TYPES OF BUSINESS:

Dental Implants
Dental Tissue Regeneration
CAD/CAM Elements & Equipment

## BRANDS/DIVISIONS/AFFILIATES:

SLActive
BoneCeramic
etkon AG
Straumann CADCAM GmbH
Ormedent spol. s.r.o.
Straumann s.r.o.
VaLiD Dental-Medical
Etkon CAD-CAM Iberica S.L.

## CONTACTS: *Note: Officers with more than one job title may be intentionally listed here more than once.*

Gilbert Achermann, CEO
Gilbert Achermann, Pres.
Beat Spalinger, CFO
Franz Maier, Exec. VP-Sales
Sandro Matter, Exec. VP-Prod. Div.
Beat Spalinger, Exec. VP-Oper.
Mark Hill, VP-Corp. Comm.
Fabian Hildbrand, VP-Investor Rel.
Beat Spalinger, Exec. VP-Finance
Oskar Ronner, Vice Chmn.
Rudolf Maag, Chmn.

| **Phone:** 41-61-965-11-11 | **Fax:** 41-61-965-11-01 |
|---|---|
| **Toll-Free:** | |
| **Address:** Peter Merian-Weg 12, Basel, CH 4052 Switzerland | |

## GROWTH PLANS/SPECIAL FEATURES:

Institut Straumann AG (Straumann) is a world leader in implant and restorative dentistry and a major provider of dental tissue regeneration products. It is a subsidiary of Straumann Holding AG, parent company of the Straumann Group. Straumann operates four geographical segments: Europe, which generated 60% of 2008 revenues; North America (21%); Asia/Pacific (12%); and the rest of the world (3%). It maintains subsidiaries and distributors in more than 60 countries, with its most important markets being Germany and the U.S. The company offers three basic products: implants, regenerative systems and computer aided design and manufacturing (CAD/CAM) technology. Implants are designed to mimic natural teeth as closely as possible, being more durable and supporting themselves better than conventional bridges. They include devices inserted at the bone or soft tissue level, surgical tools and 3D modeling software used to plan surgery. The firm also offers SLActive, an implant surface technology that decreases the healing time after surgery and increases general stability of the dental implants. Regenerative systems include products that support or repair oral structures, such as BoneCeramic, a synthetic bone graft substitute that augments the patient's jaw bone, reinforcing it to support other dental appliances, like implants. Lastly, CAD/CAM products, gained following the acquisition of etkon AG, are divided into prosthetic elements, such as crowns, inlays, overlays or bridges; and manufacturing equipment, such as software, scanners and milling units. Over the past half decade, the company has been acquiring its distributors, including most recently Ormedent spol. s.r.o. (now Straumann s.r.o.), its distributor in the Czech Republic and Slovakia, and VaLiD Dental-Medical, its Hungarian distributor. In June 2008, the firm acquired Etkon CAD-CAM Iberica S.L. and the remaining 4.6% of etkon AG, which it subsequently renamed Straumann CADCAM GmbH.

## FINANCIALS:  Sales and profits are in thousands of dollars—add 000 to get the full amount. 2008 Note: Financial information for 2008 was not available for all companies at press time.

| | | |
|---|---|---|
| 2008 Sales: $726,850 | 2008 Profits: $7,650 | **U.S. Stock Ticker:** |
| 2007 Sales: $678,000 | 2007 Profits: $167,100 | **Int'l Ticker:** STMN    Int'l Exchange: Zurich-SWX |
| 2006 Sales: $569,200 | 2006 Profits: $134,600 | Employees: 1,955 |
| 2005 Sales: $395,800 | 2005 Profits: $121,500 | Fiscal Year Ends: 12/31 |
| 2004 Sales: $327,082 | 2004 Profits: $78,210 | Parent Company: STRAUMANN HOLDING AG |

## SALARIES/BENEFITS:

| Pension Plan: | ESOP Stock Plan: | Profit Sharing: | Top Exec. Salary: $478,735 | Bonus: $109,844 |
|---|---|---|---|---|
| Savings Plan: | Stock Purch. Plan: | | Second Exec. Salary: $ | Bonus: $ |

## OTHER THOUGHTS:

**Apparent Women Officers or Directors:**
**Hot Spot for Advancement for Women/Minorities:**

## LOCATIONS: ("Y" = Yes)

| West: | Southwest: | Midwest: | Southeast: | Northeast: | International: |
|---|---|---|---|---|---|
| | | | | Y | Y |

Note: Financial information, benefits and other data can change quickly and may vary from those stated here.

# INTEGRA LIFESCIENCES HOLDINGS CORP          www.integra-ls.com

**Industry Group Code: 33911  Ranks within this company's industry group:  Sales: 9    Profits: 8**

| Drugs: | Other: | Clinical: | Computers: | Services: |
|--------|--------|-----------|------------|-----------|
| Discovery: | AgriBio: | Trials/Services: | Hardware: | Specialty Services: |
| Licensing: | Genetic Data: | Labs: | Software: | Consulting: |
| Manufacturing:  Y | Tissue Replacement: | Equipment/Supplies:  Y | Arrays: | Blood Collection: |
| Genetics: | | Research & Development Services: | Database Management: | Drug Delivery: |
| | | Diagnostics: | | Drug Distribution: |

## TYPES OF BUSINESS:

Medical Equipment Manufacturing
Implants & Biomaterials
Absorbable Medical Products
Tissue Regeneration Technology
Neurosurgery Products
Skin Replacement Products

## BRANDS/DIVISIONS/AFFILIATES:

Integra NeuroSciences
Theken Spine LLC
Theken Disc LLC
Therics LLC
Minnesota Scientific, Inc.
Omni-Tract Surgical

## CONTACTS:  *Note: Officers with more than one job title may be intentionally listed here more than once.*

Stuart M. Essig, CEO
Gerard S. Carlozzi, COO/Exec. VP
Stuart M. Essig, Pres.
John B. Henneman III, CFO/Exec. VP
Deborah A. Leonetti, Chief Mktg. Officer
Richard D. Gorelick, Sr. VP-Human Resources
Simon J. Archibald, Chief Scientific Officer
Gabrielle Wolfson, CIO
John B. Henneman, III, Exec. VP-Admin.
Richard D. Gorelick, General Counsel/Sr. VP/Sec.
James A. Oti, Sr. VP-Global Oper.
Maria Platsis, VP-Corp. Dev.
Angela Steinway, Mgr.-Investor Rel.
John B. Henneman III, Exec. VP-Finance
Nora Brennan, Treas.
Judith E. O'Grady, Sr. VP-Quality, Regulatory & Clinical Affairs
Jerry Corbin, Corp. Controller/VP
Brain Baker, Pres., Integra Pain Management
Richard E. Caruso, Chmn.
Eric Fourcault, Pres., EMEA Div.

| Phone: 609-275-0500 | Fax: 609-275-5363 |
|---|---|
| Toll-Free: 800-654-2873 | |
| Address: 311 Enterprise Dr., Plainsboro, NJ 08536 US | |

## GROWTH PLANS/SPECIAL FEATURES:

Integra Lifesciences Holdings Corporation develops, manufactures and markets medical devices for use in cranial and spinal procedures, peripheral nerve repair, small bone and joint injuries and the repair and reconstruction of soft tissue. The company divides its products into three groups: neurosciences, orthopedics and medical instruments. The neurosurgical products group include dural grafts for the repair of the dura mater; ultrasonic surgery systems for tissue ablation; cranial stabilization and brain retraction systems; systems for measurement of brain parameters; and devices that assist in accessing the cranial cavity and draining excess cerebrospinal fluid. The orthopedics group includes specialty metal implants for surgery of the extremities and spine; orthobiologics products for repair and grafting of bone; dermal regeneration products; tissue engineered wound dressings; and nerve and tendon repair products. The firm's medical instruments products include a range of specialty and general surgical and dental instruments and surgical lighting, sold primarily to hospitals, outpatient surgery centers and physician, veterinarian and dental practices. As well as marketing its products through its subsidiaries, the firm has strategic alliances with Johnson & Johnson; Medtronic; Wyeth and Zimmer; and others. Outside the U.S., Integra sells its products directly in Europe, Canada, Australia and New Zealand, and through distributors elsewhere. In August 2008, the company acquired Theken Spine LLC, Theken Disc LLC and Therics LLC (collectively Therics), spinal implant technologies developers, for $72 million. In November 2008, Integra acquired its Australia and New Zealand distributors; in December, the firm also acquired Minnesota Scientific, Inc., doing business as Omni-Tract Surgical, a developer of table mounted retractors.

Integra offers its employees medical, dental and vision plans; flexible spending accounts; an employee stock purchase plan; a 401(k) plan; and educational assistance.

## FINANCIALS:  Sales and profits are in thousands of dollars—add 000 to get the full amount. 2008 Note: Financial information for 2008 was not available for all companies at press time.

| | | |
|---|---|---|
| 2008 Sales: $654,604 | 2008 Profits: $34,933 | **U.S. Stock Ticker: IART** |
| 2007 Sales: $550,459 | 2007 Profits: $33,471 | **Int'l Ticker:**    Int'l Exchange: |
| 2006 Sales: $419,297 | 2006 Profits: $29,407 | Employees:  2,800 |
| 2005 Sales: $277,935 | 2005 Profits: $37,194 | Fiscal Year Ends: 12/31 |
| 2004 Sales: $229,825 | 2004 Profits: $17,197 | Parent Company: |

## SALARIES/BENEFITS:

| Pension Plan: | ESOP Stock Plan: | Profit Sharing: | Top Exec. Salary: $600,000 | Bonus: $ |
|---|---|---|---|---|
| Savings Plan: Y | Stock Purch. Plan: Y | | Second Exec. Salary: $444,808 | Bonus: $ |

## OTHER THOUGHTS:

**Apparent Women Officers or Directors: 7**
**Hot Spot for Advancement for Women/Minorities: Y**

## LOCATIONS: ("Y" = Yes)

| West: | Southwest: | Midwest: | Southeast: | Northeast: | International: |
|-------|-----------|----------|-----------|-----------|---------------|
| Y | | Y | | Y | Y |

# INTERMUNE INC

**www.intermune.com**

**Industry Group Code:** 325412 **Ranks within this company's industry group:** Sales: 86   Profits: 158

| Drugs: | | Other: | Clinical: | Computers: | Services: |
|---|---|---|---|---|---|
| Discovery: | Y | AgriBio: | Trials/Services: | Hardware: | Specialty Services: |
| Licensing: | Y | Genetic Data: | Labs: | Software: | Consulting: |
| Manufacturing: | | Tissue Replacement: | Equipment/Supplies: | Arrays: | Blood Collection: |
| Genetics: | | | Research & Development Services: | Database Management: | Drug Delivery: |
| | | | Diagnostics: | | Drug Distribution: |

## TYPES OF BUSINESS:
Drugs-Pulmonology & Hepatology
Infectious Disease & Cancer Treatments
Drugs-Osteopetrosis
Drugs-Chronic Granulomatous Disease

## BRANDS/DIVISIONS/AFFILIATES:
Actimmune
Pirfenidone
ITMN-191

## CONTACTS: *Note: Officers with more than one job title may be intentionally listed here more than once.*
Daniel G. Welch, CEO
Daniel G. Welch, Pres.
John Hodgman, CFO/Sr. VP
Howard Simon, Sr. VP-Human Resources & Corp. Svcs.
Williamson Bradford, Sr. VP-Clinical Science & Biometrics
Robin Steele, General Counsel/Sr. VP/Corp. Sec.
Steven Porter, Chief Medical Officer
Marianne Armstrong, Chief Medical Affairs & Regulatory Officer
Daniel G. Welch, Chmn.

| Phone: 415-466-2200 | Fax: 415-466-2300 |
|---|---|

**Toll-Free:**

**Address:** 3280 Bayshore Blvd., Brisbane, CA 94005 US

## GROWTH PLANS/SPECIAL FEATURES:

InterMune, Inc. is a biotechnology company that develops and commercializes products and therapies for pulmonary and hepatic diseases. The company's core and sole approved product is Actimmune, which is used for the treatment of severe, malignant osteopetrosis and chronic granulomatous disease. The active ingredient in Actimmune, interferon gamma-1b, consists of a bioengineered version of a human protein that acts as a potent stimulator of the immune system. The company's developing drug pipeline falls into two categories, pulmonology and hepatology. InterMune is currently developing one pulmonary therapy for the treatment of idiopathic pulmonary fibrosis (IPF), pirfenidone. Pirfenidone, currently in Phase III trials and granted fast track treatment by the FDA, is an orally available small molecule that shows potential in inhibiting collagen synthesis and decreasing fibroblast proliferation. The company, along with co-licensor KDL GmbH, bought back a former license agreement for the drug from Marnac, Inc. In hepatology, the company is developing ITMN-191 as a treatment for hepatitis C virus (HCV) infections and a protease inhibitor program. In October 2008, the company received regulatory approval to market pirfenidone for IPF in Japan. In February 2009, the firm was issued a U.S. patent for ITMN-191.

InterMune offers its employees medical, dental and vision plans; short- and long-term disability coverage; life and AD&D coverage; a 401(k) plan; tuition reimbursement; stock options; flexible spending accounts; a 529 college savings plan; and employee referral benefits.

## FINANCIALS: Sales and profits are in thousands of dollars—add 000 to get the full amount. 2008 Note: Financial information for 2008 was not available for all companies at press time.

| | | |
|---|---|---|
| 2008 Sales: $48,152 | 2008 Profits: $-97,739 | **U.S. Stock Ticker:** ITMN |
| 2007 Sales: $66,692 | 2007 Profits: $-89,602 | **Int'l Ticker:**   **Int'l Exchange:** |
| 2006 Sales: $90,784 | 2006 Profits: $-107,206 | Employees:   130 |
| 2005 Sales: $110,496 | 2005 Profits: $-5,235 | Fiscal Year Ends: 12/31 |
| 2004 Sales: $128,680 | 2004 Profits: $-59,478 | Parent Company: |

## SALARIES/BENEFITS:

| | | | | |
|---|---|---|---|---|
| Pension Plan: | ESOP Stock Plan: | Profit Sharing: | Top Exec. Salary: $631,110 | Bonus: $473,332 |
| Savings Plan: Y | Stock Purch. Plan: Y | | Second Exec. Salary: $340,026 | Bonus: $75,000 |

## OTHER THOUGHTS:
**Apparent Women Officers or Directors:** 2
**Hot Spot for Advancement for Women/Minorities:** Y

## LOCATIONS: ("Y" = Yes)

| West: | Southwest: | Midwest: | Southeast: | Northeast: | International: |
|---|---|---|---|---|---|
| Y | | | | | |

Note: Financial information, benefits and other data can change quickly and may vary from those stated here.

# INTERNATIONAL ISOTOPES INC

## www.intisoid.com

Industry Group Code: 325  Ranks within this company's industry group: Sales: 7  Profits: 7

| Drugs: | | Other: | | Clinical: | | Computers: | | Services: | |
|---|---|---|---|---|---|---|---|---|---|
| Discovery: | | AgriBio: | | Trials/Services: | | Hardware: | | Specialty Services: | Y |
| Licensing: | | Genetic Data: | | Labs: | | Software: | | Consulting: | |
| Manufacturing: | Y | Tissue Replacement: | | Equipment/Supplies: | Y | Arrays: | | Blood Collection: | |
| Genetics: | | | | Research & Development Services: | | Database Management: | | Drug Delivery: | |
| | | | | Diagnostics: | | | | Drug Distribution: | |

## TYPES OF BUSINESS:

Radioactive Isotopes Manufacturing
Nuclear Medicine Standards Publishing
Assay Analysis Services
Fluorine Gas
Depleted Uranium Oxide Products
Gemstone Processing

## BRANDS/DIVISIONS/AFFILIATES:

International Isotopes Idaho, Inc.
International Isotopes Flourine Products, Inc.
International Isotopes Transportation Svcs., Inc.
RadQual LLC

## CONTACTS: Note: Officers with more than one job title may be intentionally listed here more than once.

Steve T. Laflin, CEO
Steve T. Laflin, Pres.
Laurie A. McKenzie-Carter, CFO
Darin Lords, Mgr.-Cobalt Prod.
Chuckie Neitzel, Dir.-Admin. & Customer Svc.
John Miller, Radiation & Industrial Safety Officer
Jody Henley, Mgr.-Quality Assurance
Ralph M. Richart, Chmn.

| Phone: 208-524-5300 | Fax: 208-524-1411 |
|---|---|
| Toll-Free: 800-699-3108 | |
| Address: 4137 Commerce Cir., Idaho Falls, ID 83401 US | |

## GROWTH PLANS/SPECIAL FEATURES:

International Isotopes, Inc. (INIS) manufactures high purity fluoride gas and depleted uranium oxide products using the Fluorine Extraction Process (FEP). Subsidiaries of the firm include International Isotopes Idaho, Inc.; International Isotopes Flourine Products, Inc.; and International Isotopes Transportation Services, Inc. (IITS). The company operates in six segments: nuclear medicine standards (NMS); cobalt products; radiochemical products; fluorine products; radiological services; and transportation. The NMS segment consists of the manufacture of sources and standards associated with SPECT (single photon emission computed tomography), patient positioning and calibration or operational testing of dose measuring equipment for the nuclear pharmacy business. These items include flood sources; dose calibrators; rod sources; flexible and rigid rulers; spot markers; pen point markers; and specialty items. INIS is an exclusive manufacturer of these products for RadQual LLC. In the cobalt products segment, the firm manufactures bulk cobalt; fabricates cobalt capsules for teletherapy or irradiation devices; and recycles expended cobalt sources. In the radiochemical products segment, the company produces and distributes various isotopically pure radiochemicals for medical, industrial and research applications. The fluorine products segment uses seven patents for the FEP to produce high purity fluoride products such as germanium tetrafluoride, silicon tetrafluoride and boron trifluoride. The radiological services segment includes a variety of services, the largest of which is processing gemstones that have undergone irradiation for color enhancement. INIS has an exclusive contract with one customer for gemstone processing that accounts for most of the segment's sales. The transportation segment, through IITS, provides transportation for INIS's products and offers for-hire transportation services of hazardous and non-hazardous cargo materials. In March 2009, the company announced selecting a site in Lea County, New Mexico for the construction of its depleted uranium de-conversion and fluorine extraction processing facility, which will be one of the first commercial uranium de-conversion facilities in the U.S.

## FINANCIALS: Sales and profits are in thousands of dollars—add 000 to get the full amount. 2008 Note: Financial information for 2008 was not available for all companies at press time.

| | | |
|---|---|---|
| 2008 Sales: $5,602 | 2008 Profits: $-2,167 | U.S. Stock Ticker: INIS |
| 2007 Sales: $4,691 | 2007 Profits: $-1,719 | Int'l Ticker:    Int'l Exchange: |
| 2006 Sales: $4,470 | 2006 Profits: $-1,037 | Employees:   29 |
| 2005 Sales: $2,985 | 2005 Profits: $- 983 | Fiscal Year Ends: 12/31 |
| 2004 Sales: $2,848 | 2004 Profits: $- 845 | Parent Company: |

## SALARIES/BENEFITS:

| Pension Plan: | ESOP Stock Plan: | Profit Sharing: | Top Exec. Salary: $166,146 | Bonus: $30,000 |
|---|---|---|---|---|
| Savings Plan: | Stock Purch. Plan: | | Second Exec. Salary: $67,479 | Bonus: $10,000 |

## OTHER THOUGHTS:

Apparent Women Officers or Directors: 2
Hot Spot for Advancement for Women/Minorities:

## LOCATIONS: ("Y" = Yes)

| West: | Southwest: | Midwest: | Southeast: | Northeast: | International: |
|---|---|---|---|---|---|
| Y | | | | | |

# ISIS PHARMACEUTICALS INC

### www.isispharm.com

**Industry Group Code: 325412  Ranks within this company's industry group:** Sales: 65  Profits: 83

| Drugs: | | Other: | | Clinical: | Computers: | Services: | |
|---|---|---|---|---|---|---|---|
| Discovery: | Y | AgriBio: | | Trials/Services: | Hardware: | Specialty Services: | |
| Licensing: | Y | Genetic Data: | Y | Labs: | Software: | Consulting: | |
| Manufacturing: | Y | Tissue Replacement: | | Equipment/Supplies: | Arrays: | Blood Collection: | |
| Genetics: | | | | Research & Development Services: | Database Management: | Drug Delivery: | |
| | | | | Diagnostics: | | Drug Distribution: | |

## TYPES OF BUSINESS:

Drugs-Antisense Technology
Antisense Technology
Biosensors
Small-Molecule Drugs

## BRANDS/DIVISIONS/AFFILIATES:

Vitravene
Ibis Technologies
Mipomersen
TIGER
IBIS T-5000 Universal Pathogen Sensor
Isis Biosciences, Inc.

## CONTACTS: Note: Officers with more than one job title may be intentionally listed here more than once.

Stanley T. Crooke, CEO
B. Lynne Parshall, COO
B. Lynne Parshall, CFO/Exec. VP
C. Frank Bennett, Sr. VP-Research
B. Lynne Parshall, Sec.
Richard S. Geary, Sr. VP-Dev.
Martin P. Bedigian, Chief Medical Officer/VP
David J. Ecker, VP/Chief Scientific Officer-Ibis
Michael J. Treble, VP/Pres., Ibis Biosciences, Inc.
Stanley T. Crooke, Chmn.

| Phone: 760-931-9200 | Fax: 760-603-2700 |
|---|---|
| Toll-Free: | |
| Address: 1896 Rutherford Rd., Carlsbad, CA 92008-7208 US | |

## GROWTH PLANS/SPECIAL FEATURES:

Isis Pharmaceuticals, Inc. (Isis) is a pioneer in the area of RNA-based antisense technology, which involves direct application and interaction of gene sequence information to combat various diseases. The company successfully developed Vitravene, the first antisense drug to achieve marketing clearance, as a treatment for cytomegalovirus retinitis for patients living with AIDS. Isis subsequently licensed Vitravene to Novartis Ophthalmics. In addition to these drugs, the company has four compounds in the pre-clinical development stages and is partnered with various pharmaceutical companies for the development of 18 others, including treatments for cardiovascular diseases, diabetes, cancer, ALS (Lou Gherig's Disease), inflammatory diseases and more. Isis is the owner or exclusive licensee of approximately 1,600 RNA-based drug discovery and development patents. The company's Ibis Technologies division focuses on technologies to support the creation of biosensors to identify biological agents, and the development of small-molecule antibacterial and antiviral drugs that bind to RNA. Ibis has developed a biosensor system called TIGER (Triangulation Identification for Genetic Evaluation of Risks), designed to identify infectious organisms. Ibis Biosciences, Inc., a wholly-owned subsidiary of Isis, has developed and commercialized the Ibis T5000 Biosensor System for rapid identification and characterization of organisms. Mipomersen, a lipid-lowering cardiovascular drug developed for patients with familial hypercholesterolemia that seeks to inhibit the production of apolipoprotein B-100, a protein that carries cholesterol and triglyceride particles. In 2008, the drug was licensed to Genzyme and is in currently in Phase III testing. In January 2009, Abbot acquired an Ibis Biosciences, Inc., a subsidiary of Isis for a total of about $215 million.

Isis offers its employees medical and dental coverage; flexible spending accounts; life, long-term disability and AD&D coverage; a 401(k) plan; stock options; and a 529 college savings plan.

## FINANCIALS: Sales and profits are in thousands of dollars—add 000 to get the full amount. 2008 Note: Financial information for 2008 was not available for all companies at press time.

| | | |
|---|---|---|
| 2008 Sales: $107,190 | 2008 Profits: $-11,963 | **U.S. Stock Ticker: ISIS** |
| 2007 Sales: $58,344 | 2007 Profits: $-10,994 | **Int'l Ticker:** Int'l Exchange: |
| 2006 Sales: $14,859 | 2006 Profits: $-45,903 | Employees: 300 |
| 2005 Sales: $40,133 | 2005 Profits: $-72,401 | Fiscal Year Ends: 12/31 |
| 2004 Sales: $42,624 | 2004 Profits: $-142,864 | Parent Company: |

## SALARIES/BENEFITS:

| Pension Plan: | ESOP Stock Plan: | Profit Sharing: | Top Exec. Salary: $648,009 | Bonus: $438,009 |
|---|---|---|---|---|
| Savings Plan: Y | Stock Purch. Plan: Y | | Second Exec. Salary: $529,095 | Bonus: $334,499 |

## OTHER THOUGHTS:

**Apparent Women Officers or Directors:** 1
**Hot Spot for Advancement for Women/Minorities:**

## LOCATIONS: ("Y" = Yes)

| West: | Southwest: | Midwest: | Southeast: | Northeast: | International: |
|---|---|---|---|---|---|
| Y | | | | | |

Note: Financial information, benefits and other data can change quickly and may vary from those stated here.

# JAZZ PHARMACEUTICALS

**www.jazzpharmaceuticals.com**

Industry Group Code: 325412  Ranks within this company's industry group: Sales: 78  Profits: 166

| Drugs: | | Other: | | Clinical: | | Computers: | | Services: | |
|---|---|---|---|---|---|---|---|---|---|
| Discovery: | Y | AgriBio: | | Trials/Services: | | Hardware: | | Specialty Services: | |
| Licensing: | Y | Genetic Data: | | Labs: | | Software: | | Consulting: | |
| Manufacturing: | | Tissue Replacement: | | Equipment/Supplies: | | Arrays: | | Blood Collection: | |
| Genetics: | | | | Research & Development Services: | | Database Management: | | Drug Delivery: | |
| | | | | Diagnostics: | | | | Drug Distribution: | |

## TYPES OF BUSINESS:

Pharmaceuticals Discovery & Development
Neurological & Psychiatric Therapeutics

## BRANDS/DIVISIONS/AFFILIATES:

Xyrem
Luvox CR
JZP-6

## CONTACTS: Note: Officers with more than one job title may be intentionally listed here more than once.

Bruce C. Cozadd, CEO
Robert M. Myers, Pres.
Edwin W. Luker, VP-Sales
Heather McGaughey, VP-Human Resources
Mark G. Eller, Sr. VP-Research & Clinical Dev.
Michael DesJardin, Sr. VP-Prod. Dev.
Nandan Oza, VP-Mfg.
Carol A. Gamble, General Counsel/Sr. VP/Corp. Sec.
Joel M. Rothman, VP-Dev. Oper.
Diane R. Guinta, VP-Clinical R&D
Janne L. T. Wissel, Chief Regulatory & Compliance Officer/Sr. VP
P. J. Honerkamp, VP/Deputy General Counsel
Bruce C. Cozadd, Chmn.
Nandan Oza, VP-Supply Chain

| Phone: 650-496-3777 | Fax: 650-496-3781 |
|---|---|
| Toll-Free: | |
| Address: 3180 Porter Dr., Palo Alto, CA 94304 US | |

## GROWTH PLANS/SPECIAL FEATURES:

Jazz Pharmaceuticals, Inc. develops and commercializes products that address unmet medical needs in neurology and psychiatry. The firm's portfolio includes two marketed products, one product candidate in late Phase III clinical development and several product candidates in various stages of clinical development. Its marketed products are Xyrem and Luvox CR. Xyrem is one of the only products approved by the U.S. Food and Drug Administration (FDA) for the treatment of both excessive daytime sleepiness and cataplexy in patients with narcolepsy. Luvox CR, a once-daily extended-release formulation of selective serotonin reuptake inhibitor fluvoxamine, was approved by the FDA for the treatment of both obsessive-compulsive disorder and social anxiety disorder in February 2008. The firm shipped initial stocking orders of Luvox CR to its wholesale customers in March 2008 and began promoting the product through its specialty sales force in April 2008. Jazz is developing sodium oxybate (JZP-6), the active pharmaceutical ingredient in Xyrem, for the treatment of fibromyalgia. The firm's other product candidates are JZP-8 (intranasal clonazepam) for the treatment of recurrent acute repetitive seizures in epilepsy patients who continue to have seizures while on stable anti-epileptic regimens; JZP-4 (sodium and calcium channel antagonist) for the treatment of epilepsy and bipolar disorder; and JZP-7 (ropinirole gel) for the treatment of restless legs syndrome. In August 2008, Jazz sold its Antizol products for $5.8 million. In November 2008, the firm presented Phase III data demonstrating that JZP-6 improved pain and the core symptoms associated with fibromyalgia. In December 2008, Jazz announced plans to reduce its workforce by approximately 24% in an effort to lower operating expenses.

Jazz offers its employees medical and dependent care flexible spending accounts; a choice of Preferred Provider Organization (PPO) and Health Maintenance Organization (HMO) plans; dental and vision care insurance; and life, AD&D and disability insurance.

## FINANCIALS: Sales and profits are in thousands of dollars—add 000 to get the full amount. 2008 Note: Financial information for 2008 was not available for all companies at press time.

| | | |
|---|---|---|
| 2008 Sales: $67,514 | 2008 Profits: $-184,339 | **U.S. Stock Ticker:** JAZZ |
| 2007 Sales: $65,303 | 2007 Profits: $-138,778 | **Int'l Ticker:** Int'l Exchange: |
| 2006 Sales: $44,856 | 2006 Profits: $-59,391 | Employees: 216 |
| 2005 Sales: $21,442 | 2005 Profits: $-85,156 | Fiscal Year Ends: 12/31 |
| 2004 Sales: $ | 2004 Profits: $ | Parent Company: |

## SALARIES/BENEFITS:

| Pension Plan: | ESOP Stock Plan: | Profit Sharing: | Top Exec. Salary: $468,266 | Bonus: $ |
|---|---|---|---|---|
| Savings Plan: Y | Stock Purch. Plan: | | Second Exec. Salary: $444,096 | Bonus: $ |

## OTHER THOUGHTS:

Apparent Women Officers or Directors: 4
Hot Spot for Advancement for Women/Minorities: Y

## LOCATIONS: ("Y" = Yes)

| West: | Southwest: | Midwest: | Southeast: | Northeast: | International: |
|---|---|---|---|---|---|
| Y | | | | | |

# JOHNSON & JOHNSON

www.jnj.com

Industry Group Code: 325412 Ranks within this company's industry group: Sales: 1 Profits: 1

| Drugs: | | Other: | Clinical: | | Computers: | | Services: | |
|---|---|---|---|---|---|---|---|---|
| Discovery: | Y | AgriBio: | Trials/Services: | | Hardware: | | Specialty Services: | |
| Licensing: | | Genetic Data: | Labs: | | Software: | | Consulting: | |
| Manufacturing: | Y | Tissue Replacement: | Equipment/Supplies: | Y | Arrays: | | Blood Collection: | |
| Genetics: | | | Research & Development Services: | | Database Management: | | Drug Delivery: | |
| | | | Diagnostics: | Y | | | Drug Distribution: | |

## TYPES OF BUSINESS:

Personal Health Care & Hygiene Products
Sterilization Products
Surgical Products
Pharmaceuticals
Skin Care Products
Baby Care Products
Contact Lenses
Medical Equipment

## BRANDS/DIVISIONS/AFFILIATES:

Scios Inc
Centocor Inc
Alza Corp
Depuy Inc
Ethicon Inc
Cordis Corp
LifeScan Inc
Omrix Biopharmaceuticals Inc

## CONTACTS: Note: Officers with more than one job title may be intentionally listed here more than once.

William C. Weldon, CEO
Dominic J. Caruso, CFO
Kaye Foster-Cheek, VP-Human Resources
Russell C. Deyo, General Counsel/VP/Chief Compliance Officer
Nicholas J. Valeriani, VP-Strategy & Growth
Dominic J. Caruso, VP-Finance
Colleen Goggins, Chmn.-Consumer Group
Sheri S. McCoy, Chmn.-Pharmaceutical Group
Alex Gorsky, Chmn.-Surgical Care Group
Donald M. Casey, Jr., Chmn.-Comprehensive Care Group
William C. Weldon, Chmn.

| Phone: 732-524-0400 | Fax: 732-214-0332 |
|---|---|
| Toll-Free: | |
| Address: 1 Johnson & Johnson Plz., New Brunswick, NJ 08933 US | |

## GROWTH PLANS/SPECIAL FEATURES:

Johnson & Johnson, founded in 1886, is one of the world's most comprehensive and well-known manufacturers of health care products. The firm owns more than 250 companies in over 90 countries and markets its products in almost every country in the world. Johnson & Johnson's worldwide operations are divided into three segments: consumer, pharmaceutical and medical devices and diagnostics. The company's principal consumer goods are personal care and hygiene products, including nonprescription drugs, adult skin and hair care, baby care, oral care, first aid and sanitary protection products. Major consumer brands include Mylanta, Band-Aid, Tylenol, Aveeno and Monistat. The pharmaceutical segment covers a wide spectrum of health fields, including antifungal, anti-infective, cardiovascular, dermatology, immunology, pain management, psychotropic and women's health. Among its pharmaceutical products are Risperdal, an antipsychotic used to treat schizophrenia, and Remicade for the treatment of Crohn's disease and rheumatoid arthritis. In the medical devices and diagnostics segment, Johnson & Johnson makes a number of products including suture and mechanical wound closure products, surgical instruments, disposable contact lenses, joint reconstruction products and intravenous catheters. Subsidiaries of the company include Cordis LLC, DePuy, Inc., Diabetes Diagnostics, Inc., Ethicon Endo-Surgery, Inc., LifeScan, Inc., McNeil Healthcare LLC, Neutrogena Corporation, SurgRx, Inc. and The Tylenol Company. In November 2008, Johnson & Johnson acquired HealthMedia, Inc. In December 2008, the company acquired Omrix Biopharmaceuticals, Inc. for $438 million. In January 2009, the firm completed its acquisition of Mentor Corporation, a producer of medical products for the aesthetic specialties market. In July 2009, Johnson & Johnson announced that it would invest approximately $1.5 billion for an 18.4% stake in Elan Corporation, an Ireland-based biotechnology firm. The transaction will also give the company certain rights related to Elan's Alzheimer's drug development program.

Johnson & Johnson offers its employees benefits that include medical coverage; an employee assistance program; health assessments and health counseling; and on-site fitness centers and fitness classes at certain locations.

## FINANCIALS: Sales and profits are in thousands of dollars—add 000 to get the full amount. 2008 Note: Financial information for 2008 was not available for all companies at press time.

| | | |
|---|---|---|
| 2008 Sales: $63,747,000 | 2008 Profits: $12,949,000 | U.S. Stock Ticker: JNJ |
| 2007 Sales: $61,095,000 | 2007 Profits: $10,576,000 | Int'l Ticker: Int'l Exchange: |
| 2006 Sales: $53,324,000 | 2006 Profits: $11,053,000 | Employees: 118,700 |
| 2005 Sales: $50,514,000 | 2005 Profits: $10,060,000 | Fiscal Year Ends: 12/31 |
| 2004 Sales: $47,348,000 | 2004 Profits: $8,180,000 | Parent Company: |

## SALARIES/BENEFITS:

| Pension Plan: | ESOP Stock Plan: | Profit Sharing: | Top Exec. Salary: $1,792,019 | Bonus: $8,972,360 |
|---|---|---|---|---|
| Savings Plan: Y | Stock Purch. Plan: | | Second Exec. Salary: $1,042,404 | Bonus: $3,721,500 |

## OTHER THOUGHTS:

Apparent Women Officers or Directors: 5
Hot Spot for Advancement for Women/Minorities: Y

## LOCATIONS: ("Y" = Yes)

| West: | Southwest: | Midwest: | Southeast: | Northeast: | International: |
|---|---|---|---|---|---|
| Y | Y | Y | Y | Y | Y |

Note: Financial information, benefits and other data can change quickly and may vary from those stated here.

# JUBILANT BIOSYS LTD                                      www.jubilantbiosys.com

**Industry Group Code: 325414  Ranks within this company's industry group:** Sales:    Profits:

| Drugs: | | Other: | Clinical: | | Computers: | | Services: |
|---|---|---|---|---|---|---|---|
| Discovery: | Y | AgriBio: | Trials/Services: | | Hardware: | | Specialty Services: |
| Licensing: | | Genetic Data: | Labs: | | Software: | Y | Consulting: |
| Manufacturing: | | Tissue Replacement: | Equipment/Supplies: | | Arrays: | | Blood Collection: |
| Genetics: | | | Research & Development Services: | | Database Management: | | Drug Delivery: |
| | | | Diagnostics: | | | | Drug Distribution: |

## TYPES OF BUSINESS:
Biotech Drug Discovery & Development

## BRANDS/DIVISIONS/AFFILIATES:
Jubilant Organosys Ltd
Jubilant Group
Jubilant Discovery Research Center
iAnnotate
Legend

## CONTACTS: *Note: Officers with more than one job title may be intentionally listed here more than once.*
Sridhar (Sri) Mosur, CEO
Jonathan P. (Jon) Northrup, COO-Drug Dev.
Sridhar (Sri) Mosur, Pres.
Pankaj Garg, CFO-Global Drug Discovery & Dev.
Kankana Barua, VP/Head-Global Human Resources
Raman Govindarajan, Sr. VP-Discovery Biology & Translational Medicine
Amit Kumar Rustagi, Head-IT
Warren Stern, Sr. VP-Drug Dev.
Corey Jacklin, Dir.-Bus. Dev.
Sridhar (Sri) Mosur, CEO/Pres., Jubilant Discovery Services, Inc.
Robert W. Dougherty, Dir/Head-Invitro Biology&High-throughput Screening
V. N. Viswanadhan, Dir./Head-Computational Chemistry
Sriram Rajagopal, VP-Biology
Hari S. Bhartia, Chmn.

| Phone: 91-120-251-6601 | Fax: 91-120-251-6628 |
|---|---|
| Toll-Free: | |
| Address: 1A, Sec 16A, Noida,  201301 India | |

## GROWTH PLANS/SPECIAL FEATURES:

Jubilant Biosys Ltd. is an Indian drug discovery firm that primarily operates in collaboration with major pharmaceutical and biotech companies.   It is a subsidiary of Jubilant Organosys Ltd., the primary company in the pharmaceuticals branch of Indian conglomerate the Jubilant Group.   The company's primary facility is the Jubilant Discovery Research Center, a 125,000-square-foot lab located in Bangalore, India, which specializes in medicinal chemistry, structural biology, discovery biology, molecular modeling, pharmacology, toxicology, ADME (adsorption, distribution, metabolism and excretion) and pharmaceutical information technology (IT).   This facility is capable of assisting clients from the initial target identification of a new drug through to the investigational new drug (IND) application.   More specific examples of the facility's services include the following. Computational chemistry services primarily comprise computer modeling and analysis tools, including databases of drugs and other chemicals.   Structured biology services primarily relate to the crystallography of protein structures, including imaging proteins, purifying samples and generating expressions of protein variations.   Medicinal chemistry services include tailoring chemicals to hit specific targets, designing novel chemicals and providing biological profiles of new chemicals.   Drug metabolism and pharmacokinetics (DMPK) services include analyzing the ADME properties of a drug, such as its toxicology potential and what pathways it follows inside the body, as well as general quality assurance services.   In-vivo biology services comprise preclinical efficacy and safety testing for inflammation, cancer, pain and other therapeutic areas.   Lastly, pharmaceutical IT products services include application development and maintenance (ADM), remote IT infrastructure support and proprietary IT products specially designed for the pharmaceuticals market, including protein imaging solution iAnnotate and screening and data management program Legend.   In May 2009, the company agreed to assist AstraZeneca research new preclinical candidates, starting with the neuroscience area. In June 2009, the firm partnered with U.S.-based Endo Pharmaceuticals to develop several new oncology (cancer) treatments.

## FINANCIALS:  Sales and profits are in thousands of dollars—add 000 to get the full amount. 2008 Note: Financial information for 2008 was not available for all companies at press time.

| | | |
|---|---|---|
| 2008 Sales: $ | 2008 Profits: $ | **U.S. Stock Ticker: Subsidiary** |
| 2007 Sales: $ | 2007 Profits: $ | **Int'l Ticker:**    Int'l Exchange: |
| 2006 Sales: $ | 2006 Profits: $ | Employees: |
| 2005 Sales: $ | 2005 Profits: $ | Fiscal Year Ends: |
| 2004 Sales: $ | 2004 Profits: $ | Parent Company: JUBILANT ORGANOSYS LTD |

## SALARIES/BENEFITS:
| Pension Plan: | ESOP Stock Plan: | Profit Sharing: | Top Exec. Salary: $ | Bonus: $ |
|---|---|---|---|---|
| Savings Plan: | Stock Purch. Plan: | | Second Exec. Salary: $ | Bonus: $ |

## OTHER THOUGHTS:
**Apparent Women Officers or Directors**: 1
**Hot Spot for Advancement for Women/Minorities**:

## LOCATIONS: ("Y" = Yes)
| West: | Southwest: | Midwest: | Southeast: | Northeast: | International: |
|---|---|---|---|---|---|
| | | | | Y | Y |

# KENDLE INTERNATIONAL INC

**www.kendle.com**

Industry Group Code: 325412  Ranks within this company's industry group: Sales: 41  Profits: 45

| Drugs: | Other: | Clinical: | | Computers: | | Services: | |
|---|---|---|---|---|---|---|---|
| Discovery: | AgriBio: | Trials/Services: | Y | Hardware: | | Specialty Services: | Y |
| Licensing: | Genetic Data: | Labs: | | Software: | Y | Consulting: | Y |
| Manufacturing: | Tissue Replacement: | Equipment/Supplies: | | Arrays: | | Blood Collection: | |
| Genetics: | | Research & Development Services: | Y | Database Management: | | Drug Delivery: | |
| | | Diagnostics: | | | | Drug Distribution: | |

## TYPES OF BUSINESS:

Pharmaceutical Development-Clinical Trials
Statistical Analysis
Technical Writing
Regulatory Assistance
Consulting Services
Clinical Trial Software
Clinical Data Management
e-Learning

## BRANDS/DIVISIONS/AFFILIATES:

eKendleCollege
TrialWare
TrialWeb
TrialBase
TrialView
TriaLine
Early Stage
Late Stage

## CONTACTS: Note: Officers with more than one job title may be intentionally listed here more than once.

Candace Kendle, CEO
Christopher C. Bergen, COO
Simon Higginbotham, Pres.
Karl Brenkert, III, CFO/Sr. VP
Alan J. Boyce, Chief Mktg. Officer
Karen L. Crone, VP-Global Human Resources
Gary Wedig, CIO/VP
Karl Brenkert, III, Corp. Sec.
Lori Dorer, Dir.-Corp. Comm.
Michael Lawson, Dir.-Investor Rel.
Anthony L. Forcellini, Treas.
Martha Feller, Sr. VP-Global Clinical Dev.
Melanie A. Bruno, VP-Global Regulatory Affairs, Quality & Safety
Philip J.W. Davies, VP-Phase I
Patricia A. Steigerwald, VP-Global Late Phase
Candace Kendle, Chmn.
Ross J. Horsburgh, VP-Global Clinical Dev., Asia/Pacific

| Phone: 513-381-5550 | Fax: 513-381-5870 |
|---|---|
| Toll-Free: 800-733-1572 | |
| Address: 441 Vine St., Ste. 1200, Cincinnati, OH 45202 US | |

## GROWTH PLANS/SPECIAL FEATURES:

Kendle International, Inc. is a global clinical research organization that provides a broad range of Phase I-IV global clinical development services to the biopharmaceutical industry. The company supplements the research and development activities of biopharmaceutical companies by offering clinical research services and information technology designed to reduce the time and expense of drug development. The firm operates in two segments: early stage, which handles all Phase I testing services; and late stage, which handles all Phase II-IV services. Kendle's services include clinical trial management, clinical data management, statistical analysis, technical writing and regulatory consulting/representation. It a state-of-the-art clinical pharmacology unit in the Netherlands, where it offers Phase I clinical trials with drugs under development. The company's therapeutic expertise covers fields such as cardiovascular, dermatological, hematological, metabolic and respiratory health. The firm's proprietary TrialWare product line includes a database management system, TrialBase; a validated medical imaging system, TrialView; an interactive voice response patient randomization systsem, TriaLine; an Internet based collaborative tool, TrialWeb; a global project management system, TrialWatch; and a late phase technology system, Trial4. Additionally, the company operates Kendle College, an online e-learning division that runs seminars and training programs, focusing on the organization of clinical trials. In June 2008, Kendle International acquired Canadian company DecisionLine Clinical Research Corporation, providing Kendle with a new high-end medical facility in Toronto; the new facility has been incorporated into the firm's early stage division. In January 2009, Kendle International opened a new office in Ahmedabad, India.

The company offers its employees medical, dental and vision insurance; flexible spending accounts; life and AD&D insurance; a 401(k) plan; tuition reimbursement; and a profit sharing plan. The firm also offers opportunities for professional and personal development through eKendle College.

## FINANCIALS: Sales and profits are in thousands of dollars—add 000 to get the full amount. 2008 Note: Financial information for 2008 was not available for all companies at press time.

| | | |
|---|---|---|
| 2008 Sales: $678,581 | 2008 Profits: $29,397 | **U.S. Stock Ticker: KNDL** |
| 2007 Sales: $568,818 | 2007 Profits: $18,687 | Int'l Ticker:  Int'l Exchange: |
| 2006 Sales: $373,936 | 2006 Profits: $8,530 | Employees: 4,275 |
| 2005 Sales: $250,639 | 2005 Profits: $10,674 | Fiscal Year Ends: 12/31 |
| 2004 Sales: $215,868 | 2004 Profits: $3,572 | Parent Company: |

## SALARIES/BENEFITS:

| Pension Plan: | ESOP Stock Plan: | Profit Sharing: Y | Top Exec. Salary: $581,450 | Bonus: $84,000 |
|---|---|---|---|---|
| Savings Plan: Y | Stock Purch. Plan: | | Second Exec. Salary: $422,050 | Bonus: $48,160 |

## OTHER THOUGHTS:

**Apparent Women Officers or Directors:** 8
**Hot Spot for Advancement for Women/Minorities:** Y

## LOCATIONS: ("Y" = Yes)

| West: | Southwest: | Midwest: | Southeast: | Northeast: | International: |
|---|---|---|---|---|---|
| Y | | Y | | Y | Y |

Note: Financial information, benefits and other data can change quickly and may vary from those stated here.

# KERYX BIOPHARMACEUTICALS INC

**www.keryx.com**

**Industry Group Code:** 325412  **Ranks within this company's industry group:** Sales: 144  Profits: 141

| Drugs: | | Other: | Clinical: | Computers: | Services: |
|---|---|---|---|---|---|
| Discovery: | Y | AgriBio: | Trials/Services: | Hardware: | Specialty Services: |
| Licensing: | Y | Genetic Data: | Labs: | Software: | Consulting: |
| Manufacturing: | | Tissue Replacement: | Equipment/Supplies: | Arrays: | Blood Collection: |
| Genetics: | | | Research & Development Services: | Database Management: | Drug Delivery: |
| | | | Diagnostics: | | Drug Distribution: |

## TYPES OF BUSINESS:

Pharmaceuticals Development
Drugs-Cancer
Drugs-End Stage Renal Failure

## BRANDS/DIVISIONS/AFFILIATES:

ACCESS Oncology, Inc.
Accumin Diagnostics, Inc.
KRX-0401
Sulonex
Zerenex
Accumin
Neryx Biopharmaceuticals, Inc.

## CONTACTS: Note: Officers with more than one job title may be intentionally listed here more than once.

Ron Bentsur, CEO
James F. Oliviero, CFO
Michael P. Tarnok, Chmn.

| **Phone:** 212-531-5965 | **Fax:** 212-531-5961 |
|---|---|
| **Toll-Free:** | |
| **Address:** 750 Lexington Ave., 20th Fl., New York, NY 10022 US | |

## GROWTH PLANS/SPECIAL FEATURES:

Keryx Biopharmaceuticals, Inc. is a biopharmaceutical company focused on the development and commercialization of pharmaceutical products for the treatment of life-threatening diseases including diabetes and cancer. The company is also involved in evaluating potential compounds and companies for licensing or acquisition as well as opportunities to co-develop products with other firms. Keryx has two drugs in Phase II clinical trials. KRX-0401 (perifosine) is an oral anti-cancer drug designed to manipulate key signal pathways involved with cell growth, death, differentiation and survival. It specifically targets Akt, which is thought to be a major contribute to cancers that do not respond to other treatments. Preliminary data suggests that KRX-0401 would be most effective as a complementary treatment along with other anti-cancer agents. The drug appears to have less toxic effects than other cancer treatments, which means fewer side effects. Zerenex, a treatment for elevated phosphate levels (hyperphosphatemia) found in the end stages of renal disease (ESRD) and is currently in Phase II trials. The drug aims to be an alternative to aluminum-, calcium-, non-calcium and lanthanum-type phosphate binders which often have undesirable side effects that limit their effectiveness and viability in treatments. The company is focusing on these two drugs after discontinuing their Sulonex program, a drug designed to treat diabetic nephropathy, after it did not meet expectations in Phase III clinical trials. As a result, the firm decided to restructure and eliminate about 50% of its workforce in 2008. The company has not yet received approval for the sale of any of its drug candidates. Keryx owns three subsidiaries located in the U.S.: ACCESS Oncology, Inc.; Neryx Biopharmaceuticals, Inc.; and Accumin Diagnostics, Inc.

## FINANCIALS: Sales and profits are in thousands of dollars—add 000 to get the full amount. 2008 Note: Financial information for 2008 was not available for all companies at press time.

| | | |
|---|---|---|
| 2008 Sales: $1,283 | 2008 Profits: $-52,881 | **U.S. Stock Ticker: KERX** |
| 2007 Sales: $ 983 | 2007 Profits: $-90,062 | **Int'l Ticker:**     Int'l Exchange: |
| 2006 Sales: $ 431 | 2006 Profits: $-73,764 | Employees:   17 |
| 2005 Sales: $ 574 | 2005 Profits: $-26,895 | Fiscal Year Ends: 12/31 |
| 2004 Sales: $ 809 | 2004 Profits: $-32,943 | Parent Company: |

## SALARIES/BENEFITS:

| Pension Plan: | ESOP Stock Plan: | Profit Sharing: | Top Exec. Salary: $434,000 | Bonus: $368,900 |
|---|---|---|---|---|
| Savings Plan: | Stock Purch. Plan: | | Second Exec. Salary: $240,000 | Bonus: $25,500 |

## OTHER THOUGHTS:

**Apparent Women Officers or Directors:**
**Hot Spot for Advancement for Women/Minorities:**

## LOCATIONS: ("Y" = Yes)

| West: | Southwest: | Midwest: | Southeast: | Northeast: | International: |
|---|---|---|---|---|---|
| | | | | Y | |

# KING PHARMACEUTICALS INC
**www.kingpharm.com**

Industry Group Code: 325412  Ranks within this company's industry group: Sales: 34  Profits: 170

| Drugs: | | Other: | Clinical: | | Computers: | Services: |
|---|---|---|---|---|---|---|
| Discovery: | | AgriBio: | Trials/Services: | | Hardware: | Specialty Services: |
| Licensing: | Y | Genetic Data: | Labs: | | Software: | Consulting: |
| Manufacturing: | Y | Tissue Replacement: | Equipment/Supplies: | | Arrays: | Blood Collection: |
| Genetics: | | | Research & Development Services: | Y | Database Management: | Drug Delivery: |
| | | | Diagnostics: | | | Drug Distribution: |

## TYPES OF BUSINESS:
Pharmaceuticals Acquisition & Manufacturing
Prescription Pharmaceuticals - Diversified

## BRANDS/DIVISIONS/AFFILIATES:
Monarch Pharmaceuticals, Inc.
King Pharmaceuticals Research and Development, Inc
Meridian Medical Technologies, Inc.
Parkedale Pharmaceuticals, Inc.
King Pharmaceuticals Canada, Inc.
Monarch Pharmaceuticals Ireland Ltd.
Alpharma, Inc.

## CONTACTS: *Note: Officers with more than one job title may be intentionally listed here more than once.*
Brian A. Markison, CEO
Brian A. Markison, Pres.
Joseph Squicciarino, CFO
Eric Carter, Chief Science Officer
Eric J. Bruce, CTO
James W. Elrod, General Counsel
James E. Green, Exec. VP-Corp. Affairs
Stephen J. Andrzejewski, Chief Commercial Officer
Brian A. Markison, Chmn.

| Phone: 423-989-8000 | Fax: 423-274-8677 |
|---|---|
| Toll-Free: 800-776-3637 | |
| Address: 501 5th St., Bristol, TN 37620 US | |

## GROWTH PLANS/SPECIAL FEATURES:
King Pharmaceuticals, Inc. (KPI) develops, manufactures and markets therapies and technologies primarily in specialty-driven markets. KPI offers neuroscience, hospital, acute care and legacy products. The company also has a Meridian Auto-Injector segment, consisting of EpiPen and nerve gas antidotes, sold to the U.S. Military, and a branded pharmaceuticals segment. Branded pharmaceutical products accounted for approximately 81% of total net revenues for 2008. KPI's royalties segment develops and licenses the third-party products, Adenoscan and Adenocard, which function as imaging agents and as a converter of irregular heart rhythms to normal sinus rhythms, respectively. Key products in the company's branded pharmaceutical segment include Altace, Corzide, Corgard, Levoxyl and Cytomel for cardiovascular/metabolic therapy; Skelaxin, Avinze and Sonata for neural therapy and Thrombin-JMI, Bicillin, Synercid and Intal for hospital/acute care treatments. Notable products of KPI include Levoxyl, a thyroid hormone replacement or supplemental therapy for hypothyroidism; Avinza, an extended release formulation of morphine for severe pain and Thrombin-JMI, which controls minor bleeding during surgery. KPI's wholly-owned subsidiaries are Monarch Pharmaceuticals, Inc.; King Pharmaceuticals Research and Development, Inc.; Meridian Medical Technologies, Inc.; Parkedale Pharmaceuticals, Inc.; King Pharmaceuticals Canada Inc.; and Monarch Pharmaceuticals Ireland Limited. In January 2008, King entered into an agreement with CorePharma, LLC granting CorePharma a license to launch an authorized generic version of Skelaxin in December 2012. In April 2008, KPI and Acura Pharmaceuticals, Inc. completed the patient enrollment portion of the Phase III trials for ACUROX tablets. ACUROX is intended to relieve pain while also resisting prescription drug abuse, including intravenous injection of dissolved tablets, nasal snorting of crushed tablets and intentional swallowing of excessive numbers of tablets. In December 2008, the company acquired Alpharma, Inc. for $1.6 billion.

KPI offers its employees benefits including medical insurance; prescription coverage; disability plans; a travel assistance program; adoption benefits; and a 401(k) plan.

## FINANCIALS: Sales and profits are in thousands of dollars—add 000 to get the full amount. 2008 Note: Financial information for 2008 was not available for all companies at press time.

| | | |
|---|---|---|
| 2008 Sales: $1,565,061 | 2008 Profits: $-333,063 | U.S. Stock Ticker: KG |
| 2007 Sales: $2,136,882 | 2007 Profits: $182,981 | Int'l Ticker:  Int'l Exchange: |
| 2006 Sales: $1,988,500 | 2006 Profits: $288,949 | Employees: 3,381 |
| 2005 Sales: $1,772,881 | 2005 Profits: $117,833 | Fiscal Year Ends: 12/31 |
| 2004 Sales: $1,304,364 | 2004 Profits: $-160,288 | Parent Company: |

## SALARIES/BENEFITS:
| Pension Plan: | ESOP Stock Plan: | Profit Sharing: | Top Exec. Salary: $980,820 | Bonus: $980,820 |
|---|---|---|---|---|
| Savings Plan: Y | Stock Purch. Plan: | | Second Exec. Salary: $552,000 | Bonus: $386,400 |

## OTHER THOUGHTS:
Apparent Women Officers or Directors:
Hot Spot for Advancement for Women/Minorities:

## LOCATIONS: ("Y" = Yes)
| West: | Southwest: | Midwest: | Southeast: | Northeast: | International: |
|---|---|---|---|---|---|
| | | Y | Y | Y | |

*Note: Financial information, benefits and other data can change quickly and may vary from those stated here.*

# KV PHARMACEUTICAL CO

www.kvpharmaceutical.com

Industry Group Code: 325412A **Ranks within this company's industry group:** Sales: 7 Profits: 6

| Drugs: | | Other: | Clinical: | Computers: | Services: | |
|---|---|---|---|---|---|---|
| Discovery: | Y | AgriBio: | Trials/Services: | Hardware: | Specialty Services: | |
| Licensing: | | Genetic Data: | Labs: | Software: | Consulting: | |
| Manufacturing: | Y | Tissue Replacement: | Equipment/Supplies: | Arrays: | Blood Collection: | |
| Genetics: | Y | | Research & Development Services: | Database Management: | Drug Delivery: | Y |
| | | | Diagnostics: | | Drug Distribution: | Y |

## TYPES OF BUSINESS:

Pharmaceutical Products
Drug Delivery & Formulation Technologies
Specialty Ingredients
Taste Masking Systems
Branded Prescription Pharmaceuticals

## BRANDS/DIVISIONS/AFFILIATES:

ETHEX Corporation
Ther-Rx Corporation
Particle Dynamics, Inc.
Evamist
SITE RELEASE
Liquette
FlavorTech
Micromask

## CONTACTS: *Note: Officers with more than one job title may be intentionally listed here more than once.*

David A. Van Vliet, Interim CEO
Ronald J. Kanterman, CFO/VP
Gregory S. Bentley, General Counsel/Sr. VP
Michael S. Anderson, VP-Dev. & Industry Presence

| Phone: 314-645-6600 | Fax: 314-646-3751 |
|---|---|
| Toll-Free: | |
| Address: 2503 S. Hanley Rd., St. Louis, MO 63144 US | |

## GROWTH PLANS/SPECIAL FEATURES:

KV Pharmaceutical Co. is a pharmaceutical company that develops, acquires, manufactures and markets technologically branded and generic prescription pharmaceutical products. The company develops a wide variety of drug delivery and formulation technologies, which are primarily focused in four areas: SITE RELEASE bioadhesives, tastemasking, oral controlled release and oral quick dissolving tablets. The firm incorporates these technologies in the products it markets to control and improve the absorption and utilization of active pharmaceutical compounds. KV has a broad range of dosage form capabilities including tablets, capsules, creams, liquids and ointments. The company manufactures and markets these specialty pharmaceutical products through three wholly-owned subsidiaries: ETHEX Corporation, which targets generic and non-branded market segments; Ther-Rx Corporation, which focuses on women's healthcare; and Particle Dynamics, Inc., a specialty ingredient products company. KV's operating units currently feature some 15 drug delivery technologies. The company's site-specific treatments isolate drugs to a specific area of the body to increase their effectiveness. Taste-masking systems and quick-dissolving tablets are aimed at improving the taste of drugs and are marketed under names including Liquette, FlavorTech and Micromask. ETHEX, the company's major subsidiary, offers more than 100 products in four major categories: cardiovascular, women's health, pain management and respiratory/cough/cold. Ther-Rx markets Evamist, an estrogen replacement therapy treatment. Particle Dynamics develops, manufactures and markets technically advanced, value-added specialty ingredient products to the pharmaceutical, nutritional and personal care industries. In February 2009, the firm announced plans to reduce its workforce by approximately 700 positions as part of the firm's overall cost-cutting plan.

The firm offers employees health, dental and life insurance; an employee stock option plan; a retirement plan; a 401(k) plan; educational assistance; and an employee assistance program.

## FINANCIALS: Sales and profits are in thousands of dollars—add 000 to get the full amount. 2008 Note: Financial information for 2008 was not available for all companies at press time.

| | | |
|---|---|---|
| 2008 Sales: $601,896 | 2008 Profits: $88,354 | **U.S. Stock Ticker:** KV |
| 2007 Sales: $443,627 | 2007 Profits: $58,090 | **Int'l Ticker:** Int'l Exchange: |
| 2006 Sales: $367,618 | 2006 Profits: $11,416 | Employees: 1,590 |
| 2005 Sales: $303,493 | 2005 Profits: $33,269 | Fiscal Year Ends: 3/31 |
| 2004 Sales: $283,900 | 2004 Profits: $45,800 | Parent Company: |

## SALARIES/BENEFITS:

| Pension Plan: | ESOP Stock Plan: | Profit Sharing: Y | Top Exec. Salary: $1,383,847 | Bonus: $4,276,735 |
|---|---|---|---|---|
| Savings Plan: Y | Stock Purch. Plan: Y | | Second Exec. Salary: $371,035 | Bonus: $117,128 |

## OTHER THOUGHTS:

Apparent Women Officers or Directors:
Hot Spot for Advancement for Women/Minorities:

## LOCATIONS: ("Y" = Yes)

| West: | Southwest: | Midwest: | Southeast: | Northeast: | International: |
|---|---|---|---|---|---|
| | | Y | | | |

# LA JOLLA PHARMACEUTICAL

**www.ljpc.com**

**Industry Group Code: 325412  Ranks within this company's industry group:** Sales:   Profits: 147

| Drugs: | | Other: | Clinical: | Computers: | Services: |
|---|---|---|---|---|---|
| Discovery: | Y | AgriBio: | Trials/Services: | Hardware: | Specialty Services: |
| Licensing: | | Genetic Data: | Labs: | Software: | Consulting: |
| Manufacturing: | | Tissue Replacement: | Equipment/Supplies: | Arrays: | Blood Collection: |
| Genetics: | | | Research & Development Services: | Database Management: | Drug Delivery: |
| | | | Diagnostics: | | Drug Distribution: |

## TYPES OF BUSINESS:

Drugs-Autoimmune Diseases
Lupus Treatments
Small-Molecule Therapeutics

## BRANDS/DIVISIONS/AFFILIATES:

Riquent

## CONTACTS: Note: Officers with more than one job title may be intentionally listed here more than once.

Deirdre Gillespie, CEO
Deirdre Gillespie, Pres.
Gail A. Sloan, Corp. Sec.
Gail A. Sloan, VP-Finance
Craig R. Smith, Chmn.

| **Phone:** 858-452-6600 | **Fax:** 858-626-2851 |
|---|---|
| **Toll-Free:** | |
| **Address:** 6455 Nancy Ridge Dr., San Diego, CA 92121-2249 US | |

## GROWTH PLANS/SPECIAL FEATURES:

La Jolla Pharmaceutical Company (LJPC) is a biopharmaceutical company focused on the research and development of therapeutic products for the treatment of certain life-threatening antibody-mediated diseases, especially autoimmune conditions such as lupus. The company was previously involved in the development of Riquent, a clinical drug candidate for the treatment of lupus kidney disease. The firm signed a commercialization agreement with BioMarin CF Limited, a subsidiary of BioMarin Pharmaceuticals, Inc., for global sales rights. However, Phase III trials indicated that the drug produced unsafe side effects, and in March 2009 the company decided to cease development of Riquent. The firm's other asset is a preclinical stage study on the use of SSAO (semicarbazide-sensitive amine oxidase) inhibitors for use in the treatment of stroke, ulcerative colitis, and other autoimmune disorders. The company has stated that this asset would likely be out-licensed as it does not currently have the resources to devote to the product's development. In February 2009, La Jolla laid off 75 employees, or most of its staff. In March 2009, BioMarin Pharma returned the rights for Riquent to the company.

## FINANCIALS: Sales and profits are in thousands of dollars—add 000 to get the full amount. 2008 Note: Financial information for 2008 was not available for all companies at press time.

| 2008 Sales: $ | 2008 Profits: $-62,854 | **U.S. Stock Ticker: LJPC** |
|---|---|---|
| 2007 Sales: $ | 2007 Profits: $-53,076 | **Int'l Ticker:**   Int'l Exchange: |
| 2006 Sales: $ | 2006 Profits: $-39,445 | Employees:   94 |
| 2005 Sales: $ | 2005 Profits: $-27,363 | Fiscal Year Ends: 12/31 |
| 2004 Sales: $ | 2004 Profits: $-40,544 | Parent Company: |

## SALARIES/BENEFITS:

| Pension Plan: | ESOP Stock Plan: Y | Profit Sharing: | Top Exec. Salary: $387,115 | Bonus: $73,125 |
|---|---|---|---|---|
| Savings Plan: Y | Stock Purch. Plan: | | Second Exec. Salary: $325,000 | Bonus: $48,750 |

## OTHER THOUGHTS:

**Apparent Women Officers or Directors:** 2
**Hot Spot for Advancement for Women/Minorities:** Y

## LOCATIONS: ("Y" = Yes)

| West: | Southwest: | Midwest: | Southeast: | Northeast: | International: |
|---|---|---|---|---|---|
| Y | | | | | Y |

Note: Financial information, benefits and other data can change quickly and may vary from those stated here.

# LANNETT COMPANY INC

**www.lannett.com**

Industry Group Code: 325412A   **Ranks within this company's industry group:**   Sales: 15    Profits: 10

| Drugs: | Other: | Clinical: | Computers: | Services: |
|---|---|---|---|---|
| Discovery: | AgriBio: | Trials/Services: | Hardware: | Specialty Services: |
| Licensing: | Genetic Data: | Labs: | Software: | Consulting: |
| Manufacturing: Y | Tissue Replacement: | Equipment/Supplies: | Arrays: | Blood Collection: |
| Genetics: Y | | Research & Development Services: | Database Management: | Drug Delivery: |
| | | Diagnostics: | | Drug Distribution: Y |

## TYPES OF BUSINESS:

Drugs-Generic
Drug Delivery System Development

## BRANDS/DIVISIONS/AFFILIATES:

Acetazolamide
Butalbital
Digoxin
Unithroid
Levothyroxine Sodium
Primidone
Hydromorphone
Phentermine

## CONTACTS: *Note: Officers with more than one job title may be intentionally listed here more than once.*

Arthur P. Bedrosian, CEO
Arthur P. Bedrosian, Pres.
Keith Ruck, Interim CFO
Kevin Smith, VP-Mktg. & Sales
Bernard Sandiford, VP-Oper.
Ernest J. Sabo, VP-Regulatory Affairs & Corp. Compliance
Jeffrey Farber, Interim Chmn.
William Schreck, VP-Logistics

| Phone: 215-333-9000 | Fax: 215-333-9004 |
|---|---|
| Toll-Free: 800-325-9994 | |
| Address: 9000 State Rd., Philadelphia, PA 19136 US | |

## GROWTH PLANS/SPECIAL FEATURES:

Lannett Company, Inc. develops, manufactures, markets and distributes pharmaceutical products sold under generic names, marketing them primarily to drug wholesalers, retail drug chains, repackagers, distributors and government agencies. The company's extensive range of products includes Acetazolamide for glaucoma; Butalbital with aspirin, caffeine and codeine for migraines; Clindamycin, an antibiotic; Digoxin for congestive heart failure; Dicyclomine, a treatment for irritable bowels; Levothyroxine Sodium for thyroid deficiency; Phentermine, a pill for weight loss; Primidone for epilepsy; Terbutaline, a medication for bronchospasms; and Unithroid, a branded thyroid medication. All of the products currently manufactured and sold by the company are prescription products. The company's key products are Butalbital, Digoxin, Primidone and Levothyroxine Sodium, contributing more than 76% of 2008 revenue. Sales of Sulfamethoxazole with Trimethoprim grew from 3% of sales in 2006 to 9% of sales in 2008. Lannett has a wholly-owned subsidiary, Cody Laboratories, Inc., in Cody, Wyoming. Lannett's services also include granulation, blending, encapsulation, compression, coating and packaging. In August 2008, the company received FDA approval for Doxycycline, any antibacterial for use in urinary tract infections, acne and other indications. In December 2008, the firm launched two new pain management products, Amantadine HCl and a topical anesthetic. In May 2009, it released Pilocarpine HCl tablets, which are used to treat symptoms of dry mouth.

Lannett offers its employees medical, dental, vision and prescription drug coverage; a 401(k) plan; an employee stock purchase plan; and tuition reimbursement.

## FINANCIALS: Sales and profits are in thousands of dollars—add 000 to get the full amount. 2008 Note: Financial information for 2008 was not available for all companies at press time.

| | | |
|---|---|---|
| 2008 Sales: $72,403 | 2008 Profits: $-2,318 | U.S. Stock Ticker: LCI |
| 2007 Sales: $82,578 | 2007 Profits: $-6,929 | Int'l Ticker:    Int'l Exchange: |
| 2006 Sales: $64,060 | 2006 Profits: $4,969 | Employees: 187 |
| 2005 Sales: $44,902 | 2005 Profits: $-32,780 | Fiscal Year Ends: 6/30 |
| 2004 Sales: $63,781 | 2004 Profits: $13,215 | Parent Company: |

## SALARIES/BENEFITS:

| | | | | |
|---|---|---|---|---|
| Pension Plan: | ESOP Stock Plan: | Profit Sharing: | Top Exec. Salary: $324,825 | Bonus: $ |
| Savings Plan: Y | Stock Purch. Plan: Y | | Second Exec. Salary: $210,361 | Bonus: $ |

## OTHER THOUGHTS:

**Apparent Women Officers or Directors:**
**Hot Spot for Advancement for Women/Minorities:**

## LOCATIONS: ("Y" = Yes)

| West: | Southwest: | Midwest: | Southeast: | Northeast: | International: |
|---|---|---|---|---|---|
| Y | | | | Y | |

# LEXICON PHARMACEUTICALS INC

**www.lexgen.com**

**Industry Group Code: 325412  Ranks within this company's industry group:** Sales: 97    Profits: 153

| Drugs: | | Other: | | Clinical: | | Computers: | | Services: | |
|---|---|---|---|---|---|---|---|---|---|
| Discovery: | Y | AgriBio: | | Trials/Services: | | Hardware: | | Specialty Services: | |
| Licensing: | | Genetic Data: | Y | Labs: | | Software: | | Consulting: | |
| Manufacturing: | | Tissue Replacement: | | Equipment/Supplies: | | Arrays: | | Blood Collection: | |
| Genetics: | | | | Research & Development Services: | Y | Database Management: | Y | Drug Delivery: | |
| | | | | Diagnostics: | | | | Drug Distribution: | |

## TYPES OF BUSINESS:

Drug Discovery & Development
Genetic Databases
Research & Development-Genetics

## BRANDS/DIVISIONS/AFFILIATES:

Lexicon Genetics Incorporated
LX2931
LX1031
LX1032
LX7101
Genentech Inc
NV Organon
OmniBank

## CONTACTS: *Note: Officers with more than one job title may be intentionally listed here more than once.*

Arthur T. Sands, CEO
Arthur T. Sands, Pres.
Brian P. Zambrowicz, Chief Scientific Officer/Exec. VP
Jeffrey L. Wade, General Counsel/Exec. VP
James Tessmer, VP-Finance & Acct.
Lance K. Ishimoto, Sr. VP-Intellectual Property
Alan J. Main, Exec. VP-Pharmaceutical Research
Philip M. Brown, VP-Clinical Dev.
Samuel L. Barker, Chmn.

| **Phone:** 281-863-3000 | **Fax:** 281-863-8088 |
|---|---|
| **Toll-Free:** | |
| **Address:** 8800 Technology Forest Pl., The Woodlands, TX 77381-1160 US | |

## GROWTH PLANS/SPECIAL FEATURES:

Lexicon Pharmaceuticals, Inc., formerly Lexicon Genetics Incorporated, is a biopharmaceutical company that focuses on the discovery and development of new medical treatments for human diseases using its proprietary gene knockout technology, which disrupts the function of genes in mice in order to determine the proper pharmaceutical treatment for the physiological and behavioral functions of each gene. Lexicon has advanced four drug candidates into human clinical trials, with one additional candidate in preclinical development and compounds from a number of additional programs in various stages of preclinical research. Lexicon is conducting a Phase 2 trial of LX1031, an orally-delivered small molecule compound for the treatment of gastrointestinal disorders such as irritable bowel syndrome. The company is conducting Phase 1 trials of two other candidates: LX2931 for the treatment of autoimmune diseases such as rheumatoid arthritis and LX4211 preclinical for the treatment of Type 2 diabetes.  LX1032 for the treatment of the symptoms of carcinoid syndrome has successfully completed Phase 1 trials.  LX7101 for the treatment of Glaucoma is currently in the preclinical stages of development.  Lexicon has identified and validated in living animals more than 100 targets with promising profiles for drug discovery in the areas of diabetes and obesity; cardiovascular disease; gastrointestinal disorders; psychiatric and neurological disorders; cancer; immune system disorders; and ophthalmic disease.  The firm also collaborates with such companies as Bristol-Myers Squibb for development in the neuroscience field, and Genentech and N.V. Organon for the development of antibody targets. Lexicon's OmniBank contains over 270,000 frozen gene knockout embryonic mouse stem cell clones.

Employees are offered medical, dental, vision and life insurance, as well as an educational assistance program.

## FINANCIALS: Sales and profits are in thousands of dollars—add 000 to get the full amount. 2008 Note: Financial information for 2008 was not available for all companies at press time.

| | | | |
|---|---|---|---|
| 2008 Sales: $32,321 | 2008 Profits: $-76,860 | **U.S. Stock Ticker:** LXRX | |
| 2007 Sales: $50,118 | 2007 Profits: $-58,794 | **Int'l Ticker:** Int'l Exchange: | |
| 2006 Sales: $72,798 | 2006 Profits: $-54,311 | Employees: 347 | |
| 2005 Sales: $75,680 | 2005 Profits: $-36,315 | Fiscal Year Ends: 12/31 | |
| 2004 Sales: $61,740 | 2004 Profits: $-47,172 | Parent Company: | |

## SALARIES/BENEFITS:

| Pension Plan: | ESOP Stock Plan: | Profit Sharing: | Top Exec. Salary: $557,500 | Bonus: $ |
|---|---|---|---|---|
| Savings Plan: Y | Stock Purch. Plan: | | Second Exec. Salary: $363,333 | Bonus: $ |

## OTHER THOUGHTS:

**Apparent Women Officers or Directors:** 2
**Hot Spot for Advancement for Women/Minorities:** Y

## LOCATIONS: ("Y" = Yes)

| West: | Southwest: | Midwest: | Southeast: | Northeast: | International: |
|---|---|---|---|---|---|
| | Y | | | Y | |

# LIFE SCIENCES RESEARCH INC

## www.lsrinc.net

**Industry Group Code: 541712  Ranks within this company's industry group:** Sales: 7   Profits: 7

| Drugs: | Other: | Clinical: | | Computers: | | Services: |
|---|---|---|---|---|---|---|
| Discovery: | AgriBio: | Trials/Services: | Y | Hardware: | | Specialty Services: |
| Licensing: | Genetic Data: | Labs: | Y | Software: | | Consulting: |
| Manufacturing: | Tissue Replacement: | Equipment/Supplies: | | Arrays: | | Blood Collection: |
| Genetics: | | Research & Development Services: | Y | Database Management: | | Drug Delivery: |
| | | Diagnostics: | | | | Drug Distribution: |

## TYPES OF BUSINESS:

Contract Research
Drug Development Services
Non-Clinical Safety Testing

## BRANDS/DIVISIONS/AFFILIATES:

Huntingdon Life Sciences Group plc

## CONTACTS: *Note: Officers with more than one job title may be intentionally listed here more than once.*

Andrew H. Baker, CEO
Brian Cass, Pres.
Richard A. Michaelson, CFO
Mark L. Bibi, General Counsel/Corp. Sec.
Julian T. Griffiths, Dir.-Oper.
Andrew H. Baker, Chmn.

| **Phone:** 732-649-9961 | **Fax:** 732-649-0021 |
|---|---|
| **Toll-Free:** | |
| **Address:** Mettlers Rd., P.O. Box 2360, East Millstone, NJ 08875 US | |

## GROWTH PLANS/SPECIAL FEATURES:

Life Sciences Research, Inc. (LSR) is a global contract research organization providing pre-clinical and non-clinical testing services for biological safety evaluation research to pharmaceutical, biotechnology, agrochemical and industrial chemical companies.  LSR serves the regulatory and commercial requirements to perform safety evaluations on new pharmaceutical compounds and chemical compounds contained within the products that humans use, eat and are otherwise exposed to.  The company also tests the effect of such compounds on the environment and assesses the safety and efficacy of veterinary products.  Pre-clinical testing helps evaluate both how the drug affects the body and how the body affects the drug in order to assess safe and appropriate dose regimens.  Non-clinical testing can focus on identifying and avoiding the longer-term cancer implications of exposure to the compound, the potential for possible reproductive implications and the stability of pharmaceuticals under a variety of storage conditions.  Over 85% of the company's orders are derived from the bio-pharmaceutical industry.  LSR has also actively pursued opportunities to extend its range of capabilities supporting late stage drug discovery, focused around in vitro and in vivo models for lead candidate drug characterization and optimization.  LSR was formed specifically to acquire the former Huntingdon Life Sciences Group plc in order to relocate the company to the U.S.  The firm operates one research facilities in the U.S. and two in the U.K., and its services are designed to address the regulatory requirements of governments around the world.  The U.K. and Europe account for approximately 59% of revenues, the U.S. for 23%, Japan for 12% and the rest of the world for the remaining 6%.

## FINANCIALS: Sales and profits are in thousands of dollars—add 000 to get the full amount. 2008 Note: Financial information for 2008 was not available for all companies at press time.

| | | |
|---|---|---|
| 2008 Sales: $242,422 | 2008 Profits: $10,418 | **U.S. Stock Ticker: LSR** |
| 2007 Sales: $236,800 | 2007 Profits: $-13,974 | **Int'l Ticker:**    Int'l Exchange: |
| 2006 Sales: $192,217 | 2006 Profits: $-14,872 | Employees:  1,648 |
| 2005 Sales: $172,013 | 2005 Profits: $1,491 | Fiscal Year Ends: 12/31 |
| 2004 Sales: $157,551 | 2004 Profits: $17,594 | Parent Company: |

## SALARIES/BENEFITS:

| Pension Plan: | ESOP Stock Plan: | Profit Sharing: | Top Exec. Salary: $636,111 | Bonus: $ |
|---|---|---|---|---|
| Savings Plan: Y | Stock Purch. Plan: | | Second Exec. Salary: $636,111 | Bonus: $ |

## OTHER THOUGHTS:

**Apparent Women Officers or Directors:**
**Hot Spot for Advancement for Women/Minorities:**

## LOCATIONS: ("Y" = Yes)

| West: | Southwest: | Midwest: | Southeast: | Northeast: | International: |
|---|---|---|---|---|---|
| | | | | Y | Y |

# LIFE TECHNOLOGIES CORP

**www.lifetechnologies.com**

**Industry Group Code:** 325413 **Ranks within this company's industry group:** Sales: 1 Profits: 5

| Drugs: | Other: | | Clinical: | | Computers: | | Services: | |
|---|---|---|---|---|---|---|---|---|
| Discovery: | AgriBio: | | Trials/Services: | | Hardware: | | Specialty Services: | |
| Licensing: | Genetic Data: | Y | Labs: | | Software: | | Consulting: | |
| Manufacturing: | Tissue Replacement: | | Equipment/Supplies: | Y | Arrays: | Y | Blood Collection: | |
| Genetics: | | | Research & Development Services: | Y | Database Management: | | Drug Delivery: | |
| | | | Diagnostics: | | | | Drug Distribution: | |

## TYPES OF BUSINESS:

Equipment-Gene Cloning Kits
Microarrays
Reagents
RNA & DNA Libraries
Mass Spectrometry

## BRANDS/DIVISIONS/AFFILIATES:

Cell Culture Systems (CCS)
BioDiscovery
CellzDirect, Inc.
Applied Biosystems, Inc.
Invitrogen Corp.

## CONTACTS: *Note: Officers with more than one job title may be intentionally listed here more than once.*

Gregory T. Lucier, CEO
Mark P. Stevenson, COO
Mark P. Stevenson, Pres.
David F. Hoffmeister, CFO/Sr. VP
Peter Leddy, Sr. VP-Global Human Resources
Claude D. Benchimol, Sr. VP-R&D
Joe Beery, CIO/Sr. VP
John A. Cottingham, General Counsel/Sr. VP/Corp. Sec.
Mark O'Donnell, Sr. VP-Global Oper. & Svcs.
Paul Grossman, Sr. VP-Strategy & Corp. Dev.
Amanda Clardy, VP-Corp. Comm.
Amanda Clardy, VP-Investor Rel.
Kelli Richard, VP-Finance/Chief Acct. Officer
Nicolas M. Barthelemy, Pres., Cell Systems
John Miller, Pres., Genetic Systems
Peter Dansky, Pres., Molecular Biology Systems
Laura Lauman, Pres., Mass Spectrometry
Gregory T. Lucier, Chmn.

| Phone: 760-603-7200 | Fax: 760-603-6500 |
|---|---|
| Toll-Free: 800-955-6288 | |
| Address: 5791 Van Allen Way, Carlsbad, CA 92008 US | |

## GROWTH PLANS/SPECIAL FEATURES:

Life Technologies Corp., formerly Invitrogen Corporation, is a global biotechnology tools company with operations in over 100 companies. Life Technologies was formed through the merger of Invitrogen and Applied Biosystems, Inc. in November 2008. The company's product portfolio includes technologies for capillary electrophoresis based sequencing, next generation sequencing, mass spectrometry, sample preparation, cell culture, RNA interference analysis, functional genomics research, proteomics and cell biology applications, as well as clinical diagnostic applications and water testing analysis. The company divides its operations into three broad segments: Biodiscovery (BD), cell systems (CS) and Applied Biosystems (AB). The BD segment includes molecular biology, cell biology and drug discovery product lines. Molecular biology products include the research tools used in reagent and kit form that simplify and improve gene acquisition, gene cloning, gene expression and gene analysis techniques. This segment also includes a full range of enzymes, nucleic acids, other biochemicals and reagents. The CS segment includes cell culture products and services. Products include sera, cell and tissue culture media, reagents used in both life sciences research and in processes to grow cells in the laboratory and to produce biopharmaceuticals and other end products made through cultured cells. The AB segment's products include complete instrument-reagent systems, such as PCR and Real-Time PCR systems, capillary electrophoresis sequencing systems and next-generation DNA sequencing systems. Additional products include mass spectrometry systems, Ambion RNA reagents and specialized applied markets products and services. In early 2008, the company agreed to acquire CellzDirect, Inc. for about $57 million. In November 2008, Invitrogen Corporation and Applied Biosystems, Inc. completed their merger and the formation of Life Technologies Corporation.

The company offers its employees medical, dental and vision insurance; a 401(k) plan; life insurance; short- and long-term disability benefits; an employee stock purchase plan; educational reimbursement; and an employee assistance plan.

## FINANCIALS: Sales and profits are in thousands of dollars—add 000 to get the full amount. 2008 Note: Financial information for 2008 was not available for all companies at press time.

| | | |
|---|---|---|
| 2008 Sales: $1,620,323 | 2008 Profits: $31,321 | **U.S. Stock Ticker:** LIFE |
| 2007 Sales: $1,281,747 | 2007 Profits: $143,190 | **Int'l Ticker:** Int'l Exchange: |
| 2006 Sales: $1,151,175 | 2006 Profits: $-191,049 | **Employees:** 9,700 |
| 2005 Sales: $1,198,452 | 2005 Profits: $132,046 | **Fiscal Year Ends:** 12/31 |
| 2004 Sales: $1,023,851 | 2004 Profits: $88,825 | **Parent Company:** |

## SALARIES/BENEFITS:

| Pension Plan: | ESOP Stock Plan: | Profit Sharing: | Top Exec. Salary: $978,404 | Bonus: $2,050,000 |
|---|---|---|---|---|
| Savings Plan: Y | Stock Purch. Plan: Y | | Second Exec. Salary: $475,192 | Bonus: $445,315 |

## OTHER THOUGHTS:

**Apparent Women Officers or Directors:** 4
**Hot Spot for Advancement for Women/Minorities:** Y

## LOCATIONS: ("Y" = Yes)

| West: | Southwest: | Midwest: | Southeast: | Northeast: | International: |
|---|---|---|---|---|---|
| Y | | | | | Y |

Note: Financial information, benefits and other data can change quickly and may vary from those stated here.

# LIFECELL CORPORATION

**www.lifecell.com**

**Industry Group Code:** 325414 **Ranks within this company's industry group:** Sales: Profits:

| Drugs: | | Other: | | Clinical: | Computers: | | Services: | |
|---|---|---|---|---|---|---|---|---|
| Discovery: | Y | AgriBio: | | Trials/Services: | Hardware: | | Specialty Services: | |
| Licensing: | | Genetic Data: | | Labs: | Software: | | Consulting: | |
| Manufacturing: | Y | Tissue Replacement: | Y | Equipment/Supplies: | Arrays: | | Blood Collection: | |
| Genetics: | | | | Research & Development Services: | Database Management: | | Drug Delivery: | |
| | | | | Diagnostics: | | | Drug Distribution: | |

## TYPES OF BUSINESS:

Tissue Replacement Products
Skin Replacement Technology
Regenerative Medicine

## BRANDS/DIVISIONS/AFFILIATES:

Kinetic Concepts Inc
AlloDerm
Strattice

## CONTACTS: Note: Officers with more than one job title may be intentionally listed here more than once.

Lisa Colleran, Pres.
Catherine M. Burzik, CEO-Kinetic Concepts

| **Phone:** 908-947-1100 | **Fax:** |
|---|---|
| **Toll-Free:** | |
| **Address:** 1 Millennium Way, Branchburg, NJ 08876 US | |

## GROWTH PLANS/SPECIAL FEATURES:

LifeCell Corporation, a wholly-owned subsidiary of Kinetic Concepts, Inc. (KCI), specializes in regenerative medicine, developing and manufacturing products geared toward the repair, replacement and preservation of human tissues. The firm's products are used in a variety of reconstructive, orthopedic and urogynecologic surgical procedures. The company has many different products. AlloDerm is a tissue matrix derived from human skin. Originally used as a skin graft for deep second and third degree burns, it is now commonly used as a soft tissue replacement, including for plastic, reconstructive, general surgical, burn and periodontal procedures. Some applications include abdominal wall reconstruction, post-mastectomy breast reconstruction, ENT/head and neck plastic reconstruction and grafting. Strattice Reconstructive Tissue Matrix is a sterile reconstructive tissue matrix that supports tissue regeneration. Strattice is used for soft tissue reinforcement in reconstructive and general surgical procedures, including hernia repair and breast reconstruction or plastic surgery. In May 2008, the firm was acquired by KCI for $1.59 billion.

## FINANCIALS: Sales and profits are in thousands of dollars—add 000 to get the full amount. 2008 Note: Financial information for 2008 was not available for all companies at press time.

| | | |
|---|---|---|
| 2008 Sales: $ | 2008 Profits: $ | **U.S. Stock Ticker:** Subsidiary |
| 2007 Sales: $191,130 | 2007 Profits: $26,883 | **Int'l Ticker:** Int'l Exchange: |
| 2006 Sales: $141,680 | 2006 Profits: $20,469 | Employees: 443 |
| 2005 Sales: $94,398 | 2005 Profits: $12,044 | Fiscal Year Ends: 12/31 |
| 2004 Sales: $61,127 | 2004 Profits: $7,184 | Parent Company: KINETIC CONCEPTS INC |

## SALARIES/BENEFITS:

| Pension Plan: | ESOP Stock Plan: | Profit Sharing: | Top Exec. Salary: $500,000 | Bonus: $397,500 |
|---|---|---|---|---|
| Savings Plan: | Stock Purch. Plan: | | Second Exec. Salary: $285,850 | Bonus: $130,162 |

## OTHER THOUGHTS:

Apparent Women Officers or Directors: 2
Hot Spot for Advancement for Women/Minorities:

## LOCATIONS: ("Y" = Yes)

| West: | Southwest: | Midwest: | Southeast: | Northeast: | International: |
|---|---|---|---|---|---|
| | | | | Y | |

# LIFECORE BIOMEDICAL INC

**www.lifecore.com**

**Industry Group Code: 33911 Ranks within this company's industry group:** Sales:    Profits:

| Drugs: | | Other: | | Clinical: | | Computers: | | Services: |
|---|---|---|---|---|---|---|---|---|
| Discovery: | | AgriBio: | | Trials/Services: | | Hardware: | | Specialty Services: |
| Licensing: | | Genetic Data: | | Labs: | | Software: | | Consulting: |
| Manufacturing: | Y | Tissue Replacement: | Y | Equipment/Supplies: | Y | Arrays: | | Blood Collection: |
| Genetics: | | | | Research & Development Services: | | Database Management: | | Drug Delivery: |
| | | | | Diagnostics: | | | | Drug Distribution: |

## TYPES OF BUSINESS:

Biomaterials & Medical Device Manufacturing
Bone Regeneration Products
Surgical Devices

## BRANDS/DIVISIONS/AFFILIATES:

Warburg Pincus LLC
Support Plus
Corgel BioHydrogel

## CONTACTS: Note: Officers with more than one job title may be intentionally listed here more than once.

Dennis J. Allingham, CEO
Dennis J. Allingham, Pres.
David M. Noel, CFO
James G. Hall, VP-Technical Oper.
Kipling Thacker, VP-New Bus. Dev.
David M. Noel, VP-Finance
Larry D. Hiebert, VP/Gen. Mgr.-Hyaluronan Div.

| Phone: 952-368-4300 | Fax: 952-368-3411 |
|---|---|
| Toll-Free: | |
| Address: 3515 Lyman Blvd., Chaska, MN 55318 US | |

## GROWTH PLANS/SPECIAL FEATURES:

Lifecore Biomedical, Inc. develops and manufactures biomaterials and medical devices with applications in various surgical markets. The company's hyaluronan division is principally involved in the development and manufacture of products utilizing hyaluronan, a naturally occurring polysaccharide that is widely distributed in the extracellar matrix of connective tissues in both animals and humans. This division sells primarily to three medical segments: ophthalmic, orthopedic and veterinary. Lifecore also supplies hyaluronan to customers pursuing other medical applications, such as wound care, aesthetic surgery, medical device coatings, tissue engineering, drug delivery and pharmaceuticals. In 2008, private equity firm Warburg Pincus acquired Lifecore, taking the company private. Recently, Warburg Pincus merged Lifecore Biomedical Inc.'s dental division into its's subisidary, Keystone Dental, Inc. In May 2009, the firm commercially launched its Corgel BioHydrogel research kits; Corgel hydrogel serves as a tissue bulking agent and drug delivery matrix, and its biocompatibility allows for cells or bioactive agents to be included directly in the gel.

## FINANCIALS: Sales and profits are in thousands of dollars—add 000 to get the full amount. 2008 Note: Financial information for 2008 was not available for all companies at press time.

| | | | |
|---|---|---|---|
| 2008 Sales: $ | 2008 Profits: $ | **U.S. Stock Ticker: Private** | |
| 2007 Sales: $69,629 | 2007 Profits: $7,719 | **Int'l Ticker:**    Int'l Exchange: | |
| 2006 Sales: $63,097 | 2006 Profits: $7,040 | Employees: 241 | |
| 2005 Sales: $55,695 | 2005 Profits: $17,511 | Fiscal Year Ends: 6/30 | |
| 2004 Sales: $47,036 | 2004 Profits: $ 707 | Parent Company: WARBURG PINCUS LLC | |

## SALARIES/BENEFITS:

| Pension Plan: | ESOP Stock Plan: | Profit Sharing: | Top Exec. Salary: $312,000 | Bonus: $81,120 |
|---|---|---|---|---|
| Savings Plan: | Stock Purch. Plan: Y | | Second Exec. Salary: $160,000 | Bonus: $27,200 |

## OTHER THOUGHTS:

**Apparent Women Officers or Directors:**
**Hot Spot for Advancement for Women/Minorities:**

## LOCATIONS: ("Y" = Yes)

| West: | Southwest: | Midwest: | Southeast: | Northeast: | International: |
|---|---|---|---|---|---|
| | | Y | | | Y |

Note: Financial information, benefits and other data can change quickly and may vary from those stated here.

# LIGAND PHARMACEUTICALS INC

**www.ligand.com**

Industry Group Code: 325412  Ranks within this company's industry group: Sales: 100   Profits: 159

| Drugs: | | Other: | Clinical: | Computers: | Services: |
|---|---|---|---|---|---|
| Discovery: | Y | AgriBio: | Trials/Services: | Hardware: | Specialty Services: |
| Licensing: | Y | Genetic Data: | Labs: | Software: | Consulting: |
| Manufacturing: | | Tissue Replacement: | Equipment/Supplies: | Arrays: | Blood Collection: |
| Genetics: | | | Research & Development Services: | Database Management: | Drug Delivery: |
| | | | Diagnostics: | | Drug Distribution: |

## TYPES OF BUSINESS:

Drugs-Diversified
Small-Molecule Drugs

## BRANDS/DIVISIONS/AFFILIATES:

Avinza
Eltrombopag
Bazedoxifene
Lasofoxifene
Promacta
Viviant
Aprela
Pharmacopeia

## CONTACTS: Note: Officers with more than one job title may be intentionally listed here more than once.

John L. Higgins, CEO
John L. Higgins, Pres.
John P. Sharp, CFO
Audrey Warfield-Graham, VP-Human Resources
Martin D. Meglasson, VP-Discovery Research
Charles Berkman, General Counsel/VP/Sec.
Syed Kazmi, VP-Bus. Dev. & Strategic Planning
John P. Sharp, VP-Finance
John W. Kozarich, Chmn.

| Phone: 858-550-7500 | Fax: 858-550-7506 |
|---|---|
| Toll-Free: | |
| Address: 10275 Science Center Dr., San Diego, CA 92121-1117 US | |

## GROWTH PLANS/SPECIAL FEATURES:

Ligand Pharmaceuticals, Inc. develops drugs to address a variety of medical needs, in areas including thrombocytopenia, anemia, cancer, hormone-related diseases, osteoporosis and inflammatory diseases. The firm uses intracellular receptor (IR) technology, which was developed by its work in the field of gene transcription. The company's products target intracellular receptors in order to change cell function by selectively turning specific genes on or off. Ligand uses IR technology in seven programs: the dual acting receptor agonist (DARA) program for high blood pressure; the selective androgen receptor modulators (SARM) program for frailty, osteoporosis, sexual dysfunction, hypogonadism and more; the erythropoiein (EPO) program for anemia; the androgen-independent prostate cancer (AiPC) program, the selective glucocorticoid receptor modulators (SGRM) program for inflammation, cancer and other indications; the chemokine receptor (CCR1) program for inflammatory and autoimmune diseases; and the JAK-3 kinase inhibitor program developed with Wyeth for rheumatoid arthritis, organ transplantation and psoriasis. In addition to these programs, the company conducts several programs in collaboration with GlaxoSmithKline; Wyeth; Pfizer; Bristol-Meyers Squibb; Schering-Plough; Celgene; Cephalon; and King Pharmaceuticals. These programs further develop drug candidates based IR technology as well as selective estrogen receptor modulators. Lead collaborative drug candidates include Eltrombopag (PROMACTA) for chronic immune thrombocytopenic purpura; AVINZA for pain; Bazedoxifene (VIVANT and APRELA) for menopausal symptoms; Lasofoxifene (FABLYN) for osteoporosis; LGD-2941; and a JAK-3 inhibitor. In December 2008, Ligand acquired Pharmacopeia and access to its royalty partnerships. In early 2009, the company's collaboration drugs Fablyn and Conbriza received approval in Europe.

## FINANCIALS: Sales and profits are in thousands of dollars—add 000 to get the full amount. 2008 Note: Financial information for 2008 was not available for all companies at press time.

| | | |
|---|---|---|
| 2008 Sales: $27,315 | 2008 Profits: $-98,114 | **U.S. Stock Ticker: LGND** |
| 2007 Sales: $12,894 | 2007 Profits: $281,688 | **Int'l Ticker:**   Int'l Exchange: |
| 2006 Sales: $140,960 | 2006 Profits: $-31,743 | Employees:   96 |
| 2005 Sales: $123,010 | 2005 Profits: $-36,399 | Fiscal Year Ends: 12/31 |
| 2004 Sales: $112,112 | 2004 Profits: $-45,141 | Parent Company: |

## SALARIES/BENEFITS:

| Pension Plan: | ESOP Stock Plan: | Profit Sharing: | Top Exec. Salary: $418,333 | Bonus: $188,250 |
|---|---|---|---|---|
| Savings Plan: | Stock Purch. Plan: Y | | Second Exec. Salary: $332,800 | Bonus: $99,840 |

## OTHER THOUGHTS:

**Apparent Women Officers or Directors**: 1
**Hot Spot for Advancement for Women/Minorities**: Y

## LOCATIONS: ("Y" = Yes)

| West: | Southwest: | Midwest: | Southeast: | Northeast: | International: |
|---|---|---|---|---|---|
| Y | | | | Y | |

# LONZA GROUP

www.lonzagroup.com

**Industry Group Code: 325  Ranks within this company's industry group:** Sales: 5   Profits: 5

| Drugs: | | Other: | | Clinical: | | Computers: | | Services: | |
|---|---|---|---|---|---|---|---|---|---|
| Discovery: | | AgriBio: | Y | Trials/Services: | | Hardware: | | Specialty Services: | Y |
| Licensing: | | Genetic Data: | | Labs: | | Software: | | Consulting: | |
| Manufacturing: | Y | Tissue Replacement: | | Equipment/Supplies: | Y | Arrays: | | Blood Collection: | |
| Genetics: | | | | Research & Development Services: | Y | Database Management: | | Drug Delivery: | |
| | | | | Diagnostics: | | | | Drug Distribution: | |

## TYPES OF BUSINESS:
Chemicals Manufacturing
Fine Chemicals & Pharmaceutical Intermediates
Biocides
Materials Research
Active Drug Ingredients
Monoclonal Antibody Drugs

## BRANDS/DIVISIONS/AFFILIATES:
Alusuisse-Lonza Group
Lonza Biotec
Lonza Biologics
Cambrex Corporation
Lonza Custom Manufacturing
Lonza Singapore Pte Ltd.
amaxa

## CONTACTS: Note: Officers with more than one job title may be intentionally listed here more than once.
Stefan Borgas, CEO
Toralf Haag, CFO
Marcela Cechova, Head-Global Human Resources
Dominick Werner, Head-Media Rel.
Alexandre Pasini, Mgr.-Investor Rel.
Uwe H. Bohlke, Head-Exclusive Synthesis
Lukas Utiger, Head-Life Science Ingredients
Stephan Kutzer, Head-Biopharmaceuticals
Anja Fiedler, Head-Lonza Bioscience
Rolf Soiron, Chmn.

| **Phone:** 41-61-316-8111 | **Fax:** 41-61-316-9111 |
|---|---|
| **Toll-Free:** | |
| **Address:** Muenchensteinerstrasse 38, Basel, 4002 Switzerland | |

## GROWTH PLANS/SPECIAL FEATURES:

The Lonza Group brings together a global portfolio of companies engaged in the production and supply of active chemical ingredients, intermediates and biotechnology solutions for a range of clients in the pharmaceutical and agrochemical industries. The group structure was established in 1999 through the de-merger of its core business units from Swiss aluminum and industrial conglomerate Alusuisse-Lonza Group. Lonza Group is focused on chemical production and materials research for pharmaceutical applications, and its exclusive synthesis unit is a leading contract manufacturer of the advanced intermediates and active ingredients used in drugs. The company operates in three segments: Life Science Ingredients; Exclusive Synthesis and Biopharmaceuticals; and Bioscience. The Life Science Ingredients division is focused on ingredients used in nutrition, microbial control and select industrial markets. This division creates products and solutions for the nutrition, hygiene and personal care and wood and water treatment markets. The Exclusive Synthesis and Biopharmaceuticals segment is a custom manufacturer of intermediates and active ingredients for healthcare companies, ultimately used in critical drugs to treat cardiovascular disease, cancer, neurological and infectious diseases. The Bioscience segment supports pharmaceutical and biotechnology companies in the research, development and commercialization of human therapeutics. Lonza Group business units will oversee production through to market commercialization. The group's customers include companies in the consumer products, health products, distribution, formulation, pharmaceutical, biotechnology and service industries. In April 2008, Lonza announced plans for a new vitamin B3 manufacturing plant. That same month, the company sold its remaining 3.37% stake in Polynt S.p.A., completely divesting the company. In May 2008, Lonza Group acquired amaxa, a leading provider of nucleic acid transfection systems and consumables, which will strengthen the company's Cell Discovery business unit.

## FINANCIALS: Sales and profits are in thousands of dollars—add 000 to get the full amount. 2008 Note: Financial information for 2008 was not available for all companies at press time.

| | | |
|---|---|---|
| 2008 Sales: $2,712,290 | 2008 Profits: $386,940 | **U.S. Stock Ticker:** |
| 2007 Sales: $2,650,410 | 2007 Profits: $277,970 | **Int'l Ticker:** LONN   Int'l Exchange: Zurich-SWX |
| 2006 Sales: $2,193,600 | 2006 Profits: $213,100 | Employees: 7,711 |
| 2005 Sales: $1,818,200 | 2005 Profits: $180,500 | Fiscal Year Ends: 12/31 |
| 2004 Sales: $1,927,800 | 2004 Profits: $121,900 | Parent Company: |

## SALARIES/BENEFITS:

| Pension Plan: Y | ESOP Stock Plan: | Profit Sharing: | Top Exec. Salary: $ | Bonus: $ |
|---|---|---|---|---|
| Savings Plan: | Stock Purch. Plan: Y | | Second Exec. Salary: $ | Bonus: $ |

## OTHER THOUGHTS:
**Apparent Women Officers or Directors:** 1
**Hot Spot for Advancement for Women/Minorities:** Y

## LOCATIONS: ("Y" = Yes)

| West: | Southwest: | Midwest: | Southeast: | Northeast: | International: |
|---|---|---|---|---|---|
| | | Y | | Y | Y |

Note: Financial information, benefits and other data can change quickly and may vary from those stated here.

# LORUS THERAPEUTICS INC

**www.lorusthera.com**

Industry Group Code: 325412  Ranks within this company's industry group:  Sales: 157    Profits: 82

| Drugs: | | Other: | Clinical: | Computers: | Services: |
|---|---|---|---|---|---|
| Discovery: | Y | AgriBio: | Trials/Services: | Hardware: | Specialty Services: |
| Licensing: | | Genetic Data: | Labs: | Software: | Consulting: |
| Manufacturing: | Y | Tissue Replacement: | Equipment/Supplies: | Arrays: | Blood Collection: |
| Genetics: | | | Research & Development Services: | Database Management: | Drug Delivery: |
| | | | Diagnostics: | | Drug Distribution: |

## TYPES OF BUSINESS:

Drugs-Cancer
Antisense Compounds
Low-Molecular-Weight Compounds

## BRANDS/DIVISIONS/AFFILIATES:

Pharma Immune, Inc.
GeneSense Technologies, Inc.
Virulizin
Zor Pharmaceuticals LLC
LOR-2040
NuChem Pharmaceuticals, Inc.

## CONTACTS: *Note: Officers with more than one job title may be intentionally listed here more than once.*

Aiping H. Young, CEO
Aiping H. Young, Pres.
Elizabeth Williams, Acting CFO
Yoon Lee, VP-Research
Elizabeth Williams, Dir.-Admin
Saeid Babaei, VP-Bus. Dev.
Elizabeth Williams, Dir.-Finance
Peter Murray, Dir.-Clinical Dev.
Denis R. Burger, Chmn.

| Phone: 416-798-1200 | Fax: 416-798-2200 |
|---|---|
| Toll-Free: | |
| Address: 2 Meridian Rd., Toronto, ON M9W 4Z7 Canada | |

## GROWTH PLANS/SPECIAL FEATURES:

Lorus Therapeutics, Inc. is a life sciences company focused on the discovery, research and development of effective anticancer therapies. Lorus has product candidates in three classes of anticancer therapies: RNA-targeted therapies (antisense and siRNA), based on synthetic segments of DNA or RNA designed to bind to the messenger RNA that is responsible for the production of proteins over-expressed in cancer cells; small molecule therapies, based on anti-angiogenic, anti-proliferative and anti-metastatic agents; and immunotherapy, based on biological response modifiers that stimulate anticancer properties of the immune system. LOR-2040 (formerly GTI-2040) is Lorus's lead targeted therapy currently in Phase II clinical development. LOR-2040 is an RNA-targeting agent that decreases expression of the R2 subunit of ribonucleotide reductase (RNR). The firm's other RNA-targeted therapy is LOR-1284, which is based on siRNA-mediated inhibition of R2 expression and is in preclinical development. The company's lead immunotherapy agent is Virulizin, which stimulates the body's immune system through several mechanisms, including the activation of macrophages and the infiltration of natural killer cells into tumors. Some of Lorus's small molecule therapies include LOR-253 and LOR-220. Lorus's subsidiaries include a GeneSense Technologies, Inc. and Pharma Immune, Inc. The company also holds an 80% interest in NuChem Pharmaceuticals, Inc. In April 2008, GeneSense Technologies, Inc., a subsidiary of Lorus, signed an exclusive multinational license agreement with Zor Pharmaceuticals LLC to further develop and commercialize Virulizin for human therapeutic applications. In June 2008, LOR-2040 was granted Orphan Drug status for the treatment of Acute Myeloid Leukemia (AML) by the Committee for Orphan Medicinal Products (COMP) of the European Medicines Agency (EMEA). In May 2009, Lorus announced a cooperative research program with the U.S. National Cancer Institute concerning the use of LOR-2501, LOR-2040 and LOR-1284 in drug cocktails for the treatment of renal cell carcinoma tumors.

## FINANCIALS:  Sales and profits are in thousands of dollars—add 000 to get the full amount. 2008 Note: Financial information for 2008 was not available for all companies at press time.

| | | |
|---|---|---|
| 2008 Sales: $ 40 | 2008 Profits: $-11,680 | **U.S. Stock Ticker: LRUSF** |
| 2007 Sales: $ 100 | 2007 Profits: $-8,910 | **Int'l Ticker: LOR**    Int'l Exchange: Toronto-TSX |
| 2006 Sales: $ | 2006 Profits: $-17,900 | Employees:    52 |
| 2005 Sales: $ | 2005 Profits: $-17,600 | Fiscal Year Ends: 5/31 |
| 2004 Sales: $ 400 | 2004 Profits: $-22,300 | Parent Company: |

## SALARIES/BENEFITS:

| Pension Plan: | ESOP Stock Plan: | Profit Sharing: | Top Exec. Salary: $287,965 | Bonus: $104,686 |
|---|---|---|---|---|
| Savings Plan: | Stock Purch. Plan: | | Second Exec. Salary: $112,671 | Bonus: $8,963 |

## OTHER THOUGHTS:

**Apparent Women Officers or Directors**: 2
**Hot Spot for Advancement for Women/Minorities**: Y

## LOCATIONS: ("Y" = Yes)

| West: | Southwest: | Midwest: | Southeast: | Northeast: | International: |
|---|---|---|---|---|---|
| | | | | | Y |

# LUMINEX CORPORATION

**www.luminexcorp.com**

**Industry Group Code: 325413 Ranks within this company's industry group: Sales: 12 Profits: 9**

| Drugs: | Other: | Clinical: | | Computers: | Services: |
|---|---|---|---|---|---|
| Discovery: | AgriBio: | Trials/Services: | | Hardware: | Specialty Services: |
| Licensing: | Genetic Data: | Labs: | | Software: | Consulting: |
| Manufacturing: | Tissue Replacement: | Equipment/Supplies: | Y | Arrays: | Blood Collection: |
| Genetics: | | Research & Development Services: | | Database Management: | Drug Delivery: |
| | | Diagnostics: | Y | | Drug Distribution: |

## TYPES OF BUSINESS:

Medical Diagnostics
Bioassays
Software
xMAP Testing

## BRANDS/DIVISIONS/AFFILIATES:

xMAP
Luminex 100 IS System
Luminex HTS System
Luminex 200 System
xPONENT
MagPlex Magnetic Microspheres
xTAG
Luminex Molecular Diagnostics

## CONTACTS: Note: Officers with more than one job title may be intentionally listed here more than once.

Patrick J. Balthrop, CEO
Douglas C. Bryant, COO/Exec. VP
Patrick J. Balthrop, Pres.
Harriss T. Currie, CFO
Darin Leigh, VP-Mktg. & Sales
Steve Back, VP-Mfg.
David S. Reiter, General Counsel/VP/Corp. Sec.
Russell W. Bradley, VP-Bus. Dev. & Strategic Planning
Harriss T. Currie, VP-Finance/Treas.
Andrew D. Ewing, VP-Luminex Tech Oper.
Gregory J. Gosch, VP-Luminex Bioscience Group
Jeremy Bridge-Cook, VP-Luminex Molecular Diagnostics
Oliver H. Meek, VP-Quality Assurance & Regulatory Affairs
G. Walter Loewenbaum, Chmn.

| **Phone:** 512-219-8020 | **Fax:** 512-219-5195 |
|---|---|
| **Toll-Free:** 888-219-8020 | |
| **Address:** 12212 Technology Blvd., Austin, TX 78727 US | |

## GROWTH PLANS/SPECIAL FEATURES:

Luminex Corporation manufactures and markets biological testing technologies for the life sciences industry. The firm's technologies enable biological tests (bioassays) to detect biochemicals, proteins and genes in samples for the purpose of protein expression profiling, genomic research, genetic disease, immunodiagnostics and biodefense/environmental testing. The company's xMAP system can perform up to 100 bioassays on a single drop of fluid. The xMAP system makes use of microspheres (microscopic polystyrene beads), lasers, digital signal processing and traditional chemistry in order to run various diagnostic tests. The system is an industry-leading technology because of the rapid and precise results it produces from a relatively small sample. Products, xPONENT software and MagPlex magnetic microspheres enhance ease-of-use and automation capabilities in the xMAP technologies: The firm licenses its xMAP technology to various partners in several industries, who then develop marketable products and services for end-users. Partners include companies within the research/drug discovery fields, such as Radix BioSolutions and Cayman Chemical Company, as well as companies within clinical diagnostics fields, such as Inverness Medical and Bayer HealthCare. The firm's xTAG technology, developed by Luminex Molecular Diagnostics, consists of several components including multiplexed PCR or target identification primers, DNA Tags, xMAP microspheres and data analysis software. The technology permits the development of molecular diagnostics assays for clinical use by hospitals and reference laboratories. Products designed for use in the Luminex Systems include the Luminex 100 IS System, an analyzer based on the principles of flow chemistry; Luminex HTS System, a high-throughput screening system which performs thousands of bioassays daily; the Luminex 200 System, used by clinical and research laboratory professionals; and various microspheres.

## FINANCIALS: Sales and profits are in thousands of dollars—add 000 to get the full amount. 2008 Note: Financial information for 2008 was not available for all companies at press time.

| | | |
|---|---|---|
| 2008 Sales: $104,447 | 2008 Profits: $3,057 | **U.S. Stock Ticker: LMNX** |
| 2007 Sales: $75,010 | 2007 Profits: $-2,711 | **Int'l Ticker:** Int'l Exchange: |
| 2006 Sales: $52,989 | 2006 Profits: $1,507 | Employees: 344 |
| 2005 Sales: $42,313 | 2005 Profits: $-2,666 | Fiscal Year Ends: 12/31 |
| 2004 Sales: $35,880 | 2004 Profits: $-3,605 | Parent Company: |

## SALARIES/BENEFITS:

| Pension Plan: | ESOP Stock Plan: | Profit Sharing: | Top Exec. Salary: $445,500 | Bonus: $470,542 |
|---|---|---|---|---|
| Savings Plan: Y | Stock Purch. Plan: | | Second Exec. Salary: $325,600 | Bonus: $257,253 |

## OTHER THOUGHTS:

**Apparent Women Officers or Directors:**
**Hot Spot for Advancement for Women/Minorities:**

## LOCATIONS: ("Y" = Yes)

| West: | Southwest: | Midwest: | Southeast: | Northeast: | International: |
|---|---|---|---|---|---|
| | Y | | | | Y |

Note: Financial information, benefits and other data can change quickly and may vary from those stated here.

# MALLINCKRODT INC

www.mallinckrodt.com

Industry Group Code: 325413  Ranks within this company's industry group: Sales:    Profits:

| Drugs: | | Other: | | Clinical: | | Computers: | | Services: | |
|---|---|---|---|---|---|---|---|---|---|
| Discovery: | | AgriBio: | | Trials/Services: | | Hardware: | | Specialty Services: | |
| Licensing: | | Genetic Data: | | Labs: | | Software: | | Consulting: | |
| Manufacturing: | Y | Tissue Replacement: | | Equipment/Supplies: | Y | Arrays: | | Blood Collection: | |
| Genetics: | Y | | | Research & Development Services: | | Database Management: | | Drug Delivery: | |
| | | | | Diagnostics: | Y | | | Drug Distribution: | |

## TYPES OF BUSINESS:

Imaging Agents & Radiopharmaceuticals
Diagnostics Products
Bulk Analgesic Pharmaceuticals
Generic Pharmaceuticals
Active Pharmaceutical Ingredients
Medical Devices
Respiratory Products

## BRANDS/DIVISIONS/AFFILIATES:

Puritan Bennett
Nellcor
OptiMax
Covidien Ltd.
Brand Pharmaceuticals
Mallinckrodt Pharmaceuticals Generics

## CONTACTS: Note: Officers with more than one job title may be intentionally listed here more than once.

Matthew K. Harbaugh, CFO/VP-Shared Svcs.
Lisa Britt, VP-Human Resources
Herbert Neuman, Chief Medical Officer
Charles Bramlage, Pres., Pharmaceutical Prod.
Michael J. Giuliani, VP-R&D-Imaging Solutions & Pharmaceutical Prod.

| Phone: 314-654-2000 | Fax: |
|---|---|
| Toll-Free: | |
| Address: 675 McDonnell Blvd., Hazelwood, MO 63042 US | |

## GROWTH PLANS/SPECIAL FEATURES:

Mallinckrodt, Inc., a subsidiary of Covidien Ltd. (formerly Tyco Healthcare Group), develops and manufactures a wide range of medical products and devices, primarily used by hospitals, freestanding imaging centers and radiopharmacies for diagnostic and treatment purposes. The company operates through three main divisions: imaging, respiratory and pharmaceuticals. Mallinckrodt's imaging group produces a full line of urology imaging systems, contrast media and delivery systems and radiopharmaceuticals, including contrast media, MRI infusion pump systems, guide-wires, x-ray contrast media, nuclear medicine products and injection systems for a variety of packages, including x-ray, magnetic resonance and cardiology applications. Brands operating under the imaging group include Liebel-Flarsheim, Baro-Cat and Medescan. The company's respiratory segment operates through two businesses: Nellcor and Puritan Bennett. Nellcor focuses on pulse oximetry and critical care solutions, and its products include airway management tubes and other equipment; temperature management products, including sensors and probes; catheters and feeding tubes; pulse oximetry equipment; and the OxiMax Pulse Oximetry System, a digitally interfaced pulse monitor and sensor. Puritan Bennett focuses on respiratory support systems, including ventilators, sleep therapy devices, sleep diagnostics, oxygen therapy units and spirometers. This business also offers Clinivision Mobile Patient Charting clinical information software. Mallinckrodt's pharmaceuticals division provides specialty chemicals and prescription pharmaceuticals and is focused on providing pain relief and addiction therapy, with a product line that includes acetaminophen, methadone and naltrexone (for alcohol addiction). There are four complementary platforms operating under the firm's pharmaceuticals division: Brand Pharmaceuticals, focusing on sleep-aid and anti-depressant products; Mallinckrodt Pharmaceuticals Generics, offering generic analgesics; Addiction Treatment therapies; and the Outsourcing segment, which supplies raw materials and active pharmaceutical ingredients, and provides contract manufacturing services.

## FINANCIALS: Sales and profits are in thousands of dollars—add 000 to get the full amount. 2008 Note: Financial information for 2008 was not available for all companies at press time.

| | | U.S. Stock Ticker: Subsidiary |
|---|---|---|
| 2008 Sales: $ | 2008 Profits: $ | Int'l Ticker:    Int'l Exchange: |
| 2007 Sales: $ | 2007 Profits: $ | Employees:  11,000 |
| 2006 Sales: $ | 2006 Profits: $ | Fiscal Year Ends: 9/30 |
| 2005 Sales: $ | 2005 Profits: $ | Parent Company: COVIDIEN LTD |
| 2004 Sales: $ | 2004 Profits: $ | |

## SALARIES/BENEFITS:

| Pension Plan: | ESOP Stock Plan: | Profit Sharing: | Top Exec. Salary: $732,840 | Bonus: $4,072,400 |
|---|---|---|---|---|
| Savings Plan: | Stock Purch. Plan: | | Second Exec. Salary: $322,793 | Bonus: $1,095,600 |

## OTHER THOUGHTS:

Apparent Women Officers or Directors: 1
Hot Spot for Advancement for Women/Minorities:

## LOCATIONS: ("Y" = Yes)

| West: | Southwest: | Midwest: | Southeast: | Northeast: | International: |
|---|---|---|---|---|---|
| Y | | Y | | Y | Y |

# MANHATTAN PHARMACEUTICALS INC

www.manhattanpharma.com

**Industry Group Code: 325412  Ranks within this company's industry group:** Sales:   Profits: 67

| Drugs: | | Other: | Clinical: | | Computers: | Services: | |
|---|---|---|---|---|---|---|---|
| Discovery: | Y | AgriBio: | Trials/Services: | | Hardware: | Specialty Services: | |
| Licensing: | | Genetic Data: | Labs: | | Software: | Consulting: | |
| Manufacturing: | | Tissue Replacement: | Equipment/Supplies: | | Arrays: | Blood Collection: | |
| Genetics: | | | Research & Development Services: | | Database Management: | Drug Delivery: | Y |
| | | | Diagnostics: | | | Drug Distribution: | |

## TYPES OF BUSINESS:

Pharmaceuticals Discovery & Development
Drug Delivery Systems

## BRANDS/DIVISIONS/AFFILIATES:

Hedrin
PTH (1-34)
Nordic Biotech Advisors ApS

## CONTACTS: *Note: Officers with more than one job title may be intentionally listed here more than once.*

Douglas Abel, CEO
Michael G. McGuinness, COO
Douglas Abel, Pres.
Michael G. McGuinness, CFO
Michelle Carroll, Dir.-Strategic Dev.
Michelle Carroll, Dir.-Investor Rel.
Mary C. Spellman, Head-Dermatology & Drug Dev.

| Phone: 212-582-3950 | Fax: 212-582-3957 |
|---|---|

**Toll-Free:**

**Address:** 48 Wall St., Ste. 1100, New York, NY 10005 US

## GROWTH PLANS/SPECIAL FEATURES:

Manhattan Pharmaceuticals, Inc. is a clinical-stage pharmaceutical company developing novel drug candidates primarily in the areas of dermatologic and immune disorders. With a pipeline consisting of two clinical-stage product candidates, the company is developing potential therapeutics for large, underserved patient populations. Product candidates include Hedrin, which the company is developing through a joint venture; and a topical product for the treatment of psoriasis, also called topical PTH (1-34). PTH (1-34) is a peptide that regulates epidermal cell growth and differentiation, currently under Phase II development as a topical treatment for psoriasis and additional hyperproliferative skin disorders. Hedrin is a novel, non-insecticide treatment for head lice and is in preclinical development in the U.S. and is on the market in Europe and the U.K. In early 2008, Manhattan announced its entry into a joint venture with Nordic Biotech Advisors ApS, which created a separate entity to develop and commercialize as well as to secure a commercialization partner for Hedrin in North America. Manhattan will manage the joint venture company. In January 2009, the U.S. Food and Drug Administration (FDA) Center for Devices and Radiological Health (CDRH) classified Hedrin as a Class III medical device.

## FINANCIALS: Sales and profits are in thousands of dollars—add 000 to get the full amount. 2008 Note: Financial information for 2008 was not available for all companies at press time.

| | | | |
|---|---|---|---|
| 2008 Sales: $ | 2008 Profits: $-4,269 | **U.S. Stock Ticker: MHAN** | |
| 2007 Sales: $ | 2007 Profits: $-12,032 | **Int'l Ticker:** Int'l Exchange: | |
| 2006 Sales: $ | 2006 Profits: $-9,695 | Employees: 4 | |
| 2005 Sales: $ | 2005 Profits: $-19,141 | Fiscal Year Ends: 12/31 | |
| 2004 Sales: $ | 2004 Profits: $-5,896 | Parent Company: | |

## SALARIES/BENEFITS:

| Pension Plan: | ESOP Stock Plan: | Profit Sharing: | Top Exec. Salary: $345,000 | Bonus: $180,000 |
|---|---|---|---|---|
| Savings Plan: | Stock Purch. Plan: | | Second Exec. Salary: $288,333 | Bonus: $ |

## OTHER THOUGHTS:

**Apparent Women Officers or Directors: 1**
**Hot Spot for Advancement for Women/Minorities:**

## LOCATIONS: ("Y" = Yes)

| West: | Southwest: | Midwest: | Southeast: | Northeast: | International: |
|---|---|---|---|---|---|
| | | | | Y | |

# MAXYGEN INC

**www.maxygen.com**

Industry Group Code: 325412  **Ranks within this company's industry group:** Sales: 67   Profits: 44

| Drugs: | Other: | | Clinical: | Computers: | Services: |
|---|---|---|---|---|---|
| Discovery: | AgriBio: | | Trials/Services: | Hardware: | Specialty Services: |
| Licensing: | Genetic Data: | Y | Labs: | Software: | Consulting: |
| Manufacturing: Y | Tissue Replacement: | | Equipment/Supplies: | Arrays: | Blood Collection: |
| Genetics: | | | Research & Development Services: | Database Management: | Drug Delivery: |
| | | | Diagnostics: | | Drug Distribution: |

## TYPES OF BUSINESS:

Drug Discovery & Development
Improved & Novel Pharmaceuticals
Research Services
Chemicals
Research & Development-Molecular Evolution

## BRANDS/DIVISIONS/AFFILIATES:

MolecularBreeding
DNAShuffling
MaxyScan
Maxy-G34
MAXY-4
Codexis

## CONTACTS: Note: Officers with more than one job title may be intentionally listed here more than once.

Russell J. Howard, CEO
Elliot Goldstein, COO
Lawrence W. Briscoe, CFO/Sr. VP
Elliot Goldstein, Chief Medical Officer
John Borkholder, Chief Corp. Counsel/Corp. Sec.
Grant Yonehiro, Chief Bus. Officer/Sr. VP
Michele Boudreau, Dir.-Public Rel.
Michele Boudreau, Dir.-Investor Rel.
Isaac Stein, Chmn.

| **Phone:** 650-298-5300 | **Fax:** 650-364-2715 |
|---|---|
| **Toll-Free:** | |
| **Address:** 301 Galveston Dr., Redwood City, CA 94063 US | |

## GROWTH PLANS/SPECIAL FEATURES:

Maxygen, Inc. is a biotechnology company that works on the discovery and development of improved protein pharmaceuticals for the treatment of diseases and serious medical conditions. Technologies developed by Maxygen include MolecularBreeding, a process that mimics the natural events of evolution using a recombination process called DNAShuffling that generates a diverse library of DNA sequences. MolecularBreeding allows the company to rapidly move from product concept to IND (investigational new drug)-ready drug candidate, lowering costs associated with research and expanding the potential for discovery. The company's MaxyScan screening system selects individual proteins with desired characteristics from gene variants within the library for additional experimentation. Maxygen currently has two product candidates in different clinical study phases: Maxy-G34, a neutropenia treatment; and MAXY-4, a rheumatoid arthritis and other immune or autoimmune diseases treatment. Maxygen additionally has an HIV vaccine research program and a minority investment in Codexis, a biotechnology company that improves the manufacturing process of small molecular pharmaceutical products. In July 2008, the company sold its hemophilia program assets, including Maxy VII, a hemophilia and possibly an acute bleeding conditions treatment, to Bayer Healthcare for $90 million. In August of the same year, the firm received a 2 year, $3.4 million grant from the U.S. Department of Defense for the development of advanced vaccine technology. Maxygen will work in collaboration with Aldevron LLC. In September 2008, the company granted Astellas Pharma Inc. the rights to commercialize MAXY-4 lead candidates for transplant rejection and autoimmune diseases.

Employees are offered medical and dental insurance; group life and AD&D Insurance; flexible spending accounts; health club reimbursements; an on-site workout room; free flu shots; disability coverage; financial planning services; emergency child and elder care; tuition reimbursement; commuter voucher programs; concierge services; and credit union membership.

## FINANCIALS: Sales and profits are in thousands of dollars—add 000 to get the full amount. 2008 Note: Financial information for 2008 was not available for all companies at press time.

| | | |
|---|---|---|
| 2008 Sales: $100,709 | 2008 Profits: $30,325 | **U.S. Stock Ticker:** MAXY |
| 2007 Sales: $23,157 | 2007 Profits: $-49,315 | **Int'l Ticker:**   Int'l Exchange: |
| 2006 Sales: $25,021 | 2006 Profits: $-16,482 | Employees:   96 |
| 2005 Sales: $14,501 | 2005 Profits: $-18,436 | Fiscal Year Ends: 12/31 |
| 2004 Sales: $16,275 | 2004 Profits: $9,342 | Parent Company: |

## SALARIES/BENEFITS:

| Pension Plan: | ESOP Stock Plan: | Profit Sharing: | Top Exec. Salary: $500,500 | Bonus: $236,100 |
|---|---|---|---|---|
| Savings Plan: Y | Stock Purch. Plan: Y | | Second Exec. Salary: $457,449 | Bonus: $225,225 |

## OTHER THOUGHTS:

**Apparent Women Officers or Directors:** 1
**Hot Spot for Advancement for Women/Minorities:**

## LOCATIONS: ("Y" = Yes)

| West: | Southwest: | Midwest: | Southeast: | Northeast: | International: |
|---|---|---|---|---|---|
| Y | | | | | Y |

# MDRNA INC

www.mdrnainc.com

**Industry Group Code:** 325412A **Ranks within this company's industry group:** Sales: 23 Profits: 28

| Drugs: | | Other: | | Clinical: | | Computers: | | Services: | |
|---|---|---|---|---|---|---|---|---|---|
| Discovery: | Y | AgriBio: | | Trials/Services: | | Hardware: | | Specialty Services: | |
| Licensing: | Y | Genetic Data: | | Labs: | | Software: | | Consulting: | |
| Manufacturing: | Y | Tissue Replacement: | | Equipment/Supplies: | | Arrays: | | Blood Collection: | |
| Genetics: | | | | Research & Development Services: | | Database Management: | | Drug Delivery: | Y |
| | | | | Diagnostics: | | | | Drug Distribution: | |

## TYPES OF BUSINESS:

Drug Delivery Systems
Drugs-Nasally Administered
Tight Junction Biology
RNA Interference Technology

## BRANDS/DIVISIONS/AFFILIATES:

Calcitonin Nasal Spray

## CONTACTS: *Note: Officers with more than one job title may be intentionally listed here more than once.*

J. Michael French, CEO
J. Michael French, Pres.
Bruce R. York, CFO
Barry Polinsky, Chief Scientific Officer
Bruce R. York, Sec.
Bruce R. Thaw, Chmn.

| Phone: 425-908-3600 | Fax: 425-908-3650 |
|---|---|
| Toll-Free: | |
| Address: 3830 Monte Villa Pkwy., Bothell, WA 98021 US | |

## GROWTH PLANS/SPECIAL FEATURES:

MDRNA Inc., formerly Nastech Pharmaceutical Company, Inc. is chiefly involved in the development and commercialization of therapies using RNA interference (RNAi) to down-regulate specific protein expressions that lead to disease without altering the DNA itself. This technology may have applications for potential treatments in oncology, inflammation, metabolic disorders and viral infections. The company plans to combine this research with optimized delivery systems that will promote better uptake, faster results, fewer side effects, lower costs and improved absorption of targeted therapies. A challenge involved with RNAi technology is that specific targeting is extremely important, and off-target deliveries can result in undesirable side effects. MDRNA's research is designed to eliminate the chance for off-target effects by using a specific type of siRNA (named for its silencing functioning). It is currently focusing its developmental research on liver diseases, specifically hepatocellular carcinoma. The firm collaborates with a number of partners for research, including the University of Helsinki, Quebec's Universite Laval and the Centers for Disease Control. The company also licenses its technology to Hoffman-La Roche, Ltd. and Novartis. The company was previously developing a generic calcitonin-salmon nasal spray for the treatment of osteoporosis. In April 2009, MDRNA decided to sell these assets, along with their New York manufacturing facility, to Par Pharmaceuticals. This move was part of the firm's plan to shift exclusively to RNAi operations, although it will continue to receive royalties from sales of the drug.

## FINANCIALS: Sales and profits are in thousands of dollars—add 000 to get the full amount. 2008 Note: Financial information for 2008 was not available for all companies at press time.

| | | | |
|---|---|---|---|
| 2008 Sales: $2,609 | 2008 Profits: $-59,220 | U.S. Stock Ticker: MRNA | |
| 2007 Sales: $18,137 | 2007 Profits: $-52,372 | Int'l Ticker: | Int'l Exchange: |
| 2006 Sales: $28,490 | 2006 Profits: $-26,877 | Employees: 44 | |
| 2005 Sales: $7,449 | 2005 Profits: $-32,163 | Fiscal Year Ends: 12/31 | |
| 2004 Sales: $1,847 | 2004 Profits: $-28,609 | Parent Company: | |

## SALARIES/BENEFITS:

| Pension Plan: | ESOP Stock Plan: | Profit Sharing: | Top Exec. Salary: $511,164 | Bonus: $ |
|---|---|---|---|---|
| Savings Plan: Y | Stock Purch. Plan: Y | | Second Exec. Salary: $318,752 | Bonus: $ |

## OTHER THOUGHTS:

**Apparent Women Officers or Directors:**
**Hot Spot for Advancement for Women/Minorities:**

## LOCATIONS: ("Y" = Yes)

| West: | Southwest: | Midwest: | Southeast: | Northeast: | International: |
|---|---|---|---|---|---|
| Y | | | | Y | |

*Note: Financial information, benefits and other data can change quickly and may vary from those stated here.*

# MDS INC

**www.mdsintl.com**

**Industry Group Code: 6215  Ranks within this company's industry group:** Sales: 2  Profits: 5

| Drugs: | | Other: | | Clinical: | | Computers: | | Services: | |
|---|---|---|---|---|---|---|---|---|---|
| Discovery: | Y | AgriBio: | | Trials/Services: | Y | Hardware: | | Specialty Services: | Y |
| Licensing: | | Genetic Data: | Y | Labs: | Y | Software: | | Consulting: | Y |
| Manufacturing: | | Tissue Replacement: | | Equipment/Supplies: | Y | Arrays: | | Blood Collection: | |
| Genetics: | | | | Research & Development Services: | Y | Database Management: | | Drug Delivery: | |
| | | | | Diagnostics: | Y | | | Drug Distribution: | Y |

## TYPES OF BUSINESS:

Drug Discovery & Development Services
Analytic Instruments & Technology
Imaging Agents
Medical & Surgical Supplies
Irradiation Systems
Health Care Product Distribution

## BRANDS/DIVISIONS/AFFILIATES:

MDS Pharma Services
MDS Analytical Technologies
MDS Nordion
Blueshift Biotechnologies
Compendia Bioscience, Inc.
OncoPredictor

## CONTACTS: *Note: Officers with more than one job title may be intentionally listed here more than once.*

Stephen P. DeFalco, CEO
Stephen P. DeFalco, Pres.
Douglas S. Prince, CFO
Mary E. Federau, Exec. VP-Global Human Resources
Thomas E. Gernon, CIO/Exec. VP-IT
Kenneth L. Horton, General Counsel
Kenneth L. Horton, Exec. VP-Corp. Dev.
Janet Ko, Sr. VP-Comm.
Kim Lee, Sr. Dir.-Investor Rel.
Douglas S. Prince, Exec. VP-Finance
Steve West, Pres., MDS Nordion
David Spaight, Pres., MDS Pharma Svcs.
Andrew W. Boorn, Pres., MDS Analytical Technologies
James S.A. MacDonald, Chmn.

| Phone: 905-267-4222 | Fax: |
|---|---|
| **Toll-Free:** | |
| **Address:** 2810 Matheson Blvd. E., Ste. 500, Mississauga, ON L4W 4X7 Canada | |

## GROWTH PLANS/SPECIAL FEATURES:

MDS, Inc., founded in 1969 as Medical Data Sciences Limited, is a global biotechnology firm providing products and services for drug development and disease treatment and diagnosis. Its customers include pharmaceutical and biotechnology companies, hospitals and health care professionals. The firm operates through three business units: MDS Pharma Services, which generated 40% of 2008 revenue; MDS Analytical Technologies, 38%; and MDS Nordion, 24%. Pharma Services provides contract research services to pharmaceutical manufacturers and biotechnology companies. Its solutions include early-stage, discovery and pre-clinical research, Phase I clinical trial, bioanalytical, late-stage, Phases II-IV clinical trial and Central Labs services. The Pharma Services segment has 38 centers of operation in 29 countries. Analytical Technologies, the technological section of the company, offers research, design, manufacture and marketing solutions for mass spectrometry, drug discovery and bioresearch. Nordion's technologies are used in medical imaging, including cancer, cardiac conditions, thyroid and neurological problem diagnosis and treatment; radiotherapeutics for liver and brain cancer; and the irradiation and sterilization disposable medical supplies, plasma, surgical implements, serums, contact lenses and cosmetics. The firm has operations in 29 countries around the world, including locations in North and South America, Europe, Asia and Africa. In 2008, MDS launched 16 new products and services. In June 2008, MDS Analytical Technologies acquired Blueshift Biotechnologies. In April 2009, MDS Pharma Services announced a strategic collaboration with Compendia Bioscience, Inc. to develop OncoPredictor, designed to improve the development of cancer treatment drugs. In May 2009, Nordion opened a new radiopharmaceutical facility in Belgium.

## FINANCIALS: Sales and profits are in thousands of dollars—add 000 to get the full amount. 2008 Note: Financial information for 2008 was not available for all companies at press time.

| | | |
|---|---|---|
| 2008 Sales: $1,315,000 | 2008 Profits: $-553,000 | **U.S. Stock Ticker: MDZ** |
| 2007 Sales: $1,210,000 | 2007 Profits: $773,000 | **Int'l Ticker: MDS**  Int'l Exchange: Toronto-TSX |
| 2006 Sales: $1,060,000 | 2006 Profits: $120,000 | Employees: 5,000 |
| 2005 Sales: $1,296,908 | 2005 Profits: $26,999 | Fiscal Year Ends: 10/31 |
| 2004 Sales: $1,447,000 | 2004 Profits: $42,000 | Parent Company: |

## SALARIES/BENEFITS:

| Pension Plan: | ESOP Stock Plan: | Profit Sharing: | Top Exec. Salary: $ | Bonus: $ |
|---|---|---|---|---|
| Savings Plan: | Stock Purch. Plan: | | Second Exec. Salary: $ | Bonus: $ |

## OTHER THOUGHTS:

**Apparent Women Officers or Directors:** 3
**Hot Spot for Advancement for Women/Minorities:** Y

## LOCATIONS: ("Y" = Yes)

| West: | Southwest: | Midwest: | Southeast: | Northeast: | International: |
|---|---|---|---|---|---|
| Y | Y | Y | | Y | Y |

# MEDAREX INC

www.medarex.com

Industry Group Code: 325412  Ranks within this company's industry group: Sales: 83  Profits: 129

| Drugs: | | Other: | | Clinical: | Computers: | Services: |
|---|---|---|---|---|---|---|
| Discovery: | Y | AgriBio: | | Trials/Services: | Hardware: | Specialty Services: |
| Licensing: | Y | Genetic Data: | Y | Labs: | Software: | Consulting: |
| Manufacturing: | | Tissue Replacement: | | Equipment/Supplies: | Arrays: | Blood Collection: |
| Genetics: | | | | Research & Development Services: | Database Management: | Drug Delivery: |
| | | | | Diagnostics: | | Drug Distribution: |

## TYPES OF BUSINESS:

Drugs-Human Monoclonal Antibodies
Transgenic Mouse Technology
Drugs-Cancer
Drugs-Autoimmune Disease

## BRANDS/DIVISIONS/AFFILIATES:

UltiMAb Human Antibody Development System
HuMAb-Mouse
KM-Mouse
TC-Mouse
MDX-1337

## CONTACTS: Note: Officers with more than one job title may be intentionally listed here more than once.

Howard H. Pien, CEO
Howard H. Pien, Pres.
Christian S. Schade, CFO
Deanna Dietl, VP-Human Resources
Nils Lonberg, Sr. VP/Dir.-Scientific
Geoffrey M. Nichol, Sr. VP-Prod. Dev.
Christian S. Schade, Sr. VP-Admin.
Ursula B. Bartels, General Counsel/Sec./Sr. VP
Ronald A. Pepin, Sr. VP-Bus. Dev.
Nichol Harber, Corp. Comm.
Christian S. Schade, Sr. VP-Finance
Howard H. Pien, Chmn.

| Phone: 609-430-2880 | Fax: 609-430-2850 |
|---|---|
| Toll-Free: | |
| Address: 707 State Rd., Princeton, NJ 08540-1437 US | |

## GROWTH PLANS/SPECIAL FEATURES:

Medarex, Inc. is a biopharmaceutical company devoted to the production of fully human antibody-based therapeutic product candidates to fight cancer, inflammation, autoimmune disorders and other diseases. The firm's UltiMAb Human Antibody Development System uses transgenic mice, specifically Medarex's HuMAb-Mouse, Kirin Brewing Co., Ltd.'s TC Mouse and the KM-Mouse, a hybrid of the former two mice. In this system, mouse-derived antibody gene expression is suppressed and effectively replaced with human antibody gene expression. With UltiMAb, the company can create fully human antibodies with no mouse proteins. Consequently, UltiMAb antibodies are less likely to be rejected by patients, have more favorable safety profiles and may be eliminated by the human body more slowly, potentially reducing the required dosing. A number of product candidates use UltiMAb technology, developed in-house or through collaborations, that are currently in clinical testing stages, including treatments for melanoma; lymphoma; breast, prostate, kidney and other cancers; rheumatoid arthritis; and inflammatory diseases. Seven of the most advanced candidates are in Phase II/III clinical trials or are the subject of regulatory applications for marketing authorization. The UltiMAb technology is licensed to pharmaceutical and biotechnology companies that also pay royalties on commercial sales of their products. Over 35 pharmaceutical and biotechnology companies have collaborative or licensing agreements with Medarex to jointly develop opportunities for new antibodies and to commercialize products, including Amgen, Bristol-Myers Squibb, Centocor Ortho Biotech, Eli Lilly, Novartis Pharma, Genmab, GlaxoSmithKline and Merck. The company claims 78 issued patents in the U.S. and 359 abroad. In May 2009, the FDA granted Medarex an allowance of an investigational new drug application for the company's wholly-owned treatment for acute myelogenous leukemia, MDX-1338. In July 2009, the firm agreed to be acquired by Bristol-Myers Squibb

## FINANCIALS: Sales and profits are in thousands of dollars—add 000 to get the full amount. 2008 Note: Financial information for 2008 was not available for all companies at press time.

| | | |
|---|---|---|
| 2008 Sales: $52,292 | 2008 Profits: $-38,465 | U.S. Stock Ticker: MEDX |
| 2007 Sales: $56,258 | 2007 Profits: $-27,055 | Int'l Ticker:    Int'l Exchange: |
| 2006 Sales: $48,646 | 2006 Profits: $-181,701 | Employees:   488 |
| 2005 Sales: $51,455 | 2005 Profits: $-148,012 | Fiscal Year Ends: 12/31 |
| 2004 Sales: $12,474 | 2004 Profits: $-186,392 | Parent Company: |

## SALARIES/BENEFITS:

| Pension Plan: | ESOP Stock Plan: | Profit Sharing: | Top Exec. Salary: $750,000 | Bonus: $712,500 |
|---|---|---|---|---|
| Savings Plan: Y | Stock Purch. Plan: Y | | Second Exec. Salary: $414,000 | Bonus: $192,000 |

## OTHER THOUGHTS:

Apparent Women Officers or Directors: 2
Hot Spot for Advancement for Women/Minorities: Y

## LOCATIONS: ("Y" = Yes)

| West: | Southwest: | Midwest: | Southeast: | Northeast: | International: |
|---|---|---|---|---|---|
| Y | | | | Y | |

Note: Financial information, benefits and other data can change quickly and may vary from those stated here.

# MEDICINES CO (THE)

### www.themedicinescompany.com

**Industry Group Code: 325412  Ranks within this company's industry group:** Sales: 47   Profits: 75

| Drugs: | | Other: | Clinical: | Computers: | Services: |
|---|---|---|---|---|---|
| Discovery: | | AgriBio: | Trials/Services: | Hardware: | Specialty Services: |
| Licensing: | Y | Genetic Data: | Labs: | Software: | Consulting: |
| Manufacturing: | Y | Tissue Replacement: | Equipment/Supplies: | Arrays: | Blood Collection: |
| Genetics: | | | Research & Development Services: | Database Management: | Drug Delivery: |
| | | | Diagnostics: | | Drug Distribution: |

## TYPES OF BUSINESS:

Pharmaceuticals Acquisition & Development
Acute Care Hospital Products
Anticoagulants
Blood Pressure Control

## BRANDS/DIVISIONS/AFFILIATES:

Angiomax
Cleviprex (Clevidine)
Cangrelor
Oritavancin
CU-2010
Curacyte Discovery GmbH
Targanta Therapeutics Corporation

## CONTACTS: *Note: Officers with more than one job title may be intentionally listed here more than once.*

Clive A. Meanwell, CEO
John P. Kelley, COO
John P. Kelley, Pres.
Glenn Sblendorio, CFO/Exec. VP
Paul M. Antinori, General Counsel/Sr. VP
Kelli Watson, Sr. VP-Global Comm.
Bill O'Connor, Chief Acct. Officer
Clive A. Meanwell, Chmn.

| **Phone:** 973-290-6000 | **Fax:** 973-656-9898 |
|---|---|
| **Toll-Free:** 800-388-1183 | |
| **Address:** 8 Sylvan Way, Parsippany, NJ 07054 US | |

## GROWTH PLANS/SPECIAL FEATURES:

The Medicines Co. (TMC) is a global pharmaceutical company specializing in acute care hospital products. The firm acquires, develops and commercializes pharmaceutical products in late stages of development. The company has two marketed products: Angiomax and Cleviprex. Angiomax is an intravenous direct thrombin inhibitor approved for use in patients undergoing percutaneous coronary intervention and, in Europe, for adult patients with acute coronary syndrome. Cleviprex is an intravenous drug intended for the control of blood pressure in anesthesiology and surgery, critical care and emergency conditions. The firm is currently developing two additional late-stage pharmaceutical products and one compound as potential acute care hospital products. The company's first potential product, cangrelor, is an intravenous antiplatelet agent that prevents platelet activation and aggregation, which is believed to have potential advantages in the treatment of vascular disease. The second product, oritavancin, is an intravenous antibiotic being developed for the treatment of gram-positive bacterial infections. Phase III trials for oritavancin are expected to start in 2009. The third product candidate, CU-2010, is a small molecule serine protease inhibitor being developed for the prevention of blood loss during surgery. CU-2010 is scheduled to begin Phase I trials in 2009. In August 2008, the company acquired Germany-based Curacyte Discovery GmbH and its lead compound, CU-2010. In February 2009, the firm completed the acquisition of Targanta Therapeutics Corporation. In connection with this acquisition, TMC acquired oritavancin, an intravenous antibiotic. In May 2009, TMC announced plans to discontinue its Phase III CHAMPION clinical trials of patients undergoing percutaneous coronary intervention.

TMC offers its employees medical, dental, prescription and vision insurance; an employee assistance plan; tuition reimbursement; an employee stock purchase plan and associate stock options; and credit union membership.

## FINANCIALS: Sales and profits are in thousands of dollars—add 000 to get the full amount. 2008 Note: Financial information for 2008 was not available for all companies at press time.

| | | |
|---|---|---|
| 2008 Sales: $348,157 | 2008 Profits: $-8,504 | **U.S. Stock Ticker:** MDCO |
| 2007 Sales: $257,534 | 2007 Profits: $-18,272 | **Int'l Ticker:**   Int'l Exchange: |
| 2006 Sales: $213,952 | 2006 Profits: $63,726 | Employees:   305 |
| 2005 Sales: $150,207 | 2005 Profits: $-7,753 | Fiscal Year Ends: 12/31 |
| 2004 Sales: $144,251 | 2004 Profits: $16,999 | Parent Company: |

## SALARIES/BENEFITS:

| Pension Plan: | ESOP Stock Plan: | Profit Sharing: | Top Exec. Salary: $588,640 | Bonus: $357,599 |
|---|---|---|---|---|
| Savings Plan: Y | Stock Purch. Plan: Y | | Second Exec. Salary: $463,500 | Bonus: $173,813 |

## OTHER THOUGHTS:

Apparent Women Officers or Directors: 2
Hot Spot for Advancement for Women/Minorities:

## LOCATIONS: ("Y" = Yes)

| West: | Southwest: | Midwest: | Southeast: | Northeast: | International: |
|---|---|---|---|---|---|
| | | | | Y | Y |

# MEDICIS PHARMACEUTICAL CORP

www.medicis.com

**Industry Group Code: 325412  Ranks within this company's industry group: Sales: 43    Profits: 50**

| Drugs: | Other: | Clinical: | Computers: | Services: |
|---|---|---|---|---|
| Discovery: | AgriBio: | Trials/Services: | Hardware: | Specialty Services: |
| Licensing: | Genetic Data: | Labs: | Software: | Consulting: |
| Manufacturing:  Y | Tissue Replacement: | Equipment/Supplies: | Arrays: | Blood Collection: |
| Genetics: | | Research & Development Services: | Database Management: | Drug Delivery: |
| | | Diagnostics: | | Drug Distribution: |

## TYPES OF BUSINESS:

Dermatological, Aesthetic & Podiatric Conditions Drugs
Acne Treatment
Topical Creams
Wrinkle Treatment

## BRANDS/DIVISIONS/AFFILIATES:

Perlane
Restylane
Solodyn
Triaz
Vanos
Ziana
LipoSonix, Inc.
McKesson Corp.

## CONTACTS: Note: Officers with more than one job title may be intentionally listed here more than once.

Jonah Shacknai, CEO
Mark A. Prygocki, Sr., COO/Exec. VP
Richard D. Peterson, CFO/Exec. VP
Vincent Ippolito, Exec. VP-Sales & Mktg.
Mitchell S. Wortzman, Chief Scientific Officer/Exec. VP
Joseph P. Cooper, Exec. VP-Prod. Dev.
Jason Hanson, General Counsel/Exec. VP/Corp. Sec.
Joseph P. Cooper, Exec. VP-Corp. Dev.
Richard D. Peterson, Treas.
Jonah Shacknai, Chmn.

| Phone: 602-808-8800 | Fax: 602-808-0822 |
|---|---|
| Toll-Free: | |
| Address: 7720 North Dobson Rd., Scottsdale, AZ 85256 US | |

## GROWTH PLANS/SPECIAL FEATURES:

Medicis Pharmaceutical Corp. is a specialty pharmaceutical company that develops and markets products for treatment of dermatological, aesthetic and podiatric conditions. The company offers a range of products addressing various conditions or aesthetic improvement, including facial wrinkles; acne; fungal infections; rosacea; hyperpigmentation; photoaging; psoriasis; skin and skin-structure infections; seborrheic dermatitis; and cosmesis (improvement in the texture and appearance of skin). These products are marketed under the firm's 18 branded lines, which consist of secondary products plus six primary brands. The primary brands include Perlane, an injectable gel for implantation into the deep dermis to superficial subcutis for the correction of facial wrinkles; Restylane, an injectable gel for treatment of facial wrinkles and folds such as nasolabial folds; Solodyn, a daily dosage for the treatment of inflammatory lesions of non-nodular acne vulgaris; Triaz, a topical patented gel and cleanser and patent-pending pad treatments for acne; Vanos, a high potency topical corticosteroid indicated for the relief of inflammatory and pruritic manifestations of corticosteroid responsive dermatoses; and Ziana, a once-daily topical gel treatment for acne vulgaris. Medicis customers include wholesale pharmaceutical distributors such as Cardinal Health, Inc. (which generated 21.2% of 2008 revenues); McKesson Corp. (45.8% of 2008 revenues); and other major drug chains. McKesson is the company's sole distributor of Restylane and Perlane products in the U.S. and Canada. In June 2008, Medicis acquired LipoSonix, Inc., an independent company that specializes in body contouring technology.

The firm offers employees medical, dental and vision coverage; a prescription plan; educational assistance; a 401(k) plan; and flexible spending accounts.

## FINANCIALS: Sales and profits are in thousands of dollars—add 000 to get the full amount. 2008 Note: Financial information for 2008 was not available for all companies at press time.

| | | |
|---|---|---|
| 2008 Sales: $517,750 | 2008 Profits: $10,276 | U.S. Stock Ticker: MRX |
| 2007 Sales: $457,394 | 2007 Profits: $70,436 | Int'l Ticker:     Int'l Exchange: |
| 2006 Sales: $393,165 | 2006 Profits: $-48,152 | Employees:  578 |
| 2005 Sales: $376,899 | 2005 Profits: $64,990 | Fiscal Year Ends: 12/31 |
| 2004 Sales: $303,722 | 2004 Profits: $30,840 | Parent Company: |

## SALARIES/BENEFITS:

| Pension Plan: | ESOP Stock Plan: | Profit Sharing: | Top Exec. Salary: $1,100,000 | Bonus: $1,014,750 |
|---|---|---|---|---|
| Savings Plan: Y | Stock Purch. Plan: | | Second Exec. Salary: $735,516 | Bonus: $ |

## OTHER THOUGHTS:

**Apparent Women Officers or Directors:**
**Hot Spot for Advancement for Women/Minorities:**

## LOCATIONS: ("Y" = Yes)

| West: | Southwest: | Midwest: | Southeast: | Northeast: | International: |
|---|---|---|---|---|---|
| Y | Y | Y | Y | Y | |

Note: Financial information, benefits and other data can change quickly and may vary from those stated here.

# MEDIDATA SOLUTIONS INC

**www.mdsol.com**

Industry Group Code: 511210D  **Ranks within this company's industry group:** Sales:    Profits:

| Drugs: | Other: | | Clinical: | Computers: | | Services: | |
|---|---|---|---|---|---|---|---|
| Discovery: | AgriBio: | | Trials/Services: | Hardware: | | Specialty Services: | Y |
| Licensing: | Genetic Data: | | Labs: | Software: | Y | Consulting: | |
| Manufacturing: | Tissue Replacement: | | Equipment/Supplies: | Arrays: | | Blood Collection: | |
| Genetics: | | | Research & Development Services: | Database Management: | Y | Drug Delivery: | |
| | | | Diagnostics: | | | Drug Distribution: | |

## TYPES OF BUSINESS:

Clinical Trial Software, Online

## BRANDS/DIVISIONS/AFFILIATES:

Medidata Rave
Medidata Grants Manager
Medidata CRO Contractor
Medidata Designer
Fast Track Systems, Inc.

## CONTACTS: Note: Officers with more than one job title may be intentionally listed here more than once.

Tarek A. Sherif, CEO
Glen M. de Vries, Pres.
Bruce D. Dalziel, CFO
Steven I. Hirschfeld, Exec. VP-Global Sales & Alliances
Arden Schneider, Sr. VP-Human Resources
Glenn Watt, VP-Global Info. Security & Privacy
Richard J. Piazza, VP-New Prod.
Michael (Mike) Otner, General Counsel
Joe Tyers, Sr. VP-Sales Oper. & Knowledge Mgmt.
Keith Howells, Sr. VP-Dev.
Lineene N. Krasnow, Exec. VP-Prod. & Mktg.
Earl Hulihan, Sr. VP-Regulatory Compliance
Vik Shah, Sr. VP-Svcs.
Shih-Yin Ho, VP-Corp. Strategy
Tarek A. Sherif, Chmn.
Steve Heath, VP/Head-EMEA & Australia

| Phone: 212-918-1800 | Fax: 212-918-1818 |
|---|---|
| Toll-Free: 877-511-4200 | |
| Address: 79 5th Ave., 8th Fl., New York, NY 10003 US | |

## GROWTH PLANS/SPECIAL FEATURES:

Medidata Solutions, Inc. offers hosted clinical trial software used by medical research and development firms worldwide. Its customers include pharmaceutical, biotechnology and medical device companies, as well as academic institutions, contract research organizations (CROs) and other organizations engaged in clinical trials. Included among these customers are some of the biggest names in their respective fields, such as Johnson & Johnson; AstraZeneca; Astellas Pharma; Amgen; and Takeda Pharmaceutical. Its principal product is Medidata Rave, a comprehensive platform that integrates electronic data capture (EDC) with a clinical data management system (CDMS) in a single scalable solution that replaces traditional paper-based methods of capturing and managing clinical data. Medidata Rave allows users to post real-time updates to clinical data and input and retrieve data in multiple languages simultaneously. Additionally, because it is an online program, customers can easily coordinate data from around the world in a single platform. Most of the revenue generated by this product derives from multi-study agreements made for a pre-determined number of studies. The company's other products similarly strive to help customers streamline the design, planning and management of key aspects of the clinical development process, including protocol development, CRO negotiation, investigator contracting, the capture and management of clinical trial data and the analysis and reporting of that data on a worldwide basis. These other products include Medidata Grants Manager, Medidata CRO Contractor and Medidata Designer (for designing protocols). Besides its products, Medidata also offers professional services, including global consulting, implementation, technical support and training for customers and investigators. Application services (those relating to software) generated 70% of 2008 revenues, with professional services generating the remainder. In March 2008, Medidata acquired clinical trial planning software developer Fast Track Systems, Inc. for $18.1 million.

## FINANCIALS: Sales and profits are in thousands of dollars—add 000 to get the full amount. 2008 Note: Financial information for 2008 was not available for all companies at press time.

| | | |
|---|---|---|
| 2008 Sales: $ | 2008 Profits: $ | U.S. Stock Ticker: MDSO |
| 2007 Sales: $ | 2007 Profits: $ | Int'l Ticker:     Int'l Exchange: |
| 2006 Sales: $ | 2006 Profits: $ | Employees: |
| 2005 Sales: $ | 2005 Profits: $ | Fiscal Year Ends: 12/31 |
| 2004 Sales: $ | 2004 Profits: $ | Parent Company: |

## SALARIES/BENEFITS:

| Pension Plan: | ESOP Stock Plan: | Profit Sharing: | Top Exec. Salary: $360,000 | Bonus: $448,000 |
|---|---|---|---|---|
| Savings Plan: | Stock Purch. Plan: | | Second Exec. Salary: $360,000 | Bonus: $448,000 |

## OTHER THOUGHTS:

**Apparent Women Officers or Directors**: 5
**Hot Spot for Advancement for Women/Minorities**: Y

## LOCATIONS: ("Y" = Yes)

| West: | Southwest: | Midwest: | Southeast: | Northeast: | International: |
|---|---|---|---|---|---|
| | | | | Y | Y |

# MEDIMMUNE INC

www.medimmune.com

**Industry Group Code: 325412 Ranks within this company's industry group: Sales: Profits:**

| Drugs: | Other: | | Clinical: | Computers: | Services: |
|--------|--------|---|-----------|------------|-----------|
| Discovery: | AgriBio: | | Trials/Services: | Hardware: | Specialty Services: |
| Licensing: | Genetic Data: | | Labs: | Software: | Consulting: |
| Manufacturing: | Tissue Replacement: | Y | Equipment/Supplies: | Arrays: | Blood Collection: |
| Genetics: | | | Research & Development Services: | Database Management: | Drug Delivery: |
| | | | Diagnostics: | | Drug Distribution: |

## TYPES OF BUSINESS:

Pharmaceuticals Development & Manufacturing
Drugs-Cancer, Infectious & Immune Diseases

## BRANDS/DIVISIONS/AFFILIATES:

MedImmune Ventures, Inc.
Ethyol
Synagis
FluMist

## CONTACTS: *Note: Officers with more than one job title may be intentionally listed here more than once.*

Tony Zook, Pres.
Tim Pearson, CFO/Sr. VP
Peter Greenleaf, Sr. VP-Mktg. & Sales
Max Donley, VP-Human Resources
Peter A. Kiener, Exec. VP-R&D
William C. Bertrand, Jr., General Counsel/Exec. VP-Legal Affairs
Bernardus N. M. Machielse, Exec. VP-Oper.
Peter Greenleaf, Sr. VP-Corp. Dev.& Strategy
Frank Malinoski, Sr. VP-Medical & Scientific Affairs
Alexander A. Zukiwski, Sr. VP-Clinical Research/Chief Medical Officer

| Phone: 301-398-0000 | Fax: 301-398-9000 |
|---|---|
| Toll-Free: 877-633-4411 | |
| Address: 1 MedImmune Way, Gaithersburg, MD 20878 US | |

## GROWTH PLANS/SPECIAL FEATURES:

MedImmune, Inc., a subsidiary of AstraZeneca plc, is a biotechnology company engaged in the development and production of pharmaceuticals for infectious diseases, cancer and inflammatory disease. The company markets three products: Synagis, Ethyol and FluMist. Synagis is a treatment for respiratory syncytial virus (RSV). The drug was the first monoclonal antibody approved for an infectious disease and has become a primary pediatric product for the prevention of RSV, a major cause of viral pneumonia and bronchiolitis in infants and children. Ethyol is marketed as a treatment for the reduction of toxicities from cancer chemotherapy and radiotherapy. FluMist is a nasally administered flu vaccine for influenzas A and B. In its product candidate pipeline, MedImmune focuses its research and development efforts in the therapeutic areas of infectious disease, inflammatory diseases and cancer. The company has 11 candidate drugs in the infectious disease category, eight in the inflammatory category and 10 in the oncology category. In recent news, AstraZeneca integrated its subsidiary, Cambridge Antibody Technology, into MedImmune, bolstering its worldwide biologics division. In October 2008, the company opened a biologics research and development facility in Cambridge, the United Kingdom.

## FINANCIALS: Sales and profits are in thousands of dollars—add 000 to get the full amount. 2008 Note: Financial information for 2008 was not available for all companies at press time.

| 2008 Sales: $ | 2008 Profits: $ | U.S. Stock Ticker: Subsidiary |
|---|---|---|
| 2007 Sales: $ | 2007 Profits: $ | Int'l Ticker: Int'l Exchange: |
| 2006 Sales: $1,276,800 | 2006 Profits: $48,700 | Employees: 2,538 |
| 2005 Sales: $1,243,900 | 2005 Profits: $-16,600 | Fiscal Year Ends: 12/31 |
| 2004 Sales: $1,141,100 | 2004 Profits: $-3,800 | Parent Company: ASTRAZENECA PLC |

## SALARIES/BENEFITS:

| Pension Plan: | ESOP Stock Plan: | Profit Sharing: | Top Exec. Salary: $991,667 | Bonus: $1,300,000 |
|---|---|---|---|---|
| Savings Plan: | Stock Purch. Plan: | | Second Exec. Salary: $570,833 | Bonus: $480,000 |

## OTHER THOUGHTS:

**Apparent Women Officers or Directors:**
**Hot Spot for Advancement for Women/Minorities:**

## LOCATIONS: ("Y" = Yes)

| West: | Southwest: | Midwest: | Southeast: | Northeast: | International: |
|---|---|---|---|---|---|
| Y | | | | Y | Y |

# MEDTOX SCIENTIFIC INC

### www.medtox.com

**Industry Group Code:** 6215 **Ranks within this company's industry group:** Sales: 4 Profits: 3

| Drugs: | Other: | Clinical: | | Computers: | | Services: | |
|---|---|---|---|---|---|---|---|
| Discovery: | AgriBio: | Trials/Services: | Y | Hardware: | | Specialty Services: | Y |
| Licensing: | Genetic Data: | Labs: | Y | Software: | | Consulting: | |
| Manufacturing: | Tissue Replacement: | Equipment/Supplies: | Y | Arrays: | | Blood Collection: | |
| Genetics: | | Research & Development Services: | | Database Management: | | Drug Delivery: | |
| | | Diagnostics: | Y | | | Drug Distribution: | |

## TYPES OF BUSINESS:

Diagnostic Device Manufacturing
Forensic & Clinical Lab Services
Diagnostic Drug Screening Devices

## BRANDS/DIVISIONS/AFFILIATES:

MEDTOX Diagnostics, Inc.
MEDTOX Laboratories, Inc.
PROFILE-II
VERDICT-II
ClearCourse
SURE-SCREEN
Drug Abuse Recognition System (DARS)
EZ-SCREEN

## CONTACTS: *Note: Officers with more than one job title may be intentionally listed here more than once.*

Richard J. Braun, CEO
Richard J. Braun, Pres.
Kevin J. Wiersma, CFO
James A. Schoonover, Chief Mktg. Officer/VP-Sales & Mktg.
Susan E. Puskas, VP-Human Resources
Kevin J. Wiersma, VP/COO-Laboratory Div.
Steven J. Schmidt, VP-Finance
Susan E. Puskas, VP-Quality & Regulatory Affairs
B. Mitchell Owens, VP/COO-Diagnostics Div.
Richard J. Braun, Chmn.

| **Phone:** 651-636-7466 | **Fax:** 651-636-5351 |
|---|---|
| **Toll-Free:** 800-832-3244 | |
| **Address:** 402 W. County Rd. D, St. Paul, MN 55112 US | |

## GROWTH PLANS/SPECIAL FEATURES:

MEDTOX Scientific, Inc. manufactures and distributes diagnostic devices and provides forensic and clinical laboratory services. The firm operates through two divisions, laboratory services and product sales. The laboratory services division is operated by the company's subsidiary MEDTOX Laboratories, Inc., which provides laboratory drug testing services for corporations, medical facilities and the federal government. MEDTOX Laboratories derives the majority of its revenues, roughly 61%, from workplace drugs-of-abuse testing. It also offers specialty services, accounting for 39% of revenues, including clinical toxicology; medical diagnostics; general clinical testing for the pharmaceutical industry; heavy metal, trace element and solvent analysis; it also offers logistics, data and program management services through its WEBTOX and eChain online services. The firm's other subsidiary, MEDTOX Diagnostics, Inc., which operates the product sales division, is based in Burlington, North Carolina and manufactures on-site drug screening products for the corporate, health care, criminal justice, temporary service and drug rehabilitation markets. MEDTOX Diagnostics' point-of-collection testing (POCT) products, which account for 91% of its sales, include the PROFILE line and MEDTOXScan, which are sold to hospitals, and the VERDICT-II and SURE-SCREEN lines, which are sold within the criminal justice and drug rehabilitation markets. These products test for a variety of drugs including amphetamines, marijuana, cocaine, opiates, phencyclidine (PCP), benzoylecgonine, morphine and methamphetamine. The company also offers its ClearCourse comprehensive drug testing program, which consists of Drug Abuse Recognition System (DARS) training; SURE-SCREEN drug testing devices; WEBTOX online data management; and MEDTOX's own laboratory assistance for confirmation testing. Among this company's other products are the EZ-SCREEN breath alcohol test and agricultural diagnostics products used to detect antibiotic residues. This division also offers contract manufacturing services. In February 2009, MEDTOX received FDA clearance to market MEDTOXScan, a drug abuse testing system.

MEDTOX offers its employees medical, dental and life insurance; tuition reimbursement; and flexible spending accounts.

## FINANCIALS: Sales and profits are in thousands of dollars—add 000 to get the full amount. 2008 Note: Financial information for 2008 was not available for all companies at press time.

| | | |
|---|---|---|
| 2008 Sales: $85,813 | 2008 Profits: $5,572 | **U.S. Stock Ticker:** MTOX |
| 2007 Sales: $80,285 | 2007 Profits: $6,690 | **Int'l Ticker:** Int'l Exchange: |
| 2006 Sales: $69,804 | 2006 Profits: $4,548 | Employees: 523 |
| 2005 Sales: $63,047 | 2005 Profits: $3,318 | Fiscal Year Ends: 12/31 |
| 2004 Sales: $56,736 | 2004 Profits: $1,821 | Parent Company: |

## SALARIES/BENEFITS:

| Pension Plan: | ESOP Stock Plan: | Profit Sharing: | Top Exec. Salary: $350,500 | Bonus: $294,840 |
|---|---|---|---|---|
| Savings Plan: Y | Stock Purch. Plan: | | Second Exec. Salary: $216,782 | Bonus: $92,248 |

## OTHER THOUGHTS:

**Apparent Women Officers or Directors:** 1
**Hot Spot for Advancement for Women/Minorities:**

## LOCATIONS: ("Y" = Yes)

| West: | Southwest: | Midwest: | Southeast: | Northeast: | International: |
|---|---|---|---|---|---|
| | | Y | | Y | |

# MERCK & CO INC

**www.merck.com**

Industry Group Code: 325412  **Ranks within this company's industry group:** Sales: 9   Profits: 5

| Drugs: | | Other: | | Clinical: | | Computers: | | Services: | |
|---|---|---|---|---|---|---|---|---|---|
| Discovery: | Y | AgriBio: | | Trials/Services: | | Hardware: | | Specialty Services: | Y |
| Licensing: | | Genetic Data: | | Labs: | | Software: | | Consulting: | |
| Manufacturing: | Y | Tissue Replacement: | | Equipment/Supplies: | | Arrays: | | Blood Collection: | |
| Genetics: | | | | Research & Development Services: | | Database Management: | | Drug Delivery: | |
| | | | | Diagnostics: | | | | Drug Distribution: | |

## TYPES OF BUSINESS:

Pharmaceuticals Development & Manufacturing
Cholesterol Drugs
Hypertension Drugs
Heart Failure Drugs
Allergy & Asthma Drugs
Animal Health Products
Vaccines
Preventative Drugs

## BRANDS/DIVISIONS/AFFILIATES:

Merck Institute for Science Education
Sirna Therapeutics, Inc.
Singulair
Propecia
Merck BioVentures
Fosamax
Gardasil
EMEND

## CONTACTS: *Note: Officers with more than one job title may be intentionally listed here more than once.*

Richard Clark, CEO
Peter N. Kellogg, CFO/Exec. VP
Wendy Yarno, Chief Mktg. Officer
Mirian Graddick-Weir, Exec. VP-Human Resources
Peter S. Kim, Pres., Research Laboratories
J. Chris Scalet, CIO/Exec. VP-Global Svcs.
Willie Deese, Pres., Mfg. Div.
Bruce N. Kuhlik, General Counsel/Exec. VP
Adele Ambrose, Chief Comm. Officer
John Canan, Controller/Sr. VP
Kenneth C. Frazier, Pres., Global Human Health
Margaret McGlynn, Pres., Merck Vaccines & Infectious Diseases
Celia Colbert, Sr. VP/Sec.
Mark McDonough, VP/Treas.
Richard Clark, Chmn.
Stefan Oschmann, Pres., EMEA & Canada

| Phone: 908-423-1000 | Fax: 908-735-1253 |
|---|---|
| Toll-Free: | |
| Address: 1 Merck Dr., Whitehouse Station, NJ 08889-0100 US | |

## GROWTH PLANS/SPECIAL FEATURES:

Merck & Co., Inc. is a leading research-driven pharmaceutical company that manufactures a broad range of products sold in approximately 150 countries. These products include therapeutic and preventative drugs generally sold by prescription and medications used to control and alleviate disease. The company operates in two segments. The Pharmaceutical segment includes human health pharmaceutical products consisting of therapeutic and preventative agents, sold by prescription. Medications include Fosamax, for the prevention of osteoporosis; Zocor, for lowering cholesterol; and Singulair, a seasonal allergy and asthma medication. The Vaccines and Infectious Diseases segment includes preventative vaccines such as Rotateq, designed to prevent gastroenteritis in infants and children; and Gardasil, a vaccine for the prevention of HPV. The segment also produces therapeutic agents for the treatment of infections such as Invanz and Cancidas, an anti-fungal product. The company also manufactures Propecia, a popular treatment for male pattern baldness. Recent FDA approved products include EMEND, a drug that prevents chemotherapy-induced nausea and ISENTRESS (raltegravir) tablets, which treats patients with the HIV-1 infection. In late 2008, Merck established a new unit, Merck BioVentures. This new group will focus on biotech drugs, including generic biotech drugs know as follow-on biologics. Also, the company will continue to focus on growing sales in emerging markets, particularly China and India. In February 2009, Merck agreed to acquire the portfolio of follow-on biologic therapeutic candidates, as well as the commercial manufacturing facilities in Boulder Colorado, of Insmed Inc. In August 2009, the company agreed to merge with Schering-Plough.

Employees are offered medical, vision and dental insurance; health and dependent care accounts; a 401(k) plan; a pension plan; financial planning services; life insurance; business travel accident insurance; credit union membership; educational assistance and scholarship programs; and corporate discounts on cars, travel, entertainment and electronics.

## FINANCIALS: Sales and profits are in thousands of dollars—add 000 to get the full amount. 2008 Note: Financial information for 2008 was not available for all companies at press time.

| | | |
|---|---|---|
| 2008 Sales: $23,850,300 | 2008 Profits: $7,808,400 | **U.S. Stock Ticker:** MRK |
| 2007 Sales: $24,197,700 | 2007 Profits: $3,275,400 | **Int'l Ticker:**    Int'l Exchange: |
| 2006 Sales: $22,636,000 | 2006 Profits: $4,433,800 | Employees:  55,200 |
| 2005 Sales: $22,011,900 | 2005 Profits: $4,631,300 | Fiscal Year Ends: 12/31 |
| 2004 Sales: $22,938,600 | 2004 Profits: $5,813,400 | Parent Company: |

## SALARIES/BENEFITS:

| Pension Plan: Y | ESOP Stock Plan: | Profit Sharing: | Top Exec. Salary: $178,334 | Bonus: $2,244,510 |
|---|---|---|---|---|
| Savings Plan: Y | Stock Purch. Plan: | | Second Exec. Salary: $1,033,338 | Bonus: $875,023 |

## OTHER THOUGHTS:

**Apparent Women Officers or Directors:** 7
**Hot Spot for Advancement for Women/Minorities:** Y

## LOCATIONS: ("Y" = Yes)

| West: | Southwest: | Midwest: | Southeast: | Northeast: | International: |
|---|---|---|---|---|---|
| Y | Y | Y | Y | Y | Y |

Note: Financial information, benefits and other data can change quickly and may vary from those stated here.

# MERCK KGAA

**www.merck.de**

**Industry Group Code: 325412  Ranks within this company's industry group:** Sales: 18  Profits: 25

| Drugs: | | Other: | | Clinical: | Computers: | | Services: | |
|---|---|---|---|---|---|---|---|---|
| Discovery: | Y | AgriBio: | | Trials/Services: | Hardware: | | Specialty Services: | |
| Licensing: | Y | Genetic Data: | | Labs: | Software: | | Consulting: | |
| Manufacturing: | Y | Tissue Replacement: | | Equipment/Supplies: | Arrays: | | Blood Collection: | |
| Genetics: | Y | | | Research & Development Services: | Database Management: | | Drug Delivery: | |
| | | | | Diagnostics: | | | Drug Distribution: | |

## TYPES OF BUSINESS:

Pharmaceuticals
Over-the-Counter Drugs & Vitamins
Generic Drugs
Chemicals
LCD Components
Reagents & Diagnostics
Nanotechnology Research

## BRANDS/DIVISIONS/AFFILIATES:

Merck Serono S.A.
Erbitux
Nasivin
Merck4Biosciences
Merck4Cosmetics
Merck4Food
Merck4LCDs & Emerging Technologies
Nano Terra LLC

## CONTACTS: *Note: Officers with more than one job title may be intentionally listed here more than once.*

Karl-Ludwig Kley, Chmn.-Exec. Board
Michael Becker, CFO
Karl-Ludwig Kley, Dir.-Human Resources
Bernd Reckmann, Dir.-Corp. Info. Svcs.
Bernd Reckmann, Dir.-Prod.
Bernd Reckmann, Dir.-Eng.
Karl-Ludwig Kley, Dir.-Legal
Karl-Ludwig Kley, Dir.-Strategic Planning
Karl-Ludwig Kley, Dir.-Corp. Comm.
Michael Becker, Dir.-Acct., Finance, Controlling & Tax
Bernd Reckmann, Mgr.-Darmstadt & Gernsheim Sites
Elmar Schnee, Dir.-Pharmaceuticals Bus. Sector
Bernd Reckmann, Dir.-Chemicals Bus. Sector
Wilhelm Simson, Chmn.-Supervisory Board

| Phone: 49-6151-72-0 | Fax: 49-6151-72-2000 |
|---|---|
| Toll-Free: | |
| Address: Frankfurter St. 250, Darmstadt, 64293 Germany | |

## GROWTH PLANS/SPECIAL FEATURES:

Merck KGaA is a global pharmaceuticals and chemicals company with 192 companies in 61 countries. Outside Germany, Merck maintains research sites in France, Spain, the U.K., the U.S. and Japan. It has two main businesses: Pharmaceuticals and Chemicals. The Pharmaceuticals business manufactures prescription drugs and over-the-counter products through two divisions, Merck Serano and Consumer Health Care. Merck Serono's products target oncology, neuro-degenerative diseases, fertility, endocrinology, cardio metabolic care and new specialist therapies. Products include colorectal cancer drug Erbitux; multiple sclerosis treatment Rebif; Type 2 diabetes treatment Glucophage; and Raptiva, a new type of treatment for psoriasis. The Consumer Health Care division offers products such as cough and cold; naturals-plant sourced; naturals-marine sourced; vitamins and minerals; dermatology; and other products. Specific products include nasal decongestant Nasivin; Mediflor medicinal teas; Cebion multivitamins; and bioManan slimming aids. The Chemical business has six segments. The biosciences segment offers life sciences research products including proteins, enzymes, reagents, kits and other supplies. The cosmetics segment supplies insect repellent and basic supplies for cosmetics including active ingredients, pigments and UV filters. The food segment offers quality control products; monitoring products for cleaning, hygiene, process water and wastewater; packaging design pigments; and raw materials, including mineral salts. The liquid crystals and emerging technologies segment manufactures licristal, licrivue and lisicon brand liquid crystals for LCD TVs, optical films and organic semiconductors; isitag brand printable semiconducting polymers for RFID tags; and isishape brand photovoltaic materials. The pharmaceutical segment supplies chemicals for all areas of pharmaceuticals manufacturing. Lastly, the printing, plastics and coatings segment offers numerous pigments. In February 2009, the company agreed to acquire MediCult for $55 million.

## FINANCIALS: Sales and profits are in thousands of dollars—add 000 to get the full amount. 2008 Note: Financial information for 2008 was not available for all companies at press time.

| | | |
|---|---|---|
| 2008 Sales: $10,000,300 | 2008 Profits: $501,600 | **U.S. Stock Ticker:** |
| 2007 Sales: $9,337,530 | 2007 Profits: $4,657,720 | **Int'l Ticker: MRK**  Int'l Exchange: Frankfurt-Euronext |
| 2006 Sales: $8,352,950 | 2006 Profits: $1,335,880 | Employees: |
| 2005 Sales: $7,697,680 | 2005 Profits: $898,150 | Fiscal Year Ends: 12/31 |
| 2004 Sales: $ | 2004 Profits: $ | Parent Company: |

## SALARIES/BENEFITS:

| Pension Plan: | ESOP Stock Plan: | Profit Sharing: | Top Exec. Salary: $ | Bonus: $ |
|---|---|---|---|---|
| Savings Plan: | Stock Purch. Plan: | | Second Exec. Salary: $ | Bonus: $ |

## OTHER THOUGHTS:

**Apparent Women Officers or Directors:** 2
**Hot Spot for Advancement for Women/Minorities:**

## LOCATIONS: ("Y" = Yes)

| West: | Southwest: | Midwest: | Southeast: | Northeast: | International: |
|---|---|---|---|---|---|
| Y | | Y | Y | Y | Y |

# MERCK SERONO SA

www.merckserono.net

**Industry Group Code: 325412 Ranks within this company's industry group: Sales: Profits:**

| Drugs: | | Other: | | Clinical: | Computers: | | Services: | |
|---|---|---|---|---|---|---|---|---|
| Discovery: | Y | AgriBio: | | Trials/Services: | Hardware: | | Specialty Services: | |
| Licensing: | | Genetic Data: | Y | Labs: | Software: | | Consulting: | |
| Manufacturing: | Y | Tissue Replacement: | | Equipment/Supplies: | Arrays: | | Blood Collection: | |
| Genetics: | | | | Research & Development Services: | Database Management: | | Drug Delivery: | Y |
| | | | | Diagnostics: | | | Drug Distribution: | |

## TYPES OF BUSINESS:

Pharmaceuticals Development
Fertility Drugs
Neurology Drugs
Growth & Metabolism Drugs
Dermatology Drugs
Oncology Research

## BRANDS/DIVISIONS/AFFILIATES:

GONAL-f
Ovidrel/Ovitrelle
Luveris
Crinone
Cetrotide
Rebif
Saizen
EMD Serono Inc

## CONTACTS: Note: Officers with more than one job title may be intentionally listed here more than once.

Elmar Schnee, Pres.
Richard Douge, Head-Global Mktg.
Francois Naef, VP-Human Resources
Bernhard Kirschbaum, Exec. VP-Global R&D
Hanns-Eberhard Erle, Head-Tech. Oper.
Francois Naef, Chief Admin. Officer
Francois Naef, VP-Legal
Vincent Aurentz, Head-Portfolio Mgmt. & Bus. Dev.
Francois Naef, Head-Comm.
Dorothea Wenzel, Head-Controlling
Fereydoun Firouz, Head-Commercial U.S.
Roberto Gradnik, Head-Commercial Europe
Wolfgang Wein, Head-Oncology
Franck Latrille, Head-Commercial Int'l

| Phone: 41-22-414-3000 | Fax: 41-22-414-2179 |
|---|---|

**Toll-Free:**

**Address:** 9, chemin des Mines, Case postale 54, Geneva 20, CH-1211 Switzerland

## GROWTH PLANS/SPECIAL FEATURES:

Merck Serono SA, formerly Serono SA, is a global biotechnology company that focuses on six therapeutic areas: oncology, neurodegenerative diseases, fertility, endocrinology, cardio-metabolic care and new specialist therapies. The company was formed in January 2007 when Merck KGaA acquired Serono in a $13 billion transaction. The company researches both traditional pharmaceutical drugs, consisting of small molecules treating the symptoms of a disease, and biopharmaceuticals, comprised of large molecules such as proteins that target the underlying mechanism of a disease. For colorectal cancer, Merck Serono has developed Erbitux, a monoclonal antibody also sold for the treatment of head and neck cancer, and UFT, an oral 5-FU (5-fluorouracil) therapy. Rebif is Merck Serono's disease modifying drug, used to treat relapsing forms of multiple sclerosis. Infertility treatments developed by the company include GONAL-f, Pergoveris, Luveris, Ovidrel/Ovitrelle, Crinone and Cetrotide. Merck Serono offers a range of specialized endocrinology products. For childhood and adult growth hormone deficiency, as well as for children born small for gestational age (SGA), or with chronic renal failure or Turner's syndrome, the company has developed Saizen. Serostim, meanwhile, is used to treat HIV-associated wasting, and Zorbtive has been developed for the treatment of short bowel syndrome. Cardio-metabolic care treatments developed by Merck Serono include Glucophage/GlucophageXR and Glucovance for diabetes; Concor/ConcorCor for cardiovascular disease; and Euthyrox for hypothyroidism, euthyroid goiter and suppressive therapy of differentiated thyroid cancer. As part of its effort to develop products in new specialist areas with unmet medical needs, the company partnered with Genentech to develop and commercialize the psoriasis drug Raptiva. Merck Serono also has a number of products in development for the treatment of such diseases as systemic lupus erythematosus, rheumatoid arthritis and Crohn's disease. In Canada and the U.S., the Merck Serono division operates as EMD Serono, Inc.

## FINANCIALS: Sales and profits are in thousands of dollars—add 000 to get the full amount. 2008 Note: Financial information for 2008 was not available for all companies at press time.

| | | |
|---|---|---|
| 2008 Sales: $ | 2008 Profits: $ | **U.S. Stock Ticker: Subsidiary** |
| 2007 Sales: $ | 2007 Profits: $ | **Int'l Ticker:** Int'l Exchange: |
| 2006 Sales: $2,804,900 | 2006 Profits: $735,400 | Employees: 4,775 |
| 2005 Sales: $2,586,400 | 2005 Profits: $-105,300 | Fiscal Year Ends: 12/31 |
| 2004 Sales: $2,177,900 | 2004 Profits: $494,200 | Parent Company: MERCK KGAA |

## SALARIES/BENEFITS:

| | | | | |
|---|---|---|---|---|
| Pension Plan: | ESOP Stock Plan: | Profit Sharing: | Top Exec. Salary: $ | Bonus: $ |
| Savings Plan: | Stock Purch. Plan: | | Second Exec. Salary: $ | Bonus: $ |

## OTHER THOUGHTS:

**Apparent Women Officers or Directors**: 1
**Hot Spot for Advancement for Women/Minorities**:

## LOCATIONS: ("Y" = Yes)

| West: | Southwest: | Midwest: | Southeast: | Northeast: | International: |
|---|---|---|---|---|---|
| | | | | Y | Y |

# MERIDIAN BIOSCIENCE INC

### www.meridianbioscience.com

**Industry Group Code:** 325413  **Ranks within this company's industry group:** Sales: 9   Profits: 6

| Drugs: | Other: | Clinical: | | Computers: | Services: |
|---|---|---|---|---|---|
| Discovery: | AgriBio: | Trials/Services: | | Hardware: | Specialty Services: |
| Licensing: | Genetic Data: | Labs: | | Software: | Consulting: |
| Manufacturing: Y | Tissue Replacement: | Equipment/Supplies: | Y | Arrays: | Blood Collection: |
| Genetics: | | Research & Development Services: | Y | Database Management: | Drug Delivery: |
| | | Diagnostics: | Y | | Drug Distribution: |

## TYPES OF BUSINESS:

Diagnostic Test Kits
Contract Manufacturing
Bulk Antigens, Antibodies & Reagents

## BRANDS/DIVISIONS/AFFILIATES:

Biodesign
OEM Concepts
Viral Antigens
ImmunoCard STAT Campy
Meridian Biologics

## CONTACTS: *Note: Officers with more than one job title may be intentionally listed here more than once.*

John A. Kraeutler, CEO
Melissa A. Lueke, CFO/VP
Todd W. Motto, VP-Mktg. & Sales
Grady Barnes, VP-R&D
Lawrence J. Baldini, Exec. VP-Info. Systems
Melissa A. Leuke, Sec.
Lawrence J. Baldini, Exec. VP-Oper.
Susan D. Rolih, VP-Regulatory Affairs & Quality Systems
Richard L. Eberly, Exec. VP/Pres., Meridian Life Science
William J. Motto, Chmn.
Antonio A. Interno, Sr. VP/Pres./Managing Dir.-Europe

| **Phone:** 513-271-3700 | **Fax:** 513-271-3762 |
|---|---|
| **Toll-Free:** 800-543-1980 | |
| **Address:** 3471 River Hills Dr., Cincinnati, OH 45244 US | |

## GROWTH PLANS/SPECIAL FEATURES:

Meridian Biosciences, Inc. is a life science company that develops, manufactures, sells and distributes diagnostic test kits, primarily for respiratory, gastrointestinal, viral and parasitic infectious diseases.   It also manufactures and distributes bulk antigens, antibodies and reagents; and contract manufactures proteins and other biologicals.   The company operates in three segments: U.S. diagnostics, European diagnostics and life science.  The U.S. diagnostics segment focuses on the development, manufacture, sale and distribution of diagnostic test kits, which utilize immunodiagnostic technologies that test samples of body fluids or tissue for the presence of antigens and antibodies of specific infectious diseases.  Products also include transport media that store and preserve specimen samples from patient collection to laboratory testing.   The European diagnostics segment focuses on the sale and distribution of diagnostic test kits.   Its sales and distribution network consists of direct sales forces in Belgium, France, Holland and Italy, and independent distributors in other European, African and Middle Eastern countries.   The life sciences segment focuses on the development, manufacture, sale and distribution of bulk antigens, antibodies and reagents, as well as contract development and manufacturing services.  The segment is represented by four product-line brands: Biodesign, which represents monoclonal and polyclonal antibodies and assay reagents; OEM (original equipment manufacturer) Concepts, which represents contract ascites and antibody production services; Viral Antigens, which represents viral proteins; and Meridian Biologics, which represents contract development and manufacturing services for drug and vaccine discovery and development.  Meridian's core diagnostic products generated 83% of revenue in 2008. In June 2009, the company received FDA clearance to market the ImmunoCard STAT! Campy, a new rapid test for detecting the Campylobacter bacteria.  This offering comes in addition to the firm's February 2009 PREMIER CAMPY culture test.

Meridian offers its employees medical, dental and life insurance; short- and long-term disability; stock options; a 401(k) plan; profit sharing; and tuition reimbursement.

## FINANCIALS:  Sales and profits are in thousands of dollars—add 000 to get the full amount. 2008 Note: Financial information for 2008 was not available for all companies at press time.

| | | |
|---|---|---|
| 2008 Sales: $139,639 | 2008 Profits: $30,202 | **U.S. Stock Ticker:** VIVO |
| 2007 Sales: $122,963 | 2007 Profits: $26,721 | **Int'l Ticker:**    Int'l Exchange: |
| 2006 Sales: $108,413 | 2006 Profits: $18,333 | Employees:   363 |
| 2005 Sales: $92,965 | 2005 Profits: $12,565 | Fiscal Year Ends: 9/30 |
| 2004 Sales: $79,606 | 2004 Profits: $9,185 | Parent Company: |

## SALARIES/BENEFITS:

| Pension Plan: | ESOP Stock Plan: | Profit Sharing: Y | Top Exec. Salary: $519,615 | Bonus: $196,875 |
|---|---|---|---|---|
| Savings Plan: Y | Stock Purch. Plan: Y | | Second Exec. Salary: $450,579 | Bonus: $178,125 |

## OTHER THOUGHTS:

**Apparent Women Officers or Directors:** 2
**Hot Spot for Advancement for Women/Minorities:** Y

## LOCATIONS: ("Y" = Yes)

| West: | Southwest: | Midwest: | Southeast: | Northeast: | International: |
|---|---|---|---|---|---|
| | | Y | Y | Y | Y |

# MGI PHARMA INC

www.mgipharma.com

**Industry Group Code: 325412  Ranks within this company's industry group:** Sales:     Profits:

| Drugs: | | Other: | Clinical: | Computers: | Services: |
|---|---|---|---|---|---|
| Discovery: | | AgriBio: | Trials/Services: | Hardware: | Specialty Services: |
| Licensing: | Y | Genetic Data: | Labs: | Software: | Consulting: |
| Manufacturing: | | Tissue Replacement: | Equipment/Supplies: | Arrays: | Blood Collection: |
| Genetics: | | | Research & Development Services: | Database Management: | Drug Delivery: |
| | | | Diagnostics: | | Drug Distribution: |

## TYPES OF BUSINESS:

Pharmaceuticals Acquisition & Development
Drugs-Oncology & Rheumatology

## BRANDS/DIVISIONS/AFFILIATES:

Aloxi Injection
Gliadel
Salagen
Hexalen
Aquavan
Dacogen
Saforis
Eisai Corporation of North America

## CONTACTS: *Note: Officers with more than one job title may be intentionally listed here more than once.*

Lonnie Moulder, Jr., CEO
Eric Loukas, COO/Exec. VP
Lonnie Moulder, Jr., Pres.
Mary L. Hedley, Chief Scientific Officer/Exec. VP

| **Phone:** 952-346-4700 | **Fax:** 952-346-4800 |
|---|---|
| **Toll-Free:** 800-562-0679 | |
| **Address:** 5775 W. Old Shakopee Rd., Ste. 100, Bloomington, MN 55437-3174 US | |

## GROWTH PLANS/SPECIAL FEATURES:

MGI Pharma, Inc., along with its subsidiaries, is a biopharmaceutical company that acquires, researches, develops and commercializes pharmaceutical products in the oncology and acute care arenas. The company acquires intellectual property or product rights from others after the basic research is completed to discover the compounds that will become the company's product candidates or marketable products. It has facilities in Bloomington, Minnesota; Lexington, Massachusetts; and Baltimore, Maryland. The company's marketable and currently marketed products include Aloxi, an injection for chemotherapy-induced nausea and vomiting; Dacogen, an injection to treat myelodysplastic syndrome; Gliadel wafers for malignant glioma at time of initial surgery and glioblastoma multiforme; Salagen tablets for symptoms of radiation-induced dry mouth in head and neck cancer patients, as well as dry mouth and dry eyes in Sjogren's syndrome patients outside the U.S.; and Hexalen capsules for ovarian cancer. Its product pipeline includes Aquavan injection, for minimal to moderate sedation of patients undergoing brief diagnostic or surgical procedures and Saforis powder oral mucositis, both in Phase III trials, as well as several others in Phase I and II and two preclinical candidates. In January 2008, MGI Pharma and its subsidiaries were acquired by Eisai Corporation of North America, a subsidiary of the Tokyo-based pharmaceutical company, Eisai Co., Ltd., for $3.9 billion.

## FINANCIALS: Sales and profits are in thousands of dollars—add 000 to get the full amount. 2008 Note: Financial information for 2008 was not available for all companies at press time.

| | | |
|---|---|---|
| 2008 Sales: $ | 2008 Profits: $ | **U.S. Stock Ticker:** Subsidiary |
| 2007 Sales: $ | 2007 Profits: $ | **Int'l Ticker:**     Int'l Exchange: |
| 2006 Sales: $342,788 | 2006 Profits: $-40,161 | Employees:    540 |
| 2005 Sales: $279,362 | 2005 Profits: $-132,410 | Fiscal Year Ends: 12/31 |
| 2004 Sales: $195,667 | 2004 Profits: $-85,723 | Parent Company: EISAI CORPORATION OF NORTH AMERICA |

## SALARIES/BENEFITS:

| Pension Plan: | ESOP Stock Plan: | Profit Sharing: | Top Exec. Salary: $470,417 | Bonus: $ |
|---|---|---|---|---|
| Savings Plan: Y | Stock Purch. Plan: | | Second Exec. Salary: $344,167 | Bonus: $ |

## OTHER THOUGHTS:

**Apparent Women Officers or Directors:** 1
**Hot Spot for Advancement for Women/Minorities:**

## LOCATIONS: ("Y" = Yes)

| West: | Southwest: | Midwest: | Southeast: | Northeast: | International: |
|---|---|---|---|---|---|
| | | Y | | Y | |

# MIGENIX INC

**www.migenix.com**

**Industry Group Code: 325412  Ranks within this company's industry group:** Sales: 154  Profits: 60

| Drugs: | | Other: | Clinical: | Computers: | Services: |
|---|---|---|---|---|---|
| Discovery: | Y | AgriBio: | Trials/Services: | Hardware: | Specialty Services: |
| Licensing: | | Genetic Data: | Labs: | Software: | Consulting: |
| Manufacturing: | | Tissue Replacement: | Equipment/Supplies: | Arrays: | Blood Collection: |
| Genetics: | | | Research & Development Services: | Database Management: | Drug Delivery: |
| | | | Diagnostics: | | Drug Distribution: |

## TYPES OF BUSINESS:

Pharmaceuticals Discovery & Development
Pre-Clinical Research & Development
Acne Treatment
Antivirals

## BRANDS/DIVISIONS/AFFILIATES:

Micrologix Biotech, Inc.
Omigard
Celgosivir

## CONTACTS: *Note: Officers with more than one job title may be intentionally listed here more than once.*

Bruce Schmidt, Interim CEO
Bruce Schmidt, Interim Pres.
Arthur J. Ayers, CFO
C. Robert Cory, VP-Bus. Dev.
Arthur J. Ayers, Sr. VP-Finance
Douglas Johnson, Chmn.

| Phone: 604-221-9666 | Fax: 604-221-9688 |
|---|---|
| **Toll-Free:** | |
| **Address:** 102, 2389 Health Sciences Mall, Vancouver, BC V6T 1Z3 Canada | |

## GROWTH PLANS/SPECIAL FEATURES:

Migenix, Inc., formerly Micrologix Biotech, Inc., is a biopharmaceutical company focusing on the discovery and development of drugs to treat and prevent infectious, degenerative and metabolic diseases. Its drug candidates are based on the development of improved derivatives of naturally occurring peptide compounds found in the defense systems of virtually all life forms. Currently, the company has no products approved for marketing and is involved only in the aspects of research and development of pharmaceutical drugs. It has three product candidates in human clinical development and three products in pre-clinical development. Currently, Omigard is the firm's most advanced drug candidate. It is a leading antimicrobial topical drug product for the reduction of life-threatening bacterial and fungal bloodstream infections related to the use of central venous catheters in seriously ill patients. The drug has completed Phase III clinical trials. Migenix has also completed a Phase II clinical trial for CLS-001, a drug candidate for the treatment of rosacea and acne. Migenix plans to initiate a Phase III trial for CLS-001 in 2009. The firm has acquired the global rights to Celgosivir, or MX-3253, a Phase II clinical-stage compound. The drug is an orally administered antiviral agent that can inhibit the replication of a broad range of enveloped viruses, including Hepatitis C virus (HCV). MX-2401 is an intravenous treatment for serious gram positive bacterial infections, currently in pre-clinical development. In May 2008, Migenix announced it would start streamlining its operations, including cutting executive salaries by an average 31%, in order to ensure it has enough cash resources to push Omigard through Phase III testing; advance MX-2401 development; and carry the company through 2009.

## FINANCIALS: Sales and profits are in thousands of dollars—add 000 to get the full amount. 2008 Note: Financial information for 2008 was not available for all companies at press time.

| | | |
|---|---|---|
| 2008 Sales: $ 180 | 2008 Profits: $- 520 | **U.S. Stock Ticker: MGIFF** |
| 2007 Sales: $ 17 | 2007 Profits: $-11,302 | **Int'l Ticker: MGI**  Int'l Exchange: Toronto-TSX |
| 2006 Sales: $ 538 | 2006 Profits: $-10,720 | Employees:  48 |
| 2005 Sales: $2,200 | 2005 Profits: $-9,463 | Fiscal Year Ends: 4/30 |
| 2004 Sales: $2,667 | 2004 Profits: $-11,215 | Parent Company: |

## SALARIES/BENEFITS:

| Pension Plan: | ESOP Stock Plan: | Profit Sharing: | Top Exec. Salary: $313,283 | Bonus: $ 186 |
|---|---|---|---|---|
| Savings Plan: | Stock Purch. Plan: | | Second Exec. Salary: $290,199 | Bonus: $ 151 |

## OTHER THOUGHTS:

**Apparent Women Officers or Directors:**
**Hot Spot for Advancement for Women/Minorities:**

## LOCATIONS: ("Y" = Yes)

| West: | Southwest: | Midwest: | Southeast: | Northeast: | International: Y |
|---|---|---|---|---|---|

Note: Financial information, benefits and other data can change quickly and may vary from those stated here.

# MILLENNIUM PHARMACEUTICALS INC

**www.mlnm.com**

**Industry Group Code: 325412  Ranks within this company's industry group:** Sales:    Profits:

| Drugs: | | Other: | | Clinical: | Computers: | | Services: | |
|---|---|---|---|---|---|---|---|---|
| Discovery: | Y | AgriBio: | | Trials/Services: | Hardware: | | Specialty Services: | |
| Licensing: | | Genetic Data: | Y | Labs: | Software: | | Consulting: | |
| Manufacturing: | Y | Tissue Replacement: | | Equipment/Supplies: | Arrays: | | Blood Collection: | |
| Genetics: | | | | Research & Development Services: | Database Management: | | Drug Delivery: | |
| | | | | Diagnostics: | | | Drug Distribution: | |

## TYPES OF BUSINESS:

Pharmaceuticals Discovery & Development
Gene-Based Drug Discovery Platform
Small-Molecule Drugs

## BRANDS/DIVISIONS/AFFILIATES:

Takeda Pharmaceutical Company Ltd
VELCADE

## CONTACTS: *Note: Officers with more than one job title may be intentionally listed here more than once.*

Deborah Dunsire, CEO
Deborah Dunsire, Pres.
Marsha Fanucci, CFO/Sr. VP
Stephen M. Gansler, Sr. VP-Human Resources
Joseph B. Bolen, Chief Scientific Officer
Laurie B. Keating, General Counsel/Sr. VP
Anna Protopapas, Sr. VP-Corp. Dev.
Lisa Adler, VP-Corp. Comm.
Todd Shegog, Sr. VP-Finance
Christophe Bianchi, Exec. VP
Peter F. Smith, Sr. VP-Non-Clinical Dev. Sciences
Nancy Simonian, Chief Medical Officer
Todd Shegog, Treas.

| **Phone:** 617-679-7000 | **Fax:** 617-374-7788 |
|---|---|
| **Toll-Free:** 800-390-5663 | |
| **Address:** 40 Landsdowne St., Cambridge, MA 02139 US | |

## GROWTH PLANS/SPECIAL FEATURES:

Millennium Pharmaceuticals, a subsidiary of Takeda Pharmaceuticals Company, Ltd., researches and manufactures therapeutic products for the treatment of patients with cancer and inflammatory diseases. The company's development platform focuses on combing knowledge of genomics and protein homeostasis, a set of particular molecular pathways that affect the establishment and progression of diseases, in order to produce drug candidates. The firm currently has one marketed product called VELCADE. VELCADE is an oncology drug that is injected into patients for the treatment of multiple myeloma, a cancer of the blood, and mantle cell lymphoma. VELCADE is also being tested in Phase II and III trials for the treatment of follicular B-cell non-Hodgkin's lymphoma, hematologic malignancies and solid tumors including lung, breast, prostate and ovarian cancers. In early 2008, the firm signed an agreement to collaborate with Harvard Medical School's Office of Technology Development on a research program in the area of protein homeostasis. In May 2008, Millennium was acquired by Takeda Pharmaceutical Company, Ltd. and became its wholly-owned subsidiary.

Millennium offers employees medical, dental and vision coverage; a 401(k); flexible spending accounts; life and disability insurance; and education and transportation assistance.

## FINANCIALS: Sales and profits are in thousands of dollars—add 000 to get the full amount. 2008 Note: Financial information for 2008 was not available for all companies at press time.

| | | |
|---|---|---|
| 2008 Sales: $ | 2008 Profits: $ | **U.S. Stock Ticker:** Subsidiary |
| 2007 Sales: $527,525 | 2007 Profits: $14,909 | **Int'l Ticker:**     Int'l Exchange: |
| 2006 Sales: $486,830 | 2006 Profits: $-43,953 | Employees:   966 |
| 2005 Sales: $558,308 | 2005 Profits: $-198,249 | Fiscal Year Ends: 12/31 |
| 2004 Sales: $448,206 | 2004 Profits: $-252,297 | Parent Company: TAKEDA PHARMACEUTICAL COMPANY LTD |

## SALARIES/BENEFITS:

| Pension Plan: | ESOP Stock Plan: | Profit Sharing: | Top Exec. Salary: $871,667 | Bonus: $1,540,000 |
|---|---|---|---|---|
| Savings Plan: Y | Stock Purch. Plan: Y | | Second Exec. Salary: $450,224 | Bonus: $315,000 |

## OTHER THOUGHTS:

**Apparent Women Officers or Directors:** 5
**Hot Spot for Advancement for Women/Minorities:** Y

## LOCATIONS: ("Y" = Yes)

| West: | Southwest: | Midwest: | Southeast: | Northeast: | International: |
|---|---|---|---|---|---|
| | | | | Y | |

Note: Financial information, benefits and other data can change quickly and may vary from those stated here.

# MILLIPORE CORP

**www.millipore.com**

**Industry Group Code: 3345   Ranks within this company's industry group:   Sales: 2   Profits: 2**

| Drugs: | Other: | Clinical: | | Computers: | | Services: | |
|---|---|---|---|---|---|---|---|
| Discovery: | AgriBio: | Trials/Services: | | Hardware: | | Specialty Services: | Y |
| Licensing: | Genetic Data: | Labs: | Y | Software: | | Consulting: | |
| Manufacturing: | Tissue Replacement: | Equipment/Supplies: | Y | Arrays: | | Blood Collection: | |
| Genetics: | | Research & Development Services: | | Database Management: | | Drug Delivery: | |
| | | Diagnostics: | | | | Drug Distribution: | |

## TYPES OF BUSINESS:

Biotechnology Instruments
Fluid Analysis, Identification & Purification Equipment
Chromatography Technologies

## BRANDS/DIVISIONS/AFFILIATES:

Direct-Q 3
Lynx S2S
MicroSafe, B.V.
Newport Bio Systems
Serologicals Corporation
MilliPROBE

## CONTACTS: Note: Officers with more than one job title may be intentionally listed here more than once.

Martin D. Madaus, CEO
Martin D. Madaus, Pres.
Charles Wagner, CFO/VP
Bruce Bonnevier, VP-Worldwide Human Resources
Dennis W. Harris, Chief Scientific Officer/VP
Peter C. Kershaw, VP-Worldwide Mfg. Oper.
Jeffrey Rudin, General Counsel/VP/Corp. Sec.
Wei Zhang, VP-Strategic & Corp. Dev.
Karen Marinella Hall, Dir.-Corp. Comm.
Joshua S. Young, Dir.-Investor Rel.
Jon DiVincenzo, VP/Pres., Bioscience Div.
Gregory J. Sam, VP-Quality
Jean-Paul Mangeolle, VP/Pres., Bioprocess Div.
Martin D. Madaus, Chmn.
Geoffrey F. Ide, VP-Millipore Int'l
Peter C. Kershaw, VP-Global Supply Chain

| Phone: 978-715-4321 | Fax: 800-645-5439 |
|---|---|
| Toll-Free: 800-645-5476 | |
| Address: 290 Concord Rd., Billerica, MA 01821 US | |

## GROWTH PLANS/SPECIAL FEATURES:

Millipore Corp. is a multinational bioscience company that provides technologies, tools and services for research, development and production. The company's products and services are based on technologies such as filtration, chromatography, cell culture supplements, antibodies and cell lines. The firm's products are offered through its two segments, the Bioscience division and the Bioprocess division. Millipore's Bioscience division, which accounted for 45% of 2008 revenue, is organized around four specific market segments: biotools for the separation, isolation and purification of biological samples; research reagents such as antibodies, dyes and biochemical reagents; drug discovery reagent for the analysis of drug candidates; and laboratory water purification systems that remove contaminants for critical laboratory analysis. The Bioprocess division, which accounted for 55% of 2008 revenue, provides bio-products and technologies for the manufacturing of biologic drugs in mammalian cell cultures; filtration, purification and chromatography technologies to clarify, concentrate, purify and remove viruses; process monitoring tools for the sampling and testing of drugs and intermediate products and advanced manufacturing systems for use in sterile biomanufacturing environments. The firm operates 11 manufacturing sites in Massachusetts, New Hampshire, Missouri, Illinois, California, France and the U.K., and 47 offices worldwide. In 2008, the firm entered into a license agreement with Bayer HeathCare AG. In January 2008, Millipore and Gen-Probe created the MilliPROBE system for real time tests of contaminants in manufacturing. In February 2009, the firm acquired Epitome Biosystems' EpiTag technology as well as Guava Technologies. Also in February, Millipore opened a new biomanufacturing sciences and training facility in Singapore. In June 2009, the firm opened a new bioprocess manufacturing facility in Massachusetts.

Millipore offers employees flexible spending accounts, tuition reimbursement, employee assistance programs and adoption assistance.

## FINANCIALS: Sales and profits are in thousands of dollars—add 000 to get the full amount. 2008 Note: Financial information for 2008 was not available for all companies at press time.

| | | |
|---|---|---|
| 2008 Sales: $1,602,138 | 2008 Profits: $145,801 | **U.S. Stock Ticker:** MIL |
| 2007 Sales: $1,531,555 | 2007 Profits: $136,472 | **Int'l Ticker:**   Int'l Exchange: |
| 2006 Sales: $1,255,371 | 2006 Profits: $96,984 | Employees: 5,900 |
| 2005 Sales: $991,031 | 2005 Profits: $80,168 | Fiscal Year Ends: 12/31 |
| 2004 Sales: $883,263 | 2004 Profits: $105,556 | Parent Company: |

## SALARIES/BENEFITS:

| Pension Plan: | ESOP Stock Plan: | Profit Sharing: | Top Exec. Salary: $742,307 | Bonus: $488,000 |
|---|---|---|---|---|
| Savings Plan: Y | Stock Purch. Plan: | | Second Exec. Salary: $339,612 | Bonus: $132,002 |

## OTHER THOUGHTS:

**Apparent Women Officers or Directors**: 3
**Hot Spot for Advancement for Women/Minorities**: Y

## LOCATIONS: ("Y" = Yes)

| West: | Southwest: | Midwest: | Southeast: | Northeast: | International: |
|---|---|---|---|---|---|
| Y | Y | Y | Y | Y | Y |

# MONSANTO CO

**www.monsanto.com**

Industry Group Code: 11511 Ranks within this company's industry group: Sales: 2 Profits: 1

| Drugs: | Other: | Clinical: | Computers: | Services: |
|---|---|---|---|---|
| Discovery: | AgriBio: Y | Trials/Services: | Hardware: | Specialty Services: |
| Licensing: | Genetic Data: | Labs: | Software: | Consulting: |
| Manufacturing: | Tissue Replacement: | Equipment/Supplies: | Arrays: | Blood Collection: |
| Genetics: | | Research & Development Services: | Database Management: | Drug Delivery: |
| | | Diagnostics: | | Drug Distribution: |

## TYPES OF BUSINESS:

Agricultural Biotechnology Products & Chemicals Manufacturing
Herbicides
Seeds
Genetic Products
Lawn & Garden Products

## BRANDS/DIVISIONS/AFFILIATES:

Asgrow
Roundup Ready
Agroeste Sementes
Delta and Pine Land Company
Seminis
De Ruiter Seeds Group BV
Marmot SA
Aly Participacoes Ltda

## CONTACTS: Note: Officers with more than one job title may be intentionally listed here more than once.

Hugh Grant, CEO
Hugh Grant, Pres.
Terrell K. Crews, CFO/Exec. VP
Steven C. Mizell, Exec. VP-Human Resources
Robert T. Fraley, CTO/Exec. VP
Mark J. Leidy, Exec. VP-Mfg.
Janet M. Holloway, Chief of Staff/VP
David F. Snively, General Counsel/Sr. VP/Sec.
Carl M. Casale, Exec. VP-Oper. & Strategy
Cheryl Morley, Sr. VP-Corp. Strategy
Janet M. Holloway, Sr. VP-Comm. Rel.
Scarlett Lee Foster, VP-Investor Rel.
Richard B. Clark, VP/Controller
Robert A. Paley, VP/Treas.
Nicole M. Ringenberg, VP-Finance & Oper./Global Commercial
Terrell K. Crews, CEO-Seminis
Gerald A. Steiner, Exec. VP-Corp. Affairs & Sustainability
Hugh Grant, Chmn.
Kerry J. Preete, Exec. VP-Int'l Commercial

| Phone: 314-694-1000 | Fax: 314-694-8394 |
|---|---|
| Toll-Free: | |
| Address: 800 N. Lindbergh Blvd., St. Louis, MO 63167 US | |

## GROWTH PLANS/SPECIAL FEATURES:

Monsanto Co. is a global provider of agricultural products for farmers. The company operates in two principal business segments: Seeds and Genomics; and Agricultural Productivity. The Seeds and Genomics segment is responsible for producing seed brands and patenting genetic traits that enable seeds to resist insects, disease, drought and weeds. Major seed brands produced by Monsanto include Agroceres, Asgrow, DEKALB, Stoneville, Vistive, Monsoy, Holden's Foundation Seeds, American Seeds, Inc., Seminis, Royal Sluis and Petoseed. The company's genetic trait products include Roundup Ready traits in soybeans, corn, canola and cotton; Bollgard and Bollgard II traits in cotton; and YieldGard Corn Borer and YieldGard Rootworm traits in corn. The Agricultural Productivity segment produces herbicide products. The firm's branded herbicides include Roundup, Harness, Degree, Machete, Maverick, Certainty, Outrider and Monitor. Monsanto market its seeds and commercial herbicides through a variety of channels and directly to farmers. Residential herbicides are marketed through the Scotts Miracle-Gro Company. Subsidiaries include Delta and Pine Land Company, a developer of cotton and soybean seeds; and Agroeste Sementes, a Brazilian corn seed company. In 2008, the company acquired the Dutch holding company, De Ruiter Seeds Group B.V. for $850 million; Marmot, S.A. which operates the private Guatemalan seed company Semillas Cristiani Burkhard; and Aly Participacoes Ltda., which operates sugarcane technology companies, CanaVialis S.A. and Alellyx S.A. Also in 2008, the company sold its POSILAC bovine somatotropin brand and related business to Eli Lilly and Company for $300 million.

Employees are offered medical, dental, and vision insurance; life insurance; disability coverage; a pension plan; a 401(k) plan; a stock purchase plan; long term care insurance; adoption assistance; group auto and home insurance; an employee assistance program; and relocation assistance.

## FINANCIALS: Sales and profits are in thousands of dollars—add 000 to get the full amount. 2008 Note: Financial information for 2008 was not available for all companies at press time.

| | | |
|---|---|---|
| 2008 Sales: $11,365,000 | 2008 Profits: $2,024,000 | U.S. Stock Ticker: MON |
| 2007 Sales: $8,563,000 | 2007 Profits: $993,000 | Int'l Ticker: Int'l Exchange: |
| 2006 Sales: $7,294,000 | 2006 Profits: $689,000 | Employees: 21,700 |
| 2005 Sales: $6,275,000 | 2005 Profits: $255,000 | Fiscal Year Ends: 8/31 |
| 2004 Sales: $5,457,000 | 2004 Profits: $267,000 | Parent Company: |

## SALARIES/BENEFITS:

| Pension Plan: Y | ESOP Stock Plan: | Profit Sharing: | Top Exec. Salary: $1,286,019 | Bonus: $3,326,796 |
|---|---|---|---|---|
| Savings Plan: Y | Stock Purch. Plan: | | Second Exec. Salary: $566,827 | Bonus: $840,000 |

## OTHER THOUGHTS:

**Apparent Women Officers or Directors:** 5
**Hot Spot for Advancement for Women/Minorities:** Y

## LOCATIONS: ("Y" = Yes)

| West: | Southwest: | Midwest: | Southeast: | Northeast: | International: |
|---|---|---|---|---|---|
| Y | Y | Y | Y | Y | Y |

Note: Financial information, benefits and other data can change quickly and may vary from those stated here.

# MYLAN INC

**Industry Group Code: 325412A  Ranks within this company's industry group:** Sales: 2  Profits: 29

| Drugs: | | Other: | Clinical: | Computers: | Services: |
|---|---|---|---|---|---|
| Discovery: | Y | AgriBio: | Trials/Services: | Hardware: | Specialty Services: |
| Licensing: | | Genetic Data: | Labs: | Software: | Consulting: |
| Manufacturing: | Y | Tissue Replacement: | Equipment/Supplies: | Arrays: | Blood Collection: |
| Genetics: | Y | | Research & Development Services: | Database Management: | Drug Delivery: |
| | | | Diagnostics: | | Drug Distribution: |

## TYPES OF BUSINESS:

Drugs-Generic
Generic Pharmaceuticals
Active Pharmaceutical Ingredients

## BRANDS/DIVISIONS/AFFILIATES:

UDL Laboratories, Inc.
Mylan Pharmaceuticals, Inc.
Mylan Technologies, Inc.
Matrix Laboratories Limited
Docpharma
Dey
Genpharm ULC
Somerset Pharmaceuticals, Inc.

## CONTACTS: Note: Officers with more than one job title may be intentionally listed here more than once.

Robert J. Coury, CEO
Heather Bresch, COO/Exec. VP
Jolene Varney, CFO/Exec. VP
David A. Lillback, Sr. VP/Global Head-Human Resource
Gregory L. Sheldon, Global CIO/Sr. VP
Rajiv Malik, Exec. VP/Head-Global Tech. Oper.
Joseph F. Haggerty, Global General Counsel/Sr. VP
Andrew G. Cuneo, VP-Global Bus. Dev.
Michael Laffin, Sr. VP-Global Public Affairs
Daniel E. Crookshank, VP-Investor Rel.
Daniel C. Rizzo Jr., Controller/Chief Acct. Officer/Sr. VP
Carolyn Meyers, Pres., Specialty
Harry A. Korman, Pres., North America
Brian Byala, Treas./Sr. VP
John Montgomery, Pres., APAC
Robert J. Coury, Chmn.
Didier Barret, Pres., EMEA

| Phone: 724-514-1800 | Fax: 724-514-1870 |
|---|---|
| Toll-Free: | |
| Address: 1500 Corporate Dr., Canonsburg, PA 15317 US | |

## GROWTH PLANS/SPECIAL FEATURES:

Mylan, Inc., formerly Mylan Laboratories, Inc., develops, licenses, manufactures, distributes and markets generic/branded generic pharmaceutical products and active pharmaceutical ingredients (APIs). The company reports as three segments: Generics, Matrix and Specialty. The generics segment operates in the U.S., Canada, Europe, the Middle East, Africa, Australia, Japan and New Zealand. In the U.S., this segment markets over 200 products through four subsidiaries: Mylan Pharmaceuticals Inc. (MPI), which sells solid oral dosage products; UDL Laboratories, Inc., which re-packages and markets products obtained from MPI and third parties; Genpharm ULC, which sells generic prescriptions for all major therapeutic needs; and Mylan Technologies, Inc., which develops and markets transdermal patches. Generic international operations primarily consist of subsidiaries that the firm acquired as part of the Merck Generics acquisition. All together, the generics business derives roughly 31% of its revenues from calcium channel blockers and 29% from narcotic agonist analgesics. Mylan's Matrix segment, operated by its 71.5 %-owned subsidiary Matrix Laboratories Limited, is a leading manufacturer of APIs. Matrix currently has over 200 APIs in the market or under development, including anti-bacterials; central nervous system agents; anti-histamines; cardiovasculars; anti-virals; anti-diabetics; anti-fungals; proton pump inhibitors; pain management drugs; and anti-retrovirals used in the treatment of HIV. These products are used in Mylan pharmaceuticals or sold to third parties. This segment also includes the Docpharma subsidiary, which is a distributor of pharmaceutical products in the Benelux region of Europe. Lastly, the specialty segment is conducted by Dey, which focuses on the respiratory and severe allergy markets. Its principal products are EpiPen, for severe allergic reactions; and Perforomist Inhalation Solution, a formoterol fumarate inhalation solution. In July 2008, Mylan bought Watson Pharmaceuticals' 50% interest in joint venture Somerset Pharmaceuticals, Inc. (now a wholly-owned subsidiary of Mylan).

## FINANCIALS: Sales and profits are in thousands of dollars—add 000 to get the full amount. 2008 Note: Financial information for 2008 was not available for all companies at press time.

| | | |
|---|---|---|
| 2008 Sales: $5,137,585 | 2008 Profits: $-181,215 | **U.S. Stock Ticker:** MYL |
| 2007 Sales: $1,611,819 | 2007 Profits: $217,284 | **Int'l Ticker:**  Int'l Exchange: |
| 2006 Sales: $1,257,164 | 2006 Profits: $184,542 | Employees: 15,000 |
| 2005 Sales: $1,253,374 | 2005 Profits: $203,592 | Fiscal Year Ends: 12/31 |
| 2004 Sales: $1,374,617 | 2004 Profits: $334,609 | Parent Company: |

## SALARIES/BENEFITS:

| Pension Plan: | ESOP Stock Plan: | Profit Sharing: | Top Exec. Salary: $1,500,000 | Bonus: $3,750,000 |
|---|---|---|---|---|
| Savings Plan: Y | Stock Purch. Plan: | | Second Exec. Salary: $500,000 | Bonus: $1,000,000 |

## OTHER THOUGHTS:

**Apparent Women Officers or Directors:** 4
**Hot Spot for Advancement for Women/Minorities:** Y

## LOCATIONS: ("Y" = Yes)

| West: | Southwest: | Midwest: | Southeast: | Northeast: | International: |
|---|---|---|---|---|---|
| | Y | Y | | Y | Y |

Note: Financial information, benefits and other data can change quickly and may vary from those stated here.

# MYRIAD GENETICS INC

www.myriad.com

**Industry Group Code: 325412 Ranks within this company's industry group:** Sales: 48 Profits: 39

| Drugs: | | Other: | | Clinical: | | Computers: | | Services: | |
|---|---|---|---|---|---|---|---|---|---|
| Discovery: | Y | AgriBio: | | Trials/Services: | | Hardware: | | Specialty Services: | |
| Licensing: | | Genetic Data: | Y | Labs: | | Software: | | Consulting: | |
| Manufacturing: | | Tissue Replacement: | | Equipment/Supplies: | | Arrays: | | Blood Collection: | |
| Genetics: | | | | Research & Development Services: | | Database Management: | | Drug Delivery: | |
| | | | | Diagnostics: | Y | | | Drug Distribution: | |

## TYPES OF BUSINESS:

Pharmaceuticals Discovery & Development
Cancer Treatments
HIV Treatments
Cancer Diagnostics

## BRANDS/DIVISIONS/AFFILIATES:

Myriad Pharmaceuticals, Inc.
TheraGuide 5-FU
Prezeon
BRACAnalysis
COLARIS
MELARIS
COLARIS AP
Azixa

## CONTACTS: Note: Officers with more than one job title may be intentionally listed here more than once.

Peter D. Meldrum, CEO
Peter D. Meldrum, Pres.
James S. Evans, CFO
Jerry Lanchbury, Exec. VP-Research
Robert G. Harrison, CIO
Richard Marsh, General Counsel/Exec. VP/Corp. Sec.
William A. Hockett, III, Exec. VP-Corp. Comm.
Gregory C. Critchfield, Pres., Myriad Genetic Laboratories, Inc.
Adrian N. Hobden, Pres., Myriad Pharmaceuticals, Inc.
Mark H. Skolnick, Chief Scientific Officer
Mark C. Capone, COO-Myriad Genetic Laboratories, Inc.
John T. Henderson, Chmn.

| Phone: 801-584-3600 | Fax: 801-584-3640 |
|---|---|
| Toll-Free: | |
| Address: 320 Wakara Way, Salt Lake City, UT 84108 US | |

## GROWTH PLANS/SPECIAL FEATURES:

Myriad Genetics, Inc. is a biopharmaceutical company that develops and markets novel therapeutic and molecular diagnostic products. The company develops a number of proprietary technologies that are aimed at understanding the genetic basis of human disease in order to effectively treat them. Myriad's molecular diagnostic business is focused on both predictive and personalized medicine. Predictive medicine analyzes genes to assess an individual's potential risk of developing while personalized medicine works to assess a patient's risk in disease progression, recurrence and drug response. The company currently owns seven commercial predictive medicine products: BRACAnalysis for breast and ovarian cancer; COLARIS for colon and uterine cancer; COLARIS AP for polyp-forming syndromes of colon cancer; MELARIS for melanoma; TheraGuide 5-FU for determining the risk of toxic reaction to 5-FU chemotherapy; Prezeon for assessing PTEN function in cancer patients; and OnDose for optimizing 5-FU chemotherapy dosage. Prezeon and OnDose are relatively new additions to the company's product line and were launched in December 2008 and April 2009, respectively. The company works with its wholly-owned subsidiary Myriad Pharmaceuticals, Inc. to develop novel small molecule drugs. The lead product pipeline includes Azixa for the treatment of melanoma, glioblastoma and anaplastic glioma; MPC-3100 for inhibiting Heat shock protein 90 (Hsp90) in cancer patients; MPC-4326 and MPC-9055 for HIV; as well as other preclinical treatments for cancer and HIV. In October 2008, the company announced plans to spin off Myriad Pharmaceuticals, Inc. and focus exclusively on its molecular diagnostics business. In January 2009, the firm acquired the HIV drug Bevirimat from Panacos Pharmaceuticals, and assigned it to Myriad Pharmaceuticals.

Myriad offers its employees medical, dental, life and AD&D insurance; an employee stock purchase plan; short- and long-term disability coverage; a 401(k) plan; and tax-free reimbursements.

## FINANCIALS: Sales and profits are in thousands of dollars—add 000 to get the full amount. 2008 Note: Financial information for 2008 was not available for all companies at press time.

| | | |
|---|---|---|
| 2008 Sales: $334,000 | 2008 Profits: $47,845 | **U.S. Stock Ticker: MYGN** |
| 2007 Sales: $157,126 | 2007 Profits: $-34,962 | **Int'l Ticker:** Int'l Exchange: |
| 2006 Sales: $114,279 | 2006 Profits: $-38,189 | Employees: 994 |
| 2005 Sales: $82,406 | 2005 Profits: $-39,978 | Fiscal Year Ends: 6/30 |
| 2004 Sales: $56,648 | 2004 Profits: $-40,620 | Parent Company: |

## SALARIES/BENEFITS:

| Pension Plan: | ESOP Stock Plan: | Profit Sharing: | Top Exec. Salary: $727,552 | Bonus: $725,812 |
|---|---|---|---|---|
| Savings Plan: Y | Stock Purch. Plan: Y | | Second Exec. Salary: $500,552 | Bonus: $400,812 |

## OTHER THOUGHTS:

**Apparent Women Officers or Directors:** 1
**Hot Spot for Advancement for Women/Minorities:**

## LOCATIONS: ("Y" = Yes)

| West: | Southwest: | Midwest: | Southeast: | Northeast: | International: |
|---|---|---|---|---|---|
| Y | | | | | |

Note: Financial information, benefits and other data can change quickly and may vary from those stated here.

# NABI BIOPHARMACEUTICALS

www.nabi.com

Industry Group Code: 325412  Ranks within this company's industry group:  Sales:      Profits: 81

| Drugs: | | Other: | Clinical: | Computers: | Services: |
|---|---|---|---|---|---|
| Discovery: | Y | AgriBio: | Trials/Services: | Hardware: | Specialty Services: |
| Licensing: | | Genetic Data: | Labs: | Software: | Consulting: |
| Manufacturing: | | Tissue Replacement: | Equipment/Supplies: | Arrays: | Blood Collection: |
| Genetics: | | | Research & Development Services: | Database Management: | Drug Delivery: |
| | | | Diagnostics: | | Drug Distribution: |

## TYPES OF BUSINESS:

Drugs-Infectious Disease & Autoimmune Disorders
Vaccines
Addiction Treatments

## BRANDS/DIVISIONS/AFFILIATES:

PentaStaph
NicVAX
EnteroVAX

## CONTACTS: *Note: Officers with more than one job title may be intentionally listed here more than once.*

Raafat E. F. Fahim, CEO
Raafat E. F. Fahim, Pres.
Raafat E. Fahim, Acting CFO
Ali E. Fattom, VP-R&D
Matthew Kalnik, VP-Strategic Planning & Bus. Oper.
Paul Kessler, Sr. VP-Clinical, Medical & Regulatory Affairs
Geoffrey F. Cox, Chmn.

| Phone: 301-770-3099 | Fax: 301-770-3097 |
|---|---|
| Toll-Free: 800-685-5579 | |
| Address: 12276 Wilkins Ave., Rockville, MD 20852 US | |

## GROWTH PLANS/SPECIAL FEATURES:

Nabi Biopharmaceuticals is a development level biopharmaceutical company that focuses on vaccines and therapies for the treatment and prevention of infectious diseases and nicotine addiction.  The firm's two leading product in development are NicVAX, a vaccine to treat nicotine addiction, and PentaStaph, a pentavalent vaccine designed to prevent staphylococcus aureus infections.  The NicVAX vaccine is designed to work by stimulating the immune system to produce antibodies that bind to nicotine, effectively keeping nicotine from entering the brain and triggering pleasure sensations.  This vaccine is currently in Phase II clinical trials.  The firm's other products in development include EnteroVAX, a vaccine in development for the prevention of enterococcal infections; and the RENS (Ring Expanded Nucleosides and Nucleotides) platform designed to develop analogs used to treat viral infections and cancer.  Possibilities for this research include treatments for hepatitis B and C, respiratory syncytial virus, Epstein-Barr virus, West Nile virus and rhinovirus, as well as numerous cancers.  The company previously marketed Nabi-HB, a human polyclonal antibody drug that prevents hepatitis B infection; Aloprima, a treatment for gout; and antibody plasma products.  These products were recently sold to Biotest AG $185 million.

Nabi offers its employees medical, dental, vision, prescription drug and life insurance; short- and long-term disability; a 401(k) plan; an employee assistance program; a flexible spending account; a stock purchase plan; tuition assistance; access to a credit union; and employee referral incentives.

## FINANCIALS:  Sales and profits are in thousands of dollars—add 000 to get the full amount. 2008 Note: Financial information for 2008 was not available for all companies at press time.

| | | |
|---|---|---|
| 2008 Sales: $ | 2008 Profits: $-10,716 | U.S. Stock Ticker: NABI |
| 2007 Sales: $80,855 | 2007 Profits: $47,069 | Int'l Ticker:     Int'l Exchange: |
| 2006 Sales: $117,852 | 2006 Profits: $-58,703 | Employees:    40 |
| 2005 Sales: $94,149 | 2005 Profits: $-128,449 | Fiscal Year Ends: 12/31 |
| 2004 Sales: $142,183 | 2004 Profits: $-50,390 | Parent Company: |

## SALARIES/BENEFITS:

| | | | | |
|---|---|---|---|---|
| Pension Plan: | ESOP Stock Plan: | Profit Sharing: | Top Exec. Salary: $458,301 | Bonus: $364,000 |
| Savings Plan: Y | Stock Purch. Plan: Y | | Second Exec. Salary: $297,808 | Bonus: $158,065 |

## OTHER THOUGHTS:

**Apparent Women Officers or Directors**: 1
**Hot Spot for Advancement for Women/Minorities**:

## LOCATIONS: ("Y" = Yes)

| West: | Southwest: | Midwest: | Southeast: | Northeast: | International: |
|---|---|---|---|---|---|
| | | | | Y | |

# NANOBIO CORPORATION

www.nanobio.com

**Industry Group Code: 325412A  Ranks within this company's industry group: Sales:     Profits: 30**

| Drugs: | | Other: | Clinical: | Computers: | Services: |
|--------|--|--------|-----------|-----------|-----------|
| Discovery: | Y | AgriBio: | Trials/Services: | Hardware: | Specialty Services: |
| Licensing: | | Genetic Data: | Labs: | Software: | Consulting: |
| Manufacturing: | | Tissue Replacement: | Equipment/Supplies: | Arrays: | Blood Collection: |
| Genetics: | | | Research & Development Services: | Database Management: | Drug Delivery: |
| | | | Diagnostics: | | Drug Distribution: |

## TYPES OF BUSINESS:

Drug Delivery Systems
Nano-Emulsion Technology
Cold Sore Treatments
Drugs-Antimicrobial

## BRANDS/DIVISIONS/AFFILIATES:

NanoStat
NanoHPX
NanoTxt

## CONTACTS: *Note: Officers with more than one job title may be intentionally listed here more than once.*

James R. Baker, CEO
David Peralta, COO
David Peralta, CFO/VP
James Baker, Chief Scientific Officer
John Coffey, VP-Bus. Dev.
Mary R. Flack, VP-Clinical Research
Joyce Sutcliffe, VP-Research
Lyou-fu Ma, Sr. VP-Dev.
Stephen Gracon, VP-Regulatory Affairs
James R. Baker, Chmn.

| Phone: 734-302-4000 | Fax: 734-302-9150 |
|---------------------|--------------------|
| Toll-Free: | |
| Address: 2311 Green Rd., Ste. A, Ann Arbor, MI 48105 US | |

## GROWTH PLANS/SPECIAL FEATURES:

NanoBio Corporation is a privately-held biopharmaceutical company that develops and markets anti-infection products and mucosal vaccines.  The company's main technology platform is NanoStat, which implements oil-in-water emulsions that are roughly 150-400 nanometers in size. Nanoemulsion particles can rapidly penetrate the skin through pores and hair shafts to the site of an infection, where it physically kills lipid-containing organisms by penetrating the outer membrane of pathogenic organisms. NanoStat emulsions have the added benefit of being selectively lethal to targeted microbes without affecting any surrounding skin and mucous membranes.  NanoBio's current commercial product pipeline consists of five anti-infective pharmaceuticals and two mucosal vaccines.  The firm's two leading anti-infective products are NB-001 (Phase III), a treatment for cold sores, and NB-002 (Phase II) for nail fungus, which are both in the clinical stages of development. NanoStat mucosal vaccines are designed to treat influenza, H5N1, hepatitis B, pneumonia, tuberculosis, small pox anthrax and other various viral and bacterial diseases. NanoBio also holds exclusive intellectual property rights for virucidal, fungicidal, sporicidal and bactericidal applications in a wide spectrum of products that include personal care products, medical products and anti-bioterrorism applications.  NanoBio also maintains a research team to discover new applications and methods relating to Varicella zoster (a virus that causes shingles), vaginitis, ocular infections, influenza prophylaxis and drug delivery systems.

NanoBio offers employees medical and dental coverage; medical and child care flexible spending accounts; a 401(k) plan; life insurance; disability income protection; as well as annual performance bonuses.

## FINANCIALS: Sales and profits are in thousands of dollars—add 000 to get the full amount. 2008 Note: Financial information for 2008 was not available for all companies at press time.

| | | |
|--|--|--|
| 2008 Sales: $ | 2008 Profits: $ | U.S. Stock Ticker: Private |
| 2007 Sales: $ | 2007 Profits: $ | Int'l Ticker:     Int'l Exchange: |
| 2006 Sales: $ | 2006 Profits: $ | Employees: |
| 2005 Sales: $ | 2005 Profits: $ | Fiscal Year Ends: 12/31 |
| 2004 Sales: $ | 2004 Profits: $ | Parent Company: |

## SALARIES/BENEFITS:

| Pension Plan: | ESOP Stock Plan: | Profit Sharing: Y | Top Exec. Salary: $ | Bonus: $ |
|---------------|------------------|-------------------|---------------------|----------|
| Savings Plan: Y | Stock Purch. Plan: | | Second Exec. Salary: $ | Bonus: $ |

## OTHER THOUGHTS:

**Apparent Women Officers or Directors:**
**Hot Spot for Advancement for Women/Minorities:**

## LOCATIONS: ("Y" = Yes)

| West: | Southwest: | Midwest: | Southeast: | Northeast: | International: |
|-------|------------|----------|------------|------------|---------------|
| | | Y | | | |

Note: Financial information, benefits and other data can change quickly and may vary from those stated here.

# NANOGEN INC

www.nanogen.com

**Industry Group Code: 325413  Ranks within this company's industry group:** Sales: 17   Profits: 18

| Drugs: | Other: | | Clinical: | | Computers: | Services: |
|---|---|---|---|---|---|---|
| Discovery: | AgriBio: | | Trials/Services: | | Hardware: | Specialty Services: |
| Licensing: | Genetic Data: | Y | Labs: | | Software: | Consulting: |
| Manufacturing: | Tissue Replacement: | | Equipment/Supplies: | | Arrays: | Blood Collection: |
| Genetics: | | | Research & Development Services: | | Database Management: | Drug Delivery: |
| | | | Diagnostics: | Y | | Drug Distribution: |

## TYPES OF BUSINESS:

Molecular Diagnostic Tests
Genomics Research
PCR Molecular Reagents
Chips-DNA Analysis
Point-of-Care Tests
Microarrays

## BRANDS/DIVISIONS/AFFILIATES:

DXpress Reader
MGB Alert
Q-PCR Alert
Cardiac STATus
Decision Point
Vyvent

## CONTACTS: *Note: Officers with more than one job title may be intentionally listed here more than once.*

Howard C. Birndorf, CEO
David Ludvigson, COO
David Ludvigson, Pres.
Nick Venuto, CFO
Robert Proulx, VP-Mktg. & Sales
Graham Lidgard, Sr. VP-R&D
William L. Respess, General Counsel/Sr. VP/Corp. Sec.
David Boudreau, VP-Oper.
Merl F. Hoekstra, VP-Bus. Dev.
Kelly Gann, Media Rel.
Leanne M. Kiviharju, VP-Clinical, Regulatory & Quality Assurance
Walt Mahoney, VP-R&D
Howard C. Birndorf, Chmn.

| **Phone:** 858-410-4600 | **Fax:** 858-410-4952 |
|---|---|
| **Toll-Free:** 877-626-6436 | |
| **Address:** 10398 Pacific Center Ct., San Diego, CA 92121 US | |

## GROWTH PLANS/SPECIAL FEATURES:

Nanogen, Inc. provides advanced diagnostic technologies and products that investigate, predict and diagnose diseases for research labs, hospitals, urgent care centers and universities. The company's product line includes DXpress Reader; MGB Alert; Q-PCR Alert; Cardiac STATus; and Decision Point. DXpress Reader is an instrument used to interpret the results of quantitative in vitro immunodiagnostic assays. MGB Alert, utilized in molecular labs, is a group of detection reagents that uses probe technology to detect disease carrying organisms. Q-PCR Alert developed by Nanogen Advanced Diagnostics research and development, is used for in vitro diagnostics. Both MGB Alert and Q-PCR Alert are used in conjunction with the firm's MGB Probes, which increases diagnostic sensitivity and precision. Cardiac STATus is a rapid biomarker test that is used for the point of care diagnosis of cardiovascular disease and myocardial infarction. Decision Point is another biomarker test, used on patients experiencing chest pain to diagnose acute myocardial infarction. Nanogen has also developed Vyvent, a blood/plasma urgent care test; the company expects FDA clearance for this test in the near future. In April 2008, the firm entered into a partnership with Thermo Fisher Scientific, Inc., who will be Nanogen's exclusive provider of products used in gene-expression experiments. These products are based on Nanogen's second-generation proprietary probe technology. In May 2008, the company announced plans to consolidate its Toronto point of care manufacturing facility into its San Diego operation. In August 2008, the firm entered into definitive agreements to merge with French diagnostics company The Elitech Group; however, in January 2009, Nanogen was permitted to seek alternatives to this merger. In November 2008, the company agreed to partner with A. Menarini Diagnostics S.r.l., of Italian diagnostics/ pharmaceuticals firm The Menarini Group, to provide MGB technology to The Menarini Group's European market base.

## FINANCIALS: Sales and profits are in thousands of dollars—add 000 to get the full amount. 2008 Note: Financial information for 2008 was not available for all companies at press time.

| | | |
|---|---|---|
| 2008 Sales: $46,920 | 2008 Profits: $-38,107 | **U.S. Stock Ticker: NGEN** |
| 2007 Sales: $38,183 | 2007 Profits: $-33,945 | **Int'l Ticker:**   Int'l Exchange: |
| 2006 Sales: $26,852 | 2006 Profits: $-46,732 | Employees:  248 |
| 2005 Sales: $12,544 | 2005 Profits: $-104,779 | Fiscal Year Ends: 12/31 |
| 2004 Sales: $5,374 | 2004 Profits: $-38,907 | Parent Company: |

## SALARIES/BENEFITS:

| Pension Plan: | ESOP Stock Plan: Y | Profit Sharing: | Top Exec. Salary: $520,150 | Bonus: $46,814 |
|---|---|---|---|---|
| Savings Plan: Y | Stock Purch. Plan: Y | | Second Exec. Salary: $360,000 | Bonus: $27,038 |

## OTHER THOUGHTS:

**Apparent Women Officers or Directors:** 2
**Hot Spot for Advancement for Women/Minorities:** Y

## LOCATIONS: ("Y" = Yes)

| West: | Southwest: | Midwest: | Southeast: | Northeast: | International: |
|---|---|---|---|---|---|
| Y | | | | | Y |

Note: Financial information, benefits and other data can change quickly and may vary from those stated here.

# NEKTAR THERAPEUTICS

**www.nektar.com**

Industry Group Code: 325412A  **Ranks within this company's industry group:** Sales: 14   Profits: 22

| Drugs: | | Other: | | Clinical: | | Computers: | | Services: | |
|---|---|---|---|---|---|---|---|---|---|
| Discovery: | | AgriBio: | | Trials/Services: | | Hardware: | | Specialty Services: | |
| Licensing: | Y | Genetic Data: | | Labs: | | Software: | | Consulting: | |
| Manufacturing: | | Tissue Replacement: | | Equipment/Supplies: | Y | Arrays: | | Blood Collection: | |
| Genetics: | | | | Research & Development Services: | Y | Database Management: | | Drug Delivery: | Y |
| | | | | Diagnostics: | | | | Drug Distribution: | |

## TYPES OF BUSINESS:

Drug Delivery Systems
PEG-Based Delivery Systems
Molecular & Particle Engineering
Equipment-Inhalers

## BRANDS/DIVISIONS/AFFILIATES:

Nektar PEGylation Technology
Nektar Pulmonary Technology
Macugen
PEGASYS
DuraSeal
Neulasta
Somavert
PEG-INTRON

## CONTACTS: *Note: Officers with more than one job title may be intentionally listed here more than once.*

Howard W. Robin, CEO
Bharatt Chowrira, COO
Howard W. Robin, Pres.
John Nicholson, CFO/Sr. VP
Dorian Rinella, VP-Human Resources & Facilities
Timothay A. Riley, Sr. VP-Global Research
Gil M. Labrucherie, General Counsel/Sec.
Rinko Ghosh, Sr. VP-Bus. Dev. & Alliance Mgmt.
Jillian B Thomsen, VP-Finance/Chief Acct. Officer
Nevan Elam, Sr. VP/Head-Pulmonary Bus. Unit
John S. Patton, Chief Research Fellow
Lorianne Masuoka, Chief Medical Officer
Robert B. Chess, Chmn.

| Phone: 650-631-3100 | Fax: 650-631-3150 |
|---|---|
| Toll-Free: | |
| Address: 201 Industrial Rd., San Carlos, CA 94070 US | |

## GROWTH PLANS/SPECIAL FEATURES:

Nektar Therapeutics is a biopharmaceutical company focused on developing technology for unmet medical needs. The company, which recently restructured to focus on its therapeutic drug development operations, develops and enables differentiated therapeutics through its technology platforms. The firm aims to create differentiated, innovative products by applying its platform technologies to established or novel medicine. Nektar Therapeutics creates potential breakthrough products in two ways: by developing products in collaboration with pharmaceutical and biotechnology companies that seek to improve and differentiate their products; and by applying its technologies to already approved drugs. The company's leading technology platforms are Pulmonary Technology and PEGylation Technology. Pulmonary Technology makes drugs inhaleable to deliver them to and through the lungs for both systemic and local lung applications. PEGylation Technology is a chemical process designed to enhance the performance of most drug classes with the potential to improve solubility and stability; increase drug half-life; reduce immune responses to an active drug; and improve the efficacy and/or safety of a molecule in certain instances. Leading products approved by the U.S. Food and Drug Administration (FDA) include Macugen treatment for macular degeneration; PEGASYS for chronic Hepatitis C; Somavert human growth hormone receptor antagonist; Exubera inhaleable insulin; PEG-INTRON for treating Hepatitis C; and Neulasta for the treatment of neutropenia, a condition where the body produces too few white blood cells. Most of these have been developed in collaboration with other pharmaceutical companies, including Amgen; Novartis Pharma; Bayer; Pfizer; and Schering-Plough. In December 2008, Novartis AG acquired certain pulmonary technology, delivery assets and intellectual property for $115 million.

Nektar offers its employees medical, dental and vision plans; accident and life insurance; retirement and investment plans; and life enrichment programs, including tuition reimbursement and fitness membership discounts.

## FINANCIALS: Sales and profits are in thousands of dollars—add 000 to get the full amount. 2008 Note: Financial information for 2008 was not available for all companies at press time.

| | | |
|---|---|---|
| 2008 Sales: $90,185 | 2008 Profits: $-34,336 | U.S. Stock Ticker: NKTR |
| 2007 Sales: $273,027 | 2007 Profits: $-32,761 | Int'l Ticker:    Int'l Exchange: |
| 2006 Sales: $217,718 | 2006 Profits: $-154,761 | Employees:  338 |
| 2005 Sales: $126,279 | 2005 Profits: $-185,111 | Fiscal Year Ends: 12/31 |
| 2004 Sales: $114,270 | 2004 Profits: $-101,886 | Parent Company: |

## SALARIES/BENEFITS:

| Pension Plan: | ESOP Stock Plan: | Profit Sharing: | Top Exec. Salary: $73,500 | Bonus: $ |
|---|---|---|---|---|
| Savings Plan: Y | Stock Purch. Plan: | | Second Exec. Salary: $65,500 | Bonus: $ |

## OTHER THOUGHTS:

**Apparent Women Officers or Directors:** 1
**Hot Spot for Advancement for Women/Minorities:** Y

## LOCATIONS: ("Y" = Yes)

| West: | Southwest: | Midwest: | Southeast: | Northeast: | International: |
|---|---|---|---|---|---|
| Y | | | Y | | Y |

Note: Financial information, benefits and other data can change quickly and may vary from those stated here.

# NEOGEN CORPORATION

**www.neogen.com**

Industry Group Code: 325413  **Ranks within this company's industry group:**  Sales: 13   Profits: 7

| Drugs: | Other: | | Clinical: | | Computers: | | Services: | |
|---|---|---|---|---|---|---|---|---|
| Discovery: | AgriBio: | Y | Trials/Services: | | Hardware: | | Specialty Services: | |
| Licensing: | Genetic Data: | | Labs: | | Software: | | Consulting: | |
| Manufacturing: Y | Tissue Replacement: | | Equipment/Supplies: | Y | Arrays: | | Blood Collection: | |
| Genetics: | | | Research & Development Services: | | Database Management: | | Drug Delivery: | |
| | | | Diagnostics: | Y | | | Drug Distribution: | |

## TYPES OF BUSINESS:

Sanitary & Livestock Diagnostic Products
Food Safety Test Kits
Animal Health Test Kits
Pharmacology Test Kits
Agricultural Test Kits
Veterinary Instruments
Veterinary Pharmaceuticals

## BRANDS/DIVISIONS/AFFILIATES:

Neogen Latino America SPA
International Diagnostic Systems Corp.

## CONTACTS: Note: Officers with more than one job title may be intentionally listed here more than once.

James L. Herbert, CEO
Lon M. Bohannon, COO
Lon M. Bohannon, Pres.
Richard R. Current, CFO/VP
Jennifer Rice, Sr. Science Officer
Kenneth V. Kodilla, VP-Mfg.
Richard R. Current, Sec.
Edward L. Bradley, VP-Food Safety Oper.
Anthony E. Maltese, VP-Corp. Dev.
Terri A. Morrical, VP-Animal Safety Oper.
Joseph M. Madden, VP-Scientific Affairs
Mark A. Mozola, VP-R&D
Paul S. Satoh, VP-Basic & Exploratory Research
James L. Herbert, Chmn.

| Phone: 517-372-9200 | Fax: 517-372-2006 |
|---|---|
| Toll-Free: 800-234-5333 | |
| Address: 620 Lesher Pl., Lansing, MI 48912 US | |

## GROWTH PLANS/SPECIAL FEATURES:

Neogen Corporation and its subsidiaries develop, manufacture and market a diverse line of products for food and animal safety. The company's food safety segment primarily sells diagnostic test kits and complementary products marketed to food and feed producers and processors to detect food-borne pathogens, spoilage organisms, natural toxins, food allergens, genetic modifications, ruminant by-products, drug residues, pesticide residues and general sanitation concerns. Revenue from the food safety segment generated 56.3% of the firm's total 2008 revenue. Many of its food safety test kits use immunoassay technology to rapidly detect target substances. Neogen's animal safety segment, which generated approximately 43.7% of its 2008 revenue, primarily develops, manufactures and markets pharmaceuticals, rodenticides, disinfectants, vaccines, veterinary instruments, topicals and diagnostic products to the worldwide animal safety market. Its line of approximately 100 drug detection immunoassay test kits are sold worldwide for the detection of approximately 300 abused and therapeutic drugs in racing animals, such as horses, greyhounds and camels, as well as for testing farm animals. The company has a sales network of over 120 distributors in 100 countries, with international sales generating 38.4% of the firm's 2008 revenue. The firm manufactures its products in Lansing, Michigan; Lexington, Kentucky; Randolph, Wisconsin; and Ayr, Scotland. In June 2008, the company formed a new subsidiary, Neogen Latino America SPA, headquartered in Mexico City, to distribute the company's products throughout Mexico. In July 2008, Neogen acquired a product line of 14 different product formulations from DuPont Animal Health Solutions. In May 2009, the firm acquired International Diagnostic Systems Corp., a developer, manufacturer and marketer of test kits to detect drug residues in food and animal feed, as well as drugs in forensic and animal samples.

## FINANCIALS: Sales and profits are in thousands of dollars—add 000 to get the full amount. 2008 Note: Financial information for 2008 was not available for all companies at press time.

| | | |
|---|---|---|
| 2008 Sales: $102,418 | 2008 Profits: $12,098 | U.S. Stock Ticker: NEOG |
| 2007 Sales: $86,138 | 2007 Profits: $9,125 | Int'l Ticker:     Int'l Exchange: |
| 2006 Sales: $72,433 | 2006 Profits: $7,941 | Employees:   447 |
| 2005 Sales: $62,756 | 2005 Profits: $5,916 | Fiscal Year Ends: 5/31 |
| 2004 Sales: $55,498 | 2004 Profits: $5,099 | Parent Company: |

## SALARIES/BENEFITS:

| Pension Plan: | ESOP Stock Plan: | Profit Sharing: | Top Exec. Salary: $290,000 | Bonus: $150,000 |
|---|---|---|---|---|
| Savings Plan: Y | Stock Purch. Plan: | | Second Exec. Salary: $205,000 | Bonus: $75,000 |

## OTHER THOUGHTS:

**Apparent Women Officers or Directors:** 2
**Hot Spot for Advancement for Women/Minorities:**

## LOCATIONS: ("Y" = Yes)

| West: | Southwest: | Midwest: | Southeast: | Northeast: | International: |
|---|---|---|---|---|---|
| | | Y | | | Y |

# NEOPHARM INC

## www.neopharm.com

Industry Group Code: 325412A  Ranks within this company's industry group: Sales:      Profits: 15

| Drugs: | | Other: | Clinical: | Computers: | Services: | |
|---|---|---|---|---|---|---|
| Discovery: | Y | AgriBio: | Trials/Services: | Hardware: | Specialty Services: | |
| Licensing: | | Genetic Data: | Labs: | Software: | Consulting: | |
| Manufacturing: | | Tissue Replacement: | Equipment/Supplies: | Arrays: | Blood Collection: | |
| Genetics: | | | Research & Development Services: | Database Management: | Drug Delivery: | Y |
| | | | Diagnostics: | | Drug Distribution: | |

## TYPES OF BUSINESS:

Drug Delivery Systems
Liposomal Drug Delivery System
Tumor-Targeting Toxins
Drugs-Cancer
Drugs-Autoimmune Diseases

## BRANDS/DIVISIONS/AFFILIATES:

NeoLipid
IL13-PE38QQR
LEP-ETU
LE-DT
LE-rafAON

## CONTACTS: Note: Officers with more than one job title may be intentionally listed here more than once.

Laurence P. Birch, CEO
Laurence P. Birch, Pres.
Laurence P. Birch, Acting CFO
Shahid Ali, Exec. VP-R&D
Aquilur Rahman, Chief Scientific Officer
John N. Kapoor, Chmn.

| Phone: 847-887-0800 | Fax: 847-406-1764 |
|---|---|
| Toll-Free: | |
| Address: 101 Waukegan Rd., Ste. 970, Lake Bluff, IL 60044 US | |

## GROWTH PLANS/SPECIAL FEATURES:

NeoPharm, Inc. is a biopharmaceutical company engaged in the research, development and commercialization of drugs for the treatment of cancer and autoimmune diseases. The company has built its drug portfolio based on two novel proprietary technology platforms: The NeoLipid liposomal drug delivery system and a tumor-targeting toxin platform. NeoPharm's lead drug candidate is IL13-PE38QQR (cintredekin besudotox), a tumor-targeting toxin that has completed Phase III trials for the treatment of brain cancer. Currently, patients diagnosed with glioblastoma multiformes have few effective treatment options due to the physical layout of the brain, specifically the blood-brain barrier. IL13-PE38QQR is designed with an alternative delivery system to overcome these obstacles and selectively target cancer cells in the brain without destroying normal tissue. Idiopathic Pulmonary Fibrosis, another possible indication for use with IL13-PE38QQR, also presents conditions that may be effectively treated with the drug's selective targeting. The company's other product candidates are based on its NeoLipid drug delivery platform, which combines drugs or other compounds with NeoPharm's proprietary lipids and allows for the creation of a stable liposome. This physical property is especially important during drug storage after the medication has been administered intravenously to a patient. These products include LEP-ETU, a liposomal formulation of the widely used cancer drug paclitaxel (approved and sold in the U.S. as Taxol); LE-DT, a metastatic solid cancer treatment based on docetaxel (Taxotere); LE-SN38, which uses the active ingredient in Pfizer's Camptosar for colorectal cancer; and LE-rafAON, for pancreatic and prostate cancer. In October 2008, NeoPharm signed a Cooperative Research and Development Agreement with the National Institute of Neurological Diseases and Stroke to study IL13-PE38QQR and its effects on untreatable brain diseases.

## FINANCIALS: Sales and profits are in thousands of dollars—add 000 to get the full amount. 2008 Note: Financial information for 2008 was not available for all companies at press time.

| | | |
|---|---|---|
| 2008 Sales: $ | 2008 Profits: $-8,218 | U.S. Stock Ticker: NEOL |
| 2007 Sales: $ | 2007 Profits: $-11,001 | Int'l Ticker:      Int'l Exchange: |
| 2006 Sales: $  11 | 2006 Profits: $-33,208 | Employees:     20 |
| 2005 Sales: $ 543 | 2005 Profits: $-38,724 | Fiscal Year Ends: 12/31 |
| 2004 Sales: $ 157 | 2004 Profits: $-57,609 | Parent Company: |

## SALARIES/BENEFITS:

| Pension Plan: | ESOP Stock Plan: | Profit Sharing: | Top Exec. Salary: $212,596 | Bonus: $137,500 |
|---|---|---|---|---|
| Savings Plan: Y | Stock Purch. Plan: Y | | Second Exec. Salary: $160,370 | Bonus: $42,863 |

## OTHER THOUGHTS:

Apparent Women Officers or Directors:
Hot Spot for Advancement for Women/Minorities:

## LOCATIONS: ("Y" = Yes)

| West: | Southwest: | Midwest: | Southeast: | Northeast: | International: |
|---|---|---|---|---|---|
| | | Y | | | |

Note: Financial information, benefits and other data can change quickly and may vary from those stated here.

# NEOPROBE CORPORATION

**www.neoprobe.com**

Industry Group Code: 325413  **Ranks within this company's industry group:** Sales: 19   Profits: 12

| Drugs: | Other: | Clinical: | | Computers: | Services: |
|---|---|---|---|---|---|
| Discovery: | AgriBio: | Trials/Services: | | Hardware: | Specialty Services: |
| Licensing: | Genetic Data: | Labs: | | Software: | Consulting: |
| Manufacturing: | Tissue Replacement: | Equipment/Supplies: | Y | Arrays: | Blood Collection: |
| Genetics: | | Research & Development Services: | | Database Management: | Drug Delivery: |
| | | Diagnostics: | Y | | Drug Distribution: |

## TYPES OF BUSINESS:

Supplies-Intraoperative Cancer Diagnosis
Gamma-Guided Surgical Instruments
Blood Flow Measurement Devices
Targeting Agents

## BRANDS/DIVISIONS/AFFILIATES:

Cardiosonix, Ltd.
Activated Cellular Therapy
neo2000 Gamma Detection Systems
Lymphoseek
Quantix
CIRA Biosciences, Inc.
RIGScan CR

## CONTACTS: *Note: Officers with more than one job title may be intentionally listed here more than once.*

David C. Bupp, CEO
David C. Bupp, Pres.
Brent L. Larson, CFO
Douglas Rash, VP-Mktg.
Frederick O. Cope, VP-Pharmaceutical Research & Clinical Dev.
Anthony K. Blair, VP-Mfg. Oper.
Brent L. Larson, Sec.
Brent L. Larson, VP-Finance/Treas.
Rodger A. Brown, VP-Regulatory Affairs & Quality Assurance
Julius R. Krevans, Chmn.

| **Phone:** 614-793-7500 | **Fax:** 614-793-7520 |
|---|---|
| **Toll-Free:** 800-793-0079 | |
| **Address:** 425 Metro Pl. N., Ste. 300, Dublin, OH 43017 US | |

## GROWTH PLANS/SPECIAL FEATURES:

Neoprobe Corporation is focused on developing and commercializing surgical-supportive products. The company focuses on improving cancer surgery outcomes using its gamma detection device in conjunction with radiopharmaceutical detection agents. Its lead product is the neo2000 gamma detection device, which uses its detector probes in Intraoperative Lymphmatic Mapping (ILM), enabling physicians to assess the potential spread of cancer to lymph node tissue or vital organs. Cancer utilizes the lymphatic system to rapidly spread through a patients system, but the neo2000 allows surgeons and oncologists to biopsy sentinel nodes that are possible pathway locations, which makes it easier to evaluate if the cancer has spread or not. Other possible medical applications include thyroid function evaluation and parathyroid surgery. The company, through its wholly-owned subsidiary Cardiosonix, Ltd., is also marketing its Quantix line of products, which aims to measure and evaluate blood flow using Doppler and trigonomic measures rather than invasive imaging systems. The line consists of two products: Quantix OR is designed to measure intraoperative volume blood flow in specific blood vessels within seconds; and the Quantix ND is used in neurovascular cases to measure cerebral blood flow in real time. Neoprobe also has another majority-owned subsidiary, Cira Biosciences, Inc., focused on the development and commercialization of an activated cellular therapy technology that has shown promising early stage patient-specific treatment potential in oncology, viral (HIV/AIDS and hepatitis) and autoimmune diseases. Neoprobe's investigational activities are currently focused on three particular developmental product initiatives: Lymphoseek, RIGScan CR and Activated Cellular Therapy (ACT). In March 2009, the company released a new high energy probe to detect isotopes such as Fludeoxyglucose F 18. In May 2009, the firm began a Phase III study to determine Lymphoseek's efficacy in patients with Head and Neck Squamous cell Carcinoma.

## FINANCIALS: Sales and profits are in thousands of dollars—add 000 to get the full amount. 2008 Note: Financial information for 2008 was not available for all companies at press time.

| | | |
|---|---|---|
| 2008 Sales: $7,886 | 2008 Profits: $-5,166 | **U.S. Stock Ticker:** NEOP |
| 2007 Sales: $7,125 | 2007 Profits: $-5,088 | **Int'l Ticker:**      Int'l Exchange: |
| 2006 Sales: $6,051 | 2006 Profits: $-4,741 | Employees:    25 |
| 2005 Sales: $5,919 | 2005 Profits: $-4,929 | Fiscal Year Ends: 12/31 |
| 2004 Sales: $5,953 | 2004 Profits: $-3,541 | Parent Company: |

## SALARIES/BENEFITS:

| Pension Plan: | ESOP Stock Plan: | Profit Sharing: | Top Exec. Salary: $325,000 | Bonus: $40,000 |
|---|---|---|---|---|
| Savings Plan: Y | Stock Purch. Plan: | | Second Exec. Salary: $177,000 | Bonus: $15,000 |

## OTHER THOUGHTS:

**Apparent Women Officers or Directors:**
**Hot Spot for Advancement for Women/Minorities:**

## LOCATIONS: ("Y" = Yes)

| West: | Southwest: | Midwest: | Southeast: | Northeast: | International: |
|---|---|---|---|---|---|
| | | Y | | | |

# NEUROBIOLOGICAL TECHNOLOGIES INC
### www.ntii.com

**Industry Group Code: 325412  Ranks within this company's industry group:** Sales: 109  Profits: 95

| Drugs: | | Other: | Clinical: | Computers: | Services: |
|---|---|---|---|---|---|
| Discovery: | Y | AgriBio: | Trials/Services: | Hardware: | Specialty Services: |
| Licensing: | Y | Genetic Data: | Labs: | Software: | Consulting: |
| Manufacturing: | | Tissue Replacement: | Equipment/Supplies: | Arrays: | Blood Collection: |
| Genetics: | | | Research & Development Services: | Database Management: | Drug Delivery: |
| | | | Diagnostics: | | Drug Distribution: |

## TYPES OF BUSINESS:
Pharmaceuticals Acquisition & Development
Drugs-Neurological Disease

## BRANDS/DIVISIONS/AFFILIATES:
Namenda
Viprinex
XERECEPT

## CONTACTS: *Note: Officers with more than one job title may be intentionally listed here more than once.*
Willaim A. Fletcher, Interim CEO
Matthew M. Loar, CFO/VP
Warren W. Wasiewski, Chief Medical Officer/VP
David E. Levy, VP-Clinical Dev.
Karl G. Trass, VP-Regulatory Affairs & Quality Assurance
Abraham E. Cohen, Chmn.

| Phone: 510-595-6000 | Fax: 510-595-6006 |
|---|---|
| Toll-Free: | |
| Address: 2000 Powell St., Ste. 800, Emeryville, CA 94608 US | |

## GROWTH PLANS/SPECIAL FEATURES:
Neurobiological Technologies, Inc. (NTI), is a biotechnology company engaged in the business of acquiring and developing central nervous system (CNS) related drug candidates. The company is focused on therapies for neurological conditions that occur in connection with Alzheimer's disease, dementia, neuropathic pain, ischemic stroke and brain cancer. NTI's strategy is to in-license and develop late-stage drug candidates that target major medical needs and that can be rapidly commercialized. The company's first marketable product, Memantine (created with partners Merz Pharmaceuticals GmbH of Frankfurt, Germany, and Children's Medical Center Corporation) is an oral treatment for Alzheimer's disease. Memantine is also being evaluated in the treatment of chronic conditions of neuropathic pain, dementia and acute conditions of traumatic brain injury and stroke. It is being marketed by Forest Laboratories, Inc., in the U.S and Merz & Co. in Europe, under the trade name Namenda. The firm also hopes to derive revenues from contract development services in relation to the drug XERECEPT. This drug was originally being developed by the firm prior to its sale to partner Celtic Pharmaceuticals Holdings L.P. NTI is currently evaluating XERECEPT for the treatment of peritumoral brain edema (swelling of brain tissue due to tumor), in Phase III clinical testing for Celtic Pharmaceuticals. In 2008, the firm entered into a collaboration agreement with the Buck Institute for Age Research in order to develop a new therapy for Alzheimer's disease. In January 2009, NTI discontinued the development of Viprinex, a reperfusion agent which was in Phase III trials for the treatment of acute ischemic stroke. Development was discontinued after no improvement was seen in stroke patients in clinical trials.

## FINANCIALS: Sales and profits are in thousands of dollars—add 000 to get the full amount. 2008 Note: Financial information for 2008 was not available for all companies at press time.

| | | | |
|---|---|---|---|
| 2008 Sales: $14,760 | 2008 Profits: $-16,322 | **U.S. Stock Ticker: NTII** | |
| 2007 Sales: $17,673 | 2007 Profits: $-14,127 | **Int'l Ticker:** Int'l Exchange: | |
| 2006 Sales: $12,339 | 2006 Profits: $-27,839 | Employees: 33 | |
| 2005 Sales: $3,100 | 2005 Profits: $-17,322 | Fiscal Year Ends: 6/30 | |
| 2004 Sales: $2,786 | 2004 Profits: $-1,808 | Parent Company: | |

## SALARIES/BENEFITS:

| | | | | |
|---|---|---|---|---|
| Pension Plan: | ESOP Stock Plan: | Profit Sharing: | Top Exec. Salary: $400,000 | Bonus: $100,000 |
| Savings Plan: Y | Stock Purch. Plan: Y | | Second Exec. Salary: $295,000 | Bonus: $76,250 |

## OTHER THOUGHTS:
**Apparent Women Officers or Directors:**
**Hot Spot for Advancement for Women/Minorities:**

## LOCATIONS: ("Y" = Yes)

| West: | Southwest: | Midwest: | Southeast: | Northeast: | International: |
|---|---|---|---|---|---|
| Y | | | | Y | |

Note: Financial information, benefits and other data can change quickly and may vary from those stated here.

# NEUROCRINE BIOSCIENCES INC

**www.neurocrine.com**

Industry Group Code: 325412  **Ranks within this company's industry group:** Sales: 131    Profits: 157

| Drugs: | | Other: | Clinical: | Computers: | Services: |
|---|---|---|---|---|---|
| Discovery: | Y | AgriBio: | Trials/Services: | Hardware: | Specialty Services: |
| Licensing: | Y | Genetic Data: | Labs: | Software: | Consulting: |
| Manufacturing: | | Tissue Replacement: | Equipment/Supplies: | Arrays: | Blood Collection: |
| Genetics: | | | Research & Development Services: | Database Management: | Drug Delivery: |
| | | | Diagnostics: | | Drug Distribution: |

## TYPES OF BUSINESS:

Pharmaceuticals Discovery & Development
Drugs-Neurological & Immune Disorder

## BRANDS/DIVISIONS/AFFILIATES:

Indiplon
Urocortin 2
GnRH Antagonist
CRF-R Antagonist
Elagolix

## CONTACTS: *Note: Officers with more than one job title may be intentionally listed here more than once.*

Kevin C. Gormon, CEO
Kevin C. Gormon, Pres.
Tim Coughlin, CFO/VP
Dimitri E. Grigoriadis, VP-Research
Bill Wilson, VP-IT
Margaret E. Valeur-Jensen, General Counsel/Exec. VP/Corp. Sec.
Chris O'Brien, Chief Medical Officer
Haig Bozigian, Sr. VP-Pharmaceutical & Preclinical Dev.
Joseph A. Mollica, Chmn.

| Phone: 858-617-7600 | Fax: 858-617-7601 |
|---|---|
| Toll-Free: | |
| Address: 12780 El Camino Real, San Diego, CA 92130 US | |

## GROWTH PLANS/SPECIAL FEATURES:

Neurocrine Biosciences, Inc. (NBI) is a neuroimmunology company focused on the discovery and development of therapeutics to treat diseases and disorders of the central nervous and immune systems. The company's neuroscience, endocrine and immunology research segments provide a biological understanding of the molecular interactions of those respective systems. The firm's leading drug candidate is elagolix, an orally active Gonadotropin-Releasing Hormone (GnRH) antagonist for the treatment of endometriosis, uterine fibroids and prostate cancer. The oral delivery system is preferable over injection techniques currently available, as there are fewer local side effects. The oral method also enables patients the ability to cease treatment immediately, which is not possible using injections. The drug also has possible applications in benign prostatic hyperplasia (BHP), which causes the prostate gland to enlarge. GnRH antagonists are able to limit the development of dihydrotestosterone, a major cause of BHP. Neurocrine is also developing corticotrophin-releasing factor (CRF) receptor antagonists with GlaxoSmithKline for the treatment of anxiety and depression; Urocortin 2, a CRF peptide agonist for cardiovascular issues such as congestive heart failure; a vesicular monoamine transporter 2 inhibitor for movement disorders; glucose dependent insulin secretagogues for type II diabetes; and anticonvulsants for use in bipolar disorder. The company's previous lead drug was indiplon, a treatment for insomnia, but it is currently awaiting FDA review after the organization cited safety issues in clinical trials. In May 2009, Neurocrine announced it was laying off about 50% of its staff and shifting focus to its most advanced programs.

NBI offers medical and dental coverage; health and dependent care flexible spending accounts and long term care programs; a 401(k) plan; stock options and bonuses based on performance; a 24-hour-a-day on-site workout room; and massage therapy and yoga sessions.

## FINANCIALS: Sales and profits are in thousands of dollars—add 000 to get the full amount. 2008 Note: Financial information for 2008 was not available for all companies at press time.

| | | |
|---|---|---|
| 2008 Sales: $3,975 | 2008 Profits: $-88,613 | **U.S. Stock Ticker: NBIX** |
| 2007 Sales: $1,224 | 2007 Profits: $-207,299 | **Int'l Ticker:**    Int'l Exchange: |
| 2006 Sales: $39,234 | 2006 Profits: $-107,205 | Employees:   125 |
| 2005 Sales: $123,889 | 2005 Profits: $-22,191 | Fiscal Year Ends: 12/31 |
| 2004 Sales: $85,176 | 2004 Profits: $-45,773 | Parent Company: |

## SALARIES/BENEFITS:

| Pension Plan: | ESOP Stock Plan: | Profit Sharing: | Top Exec. Salary: $440,000 | Bonus: $240,000 |
|---|---|---|---|---|
| Savings Plan: Y | Stock Purch. Plan: Y | | Second Exec. Salary: $395,000 | Bonus: $190,000 |

## OTHER THOUGHTS:

**Apparent Women Officers or Directors:** 1
**Hot Spot for Advancement for Women/Minorities:** Y

## LOCATIONS: ("Y" = Yes)

| West: | Southwest: | Midwest: | Southeast: | Northeast: | International: |
|---|---|---|---|---|---|
| Y | | | | | |

# NEUROGEN CORP

www.neurogen.com

**Industry Group Code: 325412 Ranks within this company's industry group: Sales: 137 Profits: 127**

| Drugs: | | Other: | Clinical: | Computers: | | Services: | |
|---|---|---|---|---|---|---|---|
| Discovery: | Y | AgriBio: | Trials/Services: | Hardware: | | Specialty Services: | |
| Licensing: | | Genetic Data: | Labs: | Software: | | Consulting: | |
| Manufacturing: | | Tissue Replacement: | Equipment/Supplies: | Arrays: | | Blood Collection: | |
| Genetics: | | | Research & Development Services: | Database Management: | Y | Drug Delivery: | |
| | | | Diagnostics: | | | Drug Distribution: | |

## TYPES OF BUSINESS:

Small-Molecule Drug Development
Drugs-Neurological & Psychiatric
Drugs-Pain & Inflammatory

## BRANDS/DIVISIONS/AFFILIATES:

Aplindore
Adipiplon

## CONTACTS: Note: Officers with more than one job title may be intentionally listed here more than once.

Stephen R. Davis, CEO
Stephen R. Davis, Pres.
Thomas A. Pitler, CFO
Bertrand L. Chenard, Sr. VP-Chemistry & Process Research
Jeffrey Dill, General Counsel/VP/Corp. Sec.
Kenneth J. Sprenger, VP-Clinical Dev. & Oper.
Diane Hoffman, Dir.-Bus. Dev.
Elaine Dodge, Dir.-Public Rel.
Diane Hoffman, Dir.-Investor Rel.
Thomas A. Pitler, Chief Bus. Officer
George D. Maynard, VP-Early Dev.
Craig Saxton, Chmn.

| Phone: 203-488-8201 | Fax: 203-481-4863 |
|---|---|

**Toll-Free:**

**Address:** 45 N.E. Industrial Rd., Branford, CT 06405 US

## GROWTH PLANS/SPECIAL FEATURES:

Neurogen Corp. is a small molecule drug discovery and development company. The firm focuses on new drug candidates to improve the lives of patients suffering from neurological and psychiatric disorders, such as Parkinson's disease and Restless Leg Syndrome (RLS). Due to 2008 restructuring efforts, Neurogen Corp. is focused on the development of Aplindore, the firm's lead clinical candidate. Aplindore is a D2 partial agonist, for the treatment of Parkinson's disease and RLS. The drug is in Phase II development. In addition, the company and development partner Merck & Co., Inc., have developed several vanilloid receptor-1 (VR1) receptor antagonists that are in Phase II; these are drug candidates for the treatment of cough and various pain disorders. The company has several other drug development programs, including Adipiplon, a GABA receptor partial agonist used to treat anxiety and insomnia; however, all programs other than Aplindore and that of the VR1 receptor antagonists have been indefinitely suspended due to financial concerns. In February 2008, as part of the afore mentioned restructuring, Neurogen Corp. laid off 70 employees, which represented approximately 50% of its total workforce. In April 2008, the firm laid off an additional 45 employees, or approximately 60% of the remaining workers. In November 2008, the firm agreed to sell its chemical library $3 million. In December 2008, the company agreed to sell its C5a patent estate and related assets for $2.25 million.

## FINANCIALS: Sales and profits are in thousands of dollars—add 000 to get the full amount. 2008 Note: Financial information for 2008 was not available for all companies at press time.

| | | |
|---|---|---|
| 2008 Sales: $3,000 | 2008 Profits: $-34,514 | **U.S. Stock Ticker:** NRGN |
| 2007 Sales: $15,437 | 2007 Profits: $-55,706 | **Int'l Ticker:** Int'l Exchange: |
| 2006 Sales: $9,813 | 2006 Profits: $-53,776 | Employees: 30 |
| 2005 Sales: $7,558 | 2005 Profits: $-37,120 | Fiscal Year Ends: 12/31 |
| 2004 Sales: $19,180 | 2004 Profits: $-18,593 | Parent Company: |

## SALARIES/BENEFITS:

| Pension Plan: | ESOP Stock Plan: | Profit Sharing: | Top Exec. Salary: $421,566 | Bonus: $164,900 |
|---|---|---|---|---|
| Savings Plan: Y | Stock Purch. Plan: | | Second Exec. Salary: $273,648 | Bonus: $45,758 |

## OTHER THOUGHTS:

**Apparent Women Officers or Directors**: 2
**Hot Spot for Advancement for Women/Minorities**:

## LOCATIONS: ("Y" = Yes)

| West: | Southwest: | Midwest: | Southeast: | Northeast: | International: |
|---|---|---|---|---|---|
| | | | | Y | |

# NEXMED INC

**www.nexmed.com**

Industry Group Code: 325412A  **Ranks within this company's industry group:** Sales: 22  Profits: 14

| Drugs: | | Other: | | Clinical: | | Computers: | | Services: | |
|---|---|---|---|---|---|---|---|---|---|
| Discovery: | | AgriBio: | | Trials/Services: | | Hardware: | | Specialty Services: | |
| Licensing: | Y | Genetic Data: | | Labs: | | Software: | | Consulting: | |
| Manufacturing: | Y | Tissue Replacement: | | Equipment/Supplies: | Y | Arrays: | | Blood Collection: | |
| Genetics: | | | | Research & Development Services: | | Database Management: | | Drug Delivery: | Y |
| | | | | Diagnostics: | | | | Drug Distribution: | |

## TYPES OF BUSINESS:

Drug Delivery Systems
Transdermal Drug Delivery Systems
Sexual Dysfunction Products
Medical Devices

## BRANDS/DIVISIONS/AFFILIATES:

NexACT
Befar
Femprox
Alprox-TD

## CONTACTS: Note: Officers with more than one job title may be intentionally listed here more than once.

Vivian Liu, CEO
Hem Pandya, COO/VP
Vivian Liu, Pres.
Mark Westgate, CFO/VP
Vivian Liu, Corp. Sec.
Mark Westgate, Treas.
Richard J. Berman, Chmn.

| **Phone:** 609-208-9688 | **Fax:** 609-208-1868 |
|---|---|
| **Toll-Free:** | |
| **Address:** 350 Corporate Blvd., Robbinsville, NJ 08691 US | |

## GROWTH PLANS/SPECIAL FEATURES:

NexMed, Inc., is a pharmaceutical and medical technology company with a focus on developing and commercializing therapeutic products based on proprietary delivery systems. The company currently focuses on new and patented topical pharmaceutical products based on a penetration enhancement drug delivery technology called NexACT. NexACT facilitates an active drug's absorption through the skin. It develops topical treatments in forms including cream, gel, patch and tape, and has 13 U.S. patents in connection with its NexACT technology. NexMed's principal product candidates include Vitaros (formerly Alprox-TD) for the treatment of male erectile dysfunction (ED); and Femprox cream for female sexual arousal disorder. NexMed, in partnership with Novartis International Pharmaceutical Ltd., is in Phase III trials for developing a topical treatment for nail fungus currently called NM100060. In January 2009, the firm signed product development agreement with Pharmaceutics International, Inc. (PII), giving PII the right to use NexACT technology on some of its products. In February 2009, NexMed sold the exclusive U.S. rights to Vitaros to Warner Chilcott Company, Inc., a subsidiary of former Vitaros licensing partner Warner Chilcott, Ltd.

The company offers employees medical, dental and vision coverage; a 401(k); life, AD&D, and disability insurance; and an employee stock plan.

## FINANCIALS: Sales and profits are in thousands of dollars—add 000 to get the full amount. 2008 Note: Financial information for 2008 was not available for all companies at press time.

| | | |
|---|---|---|
| 2008 Sales: $5,957 | 2008 Profits: $-5,171 | **U.S. Stock Ticker: NEXM** |
| 2007 Sales: $1,270 | 2007 Profits: $-8,787 | **Int'l Ticker:**    Int'l Exchange: |
| 2006 Sales: $1,867 | 2006 Profits: $-8,043 | Employees:    13 |
| 2005 Sales: $2,399 | 2005 Profits: $-15,442 | Fiscal Year Ends: 12/31 |
| 2004 Sales: $ 359 | 2004 Profits: $-17,024 | Parent Company: |

## SALARIES/BENEFITS:

| Pension Plan: | ESOP Stock Plan: | Profit Sharing: | Top Exec. Salary: $300,000 | Bonus: $ |
|---|---|---|---|---|
| Savings Plan: Y | Stock Purch. Plan: | | Second Exec. Salary: $235,000 | Bonus: $ |

## OTHER THOUGHTS:

**Apparent Women Officers or Directors:** 1
**Hot Spot for Advancement for Women/Minorities:**

## LOCATIONS: ("Y" = Yes)

| West: | Southwest: | Midwest: | Southeast: | Northeast: | International: |
|---|---|---|---|---|---|
| | | | | Y | Y |

# NORTH AMERICAN SCIENTIFIC

## www.nasmedical.com

**Industry Group Code: 33911  Ranks within this company's industry group:** Sales: 15  Profits: 15

| Drugs: | Other: | Clinical: | | Computers: | | Services: |
|---|---|---|---|---|---|---|
| Discovery: | AgriBio: | Trials/Services: | | Hardware: | | Specialty Services: |
| Licensing: | Genetic Data: | Labs: | | Software: | Y | Consulting: |
| Manufacturing: | Tissue Replacement: | Equipment/Supplies: | Y | Arrays: | | Blood Collection: |
| Genetics: | | Research & Development Services: | | Database Management: | | Drug Delivery: |
| | | Diagnostics: | Y | | | Drug Distribution: |

## TYPES OF BUSINESS:

Radiation Therapy Products
Radioisotopic Products
Brachytherapy Products
Radiation Therapy Software
NAS Medical

## BRANDS/DIVISIONS/AFFILIATES:

Prospera
SurTRAK
STP-110 Precision Stepper
RTP-6000 Precision Stabilizer
Horizontal Needle Box
ClearPath

## CONTACTS: *Note: Officers with more than one job title may be intentionally listed here more than once.*

John B. Rush, CEO
Troy Barring, COO/Sr. VP
John B. Rush, Pres.
Brett L. Scott, CFO
Robin Gibbs, VP-Mktg.
L. Michael Cutrer, CTO/Exec. VP
Gary N. Wilner, Chmn.

| **Phone:** 818-734-8600 | **Fax:** 818-734-5200 |
|---|---|
| **Toll-Free:** 866-589-6274 | |
| **Address:** 20200 Sunburst St., Chatsworth, CA 91311 US | |

## GROWTH PLANS/SPECIAL FEATURES:

North American Scientific, Inc. (NASI), which does business as NAS Medical, designs, develops, produces and sells products for radiation therapy, focusing primarily on localized delivery radiation called brachytherapy. Brachytherapy is a minimally invasive medical procedure in which high-dose-rate (HDR) or low-dose-rate (LDR) sealed radioactive sources are temporarily or permanently implanted into cancerous tissue, delivering a therapeutically prescribed dose of radiation that is lethal to the cancerous tissue. NASI manufactures both iodine- and palladium-based brachytherapy seeds, the two most commonly used seeds for the treatment of prostate cancer. The brachytherapy seeds include Prospera I-125, Prospera Pd-103 and SurTRAK. The SurTRAK family of products includes the SurTRAK pre-plugged needle with a synthetic, micro-angled insert and the SurTRAK strand which, used in conjunction with the needle, is bio-resorbable and designed to hold the seeds at predetermined distances. NASI's products also include the STP-110 Precision Stepper and RTP-6000 Precision Stabilizer, which positions and holds the trans-rectal ultrasound probe during the LDR brachytherapy procedure. Other products include radiation shielding and needle loading accessories such as the Horizontal Needle Box, Needle Loading Shield, Needle Loading Box and Needle Loading Carousel. The company's brand Prospera is used in the treatment of ocular melanoma and other solid tumor applications. NASI directly markets this product line to ophthalmologists and medical physicists. The ClearPath system is a treatment that is placed through a single incision and is designed to conform to the resection cavity, allowing for more conformal therapeutic radiation dose distribution following lumpectomy. NASI also offers non-therapeutic products such as radiation calibration and reference source products, which it sells to certain government agency contractors and laboratories. In September 2008, the company sold its non-therapeutic product line to Eckert & Ziegler Isotope Products, Inc. In February 2009, the firm agreed to sell its prostate brachytherapy product line to Best Theratronics, Ltd.

## FINANCIALS: Sales and profits are in thousands of dollars—add 000 to get the full amount. 2008 Note: Financial information for 2008 was not available for all companies at press time.

| | | |
|---|---|---|
| 2008 Sales: $13,927 | 2008 Profits: $-10,945 | **U.S. Stock Ticker:** NASI |
| 2007 Sales: $15,317 | 2007 Profits: $-20,998 | **Int'l Ticker:**  Int'l Exchange: |
| 2006 Sales: $12,594 | 2006 Profits: $-17,130 | **Employees:** 84 |
| 2005 Sales: $12,032 | 2005 Profits: $-55,513 | **Fiscal Year Ends:** 10/31 |
| 2004 Sales: $24,737 | 2004 Profits: $-36,307 | **Parent Company:** |

## SALARIES/BENEFITS:

| Pension Plan: | ESOP Stock Plan: | Profit Sharing: | Top Exec. Salary: $310,708 | Bonus: $50,366 |
|---|---|---|---|---|
| Savings Plan: Y | Stock Purch. Plan: Y | | Second Exec. Salary: $279,059 | Bonus: $ |

## OTHER THOUGHTS:

**Apparent Women Officers or Directors:**
**Hot Spot for Advancement for Women/Minorities:**

## LOCATIONS: ("Y" = Yes)

| West: | Southwest: | Midwest: | Southeast: | Northeast: | International: |
|---|---|---|---|---|---|
| Y | | | | Y | Y |

Note: Financial information, benefits and other data can change quickly and may vary from those stated here.

# NOVARTIS AG

**www.novartis.com**

**Industry Group Code:** 325412 **Ranks within this company's industry group:** Sales: 4 Profits: 3

| Drugs: | | Other: | Clinical: | Computers: | Services: | |
|---|---|---|---|---|---|---|
| Discovery: | Y | AgriBio: | Trials/Services: | Hardware: | Specialty Services: | |
| Licensing: | Y | Genetic Data: | Labs: | Software: | Consulting: | |
| Manufacturing: | Y | Tissue Replacement: | Equipment/Supplies: | Arrays: | Blood Collection: | |
| Genetics: | Y | | Research & Development Services: | Database Management: | Drug Delivery: | |
| | | | Diagnostics: | | Drug Distribution: | |

## TYPES OF BUSINESS:

Drugs-Diversified
Therapeutic Drug Discovery
Therapeutic Drug Manufacturing
Generic Drugs
Over-the-Counter Drugs
Ophthalmic Products
Nutritional Products
Veterinary Products

## BRANDS/DIVISIONS/AFFILIATES:

CIBA Vision
Chiron Corp
Sandoz
Novartis Institute for Biomedical Research Inc
Novartis Oncology
Alcon Inc

## CONTACTS: *Note: Officers with more than one job title may be intentionally listed here more than once.*

Daniel Vasella, CEO
Joerg Reinhardt, COO
Daniel Vasella, Pres.
Raymund Breu, CFO
Jurgen Brokatzky-Geiger, Head-Human Resources
Paul Herrling, Head-Corp. Research
Leon V. Schumacher, CIO
Thomas Werlen, General Counsel
Ann Bailey, Head-Corp. Comm.
George Gunn, CEO-Consumer Health
Joseph Jimenez, CEO-Pharmaceuticals
Mark Fishman, Pres., Novartis Institute for Biomedical Research
Jeffrey George, CEO-Sandoz
Daniel Vasella, Chmn.

| **Phone:** 41-61-324-1111 | **Fax:** 41-61-324-8001 |
|---|---|
| **Toll-Free:** | |
| **Address:** Lichtstrasse 35, Basel, 4056 Switzerland | |

## GROWTH PLANS/SPECIAL FEATURES:

Novartis AG researches and develops pharmaceuticals as well as a large number of consumer and animal health products. It operates in four segments: pharmaceuticals; vaccines and diagnostics; consumer health; and Sandoz. The pharmaceuticals division, which accounts for 64% of sales, develops, manufactures, distributes and sells prescription medications in a variety of areas, which include cardiovascular and metabolism, oncology and hematology, neuroscience and ophthalmics, respiratory, immunology and infectious diseases. The segment is organized into global business franchises responsible for the marketing of various products as well as a business unit called Novartis Oncology, responsible for the global development and marketing of oncology products. The vaccines and diagnostics division is focused on the development of preventive vaccine treatments and diagnostic tools. It has two activities: Novartis Vaccines, whose key products include meningococcal and travel vaccines; and Chiron, a blood testing and molecular diagnostics activity dedicated to preventing the spread of infectious diseases through the development of blood-screening tools that protect the world's blood supply. The Sandoz division is a global generic pharmaceuticals company that develops, produces and markets drugs along with pharmaceutical and biotechnological active substances. The segment has activities in retail generics, anti-infectives and biopharmaceuticals. Sandoz offers some 950 compounds in more than 5,000 forms in 130 countries. The most important product groups include antibiotics, treatments for the central nervous system disorders, gastrointestinal medicines, cardiovascular treatments and hormone therapies. The consumer health division consists of three business units: over-the-counter medicines; animal health, which provides veterinary products for farm and companion animals; and CIBA Vision, which markets contact lenses and lens care products. In July 2008, the company raised its stake in Switzerland-based eye car firm Alcon, Inc. to 48%, for approximately $10.4 billion. Novartis has the option to acquire the remaining 52% of Alcon, from Nestle SA, during 2010 or 2011.

## FINANCIALS: Sales and profits are in thousands of dollars—add 000 to get the full amount. 2008 Note: Financial information for 2008 was not available for all companies at press time.

| | | |
|---|---|---|
| 2008 Sales: $41,459,000 | 2008 Profits: $8,233,000 | **U.S. Stock Ticker:** NVS |
| 2007 Sales: $38,072,000 | 2007 Profits: $11,968,000 | **Int'l Ticker:** NOVN   Int'l Exchange: Zurich-SWX |
| 2006 Sales: $34,393,000 | 2006 Profits: $7,202,000 | Employees: 96,717 |
| 2005 Sales: $31,005,000 | 2005 Profits: $6,141,000 | Fiscal Year Ends: 12/31 |
| 2004 Sales: $27,126,000 | 2004 Profits: $5,601,000 | Parent Company: |

## SALARIES/BENEFITS:

| Pension Plan: Y | ESOP Stock Plan: | Profit Sharing: | Top Exec. Salary: $ | Bonus: $ |
|---|---|---|---|---|
| Savings Plan: | Stock Purch. Plan: | | Second Exec. Salary: $ | Bonus: $ |

## OTHER THOUGHTS:

**Apparent Women Officers or Directors:** 4
**Hot Spot for Advancement for Women/Minorities:** Y

## LOCATIONS: ("Y" = Yes)

| West: | Southwest: | Midwest: | Southeast: | Northeast: | International: |
|---|---|---|---|---|---|
| Y | | | | Y | Y |

# NOVAVAX INC

www.novavax.com

**Industry Group Code: 325412A  Ranks within this company's industry group:**  Sales: 25  Profits: 23

| Drugs: | | Other: | | Clinical: | | Computers: | | Services: | |
|---|---|---|---|---|---|---|---|---|---|
| Discovery: | Y | AgriBio: | | Trials/Services: | | Hardware: | | Specialty Services: | |
| Licensing: | | Genetic Data: | | Labs: | | Software: | | Consulting: | |
| Manufacturing: | Y | Tissue Replacement: | | Equipment/Supplies: | | Arrays: | | Blood Collection: | |
| Genetics: | | | | Research & Development Services: | Y | Database Management: | | Drug Delivery: | Y |
| | | | | Diagnostics: | | | | Drug Distribution: | |

## TYPES OF BUSINESS:

Drug Delivery Systems
Drugs-Bacterial & Viral Infection
Hormone Replacement Therapies
Contract Research Services

## BRANDS/DIVISIONS/AFFILIATES:

Estrasorb
Androsorb
Micellar Nanoparticle Technology
VLP Technology

## CONTACTS: *Note: Officers with more than one job title may be intentionally listed here more than once.*

Rahul Singhvi, CEO
Rahul Singhvi, Pres.
Gale Smith, VP-Vaccine Dev.
James Robinson, VP-Tech. & Quality Oper.
Thomas S. Johnston, VP-Strategy
Raymond J. Hage, Jr., Sr. VP-Commercial Oper.
Penny M. Heaton, Chief Medical Officer/VP
John Lambert, Chmn.

| Phone: 240-268-2000 | Fax: |
|---|---|
| Toll-Free: | |
| Address: 9920 Belward Campus Dr., Rockville, MD 20850 US | |

## GROWTH PLANS/SPECIAL FEATURES:

Novavax, Inc. is a clinical-stage biopharmaceutical company focused on creating vaccines that improve upon current preventive options for a range of infectious diseases. These vaccines leverage the company's virus-like particle (VLP) platform technology. VLPs imitate the three-dimensional structures of viruses but are composed of recombinant proteins believed to be incapable of causing infection and disease. The company is initially focused on the pandemic influenza virus and seasonal flu development programs. The firm also has several additional products in various stages of development including seasonal influenza vaccine; varicella zoster (also known as shingles) vaccine; and undisclosed disease target vaccines. The company's research and development efforts are advancing industry standards for vaccine and drug delivery systems. The firm's drug delivery platform based on micellar nanoparticles (MNPs), proprietary oil and water nanoemulsions used for the topical delivery of drugs. The MNP technology uses oil and water nanoemulsions to encapsulate alcohol-soluble drugs and deliver them directly into the bloodstream through topical lotions. This same transdermal technology is being used for future products including Androsorb, a testosterone lotion used to treat female sexual dysfunction. This patented delivery platform is envisioned to support a range of drugs and therapeutics, including hormones, antibacterials and antiviral products, as well as central nervous system drugs, anti-inflammatory agents and advanced topical analgesics. The MNP technology was the basis for Novavax's first FDA-approved estrogen replacement product, Estrasorb. The drug offers a high level of efficacy in treating hot flashes, with additional advantages including lower incidence of irritation and nausea. Novavax is also researching virus-like particles, employing non-infectious protein particulate structures to elicit immune system responses. These systems are being explored for potential treatment of infectious diseases including HIV, SARS and influenza.

The firm offers employees medical, dental and vision coverage; a 401(k) plan; flexible spending accounts; stock options; and an employee referral program.

## FINANCIALS:  Sales and profits are in thousands of dollars—add 000 to get the full amount. 2008 Note: Financial information for 2008 was not available for all companies at press time.

| | | | |
|---|---|---|---|
| 2008 Sales: $1,064 | 2008 Profits: $-36,049 | **U.S. Stock Ticker: NVAX** | |
| 2007 Sales: $1,513 | 2007 Profits: $-34,765 | **Int'l Ticker:**  Int'l Exchange: | |
| 2006 Sales: $1,738 | 2006 Profits: $-23,068 | Employees:  70 | |
| 2005 Sales: $5,343 | 2005 Profits: $-11,174 | Fiscal Year Ends: 12/31 | |
| 2004 Sales: $8,260 | 2004 Profits: $-25,920 | Parent Company: | |

## SALARIES/BENEFITS:

| Pension Plan: | ESOP Stock Plan: | Profit Sharing: | Top Exec. Salary: $406,253 | Bonus: $ |
|---|---|---|---|---|
| Savings Plan: Y | Stock Purch. Plan: Y | | Second Exec. Salary: $285,405 | Bonus: $ |

## OTHER THOUGHTS:

**Apparent Women Officers or Directors:** 1
**Hot Spot for Advancement for Women/Minorities:**

## LOCATIONS: ("Y" = Yes)

| West: | Southwest: | Midwest: | Southeast: | Northeast: | International: |
|---|---|---|---|---|---|
| Y | | | | Y | |

Note: Financial information, benefits and other data can change quickly and may vary from those stated here.

# NOVEN PHARMACEUTICALS

www.noven.com

Industry Group Code: 325412A  Ranks within this company's industry group:  Sales: 12    Profits: 8

| Drugs: | | Other: | Clinical: | Computers: | Services: | |
|---|---|---|---|---|---|---|
| Discovery: | Y | AgriBio: | Trials/Services: | Hardware: | Specialty Services: | |
| Licensing: | | Genetic Data: | Labs: | Software: | Consulting: | |
| Manufacturing: | Y | Tissue Replacement: | Equipment/Supplies: | Arrays: | Blood Collection: | |
| Genetics: | | | Research & Development Services: | Database Management: | Drug Delivery: | Y |
| | | | Diagnostics: | | Drug Distribution: | |

## TYPES OF BUSINESS:

Drug Delivery Systems
Hormone Replacement Products
Pain Management Products
Central Nervous System Products
Transdermal Drug Delivery Systems

## BRANDS/DIVISIONS/AFFILIATES:

Vivelle
Menorest
Daytrana
DOT Matrix
Estalis
Noven Therapeutics
Estradot
Stavzor

## CONTACTS: Note: Officers with more than one job title may be intentionally listed here more than once.

Peter C. Brandt, CEO
Peter C. Brandt, Pres.
Michael D. Price, CFO/VP
Anthony Venditti, VP-Mktg. & Sales
Carolyn Donaldson, VP-Human Resources
Steven F. Dinh, Chief Scientific Officer/VP
Juan A. Mantelle, CTO/VP
Joel S. Lippman, VP-Clinical Dev./Chief Medical Officer
Jeff T. Mihm, General Counsel/VP
Richard Gilbert, VP-Oper.
Paven Handa, VP-Bus. Dev.
Joseph C. Jones, VP-Corp. Affairs
Jeffrey F. Eisenberg, Exec. VP
Peter G. Amanatides, VP-Quality Assurance & Quality Control
Wayne P. Yetter, Chmn.

| Phone: 305-253-5099 | Fax: 305-251-1887 |
|---|---|
| Toll-Free: | |
| Address: 11960 S.W. 144th St., Miami, FL 33186 US | |

## GROWTH PLANS/SPECIAL FEATURES:

Noven Pharmaceuticals, Inc. develops and manufactures advanced transdermal drug delivery systems and prescription transdermal products. Its principal commercialized products are transdermal drug delivery systems designed with its DOT Matrix technology for use in hormone replacement therapy. The firm's first product was an estrogen patch for the treatment of menopausal symptoms, marketed under the name Vivelle in the U.S. and Canada, and under the name Menorest in Europe and other markets. The company also launched the smallest transdermal estrogen patch ever approved by the U.S. FDA, Vivelle-Dot, and markets the product in several foreign countries under the name Estradot. The firm also markets its Lidocaine/DentiPatch for dental pain associated with dental procedures. Other Noven products include Pexeva for depression, panic disorder, OCD and GAD; Lithobid for bipolar disorder; Stavzor for bipolar disorder, migraine and epilepsy; and Daytrana for ADHD. The company also has four products in development: Lithium QD for bipolar disorder; Stavzor ER (extended release) for bipolar disorder, migraines and epilepsy; Mesafem for hot flashes; and a treatment for ADHD. Another transdermal for testosterone is currently on hold. Noven has partial ownership of a joint venture formed in partnership with Novartis Pharmaceuticals, Vivelle Ventures, doing business under the name Novogyne Pharmaceuticals. Novogyne markets Vivelle, Vivelle-Dot and CombiPatch (a combination estrogen/progestin transdermal patch for the treatment of menopausal symptoms) in the U.S. Beyond Novartis, Noven collaborates with a number of other pharmaceutical companies, such as Aventis, Shire, and Procter & Gamble.

## FINANCIALS: Sales and profits are in thousands of dollars—add 000 to get the full amount. 2008 Note: Financial information for 2008 was not available for all companies at press time.

| | | |
|---|---|---|
| 2008 Sales: $108,175 | 2008 Profits: $21,412 | U.S. Stock Ticker: NOVN |
| 2007 Sales: $83,161 | 2007 Profits: $-45,376 | Int'l Ticker:    Int'l Exchange: |
| 2006 Sales: $60,689 | 2006 Profits: $15,988 | Employees:   610 |
| 2005 Sales: $52,532 | 2005 Profits: $9,972 | Fiscal Year Ends: 12/31 |
| 2004 Sales: $45,891 | 2004 Profits: $11,224 | Parent Company: |

## SALARIES/BENEFITS:

| Pension Plan: | ESOP Stock Plan: | Profit Sharing: | Top Exec. Salary: $483,524 | Bonus: $182,500 |
|---|---|---|---|---|
| Savings Plan: Y | Stock Purch. Plan: | | Second Exec. Salary: $422,500 | Bonus: $365,625 |

## OTHER THOUGHTS:

Apparent Women Officers or Directors: 1
Hot Spot for Advancement for Women/Minorities: Y

## LOCATIONS: ("Y" = Yes)

| West: | Southwest: | Midwest: | Southeast: | Northeast: | International: |
|---|---|---|---|---|---|
| | | | Y | | |

# NOVO-NORDISK AS
## www.novonordisk.com

**Industry Group Code: 325412 Ranks within this company's industry group: Sales: 21    Profits: 19**

| Drugs: | | Other: | Clinical: | Computers: | Services: | |
|---|---|---|---|---|---|---|
| Discovery: | Y | AgriBio: | Trials/Services: | Hardware: | Specialty Services: | |
| Licensing: | | Genetic Data: | Labs: | Software: | Consulting: | |
| Manufacturing: | Y | Tissue Replacement: | Equipment/Supplies: | Arrays: | Blood Collection: | |
| Genetics: | | | Research & Development Services: | Database Management: | Drug Delivery: | Y |
| | | | Diagnostics: | | Drug Distribution: | |

## TYPES OF BUSINESS:
Drugs-Diabetes
Hormone Replacement Therapy
Growth Hormone Drugs
Hemophilia Drugs
Insulin Delivery Systems
Educational & Training Services

## BRANDS/DIVISIONS/AFFILIATES:
Levemir
Insulatard
NovoNorm
FlexPen
Norditropin
NovoSeven
Estrofem

## CONTACTS: *Note: Officers with more than one job title may be intentionally listed here more than once.*
Lars R. Sorensen, CEO
Kare Schultz, COO/Exec. VP
Lars R. Sorensen, Pres.
Jesper Brandgaard, CFO/Exec. VP
Lise Kingo, Chief of Staff/Exec. VP
Mads K. Thomsen, Chief Science Officer/Exec. VP
Mads Veggerby Lausten, Head-Investor Rel.
Goran A. Ando, Vice Chmn.
Sten Scheibye, Chmn.
Hans Rommer, Dir.-Investor Rel., North America

| Phone: 45-4444-8888 | Fax: 45-4449-0555 |
|---|---|
| Toll-Free: | |
| Address: Novo Alle, Bagsvaerd, 2880 Denmark | |

## GROWTH PLANS/SPECIAL FEATURES:
Novo Nordisk A/S is a healthcare company that focuses on developing treatments for diabetes, hemostasis management and hormone therapy. It has two segments: Diabetes care, which generated 73% of 2008 sales; and biopharmaceuticals, 27%. The diabetes care segment manages the firm's insulin franchise, including modern insulins, human insulins, insulin-related sales and oral antidiabetic drugs. Specific products include Levemir and NovoRapid modern insulin; Insulatard and Actrapid human insulin; NovoNorm, an oral antidiabetic drug; NovoPen 4, FlexPen and Innolet insulin injectors; and GlucaGen and NovoFine (needles) diabetic devices. The biopharmaceuticals segment covers hemostasis management, growth hormone therapy, hormone replacement therapy, inflammation therapy and other therapy areas. Specific products include the following. Norditropin is a premixed liquid growth hormone designed to provide a very flexible and accurate dosing, while NordiFlex, NordiFlex PenMate and NordiLet are human growth hormone injection systems. NovoSeven is a hemostasis management product, a hemophilia treatment consisting of a recombinant coagulation factor that enables coagulation to proceed in the absence of natural blood factors. Lastly, the firm's post-menopausal hormone replacement therapy products include Activelle, Novofem, Estrofem and Vagifem. Novo Nordisk has employees in 81 countries and its products are marketed in 179 countries. Besides its products, the company offers educational services and training materials for both patients and health care professionals. Novo Nordisk owns and operates dedicated research centers in the U.S., Denmark and China, as well as clinical development centers in Zurich, Switzerland; Beijing, China; Tokyo, Japan; and in the U.S. in Princeton, New Jersey and Seattle, Washington. In November 2008, the company announced plans to invest $400 million in a new insulin production plant in Tianjin, China, expected to be operational by 2012.

Novo Nordisk offers its U.S. employees health, life, dental and supplemental insurance, as well as tuition reimbursement, among other benefits.

## FINANCIALS: Sales and profits are in thousands of dollars—add 000 to get the full amount. 2008 Note: Financial information for 2008 was not available for all companies at press time.
| | | |
|---|---|---|
| 2008 Sales: $8,239,030 | 2008 Profits: $1,744,460 | **U.S. Stock Ticker: NVO** |
| 2007 Sales: $8,190,000 | 2007 Profits: $1,670,000 | **Int'l Ticker: NOVO B**   Int'l Exchange: Copenhagen-CSE |
| 2006 Sales: $6,913,700 | 2006 Profits: $1,126,020 | Employees: 26,000 |
| 2005 Sales: $5,446,472 | 2005 Profits: $946,073 | Fiscal Year Ends: 12/31 |
| 2004 Sales: $5,324,285 | 2004 Profits: $859,229 | Parent Company: |

## SALARIES/BENEFITS:
| Pension Plan: | ESOP Stock Plan: | Profit Sharing: | Top Exec. Salary: $ | Bonus: $ |
|---|---|---|---|---|
| Savings Plan: Y | Stock Purch. Plan: | | Second Exec. Salary: $ | Bonus: $ |

## OTHER THOUGHTS:
**Apparent Women Officers or Directors:** 3
**Hot Spot for Advancement for Women/Minorities:** Y

## LOCATIONS: ("Y" = Yes)
| West: | Southwest: | Midwest: | Southeast: | Northeast: | International: |
|---|---|---|---|---|---|
| Y | | | | Y | Y |

Note: Financial information, benefits and other data can change quickly and may vary from those stated here.

# NOVOZYMES

**www.novozymes.com**

Industry Group Code: 325414  **Ranks within this company's industry group:** Sales: 3  Profits: 2

| Drugs: | Other: | | Clinical: | Computers: | Services: |
|---|---|---|---|---|---|
| Discovery: | AgriBio: | Y | Trials/Services: | Hardware: | Specialty Services: |
| Licensing: | Genetic Data: | Y | Labs: | Software: | Consulting: |
| Manufacturing: | Tissue Replacement: | | Equipment/Supplies: | Arrays: | Blood Collection: |
| Genetics: | | | Research & Development Services: | Database Management: | Drug Delivery: |
| | | | Diagnostics: | | Drug Distribution: |

## TYPES OF BUSINESS:

Industrial Enzyme & Microorganism Production
Biopharmaceuticals
Enzymes
Microbiology

## BRANDS/DIVISIONS/AFFILIATES:

Mannaway
Stainzyme Plus
Spirizyme Ultra
Sucrozyme
Acrylaway
Saczyme
CellPrimeTM rTransferrin AF
RONOZYME ProAct

## CONTACTS: Note: Officers with more than one job title may be intentionally listed here more than once.

Steen Riisgaard, CEO
Steen Riisgaard, Pres.
Benny Loft, CFO/Exec. VP
Henrik Meyer, VP-Mktg.
Per Falholt, Chief Scientific Officer/Exec. VP
Kristian Merser, VP-Legal Affairs
Rasmus von Gottberg, VP-Bus. Dev. & Acquisitions
Thomas Nagy, Exec. VP-Stakeholder Rel.
Mads Bodenhoff, VP-Finance
Thomas Videbaek, Exec. VP-Bio Bus.
Peder Holk Nielsen, Exec. VP/Head-Enzyme Bus.
Michael Fredskov Christiansen, Regional Pres., China
Erik Gormsen, VP-R&D
Henrik Gurtler, Chmn.
Pedro Luiz Fernandes, Regional Pres., Brazil
Anders Spohr, VP-Supply Chain Oper.

| Phone: 45-44-46-00-00 | Fax: 45-44-46-99-99 |
|---|---|
| Toll-Free: | |
| Address: Krogshoejvej 36, Bagsvaerd, 2880 Denmark | |

## GROWTH PLANS/SPECIAL FEATURES:

Novozymes is a biotechnology company that specializes in microbiology and enzymes. The firm currently sells over 700 products in 130 countries worldwide. It splits its business into two main areas: Enzyme Business and BioBusiness. The Enzyme Business, which accounts for over 90% of sales, is split into technical, food and feed enzymes. The company's technical enzymes are used in detergents, such as Mannaway and Stainzyme Plus enzymes; transform starch in sugar for starch and fuel industries; and are applied in the textile, leather, forestry and alcohol industries. Technical enzymes include the brand name enzymes Spirizyme Ultra and Sucrozyme. The company's food enzymes increase the quality or production efficiency in the production of food products such as bread, wine, juice, beer, noodles, alcohol and pasta. Food enzyme products include Acrylaway, Saczyme and Viscoferm. Feed enzymes, such as Ronozyme NP, are designed to increase the nutritional value of feed and improve phosphorus absorption in animals. This leads to faster growth of animals, while improving the environment by decreasing the phosphorus released through manure. The firm's BioBusiness includes microorganisms and biopharmaceutical ingredients. Microorganisms have three main applications: wastewater treatment, cleaning products and natural growth enhancements for plants and turf grass. Biopharmaceutical products, such as proteins, are provided by the company as a replacement of human/animal substances, lowering the risk of disease transfer in pharmaceutical products. In June 2008, the firm announced plans to build a new enzyme production facility in Nebraska. In October 2008, Novozymes and DSM Nutritional Products launched RONOZYME ProAct, a pure protease for poultry consumption. In November 2008, the firm expanded its Chinese enzyme fermentation facility. In April 2009, Novozymes and Millipore launched CellPrimeTM rTransferrin AF , an animal-free cell culture supplement. In May 2009, the company opened two new buildings at its research and development facility in California.

## FINANCIALS: Sales and profits are in thousands of dollars—add 000 to get the full amount. 2008 Note: Financial information for 2008 was not available for all companies at press time.

| | | |
|---|---|---|
| 2008 Sales: $1,453,610 | 2008 Profits: $189,510 | **U.S. Stock Ticker:** |
| 2007 Sales: $1,327,270 | 2007 Profits: $185,940 | **Int'l Ticker: NZYM**  Int'l Exchange: Copenhagen-CSE |
| 2006 Sales: $1,251,490 | 2006 Profits: $167,610 | Employees:  5,000 |
| 2005 Sales: $1,079,840 | 2005 Profits: $148,025 | Fiscal Year Ends: 12/31 |
| 2004 Sales: $1,029,470 | 2004 Profits: $133,240 | Parent Company: |

## SALARIES/BENEFITS:

| Pension Plan: Y | ESOP Stock Plan: | Profit Sharing: | Top Exec. Salary: $ | Bonus: $ |
|---|---|---|---|---|
| Savings Plan: Y | Stock Purch. Plan: | | Second Exec. Salary: $ | Bonus: $ |

## OTHER THOUGHTS:

**Apparent Women Officers or Directors:** 1
**Hot Spot for Advancement for Women/Minorities:** Y

## LOCATIONS: ("Y" = Yes)

| West: | Southwest: | Midwest: | Southeast: | Northeast: | International: |
|---|---|---|---|---|---|
| Y | | | | Y | Y |

# NPS PHARMACEUTICALS INC

www.npsp.com

**Industry Group Code: 325412  Ranks within this company's industry group:** Sales: 66  Profits: 124

| Drugs: | | Other: | | Clinical: | Computers: | Services: |
|---|---|---|---|---|---|---|
| Discovery: | Y | AgriBio: | | Trials/Services: | Hardware: | Specialty Services: |
| Licensing: | Y | Genetic Data: | | Labs: | Software: | Consulting: |
| Manufacturing: | Y | Tissue Replacement: | | Equipment/Supplies: | Arrays: | Blood Collection: |
| Genetics: | | | | Research & Development Services: | Database Management: | Drug Delivery: |
| | | | | Diagnostics: | | Drug Distribution: |

## TYPES OF BUSINESS:

Small Molecule Drugs & Recombinant Proteins
Drugs-Gastrointestinal Diseases
Drugs-Hyperparathyroidism

## BRANDS/DIVISIONS/AFFILIATES:

Preotact
Preos
Sensipar
Mimpara
GATTEX
NPSP558
NPSP156
REGPARA

## CONTACTS: Note: Officers with more than one job title may be intentionally listed here more than once.

Francois Nader, CEO
Francois Nader, Pres.
Luke M. Beshar, CFO/Sr. VP
Roger J. Garceau, Chief Medical Officer/Sr. VP
Andrew Rackear, General Counsel/Sr. VP-Legal Affairs
Sandra C. Cottrell, Sr. VP-Regulatory Affairs & Drug Safety
Peter G. Tombros, Chmn.

| Phone: 908-450-5300 | Fax: 908-450-5351 |
|---|---|
| Toll-Free: | |
| Address: 550 Hills Drive, 3rd Fl., Bedminster, NJ 07921 US | |

## GROWTH PLANS/SPECIAL FEATURES:

NPS Pharmaceuticals, Inc. is a biopharmaceutical company focused on the development and commercialization of small molecule drugs and recombinant proteins for the treatment of bone and mineral disorders; gastrointestinal disorders; and central nervous system disorders.  The company's lead pipeline products include GATTEX (teduglutide), which is in Phase III trials for the treatment of the short bowel syndrome (SBS), Phase II for Crohn's Disease and preclinical trials for Necrotizing Enterocolitis and GI Mucositis; NPSP558, which is in Phase II trials as a treatment for hypoparathyroidism; and a glycine reuptake inhibitor.  Other products include Preos, for the treatment of post-menopausal osteoporosis, which is approved for marketing in Europe under the name Preotact and is awaiting FDA approval for commercialization in the U.S., and NPSP156, in preclinical trials for central nervous system disorders.  The firm's licensees, Amgen and Kyowa Kirin, developed the FDA-approved cinacalcet HCl for the treatment of hyperparathyroidism, marketed under the trademark Sensipar in the U.S., Mimpara in Europe and REGPARA in Asia.  Furthermore, Nycomed licenses Preotact for distribution outside of the U.S., excluding Japan and Israel.  NPS Pharmaceuticals has collaborative research and license agreements with several companies including Sensipar with Amgen; Ronacaleret for osteoporosis with GlaxoSmithKline; REGPARA with Kirin; Tapentadol with Ortho-McNeil; and Preotact and GATTEX with Nycomed.  The company has been issued roughly 188 patents in the U.S. as well as hundreds of patents in other countries.  In December 2008, NPS began a Phase III study for the use of NPSP558 in Hypoparathyroidism and GATTEX in Short Bowel Syndrome.

## FINANCIALS: Sales and profits are in thousands of dollars—add 000 to get the full amount. 2008 Note: Financial information for 2008 was not available for all companies at press time.

| | | |
|---|---|---|
| 2008 Sales: $102,279 | 2008 Profits: $-31,726 | U.S. Stock Ticker: NPSP |
| 2007 Sales: $86,248 | 2007 Profits: $- 657 | Int'l Ticker:   Int'l Exchange: |
| 2006 Sales: $48,502 | 2006 Profits: $-112,668 | Employees:   47 |
| 2005 Sales: $12,825 | 2005 Profits: $-169,723 | Fiscal Year Ends: 12/31 |
| 2004 Sales: $14,237 | 2004 Profits: $-168,251 | Parent Company: |

## SALARIES/BENEFITS:

| Pension Plan: | ESOP Stock Plan: | Profit Sharing: | Top Exec. Salary: $468,634 | Bonus: $260,531 |
|---|---|---|---|---|
| Savings Plan: Y | Stock Purch. Plan: | | Second Exec. Salary: $302,769 | Bonus: $ |

## OTHER THOUGHTS:

**Apparent Women Officers or Directors**: 2
**Hot Spot for Advancement for Women/Minorities**:

## LOCATIONS: ("Y" = Yes)

| West: | Southwest: | Midwest: | Southeast: | Northeast: | International: |
|---|---|---|---|---|---|
| | | | | Y | |

# NYCOMED

**www.nycomed.com**

**Industry Group Code:** 325412  **Ranks within this company's industry group:** Sales: 26  Profits: 160

| Drugs: | | Other: | Clinical: | Computers: | Services: |
|---|---|---|---|---|---|
| Discovery: | Y | AgriBio: | Trials/Services: | Hardware: | Specialty Services: |
| Licensing: | Y | Genetic Data: | Labs: | Software: | Consulting: |
| Manufacturing: | Y | Tissue Replacement: | Equipment/Supplies: | Arrays: | Blood Collection: |
| Genetics: | | | Research & Development Services: | Database Management: | Drug Delivery: |
| | | | Diagnostics: | | Drug Distribution: |

## TYPES OF BUSINESS:

Pharmaceuticals

## BRANDS/DIVISIONS/AFFILIATES:

Alvesco
Curosurf
Angiox
Preotact
Matrifen
Beriplast
Altana Pharma
Bradley Pharmaceuticals

## CONTACTS: *Note: Officers with more than one job title may be intentionally listed here more than once.*

Hakan Bjorklund, CEO
Runar Bjorklund, CFO
Dick Soderberg, Exec. VP-Mktg.
Charles Depasse, Exec. VP-Human Resources
Anders Ullman, Exec. VP-R&D
Charles Depasse, Exec. VP-IT
Thomas Redemann, Dir.-Global Contract Mfg.
Michael Kuner, Exec. VP-Legal
Barthold Piening, Exec. VP-Oper.
Kerstin Valinder, Exec. VP-Bus. Dev.
Tobias Cottmann, Dir.-External Comm.
Christian B. Seidelin, VP-Controlling, Treasury & Insurance
Christian Kanzelmeyer, VP-Licensing
Philippe de Lavenne, Sr. VP-Latin America
Poul Haukrog Moller, VP-Int'l Mktg.

| **Phone:** 41-44-55-51-000 | **Fax:** 41-44-55-51-001 |
|---|---|
| **Toll-Free:** | |
| **Address:** Leutschenbachstr 95, Zurich, CH-8050 Switzerland | |

## GROWTH PLANS/SPECIAL FEATURES:

Nycomed is a European-based pharmaceutical company engaged in the research, licensing, manufacturing and marketing a wide array of products in the therapeutic areas of cardiology, gastroenterology, osteoporosis, respiratory, pain and tissue management. Products include Alvesco, Curosurf and OMNARIS/SOMNAIR for respiratory problems; Angiox and Ebrantil i.v. for cardiology problems; Beriplast, and TachoSil for tissue management; CalciChew and Preotact for treatment of osteoporosis; Matrifen, Neosaldina and Xefo Rapid for pain management; and Pantoprazole and Riopan for gastroenterology problems. The company is privately-owned and has operates in over 50 countries, including Austria, Brazil, Ireland, Mexico, Estonia, Denmark, Germany, Finland, India, Norway, Belgium, Poland and the U.S. The firm's research and development program is led out of Konstanz, Germany with additional sites in Denmark, India and the U.S. Its subsidiary, Altana Pharma, researches therapeutic drugs for the treatment of gastrointestinal and respiratory diseases. It markets Alvesco, a respiratory drug, in 26 countries. Early in 2008, Nycomed acquired Bradley Pharmaceuticals, a company focused on niche therapeutic markets in the U.S. Nycomed also recently announced plans to establish a marketing and sales subsidiary in Venezuela in an effort to strengthen its Latin American presence in pharmaceutical markets in Mexico, Brazil, Venezuela and Argentina.

## FINANCIALS: Sales and profits are in thousands of dollars—add 000 to get the full amount. 2008 Note: Financial information for 2008 was not available for all companies at press time.

| | | |
|---|---|---|
| 2008 Sales: $4,540,310 | 2008 Profits: $-110,860 | **U.S. Stock Ticker:** Private |
| 2007 Sales: $4,908,150 | 2007 Profits: $334,900 | **Int'l Ticker:**    Int'l Exchange: |
| 2006 Sales: $ | 2006 Profits: $ | Employees:  12,000 |
| 2005 Sales: $ | 2005 Profits: $ | Fiscal Year Ends: 12/31 |
| 2004 Sales: $ | 2004 Profits: $ | Parent Company: |

## SALARIES/BENEFITS:

| Pension Plan: | ESOP Stock Plan: | Profit Sharing: | Top Exec. Salary: $ | Bonus: $ |
|---|---|---|---|---|
| Savings Plan: | Stock Purch. Plan: | | Second Exec. Salary: $ | Bonus: $ |

## OTHER THOUGHTS:

**Apparent Women Officers or Directors:** 1
**Hot Spot for Advancement for Women/Minorities:** Y

## LOCATIONS: ("Y" = Yes)

| West: | Southwest: | Midwest: | Southeast: | Northeast: | International: |
|---|---|---|---|---|---|
| | | | | | Y |

# ONCOGENEX PHARMACEUTICALS

**www.oncogenex.com**

Industry Group Code: 325412A  Ranks within this company's industry group: Sales:  Profits: 12

| Drugs: | | Other: | Clinical: | Computers: | Services: | |
|---|---|---|---|---|---|---|
| Discovery: | Y | AgriBio: | Trials/Services: | Hardware: | Specialty Services: | |
| Licensing: | | Genetic Data: | Labs: | Software: | Consulting: | |
| Manufacturing: | | Tissue Replacement: | Equipment/Supplies: | Arrays: | Blood Collection: | |
| Genetics: | | | Research & Development Services: | Database Management: | Drug Delivery: | Y |
| | | | Diagnostics: | | Drug Distribution: | |

## TYPES OF BUSINESS:

Drug Delivery Systems
Drugs-Cancer

## BRANDS/DIVISIONS/AFFILIATES:

SN2310 Injectable Emulsion
TOCOSOL Paclitaxel

## CONTACTS: Note: Officers with more than one job title may be intentionally listed here more than once.

Scott Cormac, CEO
Alan Fuhrman, CFO/Sr. VP
Dean R. Kessler, VP-Preclinical Dev.
Elaine Waller, VP-Regulatory Affairs & Quality Assurance
Robert E. Ivy, Chmn.

| Phone: 425-487-9500 | Fax: 425-489-0626 |
|---|---|
| Toll-Free: | |
| Address: 1522 217th Place S.E., Ste. 100, Bothell, WA 98021 US | |

## GROWTH PLANS/SPECIAL FEATURES:

OncoGenex Pharmaceuticals, formerly Sonus Pharmaceuticals, Inc., is a subsidiary of OncoGenex Technologies and is focused on the development of small molecule drugs that may offer improved effectiveness, safety, tolerability and administration for the treatment of cancer and related therapies. The firm has five product candidates in development: OGX-011, or custirsen sodium, inhibits the production of clusterin, a protein that is associated with treatment resistance in a number of solid tumors, including prostate, breast, non-small cell lung, ovarian and bladder cancers; OGX-427, which reduces production of Hsp27, a protein that is over-produced in response to many cancer treatments including hormone ablation therapy, chemotherapy and radiation therapy; SN2310, which is an Injectable Emulsion and camptothecin derivative intended to provide enhanced anti-tumor activity and improved tolerability compared to competitive camptothecin-based products; CSP-9222, which is a caspase activator and can activate cell death in tumor cells; and OGX-225, which reduces the production of Insulin-Like Growth Factor Binding Protein -2 and Insulin-Like Growth Factor Binding Protein -5, which in turn make an alternate hormone, IGF-1, that increases tumor growth. OncoGenex's OGX-011, OGX-427 and OGX-225 are currently in clinical development. The firm's lead product, OGX-011, in five Phase II clinical trials and a Phase III trial is currently being designed; both OXG-427 and OGX-225 are in Phase I trials. In May 2008, the firm merged with OncoGenex Technologies and changed its name to OncoGenex Pharmaceuticals. In August 2008, the Federal Drug Administration granted fast-track designation to OGX-011.

## FINANCIALS: Sales and profits are in thousands of dollars—add 000 to get the full amount. 2008 Note: Financial information for 2008 was not available for all companies at press time.

| | | |
|---|---|---|
| 2008 Sales: $ | 2008 Profits: $-4,204 | U.S. Stock Ticker: Subsidiary |
| 2007 Sales: $20,131 | 2007 Profits: $-8,536 | Int'l Ticker:  Int'l Exchange: |
| 2006 Sales: $22,392 | 2006 Profits: $-11,594 | Employees:  26 |
| 2005 Sales: $8,254 | 2005 Profits: $-21,097 | Fiscal Year Ends: 12/31 |
| 2004 Sales: $ | 2004 Profits: $-16,311 | Parent Company: ONCOGENEX TECHNOLOGIES |

## SALARIES/BENEFITS:

| Pension Plan: | ESOP Stock Plan: | Profit Sharing: | Top Exec. Salary: $374,040 | Bonus: $ |
|---|---|---|---|---|
| Savings Plan: | Stock Purch. Plan: | | Second Exec. Salary: $245,699 | Bonus: $ |

## OTHER THOUGHTS:

**Apparent Women Officers or Directors**: 1
**Hot Spot for Advancement for Women/Minorities**: Y

## LOCATIONS: ("Y" = Yes)

| West: | Southwest: | Midwest: | Southeast: | Northeast: | International: |
|---|---|---|---|---|---|
| Y | | | | | |

Note: Financial information, benefits and other data can change quickly and may vary from those stated here.

# ONCOTHYREON INC

## www.oncothyreon.com

Industry Group Code: 325412  Ranks within this company's industry group: Sales: 91   Profits: 53

| Drugs: | | Other: | Clinical: | Computers: | Services: |
|---|---|---|---|---|---|
| Discovery: | Y | AgriBio: | Trials/Services: | Hardware: | Specialty Services: |
| Licensing: | Y | Genetic Data: | Labs: | Software: | Consulting: |
| Manufacturing: | | Tissue Replacement: | Equipment/Supplies: | Arrays: | Blood Collection: |
| Genetics: | | | Research & Development Services: | Database Management: | Drug Delivery: |
| | | | Diagnostics: | | Drug Distribution: |

## TYPES OF BUSINESS:

Drugs-Cancer
Cancer Vaccines
Drugs-Immunological

## BRANDS/DIVISIONS/AFFILIATES:

Stimuvax
PX-12
PX-478
PX-866
BGLP40

## CONTACTS: Note: Officers with more than one job title may be intentionally listed here more than once.

Robert L. Kirkman, CEO
Gary Christianson, COO
Robert L. Kirkman, Pres.
Julie Rathbun, Media Rel. Contact
Julie Rathbun, Investor Rel. Contact
Shashi Karan, Corp. Controller
Christopher S. Henney, Chmn.

| Phone: 206-801-2100 | Fax: 206-801-2101 |
|---|---|
| Toll-Free: | |
| Address: 2601 4th Ave., Ste. 500, Seattle, WA 98121 US | |

## GROWTH PLANS/SPECIAL FEATURES:

Oncothyreon, Inc., based in Washington, focuses its biotechnology expertise on the development of cancer therapeutics. The company is seeking to develop vaccine-like treatments for cancer that will stimulate a patient's immune system to recognize and fight malignant cells. Oncothyreon's BGLP40 liposome vaccine has the potential for use in multiple forms of cancer and is in preclinical development. The company is also developing two cancer-fighting inhibitors, PX-12 and PX-478, which treat patients with advanced pancreatic cancer and those with advanced metastatic cancer, respectively. In June 2008, Another metastatic cancer inhibitor, PX-866, entered a Phase 1 trial. In December 2008, the firm sold the manufacturing rights to Stimuvax, a liposome vaccine and Oncothyreon's former leading product candidate, to German company Merck KGaA for $13 million. Stimuvax, which is being developed to generate a cellular immune response to the tumor-associated antigen mucin MUC, is currently in Phase 3 development for stage III NSCLC (non-small cell lung cancer). As a result of this transaction, the firm also transferred its rights to its Stimuvax inventory and assets, as well as its manufacturing facility in Edmonton, Canada, to Merck KGaA affiliate EMD Serono Canada, Inc.

The company offers employees medical, dental and vision benefits; life and long-term disability insurance; and a matching 401(k).

## FINANCIALS: Sales and profits are in thousands of dollars—add 000 to get the full amount. 2008 Note: Financial information for 2008 was not available for all companies at press time.

| | | |
|---|---|---|
| 2008 Sales: $39,998 | 2008 Profits: $7,125 | U.S. Stock Ticker: ONTY |
| 2007 Sales: $3,798 | 2007 Profits: $-20,340 | Int'l Ticker: BRA   Int'l Exchange: Toronto-TSX |
| 2006 Sales: $4,199 | 2006 Profits: $-16,591 | Employees:   18 |
| 2005 Sales: $3,800 | 2005 Profits: $-16,400 | Fiscal Year Ends: 12/31 |
| 2004 Sales: $7,500 | 2004 Profits: $-10,200 | Parent Company: |

## SALARIES/BENEFITS:

| Pension Plan: | ESOP Stock Plan: | Profit Sharing: | Top Exec. Salary: $375,000 | Bonus: $176,250 |
|---|---|---|---|---|
| Savings Plan: Y | Stock Purch. Plan: | | Second Exec. Salary: $270,947 | Bonus: $ |

## OTHER THOUGHTS:

Apparent Women Officers or Directors: 1
Hot Spot for Advancement for Women/Minorities: Y

## LOCATIONS: ("Y" = Yes)

| West: | Southwest: | Midwest: | Southeast: | Northeast: | International: |
|---|---|---|---|---|---|
| Y | Y | | | | |

# ONYX PHARMACEUTICALS INC                 www.onyx-pharm.com

**Industry Group Code: 325412  Ranks within this company's industry group:** Sales: 59    Profits: 56

| Drugs: | | Other: | Clinical: | Computers: | Services: |
|---|---|---|---|---|---|
| Discovery: | Y | AgriBio: | Trials/Services: | Hardware: | Specialty Services: |
| Licensing: | | Genetic Data: | Labs: | Software: | Consulting: |
| Manufacturing: | Y | Tissue Replacement: | Equipment/Supplies: | Arrays: | Blood Collection: |
| Genetics: | | | Research & Development Services: | Database Management: | Drug Delivery: |
| | | | Diagnostics: | | Drug Distribution: |

## TYPES OF BUSINESS:
Pharmaceuticals Discovery & Development
Small-Molecule Drugs
Cancer Treatments

## BRANDS/DIVISIONS/AFFILIATES:
Nexavar
ONX 0801
ONX 0803
ONX 0805

## CONTACTS: *Note: Officers with more than one job title may be intentionally listed here more than once.*
N. Anthony Coles, CEO
Laura A. Brege, COO/Exec. VP
N. Anthony Coles, Pres.
Matthew K. Fust, CFO/Exec. VP
Randy A. Kelley, Sr. VP-Sales & Mktg.
Judy Batlin, VP-Human Resources
Jeffrey D. Bloss, VP-Clinical Dev.
Juergen Lasowski, Sr. VP-Corp. Dev.
Julianna Wood, VP-Corp. Comm.
Julianna Wood, VP-Investor Rel.

| Phone: 510-597-6500 | Fax: 510-597-6600 |
|---|---|
| Toll-Free: | |
| Address: 2100 Powell St., Emeryville, CA 94608 US | |

## GROWTH PLANS/SPECIAL FEATURES:
Onyx Pharmaceuticals, Inc. is engaged in the discovery, development and commercialization of innovative products that target oncological molecular mechanisms. The company's flagship drug, Nexavar, is the result of a partnership with Bayer Pharmaceuticals and is aimed at blocking inappropriate growth signals in tumor cells by inhibiting the responsible active enzymes that induce cancer cell growth. Nexavar is the first and only oral targeted therapy to significantly improve overall survival in patients with liver and kidney cancer by inhibiting both tumor cell proliferation and angiogenesis. The drug is also approved in Europe for the treatment of advanced renal cell carcinoma in patients that have failed to respond or are unsuited for other forms of therapy. Nexavar is also being tested in Phase II and Phase III clinical trials in combination with standard chemotherapeutic and other anticancer agents for additional treatments of liver, lung, kidney, breast, colorectal and ovarian cancers. Onyx is also developing a cell cycle kinase inhibitor with Warner-Lambert Company, which regulates a cell's ability to replicate itself and case cancer. ONX 0801 is a preclinical anti-cancer development drug aimed at receptor-mediated targeting of tumor cells and the inhibition of thymidylate synthase. ONX 0803 and ONX 0805, in collaborational development with S*BIO, are drugs that are designed to inhibit JAK2, a major factor in solid tumors and rheumatoid arthritis. In May 2009, the company announced that Nexavar has been approved for the treatment of advanced liver cancer in Japan.

Onyx offers its employees medical, dental, vision, life and AD&D insurance; medical and dependent flexible spending accounts; incentive stock options; an employee stock purchase plan; a 401(k) plan; tuition reimbursement; and employee assistance and referral awards.

## FINANCIALS: Sales and profits are in thousands of dollars—add 000 to get the full amount. 2008 Note: Financial information for 2008 was not available for all companies at press time.

| | | |
|---|---|---|
| 2008 Sales: $194,343 | 2008 Profits: $1,948 | U.S. Stock Ticker: ONXX |
| 2007 Sales: $90,429 | 2007 Profits: $-34,167 | Int'l Ticker:    Int'l Exchange: |
| 2006 Sales: $29,524 | 2006 Profits: $-92,681 | Employees:   197 |
| 2005 Sales: $1,000 | 2005 Profits: $-95,174 | Fiscal Year Ends: 12/31 |
| 2004 Sales: $ 500 | 2004 Profits: $-46,756 | Parent Company: |

## SALARIES/BENEFITS:
| Pension Plan: | ESOP Stock Plan: | Profit Sharing: | Top Exec. Salary: $471,134 | Bonus: $825,000 |
|---|---|---|---|---|
| Savings Plan: Y | Stock Purch. Plan: Y | | Second Exec. Salary: $425,000 | Bonus: $187,425 |

## OTHER THOUGHTS:
**Apparent Women Officers or Directors:** 4
**Hot Spot for Advancement for Women/Minorities:** Y

## LOCATIONS: ("Y" = Yes)
| West: | Southwest: | Midwest: | Southeast: | Northeast: | International: |
|---|---|---|---|---|---|
| Y | | | | | |

# ORASURE TECHNOLOGIES INC

www.orasure.com

**Industry Group Code: 325413 Ranks within this company's industry group:** Sales: 14   Profits: 16

| Drugs: | Other: | Clinical: | | Computers: | Services: |
|---|---|---|---|---|---|
| Discovery: | AgriBio: | Trials/Services: | | Hardware: | Specialty Services: |
| Licensing: | Genetic Data: | Labs: | | Software: | Consulting: |
| Manufacturing: | Tissue Replacement: | Equipment/Supplies: | Y | Arrays: | Blood Collection: |
| Genetics: | | Research & Development Services: | Y | Database Management: | Drug Delivery: |
| | | Diagnostics: | Y | | Drug Distribution: |

## TYPES OF BUSINESS:

Medical Devices Manufacturing
Oral Fluid Collection Devices
Cryosurgical Products

## BRANDS/DIVISIONS/AFFILIATES:

OraQuick
OraSure
Histofreezer
Intercept
Q.E.D.
Micro-Plate & Auto-Lite

## CONTACTS: *Note: Officers with more than one job title may be intentionally listed here more than once.*

Douglas A. Michels, CEO
Ronald H. Spair, COO
Douglas A. Michels, Pres.
Ronald H. Spair, CFO
Joseph E. Zack, Exec. VP-Sales & Mktg.
Henry B. Cohen, Sr. VP-Human Resources
Stephen Lee, Chief Science Officer/Exec. VP
Jack E. Jerrett, General Counsel/Sr. VP/Sec.
P. Michael Formica, Exec. VP-Oper.
Mark Kirtland, Sr. VP-Bus. Dev.
Ron Ticho, VP-Corp. Comm.
Mark K. Luna, Sr. VP-Finance/Controller
P. Michael Formica, Gen. Mgr.-Cryosurgical Systems Div.
Debra Y. Fraser-Howze, VP-Gov't & External Affairs
Douglas Watson, Chmn.

| Phone: 610-882-1820 | Fax: 610-882-1830 |
|---|---|
| Toll-Free: | |
| Address: 220 E. First St., Bethlehem, PA 18015 US | |

## GROWTH PLANS/SPECIAL FEATURES:

OraSure Technologies, Inc. develops, manufactures, markets and sells oral fluid specimen collection devices using proprietary oral fluid technologies. The company is also interested in additional diagnostic products including immunoassays and in-vitro diagnostic tests that are used on other specimen types and medical devices. The company's diagnostic products include tests that are processed in a laboratory on a rapid basis at the point of care. The firm's primary products include the OraQuick and OraSure tests for HIV and other infectious diseases; the Intercept and Q.E.D. tests for drugs and alcohol; the Histofreezer portable cryosurgical system for various skin issues; the OraSure Oral Specimen Vial, Micro-Plate and Auto-Lyte kits for insurance markets and forensic toxicology; and over-the-counter (OTC) treatments for wart removal. OraSure is also developing an oral test for Hepatitis C; OTC HIV antibody tests; and fully automated drug testing using OraQuick, currently in preclinical and clinical trials. The company collaborates with other pharmaceutical companies, including Schering-Plough and Roche Diagnostics, to bring new products to market. OraSure Technologies' products are sold in the U.S. and internationally to various clinical laboratories, hospitals, clinics, community-based organizations and other public health organizations, distributors, government agencies, physicians' offices and commercial and industrial entities, with international sales generating 19% of revenue in 2008.

The company offers its employees health and dental coverage; life insurance; short- and long-term disability; flexible spending accounts; a 401(k) plan; a stock award plan; service recognition; employee referrals; free will preparation; discount plans; and employee assistance.

## FINANCIALS: Sales and profits are in thousands of dollars—add 000 to get the full amount. 2008 Note: Financial information for 2008 was not available for all companies at press time.

| | | |
|---|---|---|
| 2008 Sales: $71,104 | 2008 Profits: $-31,275 | **U.S. Stock Ticker:** OSUR |
| 2007 Sales: $82,686 | 2007 Profits: $2,473 | **Int'l Ticker:**   Int'l Exchange: |
| 2006 Sales: $68,155 | 2006 Profits: $5,268 | Employees:  282 |
| 2005 Sales: $69,366 | 2005 Profits: $27,448 | Fiscal Year Ends: 12/31 |
| 2004 Sales: $54,008 | 2004 Profits: $- 560 | Parent Company: |

## SALARIES/BENEFITS:

| Pension Plan: | ESOP Stock Plan: | Profit Sharing: | Top Exec. Salary: $474,884 | Bonus: $88,450 |
|---|---|---|---|---|
| Savings Plan: Y | Stock Purch. Plan: | | Second Exec. Salary: $380,119 | Bonus: $58,975 |

## OTHER THOUGHTS:

**Apparent Women Officers or Directors**: 1
**Hot Spot for Advancement for Women/Minorities**:

## LOCATIONS: ("Y" = Yes)

| West: | Southwest: | Midwest: | Southeast: | Northeast: | International: |
|---|---|---|---|---|---|
| | | | | Y | |

# ORCHID CELLMARK INC
## www.orchid.com

**Industry Group Code: 6215  Ranks within this company's industry group: Sales: 5    Profits: 4**

| Drugs: | Other: | | Clinical: | | Computers: | | Services: | |
|---|---|---|---|---|---|---|---|---|
| Discovery: | AgriBio: | | Trials/Services: | | Hardware: | | Specialty Services: | Y |
| Licensing: | Genetic Data: | Y | Labs: | | Software: | | Consulting: | |
| Manufacturing: | Tissue Replacement: | | Equipment/Supplies: | | Arrays: | | Blood Collection: | |
| Genetics: | | | Research & Development Services: | | Database Management: | | Drug Delivery: | |
| | | | Diagnostics: | Y | | | Drug Distribution: | |

## TYPES OF BUSINESS:

Research-Bioinformatics
Genomics Services
Diagnostic Products & Services
Genetic Databases
DNA Testing
Forensic Testing

## BRANDS/DIVISIONS/AFFILIATES:

Orchid Cellmark
STRs
SNPs
OrchidCellMark.com

## CONTACTS: Note: Officers with more than one job title may be intentionally listed here more than once.

Thomas A. Bologna, CEO
Thomas A. Bologna, Pres.
James F. Smith, CFO/VP
William J. Thomas, General Counsel/VP
Jeffrey S. Boschwitz, VP-North America Sales & Mktg.
George H. Poste, Chmn.

| Phone: 609-750-2200 | Fax: 609-750-6405 |
|---|---|
| Toll-Free: | |
| Address: 4390 US Route 1, Princeton, NJ 08540 US | |

## GROWTH PLANS/SPECIAL FEATURES:

Orchid Cellmark, Inc. (Orchid), formerly Orchid BioSciences, Inc., provides identity genomics services for the forensic and public health markets as well as paternity DNA testing and animal DNA testing. Forensic DNA testing is primarily used to establish and maintain DNA profile databases of individuals arrested or convicted of crimes, to analyze and compare evidence from crime scenes or to determine if a man has fathered a particular child in paternity cases. In agricultural applications, DNA testing services are available for selective trait breeding and traceability applications. Technologies utilized by Orchid include short tandem repeats (STRs) for forensic and paternal testing and single nucleotide polymorphisms (SNPs) for DNA agricultural applications. Agricultural projects currently pursued by the company include scrapie genotyping, a U.K. government project designed to reduce the animal disease, scrapie, on sheep farms. In addition to casework testing, the company provides DNA identification profiles of individuals for national, state and local criminal DNA databases. Orchid's services have been selected by major police departments in the U.S. and London's Metropolitan Police Force (Scotland Yard). The company has conducted DNA testing for such notable cases as those of O.J. Simpson, Jon Benet Ramsey and the Unabomber; in addition, its SNP technology was used to identify a large number of previously unidentified World Trade Center victims. Orchid Cellmark owns or has exclusive licenses to 124 patents (57 U.S., 67 international), has received allowance for two more and applied for an additional 54 (12 U.S., 42 international). In April 2008, the company agreed to open a new forensic laboratory in Northwest England. In June 2008, Orchid Cellmark launched a new web site, www.orchidcellmark.com, which offers customers at-home DNA test kits and court ready paternity tests.

The company offers employees medical, dental and vision insurance; disability, life and AD&D insurance; a 401(k) plan; an employee assistance program; and flexible spending accounts.

## FINANCIALS: Sales and profits are in thousands of dollars—add 000 to get the full amount. 2008 Note: Financial information for 2008 was not available for all companies at press time.

| | | |
|---|---|---|
| 2008 Sales: $57,595 | 2008 Profits: $-4,481 | **U.S. Stock Ticker: ORCH** |
| 2007 Sales: $60,303 | 2007 Profits: $-2,967 | **Int'l Ticker:**    Int'l Exchange: |
| 2006 Sales: $56,854 | 2006 Profits: $-11,271 | Employees:  427 |
| 2005 Sales: $61,609 | 2005 Profits: $-9,439 | Fiscal Year Ends: 12/31 |
| 2004 Sales: $62,499 | 2004 Profits: $-8,812 | Parent Company: |

## SALARIES/BENEFITS:

| Pension Plan: | ESOP Stock Plan: | Profit Sharing: | Top Exec. Salary: $560,250 | Bonus: $ |
|---|---|---|---|---|
| Savings Plan: Y | Stock Purch. Plan: | | Second Exec. Salary: $246,575 | Bonus: $ |

## OTHER THOUGHTS:

**Apparent Women Officers or Directors:** 1
**Hot Spot for Advancement for Women/Minorities:**

## LOCATIONS: ("Y" = Yes)

| West: | Southwest: | Midwest: | Southeast: | Northeast: | International: |
|---|---|---|---|---|---|
| | Y | Y | Y | Y | Y |

# ORE PHARMACEUTICALS INC

## www.orepharma.com

**Industry Group Code: 541712  Ranks within this company's industry group:** Sales: 19　　Profits: 12

| Drugs: | Other: | | Clinical: | | Computers: | | Services: | |
|---|---|---|---|---|---|---|---|---|
| Discovery: | AgriBio: | | Trials/Services: | | Hardware: | | Specialty Services: | |
| Licensing: | Genetic Data: | Y | Labs: | | Software: | | Consulting: | |
| Manufacturing: | Tissue Replacement: | | Equipment/Supplies: | | Arrays: | | Blood Collection: | |
| Genetics: | | | Research & Development Services: | Y | Database Management: | Y | Drug Delivery: | |
| | | | Diagnostics: | | | | Drug Distribution: | |

## TYPES OF BUSINESS:

Drug Repositioning
Drug Discovery

## BRANDS/DIVISIONS/AFFILIATES:

Gene Logic Inc
Ocimum Biosolutions Ltd.

## CONTACTS: *Note: Officers with more than one job title may be intentionally listed here more than once.*

Charles L. Dimmler, III, CEO
Charles L. Dimmler, III, Pres.
Philip L. Rohrer, Jr., CFO
Harlene Stanchfield, Sr. VP-Human Resources
F. Dudley Staples, Jr., General Counsel/Sr. VP/Corp. Sec.
Larry Tiffany, Interim Head-Commercial Oper.
Deborah Barbara, VP-Bus. Dev.
Christopher Culotta, Sr. Dir.-Strategic Comm.
J. Stark Thompson, Chmn.

| Phone: 617-649-2001 | Fax: |
|---|---|
| Toll-Free: 877-673-7476 | |
| Address: 610 Professional Dr., Gaithersburg, MD 20879 US | |

## GROWTH PLANS/SPECIAL FEATURES:

Ore Pharmaceuticals, Inc., formerly Gene Logic, Inc., is a drug repositioning and development company. Ore applies its experience in integrative pharmacology to identify potential new uses for drug candidates that have failed clinical development for reasons other than safety. To date the company has initiated the evaluation of over 100 compounds provided by its pharmaceutical company partners. It has been able to propose alternative indications for approximately one-third of the compounds for which the full evaluations have been completed. Some of the compounds for which Ore has found new indications have or will progress to in vivo validation or efficacy studies to evaluate and support their suitability for re-entering clinical trials for the proposed new indications. The validation study is the study Ore has agreed to perform to confirm its hypothesis. For partnered compounds, the company typically agrees on the design of the study with its partner and pays for the validation studies, which are then subject to reimbursement from its pharmaceutical partners. After completion of the initial validation study, a partner may wish to conduct additional studies to learn more about the efficacy of the compound for the new indication before deciding whether to reinstate the compound into its development pipeline. Generally, Ore's partners would conduct such in vivo efficacy studies at their own expense. No compound has yet been selected by the firm's partners to re-enter clinical trials. The technologies used by Ore to develop hypotheses about which diseases a compound may treat include real-time in vivo imaging, ex vivo multiplex bioanalytics, in vitro molecular pharmacology and in silico biology using the genomics database developed by its former genomics division. In September 2008, the company agreed to sell its DioGenix division to Nerveda, Inc.

## FINANCIALS: Sales and profits are in thousands of dollars—add 000 to get the full amount. 2008 Note: Financial information for 2008 was not available for all companies at press time.

| | | |
|---|---|---|
| 2008 Sales: $1,950 | 2008 Profits: $-22,461 | **U.S. Stock Ticker: ORXE** |
| 2007 Sales: $1,596 | 2007 Profits: $-34,688 | **Int'l Ticker:**　Int'l Exchange: |
| 2006 Sales: $24,346 | 2006 Profits: $-54,710 | Employees:　14 |
| 2005 Sales: $57,190 | 2005 Profits: $-48,304 | Fiscal Year Ends: 12/31 |
| 2004 Sales: $52,171 | 2004 Profits: $-28,520 | Parent Company: |

## SALARIES/BENEFITS:

| Pension Plan: | ESOP Stock Plan: | Profit Sharing: | Top Exec. Salary: $425,000 | Bonus: $59,500 |
|---|---|---|---|---|
| Savings Plan: Y | Stock Purch. Plan: | | Second Exec. Salary: $275,000 | Bonus: $137,500 |

## OTHER THOUGHTS:

**Apparent Women Officers or Directors:** 2
**Hot Spot for Advancement for Women/Minorities:**

## LOCATIONS: ("Y" = Yes)

| West: | Southwest: | Midwest: | Southeast: | Northeast: | International: |
|---|---|---|---|---|---|
| | | | | Y | |

# ORGANOGENESIS INC

www.organogenesis.com

Industry Group Code: 325414   Ranks within this company's industry group:   Sales:      Profits:

| Drugs: | Other: | | Clinical: | | Computers: | | Services: | |
|---|---|---|---|---|---|---|---|---|
| Discovery: | AgriBio: | | Trials/Services: | | Hardware: | | Specialty Services: | |
| Licensing: | Genetic Data: | | Labs: | | Software: | | Consulting: | |
| Manufacturing: Y | Tissue Replacement: | Y | Equipment/Supplies: | Y | Arrays: | | Blood Collection: | |
| Genetics: | | | Research & Development Services: | Y | Database Management: | | Drug Delivery: | |
| | | | Diagnostics: | | | | Drug Distribution: | |

## TYPES OF BUSINESS:

Tissue Replacement Products
Wound Dressing Products

## BRANDS/DIVISIONS/AFFILIATES:

Apligraf
BioSTAR
FortaPerm
FortaGen
CuffPatch
CelTx
NanoMatrix, Inc.

## CONTACTS: Note: Officers with more than one job title may be intentionally listed here more than once.

Geoff MacKay, CEO
Gary S. Gillheeney, Sr., COO/Exec. VP
Geoff MacKay, Pres.
Gary S. Gillheeney, Sr., CFO/Exec. VP
Santino Costanzo, VP-Sales, Bio-active Wound Healing
Houda Samaha, Dir.-Human Resources
Vincent Ronfard, VP-Research
Phillip Nolan, VP-Mfg. Oper.
Richard Shaw, VP-Finance
Dario Eklund, VP-Bio-Surgery & Oral Regeneration
Stephen Farrell, VP-Bio-Active Wound Healing
Damien Bates, Chief Medical Officer
Patrick Bilbo, VP-Regulatory

| Phone: 781-575-0775 | Fax: 781-575-0440 |
|---|---|
| Toll-Free: | |
| Address: 150 Dan Rd., Canton, MA 02021 US | |

## GROWTH PLANS/SPECIAL FEATURES:

Organogenesis, Inc., is a regenerative medicine firm that designs, develops and manufactures products containing living cells or natural connective tissue. It specializes in bio-active wound healing, oral regeneration and bio-surgery. The firm's bio-active wound healing products include: Apligraf, which has been used by over 250,000 patients, is designed for the treatment of venous leg ulcers and diabetic foot ulcers; and VCT01, a bio-engineered skin substitute currently in late stage development. Organogenesis is testing Apligraf in clinical studies for the treatment of burns and epidermolysis bullosa, a rare genetic disease characterized by extremely fragile skin. CelTx, the company's oral soft tissue regeneration product, has completed Phase III testing. Bio-surgery products include: CuffPatch for rotator cuff surgery, which repairs the tendons that connect the upper arm bones to the shoulder blade; FortaGen, used for tissue repair in cases of vaginal prolapse, cystocele, rectocele and abdominal hernias; FortaPerm, used as a sling to raise a dropped bladder neck back into place, fixing stress urinary incontinence; and BioSTAR, an implant technology for the treatment of cardiac sources of migraine headaches, strokes and other potential brain attacks. Organogenesis is also researching the potential connection between a common heart defect called patent foramen ovale and the aforementioned brain attacks. In recent years, the company acquired NanoMatrix, Inc., a firm which uses electrospinning to create designer scaffolds for the purposes of regenerative medicine. In January 2009, the firm received a $7.4 million grant from The Massachusetts Life Sciences Center; Organogenesis plans to use this funding to build a new medical research, development and manufacturing plant. The company plans to add 280 employees with this expansion.

The company offers employees benefits that include a 401(k).

## FINANCIALS: Sales and profits are in thousands of dollars—add 000 to get the full amount. 2008 Note: Financial information for 2008 was not available for all companies at press time.

| | | | |
|---|---|---|---|
| 2008 Sales: $ | 2008 Profits: $ | U.S. Stock Ticker: Private | |
| 2007 Sales: $ | 2007 Profits: $ | Int'l Ticker:    Int'l Exchange: | |
| 2006 Sales: $ | 2006 Profits: $ | Employees:   182 | |
| 2005 Sales: $ | 2005 Profits: $ | Fiscal Year Ends: 12/31 | |
| 2004 Sales: $ | 2004 Profits: $ | Parent Company: | |

## SALARIES/BENEFITS:

| Pension Plan: | ESOP Stock Plan: | Profit Sharing: | Top Exec. Salary: $277,420 | Bonus: $56,000 |
|---|---|---|---|---|
| Savings Plan: Y | Stock Purch. Plan: | | Second Exec. Salary: $229,836 | Bonus: $30,000 |

## OTHER THOUGHTS:

Apparent Women Officers or Directors: 3
Hot Spot for Advancement for Women/Minorities: Y

## LOCATIONS: ("Y" = Yes)

| West: | Southwest: | Midwest: | Southeast: | Northeast: | International: |
|---|---|---|---|---|---|
| | | | | Y | Y |

Note: Financial information, benefits and other data can change quickly and may vary from those stated here.

# OSCIENT PHARMACEUTICALS CORPORATION          www.oscient.com

**Industry Group Code: 325412  Ranks within this company's industry group:  Sales: 72    Profits: 148**

| Drugs: | Other: | Clinical: | Computers: | Services: |
|---|---|---|---|---|
| Discovery: | AgriBio: | Trials/Services: | Hardware: | Specialty Services: |
| Licensing:          Y | Genetic Data: | Labs: | Software: | Consulting: |
| Manufacturing: | Tissue Replacement: | Equipment/Supplies: | Arrays: | Blood Collection: |
| Genetics: | | Research & Development Services: | Database Management: | Drug Delivery: |
| | | Diagnostics: | | Drug Distribution: |

## TYPES OF BUSINESS:

Pharmaceuticals Commercialization
Pharmaceuticals Development
Antibiotics

## BRANDS/DIVISIONS/AFFILIATES:

FACTIVE
ANTARA
Ramoplanin

## CONTACTS: *Note: Officers with more than one job title may be intentionally listed here more than once.*

Steven M. Rauscher, CEO
Steven M. Rauscher, Pres.
Philippe M. Maitre, CFO/Exec. VP
Mark Glickman, Sr. VP-Sales & Mktg.
Joseph A. Pane, VP-Human Resources
Inder Kaul, VP-Clinical Dev., Medical & Regulatory Affairs
Diane McGuire, VP-Oper.
Christopher J.M. Taylor, VP-Corp. Comm.
Christopher J.M. Taylor, VP-Investor Rel.
David K. Stone, Chmn.

| Phone: 781-398-2300 | Fax: 781-893-9535 |
|---|---|
| Toll-Free: | |
| Address: 1000 Winter St., Ste. 2200, Waltham, MA 02451 US | |

## GROWTH PLANS/SPECIAL FEATURES:

Oscient Pharmaceuticals Corporation primarily engages in the commercialization of FDA approved products, which the firm obtains through drug acquisitions, in-licensing and co-promotion. It markets pharmaceuticals through a national sales force that targets primary care physicians, cardiologists, endocrinologists and pulmonologists. The company currently sells two approved products: Factive and Antara. Factive (gemifloxacin mesylate) is a fluoroquinolone antibiotic that has been approved in tablet formulation for two indications: community-acquired pneumonia of mild to moderate severity and acute bacterial exacerbations of chronic bronchitis. It works by inhibiting bacterial DNA synthesis through the blocking of both DNA gyrase and topoisomerase IV, enzymes needed for bacterial growth and survival. The firm licenses the rights to gemifloxacin (Factive's active ingredient) from LG Life Sciences and has in turn sublicensed the rights to market Factive tablets to several other companies: Pfizer, S.A. de C.V. in Mexico; Abbott Laboratories Ltd. in Canada; and Menarini International Operation Luxembourg S.A. in areas of Europe. Antara (fenofibrate) is indicated for the adjunct treatment of hypercholesterolemia (high blood cholesterol) and hypertriglyceridemia (high triglycerides) in combination with diet. It works by activating lipoprotein lipase and reducing lipoprotein lipase inhibitors in order to eliminate triglyceride-rich particles from plasma. In addition to its marketed drugs, the company is engaged in advanced clinical development of a novel antibiotic candidate, Ramoplanin, for the treatment of Clostridium difficile-associated disease. In February 2009, the firm announced plans to reduce its workforce by 32%.

Employees are offered medical, dental and vision insurance; flexible spending accounts; life insurance; a 401(k) plan; short- and long-term disability; tuition reimbursement; adoption assistance; a travel assistance program; an employee assistance program; and a 529 college savings program.

## FINANCIALS:  Sales and profits are in thousands of dollars—add 000 to get the full amount. 2008 Note: Financial information for 2008 was not available for all companies at press time.

| | | |
|---|---|---|
| 2008 Sales: $86,325 | 2008 Profits: $-64,755 | **U.S. Stock Ticker: OSCI** |
| 2007 Sales: $79,969 | 2007 Profits: $-29,853 | **Int'l Ticker:**      Int'l Exchange: |
| 2006 Sales: $46,152 | 2006 Profits: $-78,477 | Employees:   316 |
| 2005 Sales: $23,609 | 2005 Profits: $-88,593 | Fiscal Year Ends: 12/31 |
| 2004 Sales: $6,613 | 2004 Profits: $-93,271 | Parent Company: |

## SALARIES/BENEFITS:

| Pension Plan: | ESOP Stock Plan: | Profit Sharing: | Top Exec. Salary: $432,600 | Bonus: $ |
|---|---|---|---|---|
| Savings Plan: Y | Stock Purch. Plan: | | Second Exec. Salary: $351,298 | Bonus: $ |

## OTHER THOUGHTS:

**Apparent Women Officers or Directors**: 1
**Hot Spot for Advancement for Women/Minorities**:

## LOCATIONS: ("Y" = Yes)

| West: | Southwest: | Midwest: | Southeast: | Northeast: | International: |
|---|---|---|---|---|---|
| | | | | Y | |

# OSI PHARMACEUTICALS INC

www.osip.com

**Industry Group Code: 325412 Ranks within this company's industry group: Sales: 46 Profits: 26**

| Drugs: | | Other: | Clinical: | Computers: | Services: |
|---|---|---|---|---|---|
| Discovery: | Y | AgriBio: | Trials/Services: | Hardware: | Specialty Services: |
| Licensing: | | Genetic Data: | Labs: | Software: | Consulting: |
| Manufacturing: | Y | Tissue Replacement: | Equipment/Supplies: | Arrays: | Blood Collection: |
| Genetics: | | | Research & Development Services: | Database Management: | Drug Delivery: |
| | | | Diagnostics: | | Drug Distribution: |

## TYPES OF BUSINESS:

Drugs-Cancer
Drugs-Small-Molecule
Drugs-Diabetes

## BRANDS/DIVISIONS/AFFILIATES:

Tarceva
OSI Oncology
OSI Prosidion
PSN-821
OSI-906
OSI-027
OSI-930
PSN-602

## CONTACTS: Note: Officers with more than one job title may be intentionally listed here more than once.

Colin Goddard, CEO
Michael G. Atieh, CFO/Exec. VP
Pierre Legault, Sr. VP-Human Resources
Robert L. Simon, Exec. VP-Pharmaceutical Dev. & Mfg.
Barbara A. Wood, General Counsel/Sec./Sr. VP
Kim Wittig, Dir.-Media Rel.
Kathy Galante, Sr. Dir.-Investor Rel.
Pierre Legault, Treas.
Gabriel Leung, Pres., OSI Oncology
Anker Lundemose, Exec. VP/Pres., OSI Prosidion
Robert A. Ingram, Chmn.

| Phone: 631-962-2000 | Fax: 631-752-3880 |
|---|---|
| Toll-Free: | |
| Address: 41 Pinelawn Rd., Melville, NY 11747 US | |

## GROWTH PLANS/SPECIAL FEATURES:

OSI Pharmaceuticals, Inc. is a biotechnology company that discovers, develops and commercializes molecular targeted therapies addressing the areas of oncology, obesity and diabetes. The firm's largest area of focus is oncology where its leading product is Tarceva, a small-molecule inhibitor of the epidermal growth factor receptor, which plays a role in the abnormal growth of many cancer cells. The drug is an oral, once-a-day pharmaceutical with approved indication for advanced non-small cell lung cancer (NSCLC) and for locally advanced and metastatic pancreatic cancer. It has been approved for treatment for advanced NSCLC after failed chemotherapy treatments in 94 countries, and has been approved in 70 countries for the treatment of pancreatic cancer. Tarceva is also in trials for the treatment of other tumor types, including hepatocellular carcinoma and ovarian and colorectal cancers. The firm collaborates with Genentech, Inc., and Roche for the continued development and commercialization of Tarceva. The company's other oncology products are all in clinical or late-stage pre-clinical development; they include OSI-906, an oral small molecule IGF-1 receptor inhibitor, which shows potential in treating NSCLC, breast, pancreas, prostate and colorectal cancers; OSI-027 a small molecule TORC1/TORC2 inhibitor; and OSI-930, which is designed to target both cancer cell proliferation and blood vessel growth in selected tumors. The firm's OSI Prosidion business unit focuses on the remaining drug candidates, including diabetes and obesity products. It has two products in clinical trials: PSN-602, a novel dual serotonin and noradrenaline reuptake inhibitor for the long-term treatment of obesity; and PSN-821, an agonist with potential anti-diabetic and appetite suppressing effects.

OSI offers its employees stock options; life and dependent life insurance; tuition reimbursement; a 401(K); and an employee assistance program.

## FINANCIALS: Sales and profits are in thousands of dollars—add 000 to get the full amount. 2008 Note: Financial information for 2008 was not available for all companies at press time.

| | | |
|---|---|---|
| 2008 Sales: $379,388 | 2008 Profits: $471,485 | **U.S. Stock Ticker: OSIP** |
| 2007 Sales: $341,030 | 2007 Profits: $66,319 | Int'l Ticker: Int'l Exchange: |
| 2006 Sales: $241,037 | 2006 Profits: $-582,184 | Employees: 514 |
| 2005 Sales: $174,194 | 2005 Profits: $-157,123 | Fiscal Year Ends: 12/31 |
| 2004 Sales: $42,800 | 2004 Profits: $-260,371 | Parent Company: |

## SALARIES/BENEFITS:

| Pension Plan: | ESOP Stock Plan: | Profit Sharing: | Top Exec. Salary: $652,214 | Bonus: $705,000 |
|---|---|---|---|---|
| Savings Plan: Y | Stock Purch. Plan: Y | | Second Exec. Salary: $438,300 | Bonus: $235,000 |

## OTHER THOUGHTS:

**Apparent Women Officers or Directors:** 5
**Hot Spot for Advancement for Women/Minorities:** Y

## LOCATIONS: ("Y" = Yes)

| West: | Southwest: | Midwest: | Southeast: | Northeast: | International: |
|---|---|---|---|---|---|
| Y | | | | Y | Y |

Note: Financial information, benefits and other data can change quickly and may vary from those stated here.

# OXIGENE INC

**www.oxigene.com**

**Industry Group Code:** 325412 **Ranks within this company's industry group:** Sales: 158 Profits: 106

| Drugs: | | Other: | Clinical: | Computers: | Services: |
|---|---|---|---|---|---|
| Discovery: | Y | AgriBio: | Trials/Services: | Hardware: | Specialty Services: |
| Licensing: | Y | Genetic Data: | Labs: | Software: | Consulting: |
| Manufacturing: | | Tissue Replacement: | Equipment/Supplies: | Arrays: | Blood Collection: |
| Genetics: | | | Research & Development Services: | Database Management: | Drug Delivery: |
| | | | Diagnostics: | | Drug Distribution: |

## TYPES OF BUSINESS:

Pharmaceuticals Acquisition & Development
Drugs-Cancer
Anti-Inflammatory Agents
Ocular Disease Treatments

## BRANDS/DIVISIONS/AFFILIATES:

Zybrestat
OXi4503

## CONTACTS: Note: Officers with more than one job title may be intentionally listed here more than once.

John A. Kollins, CEO
James B. Murphy, CFO/VP
David (Dai) Chaplin, Chief Scientific Officer/VP/Head-R&D
Patricia Walicke, Chief Medical Officer/VP
William Shiebler, Chmn.

| Phone: 781-547-5900 | Fax: 781-547-6800 |
|---|---|
| Toll-Free: | |
| Address: 230 3rd Ave., Waltham, MA 02451 US | |

## GROWTH PLANS/SPECIAL FEATURES:

OXiGENE, Inc. is a clinical stage international biopharmaceutical company that researches and develops products to treat cancer and certain ocular diseases. It in-licenses complementary compounds from academic institutions in order to lead them through early-stage clinical trials and negotiate contracts with pharmaceutical companies to develop, market and manufacture resulting commercial drugs and clinical products. OXiGENE's primary drug development programs are based on a series of natural products called Combretastatins, which were originally isolated from the African bush willow tree (Combretum caffrum) by researchers at Arizona State University (ASU). ASU has granted the company a worldwide license for the use of the Combretastatins. OXiGENE has developed its primary technologies based on Combretastatins, called vascular disrupting agents (VDAs). The company's VDA compound, ZYBRESTAT (fosbretabulin, formerly known as combretastatin A4 phosphate or CA4P), attacks certain solid tumors and other diseases by selectively destroying their characteristically abnormal blood vessels. ZYBRESTAT is currently in Phase III clinical testing for anaplastic thyroid cancer; near the end of Phase II testing for platinum resistant ovarian cancer; and in the middle of Phase II testing for non-small cell lung cancer in combination with Avastin (bevacizumab), an anti-angiogenic agent developed by third-party Genentech, Inc. Additionally, the firm is developing OXi4503, which is a structural analog to fosbretabulin, but differs in that it also possesses direct cytotoxic activity. This drug is currently in Phase I testing for solid and hepatic tumors. The company is also developing ZYBRESTAT for topical use in ophthalmological conditions for wet age-related macular degeneration (ARMD). This delivery method is preferable to current treatments, which require direct injection into the eye.

OXiGENE offers its employees medical and dental coverage, flexible spending accounts and a 401(k) plan.

## FINANCIALS: Sales and profits are in thousands of dollars—add 000 to get the full amount. 2008 Note: Financial information for 2008 was not available for all companies at press time.

| | | U.S. Stock Ticker: OXGN |
|---|---|---|
| 2008 Sales: $ 12 | 2008 Profits: $-21,401 | Int'l Ticker: Int'l Exchange: |
| 2007 Sales: $ 12 | 2007 Profits: $-20,389 | Employees: 34 |
| 2006 Sales: $ | 2006 Profits: $-15,457 | Fiscal Year Ends: 12/31 |
| 2005 Sales: $ 1 | 2005 Profits: $-11,909 | Parent Company: |
| 2004 Sales: $ 7 | 2004 Profits: $-10,024 | |

## SALARIES/BENEFITS:

| Pension Plan: | ESOP Stock Plan: | Profit Sharing: | Top Exec. Salary: $334,409 | Bonus: $ |
|---|---|---|---|---|
| Savings Plan: Y | Stock Purch. Plan: | | Second Exec. Salary: $327,304 | Bonus: $ |

## OTHER THOUGHTS:

**Apparent Women Officers or Directors:** 1
**Hot Spot for Advancement for Women/Minorities:**

## LOCATIONS: ("Y" = Yes)

| West: | Southwest: | Midwest: | Southeast: | Northeast: | International: |
|---|---|---|---|---|---|
| Y | | | | Y | Y |

# PACIFIC BIOMETRICS INC

**www.pacbio.com**

**Industry Group Code:** 541712  **Ranks within this company's industry group:** Sales: 18   Profits: 8

| Drugs: | Other: | Clinical: | | Computers: | | Services: | |
|---|---|---|---|---|---|---|---|
| Discovery: | AgriBio: | Trials/Services: | Y | Hardware: | | Specialty Services: | |
| Licensing: | Genetic Data: | Labs: | Y | Software: | | Consulting: | Y |
| Manufacturing: | Tissue Replacement: | Equipment/Supplies: | | Arrays: | | Blood Collection: | |
| Genetics: | | Research & Development Services: | Y | Database Management: | | Drug Delivery: | |
| | | Diagnostics: | Y | | | Drug Distribution: | |

## TYPES OF BUSINESS:

Clinical Trials
Laboratory Services
Contract Research & Development
Diagnostic Tests
DNA Amplification Systems

## BRANDS/DIVISIONS/AFFILIATES:

PBI Technology Inc
Pacific Biomarkers Inc
Cholesterol Reference Method Laboratory Network

## CONTACTS: *Note: Officers with more than one job title may be intentionally listed here more than once.*

Ronald R. Helm, CEO
Elizabeth Teng Leary, Chief Scientific Officer
Michael Murphy, Sr. VP-Oper.
Patrick Jackle, Dir.-Bus. Dev.
Kari Charbonnel, Investor Rel.
John Jensen, Controller/VP-Finance
Mario Ehlers, CEO/Dir.-Pacific Biomarker's, Inc.
Kristin Walsh, Mgr.-Oper.
Timothy Carlson, Dir.-Laboratory
Tonya Aggoune, Dir.-Project Svcs.
Ronald R. Helm, Chmn.

| **Phone:** 206-298-0068 | **Fax:** 206-298-9838 |
|---|---|
| **Toll-Free:** | |
| **Address:** 220 W. Harrison St., Seattle, WA 98119 US | |

## GROWTH PLANS/SPECIAL FEATURES:

Pacific Biometrics, Inc. (PBI) provides specialty laboratory and clinical research services to pharmaceutical, biotechnology and laboratory manufacturers. Laboratory services are offered primarily in support of clinical trials and diagnostic product development. In clinical trial support, tailored databases are customized for each clinical research study protocol while a data analyst facilitates adapted data management plans for all clients. Diagnostic product development services include the development and improvement of reagents and point-of-care devices for diagnostic companies. PBI's areas of specialty include cardiovascular disease, diabetes and bone and joint diseases. Services offered include the measurement of cardiovascular disease markers through lipoprotein components, cholesterol, triglycerides, phospholipids and apolipoproteins; testing for diabetes markers such as glucose, HbA1c, microalbumin and non-esterified fatty acids; measurements of hormone and biochemical markers such as pyridonolines, procollagens and osteocalcin; and bone-specific alkaline phosphatase and cartilage oligomeric matrix protein for osteoporosis, bone and cartilage metabolism. PBI's wholly-owned subsidiary, PBI Technology, Inc., develops and commercializes molecular diagnostic technologies, non-invasive diagnostic devices and early-stage drug candidates. PBI Technology also owns DNA-based proprietary technologies, processes and equipment, including a proprietary isothermal DNA amplification method (LIDA) and a genetic method for distinguishing live cells from dead cells (Cell Viability). The Pacific Biometrics Research Foundation, an affiliated non-profit organization, is certified by the U.S. Centers for Disease Control and Prevention (CDC) as part of the Cholesterol Reference Method Laboratory Network (CRMLN), enabling it to perform official testing related to cholesterol measurement. During 2008, PBI formed a new subsidiary, Pacific Biomarkers, Inc., to focus on the delivery of clinical biomarker services to pharmaceutical and biotechnology clients. Pacific Biomarkers provides assay development services for novel biomarkers, as well as custom assay services for immunogenicity testing and multiplex testing.

PBI offers its employees medical, dental and vision plans, flexible spending accounts and a 401(k) savings plan, among other benefits.

## FINANCIALS: Sales and profits are in thousands of dollars—add 000 to get the full amount. 2008 Note: Financial information for 2008 was not available for all companies at press time.

| | | |
|---|---|---|
| 2008 Sales: $8,265 | 2008 Profits: $- 571 | **U.S. Stock Ticker:** PBME |
| 2007 Sales: $8,480 | 2007 Profits: $-1,213 | **Int'l Ticker:**   Int'l Exchange: |
| 2006 Sales: $10,750 | 2006 Profits: $ 179 | Employees:   56 |
| 2005 Sales: $3,230 | 2005 Profits: $-2,993 | Fiscal Year Ends: 6/30 |
| 2004 Sales: $4,801 | 2004 Profits: $-1,884 | Parent Company: |

## SALARIES/BENEFITS:

| Pension Plan: | ESOP Stock Plan: | Profit Sharing: | Top Exec. Salary: $239,000 | Bonus: $24,000 |
|---|---|---|---|---|
| Savings Plan: Y | Stock Purch. Plan: | | Second Exec. Salary: $206,250 | Bonus: $10,000 |

## OTHER THOUGHTS:

**Apparent Women Officers or Directors**: 3
**Hot Spot for Advancement for Women/Minorities**: Y

## LOCATIONS: ("Y" = Yes)

| West: | Southwest: | Midwest: | Southeast: | Northeast: | International: |
|---|---|---|---|---|---|
| Y | | | | | |

Note: Financial information, benefits and other data can change quickly and may vary from those stated here.

# PAIN THERAPEUTICS INC

### www.paintrials.com

**Industry Group Code:** 325412  **Ranks within this company's industry group:** Sales: 80   Profits: 49

| Drugs: | | Other: | Clinical: | Computers: | Services: | |
|---|---|---|---|---|---|---|
| Discovery: | Y | AgriBio: | Trials/Services: | Hardware: | Specialty Services: | |
| Licensing: | Y | Genetic Data: | Labs: | Software: | Consulting: | |
| Manufacturing: | | Tissue Replacement: | Equipment/Supplies: | Arrays: | Blood Collection: | |
| Genetics: | | | Research & Development Services: | Database Management: | Drug Delivery: | Y |
| | | | Diagnostics: | | Drug Distribution: | |

## TYPES OF BUSINESS:

Drugs, Opioids
Abuse-Resistant Drug Delivery
Drugs-Metastatic Melanoma
Drugs-Hemophilia

## BRANDS/DIVISIONS/AFFILIATES:

Oxytrex
Remoxy
PTI-202
PTI-721
PTI-188
King Pharmaceuticals, Inc.

## CONTACTS: *Note: Officers with more than one job title may be intentionally listed here more than once.*

Remi Barbier, CEO
Nadav Friedmann, COO
Remi Barbier, Pres.
Peter S. Roddy, CFO/VP
Grant L. Schoenhard, Chief Scientific Officer
George Ben Thornton, Sr. VP-Tech.
Roger Fu, VP-Pharmaceutical Dev.
Michael J. O'Donnell, Sec.
Peter Butera, VP-Clinical Oper.
Nadav Friedmann, Chief Medical Officer
Michael Zamloot, Sr. VP-Tech. Oper.
Michael Marsman, VP-Regulatory Affairs
Annelies de Kater, VP-Nonclinical Dev.
Remi Barbier, Chmn.

| **Phone:** 650-624-8200 | **Fax:** 650-624-8222 |
|---|---|
| **Toll-Free:** | |
| **Address:** 2211 Bridgepointe Pkwy., Ste. 500, San Mateo, CA 94404 US | |

## GROWTH PLANS/SPECIAL FEATURES:

Pain Therapeutics, Inc. is a biopharmaceutical company that develops novel drugs, mainly for oncology (cancer) and severe pain, particularly opioids. The company's lead drug candidate is Remoxy, which is currently in Phase III trials. Remoxy is an abuse-deterrent long-acting oral version of oxycodone designed to foil abusers who attempt to use the drug recreationally and to prevent accidental overdose by releasing just a small amount of oxycodone over time. Pain Therapeutics has a collaboration agreement with King Pharmaceuticals, Inc. to develop and commercialize Remoxy and other abuse-resistant opioid painkillers. The firm has three drugs in Phase I clinical studies: PTI-202 and PTI-721, which are also abuse-resistant opioid painkillers; and PTI-188, a novel radio-labeled monoclonal antibody for metastatic melanoma, a rare but deadly form of skin cancer, the rights of which were obtained from the Albert Einstein College of Medicine. It is also working on a Factor IX replacement for hemophilia, which is in pre-clinical study. Pain Therapeutics was previously working on Oxytrex, another non-addictive oral opioid, but work on this project was discontinued to focus on other developmental biotech products. In December 2008, the FDA declined to approve Remoxy, citing the ease of possible abuse of Remoxy's active ingredient. The decision is likely to delay any possible launch of the drug by months. In May 2009, the company launched a second Phase I study for PTI-188.

## FINANCIALS: Sales and profits are in thousands of dollars—add 000 to get the full amount. 2008 Note: Financial information for 2008 was not available for all companies at press time.

| | | |
|---|---|---|
| 2008 Sales: $63,725 | 2008 Profits: $15,347 | **U.S. Stock Ticker:** PTIE |
| 2007 Sales: $65,984 | 2007 Profits: $20,305 | **Int'l Ticker:**     Int'l Exchange: |
| 2006 Sales: $53,918 | 2006 Profits: $6,188 | Employees:    48 |
| 2005 Sales: $5,080 | 2005 Profits: $-30,670 | Fiscal Year Ends: 12/31 |
| 2004 Sales: $ | 2004 Profits: $-37,776 | Parent Company: |

## SALARIES/BENEFITS:

| Pension Plan: | ESOP Stock Plan: | Profit Sharing: | Top Exec. Salary: $566,250 | Bonus: $425,000 |
|---|---|---|---|---|
| Savings Plan: Y | Stock Purch. Plan: Y | | Second Exec. Salary: $438,542 | Bonus: $320,000 |

## OTHER THOUGHTS:

Apparent Women Officers or Directors: 1
**Hot Spot for Advancement for Women/Minorities:**

## LOCATIONS: ("Y" = Yes)

| West: | Southwest: | Midwest: | Southeast: | Northeast: | International: |
|---|---|---|---|---|---|
| Y | | | | | |

# PALATIN TECHNOLOGIES INC

**www.palatin.com**

**Industry Group Code:** 325412 **Ranks within this company's industry group:** Sales: 113 Profits: 90

| Drugs: | | Other: | | Clinical: | | Computers: | | Services: | |
|---|---|---|---|---|---|---|---|---|---|
| Discovery: | Y | AgriBio: | | Trials/Services: | | Hardware: | | Specialty Services: | Y |
| Licensing: | | Genetic Data: | Y | Labs: | | Software: | | Consulting: | |
| Manufacturing: | | Tissue Replacement: | | Equipment/Supplies: | | Arrays: | | Blood Collection: | |
| Genetics: | | | | Research & Development Services: | | Database Management: | | Drug Delivery: | |
| | | | | Diagnostics: | Y | | | Drug Distribution: | |

## TYPES OF BUSINESS:

Drugs-Diversified
Sexual Dysfunction Drugs
Inflammation Drugs
Peptide Technology
Diagnostic Imaging Products
Obesity Treatments

## BRANDS/DIVISIONS/AFFILIATES:

NeutroSpec
MIDAS
Bremelanotide

## CONTACTS: Note: Officers with more than one job title may be intentionally listed here more than once.

Carl Spana, CEO
Carl Spana, Pres.
Stephen T. Wills, CFO
Trevor Hallam, Exec. VP-R&D
Stephen T. Wills, Sec.
Stephen T. Wills, Exec. VP-Oper.
John Prendergast, Chmn.

| **Phone:** 609-495-2200 | **Fax:** 609-495-2201 |
|---|---|
| **Toll-Free:** | |
| **Address:** 4-C Cedar Brook Dr., Cranbury, NJ 08512 US | |

## GROWTH PLANS/SPECIAL FEATURES:

Palatin Technologies, Inc. is a development-stage biopharmaceutical company committed to the discovery, development and commercialization of novel therapeutics. Its primary focus is discovering and developing melanocortin (MC)-based therapeutics. The MC family of receptors has been identified with a variety of conditions and diseases, including sexual dysfunction; obesity; cachexia (extreme wasting, generally secondary to a chronic disease); and inflammation. Palatin's patented MIDAS (metal ion-induced distinctive array of structures) platform allows the company to synthesize pharmaceuticals that mimic the activity of peptides. MIDAS can generate both receptor antagonists and agonists, to either block or promote metabolic responses. The company is engaged in research and development using MIDAS to diagnose infections and treat cancer, sexual dysfunction, cachexia, congestive heart failure, obesity and inflammation. Products in development consist of PL-3994, a peptide receptor agonist for the treatment of hearth failure and hypertension; Bremelanotide, a peptide melanocortin receptor agonist for prevention of organ damage secondary to cardiac surgery; PL-6983, a peptide melanocortin receptor agonist for the treatment of female sexual dysfunction; NeutroSpec, a radiolabeled monoclonal antibody product for imaging and diagnosing infection. The ongoing clinical trials and regulatory approvals for NeutroSpec have been suspended while the company evaluates its future development and marketing activities with the Mallinckrodt division of Covidien Ltd. The firm also has a strategic alliance with AstraZeneca AB for the research of melanocortin receptor-based compounds for the treatment of obesity.

Employees are offered medical, vision and dental insurance; life and disability insurance; a 401(k) savings plan; incentive stock option plans; educational assistance; and tuition reimbursement.

## FINANCIALS: Sales and profits are in thousands of dollars—add 000 to get the full amount. 2008 Note: Financial information for 2008 was not available for all companies at press time.

| | | |
|---|---|---|
| 2008 Sales: $11,483 | 2008 Profits: $-14,384 | **U.S. Stock Ticker:** PTN |
| 2007 Sales: $14,406 | 2007 Profits: $-27,752 | **Int'l Ticker:** Int'l Exchange: |
| 2006 Sales: $19,749 | 2006 Profits: $-28,959 | Employees: 46 |
| 2005 Sales: $17,957 | 2005 Profits: $-14,358 | Fiscal Year Ends: 6/30 |
| 2004 Sales: $2,315 | 2004 Profits: $-26,318 | Parent Company: |

## SALARIES/BENEFITS:

| Pension Plan: | ESOP Stock Plan: | Profit Sharing: | Top Exec. Salary: $390,000 | Bonus: $ |
|---|---|---|---|---|
| Savings Plan: Y | Stock Purch. Plan: Y | | Second Exec. Salary: $321,000 | Bonus: $ |

## OTHER THOUGHTS:

**Apparent Women Officers or Directors:**
**Hot Spot for Advancement for Women/Minorities:**

## LOCATIONS: ("Y" = Yes)

| West: | Southwest: | Midwest: | Southeast: | Northeast: | International: |
|---|---|---|---|---|---|
| | | | | Y | |

# PAR PHARMACEUTICAL COMPANIES INC

www.parpharm.com

**Industry Group Code:** 325412A **Ranks within this company's industry group:** Sales: 8 Profits: 27

| Drugs: | | Other: | | Clinical: | | Computers: | | Services: | |
|---|---|---|---|---|---|---|---|---|---|
| Discovery: | Y | AgriBio: | | Trials/Services: | | Hardware: | | Specialty Services: | |
| Licensing: | | Genetic Data: | | Labs: | | Software: | | Consulting: | |
| Manufacturing: | Y | Tissue Replacement: | | Equipment/Supplies: | | Arrays: | | Blood Collection: | |
| Genetics: | Y | | | Research & Development Services: | | Database Management: | | Drug Delivery: | |
| | | | | Diagnostics: | | | | Drug Distribution: | |

## TYPES OF BUSINESS:

Drugs-Generic & Branded
Pharmaceutical Intermediates

## BRANDS/DIVISIONS/AFFILIATES:

Pharmaceutical Resources, Inc.
Par Pharmaceutical, Inc.
Megace ES
Nascobal
Strativa Pharmaceuticals
AstraZeneca
GlaxoSmithKline plc
Bristol-Meyers Squibb Company

## CONTACTS: *Note: Officers with more than one job title may be intentionally listed here more than once.*

Patrick G. LePore, CEO
Gerard A. Martino, COO/Exec. VP
Patrick G. LePore, Pres.
Lawrence A. Kenyon, CFO/Exec. VP
Thomas Haughey, General Counsel/Corp. Sec./Exec. VP
Paul V. Campanelli, Exec. VP/Pres., Generics Div.
John A. MacPhee, Exec. VP/Pres., Branded Products Div.
Patrick G. LePore, Chmn.

| | |
|---|---|
| **Phone:** 201-802-4000 | **Fax:** 201-802-4600 |
| **Toll-Free:** | |
| **Address:** 300 Tice Blvd., Woodcliff Lake, NJ 07677 US | |

## GROWTH PLANS/SPECIAL FEATURES:

Par Pharmaceutical Companies, Inc. (formerly Pharmaceutical Resources, Inc.) develops, manufactures and markets branded and generic pharmaceuticals through its principal subsidiary, Par Pharmaceutical, Inc. Par operates in two segments, generic pharmaceuticals and brand pharmaceuticals. In the generic segment, the company's product line includes approximately 244 products. These are manufactured principally in the solid oral dosage form (tablet, caplet and two-piece hard shell capsule). Some products are the result of license agreements with the branded drug's manufacturer, including generics of Glucophage and Glucovance through Bristol-Meyers Squibb Company; Flonase and Zantac through GlaxoSmithKline plc; and Toprol XL through AstraZeneca. Par markets its generic products primarily to wholesalers, drug store chains, supermarket chains, mass merchandisers, distributors, managed health care organizations, mail order accounts and government, principally through its internal staff. Brand products are marketed by wholly-owned subsidiary, Strativa Pharmaceuticals. Par's only product in the branded segment is Megace ES, which is approved for the treatment of anorexia or an unexplained, significant weight loss in patients with a diagnosis of AIDS. Par's research and development activities consist principally of identifying and conducting patent and market research on brand name drugs for which patent protection has expired or will soon expire; identifying and conducting patent and market research on brand name drugs for which Par believes the patents are invalid or Par can develop a non-infringing formulation; researching and developing new product formulations based upon such drugs; and introducing technology to improve production efficiency and enhance product quality. In April 2009, subsidiary Strativa Pharmaceuticals acquired the worldwide rights to Nascobal from QOL Medical LLC. Nascobal is a vitamin B12 nasal spray that offers an alternative to B12 injections.

The firm offers employees medical and dental coverage; life insurance; flexible spending plans; a 401(k) plan; the 529 college savings plan; a stock purchase plan; and bonus programs.

## FINANCIALS: **Sales and profits are in thousands of dollars—add 000 to get the full amount. 2008 Note: Financial information for 2008 was not available for all companies at press time.**

| | | |
|---|---|---|
| 2008 Sales: $578,115 | 2008 Profits: $-47,768 | **U.S. Stock Ticker: PRX** |
| 2007 Sales: $769,666 | 2007 Profits: $49,898 | **Int'l Ticker:** Int'l Exchange: |
| 2006 Sales: $725,168 | 2006 Profits: $5,847 | Employees: 654 |
| 2005 Sales: $432,256 | 2005 Profits: $-15,309 | Fiscal Year Ends: 9/30 |
| 2004 Sales: $647,975 | 2004 Profits: $7,558 | Parent Company: |

## SALARIES/BENEFITS:

| | | | | |
|---|---|---|---|---|
| Pension Plan: | ESOP Stock Plan: | Profit Sharing: | Top Exec. Salary: $800,000 | Bonus: $ |
| Savings Plan: Y | Stock Purch. Plan: Y | | Second Exec. Salary: $470,000 | Bonus: $ |

## OTHER THOUGHTS:

**Apparent Women Officers or Directors:** 1
**Hot Spot for Advancement for Women/Minorities:**

## LOCATIONS: ("Y" = Yes)

| West: | Southwest: | Midwest: | Southeast: | Northeast: | International: |
|---|---|---|---|---|---|
| | | | | Y | |

# PAREXEL INTERNATIONAL CORP

**www.parexel.com**

Industry Group Code: 541712  Ranks within this company's industry group: Sales: 4  Profits: 4

| Drugs: | Other: | Clinical: | | Computers: | | Services: | |
|---|---|---|---|---|---|---|---|
| Discovery: | AgriBio: | Trials/Services: | Y | Hardware: | Y | Specialty Services: | Y |
| Licensing: | Genetic Data: | Labs: | | Software: | Y | Consulting: | Y |
| Manufacturing: | Tissue Replacement: | Equipment/Supplies: | | Arrays: | | Blood Collection: | |
| Genetics: | | Research & Development Services: | Y | Database Management: | Y | Drug Delivery: | |
| | | Diagnostics: | | | | Drug Distribution: | |

## TYPES OF BUSINESS:

Clinical Trial & Data Management
Biostatistical Analysis & Reporting
Medical Communications Services
Clinical Pharmacology Services
Consulting Services

## BRANDS/DIVISIONS/AFFILIATES:

Clinical Research Services
PAREXEL Consulting & Medical Communications Svcs.
Perceptive Informatics, Inc.
Synchron Research
ClinPhone plc
Safe Implementation of Treatments in Stroke

## CONTACTS: *Note: Officers with more than one job title may be intentionally listed here more than once.*

Josef H. von Rickenbach, CEO
Mark A. Goldberg, COO
James F. Winschel, Jr., CFO/Sr. VP
Ulf Schneider, Chief Admin. Officer/Sr. VP
Douglas A. Batt, General Counsel/Corp. Sec./Sr. VP
Jill L. Baker, VP-Investor Rel.
Kurt A. Brykman, Pres., PAREXEL Consulting & Medical Comm. Svcs.
Josef H. von Rickenbach, Chmn.

| | |
|---|---|
| **Phone:** 781-487-9900 | **Fax:** 781-768-5512 |
| **Toll-Free:** | |
| **Address:** 200 West St., Waltham, MA 02451 US | |

## GROWTH PLANS/SPECIAL FEATURES:

PAREXEL International is a biopharmaceutical services company providing clinical research, medical communications services, consulting and informatics and advanced technology products and services to the worldwide pharmaceutical, biotechnology and medical device industries. Operating in 71 locations throughout 52 countries, PAREXEL has three business segments: Clinical Research Services (CRS); PAREXEL Consulting and Medical Communications Services (PCMS); and Perceptive Informatics, Inc. PAREXEL's core business, CRS, provides clinical trials management and biostatistics; data management; clinical pharmacology; and related medical advisory, patient recruitment and investigator site services. PCMS provides technical expertise and advice for drug development, regulatory affairs and biopharmaceutical process consulting; offers product launch support, including market development, product development and targeted communications services; identifies alternatives and solutions regarding product development, registration and commercialization; and provides health policy consulting and strategic reimbursement services. Lastly, Perceptive provides information technology designed to improve clients' product development processes, including medical imaging services, IVRS, CTMS, web-based portals, systems integration, and patient diary applications. In March 2008, the company sold a bioanalytical laboratory in Poitiers, France, to a subsidiary of Synchron Research. Also in March 2008, PAREXEL increased its minority stake in clinical pharmacology business of Synchron from 19.5% to 31%. In August 2008, the firm acquired ClinPhone plc, a clinical technology organization. In November 2008, PAREXEL opened an office in Lima, Peru. In December 2008, the company established an alliance with Safe Implementation of Treatments in Stroke (SITS) International to provide PAREXEL clients access to SITS' 800 investigator sites in 40 countries.

## FINANCIALS: Sales and profits are in thousands of dollars—add 000 to get the full amount. 2008 Note: Financial information for 2008 was not available for all companies at press time.

| | | |
|---|---|---|
| 2008 Sales: $964,283 | 2008 Profits: $64,640 | **U.S. Stock Ticker: PRXL** |
| 2007 Sales: $741,955 | 2007 Profits: $37,289 | **Int'l Ticker:** Int'l Exchange: |
| 2006 Sales: $614,947 | 2006 Profits: $23,544 | Employees: 8,050 |
| 2005 Sales: $544,726 | 2005 Profits: $-35,177 | Fiscal Year Ends: 6/30 |
| 2004 Sales: $540,983 | 2004 Profits: $13,791 | Parent Company: |

## SALARIES/BENEFITS:

| | | | | |
|---|---|---|---|---|
| Pension Plan: | ESOP Stock Plan: | Profit Sharing: | Top Exec. Salary: $550,000 | Bonus: $633,600 |
| Savings Plan: | Stock Purch. Plan: | | Second Exec. Salary: $409,679 | Bonus: $178,882 |

## OTHER THOUGHTS:

**Apparent Women Officers or Directors:** 1
**Hot Spot for Advancement for Women/Minorities:**

## LOCATIONS: ("Y" = Yes)

| West: | Southwest: | Midwest: | Southeast: | Northeast: | International: |
|---|---|---|---|---|---|
| Y | | | | Y | Y |

# PDL BIOPHARMA

**www.pdl.com**

**Industry Group Code: 325412  Ranks within this company's industry group:  Sales: 50    Profits: 36**

| Drugs: | | Other: | Clinical: | Computers: | Services: |
|---|---|---|---|---|---|
| Discovery: | Y | AgriBio: | Trials/Services: | Hardware: | Specialty Services: |
| Licensing: | Y | Genetic Data: | Labs: | Software: | Consulting: |
| Manufacturing: | | Tissue Replacement: | Equipment/Supplies: | Arrays: | Blood Collection: |
| Genetics: | | | Research & Development Services: | Database Management: | Drug Delivery: |
| | | | Diagnostics: | | Drug Distribution: |

## TYPES OF BUSINESS:

Pharmaceuticals Development
Drugs-Kidney Transplant Rejection
Humanized Monoclonal Antibodies
Oncology Drugs
Asthma Drugs
Autoimmune Disease Drugs

## BRANDS/DIVISIONS/AFFILIATES:

Protein Design Labs, Inc.
HuLuc63
Zenapax
Cardene IV
Retavase
Daclizumab
Volociximab
PDL192

## CONTACTS: *Note: Officers with more than one job title may be intentionally listed here more than once.*

John McLaughlin, CEO
John McLaughlin, Pres.
Christine Larson, CFO/VP
Richard Murray, Chief Scientific Officer/Exec. VP
Behrooz Najafi, VP-IT
Christopher L. Stone, General Counsel/VP/Sec.
Kathleen Rinehart, Corp. Rel.
Jean Suzuki, Investor Rel.
Karen J. Wilson, VP-Finance/Principal Acct. Officer
Mark McCamish, Chief Medical Officer/Sr. VP
Jaisim Shah, Chief Bus. Officer/Sr. VP
Maninder Hora, VP-Product Oper.
Julie Badillo, VP-Program & Alliance Mgmt.
Brad Goodwin, Chmn.

| Phone: 650-454-1000 | Fax: 650-454-2000 |
|---|---|
| Toll-Free: | |
| Address: 1400 Seaport Blvd, Redwood City, CA 94063 US | |

## GROWTH PLANS/SPECIAL FEATURES:

PDL BioPharma is a biopharmaceutical company focused on discovery and development of novel antibodies in oncology and immunologic diseases. PDL also receives royalties and other revenues through licensing agreements with numerous biotechnology and pharmaceutical companies based on its proprietary antibody-based platform. These licensing agreements have contributed to the development by its licensees of nine marketed products and cover several antibodies in clinical development. PDL currently has several investigational compounds in clinical development for severe or life-threatening diseases, some in collaboration. PDL's Zenapax was developed for the prevention of kidney transplant rejection. The product is currently licensed to Hoffman-La Roche, Inc. for marketing in the U.S., Europe and other countries. In the product pipeline, the company is developing Daclizumab, for asthma, multiple sclerosis and transplant management; Volociximab and PDL192 for the treatment of solid tumors; and HuLuc63 for the treatment of multiple myeloma. Daclizumab and Volociximab are being developed in collaboration with Biogen Idec. In recent news, the company sought to sell itself entirely, but in March 2008 ended the sale process and instead turned to restructuring itself, including a significant reduction in workforce. Also in March, the firm sold all rights to IV Busulfex to Otsuka Pharmaceutical; all rights to its cardiovascular products to EKR Therapeutics; and its antibody manufacturing facility in Minnesota to Genmab. In December 2008, PDL completed the spin-off of its biotechnology assets.

PDL offers employees tuition reimbursement, professional seminars and memberships, social events, adoption assistance, an employee assistance program, pet insurance and local attorney access.

## FINANCIALS: Sales and profits are in thousands of dollars—add 000 to get the full amount. 2008 Note: Financial information for 2008 was not available for all companies at press time.

| | | |
|---|---|---|
| 2008 Sales: $294,270 | 2008 Profits: $68,387 | **U.S. Stock Ticker: PDLI** |
| 2007 Sales: $224,913 | 2007 Profits: $-21,061 | **Int'l Ticker:**    Int'l Exchange: |
| 2006 Sales: $187,343 | 2006 Profits: $-130,020 | **Employees:**    887 |
| 2005 Sales: $280,569 | 2005 Profits: $-166,577 | **Fiscal Year Ends:** 12/31 |
| 2004 Sales: $96,024 | 2004 Profits: $-53,241 | **Parent Company:** |

## SALARIES/BENEFITS:

| Pension Plan: | ESOP Stock Plan: | Profit Sharing: | Top Exec. Salary: $373,277 | Bonus: $274,874 |
|---|---|---|---|---|
| Savings Plan: Y | Stock Purch. Plan: Y | | Second Exec. Salary: $345,804 | Bonus: $296,000 |

## OTHER THOUGHTS:

**Apparent Women Officers or Directors:** 4
**Hot Spot for Advancement for Women/Minorities:** Y

## LOCATIONS: ("Y" = Yes)

| West: | Southwest: | Midwest: | Southeast: | Northeast: | International: |
|---|---|---|---|---|---|
| Y | | | | | Y |

Note: Financial information, benefits and other data can change quickly and may vary from those stated here.

# PENWEST PHARMACEUTICALS CO

## www.penw.com

**Industry Group Code: 325412A  Ranks within this company's industry group: Sales: 21   Profits: 21**

| Drugs: | | Other: | | Clinical: | | Computers: | | Services: | |
|---|---|---|---|---|---|---|---|---|---|
| Discovery: | Y | AgriBio: | | Trials/Services: | | Hardware: | | Specialty Services: | |
| Licensing: | Y | Genetic Data: | | Labs: | | Software: | | Consulting: | |
| Manufacturing: | Y | Tissue Replacement: | | Equipment/Supplies: | | Arrays: | | Blood Collection: | |
| Genetics: | | | | Research & Development Services: | | Database Management: | | Drug Delivery: | Y |
| | | | | Diagnostics: | | | | Drug Distribution: | |

## TYPES OF BUSINESS:

Drug Delivery Systems
Nervous System Disorders Drugs

## BRANDS/DIVISIONS/AFFILIATES:

TIMERx
Geminex
SyncroDose
GastroDose
Opana ER
Procardia XL
PW 4153
A0001

## CONTACTS: *Note: Officers with more than one job title may be intentionally listed here more than once.*

Jennifer L. Good, CEO
Jennifer L. Good, Pres.
Paul Hayes, Sr. VP-Strategic Mktg.
Anand R. Baichwal, Chief Scientific Officer/Sr. VP-Licensing
Amale Hawi, Sr. VP-Pharmaceutical Dev.
Frank P. Muscolo, Chief Acct. Officer/Controller
Thomas R. Sciascia, Chief Medical Officer/Sr. VP
Paula D. Buckley, VP-QA/QC
Paul E. Freiman, Chmn.

| Phone: 203-796-3700 | Fax: 203-794-1393 |
|---|---|
| Toll-Free: | |
| Address: 39 Old Ridgebury Rd., Ste. 11, Danbury, CT 06810 US | |

## GROWTH PLANS/SPECIAL FEATURES:

Penwest Pharmaceuticals Co. develops pharmaceutical products based on proprietary drug delivery technologies with a focus on products that address disorders of the nervous system. The firm's products are based on oral drug delivery technologies. Opana ER is an extended release formulation of oxymorphone hydrochloride that the company developed with Endo Pharmaceuticals, Inc. using the TIMERx drug delivery technology. The drug is an oral extended release opioid analgesic approved for twice-a-day dosing in patients with moderate to severe pain requiring extended, continuous opioid treatment. Other products that utilize Timerx technology include Procardia XL, Slofedipine and Cronodipin for Hypertension and Cystrin CR for Urinary Incontinence. Products in clinical development include nalbuphine ER, a controlled release formulation of nalbuphine hydrochloride, for the treatment of pain; and A0001, a second-generation coenzyme Q10 analog for respiratory chain diseases. Penwest currently has four proprietary drug delivery technologies: TIMERx, a controlled-release technology; Geminex, a technology enabling drug release at two different rates; SyncroDose, a technology enabling controlled release at the appropriate site in the body; and the GastroDose system, a technology enabling drug delivery to the upper gastrointestinal tract. The company has more than 32 U.S. and 96 foreign patents. The firm commenced clinical trials in July 2008 for PW 4153, an extended-release drug that is administered several times per day for the treatment of Parkinson's disease. In February 2009, the Penwest began Phrase Ib clinical trial of A0001.

Employees are offered medical, dental and vision insurance; disability benefits; life and accident insurance; flexible spending accounts; education reimbursement; a 401(k) savings plan; and an employee stock purchase program.

## FINANCIALS: Sales and profits are in thousands of dollars—add 000 to get the full amount. 2008 Note: Financial information for 2008 was not available for all companies at press time.

| | | |
|---|---|---|
| 2008 Sales: $8,534 | 2008 Profits: $-26,734 | **U.S. Stock Ticker: PPCO** |
| 2007 Sales: $3,308 | 2007 Profits: $-34,465 | **Int'l Ticker:**  Int'l Exchange: |
| 2006 Sales: $3,499 | 2006 Profits: $-31,312 | Employees:  48 |
| 2005 Sales: $6,213 | 2005 Profits: $-22,898 | Fiscal Year Ends: 12/31 |
| 2004 Sales: $5,108 | 2004 Profits: $-23,785 | Parent Company: |

## SALARIES/BENEFITS:

| Pension Plan: | ESOP Stock Plan: | Profit Sharing: | Top Exec. Salary: $381,374 | Bonus: $92,880 |
|---|---|---|---|---|
| Savings Plan: Y | Stock Purch. Plan: Y | | Second Exec. Salary: $305,086 | Bonus: $55,800 |

## OTHER THOUGHTS:

**Apparent Women Officers or Directors**: 3
**Hot Spot for Advancement for Women/Minorities**: Y

## LOCATIONS: ("Y" = Yes)

| West: | Southwest: | Midwest: | Southeast: | Northeast: | International: |
|---|---|---|---|---|---|
| | | | | Y | |

# PEREGRINE PHARMACEUTICALS INC                      www.peregrineinc.com

**Industry Group Code: 325412  Ranks within this company's industry group:** Sales: 127    Profits: 109

| Drugs: | | Other: | Clinical: | | Computers: | | Services: |
|---|---|---|---|---|---|---|---|
| Discovery: | Y | AgriBio: | Trials/Services: | | Hardware: | | Specialty Services: |
| Licensing: | | Genetic Data: | Labs: | | Software: | | Consulting: |
| Manufacturing: | Y | Tissue Replacement: | Equipment/Supplies: | | Arrays: | | Blood Collection: |
| Genetics: | | | Research & Development Services: | Y | Database Management: | | Drug Delivery: |
| | | | Diagnostics: | | | | Drug Distribution: |

## TYPES OF BUSINESS:

Cancer & Hepatitis C Drugs
Contract Manufacturing Services

## BRANDS/DIVISIONS/AFFILIATES:

Avid Bioservices, Inc.
Cotara
Peregrine Beijing Pharmaceuticals Technology Dev.
Bavituximab

## CONTACTS: *Note: Officers with more than one job title may be intentionally listed here more than once.*

Steven W. King, CEO
Steven W. King, Pres.
Paul J. Lytle, CFO
Richard Richieri, Sr. VP-Mfg. & BioProcess Dev.
F. David King, VP-Bus. Dev.
Joseph Shan, Exec. Dir.-Clinical & Regulatory Affairs
Shelley Fussey, VP-Intellectual Property
John Quick, Head-Quality Systems
Thomas A. Waltz, Chmn.
Mary J. Boyd, Head-Bus. Dev., Asia & Europe

| Phone: 714-508-6000 | Fax: 714-838-5817 |
|---|---|
| Toll-Free: | |
| Address: 14272 Franklin Ave., Ste. 100, Tustin, CA 92780 US | |

## GROWTH PLANS/SPECIAL FEATURES:

Peregrine Pharmaceuticals, Inc. is a biopharmaceutical company with a portfolio of clinical stage and preclinical product candidates using monoclonal antibodies for the treatment of cancer and viral diseases. The company's products fall under two technology platforms: Anti-phosphatidylserine (anti-PS) and tumor necrosis therapy (TNT). The anti-PS immunotherapeutics are monoclonal antibodies that target and bind to components of cells normally found only on the inner surface of the cell membrane. The TNT technology uses monoclonal antibodies that target and bind DNA and associated histone proteins accessible in dead and dying cells found at the core of solid tumors. The firm's lead anti-PS product, bavituximab, is in separate clinical trials for the treatment of solid cancers and hepatitis C virus infection. Anti-PS therapy targets phospholipids on the external walls and penetrates those cells, leaving the healthy cells undisturbed. Under the firm's TNT technology, the lead product candidate is Cotara, for treatment of brain cancer. Additionally, the firm is pursuing three primary preclinical stage technologies: Anti-Angiogenesis Agents, which operate by inhibiting Vascular Endothelial Growth Factor to halt formation of blood vessels in tumors; Vasopermeation Enhancement Agents, which improve the uptake of chemotherapy drugs by tumors; and Vascular Targeting Agents, which connect with tumor blood vessels to inhibit blood flow. Peregrine operates a wholly-owned contact manufacturing subsidiary, Avid Bioservices, Inc. Avid provides several functions for Peregrine such as the manufacturing of all clinical supplies, commercial scale-up of products in clinical trials and assisting with the advancement of new clinical candidates. Avid also provides contract manufacturing services for outside biotechnology and biopharmaceutical companies on a fee-for-service basis. The firm also has a wholly-owned subsidiary in China called Peregrine Beijing Pharmaceuticals Technology Development, Ltd.

## FINANCIALS: Sales and profits are in thousands of dollars—add 000 to get the full amount. 2008 Note: Financial information for 2008 was not available for all companies at press time.

| | | |
|---|---|---|
| 2008 Sales: $6,093 | 2008 Profits: $-23,176 | **U.S. Stock Ticker:** PPHM |
| 2007 Sales: $3,708 | 2007 Profits: $-20,796 | **Int'l Ticker:**      Int'l Exchange: |
| 2006 Sales: $3,193 | 2006 Profits: $-17,061 | Employees:   138 |
| 2005 Sales: $4,959 | 2005 Profits: $-15,452 | Fiscal Year Ends: 4/30 |
| 2004 Sales: $3,314 | 2004 Profits: $-14,345 | Parent Company: |

## SALARIES/BENEFITS:

| Pension Plan: | ESOP Stock Plan: | Profit Sharing: | Top Exec. Salary: $384,534 | Bonus: $ |
|---|---|---|---|---|
| Savings Plan: Y | Stock Purch. Plan: | | Second Exec. Salary: $306,881 | Bonus: $ |

## OTHER THOUGHTS:

**Apparent Women Officers or Directors:** 2
**Hot Spot for Advancement for Women/Minorities:** Y

## LOCATIONS: ("Y" = Yes)

| West: | Southwest: | Midwest: | Southeast: | Northeast: | International: |
|---|---|---|---|---|---|
| Y | | | | | Y |

# PERRIGO CO

**www.perrigo.com**

Industry Group Code: 325412A  **Ranks within this company's industry group:** Sales: 4    Profits: 5

| Drugs: | | Other: | | Clinical: | | Computers: | | Services: | |
|---|---|---|---|---|---|---|---|---|---|
| Discovery: | | AgriBio: | | Trials/Services: | | Hardware: | | Specialty Services: | Y |
| Licensing: | | Genetic Data: | | Labs: | | Software: | | Consulting: | |
| Manufacturing: | Y | Tissue Replacement: | | Equipment/Supplies: | | Arrays: | | Blood Collection: | |
| Genetics: | Y | | | Research & Development Services: | | Database Management: | | Drug Delivery: | |
| | | | | Diagnostics: | | | | Drug Distribution: | |

## TYPES OF BUSINESS:

Generic Prescription Drugs
Over-the-Counter Pharmaceuticals
Nutritional Products
Active Pharmaceutical Ingredients
Consumer Products

## BRANDS/DIVISIONS/AFFILIATES:

Chemagis
Neca
Natural Formula
Perrigo New York, Inc.
Perrigo Israel Pharmaceuticals, Ltd.
JB Laboratories
Laboratorios Diba, S.A.
Unico Holdings

## CONTACTS: Note: Officers with more than one job title may be intentionally listed here more than once.

Joseph C. Papa, CEO
Joseph C. Papa, Pres.
Judy L. Brown, CFO/Exec. VP
James Tomshack, Sr. VP-Consumer Healthcare Sales
Michael Stewart, Sr. VP-Global Human Resources
Jatin Shah, Chief Scientific Officer/Sr. VP
Thomas M. Farrington, CIO
Todd W. Kingma, General Counsel/Exec. VP/Sec.
John T. Hendrickson, Exec. VP-Global Oper.
Jeffrey R. Needham, Sr. VP-Global Bus. Dev.
Arthur J. Shannon, VP-Corp. Comm.
Arthur J. Shannon, VP-Investor Rel.
Sharon Kochan, Exec. VP-U.S. Generics
Louis Yu, Sr. VP-Global Quality & Compliance
David T. Gibbons, Chmn.
Refael Lebel, Exec. VP/Gen. Mgr.-Perrigo Israel
John T. Hendrickson, Exec. VP-Supply Chain

| Phone: 269-673-8451 | Fax: 269-673-7534 |
|---|---|
| Toll-Free: | |
| Address: 515 Eastern Ave., Allegan, MI 49010 US | |

## GROWTH PLANS/SPECIAL FEATURES:

Perrigo Co. is a global healthcare supplier and one of the world's largest manufacturers of over-the-counter pharmaceutical and nutritional products for the store brand market. The company also develops and manufactures generic prescription drugs, active pharmaceutical ingredients (API) and consumer products. The firm operates through three segments: consumer healthcare, prescription pharmaceuticals and API. The consumer healthcare segment makes a broad line of products including analgesics, cough/cold/allergy/sinus, gastrointestinal, smoking cessation, first aid, vitamin and nutritional supplement products. The pharmaceuticals segment's primary activity is the development, manufacture and sale of generic prescription drug products, generally for the U.S. market. The company currently markets roughly 250 generic prescription products to approximately 110 customers. The API segment, through subsidiary Chemagis, develops, manufactures and markets API for the drug industry and branded pharmaceutical companies. In addition, Perrigo's operations also include the Israel consumer products segment, which consist of cosmetics, toiletries and detergents generally sold under brands such as Careline, Neca and Natural Formula, and the Israel pharmaceutical and diagnostic products segment, which includes the marketing and manufacturing of branded prescription drugs under long-term exclusive licenses and the importation of pharmaceutical, diagnostics and other medical products into Israel based on exclusive agreement with the manufacturers. Perrigo operates through several wholly-owned subsidiaries. In the U.S., these subsidiaries consist primarily of L. Perrigo Co.; Perrigo Co. of South Carolina; and Perrigo New York, Inc. Outside the U.S., the subsidiaries consist primarily of Perrigo Israel Pharmaceuticals, Ltd.; Quimica Y Farmacia S.A. de C.V.; Wrafton Laboratories, Ltd.; and Perrigo U.K., Ltd. In 2008, the company acquired JB Laboratories; the Mexican Manufacturer, Laboratorios Diba, S.A.; and Unico Holdings. In March 2009, the firm received FDA approval to market Sulfacetamide Sodium Topical Suspension, 10%, a topical acne treatment.

## FINANCIALS: Sales and profits are in thousands of dollars—add 000 to get the full amount. 2008 Note: Financial information for 2008 was not available for all companies at press time.

| | | |
|---|---|---|
| 2008 Sales: $1,822,131 | 2008 Profits: $135,773 | **U.S. Stock Ticker:** PRGO |
| 2007 Sales: $1,447,428 | 2007 Profits: $73,797 | **Int'l Ticker:**    Int'l Exchange: |
| 2006 Sales: $1,366,821 | 2006 Profits: $71,400 | Employees: 6,200 |
| 2005 Sales: $1,024,098 | 2005 Profits: $-325,983 | Fiscal Year Ends: 6/30 |
| 2004 Sales: $898,204 | 2004 Profits: $80,567 | Parent Company: |

## SALARIES/BENEFITS:

| Pension Plan: | ESOP Stock Plan: | Profit Sharing: | Top Exec. Salary: $775,000 | Bonus: $1,658,500 |
|---|---|---|---|---|
| Savings Plan: | Stock Purch. Plan: | | Second Exec. Salary: $389,275 | Bonus: $487,920 |

## OTHER THOUGHTS:

**Apparent Women Officers or Directors:** 3
**Hot Spot for Advancement for Women/Minorities:** Y

## LOCATIONS: ("Y" = Yes)

| West: | Southwest: | Midwest: | Southeast: | Northeast: | International: |
|---|---|---|---|---|---|
| | | | | Y | Y |

Note: Financial information, benefits and other data can change quickly and may vary from those stated here.

# PFIZER INC

**www.pfizer.com**

**Industry Group Code: 325412  Ranks within this company's industry group:** Sales: 2   Profits: 4

| Drugs: | | Other: | Clinical: | Computers: | Services: |
|---|---|---|---|---|---|
| Discovery: | Y | AgriBio: | Trials/Services: | Hardware: | Specialty Services: |
| Licensing: | | Genetic Data: | Labs: | Software: | Consulting: |
| Manufacturing: | Y | Tissue Replacement: | Equipment/Supplies: | Arrays: | Blood Collection: |
| Genetics: | | | Research & Development Services: | Database Management: | Drug Delivery: |
| | | | Diagnostics: | | Drug Distribution: |

## TYPES OF BUSINESS:

Pharmaceutical Drugs
Prescription Pharmaceuticals
Veterinary Pharmaceuticals

## BRANDS/DIVISIONS/AFFILIATES:

Norvasc
Viagra
Zoloft
Revolution/Stronghold
Sutent
Lipitor
Chantix
Rimadyl

## CONTACTS: *Note: Officers with more than one job title may be intentionally listed here more than once.*

Jeffrey B. Kindler, CEO
Frank D'Amelio, CFO
Mary S. McLeod, Sr. VP-Worldwide Human Resources
Martin Mackay, Pres., Pfizer Global R&D
Natale Ricciardi, Pres./Team Leader-Pfizer Global Mfg.
Amy Schulman, General Counsel/Corp. Sec./Sr. VP
William Ringo, Sr. VP-Strategy & Bus. Dev.
Sally Susman, Chief Comm. Officer/Sr. VP
Charles E. Triano, Sr. VP-Investor Rel.
Joe Feckzo, Chief Medical Officer
Corey Goodman, Pres., Biotherapeutics & Bioinnovation Center
Ian Read, Sr. VP/Pres., Worldwide Pharmaceutical Oper.
Jeffrey B. Kindler, Chmn.

| Phone: 212-573-2323 | Fax: |
|---|---|
| Toll-Free: | |
| Address: 235 E. 42nd St., New York, NY 10017 US | |

## GROWTH PLANS/SPECIAL FEATURES:

Pfizer, Inc. is a research-based, global pharmaceutical company. It discovers, develops, manufactures and markets prescription medicines for humans and animals. The company operates in two segments: pharmaceutical and animal health. The pharmaceutical business is one of the largest in the world, with medicines across 11 therapeutic areas: cardiovascular and metabolic diseases; central nervous system disorders; arthritis and pain; infectious and respiratory diseases; urology; oncology; ophthalmology; and endocrine disorders. Major pharmaceutical products include Lipitor, for the treatment of LDL-cholesterol levels in the blood; Norvasc, for treating hypertension; Zoloft, for the treatment of major depressive disorder and other conditions; and Viagra, a treatment for erectile dysfunction. Pfizer's latest drugs include Chantix, an anti-smoking agent; and Sutent, which kills cancer-infected cells and prevents blood flow from reaching existing tumors. The animal health segment develops and sells products for the prevention and treatment of diseases in livestock/companion animals. Among the products it markets are parasiticides, anti-inflammatories, antibiotics, vaccines, and anti-obesity agents. Brands include Revolution/Stronghold, a parasiticide for dogs and cats; Rimadyl, an arthritis pain medication; and Draxxin, an antibiotic used to treat infections in cattle and swine. In March 2008, the firm agreed to acquire Serenex, Inc. In June 2008, the company acquired Encysive Pharmaceuticals, Inc. In September 2008, Pfizer partnered with Medivation, Inc., to market Dimebon, potential inhibitor of Alzheimer's and Huntington's disease; and bought several European animal product franchises from Schering-Plough Corporation. In January 2009, the company agreed to acquire pharmaceuticals company Wyeth. In March 2009, Pfizer signed licensing agreements with India-based pharmaceutical company Aurobindo Pharma Ltd., expanding its generic medicines market; the firm also partnered with Bausch & Lomb to promote opthamalic pharmaceuticals. Also in March 2009, Pfizer agreed to sell its German insulin plant and assets to MannKind Corporation, and opened a new development plant in Dalian, Liaoning, China.

## FINANCIALS:  Sales and profits are in thousands of dollars—add 000 to get the full amount. 2008 Note: Financial information for 2008 was not available for all companies at press time.

| | | |
|---|---|---|
| 2008 Sales: $48,296,000 | 2008 Profits: $8,104,000 | **U.S. Stock Ticker: PFE** |
| 2007 Sales: $48,418,000 | 2007 Profits: $8,144,000 | **Int'l Ticker:**   Int'l Exchange: |
| 2006 Sales: $48,371,000 | 2006 Profits: $19,337,000 | Employees:  81,800 |
| 2005 Sales: $47,405,000 | 2005 Profits: $8,085,000 | Fiscal Year Ends: 12/31 |
| 2004 Sales: $48,988,000 | 2004 Profits: $11,361,000 | Parent Company: |

## SALARIES/BENEFITS:

| Pension Plan: Y | ESOP Stock Plan: | Profit Sharing: | Top Exec. Salary: $1,575,000 | Bonus: $ |
|---|---|---|---|---|
| Savings Plan: Y | Stock Purch. Plan: | | Second Exec. Salary: $1,051,500 | Bonus: $ |

## OTHER THOUGHTS:

**Apparent Women Officers or Directors:** 4
**Hot Spot for Advancement for Women/Minorities:** Y

## LOCATIONS: ("Y" = Yes)

| West: | Southwest: | Midwest: | Southeast: | Northeast: | International: |
|---|---|---|---|---|---|
| Y | Y | Y | Y | Y | Y |

Note: Financial information, benefits and other data can change quickly and may vary from those stated here.

# PHARMACEUTICAL PRODUCT DEVELOPMENT INC www.ppdi.com

**Industry Group Code: 541712  Ranks within this company's industry group: Sales: 2  Profits: 2**

| Drugs: | Other: | Clinical: | | Computers: | | Services: | |
|---|---|---|---|---|---|---|---|
| Discovery: | AgriBio: | Trials/Services: | | Hardware: | | Specialty Services: | Y |
| Licensing: | Genetic Data: | Labs: | | Software: | Y | Consulting: | Y |
| Manufacturing: | Tissue Replacement: | Equipment/Supplies: | Y | Arrays: | | Blood Collection: | |
| Genetics: | | Research & Development Services: | Y | Database Management: | | Drug Delivery: | |
| | | Diagnostics: | | | | Drug Distribution: | |

## TYPES OF BUSINESS:

Contract Research
Drug Discovery & Development Services
Clinical Data Consulting Services
Medical Marketing & Information Support Services
Drug Development Software
Medical Device Development

## BRANDS/DIVISIONS/AFFILIATES:

PPD Discovery Sciences
PPD Development
CSS Informatics
PPD Medical Communications
PPD Virtual
InnoPharm Ltd

## CONTACTS: *Note: Officers with more than one job title may be intentionally listed here more than once.*

Fred N. Eshelman, CEO
William J. Sharbaugh, COO
Daniel Darazsdi, CFO
Christine A. Dingivan, Chief Medical Officer/Exec. VP
B. Judd Hartman, General Counsel/Corp. Sec.
William W. Richardson, Sr. VP-Global Bus. Dev.
Louise Caudle, Dir.-Corp. Comm.
Luke Heagle, Dir.-Investor Rel.
Sue Ann Pentecost, Mgr.-Corp. Comm.
Michael O. Wilkinson, Exec. VP-Global Clinical Dev.
Ernest Mario, Chmn.

| **Phone:** 910-251-0081 | **Fax:** 910-762-5820 |
|---|---|
| **Toll-Free:** | |
| **Address:** 929 N. Front St., Wilmington, NC 28401-3331 US | |

## GROWTH PLANS/SPECIAL FEATURES:

Pharmaceutical Product Development, Inc. (PPD) provides drug discovery and development services to pharmaceutical and biotechnology companies as well as academic and government organizations. PPD's services are primarily divided into two company segments: Discovery Sciences and Development. Through the combined services of these segments, PPD helps pharmaceutical companies through all stages of clinical testing. The stages of testing can be specifically divided into preclinical, phase I, phase II-IIIb and post-approval. In the preclinical stages of drug testing, PPD provides information concerning the pharmaceutical composition of a new drug, its safety, its formulaic design and how it will be administered to children and adults. During phase I of testing, PPD conducts healthy volunteer clinics, provides data management services and guides companies/laboratories through regulatory affairs. In phase II and III tests, PPD oversees the later stages of product development and government approval, providing project management and clinical monitoring. In the post-approval stage of a drug's development, PPD provides technology and marketing services aimed to maximize the new drug's lifecycle. PPD has experience conducting research and drug development in the areas of antiviral studies, cardiovascular diseases, critical care studies, endocrine/metabolic studies, vaccine development, hematology/oncology studies, immunology studies and ophthalmology studies. The firm conducts regional, national and global studies and research projects through offices in 33 countries worldwide. In June 2008, the company opened an office in Istanbul. In October of the same year, PPD acquired Russia-based independent contract research organization, InnoPharm. In February 2009, the firm agreed to acquire AbC.R.O., Inc., a contract research organization that serves Central and Eastern Europe.

Employees are offered health insurance; life insurance; wellness programs; and a 401(k) savings plan.

## FINANCIALS: Sales and profits are in thousands of dollars—add 000 to get the full amount. 2008 Note: Financial information for 2008 was not available for all companies at press time.

| | | |
|---|---|---|
| 2008 Sales: $1,569,901 | 2008 Profits: $187,519 | **U.S. Stock Ticker:** PPDI |
| 2007 Sales: $1,414,465 | 2007 Profits: $163,401 | **Int'l Ticker:** **Int'l Exchange:** |
| 2006 Sales: $1,247,682 | 2006 Profits: $156,652 | Employees: 10,500 |
| 2005 Sales: $1,037,090 | 2005 Profits: $119,897 | Fiscal Year Ends: 12/31 |
| 2004 Sales: $841,256 | 2004 Profits: $91,684 | Parent Company: |

## SALARIES/BENEFITS:

| Pension Plan: | ESOP Stock Plan: | Profit Sharing: | Top Exec. Salary: $729,167 | Bonus: $255,208 |
|---|---|---|---|---|
| Savings Plan: Y | Stock Purch. Plan: | | Second Exec. Salary: $366,825 | Bonus: $91,706 |

## OTHER THOUGHTS:

**Apparent Women Officers or Directors:** 4
**Hot Spot for Advancement for Women/Minorities:** Y

## LOCATIONS: ("Y" = Yes)

| West: | Southwest: | Midwest: | Southeast: | Northeast: | International: |
|---|---|---|---|---|---|
| Y | Y | Y | Y | Y | Y |

Note: Financial information, benefits and other data can change quickly and may vary from those stated here.

# PHARMACYCLICS INC

**www.pharmacyclics.com**

Industry Group Code: 325412   **Ranks within this company's industry group:** Sales:    Profits: 113

| Drugs: | | Other: | Clinical: | Computers: | Services: |
|---|---|---|---|---|---|
| Discovery: | Y | AgriBio: | Trials/Services: | Hardware: | Specialty Services: |
| Licensing: | Y | Genetic Data: | Labs: | Software: | Consulting: |
| Manufacturing: | | Tissue Replacement: | Equipment/Supplies: | Arrays: | Blood Collection: |
| Genetics: | | | Research & Development Services: | Database Management: | Drug Delivery: |
| | | | Diagnostics: | | Drug Distribution: |

## TYPES OF BUSINESS:

Pharmaceuticals Development
Drugs-Cancer & Cardiovascular Disease
Texaphyrins

## BRANDS/DIVISIONS/AFFILIATES:

Xcytrin
PCI-24781

## CONTACTS: *Note: Officers with more than one job title may be intentionally listed here more than once.*

Robert W. Duggan, CEO
Glenn C. Rice, COO
Glenn C. Rice, Pres.
Maky Zanganeh, VP-Mktg.
Joseph J. Buggy, VP-Research
Maky Zanganeh, VP-Bus. Dev.
Ramses Erdtmann, VP-Finance
Ahmed Hamdy, Chief Medical Officer
Gregory Hemmi, VP-Chemical Oper.
David Loury, VP-Preclinical Sciences
Robert W. Duggan, Chmn.

| Phone: 408-774-0330 | Fax: 408-774-0340 |
|---|---|
| **Toll-Free:** | |
| **Address:** 995 E. Arques Ave., Sunnyvale, CA 94085-4521 US | |

## GROWTH PLANS/SPECIAL FEATURES:

Pharmacyclics, Inc. is a pharmaceutical company developing innovative treatments for cancer and other diseases. The company is working on a range of clinical and laboratory programs and is pursuing the development of novel drugs aimed at specific pathways. The company's most advanced product candidate is Xcytrin (motexafin gadolinium), a rationally designed small molecule, the first in a class of drugs that concentrates on cancer cells and disrupts cellular metabolism by a unique mechanism of action. Xcytrin is in late-stage human clinical trials for lung cancer and other oncology indications. Pharmacyclics is also developing other product candidates derived from structure-based research. A series of histone deacetylase (HDAC) inhibitors appear to inhibit tumor growth by interfering with cancer cell DNA transcription. Pharmacyclics' lead HDAC inhibitor is PCI-24781, which is in Phase II clinical trials. The company has also developed a small molecule inhibitor of Factor VIIa, which appears to inhibit tumor growth and angiogenesis when complexed with intrinsic tissue factor. Pharmacyclics, through its Bruton's Tyrosine Kinase (BTK) Inhibitors program, has developed drugs that can inhibit an enzyme required for early B-cells to mature into fully functioning cells. Tyrosine kinase inhibitors prevent B-cell maturation and are undergoing investigation for the treatment of lymphomas and autoimmune disorders. In April 2009, the company announced a strategic relationship with Les Labratories Servier, a French pharmaceutical company that works with HDAC inhibitors. Pharmacyclics also began a Phase I clinical trial for a BTK inhibitor that targets indications of Non-Hodgkin's Lymphoma.

## FINANCIALS: Sales and profits are in thousands of dollars—add 000 to get the full amount. 2008 Note: Financial information for 2008 was not available for all companies at press time.

| | | |
|---|---|---|
| 2008 Sales: $ | 2008 Profits: $-24,298 | U.S. Stock Ticker: PCYC |
| 2007 Sales: $ 126 | 2007 Profits: $-26,217 | Int'l Ticker:   Int'l Exchange: |
| 2006 Sales: $ 181 | 2006 Profits: $-42,158 | Employees: 47 |
| 2005 Sales: $ | 2005 Profits: $-31,048 | Fiscal Year Ends: 6/30 |
| 2004 Sales: $ | 2004 Profits: $-29,165 | Parent Company: |

## SALARIES/BENEFITS:

| Pension Plan: | ESOP Stock Plan: | Profit Sharing: | Top Exec. Salary: $438,973 | Bonus: $ |
|---|---|---|---|---|
| Savings Plan: Y | Stock Purch. Plan: Y | | Second Exec. Salary: $257,584 | Bonus: $ |

## OTHER THOUGHTS:

**Apparent Women Officers or Directors:** 2
**Hot Spot for Advancement for Women/Minorities:**

## LOCATIONS: ("Y" = Yes)

| West: | Southwest: | Midwest: | Southeast: | Northeast: | International: |
|---|---|---|---|---|---|
| Y | | | | | |

# PHARMANET DEVELOPMENT GROUP INC
## www.pharmanet.com

Industry Group Code: 541712 Ranks within this company's industry group: Sales: 6 Profits: 19

| Drugs: | Other: | Clinical: | | Computers: | | Services: | |
|---|---|---|---|---|---|---|---|
| Discovery: | AgriBio: | Trials/Services: | Y | Hardware: | | Specialty Services: | |
| Licensing: | Genetic Data: | Labs: | Y | Software: | Y | Consulting: | |
| Manufacturing: | Tissue Replacement: | Equipment/Supplies: | | Arrays: | | Blood Collection: | |
| Genetics: | | Research & Development Services: | Y | Database Management: | | Drug Delivery: | |
| | | Diagnostics: | | | | Drug Distribution: | |

## TYPES OF BUSINESS:

Contract Research
Clinical Trial Services
Bioanalytical Laboratory Services
Clinical Trial Software

## BRANDS/DIVISIONS/AFFILIATES:

SFBC International, Inc.
PharmaNet, Inc.
Anapharm, Inc.
PharmaSoft 2007

## CONTACTS: Note: Officers with more than one job title may be intentionally listed here more than once.

Jeffry P. McMullen, CEO
Steven A. George, COO/Sr. Exec. VP-Late Stage Dev.
Jeffrey P. McMullen, Pres.
John P. Hamill, CFO/Exec. VP
Robin C. Sheldrick, Sr. VP-Human Resources
Steven A. George, Sr. VP-IT
Mark DiIanni, Exec. VP-Strategic Initiatives
Pablo Fernandez, Sr. VP-Medical Affairs
Johane Boucher-Champagne, Exec. VP-Early Clinical Dev./Pres., Anapharm
Mark DiIanni, Pres., Early Stage Dev.
Thomas J. Newman, Exec. VP-Late Stage Dev.
Peter G. Tombros, Chmn.
Michael E. Laird, Sr. VP-Global Bus. Dev.

| Phone: 609-951-6800 | Fax: 609-514-0390 |
|---|---|
| Toll-Free: | |
| Address: 504 Carnegie Ctr., Princeton, NJ 08540 US | |

## GROWTH PLANS/SPECIAL FEATURES:

PharmaNet Development Group, Inc. (PharmaNet), formerly SFBC International, Inc., is a global drug development services company, which provides a broad range of both early and late stage clinical drug development services to branded pharmaceutical, biotechnology and generic drug and medical device companies around the world. PharmaNet also offers early clinical pharmacology research; biostatistics and data management; scientific and medical writing; and regulatory audit planning. The company reports in two operating segments: early stage and late stage clinical development. Early stage clinical development services consists of Phase I clinical trials and bioanalytical laboratory services. The company operates four early stage clinical trial facilities in Canada and bioanalytical services through five laboratories in the U.S., Canada and Spain. PharmaNet has developed and currently maintains extensive databases of available individuals who have indicated an interest in participating in future early stage clinical trials. PharmaNet provides late stage clinical development, including Phase II through Phase IV clinical trials, and related services through a network of 36 offices worldwide. It uses a full line of proprietary software products specifically designed to support clinical development activities. These web-based products facilitate the collection, management and reporting of clinical trial information. The firm has acquired a number of subsidiaries, which include PharmaNet, Inc., a late stage development subsidiary, and Anapharm, Inc., which is involved with early stage clinical trials and bioanalytical research. In January 2008, the firm purchased chemistry and immunoassay capabilities from Princeton Bioanalytical Laboratory, LLC. In November 2008, PharmaNet expanded the immunochemistry capabilities of its subsidiary Taylor Technology, Inc. In February 2009, the firm agreed to be acquired by JLL Partners. Also in February, the firm opened an office in Brazil.

## FINANCIALS: Sales and profits are in thousands of dollars—add 000 to get the full amount. 2008 Note: Financial information for 2008 was not available for all companies at press time.

| | | |
|---|---|---|
| 2008 Sales: $451,453 | 2008 Profits: $-251,093 | U.S. Stock Ticker: PDGI |
| 2007 Sales: $470,257 | 2007 Profits: $12,916 | Int'l Ticker:   Int'l Exchange: |
| 2006 Sales: $406,956 | 2006 Profits: $-36,025 | Employees: 2,400 |
| 2005 Sales: $361,506 | 2005 Profits: $4,779 | Fiscal Year Ends: 12/31 |
| 2004 Sales: $111,894 | 2004 Profits: $19,659 | Parent Company: |

## SALARIES/BENEFITS:

| Pension Plan: | ESOP Stock Plan: | Profit Sharing: | Top Exec. Salary: $741,932 | Bonus: $ |
|---|---|---|---|---|
| Savings Plan: Y | Stock Purch. Plan: | | Second Exec. Salary: $534,867 | Bonus: $ |

## OTHER THOUGHTS:

**Apparent Women Officers or Directors**: 4
**Hot Spot for Advancement for Women/Minorities**: Y

## LOCATIONS: ("Y" = Yes)

| West: | Southwest: | Midwest: | Southeast: | Northeast: | International: |
|---|---|---|---|---|---|
| Y | | Y | Y | Y | Y |

Note: Financial information, benefits and other data can change quickly and may vary from those stated here.

# PHARMOS CORP

**www.pharmoscorp.com**

**Industry Group Code: 325412  Ranks within this company's industry group:  Sales:      Profits: 78**

| Drugs: | | Other: | Clinical: | Computers: | Services: | |
|---|---|---|---|---|---|---|
| Discovery: | Y | AgriBio: | Trials/Services: | Hardware: | Specialty Services: | |
| Licensing: | | Genetic Data: | Labs: | Software: | Consulting: | |
| Manufacturing: | | Tissue Replacement: | Equipment/Supplies: | Arrays: | Blood Collection: | |
| Genetics: | | | Research & Development Services: | Database Management: | Drug Delivery: | Y |
| | | | Diagnostics: | | Drug Distribution: | |

## TYPES OF BUSINESS:

Neurological Diseases Drugs
Drug Delivery Systems
Synthetic Cannabinoids
Cannabinoids

## BRANDS/DIVISIONS/AFFILIATES:

Dextofisopam
Cannabinor
PRS-639,058
Tianeptine

## CONTACTS: Note: Officers with more than one job title may be intentionally listed here more than once.

S. Colin Neill, Pres.
S. Colin Neill, CFO
S. Colin Neill, Sec.
S. Colin Neill, Treas.
Robert F. Johnston, Chmn.

| **Phone:** 732-452-9556 | **Fax:** 732-452-9557 |
|---|---|
| **Toll-Free:** | |
| **Address:** 99 Wood Ave. S., Ste. 311, Iselin, NJ 08830 US | |

## GROWTH PLANS/SPECIAL FEATURES:

Pharmos Corp. is a biopharmaceutical company that discovers and develops novel therapeutics focusing on specific diseases of the nervous system including disorders of the brain-gut axis (GI/IBS), pain/inflammation and autoimmune disorders.  The company's lead product, Dextofisopam, positively completed a Phase IIa trial, and has initiated a Phase IIb trial for treating irritable bowel syndrome (IBS) in women.  Dextofisopam is currently the only drug under active development.  Pharmos' past discovery efforts were focused primarily on CB2-selective compounds, which are small molecule cannabinoid receptor agonists that bind preferentially to CB2 receptors found primarily in peripheral immune cells and peripheral nervous system.  Cannabinor completed two Phase IIa tests for pain indications with an intravenous formulation, but failed to meet its endpoints. Another CB2-selective compound, PRS-639,058, was able to overcome some of the difficulties commonly occurring in cannabinoid drugs, including low solubility and metabolism, poor bioavailability and side effects.  Pharmos also worked to develop Tianeptine, a possible follow-up to Dextofisopam in IBS, but this program is also currently inactive.  In October 2008, the firm announced the closure of its Israeli operations.  In February 2009, the company agreed to sell some of its CB2 selective agonists, including PRS-639,058, to the Israel-based firm Reperio Pharmaceuticals, Ltd, which was established by two former Pharmos scientists.  Pharmos is currently focusing its efforts on developing Dextofisopam.

## FINANCIALS:  Sales and profits are in thousands of dollars—add 000 to get the full amount. 2008 Note: Financial information for 2008 was not available for all companies at press time.

| | | | |
|---|---|---|---|
| 2008 Sales: $ | 2008 Profits: $-10,089 | **U.S. Stock Ticker:** PARS | |
| 2007 Sales: $ | 2007 Profits: $-15,626 | **Int'l Ticker:**      Int'l Exchange: | |
| 2006 Sales: $ | 2006 Profits: $-35,137 | Employees:     5 | |
| 2005 Sales: $ | 2005 Profits: $-2,930 | Fiscal Year Ends: 12/31 | |
| 2004 Sales: $ | 2004 Profits: $-21,968 | Parent Company: | |

## SALARIES/BENEFITS:

| Pension Plan: | ESOP Stock Plan: | Profit Sharing: | Top Exec. Salary: $300,000 | Bonus: $ |
|---|---|---|---|---|
| Savings Plan: Y | Stock Purch. Plan: Y | | Second Exec. Salary: $324,450 | Bonus: $ |

## OTHER THOUGHTS:

**Apparent Women Officers or Directors:**
**Hot Spot for Advancement for Women/Minorities:**

**LOCATIONS: ("Y" = Yes)**

| West: | Southwest: | Midwest: | Southeast: | Northeast: | International: |
|---|---|---|---|---|---|
| | | | | Y | |

# POLYDEX PHARMACEUTICALS

**www.polydex.com**

Industry Group Code: 325414  Ranks within this company's industry group: Sales: 7  Profits: 4

| Drugs: | | Other: | Clinical: | | Computers: | Services: | |
|---|---|---|---|---|---|---|---|
| Discovery: | | AgriBio: | Trials/Services: | | Hardware: | Specialty Services: | |
| Licensing: | | Genetic Data: | Labs: | | Software: | Consulting: | |
| Manufacturing: | Y | Tissue Replacement: | Equipment/Supplies: | | Arrays: | Blood Collection: | |
| Genetics: | | | Research & Development Services: | Y | Database Management: | Drug Delivery: | Y |
| | | | Diagnostics: | | | Drug Distribution: | Y |

## TYPES OF BUSINESS:

Drugs-Raw Ingredients
Contraceptive Development
HIV & STD Preventatives
Anemia Prevention-Veterinary

## BRANDS/DIVISIONS/AFFILIATES:

Dextran Products Limited
Chemdex Inc.
Iron Dextran
Dextran Sulphate
DEAE Dextran
Ushercell

## CONTACTS: *Note: Officers with more than one job title may be intentionally listed here more than once.*

George G. Usher, CEO
Sharon L. Wardlaw, COO
George G. Usher, Pres.
John A. Luce, CFO
Sharon L. Wardlaw, Sec.
Sharon L. Wardlaw, Treas.
George G. Usher, Chmn.

| Phone: 416-755-2231 | Fax: |
|---|---|
| Toll-Free: | |
| Address: 421 Comstock Rd., Toronto, ON M1L 2H5 Canada | |

## GROWTH PLANS/SPECIAL FEATURES:

Polydex Pharmaceuticals, Ltd., is engaged in the research, development, manufacture and marketing of biotechnology-based products for the human pharmaceutical market and also manufactures bulk pharmaceutical intermediates for the global veterinary industry.   Polydex conduct business through two wholly-owned subsidiaries: Dextran Products Limited, based in Canada, and Chemdex, Inc., based in Kansas.  The company focuses on the manufacture and sale of dextran and derivative products, including iron dextran, dextran sulfate and other specialty chemicals.  Dextran is the generic name applied to polymer compounds formed by bacterial growth on sucrose.  Iron dextran is a derivative of dextran produced by complexing iron with dextran and is injected into most pigs at birth as a treatment for anemia.  Polydex sells iron dextran to independent wholesalers primarily in Europe, the U.S. and Canada, with increasing sales in China.   Chemdex has U.S. Food and Drug Administration (FDA) approval for the manufacture and sale of iron dextran for veterinary use.   Dextran sulfate is a specialty chemical derivative of dextran used in research applications by the pharmaceutical industry and other centers of chemical research for DNA hybridization, ulcer treatment, blood coagulation, hyperlipidemia treatment and as a photographic emulsion additive.  Additional dextran derivatives produced by the company include DEAE (diethylaminoethyl) dextran, a treatment for obesity, hyperlipemia and hypercholestremia in adults; and low-molecular weight dextran, studied for a prevention of bacteria colonization in the mouth due to cystic fibrosis.  Polydex is developing its Ushercell product for use as a contraceptive microbial agent protecting against HIV and other sexually transmitted diseases.   The company has temporarily suspended two Phase III clinical trials of Ushercell and will reevaluate continuing the trials in 2010.  The company is also researching potential development of a human iron supplement.  In addition to its Dextran-based products, the firm also owns a patented method for the production of cellulose sulfate.

## FINANCIALS:  Sales and profits are in thousands of dollars—add 000 to get the full amount. 2008 Note: Financial information for 2008 was not available for all companies at press time.

| | | |
|---|---|---|
| 2008 Sales: $5,735 | 2008 Profits: $- 885 | U.S. Stock Ticker: POLXF |
| 2007 Sales: $6,499 | 2007 Profits: $- 261 | Int'l Ticker:      Int'l Exchange: |
| 2006 Sales: $5,265 | 2006 Profits: $-1,489 | Employees:   20 |
| 2005 Sales: $6,372 | 2005 Profits: $1,139 | Fiscal Year Ends: 1/31 |
| 2004 Sales: $14,100 | 2004 Profits: $ 100 | Parent Company: |

## SALARIES/BENEFITS:

| Pension Plan: | ESOP Stock Plan: | Profit Sharing: | Top Exec. Salary: $281,000 | Bonus: $ |
|---|---|---|---|---|
| Savings Plan: | Stock Purch. Plan: | | Second Exec. Salary: $111,000 | Bonus: $ |

## OTHER THOUGHTS:

**Apparent Women Officers or Directors**: 1
**Hot Spot for Advancement for Women/Minorities:**

## LOCATIONS: ("Y" = Yes)

| West: | Southwest: | Midwest: | Southeast: | Northeast: | International: |
|---|---|---|---|---|---|
| | | | | | Y |

Note: Financial information, benefits and other data can change quickly and may vary from those stated here.

# PONIARD PHARMACEUTICALS INC

**www.poniard.com**

Industry Group Code: 325412 **Ranks within this company's industry group:** Sales:      Profits: 137

| Drugs: | | Other: | Clinical: | Computers: | Services: |
|---|---|---|---|---|---|
| Discovery: | Y | AgriBio: | Trials/Services: | Hardware: | Specialty Services: |
| Licensing: | Y | Genetic Data: | Labs: | Software: | Consulting: |
| Manufacturing: | | Tissue Replacement: | Equipment/Supplies: | Arrays: | Blood Collection: |
| Genetics: | | | Research & Development Services: | Database Management: | Drug Delivery: |
| | | | Diagnostics: | | Drug Distribution: |

## TYPES OF BUSINESS:

Drugs-Cancer

## BRANDS/DIVISIONS/AFFILIATES:

Picoplatin

## CONTACTS: Note: Officers with more than one job title may be intentionally listed here more than once.

Jerry McMahon, CEO
Ronald A. Martell, COO
Ronald A. Martell, Pres.
Greg Weaver, CFO/Sr. VP
Robert De Jager, Chief Medical Officer
Anna L. Wight, VP-Legal
Cheni Kwok, Sr. VP-Corp. Dev.
Janet R. Rea, VP-Regulatory Affairs & Quality
Jerry McMahon, Chmn.

| Phone: 650-583-3774 | Fax: 650-583-3789 |
|---|---|
| **Toll-Free:** | |
| **Address:** 7000 Shoreline Ct., Ste. 270, South San Francisco, CA 94080 US | |

## GROWTH PLANS/SPECIAL FEATURES:

Poniard Pharmaceuticals, Inc. develops therapeutic biopharmaceuticals for the treatment of cancer. In particular, the company is currently focused on platinum based cancer treatment agents. Poniards lead product candidate is picoplatin, a cytotoxic platinum chemotherapeutic agent candidate with an improved safety profile relative to existing platinum based cancer therapies. It targets solid tumors by binding to DNA and interfering with its replication and transcription processes, thereby causing apoptosis (cell death). Potential benefits of Picoplatin are that it is broadly applicable for the treatment of solid tumors; the compound may have less severe side effects than other platinum treatments, and it may be possible to use this product as treatment for platinum sensitive, resistant and refractory disease. Picoplatin has received orphan drug designation from the FDA and is in Phase III trials for the treatment of small cell lung cancer (SCLC). The drug is also being assessed in Phase II trials for patients with metastatic colorectal cancer and patients with hormone refractory prostate cancer. The company is also initiating a Phase I trial of an oral version of the drug. In November 2008, Poniard announced an agreement with Baxter Oncology GmbH to supply the company with the injectable form of picoplatin. In March 2009, the company reduced its research staff by about 12%.

## FINANCIALS: Sales and profits are in thousands of dollars—add 000 to get the full amount. 2008 Note: Financial information for 2008 was not available for all companies at press time.

| | | |
|---|---|---|
| 2008 Sales: $ | 2008 Profits: $-48,565 | **U.S. Stock Ticker: PARD** |
| 2007 Sales: $ | 2007 Profits: $-32,782 | **Int'l Ticker:** Int'l Exchange: |
| 2006 Sales: $ | 2006 Profits: $-23,294 | Employees: 59 |
| 2005 Sales: $ 15 | 2005 Profits: $-20,997 | Fiscal Year Ends: 12/31 |
| 2004 Sales: $1,015 | 2004 Profits: $-19,371 | Parent Company: |

## SALARIES/BENEFITS:

| Pension Plan: | ESOP Stock Plan: | Profit Sharing: | Top Exec. Salary: $433,691 | Bonus: $173,476 |
|---|---|---|---|---|
| Savings Plan: Y | Stock Purch. Plan: | | Second Exec. Salary: $332,800 | Bonus: $93,184 |

## OTHER THOUGHTS:

**Apparent Women Officers or Directors**: 3
**Hot Spot for Advancement for Women/Minorities**: Y

## LOCATIONS: ("Y" = Yes)

| West: | Southwest: | Midwest: | Southeast: | Northeast: | International: |
|---|---|---|---|---|---|
| Y | | | | | |

# POZEN INC

www.pozen.com

**Industry Group Code:** 325412 **Ranks within this company's industry group:** Sales: 79 Profits: 69

| Drugs: | | Other: | Clinical: | Computers: | Services: |
|---|---|---|---|---|---|
| Discovery: | Y | AgriBio: | Trials/Services: | Hardware: | Specialty Services: |
| Licensing: | | Genetic Data: | Labs: | Software: | Consulting: |
| Manufacturing: | | Tissue Replacement: | Equipment/Supplies: | Arrays: | Blood Collection: |
| Genetics: | | | Research & Development Services: | Database Management: | Drug Delivery: |
| | | | Diagnostics: | | Drug Distribution: |

## TYPES OF BUSINESS:

Drugs, Analgesic
Migraine Therapy

## BRANDS/DIVISIONS/AFFILIATES:

Treximet
PN 400

## CONTACTS: Note: Officers with more than one job title may be intentionally listed here more than once.

John R. Plachetka, CEO
John R. Plachetka, Pres.
William L. Hodges, CFO
John G. Fort, Chief Medical Officer
Everadus Orlemans, Exec. VP-Prod. Dev.
William L. Hodges, Sr. VP-Admin.
Gilda M. Thomas, General Counsel/Sr. VP
William L. Hodges, Sr. VP-Finance
John R. Plachetka, Chmn.

| Phone: 919-913-1030 | Fax: 919-913-1039 |
|---|---|
| Toll-Free: | |
| Address: 1414 Raleigh Rd., Chapel Hill, NC 27517 US | |

## GROWTH PLANS/SPECIAL FEATURES:

Pozen, Inc. is a pharmaceutical company focused primarily on products that can provide improved efficacy, safety or patient convenience in the treatment of migraine, acute and chronic pain and other pain-related indications. Pozen's primary product is Treximet, developed in collaboration with GlaxoSmithKline, for the treatment of acute migraines. Treximet is a single table containing sumatriptan succinate, a 5-HT 1B/1D agonist formulated with GSK's RT Technology, and naproxen sodium, a non-steroidal anti-inflammatory drug (NSAID). The company also has a handful of other drugs in development. Under Pozen's PN program, the company has completed formulation development and clinical studies for combining a proton pump inhibitor (PPI) and an NSAID into a single tablet, meant to manage pain associated with chronic osteoarthritis, rheumatoid arthritis and ankylosing spondylitis in patients who are prone to developing gastric ulcers caused by NSAIDs. The drug is currently being developed in collaboration with AstraZeneca. Similarly, the company's PA program combines aspirin with a PPI to reduce the risk of colorectal cancer and adenomas with fewer gastrointestinal side effects. Pozen is also examining other indications in which it will develop drugs, including exploratory initiatives in cardiovascular and oncology areas and proof of concept trials in gastroenterology and cardiovascular studies. In May 2008, the company announced it will seek FDA approval for the use of PN 400 in treating chronic pain.

## FINANCIALS: Sales and profits are in thousands of dollars—add 000 to get the full amount. 2008 Note: Financial information for 2008 was not available for all companies at press time.

| | | |
|---|---|---|
| 2008 Sales: $66,133 | 2008 Profits: $-5,976 | **U.S. Stock Ticker:** POZN |
| 2007 Sales: $53,444 | 2007 Profits: $4,666 | **Int'l Ticker:** Int'l Exchange: |
| 2006 Sales: $13,517 | 2006 Profits: $-19,310 | Employees: 34 |
| 2005 Sales: $28,647 | 2005 Profits: $1,959 | Fiscal Year Ends: 12/31 |
| 2004 Sales: $23,088 | 2004 Profits: $-5,261 | Parent Company: |

## SALARIES/BENEFITS:

| | | | | |
|---|---|---|---|---|
| Pension Plan: | ESOP Stock Plan: | Profit Sharing: | Top Exec. Salary: $511,156 | Bonus: $762,367 |
| Savings Plan: Y | Stock Purch. Plan: | | Second Exec. Salary: $324,584 | Bonus: $44,700 |

## OTHER THOUGHTS:

**Apparent Women Officers or Directors:** 1
**Hot Spot for Advancement for Women/Minorities:**

## LOCATIONS: ("Y" = Yes)

| West: | Southwest: | Midwest: | Southeast: | Northeast: | International: |
|---|---|---|---|---|---|
| | | | | Y | |

# PRA INTERNATIONAL

**www.prainternational.com**

Industry Group Code: 541712  Ranks within this company's industry group:  Sales:  Profits:

| Drugs: | Other: | Clinical: | | Computers: | | Services: | |
|---|---|---|---|---|---|---|---|
| Discovery: | AgriBio: | Trials/Services: | Y | Hardware: | | Specialty Services: | |
| Licensing: | Genetic Data: | Labs: | | Software: | | Consulting: | |
| Manufacturing: | Tissue Replacement: | Equipment/Supplies: | | Arrays: | | Blood Collection: | |
| Genetics: | | Research & Development Services: | Y | Database Management: | Y | Drug Delivery: | |
| | | Diagnostics: | | | | Drug Distribution: | |

## TYPES OF BUSINESS:

Clinical Research & Testing Services
Clinical Development Services
Clinical Trials
Data Management Services

## BRANDS/DIVISIONS/AFFILIATES:

Pharmaceutical Research Associates, Inc.
PRA E-TMF
Pharmacon
Genstar Capital, LLC
Flex DMA

## CONTACTS: Note: Officers with more than one job title may be intentionally listed here more than once.

Terry Bieker, CEO
Colin Shannon, COO
Colin Shannon, Pres.
Linda Baddour, CFO/Exec. VP
Roger Boutin, Dir.-Mktg. Sales & Support
Jeff Chambers, Sr. VP-Human Resources
Willem Jan Drijfhout, Sr. VP-Early Dev. Svcs.
Tami Klerr-Naivar, Sr.  VP-Bus. Dev.
Tim Grace, Financial Rel.
Bucky Walsh, Exec. VP-Corp. Svcs.
Susan C. Stansfield, Exec. VP-Product Registration, Europe
Kent Thoelke, Sr. VP/Head-Scientific & Medical Affairs
Bruce Teplitzky, Exec. VP-Strategic Bus. Dev.
Melvin Booth, Chmn.

| Phone: 919-786-8200 | Fax: 919-786-8201 |
|---|---|
| Toll-Free: | |
| Address: 4130 Park Lake Ave.  Ste. 400, Raleigh, NC 27612 US | |

## GROWTH PLANS/SPECIAL FEATURES:

PRA International is a contract research organization (CRO) that provides clinical drug development services to pharmaceutical and biotechnology companies around the world.  CROs typically assist companies in developing drug compounds, biologics, drug delivery devices and the attainment of certain regulatory approvals necessary to market these technologies.  PRA International specializes in oncology, CNS, respiratory/allergy, cardiovascular and infectious diseases.  In addition, PRA provides a broad array of services in clinical development programs, including the creation of drug development and regulatory strategy plans, the utilization of bioanalytical laboratory testing and the development of integrated global clinical databases.  Bioanalytical sample testing includes the use of LC-MS Machines, the Packard Robotic System, the HPLL System and the Immuno-Analysis suite.  PRA also provides data management services such as electronic data capture, data monitoring and database development.  The company supports its data services through the use of its proprietary PRA E-TMF enterprise-wide electronic document management system, which provides clients with faster document handling and archival services, and Flex DMA, which allows direct entry of clinical trial data by investigational sites.  Clinical trials in the U.S. are largely centered around PRA's facilities in Kansas. The company performs studies in 60 countries on 6 continents.  In recent years, PRA International was acquired by Genstar Capital, LLC, a middle market private equity firm, for $797 million. In December 2008, the company opened an office Munich, the company's third in Germany, and one in Mexico City, Mexico.  In May 2009, the firm opened a Drug Safety Center in its Sao Paulo, Brazil office.

PRA International offers employees comprehensive medial benefits, life insurance, retirements programs, holidays and paid time off, tuition advance payment programs and a scholarship program for employee dependents.

## FINANCIALS: Sales and profits are in thousands of dollars—add 000 to get the full amount. 2008 Note: Financial information for 2008 was not available for all companies at press time.

| | | |
|---|---|---|
| 2008 Sales: $ | 2008 Profits: $ | **U.S. Stock Ticker: Private** |
| 2007 Sales: $ | 2007 Profits: $ | **Int'l Ticker:**  Int'l Exchange: |
| 2006 Sales: $338,166 | 2006 Profits: $26,845 | Employees:  2,700 |
| 2005 Sales: $326,244 | 2005 Profits: $32,223 | Fiscal Year Ends: 12/31 |
| 2004 Sales: $307,644 | 2004 Profits: $20,749 | Parent Company: GENSTAR CAPITAL LLC |

## SALARIES/BENEFITS:

| Pension Plan: | ESOP Stock Plan: | Profit Sharing: | Top Exec. Salary: $419,583 | Bonus: $ |
|---|---|---|---|---|
| Savings Plan: Y | Stock Purch. Plan: Y | | Second Exec. Salary: $293,917 | Bonus: $ |

## OTHER THOUGHTS:

**Apparent Women Officers or Directors:** 3
**Hot Spot for Advancement for Women/Minorities:** Y

## LOCATIONS: ("Y" = Yes)

| West: | Southwest: | Midwest: | Southeast: | Northeast: | International: |
|---|---|---|---|---|---|
| Y | Y | Y | | Y | Y |

# PROGENICS PHARMACEUTICALS
www.progenics.com

**Industry Group Code: 325412  Ranks within this company's industry group:** Sales: 77  Profits: 132

| Drugs: | | Other: | | Clinical: | Computers: | Services: |
|---|---|---|---|---|---|---|
| Discovery: | Y | AgriBio: | Y | Trials/Services: | Hardware: | Specialty Services: |
| Licensing: | Y | Genetic Data: | | Labs: | Software: | Consulting: |
| Manufacturing: | Y | Tissue Replacement: | | Equipment/Supplies: | Arrays: | Blood Collection: |
| Genetics: | | | | Research & Development Services: | Database Management: | Drug Delivery: |
| | | | | Diagnostics: | | Drug Distribution: |

## TYPES OF BUSINESS:
Pharmaceuticals Development
Drugs-HIV
Drugs-Cancer
Small-Molecule Drugs
Viral Entry Inhibitors
Vaccines

## BRANDS/DIVISIONS/AFFILIATES:
RELISTOR
PRO 140

## CONTACTS: Note: Officers with more than one job title may be intentionally listed here more than once.
Paul J. Maddon, CEO
Robert A. McKinney, CFO
William C. Olson, VP-R&D
Thomas A. Boyd, Sr. VP-Prod. Dev.
Nitya G. Ray, VP-Mfg.
Mark R. Baker, General Counsel/Exec. VP
Walter M. Capone, VP-Oper.
Walter M. Capone, VP-Commercial Dev.
Richard W. Krawiec, VP-Corp. Affairs
Robert A. McKinney, Sr. VP-Finance/Treas.
Paul J. Maddon, Chief Science Officer
Robert J. Isreal, Sr. VP-Medical Affairs
Benedict Osorio, VP-Quality
Ann Marie Assumma, VP-Regulatory Affairs
Kurt W. Briner, Chmn.

| Phone: 914-789-2800 | Fax: 914-789-2817 |
|---|---|
| Toll-Free: | |
| Address: 777 Old Saw Mill River Rd., Tarrytown, NY 10591 US | |

## GROWTH PLANS/SPECIAL FEATURES:
Progenics Pharmaceuticals, Inc. develops and commercializes therapeutic products to treat patients with debilitating conditions and life-threatening diseases. The firm's principal programs are directed toward symptom management and supportive care in the treatment of virology, oncology and gastroenterology. The firm's lead product in the area of symptom management and supportive care is RELISTOR, which is designed to reverse the side effects of opioid pain medications while maintaining pain relief. RELISTOR, created in collaboration with Wyeth Pharmaceuticals, is approved for use in its subcutaneous form in the U.S., Australia, Canada and E.U. countries. The company is also developing intravenous and oral delivery methods for RELISTOR. In the area of virology, the firm has developed viral entry inhibitors, which are molecules designed to inhibit a virus' ability to enter certain types of immune system cells. The company is currently developing Phase II trials for the PRO 140 viral entry inhibitor, a potential treatment for HIV infection. The firm's virology division is also focusing on discovering treatments that block entry of the Hepatitis C virus into cells. In addition, Progenics is developing immunotherapies for prostate cancer, including monoclonal antibodies directed against prostate specific membrane antigen (PSMA), a protein found on the surface of prostate cancer cells. Progenics is also creating a therapeutic prostate cancer vaccine designed to use the body's own defense mechanisms to identify and destroy prostate cancer cells. In October 2008, the company announced an agreement in which Ono Pharmaceuticals acquired the exclusive rights to develop and sell RELISTOR in Japan. In May 2009, the firm finalized its plans to cut 10% of its staff in order to streamline production.

Progenics offers employees medical, vision, dental and life insurance; a 401(k) plan; an employee stock options plan; an employee stock purchase plan; flexible spending accounts; short and long-term disability; tuition reimbursement; and an employee assistance plan.

## FINANCIALS: Sales and profits are in thousands of dollars—add 000 to get the full amount. 2008 Note: Financial information for 2008 was not available for all companies at press time.

| | | |
|---|---|---|
| 2008 Sales: $67,671 | 2008 Profits: $-44,672 | **U.S. Stock Ticker: PGNX** |
| 2007 Sales: $75,646 | 2007 Profits: $-43,688 | **Int'l Ticker:**  Int'l Exchange: |
| 2006 Sales: $69,906 | 2006 Profits: $-21,618 | Employees:  244 |
| 2005 Sales: $9,486 | 2005 Profits: $-69,429 | Fiscal Year Ends: 12/31 |
| 2004 Sales: $9,576 | 2004 Profits: $-42,018 | Parent Company: |

## SALARIES/BENEFITS:
| Pension Plan: | ESOP Stock Plan: Y | Profit Sharing: | Top Exec. Salary: $618,000 | Bonus: $330,000 |
|---|---|---|---|---|
| Savings Plan: Y | Stock Purch. Plan: Y | | Second Exec. Salary: $385,000 | Bonus: $160,000 |

## OTHER THOUGHTS:
**Apparent Women Officers or Directors: 1**
**Hot Spot for Advancement for Women/Minorities:**

## LOCATIONS: ("Y" = Yes)
| West: | Southwest: | Midwest: | Southeast: | Northeast: | International: |
|---|---|---|---|---|---|
| | | | | Y | |

Note: Financial information, benefits and other data can change quickly and may vary from those stated here.

# PROMEGA CORP

www.promega.com

**Industry Group Code: 325413  Ranks within this company's industry group:  Sales:    Profits:**

| Drugs: | Other: | Clinical: | | Computers: | | Services: | |
|---|---|---|---|---|---|---|---|
| Discovery: | AgriBio: | Trials/Services: | | Hardware: | | Specialty Services: | Y |
| Licensing: | Genetic Data: | Labs: | | Software: | | Consulting: | |
| Manufacturing: | Tissue Replacement: | Equipment/Supplies: | Y | Arrays: | | Blood Collection: | |
| Genetics: | | Research & Development Services: | | Database Management: | | Drug Delivery: | |
| | | Diagnostics: | | | | Drug Distribution: | |

## TYPES OF BUSINESS:

Life Sciences Solutions & Technical Support
Specialty Biochemicals

## BRANDS/DIVISIONS/AFFILIATES:

Terso Solutions
Maxwell 16 System
ProteasMAX

## CONTACTS: *Note: Officers with more than one job title may be intentionally listed here more than once.*

William A. Linton, CEO
William A. Linton, Pres.
Laura Francis, CFO
Randy Dimond, CTO/VP
Laura Francis, VP-Finance
William A. Linton, Chmn.

| | |
|---|---|
| **Phone:** 608-274-4330 | **Fax:** 608-277-2601 |
| **Toll-Free:** 800-356-9526 | |
| **Address:** 2800 Woods Hollow Rd., Madison, WI 53711 US | |

## GROWTH PLANS/SPECIAL FEATURES:

Promega Corp. provides solutions and technical support to the life sciences industry.  The company's over 2,000 products aid scientists in life science research, especially genomics, proteomics and cellular analysis.  The firm's products are comprised of kits and reagents, as well as integrated solutions for life sciences research and drug discovery.  Reagents include DNA and RNA purification; genotype analysis; protein expression and analysis; and DNA sequencing.  Integrated solutions provide customers with personal automation tools that combine instrumentation and reagents into one system.  Promega's Terso Solutions subsidiary focuses on leveraging RFID technology to solve customers' needs for complete inventory management solutions.  The company has U.S. and foreign patents in areas including nucleic acid purification, human identification and bioluminescence, coupled with in vitro transcription and translation and cell biology.  The firm has branches in 12 countries and sells its products through over 50 global distributors.  The firm has 200 patents in nucleic acid purification, human identification, bioluminescence, coupled in-vitro transcription/translation and cell biology.  In May 2008, the firm's Maxwell 16 system for DNA-based tests received the CE Mark, meaning it can be marketed for diagnostic use throughout the European Union.  The Maxwell 16 system purifies up to 16 DNA samples simultaneously in 30 minutes.  Also in 2008, the firm launched ProteasMAX, a surfactant which provides improved protein characterization.  In 2009, Promega launched its StemElite ID system, which improves human cell line authentication; GloMax-Multi+ detection system; a 16-locus STR system to profile inhibited DNA samples; and the pmirGLO Dual-Luciferase miRNA Target Expression Vector for microRNA study.

The company offers its employees health, life, disability, long-term care and dental insurance; flexible spending accounts; tuition assistance; and an employee assistance program.  Additional perks include an on-campus Farmer's Market, convenience services and wellness programs, which include on-site fitness classes, exercise facilities, yoga classes and massage therapy.

## FINANCIALS:  Sales and profits are in thousands of dollars—add 000 to get the full amount. 2008 Note: Financial information for 2008 was not available for all companies at press time.

| | | |
|---|---|---|
| 2008 Sales: $ | 2008 Profits: $ | **U.S. Stock Ticker:** Private |
| 2007 Sales: $ | 2007 Profits: $ | **Int'l Ticker:**    Int'l Exchange: |
| 2006 Sales: $200,000 | 2006 Profits: $ | Employees:  911 |
| 2005 Sales: $ | 2005 Profits: $ | Fiscal Year Ends: 12/31 |
| 2004 Sales: $169,500 | 2004 Profits: $ | Parent Company: |

## SALARIES/BENEFITS:

| | | | | |
|---|---|---|---|---|
| Pension Plan: | ESOP Stock Plan: | Profit Sharing: | Top Exec. Salary: $ | Bonus: $ |
| Savings Plan: Y | Stock Purch. Plan: | | Second Exec. Salary: $ | Bonus: $ |

## OTHER THOUGHTS:

Apparent Women Officers or Directors: 1
**Hot Spot for Advancement for Women/Minorities:**

## LOCATIONS: ("Y" = Yes)

| West: | Southwest: | Midwest: | Southeast: | Northeast: | International: |
|---|---|---|---|---|---|
| Y | | Y | | Y | Y |

# PROTEIN POLYMER TECHNOLOGIES

**www.ppti.com**

**Industry Group Code: 325411  Ranks within this company's industry group: Sales: 1    Profits: 1**

| Drugs: | | Other: | | Clinical: | | Computers: | | Services: | |
|---|---|---|---|---|---|---|---|---|---|
| Discovery: | | AgriBio: | | Trials/Services: | | Hardware: | | Specialty Services: | |
| Licensing: | | Genetic Data: | | Labs: | | Software: | | Consulting: | |
| Manufacturing: | Y | Tissue Replacement: | Y | Equipment/Supplies: | | Arrays: | | Blood Collection: | |
| Genetics: | | | | Research & Development Services: | Y | Database Management: | | Drug Delivery: | Y |
| | | | | Diagnostics: | | | | Drug Distribution: | |

## TYPES OF BUSINESS:

Medical Products Development
Medical Products-Bodily Repair
Drug Delivery Devices
Protein Polymers

## BRANDS/DIVISIONS/AFFILIATES:

Spine Wave, Inc.
NuCore
Polymer 47K

## CONTACTS: *Note: Officers with more than one job title may be intentionally listed here more than once.*

James B. McCarthy, CEO
James B. McCarthy, Interim Pres.
Joseph Cappello, VP-R&D
Joseph Cappello, CTO
James B. McCarthy, Interim Chief Acct. Officer
Franco A. Ferrari, VP-Laboratory Oper. & Polymer Prod.
James B. McCarthy, Chmn.

| Phone: 858-558-6064 | Fax: 858-558-6477 |
|---|---|

**Toll-Free:**

**Address:** 10655 Sorrento Valley Rd., San Diego, CA 92121 US

## GROWTH PLANS/SPECIAL FEATURES:

Protein Polymer Technologies, Inc. (PPT) is a development-stage company that uses its proprietary protein-based biomaterials technology to research, produce and clinically test medical products that aid in the natural process of bodily repair.  Its product focus includes surgical adhesives and sealants, soft-tissue augmentation products, wound-healing matrices, drug delivery devices and surgical adhesion barriers.  PPT uses biomaterials that can mimic the natural properties and functions of proteins and peptides, and interact with a cell's enzymatic reactions to spur protein formation.  The company has developed surgical tissue sealants which combine the biocompatibility of fibrin glues, without the risks associated with use of blood-derived products, with high strength and fast setting times.  The product adheres well to tissue to seal gas and fluid leaks and comes in resorbable and non-resorbable formulations.  PPT has developed protein polymers for use on dermal wounds, particularly chronic wounds such as decubitous ulcers, which assist the tissue to heal in such a way that it regenerates as functional tissue instead of scar tissue.  Polymer 47K is the company's urethral bulking agent for the treatment of female stress urinary incontinence and is given as an injection.

The FDA approved the company's Investigational Device Exemption (IDE), which allows the testing of the safety and effectiveness of the incontinence product in women over the age of 40 who have become incontinent due to the shifting of their bladder or the weakening of the muscle at its base.  Another application for the firm's bodily repair technology is spine applications.  PPT used the company's patented tissue adhesive technology to create Spine Wave's NuCore intervertebral disc repair material.  PPT is a technical partner of Spine Wave, Inc., and the company manufactured the NuCore material for Spine Wave's clinical trials.

## FINANCIALS: Sales and profits are in thousands of dollars—add 000 to get the full amount. 2008 Note: Financial information for 2008 was not available for all companies at press time.

| | | | |
|---|---|---|---|
| 2008 Sales: $ 25 | 2008 Profits: $-2,938 | **U.S. Stock Ticker: PPTI.OB** | |
| 2007 Sales: $ 287 | 2007 Profits: $-3,252 | **Int'l Ticker:** | Int'l Exchange: |
| 2006 Sales: $ 605 | 2006 Profits: $-7,878 | Employees:    2 | |
| 2005 Sales: $ 867 | 2005 Profits: $-5,822 | Fiscal Year Ends: 12/31 | |
| 2004 Sales: $ 453 | 2004 Profits: $-3,565 | Parent Company: | |

## SALARIES/BENEFITS:

| Pension Plan: | ESOP Stock Plan: | Profit Sharing: | Top Exec. Salary: $300,000 | Bonus: $ |
|---|---|---|---|---|
| Savings Plan: Y | Stock Purch. Plan: | | Second Exec. Salary: $280,000 | Bonus: $ |

## OTHER THOUGHTS:

**Apparent Women Officers or Directors:**
**Hot Spot for Advancement for Women/Minorities:**

## LOCATIONS: ("Y" = Yes)

| West: | Southwest: | Midwest: | Southeast: | Northeast: | International: |
|---|---|---|---|---|---|
| Y | | | | | |

# QIAGEN NV

**www.qiagen.com**

Industry Group Code: 325413   Ranks within this company's industry group: Sales: 3   Profits: 3

| Drugs: | Other: | | Clinical: | | Computers: | | Services: | |
|--------|--------|---|-----------|---|-----------|---|-----------|---|
| Discovery: | AgriBio: | | Trials/Services: | | Hardware: | | Specialty Services: | Y |
| Licensing: | Genetic Data: | Y | Labs: | | Software: | | Consulting: | |
| Manufacturing: | Tissue Replacement: | | Equipment/Supplies: | Y | Arrays: | | Blood Collection: | |
| Genetics: | | | Research & Development Services: | | Database Management: | | Drug Delivery: | |
| | | | Diagnostics: | Y | | | Drug Distribution: | |

## TYPES OF BUSINESS:

Supplies-Plasmid Purification
Genomics Analysis Products
Diagnostic Products

## BRANDS/DIVISIONS/AFFILIATES:

Qiagen, GmbH
Tianwei Times
Nextal Biotechnology, Inc.
artus GmbH
Shenzhen PG Biotech Co. Ltd.
Gentra Systems, Inc.
QIA Symphony SP
Corbett Life Science Pty Ltd

## CONTACTS: *Note: Officers with more than one job title may be intentionally listed here more than once.*

Peer Schatz, CEO
Roland Sackers, CFO
Bernd Uder, Sr. VP-Global Sales
Gisela Orth, VP-Global Human Resources
Joachim Schorr, Sr. VP-Global Research & Dev.
Douglas Liu, VP-Global Oper.
Ulrich Schriek, VP-Corp. Bus. Dev.
Solveigh Mahler, Dir.-Public Rel.
Solveigh Mahler, Dir.-Investor Rel.
Michael Collasius, VP-Automated Systems
Thomas Schweins, VP-Mktg. & Strategy
Detlev H. Riesner, Chmn.

| Phone: 31-77-320-8400 | Fax: 31-77-320-8409 |
|---|---|
| Toll-Free: | |
| Address: Spoorstraat 50, Venlo, 5911 KJ The Netherlands | |

## GROWTH PLANS/SPECIAL FEATURES:

Qiagen NV provides technologies and products for the separation and purification of nucleic acids, used in genomics, molecular diagnostics and genetic vaccination and gene therapy. Through its many subsidiaries, Qiagen offers more than 500 products in over 40 countries worldwide. The firm's products and services find application in animal and veterinary research, biomedical research, biosecurity and biodefense, epigenetics, genetic identity and forensics, gene expression analysis, gene silencing, influenza research, molecular diagnostics, pharmacogenomics, plant research and protein science fields. Qiagen owns Tianwei Times; Nextal Biotechnology, Inc.; Artus GmbH; and Shenzhen PG Biotech Co. Ltd. Tianwei, based in Beijing, develops, manufactures and supplies nucleic acid sample preparation consumables in China. Nextal provides proprietary sample preparation tools that make protein crystallization more accessible. Artus GmbH is a leader in PCR molecular diagnostic systems. Shenzhen PG Biotech develops, manufactures and supplies PCR based molecular diagnostic kits in China. In 2008, the firm launched the QIA Symphony SP, a modular processing platform system that automates laboratory workflows. In June 2008, Qiagen launched the following gene expression analysis products: The RNeasy Plus 96 Kit, the QuantiFast Multiplex PCR Kit and the AllPrep DNA/RNA 96 Kit. In July 2008, the company acquired Corbett Life Science Pty. Ltd., an Australia-based developer of life sciences instruments, for $66 million. In October 2008, the firm acquired the Biosystems business from Biotage AB.

Employees are offered medical, dental and vision insurance; health and dependent care flexible spending accounts; life insurance; short-and long-term disability coverage; home and auto insurance discounts; tuition assistance; an employee assistance program; and an adoption assistance program.

## FINANCIALS: Sales and profits are in thousands of dollars—add 000 to get the full amount. 2008 Note: Financial information for 2008 was not available for all companies at press time.

| | | | |
|---|---|---|---|
| 2008 Sales: $892,975 | 2008 Profits: $89,033 | **U.S. Stock Ticker:** QGEN | |
| 2007 Sales: $649,774 | 2007 Profits: $50,122 | **Int'l Ticker:** QIA   Int'l Exchange: Frankfurt-Euronext | |
| 2006 Sales: $465,778 | 2006 Profits: $70,539 | Employees: 2,662 | |
| 2005 Sales: $298,395 | 2005 Profits: $62,225 | Fiscal Year Ends: 12/31 | |
| 2004 Sales: $380,629 | 2004 Profits: $48,705 | Parent Company: | |

## SALARIES/BENEFITS:

| Pension Plan: | ESOP Stock Plan: | Profit Sharing: | Top Exec. Salary: $ | Bonus: $ |
|---|---|---|---|---|
| Savings Plan: Y | Stock Purch. Plan: | | Second Exec. Salary: $ | Bonus: $ |

## OTHER THOUGHTS:

**Apparent Women Officers or Directors**: 2
**Hot Spot for Advancement for Women/Minorities**: Y

## LOCATIONS: ("Y" = Yes)

| West: | Southwest: | Midwest: | Southeast: | Northeast: | International: |
|---|---|---|---|---|---|
| Y | | | | Y | Y |

# QLT INC

**www.qltinc.com**

Industry Group Code: 325412  **Ranks within this company's industry group:** Sales: 63  Profits: 34

| Drugs: | | Other: | | Clinical: | Computers: | Services: | |
|---|---|---|---|---|---|---|---|
| Discovery: | Y | AgriBio: | | Trials/Services: | Hardware: | Specialty Services: | |
| Licensing: | Y | Genetic Data: | | Labs: | Software: | Consulting: | |
| Manufacturing: | Y | Tissue Replacement: | | Equipment/Supplies: | Arrays: | Blood Collection: | |
| Genetics: | | | | Research & Development Services: | Database Management: | Drug Delivery: | Y |
| | | | | Diagnostics: | | Drug Distribution: | |

## TYPES OF BUSINESS:

Drugs-Photodynamic
Cancer Treatments
Eye Disease Treatments
Drug Delivery-Controlled & Extended Release

## BRANDS/DIVISIONS/AFFILIATES:

QLT USA, Inc.
Visudyne
Eligard
Atrigel
Novartis Pharma AG

## CONTACTS: *Note: Officers with more than one job title may be intentionally listed here more than once.*

Robert L. Butchofsky, CEO
Robert L. Butchofsky, Pres.
Cameron Nelson, CFO
Linda Lupini, Sr. VP-Human Resources
Alexander R. Lussow, Sr. VP-Bus. Dev. & Commercial Oper.
Therese Hayes, VP-Corp. Comm.
Therese Hayes, VP-Investor Rel.
Cameron Nelson, VP-Finance
Linda Lupini, Sr. VP-Organizational Dev.
C. Boyd Clarke, Chmn.

| | |
|---|---|
| **Phone:** 604-707-7000 | **Fax:** 604-707-7001 |
| **Toll-Free:** 800-663-5486 | |
| **Address:** 887 Great Northern Way, Vancouver, BC V5T 4T5 Canada | |

## GROWTH PLANS/SPECIAL FEATURES:

QLT, Inc. is a global biopharmaceutical company that focuses on the discovery, development and commercialization of pharmaceutical products in the fields of ophthalmology and dermatology. The firm uses four technology platforms to create two products, Visudyne and Eligard. The photodynamic therapy platform stifles and kills abnormally growing tissue by administering drugs to target proteins then activating the drugs through externally provided light. The firm's photosensitizer, Visudyne, is one of the few approved treatments for wet age-related macular degeneration (AMD), the leading cause of blindness in people over the age of 55. Visudyne, which was co-developed by Swiss firm Novartis Pharma AG, is also approved for treatment of subfoveal choroidal neovascularization (CNV) and is sold in more than 80 countries. The Atrigel platform is a system consisting of biodegradable polymers, which are injected into tissue sites, containing controlled-release pharmaceuticals. The company's Eligard drug is a treatment for prostate cancer that utilizes the Atrigel system in its delivery. QLT's final drug platform is the punctal plug drug delivery technology, a minimally invasive system that delivers a variety of drugs to the eye through controlled sustained release to the tear film. The firm is researching the use of this platform for a broad range of ocular diseases that are currently being treated with eye drops. In early 2008, the firm laid off 115 employees, approximately 45% of its workforce. In July 2008, the company sold the rights to Aczone, an acne medication the firm developed, to Allergan Sales, LLC, for $150 million. In August 2008, subsidiary QLT USA, Inc., signed an agreement giving the majority of the rights to the Atrigel platform to Reckitt Benckiser Pharmaceuticals Inc. for $25 million. In September 2008, the company sold its corporate headquarters and surrounding property to Discovery Parks Holdings Ltd. for $58.4 million.

## FINANCIALS: Sales and profits are in thousands of dollars—add 000 to get the full amount. 2008 Note: Financial information for 2008 was not available for all companies at press time.

| | | |
|---|---|---|
| 2008 Sales: $124,140 | 2008 Profits: $134,891 | **U.S. Stock Ticker:** |
| 2007 Sales: $127,606 | 2007 Profits: $-109,997 | **Int'l Ticker:** QLT    Int'l Exchange: Toronto-TSX |
| 2006 Sales: $174,136 | 2006 Profits: $-101,605 | Employees:  135 |
| 2005 Sales: $241,973 | 2005 Profits: $-325,412 | Fiscal Year Ends: 12/31 |
| 2004 Sales: $186,072 | 2004 Profits: $-165,709 | Parent Company: |

## SALARIES/BENEFITS:

| | | | | |
|---|---|---|---|---|
| Pension Plan: Y | ESOP Stock Plan: | Profit Sharing: | Top Exec. Salary: $460,976 | Bonus: $374,543 |
| Savings Plan: | Stock Purch. Plan: | | Second Exec. Salary: $320,122 | Bonus: $129,587 |

## OTHER THOUGHTS:

**Apparent Women Officers or Directors:** 3
**Hot Spot for Advancement for Women/Minorities:** Y

## LOCATIONS: ("Y" = Yes)

| West: | Southwest: | Midwest: | Southeast: | Northeast: | International: |
|---|---|---|---|---|---|
| Y | | | | | Y |

# QUEST DIAGNOSTICS INC

**www.questdiagnostics.com**

Industry Group Code: 6215  Ranks within this company's industry group:  Sales: 1   Profits: 1

| Drugs: | Other: | Clinical: | | Computers: | Services: |
|---|---|---|---|---|---|
| Discovery: | AgriBio: | Trials/Services: | | Hardware: | Specialty Services: |
| Licensing: | Genetic Data: | Labs: | Y | Software: | Consulting: |
| Manufacturing: | Tissue Replacement: | Equipment/Supplies: | | Arrays: | Blood Collection: |
| Genetics: | | Research & Development Services: | | Database Management: | Drug Delivery: |
| | | Diagnostics: | Y | | Drug Distribution: |

## TYPES OF BUSINESS:

Services-Testing & Diagnostics
Clinical Laboratory Testing
Clinical Trials Testing
Esoteric Testing Laboratories

## BRANDS/DIVISIONS/AFFILIATES:

Cardio CRP
HEPTIMAX
Bio-Intact PTH
Nichols Institute

## CONTACTS: Note: Officers with more than one job title may be intentionally listed here more than once.

Surya N. Mohapatra, CEO
Robert A. Hagemann, CFO/Sr. VP
Jon R. Cohen, Chief Medical Officer/Sr. VP
Michael E. Prevoznik, General Counsel/VP-Legal & Compliance
Wayne R. Simmons, VP-Oper.
Laura Park, VP-Comm.
Laura Park, VP-Investor Rel.
Joan E. Miller, Sr. VP-Pathology & Hospital Svcs.
Surya N. Mohapatra, Chmn.

| Phone: 201-393-5000 | Fax: 201-729-8920 |
|---|---|
| Toll-Free: 800-222-0446 | |
| Address: 3 Giralda Farms, Madison, NJ 07940 US | |

## GROWTH PLANS/SPECIAL FEATURES:

Quest Diagnostics, Inc. is a U.S. clinical laboratory testing company, offering diagnostic testing and related services to the health care industry. The firm's operations consist of routine, esoteric and clinical trials testing. Quest operates through its national network of over 2,000 patient service centers, principal laboratories in more than 30 major metropolitan areas, approximately 150 rapid-response laboratories, outpatient anatomic pathology centers, hospital-based laboratories and esoteric testing laboratories on both coasts. Routine tests measure various important bodily health parameters. Tests in this category include blood cholesterol level tests, complete blood cell counts, pap smears, HIV-related tests, urinalyses, pregnancy and prenatal tests and substance-abuse tests. Esoteric tests require more sophisticated technology and highly skilled personnel. The firm's tests in this field include Cardio CRP and HEPTIMAX. Quest's two esoteric testing laboratories, comprising the Nichols Institute, are among the leading esoteric clinical testing laboratories in the world. Esoteric tests involve endocrinology, genetics, immunology, microbiology, oncology, serology and special chemistry. Clinical trial testing primarily involves assessing the safety and efficacy of new drugs to meet FDA requirements, with services including Bio-Intact PTH. In May 2009, the company launched the first commercial laboratory test for the H1N1 swine flu virus. In June of the same year, the firm introduced the EGFR Pathway test, designed to identify genetic mutations in KRAS, NRAS and BRAF genes.

Employees are offered health and dental insurance; a 401(k) plan; an employee stock purchase plan; short- and long-term disability; educational assistance; flexible spending accounts; life insurance; and employee assistance program; credit union membership; veterinary pet care; legal services; auto and homeowners/renters insurance; financial management consulting services; banking services; adoption assistance; and discounted purchase programs.

## FINANCIALS: Sales and profits are in thousands of dollars—add 000 to get the full amount. 2008 Note: Financial information for 2008 was not available for all companies at press time.

| | | |
|---|---|---|
| 2008 Sales: $7,249,447 | 2008 Profits: $581,490 | U.S. Stock Ticker: DGX |
| 2007 Sales: $6,704,907 | 2007 Profits: $339,939 | Int'l Ticker:      Int'l Exchange: |
| 2006 Sales: $6,268,659 | 2006 Profits: $586,421 | Employees: 43,500 |
| 2005 Sales: $5,456,726 | 2005 Profits: $546,277 | Fiscal Year Ends: 12/31 |
| 2004 Sales: $5,066,986 | 2004 Profits: $499,195 | Parent Company: |

## SALARIES/BENEFITS:

| Pension Plan: | ESOP Stock Plan: | Profit Sharing: | Top Exec. Salary: $1,143,868 | Bonus: $1,915,693 |
|---|---|---|---|---|
| Savings Plan: Y | Stock Purch. Plan: Y | | Second Exec. Salary: $517,713 | Bonus: $520,224 |

## OTHER THOUGHTS:

**Apparent Women Officers or Directors:** 4
**Hot Spot for Advancement for Women/Minorities:** Y

## LOCATIONS: ("Y" = Yes)

| West: | Southwest: | Midwest: | Southeast: | Northeast: | International: |
|---|---|---|---|---|---|
| Y | Y | Y | Y | Y | Y |

Note: Financial information, benefits and other data can change quickly and may vary from those stated here.

# QUESTCOR PHARMACEUTICALS

**www.questcor.com**

Industry Group Code: 325412  Ranks within this company's industry group: Sales: 68  Profits: 40

| Drugs: | | Other: | Clinical: | Computers: | Services: | |
|---|---|---|---|---|---|---|
| Discovery: | Y | AgriBio: | Trials/Services: | Hardware: | Specialty Services: | |
| Licensing: | | Genetic Data: | Labs: | Software: | Consulting: | |
| Manufacturing: | | Tissue Replacement: | Equipment/Supplies: | Arrays: | Blood Collection: | |
| Genetics: | | | Research & Development Services: | Database Management: | Drug Delivery: | |
| | | | Diagnostics: | | Drug Distribution: | Y |

## TYPES OF BUSINESS:

Drugs-Neurology
Multiple Sclerosis Treatment

## BRANDS/DIVISIONS/AFFILIATES:

HP Acthar Gel
Doral
QSC-001
Acthar

## CONTACTS: *Note: Officers with more than one job title may be intentionally listed here more than once.*

Don M. Baily, CEO
Don M. Baily, Pres.
Gary Sawka, CFO
Timothy O'Neill, VP-Contract Mfg.
Michael H. Mulroy, Sec.
Steve Cartt, Exec. VP-Corp. Dev.
Gary Sawka, Sr. VP-Finance
Steven Halladay, Sr. VP-Clinical & Regulatory Affairs
Eldon Mayer, VP-Commercial Oper.
Dave Medieros, Sr. VP-Pharmaceutical Oper.
Virgil D. Thompson, Chmn.

| Phone: 510-400-0700 | Fax: 510-400-0799 |
|---|---|
| **Toll-Free:** | |
| **Address:** 3260 Whipple Rd., Union City, CA 94587 US | |

## GROWTH PLANS/SPECIAL FEATURES:

Questcor Pharmaceuticals, Inc. is an integrated specialty pharmaceutical company that focuses on novel therapeutics for the treatment of diseases and disorders of the central nervous system (CNS). The company currently owns and markets two commercial products: H.P. Acthar Gel (Acthar) and Doral. Acthar is a natural source, highly purified preparation of the adrenal corticotropin hormone, which is specially formulated to provide prolonged release after intramuscular or subcutaneous injection. The product is indicated for use in acute exacerbations of multiple sclerosis (MS) and is prescribed for patients that have MS and experience painful, episodic flares. The company is also in the process of getting Acthar approved for infantile spasms, a condition for which no other drug has been approved. Doral (quazepam) is a non-narcotic, selective benzodiazepine receptor agonist that is indicated for the treatment of insomnia. Sleep disturbance and insomnia are very common side effects of many neurological diseases and disorders such as MS, Epilepsy, Parkinson's disease and Alzheimer's disease. The company is developing QSC-001, in conjunction with Eurand, which is an orally dissolving tablet for severe pain in patients with swallowing difficulties. In October 2008, the company announced plans for evaluating Achtar for the treatment of the kidney disorder, nephrotic syndrome.

Employees are offered medical, dental and vision benefits; short- and long-term disability plans; life insurance; flexible spending accounts; an employee assistance program; a 401(k) plan; and an employee stock purchase plan.

## FINANCIALS: Sales and profits are in thousands of dollars—add 000 to get the full amount. 2008 Note: Financial information for 2008 was not available for all companies at press time.

| | | |
|---|---|---|
| 2008 Sales: $95,248 | 2008 Profits: $40,532 | **U.S. Stock Ticker:** QSC |
| 2007 Sales: $49,768 | 2007 Profits: $36,449 | **Int'l Ticker:**  Int'l Exchange: |
| 2006 Sales: $12,788 | 2006 Profits: $-10,109 | Employees:  46 |
| 2005 Sales: $14,162 | 2005 Profits: $7,392 | Fiscal Year Ends: 12/31 |
| 2004 Sales: $18,404 | 2004 Profits: $- 832 | Parent Company: |

## SALARIES/BENEFITS:

| Pension Plan: | ESOP Stock Plan: | Profit Sharing: | Top Exec. Salary: $525,000 | Bonus: $429,188 |
|---|---|---|---|---|
| Savings Plan: Y | Stock Purch. Plan: Y | | Second Exec. Salary: $350,000 | Bonus: $252,500 |

## OTHER THOUGHTS:

**Apparent Women Officers or Directors:**
**Hot Spot for Advancement for Women/Minorities:**

## LOCATIONS: ("Y" = Yes)

| West: | Southwest: | Midwest: | Southeast: | Northeast: | International: |
|---|---|---|---|---|---|
| Y | | | | | |

Note: Financial information, benefits and other data can change quickly and may vary from those stated here.

# QUINTILES TRANSNATIONAL CORP

**www.quintiles.com**

Industry Group Code: 541712  **Ranks within this company's industry group:** Sales:    Profits:

| Drugs: | | Other: | Clinical: | | Computers: | | Services: | |
|---|---|---|---|---|---|---|---|---|
| Discovery: | | AgriBio: | Trials/Services: | Y | Hardware: | | Specialty Services: | Y |
| Licensing: | Y | Genetic Data: | Labs: | Y | Software: | | Consulting: | Y |
| Manufacturing: | | Tissue Replacement: | Equipment/Supplies: | | Arrays: | | Blood Collection: | |
| Genetics: | | | Research & Development Services: | Y | Database Management: | | Drug Delivery: | |
| | | | Diagnostics: | | | | Drug Distribution: | |

## TYPES OF BUSINESS:

Contract Research
Pharmaceutical, Biotech & Medical Device Research
Consulting & Training Services
Sales & Marketing Services

## BRANDS/DIVISIONS/AFFILIATES:

Novaquest
Innovex
Eidetics

## CONTACTS: *Note: Officers with more than one job title may be intentionally listed here more than once.*

Dennis Gillings, CEO
John Ratliff, COO
Mike Troullis, CFO
Christopher Cabell, Chief Medical & Scientific Officer
William R. Deam, CIO/Exec. VP
Ron Wooten, Exec. VP-Corp. Dev.
Oren Cohen, Sr. VP-Clinical Research Strategies
Dennis Gillings, Chmn.

| Phone: 919-998-2000 | Fax: 919-998-2094 |
|---|---|
| Toll-Free: | |
| Address: 4709 Creekstone Dr., Ste. 200, Durham, NC 27703 US | |

## GROWTH PLANS/SPECIAL FEATURES:

Quintiles Transnational Corp. provides full-service contract research, sales and marketing services to the global pharmaceutical, biotechnology and medical device industries. The company is one of the world's top contract research organizations, and it provides a broad range of contract services to speed the process from development to peak sales of a new drug or medical device. The firm operates through offices in 50 countries, organized in three primary business segments, including the product development group, the Innovex commercialization group and the Novaquest strategic partnering solutions group. The product development group provides a full range of drug development services from strategic planning and preclinical services to regulatory submission and approval. The commercial services group, which operates under the Innovex brand, engages in sales solutions such as recruitment, training and deployment of Innovex managed sales teams; and medical communications, which provides and promotes physician education. The Novaquest strategic partnering solutions segment attempts to optimize portfolio development, company growth and profits for pharmaceutical and biotech companies through a variety of solutions such as structured finance, strategic resourcing and eBio. In January 2009, the firm announced plans to open a Phase I clinical trial unit in India through a partnership with Apollo Hospitals Group. In May 2008, the firm agreed to acquire Eidetics, a decision-analytics and market research firm.

The company offers its employees a comprehensive benefits package including on-the-job training, recreational activities and community support activities.

## FINANCIALS:  Sales and profits are in thousands of dollars—add 000 to get the full amount. 2008 Note: Financial information for 2008 was not available for all companies at press time.

| | | |
|---|---|---|
| 2008 Sales: $ | 2008 Profits: $ | U.S. Stock Ticker: Private |
| 2007 Sales: $ | 2007 Profits: $ | Int'l Ticker:    Int'l Exchange: |
| 2006 Sales: $ | 2006 Profits: $ | Employees:  16,000 |
| 2005 Sales: $2,398,583 | 2005 Profits: $ 648 | Fiscal Year Ends: 12/31 |
| 2004 Sales: $1,956,254 | 2004 Profits: $-7,427 | Parent Company: |

## SALARIES/BENEFITS:

| Pension Plan: | ESOP Stock Plan: | Profit Sharing: | Top Exec. Salary: $706,061 | Bonus: $825,000 |
|---|---|---|---|---|
| Savings Plan: | Stock Purch. Plan: | | Second Exec. Salary: $471,224 | Bonus: $709,000 |

## OTHER THOUGHTS:

**Apparent Women Officers or Directors:**
**Hot Spot for Advancement for Women/Minorities:**

## LOCATIONS: ("Y" = Yes)

| West: | Southwest: | Midwest: | Southeast: | Northeast: | International: |
|---|---|---|---|---|---|
| Y | Y | Y | Y | Y | Y |

# RANBAXY LABORATORIES LIMITED

www.ranbaxy.com

**Industry Group Code: 325412A  Ranks within this company's industry group: Sales: 5    Profits: 30**

| Drugs: | | Other: | | Clinical: | | Computers: | | Services: | |
|---|---|---|---|---|---|---|---|---|---|
| Discovery: | | AgriBio: | | Trials/Services: | | Hardware: | | Specialty Services: | |
| Licensing: | Y | Genetic Data: | | Labs: | | Software: | | Consulting: | |
| Manufacturing: | Y | Tissue Replacement: | | Equipment/Supplies: | | Arrays: | | Blood Collection: | |
| Genetics: | Y | | | Research & Development Services: | | Database Management: | | Drug Delivery: | Y |
| | | | | Diagnostics: | Y | | | Drug Distribution: | |

## TYPES OF BUSINESS:

Generic Pharmaceuticals
Active Pharmaceutical Ingredients
Drugs-Anti-Retroviral & HIV/AIDS
Drug Delivery Systems
Veterinary Pharmaceuticals
Specialty Chemicals
Diagnostics

## BRANDS/DIVISIONS/AFFILIATES:

Ranbaxy Pharmaceuticals, Inc.
Terapia
Revital
Pepfiz
Gesdyp
Garlic Pearls
Daiichi Sankyo

## CONTACTS: *Note: Officers with more than one job title may be intentionally listed here more than once.*

Atul Sobti, CEO
Omesh Sethi, CFO
Sudarshan K. Arora, Pres., R&D
Arun Sawhney, Pres., Global Mfg.
Dipak Chattaraj, Pres., Corp. Dev. & Strategy
Ramesh L. Adige, Pres., Corp. Affairs & Global Corp. Comm.
Tsutomu Une, Chmn.
Arun Sawhney, Pres.-Supply Chain

| | |
|---|---|
| **Phone:** 91-124-413-5000 | **Fax:** 91-124-4106490 |
| **Toll-Free:** | |
| **Address:** Plot 90, Sector 32, Gurgaon, 122001 India | |

## GROWTH PLANS/SPECIAL FEATURES:

Ranbaxy Laboratories Limited is a leading Indian manufacturer and marketer of generic pharmaceuticals, branded generics, over-the-counter medications and active pharmaceutical ingredients. It has manufacturing operations in 11 countries, a ground presence in 49 countries and its products are available in over 125 nations worldwide. Ranbaxy operates in the U.S. through Ranbaxy Pharmaceuticals, Inc.; Ranbaxy Laboratories, Inc.; and Ohm Laboratories, Inc. With the purchase of RPG (Aventis) SA, Ranbaxy established itself as one of the leading generic drug companies in France and continues to pursue European and international expansion. The company's research and development activities focus on select infectious diseases, metabolic diseases, inflammatory/respiratory disease and oncology. Major products include Simvastatin, CoAmoxyclav, Amoxycilin, Ciprofloxacin, Isotretinon and Cephalexin. Rambaxy conducts major research collaborations with GlaxoSmithKline (GSK). The company's Active Pharmaceutical Ingredient (API) business has over 50 products in its portfolio and supplies APIs to other generic manufacturing businesses worldwide. The company also has a global consumer healthcare segment, which includes the brands Revital, Pepfiz, Gesdyp and Garlic Pearls. This segment also uses a communication platform called The Rainbow Coalition, a program that targets both physicians and consumers, to assist in providing knowledge about the company's products to the public. The company is currently looking to build collaborative partnerships to develop drugs in the areas of anti-infectives, inflammation, metabolic diseases and oncology. In November 2008, the Japanese pharmaceutical firm Daiichi Sankyo acquired a controlling stake in the company.

Ranbaxy offers employees group life insurance and medical insurance. Rambaxy's employees come from 51 nations.

## FINANCIALS: Sales and profits are in thousands of dollars—add 000 to get the full amount. 2008 Note: Financial information for 2008 was not available for all companies at press time.

| | | |
|---|---|---|
| 2008 Sales: $1,531,700 | 2008 Profits: $-196,320 | **U.S. Stock Ticker: RBXLF.PK** |
| 2007 Sales: $1,399,530 | 2007 Profits: $159,870 | **Int'l Ticker: 500359**    Int'l Exchange: Bombay-BSE |
| 2006 Sales: $1,444,400 | 2006 Profits: $123,710 | Employees:  12,174 |
| 2005 Sales: $1,244,190 | 2005 Profits: $60,200 | Fiscal Year Ends: 12/31 |
| 2004 Sales: $1,265,200 | 2004 Profits: $160,000 | Parent Company: |

## SALARIES/BENEFITS:

| | | | | |
|---|---|---|---|---|
| Pension Plan: Y | ESOP Stock Plan: | Profit Sharing: | Top Exec. Salary: $ | Bonus: $ |
| Savings Plan: | Stock Purch. Plan: | | Second Exec. Salary: $ | Bonus: $ |

## OTHER THOUGHTS:

**Apparent Women Officers or Directors:**
**Hot Spot for Advancement for Women/Minorities:**

## LOCATIONS: ("Y" = Yes)

| West: | Southwest: | Midwest: | Southeast: | Northeast: | International: |
|---|---|---|---|---|---|
| | | | Y | Y | Y |

# REGENERON PHARMACEUTICALS INC
## www.regeneron.com

**Industry Group Code: 325412   Ranks within this company's industry group: Sales: 55   Profits: 154**

| Drugs: | | Other: | | Clinical: | | Computers: | | Services: | |
|---|---|---|---|---|---|---|---|---|---|
| Discovery: | Y | AgriBio: | | Trials/Services: | | Hardware: | | Specialty Services: | |
| Licensing: | Y | Genetic Data: | Y | Labs: | | Software: | | Consulting: | |
| Manufacturing: | Y | Tissue Replacement: | | Equipment/Supplies: | | Arrays: | | Blood Collection: | |
| Genetics: | | | | Research & Development Services: | | Database Management: | | Drug Delivery: | |
| | | | | Diagnostics: | | | | Drug Distribution: | |

## TYPES OF BUSINESS:

Drugs-Diversified
Protein-Based Drugs
Small-Molecule Drugs
Genetics & Transgenic Mouse Technologies

## BRANDS/DIVISIONS/AFFILIATES:

VelocImmune
VelociGene
VelociMouse
VEGF Trap
VEGF Trap-Eye
IL-1 Trap
ARCALYST
REGN88

## CONTACTS: *Note: Officers with more than one job title may be intentionally listed here more than once.*

Leonard S. Schleifer, CEO
Leonard S. Schleifer, Pres.
Murray A. Goldberg, CFO
George D. Yancopoulos, Chief Scientific Officer/Exec. VP
Murray A. Goldberg, Sr. VP-Admin.
Stuart A. Kolinski, General Counsel/Sr. VP/Sec.
Murray A. Goldberg, Sr. VP-Finance/Treas.
George D. Yancopoulos, Pres., Regeneron Research Laboratories
Neil Stahl, Sr. VP-R&D Sciences
Peter Powchik, Sr. VP-Clinical Dev.
Robert J. Terifay, Sr. VP-Commercial
P. Roy Vagelos, Chmn.

| Phone: 914-345-7400 | Fax: 914-347-2113 |
|---|---|
| Toll-Free: | |
| Address: 777 Old Saw Mill River Rd., Tarrytown, NY 10591-6707 US | |

## GROWTH PLANS/SPECIAL FEATURES:

Regeneron Pharmaceuticals, Inc. is a biopharmaceutical company that discovers, develops and commercializes pharmaceutical drugs for the treatment of serious medical conditions. The company's sole marketed product is Arcalyst, a therapy for the treatment of Cryopyrin-Associated Periodic Syndromes (CAPS), a rare, inherited inflammatory condition. The firm currently has three late-stage clinical development programs: aflibercept (VEGF Trap) in oncology; VEGF Trap eye formulation (VEGF Trap-Eye) in eye diseases using intraocular delivery; and rilonacept (IL-1 Trap) in gout. The VEGF Trap oncology development program is being developed jointly with the sanofi-aventis Group through a 2003 agreement. Also in collaboration with sanofi-aventis is REGN88, an antibody to the Interleukin-6 receptor (IL-6R) for treatment of rheumatoid arthritis, developed using VelocImmune technologies. Regeneron's preclinical research programs are in the areas of oncology, angiogenesis, ophthalmology, metabolic and related diseases, muscle diseases and disorders, inflammation and immune diseases, bone and cartilage, pain, metabolic diseases, infectious diseases and cardiovascular diseases. The company expects that its next generation of product candidates will be based on its proprietary technologies for developing Traps and Human Monoclonal Antibodies. Regeneron's proprietary technologies include VelociGene, VelociMouse and VelocImmune, among others. The VelociGene technology allows precise DNA manipulation and gene staining, helping to identify where a particular gene is active in the body. VelociMouse technology allows for the direct and immediate generation of genetically altered mice from ES cells, avoiding the lengthy process involved in generating and breeding knock-out mice from chimeras. VelocImmune is a novel mouse technology platform for producing fully human monoclonal antibodies. Additionally, the VelociMab suite of technologies allows rapid screenings of therapeutic antibodies and eliminates the need for slower development techniques. In August 2008, the company signed an agreement with sanofi-aventis to license its VelociGene technology.

Regeneron employee benefits include medical, dental, vision, prescription, life and AD&D insurance; short- and long-term disability; flexible spending accounts, and tuition reimbursement.

## FINANCIALS: Sales and profits are in thousands of dollars—add 000 to get the full amount. 2008 Note: Financial information for 2008 was not available for all companies at press time.

| | | | |
|---|---|---|---|
| 2008 Sales: $238,457 | 2008 Profits: $-82,710 | **U.S. Stock Ticker:** REGN | |
| 2007 Sales: $125,024 | 2007 Profits: $-105,600 | **Int'l Ticker:** Int'l Exchange: | |
| 2006 Sales: $63,447 | 2006 Profits: $-102,337 | Employees: 919 | |
| 2005 Sales: $66,193 | 2005 Profits: $-95,446 | Fiscal Year Ends: 12/31 | |
| 2004 Sales: $174,017 | 2004 Profits: $41,699 | Parent Company: | |

## SALARIES/BENEFITS:

| Pension Plan: | ESOP Stock Plan: | Profit Sharing: | Top Exec. Salary: $713,000 | Bonus: $470,580 |
|---|---|---|---|---|
| Savings Plan: Y | Stock Purch. Plan: | | Second Exec. Salary: $592,100 | Bonus: $325,655 |

## OTHER THOUGHTS:

Apparent Women Officers or Directors:
Hot Spot for Advancement for Women/Minorities:

## LOCATIONS: ("Y" = Yes)

| West: | Southwest: | Midwest: | Southeast: | Northeast: | International: |
|---|---|---|---|---|---|
| | | | | Y | |

Note: Financial information, benefits and other data can change quickly and may vary from those stated here.

# REPLIGEN CORPORATION

www.repligen.com

**Industry Group Code: 325412  Ranks within this company's industry group:** Sales: 104   Profits: 41

| Drugs: | | Other: | | Clinical: | | Computers: | | Services: | |
|---|---|---|---|---|---|---|---|---|---|
| Discovery: | Y | AgriBio: | | Trials/Services: | | Hardware: | | Specialty Services: | |
| Licensing: | | Genetic Data: | | Labs: | | Software: | | Consulting: | |
| Manufacturing: | Y | Tissue Replacement: | | Equipment/Supplies: | | Arrays: | | Blood Collection: | |
| Genetics: | | | | Research & Development Services: | | Database Management: | | Drug Delivery: | |
| | | | | Diagnostics: | Y | | | Drug Distribution: | |

## TYPES OF BUSINESS:

Neuropsychiatric Drugs
Protein A Drugs

## BRANDS/DIVISIONS/AFFILIATES:

SecreFlo
RG2417
RPA50
IPA300
IPA400
RG1068
CTLA4-Ig

## CONTACTS: Note: Officers with more than one job title may be intentionally listed here more than once.

Walter C. Herlihy, CEO
Walter C. Herlihy, Pres.
William J. Kelly, CFO
Laura Whitehouse Pew, VP-Market Dev.
James R. Rusche, Sr. VP-R&D
William J. Kelly, VP-Admin.
William J. Kelly, Sec.
Daniel P. Witt, VP-Oper.
Howard Benjamin, VP-Bus. Dev.
William J. Kelly, VP-Finance
Stephen Tingley, VP-Bioprocessing Bus. Dev.
Alexander Rich, Chmn.

| **Phone:** 781-250-0111 | **Fax:** 781-250-0115 |
|---|---|

**Toll-Free:** 800-622-2259

**Address:** 41 Seyon St., Bldg. 1, Ste. 100, Waltham, MA 02453 US

## GROWTH PLANS/SPECIAL FEATURES:

Repligen Corp. develops therapeutics for the treatment of diseases of the central nervous system. The company also manufactures a line of products based on protein A, which is used in the production of many therapeutic monoclonal antibodies. Its protein A products include recombinant protein A (RPA50), immobilized protein A (IPA300, IPA 400 and IPA400HC) and protein A ELISA kits. The firm sells these products to the biotechnology and pharmaceutical companies, including GE Healthcare and Applied Biosystems, Inc. Protein A products have a wide variety of end-uses including applications in drugs for colon cancer, infection, Crohn's disease and arthritis. In addition to protein A products, the firm market SecreFlo, a synthetic form of the hormone secretin, which is used as an aid in the diagnosis chronic pancreatitis and gastrinoma. Repligen also has products in the development stage for neuropsychiatric disorders and other applications: RG1068 (secretin), a hormone produced in the small intestine that regulates the function of the pancreas as part of the process of digestion, evaluated by the company for improvement of MRI imaging of the pancreas; RG2417 (uridine), a biological compound essential for the synthesis of DNA and RNA, being tested by the firm under an oral formulation for the treatment of Bipolar Depression; and HDAC inhibitors for Friedreich's ataxia and Huntington's disease. The firm also owns the exclusive rights for the use of CTLA4-Ig, a novel autoimmune therapy marketed by Bristol-Myers Squibb under the name Orencia.

Repligen offers its employees benefits that include health and dental insurance; life and long-term disability coverage; a 401(k) plan; and equity participation.

## FINANCIALS: Sales and profits are in thousands of dollars—add 000 to get the full amount. 2008 Note: Financial information for 2008 was not available for all companies at press time.

| | | |
|---|---|---|
| 2008 Sales: $19,296 | 2008 Profits: $37,107 | **U.S. Stock Ticker: RGEN** |
| 2007 Sales: $14,074 | 2007 Profits: $- 889 | **Int'l Ticker:**    Int'l Exchange: |
| 2006 Sales: $12,911 | 2006 Profits: $ 697 | Employees:   69 |
| 2005 Sales: $9,360 | 2005 Profits: $-2,984 | Fiscal Year Ends: 3/31 |
| 2004 Sales: $6,914 | 2004 Profits: $-9,551 | Parent Company: |

## SALARIES/BENEFITS:

| Pension Plan: | ESOP Stock Plan: Y | Profit Sharing: | Top Exec. Salary: $344,000 | Bonus: $103,349 |
|---|---|---|---|---|
| Savings Plan: Y | Stock Purch. Plan: | | Second Exec. Salary: $234,000 | Bonus: $46,818 |

## OTHER THOUGHTS:

**Apparent Women Officers or Directors:** 2
**Hot Spot for Advancement for Women/Minorities:**

## LOCATIONS: ("Y" = Yes)

| West: | Southwest: | Midwest: | Southeast: | Northeast: | International: |
|---|---|---|---|---|---|
| | | | | Y | |

Note: Financial information, benefits and other data can change quickly and may vary from those stated here.

# REPROS THERAPEUTICS INC

**www.reprosrx.com**

Industry Group Code: 325412 **Ranks within this company's industry group:** Sales: 146   Profits: 117

| Drugs: | | Other: | Clinical: | Computers: | Services: |
|---|---|---|---|---|---|
| Discovery: | Y | AgriBio: | Trials/Services: | Hardware: | Specialty Services: |
| Licensing: | Y | Genetic Data: | Labs: | Software: | Consulting: |
| Manufacturing: | | Tissue Replacement: | Equipment/Supplies: | Arrays: | Blood Collection: |
| Genetics: | | | Research & Development Services: | Database Management: | Drug Delivery: |
| | | | Diagnostics: | | Drug Distribution: |

## TYPES OF BUSINESS:

Drugs-Fertility & Sexual Dysfunction
Reproductive System Disorder Treatment
Hormonal Disorder Treatment

## BRANDS/DIVISIONS/AFFILIATES:

Proellex
Androxal
VASOMAX

## CONTACTS: Note: Officers with more than one job title may be intentionally listed here more than once.

Joseph S. Podolski, CEO
Paul Lammers, Pres.
Louis Ploth, CFO
Ronald Wiehle, VP-R&D
Louis Ploth, Corp. Sec.
Louis Ploth, VP-Bus. Dev.
Andre van As, Sr. VP-Clinical & Regulatory Affairs
Andre van As, Chief Medical Officer
Mark Lappe, Chmn.

| Phone: 281-719-3400 | Fax: 281-719-3446 |
|---|---|
| Toll-Free: | |
| Address: 2408 Timberloch Pl., Ste. B-7, The Woodlands, TX 77380 US | |

## GROWTH PLANS/SPECIAL FEATURES:

Repros Therapeutics, Inc., formerly Zonagen, Inc., develops treatments of hormonal and reproductive system disorders. Proellex, its leading product candidate, is an orally active small molecule compound selective progesterone receptor blocker for the treatment of uterine fibroids and the treatment of endometriosis. The National Uterine Fibroid Foundation estimates that as many as 80% of all women in the U.S. have uterine fibroids, and one in four of these women require treatment. According to The Endometriosis Association, endometriosis affects 5.5 million women in the U.S. and Canada. As a new chemical entity, Proellex is required to undergo the full regulatory approval process, including a two-year carcinogenicity study, which the company completed in 2008. For the treatment of anemia and uterine fibroids, Proellex is undergoing Phase III clinical trials. For the treatment of endometriosis, Proellex is undergoing a Phase II trial. The company's other product candidate, Androxal, is an orally available small molecule compound. Androxal targets adult-onset idiopathic hypogonadotrophic hypogonadism with concomitant plasma glucose and lipid elevations; and improves or maintains fertility and/or sperm function in men being treated for low testosterone. The drug is also being evaluated for use in type II diabetes. Androxal is currently in Phase IIb testing in the U.S. The company has a third, phentolamine-based product, VASOMAX, which is out-licensed in Mexico and Brazil for the treatment of male erectile dysfunction (ED). VASOMAX is currently on partial clinical hold in the U.S. In January 2009, the firm reported positive results in its Proellex endometriosis trials, with statistically significant pain reductions compared to a placebo.

## FINANCIALS: Sales and profits are in thousands of dollars—add 000 to get the full amount. 2008 Note: Financial information for 2008 was not available for all companies at press time.

| | | |
|---|---|---|
| 2008 Sales: $ 433 | 2008 Profits: $-25,202 | **U.S. Stock Ticker:** RPRX |
| 2007 Sales: $1,508 | 2007 Profits: $-13,700 | **Int'l Ticker:**   Int'l Exchange: |
| 2006 Sales: $ 596 | 2006 Profits: $-14,195 | Employees:   10 |
| 2005 Sales: $ 634 | 2005 Profits: $-7,391 | Fiscal Year Ends: 12/31 |
| 2004 Sales: $ 257 | 2004 Profits: $-3,697 | Parent Company: |

## SALARIES/BENEFITS:

| Pension Plan: | ESOP Stock Plan: | Profit Sharing: | Top Exec. Salary: $424,684 | Bonus: $84,087 |
|---|---|---|---|---|
| Savings Plan: | Stock Purch. Plan: | | Second Exec. Salary: $276,225 | Bonus: $ |

## OTHER THOUGHTS:

**Apparent Women Officers or Directors**: 1
**Hot Spot for Advancement for Women/Minorities**:

## LOCATIONS: ("Y" = Yes)

| West: | Southwest: | Midwest: | Southeast: | Northeast: | International: |
|---|---|---|---|---|---|
| | Y | | | | |

# RIGEL PHARMACEUTICALS INC

www.rigel.com

**Industry Group Code: 325412 Ranks within this company's industry group: Sales:    Profits: 162**

| Drugs: | | Other: | Clinical: | Computers: | Services: |
|---|---|---|---|---|---|
| Discovery: | Y | AgriBio: | Trials/Services: | Hardware: | Specialty Services: |
| Licensing: | Y | Genetic Data: | Labs: | Software: | Consulting: |
| Manufacturing: | | Tissue Replacement: | Equipment/Supplies: | Arrays: | Blood Collection: |
| Genetics: | | | Research & Development Services: | Database Management: | Drug Delivery: |
| | | | Diagnostics: | | Drug Distribution: |

## TYPES OF BUSINESS:

Biopharmaceuticals Development
Small-Molecule Drugs
Drugs-Cancer & Inflammatory Diseases
Drugs-Viral Diseases
Drugs-Autoimmune Diseases

## BRANDS/DIVISIONS/AFFILIATES:

R788
R763
R348
R343

## CONTACTS: Note: Officers with more than one job title may be intentionally listed here more than once.

James M. Gower, CEO
Raul R. Rodriguez, COO/Exec. VP
Ryan Maynard, CFO/VP
Donald G. Payan, Chief Scientific Officer
Dolly Vance, General Counsel/Sr. VP/Corp. Sec.
Donald G. Payan, Exec. VP-Research & Discovery
Elliott B. Grossbard, Chief Medical Officer/Exec. VP
James M. Gower, Chmn.

| Phone: 650-624-1100 | Fax: 650-624-1101 |
|---|---|

**Toll-Free:**

**Address:** 1180 Veterans Blvd., South San Francisco, CA 94080 US

## GROWTH PLANS/SPECIAL FEATURES:

Rigel Pharmaceuticals, Inc. is a biotechnology company focused on developing small-molecule drugs in the fields of inflammatory diseases, cancer, autoimmune diseases and viral diseases. The firm's lead candidate, R788, a potential drug for the treatment of rheumatoid arthritis functions by inhibiting IgG receptor signaling in macrophages and B-cells and is in Phase II of the development process. R788 is also being tested by the firm in various stages of clinical and preclinical trials for indications of B-Cell lymphoma, lupus and immune thrombocytopenia purpura, a blood disorder in which the immune system destroys platelets in the blood. The company's cancer treatment candidate, R763, is a specific inhibitor of Aurora kinase, shown to block proliferation of trigger apoptosis in several tumor cell lines. Rigel develops R763 in partnership with Merck Serono, and the drug is currently in Phase I clinical trials for the treatment of patients with refractory solid tumors and hematological malignancies. R343, an oral Syk kinase inhibitor and the company's asthma and allergy treatment candidate, functions by inhibiting the IgE receptor signaling in respiratory tract mast cells. The company holds a partnership with Pfizer for the development of R343, which is currently in Phase I development. The company's final drug in clinical development stages is R348, an oral Janus Tyrosine Kinase 3 (JAK3) inhibitor. R348 is currently in Phase I trials and has shown potential for the treatment of rheumatoid arthritis, psoriasis and transplant rejection. The firm has several other research projects in the preclinical development stages, including programs studying HIV and ligase inhibitors. In February 2009, the company announced it would cut 20% of its workforce as it looks for a partner for its arthritis drug R788.

Rigel offers employees medical, dental and vision coverage; education reimbursement; a 401(k) plan; subsidized lunches; an employee assistance program; and wellness benefits.

## FINANCIALS: Sales and profits are in thousands of dollars—add 000 to get the full amount. 2008 Note: Financial information for 2008 was not available for all companies at press time.

| | | |
|---|---|---|
| 2008 Sales: $ | 2008 Profits: $-132,435 | **U.S. Stock Ticker: RIGL** |
| 2007 Sales: $12,600 | 2007 Profits: $-74,272 | **Int'l Ticker:** Int'l Exchange: |
| 2006 Sales: $33,473 | 2006 Profits: $-37,637 | Employees: 181 |
| 2005 Sales: $16,526 | 2005 Profits: $-45,256 | Fiscal Year Ends: 12/31 |
| 2004 Sales: $4,733 | 2004 Profits: $-56,255 | Parent Company: |

## SALARIES/BENEFITS:

| Pension Plan: | ESOP Stock Plan: | Profit Sharing: | Top Exec. Salary: $600,000 | Bonus: $ |
|---|---|---|---|---|
| Savings Plan: | Stock Purch. Plan: Y | | Second Exec. Salary: $483,000 | Bonus: $ |

## OTHER THOUGHTS:

**Apparent Women Officers or Directors:** 1
**Hot Spot for Advancement for Women/Minorities:**

## LOCATIONS: ("Y" = Yes)

| West: | Southwest: | Midwest: | Southeast: | Northeast: | International: |
|---|---|---|---|---|---|
| Y | | | | | |

# ROCHE HOLDING LTD

**www.roche.com**

**Industry Group Code: 325412  Ranks within this company's industry group:** Sales: 3    Profits: 6

| Drugs: | | Other: | | Clinical: | | Computers: | | Services: | |
|---|---|---|---|---|---|---|---|---|---|
| Discovery: | Y | AgriBio: | | Trials/Services: | | Hardware: | | Specialty Services: | |
| Licensing: | | Genetic Data: | | Labs: | | Software: | | Consulting: | |
| Manufacturing: | Y | Tissue Replacement: | | Equipment/Supplies: | | Arrays: | | Blood Collection: | |
| Genetics: | | | | Research & Development Services: | | Database Management: | | Drug Delivery: | |
| | | | | Diagnostics: | Y | | | Drug Distribution: | |

## TYPES OF BUSINESS:

Pharmaceuticals Manufacturing
Antibiotics
Diagnostics
Cancer Drugs
Virology Products
HIV/AIDS Treatments
Transplant Drugs

## BRANDS/DIVISIONS/AFFILIATES:

Genentech Inc
454 Life Sciences
NimbleGen Systems Inc
Therapeutic Human Polyclonals Inc
Ventana Medical Systems Inc
Chugai Pharmaceuticals
Gilead Sciences
Memory Pharmaceuticals Corp

## CONTACTS: *Note: Officers with more than one job title may be intentionally listed here more than once.*

Severin Schwan, CEO
Erich Hunziker, CFO
Sylvia Ayyoubi, Head-Human Resources
Jonathan Knowles, Head-Group Research
Gottlieb Keller, General Counsel/Head-Corp. Svcs.
Per-Olof Attinger, Head-Global Corp. Comm.
William M. Burns, CEO-Roche Pharmaceuticals
Jurgen Schwiezer, CEO-Roche Diagnostics
Osamu Nagayama, Pres./CEO-Chugai
Pascal Soriot, Head-Commercial Oper., Pharmaceuticals
Franz B. Humer, Chmn.

| Phone: 41-61-688-1111 | Fax: 41-61-691-9391 |
|---|---|
| Toll-Free: | |
| Address: Grenzacherstrasse 124, Basel,  4070 Switzerland | |

## GROWTH PLANS/SPECIAL FEATURES:

Roche Holding, Ltd. is one of the world's largest health care companies, occupying an industry-leading position in the global diagnostics market and ranking as one of the top producers of pharmaceuticals, with particular recognition in the areas of cancer drugs, autoimmune disease and metabolic disorder treatments, virology and transplantation medicine.   The company's operations currently extend to over 150 countries, with additional alliances and research and development agreements with corporate and institutional partners, furthering Roche's collective reach.   Among the company's related corporate interests are majority ownership holdings in U.S.-based Genentech and Japanese pharmaceutical firm Chugai.  The firm operates through two divisions, Pharmaceuticals and Diagnostics, and its products include the cancer drugs Avastin, Bondronat, Xeloda, Herceptin and Tarceva; the antibiotic Rocephin; the HIV/AIDS treatments Viracept, Invirase and Fuzeon; and Tamiflu, which is used to prevent and treat influenza.  Roche companies control proprietary diagnostic technologies across a range of areas, including advanced DNA tests, leading consumer diabetes monitoring devices and applied sciences methodologies for laboratory research.   As part of the mobilization of Tamiflu, the company has maintained a long-term strategic partnership with Gilead Sciences to coordinate the licensing and manufacture of the drug, important in the case of a flu pandemic.  In January 2008, the firm agreed to acquire Ventana Medical Systems, Inc. for $3.4 billion.  In April 2008, Roche's Actemra, a treatment for adult rheumatoid arthritis and juvenile idiopathic arthritis, received approval for use in Japan.  In July 2008, the firm announced that it hopes to acquire the remaining shares of biotech giant Genentech.  In November 2008, Roche announced that it would acquire Memory Pharmaceuticals Corp., a developer of drug treatments for disorders of the central nervous system such as Alzheimer's, for approximately $50 million.

## FINANCIALS: Sales and profits are in thousands of dollars—add 000 to get the full amount. 2008 Note: Financial information for 2008 was not available for all companies at press time.

| | | |
|---|---|---|
| 2008 Sales: $41,676,500 | 2008 Profits: $7,803,000 | **U.S. Stock Ticker: RHHBY** |
| 2007 Sales: $40,650,000 | 2007 Profits: $8,600,000 | **Int'l Ticker: RO**    Int'l Exchange: Zurich-SWX |
| 2006 Sales: $34,851,500 | 2006 Profits: $7,116,030 | Employees:  68,218 |
| 2005 Sales: $27,385,668 | 2005 Profits: $5,189,777 | Fiscal Year Ends: 12/31 |
| 2004 Sales: $22,767,021 | 2004 Profits: $5,446,567 | Parent Company: |

## SALARIES/BENEFITS:

| Pension Plan: Y | ESOP Stock Plan: | Profit Sharing: | Top Exec. Salary: $ | Bonus: $ |
|---|---|---|---|---|
| Savings Plan: | Stock Purch. Plan: | | Second Exec. Salary: $ | Bonus: $ |

## OTHER THOUGHTS:

**Apparent Women Officers or Directors:** 4
**Hot Spot for Advancement for Women/Minorities:** Y

## LOCATIONS: ("Y" = Yes)

| West: | Southwest: | Midwest: | Southeast: | Northeast: | International: |
|---|---|---|---|---|---|
| Y | Y | Y | Y | Y | Y |

# ROSETTA INPHARMATICS LLC

## www.rii.com

Industry Group Code: 541712  Ranks within this company's industry group:  Sales:    Profits:

| Drugs: | Other: | Clinical: | | Computers: | | Services: | |
|---|---|---|---|---|---|---|---|
| Discovery: | AgriBio: | Trials/Services: | | Hardware: | | Specialty Services: | |
| Licensing: | Genetic Data: | Labs: | | Software: | Y | Consulting: | |
| Manufacturing: | Tissue Replacement: | Equipment/Supplies: | | Arrays: | Y | Blood Collection: | |
| Genetics: | | Research & Development Services: | Y | Database Management: | | Drug Delivery: | |
| | | Diagnostics: | | | | Drug Distribution: | |

## TYPES OF BUSINESS:

Biotechnology Research
Bioinformatics Software
Gene Expression Research
DNA Microarrays

## BRANDS/DIVISIONS/AFFILIATES:

Rosetta Biosoftware
Rosetta Resolver
Rosetta Syllego
Rosetta Elucidator

## CONTACTS: Note: Officers with more than one job title may be intentionally listed here more than once.

S.J. Rupert Vessey, Site Head
Douglas E. Bassett, Jr., Co-Site Head
Stephen H. Friend, Pres.
Eric Schadt, Exec. Scientific Dir.-Genetics
S.J. Rupert Vessey, VP-Molecular Profiling
Douglas E. Bassett, Jr., Exec. Dir.-Molecular Profiling

| Phone: 206-802-7000 | Fax: 206-802-6501 |
|---|---|
| Toll-Free: | |
| Address: 401 Terry Ave. N., Seattle, WA 98109 US | |

## GROWTH PLANS/SPECIAL FEATURES:

Rosetta Inpharmatics LLC, a wholly-owned subsidiary of Merck and Co., uses gene expression research and DNA microarray technologies to support Merck's drug discovery efforts and to enhance drug development activities. Through its Rosetta Biosoftware business unit, the company continues the commercial release of the Rosetta Resolver gene expression data analysis system, an enterprise-level bioinformatics software package launched in 2000 and currently in its seventh version. The software remains the firm's flagship product, licensed to many of the world's leading academic research institutions and life sciences corporations. The firm also offers two other software packages: Rosetta Syllego, for genetic data management; and Rosetta Elucidator, for protein expression analysis. The company's expertise has led to new applications of gene expression technologies in areas including molecular toxicology, biomarker discovery and disease classification. Rosetta Inpharmatics is committed to using genomic research and data analysis to enable more accurate selection of drug targets and more efficient drug development, while also developing new tools to extend the analysis of gene expression data generated by DNA microarrays. These developments not only support the pharmaceutical industry, but also bring new capabilities to other sectors, including agrochemical research and biotechnology. In June 2008, the firm released version 2.0 of its Rosetta Syllego software, with an expanded set of data analysis tools. In December 2008, Rosetta released version 3.2 of its Rosetta Elucidator software.

Rosetta's work atmosphere includes clubs for rock climbers and theater goers, as well as a post-doctorate program to facilitate the transition of fellows to independent scientific investigators in a rigorous research setting. Employees benefits include education assistance; scholarship programs; paid vacation; wellness centers; daycare centers; flexible work arrangements; a 401(k) plan and a pension plan; flexible spending accounts; and medical, dental, prescription, and vision insurance.

## FINANCIALS: Sales and profits are in thousands of dollars—add 000 to get the full amount. 2008 Note: Financial information for 2008 was not available for all companies at press time.

| | | |
|---|---|---|
| 2008 Sales: $ | 2008 Profits: $ | U.S. Stock Ticker: Subsidiary |
| 2007 Sales: $ | 2007 Profits: $ | Int'l Ticker:    Int'l Exchange: |
| 2006 Sales: $ | 2006 Profits: $ | Employees:  266 |
| 2005 Sales: $ | 2005 Profits: $ | Fiscal Year Ends: 12/31 |
| 2004 Sales: $ | 2004 Profits: $ | Parent Company: MERCK & CO INC |

## SALARIES/BENEFITS:

| Pension Plan: Y | ESOP Stock Plan: | Profit Sharing: | Top Exec. Salary: $ | Bonus: $ |
|---|---|---|---|---|
| Savings Plan: Y | Stock Purch. Plan: | | Second Exec. Salary: $ | Bonus: $ |

## OTHER THOUGHTS:

Apparent Women Officers or Directors:
Hot Spot for Advancement for Women/Minorities:

## LOCATIONS: ("Y" = Yes)

| West: | Southwest: | Midwest: | Southeast: | Northeast: | International: |
|---|---|---|---|---|---|
| Y | | | | | |

# SALIX PHARMACEUTICALS

**www.salix.com**

Industry Group Code: 325412  Ranks within this company's industry group:  Sales: 60    Profits: 135

| Drugs: | | Other: | Clinical: | | Computers: | | Services: |
|---|---|---|---|---|---|---|---|
| Discovery: | | AgriBio: | Trials/Services: | | Hardware: | | Specialty Services: |
| Licensing: | | Genetic Data: | Labs: | | Software: | | Consulting: |
| Manufacturing: | Y | Tissue Replacement: | Equipment/Supplies: | | Arrays: | | Blood Collection: |
| Genetics: | | | Research & Development Services: | | Database Management: | | Drug Delivery: |
| | | | Diagnostics: | | | | Drug Distribution: |

## TYPES OF BUSINESS:

Pharmaceuticals Development & Manufacturing
Drugs-Gastroenterology

## BRANDS/DIVISIONS/AFFILIATES:

Colazal
Azasan
Proctocort
Anusol-HC
OsmoPrep
Xifaxan
Visicol
DIACOL

## CONTACTS: *Note: Officers with more than one job title may be intentionally listed here more than once.*

Carolyn J. Logan, CEO
Carolyn J. Logan, Pres.
Adam C. Derbyshire, CFO
William P. Forbes, VP-R&D
Adam C. Derbyshire, Sr. VP-Admin.
William P. Forbes, Chief Dev. Officer
Adam C. Derbyshire, Sr. VP-Finance
John F. Chappell, Chmn.

| Phone: 919-862-1000 | Fax: 919-862-1095 |
|---|---|
| Toll-Free: 888-802-9956 | |
| Address: 1700 Perimeter Park Dr., Morrisville, NC 27560 US | |

## GROWTH PLANS/SPECIAL FEATURES:

Salix Pharmaceuticals is a specialty pharmaceutical company dedicated to acquiring, developing and commercializing prescription drugs used in the treatment of a variety of gastrointestinal diseases. The company seeks to identify late-stage or approved proprietary therapeutics for in-licensing, which subsequently advances the new drugs through regulatory procedures and final product development stages. The firm's Colazal (balsalazide disodium) treats ulcerative colitis. Azasan (azathioprine tablets), initially intended to suppress immune response in organ transplant recipients, is also marketed by Salix as a treatment for rheumatoid arthritis. The company also sells Xifaxam (rifaximin), a gastrointestinal-specific oral antibiotic; Visicol, a product indicated for cleansing of the bowel as a preparation for colonoscopy; OsmoPrep tablets; MoviPrep oral solution; Anusol-HC rectal suppositories; and Proctocort, which is available in a cream form that is indicated for the relief of the inflammatory and pruritic manifestations of corticosteroid-responsive dermatoses, and in a suppository form, which is indicated for use in inflamed hemorrhoids and postirradiation proclitis. Salix recently added Pepcid Oral Suspension and Diuril Oral Suspension to its line of products by acquiring the rights to them from Merck & Co., Inc. The primary product candidates Salix is developing are balsalazide disodium tablets, which the company intends to sell for the treatment of ulcerative colitis; a patented, granulated formula of mesalamine, which it intends to sell for the treatment of ulcerative colitis; rifaximin for various additional gastrointestinal issues; Metoclopramide, for short-term therapy following gastroesophageal reflux; and Vapreotide Acetate Powder, for the treatment of acute esophageal variceal bleeding.

## FINANCIALS:  Sales and profits are in thousands of dollars—add 000 to get the full amount. 2008 Note: Financial information for 2008 was not available for all companies at press time.

| | | |
|---|---|---|
| 2008 Sales: $178,766 | 2008 Profits: $-47,037 | **U.S. Stock Ticker: SLXP** |
| 2007 Sales: $235,792 | 2007 Profits: $8,225 | **Int'l Ticker:**    Int'l Exchange: |
| 2006 Sales: $208,533 | 2006 Profits: $31,510 | Employees:   280 |
| 2005 Sales: $154,903 | 2005 Profits: $-60,585 | Fiscal Year Ends: 12/31 |
| 2004 Sales: $105,496 | 2004 Profits: $6,839 | Parent Company: |

## SALARIES/BENEFITS:

| Pension Plan: Y | ESOP Stock Plan: | Profit Sharing: | Top Exec. Salary: $713,000 | Bonus: $320,850 |
|---|---|---|---|---|
| Savings Plan: Y | Stock Purch. Plan: | | Second Exec. Salary: $352,000 | Bonus: $175,000 |

## OTHER THOUGHTS:

**Apparent Women Officers or Directors:** 1
**Hot Spot for Advancement for Women/Minorities:**

## LOCATIONS: ("Y" = Yes)

| West: | Southwest: | Midwest: | Southeast: | Northeast: | International: |
|---|---|---|---|---|---|
| | | | | Y | |

# SANGAMO BIOSCIENCES INC

**www.sangamo.com**

**Industry Group Code: 541712 Ranks within this company's industry group: Sales: 16 Profits: 13**

| Drugs: | | Other: | | Clinical: | | Computers: | | Services: | |
|---|---|---|---|---|---|---|---|---|---|
| Discovery: | Y | AgriBio: | | Trials/Services: | | Hardware: | | Specialty Services: | |
| Licensing: | | Genetic Data: | Y | Labs: | | Software: | | Consulting: | |
| Manufacturing: | | Tissue Replacement: | Y | Equipment/Supplies: | | Arrays: | | Blood Collection: | |
| Genetics: | | | | Research & Development Services: | Y | Database Management: | | Drug Delivery: | |
| | | | | Diagnostics: | | | | Drug Distribution: | |

## TYPES OF BUSINESS:

Drug Research & Development
Gene Expression Regulation Therapies
Transcription Factor Technology

## BRANDS/DIVISIONS/AFFILIATES:

ZFP Therapeutic
ZFP Transcription Factors (ZFP TFs)
ZFP Nucleases (ZFNs)
Zinc Finger DNA-Binding Proteins
SB-509
SB-728-T

## CONTACTS: *Note: Officers with more than one job title may be intentionally listed here more than once.*

Edward O. Lanphier, II, CEO
Edward O. Lanphier, II, Pres.
H. Ward Wolff, CFO/Exec. VP
Phillip D. Gregory, VP-Research
Edward J. Rebar, VP-Tech.
Gregory S. Zante, VP-Admin.
David G. Ichikawa, Sr. VP-Bus. Dev.
Gregory S. Zante, VP-Finance
Ely Benaim, VP-Clinical Affairs
Dale G. Ando, Chief Medical Officer/VP-Therapeutic Dev.

| Phone: 510-970-6000 | Fax: 510-236-8951 |
|---|---|
| Toll-Free: | |
| Address: 501 Canal Blvd., Ste. A100, Richmond, CA 94804 US | |

## GROWTH PLANS/SPECIAL FEATURES:

Sangamo BioSciences, Inc. is a biotechnology company that develops, researches and commercializes zinc finger DNA-binding proteins (ZFPs), a naturally occurring class of proteins. ZFPs can be engineered to make ZFP transcription factors (ZFP TFs), which can be used to turn genes on or off, and ZFP nucleases (ZFNs), which enable the modification of DNA sequences in a variety of ways. Sangamo's lead ZFP Therapeutic, SB-509, is a plasmid formulation of a ZFP TF activator of the vascular endothelial growth factor-A gene, and is currently in three Phase 2 clinical trials for the treatment of diabetic neuropathy (DN) and one Phase 2 trial for amyotrophic lateral sclerosis (ALS), or Lou Gehrig's Disease. Sangamo expects to have clinical data from its Phase 2 trials in DN in 2009, as well as to complete enrollment and treatment in its Phase 2 study for ALS. The firm has additional research-stage programs in X-linked severe combined immunodeficiency (X-linked SCID), hemophilia and hemoglobinopathies. The company additionally expects to expect to file an Investigational New Drug (IND) application for a Phase 1 trial to evaluate a ZFN-based therapeutic for the treatment of glioblastoma multiforme, a type of brain cancer. As ZFPs act at the DNA level, they have broad potential applications in several areas including human therapeutics, plant agriculture, research reagents and cell-line engineering. Sangamo has attempted to capitalize on the broad potential commercial applications of its ZFPs by facilitating the sale and licensing of its ZFP TFs and ZFNs to companies working in fields outside of human therapeutics, including Dow AgroSciences; Pfizer; Sigma-Aldrich; Medarex; Novo Nordisk; Novartis A/G; Kirin Brewery Company; and Genentech. In 2008, the company initiated a Phase 1 clinical trial to evaluate SB-728-T for the treatment of HIV/AIDS.

## FINANCIALS: Sales and profits are in thousands of dollars—add 000 to get the full amount. 2008 Note: Financial information for 2008 was not available for all companies at press time.

| | | |
|---|---|---|
| 2008 Sales: $16,186 | 2008 Profits: $-24,302 | U.S. Stock Ticker: SGMO |
| 2007 Sales: $9,098 | 2007 Profits: $-21,480 | Int'l Ticker: Int'l Exchange: |
| 2006 Sales: $7,885 | 2006 Profits: $-17,864 | Employees: 77 |
| 2005 Sales: $2,484 | 2005 Profits: $-13,293 | Fiscal Year Ends: 12/31 |
| 2004 Sales: $1,315 | 2004 Profits: $-13,818 | Parent Company: |

## SALARIES/BENEFITS:

| Pension Plan: | ESOP Stock Plan: | Profit Sharing: | Top Exec. Salary: $510,000 | Bonus: $ |
|---|---|---|---|---|
| Savings Plan: Y | Stock Purch. Plan: Y | | Second Exec. Salary: $385,000 | Bonus: $ |

## OTHER THOUGHTS:

**Apparent Women Officers or Directors: 1**
**Hot Spot for Advancement for Women/Minorities: Y**

## LOCATIONS: ("Y" = Yes)

| West: | Southwest: | Midwest: | Southeast: | Northeast: | International: |
|---|---|---|---|---|---|
| Y | | | | | |

# SANOFI-AVENTIS SA

**www.en.sanofi-aventis.com**

**Industry Group Code: 325412  Ranks within this company's industry group: Sales: 5   Profits: 8**

| Drugs: | | Other: | Clinical: | Computers: | Services: |
|---|---|---|---|---|---|
| Discovery: | Y | AgriBio: | Trials/Services: | Hardware: | Specialty Services: |
| Licensing: | | Genetic Data: | Labs: | Software: | Consulting: |
| Manufacturing: | Y | Tissue Replacement: | Equipment/Supplies: | Arrays: | Blood Collection: |
| Genetics: | | | Research & Development Services: | Database Management: | Drug Delivery: |
| | | | Diagnostics: | | Drug Distribution: |

## TYPES OF BUSINESS:

Pharmaceuticals Development & Manufacturing
Over-the-Counter Drugs
Cardiovascular Drugs
CNS Drugs
Oncology Drugs
Diabetes Drugs
Generics
Vaccines

## BRANDS/DIVISIONS/AFFILIATES:

Aprovel
Plavix
Allegra
Depakine
Stilnox
Sanofi Pasteur
Eloxatin
Lantus

## CONTACTS: *Note: Officers with more than one job title may be intentionally listed here more than once.*

Chris Viehbacher, CEO
Laurence Debroux, CFO/Sr. VP/Chief Strategic Officer
Pierre Chancel, Sr. VP-Global Mktg.
Gilles Lhernould, Sr. VP-Human Resources
Marc Cluzel, Sr. VP-R&D
Karen Linehan, General Counsel/Sr. VP-Legal Affairs
Hanspeter Spek, Exec. VP-Pharmaceutical Oper.
Philippe Fauchet, Sr. VP-Bus. Dev.
Michel Labie, Sr. VP-Comm./VP-Institutional & Professional Rel.
Philippe Peyre, Sr. VP-Corp. Affairs
Olivier Charmeil, Sr. VP-Pharmaceutical Oper., Asia-Pacific & Japan
Philippe Luscan, Sr. VP-Industrial Affairs
Gregory Irace, Sr. VP-Pharmaceutical Oper., U.S.
Jean-Francois Dehecq, Chmn.
Antoine Ortoli, Sr. VP-Pharmaceuticals Oper., Int'l

| Phone: 33-1-53-77-4000 | Fax: |
|---|---|
| Toll-Free: | |
| Address: 174 Ave. de France, Paris,  75365 France | |

## GROWTH PLANS/SPECIAL FEATURES:

Sanofi-Aventis SA is an international pharmaceutical group engaged in the research, development, manufacturing and marketing of primarily prescription pharmaceutical products. The firm conducts research and produces pharmaceutical products in seven therapeutic areas: cardiovascular diseases, thrombosis, metabolic disorders (especially diabetes), oncology, central nervous system (CNS) disorders, internal medicine and vaccines. Cardiovascular medications include the blood pressure medication, Aprovel. One of the company's thrombosis medications is the anti-clotting agent, Plavix. In the field of oncology, Sanofi-Aventis manufactures Eloxatin, a treatment for colon-rectal cancer, and Taxotere, a medication for breast cancer patients. CNS medications include Stilnox, an insomnia medication, and Depakine, a treatment for epilepsy. Products in the internal medicine sector include the antihistamine, Allegra, as well as Xatral, a treatment for enlarged prostrates. The firm has three metabolism/diabetes treatments: Amaryl, Apidra and Lantus. Subsidiary Sanofi Pasteur produces vaccines that fight 20 diseases, immunizing over 500 million people annually. In September 2008, the company acquired Symbion Consumer, Australian distributor of vitamins and mineral supplements, from Primary Health Care Limited for approximately $363 million. In February 2009, the firm agreed to partner with the Salk Institute for Biological Studies, forming the Sanofi-Aventis Regenerative Medicine Program. In June 2009, Sanofi-Aventis announced plans to increase its focus on generic drug operations in emerging markets, as well as to reduce research and development spending across several drug areas. The moves are addressed, at least in part, toward developing new areas of revenue before the 2011 loss of U.S. patent protection for Plavix, which will open up the market for generic versions of the anti-clotting drug.

## FINANCIALS: Sales and profits are in thousands of dollars—add 000 to get the full amount. 2008 Note: Financial information for 2008 was not available for all companies at press time.

| | | |
|---|---|---|
| 2008 Sales: $36,751,700 | 2008 Profits: $5,721,790 | **U.S. Stock Ticker: SNY** |
| 2007 Sales: $37,397,000 | 2007 Profits: $7,574,840 | **Int'l Ticker: SAN**    Int'l Exchange: Paris-Euronext |
| 2006 Sales: $38,722,100 | 2006 Profits: $6,003,540 | Employees:  98,213 |
| 2005 Sales: $37,272,700 | 2005 Profits: $3,538,800 | Fiscal Year Ends: 12/31 |
| 2004 Sales: $20,377,000 | 2004 Profits: $-4,890,000 | Parent Company: |

## SALARIES/BENEFITS:

| Pension Plan: | ESOP Stock Plan: | Profit Sharing: | Top Exec. Salary: $ | Bonus: $ |
|---|---|---|---|---|
| Savings Plan: | Stock Purch. Plan: | | Second Exec. Salary: $ | Bonus: $ |

## OTHER THOUGHTS:

**Apparent Women Officers or Directors: 4**
**Hot Spot for Advancement for Women/Minorities: Y**

## LOCATIONS: ("Y" = Yes)

| West: | Southwest: | Midwest: | Southeast: | Northeast: | International: |
|---|---|---|---|---|---|
| Y | Y | Y | Y | Y | Y |

Note: Financial information, benefits and other data can change quickly and may vary from those stated here.

# SAVIENT PHARMACEUTICALS INC

### www.savientpharma.com

**Industry Group Code: 325412  Ranks within this company's industry group:  Sales: 133    Profits: 155**

| Drugs: | | Other: | | Clinical: | | Computers: | | Services: | |
|---|---|---|---|---|---|---|---|---|---|
| Discovery: | Y | AgriBio: | | Trials/Services: | | Hardware: | | Specialty Services: | |
| Licensing: | | Genetic Data: | | Labs: | | Software: | | Consulting: | |
| Manufacturing: | Y | Tissue Replacement: | | Equipment/Supplies: | | Arrays: | | Blood Collection: | |
| Genetics: | Y | | | Research & Development Services: | | Database Management: | | Drug Delivery: | |
| | | | | Diagnostics: | | | | Drug Distribution: | |

## TYPES OF BUSINESS:

Pharmaceuticals Discovery & Development
Weight Gain Products
Hormone Therapy

## BRANDS/DIVISIONS/AFFILIATES:

Oxandrin
Puricase
Oxandrolone
Pegloticase

## CONTACTS: Note: Officers with more than one job title may be intentionally listed here more than once.

Paul Hamelin, Pres.
David Gionco, CFO/Sr. VP
Zeb Horowitz, Chief Medical Officer/Sr. VP
Philip K. Yachmetz, General Counsel/Sr. VP/Sec.
David Gionco, Treas.
Robert Lamm, Sr. VP-Quality & Regulatory Affairs
Stephen Jaeger, Chmn.

| Phone: 732-418-9300 | Fax: 732-418-0570 |
|---|---|
| Toll-Free: | |
| Address: 1 Tower Ctr., 14th Fl., E. Brunswick, NJ 08816 US | |

## GROWTH PLANS/SPECIAL FEATURES:

Savient Pharmaceuticals, Inc. is engaged in the development, manufacture and marketing of both genetically engineered and niche-focused specialty pharmaceutical products. Through a combination of internal research and development, acquisitions, collaborative relationships and licensing arrangements, the company has developed a number of therapeutics. Currently, the primary product marketed worldwide by Savient is Oxandrin, an oral anabolic agent primarily used to promote weight gain following involuntary weight loss. Oxandrin is currently used for patients suffering from weight loss due to a number of medical conditions, including HIV/AIDS, surgery recovery, cancer and chronic diseases. The company is also considering employing the drug's applications on frail, elderly patients. Oxandrin is also beneficial in the relief of bone pain for osteoporosis patients. Savient distributes a generic version of the same drug oxandrolone. The firm is currently developing the drug Puricase, also referred to as Pegloticase, an infused genetically engineered enzyme conjugate being studied for the elimination of excess uric acid in individuals suffering from severe gout. The rights to this technology were purchased from Duke University and Mountain View Pharmaceuticals, Inc. Current treatments and research endeavors aim to block uric acid production, but do little to reduce levels present in the body. As a result, these therapies sometimes fail to help tens of thousands of people suffering from the disease. Puricase is designed to convert the acid to allantoin, which the body can remove by itself. The drug is currently engaged in the late stages of Phase III development. In December 2008, the FDA accepted a priority review of Savient's Biologics License Application for Puricase.

Savient offers its employees medical, vision, dental, life and AD&D insurance; a flexible spending account; tuition reimbursement; short- and long-term disability; a 401(k) plan; and an employee referral program.

## FINANCIALS: Sales and profits are in thousands of dollars—add 000 to get the full amount. 2008 Note: Financial information for 2008 was not available for all companies at press time.

| | | |
|---|---|---|
| 2008 Sales: $3,181 | 2008 Profits: $-84,169 | **U.S. Stock Ticker: SVNT** |
| 2007 Sales: $14,024 | 2007 Profits: $-48,668 | **Int'l Ticker:**    Int'l Exchange: |
| 2006 Sales: $47,514 | 2006 Profits: $60,325 | Employees:   73 |
| 2005 Sales: $49,495 | 2005 Profits: $5,968 | Fiscal Year Ends: 12/31 |
| 2004 Sales: $62,353 | 2004 Profits: $-27,515 | Parent Company: |

## SALARIES/BENEFITS:

| Pension Plan: | ESOP Stock Plan: | Profit Sharing: | Top Exec. Salary: $517,274 | Bonus: $310,365 |
|---|---|---|---|---|
| Savings Plan: Y | Stock Purch. Plan: Y | | Second Exec. Salary: $344,150 | Bonus: $ |

## OTHER THOUGHTS:

**Apparent Women Officers or Directors:**
**Hot Spot for Advancement for Women/Minorities:**

## LOCATIONS: ("Y" = Yes)

| West: | Southwest: | Midwest: | Southeast: | Northeast: | International: |
|---|---|---|---|---|---|
| | | | | Y | |

Note: Financial information, benefits and other data can change quickly and may vary from those stated here.

# SCHERING-PLOUGH CORP

www.schering-plough.com

**Industry Group Code:** 325412 **Ranks within this company's industry group:** Sales: 13 Profits: 17

| Drugs: | | Other: | Clinical: | Computers: | Services: |
|---|---|---|---|---|---|
| Discovery: | Y | AgriBio: | Trials/Services: | Hardware: | Specialty Services: |
| Licensing: | | Genetic Data: | Labs: | Software: | Consulting: |
| Manufacturing: | Y | Tissue Replacement: | Equipment/Supplies: | Arrays: | Blood Collection: |
| Genetics: | | | Research & Development Services: | Database Management: | Drug Delivery: |
| | | | Diagnostics: | | Drug Distribution: |

## TYPES OF BUSINESS:

Drugs-Diversified
Anti-Infective & Anti-Cancer Drugs
Dermatologicals
Cardiovascular Drugs
Animal Health Products
Over-the-Counter Drugs
Foot & Sun Care Products

## BRANDS/DIVISIONS/AFFILIATES:

Bovilis/Vista
HOMEAGAIN
Dr. Scholl's
Coppertone
Nuflor
Livial
Vytorin
Lotrimin

## CONTACTS: *Note: Officers with more than one job title may be intentionally listed here more than once.*

Fred Hassan, CEO
Robert J. Bertolini, CFO/Exec. VP
C. Ron Cheeley, Sr. VP-Global Human Resources
Thomas P. Koestler, Exec. VP/Pres., Schering-Plough Research Institute
Thomas Sabatino, Jr., General Counsel/Exec. VP
Janet M. Barth, VP-Investor Rel.
Steven H. Koehler, Chief Acct. Officer/Controller/VP
Richard S. Bowles III, Sr. VP-Global Quality Oper.
Carrie S. Cox, Exec. VP/Pres., Global Pharmaceuticals
Lori Queisser, Sr. VP-Global Compliance & Bus. Practices
Brent Saunders, Sr. VP/Pres., Consumer Health Care
Fred Hassan, Chmn.
Ian McInnes, Sr. VP/Pres., Global Supply Chain

| Phone: 908-298-4000 | Fax: 908-298-7653 |
|---|---|
| Toll-Free: | |
| Address: 2000 Galloping Hill Rd., Kenilworth, NJ 07033 US | |

## GROWTH PLANS/SPECIAL FEATURES:

Schering-Plough Corp. is a science-centered global heath care company. The company operates in three segments: human prescription pharmaceuticals, animal health and consumer health care. The human prescription pharmaceuticals segment discovers, develops, manufactures and markets human pharmaceutical products. Within the segment, the firm has a broad range of research projects and marketed products in six therapeutic areas: cardiovascular, central nervous system, immunology and infectious disease, oncology, respiratory and women's health. Marketed products include Vytorin, a cholesterol-lowering tablet; Remeron, an antidepressant; Livial, a menopausal therapy; and Temodar capsules for certain types of brain tumors including newly diagnosed glioblastoma multiforme. The animal health segment discovers, develops, manufactures and markets animal health products, including vaccines. Principal marketed products in this segment include Nuflor fish, bovine and swine antibiotics; Bovilis/Vista vaccine lines for infectious diseases in cattle; Innovax ND-SB vaccine for poultry; Otomax, a canine ear treatment; and HOMEAGAIN, which identifies pet cats and dogs and makes them easier to recover if lost. The consumer health care segment develops, manufactures and markets various over-the-counter, foot care and sun care products. Principal products in this division include Claritin non-sedating antihistamines; Dr. Scholl's foot care products; Lotrimin topical antifungal products; and Coppertone sun care lotions, sprays, dry oils, lip protection products and sunless tanning products. Throughout 2008, the firm continued the integration of its 2007 acquisition, Organon BioSciences, into its operations. In February 2008, the company partnered with OraSure Technologies, Inc., to develop a rapid oral hepatitis C test. In April 2008, Schering-Plough agreed to sell a combined 12 European animal product franchises to Virbac, S.A., and Pfizer Animal Health. In August 2008, the firm founded Shanghai Schering-Plough Pharmaceutical Co. Ltd., to expand its Chinese allergy and skincare product market. In August 2009, the firm approved a merger agreement with Merck & Co.

## FINANCIALS: Sales and profits are in thousands of dollars—add 000 to get the full amount. 2008 Note: Financial information for 2008 was not available for all companies at press time.

| | | |
|---|---|---|
| 2008 Sales: $18,502,000 | 2008 Profits: $1,903,000 | **U.S. Stock Ticker:** SGP |
| 2007 Sales: $12,690,000 | 2007 Profits: $-1,473,000 | **Int'l Ticker:** Int'l Exchange: |
| 2006 Sales: $10,594,000 | 2006 Profits: $1,143,000 | **Employees:** 51,000 |
| 2005 Sales: $9,508,000 | 2005 Profits: $269,000 | **Fiscal Year Ends:** 12/31 |
| 2004 Sales: $8,272,000 | 2004 Profits: $-947,000 | **Parent Company:** |

## SALARIES/BENEFITS:

| Pension Plan: | ESOP Stock Plan: | Profit Sharing: | Top Exec. Salary: $1,720,250 | Bonus: $3,387,150 |
|---|---|---|---|---|
| Savings Plan: | Stock Purch. Plan: Y | | Second Exec. Salary: $1,089,000 | Bonus: $1,146,080 |

## OTHER THOUGHTS:

**Apparent Women Officers or Directors:** 8
**Hot Spot for Advancement for Women/Minorities:** Y

## LOCATIONS: ("Y" = Yes)

| West: | Southwest: | Midwest: | Southeast: | Northeast: | International: |
|---|---|---|---|---|---|
| Y | Y | Y | Y | Y | Y |

# SCICLONE PHARMACEUTICALS

**www.sciclone.com**

Industry Group Code: 325412 Ranks within this company's industry group: Sales: 82 Profits: 74

| Drugs: | | Other: | Clinical: | Computers: | Services: |
|---|---|---|---|---|---|
| Discovery: | Y | AgriBio: | Trials/Services: | Hardware: | Specialty Services: |
| Licensing: | Y | Genetic Data: | Labs: | Software: | Consulting: |
| Manufacturing: | Y | Tissue Replacement: | Equipment/Supplies: | Arrays: | Blood Collection: |
| Genetics: | | | Research & Development Services: | Database Management: | Drug Delivery: |
| | | | Diagnostics: | | Drug Distribution: |

## TYPES OF BUSINESS:

Pharmaceuticals Acquisition & Development
Immune System Enhancers
Infectious Disease Therapies
Cancer Therapies

## BRANDS/DIVISIONS/AFFILIATES:

SciClone Pharmaceuticals International, Ltd.
ZADAXIN
SCV-07
RP101
DC Bead

## CONTACTS: Note: Officers with more than one job title may be intentionally listed here more than once.

Friedhelm Blobel, CEO
Friedhelm Blobel, Pres.
Gary Titus, CFO
Randy McBeath, VP-Mktg.
Cynthia W. Tuthill, VP-Scientific Affairs/Chief Science Officer
Eric J. Hoechstetter, VP-Legal Affairs
Jeffery Lange, VP-Bus. Dev.
Gary Titus, Sr. VP-Finance
Israel Rios, Chief Medical Officer/Sr. VP-Medical Affairs
Craig Halverson, VP-Regulatory Affairs & Quality Assurance
Dean Woodman, Chmn.
Hans P. Schmid, Pres., SciClone Pharmaceuticals Int'l, Ltd.

| Phone: 650-358-3456 | Fax: 650-358-3469 |
|---|---|
| Toll-Free: 800-724-2566 | |
| Address: 950 Tower Ln., Ste. 900, Foster City, CA 94404 US | |

## GROWTH PLANS/SPECIAL FEATURES:

SciClone Pharmaceuticals, Inc. develops and commercializes pharmaceutical and biological therapeutic compounds for oncology and infectious diseases. The firm's lead product ZADAXIN, synthetic preparation of thymosin alpha 1, is being evaluated in two Phase III hepatitis C virus (HCV) clinical trials in the U.S. The company is also planning a Phase III clinical trial to evaluate the drug in treating malignant melanoma. The drug is also being evaluated in late-stage clinical trials for the treatment of hepatitis B virus and certain cancers. ZADAXIN is approved for sale in over 34 countries, primarily in Asia, the Middle East and Latin America, with 92% of the company's sales of ZADAXIN coming from China through SciClone China. The company also markets DC Bead, an injected embolic agent that blocks blood delivery to tumors and can deliver chemotherapeutic or other medications to targeted sites. The firm obtained the Chinese marketing rights for the treatment from Biocompatibles International PLC and is investigating the drug's efficacy and safety in Phase II trials for the U.S. market. SciClone's other proprietary drug development candidates are SCV-07 for oral mucositis in head and neck cancer and RP101 for pancreatic cancer. SCV-07 is a synthetic dipeptide that has demonstrated immunomodulatory activity by increasing T-cell differentiation and function, biological processes that are necessary for the body to fight infection. The company acquired exclusive worldwide rights, outside of Russia, to SCV-07 from Verta, Ltd. RP101 is a nucleoside analog which may enhance the effectiveness of chemotherapy, and has been granted Orphan Drug Designation by the FDA for the adjunct treatment of pancreatic cancer. SciClone acquired development rights for RP101 in the U.S. and Canada from Resistys, Inc.

SciClone offers employees medical, dental, vision and life insurance; short- and long-term disability coverage; flexible spending accounts; a 401(k) plan; and an employee stock purchase program.

## FINANCIALS: Sales and profits are in thousands of dollars—add 000 to get the full amount. 2008 Note: Financial information for 2008 was not available for all companies at press time.

| | | |
|---|---|---|
| 2008 Sales: $54,113 | 2008 Profits: $-8,348 | **U.S. Stock Ticker: SCLN** |
| 2007 Sales: $37,058 | 2007 Profits: $-9,948 | **Int'l Ticker:** Int'l Exchange: |
| 2006 Sales: $32,662 | 2006 Profits: $ 727 | Employees: 227 |
| 2005 Sales: $28,334 | 2005 Profits: $-7,713 | Fiscal Year Ends: 12/31 |
| 2004 Sales: $24,396 | 2004 Profits: $-13,278 | Parent Company: |

## SALARIES/BENEFITS:

| Pension Plan: | ESOP Stock Plan: | Profit Sharing: | Top Exec. Salary: $442,000 | Bonus: $439,470 |
|---|---|---|---|---|
| Savings Plan: Y | Stock Purch. Plan: Y | | Second Exec. Salary: $318,000 | Bonus: $140,000 |

## OTHER THOUGHTS:

**Apparent Women Officers or Directors:** 1
**Hot Spot for Advancement for Women/Minorities:**

## LOCATIONS: ("Y" = Yes)

| West: | Southwest: | Midwest: | Southeast: | Northeast: | International: |
|---|---|---|---|---|---|
| Y | | | | | Y |

# SCIELE PHARMA INC

**www.sciele.com**

Industry Group Code: 325412  Ranks within this company's industry group:  Sales:    Profits:

| Drugs: | | Other: | Clinical: | Computers: | Services: |
|---|---|---|---|---|---|
| Discovery: | | AgriBio: | Trials/Services: | Hardware: | Specialty Services: |
| Licensing: | Y | Genetic Data: | Labs: | Software: | Consulting: |
| Manufacturing: | | Tissue Replacement: | Equipment/Supplies: | Arrays: | Blood Collection: |
| Genetics: | | | Research & Development Services: | Database Management: | Drug Delivery: |
| | | | Diagnostics: | | Drug Distribution: |

## TYPES OF BUSINESS:

Drugs-Acquisition & Licensing
Prescription Drug Sales & Marketing
Prescription Drug Research & Development

## BRANDS/DIVISIONS/AFFILIATES:

Shionogi & Co., Ltd.
Prenatal Elite
Fortamet
Altoprev
Triglide
Nitrolingual
Prenate DHA
Allegra

## CONTACTS: *Note: Officers with more than one job title may be intentionally listed here more than once.*

Patrick P. Fourteau, CEO
Edward Schutter, COO
Edward Schutter, Pres.
Darrell Borne, CFO/Exec. VP
Darrell Borne, Corp. Sec.
Darrell Borne, Treas.
Joseph J. Ciaffoni, Chief Commercial Officer
Larry M. Dillaha, Chief Medical Officer/Exec. VP
Leslie Zacks, Chief Legal & Compliance Officer/Exec. VP

| Phone: 770-442-9707 | Fax: 678-341-1470 |
|---|---|
| Toll-Free: 800-461-3696 | |
| Address: 5 Concourse Pkwy., Ste. 1800, Atlanta, GA 30328 US | |

## GROWTH PLANS/SPECIAL FEATURES:

Sciele Pharma Inc., develops, markets and sells brand-name prescription products, focusing on cardiology, diabetes, women's health, and pediatric treatment.  It both develops drug candidates and acquires or licenses pharmaceutical products that have high sales growth potential and complement its existing products.  In total, the company sells 22 products, including Sular for hypertension; Fortamet, an adjunct to diet and exercise that lowers blood glucose in type 2 diabetes patients; Altoprev, for cholesterol reduction and coronary heart disease; Triglide, for hyperchlesterolemia and hypertriglyceridemia; Nitrolingual, a pumpspray that provides acute relief during attacks of angina pectoris due to coronary artery disease; Prenate Elite and Prenate DHA, prenatal vitamins; Zovirax, a topical herpes medication; Allegra, a pediatric allergy and chronic idiopathic urticaria treatment; Methylin chewable tablets, for attention deficit/hyperactivity disorder; and Orapred, whose applications include severe allergy relief for patients with asthma.  Sciele's other products include treatments for swimmer's ear infection, tension headaches, peptic ulcers, dementia, urinary tract infections and seasonal allergies.  In addition to its marketed drugs, the company has products under development for a range of indications including chronic drooling, premature ejaculation, head lice, diabetes and hypertension.  The firm enlists third-party manufacturers for all its products.  In March 2008, the company acquired the Twinject epinephrine auto-injector, which treats severe allergic reactions and anaphylaxis, from Verus Pharmaceuticals, Inc.  In October 2008, the firm was acquired by Japanese drug maker Shionogi & Co., Ltd.; Sciele Pharma is now an indirect, wholly-owned subsidiary of Shionogi.

## FINANCIALS: Sales and profits are in thousands of dollars—add 000 to get the full amount. 2008 Note: Financial information for 2008 was not available for all companies at press time.

| | | |
|---|---|---|
| 2008 Sales: $ | 2008 Profits: $ | U.S. Stock Ticker: Subsidiary |
| 2007 Sales: $382,255 | 2007 Profits: $45,407 | Int'l Ticker:    Int'l Exchange: |
| 2006 Sales: $293,181 | 2006 Profits: $45,244 | Employees:   920 |
| 2005 Sales: $216,358 | 2005 Profits: $39,209 | Fiscal Year Ends: 12/31 |
| 2004 Sales: $151,967 | 2004 Profits: $26,554 | Parent Company: SHIONOGI & CO., LTD. |

## SALARIES/BENEFITS:

| Pension Plan: | ESOP Stock Plan: Y | Profit Sharing: | Top Exec. Salary: $375,000 | Bonus: $562,500 |
|---|---|---|---|---|
| Savings Plan: Y | Stock Purch. Plan: | | Second Exec. Salary: $280,000 | Bonus: $294,000 |

## OTHER THOUGHTS:

**Apparent Women Officers or Directors:**
**Hot Spot for Advancement for Women/Minorities:**

## LOCATIONS: ("Y" = Yes)

| West: | Southwest: | Midwest: | Southeast: | Northeast: | International: |
|---|---|---|---|---|---|
| | | | Y | | |

# SCIOS INC

www.sciosinc.com

**Industry Group Code: 325412  Ranks within this company's industry group: Sales:    Profits:**

| Drugs: | | Other: | Clinical: | Computers: | | Services: |
|---|---|---|---|---|---|---|
| Discovery: | Y | AgriBio: | Trials/Services: | Hardware: | | Specialty Services: |
| Licensing: | | Genetic Data: | Labs: | Software: | | Consulting: |
| Manufacturing: | Y | Tissue Replacement: | Equipment/Supplies: | Arrays: | | Blood Collection: |
| Genetics: | | | Research & Development Services: | Database Management: | Y | Drug Delivery: |
| | | | Diagnostics: | | | Drug Distribution: |

## TYPES OF BUSINESS:

Cardiovascular Drugs

## GROWTH PLANS/SPECIAL FEATURES:

Scios, Inc., a subsidiary of Johnson & Johnson, is a biopharmaceutical company developing treatments for cardiovascular diseases. The firm's technology platform fuses classical medicinal chemistry with the most recent advances in disease-based gene array, bioinformatics and computational chemistry. The company's lead product, Natrecor, is an intravenous cardiovascular drug approved to treat acutely decompensated congestive heart failure (ADHF) for patients who have dyspnea (shortness of breath) at rest or with minimal activity, such as talking, eating or bathing. Natrecor is a recombinant form of the human B-type natriuretic peptide (hBNP), which is normally produced by the heart. The firm's current research and development program focuses on the discovery of new therapeutics for other cardiovascular diseases. Scios also has a program call ADHERE, Acute Decompensated Heart Failure National Registry, which is an observational registry that lists data on heart failure patient treatment and associated outcomes.

## BRANDS/DIVISIONS/AFFILIATES:

Natrecor
ADHERE

## CONTACTS: Note: Officers with more than one job title may be intentionally listed here more than once.

James R. Mitchell, Pres.
Tao Fu, VP-Corp. Dev. & Planning
Jim Barr, VP-Finance
William C. Weldon, CEO-Johnson & Johnson

| **Phone:** 650-564-5000 | **Fax:** 650-564-7070 |
|---|---|
| **Toll-Free:** 877-462-873267 | |
| **Address:** 1900 Charleston Rd., Mountain View, CA 94039 US | |

## FINANCIALS: Sales and profits are in thousands of dollars—add 000 to get the full amount. 2008 Note: Financial information for 2008 was not available for all companies at press time.

| | | |
|---|---|---|
| 2008 Sales: $ | 2008 Profits: $ | **U.S. Stock Ticker: Subsidiary** |
| 2007 Sales: $98,600 | 2007 Profits: $ | **Int'l Ticker:**    Int'l Exchange: |
| 2006 Sales: $ | 2006 Profits: $ | Employees:   777 |
| 2005 Sales: $ | 2005 Profits: $ | Fiscal Year Ends: 12/31 |
| 2004 Sales: $ | 2004 Profits: $ | Parent Company: JOHNSON & JOHNSON |

## SALARIES/BENEFITS:

| Pension Plan: | ESOP Stock Plan: | Profit Sharing: | Top Exec. Salary: $460,000 | Bonus: $500,000 |
|---|---|---|---|---|
| Savings Plan: | Stock Purch. Plan: | | Second Exec. Salary: $267,750 | Bonus: $180,000 |

## OTHER THOUGHTS:

Apparent Women Officers or Directors:
Hot Spot for Advancement for Women/Minorities:

## LOCATIONS: ("Y" = Yes)

| West: | Southwest: | Midwest: | Southeast: | Northeast: | International: |
|---|---|---|---|---|---|
| | | | | Y | |

# SEATTLE GENETICS

**www.seattlegenetics.com**

Industry Group Code: 325412  **Ranks within this company's industry group:** Sales: 96   Profits: 156

| Drugs: | | Other: | Clinical: | Computers: | Services: |
|---|---|---|---|---|---|
| Discovery: | Y | AgriBio: | Trials/Services: | Hardware: | Specialty Services: |
| Licensing: | Y | Genetic Data: | Labs: | Software: | Consulting: |
| Manufacturing: | | Tissue Replacement: | Equipment/Supplies: | Arrays: | Blood Collection: |
| Genetics: | | | Research & Development Services: | Database Management: | Drug Delivery: |
| | | | Diagnostics: | | Drug Distribution: |

## TYPES OF BUSINESS:

Biopharmaceuticals Development
Cancer Treatments
Monoclonal Antibodies
Autoimmune Diseases

## BRANDS/DIVISIONS/AFFILIATES:

SGN-40
SGN-75
SGN-33
SGN-35
SGN-70
Antibody-Drug Conjugates (ADCs)

## CONTACTS:
*Note: Officers with more than one job title may be intentionally listed here more than once.*

Clay B. Siegall, CEO
Clay B. Siegall, Pres.
Todd Simpson, CFO
Christopher Pawlowicz, Sr. VP-Human Resources
Vaughn B. Himes, Exec. VP-Tech. Oper.
Morris Z. Rosenberg, Sr. VP-Dev.
Thomas C. Reynolds, Chief Medical Officer
Iqbal S. Grewal, VP-Preclinical Therapeutics
Eric L. Dobmeier, Chief Bus. Officer
Peter D. Senter, VP-Chemistry
Clay B. Siegall, Chmn.

| Phone: 425-527-4000 | Fax: 425-527-4001 |
|---|---|
| Toll-Free: | |
| Address: 21823 30th Dr. S.E., Bothell, WA 98021 US | |

## GROWTH PLANS/SPECIAL FEATURES:

Seattle Genetics develops monoclonal antibody (mAb)-based therapies for the treatment of cancer and autoimmune diseases. The company has an exclusive, worldwide license agreement with Genentech to develop and commercialize its lead product candidate SGN-40 (dacetuzumab). Its research and development activities focus on mAb-based (monoclonal antibodies) therapies for human cancers including lung, renal cell, Hodgkin's disease, non-Hodgkin's lymphoma, multiple myeloma and melanoma. Products in the developmental stage include SGN-33 (lintuzumab), in testing for acute myeloid leukemia (AML) and myelodysplastic syndromes (MDS); SGN-40, in testing for multiple myeloma, diffuse large B-cell lymphoma and non-Hodgkin's lymphoma; SGN-35, in testing for relapsed Hodgkin's lymphoma and systemic anaplastic large cell lymphoma; SGN-70, an IND for autoimmune diseases; SGN-75, a preclinical candidate to treat solid tumors; and SGN-19A, a preclinical candidate for cancer. These product candidates represent applications of Seattle Genetics' primary platform technologies: genetically engineered monoclonal antibodies and antibody-drug conjugates (ADC). Each of these platforms is designed to support therapies that are able to identify and kill cancer cells while limiting damage to normal tissue. Seattle Genetics currently has license agreements for its proprietary ADC technology with Genentech; Bayer Pharmaceuticals, Inc.; CuraGen; Progenics; Daiichi Sankyo; and MedImmune, Inc. The company has a co-development agreement is Agensys to research and develop ADC products. In March 2009, the FDA granted fast track designation for SGN-35 for use in Hodgkin's lymphoma.

Seattle Genetics offers its employees medical, dental and vision coverage; a 401(k) plan; a flexible spending plan; short- and long-term disability; life and AD&D coverage; and an employee stock purchase plan.

## FINANCIALS:
**Sales and profits are in thousands of dollars—add 000 to get the full amount. 2008 Note: Financial information for 2008 was not available for all companies at press time.**

| | | |
|---|---|---|
| 2008 Sales: $35,236 | 2008 Profits: $-85,501 | **U.S. Stock Ticker: SGEN** |
| 2007 Sales: $22,420 | 2007 Profits: $-48,932 | **Int'l Ticker:** Int'l Exchange: |
| 2006 Sales: $10,005 | 2006 Profits: $-36,015 | Employees: 189 |
| 2005 Sales: $9,757 | 2005 Profits: $-29,433 | Fiscal Year Ends: 12/31 |
| 2004 Sales: $6,701 | 2004 Profits: $-35,439 | Parent Company: |

## SALARIES/BENEFITS:

| Pension Plan: | ESOP Stock Plan: | Profit Sharing: | Top Exec. Salary: $567,958 | Bonus: $360,875 |
|---|---|---|---|---|
| Savings Plan: Y | Stock Purch. Plan: Y | | Second Exec. Salary: $352,876 | Bonus: $179,439 |

## OTHER THOUGHTS:

**Apparent Women Officers or Directors:**
**Hot Spot for Advancement for Women/Minorities:**

## LOCATIONS: ("Y" = Yes)

| West: | Southwest: | Midwest: | Southeast: | Northeast: | International: |
|---|---|---|---|---|---|
| Y | | | | | |

# SENETEK PLC

**www.senetekplc.com**

**Industry Group Code: 325412  Ranks within this company's industry group: Sales: 141  Profits: 65**

| Drugs: | | Other: | Clinical: | Computers: | Services: |
|---|---|---|---|---|---|
| Discovery: | Y | AgriBio: | Trials/Services: | Hardware: | Specialty Services: |
| Licensing: | Y | Genetic Data: | Labs: | Software: | Consulting: |
| Manufacturing: | Y | Tissue Replacement: | Equipment/Supplies: | Arrays: | Blood Collection: |
| Genetics: | | | Research & Development Services: | Database Management: | Drug Delivery: |
| | | | Diagnostics: | | Drug Distribution: |

## TYPES OF BUSINESS:

Drugs-Sexual Dysfunction
Anti-Aging Products
Skin Care Drugs

## BRANDS/DIVISIONS/AFFILIATES:

Invicorp
Kinetin
Zeatin
Pyratine-6
Reliaject
4HBAP
Pyratine XR

## CONTACTS: Note: Officers with more than one job title may be intentionally listed here more than once.

Frank J. Massino, CEO
Phillip J. Rose, COO
William O'Kelly, CFO
Brian Clark, Chief Scientific Officer
William F. O'Kelly, Sec.
Jan-Elo Jorgensen, Dir.-Research/Managing Dir.-Senetek Denmark ApS
Frank J. Massino, Chmn.

| Phone: 707-226-3900 | Fax: 707226-3999 |
|---|---|
| **Toll-Free:** | |
| **Address:** 831A Latour Ct., Napa, CA 94558 US | |

## GROWTH PLANS/SPECIAL FEATURES:

Senetek PLC develops and markets proprietary skin treatments, anti-aging products and products for the treatment of sexual dysfunction. In the anti-aging market, the company offers Kinetin, an antioxidant skin cream that has been shown to reduce fine lines and blotchiness and increase the amount of moisture retained by the skin. Senetek developed an entire line of Kinetin products, as well as Zeatin, an analog of Kinetin that is indicated for dermatological applications including the treatment of psoriasis. Kinetin products are marketed worldwide by Valeant Pharmaceuticals International. The first of Senetek's group of second-generation cytokinins to complete clinical testing, Pyratine-6, has been successfully evaluated for the treatment of the signs of aging and its effect on reducing acne lesions and erythema. Senetek also has a range of other second generation cytokinin treatments in development, and recently concluded clinical trials for 4HBAP, a proprietary aromatic cytokinin for use with photodamaged skin. For erectile dysfunction (ED), the company markets Invicorp, a self-administered injection that delivers a combination therapy of two specific drugs, vasoactive intestinal polypeptide and phentolamine mesylate. Invicorp is meant to be a second-line treatment, after oral therapies have failed. Plethora Solutions recently signed a collaborative marketing agreement with Senetek to market Invicorp in North America. Aside from drugs, Reliaject, still under going clinical studies, is an auto-injector developed by Senetek that employs an ultra-fine gauge needle, preset to achieve the appropriate penetration before drug flow occurs, thereby reducing reliance on the patient's technique for accuracy and safe delivery. In March 2009, the company launched Pyratine XR for the treatment of rosacea.

## FINANCIALS: Sales and profits are in thousands of dollars—add 000 to get the full amount. 2008 Note: Financial information for 2008 was not available for all companies at press time.

| | | |
|---|---|---|
| 2008 Sales: $1,845 | 2008 Profits: $-3,759 | **U.S. Stock Ticker: SNTK** |
| 2007 Sales: $26,471 | 2007 Profits: $18,632 | **Int'l Ticker:** Int'l Exchange: |
| 2006 Sales: $8,431 | 2006 Profits: $1,883 | Employees: 8 |
| 2005 Sales: $5,871 | 2005 Profits: $-1,739 | Fiscal Year Ends: 12/31 |
| 2004 Sales: $7,550 | 2004 Profits: $ 566 | Parent Company: |

## SALARIES/BENEFITS:

| Pension Plan: | ESOP Stock Plan: | Profit Sharing: | Top Exec. Salary: $340,000 | Bonus: $93,750 |
|---|---|---|---|---|
| Savings Plan: | Stock Purch. Plan: | | Second Exec. Salary: $142,500 | Bonus: $25,000 |

## OTHER THOUGHTS:

**Apparent Women Officers or Directors:**
**Hot Spot for Advancement for Women/Minorities:**

## LOCATIONS: ("Y" = Yes)

| West: | Southwest: | Midwest: | Southeast: | Northeast: | International: |
|---|---|---|---|---|---|
| Y | | | | | Y |

Note: Financial information, benefits and other data can change quickly and may vary from those stated here.

# SEPRACOR INC

**www.sepracor.com**

**Industry Group Code: 325412  Ranks within this company's industry group:** Sales: 35    Profits: 24

| Drugs: | | Other: | | Clinical: | Computers: | Services: | |
|---|---|---|---|---|---|---|---|
| Discovery: | Y | AgriBio: | | Trials/Services: | Hardware: | Specialty Services: | |
| Licensing: | Y | Genetic Data: | | Labs: | Software: | Consulting: | |
| Manufacturing: | Y | Tissue Replacement: | | Equipment/Supplies: | Arrays: | Blood Collection: | |
| Genetics: | | | | Research & Development Services: | Database Management: | Drug Delivery: | Y |
| | | | | Diagnostics: | | Drug Distribution: | |

## TYPES OF BUSINESS:

Pharmaceuticals Discovery & Development
Respiratory Treatments
Central Nervous System Disorder Treatments
Insomnia

## BRANDS/DIVISIONS/AFFILIATES:

LUNESTA
XOPENEX
STEDESA
Oryx Pharmaceuticals, Inc.
ALVESCO
BROVANA
OMNARIS
XOPENEX

## CONTACTS: Note: Officers with more than one job title may be intentionally listed here more than once.

Adrian Adams, CEO
Adrian Adams, Pres.
Robert F. Scumaci, CFO/Exec. VP
Richard Ranieri, Exec. VP-Human Resources
Mark H. N. Corrigan, Exec. VP-R&D
Richard Ranieri, Exec. VP-Admin.
Andrew I. Koven, General Counsel/Corp. Sec./Exec. VP
Mark Iwicki, Chief Commercial Officer/Exec. VP
Timothy J. Barberich, Exec. Chmn.

| Phone: 508-481-6700 | Fax: 508-357-7499 |
|---|---|
| Toll-Free: | |
| Address: 84 Waterford Dr., Marlborough, MA 01752 US | |

## GROWTH PLANS/SPECIAL FEATURES:

Sepracor, Inc. is a research-based pharmaceutical company whose goal is to discover, develop and market products that are directed toward the treatment of respiratory and central nervous system (CNS) disorders. Serpacor's lead product is adult insomnia treatment Lunesta (eszopiclone). It also markets Xopenex Inhalation Solution for use in nebulizer machines, which turn liquids into fine sprays; it is indicated for patients with asthma or chronic obstructive pulmonary disease (COPD). These two drugs alone accounted for 81% of Sepracor's 2008 product sales. The company also markets Brovana for COPD; Omnaris for allergic rhinitis; and Alvesco for asthma. Sepracor markets its own and other companies' products through its sales force, co-promotion agreements and out-licensing partnerships. Due to its patents relating to the chemicals desloratadine, fexofenadine and levocetirizine, the company has out-licensing agreements for various allergy medications, including Schering-Plough for Clarinex (desloratadine); sanofi-aventis for Allegra (fexofenadine); and UCB Farchim SA for its Xusal/Xyzal products (levocetirizine). Beyond these, the firm is also developing new treatments for epilepsy, depression, insomnia, ADHD, neuropathic pain, COPD and asthma. In June 2008, the company acquired Oryx Pharmaceuticals Inc., and expanded its presence in Canada and its specialty pharmaceutical market. In January 2009, Sepracor announced it would cut 20% of its workforce, or about 530 jobs. In March 2009, the firm formally submitted a new drug application for Stedesa, a treatment for epilepsy, to the FDA. In May 2009, the company withdrew plans to market Lunesta in Europe after the European Medical Agency declined to grant the drug a new active substance status.

Sepracor offers its employees medical, dental and vision coverage; short- and long-term disability; life and AD&D insurance; flexible spending accounts; a 401(k) plan; tuition assistance; and childcare services.

## FINANCIALS: Sales and profits are in thousands of dollars—add 000 to get the full amount. 2008 Note: Financial information for 2008 was not available for all companies at press time.

| | | |
|---|---|---|
| 2008 Sales: $1,292,289 | 2008 Profits: $515,110 | **U.S. Stock Ticker:** SEPR |
| 2007 Sales: $1,225,230 | 2007 Profits: $58,333 | **Int'l Ticker:**    Int'l Exchange: |
| 2006 Sales: $1,183,133 | 2006 Profits: $171,161 | Employees: 2,277 |
| 2005 Sales: $820,928 | 2005 Profits: $3,927 | Fiscal Year Ends: 12/31 |
| 2004 Sales: $380,877 | 2004 Profits: $-296,910 | Parent Company: |

## SALARIES/BENEFITS:

| Pension Plan: | ESOP Stock Plan: | Profit Sharing: | Top Exec. Salary: $1,050,000 | Bonus: $682,500 |
|---|---|---|---|---|
| Savings Plan: Y | Stock Purch. Plan: Y | | Second Exec. Salary: $648,442 | Bonus: $ |

## OTHER THOUGHTS:

**Apparent Women Officers or Directors:** 1
**Hot Spot for Advancement for Women/Minorities:**

## LOCATIONS: ("Y" = Yes)

| West: | Southwest: | Midwest: | Southeast: | Northeast: | International: |
|---|---|---|---|---|---|
| | | | | Y | Y |

# SEQUENOM INC

**www.sequenom.com**

**Industry Group Code: 325413  Ranks within this company's industry group: Sales: 16  Profits: 19**

| Drugs: | Other: | | Clinical: | Computers: | | Services: | |
|---|---|---|---|---|---|---|---|
| Discovery: | AgriBio: | | Trials/Services: | Hardware: | | Specialty Services: | |
| Licensing: | Genetic Data: | Y | Labs: | Software: | Y | Consulting: | |
| Manufacturing: | Tissue Replacement: | | Equipment/Supplies: | Arrays: | Y | Blood Collection: | |
| Genetics: | | | Research & Development Services: | Database Management: | Y | Drug Delivery: | |
| | | | Diagnostics: | | | Drug Distribution: | |

## TYPES OF BUSINESS:

Chips-DNA Arrays
Genotype Analysis Software
Cell Research Database

## BRANDS/DIVISIONS/AFFILIATES:

MassARRAY
iPLEX Gold
SEQureDx
Center for Molecular Medicine
AttoSense
SensiGen LLC

## CONTACTS: Note: Officers with more than one job title may be intentionally listed here more than once.

Harry Stylli, CEO
Harry Stylli, Pres.
Paul W. Hawran, CFO
Michael Monko, Sr. VP-Sales & Mktg.
Alisa Judge, VP-Human Resources
Charles R. Cantor, Chief Scientific Officer
Clarke Neumann, General Counsel/VP
Larry Myres, VP-Oper.
Elizabeth Dragon, Sr. VP-R&D
Steven Owings, VP-Commercial Dev., Prenatal Diagnostics
Robert M. Di Tullio, VP-Regulatory Affairs/Quality Sequenom, Inc.
Allan T. Lombard, Chief Medical Officer

| Phone: 858-202-9000 | Fax: 858-202-9001 |
|---|---|
| **Toll-Free:** | |
| **Address:** 3595 John Hopkins Ct., San Diego, CA 92121-1331 US | |

## GROWTH PLANS/SPECIAL FEATURES:

Sequenom, Inc. is a genetics and molecular diagnostic company providing genetic analysis products, services and diagnostic applications for the development of noninvasive diagnostics in prenatal, oncology and infectious diseases and other disorders. The company's main source of revenue is derived from MassARRAY, a hardware and software application with consumable chips and reagents that analyzes high performance nucleic acids and quantitatively measures genetic target material and variations. The MassARRAY system offers cost-effective methods for numerous types of DNA analysis applications, which can range from SNP genotyping and allelotyping; quantitative gene expression analysis; quantitative methylation marker analysis; epigenomics; and pathogen typing. Sequenom's iPLEX Gold assay provides for multiplexed DNA sample analysis that enables the user to perform multiple sample genotyping analyses using a similar amount of reagents and chip surface area as used for a single DNA sample analysis. Customers of MassARRAY include clinical research laboratories, biotechnology companies, academic institutions and government agencies in North America, Europe, India, Australia, France, South Korea, New Zealand, Singapore, Taiwan and Turkey. Sequenom is developing various molecular diagnostic tests in prenatal genetic disorders, oncology and infectious diseases under the brand name SEQureDx. In the agricultural market, MassARRAY is utilized to provide farm-of-origin verification, country-of-origin verification, age verification and national ID programs for traceability analysis of livestock. In July 2008, Sequenom formed a global alliance with SensiGen LLC to develop and market advanced diagnostic tests and systems. In November 2008, the company acquired the Center for Molecular Medicine in Grand Rapids, Michigan, a Clinical Laboratory Improvement Act (CLIA) certified clinical diagnostics laboratory. In January 2009, Sequenom agreed to acquire the complete AttoSense portfolio of tests from privately held biotechnology company SensiGen LLC.

Sequenom offers its employees a Section 529 College Savings Plan; an employee assistance program; flexible spending accounts; and medical, dental, vision and disability insurance.

## FINANCIALS: Sales and profits are in thousands of dollars—add 000 to get the full amount. 2008 Note: Financial information for 2008 was not available for all companies at press time.

| | | |
|---|---|---|
| 2008 Sales: $47,149 | 2008 Profits: $-44,154 | **U.S. Stock Ticker: SQNM** |
| 2007 Sales: $41,002 | 2007 Profits: $-21,983 | **Int'l Ticker:**  Int'l Exchange: |
| 2006 Sales: $28,496 | 2006 Profits: $-17,577 | Employees:  192 |
| 2005 Sales: $19,421 | 2005 Profits: $-26,537 | Fiscal Year Ends: 12/31 |
| 2004 Sales: $22,449 | 2004 Profits: $-34,625 | Parent Company: |

## SALARIES/BENEFITS:

| Pension Plan: | ESOP Stock Plan: Y | Profit Sharing: | Top Exec. Salary: $441,000 | Bonus: $181,912 |
|---|---|---|---|---|
| Savings Plan: Y | Stock Purch. Plan: | | Second Exec. Salary: $319,020 | Bonus: $82,247 |

## OTHER THOUGHTS:

**Apparent Women Officers or Directors:** 2
**Hot Spot for Advancement for Women/Minorities:** Y

## LOCATIONS: ("Y" = Yes)

| West: | Southwest: | Midwest: | Southeast: | Northeast: | International: |
|---|---|---|---|---|---|
| Y | | | | Y | Y |

Note: Financial information, benefits and other data can change quickly and may vary from those stated here.

# SERACARE LIFE SCIENCES INC

**www.seracare.com**

**Industry Group Code:** 325414　**Ranks within this company's industry group:** Sales: 4　Profits: 5

| Drugs: | Other: | Clinical: | Computers: | Services: |
|---|---|---|---|---|
| Discovery: | AgriBio: | Trials/Services: | Hardware: | Specialty Services: Y |
| Licensing: | Genetic Data: | Labs: | Software: Y | Consulting: |
| Manufacturing: Y | Tissue Replacement: | Equipment/Supplies: Y | Arrays: Y | Blood Collection: Y |
| Genetics: | | Research & Development Services: Y | Database Management: | Drug Delivery: |
| | | Diagnostics: Y | | Drug Distribution: |

## TYPES OF BUSINESS:

Diagnostics Products
Blood & Plasma Collection & Processing
Research Support Services
Plasma-based Products
Biological Product Database

## BRANDS/DIVISIONS/AFFILIATES:

ACCURUN

## CONTACTS: *Note: Officers with more than one job title may be intentionally listed here more than once.*

Susan L. N. Vogt, CEO
Susan L. N. Vogt, Pres.
Gregory A. Gould, CFO
Bill Smutny, VP-Mktg. & Sales
Kathi Benjamin, VP-Human Resources
Mark M. Manak, Chief Scientific Officer
Ron Dilling, VP-Mfg.
Gregory A. Gould, Sec.
Michael K. Steele, VP-Bus. Dev.
Gregory A. Gould, Treas.
Kathi Shea, VP-BioServices Oper.
David M. Olsen, VP-Corp. Quality
Eugene I. Davis, Chmn.

| | |
|---|---|
| **Phone:** 508-244-6400 | **Fax:** 508-634-3394 |
| **Toll-Free:** 800-676-1881 | |
| **Address:** 37 Birch St., Milford, MA 01757 US | |

## GROWTH PLANS/SPECIAL FEATURES:

SeraCare Life Sciences, Inc. develops, manufactures and sells a broad range of biological based materials and services, which facilitate the discovery, development and production of diagnostic and therapeutic products. The company's offerings include plasma-based reagents, diagnostic controls and molecular biomarkers. SeraCare's business is divided into two segments: The diagnostic and biopharmaceutical products division and the bioservices division. The diagnostic and biopharmaceutical products division includes two product lines: reagents and bioprocessing products along with controls and panels. The company's reagents and bioprocessing products include diagnostic intermediates, which are plasma-based products; cell culture additives and media; therapeutic grade albumin; and purified viable human cells. The firm also markets 100 control and panel products for infectious diseases including HIV, hepatitis A, hepatitis b, hepatitis C, West Nile Virus, Chagas and human papillomavirus. These products are used for quality control of infectious disease testing in hospitals, clinical testing labs and blood banks as well as aiding in the development and evaluation of new tests and markers of disease. The firm's controls are primarily sold under the ACCURUN brand name. The bioservices division consists of biobanking, in which SeraCare manages and stores more than 19 million biological samples, and contract research services. This division supplies technical support and training services as well as research and clinical trial support services including assistance with collecting, cataloging, processing, transporting, cryopreservation and tracking of samples collected during research studies and clinical trials. In October 2008, the firm launched its genetic testing controls line, with ACCURUN tests for warfarin sensitivity and thrombophilia.

SeraCare offers its employees health and dental coverage; short- and long-term disability; life and travel insurance; a 401(k) plan; tuition reimbursement; a flexible spending account; and an employee assistance program.

## FINANCIALS: Sales and profits are in thousands of dollars—add 000 to get the full amount. 2008 Note: Financial information for 2008 was not available for all companies at press time.

| | | |
|---|---|---|
| 2008 Sales: $48,967 | 2008 Profits: $-11,963 | **U.S. Stock Ticker:** SRLS |
| 2007 Sales: $47,304 | 2007 Profits: $-13,165 | **Int'l Ticker:**　Int'l Exchange: |
| 2006 Sales: $49,176 | 2006 Profits: $-24,278 | Employees:　243 |
| 2005 Sales: $50,300 | 2005 Profits: $-21,097 | Fiscal Year Ends: 9/30 |
| 2004 Sales: $28,441 | 2004 Profits: $4,155 | Parent Company: |

## SALARIES/BENEFITS:

| | | | | |
|---|---|---|---|---|
| Pension Plan: | ESOP Stock Plan: | Profit Sharing: | Top Exec. Salary: $377,138 | Bonus: $222,775 |
| Savings Plan: Y | Stock Purch. Plan: | | Second Exec. Salary: $269,385 | Bonus: $159,126 |

## OTHER THOUGHTS:

**Apparent Women Officers or Directors:** 5
**Hot Spot for Advancement for Women/Minorities:** Y

## LOCATIONS: ("Y" = Yes)

| West: | Southwest: | Midwest: | Southeast: | Northeast: | International: |
|---|---|---|---|---|---|
| | | | | Y | Y |

# SHIRE BIOCHEM INC

www.shire.com

**Industry Group Code: 325412  Ranks within this company's industry group: Sales:    Profits:**

| Drugs: | | Other: | | Clinical: | Computers: | | Services: | |
|---|---|---|---|---|---|---|---|---|
| Discovery: | Y | AgriBio: | | Trials/Services: | Hardware: | | Specialty Services: | |
| Licensing: | | Genetic Data: | | Labs: | Software: | | Consulting: | |
| Manufacturing: | Y | Tissue Replacement: | | Equipment/Supplies: | Arrays: | | Blood Collection: | |
| Genetics: | | | | Research & Development Services: | Database Management: | | Drug Delivery: | |
| | | | | Diagnostics: | | | Drug Distribution: | |

## TYPES OF BUSINESS:

Pharmaceuticals Discovery & Development
Influenza Vaccine
Drugs-HIV

## BRANDS/DIVISIONS/AFFILIATES:

Shire Pharmaceuticals Group
Biochem Pharma
Shire PLC
3TC
Aderall XR
Agrylin
Alertec
Hectorol

## CONTACTS: *Note: Officers with more than one job title may be intentionally listed here more than once.*

Angus Russell, CEO
Graham Hetherington, CFO
Anita Graham, Chief Admin. Officer
Tatjana May, General Counsel/Sec./Exec. VP-Legal Affairs
Barbara Deptula, Chief Corp. Dev. Officer/Exec. VP
Sylvie Gregoire, Pres., Human Genetic Therapies
Mike Cola, Pres., Specialty Pharmaceuticals
Anita Graham, Exec. VP-Corp. Bus. Svcs.
Tatjana May, Exec. VP-Global Legal Affairs

| Phone: 514-787-2300 | Fax: 514-787-2427 |
|---|---|
| Toll-Free: | |
| Address: 275 Boul Armand-Frappier Bureau, Ste. 500, Ville Saint-Laurent, QC H4S 2C9 Canada | |

## GROWTH PLANS/SPECIAL FEATURES:

Shire BioChem, Inc. is a Canadian company that focuses on research, development and commercialization of innovative products for the prevention and treatment of human diseases.  The company is also in charge of marketing products in Canada.  The firm is a subsidiary of Shire PLC, formerly Shire Pharmaceuticals Group, a global specialty pharmaceutical company with a focus on developing and marketing products in the areas of central nervous system, gastrointestinal and renal diseases.  The parent company has operations in global pharmaceutical markets including the U.S., Canada, the U.K., France, Italy, Spain and Germany, as well as a specialist drug delivery unit in the U.S.  Shire PLC's world headquarters are in Basingstroke, U.K., with its U.S. headquarters are in Wayne, Pennsylvania.  Shire BioChem currently engages in researching cancer and infectious diseases such as HIV and influenza.  The firm's 3TC is the cornerstone of the therapies that are now standard treatments for HIV and AIDS.  3TC is available worldwide and is marketed in partnership with GlaxcoSmithKline.  Among other products Shire BioChem currently has on the market in Canada are Aderall XR, for ADHD; Agrylin, treatment of essential thrombocythaemia; Alertec; Amatine; Elaprase; Estrace; Fosrenol, reduces serum phosphate in patients that suffer from end-stage renal disease; Mezavant; and Replagal.  In May 2008, the company opened an office in Norway.  In February 2009, the company agreed to acquire the worldwide rights to Equasym IR and XL used for the treatment of ADHD.

## FINANCIALS: Sales and profits are in thousands of dollars—add 000 to get the full amount. 2008 Note: Financial information for 2008 was not available for all companies at press time.

| | | U.S. Stock Ticker: Subsidiary |
|---|---|---|
| 2008 Sales: $ | 2008 Profits: $ | Int'l Ticker:    Int'l Exchange: |
| 2007 Sales: $ | 2007 Profits: $ | Employees:   500 |
| 2006 Sales: $ | 2006 Profits: $ | Fiscal Year Ends: 12/31 |
| 2005 Sales: $ | 2005 Profits: $ | Parent Company: SHIRE PLC |
| 2004 Sales: $ | 2004 Profits: $ | |

## SALARIES/BENEFITS:

| Pension Plan: | ESOP Stock Plan: | Profit Sharing: | Top Exec. Salary: $ | Bonus: $ |
|---|---|---|---|---|
| Savings Plan: | Stock Purch. Plan: | | Second Exec. Salary: $ | Bonus: $ |

## OTHER THOUGHTS:

**Apparent Women Officers or Directors**: 4
**Hot Spot for Advancement for Women/Minorities**: Y

## LOCATIONS: ("Y" = Yes)

| West: | Southwest: | Midwest: | Southeast: | Northeast: | International: |
|---|---|---|---|---|---|
| | | | | | Y |

# SHIRE PLC

www.shire.com

**Industry Group Code: 325412  Ranks within this company's industry group:** Sales: 30  Profits: 32

| Drugs: | | Other: | | Clinical: | Computers: | | Services: | |
|---|---|---|---|---|---|---|---|---|
| Discovery: | Y | AgriBio: | | Trials/Services: | Hardware: | | Specialty Services: | |
| Licensing: | Y | Genetic Data: | | Labs: | Software: | | Consulting: | |
| Manufacturing: | Y | Tissue Replacement: | | Equipment/Supplies: | Arrays: | | Blood Collection: | |
| Genetics: | | | | Research & Development Services: | Database Management: | | Drug Delivery: | Y |
| | | | | Diagnostics: | | | Drug Distribution: | |

## TYPES OF BUSINESS:

Drugs-Diversified
Drug Delivery Technology
Small-Molecule Drugs

## BRANDS/DIVISIONS/AFFILIATES:

Adderall
Daytrana
Carbatrol
Pentasa
Colazide
Replagal
Elaprase
Jerini AG

## CONTACTS: Note: Officers with more than one job title may be intentionally listed here more than once.

Angus Russell, CEO
Graham Hetherington, CFO
Jeffrey Jonas, Sr. VP-R&D, Specialty Pharmaceuticals
Joseph Rus, Exec. VP-New Prod. Dev.
Anita Graham, Chief Admin. Officer/Exec. VP-Corp. Bus. Svcs.
Tatjana May, General Counsel/Exec. VP-Global Legal Affairs/Sec.
Barbara Deptula, Chief Corp. Dev. Officer/Exec. VP
Jessica Mann, Sr. VP-Global Corp. Comm.
Joseph Rus, Exec. VP-Market Alliance
Mike Cola, Pres., Specialty Pharmaceuticals
Sylvie Gregoire, Pres., Human Genetic Therapies
Matthew Emmens, Chmn.
Soh Fujiwara, Managing Dir.-Japanese Oper.

| Phone: 44-1256-894-000 | Fax: 44-1256-894-708 |
|---|---|
| Toll-Free: | |
| Address: Hampshire Int'l Business Park, Chineham, Basingstoke, Hampshire RG24 8EP UK | |

## GROWTH PLANS/SPECIAL FEATURES:

Shire plc, formerly Shire Ltd., is an international specialty pharmaceutical company. The firm is focused on three therapeutic areas: attention deficit and hyperactivity disorder (ADHD), human genetic therapies (HGT) and gastrointestinal diseases. Shire's products for the treatment of ADHD include Adderall XR, a once-daily treatment, and Daytrana, an FDA-approved transdermal medication for ADHD. Carbatrol is the company's epilepsy treatment product. For the treatment of ulcerative colitis, Shire markets Pentasa. Shire's HGT products include Replagal for the treatment of Fabry disease, a rare genetic disorder resulting from the deficiency of an enzyme involved in the breakdown of fats; and Elaprase for the treatment of Hunter syndrome, a rare genetic disorder interfering with the processing of certain waste substances in the body. For renal disease, Shire markets Fosrenol, a phosphate binder for use in end-stage renal failure patients receiving dialysis. For the treatment of myeloproliferative disorders, a group of diseases in which one or more blood cell types are overproduced, Shire markets Agrylin in the U.S. and Xagrid in Europe. Other products marketed by Shire include Reminyl, CalciChew, Vyvanse and Promatine. The firm markets products in the U.S., Canada, the U.K., Germany, France, Italy, Ireland, Australia and Spain. In August 2008, the firm acquired 81% of Jerini AG, a biotechnology company. In October 2008, the firm changed its name from Shire Ltd. to Shire plc. In February 2009, Shire plc opened a new office in Tokyo, Japan. Also in February 2009, the firm agreed to acquire the global rights (excluding the U.S., Canada and Barbados) to ADHD treatments Equasym IR and Equasym XL. In March 2009, the company and GlaxoSmithKline plc agreed to co-promote ADHD treatment Vyvanse.

Shire offers its employees benefits including medical, dental, life, disability, business travel accident and AD&D insurance; an assistance program; and flexible spending account arrangements.

## FINANCIALS: Sales and profits are in thousands of dollars—add 000 to get the full amount. 2008 Note: Financial information for 2008 was not available for all companies at press time.

| | | |
|---|---|---|
| 2008 Sales: $3,022,200 | 2008 Profits: $156,000 | **U.S. Stock Ticker: SHPGY** |
| 2007 Sales: $2,436,300 | 2007 Profits: $-1,451,800 | **Int'l Ticker: SHP.L**  Int'l Exchange: London-LSE |
| 2006 Sales: $1,796,500 | 2006 Profits: $278,200 | Employees: 3,769 |
| 2005 Sales: $1,599,300 | 2005 Profits: $-578,400 | Fiscal Year Ends: 12/31 |
| 2004 Sales: $1,363,200 | 2004 Profits: $263,300 | Parent Company: |

## SALARIES/BENEFITS:

| Pension Plan: | ESOP Stock Plan: | Profit Sharing: | Top Exec. Salary: $1,227,657 | Bonus: $ |
|---|---|---|---|---|
| Savings Plan: Y | Stock Purch. Plan: Y | | Second Exec. Salary: $828,754 | Bonus: $ |

## OTHER THOUGHTS:

**Apparent Women Officers or Directors:** 6
**Hot Spot for Advancement for Women/Minorities:** Y

## LOCATIONS: ("Y" = Yes)

| West: | Southwest: | Midwest: | Southeast: | Northeast: | International: |
|---|---|---|---|---|---|
| | | | | Y | Y |

# SIEMENS HEALTHCARE DIAGNOSTICS

diagnostics.siemens.com

**Industry Group Code: 325413  Ranks within this company's industry group:  Sales:    Profits:**

| Drugs: | Other: | Clinical: | | Computers: | Services: |
|---|---|---|---|---|---|
| Discovery: | AgriBio: | Trials/Services: | | Hardware: | Specialty Services: |
| Licensing: | Genetic Data: | Labs: | | Software: | Consulting: |
| Manufacturing: | Tissue Replacement: | Equipment/Supplies: | Y | Arrays: | Blood Collection: |
| Genetics: | | Research & Development Services: | | Database Management: | Drug Delivery: |
| | | Diagnostics: | Y | | Drug Distribution: |

## TYPES OF BUSINESS:

Supplies-Immunodiagnostic Kits
Nonisotopic Diagnostic Tests
Immunoassay Analyzers
Allergy Testing

## BRANDS/DIVISIONS/AFFILIATES:

IMMULITE
ADVIA
Bayer Diagnostics
RapidLab
Clinitek
DCA 2000+
Siemens AG
Diagnostic Products Corp.

## CONTACTS: *Note: Officers with more than one job title may be intentionally listed here more than once.*

Dominic M. (Donal) Quinn, CEO
Sidney A. (Sid) Aroesty, COO
Dominic M. (Donal) Quinn, Pres.
Jochen Schmitz, CFO

| Phone: 914-631-8000 | Fax: 914-524-2132 |
|---|---|
| Toll-Free: 800-431-1970 | |
| Address: 511 Benedict Ave., Tarrytown, NY 10591-5097 US | |

## GROWTH PLANS/SPECIAL FEATURES:

Siemens Medical Solutions Diagnostics (SMSD), formed from the acquisition of Diagnostic Products Corp. (DPC) and Bayer Diagnostics, is a wholly-owned subsidiary of Siemens Medical Solutions.  Its products and services offer a balance of science, technology and practicality to provide healthcare professionals with the information they need to deliver better, more personalized healthcare to patients around the world.  With the acquisition of DPC and Bayer Diagnostics, SMSD has bridged the gap between in-vivo and in-vitro diagnostics to become the first full service diagnostics company.  By bringing together medical imaging, laboratory diagnostics and healthcare and healthcare information technology, SMSD offers a unique set of solutions that provide clinical, operational and financial outcomes.  The portfolio of SMSD includes products and services designed to optimize efficiency, improve workflow, help ensure patient safety and provide the highest levels of productivity and flexibility.  Its broad spectrum of immunoassay, integrated chemistry, routine chemistry, automation, hematology, microbiology, hemostasis, molecular, urinalysis, diabetes and blood gas testing systems, in conjunction with automation, informatics and consulting solutions, serve the needs of laboratories of any size.  SMSD's proprietary products include the IMMULITE, ADVIA, RapidLab and Clinitek families of products, along with the DCA 2000+, the MicroMix 5, 3gAllergy and SMS products.  In November 2007, the firm acquired Dade Behring Holdings, Inc., a laboratory diagnostics company.

## FINANCIALS:  Sales and profits are in thousands of dollars—add 000 to get the full amount. 2008 Note: Financial information for 2008 was not available for all companies at press time.

| | | |
|---|---|---|
| 2008 Sales: $ | 2008 Profits: $ | U.S. Stock Ticker: Subsidiary |
| 2007 Sales: $ | 2007 Profits: $ | Int'l Ticker:    Int'l Exchange: |
| 2006 Sales: $ | 2006 Profits: $ | Employees:  2,495 |
| 2005 Sales: $481,100 | 2005 Profits: $67,200 | Fiscal Year Ends: 12/31 |
| 2004 Sales: $446,800 | 2004 Profits: $61,700 | Parent Company: SIEMENS AG |

## SALARIES/BENEFITS:

| Pension Plan: | ESOP Stock Plan: | Profit Sharing: | Top Exec. Salary: $600,000 | Bonus: $ |
|---|---|---|---|---|
| Savings Plan: | Stock Purch. Plan: | | Second Exec. Salary: $450,000 | Bonus: $30,000 |

## OTHER THOUGHTS:

**Apparent Women Officers or Directors:**
**Hot Spot for Advancement for Women/Minorities:**

## LOCATIONS: ("Y" = Yes)

| West: | Southwest: | Midwest: | Southeast: | Northeast: | International: |
|---|---|---|---|---|---|
| Y | | | | Y | Y |

# SIGA TECHNOLOGIES INC

**www.siga.com**

Industry Group Code: 325412  Ranks within this company's industry group:  Sales: 119    Profits: 76

| Drugs: | | Other: | Clinical: | Computers: | Services: | |
|---|---|---|---|---|---|---|
| Discovery: | Y | AgriBio: | Trials/Services: | Hardware: | Specialty Services: | |
| Licensing: | | Genetic Data: | Labs: | Software: | Consulting: | |
| Manufacturing: | | Tissue Replacement: | Equipment/Supplies: | Arrays: | Blood Collection: | |
| Genetics: | | | Research & Development Services: | Database Management: | Drug Delivery: | Y |
| | | | Diagnostics: | | Drug Distribution: | |

## TYPES OF BUSINESS:

Drugs-Infectious Diseases
Vaccines
Antibiotics
Biothreat Rapid-Response Therapeutics
Mucosal Drug Delivery

## BRANDS/DIVISIONS/AFFILIATES:

SIGA-246
ST-294
ST-193

## CONTACTS: *Note: Officers with more than one job title may be intentionally listed here more than once.*

Eric A. Rose, CEO
Ayelet Dugary, CFO
Dennis E. Hruby, Chief Scientific Officer
Ayelet Dugary, Sec.
Eric A. Rose, Chmn.

| **Phone:** 212-672-9100 | **Fax:** 212-697-3130 |
|---|---|
| **Toll-Free:** | |
| **Address:** 420 Lexington Ave., Ste. 408, New York, NY 10170 US | |

## GROWTH PLANS/SPECIAL FEATURES:

SIGA Technologies is a development-stage biotechnology company focused on the discovery, development and commercialization of vaccines, antibiotics and anti-infectives for serious infectious diseases.  The major focus of the company's activities is on products for use in defense against biological warfare agents such as smallpox and arenaviruses (hemorrhagic fevers).  Its lead product, SIGA-246, is an orally administered anti-viral drug targeting the smallpox virus that has been granted fast-track orphan drug status by the FDA.  SIGA's antiviral programs are designed to prevent or limit the replication of the viral pathogen.  Its anti-infectives programs are aimed at the increasingly serious problem of drug resistance.  Further biological warfare defense products in the company's portfolio include ST-294 and St-193, anti-arenavirus drug candidates.  SIGA also has programs against other hemorrhagic fever viruses, including Lassa virus, Dengue fever, Rift Valley fever, Lymphocytic choriomeningitis virus (LCMV) and Ebola.  The firm's other technology platforms include bacterial commensal vector vaccine delivery systems and antibiotics that target bacterial adhesion organelles.  The company has collaborations with organizations including the National Institutes of Health, the U.S. Air Force and TransTech Pharma, Inc.

## FINANCIALS:  Sales and profits are in thousands of dollars—add 000 to get the full amount. 2008 Note: Financial information for 2008 was not available for all companies at press time.

| | | |
|---|---|---|
| 2008 Sales: $8,066 | 2008 Profits: $-8,599 | **U.S. Stock Ticker: SIGA** |
| 2007 Sales: $6,699 | 2007 Profits: $-5,639 | **Int'l Ticker:**    Int'l Exchange: |
| 2006 Sales: $7,258 | 2006 Profits: $-9,899 | Employees:    43 |
| 2005 Sales: $8,477 | 2005 Profits: $-2,288 | Fiscal Year Ends: 12/31 |
| 2004 Sales: $1,839 | 2004 Profits: $-9,373 | Parent Company: |

## SALARIES/BENEFITS:

| Pension Plan: | ESOP Stock Plan: | Profit Sharing: | Top Exec. Salary: $400,000 | Bonus: $ |
|---|---|---|---|---|
| Savings Plan: Y | Stock Purch. Plan: | | Second Exec. Salary: $250,000 | Bonus: $ |

## OTHER THOUGHTS:

**Apparent Women Officers or Directors:** 1
**Hot Spot for Advancement for Women/Minorities:**

## LOCATIONS: ("Y" = Yes)

| West: | Southwest: | Midwest: | Southeast: | Northeast: | International: |
|---|---|---|---|---|---|
| Y | | | | Y | |

# SIGMA-ALDRICH CORP

**www.sigmaaldrich.com**

**Industry Group Code: 325** **Ranks within this company's industry group:** Sales: 6    Profits: 6

| Drugs: | Other: | Clinical: | | Computers: | Services: |
|--------|--------|-----------|---|-----------|-----------|
| Discovery: | AgriBio: | Trials/Services: | | Hardware: | Specialty Services: |
| Licensing: | Genetic Data: | Labs: | | Software: | Consulting: |
| Manufacturing: | Tissue Replacement: | Equipment/Supplies: | Y | Arrays: | Blood Collection: |
| Genetics: | | Research & Development Services: | | Database Management: | Drug Delivery: |
| | | Diagnostics: | Y | | Drug Distribution: |

## TYPES OF BUSINESS:

Chemicals Manufacturing
Biotechnology Equipment
Pharmaceutical Ingredients
Fine Chemicals
Chromatography Products

## BRANDS/DIVISIONS/AFFILIATES:

Research Essentials
Research Specialties
Research Biotech
SAFC
Seppro
Sigma-Aldrich Quimica Ltda

## CONTACTS: Note: Officers with more than one job title may be intentionally listed here more than once.

Jai Nagarkatti, CEO
Jai Nagarkatti, Pres.
Rakesh Sachdev, CFO/VP
Gerrit van den Dool, VP-Sales
Doug Rau, VP-Human Resources
Carl Turza, CIO
Rakesh Sachdev, Sec.
Karen Miller, VP-Strategy & Corp. Dev.
Kirk Richter, Treas.
Gilles Cottier, Pres., SAFC
Frank Wicks, Pres., Research Specialties & Essentials
David Smoller, Pres., Research Biotech
Steve Walton, VP-Quality & Safety
Jai Nagarkatti, Chmn.
Dave Julien, Pres., Supply Chain

| Phone: 314-771-5765 | Fax: 314-771-5757 |
|---|---|
| Toll-Free: | |
| Address: 3050 Spruce St., St. Louis, MO 63103 US | |

## GROWTH PLANS/SPECIAL FEATURES:

Sigma-Aldrich Corp. is a life science and technology company that develops, manufactures, purchases and distributes a broad range of biochemicals and organic chemicals. The company offers roughly 100,000 chemicals (including 46,000 chemicals manufactured in-house) and 30,000 equipment products used for scientific and genomic research; biotechnology; pharmaceutical development; disease diagnosis; and pharmaceutical and high technology manufacturing. Sigma-Aldrich is structured into four units: Research essentials, research specialties, research biotech and SAFC, a fine chemicals unit. The research essentials unit sells biological buffers; cell culture reagents; biochemicals; chemicals; solvents; and other reagents and kits. The research specialties unit provides organic chemicals, biochemicals, analytical reagents, chromatography consumables, reference materials and high-purity products. The research biotech unit supplies immunochemical, molecular biology, cell signaling and neuroscience biochemicals and kits used in biotechnology, genomic, proteomic and other life science research applications. The SAFC unit offers large-scale organic chemicals and biochemicals used in development and production by pharmaceutical, biotechnology, industrial and diagnostic companies. The company operates in 37 countries, selling its products in nearly 160 countries and servicing over 1 million customers. Customers include commercial laboratories; pharmaceutical and industrial companies; universities; diagnostics, chemical and biotechnology companies and hospitals; non-profit organizations; and governmental institutions. In January 2009, the firm acquired the Seppro affinity depletion technology and 700 avian-derived antibodies from GenWay Biotech, Inc. Also in 2009, Sigma-Aldrich acquired Sigal Ltda, its Chilean distributor and established Sigma-Aldrich Quimica Ltda.

## FINANCIALS: Sales and profits are in thousands of dollars—add 000 to get the full amount. 2008 Note: Financial information for 2008 was not available for all companies at press time.

| | | |
|---|---|---|
| 2008 Sales: $2,200,700 | 2008 Profits: $341,500 | **U.S. Stock Ticker:** SIAL |
| 2007 Sales: $2,038,700 | 2007 Profits: $311,100 | **Int'l Ticker:** Int'l Exchange: |
| 2006 Sales: $1,797,500 | 2006 Profits: $276,800 | Employees: 7,925 |
| 2005 Sales: $1,666,500 | 2005 Profits: $258,300 | Fiscal Year Ends: 12/31 |
| 2004 Sales: $1,409,200 | 2004 Profits: $232,900 | Parent Company: |

## SALARIES/BENEFITS:

| Pension Plan: Y | ESOP Stock Plan: | Profit Sharing: | Top Exec. Salary: $750,000 | Bonus: $356,625 |
|---|---|---|---|---|
| Savings Plan: Y | Stock Purch. Plan: | | Second Exec. Salary: $394,167 | Bonus: $124,951 |

## OTHER THOUGHTS:

**Apparent Women Officers or Directors:** 2
**Hot Spot for Advancement for Women/Minorities:** Y

## LOCATIONS: ("Y" = Yes)

| West: | Southwest: | Midwest: | Southeast: | Northeast: | International: |
|-------|-----------|----------|-----------|-----------|---------------|
| Y | Y | Y | Y | Y | Y |

Note: Financial information, benefits and other data can change quickly and may vary from those stated here.

# SIMCERE PHARMACEUTICAL GROUP

www.simcere.com

**Industry Group Code:** 325412  **Ranks within this company's industry group:** Sales: 57   Profits: 38

| Drugs: | | Other: | Clinical: | Computers: | Services: |
|---|---|---|---|---|---|
| Discovery: | | AgriBio: | Trials/Services: | Hardware: | Specialty Services: |
| Licensing: | | Genetic Data: | Labs: | Software: | Consulting: |
| Manufacturing: | Y | Tissue Replacement: | Equipment/Supplies: | Arrays: | Blood Collection: |
| Genetics: | Y | | Research & Development Services: | Database Management: | Drug Delivery: |
| | | | Diagnostics: | | Drug Distribution: |

## TYPES OF BUSINESS:

Branded Generic Pharmaceuticals
Drug Research

## BRANDS/DIVISIONS/AFFILIATES:

Endu
Bicun
Zailin
Yingtaiqing
Anqi
Biqi
Simcere Medgenn Bio-Pharmaceutical Co., Ltd.
Anxin

## CONTACTS: *Note: Officers with more than one job title may be intentionally listed here more than once.*

Jinsheng Ren, CEO
Frank Zhigang Zhao, CFO
Mark Peizhi Chen, VP-Mktg.
Qinseng Li, VP-Human Resources
Peng Wang, Chief Scientific Officer
Haibo Qian, Corp. Sec.
Eric Wang Lam Cheung, VP-Investor Rel.
Jindong Zhou, Exec. VP
Dazheng Sun, VP-Commercial Sales
Jialun Tian, VP-Hospital Sales
Xiaojin Yin, Sr. VP-R&D
Jinsheng Ren, Chmn.

| Phone: 86-25-8556-6666 | Fax: 85-25-8547-1729 |
|---|---|
| Toll-Free: | |
| Address: No. 699-18 Xuan Wu Ave., Xuan Wu District, Nanjing, Jiangsu, 210042 China | |

## GROWTH PLANS/SPECIAL FEATURES:

Simcere Pharmaceutical Group is a Chinese manufacturer and supplier of branded generic pharmaceuticals. The company manufactures and sells 39 pharmaceutical products and is the exclusive distributor of three additional pharmaceutical products marketed under the firm's brand name. The firm's products are used for treatment of a wide range of diseases including cancer; cerebrovascular and cardiovascular diseases; infections; arthritis; diarrhea; allergies; respiratory conditions; and urinary conditions. Simcere's products include Bicun, an anti-stroke medication and the first synthetic-free radical scavenger sold in China; Zailin, a generic amoxicillin granule antibiotic; Endu, an anticancer medication and recombinant human endostatin injection; Yingtaiqing, a generic diclofenac sodium sustained-release capsule for inflammation and pain relief; Anqi, a generic amoxicillin with clavulanate potassium antibiotic; and Biqi, an over-the-counter generic smectite powder for diarrhea. In addition, the Chinese FDA has approved the sale of more than 210 other company products. Simcere Pharmaceutical has six manufacturing bases, two nationwide sales and marketing subsidiaries, and one research and development center. The company owns 80% of Shandong Simcere Medgenn Bio-Pharmaceutical Co., Ltd. In recent years, the firm acquired Master Luck Corporation Limited, which is a majority-owner of Nanjing Chit Pharmaceutical, a quick-growing manufacturer and supplier of anti-cancer drugs in China. In April 2008, Simcere acquired a 70% stake in Wuhu Zhong Ren Pharmaceutical Co. Ltd., a Chinese drug manufacturer specializing in the production of sustained release anti-tumor implants. In May 2008, the firm received Chinese FDA approval to manufacture and market a generic Biapenem injection (under the brand name Anxin) for the treatment of serious infections. In May 2009, Simcere agreed to acquire 37.5% ownership of Jiangsu Yanshen Biological Technology Stock Co., Ltd. for approximately $28.6 million.

The company offers employee benefits that include health insurance and a housing plan.

## FINANCIALS: Sales and profits are in thousands of dollars—add 000 to get the full amount. 2008 Note: Financial information for 2008 was not available for all companies at press time.

| | | |
|---|---|---|
| 2008 Sales: $225,206 | 2008 Profits: $51,323 | **U.S. Stock Ticker:** SCR |
| 2007 Sales: $187,638 | 2007 Profits: $41,300 | **Int'l Ticker:**    Int'l Exchange: |
| 2006 Sales: $ | 2006 Profits: $ | Employees: 2,615 |
| 2005 Sales: $91,300 | 2005 Profits: $12,700 | Fiscal Year Ends: 12/31 |
| 2004 Sales: $68,100 | 2004 Profits: $5,600 | Parent Company: |

## SALARIES/BENEFITS:

| Pension Plan: | ESOP Stock Plan: | Profit Sharing: | Top Exec. Salary: $ | Bonus: $ |
|---|---|---|---|---|
| Savings Plan: | Stock Purch. Plan: | | Second Exec. Salary: $ | Bonus: $ |

## OTHER THOUGHTS:

**Apparent Women Officers or Directors:**
**Hot Spot for Advancement for Women/Minorities:**

## LOCATIONS: ("Y" = Yes)

| West: | Southwest: | Midwest: | Southeast: | Northeast: | International: |
|---|---|---|---|---|---|
| | | | | | Y |

# SKYEPHARMA PLC

www.skyepharma.com

**Industry Group Code: 325412A  Ranks within this company's industry group:  Sales: 13    Profits: 25**

| Drugs: | | Other: | | Clinical: | Computers: | | Services: | |
|---|---|---|---|---|---|---|---|---|
| Discovery: | Y | AgriBio: | | Trials/Services: | Hardware: | | Specialty Services: | |
| Licensing: | Y | Genetic Data: | | Labs: | Software: | | Consulting: | |
| Manufacturing: | Y | Tissue Replacement: | | Equipment/Supplies: | Arrays: | | Blood Collection: | |
| Genetics: | | | | Research & Development Services: | Database Management: | | Drug Delivery: | Y |
| | | | | Diagnostics: | | | Drug Distribution: | |

## TYPES OF BUSINESS:

Drug Delivery Systems
Generic Drugs

## BRANDS/DIVISIONS/AFFILIATES:

Sular
Xatral
Lodotra
ZYFLO CR
Paxil CR
Corumo
Pulmicort HFA-MDI
Solaraze

## CONTACTS: *Note: Officers with more than one job title may be intentionally listed here more than once.*

Ken Cunningham, CEO
Peter Grant, CFO
John Murphy, General Counsel/Sec.
Peter Laing, Dir.-Corp. Comm.
Jeremy Scudamore, Chmn.

| Phone: 44-20-7491-1777 | Fax: 44-20-7491-3338 |
|---|---|
| Toll-Free: | |
| Address: 105 Piccadilly, London, W1J 7NJ UK | |

## GROWTH PLANS/SPECIAL FEATURES:

SkyePharma PLC is a worldwide provider of drug delivery technologies. Its drug delivery products and new drugs are manufactured under the SkyePharma name, which is contracted to several global pharmaceutical companies, including GlaxoSmithKline and Abbott Laboratories. SkyePharma's platform consists of oral, topical and inhalation technologies. Each of these areas is supported by dedicated research facilities located primarily in Switzerland, with additional operations in Lyon, France. The company also develops generic and super-generic drugs (drugs that are difficult to replicate, or that are improved versions of off-patent originals). The firm's oral treatment products include Xatral OD/Uroxatral for genito-urinary diseases; Madopar DR for Parkinson's disease; Diclofenac-ratiopharm for inflammation; Coruno for angina; and Sular for hypertension. Two examples of the companies inhalation products are Foradil Certihaler and Pulmicort, both used to treat asthma. The firm's topical drug, Solaraze, is used to treat Actinic Keratosis. SkyePharma is currently developing oral drugs Lodotra for rheumatoid arthritis and SKP-1041 for sleep maintenance. The company is also in the process of developing Flutiform, a new inhalation product for the treatment of asthma.

## FINANCIALS: Sales and profits are in thousands of dollars—add 000 to get the full amount. 2008 Note: Financial information for 2008 was not available for all companies at press time.

| | | |
|---|---|---|
| 2008 Sales: $94,710 | 2008 Profits: $-43,700 | **U.S. Stock Ticker:** |
| 2007 Sales: $63,340 | 2007 Profits: $-36,540 | **Int'l Ticker: SKP**    Int'l Exchange: London-LSE |
| 2006 Sales: $85,100 | 2006 Profits: $-156,600 | Employees:   420 |
| 2005 Sales: $100,600 | 2005 Profits: $-100,800 | Fiscal Year Ends: 12/31 |
| 2004 Sales: $119,200 | 2004 Profits: $-46,600 | Parent Company: |

## SALARIES/BENEFITS:

| Pension Plan: Y | ESOP Stock Plan: | Profit Sharing: | Top Exec. Salary: $685,860 | Bonus: $288,190 |
|---|---|---|---|---|
| Savings Plan: | Stock Purch. Plan: Y | | Second Exec. Salary: $590,870 | Bonus: $231,840 |

## OTHER THOUGHTS:

Apparent Women Officers or Directors:
Hot Spot for Advancement for Women/Minorities:

## LOCATIONS: ("Y" = Yes)

| West: | Southwest: | Midwest: | Southeast: | Northeast: | International: |
|---|---|---|---|---|---|
| | | | | | Y |

# SPECIALTY LABORATORIES INC

www.specialtylabs.com

**Industry Group Code:** 6215 **Ranks within this company's industry group:** Sales: Profits:

| Drugs: | Other: | Clinical: | | Computers: | Services: |
|---|---|---|---|---|---|
| Discovery: | AgriBio: | Trials/Services: | | Hardware: | Specialty Services: |
| Licensing: | Genetic Data: | Labs: | Y | Software: | Consulting: |
| Manufacturing: | Tissue Replacement: | Equipment/Supplies: | | Arrays: | Blood Collection: |
| Genetics: | | Research & Development Services: | | Database Management: | Drug Delivery: |
| | | Diagnostics: | Y | | Drug Distribution: |

## TYPES OF BUSINESS:

Clinical Laboratory Tests
Assays

## BRANDS/DIVISIONS/AFFILIATES:

DataPassport MD
Quest Diagnostics, Inc.

## CONTACTS: *Note: Officers with more than one job title may be intentionally listed here more than once.*

Christopher Lockhart, Laboratory Dir.
Surya N. Mohapatra, Chmn./CEO-Quest Diagnostics

| | |
|---|---|
| **Phone:** 661-799-6543 | **Fax:** 661-799-6634 |
| **Toll-Free:** 800-421-7110 | |
| **Address:** 27027 Tourney Rd., Valencia, CA 91355 US | |

## GROWTH PLANS/SPECIAL FEATURES:

Specialty Laboratories, Inc. (SL), a subsidiary of Quest Diagnostics, Inc., is a research-based clinical laboratory, predominantly focused on developing and performing esoteric clinical laboratory tests, referred to as assays. The firm offers a comprehensive menu of thousands of assays, many of which it developed through internal research and development efforts, that are used to diagnose, evaluate and monitor patients in the areas of endocrinology; genetics; infectious diseases; neurology; pediatrics; urology; allergy and immunology; cardiology and coagulation; hepatology; microbiology; oncology; rheumatology; women's health; dermatopathology; gastroenterology; nephrology; pathology; and toxicology. Some of the company's newest assays include evaluations for vitamin D deficiency, potassium and chlorine levels and protein chemistry, and the ovarian cancer marker HE4. The company also partners with the Center for Advanced Diagnostics and the Molecular Profiling Institute, Inc. (MPI), a subsidiary of Translational Genomics Research Institute (TGen), to offer molecular-based assays. In addition, SL owns proprietary information technology that accelerates and automates test ordering and results reporting with customers. Products include DataPassportMD, a web-based laboratory order entry and resulting system. The company's primary customers are hospitals, independent clinical laboratories and physicians.

The company offers its employees health and dental plans; a 401(k) plan; a stock purchase plan; short- and long-term disability plans; flexible spending accounts; an employee assistance program; and an education reimbursement program.

## FINANCIALS: Sales and profits are in thousands of dollars—add 000 to get the full amount. 2008 Note: Financial information for 2008 was not available for all companies at press time.

| | | |
|---|---|---|
| 2008 Sales: $ | 2008 Profits: $ | **U.S. Stock Ticker: Subsidiary** |
| 2007 Sales: $ | 2007 Profits: $ | **Int'l Ticker:** Int'l Exchange: |
| 2006 Sales: $ | 2006 Profits: $ | Employees: 689 |
| 2005 Sales: $ | 2005 Profits: $ | Fiscal Year Ends: 12/31 |
| 2004 Sales: $134,803 | 2004 Profits: $-12,950 | Parent Company: QUEST DIAGNOSTICS INC |

## SALARIES/BENEFITS:

| | | | | |
|---|---|---|---|---|
| Pension Plan: | ESOP Stock Plan: | Profit Sharing: | Top Exec. Salary: $422,908 | Bonus: $ |
| Savings Plan: Y | Stock Purch. Plan: | | Second Exec. Salary: $262,000 | Bonus: $ |

## OTHER THOUGHTS:

**Apparent Women Officers or Directors:**
**Hot Spot for Advancement for Women/Minorities:**

## LOCATIONS: ("Y" = Yes)

| West: | Southwest: | Midwest: | Southeast: | Northeast: | International: |
|---|---|---|---|---|---|
| Y | | | | | |

# SPECTRAL DIAGNOSTICS INC

### www.spectraldx.com

**Industry Group Code: 325413   Ranks within this company's industry group: Sales: 22    Profits: 11**

| Drugs: | Other: | Clinical: | | Computers: | | Services: |
|---|---|---|---|---|---|---|
| Discovery: | AgriBio: | Trials/Services: | | Hardware: | | Specialty Services: |
| Licensing: | Genetic Data: | Labs: | | Software: | | Consulting: |
| Manufacturing: | Tissue Replacement: | Equipment/Supplies: | Y | Arrays: | | Blood Collection: |
| Genetics: | | Research & Development Services: | | Database Management: | | Drug Delivery: |
| | | Diagnostics: | Y | | | Drug Distribution: |

## TYPES OF BUSINESS:

Medical Diagnostics Products
West Nile Virus Diagnostics
Sepsis Diagnostics
Antibodies

## BRANDS/DIVISIONS/AFFILIATES:

RapidWN
Endotoxin Activity Assays
Toraymyxin

## CONTACTS: Note: Officers with more than one job title may be intentionally listed here more than once.

Paul M. Walker, CEO
Paul M. Walker, Pres.
Tony Businskas, CFO/Exec. VP
Nisar A. Shaikh, Sr. VP-Scientific Affairs
Robert Verhagen, VP-Bus. Dev.
Nisar A. Shaikh, Sr. VP-Corp. Quality
Debra M. Foster, Dir.-Sepsis Program

| Phone: 416-626-3233 | Fax: 416-626-7383 |
|---|---|
| Toll-Free: 888-426-4264 | |
| Address: 135 The West Mall, Toronto, ON M9C 1C2 Canada | |

## GROWTH PLANS/SPECIAL FEATURES:

Spectral Diagnostics, Inc., is a medical diagnostics company that focuses on developing, manufacturing and selling rapid assays for the identification of life-threatening illnesses. Its activities are primarily involved with diagnostics products for sepsis and West Nile Virus. Spectral makes Endotoxin Activity Assays (EAA), rapid, whole blood tests used to identify patients who are at risk for sepsis. With the development of the EAA test as a rapid indicator of endotoxin, physicians are assisted in stratifying patients into those who are at low risk of severe sepsis and those who require directed anti-sepsis or anti-infection therapy. The company's other principal product, its RapidWN West Nile Virus rapid diagnostic test, diagnoses patients potentially infected with West Nile Virus in less than an hour. Spectral has an in-house antibody generation and assay development team, which produces materials for its assay kits and for third-party customers. The company's remaining products include high performance monoclonal antibodies and recombinant proteins. Spectral has successfully engineered and produced over 30 antigens in bacteria and has created several patented novel molecules. The firm's protein, antibodies and reagents are sold to the research and development industry, as well as for use in products manufactured by other diagnostic companies. In July 2008, Spectral Diagnostics signed an exclusive agreement with BB Medical for the distribution of EAA in Russia. In March 2009, the firm has signed a license agreement with Japan-based Toray Industries, Inc.; the agreement gave Spectral the exclusive development and commercial rights in the U.S. for Toraymyxin, a therapeutic device for the treatment of sepsis.

## FINANCIALS: Sales and profits are in thousands of dollars—add 000 to get the full amount. 2008 Note: Financial information for 2008 was not available for all companies at press time.

| | | |
|---|---|---|
| 2008 Sales: $2,800 | 2008 Profits: $-1,400 | **U.S. Stock Ticker: DIAGF** |
| 2007 Sales: $2,800 | 2007 Profits: $-1,500 | **Int'l Ticker: SDI**   Int'l Exchange: Toronto-TSX |
| 2006 Sales: $2,900 | 2006 Profits: $3,800 | Employees:   70 |
| 2005 Sales: $9,700 | 2005 Profits: $-7,100 | Fiscal Year Ends: 12/31 |
| 2004 Sales: $9,000 | 2004 Profits: $-1,600 | Parent Company: |

## SALARIES/BENEFITS:

| Pension Plan: | ESOP Stock Plan: | Profit Sharing: | Top Exec. Salary: $268,252 | Bonus: $ |
|---|---|---|---|---|
| Savings Plan: | Stock Purch. Plan: | | Second Exec. Salary: $201,189 | Bonus: $ |

## OTHER THOUGHTS:

**Apparent Women Officers or Directors: 1**
**Hot Spot for Advancement for Women/Minorities:**

## LOCATIONS: ("Y" = Yes)

| West: | Southwest: | Midwest: | Southeast: | Northeast: | International: |
|---|---|---|---|---|---|
| Y | | Y | | Y | Y |

# SPECTRUM PHARMACEUTICALS INC
## www.spectrumpharm.com

**Industry Group Code: 325412 Ranks within this company's industry group: Sales: 99 Profits: 93**

| Drugs: | | Other: | Clinical: | Computers: | Services: |
|---|---|---|---|---|---|
| Discovery: | Y | AgriBio: | Trials/Services: | Hardware: | Specialty Services: |
| Licensing: | Y | Genetic Data: | Labs: | Software: | Consulting: |
| Manufacturing: | Y | Tissue Replacement: | Equipment/Supplies: | Arrays: | Blood Collection: |
| Genetics: | | | Research & Development Services: | Database Management: | Drug Delivery: |
| | | | Diagnostics: | | Drug Distribution: |

## TYPES OF BUSINESS:

Oncology & Urology Drugs
Cancer Treatments

## BRANDS/DIVISIONS/AFFILIATES:

Fusilev
Elsamitrucin
Lucanthone
EOquin
Ortataxel
Ozarelix
Satraplatin
RenaZorb

## CONTACTS: Note: Officers with more than one job title may be intentionally listed here more than once.

Rajesh C. Shrotriya, CEO
Rajesh C. Shrotriya, Pres.
George Uy, VP-Mktg. & Sales
William Pedranti, General Counsel/VP
Russell L. Skibsted, Chief Business Officer/Sr. VP
Shyam Kumaria, VP-Finance
Michael Adam, Sr. VP-Pharmaceutical Oper.
Andrew Sandler, Chief Medical Officer
Amar Singh, Chief Commercial Officer
Rajesh C. Shrotriya, Chmn.

| Phone: 949-788-6700 | Fax: 949-788-6706 |
|---|---|
| Toll-Free: | |
| Address: 157 Technology Dr., Irvine, CA 92618 US | |

## GROWTH PLANS/SPECIAL FEATURES:

Spectrum Pharmaceuticals, Inc. is a biopharmaceutical company that develops, acquires and advances a diversified portfolio of drug candidates, with a focus on oncology, urology and other critical health challenges. The company currently has 11 drugs in development, including two recently approved drugs with late stage alternative indications. Fusilev (levoleucovorin) is a novel folate analog formulation that is indicated for patients with osteosarcoma after high-dose methotrexate therapy, a common cancer treatment that can cause toxicity or inadvertent overdose of folic acid antagonists. Fusilev has also been submitted for a new drug application (NDA) for the treatment of colorectal cancer. Spectrum's other marketed drug, Zevalin, is a treatment for patients with relapsed or refractory low-grade or follicular B-cell non-Hodgkin's lymphoma (NHL). The drug has also been submitted as an NDA for first-line therapy in NHL. The firm's late-stage development product is EOquin, an anti-cancer agent for the treatment of non-invasive bladder cancer. The company's Phase I and II drug candidates include Ozarelix, an antagonist that is being investigated for indications in non muscle-invasive bladder cancer, hormone dependent prostate cancer, benign prostatic hypertrophy and endometriosis; Ortataxel, a treatment for solid tumors; Satraplatin, a treatment for non-small cell lung cancer; SPI-1620, an adjunct to chemotherapy; Elsamitrucin, an anti-tumor antibiotic; Lucanthone, a chemo sensitizer in malignant brain tumors; RenaZorb, a second-generation lanthanum-based treatment for hyperphosphatemia; and SPI-205, a drug with possible benefits in treating chemotherapy-induced pheripheral neuropathy. In August 2008, the firm launched Fusilev for post-methotrexate therapy. In October 2008, the company and Allergan, Inc. entered into a collaboration agreement to develop and commercialize EOquin. In May 2009, Spectrum obtained 100% control of RIT Oncology, LLC, a joint company formed with Cell Therapeutics, Inc. to market Zevalin.

## FINANCIALS: Sales and profits are in thousands of dollars—add 000 to get the full amount. 2008 Note: Financial information for 2008 was not available for all companies at press time.

| | | |
|---|---|---|
| 2008 Sales: $28,725 | 2008 Profits: $-15,467 | **U.S. Stock Ticker: SPPI** |
| 2007 Sales: $7,672 | 2007 Profits: $-34,036 | **Int'l Ticker:** Int'l Exchange: |
| 2006 Sales: $5,673 | 2006 Profits: $-23,284 | Employees: 84 |
| 2005 Sales: $ 577 | 2005 Profits: $-18,642 | Fiscal Year Ends: 12/31 |
| 2004 Sales: $ 258 | 2004 Profits: $-12,286 | Parent Company: |

## SALARIES/BENEFITS:

| Pension Plan: | ESOP Stock Plan: | Profit Sharing: | Top Exec. Salary: $600,000 | Bonus: $1,000,000 |
|---|---|---|---|---|
| Savings Plan: Y | Stock Purch. Plan: | | Second Exec. Salary: $288,466 | Bonus: $ |

## OTHER THOUGHTS:

**Apparent Women Officers or Directors:**
**Hot Spot for Advancement for Women/Minorities:**

## LOCATIONS: ("Y" = Yes)

| West: | Southwest: | Midwest: | Southeast: | Northeast: | International: |
|---|---|---|---|---|---|
| Y | | | | | Y |

# STEMCELLS INC

**www.stemcellsinc.com**

Industry Group Code: 325414  **Ranks within this company's industry group:** Sales: 8  Profits: 7

| Drugs: | | Other: | | Clinical: | Computers: | Services: |
|---|---|---|---|---|---|---|
| Discovery: | Y | AgriBio: | | Trials/Services: | Hardware: | Specialty Services: |
| Licensing: | Y | Genetic Data: | Y | Labs: | Software: | Consulting: |
| Manufacturing: | | Tissue Replacement: | Y | Equipment/Supplies: | Arrays: | Blood Collection: |
| Genetics: | | | | Research & Development Services: | Database Management: | Drug Delivery: |
| | | | | Diagnostics: | | Drug Distribution: |

## TYPES OF BUSINESS:

Cell-Based Therapeutics

## BRANDS/DIVISIONS/AFFILIATES:

HuCNS-SC
hLEC

## CONTACTS: *Note: Officers with more than one job title may be intentionally listed here more than once.*

Martin McGlynn, CEO
Ann Tsukamoto, COO
Martin McGlynn, Pres.
Rodney Young, CFO
Ann Tsukamoto, Exec. VP-R&D
Rodney Young, VP-Admin.
Ken Stratton, General Counsel
Stewart Craig, Sr. VP-Oper.
Stewart Craig, Sr. VP-Dev.
Rodney Young, VP-Finance
Stephen Huhn, VP/Head-CNS Program
Nobuko Uchida, VP-Stem Cell Biology
Maria Millian, VP/Head-Liver Program
John J. Schwartz, Chmn.

| Phone: 650-475-3100 | Fax: 650-475-3101 |
|---|---|
| **Toll-Free:** | |
| **Address:** 3155 Porter Dr., Palo Alto, CA 94304 US | |

## GROWTH PLANS/SPECIAL FEATURES:

StemCells, Inc. is focused on the discovery and development of cell-based therapeutics to treat damage to, or degeneration of, major organ systems including the central nervous system, liver and pancreas. The company seeks to identify and purify rare stem and progenitor cells, expand and bank them as transplantable cells and then develop therapeutic agents. StemCells is currently conducting a Phase I clinical trial to evaluate the safety and efficacy of its lead product, the human neural stem cell (HuCNS-SC), as a treatment for infantile and late infantile neuronal ceroid lipofuscinosis, two forms of a group of disorders often referred to as Batten disease. In addition, the company identified a population of cells derived from liver tissue, the human liver engrafting cells (hLEC), which, when transplanted into a mouse model of liver degeneration, show long-term engraftment, differentiate into human hepatocytes, secrete human hepatic proteins and form structural elements of the liver. Based on this data, the firm is developing the hLEC for potential therapeutic applications to liver diseases. Liver stem cells may be useful in the treatment of diseases such as hepatitis, liver failure, blood-clotting disorder, cirrhosis of the liver and liver cancer. StemCells has also identified a candidate stem cell of the pancreas. Other possible applications include spinal cord treatments, Alzheimer's Disease, amyotropic lateral sclerosis (ALS, or Lou Gherig's Disease) and epilepsy. The firm owns 50 issued U.S. patents and 150 foreign patents. In March 2009, StemCells acquired Stem Cell Sciences, PLC and their respective technologies and subsidiaries. In June 2009, the company shuttered its Australian facility, laying off nine employees.

## FINANCIALS:  Sales and profits are in thousands of dollars—add 000 to get the full amount. 2008 Note: Financial information for 2008 was not available for all companies at press time.

| | | |
|---|---|---|
| 2008 Sales: $ 232 | 2008 Profits: $-29,087 | **U.S. Stock Ticker: STEM** |
| 2007 Sales: $ 57 | 2007 Profits: $-25,023 | **Int'l Ticker:**  Int'l Exchange: |
| 2006 Sales: $ 93 | 2006 Profits: $-18,948 | Employees:  55 |
| 2005 Sales: $ 206 | 2005 Profits: $-11,738 | Fiscal Year Ends: 12/31 |
| 2004 Sales: $ 141 | 2004 Profits: $-15,330 | Parent Company: |

## SALARIES/BENEFITS:

| Pension Plan: | ESOP Stock Plan: | Profit Sharing: | Top Exec. Salary: $385,000 | Bonus: $77,000 |
|---|---|---|---|---|
| Savings Plan: Y | Stock Purch. Plan: | | Second Exec. Salary: $300,000 | Bonus: $37,500 |

## OTHER THOUGHTS:

**Apparent Women Officers or Directors**: 3
**Hot Spot for Advancement for Women/Minorities**: Y

## LOCATIONS: ("Y" = Yes)

| West: | Southwest: | Midwest: | Southeast: | Northeast: | International: |
|---|---|---|---|---|---|
| Y | | | | Y | |

# STIEFEL LABORATORIES INC

**www.stiefel.com**

Industry Group Code: 325412  Ranks within this company's industry group:  Sales:    Profits:

| Drugs: | Other: | Clinical: | Computers: | Services: |
|---|---|---|---|---|
| Discovery: | AgriBio: | Trials/Services: | Hardware: | Specialty Services:  Y |
| Licensing: | Genetic Data: | Labs: | Software: | Consulting: |
| Manufacturing:  Y | Tissue Replacement: | Equipment/Supplies: | Arrays: | Blood Collection: |
| Genetics: | | Research & Development Services: | Database Management: | Drug Delivery: |
| | | Diagnostics: | | Drug Distribution:  Y |

## TYPES OF BUSINESS:

Dermatological & Skin Care Products

## BRANDS/DIVISIONS/AFFILIATES:

Evoclin
Duac
Revaleskin
MimyX
LUXIQ Foam
Verdeso
Olux
ABR Development

## CONTACTS: Note: Officers with more than one job title may be intentionally listed here more than once.

Charles W. Stiefel, CEO
Bill Humphries, Pres.
Alfonso Ugarte, VP-Global Mktg.
Steve Karasick, Sr. VP-People
Gavin Corcoran, Chief Scientific Officer
Steve Karasick, Sr. VP-IT
Brent Stiefel, Exec. VP-Global Corp. Dev. & Prod. Portfolio
Erin Bacher, Associate Dir.-Global Public Rel.
Todd Stiefel, Exec. VP-Global Strategy
Wayne Wilson, VP-U.S. Sales
Jim Hartman, Sr. VP-U.S. Commercial Oper.
Jeff Klimaski, VP/Global Corp. Compliance Officer
Charles W. Stiefel, Chmn.
Richard MacKay, Pres., Stiefel Canada, Inc.

| Phone: 305-443-3800 | Fax: 305-443-3467 |
|---|---|
| Toll-Free: | |
| Address: 255 Alhambra Cir., Coral Gables, FL 33134-7412 US | |

## GROWTH PLANS/SPECIAL FEATURES:

Stiefel Laboratories, Inc. is a specialized pharmaceutical company that focuses on the advancement of dermatology and skin care. The company's sells hundreds of products that treat a wide range of dermatological ailments including acne, psoriasis, fungal infections, eczema, dry skin, oily skin, rosacea, seborrhea, pruritus and sun damaged and aging skin. Leading products include Evoclin and Duac, acne skin care products; Revaleskin, for younger and smoother skin; MimyX, a cream that relieves dry and waxy skin for a variety of dermatoses; LUXIQ Foam, which reduces the signs of scalp dermatoses, such as scaling, redness, and plaques; Olux, also used for scalp treatment; and Verdeso, a treatment for eczema. The firm also offers pharmaceutical contract manufacturing of gels, creams, ointments, topical solutions and liquid orals, both prescription and over-the-counter. The company has subsidiaries in more than 30 countries and its products are available in more than 100 countries. In May 2008, the firm acquired ABR Invent and ABR Development, developers of the dermal filler Atlean. In August 2008, Stiefel acquired Barrier Therapeutics, a pharmaceutical company that specializes on the development and commercialization of dermatological products. In April 2009, the company agreed to be acquired by GlaxoSmithKline, a world leader in research-based pharmaceutical and healthcare companies, for $2.9 billion.

## FINANCIALS:  Sales and profits are in thousands of dollars—add 000 to get the full amount. 2008 Note: Financial information for 2008 was not available for all companies at press time.

| | | |
|---|---|---|
| 2008 Sales: $ | 2008 Profits: $ | U.S. Stock Ticker: Private |
| 2007 Sales: $444,900 | 2007 Profits: $ | Int'l Ticker:    Int'l Exchange: |
| 2006 Sales: $ | 2006 Profits: $ | Employees:  2,000 |
| 2005 Sales: $184,264 | 2005 Profits: $33,958 | Fiscal Year Ends: |
| 2004 Sales: $144,355 | 2004 Profits: $19,015 | Parent Company: |

## SALARIES/BENEFITS:

| Pension Plan: | ESOP Stock Plan: | Profit Sharing: | Top Exec. Salary: $530,000 | Bonus: $325,000 |
|---|---|---|---|---|
| Savings Plan: | Stock Purch. Plan: | | Second Exec. Salary: $393,083 | Bonus: $190,000 |

## OTHER THOUGHTS:

Apparent Women Officers or Directors: 3
Hot Spot for Advancement for Women/Minorities: Y

## LOCATIONS: ("Y" = Yes)

| West: | Southwest: | Midwest: | Southeast: | Northeast: | International: |
|---|---|---|---|---|---|
| Y | | | Y | Y | Y |

# STRATAGENE CORP

www.stratagene.com

Industry Group Code: 325413 Ranks within this company's industry group: Sales: Profits:

| Drugs: | Other: | | Clinical: | | Computers: | | Services: | |
|---|---|---|---|---|---|---|---|---|
| Discovery: | AgriBio: | | Trials/Services: | | Hardware: | Y | Specialty Services: | |
| Licensing: | Genetic Data: | Y | Labs: | | Software: | Y | Consulting: | |
| Manufacturing: | Tissue Replacement: | | Equipment/Supplies: | Y | Arrays: | Y | Blood Collection: | |
| Genetics: | | | Research & Development Services: | | Database Management: | Y | Drug Delivery: | |
| | | | Diagnostics: | Y | | | Drug Distribution: | |

## TYPES OF BUSINESS:

Laboratory Instrument Manufacturing
Gene-Sequencing Software
Clinical Diagnostic Equipment

## BRANDS/DIVISIONS/AFFILIATES:

KOVA Microscopic Urinalysis System
PicoFuge
StrataCooler
Absolutely RNA
GeneMorph II
ArrayAssist
PathwayAssist
Hycor

## CONTACTS: Note: Officers with more than one job title may be intentionally listed here more than once.

Joseph A. Sorge, CEO
Joseph A. Sorge, Pres.
Steve Martin, CFO
John R. Pouk, Sr. VP-Global Sales
Ronni L. Sherman, General Counsel
Nelson F. Thune, Sr. VP-Oper./Gen. Mgr.
Nelson F. Thune, Sr. VP-Oper., Hycor Biomedical Inc.
Joseph A. Sorge, Chmn.

| Phone: 858-373-6300 | Fax: 858-535-0034 |
|---|---|
| Toll-Free: 800-894-1304 | |
| Address: 1011 N. Torrey Pines Rd., La Jolla, CA 92037 US | |

## GROWTH PLANS/SPECIAL FEATURES:

Stratagene Corp., a subsidiary of Agilent Technologies, Inc., develops, manufactures and markets lab instruments and materials for the molecular biology, genomics, proteomics, drug discovery and toxicology fields. The firm's products are applied to amplification; nucleic acid analysis; cell biology; protein function and analysis; cloning; quantitative PCR; and microarrays. Amplification products include AccuScript, used to synthesize cDNA. Nucleic acid analysis products offer DNA purification, gene regulation, labware and more. Cell Biology products include the Mitochondrial Membrane Potential Detection Kit and MycoSensor QPCR Gradient Cycler. With regard to protein function and analysis, Stratagene offers analysis products such as the QuikChange Lightning Sit-Directed Mutagenesis Kit. StrataClone, the firm's new cloning kit, is a less expensive alternative to Topoisomerase-based cloning. The Brilliant II FAST QPCR Master Mix is one of the company's quantitative PCR amplification products. Microarray products include Fairplay III Microarray Labeling Kit, which converts total RNA to fluorescence-labeled cDNA for hybridization to microarrays. In Spring 2008, Third Wave Technologies Inc., a developer of genetic testing products for diagnostic applications, acquired Agilent's Full Velocity portfolio of patents (developed by Stratagene), which cover nucleic acid amplification technology, for an undisclosed amount.

StrataGene offers employees an on-site workout facility and credit union membership.

## FINANCIALS: Sales and profits are in thousands of dollars—add 000 to get the full amount. 2008 Note: Financial information for 2008 was not available for all companies at press time.

| | | |
|---|---|---|
| 2008 Sales: $ | 2008 Profits: $ | U.S. Stock Ticker: Subsidiary |
| 2007 Sales: $ | 2007 Profits: $ | Int'l Ticker: Int'l Exchange: |
| 2006 Sales: $95,557 | 2006 Profits: $ 59 | Employees: 453 |
| 2005 Sales: $130,285 | 2005 Profits: $7,788 | Fiscal Year Ends: 12/31 |
| 2004 Sales: $84,813 | 2004 Profits: $7,438 | Parent Company: AGILENT TECHNOLOGIES INC |

## SALARIES/BENEFITS:

| Pension Plan: | ESOP Stock Plan: | Profit Sharing: | Top Exec. Salary: $467,608 | Bonus: $139,933 |
|---|---|---|---|---|
| Savings Plan: Y | Stock Purch. Plan: Y | | Second Exec. Salary: $328,884 | Bonus: $98,514 |

## OTHER THOUGHTS:

Apparent Women Officers or Directors:
Hot Spot for Advancement for Women/Minorities:

## LOCATIONS: ("Y" = Yes)

| West: | Southwest: | Midwest: | Southeast: | Northeast: | International: |
|---|---|---|---|---|---|
| Y | Y | | | | Y |

Note: Financial information, benefits and other data can change quickly and may vary from those stated here.

# SUPERGEN INC

**www.supergen.com**

**Industry Group Code:** 325412  **Ranks within this company's industry group:** Sales: 93    Profits: 77

| Drugs: | | Other: | Clinical: | Computers: | Services: |
|---|---|---|---|---|---|
| Discovery: | Y | AgriBio: | Trials/Services: | Hardware: | Specialty Services: |
| Licensing: | Y | Genetic Data: | Labs: | Software: | Consulting: |
| Manufacturing: | | Tissue Replacement: | Equipment/Supplies: | Arrays: | Blood Collection: |
| Genetics: | | | Research & Development Services: | Database Management: | Drug Delivery: |
| | | | Diagnostics: | | Drug Distribution: |

## TYPES OF BUSINESS:

Pharmaceuticals Acquisition & Development
Oncology Drugs

## BRANDS/DIVISIONS/AFFILIATES:

MP529
MP470
SGI-1776
S-110

## CONTACTS: Note: Officers with more than one job title may be intentionally listed here more than once.

James S. Manuso, CEO
James S. Manuso, Pres.
Michael Molkentin, CFO/Corp. Sec.
David Bearss, Chief Scientific Officer
Sanjeev Redkar, VP-Mfg. & Pre-Clinical Dev.
Shu Lee, VP-Legal Affairs & Intellectual Property
Timothy L. Enns, Sr. VP-Bus. Dev.
Timothy L. Enns, Sr. VP-Corp. Comm.
Gavin Choy, VP-Clinical Oper.
David S. Smith, VP-Regulatory & Quality Affairs
Michael McCullar, VP-Drug Discovery Oper.
James S. Manuso, Chmn.

| | |
|---|---|
| **Phone:** 925-560-0100 | **Fax:** 925-560-0101 |
| **Toll-Free:** | |
| **Address:** 4140 Dublin Blvd., Ste. 200, Dublin, CA 94568 US | |

## GROWTH PLANS/SPECIAL FEATURES:

SuperGen Inc. develops and commercializes pharmaceutical products intended to treat patients with cancer. The firm's development platform is based on its proprietary CLIMB process that merges rapid screening of compound libraries with computational chemistry and systems biology techniques to identify small-molecule drug candidates. SuperGen currently possesses a portfolio of five oncological drugs candidates that treat a variety of solid tumors and hematological malignancies. The company's leading product, MP-470, is a tyrosine kinase inhibitor/rad 51 suppressor, which is in Phase I clinical trials. MP-470 is being tested on patients with refractory solid tumors and lymphomas. SuperGen's other products, which are in preclinical stages, include MP529, an aurora A kinase inhibitor; SGI-1776, a pim kinase inhibitor; S-110, a compound which targets and blocks the mechanism by which methylation occurs, thus allowing tumor suppressor genes to function. The JAK2 kinase inhibitor is also in the company's pipeline that has not yet reached preclinical stages. In addition to drug candidates, SuperGen continues to generate royalties from the 2004 sale of its drug Dacogen to MGI Pharma Inc. MGI has worldwide rights to develop, manufacture and distribute Dacogen and pays the firm at specific regulatory and commercialization milestones. Dacogen is a therapeutic product that decreases the degree of methylation, the replacement of hydrogen with methyl, at certain DNA sites to assist patients with myelodysplastic syndrome. In August 2008, the FDA granted orphan drug designation to MP-470 for the treatment of glioblastoma multiforme, a rare form of brain cancer.

Employees are offered medical, dental and vision insurance; life insurance; short- and long-term disability coverage; flexible spending accounts; an employee assistance plan; a 401(k) plan; and an employee stock purchase plan.

## FINANCIALS: Sales and profits are in thousands of dollars—add 000 to get the full amount. 2008 Note: Financial information for 2008 was not available for all companies at press time.

| | | |
|---|---|---|
| 2008 Sales: $38,422 | 2008 Profits: $-9,111 | **U.S. Stock Ticker:** SUPG |
| 2007 Sales: $22,954 | 2007 Profits: $13,081 | **Int'l Ticker:**    Int'l Exchange: |
| 2006 Sales: $38,083 | 2006 Profits: $-16,487 | Employees:    86 |
| 2005 Sales: $30,169 | 2005 Profits: $-14,482 | Fiscal Year Ends: 12/31 |
| 2004 Sales: $31,993 | 2004 Profits: $-46,860 | Parent Company: |

## SALARIES/BENEFITS:

| | | | | |
|---|---|---|---|---|
| Pension Plan: | ESOP Stock Plan: | Profit Sharing: | Top Exec. Salary: $560,406 | Bonus: $500,000 |
| Savings Plan: Y | Stock Purch. Plan: Y | | Second Exec. Salary: $370,125 | Bonus: $91,000 |

## OTHER THOUGHTS:

Apparent Women Officers or Directors:
Hot Spot for Advancement for Women/Minorities:

## LOCATIONS: ("Y" = Yes)

| West: | Southwest: | Midwest: | Southeast: | Northeast: | International: |
|---|---|---|---|---|---|
| Y | | | | | |

# SYNBIOTICS CORP

**www.synbiotics.com**

Industry Group Code: 325412B  Ranks within this company's industry group: Sales:    Profits:

| Drugs: | Other: | Clinical: | Computers: | Services: |
|---|---|---|---|---|
| Discovery: | AgriBio: | Trials/Services: | Hardware: | Specialty Services: Y |
| Licensing: | Genetic Data: | Labs: | Software: Y | Consulting: |
| Manufacturing: Y | Tissue Replacement: | Equipment/Supplies: Y | Arrays: | Blood Collection: |
| Genetics: | | Research & Development Services: | Database Management: | Drug Delivery: |
| | | Diagnostics: Y | | Drug Distribution: |

## TYPES OF BUSINESS:

Veterinary Products Manufacturing
Animal Health Diagnostics
Animal Pregnancy Tests
Breeding Services
Flock Management Software

## BRANDS/DIVISIONS/AFFILIATES:

Synbiotics Europe
Fungassay
Ovassay
ViraCHEK
WITNESS
Profile
Tuberculin PPD
Flu Detect

## CONTACTS: *Note: Officers with more than one job title may be intentionally listed here more than once.*

Paul R. Hays, CEO
Paul R. Hays, Pres.
Dan O'Rourke, CFO/VP
Kevin Hayes, VP-Mktg. & Sales
Chinta Lamichhane, VP-R&D
Clifford J. Frank, VP-Strategic Projects

| Phone: 816-464-3500 | Fax: 816-464-3521 |
|---|---|
| Toll-Free: 800-228-4305 | |
| Address: 12200 N.W. Ambassador Dr., Ste. 101, Kansas City, MO 64163 US | |

## GROWTH PLANS/SPECIAL FEATURES:

Synbiotics Corp. develops, manufactures and markets animal diagnostic products to veterinarians and breeders. The firm produces products for poultry, cows, swine, dogs, horses, cats and non-human primates, and is able to treat animal coagulation, dermatophytes, gastrointestinal parasites, tuberculosis, heartworm, parvovirus and arthritis. Synbiotics' brand names include D-Tec to diagnose Brucella canis infections in dogs; ReproCHEK, a canine pregnancy test; TiterCHEK, a canine distemper and parvovirus antibody test; Fungassay for dermatophyte infections; Ovassay for identifying gastrointestinal parasites; CRF, a test for canine rheumatoid factor; ViraCHEK, a coronavirus identifier; and WITNESS, the company's heartworm identification kit. Synbiotics also markets domestic animal pregnancy tests and the Profile flock management software. The firm distributes its products through a number of companies in the U.S, Canada, Latin America, Asia, New Zealand and Australia. The company offers canine reproduction services that include a referral network and a freezing center that allows breeders to conduct long-distance breeding. The firm's specialized use of frozen semen allows dog breeders to preserve valuable genetic lines and to conduct long-distance breeding without shipping the dogs themselves. Synbiotics manufactures most of its products at its facilities located in Kansas City, Missouri and Lyon, France.

## FINANCIALS: Sales and profits are in thousands of dollars—add 000 to get the full amount. 2008 Note: Financial information for 2008 was not available for all companies at press time.

| | | |
|---|---|---|
| 2008 Sales: $ | 2008 Profits: $ | U.S. Stock Ticker: SYNB.PK |
| 2007 Sales: $ | 2007 Profits: $ | Int'l Ticker:    Int'l Exchange: |
| 2006 Sales: $ | 2006 Profits: $ | Employees:    93 |
| 2005 Sales: $ | 2005 Profits: $ | Fiscal Year Ends: 12/31 |
| 2004 Sales: $19,219 | 2004 Profits: $- 647 | Parent Company: |

## SALARIES/BENEFITS:

| Pension Plan: | ESOP Stock Plan: | Profit Sharing: | Top Exec. Salary: $250,000 | Bonus: $20,000 |
|---|---|---|---|---|
| Savings Plan: | Stock Purch. Plan: | | Second Exec. Salary: $188,700 | Bonus: $ |

## OTHER THOUGHTS:

**Apparent Women Officers or Directors:**
**Hot Spot for Advancement for Women/Minorities:**

## LOCATIONS: ("Y" = Yes)

| West: | Southwest: | Midwest: | Southeast: | Northeast: | International: |
|---|---|---|---|---|---|
| Y | | | | Y | Y |

# SYNGENTA AG
### www.syngenta.com

**Industry Group Code:** 11511 **Ranks within this company's industry group:** Sales: 1 Profits: 2

| Drugs: | Other: | | Clinical: | Computers: | Services: |
|---|---|---|---|---|---|
| Discovery: | AgriBio: | Y | Trials/Services: | Hardware: | Specialty Services: |
| Licensing: | Genetic Data: | | Labs: | Software: | Consulting: |
| Manufacturing: | Tissue Replacement: | | Equipment/Supplies: | Arrays: | Blood Collection: |
| Genetics: | | | Research & Development Services: | Database Management: | Drug Delivery: |
| | | | Diagnostics: | | Drug Distribution: |

## TYPES OF BUSINESS:

Agricultural Biotechnology Products & Chemicals Manufacturing
Crop Protection Products
Seeds

## BRANDS/DIVISIONS/AFFILIATES:

Dual Gold
Bicept Magnum
Acanto
Score
Syngenta Biotechnology, Inc.
Wageningen Agricultural University
Fischer Group
Zeraim Gedera Ltd.

## CONTACTS: *Note: Officers with more than one job title may be intentionally listed here more than once.*

Michael Mack, CEO
John Ramsay, CFO
David Lawrence, VP-R&D
Christoph Mader, Head-Legal & Taxes/Corp. Sec.
Mark Peacock, Head-Global Oper.
Robert Berendes, Head-Bus. Dev.
John Atkin, COO-Crop Protection
Davor Pisk, COO-Syngenta Seeds
Martin Taylor, Chmn.

| Phone: 41-61-697-1111 | Fax: 41-61-323-2324 |
|---|---|
| Toll-Free: | |
| Address: Schwarzwaldallee 215, Basel, 4058 Switzerland | |

## GROWTH PLANS/SPECIAL FEATURES:

Syngenta AG is one of the world's largest agrochemical companies and a leading worldwide supplier of conventional and bioengineered crop protection and seeds. Products designed for crop protection include herbicides, fungicides and insecticides. Additionally, the firm produces seeds for field crops, vegetables and flowers. Its leading marketed products include the following: Dual Gold and Bicept Magnum herbicides; Acanto, Score and Amistar fungicides; and Proclaim, an insecticide. The seeds that Syngenta markets are for field crops, such as corn, soybeans, sugar beets, sunflowers and oilseed rape (canola); fruits and vegetables, including tomatoes, lettuce, melons, squash, cabbages, peppers, beans and radishes; and a wide variety of garden plants such as begonias, lavender, sage and many other seasonal flowers and herbs, some of which can also be purchased as plugs or full-grown plants. The company spends a significant amount of resources on research and development, with major laboratories located in Basel and Stein, Switzerland; Jealott's Hill, in the U.K.; and Syngenta Biotechnology, Inc., in North Carolina. Spending on research and development efforts regularly exceeds $800 million annually. Syngenta's plant science division is engaged in collaborations with several companies and universities, including Wageningen Agricultural University in the Netherlands. The research and development activities at Syngenta are currently devoted to advances in genomics, crop protection, health assessment, environmental issues, genetic mapping and traits development. The company's gene technology has become so refined that single genes can be isolated from a type of plant material and transferred to the DNA of another. This process allows manipulation of such traits as nutrient composition, appearance, and even the specifics of the taste of a certain crop. In March 2007, the firm acquired the Fischer Group, a vegetative flower company, for roughly $67 million. In July 2007, Syngenta acquired Zeraim Gedera Ltd., an Israeli seeds company, for $95 million.

## FINANCIALS: Sales and profits are in thousands of dollars—add 000 to get the full amount. 2008 Note: Financial information for 2008 was not available for all companies at press time.

| | | |
|---|---|---|
| 2008 Sales: $11,624,000 | 2008 Profits: $1,385,000 | **U.S. Stock Ticker:** SYT |
| 2007 Sales: $9,240,000 | 2007 Profits: $1,109,000 | **Int'l Ticker:** SYNN    Int'l Exchange: Zurich-SWX |
| 2006 Sales: $8,046,000 | 2006 Profits: $634,000 | Employees: 24,000 |
| 2005 Sales: $8,104,000 | 2005 Profits: $622,000 | Fiscal Year Ends: 12/31 |
| 2004 Sales: $7,269,000 | 2004 Profits: $428,000 | Parent Company: |

## SALARIES/BENEFITS:

| Pension Plan: Y | ESOP Stock Plan: | Profit Sharing: | Top Exec. Salary: $ | Bonus: $ |
|---|---|---|---|---|
| Savings Plan: Y | Stock Purch. Plan: Y | | Second Exec. Salary: $ | Bonus: $ |

## OTHER THOUGHTS:

**Apparent Women Officers or Directors:** 1
**Hot Spot for Advancement for Women/Minorities:**

## LOCATIONS: ("Y" = Yes)

| West: | Southwest: | Midwest: | Southeast: | Northeast: | International: |
|---|---|---|---|---|---|
| Y | Y | Y | Y | Y | Y |

# SYNOVIS LIFE TECHNOLOGIES INC

www.synovislife.com

Industry Group Code: 33911 Ranks within this company's industry group: Sales: 12 Profits: 10

| Drugs: | Other: | Clinical: | | Computers: | | Services: |
|---|---|---|---|---|---|---|
| Discovery: | AgriBio: | Trials/Services: | | Hardware: | | Specialty Services: |
| Licensing: | Genetic Data: | Labs: | | Software: | | Consulting: |
| Manufacturing: | Tissue Replacement: | Equipment/Supplies: | Y | Arrays: | | Blood Collection: |
| Genetics: | | Research & Development Services: | Y | Database Management: | | Drug Delivery: |
| | | Diagnostics: | | | | Drug Distribution: |

## TYPES OF BUSINESS:

Surgical & Interventional Treatment Products
Implantable Biomaterials

## BRANDS/DIVISIONS/AFFILIATES:

Synovis Surgical Innovations
Synovis Micro Companies Alliance
Biover Microvascular Clamp
Microvascular Anastomotis COUPLER System
Peri-Strips
Veritas
Tissue-Guard
GEM MicroClip

## CONTACTS: Note: Officers with more than one job title may be intentionally listed here more than once.

Richard W. Kramp, CEO
Richard W. Kramp, Pres.
Brett A. Reynolds, CFO
Daniel L. Mooradian, VP-R&D
Brett A. Reynolds, Corp. Sec.
Tim Floeder, VP-Corp. Dev.
Brett A. Reynolds, VP-Finance
Michael K. Campbell, Pres., Micro Companies Alliance, Inc.
Mary L. Frick, VP-Regulatory Affairs & Quality Assurance
Mary L. Frick, VP-Clinical Affairs
David Buche, VP/COO-Synovis Surgical Innovations
Timothy M. Scanlan, Chmn.

| Phone: 651-796-7300 | Fax: 651-642-9018 |
|---|---|
| Toll-Free: 800-255-4018 | |
| Address: 2575 University Ave. W., St. Paul, MN 55114 US | |

## GROWTH PLANS/SPECIAL FEATURES:

Synovis Life Technologies, Inc. is a diversified medical device company engaged in developing, manufacturing and bringing to market products for the surgical treatment of disease. Its products are marketed for use in bariatric, cardiac, neurologic, thoracic, urogynecologic, vascular and general surgeries. The company operates in two segments: Synovis Surgical Innovations and Synovis Micro Companies Alliance. Synovis Surgical Innovations is focused on developing implantable biomaterial products designed to improve a patient's outcome in critical surgeries. The products are made from bovine pericardium, which easily assimilates into host tissue. Products offered by the firm include Peri-Strips, which are stapling buttresses used to reinforce surgical staple lines and reduce potential leakage; Tissue-Guard, which is used to repair and replace damaged tissue in cardiac, vascular, thoracic, abdominal and neuron surgeries; and Veritas, for use in pelvic floor reconstruction, stress urinary incontinence treatment, vaginal and rectal prolapse repair, hernia repair as well as soft tissue repair. Synovis Micro Companies Alliance, acquired in 2001, provides products to the niche microsurgery market. Products offered include the Microvascular Anastomotic COUPLER System, which is used in small vessel anastomases; GEM Neurotube, which is used in primary or secondary nerve repair; GEM MicroClip, a hemostatic clip; GEM03 Focus Headlight, a rechargeable medical lamp; and the Biover Microvascular Clamp, a single use clamp for arteries and veins. The company previously operated an interventional business, but this was sold to Heraeus Vadnais Inc. in 2008.

## FINANCIALS: Sales and profits are in thousands of dollars—add 000 to get the full amount. 2008 Note: Financial information for 2008 was not available for all companies at press time.

| | | |
|---|---|---|
| 2008 Sales: $49,800 | 2008 Profits: $11,485 | U.S. Stock Ticker: SYNO |
| 2007 Sales: $37,691 | 2007 Profits: $3,810 | Int'l Ticker: Int'l Exchange: |
| 2006 Sales: $27,743 | 2006 Profits: $-1,481 | Employees: 235 |
| 2005 Sales: $60,256 | 2005 Profits: $ 883 | Fiscal Year Ends: 10/31 |
| 2004 Sales: $55,044 | 2004 Profits: $1,278 | Parent Company: |

## SALARIES/BENEFITS:

| Pension Plan: | ESOP Stock Plan: | Profit Sharing: | Top Exec. Salary: $348,000 | Bonus: $152,581 |
|---|---|---|---|---|
| Savings Plan: Y | Stock Purch. Plan: | | Second Exec. Salary: $209,066 | Bonus: $72,650 |

## OTHER THOUGHTS:

Apparent Women Officers or Directors: 2
Hot Spot for Advancement for Women/Minorities: Y

## LOCATIONS: ("Y" = Yes)

| West: | Southwest: | Midwest: | Southeast: | Northeast: | International: |
|---|---|---|---|---|---|
| | | Y | Y | | |

# SYNVISTA THERAPEUTICS INC

**www.synvista.com**

Industry Group Code: 325412  Ranks within this company's industry group:  Sales:      Profits:

| Drugs: | | Other: | Clinical: | Computers: | Services: |
|---|---|---|---|---|---|
| Discovery: | Y | AgriBio: | Trials/Services: | Hardware: | Specialty Services: |
| Licensing: | Y | Genetic Data: | Labs: | Software: | Consulting: |
| Manufacturing: | | Tissue Replacement: | Equipment/Supplies: | Arrays: | Blood Collection: |
| Genetics: | | | Research & Development Services: | Database Management: | Drug Delivery: |
| | | | Diagnostics: | | Drug Distribution: |

## TYPES OF BUSINESS:

Pharmaceutical Discovery & Development
Cardiovascular Drugs
Diabetes Drugs

## BRANDS/DIVISIONS/AFFILIATES:

Haptoguard, Inc.

## CONTACTS: Note: Officers with more than one job title may be intentionally listed here more than once.

Noah Berkowitz, CEO
Noah Berkowitz, Pres.
David Tantillo, Senior Dir.-Mktg. & Sales
Jacob Victor, Exec. Dir.-Prod. Dev.
Wendy Milici, Dir.-Admin.
Wendy Milici, Dir.-Finance
Jacob Victor, Exec. Dir.-Clinical Diagnostics

| | |
|---|---|
| **Phone:** 201-934-5000 | **Fax:** 201-934-8880 |
| **Toll-Free:** | |
| **Address:** 221 W. Grand Ave., Montvale, NJ 07645 US | |

## GROWTH PLANS/SPECIAL FEATURES:

Synvista Therapeutics is a product-based biopharmaceutical company primarily engaged in the discovery and development of oral drugs to reverse or inhibit cardiovascular aging and diabetic complications. These drug candidates were developed as a result of research on the advanced glycosylation end-product (AGE) pathway, a fundamental process and consequence of aging and diabetes that causes or contributes to many medical disorders. Synvista's current research and drug development activities targeting this pathway take three directions: breaking AGE crosslinks between proteins; preventing or inhibiting AGE formation; and reducing the AGE burden through a novel class of anti-hyperglycemic agents, or glucose-lowering agents. Since its inception, Synvista has created an extensive library of compounds targeting the AGE pathway. The company's lead product candidate is alagebrium chloride, formerly ALT-711, an AGE crosslink breaker that offers the first therapeutic approach to breaking these crosslinks, which may reverse tissue damage caused by aging and diabetes, thereby restoring flexibility and function to blood vessels and organs of the body. Results suggest that alagebrium is a novel potential therapy for many conditions that are associated with diabetes and aging. Synvista is also working on its compound ALT-2074, which uses a certain form of haptoglobin which is present in 40% of the human population. With this technology, it will be able to offer directed therapy based on variability in the form of haptoglobin in the blood. The compound is currently in testing. Synvista maintains a number of licensing and collaboration agreements relating to the development and distribution of its AGE technology. HaptoGuard, a wholly-owned subsidiary, has a product, ALT-2074, being tested in diabetic patients recovering from a recent myocardial infarction or acute coronary syndrome.

## FINANCIALS: Sales and profits are in thousands of dollars—add 000 to get the full amount. 2008 Note: Financial information for 2008 was not available for all companies at press time.

| | | |
|---|---|---|
| 2008 Sales: $ | 2008 Profits: $ | **U.S. Stock Ticker: SYI** |
| 2007 Sales: $ 51 | 2007 Profits: $-16,093 | **Int'l Ticker:** Int'l Exchange: |
| 2006 Sales: $ 62 | 2006 Profits: $-17,680 | Employees: 7 |
| 2005 Sales: $ 458 | 2005 Profits: $-12,614 | Fiscal Year Ends: 12/31 |
| 2004 Sales: $ 334 | 2004 Profits: $-13,959 | Parent Company: |

## SALARIES/BENEFITS:

| | | | | |
|---|---|---|---|---|
| Pension Plan: | ESOP Stock Plan: | Profit Sharing: | Top Exec. Salary: $264,000 | Bonus: $83,160 |
| Savings Plan: Y | Stock Purch. Plan: | | Second Exec. Salary: $240,000 | Bonus: $72,000 |

## OTHER THOUGHTS:

**Apparent Women Officers or Directors:** 1
**Hot Spot for Advancement for Women/Minorities:**

## LOCATIONS: ("Y" = Yes)

| West: | Southwest: | Midwest: | Southeast: | Northeast: | International: |
|---|---|---|---|---|---|
| | | | | Y | |

# TAKEDA PHARMACEUTICAL COMPANY LTD    www.takeda.com

**Industry Group Code: 325412  Ranks within this company's industry group: Sales: 16  Profits: 13**

| Drugs: | | Other: | | Clinical: | Computers: | Services: | |
|---|---|---|---|---|---|---|---|
| Discovery: | Y | AgriBio: | | Trials/Services: | Hardware: | Specialty Services: | |
| Licensing: | Y | Genetic Data: | | Labs: | Software: | Consulting: | |
| Manufacturing: | Y | Tissue Replacement: | | Equipment/Supplies: | Arrays: | Blood Collection: | |
| Genetics: | | | | Research & Development Services: | Database Management: | Drug Delivery: | |
| | | | | Diagnostics: | | Drug Distribution: | |

## TYPES OF BUSINESS:

Pharmaceuticals Discovery & Development
Over-the-Counter Drugs
Vitamins
Chemicals
Agricultural & Food Products

## BRANDS/DIVISIONS/AFFILIATES:

Takeda America Holdings, Inc.
Takeda Research Investment, Inc.
TAP Pharmaceutical Products Inc.
Takeda Europe Research and Development Center Ltd.
Takeda Ireland Limited
Boie-Takeda Chemicals, Inc.
Paradigm Therapeutics Ltd.
Millennium Pharmaceuticals, Inc.

## CONTACTS: Note: Officers with more than one job title may be intentionally listed here more than once.

Yasuchika Hasegawa, Pres.
Tsudoi Miyoshi, Gen. Mgr.-Human Resources Dept.
Shigenori Ohkawa, Gen. Mgr.-Pharmaceutical Research Div.
Hiroshi Takahara, Gen. Mgr.-Finance & Acct.
Kanji Negi, Gen. Mgr.-Admin. Mgmt., Pharmaceutical
Hiroshi Sakiyama, Gen. Mgr.-Tokyo Branch
Teruo Sakurada, Gen. Mgr.-Osaka Branch
Hiroshi Ohtsuki, Pres., Consumer Healthcare Company
Kunio Takeda, Chmn.
Naohisa Takeda, Gen. Mgr.-Overseas Bus. Planning Dept.

| **Phone:** 81-6-6204-2111 | **Fax:** 81-6-6204-2880 |
|---|---|
| **Toll-Free:** | |
| **Address:** 1-1, Doshomachi 4-chome, Chuo-ku,, Osaka, 540-8645 Japan | |

## GROWTH PLANS/SPECIAL FEATURES:

Takeda Pharmaceutical Company Ltd., based in Japan, is an international research-based company focused on pharmaceuticals. One of the largest pharmaceutical companies in Japan, it operates research and development facilities in four countries and production facilities in five countries. Takeda discovers, develops, manufactures and markets pharmaceutical products in two categories: ethical and consumer health care drugs. Ethical drugs, as the firm denominates them, are marketed in about 90 countries worldwide. This segment includes the anti-prostatic cancer agent leuprolide acetate, marketed as Lupron Depot, Enantone, Prostap and Leuplin; the anti-peptic ulcer agent lansoprazole, marketed as Prevacid, Ogast, Takepron and other names; the anti-hypertensive agent candesartan cilexetil, marketed as Blopress, Kenzen and Amias; and the anti-diabetic agent pioglitazone hydrochloride, marketed as Actos. The company's consumer health care division focuses on the over-the-counter drug market. Takeda's main consumer brands include Alinamin, a vitamin B1 derivative; Benza, a cold remedy; and Hicee, a vitamin C preparation. Outside Japan, Takeda's subsidiaries include development, production, and sales and marketing companies, as well as holding companies in the U.S. and Europe. The company also operates subsidiaries with the agro, food, urethane chemicals and bulk vitamin businesses. Within research and development, Takeda focuses on the life-style related diseases, oncology, urologic diseases, central nervous system diseases and gastroenterology life cycle management. The company pursues alliances with other pharmaceutical manufacturers, biotechnology companies, universities and other research institutions to efficiently introduce key technologies. In May 2008, the company acquired Millennium Pharmaceuticals, Inc.

## FINANCIALS: Sales and profits are in thousands of dollars—add 000 to get the full amount. 2008 Note: Financial information for 2008 was not available for all companies at press time.

| | | |
|---|---|---|
| 2008 Sales: $13,748,020 | 2008 Profits: $3,554,540 | **U.S. Stock Ticker: TKPHF** |
| 2007 Sales: $11,060,737 | 2007 Profits: $2,845,805 | **Int'l Ticker: 4502** Int'l Exchange: Tokyo-TSE |
| 2006 Sales: $10,360,744 | 2006 Profits: $2,677,342 | Employees: 15,717 |
| 2005 Sales: $10,441,300 | 2005 Profits: $2,579,600 | Fiscal Year Ends: 3/31 |
| 2004 Sales: $10,284,200 | 2004 Profits: $2,700,300 | Parent Company: |

## SALARIES/BENEFITS:

| Pension Plan: | ESOP Stock Plan: | Profit Sharing: | Top Exec. Salary: $ | Bonus: $ |
|---|---|---|---|---|
| Savings Plan: | Stock Purch. Plan: | | Second Exec. Salary: $ | Bonus: $ |

## OTHER THOUGHTS:

**Apparent Women Officers or Directors:**
**Hot Spot for Advancement for Women/Minorities:**

## LOCATIONS: ("Y" = Yes)

| West: | Southwest: | Midwest: | Southeast: | Northeast: | International: |
|---|---|---|---|---|---|
| Y | | Y | | Y | Y |

Note: Financial information, benefits and other data can change quickly and may vary from those stated here.

# TARGETED GENETICS CORP

**www.targen.com**

**Industry Group Code:** 325412 **Ranks within this company's industry group:** Sales: 117   Profits: 104

| Drugs: | | Other: | | Clinical: | Computers: | | Services: | |
|--------|---|--------|---|-----------|------------|---|-----------|---|
| Discovery: | Y | AgriBio: | | Trials/Services: | Hardware: | | Specialty Services: | |
| Licensing: | | Genetic Data: | Y | Labs: | Software: | | Consulting: | |
| Manufacturing: | | Tissue Replacement: | | Equipment/Supplies: | Arrays: | | Blood Collection: | |
| Genetics: | | | | Research & Development Services: | Database Management: | | Drug Delivery: | Y |
| | | | | Diagnostics: | | | Drug Distribution: | |

## TYPES OF BUSINESS:

Gene Therapeutics

## BRANDS/DIVISIONS/AFFILIATES:

Adeno-Associated Viral (AAV) vectors
tgAAC94
tgAAC09

## CONTACTS: *Note: Officers with more than one job title may be intentionally listed here more than once.*

B. G. Susan Robinson, CEO
B. G. Susan Robinson, Pres.
David J. Poston, CFO
Pervin Anklesaria, VP-Prod. Dev.
Richard W. Peluso, VP-Mfg. & Process Sciences
David J. Poston, Sec.
David J. Poston, VP-Finance/Treas.
Jeremy Curnock Cook, Chmn.

| Phone: 206-623-7612 | Fax: 206-223-0288 |
|---|---|
| Toll-Free: | |
| Address: 1100 Olive Way, Ste. 100, Seattle, WA 98101 US | |

## GROWTH PLANS/SPECIAL FEATURES:

Targeted Genetics Corp. is a clinical-stage biotechnology company that develops gene therapeutics. The company's gene therapeutics consist of a delivery vehicle, called a vector, and genetic material. The role of the vector is to carry the genetic material into a target cell. Once delivered into the cell, the gene can express or direct production of the specific proteins encoded by the gene. The firm develops and manufactures adeno-associated viral (AAV) vectors, or AAV-based gene therapeutics. The company has fours products in development, including treatments for inflammatory arthritis, HIV/AIDS, congestive heart failure and Huntington's Disease. TGAAC94, the inflammatory arthritis candidate, works to administer an inhibitor of TNF-alpha (Tumor Necrosis Factor) in order to modulate inflammation. TGAAC09, a potential HIV/AIDS vaccine based on HIV subtype C, is designed to use the body's immune system to counteract the disease. The program is being developed in conjunction with the International AIDS Vaccine Initiative, Columbus Children's Research Institute and the Children's Hospital of Philadelphia. The congestive heart failure program, being developed with Celladon Corporation, uses the AAV delivery method with SERCA2a and phospholamban gene variants in order to modify the heart's ability to contract. Targeted Genetics' Huntington's Disease program was acquired from Sirna Therapeutics in 2008 and aims to use the AAV vector with siRNA (short interfering RNA) to destroy specific RNA targets. In December 2008, the company cut about 10% of its workforce and scaled back executive pay. In May 2009, the firm announced it may shut down if it is unable to raise enough capital.

Targeted Genetics offers its employees medical, dental, vision, life and AD&D insurance; short- and long-term disability; an employee assistance program; flexible spending accounts; a 401(k) plan; and stock options.

## FINANCIALS: Sales and profits are in thousands of dollars—add 000 to get the full amount. 2008 Note: Financial information for 2008 was not available for all companies at press time.

| | | | |
|---|---|---|---|
| 2008 Sales: $8,718 | 2008 Profits: $-20,720 | **U.S. Stock Ticker:** TGEN | |
| 2007 Sales: $10,332 | 2007 Profits: $-16,127 | **Int'l Ticker:**   Int'l Exchange: | |
| 2006 Sales: $9,864 | 2006 Profits: $-33,990 | Employees:   58 | |
| 2005 Sales: $6,874 | 2005 Profits: $-19,198 | Fiscal Year Ends: 12/31 | |
| 2004 Sales: $9,652 | 2004 Profits: $-14,257 | Parent Company: | |

## SALARIES/BENEFITS:

| Pension Plan: | ESOP Stock Plan: | Profit Sharing: | Top Exec. Salary: $374,000 | Bonus: $ |
|---|---|---|---|---|
| Savings Plan: Y | Stock Purch. Plan: | | Second Exec. Salary: $257,500 | Bonus: $ |

## OTHER THOUGHTS:

**Apparent Women Officers or Directors:** 1
**Hot Spot for Advancement for Women/Minorities:**

## LOCATIONS: ("Y" = Yes)

| West: | Southwest: | Midwest: | Southeast: | Northeast: | International: |
|---|---|---|---|---|---|
| Y | | | | | |

# TARO PHARMACEUTICAL INDUSTRIES

www.taro.com

**Industry Group Code:** 325412A **Ranks within this company's industry group:** Sales: Profits:

| Drugs: | | Other: | | Clinical: | | Computers: | | Services: | |
|---|---|---|---|---|---|---|---|---|---|
| Discovery: | Y | AgriBio: | | Trials/Services: | | Hardware: | | Specialty Services: | |
| Licensing: | Y | Genetic Data: | | Labs: | | Software: | | Consulting: | |
| Manufacturing: | Y | Tissue Replacement: | | Equipment/Supplies: | | Arrays: | | Blood Collection: | |
| Genetics: | Y | | | Research & Development Services: | | Database Management: | | Drug Delivery: | |
| | | | | Diagnostics: | | | | Drug Distribution: | |

## TYPES OF BUSINESS:

Drugs-Generic & Proprietary
Over-the-Counter Analgesics
Vitamins
Anti-Cancer Drugs
Dermatological Drugs

## BRANDS/DIVISIONS/AFFILIATES:

Taro Pharmaceuticals U.S.A., Inc.
Kusch Manual
RxDesktop
Lustra
Ovide
Clotrimazole/Betamethasomne
Phenytoin
Cleanse & Treat

## CONTACTS: Note: Officers with more than one job title may be intentionally listed here more than once.

Ron Kolker, CFO/VP
Zahava Rafalowicz, VP-Mktg. & Sales
Inbal Rothman, VP-Human Resources & Community Affairs
Avraham Yacobi, Sr. VP-R&D
Tzvi Tal, VP-IT
Roman Kaplan, VP-Tech. Oper. & Pharmaceuticals
Noam Shamir, VP-Industrial Eng.
Tal Levitt, Corp. Sec.
Tal Levitt, Sr. VP-Corp. Affairs/Treas.
Hannah Bayer, VP-Finance/Chief Acct. Officer
Hagai Reingold, VP-API Div.
Mariana Bacalu, VP-Quality Affairs
Ram Zajicek, VP/Mgr.-Haifa Site
Yohanan Dichter, Sr. Mgr.-Quality
Barrie Levitt, Chmn.
Noam Shamir, VP-Supply Chain

| Phone: 972-9-971-1821 | Fax: 972-9-955-7443 |
|---|---|
| Toll-Free: | |
| Address: Italy House, Euro Park, Yakum, 60972 Israel | |

## GROWTH PLANS/SPECIAL FEATURES:

Taro Pharmaceutical Industries is a multinational pharmaceutical company that discovers, develops, manufactures and markets a wide range of both proprietary and generic health care products. These range from over-the-counter and prescription analgesics to vitamins. Its products address illnesses varying from the common cold to cancer, and are mainly used in dermatology, cardiology, neurology and pediatrics. The firm operates through three entities: Taro Pharmaceutical Industries (Israel) and its subsidiaries Taro Pharmaceuticals U.S.A., Inc. and Taro Pharmaceuticals, Inc. (Canada). Taro produces more than 200 pharmaceutical products, including topical preparations such as creams, ointments, gels and solutions; oral medications such as tablets, capsules, powders and liquids; and sterile products such as injectables, ophthalmic drops and powders. Top products manufactured by Taro include Lustra, a treatment for dyschromia (discolored skin); Ovide, a lotion for the treatment of head lice; Warfarin, sodium tablets for certain heart conditions; Clotrimazole and Betamethasomne Dipropionate Creams (generic equivalents of Lotrisone), for the treatment of the topical effects of fungus; Phenytoin Oral Suspension, for the treatment of epilepsy; and Terconazole Vaginal Cream, for treating yeast infections. Taro also funds the Kusch Manual, a dermatology diagnosis manual for physicians; and RxDesktop, a program for health care professionals that contains conversation tools, dosage calculators and other information. The firm has six subsidiaries for distribution and manufacturing in the U.S., Canada, Israel and the U.K.

## FINANCIALS: Sales and profits are in thousands of dollars—add 000 to get the full amount. 2008 Note: Financial information for 2008 was not available for all companies at press time.

| | | | |
|---|---|---|---|
| 2008 Sales: $ | 2008 Profits: $ | **U.S. Stock Ticker:** TAROF.PK | |
| 2007 Sales: $ | 2007 Profits: $ | **Int'l Ticker:** Int'l Exchange: | |
| 2006 Sales: $ | 2006 Profits: $ | Employees: 1,400 | |
| 2005 Sales: $ | 2005 Profits: $ | Fiscal Year Ends: 12/31 | |
| 2004 Sales: $284,100 | 2004 Profits: $11,100 | Parent Company: | |

## SALARIES/BENEFITS:

| Pension Plan: Y | ESOP Stock Plan: | Profit Sharing: | Top Exec. Salary: $ | Bonus: $ |
|---|---|---|---|---|
| Savings Plan: | Stock Purch. Plan: Y | | Second Exec. Salary: $ | Bonus: $ |

## OTHER THOUGHTS:

**Apparent Women Officers or Directors**: 3
**Hot Spot for Advancement for Women/Minorities**: Y

## LOCATIONS: ("Y" = Yes)

| West: | Southwest: | Midwest: | Southeast: | Northeast: | International: |
|---|---|---|---|---|---|
| | | | | Y | Y |

Note: Financial information, benefits and other data can change quickly and may vary from those stated here.

# TECHNE CORP

**www.techne-corp.com**

**Industry Group Code:** 325413  **Ranks within this company's industry group:** Sales: 6   Profits: 2

| Drugs: | Other: | Clinical: | | Computers: | Services: |
|---|---|---|---|---|---|
| Discovery: | AgriBio: | Trials/Services: | | Hardware: | Specialty Services: |
| Licensing: | Genetic Data: | Labs: | | Software: | Consulting: |
| Manufacturing: | Tissue Replacement: | Equipment/Supplies: | Y | Arrays: | Blood Collection: |
| Genetics: | | Research & Development Services: | Y | Database Management: | Drug Delivery: |
| | | Diagnostics: | Y | | Drug Distribution: |

## TYPES OF BUSINESS:

Biotechnology Products
Reagents, Antibodies & Assay Kits
Hematology Products

## BRANDS/DIVISIONS/AFFILIATES:

Research and Diagnostic Systems, Inc.
R&D Systems Europe, Ltd.
R&D Systems GmbH
R&D Systems China Co. Ltd.

## CONTACTS: Note: Officers with more than one job title may be intentionally listed here more than once.

Thomas E. Oland, CEO
Thomas E. Oland, Pres.
Gergory J. Melson, CFO
Gregory J. Melson, VP-Finance
Marcel Veronneau, VP-Hematology Oper.
Thomas E. Oland, Treas.
Richard A. Krzyzek, VP-Biotech Div.
Thomas E. Oland, Chmn.
Wendy Shao, Gen. Mgr.-R&D Systems China Co. Ltd.

| Phone: 612-379-8854 | Fax: 612-379-6580 |
|---|---|
| **Toll-Free:** | |
| **Address:** 614 McKinley Pl. NE, Minneapolis, MN 55413-2610 US | |

## GROWTH PLANS/SPECIAL FEATURES:

Techne Corp. is a holding company involved in biotechnology and hematology products. The firm operates through two subsidiaries: Research and Diagnostic Systems, Inc. (R&D Systems) and R&D Systems Europe, Ltd. (R&D Europe). R&D Systems manufactures biological products in two major segments: hematology controls, which are used in clinical and hospital laboratories to monitor the accuracy of blood analysis instruments; and biotechnology products, which including purified proteins and antibodies that are sold exclusively to the research market and assay kits that are sold to the research and clinical diagnostic markets. R&D Europe distributes biotechnology products throughout Europe and operates sales offices in France and Germany through its German subsidiary, R&D Systems GmbH. In recent years, R&D Systems has also expanded its product portfolio to include enzymes and intracellular cell signaling reagents such as kinases, proteases and phosphatases; these reagents detect diseases such as cancer, Alzheimer's, arthritic, autoimmunity, diabetes, hypertension, obesity, AIDS and SARS. Techne also produces controls and calibrators for a variety of medical brands such as Abbott Diagnostics, Beckman Coulter, Siemens Healthcare Diagnostics and Sysmex. In the hematology sector, the company's Whole Blood Flow Cytometry Controls are used to identify and quantify white blood cells by their surface antigens while linearity and reportable range controls assess the linearity of hematology analyzers for white blood cells, red blood cells, platelets and reticulocytes. R&D Systems China Co. Ltd. is a wholly-owned subsidiary established to improve the level of service offered to R&D Systems' distributors and customers in Shanghai, China. The company operates as a warehouse/distribution hub for R&D Systems' cell biology research products and offers technical services and marketing support for the Chinese market.

## FINANCIALS: Sales and profits are in thousands of dollars—add 000 to get the full amount. 2008 Note: Financial information for 2008 was not available for all companies at press time.

| | | |
|---|---|---|
| 2008 Sales: $257,420 | 2008 Profits: $103,558 | **U.S. Stock Ticker: TECH** |
| 2007 Sales: $223,482 | 2007 Profits: $85,111 | **Int'l Ticker:**   Int'l Exchange: |
| 2006 Sales: $202,617 | 2006 Profits: $73,351 | Employees:  719 |
| 2005 Sales: $178,700 | 2005 Profits: $66,100 | Fiscal Year Ends: 6/30 |
| 2004 Sales: $161,257 | 2004 Profits: $52,928 | Parent Company: |

## SALARIES/BENEFITS:

| Pension Plan: | ESOP Stock Plan: | Profit Sharing: Y | Top Exec. Salary: $260,000 | Bonus: $47,320 |
|---|---|---|---|---|
| Savings Plan: Y | Stock Purch. Plan: Y | | Second Exec. Salary: $254,100 | Bonus: $ |

## OTHER THOUGHTS:

**Apparent Women Officers or Directors:** 2
**Hot Spot for Advancement for Women/Minorities:** Y

## LOCATIONS: ("Y" = Yes)

| West: | Southwest: | Midwest: | Southeast: | Northeast: | International: |
|---|---|---|---|---|---|
| | | Y | | | Y |

# TELIK INC

**www.telik.com**

Industry Group Code: 325412  **Ranks within this company's industry group:** Sales:  Profits: 125

| Drugs: | | Other: | | Clinical: | Computers: | | Services: | |
|---|---|---|---|---|---|---|---|---|
| Discovery: | Y | AgriBio: | | Trials/Services: | Hardware: | | Specialty Services: | |
| Licensing: | Y | Genetic Data: | | Labs: | Software: | | Consulting: | |
| Manufacturing: | | Tissue Replacement: | | Equipment/Supplies: | Arrays: | | Blood Collection: | |
| Genetics: | | | | Research & Development Services: | Database Management: | | Drug Delivery: | |
| | | | | Diagnostics: | | | Drug Distribution: | |

## TYPES OF BUSINESS:

Drugs-Cancer
Small-Molecule Drugs

## BRANDS/DIVISIONS/AFFILIATES:

Target-Related Affinity Profiling (TRAP)
TELCYTA
TELINTRA
TLK58747

## CONTACTS: *Note: Officers with more than one job title may be intentionally listed here more than once.*

Michael M. Wick, CEO
Cynthia M. Butitta, COO
Michael M. Wick, Pres.
Cynthia M. Butitta, CFO
Gail L. Brown, Chief Medical Officer/Sr. VP
William P. Kaplan, General Counsel/Corp. Sec./VP
Marc L. Steuer, Sr. VP-Bus. Dev.
Michael M. Wick, Chmn.

| **Phone:** 650-845-7700 | **Fax:** 650-845-7800 |
|---|---|
| **Toll-Free:** | |
| **Address:** 3165 Porter Dr., Palo Alto, CA 94304 US | |

## GROWTH PLANS/SPECIAL FEATURES:

Telik, Inc. discovers, develops and commercializes small-molecule pharmaceuticals, mainly for the treatment of specific cancers. The company's proprietary Target-Related Affinity Profiling (TRAP) chemoinformatics technology is used by Telik and its partners to rapidly identify promising chemicals for development. TRAP can select a small sample from its large compound library and effectively identify small, biologically active molecules. The company currently has two product candidates in clinical trials. Telcyta is a novel tumor-activated compound that is currently in Phase III clinical trials for use against ovarian and non-small cell lung cancers and in Phase II of development for colorectal, breast, and certain forms of lung and ovarian cancers. The drug product candidate is designed to be activated in cancer cells through binding to the GST P1-1 protein, which is elevated in many human cancers, even more so in patients who have already been treated with other standard chemotherapy drugs. Once bound to the protein inside a cancer cell, a chemical reaction occurs that releases Telcyta and causes programmed cancer death. Telintra, a small-molecule bone marrow stimulant, is in Phase II trials for myelodysplastic syndrome (MDS), a form of pre-leukemia. The product is used in the treatment of blood disorders with low blood cell levels, such as neutropenia or anemia. The company has a variety of potential products in the preclinical pipeline, including TLK58747, which has shown significant anti-tumor activity in human breast, pancreatic and colon tumors while in preclinical study. In August 2008, Telik announced that it would license compounds discovered using its proprietary TRAP technology to ReceptorBio, Inc. for the development of a diabetes drug. In February 2009, the company announced plans to reduce its work force by 44% and focus on the development of Telintra.

## FINANCIALS: Sales and profits are in thousands of dollars—add 000 to get the full amount. 2008 Note: Financial information for 2008 was not available for all companies at press time.

| | | |
|---|---|---|
| 2008 Sales: $ | 2008 Profits: $-31,763 | **U.S. Stock Ticker:** TELK |
| 2007 Sales: $ | 2007 Profits: $-55,215 | **Int'l Ticker:**  Int'l Exchange: |
| 2006 Sales: $ | 2006 Profits: $-79,624 | Employees:  52 |
| 2005 Sales: $  19 | 2005 Profits: $-75,542 | Fiscal Year Ends: 12/31 |
| 2004 Sales: $ 163 | 2004 Profits: $-69,817 | Parent Company: |

## SALARIES/BENEFITS:

| Pension Plan: | ESOP Stock Plan: | Profit Sharing: | Top Exec. Salary: $514,000 | Bonus: $ |
|---|---|---|---|---|
| Savings Plan: | Stock Purch. Plan: Y | | Second Exec. Salary: $379,000 | Bonus: $ |

## OTHER THOUGHTS:

**Apparent Women Officers or Directors:** 2
**Hot Spot for Advancement for Women/Minorities:** Y

## LOCATIONS: ("Y" = Yes)

| West: | Southwest: | Midwest: | Southeast: | Northeast: | International: |
|---|---|---|---|---|---|
| Y | | | | | |

Note: Financial information, benefits and other data can change quickly and may vary from those stated here.

# TEVA PARENTERAL MEDICINES INC

**www.tevausa.com**

Industry Group Code: 325412  **Ranks within this company's industry group:** Sales:     Profits:

| Drugs: | | Other: | | Clinical: | | Computers: | | Services: | |
|---|---|---|---|---|---|---|---|---|---|
| Discovery: | | AgriBio: | | Trials/Services: | | Hardware: | | Specialty Services: | |
| Licensing: | Y | Genetic Data: | | Labs: | Y | Software: | | Consulting: | |
| Manufacturing: | Y | Tissue Replacement: | | Equipment/Supplies: | | Arrays: | | Blood Collection: | |
| Genetics: | | | | Research & Development Services: | Y | Database Management: | | Drug Delivery: | |
| | | | | Diagnostics: | | | | Drug Distribution: | |

## TYPES OF BUSINESS:

Drugs-Diversified
Bulk Pharmaceutical Ingredients
Contract Manufacturing & Research
Injectable Pharmaceuticals

## BRANDS/DIVISIONS/AFFILIATES:

Teva Pharmaceuticals Industries, Ltd.
Teva Pharmaceuticals USA
Teva Biopharmaceuticals USA
Teva Speciality Pharmaceuticals, LLC
TAPI
Gate Pharmaceuticals
Teva Animal Health
Barr Laboratories

## CONTACTS: *Note: Officers with more than one job title may be intentionally listed here more than once.*

Frank Becker, CEO
Shlomo Yanai, CEO/Pres., Teva Pharmaceutical Industries Ltd.
Eyal Desheh, CFO-Teva Pharmaceutical Industries, Ltd.
Eli Hurvitz, Chmn., Teva Pharmaceutical Industries, Ltd.

| Phone: 949-455-4700 | Fax: 949-855-8210 |
|---|---|
| Toll-Free: | |
| Address: 19 Hughes, Irvine, CA 92618 US | |

## GROWTH PLANS/SPECIAL FEATURES:

Teva Parenteral Medicines, Inc., formerly Sicor, Inc., is a wholly-owned subsidiary of Teva Pharmaceutical Industries and a global leader in the development and marketing of generic injectable drugs. The company uses its internal research and development capabilities, together with its operational manufacturing and regulatory experience, to develop a wide range of pharmaceutical products. The firm markets oncolytic and specialty injectable products, particularly in the oncology and anesthesiology care areas. It concentrates on products and technologies that face significant barriers to entering the worldwide market. A few of the firm's products include Acyclovir powder for injection; Cholografin Meglumine injection; Leucovorin Clacium injection USP; Octreotide Acetate injection; and Sulfamethoxazole and Trimethoprim injection. Teva Parenteral Medicines has concentrated its expansion toward securing marketing approval for new drugs and marketing them first in developing nations, then in Europe and finally in the U.S. The firm is one of 11 business units controlled by Teva Pharmaceutical Industries. The business units include Teva Pharmaceuticals USA; Teva Biopharmaceuticals USA; Teva Parenteral Medicines, Inc.; TAPI; Teva Speciality Pharmaceuticals, LLC; Gate Pharmaceuticals; Teva Animal Health; Barr Laboratories; Duramed; Global Generic R&D; and Global Respiratory R&D.

The firm offers employees medical and dental coverage; life insurance; long and short term disability; a child and elder care referral program; flexible spending accounts; tuition reimbursement; an employee assistance program; a 401(k) plan; an employee stock purchase plan; and professional development programs.

## FINANCIALS: Sales and profits are in thousands of dollars—add 000 to get the full amount. 2008 Note: Financial information for 2008 was not available for all companies at press time.

| | | |
|---|---|---|
| 2008 Sales: $ | 2008 Profits: $ | U.S. Stock Ticker: Subsidiary |
| 2007 Sales: $ | 2007 Profits: $ | Int'l Ticker:     Int'l Exchange: |
| 2006 Sales: $ | 2006 Profits: $ | Employees:  1,905 |
| 2005 Sales: $ | 2005 Profits: $ | Fiscal Year Ends: |
| 2004 Sales: $ | 2004 Profits: $ | Parent Company: TEVA PHARMACEUTICAL INDUSTRIES |

## SALARIES/BENEFITS:

| Pension Plan: | ESOP Stock Plan: | Profit Sharing: Y | Top Exec. Salary: $500,000 | Bonus: $463,417 |
|---|---|---|---|---|
| Savings Plan: Y | Stock Purch. Plan: Y | | Second Exec. Salary: $368,125 | Bonus: $158,036 |

## OTHER THOUGHTS:

**Apparent Women Officers or Directors:**
**Hot Spot for Advancement for Women/Minorities:**

## LOCATIONS: ("Y" = Yes)

| West: | Southwest: | Midwest: | Southeast: | Northeast: | International: |
|---|---|---|---|---|---|
| Y | | | | | Y |

Note: Financial information, benefits and other data can change quickly and may vary from those stated here.

# TEVA PHARMACEUTICAL INDUSTRIES        www.tevapharm.com

**Industry Group Code: 325412A Ranks within this company's industry group: Sales: 1  Profits: 1**

| Drugs: | Other: | Clinical: | Computers: | Services: |
|---|---|---|---|---|
| Discovery: | AgriBio: | Trials/Services: | Hardware: | Specialty Services: |
| Licensing: | Genetic Data: | Labs: | Software: | Consulting: |
| Manufacturing: Y | Tissue Replacement: | Equipment/Supplies: Y | Arrays: | Blood Collection: |
| Genetics: Y | | Research & Development Services: | Database Management: | Drug Delivery: |
| | | Diagnostics: | | Drug Distribution: |

## TYPES OF BUSINESS:
Drugs-Generic
Active Pharmaceutical Ingredients

## BRANDS/DIVISIONS/AFFILIATES:
Teva Pharmaceuticals USA
Pharmachemie BV
Copaxone
Ivax Corporation
Bentley Pharmaceuticals, Inc.
Azilect
CoGenesys, Inc.
Barr Pharmaceuticals

## CONTACTS: Note: Officers with more than one job title may be intentionally listed here more than once.
Shlomo Yanai, CEO
Shlomo Yanai, Pres.
Eyal Desheh, CFO
Judith Vardi, VP-Israel Pharmaceutical Sales
Isaac Abravanel, Corp. VP-Human Resources
Ben-Zion Weiner, Chief R&D Officer
Uzi Karniel, Chief Legal Officer/Corp. Sec.
Itzhak Krinsky, Corp. VP-Bus. Dev.
Ayala Miller, Dir.-Corp. Comm., Teva Pharmaceutical Industries
Ron Grupel, Internal Auditor
William S. Marth, CEO/Pres., Teva North America
Amir Elstein, VP-Global Pharmaceutical Resources
Gerard Van Odlijk, Group VP-Europe
Eli Shohet, Chief Integration Officer
Eli Hurvitz, Chmn.
Chaim Hurvitz, Group VP-Int'l

**Phone:** 972-3-926-7267  **Fax:** 972-3-923-4050
**Toll-Free:**
**Address:** 5 Basel St., Petach Tikva, 49131 Israel

## GROWTH PLANS/SPECIAL FEATURES:
Teva Pharmaceutical Industries, Ltd., based in Israel, is a pharmaceutical company that produces, distributes and sells pharmaceutical products internationally. Approximately 80% of the firm's sales are generated from North America and Europe. The firm specializes in the development of generic, also known as human pharmaceuticals (HP), and active pharmaceutical ingredients (API). The HP segment produces generic drugs in all major therapeutics in a variety of dosage forms, including capsules, tablets, creams, ointments and liquids. The API segment, which accounts for 90% of total sales, distributes ingredients to manufacturers worldwide, in addition to supporting its own pharmaceutical products. Teva also manufactures innovative drugs in niche markets. Currently, the firm focuses on products for neurological disorders and auto-immune diseases such as Alzheimer's disease, Amyotrophic Lateral Sclerosis, Lupus and Parkinson's disease. Subsidiary Teva Pharmaceuticals USA is a dominant figure in the manufacture and distribution of generic drugs in the U.S. In addition, subsidiary Pharmachemie BV has the largest market share in generics in the Netherlands. The firm's two chief products are Copaxone, a branded treatment for multiple sclerosis (MS), currently marketed and sold in 42 countries including the U.S., and Azilect, a treatment for Parkinson's. In December 2008, the firm acquired Barr Pharmaceuticals and its subsidiary Pliva. In July 2008, Teva acquired Bentley Pharmaceuticals, Inc. for approximately $360 million. In February 2008, the company acquired privately-held CoGenesys, Inc., a biopharmaceutical company with a broad based biotechnology platform and focused on the development of peptide- and protein-based medicines across broad therapeutic categories.

## FINANCIALS: Sales and profits are in thousands of dollars—add 000 to get the full amount. 2008 Note: Financial information for 2008 was not available for all companies at press time.
2008 Sales: $11,085,000   2008 Profits: $635,000
2007 Sales: $9,408,000   2007 Profits: $1,952,000
2006 Sales: $8,408,000   2006 Profits: $546,000
2005 Sales: $5,250,000   2005 Profits: $1,072,000
2004 Sales: $4,799,000   2004 Profits: $-331,800

**U.S. Stock Ticker:** TEVA
**Int'l Ticker:** TEVA   Int'l Exchange: Tel Aviv-TASE
Employees: 38,307
Fiscal Year Ends: 12/31
Parent Company:

## SALARIES/BENEFITS:
Pension Plan: Y   ESOP Stock Plan:   Profit Sharing:   Top Exec. Salary: $   Bonus: $
Savings Plan: Y   Stock Purch. Plan:   Second Exec. Salary: $   Bonus: $

## OTHER THOUGHTS:
**Apparent Women Officers or Directors:** 6
**Hot Spot for Advancement for Women/Minorities:** Y

## LOCATIONS: ("Y" = Yes)
| West: | Southwest: | Midwest: | Southeast: | Northeast: | International: |
|---|---|---|---|---|---|
| Y | Y | Y | Y | Y | Y |

Note: Financial information, benefits and other data can change quickly and may vary from those stated here.

# THERAGENICS CORP

**www.theragenics.com**

**Industry Group Code:** 33911 **Ranks within this company's industry group:** Sales: 11 Profits: 17

| Drugs: | Other: | Clinical: | | Computers: | Services: |
|---|---|---|---|---|---|
| Discovery: | AgriBio: | Trials/Services: | | Hardware: | Specialty Services: |
| Licensing: | Genetic Data: | Labs: | | Software: | Consulting: |
| Manufacturing: | Tissue Replacement: | Equipment/Supplies: | Y | Arrays: | Blood Collection: |
| Genetics: | | Research & Development Services: | Y | Database Management: | Drug Delivery: |
| | | Diagnostics: | | | Drug Distribution: |

## TYPES OF BUSINESS:

Medical Devices
Surgical Products

## BRANDS/DIVISIONS/AFFILIATES:

TheraSeed
Galt Medical Corp.
CP Medical Corp.
I-seed
NeedleTech Products, Inc.

## CONTACTS: *Note: Officers with more than one job title may be intentionally listed here more than once.*

M. Christine Jacobs, CEO
M. Christine Jacobs, Pres.
Frank J. Tarallo, CFO
Bruce W. Smith, Corp. Sec.
Bruce W. Smith, Exec. VP-Strategy & Bus. Dev.
Frank J. Tarallo, Treas.
Michael F. Lang, Pres., Galt Medical
Janet E. Zeman, Pres., CP Medical
R. Michael O'Bannon, Exec. VP-Organizational Dev.
M. Christine Jacobs, Chmn.

| **Phone:** 770-271-0233 | **Fax:** |
|---|---|
| **Toll-Free:** | |
| **Address:** 5203 Bristol Industrial Way, Buford, GA 30518 US | |

## GROWTH PLANS/SPECIAL FEATURES:

Theragenics Corp. is a medical device company serving the cancer treatment and surgical markets. The company operates in two segments, the brachytherapy seed business and the surgical products business. The brachytherapy seed business segment produces, markets and sells TheraSeed, the firm's premier palladium-103 prostate cancer treatment device; I-seed, its iodone-125 based prostrate cancer treatment device; and other related products and services. Theragenics is the world's largest producer of palladium-103, the radioactive isotope that supplies the therapeutic radiation for its TheraSeed device. TheraSeed is an implant the size of a grain of rice that is used primarily in treating localized prostate cancer with a one-time, minimally invasive procedure. The implant emits radiation within the immediate prostate area, killing the tumor while sparing surrounding organs from significant radiation exposure. Physicians, hospitals and other healthcare providers, primarily located in the U.S., utilize the TheraSeed device. The majority of TheraSeed sales are channeled through one third-party distributor. The surgical products business segment consists of wound closure and vascular access products. Wound closure include sutures, needles and other surgical products with applications in, among other areas, urology, veterinary, cardiology, orthopedics, plastic surgery and dental. Vascular access includes introducers and guidewires used in the interventional radiology, interventional cardiology and vascular surgery markets. The surgical products business sells its devices and components primarily to original equipment manufacturers (OEM) and a network of distributors. Theragenics' subsidiaries, CP Medical Corp., Galt Medical Corp. and NeedleTech Products, Inc. accounted for roughly 57% of revenue in 2008. In July 2008, the company acquired NeedleTech Products, Inc., a manufacturer of specialty needles and related medical devices, for $48.7 million.

Employees are offered medical, dental and life insurance; short- and long-term disability coverage; a 401(k) plan; and a stock purchase.

## FINANCIALS: Sales and profits are in thousands of dollars—add 000 to get the full amount. 2008 Note: Financial information for 2008 was not available for all companies at press time.

| | | | |
|---|---|---|---|
| 2008 Sales: $67,358 | 2008 Profits: $-58,540 | **U.S. Stock Ticker:** TGX | |
| 2007 Sales: $62,210 | 2007 Profits: $5,635 | **Int'l Ticker:** Int'l Exchange: | |
| 2006 Sales: $54,096 | 2006 Profits: $6,865 | Employees: 338 | |
| 2005 Sales: $44,270 | 2005 Profits: $-29,006 | Fiscal Year Ends: 12/31 | |
| 2004 Sales: $33,338 | 2004 Profits: $-4,310 | Parent Company: | |

## SALARIES/BENEFITS:

| Pension Plan: | ESOP Stock Plan: | Profit Sharing: | Top Exec. Salary: $535,000 | Bonus: $204,560 |
|---|---|---|---|---|
| Savings Plan: Y | Stock Purch. Plan: Y | | Second Exec. Salary: $295,000 | Bonus: $147,000 |

## OTHER THOUGHTS:

**Apparent Women Officers or Directors**: 2
**Hot Spot for Advancement for Women/Minorities**: Y

## LOCATIONS: ("Y" = Yes)

| West: | Southwest: | Midwest: | Southeast: | Northeast: | International: |
|---|---|---|---|---|---|
| Y | Y | | Y | | |

# THERMO FISHER SCIENTIFIC INC

www.thermofisher.com

**Industry Group Code: 423450   Ranks within this company's industry group: Sales: 1   Profits: 1**

| Drugs: | | Other: | | Clinical: | | Computers: | | Services: | |
|---|---|---|---|---|---|---|---|---|---|
| Discovery: | | AgriBio: | | Trials/Services: | Y | Hardware: | Y | Specialty Services: | Y |
| Licensing: | | Genetic Data: | | Labs: | | Software: | Y | Consulting: | Y |
| Manufacturing: | Y | Tissue Replacement: | | Equipment/Supplies: | Y | Arrays: | | Blood Collection: | |
| Genetics: | | | | Research & Development Services: | Y | Database Management: | | Drug Delivery: | |
| | | | | Diagnostics: | Y | | | Drug Distribution: | |

## TYPES OF BUSINESS:

Laboratory Equipment & Supplies Distribution
Contract Manufacturing
Equipment Calibration & Repair
Clinical Trial Services
Laboratory Workstations
Clinical Consumables
Diagnostic Reagents
Custom Chemical Synthesis

## BRANDS/DIVISIONS/AFFILIATES:

Biolab
LTQ Velos
Heraeus
M-PULSe
Affinity BioReagents
FIBERLite
Open Biosystems, Inc.

## CONTACTS: *Note: Officers with more than one job title may be intentionally listed here more than once.*

Marijn E. Dekkers, CEO
Marc N. Casper, COO/Exec. VP
Marijn E. Dekkers, Pres.
Peter M. Wilver, CFO/Sr. VP
Stephen G. Sheehan, Sr. VP-Human Resources
Seth H. Hoogasian, General Counsel/Sec./Sr. VP
Peter E. Hornstra, Chief Acct. Officer/VP
Alan J. Malus, Sr. VP
Gregory J. Herrema, Sr. VP
Yuh-Geng Tsay, Sr. VP
Jim Manzi, Chmn.

| **Phone:** 781-622-1000 | **Fax:** 781-622-1207 |
|---|---|
| **Toll-Free:** 800-678-5599 | |
| **Address:** 81 Wyman St., Waltham, MA 02454 US | |

## GROWTH PLANS/SPECIAL FEATURES:

Thermo Fisher Scientific Inc. is a distributor of products and services principally to the scientific-research and clinical laboratory markets. The firm serves over 350,000 customers including biotechnology and pharmaceutical companies; colleges and universities; medical-research institutions; hospitals; reference, quality control, process-control and research and development labs in various industries; and government agencies. Thermo Fisher offers an array of products, including analytical instruments; automation and robotics; life science research supplies; chemicals; disposable laboratory supplies; diagnostic supplies; laboratory equipment; furniture; software; and custom products. The company's services include asset management, instrument standards compliance services, education, technical support and professional assistance. In 2008, the firm acquired the Analytical Technologies and Environment Instrumentation divisions of Chemito Technologies Pvt. Ltd, the leading analytical instrument provider in India; Open Biosystems Inc., a provider of RNA Interference, gene expression and protein detection products for life science research and drug discovery; Affinity BioReagents, a provider of antibodies, peptides, proteins and other reagents for research; and FIBERLite, the world's leading supplier of carbon fiber centrifuge rotors. In December 2008, Thermo Fisher opened a new clinical services facility in India in order to establish a more significant presence in that market. In April 2009, the company acquired Biolab, an Australian provider of analytical equipment and the scientific and medical division of Alesco Corp., Ltd., for $120 million, in addition to a handful of other company acquisitions.

ThermoFisher offers its employees comprehensive benefits, paid time off, tuition reimbursement, retirement plans and employee recognition awards.

## FINANCIALS: Sales and profits are in thousands of dollars—add 000 to get the full amount. 2008 Note: Financial information for 2008 was not available for all companies at press time.

| | | |
|---|---|---|
| 2008 Sales: $10,498,000 | 2008 Profits: $994,200 | **U.S. Stock Ticker:** TMO |
| 2007 Sales: $9,746,400 | 2007 Profits: $761,100 | **Int'l Ticker:**     Int'l Exchange: |
| 2006 Sales: $3,791,617 | 2006 Profits: $168,935 | Employees: 34,500 |
| 2005 Sales: $2,633,027 | 2005 Profits: $223,218 | Fiscal Year Ends: 12/31 |
| 2004 Sales: $2,205,995 | 2004 Profits: $361,837 | Parent Company: THERMO ELECTRON CORP |

## SALARIES/BENEFITS:

| Pension Plan: | ESOP Stock Plan: | Profit Sharing: | Top Exec. Salary: $1,163,750 | Bonus: $1,865,643 |
|---|---|---|---|---|
| Savings Plan: Y | Stock Purch. Plan: | | Second Exec. Salary: $701,250 | Bonus: $823,266 |

## OTHER THOUGHTS:

**Apparent Women Officers or Directors:** 2
**Hot Spot for Advancement for Women/Minorities:**

## LOCATIONS: ("Y" = Yes)

| West: | Southwest: | Midwest: | Southeast: | Northeast: | International: |
|---|---|---|---|---|---|
| Y | | Y | Y | Y | Y |

Note: Financial information, benefits and other data can change quickly and may vary from those stated here.

# THERMO SCIENTIFIC

**www.thermo.com/perbio**

Industry Group Code: 325413   **Ranks within this company's industry group:**   Sales:       Profits:

| Drugs: | Other: | Clinical: | | Computers: | Services: | |
|---|---|---|---|---|---|---|
| Discovery: | AgriBio: | Trials/Services: | | Hardware: | Specialty Services: | |
| Licensing: | Genetic Data: | Labs: | | Software: | Consulting: | |
| Manufacturing: | Tissue Replacement: | Equipment/Supplies: | Y | Arrays: | Blood Collection: | |
| Genetics: | | Research & Development Services: | | Database Management: | Drug Delivery: | |
| | | Diagnostics: | Y | | Drug Distribution: | |

## TYPES OF BUSINESS:

Research Diagnostics Products

## BRANDS/DIVISIONS/AFFILIATES:

Pierce
HyClone
Endogen
Dharmacon
Cellomics
Thermo Fisher Scientific, Inc.

## CONTACTS: Note: Officers with more than one job title may be intentionally listed here more than once.

Ragnar Lindqvist, Sec.
Mats Fischier, Chmn.

| Phone: 46-42-26-90-90 | Fax: 46-42-26-90-98 |
|---|---|
| Toll-Free: | |
| Address: Knutpunkten 34, Helsingborg,  252 78 Sweden | |

## GROWTH PLANS/SPECIAL FEATURES:

Thermo Scientific (TS) (formerly Perbio Science AB), a subsidiary of Thermo Fisher Scientific, Inc., focuses on protein production and research.  The firm specializes in products, systems and services for the study and production of proteins, cell cultures and RNA, primarily for academic and research institutes, pharmaceutical companies and diagnostics companies.  The company operates in three segments: bioresearch, cell culture and medical devices.  The bioresearch segment develops products used for research in protein chemistry and molecular biology.  Products include kits, reagents and services for identifying, quantifying, purifying and modifying proteins, amidites and nucleotides for nucleic acid-based drugs and diagnostics.  The cell culture segment offers proper nutrition for animal cells, process liquids for protein purification, sterile liquid handling systems and disposable sterile liquid processing systems.  Products include animal sera such as fetal bovine serum, cosmic calf serum and equine serum, as well as cell culture media in powder and liquid forms.  The medical devices segment provides voice prosthesis that helps laryngectomy patients regain their speech ability; devices for laryngectomized and tracheostomized patients; and devices for the treatment of excessive menstrual bleeding and obesity.  Leading brands include Pierce, a provider of research products for use in protein chemistry, sample handling, immunology and chromatography; Endogen, a provider of kits for the detection of infections, multiplex assays testing up to 16 types of antibodies and antibodies to cytokines and kinases; HyClone, a provider of cell-culture products and bioprocessing systems; Dharmacon, a provider of RNA products and technologies; and Cellomics, a provider of cellular measuring tools and screening instruments.  TS has operating companies in the U.K., France, Germany, Switzerland, the Netherlands and Belgium.  In May 2009, TS introduced a new line of Thermo Scientific DyLight-Labeled Highly Cross-Adsorbed Secondary antibodies.

## FINANCIALS:  Sales and profits are in thousands of dollars—add 000 to get the full amount. 2008 Note: Financial information for 2008 was not available for all companies at press time.

| | | |
|---|---|---|
| 2008 Sales: $ | 2008 Profits: $ | **U.S. Stock Ticker: Subsidiary** |
| 2007 Sales: $ | 2007 Profits: $ | **Int'l Ticker:**    Int'l Exchange: |
| 2006 Sales: $ | 2006 Profits: $ | Employees:  1,266 |
| 2005 Sales: $ | 2005 Profits: $ | Fiscal Year Ends: 12/31 |
| 2004 Sales: $ | 2004 Profits: $ | Parent Company: THERMO FISHER SCIENTIFIC INC |

## SALARIES/BENEFITS:

| Pension Plan: | ESOP Stock Plan: | Profit Sharing: | Top Exec. Salary: $ | Bonus: $ |
|---|---|---|---|---|
| Savings Plan: | Stock Purch. Plan: | | Second Exec. Salary: $ | Bonus: $ |

## OTHER THOUGHTS:

**Apparent Women Officers or Directors:**
**Hot Spot for Advancement for Women/Minorities:**

## LOCATIONS: ("Y" = Yes)

| West: | Southwest: | Midwest: | Southeast: | Northeast: | International: |
|---|---|---|---|---|---|
| | | | | | Y |

# THIRD WAVE TECHNOLOGIES INC

**www.twt.com**

Industry Group Code: 325413 Ranks within this company's industry group: Sales: Profits:

| Drugs: | Other: | | Clinical: | | Computers: | | Services: | |
|---|---|---|---|---|---|---|---|---|
| Discovery: | AgriBio: | | Trials/Services: | | Hardware: | | Specialty Services: | |
| Licensing: | Genetic Data: | Y | Labs: | | Software: | | Consulting: | |
| Manufacturing: | Tissue Replacement: | | Equipment/Supplies: | Y | Arrays: | | Blood Collection: | |
| Genetics: | | | Research & Development Services: | | Database Management: | | Drug Delivery: | |
| | | | Diagnostics: | Y | | | Drug Distribution: | |

## TYPES OF BUSINESS:

Genetic Analysis Products
Assays

## BRANDS/DIVISIONS/AFFILIATES:

Invader
Cleavase
Invader InPlex
Invader Plus
Universal Invader Program
qInvader

## CONTACTS: *Note: Officers with more than one job title may be intentionally listed here more than once.*

John W. Cummings, CEO-Hologic

| Phone: 608-273-8933 | Fax: 608-273-8618 |
|---|---|
| Toll-Free: 888-898-2357 | |
| Address: 502 S. Rosa Rd., Madison, WI 53719 US | |

## GROWTH PLANS/SPECIAL FEATURES:

Third Wave Technologies, Inc., a subsidiary of Hologic, Inc., develops, manufactures and markets genetic analysis products for clinical diagnostics and studies. Its patented genetic analysis platform, Invader, offers several advantages over conventional genetic analysis technologies. Invader relies upon the company's proprietary Cleavase enzyme for testing rather than using a complex copying technique known as a polymerase chain reaction (PCR). Available in ready-to-use formats, Third Wave's products are compatible with existing automation processes and detection platforms. These advantages make the Invader platform a convenient solution for genetic analysis, with applications ranging from disease discovery to patient care, including large-scale disease association studies, drug response marker profiling and molecular diagnostics. The company's Invader Plus product relies on a combination of the Invader platform with traditional PCR amplification. Developed in collaboration with 3M, the Invader InPlex product combines Invader chemistry with 3M's microfluidic technology, which helps to improve speed, efficiency and ease of use. Finally, qInvader detects and quantifies nucleic acids as a tool to understanding the development of cancer and other diseases. In addition to Invader products, Third Wave offers assays for cardiovascular disease and deep-vein thrombosis detection, as well as tests relating to animal and plant genetics. The company's headquarters are located in Madison, Wisconsin. In July 2008, the company became a wholly-owned subsidiary of Hologic, Inc. In 2008, the company acquired Agilent Technologies Inc.'s Full Velocity portfolio of patents, which cover nucleic acid amplification technology, for an undisclosed amount.

## FINANCIALS: Sales and profits are in thousands of dollars—add 000 to get the full amount. 2008 Note: Financial information for 2008 was not available for all companies at press time.

| | | |
|---|---|---|
| 2008 Sales: $ | 2008 Profits: $ | U.S. Stock Ticker: Subsidiary |
| 2007 Sales: $31,100 | 2007 Profits: $ | Int'l Ticker: Int'l Exchange: |
| 2006 Sales: $28,027 | 2006 Profits: $-18,887 | Employees: 154 |
| 2005 Sales: $23,906 | 2005 Profits: $-22,346 | Fiscal Year Ends: 12/31 |
| 2004 Sales: $46,493 | 2004 Profits: $-1,942 | Parent Company: HOLOGIC INC |

## SALARIES/BENEFITS:

| Pension Plan: | ESOP Stock Plan: | Profit Sharing: | Top Exec. Salary: $422,917 | Bonus: $671,241 |
|---|---|---|---|---|
| Savings Plan: Y | Stock Purch. Plan: | | Second Exec. Salary: $274,583 | Bonus: $360,666 |

## OTHER THOUGHTS:

**Apparent Women Officers or Directors**: 1
**Hot Spot for Advancement for Women/Minorities**:

## LOCATIONS: ("Y" = Yes)

| West: | Southwest: | Midwest: | Southeast: | Northeast: | International: |
|---|---|---|---|---|---|
| | | Y | | | |

# TITAN PHARMACEUTICALS

**www.titanpharm.com**

Industry Group Code: 325412  Ranks within this company's industry group:  Sales:      Profits:

| Drugs: | | Other: | Clinical: | Computers: | Services: | |
|---|---|---|---|---|---|---|
| Discovery: | Y | AgriBio: | Trials/Services: | Hardware: | Specialty Services: | |
| Licensing: | | Genetic Data: | Labs: | Software: | Consulting: | |
| Manufacturing: | Y | Tissue Replacement: | Equipment/Supplies: | Arrays: | Blood Collection: | |
| Genetics: | | | Research & Development Services: | Database Management: | Drug Delivery: | Y |
| | | | Diagnostics: | | Drug Distribution: | |

## TYPES OF BUSINESS:

Central Nervous System, Cardiovascular & Bone Diseases Therapeutics
Drug Delivery Systems

## BRANDS/DIVISIONS/AFFILIATES:

Fanapt
Probuphine
ProNeura

## CONTACTS: Note: Officers with more than one job title may be intentionally listed here more than once.

Sunil Bhonsle, Pres.
Kate Beebe, Sr. VP-Clinical Dev. & Medical Affairs
Sunil Bhonsle, Sec.
Mark Goldstein, Sr. VP
Marc Rubin, Exec. Chmn.

| Phone: 650-244-4990 | Fax: 650-244-4956 |
|---|---|
| Toll-Free: | |
| Address: 400 Oyster Point Blvd., Ste. 505, South San Francisco, CA 94080 US | |

## GROWTH PLANS/SPECIAL FEATURES:

Titan Pharmaceuticals, Inc. is a biopharmaceutical company that develops proprietary therapeutics for the treatment of central nervous system disorders, cardiovascular disease, bone disease and other disorders.  The firm's main product, Fanapt (formerly Iloperidone), which was developed in collaboration with Vanda Pharmaceutical, Inc., is a treatment of adult schizophrenia.  The company is focused on clinical development of Probuphine, for the treatment of opioid dependence.  Titan Pharmaceuticals utilizes the ProNeura technology in the Probuphine treatment.  ProNeura is a drug delivery system that lasts approximately six months and is placed subcutaneously in the patient.  ProNeura also has potential applications where conventional treatment is limited by poor patient compliance and variability in blood drug levels.  In May 2009, Fanapt received marketing approval from the U.S. FDA.

## FINANCIALS: Sales and profits are in thousands of dollars—add 000 to get the full amount. 2008 Note: Financial information for 2008 was not available for all companies at press time.

| | | | |
|---|---|---|---|
| 2008 Sales: $ | 2008 Profits: $ | U.S. Stock Ticker: TTP | |
| 2007 Sales: $ 24 | 2007 Profits: $-17,647 | Int'l Ticker: | Int'l Exchange: |
| 2006 Sales: $ 32 | 2006 Profits: $-15,737 | Employees: 44 | |
| 2005 Sales: $ 89 | 2005 Profits: $-22,462 | Fiscal Year Ends: 12/31 | |
| 2004 Sales: $ 31 | 2004 Profits: $-26,004 | Parent Company: | |

## SALARIES/BENEFITS:

| Pension Plan: | ESOP Stock Plan: | Profit Sharing: | Top Exec. Salary: $493,328 | Bonus: $ |
|---|---|---|---|---|
| Savings Plan: | Stock Purch. Plan: | | Second Exec. Salary: $297,583 | Bonus: $ |

## OTHER THOUGHTS:

**Apparent Women Officers or Directors:** 1
**Hot Spot for Advancement for Women/Minorities:**

## LOCATIONS: ("Y" = Yes)

| West: | Southwest: | Midwest: | Southeast: | Northeast: | International: |
|---|---|---|---|---|---|
| Y | | | | | |

# TORREYPINES THERAPEUTICS INC
www.torreypinestherapeutics.com

**Industry Group Code: 325412 Ranks within this company's industry group: Sales: 128 Profits: 108**

| Drugs: | | Other: | Clinical: | Computers: | Services: |
|---|---|---|---|---|---|
| Discovery: | Y | AgriBio: | Trials/Services: | Hardware: | Specialty Services: |
| Licensing: | Y | Genetic Data: | Labs: | Software: | Consulting: |
| Manufacturing: | | Tissue Replacement: | Equipment/Supplies: | Arrays: | Blood Collection: |
| Genetics: | | | Research & Development Services: | Database Management: | Drug Delivery: |
| | | | Diagnostics: | | Drug Distribution: |

## TYPES OF BUSINESS:

Pharmaceutical Acquisition & Development
Biopharmaceuticals
Drug Development
Central Nervous System Disorder Treatment Products
Alzheimer's Disease Treatment Products
CIAS Treatment Products
Migraine Treatment Products

## BRANDS/DIVISIONS/AFFILIATES:

AXONYX, Inc.
tezampanel
NGX426
NGX267
NGX292

## CONTACTS: Note: Officers with more than one job title may be intentionally listed here more than once.

Evelyn A. Graham, CEO
Craig Johnson, CFO/Sec.
Paul R. Schneider, General Counsel/VP
Craig Johnson, VP-Finance
Sue Mellberg, VP-Project Mgmt.
Peter Davis, Chmn.

| Phone: 858-623-5665 | Fax: 858-623-5666 |
|---|---|
| Toll-Free: | |
| **Address:** 11085 N. Torrey Pines Rd., Ste. 300, La Jolla, CA 92037 US | |

## GROWTH PLANS/SPECIAL FEATURES:

TorreyPines Therapeutics, Inc., formerly AXONYX, Inc., engages in the discovery, development, and commercialization of novel small molecules to treat diseases and disorders of the central nervous system. Its therapeutic focus on chronic diseases including dry mouth, migraine and neuropathic pain; and cognitive disorders, including cognitive impairment associated with schizophrenia and Alzheimer's disease. TorreyPines has two product candidates in clinical trials: Tezampanel, which has undergone a Phase IIb trial for the abortive treatment of migraines; and its oral prodrug, NGX426, which recently completed its Phase I clinical trial for chronic pain treatment. Tezampanel does not constrict blood vessels, as current migraine treatment options do, but instead is believed to work by selectively blocking the transmission of pain signals to the brain. Another drug candidate, NGX267, is undergoing Phase II clinical trials for the treatment of xerostomia (dry mouth) associated with Sjogren's syndrome. NGX292, currently in preclinical development, is a backup candidate for NGX267. TorreyPines has additional product candidates in development focused on the treatment of Alzheimer's disease: Phenserine, for which the company has completed Phase III clinical trials and is currently pursuing out-licensing opportunities; posiphen, which has completed Phase I clinical trials; and bisnorcymserine and NGX555, both of which are in preclinical development. In November 2008, the firm agreed to sell its Alzheimer's disease genetics research program to Eisai Co., Ltd. Also in 2008, TorreyPines licensed its Alzheimer treatment compounds, posiphen, bisnorcymcerine and phenserine to QR Pharma, Inc, in order to focus solely on its NGX426 and NGX267 developments. In February 2009, the company completed a Phase I multiple dose clinical trial of NGX426.

## FINANCIALS: Sales and profits are in thousands of dollars—add 000 to get the full amount. 2008 Note: Financial information for 2008 was not available for all companies at press time.

| | | |
|---|---|---|
| 2008 Sales: $6,071 | 2008 Profits: $-22,785 | **U.S. Stock Ticker:** TPTX |
| 2007 Sales: $9,850 | 2007 Profits: $-23,369 | **Int'l Ticker:** Int'l Exchange: |
| 2006 Sales: $9,850 | 2006 Profits: $-25,377 | Employees: 10 |
| 2005 Sales: $7,967 | 2005 Profits: $-11,542 | Fiscal Year Ends: 12/31 |
| 2004 Sales: $3,551 | 2004 Profits: $-10,356 | Parent Company: |

## SALARIES/BENEFITS:

| Pension Plan: | ESOP Stock Plan: | Profit Sharing: | Top Exec. Salary: $320,015 | Bonus: $ |
|---|---|---|---|---|
| Savings Plan: | Stock Purch. Plan: | | Second Exec. Salary: $304,667 | Bonus: $ |

## OTHER THOUGHTS:

**Apparent Women Officers or Directors:** 2
**Hot Spot for Advancement for Women/Minorities:** Y

## LOCATIONS: ("Y" = Yes)

| West: | Southwest: | Midwest: | Southeast: | Northeast: | International: |
|---|---|---|---|---|---|
| Y | | | | | Y |

Note: Financial information, benefits and other data can change quickly and may vary from those stated here.

# TOYOBO CO LTD

**www.toyobo.co.jp**

Industry Group Code: 313  Ranks within this company's industry group: Sales: 1  Profits: 1

| Drugs: | | Other: | | Clinical: | | Computers: | | Services: | |
|---|---|---|---|---|---|---|---|---|---|
| Discovery: | Y | AgriBio: | | Trials/Services: | | Hardware: | | Specialty Services: | |
| Licensing: | | Genetic Data: | | Labs: | | Software: | | Consulting: | |
| Manufacturing: | Y | Tissue Replacement: | | Equipment/Supplies: | | Arrays: | | Blood Collection: | |
| Genetics: | | | | Research & Development Services: | | Database Management: | | Drug Delivery: | |
| | | | | Diagnostics: | | | | Drug Distribution: | |

## TYPES OF BUSINESS:

Textile & Fiber Manufacturing
Advanced Materials
Biomedical Products
Industrial Textiles
Plastics & Films
Engineering
Pharmaceuticals
Water Treatment Membranes

## BRANDS/DIVISIONS/AFFILIATES:

Zylon
Dyneema
Tsunooga

## CONTACTS: Note: Officers with more than one job title may be intentionally listed here more than once.

Ryuzou Sakamoto, COO
Ryuzou Sakakmoto, Pres.
Kenji Hayashi, Controlling Supervisor-Personnel & Labor Dept.
Kazuyuki Yabuki, Sr. Gen. Mgr.-Corp. R&D
Kenji Hayashi, Controlling Supervisor-Gen. Admin. Dept.
Kenji Hayashi, Controlling Supervisor-Law Dept.
Controlling Supervisor-Bus. Dev,
Fumishige Imamura, Dir.-Audit Dept. & Finance, Acct., Control Dept.
Masahiko Hachimaru, Head-Plastics Div./Controlling Supervisor-Tokyo
Kazuo Kurita, Head-Bioscience & Medical Div.
Hiroyuki Kagawa, Head-Fibers Div.
Masaaki Sekino, Controlling Supervisor-Intellectual Property
Junji Tsumura, Chmn.
Masayuki Yoshikawa, Gen. Mgr.-Procurement Oper. Office

| Phone: 81-6-6348-3111 | Fax: 81-6-6348-3206 |
|---|---|
| Toll-Free: | |
| Address: 2-8, Dojima Hama 2-chome, Kit-ku, Osaka, 530-8230 Japan | |

## GROWTH PLANS/SPECIAL FEATURES:

Toyobo Co. Ltd., founded as a textile business in 1882, has expanded into domains that utilize its core technologies of polymerization, modification, processing and bioscience. Currently, the company operates in four segments: textiles; life science; films and functional polymers; and industrial materials. The textiles segment produces a range of fabric materials, including polyester, nylon, spandex, polynostic, cotton and wool. The company has also developed super-fibers, which combine the strength of metal with the lightness of fibers. Toyobo's life science segment produces pharmaceuticals, bioproducts, medical membranes and equipment; and water treatment membranes. The films and functional polymers segment produces polyester film for soft packaging and optical polyester film for IT equipment, as well as various plastics and resins. The industrial materials segment produces fibers for airbags and tire cords and a variety of air filter products, as well as high-performance fibers. These include Zylon, a fiber with high thermal resistance and Dyneema, a super-strong fiber that is lighter than water. In April 2009, the company developed a cut resistant, high strength melt spinning polyethylene fiber called Tsunooga.

## FINANCIALS: Sales and profits are in thousands of dollars—add 000 to get the full amount. 2008 Note: Financial information for 2008 was not available for all companies at press time.

| | | |
|---|---|---|
| 2008 Sales: $4,305,989 | 2008 Profits: $46,891 | **U.S. Stock Ticker: TYOBY** |
| 2007 Sales: $3,614,282 | 2007 Profits: $114,121 | **Int'l Ticker: 3101**   Int'l Exchange: Tokyo-TSE |
| 2006 Sales: $3,418,200 | 2006 Profits: $107,100 | Employees: 3,249 |
| 2005 Sales: $3,660,500 | 2005 Profits: $113,500 | Fiscal Year Ends: 3/31 |
| 2004 Sales: $3,531,400 | 2004 Profits: $82,900 | Parent Company: |

## SALARIES/BENEFITS:

| Pension Plan: | ESOP Stock Plan: | Profit Sharing: | Top Exec. Salary: $ | Bonus: $ |
|---|---|---|---|---|
| Savings Plan: | Stock Purch. Plan: | | Second Exec. Salary: $ | Bonus: $ |

## OTHER THOUGHTS:

Apparent Women Officers or Directors:
Hot Spot for Advancement for Women/Minorities:

## LOCATIONS: ("Y" = Yes)

| West: | Southwest: | Midwest: | Southeast: | Northeast: | International: |
|---|---|---|---|---|---|
| | | | | Y | Y |

# TRIMERIS INC

**www.trimeris.com**

**Industry Group Code: 325412  Ranks within this company's industry group: Sales: 103    Profits: 51**

| Drugs: | | Other: | Clinical: | Computers: | Services: |
|---|---|---|---|---|---|
| Discovery: | Y | AgriBio: | Trials/Services: | Hardware: | Specialty Services: |
| Licensing: | | Genetic Data: | Labs: | Software: | Consulting: |
| Manufacturing: | | Tissue Replacement: | Equipment/Supplies: | Arrays: | Blood Collection: |
| Genetics: | | | Research & Development Services: | Database Management: | Drug Delivery: |
| | | | Diagnostics: | | Drug Distribution: |

## TYPES OF BUSINESS:

Antivirals
HIV Drugs

## BRANDS/DIVISIONS/AFFILIATES:

FUZEON
TRI-1144
F. Hoffman-La Roche, Ltd.

## CONTACTS: *Note: Officers with more than one job title may be intentionally listed here more than once.*

Martin Mattingly, CEO
Andrew Graham, CFO/Sec.
Michael Alrutz, General Counsel

| Phone: 919-806-4682 | Fax: 919-806-4770 |
|---|---|
| Toll-Free: | |
| Address: 2530 Meridian Pkwy., Fl. 2, Durham, NC 27713 US | |

## GROWTH PLANS/SPECIAL FEATURES:

Trimeris, Inc. is a biopharmaceutical company primarily engaged in the development and commercialization of a class of antiviral drug treatments called fusion inhibitors. Fusion inhibitors impair viral fusion, a complex process by which viruses attach to, penetrate and infect host cells. The firm focuses on Fuzeon, an HIV fusion inhibitor that was developed in collaboration with F. Hoffman-La Roche, Ltd. (Roche). When used in combination with other anti-HIV drugs, Fuzeon has been shown to reduce the amount of HIV in the blood and increase the number of T-cells. The drug is approved for marketing by FDA in combination with other anti-HIV drugs for the treatment of HIV-1 infection in treatment-experienced patients with evidence of HIV-1 replication despite ongoing anti-HIV therapy, and by the European Agency for the Evaluation of Medicinal Products in exceptional circumstances. Trimeris currently relies on Roche for the manufacture, sales, marketing and distribution of Fuzeon. Fuzeon is available through retail and specialty pharmacies across the U.S. and Canada. The company's other product, TRI-1144, is a next-generation fusion inhibitor peptide meant to suppress HIV by raising the genetic barrier to the development of resistance. TRI-1144 has completed a Phase I clinical trial. An Investigational New Drug (IND) application for TRI-1144 was filed in January 2008. In June 2008, the firm discontinued its active research and development activities.

## FINANCIALS: Sales and profits are in thousands of dollars—add 000 to get the full amount. 2008 Note: Financial information for 2008 was not available for all companies at press time.

| | | |
|---|---|---|
| 2008 Sales: $20,613 | 2008 Profits: $8,009 | U.S. Stock Ticker: TRMS |
| 2007 Sales: $49,385 | 2007 Profits: $27,425 | Int'l Ticker:    Int'l Exchange: |
| 2006 Sales: $36,980 | 2006 Profits: $7,384 | Employees:    4 |
| 2005 Sales: $19,059 | 2005 Profits: $-8,106 | Fiscal Year Ends: 12/31 |
| 2004 Sales: $6,708 | 2004 Profits: $-40,088 | Parent Company: |

## SALARIES/BENEFITS:

| Pension Plan: | ESOP Stock Plan: | Profit Sharing: | Top Exec. Salary: $385,000 | Bonus: $231,000 |
|---|---|---|---|---|
| Savings Plan: | Stock Purch. Plan: | | Second Exec. Salary: $175,000 | Bonus: $70,000 |

## OTHER THOUGHTS:

**Apparent Women Officers or Directors:**
**Hot Spot for Advancement for Women/Minorities:**

## LOCATIONS: ("Y" = Yes)

| West: | Southwest: | Midwest: | Southeast: | Northeast: | International: |
|---|---|---|---|---|---|
| | | | | Y | |

# TRINITY BIOTECH PLC

**www.trinitybiotech.com**

**Industry Group Code: 325413  Ranks within this company's industry group:** Sales: 8  Profits: 21

| Drugs: | Other: | Clinical: | | Computers: | | Services: |
|---|---|---|---|---|---|---|
| Discovery: | AgriBio: | Trials/Services: | | Hardware: | | Specialty Services: |
| Licensing: | Genetic Data: | Labs: | | Software: | | Consulting: |
| Manufacturing: | Tissue Replacement: | Equipment/Supplies: | Y | Arrays: | | Blood Collection: |
| Genetics: | | Research & Development Services: | | Database Management: | | Drug Delivery: |
| | | Diagnostics: | Y | | | Drug Distribution: |

## TYPES OF BUSINESS:

Medical Diagnostics Products
Immunoassay Technology

## BRANDS/DIVISIONS/AFFILIATES:

GeneSys
Trini
Capillus
Destiny MAX
UniGold

## CONTACTS: Note: Officers with more than one job title may be intentionally listed here more than once.

Ronan O'Caoimh, CEO
Rory Nealon, COO
Kevin Tansley, CFO
Kevin Tansley, Sec.
Ronan O'Caoimh, Chmn.

| **Phone:** 353-1276-9800 | **Fax:** 353-1276-9888 |
|---|---|
| **Toll-Free:** | |
| **Address:** 1 Southern Cross, IDA Business Park, Bray, Ireland UK | |

## GROWTH PLANS/SPECIAL FEATURES:

Trinity Biotech plc develops, acquires, manufactures and markets medical diagnostic products for the clinical laboratory and point-of-care (POC) segments of the diagnostic market. Trinity has seven manufacturing sites in Ireland, Germany, Sweden and the U.S., and sells its products in 80 countries worldwide through its own sales force, as well as international distributors and strategic partners. The company's product offerings are used for testing and diagnosis in autoimmune, infectious and sexually transmitted diseases (STDs), diabetes and disorders of the blood, liver and intestine. The firm also provides raw materials to the life sciences industry. Trinity's products can be divided into four lines, which include hemostasis, point-of-care, infectious diseases and clinical chemistry. The hemostasis product line includes test kits and instrumentation for the detection of blood coagulation and clotting disorders. It includes the products Destiny MAX and Trini. Point-of-care products primarily include tests for the presence of HIV antibodies, including the products UniGold and Capillus. The infectious disease product line tests for a wide range of issues, such as autoimmune diseases, hormonal imbalances, STDs, intestinal infections, lung/bronchial infections, cardiovascular and other diseases. Products are sold primarily in North America, Europe and Asia. Trinity's clinical chemistry products have proven performance in the diagnosis of many disease states from liver and kidney disease to G6PDH deficiency, which is an indicator of hemolytic anemia. Since its inception in 1992, Trinity has expanded through a series of acquisitions. In 2008, Trinity was named the distributor of Akers Biosciences' Heparin Allergy Test in U.S. and German markets. Also in 2008, the firm announced the launch of its new HIV Incidence Assay; the GeneSys platform for infant hemoglobinopathies; and the European launch of Destiny MAX, its high throughput hemostasis analyzer. In December 2008, as part of a cost reduction program, the company announced a 10% workforce reduction.

## FINANCIALS: Sales and profits are in thousands of dollars—add 000 to get the full amount. 2008 Note: Financial information for 2008 was not available for all companies at press time.

| | | |
|---|---|---|
| 2008 Sales: $140,139 | 2008 Profits: $-77,778 | **U.S. Stock Ticker: TRIB** |
| 2007 Sales: $143,617 | 2007 Profits: $-35,372 | **Int'l Ticker: TWU**  Int'l Exchange: Dublin-ISE |
| 2006 Sales: $118,674 | 2006 Profits: $3,276 | Employees:  711 |
| 2005 Sales: $98,560 | 2005 Profits: $5,280 | Fiscal Year Ends: 12/31 |
| 2004 Sales: $80,008 | 2004 Profits: $5,714 | Parent Company: |

## SALARIES/BENEFITS:

| Pension Plan: | ESOP Stock Plan: | Profit Sharing: | Top Exec. Salary: $656,000 | Bonus: $207,000 |
|---|---|---|---|---|
| Savings Plan: | Stock Purch. Plan: | | Second Exec. Salary: $509,000 | Bonus: $125,000 |

## OTHER THOUGHTS:

**Apparent Women Officers or Directors:**
**Hot Spot for Advancement for Women/Minorities:**

## LOCATIONS: ("Y" = Yes)

| West: | Southwest: | Midwest: | Southeast: | Northeast: | International: |
|---|---|---|---|---|---|
| Y | | Y | | Y | Y |

Note: Financial information, benefits and other data can change quickly and may vary from those stated here.

# TRIPOS INTERNATIONAL

www.tripos.com

**Industry Group Code: 511210D  Ranks within this company's industry group:** Sales:  Profits:

| Drugs: | Other: | Clinical: | Computers: | | Services: | |
|---|---|---|---|---|---|---|
| Discovery: | AgriBio: | Trials/Services: | Hardware: | | Specialty Services: | Y |
| Licensing: | Genetic Data: | Labs: | Software: | Y | Consulting: | Y |
| Manufacturing: | Tissue Replacement: | Equipment/Supplies: | Arrays: | | Blood Collection: | |
| Genetics: | | Research & Development Services: | Database Management: | | Drug Delivery: | |
| | | Diagnostics: | | | Drug Distribution: | |

## TYPES OF BUSINESS:

Biotech Software, Research Products & Services
Clinical Software
Consulting Services

## BRANDS/DIVISIONS/AFFILIATES:

SYBYL
Benchware
Tripos, Inc.
Vector Capital
Certara
Pharsight Corporation
Pantheon

## CONTACTS: *Note: Officers with more than one job title may be intentionally listed here more than once.*

Jim Hopkins, CEO
Patrick Flanagan, COO
John D. Yingling, CFO
Diana O'Rourke, Dir.-Mktg.
Richard D. Cramer III, Chief Scientific Officer
Gregory B. Smith, CTO
William Cobert, VP-European Sales, Certara

| | |
|---|---|
| **Phone:** 314-647-1099 | **Fax:** 314-647-9241 |
| **Toll-Free:** 800-323-2960 | |
| **Address:** 1699 S. Hanley Rd., St. Louis, MO 63144-2319 US | |

## GROWTH PLANS/SPECIAL FEATURES:

Tripos International, formed from the Discovery Informatics business of Tripos, Inc., provides research products and services for biotechnology, pharmaceutical and other life science enterprises. Drug development companies use the firm's informatics technologies, including computational methods and software tools, to study products at a molecular level. The firm serves over 1,000 customers in 46 countries. The company's two primary products lines are the SYBYL suite of molecular modeling applications and Benchware software products. The SYBYL program offers complete computational chemistry and molecular modeling systems that are tailored to accelerate and ease the discovery process. It comes with its Standard Base, which offers tools and functions designed to optimize, visualize and compare the attributes of molecular models and structures. SYBYL also offers two optional base modules: MOLCAD, which creates graphical images that reveal the properties of molecules; and Advanced Computation, a fast systematic conformational searching algorithm. The SIBYL suite also includes additional programs that focus on ligands, receptors or protein structures to design new compounds and drugs. The Benchware line of products, designed for laboratory scientists, digitally stores, searches and retrieves research information; supports chemical synthesis and testing decisions; and shares experimental results with project team members, including computational scientists. Additionally, Tripos International provides informatics services, including product support and training. Tripos International is owned by Vector Capital, a private equity firm that specializes in taking struggling public companies and making them private. In September 2008, the firm agreed to acquire Pharsight Corporation, a software and scientific services provider to the drug manufacturing industry, for $57 million. Upon completion of the Pharsight Corporation acquisition, Tripos International and Pharsight formed a new company, Certara, which offers the combined services of the two firms. In October 2008, the company developed Pantheon, a software platform that addresses the cheminformatics requirements of casual modelers and medicinal chemists.

## FINANCIALS: Sales and profits are in thousands of dollars—add 000 to get the full amount. 2008 Note: Financial information for 2008 was not available for all companies at press time.

| | | |
|---|---|---|
| 2008 Sales: $ | 2008 Profits: $ | **U.S. Stock Ticker:** Private |
| 2007 Sales: $ | 2007 Profits: $ | **Int'l Ticker:**   Int'l Exchange: |
| 2006 Sales: $27,384 | 2006 Profits: $-38,593 | Employees:  327 |
| 2005 Sales: $27,981 | 2005 Profits: $-4,288 | Fiscal Year Ends: 12/31 |
| 2004 Sales: $20,478 | 2004 Profits: $ 232 | Parent Company: VECTOR CAPITAL |

## SALARIES/BENEFITS:

| | | | | |
|---|---|---|---|---|
| Pension Plan: | ESOP Stock Plan: | Profit Sharing: | Top Exec. Salary: $380,000 | Bonus: $ |
| Savings Plan: | Stock Purch. Plan: | | Second Exec. Salary: $240,000 | Bonus: $ |

## OTHER THOUGHTS:

**Apparent Women Officers or Directors**: 1
**Hot Spot for Advancement for Women/Minorities:** Y

## LOCATIONS: ("Y" = Yes)

| West: | Southwest: | Midwest: | Southeast: | Northeast: | International: |
|---|---|---|---|---|---|
| | | Y | | | Y |

# UCB SA

**www.ucb-group.com**

Industry Group Code: 325412 **Ranks within this company's industry group:** Sales: 24 Profits: 143

| Drugs: | | Other: | | Clinical: | | Computers: | | Services: | |
|---|---|---|---|---|---|---|---|---|---|
| Discovery: | | AgriBio: | Y | Trials/Services: | | Hardware: | | Specialty Services: | |
| Licensing: | | Genetic Data: | | Labs: | | Software: | | Consulting: | |
| Manufacturing: | Y | Tissue Replacement: | | Equipment/Supplies: | | Arrays: | Y | Blood Collection: | |
| Genetics: | | | | Research & Development Services: | | Database Management: | | Drug Delivery: | |
| | | | | Diagnostics: | | | | Drug Distribution: | |

## TYPES OF BUSINESS:

Pharmaceuticals Development
Industrial Chemical Products
Allergy & Respiratory Treatments
Central Nervous System Disorder Treatments

## BRANDS/DIVISIONS/AFFILIATES:

Keppra
Neupro
Cimzia
Lortab
Xyzal
Zyrtec
Cyclofluidic
Tussionex

## CONTACTS: *Note: Officers with more than one job title may be intentionally listed here more than once.*

Roch Doliveux, CEO
Detlef Thielgen, CFO/Exec. VP
Fabrice Enderlin, Exec. VP-Human Resources
Michele Antonelli, Exec. VP-Tech. Oper.
Bob Trainor, General Counsel/Exec. VP
Bill Robinson, Exec. VP-Global Oper.
Mark McDade, VP-Bus. Dev. & Corp. Strategy
Antje Witte, VP-Corp. Comm.
Antje Witte, VP-Investor Rel.
Iris Low-Friedrich, Exec. VP-Dev./Chief Medical Officer
Michele Antonelli, Exec. VP-Quality Assurance
Karel Boone, Chmn.

| Phone: 32-2-559-99-99 | Fax: 32-2-599-99-00 |
|---|---|
| Toll-Free: | |
| Address: Allee de la Recherche, 60, Brussels, 1070 Belgium | |

## GROWTH PLANS/SPECIAL FEATURES:

UCB S.A. is a Belgian biopharmaceutical firm with operations in more than 40 countries. The firm's focuses on severe diseases in the fields of the central nervous system (CNS), inflammation and oncology. Keppra, the firm's lead product, is used in both monotherapy and adjunctive therapy for the treatment of epilepsy. Another CNS product is Neupro, approved in Europe for the treatment of early-stage Parkinson's disease. UCB's anti-allergics include Xyzal, an anti-allergic designed to treat and prevent persistent rhinitis in children, characterized by severe and long-lasting allergic symptoms with a tendency to evolve towards allergic asthma; Cimzia, approved in the U.S. and Switzerland for the treatment of Crohn's disease; and Zyrtec, an antihistamine approved in over 100 countries for use in children from the age of six months. Other products include Nootropil, a cerebral function regulator used to treat adults and the elderly; Lortab, an analgesic approved in the U.S. for the relief of moderate to moderately severe pain; Tussionex, a 12-hour cough suppressant approved in the U.S.; Innovair, a combination therapy for asthma; Isoket, used to treat coronary heart disease; and BUP-4, a once-daily treatment for urinary incontinence. In August 2008, the firm agreed to layoff 2,000 employees worldwide. In October 2008, UCB and Pfizer Ltd. founded a new technology company, Cyclofluidic, for the purpose of accelerating the drug discovery process. In January 2009, the company agreed to sell all the rights to Somatostatine-UCB, an anti-hemorrhagic product, to Eumedica. In February 2009, UCB agreed to sell the global rights (excluding the U.S., Barbados and Canada) to EQUASYM IR/XL, an Attention Deficit Hyperactivity Disorder (ADHD) treatment, to Shire plc.

## FINANCIALS: Sales and profits are in thousands of dollars—add 000 to get the full amount. 2008 Note: Financial information for 2008 was not available for all companies at press time.

| | | U.S. Stock Ticker: |
|---|---|---|
| 2008 Sales: $4,860,380 | 2008 Profits: $-58,040 | Int'l Ticker: UCB    Int'l Exchange: Brussels-Euronext |
| 2007 Sales: $4,894,120 | 2007 Profits: $295,590 | Employees: 11,403 |
| 2006 Sales: $2,987,500 | 2006 Profits: $501,100 | Fiscal Year Ends: 12/31 |
| 2005 Sales: $2,602,500 | 2005 Profits: $961,760 | Parent Company: |
| 2004 Sales: $2,132,440 | 2004 Profits: $419,100 | |

## SALARIES/BENEFITS:

| Pension Plan: Y | ESOP Stock Plan: | Profit Sharing: | Top Exec. Salary: $ | Bonus: $ |
|---|---|---|---|---|
| Savings Plan: | Stock Purch. Plan: | | Second Exec. Salary: $ | Bonus: $ |

## OTHER THOUGHTS:

**Apparent Women Officers or Directors:** 2
**Hot Spot for Advancement for Women/Minorities:**

## LOCATIONS: ("Y" = Yes)

| West: | Southwest: | Midwest: | Southeast: | Northeast: | International: |
|---|---|---|---|---|---|
| | | Y | Y | Y | Y |

# UNIGENE LABORATORIES

## www.unigene.com

**Industry Group Code:** 325412 **Ranks within this company's industry group:** Sales: 105 Profits: 70

| Drugs: | | Other: | Clinical: | Computers: | Services: | |
|---|---|---|---|---|---|---|
| Discovery: | Y | AgriBio: | Trials/Services: | Hardware: | Specialty Services: | |
| Licensing: | Y | Genetic Data: | Labs: | Software: | Consulting: | |
| Manufacturing: | Y | Tissue Replacement: | Equipment/Supplies: | Arrays: | Blood Collection: | |
| Genetics: | | | Research & Development Services: | Database Management: | Drug Delivery: | Y |
| | | | Diagnostics: | | Drug Distribution: | |

## TYPES OF BUSINESS:

Peptides Research & Production
Drug Delivery Systems

## BRANDS/DIVISIONS/AFFILIATES:

Fortical
Forcaltonin
EnteriPep
SecraPep
NasaPep

## CONTACTS: *Note: Officers with more than one job title may be intentionally listed here more than once.*

Warren Levy, CEO
Warren P. Levy, Pres.
Nozer M. Mehta, VP-Biological R&D
James Gilligan, VP-Prod. Dev.
Paul Shields, VP-Mfg. Oper.
Ronald Levy, Sec./Exec. VP
William Steinhauer, VP-Finance
Jay Levy, Treas.
Jay Levy, Chmn.

| **Phone:** 973-265-1100 | **Fax:** 973-335-0972 |
|---|---|
| **Toll-Free:** | |
| **Address:** 81 Fulton St., Boonton, NJ 07005 US | |

## GROWTH PLANS/SPECIAL FEATURES:

Unigene Laboratories, Inc. is a biopharmaceutical company that focuses on the research, production and delivery of small proteins, or peptides, for medical use. The company uses its proprietary peptide manufacturing and delivery technology, including SecraPrep for cost-effective production; EnteriPep for oral delivery; and NasaPep for nasal delivery. The firm's primary focus is on the development of salmon calcitonin and other peptide products, for the treatment of osteoporosis and other indications. Unigene has licensed worldwide rights to its manufacturing and delivery technologies for oral parathyroid hormone, which is used to regulate calcium and phosphate metabolism in the bones and kidneys, to GlaxoSmithKline. In the U.S., the company licensed its nasal salmon calcitonin product, trademarked Fortical, to Upsher-Smith Laboratories, Inc. Fortical, aimed at treating osteoporosis, was the company's first drug approved in the U.S. The firm licensed worldwide rights to its patented manufacturing technology for the production of calcitonin to Novartis Pharma AG. In addition to Fortical, Unigene Laboratories has an injectable calcitonin product, Forcaltonin, which is approved for sale in the E.U. for osteoporosis, but has yet to generate significant revenue. The company holds nine U.S. patents. Unigene is also developing site-directed bone-growth technology in conjunction with Yale University. Other preclinical developments include treatments for inflammatory diseases, strokes and heart attacks, metabolic disorders and hormone regulation. In October 2008, the company began a Phase I clinical study for its oral parathyroid hormone treatment, which utilizes the company's EnteriPep delivery technology. In May 2009, the firm was issued its first patent for site-directed bone growth technology.

## FINANCIALS: Sales and profits are in thousands of dollars—add 000 to get the full amount. 2008 Note: Financial information for 2008 was not available for all companies at press time.

| | | |
|---|---|---|
| 2008 Sales: $19,229 | 2008 Profits: $-6,078 | **U.S. Stock Ticker:** UGNE |
| 2007 Sales: $20,423 | 2007 Profits: $-3,448 | **Int'l Ticker:** Int'l Exchange: |
| 2006 Sales: $6,059 | 2006 Profits: $-11,784 | Employees: 103 |
| 2005 Sales: $14,276 | 2005 Profits: $- 496 | Fiscal Year Ends: 12/31 |
| 2004 Sales: $8,400 | 2004 Profits: $-5,900 | Parent Company: |

## SALARIES/BENEFITS:

| Pension Plan: | ESOP Stock Plan: | Profit Sharing: | Top Exec. Salary: $292,962 | Bonus: $ |
|---|---|---|---|---|
| Savings Plan: Y | Stock Purch. Plan: | | Second Exec. Salary: $260,308 | Bonus: $ |

## OTHER THOUGHTS:

**Apparent Women Officers or Directors:**
**Hot Spot for Advancement for Women/Minorities:**

## LOCATIONS: ("Y" = Yes)

| West: | Southwest: | Midwest: | Southeast: | Northeast: | International: |
|---|---|---|---|---|---|
| | | | | Y | |

# UNITED THERAPEUTICS CORP

**www.unither.com**

Industry Group Code: 325412  Ranks within this company's industry group:  Sales: 52   Profits: 131

| Drugs: | | Other: | Clinical: | | Computers: | | Services: | |
|---|---|---|---|---|---|---|---|---|
| Discovery: | Y | AgriBio: | Trials/Services: | | Hardware: | | Specialty Services: | |
| Licensing: | | Genetic Data: | Labs: | | Software: | | Consulting: | |
| Manufacturing: | Y | Tissue Replacement: | Equipment/Supplies: | Y | Arrays: | | Blood Collection: | |
| Genetics: | | | Research & Development Services: | | Database Management: | | Drug Delivery: | |
| | | | Diagnostics: | | | | Drug Distribution: | |

## TYPES OF BUSINESS:

Cardiovascular, Cancer & Infectious Diseases Therapeutics
Dietary Supplements
Telecardiology Products

## BRANDS/DIVISIONS/AFFILIATES:

Unither Pharma, Inc.
Medicomp, Inc.
HeartBar
Remodulin
OvaRex
Lung Rx

## CONTACTS: *Note: Officers with more than one job title may be intentionally listed here more than once.*

Martine Rothblatt, CEO
Roger Jeffs, COO
Roger Jeffs, Pres.
John Ferrari, CFO
Shola Oyewole, CIO
Paul A. Mahon, General Counsel
Paul A. Mahon, Exec. VP-Strategic Planning
John Ferrari, Treas.
Dan Balda, Pres./COO-Medicomp, Inc.
David Walsh, Exec. VP/COO-Production
Martine Rothblatt, Chmn.

| Phone: 301-608-9292 | Fax: 301-608-9291 |
|---|---|
| Toll-Free: | |
| Address: 1110 Spring St., Silver Spring, MD 20910 US | |

## GROWTH PLANS/SPECIAL FEATURES:

United Therapeutics Corp. (UTC) is a biotechnology company focused on the development and commercialization of therapeutic products for patients with cancer, cardiovascular diseases, and other infectious diseases. The company's key therapeutic platforms include: prostacyclin analogs, which are stable synthetic forms of prostacyclin, a molecule produced by the body that affects blood-vessel health and function; immunotherapeutic monoclonal antibodies, which are antibodies that activate patients' immune systems to treat cancer; and glycobiology antiviral agents, which are a class of small molecules that have shows preclinical indications of efficacy against a broad range of viruses. Remodulin, the company's lead product, is an FDA-approved, treprostinil-based compound designed for the treatment of pulmonary arterial hypertension (PAH), for patients with negative symptoms associated with exercise. UTC has commercial rights for Remodulin in most European Union countries, Israel, Australia, Mexico, Argentina, and Peru. The firm is developing inhaled and oral formulations of treprostinil-based products, in addition to Beraprost-MR, a prostacyclin analog designed to treat cardiovascular diseases. One of the firm's subsidiaries, Unither Pharma, Inc., markets the HeartBar line of products, which are arginine-enriched dietary supplements designed to help maintaining healthy circulatory function. Medicomp, Inc., a wholly-owned subsidiary, manufactures and markets a variety of telecardiology services, including cardiac Holter monitoring, event monitoring and analysis, and pacemaker monitoring. Medicomp's services are delivered through its proprietary Digital Decipher Holter recorder/analyzer and its CardioPAL family of event monitors. In June 2008, Lung Rx, a wholly-owned subsidiary focused on pulmonary medicine and pulmonary therapeutic products, submitted a New Drug Application to the FDA for marketing approval of an inhaled form of treprostinil, for the treatment of pulmonary arterial hypertension.

UTC offers its employees medical, vision, dental and prescription insurance; life and AD&D insurance; short- and long-term disability; flexible spending accounts; educational assistance; and an employee assistance program.

## FINANCIALS: Sales and profits are in thousands of dollars—add 000 to get the full amount. 2008 Note: Financial information for 2008 was not available for all companies at press time.

| | | |
|---|---|---|
| 2008 Sales: $281,497 | 2008 Profits: $-42,789 | **U.S. Stock Ticker:** UTHR |
| 2007 Sales: $210,943 | 2007 Profits: $19,859 | **Int'l Ticker:**   Int'l Exchange: |
| 2006 Sales: $159,632 | 2006 Profits: $73,965 | Employees:   320 |
| 2005 Sales: $115,915 | 2005 Profits: $65,016 | Fiscal Year Ends: 12/31 |
| 2004 Sales: $73,590 | 2004 Profits: $15,449 | Parent Company: |

## SALARIES/BENEFITS:

| Pension Plan: | ESOP Stock Plan: | Profit Sharing: | Top Exec. Salary: $796,300 | Bonus: $656,300 |
|---|---|---|---|---|
| Savings Plan: Y | Stock Purch. Plan: | | Second Exec. Salary: $701,300 | Bonus: $424,400 |

## OTHER THOUGHTS:

**Apparent Women Officers or Directors**: 1
**Hot Spot for Advancement for Women/Minorities**:

## LOCATIONS: ("Y" = Yes)

| West: | Southwest: | Midwest: | Southeast: | Northeast: | International: |
|---|---|---|---|---|---|
| | | | Y | Y | Y |

# UNITED-GUARDIAN INC

www.u-g.com

**Industry Group Code:** 325412 **Ranks within this company's industry group:** Sales: 111 Profits: 55

| Drugs: | | Other: | | Clinical: | | Computers: | | Services: | |
|---|---|---|---|---|---|---|---|---|---|
| Discovery: | Y | AgriBio: | | Trials/Services: | | Hardware: | | Specialty Services: | |
| Licensing: | | Genetic Data: | | Labs: | | Software: | | Consulting: | |
| Manufacturing: | Y | Tissue Replacement: | | Equipment/Supplies: | | Arrays: | | Blood Collection: | |
| Genetics: | | | | Research & Development Services: | Y | Database Management: | | Drug Delivery: | |
| | | | | Diagnostics: | | | | Drug Distribution: | |

## TYPES OF BUSINESS:

Cosmetic Ingredients & Pharmaceuticals
Personal & Healthcare Products
Organic Chemicals
Test Solutions
Indicators
Dyes
Reagents

## BRANDS/DIVISIONS/AFFILIATES:

Guardian Laboratories
Orchid Complex
Razoride
Lubrajel
Recidin
Clorpactin
Klensoft
Confetti Dermal Delivery Flakes

## CONTACTS: *Note: Officers with more than one job title may be intentionally listed here more than once.*

Kenneth H. Globus, CEO
Kenneth H. Globus, Pres.
Robert S. Rubinger, CFO/Exec. VP
Joseph J. Vernice, VP/Mgr.-R&D
Joseph J. Vernice, Dir.-Tech. Svcs.
Robert S. Rubinger, Dir.-Prod. Dev./Sec.
Kenneth H. Globus, General Counsel
Cecile M. Brophy, Treas./Controller
Charles W. Castanza, Sr. VP/Dir.-Plant Oper.
Peter A. Hiltunen, VP/Production Mgr.

**Phone:** 631-273-0900 **Fax:** 631-273-0858
**Toll-Free:** 800-645-5566
**Address:** 230 Marcus Blvd., Hauppauge, NY 11788 US

## GROWTH PLANS/SPECIAL FEATURES:

United-Guardian, Inc., through Guardian Laboratories, conducts research, product development, manufacturing and marketing of cosmetic ingredients and other personal care products, pharmaceuticals, medical and health care products and specialty industrial products. Guardian Laboratories' two largest marketed product lines are the Lubrajel line of cosmetic ingredients and Rencidin Irrigation. The Lubrajel line of cosmetic ingredients are nondrying water-based moisturizing and lubricating gels that have applications in the cosmetic industry primarily as a moisturizer and as a base for other cosmetic ingredients, and in the medical field as a lubricant. The Lubrajel line accounted for roughly 77% of revenue in 2008. Rencidin Irrigation, a urological prescription drug, used primarily to prevent the formation of and to dissolve calcifications in catheters implanted in the urinary bladder, accounted for roughly 18% of revenue in 2008. Other products include Clorpactin, a microbicidal product used primarily in urology and surgery as an antiseptic for treating a wide range of localized infections in the urinary bladder, the peritoneum, the abdominal cavity, the eye, ear, nose and throat, and the sinuses; Lubrislide, a powdered lubricant; Lubrajel PF, a preservative-free form of Lubrajel; Unitwix, a cosmetic additive used as a thickener for oils and oil-based liquids; Orchid Complex, base for skin creams, lotions, cleansers, and other cosmetics; Razoride, a water-based shaving product; Klensoft, a surfactant that can be used in shampoos, shower gels, makeup removers and other cosmetic formulations; and Confetti Dermal Delivery Flakes, a product line that incorporates various functional oil-soluble ingredients into colorful flakes that can be added to, and suspended in, various water-based products. The company markets its products primarily through marketing partners, distributors, wholesalers, direct advertising, mailings and trade exhibitions. In July 2008, United-Guardian's stock was added to the Russell Microcap Index.

## FINANCIALS: **Sales and profits are in thousands of dollars—add 000 to get the full amount. 2008 Note: Financial information for 2008 was not available for all companies at press time.**

| | | |
|---|---|---|
| 2008 Sales: $12,292 | 2008 Profits: $3,163 | **U.S. Stock Ticker: UG** |
| 2007 Sales: $11,889 | 2007 Profits: $3,544 | **Int'l Ticker:** Int'l Exchange: |
| 2006 Sales: $11,208 | 2006 Profits: $2,737 | Employees: 39 |
| 2005 Sales: $12,135 | 2005 Profits: $2,617 | Fiscal Year Ends: 12/31 |
| 2004 Sales: $11,123 | 2004 Profits: $2,475 | Parent Company: |

## SALARIES/BENEFITS:

| | | | | |
|---|---|---|---|---|
| Pension Plan: | ESOP Stock Plan: | Profit Sharing: | Top Exec. Salary: $240,989 | Bonus: $75,000 |
| Savings Plan: | Stock Purch. Plan: | | Second Exec. Salary: $159,410 | Bonus: $18,200 |

## OTHER THOUGHTS:

**Apparent Women Officers or Directors**: 1
**Hot Spot for Advancement for Women/Minorities**:

## LOCATIONS: ("Y" = Yes)

| West: | Southwest: | Midwest: | Southeast: | Northeast: | International: |
|---|---|---|---|---|---|
| | | | | Y | |

# URIGEN PHARMACEUTICALS INC

**www.urigen.com**

Industry Group Code: 325412　Ranks within this company's industry group:　Sales:　　Profits: 68

| Drugs: | | Other: | Clinical: | Computers: | Services: |
|---|---|---|---|---|---|
| Discovery: | | AgriBio: | Trials/Services: | Hardware: | Specialty Services: |
| Licensing: | Y | Genetic Data: | Labs: | Software: | Consulting: |
| Manufacturing: | Y | Tissue Replacement: | Equipment/Supplies: | Arrays: | Blood Collection: |
| Genetics: | | | Research & Development Services: | Database Management: | Drug Delivery: |
| | | | Diagnostics: | | Drug Distribution: |

## TYPES OF BUSINESS:

Drugs-Urological

## BRANDS/DIVISIONS/AFFILIATES:

URG101
URG301

## CONTACTS: *Note: Officers with more than one job title may be intentionally listed here more than once.*

William J. Garner, CEO
Terry M. Nida, COO
William J. Garner, Pres.
Martin E. Shmagin, CFO
Dennis H. Giesing, Chief Scientific Officer
Amie E. Franklin, Mgr.-Clinical, Regulatory & Intellectual Property
Tracy Taylor, Chmn.

| Phone: 415-781-0350 | Fax: 415-781-0385 |
|---|---|
| Toll-Free: | |
| Address: 27 Maiden Ln., Ste. 595, San Francisco, CA 94108 US | |

## GROWTH PLANS/SPECIAL FEATURES:

Urigen Pharmaceuticals, Inc. specializes in the development and commercialization of products for the treatment and diagnosis of urological disorders. The company has two products that are currently in clinical trials, and another, an intranasal testosterone, in the developmental stages. URG101, the firm's leading product, is a proprietary combination therapy of components approved by global regulatory authorities that is locally delivered to the bladder for rapid relief of pain and urgency. It is in Phase II clinical trials for painful bladder syndrome (PBS), but the company is also developing indications for radiation cystitis and dyspareunia, or painful intercourse. URG301 is in Phase II trials for both overactive bladder and urethritis. In addition, URG301 is being evaluated as a possible treatment of acute urethral discomfort. The company maintains strategic partnerships with Life Science Strategy Group and Kalium, Inc. In March 2008, the firm announces successful results from a Phase II clinical trial of URG101 in patients with Painful Bladder Syndrome/Interstitial Cystitis (PBS/IC). In April 2008, Urigen ended its Development and License Agreement with M et P Patent AG, for intranasal testosterone, in an effort to focus resources on URG101 and URG301. In April 2008, the firm announced an allowance of claims in its first patent application for URG101, to be used for the treatment of PBS. The product has since been moved into late-stage clinical development.

## FINANCIALS: Sales and profits are in thousands of dollars—add 000 to get the full amount. 2008 Note: Financial information for 2008 was not available for all companies at press time.

| | | |
|---|---|---|
| 2008 Sales: $ | 2008 Profits: $-4,844 | U.S. Stock Ticker: URGP.OB |
| 2007 Sales: $ 571 | 2007 Profits: $-3,733 | Int'l Ticker:　Int'l Exchange: |
| 2006 Sales: $ 727 | 2006 Profits: $-15,337 | Employees:　6 |
| 2005 Sales: $2,177 | 2005 Profits: $-11,083 | Fiscal Year Ends: 6/30 |
| 2004 Sales: $7,478 | 2004 Profits: $-6,484 | Parent Company: |

## SALARIES/BENEFITS:

| Pension Plan: | ESOP Stock Plan: | Profit Sharing: | Top Exec. Salary: $250,000 | Bonus: $ |
|---|---|---|---|---|
| Savings Plan: | Stock Purch. Plan: | | Second Exec. Salary: $225,000 | Bonus: $ |

## OTHER THOUGHTS:

Apparent Women Officers or Directors: 1
Hot Spot for Advancement for Women/Minorities:

## LOCATIONS: ("Y" = Yes)

| West: | Southwest: | Midwest: | Southeast: | Northeast: | International: |
|---|---|---|---|---|---|
| Y | | | | | |

# VALEANT PHARMACEUTICALS INTERNATIONAL www.valeant.com

Industry Group Code: 325412  Ranks within this company's industry group: Sales: 42    Profits: 110

| Drugs: | | Other: | | Clinical: | | Computers: | | Services: | |
|---|---|---|---|---|---|---|---|---|---|
| Discovery: | Y | AgriBio: | | Trials/Services: | | Hardware: | | Specialty Services: | |
| Licensing: | Y | Genetic Data: | | Labs: | | Software: | | Consulting: | |
| Manufacturing: | Y | Tissue Replacement: | | Equipment/Supplies: | | Arrays: | | Blood Collection: | |
| Genetics: | | | | Research & Development Services: | | Database Management: | | Drug Delivery: | |
| | | | | Diagnostics: | | | | Drug Distribution: | |

## TYPES OF BUSINESS:

Prescription & Non-Prescription Pharmaceuticals
Neurology Drugs
Dermatology Drugs
Infectious Diseases Drugs

## BRANDS/DIVISIONS/AFFILIATES:

Efudex/Efudix
Dermatix
Oxsoralen-Ultra
Cloderm
Virazole
Coria Laboratories, Ltd.
DermaTech Pty Ltd.
Dow Pharmaceuticals Sciences, Inc

## CONTACTS: Note: Officers with more than one job title may be intentionally listed here more than once.

J. Michael Pearson, CEO
Peter J. Blott, CFO/Exec. VP
Elisa A. Carlson, Chief Admin. Officer/Exec. VP
Steve T. Min, General Counsel/Exec. VP/Corp. Sec.
Richard K. Masterson, Exec. VP-Commercial Dev.
J. Michael Pearson, Chmn.
Bhaskar Chaudhuri, Pres., Valeant Pharmaceuticals Int'l

| Phone: 949-461-6000 | Fax: 949-461-6609 |
|---|---|
| Toll-Free: 800-548-5100 | |
| Address: 1 Enterprise, Aliso Viejo, CA 92656 US | |

## GROWTH PLANS/SPECIAL FEATURES:

Valeant Pharmaceuticals International is a pharmaceutical company that develops, manufactures and markets a broad spectrum of prescription and non-prescription pharmaceuticals. The company focuses in the therapeutic areas of neurology, dermatology and infectious diseases. The firm's products also treat neuromuscular disorders, cancer, cardiovascular disease, diabetes and psychiatric disorders. Valeant's products are sold through three pharmaceutical segments: Specialty Pharmaceuticals, Branded Generics-Europe and Branded Generics-Latin America. The company's specialty pharmaceuticals segment includes product revenues primarily from the U.S., Canada and Australia. Its product portfolio comprises roughly 389 branded products with approximately 982 stock keeping units. Products in the neurology field include Mestinon, Librax, Migranal, Tasmar, Zelapar and Diastat AcuDial; in the dermatology field, Efudex, Kinerase, Oxsoralen-Ultra and Cloderm. Another drug, Virazole, is for the treatment of various viral diseases. The Branded Generics-Europe segment includes revenues from products in Poland, Hungary, the Czech Republic and Slovakia. The Branded Generics-Latin America segment includes revenues from products in Mexico and Brazil. Valeant's research and development program focuses on preclinical and clinical development of identified molecules. The company is developing product candidates, including taribavirin and retigabine. The firm's marketing and promotion efforts focus on the Promoted Products, which consists of products sold in more than 100 global markets. In October 2008, the company acquired Coria Laboratories, Ltd. a private pharmaceutical company specializing in dermatology medications. In November of the same year, the firm acquired DermaTech Pty Ltd., an Australia-based company also focused on dermatology products. In 2009, the company acquired Dow Pharmaceuticals Sciences, Inc. a dermatology company that develops topical products. Also in 2009, the firm drastically cut its research budget from $100 million to $50 million.

## FINANCIALS: Sales and profits are in thousands of dollars—add 000 to get the full amount. 2008 Note: Financial information for 2008 was not available for all companies at press time.

| | | | |
|---|---|---|---|
| 2008 Sales: $656,977 | 2008 Profits: $-23,714 | U.S. Stock Ticker: VRX | |
| 2007 Sales: $689,503 | 2007 Profits: $-6,186 | Int'l Ticker:   Int'l Exchange: | |
| 2006 Sales: $685,052 | 2006 Profits: $-57,568 | Employees:  2,331 | |
| 2005 Sales: $823,886 | 2005 Profits: $-188,143 | Fiscal Year Ends: 12/31 | |
| 2004 Sales: $684,251 | 2004 Profits: $-154,653 | Parent Company: | |

## SALARIES/BENEFITS:

| Pension Plan: | ESOP Stock Plan: | Profit Sharing: | Top Exec. Salary: $916,667 | Bonus: $1,000,000 |
|---|---|---|---|---|
| Savings Plan: Y | Stock Purch. Plan: | | Second Exec. Salary: $390,000 | Bonus: $250,000 |

## OTHER THOUGHTS:

**Apparent Women Officers or Directors**: 2
**Hot Spot for Advancement for Women/Minorities**: Y

## LOCATIONS: ("Y" = Yes)

| West: | Southwest: | Midwest: | Southeast: | Northeast: | International: |
|---|---|---|---|---|---|
| Y | | | | | Y |

# VASOGEN INC

**www.vasogen.com**

**Industry Group Code:** 325412  **Ranks within this company's industry group:** Sales:    Profits: 92

| Drugs: | | Other: | Clinical: | Computers: | Services: |
|---|---|---|---|---|---|
| Discovery: | Y | AgriBio: | Trials/Services: | Hardware: | Specialty Services: |
| Licensing: | | Genetic Data: | Labs: | Software: | Consulting: |
| Manufacturing: | | Tissue Replacement: | Equipment/Supplies: | Arrays: | Blood Collection: |
| Genetics: | | | Research & Development Services: | Database Management: | Drug Delivery: |
| | | | Diagnostics: | | Drug Distribution: |

## TYPES OF BUSINESS:

Biopharmaceuticals Development
Immune Modulation Therapies

## BRANDS/DIVISIONS/AFFILIATES:

Celacade

## CONTACTS: Note: Officers with more than one job title may be intentionally listed here more than once.

Christopher J. Waddick, CEO
Christopher J. Waddick, Pres.
Graham D. Neil, CFO
Eldon R. Smith, Sr. VP-Scientific Affairs/Chief Medical Officers
Graham D. Neil, VP-Finance
Eldon R. Smith, Chmn.

| **Phone:** 905-402-9925 | **Fax:** 905-847-6270 |
|---|---|
| **Toll-Free:** | |
| **Address:** 4 Robert Speck Pkwy., 15th Fl., Mississauga, ON L4Z 1S1 Canada | |

## GROWTH PLANS/SPECIAL FEATURES:

Vasogen, Inc. researches, develops and commercializes immune modulation therapies designed to target destructive inflammatory process associated with cardiovascular and neurodegenerative disorders.   Historically, the firm has worked on two leading drug therapies, the Celacade System and the VP drug series.  Celacade is an immune modulation therapy designed to treat chronic heart failure and peripheral arterial disease.   The VP Series of drugs includes two candidates, VP015 and VP025.  The drugs are designed to target   inflammation   underlying   certain   neurological conditions.  None of these drugs are currently being actively developed.  In April 2008, Vasogen announced a corporate restructuring plan that involved the reduction of its workforce by 85%; the suspension of plans for ACCLAIM II, a clinical trial designed to support the use of Celacade in the treatment of patients with NYHA Class II heart failure; and the discontinuation   of   support   for   the   European commercialization or any further development of Celacade.

## FINANCIALS:  Sales and profits are in thousands of dollars—add 000 to get the full amount. 2008 Note: Financial information for 2008 was not available for all companies at press time.

| | | |
|---|---|---|
| 2008 Sales: $ | 2008 Profits: $-14,800 | **U.S. Stock Ticker: VSGN** |
| 2007 Sales: $ | 2007 Profits: $-26,500 | **Int'l Ticker: VAS**    Int'l Exchange: Toronto-TSX |
| 2006 Sales: $ | 2006 Profits: $-61,100 | Employees:   104 |
| 2005 Sales: $90,700 | 2005 Profits: $-90,300 | Fiscal Year Ends: 11/30 |
| 2004 Sales: $ | 2004 Profits: $-62,800 | Parent Company: |

## SALARIES/BENEFITS:

| Pension Plan: | ESOP Stock Plan: | Profit Sharing: | Top Exec. Salary: $330,000 | Bonus: $82,500 |
|---|---|---|---|---|
| Savings Plan: | Stock Purch. Plan: | | Second Exec. Salary: $157,500 | Bonus: $31,500 |

## OTHER THOUGHTS:

**Apparent Women Officers or Directors:**
**Hot Spot for Advancement for Women/Minorities:**

## LOCATIONS: ("Y" = Yes)

| West: | Southwest: | Midwest: | Southeast: | Northeast: | International: |
|---|---|---|---|---|---|
| | | | | | Y |

# VAXGEN INC

**www.vaxgen.com**

**Industry Group Code:** 325412 **Ranks within this company's industry group:** Sales: 151 Profits: 88

| Drugs: | | Other: | Clinical: | Computers: | | Services: | |
|---|---|---|---|---|---|---|---|
| Discovery: | Y | AgriBio: | Trials/Services: | Hardware: | | Specialty Services: | |
| Licensing: | | Genetic Data: | Labs: | Software: | | Consulting: | |
| Manufacturing: | | Tissue Replacement: | Equipment/Supplies: | Arrays: | Y | Blood Collection: | |
| Genetics: | | | Research & Development Services: | Database Management: | | Drug Delivery: | |
| | | | Diagnostics: | | | Drug Distribution: | |

## TYPES OF BUSINESS:

Biologic Products

## BRANDS/DIVISIONS/AFFILIATES:

## CONTACTS: *Note: Officers with more than one job title may be intentionally listed here more than once.*

James P. Panek, CEO
James P. Panek, Pres.
Piers Whitehead, Sec.
Piers Whitehead, VP-Corp. & Bus. Dev.
James P. Panek, Chief Acct. Officer

| **Phone:** 650-624-2439 | **Fax:** 650-624-4785 |
|---|---|
| **Toll-Free:** | |
| **Address:** 379 Oyster Point Blvd., Ste. 10, South San Francisco, CA 94080 US | |

## GROWTH PLANS/SPECIAL FEATURES:

VaxGen, Inc. is a biopharmaceutical company that historically developed, manufactured and commercialized biologic products for the prevention and treatment of human infectious diseases. VaxGen owns a facility in California with a 1,000-liter bioreactor that can be used to make cell culture or microbial biologic products, and capable of either microbial or mammalian fermentation. However, the company has ended all product development activities and sold or otherwise terminated its drug development programs. As a result of a proposed merger between VaxGen and Raven BioTechnologies, Inc., which was terminated, VaxGen underwent significant restructuring in April 2008, which included reduction of workforce by 75%. During August 2008, the firm further reduced its workforce by an additional 25%. Also in 2008, VaxGen sold all of its assets and rights related to its anthrax vaccine candidate to Emergent BioSolutions, Inc. for $2 million with another possible $8 million in milestone payments.

## FINANCIALS: Sales and profits are in thousands of dollars—add 000 to get the full amount. 2008 Note: Financial information for 2008 was not available for all companies at press time.

| | | |
|---|---|---|
| 2008 Sales: $ 293 | 2008 Profits: $-12,563 | **U.S. Stock Ticker:** VXGN |
| 2007 Sales: $5,011 | 2007 Profits: $-44,180 | **Int'l Ticker:** Int'l Exchange: |
| 2006 Sales: $14,836 | 2006 Profits: $37,592 | Employees: 3 |
| 2005 Sales: $29,939 | 2005 Profits: $-55,958 | Fiscal Year Ends: 12/31 |
| 2004 Sales: $ | 2004 Profits: $ | Parent Company: |

## SALARIES/BENEFITS:

| | | | | |
|---|---|---|---|---|
| Pension Plan: | ESOP Stock Plan: | Profit Sharing: | Top Exec. Salary: $390,000 | Bonus: $58,500 |
| Savings Plan: | Stock Purch. Plan: | | Second Exec. Salary: $160,417 | Bonus: $ |

## OTHER THOUGHTS:

**Apparent Women Officers or Directors:** 1
**Hot Spot for Advancement for Women/Minorities:**

## LOCATIONS: ("Y" = Yes)

| West: | Southwest: | Midwest: | Southeast: | Northeast: | International: |
|---|---|---|---|---|---|
| Y | | | | | |

# VENTRIA BIOSCIENCE

www.ventria.com

Industry Group Code: 325411 **Ranks within this company's industry group:** Sales: Profits:

| Drugs: | Other: | | Clinical: | Computers: | Services: |
|---|---|---|---|---|---|
| Discovery: | AgriBio: | Y | Trials/Services: | Hardware: | Specialty Services: |
| Licensing: | Genetic Data: | | Labs: | Software: | Consulting: |
| Manufacturing: Y | Tissue Replacement: | | Equipment/Supplies: | Arrays: | Blood Collection: |
| Genetics: | | | Research & Development Services: | Database Management: | Drug Delivery: |
| | | | Diagnostics: | | Drug Distribution: |

## TYPES OF BUSINESS:

Medical Supplies, Manufacturing
Lactoferrin Manufacturing
Lysozyme Manufacturing

## BRANDS/DIVISIONS/AFFILIATES:

ExpressPro
ExpressTec
ExpressMab
BioShare

## CONTACTS: *Note: Officers with more than one job title may be intentionally listed here more than once.*

Scott E. Deeter, CEO
Scott E. Deeter, Pres.
Ning Huang, VP-R&D
Randy Semadeni, VP-Bus. Dev.
Randy Semadeni, VP-Finance
Delia R. Bethell, VP-Clinical Dev.
Victor Hicks, VP/Commercial Dir., InVitria Division
Greg Unruh, VP/Gen. Mgr.

| Phone: 916-921-6148 | Fax: 916-921-5611 |
|---|---|
| Toll-Free: | |
| Address: 4110 N. Freeway Blvd., Sacramento, CA 95834 US | |

## GROWTH PLANS/SPECIAL FEATURES:

Ventria Bioscience focuses on human nutrition and human therapeutics products. The company manufactures two products, lactoferrin and lysozyme. Lactoferrin is a globular multifunctional protein with antimicrobial activity. Lactoferrin protein has applications in the alleviation of fungal infections gastrointestinal health, dietary management of acute diarrhea and the treatment of topical infections and inflammations. Lysozyme is an enzyme that attacks the cell walls of many bacterias and has anti-bacterial, anti-viral and anti-fungal properties. Lysozyme enzyme has applications in the alleviation of fungal infections gastrointestinal health, dietary management of acute diarrhea and the treatment of topical infections and inflammations. The materials are generally found in human breast milk, as well as tears, nasogastric secretions, saliva and bronchial secretions. Ventria's ExpressTec platform uses self-pollinating crops, specifically rice and barley, as the production host for these products. Ventria uses the ExpressTec production system as a basis for forming more specialized systems that produce specific molecules. These platforms include ExpressPro for production of proteins, ExpressTide for production of peptides and ExpressMab for production of monoclonal antibodies. The company has a program for researchers called BioShare, in which researchers submit requests for materials and are granted access to a free supply of recombinant proteins and peptides. In October 2008, Ventria expanded its lactoferrin production.

## FINANCIALS: Sales and profits are in thousands of dollars—add 000 to get the full amount. 2008 Note: Financial information for 2008 was not available for all companies at press time.

| | | |
|---|---|---|
| 2008 Sales: $ | 2008 Profits: $ | **U.S. Stock Ticker:** Private |
| 2007 Sales: $ | 2007 Profits: $ | **Int'l Ticker:** Int'l Exchange: |
| 2006 Sales: $ | 2006 Profits: $ | Employees: |
| 2005 Sales: $ | 2005 Profits: $ | Fiscal Year Ends: |
| 2004 Sales: $ | 2004 Profits: $ | Parent Company: |

## SALARIES/BENEFITS:

| Pension Plan: | ESOP Stock Plan: | Profit Sharing: | Top Exec. Salary: $ | Bonus: $ |
|---|---|---|---|---|
| Savings Plan: Y | Stock Purch. Plan: | | Second Exec. Salary: $ | Bonus: $ |

## OTHER THOUGHTS:

**Apparent Women Officers or Directors**: 1
**Hot Spot for Advancement for Women/Minorities**:

## LOCATIONS: ("Y" = Yes)

| West: | Southwest: | Midwest: | Southeast: | Northeast: | International: |
|---|---|---|---|---|---|
| Y | | Y | | | |

# VERENIUM CORPORATION

**www.verenium.com**

**Industry Group Code: 541712   Ranks within this company's industry group: Sales: 11   Profits: 18**

| Drugs: | Other: | | Clinical: | Computers: | Services: | |
|---|---|---|---|---|---|---|
| Discovery: | AgriBio: | Y | Trials/Services: | Hardware: | Specialty Services: | Y |
| Licensing: | Genetic Data: | | Labs: | Software: | Consulting: | |
| Manufacturing: | Tissue Replacement: | | Equipment/Supplies: | Arrays: | Blood Collection: | |
| Genetics: | | | Research & Development Services: | Database Management: | Drug Delivery: | |
| | | | Diagnostics: | | Drug Distribution: | |

## TYPES OF BUSINESS:

Specialty Enzyme Production
Biofuels
Cellulosic Ethanol Technology

## BRANDS/DIVISIONS/AFFILIATES:

BP
Fuelzyme-LF
Fuelzyme-CX
Purifine
Luminase
Quantum
Marubeni Corporation
Tsukishima Kikai Corporation, Ltd.

## CONTACTS: *Note: Officers with more than one job title may be intentionally listed here more than once.*

Carlos A. Riva, CEO
Carlos A. Riva, Pres.
James Levine, CFO/Exec. VP
Nell Jones, Sr. VP-Human Resources
Gregory Powers, Exec. VP-R&D
J. Chris Terajewicz, Sr. VP-Eng. & Construction
Gerald M. Haines II, Chief Legal Officer/Sr. VP
William H. Baum, Exec. VP-Bus. Dev.
Jeffrey G. Black, Chief Acct. Officer/Sr. VP
John B. Howe, VP-Public Affairs
Janet Roemer, Exec. VP-Specialty Enzymes Bus. Unit
James Cavanaugh, Chmn.

| Phone: 617-674-5300 | Fax: |
|---|---|
| Toll-Free: | |
| Address: 55 Cambridge Pkwy., Cambridge, MA 02142 US | |

## GROWTH PLANS/SPECIAL FEATURES:

Verenium Corporation, formerly Diversa Corporation, is an alternative energy and bioenzymes company. The firm operates two segments, representing Diversa's and Celunol's old businesses, respectively: Specialty Enzymes and Biofuels. It also operates a Research and Development business unit supporting both segments. The Specialty Enzymes segment manufactures chemicals for three target markets: Alternative Fuels, Specialty Industrial Processes and Animal Health & Nutrition. Its commercial enzymes for the biofuels market include Fuelzyme-LF, used to increase the efficiency of corn ethanol production and Purifine, used both to convert vegetable oils to biodiesel and oilseed into edible oils. For the Industrial market, Luminase makes pulp more susceptible to bleaching and Cottonase is designed to provide better cleaning of cotton that traditional methods like chemical scouring. For the Animal Health & Nutrition market, the Phytase enzyme helps phosphorous and other nutrients in animal feed become more digestible. The Biofuels segment focuses on cellulosic ethanol production, aiming to commercially produce ethanol fuel from almost any cellulose-based biomass, including agricultural waste, switchgrass and wood pulp. It currently has one research and development ethanol plant in Louisiana dedicated to improving Verenium's extraction process and extending it to new biomass materials; and it is nearing the production phase of a demonstration plant, also in Louisiana, which will ultimately produce around 1.4 million gallons of ethanol annually. The firm will partner with BP in the development of a massive cellulosic ethanol production plant in Florida at a cost of about $300 million. In addition, BP invested, in early 2009, $112.5 million in Verenium and received a 50% stake in Verenium's technology licensing business.

Employees are offered health insurance, a 401(k) plan and stock options.

## FINANCIALS: Sales and profits are in thousands of dollars—add 000 to get the full amount. 2008 Note: Financial information for 2008 was not available for all companies at press time.

| | | |
|---|---|---|
| 2008 Sales: $69,659 | 2008 Profits: $-185,542 | **U.S. Stock Ticker: VRNM** |
| 2007 Sales: $46,273 | 2007 Profits: $-107,585 | **Int'l Ticker:**   Int'l Exchange: |
| 2006 Sales: $49,198 | 2006 Profits: $-39,271 | Employees:   303 |
| 2005 Sales: $54,303 | 2005 Profits: $-89,718 | Fiscal Year Ends: 12/31 |
| 2004 Sales: $ | 2004 Profits: $ | Parent Company: |

## SALARIES/BENEFITS:

| Pension Plan: | ESOP Stock Plan: | Profit Sharing: | Top Exec. Salary: $495,000 | Bonus: $208,000 |
|---|---|---|---|---|
| Savings Plan: Y | Stock Purch. Plan: Y | | Second Exec. Salary: $367,422 | Bonus: $100,000 |

## OTHER THOUGHTS:

**Apparent Women Officers or Directors**: 3
**Hot Spot for Advancement for Women/Minorities**: Y

## LOCATIONS: ("Y" = Yes)

| West: | Southwest: | Midwest: | Southeast: | Northeast: | International: |
|---|---|---|---|---|---|
| Y | | | Y | Y | |

# VERMILLION INC

**www.vermillion.com**

Industry Group Code: 325413  Ranks within this company's industry group:  Sales:    Profits:

| Drugs: | Other: | | Clinical: | | Computers: | | Services: | |
|--------|--------|---|-----------|---|-----------|---|-----------|---|
| Discovery: | AgriBio: | | Trials/Services: | | Hardware: | | Specialty Services: | |
| Licensing: | Genetic Data: | Y | Labs: | Y | Software: | | Consulting: | |
| Manufacturing: | Tissue Replacement: | | Equipment/Supplies: | Y | Arrays: | | Blood Collection: | |
| Genetics: | | | Research & Development Services: | Y | Database Management: | | Drug Delivery: | |
| | | | Diagnostics: | Y | | | Drug Distribution: | |

## TYPES OF BUSINESS:

Biomarkers
Translation Proteomics
Research Services
Specialty Diagnostics

## BRANDS/DIVISIONS/AFFILIATES:

Ciphergen Biosystems, Inc.
OVA1

## CONTACTS: Note: Officers with more than one job title may be intentionally listed here more than once.

Gail S. Page, Chmn.

| Phone: 510-226-2800 | Fax: 510-226-2801 |
|---------------------|-------------------|
| Toll-Free: | |
| Address: 47350 Fremont Blvd., Fremont, CA 94538 US | |

## GROWTH PLANS/SPECIAL FEATURES:

Vermillion, Inc., formerly Ciphergen Biosystems, Inc., discovers, develops and commercializes high-value diagnostic tests in the fields of oncology, hematology, cardiology and women's health. Vermillion utilizes advanced protein separation methods to identify and resolve variants of specific biomarkers (known as translational proteomics) for developing a procedure to measure a property or concentration of an analyte (known as an assay) and commercializing novel diagnostic tests. It will also address clinical questions related to early disease detection, treatment response, monitoring of disease progression, prognosis and others through collaborations with leading academic and research institutions in addition to a strategic alliance agreement with Quest. Academic and research institutions that Vermillion has collaborated with include The Johns Hopkins University School of Medicine; The University of Texas M.D. Anderson Cancer Center; University College London; The University of Texas Medical Branch; The Katholieke Universiteit Leuven; The Ohio State University Research Foundation; and Stanford University. Together with its collaborators, Vermillion is currently conducting large-scale protein biomarker studies. The company has one diagnostic product commercially available for thrombotic thrombocytopenic purpura, a rare disorder of the blood coagulation system that is potentially life-threatening if left untreated. It also has multiple ovarian cancer diagnostic tests in development, the most advanced of which is OVA1. In March 2009, Vermillion filed for Chapter 11 bankruptcy protection. In connection with filing for bankruptcy, the company reduced its workforce.

## FINANCIALS: Sales and profits are in thousands of dollars—add 000 to get the full amount. 2008 Note: Financial information for 2008 was not available for all companies at press time.

| | | |
|---|---|---|
| 2008 Sales: $ | 2008 Profits: $ | U.S. Stock Ticker: VRML |
| 2007 Sales: $  44 | 2007 Profits: $-21,282 | Int'l Ticker:    Int'l Exchange: |
| 2006 Sales: $18,215 | 2006 Profits: $-22,066 | Employees:   30 |
| 2005 Sales: $27,246 | 2005 Profits: $-35,433 | Fiscal Year Ends: 12/31 |
| 2004 Sales: $20,181 | 2004 Profits: $-19,841 | Parent Company: |

## SALARIES/BENEFITS:

| Pension Plan: | ESOP Stock Plan: | Profit Sharing: | Top Exec. Salary: $360,958 | Bonus: $108,150 |
|---------------|------------------|-----------------|----------------------------|-----------------|
| Savings Plan: | Stock Purch. Plan: Y | | Second Exec. Salary: $215,458 | Bonus: $38,700 |

## OTHER THOUGHTS:

**Apparent Women Officers or Directors:**
**Hot Spot for Advancement for Women/Minorities:**

## LOCATIONS: ("Y" = Yes)

| West: | Southwest: | Midwest: | Southeast: | Northeast: | International: |
|-------|-----------|----------|------------|------------|---------------|
| Y | | | | | |

# VERNALIS PLC

**www.vernalis.com**

**Industry Group Code: 325412  Ranks within this company's industry group:  Sales: 71    Profits: 58**

| Drugs: | | Other: | Clinical: | Computers: | Services: | |
|---|---|---|---|---|---|---|
| Discovery: | Y | AgriBio: | Trials/Services: | Hardware: | Specialty Services: | Y |
| Licensing: | Y | Genetic Data: | Labs: | Software: | Consulting: | |
| Manufacturing: | | Tissue Replacement: | Equipment/Supplies: | Arrays: | Blood Collection: | |
| Genetics: | | | Research & Development Services: | Database Management: | Drug Delivery: | |
| | | | Diagnostics: | | Drug Distribution: | Y |

## TYPES OF BUSINESS:

Drugs-Neurology & Acute Pain
Drugs-Parkinson's Disease
Drugs-Migraine
Drugs-Obesity
Drugs-Oncology

## BRANDS/DIVISIONS/AFFILIATES:

Frovatriptan
V10153
V1512
V3381
SK Chemicals

## CONTACTS: *Note: Officers with more than one job title may be intentionally listed here more than once.*

Ian Garland, CEO
John Slater, COO
David Mackney, CFO
Nigel Clark, VP-Bus. Dev.
Peter Fellner, Exec. Chmn.

| Phone: 440-118-977-3133 | Fax: 440-118-989-9300 |
|---|---|

**Toll-Free:**

**Address:** Oakdene Ct., 613 Reading Rd., Winnersh,  RG41 5UA UK

## GROWTH PLANS/SPECIAL FEATURES:

Vernalis plc discovers, researches and develops new compounds for central nervous system disorders. The company's main two developmental franchises are for pain, including the treatment of migraines, and for neurological disorders, such as Parkinson's disease. The firm actively seeks partnerships with pharmaceutical companies to complete the development and marketing of those compounds. Vernalis currently has one marketed product and seven products in clinical development. Vernalis' marketed drug is frovatriptan, intended for the oral treatment of migraines. Frovatriptan has been approved in the U.S., Canada, Europe and seven Central American countries. The firm's products in clinical development include V10153, a dual-acting, thrombolytic protein being developed to treat acute ischaemic stroke; V1512, a combination therapy for the treatment of Parkinson's disease; and V3381, an NMDA antagonist and MAO-A inhibitor being developed for the treatment of acute neuropathic pain, such as pain from long-standing diabetes. Vernalis is investigating the potential use of certain receptor antagonists to treat Parkinson's disease and obesity at the level of the nervous system, and is working with Hsp90 in an attempt to inhibit cancer growth. In June 2008, Vernalis agreed to sell Apokyn, a treatment for immobilization due to Parkinson's disease, to Ipsen S.A., along with Vernalis's U.S. commercial operations. In June 2009, the firm's partner, SK Chemicals, was granted approval to market frovatriptan in South Korea.

## FINANCIALS: Sales and profits are in thousands of dollars—add 000 to get the full amount. 2008 Note: Financial information for 2008 was not available for all companies at press time.

| | | |
|---|---|---|
| 2008 Sales: $89,900 | 2008 Profits: $- 100 | **U.S. Stock Ticker:** |
| 2007 Sales: $32,500 | 2007 Profits: $-74,500 | **Int'l Ticker: VER**    Int'l Exchange: London-LSE |
| 2006 Sales: $20,400 | 2006 Profits: $-69,600 | Employees:    90 |
| 2005 Sales: $28,000 | 2005 Profits: $-65,000 | Fiscal Year Ends: 12/31 |
| 2004 Sales: $29,100 | 2004 Profits: $-56,000 | Parent Company: |

## SALARIES/BENEFITS:

| Pension Plan: | ESOP Stock Plan: | Profit Sharing: | Top Exec. Salary: $467,464 | Bonus: $232,365 |
|---|---|---|---|---|
| Savings Plan: | Stock Purch. Plan: | | Second Exec. Salary: $344,557 | Bonus: $106,648 |

## OTHER THOUGHTS:

**Apparent Women Officers or Directors**: 1
**Hot Spot for Advancement for Women/Minorities**: Y

## LOCATIONS: ("Y" = Yes)

| West: | Southwest: | Midwest: | Southeast: | Northeast: | International: |
|---|---|---|---|---|---|
| | | | | | Y |

# VERTEX PHARMACEUTICALS INC

www.vpharm.com

**Industry Group Code: 325412  Ranks within this company's industry group: Sales: 61  Profits: 171**

| Drugs: | | Other: | Clinical: | Computers: | Services: |
|---|---|---|---|---|---|
| Discovery: | Y | AgriBio: | Trials/Services: | Hardware: | Specialty Services: |
| Licensing: | Y | Genetic Data: | Labs: | Software: | Consulting: |
| Manufacturing: | | Tissue Replacement: | Equipment/Supplies: | Arrays: | Blood Collection: |
| Genetics: | | | Research & Development Services: | Database Management: | Drug Delivery: |
| | | | Diagnostics: | | Drug Distribution: |

## TYPES OF BUSINESS:

Small Molecule Drugs

## BRANDS/DIVISIONS/AFFILIATES:

Telaprevir
Lexiva
Telzir
MK-0457
VX-702
VX-770
VX-883
VX-409

## CONTACTS: Note: Officers with more than one job title may be intentionally listed here more than once.

Joshua Boger, CEO
Joshua Boger, Pres.
Ian F. Smith, CFO/Exec. VP
Lisa Kelly-Croswell, Sr. VP-Human Resources
Peter Mueller, Chief Scientific Officer
Kenneth S. Boger, General Counsel/Sr. VP
Kurt C. Graves, Exec. VP/Head-Strategic Dev.
Peter Mueller, Exec. VP-Drug Innovation & Realization
Richard C. Garrison, Sr. VP-Catalyst
John J. Alam, Exec. VP-Medicine Dev./Chief Medical Officer
Amit K. Sachdev, Sr. VP-Public Policy & Gov't Affairs
Charles A. Sanders, Chmn.

| Phone: 617-444-6100 | Fax: 617-444-6680 |
|---|---|
| Toll-Free: | |
| Address: 130 Waverly St., Cambridge, MA 02139 US | |

## GROWTH PLANS/SPECIAL FEATURES:

Vertex Pharmaceuticals, Inc. discovers, develops and commercializes small molecule drugs for the treatment of serious diseases. The company concentrates most of its drug development resources on four drug candidates: Telaprevir, for the treatment of hepatitis C virus infection; VX-702, for the treatment of rheumatoid arthritis and other inflammatory diseases; VX-770, for the treatment of cystic fibrosis; and VX-883, for the treatment of bacterial infection. The firm's lead product, telaprevir, is an oral hepatitis C protease inhibitor in Phase III clinical trials. The U.S. Food and Drug Administration (FDA) granted fast track designation to telaprevir. Vertex's pipeline includes several drug candidates that are being developed by its collaborators. The most advanced of these drug candidates is MK-0457, an Aurora kinase inhibitor developed by Merck & Co., Inc. for the treatment of cancer. Other collaborations include Cystic Fibrosis Foundation Therapeutics, Inc., with which it is developing VX-770; GlaxoSmithKline, for the development and commercialization of VX-409; and Kissei Pharmaceutical Co., Ltd., for the development and commercialization of VX-702. Forsamprenavir calcium, a company-discovered compound for the treatment of HIV infection, is marketed by the firm's collaborator GlaxoSmithKlein plc as Lexiva in the U.S. and Telzir in Europe. The firm began Phase III clinical trials for telaprevir in March 2008.

Employee benefits at Vertex include medical, dental and vision insurance; flexible benefits; educational assistance; commuter benefits; employee assistance; child and elder care; banking and mortgage discounts; and discount entertainment passes.

## FINANCIALS: Sales and profits are in thousands of dollars—add 000 to get the full amount. 2008 Note: Financial information for 2008 was not available for all companies at press time.

| | | |
|---|---|---|
| 2008 Sales: $175,504 | 2008 Profits: $-459,851 | **U.S. Stock Ticker: VRTX** |
| 2007 Sales: $199,012 | 2007 Profits: $-391,279 | **Int'l Ticker:** Int'l Exchange: |
| 2006 Sales: $216,356 | 2006 Profits: $-206,891 | Employees: 1,150 |
| 2005 Sales: $160,890 | 2005 Profits: $-203,417 | Fiscal Year Ends: 12/31 |
| 2004 Sales: $102,717 | 2004 Profits: $-166,247 | Parent Company: |

## SALARIES/BENEFITS:

| Pension Plan: | ESOP Stock Plan: | Profit Sharing: | Top Exec. Salary: $708,539 | Bonus: $978,750 |
|---|---|---|---|---|
| Savings Plan: Y | Stock Purch. Plan: Y | | Second Exec. Salary: $456,664 | Bonus: $385,324 |

## OTHER THOUGHTS:

**Apparent Women Officers or Directors: 2**
**Hot Spot for Advancement for Women/Minorities: Y**

## LOCATIONS: ("Y" = Yes)

| West: | Southwest: | Midwest: | Southeast: | Northeast: | International: |
|---|---|---|---|---|---|
| Y | | | | Y | Y |

# VIACELL INC

www.viacellinc.com

Industry Group Code: 325414   Ranks within this company's industry group: Sales:   Profits:

| Drugs: | Other: | Clinical: | Computers: | Services: | |
|---|---|---|---|---|---|
| Discovery: | AgriBio: | Trials/Services: | Hardware: | Specialty Services: | Y |
| Licensing: | Genetic Data: | Labs: | Software: | Consulting: | |
| Manufacturing: | Tissue Replacement: Y | Equipment/Supplies: | Arrays: | Blood Collection: | |
| Genetics: | | Research & Development Services: | Database Management: | Drug Delivery: | |
| | | Diagnostics: | | Drug Distribution: | |

## TYPES OF BUSINESS:

Human Cells-Based Medicine
Stem Cell Research

## BRANDS/DIVISIONS/AFFILIATES:

ViaCord
ViaCyte
ViaCell
ViaCord Research Institute

## CONTACTS: Note: Officers with more than one job title may be intentionally listed here more than once.

Marc D. Beer, CEO
Jim Corbett, Pres.
John F. Thero, CFO/Sr. VP
Nadia Altomare, VP-Sales, Service & Oper.
Morey Kraus, CTO
Karen Nichols, VP-Prod. Dev.
Mary T. Thistle, Sr. VP-Bus. Dev.
Karen Foster, VP-ViaCord Processing Laboratory
Karen Nichols, VP-Regulatory Affairs & Quality Systems
Christopher Stump, VP-Sales & Training
Vaughn M. Kailian, Chmn.

| Phone: 617-914-3900 | Fax: 866-565-2243 |
|---|---|
| Toll-Free: 866-668-4895 | |
| Address: 245 First St., Cambridge, MA 02142 US | |

## GROWTH PLANS/SPECIAL FEATURES:

ViaCord, Inc., formerly ViaCell, Inc., is a leading provider of neonatal screening systems. The firm's sole offering is the ViaCord cord blood banking product, which includes the collection, testing, processing and preserving of umbilical cord blood. Currently, ViaCord stores approximately 145,000 cord blood units for its customers. Additional services are offered when a customer requests cord blood banking. These services include comprehensive testing, which is conducted at the firm's laboratory and includes several tests that are necessary in the event the unit is ever needed for transplant purposes; processing, which is also done at the firm's laboratory and is designed to maximize the number of stem cells preserved; cryopreservation, which consists of the firm freezing the cord blood unit and storing it in liquid nitrogen; and provision of collection kits, which includes all required items for the collection of the newborn umbilical cord blood at the time of birth. Stem cells from umbilical cord blood are a potential treatment option for more than 70 diseases, including certain blood cancers and genetic diseases. The company's lead drug candidate, ViaCyte, is being studied for its potential to broaden reproductive choices for women through the cryopreservation of human unfertilized eggs. The company also operates the ViaCord Research Institute and concentrates on five primary areas of research: cord blood technologies, emerging stem cell therapies, genetic screening, therapeutic development and related transplants.

## FINANCIALS: Sales and profits are in thousands of dollars—add 000 to get the full amount. 2008 Note: Financial information for 2008 was not available for all companies at press time.

| | | | |
|---|---|---|---|
| 2008 Sales: $ | 2008 Profits: $ | U.S. Stock Ticker: Private | |
| 2007 Sales: $ | 2007 Profits: $ | Int'l Ticker:   Int'l Exchange: | |
| 2006 Sales: $54,426 | 2006 Profits: $-21,330 | Employees:   254 | |
| 2005 Sales: $44,443 | 2005 Profits: $-14,677 | Fiscal Year Ends: 12/31 | |
| 2004 Sales: $38,274 | 2004 Profits: $-21,097 | Parent Company: PERKINELMER INC | |

## SALARIES/BENEFITS:

| Pension Plan: | ESOP Stock Plan: | Profit Sharing: | Top Exec. Salary: $363,462 | Bonus: $137,592 |
|---|---|---|---|---|
| Savings Plan: | Stock Purch. Plan: | | Second Exec. Salary: $284,278 | Bonus: $78,650 |

## OTHER THOUGHTS:

Apparent Women Officers or Directors: 5
Hot Spot for Advancement for Women/Minorities: Y

## LOCATIONS: ("Y" = Yes)

| West: | Southwest: | Midwest: | Southeast: | Northeast: | International: |
|---|---|---|---|---|---|
| | | Y | | Y | Y |

Note: Financial information, benefits and other data can change quickly and may vary from those stated here.

# VICAL INC

www.vical.com

Industry Group Code: 325412  Ranks within this company's industry group: Sales: 120   Profits: 128

| Drugs: | | Other: | | Clinical: | | Computers: | | Services: | |
|---|---|---|---|---|---|---|---|---|---|
| Discovery: | Y | AgriBio: | | Trials/Services: | | Hardware: | | Specialty Services: | |
| Licensing: | Y | Genetic Data: | | Labs: | | Software: | | Consulting: | |
| Manufacturing: | | Tissue Replacement: | | Equipment/Supplies: | | Arrays: | | Blood Collection: | |
| Genetics: | | | | Research & Development Services: | | Database Management: | | Drug Delivery: | |
| | | | | Diagnostics: | | | | Drug Distribution: | |

## TYPES OF BUSINESS:

Drug Delivery Systems
Cancer, Infectious Disease & Metabolic Disease Drugs & Vaccines

## BRANDS/DIVISIONS/AFFILIATES:

Allovectin-7
Vaxfectin

## CONTACTS: Note: Officers with more than one job title may be intentionally listed here more than once.

Vijay B. Samant, CEO
Vijay B. Samant, Pres.
Jill M. Church, CFO/Sr. VP
Alain P. Rolland, Sr. VP-Prod. Dev.
Kevin R. Bracken, VP-Mfg.
Jill M. Church, Sec.
Robert M. Jackman, Sr. VP-Bus. Oper.
Andrew R. de Guttadauro, VP-Corp. Dev.
Larry R. Smith, VP-Vaccine Research
R. Gordon Douglas, Chmn.

| Phone: 858-646-1100 | Fax: 858-646-1150 |
|---|---|
| Toll-Free: | |
| Address: 10390 Pacific Ctr. Ct., San Diego, CA 92121 US | |

## GROWTH PLANS/SPECIAL FEATURES:

Vical, Inc. researches and develops biopharmaceutical products based on its patented DNA delivery technologies for the prevention and treatment of serious or life-threatening diseases. The company's research areas include vaccines for use in high-risk population for infectious disease targets; vaccines for general pediatric, adolescent and adult populations for infectious disease applications; and cancer vaccines or immunotherapies. The firm has five active independent development programs in the areas of infectious disease and cancer, which include a Phase III clinical trial using the Allovectin-7 immunotherapeutic in patients with metastatic melanoma; a Phase II clinical trial using the cytomegalovirus DNA vaccine in hematopoietic cell transplant patients; a Phase I clinical trial of a pandemic influenza DNA vaccine candidate using the proprietary Vaxfectin as an adjuvant; a preclinical stage candidate to prevent cytomegalovirus infection by fetal transmission prior to or in the course of a pregnancy; and a research stage study into reducing viral shedding for type II herpes. Vical has licenses its technologies to companies including Merck & Co., Inc.; the Sanofi-Aventis Groups; and AnGes. The firm also has two veterinary vaccine studies, licensed to Merial, Ltd. and Aqua Health, Ltd., for infectious diseases and cancer in household animals. In November 2008, Vical, the U.S. Navy and the U.S. Army agreed to develop a DNA vaccine for Dengue Fever in a deal that will give Vical a $1.3 million manufacturing and expertise contract. In November 2008, the company announced it would lay off about 20% of its workforce and accelerate the closure of one of its research facilities. In May 2009, the firm entered into a Cooperative Research and Development Agreement with the U.S. Navy to expedite the development of a vaccine for H1N1 influenza (swine flu).

## FINANCIALS: Sales and profits are in thousands of dollars—add 000 to get the full amount. 2008 Note: Financial information for 2008 was not available for all companies at press time.

| | | | |
|---|---|---|---|
| 2008 Sales: $7,956 | 2008 Profits: $-36,896 | U.S. Stock Ticker: VICL | |
| 2007 Sales: $5,512 | 2007 Profits: $-35,894 | Int'l Ticker: | Int'l Exchange: |
| 2006 Sales: $14,740 | 2006 Profits: $-23,148 | Employees: 118 | |
| 2005 Sales: $12,003 | 2005 Profits: $-24,357 | Fiscal Year Ends: 12/31 | |
| 2004 Sales: $14,545 | 2004 Profits: $-23,733 | Parent Company: | |

## SALARIES/BENEFITS:

| Pension Plan: | ESOP Stock Plan: | Profit Sharing: | Top Exec. Salary: $470,000 | Bonus: $190,000 |
|---|---|---|---|---|
| Savings Plan: Y | Stock Purch. Plan: | | Second Exec. Salary: $306,000 | Bonus: $80,000 |

## OTHER THOUGHTS:

**Apparent Women Officers or Directors:** 1
**Hot Spot for Advancement for Women/Minorities:**

## LOCATIONS: ("Y" = Yes)

| West: | Southwest: | Midwest: | Southeast: | Northeast: | International: |
|---|---|---|---|---|---|
| Y | | | | | |

# VION PHARMACEUTICALS INC

**www.vionpharm.com**

Industry Group Code: 325412  **Ranks within this company's industry group:** Sales: 156    Profits: 122

| Drugs: | | Other: | Clinical: | Computers: | Services: | |
|---|---|---|---|---|---|---|
| Discovery: | Y | AgriBio: | Trials/Services: | Hardware: | Specialty Services: | |
| Licensing: | | Genetic Data: | Labs: | Software: | Consulting: | |
| Manufacturing: | | Tissue Replacement: | Equipment/Supplies: | Arrays: | Blood Collection: | |
| Genetics: | | | Research & Development Services: | Database Management: | Drug Delivery: | Y |
| | | | Diagnostics: | | Drug Distribution: | |

## TYPES OF BUSINESS:

Cancer Treatment Drugs
Drug Delivery Systems

## BRANDS/DIVISIONS/AFFILIATES:

Triapine
Tumor Amplified Protein Expression Therapy (TAPET)
Cloretazine

## CONTACTS: *Note: Officers with more than one job title may be intentionally listed here more than once.*

Alan Kessman, CEO
Howard B. Johnson, Pres.
Howard B. Johnson, CFO
Ivan King, VP-R&D
James Tanguay, VP-Mfg., Chemistry & Control
Karen Schmedlin, Sec.
Karen Schmedlin, VP-Finance/Chief Acct. Officer
Ann Cahill, VP-Clinical Dev.
William Hahne, VP-Medical
Aileen Ryan, VP-Regulatory Affairs
Tanya Lewis, VP-Regulatory Affairs & Quality Assurance
William R. Miller, Chmn.

| Phone: 203-498-4210 | Fax: 203-498-4211 |
|---|---|
| Toll-Free: | |
| Address: 4 Science Park, New Haven, CT 06511 US | |

## GROWTH PLANS/SPECIAL FEATURES:

Vion Pharmaceuticals, Inc. is a pharmaceutical company engaged in the development of products for the treatment of cancer. The company's portfolio of product candidates consists of two small molecule anticancer agents in clinical development and additional small molecules in preclinical development. Vion's lead product candidate, Onrigin (formerly Cloretazine), is an alkylating (DNA-damaging) agent for the treatment of acute myelogenous leukemia (AML). The drug received two fast track designations from FDA for the treatment of relapsed AML and elderly poor-risk AML and an orphan drug designation for the treatment of AML in the U.S. and the E.U. Triapine, the second product candidate in clinical trials, is a small molecule that in preclinical models inhibits the enzyme ribonucleotide reductase and therefore prevents the replication of tumor cells by blocking a critical step in DNA synthesis. Clinical trials of Triapine are sponsored by the National Cancer Institute's (NCI) Cancer Therapy Evaluation Program under a long-term clinical trials agreement with the NCI's Division of Cancer Treatment and Diagnosis. Other preclinical products in Vion's portfolio include VNP40541, a cytotoxic (cell-damaging) compound that has been demonstrated in preclinical studies to be highly selective for hypoxic (poorly oxygenated) cells that can be difficult to treat in cancerous tumors, and Tumor Amplified Protein Expression Therapy (TAPET), a drug delivery technology for the treatment of cancer. The company is not actively developing these preclinical technologies at this time. In April 2009, the FDA accepted Vion's Onrigin new drug application for review.

The company offers its employees medical and dental insurance; life, disability and long-term care insurance; a 401(k) plan; an employee stock purchase plan; relocation assistance; and an education reimbursement.

## FINANCIALS: Sales and profits are in thousands of dollars—add 000 to get the full amount. 2008 Note: Financial information for 2008 was not available for all companies at press time.

| | | |
|---|---|---|
| 2008 Sales: $ 43 | 2008 Profits: $-29,848 | **U.S. Stock Ticker:** VION |
| 2007 Sales: $ 66 | 2007 Profits: $-33,993 | **Int'l Ticker:** Int'l Exchange: |
| 2006 Sales: $ 22 | 2006 Profits: $-25,347 | **Employees:** 43 |
| 2005 Sales: $ 23 | 2005 Profits: $-18,041 | **Fiscal Year Ends:** 12/31 |
| 2004 Sales: $ 275 | 2004 Profits: $-16,055 | **Parent Company:** |

## SALARIES/BENEFITS:

| Pension Plan: | ESOP Stock Plan: | Profit Sharing: | Top Exec. Salary: $458,988 | Bonus: $114,747 |
|---|---|---|---|---|
| Savings Plan: Y | Stock Purch. Plan: Y | | Second Exec. Salary: $297,440 | Bonus: $54,000 |

## OTHER THOUGHTS:

**Apparent Women Officers or Directors:** 3
**Hot Spot for Advancement for Women/Minorities:** Y

## LOCATIONS: ("Y" = Yes)

| West: | Southwest: | Midwest: | Southeast: | Northeast: | International: |
|---|---|---|---|---|---|
| | | | | Y | |

# VIRBAC CORP

## www.virbaccorp.com

**Industry Group Code: 325412B  Ranks within this company's industry group:** Sales:    Profits:

| Drugs: | | Other: | | Clinical: | | Computers: | | Services: | |
|---|---|---|---|---|---|---|---|---|---|
| Discovery: | Y | AgriBio: | | Trials/Services: | | Hardware: | | Specialty Services: | |
| Licensing: | | Genetic Data: | | Labs: | | Software: | | Consulting: | |
| Manufacturing: | Y | Tissue Replacement: | | Equipment/Supplies: | | Arrays: | | Blood Collection: | |
| Genetics: | | | | Research & Development Services: | | Database Management: | | Drug Delivery: | |
| | | | | Diagnostics: | | | | Drug Distribution: | |

## TYPES OF BUSINESS:

Drugs-Animal Health & Pet Care

## BRANDS/DIVISIONS/AFFILIATES:

C.E.T.
IVERHART MAX
Preventic
Soloxine
Pancrezyme
Novifit NoviSAMe
ResiKetoChlor Leave-On Lotion
Rebound OES

## CONTACTS: *Note: Officers with more than one job title may be intentionally listed here more than once.*

Erik R. Martinez, CEO
Erik R. Martinez, Pres.
Christo White, CFO
Dena Ware, Mgr.-Mktg.
George Stanly, Dir.-IT
Mary King, Dir.-Tech. Svcs.
Eric Maree, Chmn.

| **Phone:** 817-831-5030 | **Fax:** 817-831-8327 |
|---|---|
| **Toll-Free:** 800-338-3659 | |
| **Address:** 3200 Meacham Blvd., Fort Worth, TX 76137 US | |

## GROWTH PLANS/SPECIAL FEATURES:

Virbac Corporation, a subsidiary of Virbac SA, develops, manufactures and markets a variety of pet and companion animal health products.  The firm specializes in the heartworm, tick/flea, dermatology, antibiotics, endocrinology and oral hygiene markets.  The company's dermatology line markets treatments for allergic dermatitis, keratoseborrheic disorders, infectious dermatitis and otitis externa.  The firm's leading brands include the C.E.T. line of at home pet oral hygiene products, which are designed specifically for the mouth of a dog or cat; IVERHART MAX, a medication that protects against heartworm disease and roundworm, hookworm and tapeworm infections; and Preventic, a line of products that prevent flea and tick infestation, including tick collars, shampoo, dips and environmental parasite control sprays and treatments for the house and yard.  The firm also sells a line of endocrinology and urology products including Soloxine, sodium tablets, for management of canine hypothyroidism; Tumil-K tablets or gel, for potassium deficiencies; Pancrezyme tablets or powder, for the treatment for exocrine pancreatic insufficiency; and Uroeze tablets or powder and Ammonil tablets, for proper urinary pH. Additional products offered by Virbac include dewormers for dogs and natural fibers to ease constipation in dogs and cats.  In addition to veterinary products, the company operates the Virbac University online web site, an interactive resource for veterinarians and their staff members.  Virbac University provides up-to-date technical and treatment information in four major areas and provides the opportunity to earn CE Credits and Certification.  Virbac also operates a Canadian branch in St Lazare, Quebec, which markets and distributes veterinary products in Canada.  In 2008, the company added several new products to its line: Biomox for soft tissue infections; Rebound OES, a rehydrating solution; Novifit NoviSAMe, developed to enhance brain activity in aging pets; C.E.T VeggieDent dental chews; and ResiKetoChlor Leave-On Lotion, which treats dermatological infections.

## FINANCIALS:  Sales and profits are in thousands of dollars—add 000 to get the full amount. 2008 Note: Financial information for 2008 was not available for all companies at press time.

| | | |
|---|---|---|
| 2008 Sales: $ | 2008 Profits: $ | **U.S. Stock Ticker: Subsidiary** |
| 2007 Sales: $80,800 | 2007 Profits: $ | **Int'l Ticker:**    Int'l Exchange: |
| 2006 Sales: $ | 2006 Profits: $ | Employees:   269 |
| 2005 Sales: $80,778 | 2005 Profits: $3,873 | Fiscal Year Ends: 10/31 |
| 2004 Sales: $77,115 | 2004 Profits: $1,471 | Parent Company: VIRBAC SA |

## SALARIES/BENEFITS:

| Pension Plan: | ESOP Stock Plan: | Profit Sharing: | Top Exec. Salary: $305,363 | Bonus: $ |
|---|---|---|---|---|
| Savings Plan: | Stock Purch. Plan: | | Second Exec. Salary: $151,470 | Bonus: $ |

## OTHER THOUGHTS:

**Apparent Women Officers or Directors**: 2
**Hot Spot for Advancement for Women/Minorities**:

## LOCATIONS: ("Y" = Yes)

| West: | Southwest: | Midwest: | Southeast: | Northeast: | International: |
|---|---|---|---|---|---|
| | Y | Y | | | Y |

# VIROPHARMA INC

**www.viropharma.com**

**Industry Group Code: 325412  Ranks within this company's industry group:** Sales: 56   Profits: 37

| Drugs: | | Other: | Clinical: | Computers: | Services: | |
|---|---|---|---|---|---|---|
| Discovery: | | AgriBio: | Trials/Services: | Hardware: | Specialty Services: | |
| Licensing: | Y | Genetic Data: | Labs: | Software: | Consulting: | |
| Manufacturing: | | Tissue Replacement: | Equipment/Supplies: | Arrays: | Blood Collection: | |
| Genetics: | | | Research & Development Services: | Database Management: | Drug Delivery: | |
| | | | Diagnostics: | | Drug Distribution: | |

## TYPES OF BUSINESS:

Oral Antibiotics
Hepatitis C Drugs
Infectious Diseases Drugs

## BRANDS/DIVISIONS/AFFILIATES:

Vanococin HCl Capsules
Maribavir
HCV-796
Camvia
Cinryze
NTCD
Lev Pharmaceuticals

## CONTACTS: *Note: Officers with more than one job title may be intentionally listed here more than once.*

Vincent J. Milano, CEO
Daniel B. Soland, COO/VP
Vincent J. Milano, Pres.
Colin Broom, Chief Scientific Officer/VP
Thomas F. Doyle, VP-Strategic Initiatives
Michel de Rosen, Chmn.
Robert G. Pietrusko, VP-Global Regulatory Affairs & Quality

| Phone: 610-458-7300 | Fax: 610-458-7380 |
|---|---|
| Toll-Free: | |
| Address: 397 Eagleview Blvd., Exton, PA 19341 US | |

## GROWTH PLANS/SPECIAL FEATURES:

ViroPharma, Inc. is a biopharmaceutical company that develops and commercializes products that address serious infectious diseases, with a focus on products used by physician specialists or in hospital settings. The company's only marketed product is Vancocin HC1 capsules, an oral antibiotic for the treatment of antibiotic-associated pseudomembranous colitis and enterocolitis, including methicilin-resistant strains. The firm is developing Camvia (formerly maribavir), currently in Phase III, for the prevention and treatment of cytomegalovirus diseases and HCV-796, currently in Phase II, for the treatment of hepatitis C virus. ViroPharma has a licensing agreement for the rights to develop non-toxigenic strains of C. difficile (NTCD) for the treatment and prevention of CDI. The firm has licensed the U.S. and Canadian rights for an intranasal formulation of pleconaril, to Schering-Plough for the treatment of picornavirus infections. Customers of Vancocin include wholesalers, who then distribute the drug to pharmacies, hospitals and long term care facilities. Recently, the firm announced that U.S. Food and Drug Administration (FDA) granted fast track designation for HCV-796 for treatment of hepatitis C virus infection; and ViroPharma announced that it had initiated a Phase III clinical trial for Camvia for patients undergoing a liver transplant procedure. In July 2008, the firm agreed to acquire Lev Pharmaceuticals for $442.9 million; this transaction includes Lev's lead product Cinryze, currently under review as a treatment for hereditary angioedema. In February 2009, the firm discontinued its Maribavir Phase III study for liver transplant patients, and announced negative results for its application in bone marrow transplant patients. Also in February 2009, the FDA granted a priority review of the firm's Cinrynze C1 Inhibitor for angioedema.

The company offers its employees medical and dental benefits; a 401(k) plan; and stock options.

## FINANCIALS: Sales and profits are in thousands of dollars—add 000 to get the full amount. 2008 Note: Financial information for 2008 was not available for all companies at press time.

| | | U.S. Stock Ticker: VPHM |
|---|---|---|
| 2008 Sales: $232,307 | 2008 Profits: $67,617 | Int'l Ticker:    Int'l Exchange: |
| 2007 Sales: $203,770 | 2007 Profits: $95,353 | Employees: 201 |
| 2006 Sales: $167,181 | 2006 Profits: $66,666 | Fiscal Year Ends: 12/31 |
| 2005 Sales: $132,417 | 2005 Profits: $113,705 | Parent Company: |
| 2004 Sales: $22,389 | 2004 Profits: $-19,534 | |

## SALARIES/BENEFITS:

| Pension Plan: | ESOP Stock Plan: | Profit Sharing: | Top Exec. Salary: $430,000 | Bonus: $231,770 |
|---|---|---|---|---|
| Savings Plan: Y | Stock Purch. Plan: Y | | Second Exec. Salary: $337,000 | Bonus: $193,775 |

## OTHER THOUGHTS:

**Apparent Women Officers or Directors:**
**Hot Spot for Advancement for Women/Minorities:**

## LOCATIONS: ("Y" = Yes)

| West: | Southwest: | Midwest: | Southeast: | Northeast: | International: |
|---|---|---|---|---|---|
| | | | | Y | |

*Note: Financial information, benefits and other data can change quickly and may vary from those stated here.*

# VYSIS INC

**www.vysis.com**

Industry Group Code: 325413  Ranks within this company's industry group:  Sales:      Profits:

| Drugs: | Other: | | Clinical: | | Computers: | | Services: | |
|--------|--------|---|----------|---|-----------|---|----------|---|
| Discovery: | AgriBio: | | Trials/Services: | | Hardware: | Y | Specialty Services: | |
| Licensing: | Genetic Data: | Y | Labs: | | Software: | | Consulting: | |
| Manufacturing: | Tissue Replacement: | | Equipment/Supplies: | Y | Arrays: | Y | Blood Collection: | |
| Genetics: | | | Research & Development Services: | | Database Management: | | Drug Delivery: | |
| | | | Diagnostics: | Y | | | Drug Distribution: | |

## TYPES OF BUSINESS:

Medical Diagnostics Products
Reagents
Microarray Technology
DNA Probes
Genomics Workstations
Genetic Diagnostic Products

## BRANDS/DIVISIONS/AFFILIATES:

Abbott Laboratories
Fluorescence In Situ Hybridization (FISH)
GenoSensor Microarray System
PathVysion
UroVysion
AneuVysion
LSI Locus Specific Identifier DNA Probes
HYBrite Hybridization System

## CONTACTS: Note: Officers with more than one job title may be intentionally listed here more than once.

Steven A. Seeling, Div. VP
Susan Zint, Dir.-Human Resources
David Reiners, Mgr.-Mktg. Comm.

| Phone: 224-361-7000 | Fax: 224-361-7138 |
|---|---|
| Toll-Free: 800-553-7042 | |
| Address: 1300 E. Touhy, Des Plaines, IL 60018-3315 US | |

## GROWTH PLANS/SPECIAL FEATURES:

Vysis, Inc., a subsidiary of Abbott Laboratories, develops and markets genomics-based clinical products for the evaluation and management of cancer, prenatal disorders and other genetic diseases. The firm's technology platforms, which it not only uses for its own research but markets for other firms, are Fluorescence in Situ Hybridization (FISH) and the GenoSensor Microarray System, both designed to detect gene and chromosome copy numbers; the VP 2000 Processor, an integrated workstation with a variety of tools for preparing slides and specimens; and various comparative genomic hybridization agents. Vysis's developed products include PathVysion HER-2 DNA Probe Kit, which enables assessment of the HER-2/NEU breast cancer gene; the UroVysion Bladder Cancer Kit, for detecting specific chromosome abnormalities indicative of bladder cancer reoccurrence; and the AneuVysion Assay, detecting trisomy 13, 18, 21 and sex chromosome anomalies for prenatal screenings. Vysis also produces many products relating to more general DNA analysis, such as CEP Chromosome Enumeration DNA Probes, LSI Locus Specific Identifier DNA Probes and TelVysion Telomere DNA Probes. The SpectraVysion Assay provides distinct identification of the 24 human chromosomes for anomalies that are not identifiable using traditional cytogenetic banding methods. HYBrite Hybridization System provides for FISH procedures through a hands-free denaturation/hybridization system, eliminating hazardous solutions such as formamide and ethanol.

## FINANCIALS: Sales and profits are in thousands of dollars—add 000 to get the full amount. 2008 Note: Financial information for 2008 was not available for all companies at press time.

| | | |
|---|---|---|
| 2008 Sales: $ | 2008 Profits: $ | **U.S. Stock Ticker: Subsidiary** |
| 2007 Sales: $ | 2007 Profits: $ | **Int'l Ticker:** Int'l Exchange: |
| 2006 Sales: $ | 2006 Profits: $ | Employees: 133 |
| 2005 Sales: $ | 2005 Profits: $ | Fiscal Year Ends: 12/31 |
| 2004 Sales: $ | 2004 Profits: $ | Parent Company: ABBOTT LABORATORIES |

## SALARIES/BENEFITS:

| Pension Plan: | ESOP Stock Plan: | Profit Sharing: | Top Exec. Salary: $ | Bonus: $ |
|---|---|---|---|---|
| Savings Plan: | Stock Purch. Plan: | | Second Exec. Salary: $ | Bonus: $ |

## OTHER THOUGHTS:

**Apparent Women Officers or Directors:** 1
**Hot Spot for Advancement for Women/Minorities:**

## LOCATIONS: ("Y" = Yes)

| West: | Southwest: | Midwest: | Southeast: | Northeast: | International: |
|-------|-----------|----------|-----------|-----------|---------------|
| | | Y | | | Y |

# WARNER CHILCOTT PLC

www.warnerchilcott.com

**Industry Group Code:** 325412 **Ranks within this company's industry group:** Sales: Profits:

| Drugs: | Other: | Clinical: | Computers: | Services: | |
|---|---|---|---|---|---|
| Discovery: | AgriBio: | Trials/Services: | Hardware: | Specialty Services: | |
| Licensing: | Genetic Data: | Labs: | Software: | Consulting: | |
| Manufacturing: Y | Tissue Replacement: | Equipment/Supplies: | Arrays: | Blood Collection: | |
| Genetics: | | Research & Development Services: | Database Management: | Drug Delivery: | |
| | | Diagnostics: | | Drug Distribution: | |

## TYPES OF BUSINESS:

Pharmaceuticals Development & Manufacturing
Contraceptives
Hormone Therapies
Vitamins
Dermatology Treatments

## BRANDS/DIVISIONS/AFFILIATES:

Loestrin 24
Estrostep
femhrt
Ovcon
Estrace
Taclonex
Dovonex
Doryx

## CONTACTS: *Note: Officers with more than one job title may be intentionally listed here more than once.*

Roger Boissonneault, CEO
Roger Boissonneault, Pres.
Paul Herendeen, CFO/Exec. VP
Herman Ellman, Sr. VP-Clinical Dev.
Leland H. Cross, Sr. VP-Tech. Oper.
Izumi Hara, General Counsel/Sr. VP/Corp. Sec.
Anthony D.Bruno, Exec. VP-Corp. Dev.
W. Carlton Reichel, Pres., Pharmaceuticals
Alvin Howard, Sr. VP-Regulatory Affairs

| Phone: 973-442-3200 | Fax: 973-442-3283 |
|---|---|
| Toll-Free: 800-521-8813 | |
| Address: 100 Enterprise Dr., Rockaway, NJ 07866 US | |

## GROWTH PLANS/SPECIAL FEATURES:

Warner Chilcott is a specialty pharmaceutical company that focuses on developing, manufacturing and marketing branded prescription pharmaceutical products in women's healthcare and dermatology. Its franchises are comprised of complementary portfolios of established, branded, development stage and new products, including recently launched products Loestrin 24 and Taclonex. Its women's healthcare franchise is anchored by its strong presence in the hormonal contraceptive and hormone therapy categories and its dermatology franchise is built on its established positions in the markets for psoriasis and acne therapies. Recently, it launched Loestrin 24 Fe, an oral contraceptive with a novel patented 24-day dosing regimen, with the goal of growing the market share position it achieved with its Ovcon and Estrostep products in the hormonal contraceptive market. The company also has a significant presence in the hormone therapy market, primarily through its products femhrt and Estrace Cream. In dermatology, its psoriasis product Dovonex enjoys the leading position in the U.S. for the non-steroidal topical treatment of psoriasis. Warner Chilcott strengthened and extended its position in the market for psoriasis therapies with the launch of Taclonex, the first once-a-day topical psoriasis treatment that combines betamethasone dipropionate, a corticosteroid, with calcipotriene, the active ingredient in Dovonex. The company's product Doryx is the leading branded oral tetracycline in the U.S. for the treatment of acne. In 2008, the company announced the approval of Doryx 150 mg delayed-release tablets. Warner Chilcott has collaborations with various entities, including: Foamix; Watson Pharmaceuticals; LEO Pharma; and Paratek. In February 2009, the company agreed to acquire the U.S. rights to a topical cream for erectile dysfunction from NexMed, Inc. In May 2009, the firm agreed to move its operations from Bermuda to Ireland; upon doing so, Warner Chilcott will become a new public holding company, Warner Chilcott plc.

## FINANCIALS: Sales and profits are in thousands of dollars—add 000 to get the full amount. 2008 Note: Financial information for 2008 was not available for all companies at press time.

| | | |
|---|---|---|
| 2008 Sales: $ | 2008 Profits: $ | U.S. Stock Ticker: Private |
| 2007 Sales: $ | 2007 Profits: $ | Int'l Ticker: Int'l Exchange: |
| 2006 Sales: $ | 2006 Profits: $ | Employees: 960 |
| 2005 Sales: $ | 2005 Profits: $ | Fiscal Year Ends: 12/31 |
| 2004 Sales: $ | 2004 Profits: $ | Parent Company: |

## SALARIES/BENEFITS:

| Pension Plan: | ESOP Stock Plan: | Profit Sharing: | Top Exec. Salary: $ | Bonus: $ |
|---|---|---|---|---|
| Savings Plan: | Stock Purch. Plan: | | Second Exec. Salary: $ | Bonus: $ |

## OTHER THOUGHTS:

Apparent Women Officers or Directors:
Hot Spot for Advancement for Women/Minorities:

## LOCATIONS: ("Y" = Yes)

| West: | Southwest: | Midwest: | Southeast: | Northeast: | International: |
|---|---|---|---|---|---|
| | | | | Y | Y |

# WATSON PHARMACEUTICALS INC                         www.watson.com

**Industry Group Code: 325412A  Ranks within this company's industry group:** Sales: 3   Profits: 2

| Drugs: | | Other: | Clinical: | Computers: | Services: |
|---|---|---|---|---|---|
| Discovery: | Y | AgriBio: | Trials/Services: | Hardware: | Specialty Services: |
| Licensing: | | Genetic Data: | Labs: | Software: | Consulting: |
| Manufacturing: | Y | Tissue Replacement: | Equipment/Supplies: | Arrays: | Blood Collection: |
| Genetics: | Y | | Research & Development Services: | Database Management: | Drug Delivery: |
| | | | Diagnostics: | | Drug Distribution: Y |

## TYPES OF BUSINESS:

Generic Pharmaceuticals
Branded Drugs
Urology Drugs
Anti-Hypertensive Drugs
Nephrology Drugs
Anti-Inflammatory Drugs
Oral Contraceptive Drugs
Pain Management Drugs

## BRANDS/DIVISIONS/AFFILIATES:

Watson Laboratories
Anda
Anda Pharmaceuticals
Andrx Corp.
Trelstar Depot
Valmed
Gelnique
Arrow Group

## CONTACTS: Note: Officers with more than one job title may be intentionally listed here more than once.

Paul M. Bisaro, CEO
Paul M. Bisaro, Pres.
Mark Durand, CFO/Sr. VP
Clare Carmichael, Sr. VP-Human Resources
Charles D. Ebert, Sr. VP-R&D
Thomas R. Giordano, CIO/Sr. VP
David A. Buchen, General Counsel/Sr. VP/Sec.
Patricia L. Eisenhauer, VP-Corp. Comm.
Patricia L. Eisenhauer, VP-Investor Rel.
Edward F. Heimers, Jr., Exec. VP/Pres., Brand Division
Albert Paonessa III, Exec. VP/COO-Anda Inc.
Gordon Munro, Sr. VP-Quality Assurance
Thomas R. Russillo, Exec. VP/Pres., U.S. Generics Division
Andrew L. Turner, Chmn.

| Phone: 951-493-5300 | Fax: |
|---|---|
| Toll-Free: | |
| Address: 311 Bonnie Cir., Corona, CA 92880 US | |

## GROWTH PLANS/SPECIAL FEATURES:

Watson Pharmaceuticals, Inc. develops, manufactures, markets, sells and distributes over 27 branded and over 150 generic pharmaceutical products. The firm operates through three segments: generic, brand and distribution. The generic segment includes pharmaceutical products that are therapeutically equivalent to proprietary products. These generic products address the therapeutic areas of antibiotics, anti-inflammatories, depression, hypertension, oral contraceptives, pain management and smoking cessation. Generic products accounted for roughly 60% of net revenues in 2008. The brand segment develops, manufactures, markets, sells and distributes products primarily through two core areas: specialty products and nephrology. The specialty products include urology and a number of non-promoted products. The nephrology product line concerns products for the treatment of iron deficiency anemia. Brands include Trelstar Depot and Trelstar LA, treatments for advanced prostate cancer; Gelnique, a treatment for overactive bladder symptoms; Rapaflo, a treatment for symptoms of benign prostatic hyperplasia; and Ferrlecit and INFeD, iron replacement therapies for patients with iron deficiency anemia. The company markets its brand products through 380 sales professionals. Brand products accounted for roughly 18% of total revenue in 2008. The distribution segment distributes generic products and select brand products to independent pharmacies, pharmacy chains, pharmacy buying groups, alternative care facilities, including long-term care pharmacies, and physicians' offices in the U.S. The company's distribution business subsidiaries include Anda, Anda Pharmaceuticals and Valmed. In November 2008, the company agreed to acquire a portfolio of generic pharmaceutical products conditional to the merger of Teva Pharmaceutical Industries, Ltd., and Barr Pharmaceuticals, Inc. In June 2009, Watson announced that it would acquire London-based Arrow Group for approximately $1.75 billion.

The company offers employees medical, dental and vision insurance; a 401(k) plan; life and AD&D insurance; domestic partner coverage; flexible spending accounts; business travel accident insurance; short- and long-term disability; pet insurance; and tuition reimbursement.

## FINANCIALS: Sales and profits are in thousands of dollars—add 000 to get the full amount. 2008 Note: Financial information for 2008 was not available for all companies at press time.

| | | |
|---|---|---|
| 2008 Sales: $2,535,501 | 2008 Profits: $238,379 | **U.S. Stock Ticker: WPI** |
| 2007 Sales: $2,496,651 | 2007 Profits: $141,030 | **Int'l Ticker:**  Int'l Exchange: |
| 2006 Sales: $1,979,244 | 2006 Profits: $-445,005 | Employees: 5,070 |
| 2005 Sales: $1,646,203 | 2005 Profits: $138,557 | Fiscal Year Ends: 12/31 |
| 2004 Sales: $1,640,551 | 2004 Profits: $150,018 | Parent Company: |

## SALARIES/BENEFITS:

| Pension Plan: | ESOP Stock Plan: | Profit Sharing: | Top Exec. Salary: $1,000,000 | Bonus: $997,200 |
|---|---|---|---|---|
| Savings Plan: Y | Stock Purch. Plan: | | Second Exec. Salary: $790,608 | Bonus: $430,486 |

## OTHER THOUGHTS:

**Apparent Women Officers or Directors**: 2
**Hot Spot for Advancement for Women/Minorities**: Y

## LOCATIONS: ("Y" = Yes)

| West: | Southwest: | Midwest: | Southeast: | Northeast: | International: |
|---|---|---|---|---|---|
| Y | | Y | Y | Y | |

Note: Financial information, benefits and other data can change quickly and may vary from those stated here.

# WHATMAN PLC

**www.whatman.com**

Industry Group Code: 3345 Ranks within this company's industry group: Sales: Profits:

| Drugs: | Other: | Clinical: | | Computers: | Services: |
|---|---|---|---|---|---|
| Discovery: | AgriBio: | Trials/Services: | | Hardware: | Specialty Services: |
| Licensing: | Genetic Data: | Labs: | | Software: | Consulting: |
| Manufacturing: | Tissue Replacement: | Equipment/Supplies: | Y | Arrays: | Blood Collection: |
| Genetics: | | Research & Development Services: | | Database Management: | Drug Delivery: |
| | | Diagnostics: | Y | | Drug Distribution: |

## TYPES OF BUSINESS:

Equipment-Filtration Systems
Diagnostic Supplies

## BRANDS/DIVISIONS/AFFILIATES:

Multiwell
UNIFILTER
VigeneTech
Whatman FAST Quant
GE Healthcare

## CONTACTS: *Note: Officers with more than one job title may be intentionally listed here more than once.*

Kieran May, CEO
Helen Evans, VP-Mktg.
Michael Harvey, Dir.-R&D
Shari States, Group Mgr.-Mktg. Comm.
Joe Hogan, CEO/Pres., GE Healthcare
LiLi Lee, Prod. Mgr.
Michael Harper, Chmn.

| Phone: 440-162-267-6670 | Fax: 440-162-269-1425 |
|---|---|
| Toll-Free: 800-942-8626 | |
| Address: Springfield Mill, James Whatman Way, Maidstone, Kent ME14 2LE UK | |

## GROWTH PLANS/SPECIAL FEATURES:

Whatman plc, operating under General Electric Co.'s GE Healthcare division, is a global leader in filtration technology and manufactures and distributes separation and filtration products used in laboratories, health care facilities and bioscience research. The firm offers over 100 different products for filtration including student kits, DNA purification and pH testers. Whatman has developed total sample preparation solutions through its robust line of filtration devices and membranes. The company's breakthrough protein array technology and FTA technology to capture, archive and purify DNA at room temperature enables it to provide novel solutions for the analytical, healthcare and bioscience markets. The firm's subsidiary companies are expansions to different countries and manufacturing plants, and are located in Asia Pacific, Germany, Ireland, the U.S., Japan, Canada and Belgium. Whatman products serve three markets: LabSciences, MedTech and BioScience. The LabSciences unit produces chromatography products used for purification, extraction products, filter papers, filtration devices, membrane filters and specialty products. The MedTech sector makes diagnostic components and medical devices such as purification and sterilization tools used by original equipment manufacturers. The BioScience unit produces the Multiwell and UNIFILTER lines of filter plates and products used for nucleic acid sample preparation. In 2008, the company was acquired by GE Healthcare. Also in 2008, the company, along with VigeneTech, announced the launch of custom software to analyze Whatman FAST Quant samples. Whatman FAST Quant kits allow researchers to accurately determine the concentration of several cytokines in dozens of biological samples simultaneously.

## FINANCIALS: Sales and profits are in thousands of dollars—add 000 to get the full amount. 2008 Note: Financial information for 2008 was not available for all companies at press time.

| | | |
|---|---|---|
| 2008 Sales: $ | 2008 Profits: $ | **U.S. Stock Ticker: Subsidiary** |
| 2007 Sales: $ | 2007 Profits: $ | **Int'l Ticker: WHM**  Int'l Exchange: London-LSE |
| 2006 Sales: $ | 2006 Profits: $ | Employees: 812 |
| 2005 Sales: $ | 2005 Profits: $ | Fiscal Year Ends: 12/31 |
| 2004 Sales: $159,500 | 2004 Profits: $-1,000 | Parent Company: GENERAL ELECTRIC CO (GE) |

## SALARIES/BENEFITS:

| Pension Plan: | ESOP Stock Plan: | Profit Sharing: | Top Exec. Salary: $ | Bonus: $ |
|---|---|---|---|---|
| Savings Plan: | Stock Purch. Plan: | | Second Exec. Salary: $ | Bonus: $ |

## OTHER THOUGHTS:

**Apparent Women Officers or Directors:**
**Hot Spot for Advancement for Women/Minorities:**

## LOCATIONS: ("Y" = Yes)

| West: | Southwest: | Midwest: | Southeast: | Northeast: | International: |
|---|---|---|---|---|---|
| | | | | Y | Y |

Note: Financial information, benefits and other data can change quickly and may vary from those stated here.

# WYETH
**www.wyeth.com**

**Industry Group Code: 325412  Ranks within this company's industry group:** Sales: 10  Profits: 11

| Drugs: | | Other: | | Clinical: | | Computers: | | Services: | |
|---|---|---|---|---|---|---|---|---|---|
| Discovery: | Y | AgriBio: | | Trials/Services: | | Hardware: | | Specialty Services: | |
| Licensing: | Y | Genetic Data: | Y | Labs: | | Software: | | Consulting: | |
| Manufacturing: | Y | Tissue Replacement: | | Equipment/Supplies: | | Arrays: | | Blood Collection: | |
| Genetics: | | | | Research & Development Services: | | Database Management: | | Drug Delivery: | |
| | | | | Diagnostics: | | | | Drug Distribution: | |

## TYPES OF BUSINESS:
Drugs-Diversified
Wholesale Pharmaceuticals
Animal Health Care Products
Biologicals
Vaccines
Over-the-Counter Drugs
Women's Health Care Products
Nutritional Supplements

## BRANDS/DIVISIONS/AFFILIATES:
Chap Stick
Premarin
Dimetapp
Advil
Robitussin
Preparation H
Thiakis Limited
Fort Dodge Animal Health

## CONTACTS: *Note: Officers with more than one job title may be intentionally listed here more than once.*
Bernard Poussot, CEO
Bernard Poussot, Pres.
Gregory Norden, CFO/Sr. VP
Denise Peppard, Sr. VP-Human Resources
Jeffrey E. Keisling, CIO/VP-Corp. Info. Svcs.
Lawrence V. Stein, General Counsel/Sr. VP
Thomas Hofstaetter, Sr. VP-Corp. Bus. Dev.
Timothy P. Cost, Sr. VP-Corp. Affairs
Justin R. Victoria, VP-Investor Rel.
Mary K. Wold, Sr. VP-Finance
Mikael Dolsten, Sr. VP
Joseph M. Mahady, Sr. VP
Leo C. Jardot, VP-Gov't Rel.
Andrew F. Davidson, VP-Internal Audit
Bernard Poussot, Chmn.

| Phone: 973-660-5000 | Fax: 973-660-7026 |
|---|---|
| **Toll-Free:** | |
| **Address:** 5 Giralda Farms, Madison, NJ 07940 US | |

## GROWTH PLANS/SPECIAL FEATURES:
Wyeth is a global leader in pharmaceuticals, consumer health care products and animal health care products. The firm discovers, develops, manufactures, distributes and sells a diversified line of products arising from three divisions: Wyeth Pharmaceuticals, Wyeth Consumer Health care and Fort Dodge Animal Health. The pharmaceuticals segment is itself divided into women's health care, neuroscience, vaccines and infectious disease, musculoskeletal, internal medicine, hemophilia and immunology and oncology. The division sells branded and generic pharmaceuticals, biological and nutraceutical products as well as animal biological products and pharmaceuticals. Its branded products include Advil, Dimetapp, Premarin, Prempro, Premphase, Triphasil, Ativan, Effexor, Altace, Inderal, Zoton, Protonix and Enbrel. The consumer health care segment's products include analgesics, cough/cold/allergy remedies, nutritional supplements, lip balm and hemorrhoidal, antacid, asthma and other relief items sold over-the-counter. The segment's well-known over-the-counter products include Advil, cold medicines Robitussin and Dimetapp and nutritional supplement Centrum, as well as Chap Stick, Caltrate, Preparation H and Solgar. The company's animal health care products include vaccines, pharmaceuticals, endectocides (dewormers that control both internal and external parasites) and growth implants under the brand names LymeVax, Duramune and Fel-O-Vax. In December 2008, Wyeth acquired Thiakis Limited, a U.K.-based private biotechnology company for approximately $30 million. In January 2009, the firm announced its agreement to be acquired by Pfizer for $68 billion.

Employees are offered a pension plan; a 401(k) plan; child care subsidies; and educational assistance.

## FINANCIALS: Sales and profits are in thousands of dollars—add 000 to get the full amount. 2008 Note: Financial information for 2008 was not available for all companies at press time.

| | | |
|---|---|---|
| 2008 Sales: $22,833,908 | 2008 Profits: $4,417,833 | **U.S. Stock Ticker:** WYE |
| 2007 Sales: $22,399,798 | 2007 Profits: $4,615,960 | **Int'l Ticker:** Int'l Exchange: |
| 2006 Sales: $20,350,655 | 2006 Profits: $4,196,706 | Employees: 50,527 |
| 2005 Sales: $18,755,790 | 2005 Profits: $3,656,298 | Fiscal Year Ends: 12/31 |
| 2004 Sales: $17,358,028 | 2004 Profits: $1,233,997 | Parent Company: |

## SALARIES/BENEFITS:

| Pension Plan: Y | ESOP Stock Plan: | Profit Sharing: | Top Exec. Salary: $1,450,000 | Bonus: $2,750,000 |
|---|---|---|---|---|
| Savings Plan: Y | Stock Purch. Plan: | | Second Exec. Salary: $925,000 | Bonus: $1,341,000 |

## OTHER THOUGHTS:
**Apparent Women Officers or Directors**: 5
**Hot Spot for Advancement for Women/Minorities**: Y

## LOCATIONS: ("Y" = Yes)

| West: | Southwest: | Midwest: | Southeast: | Northeast: | International: |
|---|---|---|---|---|---|
| Y | Y | Y | Y | Y | Y |

---

Note: Financial information, benefits and other data can change quickly and may vary from those stated here.

# XECHEM INTERNATIONAL                               www.xechem.com

**Industry Group Code: 325412  Ranks within this company's industry group:  Sales:     Profits:**

| Drugs: | | Other: | Clinical: | | Computers: | | Services: | |
|---|---|---|---|---|---|---|---|---|
| Discovery: | Y | AgriBio: | Trials/Services: | | Hardware: | | Specialty Services: | |
| Licensing: | | Genetic Data: | Labs: | | Software: | | Consulting: | |
| Manufacturing: | Y | Tissue Replacement: | Equipment/Supplies: | | Arrays: | | Blood Collection: | |
| Genetics: | Y | | Research & Development Services: | Y | Database Management: | | Drug Delivery: | |
| | | | Diagnostics: | | | | Drug Distribution: | |

## TYPES OF BUSINESS:

Pharmaceuticals Development & Manufacturing
Drugs-Proprietary
Drugs-Generic
Research & Development Services
Fine Chemicals
Nutraceuticals

## BRANDS/DIVISIONS/AFFILIATES:

Xechem Nigeria Pharmaceuticals, Ltd.
NICOSAN
5-HMF

## CONTACTS: *Note: Officers with more than one job title may be intentionally listed here more than once.*

Robert Swift, Interim Pres.
Robert Swift, Chief Oversight Officer
Robert Swift, Interim Chmn.

| **Phone:** 732-205-0500 | **Fax:** 732-2474090 |
|---|---|
| **Toll-Free:** | |
| **Address:** 379 Thornall, Edison, NJ 08818 US | |

## GROWTH PLANS/SPECIAL FEATURES:

Xechem International is a development stage biopharmaceutical company working on sickle cell disease. Recently, its focus has been the development of NICOSAN, as well as another sickle cell compound, 5-HMF, which it licensed from Virginia Commonwealth University. Xechem is restructuring its U.S. operations as well as the oversight of its Nigerian subsidiary, Xechem Nigeria Pharmaceuticals, Ltd., which produces and sells NICOSAN. As a part of this restructuring, the company recently closed its operations in New Brunswick, New Jersey; this included termination of personnel and the transfer of over $2 million in equipment to Nigeria. The company still maintains its Edison, New Jersey headquarters. Xechem, in partnership with Rutgers University, has developed a formulation of NICOSAN that can be made into tablets for sale in the U.S. as a nutraceutical. In November 2008, Xechem International and one of its subsidiaries, Xechem, Inc., filed for Chapter 11 protection; the firms are continuing restructuring efforts with the intent to emerge from bankruptcy.

## FINANCIALS: Sales and profits are in thousands of dollars—add 000 to get the full amount. 2008 Note: Financial information for 2008 was not available for all companies at press time.

| | | |
|---|---|---|
| 2008 Sales: $ | 2008 Profits: $ | **U.S. Stock Ticker: XKEMQ** |
| 2007 Sales: $ | 2007 Profits: $ | **Int'l Ticker:** Int'l Exchange: |
| 2006 Sales: $ 202 | 2006 Profits: $-11,130 | Employees: 80 |
| 2005 Sales: $ 6 | 2005 Profits: $-10,039 | Fiscal Year Ends: 12/31 |
| 2004 Sales: $ 168 | 2004 Profits: $-17,606 | Parent Company: |

## SALARIES/BENEFITS:

| Pension Plan: | ESOP Stock Plan: | Profit Sharing: | Top Exec. Salary: $324,167 | Bonus: $ |
|---|---|---|---|---|
| Savings Plan: | Stock Purch. Plan: Y | | Second Exec. Salary: $71,500 | Bonus: $ |

## OTHER THOUGHTS:

**Apparent Women Officers or Directors:**
**Hot Spot for Advancement for Women/Minorities:**

## LOCATIONS: ("Y" = Yes)

| West: | Southwest: | Midwest: | Southeast: | Northeast: | International: |
|---|---|---|---|---|---|
| | | | | Y | Y |

# XENOPORT INC

**www.xenoport.com**

Industry Group Code: 325412  **Ranks within this company's industry group:** Sales: 89    Profits: 146

| Drugs: | | Other: | | Clinical: | | Computers: | | Services: | |
|---|---|---|---|---|---|---|---|---|---|
| Discovery: | Y | AgriBio: | | Trials/Services: | | Hardware: | | Specialty Services: | |
| Licensing: | Y | Genetic Data: | | Labs: | | Software: | | Consulting: | |
| Manufacturing: | | Tissue Replacement: | | Equipment/Supplies: | | Arrays: | | Blood Collection: | |
| Genetics: | | | | Research & Development Services: | | Database Management: | | Drug Delivery: | Y |
| | | | | Diagnostics: | | | | Drug Distribution: | |

## TYPES OF BUSINESS:

Drug Development

## BRANDS/DIVISIONS/AFFILIATES:

Xanodyne Pharmaceuticals, Inc.
Astellas Pharma Inc
GlaxoSmithKline plc
XP13512
XP19986
XP21279
XP20925
XP21510

## CONTACTS: *Note: Officers with more than one job title may be intentionally listed here more than once.*

Ronald W. Barrett, CEO
William J. Rieflin, Pres.
William G. Harris, CFO
Mark A. Gallop, Sr. VP-Research
Kenneth C. Cundy, Sr. VP-Preclinical Dev.
Gianna M. Bosko, Sec.
David R. Savello, Sr. VP-Dev.
William G. Harris, Sr. VP-Finance
David A. Stamler, Chief Medical Officer/Sr. VP
Vincent J. Angotti, Chief Commercialization Officer/Sr. VP

| Phone: 408-616-7200 | Fax: |
|---|---|
| Toll-Free: | |
| Address: 3410 Central Expwy., Santa Clara, CA 95051 US | |

## GROWTH PLANS/SPECIAL FEATURES:

XenoPort, Inc. develops and commercializes potential treatments of central nervous system (CNS) disorders. Its drug candidates, which it calls Transported Prodrugs, are designed to modify the chemical structure of currently marketed drugs, called the parent drug, in order to correct deficiencies in oral absorption, distribution and/or metabolism by utilizing the body's natural nutrient transporter mechanisms. The firm's most advanced product candidate, XP13512 (parent drug: gabapentin), passed its Phase III clinical trial for moderate to severe restless leg syndrome (RLS) in 2008. It has already passed Phase II trials for the treatment of post-herpetic neuralgia (PHN), a type of nerve damage associated with certain strains of the herpes virus. Partner company Astellas Pharma, Inc. is conducting two Phase II trials in Japan evaluating XP13512 for the treatment of painful diabetic neuropathy (PDN) and RLS; another partner, GlaxoSmithKline plc (GSK), plans to evaluate XP13512 for PHN, PDN and migraine prophylaxis. Astellas has the exclusive right to develop and commercialize XP13512 in Japan, Korea, the Philippines, Indonesia, Thailand and Taiwan; GSK has exclusive development and commercialization rights for all other territories. The company's second product candidate, XP19986 (baclofen), is in separate Phase II trials for two indications: the potential treatment of gastroesophageal reflux disease (GERD), sometimes called acid reflux, and for the treatment of spasticity. The firm's third candidate, XP21279 (levodopa), is in Phase I trials for the treatment of Parkinson's disease. XenoPort has other drugs in development or already commercialized, including XP21520 (tranexamic acid), in use in Europe and Asia to treat women with menorrhagia (heavy menstrual flow), of which Xanodyne owns the rights for commercialization inside the U.S., and XP20925 for migraines. In March 2009, the FDA accepted for review a new drug application for XenoPort and GSK's RLS drug, Solzira.

## FINANCIALS: Sales and profits are in thousands of dollars—add 000 to get the full amount. 2008 Note: Financial information for 2008 was not available for all companies at press time.

| | | |
|---|---|---|
| 2008 Sales: $41,996 | 2008 Profits: $-62,540 | U.S. Stock Ticker: XNPT |
| 2007 Sales: $113,822 | 2007 Profits: $28,193 | Int'l Ticker:    Int'l Exchange: |
| 2006 Sales: $10,606 | 2006 Profits: $-64,313 | Employees:   222 |
| 2005 Sales: $4,753 | 2005 Profits: $-42,909 | Fiscal Year Ends: 12/31 |
| 2004 Sales: $ | 2004 Profits: $ | Parent Company: |

## SALARIES/BENEFITS:

| Pension Plan: | ESOP Stock Plan: | Profit Sharing: | Top Exec. Salary: $500,000 | Bonus: $266,250 |
|---|---|---|---|---|
| Savings Plan: | Stock Purch. Plan: | | Second Exec. Salary: $375,000 | Bonus: $150,820 |

## OTHER THOUGHTS:

**Apparent Women Officers or Directors**: 2
**Hot Spot for Advancement for Women/Minorities**:

## LOCATIONS: ("Y" = Yes)

| West: | Southwest: | Midwest: | Southeast: | Northeast: | International: |
|---|---|---|---|---|---|
| Y | | | | | |

# XOMA LTD

**www.xoma.com**

Industry Group Code: 325412  Ranks within this company's industry group: Sales: 76  Profits: 133

| Drugs: | | Other: | | Clinical: | Computers: | Services: |
|---|---|---|---|---|---|---|
| Discovery: | Y | AgriBio: | | Trials/Services: | Hardware: | Specialty Services: |
| Licensing: | Y | Genetic Data: | Y | Labs: | Software: | Consulting: |
| Manufacturing: | Y | Tissue Replacement: | | Equipment/Supplies: | Arrays: | Blood Collection: |
| Genetics: | | | | Research & Development Services: | Database Management: | Drug Delivery: |
| | | | | Diagnostics: | | Drug Distribution: |

## TYPES OF BUSINESS:

Therapeutic Antibodies

## BRANDS/DIVISIONS/AFFILIATES:

Raptiva
Lucentis
CIMZIA
XOMA 3AB
XOMA 052
Takeda Pharmaceutical Company Ltd.

## CONTACTS: *Note: Officers with more than one job title may be intentionally listed here more than once.*

Steven B. Engle, CEO
Steven B. Engle, Pres.
Fred Kurland, CFO
Charles C. Wells, VP-Human Resources
Stephen K. Doberstein, VP-Research
Charles C. Wells, VP-IT
Christopher J. Margolin, General Counsel/VP/Sec.
Robert S. Tenerowicz, VP-Oper.
Mary L. Anderson, VP-Bus. Dev.
Fred Kurland, VP-Finance
Patrick J. Scannon, Chief Medical Officer/Exec. VP
Daniel P. Cafaro, VP-Regulatory Affairs & Compliance
Calvin L. McGoogan, VP-Quality & Facilities
Steven B. Engle, Chmn.

| Phone: 510-204-7200 | Fax: 510-644-2011 |
|---|---|
| Toll-Free: | |
| Address: 2910 7th St., Berkeley, CA 94710 US | |

## GROWTH PLANS/SPECIAL FEATURES:

Xoma, Ltd., is a biopharmaceutical company that discovers, develops and manufactures therapeutic antibodies and other agents designed to treat inflammatory, autoimmune, infectious and oncological diseases. The company has interests in therapeutic antibody candidates being developed by others as a result of licensing technologies, and royalty interests in two products, Lucentis and CIMZIA. Lucentis, by Genentech, is an antibody fragment against vascular endothelial growth factor for the treatment of age-related macular degeneration. Another product, CIMZIA, is approved in the U.S. and Switzerland for the treatment of Crohn's disease. Formerly, Xoma had a royalty interest in antibody product Raptiva. Raptiva is a humanized therapeutic monoclonal antibody developed to treat immune system disorders. Xoma began its voluntary removal of Raptiva from the U.S. market in April 2009, due to the association of the drug with increased risk of progressive multifocal leukoencephalopathy (PML). Xoma's product development pipeline includes proprietary products and collaborative programs at various stages of preclinical and clinical development. Products in the development pipeline include XOMA 052, an anti-IL-1 beta antibody; XOMA 3AB, a biodefense anti-botulism antibody candidate; and five antibodies in preclinical development. The firm possesses a broad technology platform for the discovery, optimization and manufacture of therapeutic antibodies. Xoma has collaborations with several companies, including Novartis AG; Takeda Pharmaceutical Company Limited; Schering-Plough Research Institute; and Pfizer, Inc. In January 2009, Xoma announced a workforce reduction of about 42%, or 144 employees. In February 2009, the company expanded its collaboration agreement with Takeda to provide access to multiple antibody technologies. In April 2009, the firm began voluntary removal of Raptiva from the U.S. market.

Xoma offers its employees medical, dental and vision plans; flexible spending accounts; an employee assistance plan; educational assistance; credit union eligibility; a commuter checks program; share option plans; and an employee share purchase plan.

## FINANCIALS: Sales and profits are in thousands of dollars—add 000 to get the full amount. 2008 Note: Financial information for 2008 was not available for all companies at press time.

| | | |
|---|---|---|
| 2008 Sales: $67,987 | 2008 Profits: $-45,245 | U.S. Stock Ticker: XOMA |
| 2007 Sales: $84,252 | 2007 Profits: $-12,326 | Int'l Ticker:  Int'l Exchange: |
| 2006 Sales: $29,498 | 2006 Profits: $-51,841 | Employees:  335 |
| 2005 Sales: $18,669 | 2005 Profits: $2,779 | Fiscal Year Ends: 12/31 |
| 2004 Sales: $3,665 | 2004 Profits: $-78,942 | Parent Company: |

## SALARIES/BENEFITS:

| Pension Plan: | ESOP Stock Plan: | Profit Sharing: | Top Exec. Salary: $515,000 | Bonus: $ |
|---|---|---|---|---|
| Savings Plan: Y | Stock Purch. Plan: Y | | Second Exec. Salary: $370,800 | Bonus: $ |

## OTHER THOUGHTS:

**Apparent Women Officers or Directors**: 1
**Hot Spot for Advancement for Women/Minorities**:

## LOCATIONS: ("Y" = Yes)

| West: | Southwest: | Midwest: | Southeast: | Northeast: | International: |
|---|---|---|---|---|---|
| Y | | | | | |

Note: Financial information, benefits and other data can change quickly and may vary from those stated here.

# ZILA INC

www.zila.com

**Industry Group Code:** 325412 **Ranks within this company's industry group:** Sales: 87 Profits: 96

| Drugs: | | Other: | | Clinical: | | Computers: | | Services: | |
|---|---|---|---|---|---|---|---|---|---|
| Discovery: | Y | AgriBio: | | Trials/Services: | | Hardware: | | Specialty Services: | |
| Licensing: | Y | Genetic Data: | | Labs: | | Software: | | Consulting: | |
| Manufacturing: | Y | Tissue Replacement: | | Equipment/Supplies: | Y | Arrays: | | Blood Collection: | |
| Genetics: | | | | Research & Development Services: | Y | Database Management: | | Drug Delivery: | |
| | | | | Diagnostics: | Y | | | Drug Distribution: | |

## TYPES OF BUSINESS:

Cancer Detection Products
Dental Products

## BRANDS/DIVISIONS/AFFILIATES:

ViziLite Plus
OraTest
Professional Dental Technologies, Inc.
Zila Pharmaceuticals, Inc.
Zila, Ltd.
Zila Technical, Inc.
Zila Biotechnology, Inc.
TBlue360

## CONTACTS: Note: Officers with more than one job title may be intentionally listed here more than once.

David R. Bethune, Interim CEO
Gary Klinefelter, General Counsel/VP
Diane E. Klein, Treas./VP-Finance
David R. Bethune, Chmn.

| Phone: 602-266-6700 | Fax: 602-234-2264 |
|---|---|
| Toll-Free: | |
| Address: 16430 N. Scottsdale Rd., Ste. 450, Scottsdale, AZ 85254 US | |

## GROWTH PLANS/SPECIAL FEATURES:

Zila, Inc. is a specialty pharmaceutical company dedicated to the prevention, detection and treatment of oral diseases, with a primary focus on oral cancer. Zila was formerly a provider of preventative healthcare technologies and products, and made the transition with the acquisition of Professional Dental Technologies, Inc., into a cancer detection company. Zila is a holding company that conducts its operations through several wholly-owned subsidiaries, including Zila Pharmaceuticals, Inc.; Professional Dental Technologies, Inc. (Pro-Dentec); Zila Biotechnology, Inc.; Zila Technical, Inc.; and Zila Limited (in the U.K.). The company's flagship product is ViziLite Plus with TBlue630, a chemiluminescent disposable light used for the illumination and marking of oral mucosal abnormalities. The firm also designs, manufactures and markets a line of periodontal products sold directly to dental professionals, which includes the Rota-dent Professional Powered Brush, the Pro-Select Platinum ultrasonic scaler and a variety of oral pharmaceutical products approved for both in-office and home use. The company also maintains a research and development segment to investigate pre-cancer/cancer detection using Zila's patented ZTC and OraTest oral cancer detection technologies. Most recently, Zila has gained approval to launch ViziLite Plus in Canada, the U.K. and Greece. In August 2008, the company announced it would be marketing its VizLite Plus product in Spain, France, Germany and Portugal through its distributor DentAid. In October 2008, the firm received marketing clearance for its OraTest product in seven European countries. In February 2009, VizLite's global market expanded to Russia, Belarus and countries in Asia.

Zila offers its employees medical, dental, vision, income protection and life insurance; flexible spending accounts; retirement and savings benefits; stock options; and time-off programs.

## FINANCIALS: Sales and profits are in thousands of dollars—add 000 to get the full amount. 2008 Note: Financial information for 2008 was not available for all companies at press time.

| | | |
|---|---|---|
| 2008 Sales: $45,061 | 2008 Profits: $-16,378 | **U.S. Stock Ticker:** ZILA |
| 2007 Sales: $28,801 | 2007 Profits: $-13,164 | **Int'l Ticker:** Int'l Exchange: |
| 2006 Sales: $2,822 | 2006 Profits: $-29,346 | Employees: 367 |
| 2005 Sales: $1,199 | 2005 Profits: $1,099 | Fiscal Year Ends: 7/31 |
| 2004 Sales: $36,682 | 2004 Profits: $-4,375 | Parent Company: |

## SALARIES/BENEFITS:

| Pension Plan: Y | ESOP Stock Plan: | Profit Sharing: | Top Exec. Salary: $370,443 | Bonus: $ |
|---|---|---|---|---|
| Savings Plan: Y | Stock Purch. Plan: Y | | Second Exec. Salary: $244,519 | Bonus: $ |

## OTHER THOUGHTS:

**Apparent Women Officers or Directors**: 2
**Hot Spot for Advancement for Women/Minorities**: Y

## LOCATIONS: ("Y" = Yes)

| West: | Southwest: | Midwest: | Southeast: | Northeast: | International: |
|---|---|---|---|---|---|
| | Y | | Y | | Y |

# ZYMOGENETICS INC

**www.zymogenetics.com**

Industry Group Code: 325412  **Ranks within this company's industry group:** Sales: 74  Profits: 161

| Drugs: | | Other: | | Clinical: | Computers: | | Services: | |
|---|---|---|---|---|---|---|---|---|
| Discovery: | Y | AgriBio: | | Trials/Services: | Hardware: | | Specialty Services: | |
| Licensing: | Y | Genetic Data: | Y | Labs: | Software: | | Consulting: | |
| Manufacturing: | Y | Tissue Replacement: | | Equipment/Supplies: | Arrays: | | Blood Collection: | |
| Genetics: | | | | Research & Development Services: | Database Management: | | Drug Delivery: | |
| | | | | Diagnostics: | | | Drug Distribution: | |

## TYPES OF BUSINESS:

Therapeutic Proteins
Hemostasis, Inflammatory & Autoimmune Diseases Drugs
Cancer & Viral Infections Drugs

## BRANDS/DIVISIONS/AFFILIATES:

rThrombin
Interleukin-21
PEG-IFN
Ataticept
rFactor XIII
Augment Bone Grafts
Bayer Schering Pharma AG
RECOTHROM

## CONTACTS: *Note: Officers with more than one job title may be intentionally listed here more than once.*

Douglas E. Williams, CEO
Stephen W. Zaruby, Pres.
James A. Johnson, CFO/Exec. VP
Darren R. Hamby, Sr. VP-Human Resources
Suzanne M. Shema, General Counsel/Sr. VP
Heather Franklin, Sr. VP-Bus. Dev.
James A. Johnson, Treas.
Eleanor L. Ramos, Chief Medical Officer/Sr. VP
Suzanne M. Shema, Sr. VP-Law & Compliance
Bruce L. A. Carter, Chmn.

| Phone: 206-442-6600 | Fax: 206-442-6608 |
|---|---|
| Toll-Free: 800-775-6686 | |
| Address: 1201 Eastlake Ave. E., Seattle, WA 98102 US | |

## GROWTH PLANS/SPECIAL FEATURES:

ZymoGenetics, Inc. discovers, develops, manufactures and commercializes therapeutic proteins for the treatment of human diseases. The company's current therapeutic focus is in the areas of hemostasis; inflammatory and autoimmune diseases; cancer; and viral infections. The firm's first internally developed product candidate, Recothrom Thrombin (also referred to as rThrombin or recombinant thrombin), was approved by the FDA in 2008 as a topical hemostat to control moderate bleeding during surgical procedures and is now being marketed in the U.S. Outside of the U.S., the company has partnered with Bayer Schering Pharma AG to develop and commercialize Recothrom. Other in-house products include Interleukin-21 (IL-21), a cytokine with potential applications for the treatment of cancer; PEG-IFN (formerly known as IL-29), a cytokine with potential applications for the treatment of viral infections; and monoclonal antibodies (mAb) for IL-21, IL-31 and FcyR1a for autoimmune diseases and dermatitis. ZymoGenetics collaborates with a number of external companies for developmental products, including Merck Serono for ataticept (formerly known as TACI-Ig), a soluble receptor with potential applications for the treatment of cancer and autoimmune diseases; with Novo Nordisk for rFactor XIII for congenital deficiency and cardiac surgery; with BioMimetic Therapeutics for Augment Bone Grafts; and many others. The company also out-licenses a number of commercial products, including treatments for diabetes, hemophilia, wound healing, periodontal defects, hypoglycemia and myocardial infarctions. In April 2009, the firm announced it would be laying off about 32% of its workforce and shift focus from oncology development to immunology.

ZymoGenetics offers its employees medical, dental and vision coverage; performance based bonuses; stock options; a 401(k) plan; retirement planning assistance; employee discounts; tuition reimbursements; life insurance; short- and long-term disability coverage; and parental leave.

## FINANCIALS: Sales and profits are in thousands of dollars—add 000 to get the full amount. 2008 Note: Financial information for 2008 was not available for all companies at press time.

| | | |
|---|---|---|
| 2008 Sales: $73,989 | 2008 Profits: $-116,241 | **U.S. Stock Ticker:** ZGEN |
| 2007 Sales: $38,477 | 2007 Profits: $-148,144 | **Int'l Ticker:**  Int'l Exchange: |
| 2006 Sales: $25,380 | 2006 Profits: $-130,002 | Employees:  570 |
| 2005 Sales: $42,909 | 2005 Profits: $-78,027 | Fiscal Year Ends: 12/31 |
| 2004 Sales: $35,694 | 2004 Profits: $-88,756 | Parent Company: |

## SALARIES/BENEFITS:

| Pension Plan: | ESOP Stock Plan: | Profit Sharing: | Top Exec. Salary: $636,681 | Bonus: $185,231 |
|---|---|---|---|---|
| Savings Plan: Y | Stock Purch. Plan: Y | | Second Exec. Salary: $482,050 | Bonus: $140,244 |

## OTHER THOUGHTS:

**Apparent Women Officers or Directors**: 4
**Hot Spot for Advancement for Women/Minorities**: Y

## LOCATIONS: ("Y" = Yes)

| West: | Southwest: | Midwest: | Southeast: | Northeast: | International: |
|---|---|---|---|---|---|
| Y | | | | | |

# ADDITIONAL INDEXES

**Contents:**

# INDEX OF FIRMS NOTED AS HOT SPOTS FOR ADVANCEMENT FOR WOMEN & MINORITIES

23ANDME
3SBIO INC
4SC AG
AASTROM BIOSCIENCES INC
ABBOTT LABORATORIES
ACAMBIS PLC
ACCELRYS INC
ADOLOR CORP
AGILENT TECHNOLOGIES INC
AKZO NOBEL NV
ALBANY MOLECULAR RESEARCH INC
ALCON INC
ALKERMES INC
ALLERGAN INC
ALPHARMA ANIMAL HEALTH
AMGEN INC
AMYLIN PHARMACEUTICALS INC
ANTIGENICS INC
APP PHARMACEUTICALS INC
APPLIED BIOSYSTEMS INC
ARIAD PHARMACEUTICALS INC
ASTRAZENECA PLC
AUTOIMMUNE INC
AXCAN PHARMA INC
BARR PHARMACEUTICALS INC
BAUSCH & LOMB INC
BAXTER INTERNATIONAL INC
BIO RAD LABORATORIES INC
BIOANALYTICAL SYSTEMS INC
BIOGEN IDEC INC
BIOHEART INC
BIOMARIN PHARMACEUTICAL INC
BIONANOMATRIX
BIOPURE CORPORATION
BIOSITE INC
BIOTECH HOLDINGS LTD
BIOVAIL CORPORATION
BRISTOL-MYERS SQUIBB CO
BURRILL & COMPANY
CALIPER LIFE SCIENCES
CAMBREX CORP
CANREG INC
CARACO PHARMACEUTICAL LABORATORIES
CARDIOME PHARMA CORP
CARDIUM THERAPEUTICS INC
CELERA CORPORATION
CELL GENESYS INC
CELL THERAPEUTICS INC
CEPHEID
CHARLES RIVER LABORATORIES INTERNATIONAL INC
COMPUGEN LTD
COVANCE INC
COVIDIEN PLC

CSL LIMITED
CUBIST PHARMACEUTICALS INC
CURAGEN CORPORATION
CYPRESS BIOSCIENCE INC
DENDREON CORPORATION
DEPOMED INC
DISCOVERY LABORATORIES INC
DURECT CORP
DYAX CORP
E I DU PONT DE NEMOURS & CO (DUPONT)
ELAN CORP PLC
ELI LILLY & COMPANY
ENCORIUM GROUP INC
ENDO PHARMACEUTICALS HOLDINGS INC
ENTREMED INC
ENZO BIOCHEM INC
EPIX PHARMACEUTICALS INC
ERESEARCH TECHNOLOGY INC
EXELIXIS INC
FLAMEL TECHNOLOGIES SA
FOREST LABORATORIES INC
GE HEALTHCARE
GENENTECH INC
GENEREX BIOTECHNOLOGY
GENOMIC HEALTH INC
GEN-PROBE INC
GENTA INC
GENZYME BIOSURGERY
GENZYME CORP
GERON CORPORATION
GILEAD SCIENCES INC
GLAXOSMITHKLINE PLC
GTC BIOTHERAPEUTICS INC
HEMISPHERX BIOPHARMA INC
HESKA CORP
HI-TECH PHARMACAL CO INC
HUMAN GENOME SCIENCES INC
ICON PLC
IDEXX LABORATORIES INC
IMCLONE SYSTEMS INC
IMMTECH PHARMACEUTICALS INC
IMMUNOGEN INC
IMMUNOMEDICS INC
IMS HEALTH INC
INCYTE CORP
INFINITY PHARMACEUTICALS INC
INSITE VISION INC
INSPIRE PHARMACEUTICALS INC
INTEGRA LIFESCIENCES HOLDINGS CORP
INTERMUNE INC
JAZZ PHARMACEUTICALS
JOHNSON & JOHNSON
KENDLE INTERNATIONAL INC
LA JOLLA PHARMACEUTICAL
LEXICON PHARMACEUTICALS INC
LIFE TECHNOLOGIES CORP
LIGAND PHARMACEUTICALS INC
LONZA GROUP

LORUS THERAPEUTICS INC
MDS INC
MEDAREX INC
MEDIDATA SOLUTIONS INC
MERCK & CO INC
MERIDIAN BIOSCIENCE INC
MILLENNIUM PHARMACEUTICALS INC
MILLIPORE CORP
MONSANTO CO
MYLAN INC
NANOGEN INC
NEKTAR THERAPEUTICS
NEUROCRINE BIOSCIENCES INC
NOVARTIS AG
NOVEN PHARMACEUTICALS
NOVO-NORDISK AS
NOVOZYMES
NYCOMED
ONCOGENEX PHARMACEUTICALS
ONCOTHYREON INC
ONYX PHARMACEUTICALS INC
ORGANOGENESIS INC
OSI PHARMACEUTICALS INC
PACIFIC BIOMETRICS INC
PDL BIOPHARMA
PENWEST PHARMACEUTICALS CO
PEREGRINE PHARMACEUTICALS INC
PERRIGO CO
PFIZER INC
PHARMACEUTICAL PRODUCT DEVELOPMENT INC
PHARMANET DEVELOPMENT GROUP INC
PONIARD PHARMACEUTICALS INC
PRA INTERNATIONAL
QIAGEN NV
QLT INC
QUEST DIAGNOSTICS INC
ROCHE HOLDING LTD
SANGAMO BIOSCIENCES INC
SANOFI-AVENTIS SA
SCHERING-PLOUGH CORP
SEQUENOM INC
SERACARE LIFE SCIENCES INC
SHIRE BIOCHEM INC
SHIRE PLC
SIGMA-ALDRICH CORP
STEMCELLS INC
STIEFEL LABORATORIES INC
SYNOVIS LIFE TECHNOLOGIES INC
TARO PHARMACEUTICAL INDUSTRIES
TECHNE CORP
TELIK INC
TEVA PHARMACEUTICAL INDUSTRIES
THERAGENICS CORP
TORREYPINES THERAPEUTICS INC
TRIPOS INTERNATIONAL
VALEANT PHARMACEUTICALS INTERNATIONAL
VERENIUM CORPORATION
VERNALIS PLC

VERTEX PHARMACEUTICALS INC
VIACELL INC
VION PHARMACEUTICALS INC
WATSON PHARMACEUTICALS INC
WYETH
ZILA INC
ZYMOGENETICS INC

# INDEX OF SUBSIDIARIES, BRAND NAMES AND AFFILIATIONS

**Brand or subsidiary, followed by the name of the related corporation**

# INDEX OF SUBSIDIARIES, BRAND NAMES AND AFFILIATIONS, CONT.

## INDEX OF SUBSIDIARIES, BRAND NAMES AND AFFILIATIONS, CONT.

Assisted Reproductive Technology; **COLUMBIA LABORATORIES INC**
Astellas Pharma Inc; **XENOPORT INC**
Astellas Pharma US, Inc.; **CARDIOME PHARMA CORP**
Astellas Venture Management, LLC; **ASTELLAS PHARMA INC**
AstraTech; **ASTRAZENECA PLC**
AstraZeneca; **PAR PHARMACEUTICAL COMPANIES INC**
AstraZeneca PLC; **ASTRAZENECA CANADA**
Atacicept; **ZYMOGENETICS INC**
ATL-104; **ALIZYME PLC**
Atra-IV; **ANTIGENICS INC**
Atria Genetics, Inc.; **CELERA CORPORATION**
Atrigel; **QLT INC**
Atripla; **GILEAD SCIENCES INC**
Atryn; **GTC BIOTHERAPEUTICS INC**
AttoSense; **SEQUENOM INC**
Augment Bone Grafts; **ZYMOGENETICS INC**
Autosuture; **COVIDIEN PLC**
AV333; **AVIGEN INC**
AV411; **AVIGEN INC**
AV650; **AVIGEN INC**
AVANT Immunotherapeutics, Inc.; **CELLEDEX THERAPEUTICS INC**
Avastin; **GENENTECH INC**
AVE1642; **IMMUNOGEN INC**
AVI-4658; **AVI BIOPHARMA INC**
AVI-6002; **AVI BIOPHARMA INC**
AVI-6003; **AVI BIOPHARMA INC**
Avid Bioservices, Inc.; **PEREGRINE PHARMACEUTICALS INC**
Avinza; **LIGAND PHARMACEUTICALS INC**
Avise MCV; **CYPRESS BIOSCIENCE INC**
Avise PG; **CYPRESS BIOSCIENCE INC**
Avista Capital Partners; **BIORELIANCE CORP**
AVONEX; **BIOGEN IDEC INC**
AXONYX, Inc.; **TORREYPINES THERAPEUTICS INC**
Axxora Life Sciences; **ENZO BIOCHEM INC**
Azactam; **ELAN CORP PLC**
Azasan; **SALIX PHARMACEUTICALS**
AzaSite; **INSITE VISION INC**
AzaSite; **INSPIRE PHARMACEUTICALS INC**
AzaSite Plus; **INSITE VISION INC**
AzaSite Xtra; **INSITE VISION INC**
Azilect; **TEVA PHARMACEUTICAL INDUSTRIES**
Azixa; **MYRIAD GENETICS INC**
BaBOOM! Energy Spray; **GENEREX BIOTECHNOLOGY**
Baclofen; **CARACO PHARMACEUTICAL LABORATORIES**
Baltimore Clinical Pharmacology Research; **BIOANALYTICAL SYSTEMS INC**

Barr Laboratories; **BARR PHARMACEUTICALS INC**
Barr Laboratories; **TEVA PARENTERAL MEDICINES INC**
Barr Pharmaceuticals; **TEVA PHARMACEUTICAL INDUSTRIES**
BASF Catalysts LLC; **BASF AG**
Basulin; **FLAMEL TECHNOLOGIES SA**
Battelle Ventures; **BIONANOMATRIX**
Bavituximab; **PEREGRINE PHARMACEUTICALS INC**
Baxter Healthcare Corporation; **ACAMBIS PLC**
Bayer; **BAYER CORP**
Bayer Business Services; **BAYER AG**
Bayer Corp; **BAYER AG**
Bayer CropScience; **BAYER AG**
Bayer CropScience, LP; **BAYER CORP**
Bayer Diagnostics; **SIEMENS HEALTHCARE DIAGNOSTICS**
Bayer HealthCare; **BAYER AG**
Bayer HealthCare AG; **BAYER CORP**
Bayer MaterialScience; **BAYER AG**
Bayer MaterialSciences, LLC; **BAYER CORP**
Bayer Schering Pharma AG; **ZYMOGENETICS INC**
Bayer Schering Pharma AG; **BAYER AG**
Bayer Technology Services; **BAYER AG**
BaySystems; **BAYER CORP**
Bazedoxifene; **LIGAND PHARMACEUTICALS INC**
BCX-4208/R3421; **BIOCRYST PHARMACEUTICALS INC**
Beacon Bioscience Inc; **ICON PLC**
Bead Block; **BIOCOMPATIBLES INTERNATIONAL PLC**
BeadArray Technology; **ILLUMINA INC**
BeadXpress Reader; **ILLUMINA INC**
Befar; **NEXMED INC**
Beijing Double-Crane Pharmaceutical Co., Ltd.; **ALLIANCE PHARMACEUTICAL CORP**
Belinostat; **CURAGEN CORPORATION**
Ben Franklin Technology Partners; **BIONANOMATRIX**
Benchware; **TRIPOS INTERNATIONAL**
Benicar; **FOREST LABORATORIES INC**
Bentley Pharmaceuticals, Inc.; **TEVA PHARMACEUTICAL INDUSTRIES**
Beriplast; **NYCOMED**
Berkeley HeartLab, Inc.; **CELERA CORPORATION**
Betaferon; **BAYER SCHERING PHARMA AG**
BEXIDEM; **IDM PHARMA INC**
BGLP40; **ONCOTHYREON INC**
BH4; **BIOMARIN PHARMACEUTICAL INC**
Bicept Magnum; **SYNGENTA AG**
Bicun; **SIMCERE PHARMACEUTICAL GROUP**
BIIB015; **IMMUNOGEN INC**
Bilastine; **INSPIRE PHARMACEUTICALS INC**
Bioadhesive Delivery System; **COLUMBIA LABORATORIES INC**

## INDEX OF SUBSIDIARIES, BRAND NAMES AND AFFILIATIONS, CONT.

## INDEX OF SUBSIDIARIES, BRAND NAMES AND AFFILIATIONS, CONT.

CardioTech International, Ltd.; **ADVANSOURCE BIOMATERIALS CORPORATION**
Cardium Biologics; **CARDIUM THERAPEUTICS INC**
Cardizem LA; **BIOVAIL CORPORATION**
Carticel; **GENZYME BIOSURGERY**
Catheter and Disposables Technology, Inc.; **ADVANSOURCE BIOMATERIALS CORPORATION**
CDX-110; **CELLEDEX THERAPEUTICS INC**
CDX-1135; **CELLEDEX THERAPEUTICS INC**
CDX-1307; **CELLEDEX THERAPEUTICS INC**
CEA; **DENDREON CORPORATION**
CEA-Scan; **IMMUNOMEDICS INC**
Cefotan; **APP PHARMACEUTICALS INC**
Cegedim Dendrite; **CEGEDIM SA**
Cegedim Strategic Data; **CEGEDIM SA**
Cegers; **CEGEDIM SA**
Cel-1000; **CEL-SCI CORPORATION**
Cel-2000; **CEL-SCI CORPORATION**
Celacade; **VASOGEN INC**
Celgosivir; **MIGENIX INC**
Cell Culture Systems (CCS); **LIFE TECHNOLOGIES CORP**
CellBeads; **BIOCOMPATIBLES INTERNATIONAL PLC**
Cellegy Pharmaceuticals, Inc.; **ADAMIS PHARMACEUTICALS CORPORATION**
CellMed; **BIOCOMPATIBLES INTERNATIONAL PLC**
Cellomics; **THERMO SCIENTIFIC**
CellPrimeTM rTransferrin AF; **NOVOZYMES**
CellzDirect, Inc.; **LIFE TECHNOLOGIES CORP**
Celsis Analytical Services; **CELSIS INTERNATIONAL PLC**
Celsis In Vitro Technologies; **CELSIS INTERNATIONAL PLC**
Celsis Rapid Detection; **CELSIS INTERNATIONAL PLC**
Celsius Control System; **CARDIUM THERAPEUTICS INC**
CelTx; **ORGANOGENESIS INC**
Center for Molecular Medicine; **SEQUENOM INC**
Centocor Inc; **JOHNSON & JOHNSON**
Cephalon, Inc.; **ALKERMES INC**
Ceravix; **GLAXOSMITHKLINE PLC**
Cerexa, Inc.; **FOREST LABORATORIES INC**
Cerezyme; **GENZYME CORP**
Certara; **TRIPOS INTERNATIONAL**
Cetaphil; **GALDERMA PHARMA SA**
Cethrin; **ALSERES PHARMACEUTICALS INC**
Cetilistat; **ALIZYME PLC**
Cetrorelix; **AETERNA ZENTARIS INC**
Cetrotide; **AETERNA ZENTARIS INC**
Cetrotide; **MERCK SERONO SA**
CGEN-25007; **COMPUGEN LTD**

Chantix; **PFIZER INC**
Chap Stick; **WYETH**
Chemagis; **PERRIGO CO**
Chemdex Inc.; **POLYDEX PHARMACEUTICALS**
ChimeriVax-Dengue; **ACAMBIS PLC**
ChimeriVax-JE; **ACAMBIS PLC**
ChimeriVax-West Nile; **ACAMBIS PLC**
China Development Industrial Bank; **AVIVA BIOSCIENCES CORP**
Chiron Corp; **NOVARTIS AG**
Chocola; **EISAI CO LTD**
CholeraGarde; **CELLEDEX THERAPEUTICS INC**
Cholesterol Reference Method Laboratory Network; **PACIFIC BIOMETRICS INC**
ChronoFlex; **ADVANSOURCE BIOMATERIALS CORPORATION**
Chugai Pharmaceuticals; **ROCHE HOLDING LTD**
CIBA Vision; **NOVARTIS AG**
Cimzia; **UCB SA**
CIMZIA; **XOMA LTD**
Cimzia; **ENZON PHARMACEUTICALS INC**
Cinryze; **VIROPHARMA INC**
Ciphergen Biosystems, Inc.; **VERMILLION INC**
CIRA Biosciences, Inc.; **NEOPROBE CORPORATION**
Cleanse & Treat; **TARO PHARMACEUTICAL INDUSTRIES**
ClearCourse; **MEDTOX SCIENTIFIC INC**
ClearPath; **NORTH AMERICAN SCIENTIFIC**
Cleavase; **THIRD WAVE TECHNOLOGIES INC**
Cleviprex (Clevidine); **MEDICINES CO (THE)**
ClindaReach; **DUSA PHARMACEUTICALS INC**
Clinical Research Services; **PAREXEL INTERNATIONAL CORP**
Clinical Trial Support Services; **BIORELIANCE CORP**
Clinitek; **SIEMENS HEALTHCARE DIAGNOSTICS**
ClinPhone plc; **PAREXEL INTERNATIONAL CORP**
Clobex; **GALDERMA PHARMA SA**
Cloderm; **VALEANT PHARMACEUTICALS INTERNATIONAL**
Clolar; **GENZYME ONCOLOGY**
Clolar; **GENZYME CORP**
Cloretazine; **VION PHARMACEUTICALS INC**
Clorpactin; **UNITED-GUARDIAN INC**
Clotrimazole/Betamethasomne; **TARO PHARMACEUTICAL INDUSTRIES**
CML Healthcare Inc.; **CML HEALTHCARE INCOME FUND**
Coag Dx; **IDEXX LABORATORIES INC**
Coastal Training Technologies Corporation; **E I DU PONT DE NEMOURS & CO (DUPONT)**
Coban; **ELI LILLY & COMPANY**
Codexis; **MAXYGEN INC**
CoGenesys, Inc.; **TEVA PHARMACEUTICAL INDUSTRIES**
COLAL; **ALIZYME PLC**

## INDEX OF SUBSIDIARIES, BRAND NAMES AND AFFILIATIONS, CONT.

## INDEX OF SUBSIDIARIES, BRAND NAMES AND AFFILIATIONS, CONT.

# INDEX OF SUBSIDIARIES, BRAND NAMES AND AFFILIATIONS, CONT.

## INDEX OF SUBSIDIARIES, BRAND NAMES AND AFFILIATIONS, CONT.

Fort Dodge Animal Health; **WYETH**
FortaGen; **ORGANOGENESIS INC**
Fortamet; **SCIELE PHARMA INC**
FortaPerm; **ORGANOGENESIS INC**
Fortical; **UNIGENE LABORATORIES**
Fosamax; **MERCK & CO INC**
Fresenius Kabi Pharmaceuticals Holding Inc; **APP PHARMACEUTICALS INC**
Frova; **ENDO PHARMACEUTICALS HOLDINGS INC**
Frovatriptan; **VERNALIS PLC**
Fuelzyme-CX; **VERENIUM CORPORATION**
Fuelzyme-LF; **VERENIUM CORPORATION**
Fujisawa Pharmaceutical Co., Ltd.; **ASTELLAS PHARMA INC**
FUMADERM; **BIOGEN IDEC INC**
Fungassay; **SYNBIOTICS CORP**
Fusilev; **SPECTRUM PHARMACEUTICALS INC**
FUZEON; **TRIMERIS INC**
Gabapentin GR; **DEPOMED INC**
Galt Medical Corp.; **THERAGENICS CORP**
Ganite; **GENTA INC**
Gardasil; **MERCK & CO INC**
Garlic Pearls; **RANBAXY LABORATORIES LIMITED**
GastroDose; **PENWEST PHARMACEUTICALS CO**
Gate Pharmaceuticals; **TEVA PARENTERAL MEDICINES INC**
GATTEX; **NPS PHARMACEUTICALS INC**
GDC-0449; **CURIS INC**
GE Healthcare; **HARVARD BIOSCIENCE INC**
GE Healthcare; **WHATMAN PLC**
GE Healthcare Bio-Sciences; **GE HEALTHCARE**
GE Healthcare Information Technologies; **GE HEALTHCARE**
GE Healthcare Technologies; **GE HEALTHCARE**
GE Technology Infrastructure; **GE HEALTHCARE**
Gelnique; **WATSON PHARMACEUTICALS INC**
GEM MicroClip; **SYNOVIS LIFE TECHNOLOGIES INC**
Geminex; **PENWEST PHARMACEUTICALS CO**
Gemzar; **ELI LILLY & COMPANY**
Genasense; **GENTA INC**
Gene Logic Inc; **ORE PHARMACEUTICALS INC**
GeneChip; **AFFYMETRIX INC**
Genelabs Technologies; **GLAXOSMITHKLINE PLC**
GeneMachines; **HARVARD BIOSCIENCE INC**
GeneMorph II; **STRATAGENE CORP**
Genentech Inc; **23ANDME**
Genentech Inc; **ROCHE HOLDING LTD**
Genentech Inc; **LEXICON PHARMACEUTICALS INC**
General Electric Co (GE); **GE HEALTHCARE**
Generx; **CARDIUM THERAPEUTICS INC**
GeneSense Technologies, Inc.; **LORUS THERAPEUTICS INC**
GeneSys; **TRINITY BIOTECH PLC**

Genetech, Inc.; **DENDREON CORPORATION**
GeneXpert; **CEPHEID**
Genome Analyzer; **ILLUMINA INC**
GenoSensor Microarray System; **VYSIS INC**
Genpharm ULC; **MYLAN INC**
Gen-Probe; **CHIRON CORP**
Genstar Capital, LLC; **PRA INTERNATIONAL**
Gensweet; **GENENCOR INTERNATIONAL INC**
Gentra Systems, Inc.; **QIAGEN NV**
Genvir; **FLAMEL TECHNOLOGIES SA**
Genzyme Corp.; **GENZYME BIOSURGERY**
Genzyme Corp.; **GENZYME ONCOLOGY**
Genzyme Corporation; **BIOMARIN PHARMACEUTICAL INC**
Gesdyp; **RANBAXY LABORATORIES LIMITED**
Gilead Sciences; **ROCHE HOLDING LTD**
GlaxoSmithKline plc; **XENOPORT INC**
GlaxoSmithKline plc; **PAR PHARMACEUTICAL COMPANIES INC**
Gliadel; **MGI PHARMA INC**
Global Pharmaceuticals; **IMPAX LABORATORIES INC**
GLP hERG Channel Drug Candidate Screening; **CELLULAR DYNAMICS INTERNATIONAL**
Glucanin; **BIOTECH HOLDINGS LTD**
Glucerna; **ABBOTT LABORATORIES**
Glucose RapidSpray; **GENEREX BIOTECHNOLOGY**
Glumetza; **DEPOMED INC**
GnRH Antagonist; **NEUROCRINE BIOSCIENCES INC**
Gold Avenue Ltd.; **IMMTECH PHARMACEUTICALS INC**
GONAL-f; **MERCK SERONO SA**
Google Inc; **23ANDME**
G-Protein Coupled Receptors (GPCRs); **ACTELION LTD**
Grafeel; **DR REDDY'S LABORATORIES LIMITED**
GRN163L; **GERON CORPORATION**
GRNCM1; **GERON CORPORATION**
GRNOPC1; **GERON CORPORATION**
GRNVAC1; **GERON CORPORATION**
Guardian Laboratories; **UNITED-GUARDIAN INC**
GVAX; **CELL GENESYS INC**
Hammerite; **AKZO NOBEL NV**
Hansa Chemie International; **BASF AG**
Haptoguard, Inc.; **SYNVISTA THERAPEUTICS INC**
Harmony; **ADVANCED BIONICS CORPORATION**
Harnal; **ASTELLAS PHARMA INC**
HCV-796; **VIROPHARMA INC**
Health Services Research Network; **IMS HEALTH INC**
HeartBar; **UNITED THERAPEUTICS CORP**
Hectorol; **SHIRE BIOCHEM INC**
Hedrin; **MANHATTAN PHARMACEUTICALS INC**
Helinx; **CERUS CORPORATION**

# INDEX OF SUBSIDIARIES, BRAND NAMES AND AFFILIATIONS, CONT.

## INDEX OF SUBSIDIARIES, BRAND NAMES AND AFFILIATIONS, CONT.

# INDEX OF SUBSIDIARIES, BRAND NAMES AND AFFILIATIONS, CONT.

# INDEX OF SUBSIDIARIES, BRAND NAMES AND AFFILIATIONS, CONT.

# INDEX OF SUBSIDIARIES, BRAND NAMES AND AFFILIATIONS, CONT.

My Study Portal; **ERESEARCH TECHNOLOGY INC**
Mylan Pharmaceuticals, Inc.; **MYLAN INC**
Mylan Technologies, Inc.; **MYLAN INC**
MyoCell; **BIOHEART INC**
MyoCell SDF-1; **BIOHEART INC**
Myonal; **EISAI CO LTD**
Myriad Pharmaceuticals, Inc.; **MYRIAD GENETICS INC**
Mytogen, Inc.; **ADVANCED CELL TECHNOLOGY INC**
Naglazyme; **BIOMARIN PHARMACEUTICAL INC**
Namenda; **NEUROBIOLOGICAL TECHNOLOGIES INC**
Namenda; **FOREST LABORATORIES INC**
Nano Terra LLC; **MERCK KGAA**
NanoCrystal; **ELAN CORP PLC**
NanoHPX; **NANOBIO CORPORATION**
NanoMatrix, Inc.; **ORGANOGENESIS INC**
NanoStat; **NANOBIO CORPORATION**
NanoTxt; **NANOBIO CORPORATION**
NasaPep; **UNIGENE LABORATORIES**
Nascobal; **PAR PHARMACEUTICAL COMPANIES INC**
Nasivin; **MERCK KGAA**
National Human Genome Research Institute; **BIONANOMATRIX**
National Institutes of Health; **BIONANOMATRIX**
Natrecor; **SCIOS INC**
Natural Formula; **PERRIGO CO**
Nebido; **ENDO PHARMACEUTICALS HOLDINGS INC**
Neca; **PERRIGO CO**
NeedleTech Products, Inc.; **THERAGENICS CORP**
Nektar PEGylation Technology; **NEKTAR THERAPEUTICS**
Nektar Pulmonary Technology; **NEKTAR THERAPEUTICS**
Nellcor; **MALLINCKRODT INC**
Nellcor; **COVIDIEN PLC**
neo2000 Gamma Detection Systems; **NEOPROBE CORPORATION**
Neogen Latino America SPA; **NEOGEN CORPORATION**
NeoLipid; **NEOPHARM INC**
Neryx Biopharmaceuticals, Inc.; **KERYX BIOPHARMACEUTICALS INC**
Nestle Corporation; **ALCON INC**
Nestle SA; **GALDERMA PHARMA SA**
Neulasta; **AMGEN INC**
Neulasta; **NEKTAR THERAPEUTICS**
NEUPOGEN; **AMGEN INC**
Neupro; **UCB SA**
NeutroSpec; **PALATIN TECHNOLOGIES INC**
Neuvenge; **DENDREON CORPORATION**
New Enterprise Associates; **23ANDME**

NewLab BioQuality AG; **CHARLES RIVER LABORATORIES INTERNATIONAL INC**
Newport Bio Systems; **MILLIPORE CORP**
NexACT; **NEXMED INC**
Nexavar; **ONYX PHARMACEUTICALS INC**
Nexera; **DOW AGROSCIENCES LLC**
Nexium; **ASTRAZENECA PLC**
Nextal Biotechnology, Inc.; **QIAGEN NV**
NFil Technology; **BIOCOMPATIBLES INTERNATIONAL PLC**
NGX267; **TORREYPINES THERAPEUTICS INC**
NGX292; **TORREYPINES THERAPEUTICS INC**
NGX426; **TORREYPINES THERAPEUTICS INC**
Niacin Receptor Agonist; **ARENA PHARMACEUTICALS INC**
Nichols Institute; **QUEST DIAGNOSTICS INC**
Nicomide; **DUSA PHARMACEUTICALS INC**
Nicomide-T; **DUSA PHARMACEUTICALS INC**
Nicorette; **GLAXOSMITHKLINE PLC**
NICOSAN; **XECHEM INTERNATIONAL**
NicVAX; **NABI BIOPHARMACEUTICALS**
NimbleGen Systems Inc; **ROCHE HOLDING LTD**
Nippon Kayaku, Kayaku Akzo Co Ltd; **AKZO NOBEL NV**
Nitrolingual; **SCIELE PHARMA INC**
Non-GLP hERG Channel Drug Candidate Screening; **CELLULAR DYNAMICS INTERNATIONAL**
Nordic Biotech Advisors ApS; **MANHATTAN PHARMACEUTICALS INC**
Norditropin; **NOVO-NORDISK AS**
Norgine B.V.; **ALIZYME PLC**
Norvasc; **PFIZER INC**
Novaquest; **QUINTILES TRANSNATIONAL CORP**
Novartis AG; **CHIRON CORP**
Novartis Institute for Biomedical Research Inc; **NOVARTIS AG**
Novartis Oncology; **NOVARTIS AG**
Novartis Pharma AG; **EMISPHERE TECHNOLOGIES INC**
Novartis Pharma AG; **QLT INC**
NovaScreen Biosciences Corp.; **CALIPER LIFE SCIENCES**
Noven Therapeutics; **NOVEN PHARMACEUTICALS**
Novifit NoviSAMe; **VIRBAC CORP**
Novo Nordisk; **ARADIGM CORPORATION**
NovoNorm; **NOVO-NORDISK AS**
NovoSeven; **NOVO-NORDISK AS**
NPSP156; **NPS PHARMACEUTICALS INC**
NPSP558; **NPS PHARMACEUTICALS INC**
NTCD; **VIROPHARMA INC**
NuChem Pharmaceuticals, Inc.; **LORUS THERAPEUTICS INC**
NuCore; **PROTEIN POLYMER TECHNOLOGIES**
Nuflor; **SCHERING-PLOUGH CORP**
NuLeusin; **3SBIO INC**

## INDEX OF SUBSIDIARIES, BRAND NAMES AND AFFILIATIONS, CONT.

# INDEX OF SUBSIDIARIES, BRAND NAMES AND AFFILIATIONS, CONT.

## INDEX OF SUBSIDIARIES, BRAND NAMES AND AFFILIATIONS, CONT.

# INDEX OF SUBSIDIARIES, BRAND NAMES AND AFFILIATIONS, CONT.

## INDEX OF SUBSIDIARIES, BRAND NAMES AND AFFILIATIONS, CONT.

## INDEX OF SUBSIDIARIES, BRAND NAMES AND AFFILIATIONS, CONT.

Takeda America Holdings, Inc.; **TAKEDA PHARMACEUTICAL COMPANY LTD**

Takeda Europe Research and Development Center Ltd.; **TAKEDA PHARMACEUTICAL COMPANY LTD**

Takeda Ireland Limited; **TAKEDA PHARMACEUTICAL COMPANY LTD**

Takeda Pharmaceutical Company Ltd; **MILLENNIUM PHARMACEUTICALS INC**

Takeda Pharmaceutical Company Ltd; **IDM PHARMA INC**

Takeda Pharmaceutical Company Ltd.; **XOMA LTD**

Takeda Research Investment, Inc.; **TAKEDA PHARMACEUTICAL COMPANY LTD**

Tamibarotene; **CYTRX CORPORATION**

TAP Pharmaceutical Products Inc.; **TAKEDA PHARMACEUTICAL COMPANY LTD**

TAPI; **TEVA PARENTERAL MEDICINES INC**

Tarceva; **OSI PHARMACEUTICALS INC**

Targanta Therapeutics Corporation; **MEDICINES CO (THE)**

Target-Related Affinity Profiling (TRAP); **TELIK INC**

Taro Pharmaceuticals U.S.A., Inc.; **TARO PHARMACEUTICAL INDUSTRIES**

Tasidotin Hydrochloride; **GENZYME ONCOLOGY**

TAXUS; **ANGIOTECH PHARMACEUTICALS**

TBlue360; **ZILA INC**

TC-Mouse; **MEDAREX INC**

Teflon; **E I DU PONT DE NEMOURS & CO (DUPONT)**

Telaprevir; **VERTEX PHARMACEUTICALS INC**

TELCYTA; **TELIK INC**

TELINTRA; **TELIK INC**

Telzir; **VERTEX PHARMACEUTICALS INC**

Tepnel Life Sciences, plc; **GEN-PROBE INC**

Terapia; **RANBAXY LABORATORIES LIMITED**

Terso Solutions; **PROMEGA CORP**

Tesetaxel; **GENTA INC**

Teva Animal Health; **TEVA PARENTERAL MEDICINES INC**

Teva Biopharmaceuticals USA; **TEVA PARENTERAL MEDICINES INC**

Teva Pharmaceutical Industries; **AUTOIMMUNE INC**

Teva Pharmaceuticals Industries, Ltd.; **TEVA PARENTERAL MEDICINES INC**

Teva Pharmaceuticals USA; **TEVA PHARMACEUTICAL INDUSTRIES**

Teva Pharmaceuticals USA; **TEVA PARENTERAL MEDICINES INC**

Teva Speciality Pharmaceuticals, LLC; **TEVA PARENTERAL MEDICINES INC**

tezampanel; **TORREYPINES THERAPEUTICS INC**

tgAAC09; **TARGETED GENETICS CORP**

tgAAC94; **TARGETED GENETICS CORP**

TGI 1200; **BIOHEART INC**

THALOMID; **CELGENE CORP**

Theken Disc LLC; **INTEGRA LIFESCIENCES HOLDINGS CORP**

Theken Spine LLC; **INTEGRA LIFESCIENCES HOLDINGS CORP**

TheraGuide 5-FU; **MYRIAD GENETICS INC**

Therapeutic Human Polyclonals Inc; **ROCHE HOLDING LTD**

TheraSeed; **THERAGENICS CORP**

TherAtoh; **GENVEC INC**

Therics LLC; **INTEGRA LIFESCIENCES HOLDINGS CORP**

Thermo Fisher Scientific, Inc.; **THERMO SCIENTIFIC**

Ther-Rx Corporation; **KV PHARMACEUTICAL CO**

Thiakis Limited; **WYETH**

Thyrogen; **GENZYME CORP**

Tianeptine; **PHARMOS CORP**

Tianwei Times; **QIAGEN NV**

Tietai Iron Sucrose Supplement; **3SBIO INC**

TIGER; **ISIS PHARMACEUTICALS INC**

TIGRIS; **GEN-PROBE INC**

TILL Photonics GmbH; **AGILENT TECHNOLOGIES INC**

TIMERx; **PENWEST PHARMACEUTICALS CO**

Tissue Genesis, Inc.; **BIOHEART INC**

Tissue Repair Cells; **AASTROM BIOSCIENCES INC**

Tissue Repair Co.; **CARDIUM THERAPEUTICS INC**

Tissue-Guard; **SYNOVIS LIFE TECHNOLOGIES INC**

TLK58747; **TELIK INC**

TNFerade; **GENVEC INC**

TNKase; **GENENTECH INC**

TOCOSOL Paclitaxel; **ONCOGENEX PHARMACEUTICALS**

Topiramate; **CARACO PHARMACEUTICAL LABORATORIES**

Toraymyxin; **SPECTRAL DIAGNOSTICS INC**

TPG (Texas Pacific Group); **AXCAN PHARMA INC**

TPIAO; **3SBIO INC**

Tracleer; **ACTELION LTD**

TRAIL Receptor Antibodies; **HUMAN GENOME SCIENCES INC**

TRANSDUR; **DURECT CORP**

Translation Suppressing Oligomers (TSO); **AVI BIOPHARMA INC**

Transtuzumab-DM1; **IMMUNOGEN INC**

Travelmin; **EISAI CO LTD**

Treanda; **CEPHALON INC**

Trelstar Depot; **WATSON PHARMACEUTICALS INC**

Treximet; **POZEN INC**

TRI-1144; **TRIMERIS INC**

Triage BNP Test; **BIOSITE INC**

Triage Cardiac Panel; **BIOSITE INC**

Triage Drugs of Abuse Panel; **BIOSITE INC**

Triage Parasite Panel; **BIOSITE INC**

Triage Profiler Panels; **BIOSITE INC**

Triage TOX Drug Screen; **BIOSITE INC**

# INDEX OF SUBSIDIARIES, BRAND NAMES AND AFFILIATIONS, CONT.

# INDEX OF SUBSIDIARIES, BRAND NAMES AND AFFILIATIONS, CONT.

**INDEX OF SUBSIDIARIES, BRAND NAMES AND AFFILIATIONS, CONT.**

# We help you understand how industries work today and the trends that will change them tomorrow.

The Industry Research Centers at Plunkett Research Online take you directly to the data you need. Each industry's data is organized in unique, easy-to-use tools:

- ▸ Market Research and Industry Trends
- ▸ Industry Statistics
- ▸ Company Profiles
- ▸ Export Company Contacts
- ▸ Industry Glossary
- ▸ Industry Associations
- ▸ Build-A-Report℠

Get instant data for business plans and competitive intelligence reports.

Filter, view, select and export contacts at thousands of industry-leading companies for the best sales prospects lists.

**NEW!** Use our exclusive Build-A-Report℠ tool to export professional-quality reports, sales guides and training tools in PDF format.

Target industry-leading employers for the most effective job searches.

**NEW!** Create a personal MyResearchProfile℠ to save searches and receive industry alerts.

 **Ask about our FREE ONLINE TRIAL at** info@plunkettresearch.c 713.932.0000

# Online access to **32** Industry Research Centers brimming with data & resources in these industries:

1. Advertising/Branding
2. Airline, Hotel & Travel
3. Alternative & Renewable Energy
4. Apparel & Textiles
5. Archives
6. Automobile
7. Banking, Mortgages & Credit
8. Biotechnology
9. Canadian Industry
10. Chemicals, Plastics & Coatings
11. Consulting
12. E-Commerce & Internet
13. Energy & Utilities
14. Engineering & Research
15. Entertainment & Media
16. Food, Beverage & Tobacco
17. Health Care
18. InfoTech, Software & Hardware
19. Insurance
20. International
21. Investment & Securities
22. Job Seekers
23. Middle Market
24. Nanotechnology & MEMs
25. Outsourcing & Offshoring
26. Private Companies
27. Real Estate & Construction
28. Retail
29. Sports
30. Telecommunications
31. Transportation & Logistics
32. Wireless, Cellular & Wi-Fi

**All data is written in-house by the Plunkett Research staff.**

# Plunkett Research Online www.plunkettresearch.com 713.932.0000

# Get this complete set of industry-specific tools from each of our 32 Industry Research Centers.

## Plunkett Research, Ltd.
Industry Statistics, Trends and In-depth Analysis of Top Companies

When you want affordable, timely industry research, contacts and business development data

**Need Help? Phone 713.932.0000**

Home | Industry Research Titles | Reviews | About Us | Contact & Support | Catalog | How to Buy

Log Out

## Plunkett's Energy Industry Research Center

[ Search input ]  Search   Save Your Searches   Historic Data

> **Market Research & Trends**
A Market overview and an analysis of major trends that are creating rapid changes in the industry today.

> **Company Profiles**
In-depth profiles of leading companies in this industry. Includes business descriptions, financial information, growth plans and executive names.

> **Export Company Contacts**
Our company contact information can be exported to Microsoft Excel or text files. Includes company name, address, phone, website and executives with job titles.

> **Export Associations/Organizations**
Our Associations/Organizations contact information can be exported to Microsoft Excel or text files. Includes organization name, address, phone, website and description.

> **Statistics**
The statistics section contains extensive data on many facets of the industry.

> **Associations/Organizations**
Industry associations, government agencies and important industry phone numbers and web sites.

> **Glossary**
An industry-specific glossary written in language that is easy-to-understand. This information is ideal to use when preparing for job interviews or sales presentations.

> **Plunkett's Build-A-Report**ᔕᴹ
Quickly build a custom report based on trends, statistics, company profiles, contacts, and glossaries.

Better Industry Data. Better User Interface. Better Tools. Better Prices.

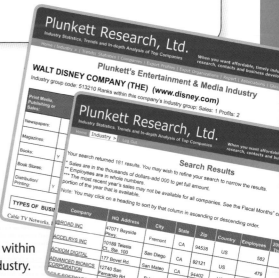

## Q&A

▸ **How does Plunkett's approach to research differ from others?**

The focus at Plunkett Research is to make it easy for the general reader to readily access and understand the most vital trends creating change within given industries – even if the reader has no current expertise in that industry.

Our philosophy is that several core sets of data should be presented in one comprehensive tool to give the reader the best total picture of an industry. That data must be user-friendly: easy-to-access, easy-to-understand, easy-to-use. That data must be affordable and represent extremely high value for the buyer/subscriber.

www.plunkettresearch.com     713.932.0000